HOLT

Environmental Science

TEACHER EDITION

Karen Arms, Ph.D., J.D.
Savannah, Georgia

Teacher Edition WALK-THROUGH

Student Edition CONTENTS IN BRIEF

HOLT, RINEHART AND WINSTON
A Harcourt Education Company
Orlando • Austin • New York • San Diego • Toronto • London

The #1 selling program in the nation

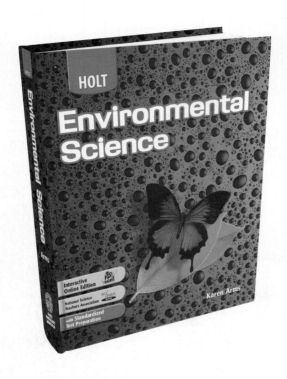

"ACTUALLY TEACHES AND DEVELOPS THE SCIENCE BEHIND THE ENVIRONMENT." —KAREN ARMS

Well-known biologist Karen Arms has written a text developed specifically for high school students. *Holt Environmental Science* presents an unbiased approach to the environment, with comprehensive content including ecology, Earth science, health, and policy topics.

STUDENTS OF ALL ABILITY LEVELS DEVELOP THE SKILLS THEY NEED FOR SUCCESS

- Math, reading, and writing skills practice are found throughout the program.
- **Standardized Test Prep** allows students to practice their test-taking skills.
- **Maps in Action** features help students build their map-reading and analytical skills.
- Activities are labeled by difficulty level in the *Teacher Edition* to help you cater classroom instruction to the abilities of your students.

A FLEXIBLE LABORATORY PROGRAM BUILDS INQUIRY AND CRITICAL-THINKING SKILLS

- The laboratory program includes in-text labs for each chapter plus additional labs found in the *Chapter Resource Files.*
- All labs are tested by teachers and rated by level of difficulty so you can be sure that they will work with your students.
- A variety of labs, from **QuickLAB** to **Design Your Own Lab,** help you meet your curriculum needs and work within the time constraints of your class schedule.

INTEGRATED TECHNOLOGY AND ONLINE RESOURCES EXPAND LEARNING BEYOND THE CLASSROOM

- Lighten the load with an interactive *Online Edition* or *CD-ROM Version* of the student text.

- **SciLinks®,** a Web service developed and maintained by the NSTA, contains current and prescreened links that engage students.

- *Holt Environmental Science Interactive Tutor CD-ROM* gives students a fun way to explore environmental science concepts at their own pace.

- All the resources you need are on the *One-Stop Planner® CD-ROM with ExamView® Test Generator,* with worksheets, customizable lesson plans, Holt Calendar Planner®, Holt PuzzlePro®, and a powerful test generator.

The Student Edition builds skills for success in science

Pre-Reading Activity includes a **FoldNote** to help students organize their ideas and improve their comprehension and retention.

RELEVANT AND MOTIVATING FEATURES

Ecofact and **Geofact** are quick tidbits of interesting, related information that motivate students to learn more.

Ecofact

Conserving Water Arthropods and vertebrates are the only two ~~ mals that have adap-~~ ~~ revent~~ dehydration so ~~ at some of them can~~ ~~ reely~~ on land on a ~~ y.~~ No other animals ~~ ptation.~~

Geofact

Minerals in Your Mouth Phosphorus is the 11th most abundant element in the Earth's crust and occurs naturally as phosphate in the mineral apatite. Apatite can exist in igneous, metamorphic, and sedimentary rocks as well as in your teeth and bones.

SciLinks® provides up-to-date Web links prescreened by NSTA.

internet connect

www.scilinks.org
Topic: Nitrogen Cycle
SciLinks code: HE4073

SCI LINKS Maintained by the National Science Teachers Association

High-interest features such as **Maps in Action** and **Science and Technology** extend content with real-world examples.

MAPS in action
WIND POWER IN THE UNITED STATES

Cross-disciplinary **Connections** show how environmental science relates to other sciences and disciplines.

Connection to Chemistry

Medicines from Plants Many of the medicines we use come from plants native to tropical rain forests. Chemists extract and test chemicals found in plants to determine if the chemicals can cure or fight diseases. Rosy periwinkle, a

MATHPRACTICE allows students to improve important math skills.

MATHPRACTICE

A Meal Fit for a Grizzly Bear Grizzly bears are omnivores that can eat up to 15 percent of their body weight per day when eating salmon and up to 33 percent of their body weight when eating fruits and

How Ecosystems Work

CHAPTER 5

1 Energy Flow in Ecosys
2 The Cycling of Materia
3 How Ecosystems Chang

PRE-READING ACTIVITY

FOLDNOTES **Double-Door Fold** Before you read this chapter, create the **FoldNote** entitled "Double-Door Fold described in the Reading a Study Skills Handbook section of the Appendix. Write "Energy flow in ecosystems on one flap of the double door and "Movement of materials in ecosystems" or the other flap. As you read the chapter, compare the two topics, and write chara teristics of each on the inside of the appropriate flap.

This green frog gets the energy needs to survive by eating other organisms, such as dragonflies.

116 Chapter 5 How Ecosystems Work

CASE STUDY Paper or Plastic?

The following question may sound familiar: Do you want paper or plastic? If you have ever stood in the checkout line of a grocery store, it probably is. Almost every grocery store today offers a choice between either paper or plastic bags for sacking grocery items. Many people make their choice based on convenience. But what is the best choice for someone who is concerned about the environment?

On the surface, it may seem that paper is the better choice. Paper comes from a renewable

Upon closer examination, however, the decision may not be as simple as it seems. Removing large numbers of trees from forests to manufacture paper can disrupt woodland ecosystems. Plus, a tremendous amount of energy is required to convert trees into pulp and then manufacture paper from the pulp.

To make the best decision about which product is better for the environment, the following questions should be considered.

► Making an educated decision at the grocery store will help ~~ solid waste~~

• How much raw material, energy,

CASE STUDY makes science relevant to students and builds critical-thinking skills.

SECTION 1
Energy Flow in Ecosystems

Just as a car cannot run without fuel, an organism cannot survive without a constant supply of energy. Where does an organism's energy come from? The answer to that question depends on the organism, but the ultimate source of energy for almost all organisms is the sun.

Life Depends on the Sun

Energy from the sun enters an ecosystem when a plant uses sunlight to make sugar molecules by a process called **photosynthesis.** During photosynthesis, plants, algae, and some bacteria capture solar energy. Solar energy drives a series of chemical reactions that require carbon dioxide and water, as shown in **Figure 1.** The result of photosynthesis is the production of sugar molecules known as *carbohydrates.* Carbohydrates are energy-rich molecules which organisms use to carry out daily activities. As organisms consume food and use energy from carbohydrates, the energy travels from one organism to another. Plants, such as the sunflowers in **Figure 2,** produce carbohydrates in their leaves. When an animal eats a plant, some energy is transferred from the plant to the animal. Organisms use this energy to move, grow, and reproduce.

Objectives

▶ Describe how energy is transferred from the sun to producers and then to consumers.
▶ Describe one way in which consumers depend on producers.
▶ List two types of consumers.
▶ Explain how energy transfer in a food web is more complex than energy transfer in a food chain.
▶ Explain why an energy pyramid is a representation of trophic levels.

Key Terms

photosynthesis
producer
consumer
decomposer
cellular respiration
food chain
food web
trophic level

Figure 1 ▶ During photosynthesis, plants use carbon dioxide, water, and solar energy to make carbohydrates and oxygen.

Figure 2 ▶ The cells in the leaves of these sunflowers contain a green chemical called *chlorophyll.* Chlorophyll helps plants trap energy from the sun to produce energy-rich carbohydrates.

Energy Flow in Ecosystems **117**

Each section begins with a list of **Objectives** and **Key Terms** that enhance student comprehension.

Accessible navigation engages students with outline-style headings and content grouped into small chunks.

The illustrations in *Holt Environmental Science* help students visualize key concepts, and many include questions so that students can apply what they have learned.

LABS AND ACTIVITIES GRAB ATTENTION

QuickLAB gives students a short, hands-on experience requiring minimal materials.

Field Activity gives students an opportunity to observe and apply a subject in a real-world setting.

QuickLAB

Sponging It Up

Procedure

1. Completely saturate **two small sponges** with **water** and allow the excess water to drain off

In-text labs, including **Skills Practice Lab, Exploration Lab,** and **Inquiry Lab,** provide more in-depth exploration of a concept using scientific methods.

REVIEW FOR TEST-READINESS

Chapter Highlights identifies key terms and concepts to help students study for the **Chapter Review.**

Each section and chapter end with **Review** questions that enable students to check their understanding and apply problem-solving skills.

Study Tip, included in the **Chapter Review,** helps students practice study and test-taking skills.

Standardized Test Prep questions help students review environmental science content in the context of a standardized test format.

✔ **STUDY TIP**

Taking Multiple-Choice Tests When you take multiple-choice tests, be sure to read all of the choices before you pick the correct answer. Be patient, and eliminate choices that are obviously incorrect.

A Teacher Edition that makes planning easy

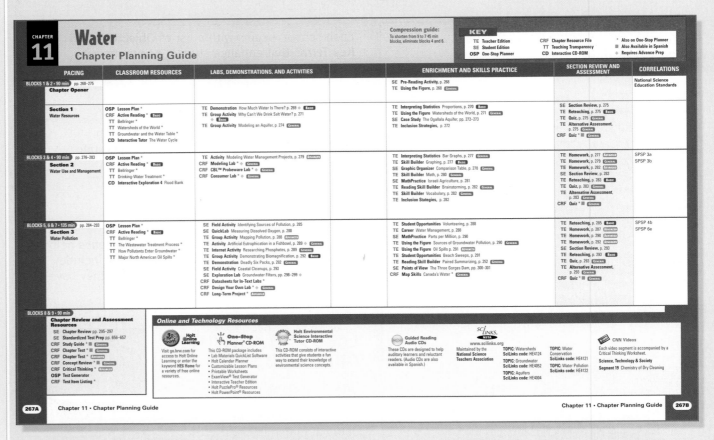

TEACHING RESOURCES DESIGNED FOR CONVENIENCE

The **Chapter Planning Guide** breaks each chapter down into flexible 45-minute blocks and offers a full listing of activities and classroom resources available for each section. Look for guidance on:
- Pacing
- Classroom Resources
- Labs, Demonstrations, and Activities
- Enrichment and Skills Practice
- Section Review and Assessment
- National Science Education Standards Correlations
- Online and Technology Resources

Chapter Enrichment, found after the **Chapter Planning Guide,** adds interesting material to extend the information in your presentation.

A **Lesson Cycle** builds structure around every lesson:

- **Focus** uses the objectives listed in the *Student Edition* to focus student attention on the upcoming content.

- **Motivate** uses demonstrations, discussions, and lively activities to get students excited about the material.

- **Teach** presents various teaching techniques including **Skill Builder, Group Activity,** and more.

- Finally, **Close** with **Reteaching,** a **Quiz,** and an **Alternative Assessment** to ensure students understand the information covered.

ACTIVITIES AND DEMONSTRATIONS FOR EVERY LEARNING LEVEL

Activities are leveled by ability level in the teacher's wrap—**Basic, General,** and **Advanced**—helping you choose the activities you think your students are ready for.

Learning styles are addressed throughout—**Interpersonal, Intrapersonal, Auditory, Kinesthetic, Logical, Visual,** and **Verbal**—so you can adapt material to different styles of learning.

- **Bellringer** activities begin each section with an activity designed to get students focused while you attend to administrative duties. Bellringer activities are also available on transparencies.

Bellringer

Tell students that human lungs have about 300 million tiny air pockets called *alveoli.* Altogether, the surface area in the lungs of humans is equivalent to the area of a tennis court. Ask students to think about why the surface area of human lungs is so large, and have them describe the effects of

- **Activity** sparks student interest and enhances understanding.

- **Group Activity** encourages collaborative effort.

- **Internet Activity** improves research skills using the Internet.

Internet Activity ———— GENERAL

Eratosthenes Experiment Have students look up a project called "Eratosthenes Experiment: A Worldwide Science and Math Experiment" on the Internet. This one-day project (which occurs twice a year during the equinox) allows students from around the globe to calculate Earth's circumference by measuring the shadow of the sun and entering the data into a simple equation. They can share data with students around the world online. **LS Logical**

CREATING RELEVANCE AND UNDERSTANDING

On almost every page you will find exciting features to help ignite class discussion and keep students thinking.

- Notable Quote
- Brain Food
- Real-Life Connections
- Student Opportunities
- Career
- Misconception Alert
- Skill Builder
- Connections
- Cultural Awareness

BRAIN FOOD

Earth's Crust The crust is thicker beneath the continents and thinner beneath the oceans. The crust is only 5 to 8 km thick beneath the oceans. The crust averages 30 km in thickness beneath the conti-

Career

Architect Architect Samuel Mockbee asked, "Does the architect have a role in addressing political or economic inequities, or transportation issues, or environmental issues? Because I think we do." In 1993, Mockbee and Dennis K. Ruth started a program called the "Rural Studio" at the University of Alabama. Students work with poor, rural residents of Hale County, Alabama, to improve their quality of life by designing affordable homes and other buildings using materials that would otherwise end up in landfills. They work with a community to address its needs from within. Mockbee decided to use his talent to combine a social commitment to the poor with an environmental focus. Have students find out more about the Rural Studio and the connections between architecture and environmental science. Students can also look for information about "Green Architecture" programs on the Internet.

REAL-LIFE ———— **CONNECTION** GENERAL

Where Does Your Water Come From? Have students find out where their tap water comes from. Ask them, "Does your community's water come mainly from surface water or groundwater? How many people can current water sources support?" If possible, plan a trip to your local water treatment plant or contact the plant administrators to see if they have a classroom resource kit.

MISCONCEPTION /// **ALERT** \\\

The Water Cycle Students may not know that plants and soil are also part of the water cycle. Plants cycle water through the water cycle as they draw liquid water from the ground and *transpire,* or release water vapor through leaf pores. You can demonstrate this for students by placing a seedling or small plant in a clear plastic cup and sealing the cup with plastic wrap. After a few hours in the sun, water from the soil and the plant will condense on the

READING SKILL BUILDER — BASIC

Reading Organizer Encourage students to make a graphic reading organizer of the problems associated with rapid population growth. Students should draw a concept map of the problems and issues based on the headings in this sec-

INCLUSION STRATEGIES MAKE MATERIAL ACCESSIBLE TO ALL

Written by a professional in the field of special needs education, **Inclusion Strategies** address many different learning exceptionalities in the classroom.

- Learning Disabled
- Developmentally Delayed
- Attention Deficit Disorder
- Gifted and Talented

INCLUSION Strategies

- *Developmentally Delayed*
- *Learning Disabled*
- *Attention Deficit Disorder*

Ask students to use **Figure 17** to make a colorful poster or bulletin board depicting

Assessment opportunities help you track students' progress

TEACHER EDITION SECTION ASSESSMENT

- **Reteaching** activities help students understand section material by presenting a concept in a different way. Use these features to customize your lesson to your student population.

- **Quiz** provides additional questions to assess student progress.

- **Alternative Assessment** provides another range of options for assessing student knowledge using a variety of modalities and methods.

Close

Reteaching — BASIC

Tectonic Processes Reinforce the idea of tectonic processes by having students build a small model of two tectonic plates with clay. Have them model divergent plates or convergent plates that result in either subduction (one plate being subsumed under the other) or uplift (mountain-building), and give an example of which plates are currently undergoing these processes. **LS Kinesthetic**

Alternative Assessment — GENERAL

The Big Event Have students write a short story in which they are scientists (either seismologists or vulcanologists). The *big event* is about to occur (either an earthquake or a volcanic eruption), and they are following its progress. Have students tell the story of what happened and how the phenomenon progressed, describing events in as much scientific detail as possible. **LS Logical** | **English Language Learners**

Quiz — GENERAL

1. What are the four components of the Earth system? (The geosphere, the hydropshere, the atmosphere, and the biosphere.)

2. What is the importance of the asthenosphere? (It is the upper layer of the mantle that is viscous enough to allow Earth's tectonic plates to move around on it.)

3. Where do earthquakes and volcanic eruptions most often occur? (They occur primarily at plate boundaries.)

STUDENT EDITION SECTION ASSESSMENT

Section Review provides a thorough review including questions that evaluate reading, writing, and critical-thinking skills.

SECTION 1 Review

1. **List** six forms of renewable energy, and compare the advantages and disadvantages of each.

2. **Describe** the differences between passive solar heating, active solar heating, and photovoltaic energy.

3. **Describe** how hydroelectric energy, geothermal energy, and geothermal heat pumps work.

4. **Explain** whether all renewable energy sources have their origin in energy from the sun.

CRITICAL THINKING

5. **Making Decisions** Which renewable energy source would be best suited to your region? Write a paragraph that explains your reasoning. **WRITING SKILLS**

6. **Identifying Trends** Identify a modern trend in hydroelectric power and in wind energy.

7. **Analyzing Relationships** Write an explanation of the differences in biomass fuel use between developed and developing countries. **WRITING SKILLS**

CHAPTER ASSESSMENT

Chapter Review presents a variety of question types such as **Key Terms, Short Answer, Concept Mapping, Critical Thinking, Interpreting Graphics, Math Skills,** and **Writing Skills** to check students' understanding. **Reading Skills** are also assessed.

Assignment Guide, found in the *Teacher Edition,* correlates **Chapter Review** questions to sections in the student textbook. You can easily customize end-of-chapter review questions to the sections you teach. This guide is a valuable resource for reteaching.

Assignment Guide

Section	Questions
1	1, 2, 6, 8, 11, 19, 26, 27, 31, 36
2	3, 4, 7, 12–16, 20, 21, 28, 34, 36
3	5, 9, 10, 17, 18, 22–25, 29, 30, 32, 33, 35, 36

General and **Advanced Chapter Tests,** found in the *Chapter Resource Files,* provide an objective-based assessment using various question formats: multiple choice, short answer, essay, and problems.

Two pages of *Standardized Test Prep* for every chapter are in the **Appendix.** Questions include multiple choice, short answer, and interpreting graphics.

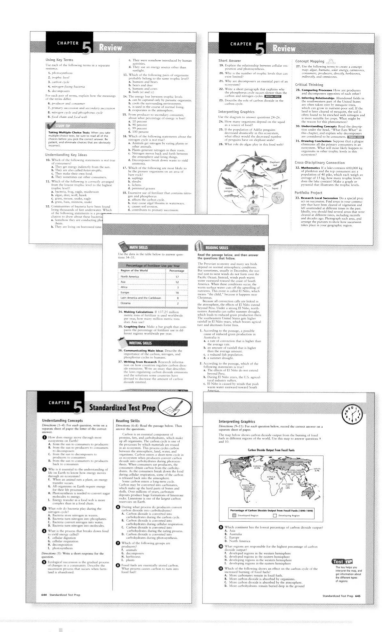

CUSTOM ASSESSMENT

One-Stop Planner® CD-ROM with ExamView® Test Generator helps you construct your own chapter tests from more than 1,000 items, correlated with text objectives. Questions can be edited and are coded by item type and level of comprehension.

Resources that make teaching easier

CHAPTER RESOURCE FILES

A **Chapter Resource File** accompanies each chapter of **Holt Environmental Science.** Everything you need to plan and manage your lessons in a convenient timesaving format is included in each chapter book. Also included is an introduction booklet, your guide to the resources in each **Chapter Resource File.** **Chapter Resource Files** include:

Skills Worksheets
- Concept Review
- Critical Thinking
- Active Reading
- Map Skills

Labs and Activities
- Datasheets for In-Text Labs
- CBL™ Probeware Labs
- Consumer Labs
- Design Your Own Labs
- Field Activities
- Long-Term Projects
- Research Activities
- Math/Graphing Labs
- Modeling Labs
- Observation Labs

Assessments
- Quizzes
- Chapter Tests

Answer Keys
- Answer Key for Skills Worksheets and Assessments
- Lab Notes and Answers

Teaching Transparency List

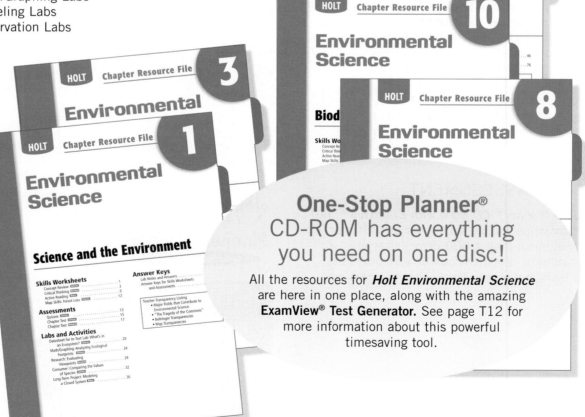

One-Stop Planner®
CD-ROM has everything you need on one disc!

All the resources for **Holt Environmental Science** are here in one place, along with the amazing **ExamView® Test Generator.** See page T12 for more information about this powerful timesaving tool.

ADDITIONAL RESOURCES

These additional resources are designed to help you reinforce and extend lessons.

- *Holt Science Skills Workshop: Reading in the Content Area* contains exercises that target key reading skills using excerpts from Holt's science textbooks.

- *Holt Science Laboratory Manager's Professional Reference* is a well-organized reference that was created to help teachers understand risk management and the hazards that can occur in the classroom.

- The *Lesson Plans* booklet provides pacing, objectives, and a suggested teaching outline with time requirements.

- *Active Reading Workbook* helps students practice essential skills for reading success as they analyze passages from the text.

- *Study Guide* contains review worksheets that reinforce the skills and concepts presented in the *Student Edition.*

- Over 100 full-color **Teaching Transparencies** use graphics directly from the text to enhance classroom presentations.

SPANISH RESOURCES BRING ENVIRONMENTAL SCIENCE TO ENGLISH-LANGUAGE LEARNERS!

- *Study Guide,* in Spanish, contains review worksheets that reinforce the skills and concepts presented in the *Student Edition.*

- *Assessments* in Spanish include: **Section Quizzes** and **General Chapter Tests.**

Technology that enhances teaching

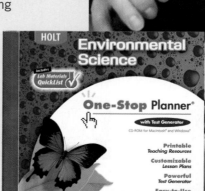

One-Stop Planner® CD-ROM
with Test Generator

Planning and managing lessons has never been easier than with this convenient, all-in-one CD-ROM. The One-Stop Planner® CD-ROM includes the following features:

PRINTABLE:
- **Teaching Resources**
- **Transparency Masters**

CUSTOMIZABLE:
- **Lesson Plans:** traditional and block-scheduling lesson plans in several word-processing formats.
- **Holt Calendar Planner®:** manage your time and resources by the day, week, month, and year.
- **Powerpoint® Resources:** Includes PowerPoint® presentations that can be customized with video resources, an image and activity bank, and standardized test preparation questions.

POWERFUL:
- **ExamView® Test Generator:** test items organized by chapter, plus hundreds of editable questions, so you can put together your own tests and quizzes. Plus you can post test questions to Holt Online Assessment for automatic grading.
- **Holt PuzzlePro®:** an easy way to create crossword puzzles and word searches that make learning vocabulary more fun.
- **Interactive Teacher Edition:** the entire teacher text, with links connecting you to the related Teaching Resources.
- **Lab Materials QuickList Software:** helps you create a customized list of the lab materials you need.

HOLT
Environmental Science
Includes Lab Materials QuickList ✓

One-Stop Planner®
with Test Generator
CD-ROM for Macintosh® and Windows®

Printable
Teaching Resources

Customizable
Lesson Plans

Powerful
Test Generator

Easy-to-Use
Lab Materials QuickList Software

GUIDED READING AUDIO CD PROGRAM

This audio program provides a direct read of **Holt Environmental Science,** making content more accessible for auditory learners and reluctant readers.

INTERACTIVE EXPLORATIONS CD-ROM

This CD-ROM lets students act as lab assistants in solving real-world environmental problems while developing inquiry, analysis, and decision-making skills.

INTERACTIVE TUTOR CD-ROM

This tutor allows students to explore environmental science concepts at their own pace through background information and 19 activities.

STUDENT EDITION, CD-ROM VERSION

Ideal for students with limited access to the Internet, but who need to lighten the load of textbooks they carry home, the entire *Student Edition* is on one easy-to-navigate CD-ROM, page-for-page.

CNN PRESENTS SCIENCE IN THE NEWS VIDEOS

Science, Technology & Society and **Earth Science Connections** videos include broadcast news segments that bring the relevance of science directly into your classroom. Each news segment showcases useful applications of science concepts, often making cross-curricular connections to industry, careers, and a variety of other areas.

NOVA VIDEOS

Engaging Nova Videos, **What's Up With the Weather** and **Fire Wars,** make environmental science even more relevant to students.

Online Resources are available anytime, anywhere!

THE *PREMIER ONLINE EDITION* IS PORTABLE, EXPANDABLE, INTERACTIVE, AND YET WEIGHS NOTHING AT ALL

The *Premier Online Edition* of **Holt Environmental Science** engages students in ways that were never before possible. And because it's all online, it's available anywhere you or your students connect to the Internet. You'll find:

- Interactive quizzes and activities
- Holt Online Assessments
- And much more!

Try it for yourself at www.hrw.com/online.

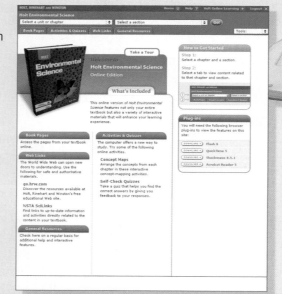

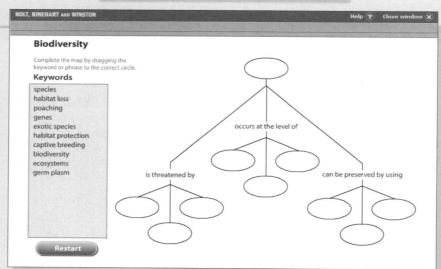

This Web service, developed and maintained by the National Science Teachers Association, contains a collection of prescreened links with current information and activities directly related to chapter topics.

- Saves you valuable time searching for relevant and up-to-date Web sites

- Sites reviewed by science context experts and educators

- **Internet Connect** boxes within each chapter offer opportunities to enrich, enhance, and extend learning

Holt's own award-winning Web site includes a variety of resources and reference materials that go along with the textbook as well as links to state resources.

A variety of labs for every purpose

Holt Environmental Science includes lab activities that meet the demands of your curriculum. This flexible laboratory program builds inquiry and critical-thinking skills.

All labs are **tested by teachers,** and **Lab Ratings** help you select the labs that suit your students' abilities.

Jennifer Seelig-Fritz
North Springs High School
Atlanta, Georgia

Lab Ratings

EASY ————————→ HARD

TEACHER PREPARATION
STUDENT SETUP
CONCEPT LEVEL
CLEANUP

BRIEF LABS

QuickLAB is an easy activity that can be completed in less than one class period.

Field Activity gives students an opportunity to observe and apply a subject in a real-world setting.

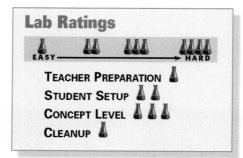

QuickLAB

Sponging It Up

Procedure

1. Completely saturate **two small sponges** with **water** and allow the excess water to drain off.

2. Measure each sponge's mass by using an **electric balance**. Record the mass.

3. Using **plastic wrap,** completely

CHAPTER 18 — **Inquiry Lab: MODELING**

Objectives

► **USING SCIENTIFIC METHODS** **Prepare** a detailed sketch of your solution to the design problem.

► **Design and build** a functional windmill that lifts a specific weight as quickly as possible.

Materials

blow-dryer, 1,500 W
dowel or smooth rod
foam board
glue, white
paper clips, large (30)
paper cup, small (1)
spools of thread, empty (2)
string, 50 cm

optional materials for windmill blades: foam board, paper plates, paper cups, or any other lightweight materials

Blowing in the Wind

MEMO

To: Division of Research and Developers

Quixote Alternative Energy Systems is accepting design proposals to develop a windmill that can be used to lift window washers to the tops of buildings. As part of the design engineering team, your division has been asked to develop a working midel of such a windmill. Your task is to design and build a prototype of a windmill that can capture energy from a 1,500 W blow-dryer. Your model must lift 30 large paper clips a vertical distance of 50 cm (approximately 2 ft) as quickly as possible.

Procedure

1. Build the base for your windmill (shown below). Begin by attaching the two spools to the foam board using the glue. Make sure the spools are parallel before you glue them.

2. Pass a dowel or a smooth rod through the center of the spools. The dowel should rotate freely. Attach one end of the string securely to the dowel between the two spools.

3. Poke a hole through the middle of the foam board to allow the string to pass through.

4. Attach the cup to the end of the string. You will use the cup to lift the paper clips.

5. Place your windmill base between two lab tables or in any other area that will allow the string to hang freely.

► **Windmill Base** Your windmill base should allow the dowel to spin as freely as possible. The pinwheel shown at the end of the dowel is a suggested design for your windmill blades.

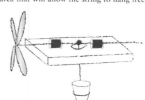

FIELD ACTIVITY

Is It Really Recyclable?

Conduct a survey of the plastic containers in your household that recyclable. Note the number plastic containers found in your sehold. Now look at the num- printed on the bottom of container. The plastics indus- has established a system of gnating which plastics are cyclable. Types 1 and 2 are st commonly recycled by most munities. Type 4 is less com- nly recycled, and types 3, 5, 6, 7 are most likely not to be

6. Prepare a sketch of your prototype windmill blades based on the objectives for this lab. Include a list of the materials that you will use and safety precautions (if necessary).

7. Have your teacher approve your design before you begin construction.

8. Construct a working prototype of your windmill blades. Test your model several times to collect data on the speed at which it lifts the paper clips. Record your data for each trial.

9. Vary the type of material used for construction of your windmill blades. Test the various blades to determine whether they improve the original plan.

10. Vary the number and size of the blades on your windmill. Test each design to determine whether the change improves the original plan.

Analysis

1. **Summarize Results** Create a data table that lists the speed for each lift for several trials. Include an average speed.

2. **Graphing Data** Prepare a bar graph that shows your results for each blade design.

Conclusions

3. **Evaluating Methods** After you observe all of the designs, decide which ones you think best solve the problem and explain why.

4. **Evaluating Models** Which change improved your windmill the most—varying the materials for the blades, varying the number of blades, or varying the size of the blades? Would you change your design further? If so, how?

Extension

1. **Research** Windmills have been used for more than 2,000 years. Research the three basic types of vertical-axis machines and the applications in which they are used. Prepare a report of your findings.

2. **Making Models** Adapt your design to make a water wheel. You'll find that water wheels can pull much more weight than a windmill can. Find designs on the Internet for micro-hydropower water wheels such as the Pelton wheel, and use the designs as inspiration for your models. You can even design your own dam and reservoir.

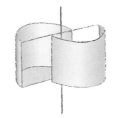

▶ **Sample Windmill Blade Designs**

EXTENDED LABS

More extensive investigation is offered with these two-page labs:

- **Skills Practice Lab** gives students a hands-on experience by guiding them through each procedural step.

- **Exploration Lab** engages students to explore a situation or phenomenon to improve their understanding.

- **Inquiry Lab** allows students to develop and perform their own procedure, often using a real-life example.

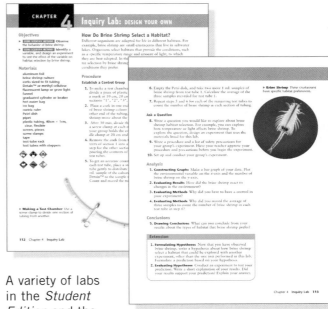

A variety of labs in the *Student Edition* and the *Chapter Resource Files* provide you with maximum flexibility in planning a complete and meaningful laboratory experience for all your students.

- CBL™ Probeware Labs
- Consumer Labs
- Field Activities
- Modeling Labs
- Math/Graphing Labs
- Observation Labs
- Research Labs
- Long-Term Projects

TEACHER EDITION LAB ACTIVITIES

Additional activities found in the *Teacher Edition* also help illustrate science concepts. Examples include **Demonstration, Activity,** and **Group Activity.**

Motivate

Demonstration — BASIC

How Much Water Is There?
Begin with 4 L of water. Add a few drops of food coloring so th students can see the water volum easily. Explain that this volume

Group Activity — GENERAL

Teaching To the Class Have students create teaching materials on probability, statistics, and risk for other students. Split the class into three groups. One group should devise a lesson on probability, another on statistics, and a third on risk. Ask students to present the

Meeting Individual Needs

Students have a wide range of abilities and learning exceptionalities. These pages show you how *Holt Environmental Science* provides resources and strategies to help you tailor your instruction to engage every student in your classroom.

Learning exceptionality	Resources and strategies	
Learning Disabled and Slow Learners Students who have dyslexia or dysgraphia, students reading below grade level, students having difficulty understanding abstract or complex concepts, and slow learners	• Inclusion Strategies labeled *Learning Disabled* • Activities labeled *Basic* • *Reteaching* activities	• Activities labeled *Visual, Kinesthetic,* or *Auditory* • Hands-on activities or projects • Oral presentations instead of written tests or assignments
Developmental Delays Students who are functioning far below grade level because of mental retardation, autism, or brain injury; goals are to learn or retain basic concepts	• Inclusion Strategies labeled *Developmentally Delayed* • Activities labeled *Basic*	• *Reteaching* activities • Project-based activities
Attention Deficit Disorders Students experiencing difficulty completing a task that has multiple steps, difficulty handling long assignments, or difficulty concentrating without sensory input from physical activity	• Inclusion Strategies labeled *Attention Deficit Disorder* • Activities labeled *Basic* • *Reteaching* activities • Activities labeled *Co-op Learning*	• Activities labeled *Visual, Kinesthetic,* or *Auditory* • Concepts broken into small chunks • Oral presentations instead of written tests or assignments
English as a Second Language Students learning English	• Activities labeled *English Language Learners* • Activities labeled *Basic*	• *Reteaching* activities • Activities labeled *Visual*
Gifted and Talented Students who are performing above grade level and demonstrate aptitude in crosscurricular assignments	• Inclusion Strategies labeled *Gifted and Talented* • Activities labeled *Advanced* • *Connection* activities	• Activities that involve multiple tasks, a strong degree of independence, and student initiative

General Strategies The following strategies can help you modify instruction to help students who struggle with common classroom difficulties.

A student experiencing difficulty with...	May benefit if you . . .	
Beginning assignments	• Assign work in small amounts • Have the student use cooperative or paired learning • Provide varied and interesting activities	• Allow choice in assignments or projects • Reinforce participation • Seat the student closer to you
Following directions	• Gain the student's attention before giving directions • Break up the task into small steps • Give written directions rather than oral directions • Use short, simple phrases • Stand near the student when you are giving directions	• Have the student repeat directions to you • Prepare the student for changes in activity • Give visual cues by posting general routines • Reinforce improvement in or approximation of following directions
Keeping track of assignments	• Have the student use folders for assignments • Have the student use assignment notebooks	• Have the student keep a checklist of assignments and highlight assignments when they are turned in
Reading the textbook	• Provide outlines of the textbook content • Reduce the length of required reading • Allow extra time for reading • Have the students read aloud in small groups	• Have the student use peer or mentor readers • Have the student use books on tape or CD • Discuss the content of the textbook in class after reading
Staying on task	• Reduce distracting elements in the classroom • Provide a task-completion checklist • Seat the student near you	• Provide alternative ways to complete assignments, such as oral projects taped with a buddy
Behavioral or social skills	• Model the appropriate behaviors • Establish class rules, and reiterate them often • Reinforce positive behavior • Assign a mentor as a positive role model to the student • Contract with the student for expected behaviors • Reinforce the desired behaviors or any steps toward improvement	• Separate the student from any peer who stimulates the inappropriate behavior • Provide a "cooling off" period before talking with the student • Address academic/instructional problems that may contribute to disruptive behaviors • Include parents in the problem-solving process through conferences, home visits, and frequent communication
Attendance	• Recognize and reinforce attendance by giving incentives or verbal praise • Emphasize the importance of attendance by letting the student know that he or she was missed when he or she was absent	• Encourage the student's desire to be in school by planning activities that are likely to be enjoyable, giving the student a preferred responsibility to be performed in class, and involving the student in extracurricular activities • Schedule problem-solving meeting with parents, faculty, or both
Test-taking skills	• Prepare the student for testing by teaching ways to study in pairs, such as using flashcards, practice tests, and study guides, and by promoting adequate sleep, nourishment, and exercise • During testing, allow the student to respond orally on tape or to respond using a computer; to use	notes; to take breaks; to take the test in another location; to work without time constraints; or to take the test in several short sessions • Decrease visual distraction by improving the visual design of the test through use of larger type, spacing, consistent layout, and shorter sentences

Build critical reading skills

FEATURES HELP STUDENTS UNDERSTAND WHAT THEY READ

SECTION 1
Energy Flow in Ecosystems

Just as a car cannot run without fuel, an organism cannot survive without a constant supply of energy. Where does an organism's energy come from? The answer to that question depends on the organism, but the ultimate source of energy for almost all organisms is the sun.

Life Depends on the Sun

Energy from the sun enters an ecosystem when a plant uses sunlight to make sugar molecules by a process called **photosynthesis.** During photosynthesis, plants, algae, and some bacteria capture solar energy. Solar energy drives a series of chemical reactions that require carbon dioxide and water, as shown in **Figure 1.** The result of photosynthesis is the production of sugar molecules known as *carbohydrates.* Carbohydrates are energy-rich molecules which organisms use to carry out daily activities. As organisms consume food and use energy from carbohydrates, the energy travels from one organism to another. Plants, such as the sunflowers in **Figure 2,** produce carbohydrates in their leaves. When an animal eats a plant, some energy is transferred from the plant to the animal. Organisms use this energy to move, grow, and reproduce.

Objectives
- ▶ Describe how energy is transferred from the sun to producers and then to consumers.
- ▶ Describe one way in which consumers depend on producers.
- ▶ List two types of consumers.
- ▶ Explain how energy transfer in a food web is more complex than energy transfer in a food chain.
- ▶ Explain why an energy pyramid is a representation of trophic levels.

Key Terms
photosynthesis
producer
consumer
decomposer
cellular respiration
food chain
food web
trophic level

Each section begins with a series of **Objectives** and a list of **Key Terms** that help focus student attention on the material they are about to read.

Vocabulary words are highlighted in the text. A glossary at the back of the textbook includes all key terms.

Reading Skill Builder in the *Teacher Edition* helps students acquire those important skills that build confidence in reading.

READING SKILL BUILDER — BASIC

Reading Organizer Encourage students to make a graphic reading organizer of the problems associated with rapid population growth. Students should draw a concept map of the problems and issues based on the headings in this section. After they read the section, encourage students to look over their concep...

Reading Warm-Up and **Reading Follow-Up** check to see how students' understanding has improved.

Critical Thinking

24. Making Comparisons Read the description of energy efficiency and energy conservation in this chapter. How are the two concepts related? Give several examples. **READING SKILLS**

25. Analyzing Ideas Does the energy used by fuel cells come from the sun? Explain your answer.

26. Analyzing Ideas Explain whether you think the most important advances of the 21st century will be new sources of energy or more...

PRE-READING ACTIVITY

FOLDNOTES — **Double-Door Fold**
Before you read this chapter, create the **FoldNote** entitled "Double-Door Fold" described in the Reading and Study Skills Handbook section of the Appendix. Write "Energy flow in ecosystems"...

Reading Skills presents reading passages for student analysis, followed by questions that develop test-taking and critical-thinking skills.

READING SKILLS

Read the passage below, and the questions that follow.

The Peruvian economy and ma depend on normal atmospheric But sometimes, usually in Dece mal east-to-west winds do not Pacific Ocean. Instead, winds p water eastward toward the coa America. When these condition warm surface water cuts off the nutrients. This event is called E means "the child," because it h Ch...mas.

Pre-Reading Activity provides **FoldNotes** activities to help students organize information presented in the chapter. Students are encouraged to take notes and then categorize what they read. In addition, the **Appendix** provides complete instruction on how to create and use the reading strategies suggested in pre-reading activities.

ADDITIONAL RESOURCES ALSO HELP IN READING COMPREHENSION

Guided Reading Audio CD Program

Auditory learners, reluctant readers, and English-language learners will all find critical reading support with these CDs.

Holt Science Skills: Reading in the Content Area

This book targets key reading skills using excerpts from Holt's science textbooks. Students learn various strategies that help them improve their reading skills.

Activities are flexible enough to be completed by students individually, or you can use the overhead transparencies provided in the *Teacher Edition* to teach analysis skills to a large group of students.

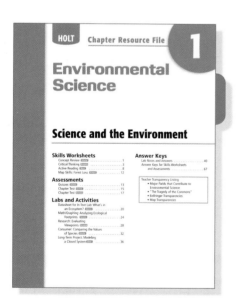

Concept Review

From straight recall to higher-order thinking, **Concept Review** in the *Chapter Resource File* helps reinforce what students have learned in each section.

Active Reading Workbook

This workbook helps students practice essential skills for reading success as they analyze passages from the text.

Writing skills help students succeed

WRITING FEATURES GET STUDENTS THINKING ABOUT CONTENT

Writing Skills questions are presented in each **Chapter Review** to help students develop a variety of written communication skills.

In addition to specific writing skills questions, opportunities to practice and build students' writing abilities are found in each **Section** and **Chapter Review**.

SECTION 3 Review

1. **Describe** how animals and angiosperms depend on each other. Write a short paragraph to explain your answer. **WRITING SKILLS**

2. **Describe** the importance of protists in the ocean.

3. **Name** the six kingdoms of life, and give two characteristics of each.

4. **Explain** the importance of bacteria and fungi in the environment.

CRITICAL THINKING

5. **Analyzing Relationships** Explain how the large number and wide distribution of angiosperm species is related to the success of insects.

6. **Understanding Concepts** Write a short paragraph that compares the reproductive structures of gymnosperms and angiosperms. **WRITING SKILLS**

WRITING SKILLS

32. **Communicating Main Ideas** Explain why scientists are working to reduce the use of the two main sources of energy people use today—fossil fuels and biomass.

33. **Writing Persuasively** Write a guide that encourages people to conserve energy and

---

CASE STUDY — Paper or Plastic?

The following question may sound familiar: Do you want paper or plastic? If you have ever stood in the checkout line of a grocery store, it probably is. Almost every grocery store today offers a choice between either paper or plastic bags for sacking grocery items. Many people make their choice based on convenience. But what is the best choice for someone who is concerned about the environment?

On the surface, it may seem that paper is the better choice. Paper comes from a renewable resource—trees—and is biodegradable. Plastic, on the other hand, comes from petroleum or natural gas, which are usually considered nonrenewable resources. In addition, plastic bags are not biodegradable.

Upon closer examination, however, the decision may not be as simple as it seems. Removing large numbers of trees from forests to manufacture paper can disrupt woodland ecosystems. Plus, a tremendous amount of energy is required to convert trees into pulp and then manufacture paper from the pulp.

To make the best decision about which product is better for the environment, the following questions should be considered.

- How much raw material, energy, and water is needed to manufacture each bag?
- What waste products will result from the manufacture of each bag, and what effect will those wastes have on water, the atmosphere, and land?

▶ Making an educated decision at the grocery store will help reduce solid waste.

- Can recycled materials be used in the manufacture of the bag? If so, to what degree will the use of recycled materials reduce the amount of raw materials, energy,

and water used and wastes produced in making the bag?
- How will the bag decompose, and what will the environmental impact be if it is incorrectly disposed of?

Although several studies have analyzed these questions, most have been conducted by parties with a vested interest, such as plastic or paper manufacturing companies. As you might expect, the studies done by plastic manufacturers conclude that plastic bags have the least environmental impact, while studies done by paper producers conclude that paper bags have the least environmental impact. Often, these researchers fail to study all of the important factors listed above.

But the plastic versus paper debate has caused both industries to improve the way their products affect the environment. For example, paper bags recently outsold plastic bags because they were considered

▶ A reusable canvas shopping bag may be the best response to the paper-or-plastic question.

stronger, better for reusing or recycling, and less harmful in a landfill.

Then, new technology allowed the plastics industry to gain a larger market share. By incorporating recycled plastic into the bags, manufacturers improved the image of plastic bags.

Therefore, the debate continues and environmentally conscious people are still wondering which is better. Right now there seems to be no right answer. However, the following are environmentally sound options.

- Carry your groceries in bags brought from home (paper, plastic, or canvas bags).
- Choose the bag you are most likely to reuse in the future.
- If you have only one or two small items do not use a bag.

CRITICAL THINKING

1. **Identifying Relationships** Explain how environmentally conscious shoppers have helped improve paper and plastic bag manufacturing in this country.

2. **Understanding Concepts** Why should a person care which bag he or she is given at the grocery store?

Case Study and end-of-chapter features such as **Society and the Environment** help increase reading comprehension and enhance writing skills.

Skill Builder for writing, found in the *Teacher Edition*, provides strategies to improve students' writing skills.

SKILL BUILDER — BASIC

Writing Have students find something of interest to them in the classroom, such as a model, a fish tank, or a globe, and write down as many observations about the object as they can in 5 minutes. Then have students switch objects with a partner and repeat the exercise. Have them compare observations and write a short summary of what they learned from this experience in their *EcoLog*.

LS Intrapersonal

English Language Learners

Support to strengthen math skills

MATH IS INTEGRATED THROUGHOUT THE STUDENT TEXT

MathPractice is found throughout the text and enhances the development of math skills. There is at least one per chapter.

Many of the illustrations in *Holt Environmental Science* are graphic representations of mathematical data. Students are often asked to interpret data by answering related questions.

MATH SKILLS

Use the table below to answer questions 33–35.

Paper Products in Municipal Solid Waste		
Product	Generation (tons)	Percentage recycled
Newspapers	13,620	56.4
Books	1,140	14.0
Magazines	2,260	20.8
Office papers	7,040	50.4

33. Evaluating Data How many tons of paper products were generated according to the table?

34. Making Calculations How many tons of newspapers were recycled? How many tons of newspapers were not recycled?

MATHPRACTICE

A Meal Fit for a Grizzly Bear Grizzly bears are omnivores that can eat up to 15 percent of their body weight per day when eating salmon and up to 33 percent of their body weight when eating fruits and other vegetation. How many pounds of salmon can a 200 lb grizzly bear eat in one day? How many pounds of fruits and other vegetation can the same bear eat in one day?

Chapter Review presents even more opportunities for students to practice their math skills by interpreting graphs and by answering specific questions in the **Math Skills** section.

ADDITIONAL MATH PROBLEMS AND HINTS ARE INCLUDED IN THE TEACHER EDITION

- **Skill Builder** for math includes activities that help improve math skills.

- **Interpreting Statistics** provides additional math practice.

SKILL BUILDER ——— GENERAL

Math According to the World Wildlife Fund, rain forests are being cleared at a rate of 26 hectares per minute. Ask students to calculate how many hectares of rainforest are being cleared in an hour, a day, and a year. (26 ha per min × 60 min per hr = 1560 ha per hr; 1560 ha per hr × 24 hr per day = 37,440 ha per day; 37,440 ha per day × 365 days per year = 13,665,600 ha per year) To help students understand the size of the area, ask them to find out the size, in hectares, of their county

Teach

Interpreting Statistics ——— GENERAL

Bar Graphs The bar graph in **Figure 5** shows how water is used in different regions of the world. ...s may think that Australia ...a use a greater volume of ...or agriculture than other ... of the world. Point out ...e graph shows percentages ...r used, not the amounts of ...sed. For each region, the ...ars do not equal 100 per-...cause of water lost to evap-... Have students explain ...

Pacing Guide

Holt Environmental Science is a flexible textbook that can be used in either a full year or one semester course. The textbook can also be adapted to match student abilities or to emphasize one area of study, such as Earth science or ecology. The following tables show how 45-minute instructional blocks can be organized to match the needs of your class.

1/2 Year Course (90 Blocks)	Basic (13 Chapters)	General (15 Chapters)	General Earth Science (14 Chapters)	General Ecology (13 Chapters)	Advanced (16 Chapters)
Chapter 1 *Science and the Environment*	7	7	6	6	4
Section 1	3	3	3	3	2
Section 2	3	3	2	2	1
Exploration Lab	1	1	1	1	1
Chapter 2 *Tools of Environmental Science*	9	9	7	7	6
Section 1	3	3	2	2	2
Section 2	3	3	2	2	2
Section 3	2	2	2	2	1
Skills Practice Lab	1	1	1	1	1
Chapter 3 *The Dynamic Earth*	–	–	11	–	–
Section 1	–	–	3	–	–
Section 2	–	–	3	–	–
Section 3	–	–	3	–	–
Inquiry Lab	–	–	2	–	–
Chapter 4 *The Organization of Life*	9	9	4	11	9
Section 1	3	3	1.5	3	3
Section 2	2	2	1	3	2
Section 3	2	2	1.5	3	2
Inquiry Lab	2	2	–	2	2
Chapter 5 *How Ecosystems Work*	8	6	–	9	7
Section 1	3	2	–	3	3
Section 2	2	2	–	3	2
Section 3	2	1	–	2	1
Exploration Lab	1	1	–	1	1
Chapter 6 *Biomes*	7	5	–	9	6
Section 1	2.5	1.5	–	3	3
Section 2	2	1.5	–	3	2
Section 3	1.5	1	–	2	1
Exploration Lab	1	1	–	1	1
Chapter 7 *Aquatic Ecosystems*	–	–	3	5	–
Section 1	–	–	1	2	–
Section 2	–	–	1.5	2	–
Skills Practice Lab	–	–	0.5	1	–
Chapter 8 *Understanding Populations*	7	7	–	8	5
Section 1	3	3	–	3	1.5
Section 2	2	2	–	2	1.5
Skills Practice Lab	2	2	–	2	2
Chapter 9 *The Human Population*	–	–	–	4	–
Section 1	–	–	–	2	–
Section 2	–	–	–	2	–
Skills Practice Lab	–	–	–	–	–
Chapter 10 *Biodiversity*	5	6	–	8	6
Section 1	1	1	–	2	1
Section 2	2	2	–	3	2
Section 3	2	1	–	3	1
Exploration Lab	–	2	–	–	2
Chapter 11 *Water*	6	7	7	–	6
Section 1	1	1	1	–	1
Section 2	2	2	2	–	2
Section 3	2	3	3	–	2
Exploration Lab	1	1	1	–	1

	Basic	General	General Earth Science	General Ecology	Advanced
	(13 Chapters)	(15 Chapters)	(14 Chapters)	(13 Chapters)	(16 Chapters)
Chapter 12 *Air*	–	–	7	–	–
Section 1	–	–	2	–	–
Section 2	–	–	2	–	–
Section 3	–	–	2	–	–
Inquiry Lab	–	–	1	–	–
Chapter 13 *Atmosphere and Climate Change*	8	5	8	5	6
Section 1	3	1.5	2	1	2
Section 2	2	1.5	2	2	2
Section 3	2	1	3	2	1
Exploration Lab	1	1	1	–	1
Chapter 14 *Land*	–	5	6	5	6
Section 1	–	1	1	1	1
Section 2	–	1.5	2	1.5	2
Section 3	–	1.5	2	2	2
Inquiry Lab	–	1	1	0.5	1
Chapter 15 *Food and Agriculture*	–	–	–	–	–
Section 1	–	–	–	–	–
Section 2	–	–	–	–	–
Section 3	–	–	–	–	–
Inquiry Lab	–	–	–	–	–
Chapters 16 *Mining and Mineral Resources*	–	–	9	–	5
Section 1	–	–	3	–	1
Section 2	–	–	3	–	2
Section 3	–	–	2	–	1
Skills Practice Lab	–	–	1	–	1
Chapter 17 *Nonrenewable Energy*	6	4	5	–	5
Section 1	3	2	3	–	2
Section 2	2	1	3	–	1
Skills Practice Lab	1	1	1	–	1
Chapter 18 *Renewable Energy*	6	5	5	–	5
Section 1	2	1.5	2	–	2
Section 2	2	1.5	1	–	1
Inquiry Lab	2	2	2	–	2
Chapter 19 *Waste*	6	5	6	–	5
Section 1	1.5	1.5	2	–	1
Section 2	2	1.5	2	–	2
Section 3	1.5	1	1	–	1
Skills Practice Lab	1	1	1	–	1
Chapter 20 *The Environment and Human Health*	–	4	–	7	4
Section 1	–	1.5	–	3	1.5
Section 2	–	1.5	–	3	1.5
Skills Practice Lab	–	1	–	1	1
Chapter 21 *Economics, Policy, and the Future*	6	6	6	6	6
Section 1	1.5	1.5	1.5	1.5	1
Section 2	1.5	1.5	1.5	1.5	2
Section 3	1	1	1	1	1
Inquiry Lab	2	2	2	2	2

T25

Pacing Guide

1 Year Course (180 Blocks)

	Basic (16 Chapters)	General (19 Chapters)	General Earth Science (17 Chapters)	General Ecology (17 Chapters)	Advanced (21 Chapters)
Chapter 1 Science and the Environment	9	8	7	7	7
Section 1	4	3.5	3	3	3
Section 2	4	3	2.5	2.5	2.5
Exploration Lab	1	1	1	1	1
Making a Difference	–	0.5	0.5	0.5	0.5
Chapter 2 Tools of Environmental Science	13	10	10	10	10
Section 1	4	3	3	3	3
Section 2	4	3	3	3	3
Section 3	3.5	2	2	2	2
Skills Practice Lab	1	1	1	1	1
Maps in Action	0.5	0.5	0.5	0.5	0.5
Science and Technology	–	0.5	0.5	0.5	0.5
Chapter 3 The Dynamic Earth	–	12	17	–	9
Section 1	–	3	5	–	2
Section 2	–	3	5	–	2
Section 3	–	3	4	–	2
Exploration Lab	–	2	2	–	2
Maps in Action	–	0.5	0.5	–	0.5
Science and Technology	–	0.5	0.5	–	0.5
Chapter 4 The Organization of Life	14	11	8	17	10
Section 1	4	3	2	5	2
Section 2	4	2.5	1.5	4.5	2.5
Section 3	4	3	2	5	3
Inquiry Lab	2	2	2	2	2
Making a Difference	–	0.5	0.5	0.5	0.5
Chapter 5 How Ecosystems Work	14	10	9	17	10
Section 1	4.5	3	2	5	3
Section 2	4	3	2	5	3
Section 3	4	2	3	5	2
Exploration Lab	1	1	1	1	1
Maps in Action	0.5	0.5	0.5	0.5	0.5
Society and the Environment	–	0.5	0.5	0.5	0.5
Chapter 6 Biomes	12	10	–	13	9.5
Section 1	4	3	–	4	3
Section 2	4	3	–	4	3
Section 3	3	2.5	–	3.5	2
Exploration Lab	1	1	–	1	1
Points of View	–	0.5	–	0.5	0.5
Chapter 7 Aquatic Ecosystems	9	7	7	10	6
Section 1	4	3	2	4	2
Section 2	3.5	2	3	4	2
Skills Practice Lab	1	1	1	1	1
Maps in Action	0.5	0.5	0.5	0.5	0.5
Science and Technology	–	0.5	0.5	0.5	0.5
Chapter 8 Understanding Populations	9	7	–	11	7
Section 1	4	3	–	4	2
Section 2	3	2	–	4.5	3
Skills Practice Lab	2	1.5	–	2	1.5
Points of View	–	0.5	–	0.5	0.5
Chapter 9 The Human Population	–	–	–	11	9
Section 1	–	–	–	4	3
Section 2	–	–	–	4	3
Skills Practice Lab	–	–	–	2	2
Maps in Action	–	–	–	0.5	0.5
Society and the Environment	–	–	–	0.5	0.5
Chapter 10 Biodiversity	11	11	–	14	9.5
Section 1	2	2	–	3	2
Section 2	4	4	–	5	2
Section 3	3	3	–	4	3
Exploration Lab	2	1.5	–	1.5	2
Making a Difference	–	0.5	–	0.5	0.5
Chapter 11 Water	12	9	12	7	9
Section 1	3	1.5	2.5	1.5	1.5
Section 2	4	3	4	2	3
Section 3	4	3	4	2	3
Exploration Lab	1	1	1	1	1
Making a Difference	–	0.5	0.5	0.5	0.5

	Basic (16 Chapters)	General (19 Chapters)	General Earth Science (17 Chapters)	General Ecology (17 Chapters)	Advanced (21 Chapters)
Chapter 12 _Air_	11	11	14	–	9
Section 1	3.5	3	4	–	2
Section 2	3	3	4	–	3
Section 3	3	3	4	–	2
Skills Practice Lab	1	1	1	–	1
Maps in Action	0.5	0.5	0.5	–	0.5
Society and the Environment	–	0.5	0.5	–	0.5
Chapter 13 _Atmosphere and Climate Change_	12	10	13	10	10
Section 1	4	3	4	3	3
Section 2	3	2.5	3.5	2.5	2.5
Section 3	4	3	4	3	3
Exploration Lab	1	1	1	1	1
Making a Difference	–	0.5	0.5	0.5	0.5
Chapter 14 _Land_	–	10	12	11	8
Section 1	–	2.5	2.5	2.5	1
Section 2	–	3	4	3	2.5
Section 3	–	3	4	4	3
Inquiry Lab	–	1	1	1	1
Making a Difference	–	0.5	0.5	0.5	0.5
Chapter 15 _Food and Agriculture_	–	–	12	–	9.5
Section 1	–	–	4	–	2
Section 2	–	–	3.5	–	3
Section 3	–	–	3	–	3
Inquiry Lab	–	–	1	–	1
Points of View	–	–	0.5	–	0.5
Chapter 16 _Mining and Mineral Resources_	12	9	15	–	9
Section 1	3.5	2	5	–	2
Section 2	4	3	4	–	3
Section 3	3	2	4	–	2
Skills Practice Lab	1	1	1	–	1
Maps in Action	0.5	0.5	0.5	–	0.5
Society and the Environment	–	0.5	0.5	–	0.5
Chapter 17 _Nonrenewable Energy_	9	7	9	7	7
Section 1	4.5	3	4	3	3
Section 2	3	2	3	2	2
Skills Practice Lab	1	1	1	1	1
Maps in Action	0.5	0.5	0.5	0.5	0.5
Science and Technology	–	0.5	0.5	0.5	0.5
Chapter 18 _Renewable Energy_	10	9	10	8	8
Section 1	4.5	3	4	3	3
Section 2	3	3	3	2	2
Inquiry Lab	2	2	2	2	2
Maps in Action	0.5	0.5	0.5	0.5	0.5
Science and Technology	–	0.5	0.5	0.5	0.5
Chapter 19 _Waste_	12	10	11	9	7.5
Section 1	4	3	3	2.5	2
Section 2	4	3	4	3	2
Section 3	3	2.5	2.5	2	2
Skills Practice Lab	1	1	1	1	1
Points of View	–	0.5	0.5	0.5	0.5
Chapter 20 _The Environment and Human Health_	–	8	6	10	8
Section 1	–	3	2	4	3
Section 2	–	3	2	4	3
Skills Practice Lab	–	1	1	1	1
Maps in Action	–	0.5	0.5	0.5	0.5
Society and the Environment	–	0.5	0.5	0.5	0.5
Chapter 21 _Economics, Policy, and the Future_	11	11	8	8	8
Section 1	3	3	2	2	2
Section 2	3	3	2	2	2
Section 3	3	2.5	1.5	1.5	1.5
Inquiry Lab	2	2	2	2	2
Making a Difference	–	0.5	0.5	0.5	0.5

Correlation to the National Science Education Standards

The following list shows the correlation of *Holt Environmental Science* with the National Science Education Standards for grades 9–12. For further details, see the teacher's wrap of each chapter opener.

UNIFYING CONCEPTS AND PROCESSES

Standard	Code	Correlation
Systems, order, and organization	UCP 1	4.1, 4.2, 4.3, 5.1, 5.2, 8.1, 8.2, 10.1, 10.2
Evidence, models, and explanation	UCP 2	2.2, 9.1, 9.2, 21.1, 21.2
Change, constancy, and measurement	UCP 3	2.1, 2.2, 8.1, 9.1, 9.2
Evolution and equilibrium	UCP 4	4.1, 4.2, 4.3, 8.1, 8.2, 10.1, 10.2
Form and function	UCP 5	8.1, 8.2

SCIENCE AS INQUIRY

Standard	Code	Correlation
Abilities necessary to do scientific inquiry	SAI 1	1.1, 1.2, 2.1, 2.2, 2.3
Understandings about scientific inquiry	SAI 2	1.2, 2.1, 2.2, 2.3

PHYSICAL SCIENCE

Standard	Code	Correlation
Structure of Atoms The nuclear forces that hold the nucleus of an atom together, at nuclear distances, are usually stronger than the electric forces that would make it fly apart. Nuclear reactions convert a fraction of the mass of interacting particles into energy, and they can release much greater amounts of energy than atomic interactions. Fission is the splitting of a large nucleus into smaller pieces. Fusion is the joining of two nuclei at extremely high temperature and pressure, and is the process responsible for the energy of the sun and other stars.	PS 1c	17.2
Chemical reactions Chemical reactions may release or consume energy. Some reactions such as the burning of fossil fuels release large amounts of energy by losing heat and by emitting light. Light can initiate many chemical reactions such as photosynthesis and the evolution of urban smog.	PS 3b	12.1, 17.1

LIFE SCIENCE

Standard	Code	Correlation
The cell		
Plant cells contain chloroplasts, the site of photosynthesis. Plants and many microorganisms use solar energy to combine molecules of carbon dioxide and water into complex, energy-rich organic compounds and release oxygen to the environment. This process of photosynthesis provides a vital connection between the sun and the energy needs of living systems.	**LS 1e**	5.1
Biological evolution		
Species evolve over time. Evolution is the consequence of the interactions of (1) the potential for a species to increase its numbers, (2) the genetic variability of offspring due to mutation and recombination of genes, (3) a finite supply of the resources required for life, and (4) the ensuing selection by the environment of those offspring better able to survive and leave offspring.	**LS 3a**	4.2, 6.2, 6.3
The great diversity of organisms is the result of more than 3.5 billion years of evolution that has filled every available niche with life forms.	**LS 3b**	4.2, 4.3, 10.1, 10.2
Biological classifications are based on how organisms are related. Organisms are classified into a hierarchy of groups and subgroups based on similarities which reflect their evolutionary relationships. Species is the most fundamental unit of classification.	**LS 3e**	4.3
Interdependence of organisms		
The atoms and molecules on the earth cycle among the living and nonliving components of the biosphere.	**LS 4a**	3.3
Energy flows through ecosystems in one direction, from photosynthetic organisms to herbivores to carnivores and decomposers.	**LS 4b**	5.1
Organisms both cooperate and compete in ecosystems. The interrelationships and interdependencies of these organisms may generate ecosystems that are stable for hundreds or thousands of years.	**LS 4c**	4.1, 6.2, 6.3, 8.1, 8.2
Living organisms have the capacity to produce populations of infinite size, but environments and resources are finite. This fundamental tension has profound effects on the interactions between organisms.	**LS 4d**	6.2, 6.3, 8.1, 8.2
Human beings live within the world's ecosystems. Increasingly, humans modify ecosystems as a result of population growth, technology, and consumption. Human destruction of habitats through direct harvesting, pollution, atmospheric changes, and other factors is threatening current global stability, and if not addressed, ecosystems will be irreversibly affected.	**LS 4e**	1.1, 7.1, 7.2, 10.1, 10.2, 15.1, 15.2, 19.1, 19.2, 19.3
Matter, energy, and organization in living systems		
As matter and energy flow through different levels of organization of living systems—cells, organs, organisms, communities—and between living systems and the physical environment, chemical elements are recombined in different ways. Each recombination results in storage and dissipation of energy into the environment as heat. Matter and energy are conserved in each change.	**LS 5f**	5.1, 5.2

LIFE SCIENCE, *continued*

Standard	Code	Correlation
The Behavior of organisms Organisms have behavioral responses to internal changes and to external stimuli. Responses to external stimuli can result from interactions with the organism's own species and others, as well as environmental changes; these responses either can be innate or learned. The broad patterns of behavior exhibited by animals have evolved to ensure reproductive success. Animals often live in unpredictable environments, and so their behavior must be flexible enough to deal with uncertainty and change. Plants also respond to stimuli.	**LS 6b**	6.2, 6.3

EARTH AND SPACE SCIENCE

Standard	Code	Correlation
Energy in the Earth System Earth systems have internal and external sources of energy, both of which create heat. The sun is the major external source of energy. Two primary sources of internal energy are the decay of radioactive isotopes and the gravitational energy from the earth's original formation.	**ES 1a**	3.1, 3.2, 3.3, 18.1
Heating of Earth's surface and atmosphere by the sun drives convection within the atmosphere and oceans, producing winds and ocean currents.	**ES 1c**	3.2, 3.3, 13.1
Global climate is determined by energy transfer from the sun at and near the earth's surface. This energy transfer is influenced by dynamic processes, such as cloud cover and the earth's rotation, and static conditions, such as the position of mountain ranges and oceans.	**ES 1d**	3.2, 3.3, 13.1, 13.2
Geochemical Cycles The Earth is a system containing essentially a fixed amount of each stable chemical atom or element. Each element can exist in several different chemical reservoirs. Each element on earth moves among reservoirs in the solid earth, oceans, atmosphere, and organisms as part of geochemical cycles.	**ES 2a**	3.1, 3.2, 3.3, 5.2
Movement of matter between reservoirs is driven by the earth's internal and external sources of energy. These movements are often accompanied by a change in the physical and chemical properties of the matter. Carbon, for example, occurs in carbonate rocks such as limestone, in the atmosphere as carbon dioxide gas, in water as dissolved carbon dioxide, and in all organisms as complex molecules that control the chemistry of life.	**ES 2b**	5.2
Origin and Evolution of the Earth System Interactions among the solid earth, the oceans, the atmosphere, and organisms have resulted in the ongoing evolution of the earth system. We can observe some changes such as earthquakes and volcanic eruptions on a human time scale, but many processes such as mountain building and plate movements take place over hundreds of millions of years.	**ES 3c**	3.1, 3.2, 3.3

SCIENCE AND TECHNOLOGY

Standard	Code	Correlation
Understandings about science and technology	ST 2a	1.1
	ST 2c	2.1

SCIENCE IN PERSONAL AND SOCIAL PERSPECTIVES

Standard	Code	Correlation
Population growth	SPSP 2a	8.1, 9.1, 19.1
	SPSP 2b	9.1, 9.2
	SPSP 2c	8.1
Natural resources	SPSP 3a	1.1, 9.2, 10.1, 10.3, 11.2, 16.1, 17.1, 17.2, 18.1, 18.2, 21.1, 21.2
	SPSP 3b	9.2, 11.2, 14.2, 14.3, 16.2, 16.3, 17.1, 17.2, 18.1, 18.2, 19.1, 19.2, 19.3
	SPSP 3c	7.1, 7.2, 19.1, 19.2, 20.1, 20.2
Environmental quality	SPSP 4a	7.1, 7.2, 12.1, 12.2, 12.3, 13.1, 13.2, 13.3, 14.1, 14.3, 15.1, 15.2, 15.3
	SPSP 4b	5.2, 11.3, 12.3, 13.1, 13.2, 13.3, 16.3
	SPSP 4c	16.3, 21.1, 21.2, 21.3
Natural and human-induced hazards	SPSP 5b	7.1, 7.2, 14.2, 14.3, 16.3
	SPSP 5c	15.1, 15.2, 15.3
	SPSP 5d	13.1, 13.2, 13.3, 20.1, 20.2
Science and technology in local, national, and global challenges	SPSP 6b	1.1, 1.2, 21.1, 21.2
	SPSP 6c	20.1, 20.2, 21.2, 21.3
	SPSP 6d	2.1, 2.2, 2.3, 19.1, 19.2, 19.3
	SPSP 6e	7.1, 7.2, 10.2, 11.3, 12.3, 14.3

HISTORY AND NATURE OF SCIENCE

Standard	Code	Correlation
Science as a human endeavor	HNS 1c	10.1, 10.3
Nature of scientific knowledge	HNS 2b	2.1
Historical perspectives	HNS 3	1.1, 2.2

Safety in Your Laboratory

Risk Assessment

MAKING YOUR LABORATORY A SAFE PLACE TO WORK AND LEARN

Concern for safety must begin before any activity in the classroom and before students enter the lab. A careful review of the facilities should be a basic part of preparation for each school term. You should investigate the physical environment, identify any safety risks, and inspect your work areas for compliance with safety regulations.

The review of the lab should be thorough, and all safety issues must be addressed immediately. Keep a file of your review, and add to the list each year. This will allow you to continue to raise the standard of safety in your lab and classroom.

Many classroom experiments, demonstrations, and other activities are classics that have been used for years. This familiarity may lead to a comfort that can obscure inherent safety concerns. Review all experiments, demonstrations, and activities for safety concerns before presenting them to the class. Identify and eliminate potential safety hazards.

1. **Identify the Risks**
   Before introducing any activity, demonstration, or experiment to the class, analyze it and consider what could possibly go wrong. Carefully review the list of materials to make sure they are safe. Inspect the equipment in your lab or classroom to make sure it is in good working order. Read the procedures to make sure they are safe. Record any hazards or concerns you identify.

2. **Evaluate the Risks**
   Minimize the risks you identified in the last step without sacrificing learning. Remember that no activity you can perform in the lab or classroom is worth risking injury. Thus, extremely hazardous activities, or those that violate your school's policies, must be eliminated. For activities that present smaller risks, analyze each risk carefully to determine its likelihood. If the pedagogical value of the activity does not outweigh the risks, the activity must be eliminated.

3. **Select Controls to Address Risks**
   Even low-risk activities require controls to eliminate or minimize the risks. Make sure that in devising controls you do not substitute an equally or more hazardous alternative. Some control methods include the following:

   - Explicit verbal and written warnings may be added or posted.

   - Equipment may be rebuilt or relocated, parts may be replaced, or the equipment may be replaced entirely by safer alternatives.

   - Risky procedures may be eliminated.

   - Activities may be changed from student activities to teacher demonstrations.

4. **Implement and Review Selected Controls**
   Controls do not help if they are forgotten or not enforced. The implementation and review of controls should be as systematic and thorough as the initial analysis of safety concerns in the lab and laboratory activities.

SAFETY RISKS AND PREVENTATIVE CONTROLS

This chart describes several possible safety hazards and controls that can be implemented to resolve them. This chart is not complete, but it can be used as a starting point to identify hazards in your laboratory.

Identified risk	Preventative control
Facilities and Equipment	
Lab tables are in disrepair, room is poorly lighted and poorly ventilated, faucets and electrical outlets do not work or are difficult to use because of their location.	Work surfaces should be level and stable. There should be adequate lighting and ventilation. Water supplies, drains, and electrical outlets should be in good working order. Any equipment in a dangerous location should not be used; it should be relocated or rendered inoperable.
Wiring, plumbing, and air circulation systems do not work or do not meet current specifications.	Specifications should be kept on file. Conduct a periodic review of all equipment, and document compliance. Damaged fixtures must be labeled as such and must be repaired as soon as possible.
Eyewash fountains and safety showers are present, but no one knows anything about their specifications.	Ensure that eyewash fountains and safety showers meet the requirements of the ANSI standard (Z358.1).
Eyewash fountains are checked and cleaned once at the beginning of each school year. No records are kept of routine checks and maintenance on the safety showers and eyewash fountains.	Flush eyewash fountains for 5 minutes every month to remove any bacteria or other organisms from pipes. Test safety showers (measure flow in gallons per min) and eyewash fountains every 6 months and keep records of the test results.
Labs are conducted in multipurpose rooms, and equipment from other courses remains accessible.	Only the items necessary for a given activity should be available to students. All equipment should be locked away when not in use.
Students are permitted to enter or work in the lab without teacher supervision.	Lock all laboratory rooms whenever a teacher is not present. Supervising teachers must be trained in lab safety and emergency procedures.
Safety equipment and emergency procedures	
Fire and other emergency drills are infrequent, and no records or measurements are made of the results of the drills.	Always carry out critical reviews of fire or other emergency drills. Be sure that plans include alternate routes. Don't wait until an emergency to find the flaws in your plans.
Emergency evacuation plans do not include instructions for securing the lab in the event of an evacuation during a lab activity.	Plan actions in case of emergency: establish what devices should be turned off, which escape route to use, where to meet outside the building.
Fire extinguishers are in out-of-the-way locations, not on the escape route.	Place fire extinguishers near escape routes so that they will be of use during an emergency.
Fire extinguishers are not maintained. Teachers are not trained to use them.	Document regular maintenance of fire extinguishers. Train supervisory personnel in the proper use of extinguishers. Instruct students not to use an extinguisher but to call for a teacher.

Identified risk	Preventative control
Safety equipment and emergency procedures, *continued*	
Teachers in labs and neighboring classrooms are not trained in CPR or first aid.	Teachers should receive training from the local chapter of the American Red Cross. Certifications should be kept current with frequent refresher courses.
Teachers are not aware of their legal responsibilities in case of an injury or accident.	Review your faculty handbook for your responsibilities regarding safety in the classroom and laboratory. Contact the legal counsel for your school district to find out the extent of their support and any rules, regulations, or procedures you must follow.
Emergency procedures are not posted. Emergency numbers are kept only at the switchboard or main office. Instructions are given verbally only at the beginning of the year.	Emergency procedures should be posted at all exits and near all safety equipment. Emergency numbers should be posted at all phones, and a script should be provided for the caller to use. Emergency procedures must be reviewed periodically, and students should be reminded of them at the beginning of each activity.
Spills are handled on a case-by-case basis and are cleaned up with whatever materials happen to be on hand.	Have the appropriate equipment and materials available for cleaning up; replace them before expiration dates. Make sure students know to alert you to spilled chemicals, blood, and broken glass.
Work habits and environment	
Safety wear is only used for activities involving chemicals or hot plates.	Aprons and goggles should be worn in the lab at all times. Long hair, loose clothing, and loose jewelry should be secured.
There is no dress code established for the laboratory; students are allowed to wear sandals or open-toed shoes.	Open-toed shoes should never be worn in the laboratory. Do not allow any footwear in the lab that does not cover feet completely.
Students are required to wear safety gear but teachers and visitors are not.	Always wear safety gear in the lab. Keep extra equipment on hand for visitors.
Safety is emphasized at the beginning of the term but is not mentioned later in the year.	Safety must be the first priority in all lab work. Students should be warned of risks and instructed in emergency procedures for each activity.
There is no assessment of students' knowledge and attitudes regarding safety.	Conduct frequent safety quizzes. Only students with perfect scores should be allowed to work in the lab.
You work alone during your preparation period to organize the day's labs.	Never work alone in a science laboratory or a storage area.
Safety inspections are conducted irregularly and are not documented. Teachers and administrators are unaware of what documentation will be necessary in case of a lawsuit.	Safety reviews should be frequent and regular. All reviews should be documented, and improvements must be implemented immediately. Contact legal counsel for your district to make sure your procedures will protect you in case of a lawsuit.

Identified risk **Purchasing, storing, and using chemicals**	Preventative control
The storeroom is too crowded, so you decide to keep some equipment on the lab benches.	Do not store reagents or equipment on lab benches. Keep shelves organized. Never place reactive chemicals (in bottles, beakers, flasks, wash bottles, etc.) near the edges of a lab bench.
You prepare solutions from concentrated stock to save money.	Reduce risks by ordering diluted instead of concentrated substances.
You purchase plenty of chemicals to be sure that you won't run out or to save money.	Purchase chemicals in class-size quantities. Do not purchase or have on hand more than one year's supply of each chemical.
You don't generally read labels on chemicals when preparing solutions for a lab, because you already know about a chemical.	Read each label to be sure it states the hazards and describes the precautions and first aid procedures (when appropriate) that apply to the contents in case someone else has to deal with that chemical in an emergency.
You never read the Material Safety Data Sheets (MSDSs) that come with your chemicals.	Always read the Material Safety Data Sheet (MSDS) for a chemical before using it. Follow the precautions described in that MSDS. File and organize MSDSs for all chemicals where they can be found easily in case of an emergency.
The main stockroom contains chemicals that haven't been used for years.	Do not leave bottles of chemicals unused on the shelves of the lab for more than one week or unused in the main stockroom for more than one year. Dispose of or use up any leftover chemicals.
No extra precautions are taken when flammable liquids are dispensed from their containers.	When transferring flammable liquids from bulk containers, ground the container; before transferring flammable liquids to a smaller metal container, ground both containers.
Students are told to put their broken glass and solid chemical wastes in the trash can.	Have separate containers for trash, for broken glass, and for different categories of hazardous chemical wastes.
You store chemicals alphabetically instead of by hazard class. Chemicals are stored without consideration of possible emergencies (fire, earthquake, flood, etc.), which could compound the hazard.	Use MSDSs to determine which chemicals are incompatible. Store chemicals by the hazard class indicated on the MSDS. Store chemicals that are incompatible with common fire-fighting media like water (such as alkali metals) or carbon dioxide (such as alkali and alkaline-earth metals) under conditions that eliminate the possibility of a reaction with water or carbon dioxide if it is necessary to fight a fire in the storage area.
Corrosives are kept above eye level, out of reach from any unauthorized person.	Always store corrosive chemicals on shelves below eye level. Remember, fumes from many corrosives can destroy metal cabinets and shelving.
Chemicals are kept on the stockroom floor on the days that they will be used so that they are easy to find.	Never store chemicals or other materials on floors or in the aisles of the laboratory or storeroom, even for a few minutes.

Master Materials List

Chemicals and Consumable Materials	Amount Needed per Lab Group	Chapter # for Lab (Q indicates QuickLab)	Ward's Ordering Code
alum (optional)	2 Tbsp	11	940 R 0006
aluminum foil	1 sheet	4	15 R 1009
bag, plastic, shopping	1	19	local
bag, plastic, garbage	3	15Q	local
bags, clear plastic, 12″ x 18″	2	12	18 R 6981
baking soda	1 tsp	19Q	37 R 5467
battery, 9V	1	18Q	14 R 5401
beans, dry	100 g (1/4 lb)	8Q	86 R 8006
bill for electric utility service	1	17	local
bottle, soft drink, 2–3 L	1	16Q	local
bottle, soft drink, 2–3 L	4	11	local
brine shrimp culture	50 mL	4	87 R 5105
bromthymol blue	5–8 drops	5Q	944 R 7106
cardboard, 25cm x 25cm	1	13Q	local
cards, index	50	21	15 R 9809
chalk, blackboard	1 stick	12Q	15 R 4637
charcoal (optional)	1 C	11	950 R 4706
compost, dry	5 g	15	local
cooking oil (optional)	100 mL	11	946 R 4206
corn syrup	200 mL per class	2	39 R 1463
cover slip for microscope slide	1	7	14 R 3555
cover slip for microscope slide	5	8	14 R 3555
cover slip for microscope slide	2	4Q	14 R 3555
cupric carbonate, basic (copper (II) carbonate), powder	25 g	16	960 R 3204
Detain™ or methyl cellulose, aqueous solution	1 mL	4	37 R 7950 or 37 R 7600
detergent (optional)	100 mL	11	37 R 2268
dirt	4 C	2Q	local
distilled water	2.25 L	7	88 R 7005
distilled water	2.5 L per class	8	88 R 7005
distilled water	1 cup	12Q	88 R 7005
egg carton	1	5	local

Chemicals and Consumable Materials, *continued*	Amount Needed per Lab Group	Chapter # for Lab (Q indicates QuickLab)	Ward's Ordering Code
Elodea plant sprig	1	5Q	86 R 7500
fertilizer, dry, household use	3 tsp	7	20 R 6020
fertilizer, dry, household use (optional)	1 Tbsp	11	20 R 6020
file folders	10	21	local
filter paper	4	15	15 R 2815
foam board, 15″ x 20″	1 piece	18	15 R 9856
foam board, 15″ x 20″ (optional extra)	1 piece	18	15 R 9856
food coloring, assorted colors	1 set	$3Q_2$, 7Q, 11	945 R 3003
glucose solution	15 mL	11	945 R 6405
glucose test paper	1 piece	11	14 R 4119
glue, white	1 bottle	18	15 R 9806
graph paper	1 pad	2, 9, 10, 14	15 R 3835
graph paper (optional)	1 pad	6	15 R 3835
grass clippings, dry, chopped	5 g	15	local
gravel	1 lb	11	21 R 1805
hay or other mulch	5,000 cm³	15Q	local
houseplants, potted, identical species	2	12	86 R 7300
ice cubes	6	2	local
ice cubes	24	13	local
incense stick	1	13	local
iron filings	1 g	16	941 R 5406
matches, safety or lighter, electric gas	1 box or 1	13	15 R 9427 or 15 R 3504
methylene blue solution, 1%	2 mL	8	946 R 1904
microscope slide	5	8	14 R 3500
microscope slide	2	4Q	14 R 3500
microscope slide	1	7	14-3500
milk carton, empty, 1 or 1/2 pint	2	3	local
owl pellet	1	5	69 R 3392
paper clip, large	30	18	15 R 9815
paper cup, small	1	2Q, 18	15 R 9830
paper cup, small (optional)	4	18	15 R 9830
paper, black	1 sheet	$3Q_1$	15 R 9825
paper, white	1 sheet	$3Q_1$, 5	local

Chemicals and Consumable Materials, *continued*	Amount Needed per Lab Group	Chapter # for Lab (Q indicates QuickLab)	Ward's Ordering Code
pebbles	1 lb or 2 kg	3	45 R 1986
pencils, assorted colors or markers, assorted colors	1 set	1, 9, 10Q, 14, 21	15 R 2576 or 15 R 4635
petroleum jelly	1 Tbsp	18Q	15 R 9832
pH paper or litmus paper	2 strips	12Q	15 R 2558 or 15 R 3100
pipet, with 0.5 mL graduations	1	4	18 R 2971
pipet, with 0.5 mL graduations	5	8	18 R 2971
plant, flowering	1	4Q	86 R 7015
plant, grass, with pollinating heads	1	4Q	local
plaster of Paris	10 lb per class	3	940 R 9507
plastic wrap, clear	1 roll	6Q, 7	15 R 9858
plate, paper (optional)	4	18	15 R 9889
pond water, with live organisms	300 mL	7	88 R 7010
poster board, 22" x 28"	1	1, 21	15 R 9856
rice	0.5 kg	16Q	local
rocks, small	2 lb or 1 kg	3	local
salt	1 Tbsp	18Q	942 R 9706
salt	1 g	20Q, 7Q	942 R 9706
sand, fine, clean	5 to 10 lb	3	942 R 5806
sand, fine, clean	1 lb	11, 2Q	942 R 5806
sawdust	5 g	15	21 R 2295
screen or mesh, fine, 4 cm^2	2	4	local
silver nitrate, aqueous solution	5 mL	20Q	37 R 9555
sod (turf grass), 1 ft x 1 ft x 1 in square	1	15Q	local
sodium nitrite	2 g	12	981 R 1706
soft drink (optional)	1 can	11	local
soil, potting or topsoil	1 lb	2Q, 11, 15Q, 16Q	local or 45 R 1980
soil, sample from ground	50 g	15	local
sponge, kitchen	2	6Q	15 R 1103
spools of thread, empty	2	18	local
straw, drinking	1	5Q	15 R 9869
string, 50 m spool	1	1, 3Q$_2$, 10, 18	15 R 9863
sugar	100 g (per class)	8	946 R 9608
sulfuric acid, 1.0 M solution	2 mL	12	970 R 9806
sulfuric acid, 1.0 M solution	150 mL	16	970 R 9306

Chemicals and Consumable Materials, *continued*	Amount Needed per Lab Group	Chapter # for Lab (Q indicates QuickLab)	Ward's Ordering Code
swab, cotton	1	4Q	14 R 5502
tape, masking	1 roll	4, 13, 3Q$_1$	15 R 9828
toothpick	1	2Q	15 R 9840
towel, paper	1 roll	19	15 R 9844
tube, cardboard, <2 cm (1") dia. x 10 cm	1	17Q	local
tubing, vinyl, 1 cm dia. (3/8" x 1/16")	40 cm	4	18 R 5083
twist tie or tape, masking	2 or 2 rolls	12	14 R 0947 or 15 R 9828
vinegar	1/2 Tbsp.	12Q	39 R 0138
vinegar	200 mL	19Q	39 R 0138
vinegar (optional)	100 mL	11	39 R 0138
wax pencil	1	7, 11	15 R 1155
wire, copper, fine-gauge	100 cm	17Q	16 R 0549
yeast, baker's	7 g per class	2	88 R 0929
yeast, baker's	20 g per class	8	88 R 0929

Equipment and Reusable Materials	Amount Needed per Lab Group	Chapter # for Lab (Q indicates QuickLab)	Ward's Ordering Code
aquarium, 15 gal, glass	1	$3Q_2$, 13	21-5242
baking tray	1	2Q	local
balance (or scale), metric, triple beam or electronic	1	6Q, 15Q, 15, 19	15-6057 or 15-6240
bar magnet	1	17Q	13-0115
beaker, 50 mL	1	12	17-4010
beaker, 100 mL	3	2, 4	17-4020
beaker, 250 mL	2	$3Q_1$, 5Q, 15	17-4040
beaker, 400 mL	3	2	17-4050
beaker, 600 mL	1	18Q, 19Q	17-4060
beaker, 800 mL	5	11	17-4070
binoculars (optional)	1	6	25-4523
block, wooden, 12″ x 1 1/2″ x 4″	1	3	local
blow-dryer, 1500 W	1	18	local
book, <1/2″ thick	1	2Q	local
bottle, glass, 4 oz or 100 mL	1	$3Q_2$	17-0315
bowl, 10 L or container, plastic, 16″ x 13″ x 4″	1	15Q	local or 20-3210
bowl, 16 oz	2	16Q	18-7204
bulb, 75 W incandescent	1	13	15-9851
bunsen burner	1	16	15-0603
calculator	1	9, 17	27-3055
calculator (optional)	1	19	27-3055
clock or stopwatch	1	2, 14Q, 15	15-1492 or 15-0512
computer with spreadsheet software (optional)	1	9	local
computer with word processing, graphing, or presentation software (optional)	1	21	local
container (tray), plastic, 16″ x 13″ x 4″	3	3, 15Q	20-3210
corks sized to fit 3/8″ tubing	2	4	15-8459
crucible (or other heat safe container)	1	15	15-3127
delivery tubes, rubber or plastic	3	2	18-7700
dissolved-oxygen test kit	1	11Q	21-0054
doorstopper	6	15Q	16-0609
dowel or smooth rod	1	18	15-0082
erlemeyer flask, 500 mL	5	8	17-2984

Equipment and Reusable Materials, continued	Amount Needed per Lab Group	Chapter # for Lab (Q indicates QuickLab)	Ward's Ordering Code
eyedropper	1	7, 15	18-2970
field guide, flora and fauna, local	1	6	local
field guide, flowers (optional)	1	1	32-2107 or 32-2108
field guide, freshwater microorganisms	1	7	32-8063
field guide, insects (optional)	1	1	32-0106
field guide, mammals, including skull illustrations	1	5	32-2114
field guide, reptiles and amphibians, including skull illustrations	1	5	32-2123
field guide, trees (optional)	1	1	32-2109 or 32-2110
floodlight, 150 W	1	$3Q_1$	36-4168
forceps	1	5	14-0512
funnel	1	15, 16	18-1300
galvanometer or battery tester	1	17Q	16-0537 or local
globe or atlas	1	6	80-5630
graduated cylinder, 100 mL	1	4, 5Q, 7	18-1730
graduated cylinder, 1000 mL	1	11, 16Q	18-1760
hand lens	1	1, 10	24-1112
hinged glass top for 15 gal aquarium	1	13	21-5307
hot plate or oven	1	15	15-7999 or local
hot-water bag	1	4	local
ice bag	1	4	local
jar or wide-mouth bottle, plastic, with lid, 250 mL	3	11Q	18-0082
jar, 1 qt (950 mL)	3	7	17-2060
lamp, goose-neck, adjustable	1	13	15-5046
light fixture, fluorescent, for growing plants	1	4, 7	20-3124
map, local area, road or topographical	1	10Q	local
meterstick	1	14Q	15-4065
meterstick or tape measure	1	1, 10	15-4065 or 15-3989
micrometer, stage type or eyepiece disc for microscope	1	8	24-0250 or 25-0230 or 25-4303
microscope, compound	1	4Q, 7, 8	24-2310
needle, dissecting	1	5	14-0950

Equipment and Reusable Materials, *continued*	Amount Needed per Lab Group	Chapter # for Lab (Q indicates QuickLab)	Ward's Ordering Code
pan, dissecting	1	5	18-3665
Petri dish	1	4	18-7100
ruler, metric	1	3, 4, 6, 11, 19	15-4650
ruler, metric	2	9	15-4650
screw clamps	3	4	15-3910
spoon, household	1	16Q	local
spoon, measuring	1	19Q	15-9870
stakes, wooden	4	1	local
stakes, wooden (optional)	4	10	local
stirring rod	1	7, 11, 15	17-6005
stoppers, no. 2, one-hole	3	2	15-8482
test tube stopper, size 00	4	4	15-8459
test tube stopper, size 5	1	5Q	15-8465
test tube, 13mm x 100mm (9 mL)	4	4	17-0610
test tube, 13mm x 100mm (9 mL)	2	7Q, 16, 18Q	17-0610
test tube, 13mm x 100mm (9 mL)	5	8	17-0610
test tube, 13mm x 100mm (9 mL)	1	20Q	17-0610
test tube, 25mm x 200mm (70ml)	3	2	17-0660
test tube, 25mm x 200mm (70ml)	1	5Q	17-0660
test-tube clamp	1	16	15-0841
test-tube rack	1	4, 16	15-2953
thermometer	1	2, $3Q_1$	15-1416
thermometer, outdoor	2	13	15-0515
tongs	1	15	14-0960

Safety Equipment	Amount Needed per Lab Group	Chapter # for Lab (Q indicates QuickLab)	Ward's Ordering Code
apron	1 per student	(all)	15-1050
gloves, disposable	1 pair per student	(all)	15-1071
gloves, heat-resistant	1 pair per student	15, 16	15-1095
safety goggles	1 pair per student	(all)	15-3046

MASTER MATERIALS LIST

ABOUT THE AUTHOR

Karen Arms, Ph.D., J.D.

Karen Arms received her Ph.D. in molecular biology from Oxford University and a doctor of law from Cornell University. She was an assistant professor of biology at Cornell University, where she taught introductory biology and courses in science and society. She also taught marine biology at the University of Georgia Marine Biology Station and introductory biology at South College in Savannah, Georgia. In addition to *Holt Environmental Science*, Dr. Arms has written several college-level biology textbooks. Her interest in and concern for the environment led her to form an ecotourism organization that introduces people to the ecosystems of the southeastern coast.

ISBN 0-03-039074-5

1 2 3 4 5 6 7   048   08 07 06 05 04

Acknowledgments

CONTRIBUTING WRITERS

Inclusion Specialist

Joan A. Altobelli
Special Education Director
Austin Independent School District
Austin, Texas

John A. Solorio
Multiple Technologies Lab Facilitator
Austin Independent School District
Austin, Texas

Feature Development and Lab Safety Consultants

Michele Benn
Science Teacher
Beaver Falls High School
Beaver Falls, Pennsylvania

Randa Flinn
Science Teacher
Northeast High School
Fort Lauderdale, Florida

Alyson M. Mike
Science Teacher
East Helena Public Schools
East Helena, Montana

Tammie Niffenegger
Science Chair and Science Teacher
Port Washington High School
Waldo, Wisconsin

Gabriell DeBear Paye
Science and Environmental Technology Lead Teacher
West Roxbury High School
West Roxbury, Massachusetts

Teresa Wilson
Science Writer
Austin, Texas

Teacher Edition Development

Kara R. Dotter
Science Writer
Austin, Texas

Lisa Parks
Science Writer
Seattle, Washington

Meredith Phillips
Science Writer
Brooklyn, New York

Catherine Podeszwa
Science Writer
Duluth, Minnesota

Teresa Wilson
Science Writer
Austin, Texas

ACADEMIC REVIEWERS

Mead Allison, Ph.D.
Associate Professor
Department of Geology and Earth Sciences
Tulane University
New Orleans, Louisiana

David M. Armstrong, Ph.D.
Professor
Environmental, Population, and Organismic Biology
University of Colorado
Boulder, Colorado

Paul D. Asimow, Ph.D.
Assistant Professor of Geology and Geochemistry
Division of Geological and Planetary Sciences
California Institute of Technology
Pasadena, California

John A. Brockhaus, Ph.D.
Director of Mapping, Charting, and Geodesy Program
Department of Geography and Environmental Engineering
United States Military Academy
West Point, New York

Gary Campbell, Ph.D.
Professor of Mineral Economics
School of Business and Economics
Michigan Technological University
Houghton, Michigan

Laura Chenault, D.V.M.
Bulverde, Texas

Marian R. Chertow, Ph.D.
Assistant Professor of Industrial Environmental Management
Yale School of Forestry and Environmental Studies
Yale University
New Haven, Connecticut

Susan B. Dickey, R.N., Ph.D.
Associate Professor
Pediatric Nursing
Temple University
Philadelphia, Pennsylvania

Dale Elifrits, Ph.D.
Professor
Department of Physics and Geology
Northern Kentucky University
Highland Heights, Kentucky

Turgay Ertekin, Ph.D.
George E. Trimble Chair in Earth and Mineral Sciences
Professor of Petroleum and Natural Gas Engineering
Engineering Department
Penn State University
University Park, Pennsylvania

Ronald A. Feldman, Ph.D.
Ruth Harris Ottman Centennial Professor for the Advancement of Social Work Education
Director, Center for the Study of Social Work Practice
Columbia University
New York, New York

Linda Gaul, Ph.D.
Epidemiologist
Texas Department of Health
Austin, Texas

John Goodge, Ph.D.
Associate Professor of Geology
Southern Methodist University
Dallas, Texas

Herbert Grossman, Ph.D.
Associate Professor of Botany and Biology
Department of Environmental Sciences
Pennsylvania State University
University Park, Pennsylvania

Acknowledgments

David Haig, Ph.D.
Associate Professor of Biology
Department of Organismic and Evolutionary Biology
Harvard University
Cambridge, Massachusetts

Vicki Hansen, Ph.D.
Professor of Geological Sciences
Department of Geology
Southern Methodist University
Dallas, Texas

Rosalind Harris, Ph.D.
Professor, Rural Agriculture
Department of Sociology
University of Kentucky
Lexington, Kentucky

Richard Hey, Ph.D.
Professor of Geophysics
School of Ocean and Earth Sciences Technology
University of Hawaii
Honolulu, Hawaii

Steven A. Jennings, Ph.D.
Associate Professor of Geography
Department of Geography and Environmental Studies
University of Colorado
Colorado Springs, Colorado

Joel Leventhal, Ph.D.
Emeritus Scientist
U.S. Geological Survey and Diversified Geochemistry
Lakewood, Colorado

Joann Mossa, Ph.D.
Associate Professor
Department of Geography
University of Florida
Gainesville, Florida

Gary Mueller, Ph.D.
Associate Professor of Nuclear Engineering
Department of Engineering
University of Missouri
Rolla, Missouri

Barbara Murck, Ph.D.
Director, Environmental Programs
University of Toronto
Mississauga, Ontario, Canada

Emily Niemeyer, Ph.D.
Assistant Professor of Chemistry
Department of Chemistry
Southwestern University
Georgetown, Texas

Eva Oberdörster, Ph.D.
Lecturer
Department of Biological Sciences
Southern Methodist University
Dallas, Texas

Hilary Olson, Ph.D.
Research Scientist
Institute of Geophysics
The University of Texas
Austin, Texas

Ken Peace, C.C.E.
Geology Supervisor
Ark Land Company
St. Louis, Missouri

Per F. Peterson, Ph.D.
Professor and Chair
Department of Nuclear Engineering
University of California
Berkeley, California

David Pimentel, Ph.D.
Professor Emeritus and Agricultural Ecologist
Department of Entomology, Systematics and Ecology
Cornell University
Ithaca, New York

Mary M. Poulton, Ph.D.
Department Head and Associate Professor of Geological Engineering
Department of Mining and Geological Engineering
University of Arizona
Tucson, Arizona

Barron Rector, Ph.D.
Associate Professor and Extension Range Specialist
Texas Agricultural Extension Service
Texas A&M University
College Station, Texas

Dork Sahagian, Ph.D.
Research Professor, Stratigraphy and Basin Analysis, Geodynamics
Global Analysis, Interpretation, and Modeling Program
University of New Hampshire
Durham, New Hampshire

Miles Silman, Ph.D.
Associate Professor of Biology
Department of Biology
Wake Forest University
Winston-Salem, North Carolina

Spencer Steinberg, Ph.D.
Associate Professor, Environmental Organic Chemistry
Chemistry Department
University of Nevada
Las Vegas, Nevada

Richard Storey, Ph.D.
Dean of the Faculty and Professor of Biology
Colorado College
Colorado Springs, Colorado

Martin VanDyke, Ph.D.
Professor of Chemistry, Emeritus
Front Range Community College
Westminster, Colorado

Judith Weis, Ph.D.
Professor of Biology
Department of Biological Sciences
Rutgers University
Newark, New Jersey

Mary Wicksten, Ph.D.
Professor of Biology
Department of Biology
Texas A&M University
College Station, Texas

TEACHER REVIEWERS

Robert Akeson
Science Teacher
Boston Latin School
Boston, Massachusetts

Dan Aude
Magnet Programs Coordinator
Montgomery Public Schools
Montgomery, Alabama

Lowell Bailey
Science Teacher
Bedford North Lawrence High School
Bedford, Indiana

Robert Baronak
Biology Teacher
Donegal High School
Mount Joy, Pennsylvania

Michele Benn
Science Teacher
Beaver Falls High School
Beaver Falls, Pennsylvania

continued on page 662

CONTENTS IN BRIEF

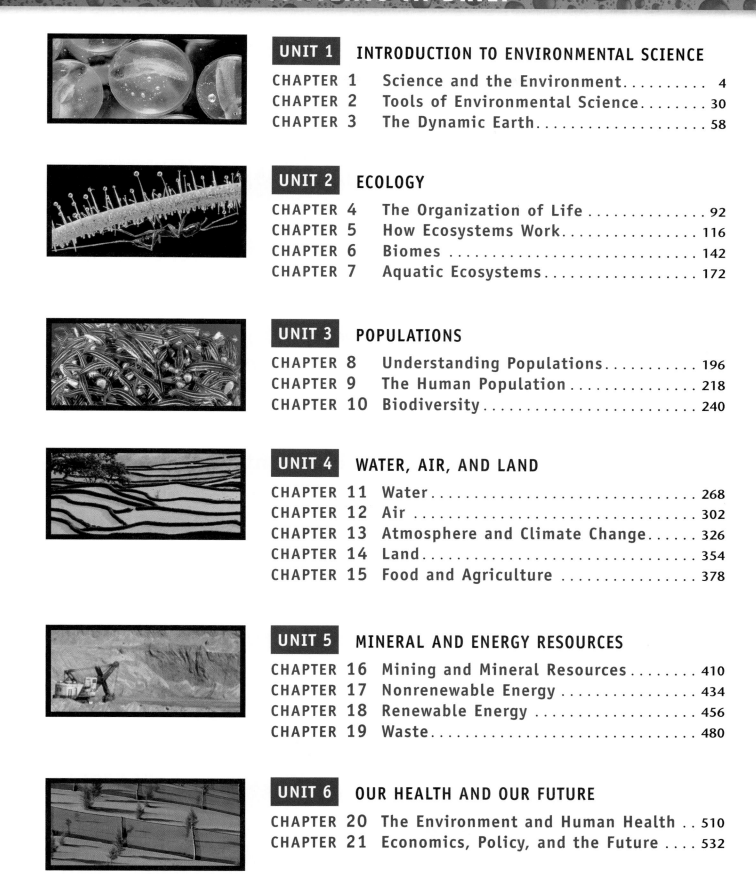

CONTENTS

INTRODUCTION TO ENVIRONMENTAL SCIENCE UNIT 1

ECOLOGY  UNIT 2

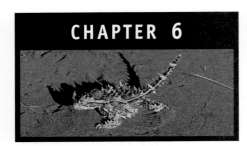

WATER, AIR, AND LAND

UNIT 4

MINERAL AND ENERGY RESOURCES

UNIT 5

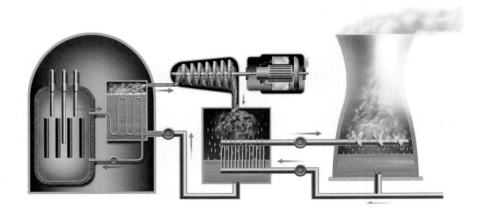

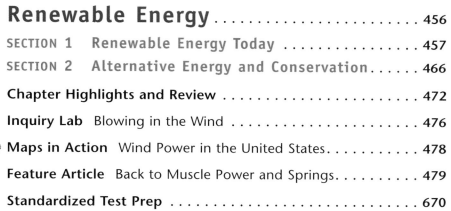

APPENDIX

LABS

Exploration Labs

Skills Practice Labs

Inquiry Labs

QuickLABS

MAPS IN ACTION

FEATURE ARTICLES

Making a Difference

Points of View

Science and Technology

Society and the Environment

CASE STUDIES

Safety Symbols

The following safety symbols will appear in this text when you are asked to perform a procedure requiring extra precautions. Once you have familiarized yourself with these safety symbols, turn to pp. 558–561 for safety guidelines to use in all of your environmental science laboratory work.

 EYE PROTECTION

- Wear safety goggles when working around chemicals, acids, bases, flames or heating devices. Contents under pressure may become projectiles and cause serious injury.
- Never look directly at the sun through any optical device or use direct sunlight to illuminate a microscope.
- Avoid wearing contact lenses in the lab.
- If any substance gets into your eyes, notify your instructor immediately and flush your eyes with running water for at least 15 minutes.

 CLOTHING PROTECTION

- Secure loose clothing and remove dangling jewelry. Do not wear open-toed shoes or sandals in the lab.
- Wear an apron or lab coat to protect your clothing when you are working with chemicals.
- If a spill gets on your clothing, rinse it off immediately with water for at least 5 minutes while notifying your instructor.

CAUSTIC SUBSTANCES

- If a chemical gets on your skin, on your clothing, or in your eyes, rinse it immediately and alert your instructor.
- If a chemical is spilled on the floor or lab bench, alert your instructor, but do not clean it up yourself unless your instructor directs you to do so.

 CHEMICAL SAFETY

- Always use caution when working with chemicals.
- Always wear appropriate protective equipment. Always wear eye goggles, gloves, and a lab apron or lab coat when you are working with any chemical or chemical solution.
- Never mix chemicals unless your instructor directs you to do so.
- Never taste, touch, or smell chemicals unless your instructor directs you to do so.
- Add an acid or base to water; never add water to an acid or base.
- Never return an unused chemical to its original container.
- Never transfer substances by sucking on a pipet or straw; use a suction bulb.
- Follow instructions for proper disposal.

 ANIMAL SAFETY

- Always obtain permission before bringing any animal to school.
- Handle animals carefully and respectfully.
- Wash your hands thoroughly after handling any animal.

 PLANT SAFETY

- Wear disposable polyethylene gloves when handling any wild plant.
- Do not eat any part of a plant or plant seed used in the lab.
- Wash hands thoroughly after handling any part of a plant.
- When outdoors, do not pick any wild plants unless your instructor directs you to do so.

 ELECTRICAL SAFETY

- Do not place electrical cords in walking areas or let cords hang over a table edge in a way that could cause equipment to fall if the cord is accidentally pulled.

- Do not use equipment that has frayed electrical cords or loose plugs.
- Be sure that equipment is in the "off" position before you plug it in.
- Never use an electrical appliance around water or with wet hands or clothing.
- Be sure to turn off and unplug electrical equipment when you are finished using it.

HEATING SAFETY

- Avoid wearing hair spray or hair gel on lab days.
- Whenever possible, use an electric hot plate instead of an open flame as a heat source.
- When heating materials in a test tube, always angle the test tube away from yourself and others.
- Glass containers used for heating should be made of heat-resistant glass.

SHARP OBJECTS

- Use knives and other sharp instruments with extreme care.
- Never cut objects while holding them in your hands. Place objects on a suitable work surface for cutting.
- Never use a double-edged razor in the lab.

HAND SAFETY

- To avoid burns, wear heat-resistant gloves whenever instructed to do so.
- Always wear protective gloves when working with an open flame, chemicals, solutions, or wild or unknown plants.
- If you do not know whether an object is hot, do not touch it.
- Use tongs when heating test tubes. Never hold a test tube in your hand to heat the test tube.

FIRE SAFETY

- Know the location of laboratory fire extinguishers and fire-safety blankets.
- Know your school's fire-evacuation routes.

GAS SAFETY

- Do not inhale any gas or vapor unless your instructor directs you to do so. Do not breathe pure gases.
- Handle materials prone to emit vapors or gases in a well-ventilated area. This work should be done in an approved chemical fume hood.

GLASSWARE SAFETY

- Check the condition of glassware before and after using it. Inform your teacher of any broken, chipped, or cracked glassware, because it should not be used.
- Do not pick up broken glass with your bare hands. Place broken glass in a specially designated disposal container.

WASTE DISPOSAL

- Clean and decontaminate all work surfaces and personal protective equipment as directed by your instructor.
- Dispose of all broken glass, contaminated sharp objects, and other contaminated materials (biological and chemical) in special containers as directed by your instructor.

HYGIENIC CARE/CLEAN HANDS

- Keep your hands away from your face and mouth.
- Always wash your hands thoroughly when you have finished with an experiment.

To the Student

Like all other sciences, environmental science is a process of satisfying our curiosity about why things are the way they are and about how things happen the way they do. For example, in studying environmental science, you may discover the answers to the following questions.

Q How could the birth of these otters indicate a healthy ecosystem?

Q How could the demise of this seemingly unimportant insect cause severe damage to the rain forest in which it lives?

Q How could the watering of this lawn affect the water quality of a nearby stream?

Q How could a population of iguanas help save a rain forest from destruction?

Q How could recycling an aluminum can help save fossil fuels and reduce both air and water pollution?

Q How could the exhaust from these cars in New York contribute to the decline of salmon in Canada?

Q How could technology used to study weather help us to understand air pollution?

Learning about Environmental Science

You may not have expected to have the questions shown on these pages answered in your study of environmental science. The answers to these questions and many more like them help define this unusual and exciting area of study. In learning about the various aspects of our environment, you will quickly come to understand how interdependent life on Earth is.

In many cases, we know so little about environmental interactions that we can't even begin to predict long-term effects. For example, it took nearly 15 years of study before we understood the relationship between the pesticide DDT and the declining populations of bald eagles. It will take even longer to understand the relationship between pollution and the climate. And what usually happens is that the answer to one question leads to a string of new questions.

Perhaps the most important question to ask at the beginning of this course is: **What do you hope to get out of this environmental science text?**

You may be interested in science and want to know more about the inner workings of our environment. Or you may be interested in learning more about human impact on the environment and what we can do to reduce the negative consequences. You may even want to know more about the environment firsthand so that you can decipher environmental issues for yourself, rather than simply accepting someone else's point of view.

Challenge

Regardless of your reasons for taking this course, my challenge to you is to think for yourself. In reading this textbook, you not only will learn a lot about science, you will learn about the complex issues facing our environment. You will explore different points of view and be exposed to a variety of differing opinions. Don't feel that you have to accept any particular opinion as your own. As your knowledge and skills in environmental science grow, so will your ability to draw your own conclusions.

Karen Arms

How to Use Your Textbook

Your Roadmap for Success with *Holt Environmental Science*

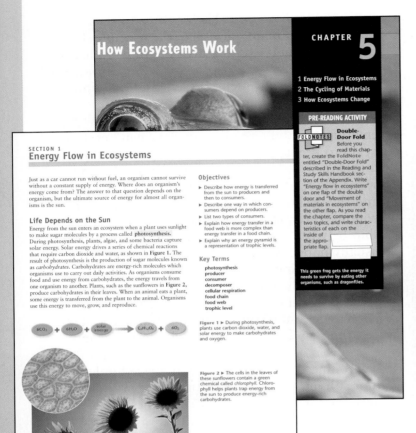

Get Organized

Do the **Pre-Reading Activity** at the beginning of each chapter to create a **FoldNote**, which is a helpful note-taking and study aid. Use the **Graphic Organizer** activity within the chapter to organize the chapter content in a way that you understand.

STUDY TIP Go to the **Reading and Study Skills** section of the Appendix for guidance on making FoldNotes and Graphic Organizers.

Read for Meaning

Read the **Objectives** at the beginning of each section, because they will tell you what you'll need to learn. **Key Terms** are also listed for each section. Each key term is also highlighted in the text. After reading each chapter, turn to the **Chapter Highlights** page and review the Key Terms and the **Main Ideas**, which are brief summaries of the chapter's key concepts. You may want to do this even before you read the chapter.

STUDY TIP If you don't understand a definition, reread the page on which the term is introduced. The surrounding text should help make the definition easier to understand.

▶ Be Resourceful, Use the Web

Internet Connect boxes in your textbook take you to resources that you can use for science projects, reports, and research papers. Go to **scilinks.org** and type in the SciLinks code to get information on a topic.

Visit go.hrw.com
Find resources and reference materials that go with your textbook at **go.hrw.com**. Enter the keyword HE6 Home to access the home page for your textbook.

Prepare for Tests

Section Reviews and **Chapter Reviews** test your knowledge of the main points of the chapter. Critical Thinking items challenge you to think about the material in different ways and in greater depth. The **standardized test prep** that is located at the end of the Appendix helps you sharpen your test-taking abilities.

STUDY TIP Reread the Section Objectives and Chapter Highlights when studying for a test to be sure you know the material.

Use the Appendix

Your **Appendix** contains a variety of resources designed to enhance your learning experience. These resources include **Reading and Study Skills**, a **Math Skills Refresher**, and a **Chemistry Refresher**, which provide helpful study aids. The appendix also contains features on **Economic Concepts**, **Environmental Careers**, and **Ecoskills**. In addition, you can find information on **SI Conversions**, **Mineral Uses**, and a variety of **Maps**.

Figure 10 ▶ **Examples of endemic species of the United States** include ❶ the cecropia moth, (declining populations), ❷ the tulip poplar tree (limited distribution), ❸ the crayfish *Cambarus mongalensis* (limited distribution), ❹ the desert pupfish (endangered), and ❺ the northern spotted owl (threatened).

Biodiversity in the United States You may notice that three of the biodiversity hotspots in **Figure 9** are partly within U.S. borders. The United States includes a wide variety of unique ecosystems, including the Florida Everglades, the California coastal region, Hawaii, the Midwestern prairies, and the forests of the Pacific Northwest. The United States holds unusually high numbers of species of freshwater fishes, mussels, snails, and crayfish. Species diversity in the United States is also high among groups of land plants such as pine trees and sunflowers. Some examples of the many unique native species are shown in **Figure 10**.

The California Floristic Province, a biodiversity hotspot, is home to 3,488 native plant species. Of these species, 2,124 are endemic and 565 are threatened or endangered. The threats to this area include the use of land for agriculture and housing, dam construction, overuse of water, destructive recreation, and mining—all stemming from local human population growth.

SECTION 2 Review

1. **Describe** four ways that species are being threatened with extinction globally.

2. **Define** and give examples of *endangered species* and *threatened species*.

3. **List** areas of the Earth that have high levels of biodiversity and many threats to species.

4. **Compare** the amount of biodiversity in the United States to that of the rest of the world.

CRITICAL THINKING

5. **Interpreting Graphics** The biodiversity hot spots shown in Figure 9 share several characteristics besides a great number of species. Look at the map, and name as many shared characteristics as you can.

6. **Expressing Opinions** Which of the various threats to biodiversity do you think will be most difficult to stop? Which are hardest to justify? Write a paragraph to explain your opinion. **WRITING SKILLS**

Section 2 **Biodiversity at Risk 251**

Visit Holt Online Learning
If your teacher gives you a special password to log onto the **Holt Online Learning** site, you'll find your complete textbook on the Web. In addition, you'll find some great learning tools and practice quizzes. You'll be able to see how well you know the material from your textbook.

SCIENCE

CHAPTER 1

Science and the Environment

CHAPTER 2

Tools of Environmental Science

CHAPTER 3

The Dynamic Earth

When they reach adulthood, amphibians, such as these un-hatched salamanders, breathe through their skin, which makes them vulnerable to pollutants in their environment. Scientists closely monitor amphibian species to determine the effects of pollution on the world's ecosystems.

Science and the Environment
Chapter Planning Guide

PACING	CLASSROOM RESOURCES	LABS, DEMONSTRATIONS, AND ACTIVITIES
BLOCKS 1, 2 & 3 · 135 min pp. 4–15		
Chapter Opener		
Section 1 Understanding Our Environment	**OSP** Lesson Plan * **CRF** Active Reading * **BASIC** **TT** Bellringer * **TT** Major Fields that Contribute to Environmental Science *	**TE** Group Activity Hold a Press Conference, p. 8 ◆ **GENERAL** **SE** Field Activity Germinating Corn, p. 10 **SE** QuickLab Classifying Resources, p. 14 **SE** Exploration Lab What's in an Ecosystem? pp. 26–27 ◆ **CRF** Datasheets for In-Text Labs * **CRF** Long-Term Project * ◆ **BASIC** **CRF** Math/Graphing Lab * ◆ **GENERAL**
BLOCKS 4 & 5 · 90 min pp. 16–21 **Section 2** The Environment and Society	**OSP** Lesson Plan * **CRF** Active Reading * **BASIC** **TT** Bellringer * **TT** "The Tragedy of the Commons" *	**TE** Group Activity "Tragedy of the Commons" Game, p. 16 ◆ **GENERAL** **TE** Internet Activity Ecological Footprints, p. 19 **GENERAL** **SE** Field Activity Critical Thinking and the News, p. 20 **TE** Group Activity Town Council Meeting, p. 20 **GENERAL** **CRF** Research Lab * ◆ **ADVANCED** **CRF** Consumer Lab * ◆ **GENERAL**

BLOCKS 6 & 7 · 90 min

Chapter Review and Assessment Resources

- **SE** Chapter Review pp. 23–25
- **SE** Standardized Test Prep pp. 636–637
- **CRF** Study Guide * ■ **GENERAL**
- **CRF** Chapter Test * ■ **GENERAL**
- **CRF** Chapter Test * **ADVANCED**
- **CRF** Concept Review * ■ **GENERAL**
- **CRF** Critical Thinking * **ADVANCED**
- **OSP** Test Generator
- **CRF** Test Item Listing *

Online and Technology Resources

Visit **go.hrw.com** for access to Holt Online Learning or enter the keyword **HE6 Home** for a variety of free online resources.

This CD-ROM package includes
- Lab Materials QuickList Software
- Holt Calendar Planner
- Customizable Lesson Plans
- Printable Worksheets
- ExamView® Test Generator
- Interactive Teacher Edition
- Holt PuzzlePro® Resources
- Holt PowerPoint® Resources

 Holt Environmental Science Interactive Tutor CD-ROM

This CD-ROM consists of interactive activities that give students a fun way to extend their knowledge of environmental science concepts.

KEY

TE	Teacher Edition	CRF	Chapter Resource File	*	Also on One-Stop Planner
SE	Student Edition	TT	Teaching Transparency	■	Also Available in Spanish
OSP	One-Stop Planner	CD	Interactive CD-ROM	◆	Requires Advance Prep

ENRICHMENT AND SKILLS PRACTICE	SECTION REVIEW AND ASSESSMENT	CORRELATIONS
SE **Pre-Reading Activity,** p. 4 TE **Using the Figure,** p. 4 `GENERAL`		**National Science Education Standards**
TE **Reading Skill Builder** Brainstorming, p. 6 `BASIC` TE **Career** Environmental Scientist, p. 6 TE **Inclusion Strategies,** p. 6 TE **Skill Builder** Vocabulary, p. 7 `GENERAL` TE **Reading Skill Builder** Reading Organizer, p. 7 `GENERAL` TE **Student Opportunities** Constructing Wildlife Habitat Spaces, p. 8 SE **Graphic Organizer** Comparison Table, p. 9 `GENERAL` TE **Reading Skill Builder** Paired Reading, p. 9 `BASIC` TE **Skill Builder** Math, p. 10 `GENERAL` TE **Skill Builder** Writing, p. 11 `ADVANCED` SE **Case Study** Lake Washington, pp. 12–13 TE **Using the Figure** Population Growth, p. 13 `GENERAL`	SE **Mid-Section Review,** p. 8 TE **Homework,** p. 11 `GENERAL` TE **Homework,** p. 14 `GENERAL` SE **Section Review,** p. 15 TE **Reteaching,** p. 15 `BASIC` TE **Quiz,** p. 15 `GENERAL` TE **Alternative Assessment,** p. 15 `GENERAL` CRF **Quiz** * ■ `GENERAL`	LS 4e ST 2a SPSP 3a SPSP 6b
SE **MathPractice** Market Equilibrium, p. 17 TE **Interpreting Statistics** Oil Prices, p. 17 `GENERAL` TE **Reading Skill Builder** K-W-L, p. 17 `BASIC` TE **Skill Builder** Writing, p. 17 `GENERAL` TE **Using the Figure** Fresh Vegetables, p. 18 `GENERAL` TE **Inclusion Strategies,** p. 18 TE **Using the Figure** Developed Vs. Developing, p. 19 `GENERAL` TE **Using the Figure** Different Footprints, p. 19 `GENERAL` TE **Student Opportunities** Set Up a Recycling Program, p. 20 SE **Making a Difference** Chicken of the Trees, pp. 28–29 CRF **Map Skills** Forest Loss * `GENERAL`	TE **Homework,** p. 18 `GENERAL` SE **Section Review,** p. 21 TE **Reteaching,** p. 21 `BASIC` TE **Quiz,** p. 21 `GENERAL` TE **Alternative Assessment,** p. 21 `GENERAL` CRF **Quiz** * ■ `GENERAL`	SPSP 6b

Guided Reading Audio CDs

These CDs are designed to help auditory learners and reluctant readers. (Audio CDs are also available in Spanish.)

www.scilinks.org

Maintained by the **National Science Teachers Association**

Topic: Careers in Environmental Science
SciLinks Code: HE4010

Topic: Solving Environmental Problems
SciLinks Code: HE4013

CNN Videos

Each video segment is accompanied by a Critical Thinking Worksheet.

Earth Science Connections

Segment 2 Why Time Zones

Chapter Enrichment

This Chapter Enrichment provides relevant and interesting information to expand and enhance your classroom instruction of the chapter material.

SECTION 1 Understanding Our Environment

Environmental Science

Environmental science is a multidisciplinary field that draws from all the sciences, as well as other fields, to help us better understand the relationship between humans and the world in which we live. Environmental science is considered an applied science. It applies the principles of pure sciences, such as chemistry and biology, to help achieve practical goals. The study of environmental science focuses on three main areas:

- conservation and protection of natural resources
- environmental education and communication
- environmental research

Environmental science was started when scientists began putting together the pieces of the environmental puzzle. They realized how aspects of ecology, biology, chemistry, geology, geography, and other fields all played a part in the environment. When these pieces were put together, the field of environmental science was born.

▶ Albatrosses, Amsterdam Island

The Extinction of North American Megafauna

North America was once populated with a variety of large mammals, known as megafauna, that would rival even the most dramatic species found on Earth today. There were dozens of species, including mammoths, camels, giant ground sloths, beavers as large as black bears, saber-toothed cats, maned lions, giant short-faced bears, and dire wolves. By about fifteen thousand years ago, all of these species, and many other megafauna, ceased to exist in North America.

▶ Hunter-gatherers, New Guinea

There are three popular theories as to how and why these animals disappeared in such a short period of time. The blitzkrieg, or overkill, theory claims that humans arrived in North America during this same period of time and hunted the animals to extinction. Another theory suggests that rapid shifts in climate associated with the end of the last ice age doomed the animals. A third theory proposes that a disease, possibly similar to influenza or rabies, was introduced by the arrival of humans and caused these species to die out. It is likely that a combination of factors contributed to the extinction of these animals.

BRAIN FOOD

Skeletal remains of hunter-gatherers from as many as 20,000 years ago show signs that these humans were responsible for the pollution of their environment and even suffered deformities as a result.

Ghosts of Evolution

The avocado plant does not belong in Mexico. No animal capable of ingesting and excreting the seed of an avocado has lived in Mexico for thousands of years. Today, people help to spread and reproduce avocado trees, but how did avocados thrive before people? Avocado trees once had partners that aided in their survival. Thousands of years ago, large animals such as mammoths and giant sloths lived throughout North America. These extinct animals called "ghosts of evolution," also supported the Kentucky coffee tree and a number of other plants that persist in modern times.

BRAIN FOOD

The large thorns of the Osage orange tree are thought to be defenses against foraging mastodons.

The Pronghorn antelope of America's Great Plains is another example of a species that reveals the influence of an extinct animal. The Pronghorn antelope can run up to 60 miles per hour, easily outrunning any predators. In fact, none of its predators, including wolves, coyotes and pumas, can approach that speed. So why does the pronghorn have to run so fast? It is thought that an animal such as the extinct American cheetah may have hunted the pronghorn thousands of years ago. The influence of this extinct predator could persist today.

SECTION
2 The Environment and Society

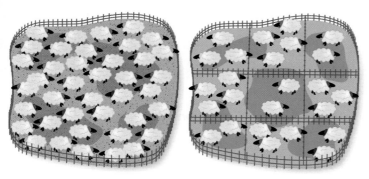

▶ When commons were replaced by enclosed fields, people tended to graze only the number of animals that the land could support

"The Tragedy of the Commons"

One of the cornerstones of environmental science was set in 1968 when Garrett Hardin published an essay titled "The Tragedy of the Commons." In his essay, he described the attitude with which people use resources. An individual may benefit from an action, but how does that action impact others? In terms of environmental science, how does one person's action affect the environment that everyone must share? A common theme in many of Hardin's works is the idea of bioethics. This is the concept of building ethics on a biological foundation. Hardin publicly debated controversial topics such as abortion, population control, foreign aid, nuclear power and immigration for over 30 years, but is best known for "The Tragedy of the Commons."

▶ Cleaning up a waterway in Oakland, California

Strategic Lawsuits Against Public Participation (SLAPPs)

The law can be used to protect, but it can also be used to intimidate. Some corporations have filed lawsuits against citizens and groups who criticize their environmental records. These lawsuits, called *strategic lawsuits against public participation* (SLAPPs), are intended to intimidate citizen groups by using up time, money, and limiting involvement. These lawsuits are usually thrown out of court, but they often consume a considerable amount of time and money. Sometimes, the group can file a counter claim, but this process takes additional time and money. Often, the corporation will offer to drop the lawsuit if the group agrees to never oppose the corporation again.

Overview

Tell students that this chapter presents an overview of environmental science and the history of humans in the environment. To solve environmental problems, we must be able to identify causes and describe solutions. The chapter also examines environmental science in a social and economic context.

Using the Figure — GENERAL

The photograph shows a French ecologist in Costa Rica's Rio Macho rain forest preserve. Tell students that many of the plants found in the rain forest canopy seem better adapted for life in the desert than in the rain forest. Ask students to hypothesize why this is true. (The rainforest canopy has trace amounts of soil, so there is nothing to hold water when it rains. This means that plants must be adapted to conserve water even though they are growing in some of the wettest places on Earth.) Some types of plants, called *bromeliads*, can catch and store their own water. Some of the largest bromeliads can hold many liters of water and can support reproducing populations of frogs. **LS Visual**

PRE-READING ACTIVITY

 You may want to assign this FoldNote activity as homework. Collect the FoldNotes to check students' understanding of the material.

For information about videos related to this chapter, go to **go.hrw.com** and type in the keyword **HE4 SENV**.

Science and the Environment

1 **Understanding Our Environment**
2 **The Environment and Society**

PRE-READING ACTIVITY

FOLDNOTES **Booklet** Before you read this chapter, create the **FoldNote** entitled "Booklet" described in the Reading and Study Skills section of the Appendix. Label each page of the booklet with a main idea from the chapter. As you read the chapter, write what you learn about each main idea on the appropriate page of the booklet.

More than 2,700 m (9,000 ft) above sea level, a forest ecologist is studying biodiversity in a Costa Rican rain forest. To ascend to the treetops, he shoots an arrow over a branch and hauls himself up with the attached rope.

Chapter Correlations *National Science Education Standards*

LS 4e Human beings live within the world's ecosystems. Increasingly, humans modify ecosystems as a result of population growth, technology, and consumption. Human destruction of habitats through direct harvesting, pollution, atmospheric changes, and other factors is threatening current global stability, and if not addressed, ecosystems will be irreversibly affected. **(Section 1)**

ST 2a Scientists in different disciplines ask different questions, use different methods of investigation, and accept different types of evidence to support their explanations. Many scientific investigations require the contributions of individuals from different disciplines, including engineering. New disciplines of science, such as geophysics and biochemistry often emerge at the interface of two older disciplines. **(Section 1)**

SPSP 3a Human populations use resources in the environment in order to maintain and improve their existence. Natural resources have been and will continue to be used to maintain human populations. **(Section 1)**

SPSP 6b Understanding basic concepts and principles of science and technology should precede active debate about the economics, policies, politics, and ethics of various science and technology related challenges. However, understanding science alone will not resolve local, national, or global challenges. **(Section 1 and Section 2)**

SECTION 1
Understanding Our Environment

When someone mentions the term *environment*, some people think of a beautiful scene, such as a stream flowing through a wilderness area or a rain-forest canopy alive with blooming flowers and howling monkeys. You might not think of your backyard or neighborhood as part of your environment. In fact, the environment is everything around us. It includes the natural world as well as things produced by humans. But the environment is also more than what you can see—it is a complex web of relationships that connects us with the world we live in.

What Is Environmental Science?

The students from Keene High School in **Figure 1** are searching the Ashuelot River in New Hampshire for dwarf wedge mussels. The mussels, which were once abundant in the river, are now in danger of disappearing completely—and the students want to know why. To find out more, the students test water samples from different parts of the river and conduct experiments. Could the problem be that sewage is contaminating the water? Or could fertilizer from a nearby golf course be causing algae in the river to grow rapidly and use up the oxygen that the mussels need to survive? Another possible explanation might be that a small dam on the river is disrupting the mussels' reproduction.

The students' efforts have been highly praised and widely recognized. Yet they hope for a more meaningful reward—the preservation of an endangered species. The students' work is just one example of a relatively new field—**environmental science,** the study of how humans interact with the environment.

Objectives

▶ Define *environmental science,* and compare environmental science with ecology.
▶ List the five major fields of study that contribute to environmental science.
▶ Describe the major environmental effects of hunter-gatherers, the agricultural revolution, and the Industrial Revolution.
▶ Distinguish between renewable and nonrenewable resources.
▶ Classify environmental problems into three major categories.

Key Terms

environmental science
ecology
agriculture
natural resource
pollution
biodiversity

Figure 1 ▶ These students are counting the number of dwarf wedge mussels in part of the Ashuelot River. They hope that the data they collect will help preserve this endangered species.

5

SECTION 1

Focus

Overview

Before beginning this section, review with your students the Objectives in the Student Edition. This section describes the fields that contribute to environmental science. It also explores the history of humans in the environment and discusses the major problems threatening the environment today.

Bellringer

Ask students, "What is the environment?" Have students brainstorm answers. As a class, establish a working definition for the term.

Motivate

Identifying Preconceptions — GENERAL

Understanding Environmental Problems Ask students to list some of the most significant environmental problems in the world today. (Answers may include global issues such as rain forest destruction or global warming.) Then ask students to name some environmental problems in your local community. Ask students to identify one of the local problems that is part of a larger global problem. For example, if smog is a problem in your area, students may point out that automobile emissions might also contribute to global warming. Then ask students to brainstorm ways that the local problem could be solved. (Ideas might include carpooling or using mass transit whenever possible.)

Chapter Resource File

• **Lesson Plan**
• **Active Reading** BASIC
• **Section Quiz** GENERAL

Transparencies

TT Bellringer

Brainstorming Before covering the passage about the fields of study that contribute to environmental science, ask students to come up with their own list similar to **Table 1.** Write their responses on the board. Then have students read the passage, revisit their list, and decide if any of the fields of study should be removed or added to the list.

LS Logical

English Language Learners

REAL-LIFE CONNECTION — GENERAL

Environmental News Journals

To help students become aware of environmental issues in your community and around the world, have them create a notebook of articles from environmental news stories in magazines, newspapers, and on the Internet. Have students make regular additions to their journals. For example, have students search for stories that relate to each section in this book. Before you introduce each section, discuss student stories with the class. Or, students could organize their journals according to your class syllabus. Students can also supplement their journals with critiques of the articles they read.

Connection to History

Rachel Carson Alarmed by the increasing levels of pesticides and other chemicals in the environment, biologist Rachel Carson published *Silent Spring* in 1962. Carson imagined a spring morning that was silent because the birds and frogs were dead after being poisoned by pesticides. Carson's carefully researched book was enthusiastically received by the public and was read by many other scientists as well as policy makers and politicians. However, many people in the chemical industry saw *Silent Spring* as a threat to their pesticide sales and launched a $250,000 campaign to discredit Carson. Carson's research prevailed, although she died in 1964—unaware that the book she had written was instrumental in the birth of the modern environmental movement.

The Goals of Environmental Science One of the major goals of environmental science is to understand and solve environmental problems. To accomplish this goal, environmental scientists study two main types of interactions between humans and their environment. One area of study focuses on how we use natural resources, such as water and plants. The other area of study focuses on how our actions alter our environment. To study these interactions, environmental scientists must gather and analyze information from many different disciplines.

Many Fields of Study Environmental science is an interdisciplinary science, which means that it involves many fields of study. One important foundation of environmental science is ecology. **Ecology** is the study of how living things interact with each other and with their nonliving environment. For example, an ecologist might study the relationship between bees and the plants bees pollinate. However, an environmental scientist might investigate how the nesting behavior of bees is influenced by human activities such as the planting of suburban landscaping.

Many sciences other than ecology also contribute to environmental science. For example, chemistry helps us understand the nature of pollutants. Geology helps us model how pollutants travel underground. Botany and zoology can provide information needed to preserve species. Paleontology, the study of fossils, can help us understand how Earth's climate has changed in the past. Using such information about the past can help us predict how future climate changes could affect life on Earth. At any given time, an environmental scientist may use information provided by other sciences. **Figure 2** shows a few examples of disciplines that contribute to environmental science.

Figure 2 ▶ Many Fields of Study

▶ This marine biologist (right) is studying a marine mammal called a *manatee*.

▶ This ornithologist (above) is studying the nesting behavior of seabirds called *albatrosses*.

6

Career

Environmental Scientist Environmental science is a broad field of study. Most environmental scientists choose to specialize in one field, such as chemistry, botany, or geology. Environmental scientists also take classes in other fields of study to understand different aspects of the field. Students who are interested in pursuing a career in environmental science should consider attending schools that offer a wide variety of sciences and interdisciplinary study programs.

INCLUSION Strategies

• *Developmentally Delayed*

Have students look in magazines to find photographs that show an environment. Have students glue each photograph to a separate index card to create postcards. On the back of the index card, students should write a short message to a friend or relative describing the environment and its plants and animals. Lower-level students may simply label the plants or animals in the photograph and dictate a story about the environment to another student or into a tape recorder.

But studying the environment also involves studying human populations, so environmental scientists may use information from the social sciences, which include economics, law, politics, and geography. Social sciences can help us answer questions such as, How do cultural attitudes affect the ways that people use the U.S. park system? or How does human migration from rural to urban areas affect the local environment? **Table 1** lists some of the major fields of study that contribute to the study of environmental science.

internet connect
www.scilinks.org
Topic: Careers in Environmental Science
SciLinks code: HE4010
SCILINKS. Maintained by the National Science Teachers Association

Table 1 ▼

Major Fields of Study That Contribute to Environmental Science	
Biology is the study of living organisms.	**Zoology** is the study of animals. **Botany** is the study of plants. **Microbiology** is the study of microorganisms. **Ecology** is the study of how organisms interact with their environment and each other.
Earth science is the study of the Earth's nonliving systems and the planet as a whole.	**Geology** is the study of the Earth's surface, interior processes, and history. **Paleontology** is the study of fossils and ancient life. **Climatology** is the study of the Earth's atmosphere and climate. **Hydrology** is the study of Earth's water resources.
Physics is the study of matter and energy.	**Engineering** is the science by which matter and energy are made useful to humans in structures, machines, and products.
Chemistry is the study of chemicals and their interactions.	**Biochemistry** is the study of the chemistry of living things. **Geochemistry**, a branch of geology, is the study of the chemistry of materials such as rocks, soil, and water.
Social sciences are the study of human populations.	**Geography** is the study of the relationship between human populations and Earth's features. **Anthropology** is the study of the interactions of the biological, cultural, geographical, and historical aspects of humankind. **Sociology** is the study of human population dynamics and statistics.

▶ This geologist is studying a volcano in Hawaii.

▶ This biologist is examining a plant that was grown in a lab from just a few cells.

READING SKILL BUILDER — GENERAL

Reading Organizer Tell students that they may find it useful to make a concept map of the fields of study that contribute to environmental science. One idea for a concept map is to have environmental science in the center and have the other fields of study as branches. Encourage students to make notes on the concept map to describe how the field of study relates to environmental science. **LS Visual**

Discussion —— GENERAL

What is Environmental Science?

Tell students that one of the basic concepts of environmental science is that everything is connected to everything else. Have students brainstorm some examples of this concept. Ask students the following focus question: "How is this basic concept related to the fact that it is often difficult for people to reach an agreement about environmental issues?" (Decisions that involve the environment often involve many different parties, and any decision or action can have many consequences. Because people have different priorities, reaching an agreement is often difficult.) **LS Verbal**

SKILL BUILDER — GENERAL

Vocabulary Many words in science have suffixes or prefixes that are based on Latin or Greek root words. Even the names of the different sciences are based on these roots. One of the first things that students may see when looking at the names of the sciences in **Table 1** is that many of the sciences have the suffix *–ology,* which means "the study of." This suffix is based on the Greek word *logos,* which means "discourse." Encourage students to find out the meaning of other common suffixes and prefixes used in science and make a table of prefixes and suffixes in their *EcoLog.* Students may want to make this list at the back of their *EcoLog* so that they may continue to add to it and refer to it throughout the course. **LS Verbal** | English Language Learners

Transparencies
TT Major Fields that Contribute to Environmental Science

Group Activity ── GENERAL

Hold a Press Conference When scientists make important discoveries, one way they announce their findings to the world is through a press conference. Have students hold a press conference about an environmental news story. Select several students to be the scientists. Ask the scientists to research a current environmental news story. They should plan a press conference to present their findings. Have the rest of the class represent the media. They should prepare questions for the scientists. After the press conferences, the media group should write a short newspaper article describing the findings. Have the groups change roles until everyone has had the opportunity to present and report on a scientific finding.

LS Interpersonal   Co-op Learning

Student Opportunities

Constructing Wildlife Habitat Spaces Habitat loss is one of the greatest threats to plants and animals in our environment. Students can help by constructing and maintaining a habitat for wildlife. Habitat projects may be as simple as planting a container garden to attract pollinators or putting up a bird feeder. If students have more space available, they could construct a small marsh or pond. Students might even consider involving their school or community in turning a vacant lot into wildlife habitat. For more information and ideas about constructing backyard habitats, visit the National Wildlife Federation Web site.

Ecofact

The Fall of Troy Environmental problems are nothing new. Nearly 3,000 years ago, the Greek poet Homer wrote about the ancient seaport of Troy, which was located beneath a wooded hillside. The Trojans cut down all the trees on the surrounding hills. Without trees to hold the soil in place, rain washed the soil into the harbor. So much silt accumulated in the harbor that large ships could not enter and Troy's economy collapsed. Today, the ruins of Troy are several miles from the sea.

Scientists as Citizens, Citizens as Scientists

Governments, businesses, and cities recognize that studying our environment is vital to maintaining a healthy and productive society. Thus, environmental scientists are often asked to share their research with the world. **Figure 3** shows scientists at a press conference that was held after a meeting on climate change.

Often, the observations of nonscientists are the first step toward addressing an environmental problem. For example, when deformed frogs started appearing in lakes in Minnesota, middle school students noticed the problem first. Likewise, the students at Dublin Scioto High School in Ohio, shown in **Figure 3**, study box turtle habitats every year. The students want to find out how these endangered turtles live and what factors affect the turtles' nesting and hibernation sites. The students track the turtles, measure the atmospheric conditions, analyze soil samples, and map the movements of the small reptiles. Why do these efforts matter? They matter because the box turtle habitat is threatened. At the end of the year, students present their findings to city planners in hopes that the most sensitive turtle habitats will be protected.

Figure 3 ▶ Environmental Science and Public Life Scientists hold a press conference on climate change (above). Students (right) are studying the movements of box turtles.

SECTION 1 Mid-Section Review

1. **Describe** the two main types of interactions that environmental scientists study. Give an example of each.

2. **Describe** the major fields of study that contribute to environmental science.

3. **Explain** why environmental science is an interdisciplinary science.

CRITICAL THINKING

4. **Making Comparisons** What is the difference between environmental science and ecology?

5. **Making Inferences** Read the Ecofact. Propose a solution to prevent the environmental problems of the seaport of Troy described in the Ecofact.
   READING SKILLS

8

Answers to Mid-Section Review

1. The two main types of interactions that environmental scientists study are how we use resources and how our actions affect the environment. A geologist might study how mineral resources are mined and used while a biologist might study insect populations to see how pesticides affect the population.

2. The five major fields of study are biology, the study of living organisms; Earth science, the study of Earth's nonliving systems and the planet; physics, the study of matter and energy; chemistry, the study of chemicals and their interactions; and social sciences, the study of human populations.

3. Answers may vary. Environmental science is a complex field and many different sciences are needed to understand the interactions of different parts of the environment.

4. Environmental science is the study of how humans interact with the environment. Ecology is the study of how living things interact with each other and the nonliving environment.

5. Answers may vary. Students might suggest that regulations limiting logging in ancient Troy could have reduced soil erosion.

Our Environment Through Time

You may think that environmental change is a modern issue, but wherever humans have hunted, grown food, or settled, they have changed the environment. For example, the land where New York City now stands was once an area where Native Americans hunted game and gathered food, as shown in **Figure 4.** The environmental change that occurred on Manhattan Island over the last 300 years was immense, yet that period of time was just a "blink" in human history.

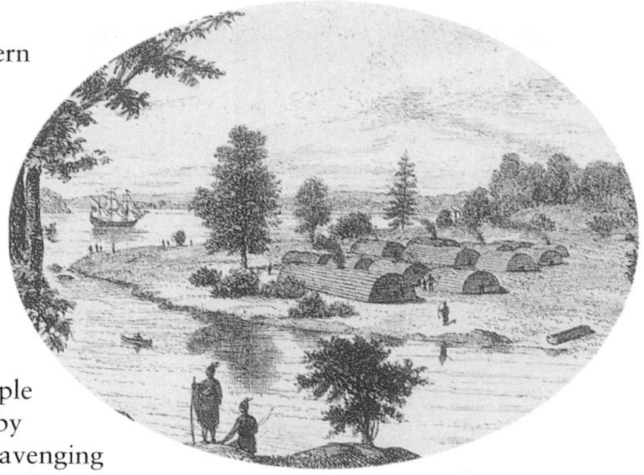

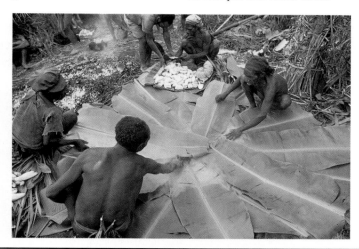

Figure 4 ▶ Three hundred years ago Manhattan was a much different place. This painting shows an area where Native Americans hunted and fished.

Hunter-Gatherers For most of human history, people were *hunter-gatherers,* or people who obtain food by collecting plants and by hunting wild animals or scavenging their remains. Early hunter-gatherer groups were small, and they migrated from place to place as different types of food became available at different times of the year. Even today there are hunter-gatherer societies in the Amazon rain forests of South America and in New Guinea, as shown in **Figure 5.**

Hunter-gatherers affect their environment in many ways. For example, some Native American tribes hunted bison, which live in grasslands. The tribes set fires to burn the prairies and prevent the growth of trees. In this way, the tribes kept the prairies as open grassland where they could hunt bison. In addition, hunter-gatherer groups probably helped spread plants to areas where the plants did not originally grow.

In North America, a combination of rapid climate changes and overhunting by hunter-gatherers may have led to the disappearance of some large mammal species. These species include giant sloths, giant bison, mastodons, cave bears, and saber-toothed cats. Huge piles of bones have been found in places where ancient hunter-gatherers drove thousands of animals into pits and killed them.

Graphic Organizer **Comparison Table**

Create the **Graphic Organizer** entitled "Comparison Table" described in the Appendix. Label the columns with "Hunter-Gatherers," "The Agricultural Revolution," and "The Industrial Revolution." Label the rows with "Characteristics" and "Effects on the Environment." Then, fill in the table with details about the characteristics and the effects on the environment of each historical period.

Figure 5 ▶ This modern hunter-gatherer group lives in New Guinea, a tropical island off the north coast of Australia.

9

MISCONCEPTION ALERT

The Purchase of Manhattan Island
According to popular legend, Manhattan Island was purchased by the Dutch in 1626 for $24 of beads. Several misconceptions are associated with the story. First, the island was purchased with various items, not beads. The value of the items that were exchanged was approximately 60 guilders. In 1847, a historian converted 60 guilders to dollars and got a value of $24. However, he did not adjust the amount for inflation. That $24 figure has persisted through the years and is probably much less than the actual sum. Secondly, it is believed that the Native American Chief who "sold" the island probably did not own the island because many different tribes lived on the island at the time.

Debate GENERAL

The Extinction of Large Mammals At the end of the last Ice Age, many large mammals became extinct in North America. Scientists are not exactly sure why the large mammals went extinct, but the extinction of these animals coincides with the arrival of humans. Divide students into two groups. Assign one group to research how people may have caused the extinction of the large mammals, and assign the other group to research other possible reasons for the extinctions. After students have gathered their information, set up a trial between the two sides. LS Verbal

READING SKILL BUILDER BASIC

Paired Reading Have students work in pairs to read this subsection. Students should alternate between reading and listening. Have the student who is listening make notes about the important ideas in the paragraph. When the pairs are finished reading, they can compile their notes in an outline. LS Verbal English Language Learners

Graphic Organizer GENERAL

Comparison Table
You may want to use this Graphic Organizer in a game to review material before the test. Divide the class into two teams. Ask students questions about material from the Comparison Table. Give points to each team that provides correct answers.

Discussion GENERAL

Selective Breeding and Domestication Selective breeding is a process in which plants and animals are bred to enhance specific characteristics. Dogs, cats, cows, horses, pigs, turkeys, and chickens have been selectively bred from wild animals. Broccoli, cauliflower, cabbage, kale, kohlrabi, and Brussels sprouts are domestic plants that are derived from the same species of wild mustard plant. Dogs are a good example of selective breeding. Some dogs, such as the Australian shepherd were selectively bred to herd animals, while the husky was selectively bred to pull a dog sled. This is why there is such a great variety of dogs, from Chihuahuas to Great Danes, which are both the same species. Have students select a domesticated plant or animal of their choice and research the wild ancestors from which it was bred. **LS** Logical

Answers to Field Activity

Answers may vary. Students should notice some grass-like features in the corn seedling.

Figure 6 ▶ This grass is thought to be a relative of the modern corn plant. Native Americans may have selectively bred a grass like this to produce corn.

FIELD ACTIVITY

Germinating Corn Many people do not realize how easy it is to grow corn plants from unpopped popcorn kernels. This ancient grass will sprout in a matter of days if it is watered frequently. Place a few popcorn kernels on a wet paper towel, and place the paper towel in a clear plastic cup so that the kernels are visible from the outside. Leave the cup on a windowsill for several days and water it frequently. As your plant grows, see if you can observe any grasslike features. Record your observations in your *EcoLog.*

Figure 7 ▶ For thousands of years humans have burned forests to create fields for agriculture. In this photo, a rain forest in Thailand is being cleared for farming.

10

The Agricultural Revolution Eventually many hunter-gatherer groups began to collect the seeds of the plants they gathered and to domesticate some of the animals in their environment. **Agriculture** is the practice of growing, breeding, and caring for plants and animals that are used for food, clothing, housing, transportation, and other purposes. The practice of agriculture started in many different parts of the world over 10,000 years ago. This change had such a dramatic impact on human societies and their environment that it is often called the *agricultural revolution.*

The agricultural revolution allowed human populations to grow at an unprecedented rate. An area of land can support up to 500 times as many people by farming as it can by hunting and gathering. As populations grew, they began to concentrate in smaller areas. These changes placed increased pressure on local environments.

The agricultural revolution also changed the food we eat. The plants we grow and eat today are descended from wild plants. During harvest season, farmers collected seeds from plants that exhibited the qualities they desired. The seeds of plants with large kernels or sweet and nutritious flesh were planted and harvested again. Over the course of many generations, the domesticated plants became very different from their wild ancestors. For example, the grass shown in **Figure 6** may be related to the grass that corn was bred from.

As grasslands, forests, and wetlands were replaced with farmland, habitat was destroyed. Slash-and-burn agriculture, shown in **Figure 7**, is one of the earliest ways that land was converted to farmland. Replacing forest with farmland on a large scale can cause soil loss, floods, and water shortages. In addition, much of this converted land was farmed poorly and is no longer fertile. The destruction of farmland had far-reaching environmental effects. For example, the early civilizations of the Tigris-Euphrates River basin collapsed, in part, because the overworked soil became waterlogged and contaminated by salts.

SKILL BUILDER GENERAL

Math According to the World Wildlife Fund, rain forests are being cleared at a rate of 26 hectares per minute. Ask students to calculate how many hectares of rainforest are being cleared in an hour, a day, and a year. (26 ha per min × 60 min per hr = 1560 ha per hr; 1560 ha per hr × 24 hr per day = 37,440 ha per day; 37,440 ha per day × 365 days per year = 13,665,600 ha per year) To help students understand the size of the area, ask them to find out the size, in hectares, of their county and their state and compare that area to the area of rain forest being cleared. **LS** Logical

BRAIN FOOD 🧠

Ancient Dogs Fossils of domesticated dogs dating from 11,000 years ago have been found in Iraq, and cultivated plants date back 10,000 years in the Americas.

Figure 8 ▶ During much of the Industrial Revolution, few limits were placed on the air pollution caused by burning fossil fuels. Locomotives such as these are powered by burning coal.

The Industrial Revolution For almost 10,000 years the tools of human societies were powered mainly by humans or animals. However, this pattern changed dramatically in the middle of the 1700s with the Industrial Revolution. The Industrial Revolution involved a shift from energy sources such as animal muscle and running water, to fossil fuels, such as coal and oil. The increased use of fossil fuels and machines, such as the steam engines shown in **Figure 8,** changed society and greatly increased the efficiency of agriculture, industry, and transportation.

During the Industrial Revolution, the large-scale production of goods in factories became less expensive than the local production of handmade goods. On the farm, machinery further reduced the amount of land and human labor needed to produce food. As fewer people grew their own food, populations in urban areas steadily grew. Fossil fuels and motorized vehicles also allowed food and other goods to be transported cheaply across great distances.

Improving Quality of Life The Industrial Revolution introduced many positive changes. Inventions such as the light bulb greatly improved our quality of life. Agricultural productivity increased, and sanitation, nutrition, and medical care vastly improved. Yet with all of these advances, the Industrial Revolution introduced many new environmental problems. As the human population grew, many environmental problems such as pollution and habitat loss became more common.

In the 1900s, modern societies increasingly began to use artificial substances in place of raw animal and plant products. Plastics, artificial pesticides and fertilizers, and many other materials are the result of this change. While many of these products have made life easier, we are now beginning to understand some of the environmental problems they present. Much of environmental science is concerned with the problems associated with the Industrial Revolution.

Figure 9 ▶ Modern communication technology, such as radios, TVs, and computers characterize the later stages of the Industrial Revolution.

11

Earth as a Living Organism In the 1960s, a NASA scientist named James Lovelock was working to design tests to search for life on Mars. Lovelock started by trying to determine how Earth was different from Mars. His search for differences led him to study the atmospheres of both planets. Earth has an atmosphere of 78 percent nitrogen, 21 percent oxygen, a small amount of carbon dioxide, and other trace gases. The atmosphere of Mars is 95 percent carbon dioxide, and has traces of oxygen and other gases. His work showed that processes on Earth regulated and maintained the atmosphere in a way similar to the manner in which an organism would maintain conditions in its body.

This discovery led Lovelock to develop the Gaia hypothesis, which stated that Earth was a living organism. Gaia was the name of the ancient Greek Earth-goddess. Lovelock published a book in 1979 called *Gaia, A New Look at Life on Earth*. Since that time, the hypothesis has been widely debated. Some people argue that the hypothesis can be interpreted literally, while others think that it is a useful analogy to understand Earth's systems as a whole. The theory is one way of approaching and trying to understand environmental problems. Discuss the Gaia hypothesis with students and have them form their own opinions. **LS** Verbal

Figure 10 ▶ This photograph was taken in 1968 by the crew of *Apollo 8*. Photographs such as this helped people realize the uniqueness of the planet we share.

Spaceship Earth

Earth has been compared to a ship traveling through space that cannot dispose of waste or take on new supplies as it travels. Earth is essentially a *closed system*—the only thing that enters Earth's atmosphere in large amounts is energy from the sun, and the only thing that leaves in large amounts is heat. A closed system of this sort has some potential problems. Some resources are limited, and as the population grows, the resources will be used more rapidly. In a closed system, there is also the chance that we will produce wastes more quickly than we can dispose of them.

Although the Earth can be thought of as a complete system, environmental problems can occur on different scales: local, regional, or global. For example, your community may be discussing where to build a new landfill, or local property owners may be arguing with environmentalists about the importance of a rare bird or insect. The drinking water in your region may be affected by a polluted river hundreds of miles away. Other environmental problems are global. For example, ozone-depleting chemicals released in Brazil may destroy the ozone layer that everyone on Earth depends on.

Lake Washington: An Environmental Success Story

Seattle is located on a narrow strip of land between two large bodies of water. To the west is the Puget Sound, which is part of the Pacific Ocean, and to the east is Lake Washington, which is a deep freshwater lake. During the 1940s and early 1950s, cities on the east side of Lake Washington built 11 sewer systems that emptied into the lake. Unlike raw sewage, this sewage was treated and was not a threat to human health. So, people were surprised by research in 1955 showing that the treated sewage was threatening their lake. Scientists working in Dr. W. T. Edmondson's lab at the University of Washington found a bacterium, *Oscillatoria rubescens,* that had never been seen in the lake before.

Dr. Edmondson knew that in several lakes in Europe, pollution from sewage had been followed by the appearance of *O. rubescens*. A short time after, the lakes deteriorated severely and became cloudy, smelly, and unable to support fish. The scientists studying Lake Washington realized that they were seeing the beginning of this process.

About this same time, Seattle set up the Metropolitan Problems Advisory Committee, chaired by James Ellis. Dr. Edmondson wrote Ellis a letter that explained what could be expected in the future if action was not taken. The best solution to the problem seemed to be to pump the sewage around the lake and empty it deep into Puget Sound. Although this solution may seem like it would save one body of water by polluting another one, it was actually a good choice. Diluting the sewage in Puget Sound is less of an environmental problem than

allowing it to build up in an enclosed lake.

Cities around the lake had to work together to connect their sewage plants to large lines that would carry the treated sewage to Puget Sound. Because there was no legal way for cities to connect plants

12

Population Growth: A Local Pressure One reason many environmental problems are so pressing today is that the agricultural revolution and the Industrial Revolution allowed the human population to grow much faster than it had ever grown before. The development of modern medicine and sanitation also helped increase the human population. As shown in **Figure 11**, the human population almost quadrupled during the 20th century. Producing enough food for such a large population has environmental consequences. In the past 50 years, nations have used vast amounts of resources to meet the world's need for food. Many of the environmental problems that affect us today such as habitat destruction and pesticide pollution are the result of feeding the world in the 20th century.

There are many different predictions of population growth for the future. But most scientists think that the human population will almost double in the 21st century before it begins to stabilize. We can expect that the pressure on the environment will continue to increase as the human population and its need for food and resources grows.

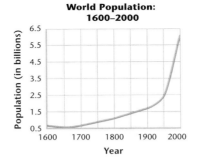

**World Population:
1600–2000**

Figure 11 ▶ The human population is now more than 10 times larger than the population of 400 years ago.

▶ Lake Washington is now clean enough for everyone to enjoy.

at the time, Ellis successfully worked for the passage of a bill in the state legislature that set up committees to handle projects of this kind. Newspaper articles and letters to the editor addressed the issue. Public forums and discussion groups were also held.

The first sewage plant was connected in 1963. Today, the lake is clearer than it has been since scientists began their studies of the lake in the 1930s. The story of Lake Washington is an example of how environmental science and public action work together to solve environmental problems. Science was essential to understanding a healthy lake ecosystem, to documenting changes that were beginning to cause problems, and to making predictions about what would happen if changes were made or if nothing was done. Engineers offered practical solutions to the problem of moving the sewage. Legislators and civic leaders addressed the legal problems. Volunteers, local media, and local activists provided public education and pressed to get the problem solved quickly. The clear, blue waters of Lake Washington stand as a monument to citizens' desires to live in a clean, healthy environment and to their ability to work together to make it happen.

CRITICAL THINKING

1. Analyzing Processes Explain how each person and group played a crucial role in the cleanup of Lake Washington.

2. Analyzing Relationships How was the scientists' work similar to the work of the Keene High School students you read about in this section?

13

Quick**LAB**

Skills Acquired:

• Classifying

• Identifying and Recognizing Patterns

Answers

1. Answers may vary.

2. Answers may vary.

Homework ——— GENERAL

Tasmanian Tigers The last known Tasmanian tiger, shown in **Figure 14,** died in captivity at the Holbart Zoo in Australia in 1936. Since that time, there have been hundreds of reported sightings but no Tasmanian tigers have been captured or photographed. It is possible that this large marsupial is still living in the forests of Tasmania. Have students find out about the status of the Tasmanian tiger.

Quick**LAB**

Classifying Resources

Procedure

1. Create a table similar to Table 2.

2. Choose five objects in your classroom, such as a **pencil,** a **notebook,** or a **chair.**

3. Observe your objects closely, and list the resources that comprise them. For example, a pencil is made of wood, graphite, paint, aluminum, rubber, and pumice.

4. Classify the resources you have observed as nonrenewable or renewable.

Analysis

1. What percentage of the resources you observed are renewable? What percentage of the resources are nonrenewable?

2. Hypothesize the origin of three of the resources you observed. If time permits, research the origin of the resources you chose to find out if you were correct.

Table 2 ▼

Renewable and Nonrenewable Resources	
Renewable	**Nonrenewable**
energy from the sun	metals such as iron, aluminum, and copper
water	
wood	nonmetallic materials such as salt, sand, and clay
soil	
air	fossil fuels

Figure 12 ▶ More than 12 million tons of copper have been mined from the Bingham mine in Utah. Once all of the copper that can be profitably extracted is used up, the copper in this mine will be depleted.

14

What Are Our Main Environmental Problems?

You may feel as though the world has an unlimited variety of environmental problems. But we can generally group environmental problems into three categories: resource depletion, pollution, or loss of biodiversity.

Resource Depletion Any natural material that is used by humans is called a **natural resource.** Natural resources can be classified as renewable and nonrenewable as shown in **Table 2.** A *renewable resource* is a resource that can be replaced relatively quickly by natural processes. Fresh water, air, soil, trees, and crops are all resources that can be renewed. Energy from the sun is also a renewable resource. A *nonrenewable resource* is a resource that forms at a much slower rate than the rate that it is consumed. The most common nonrenewable resources are minerals and fossil fuels. Once the supply of a nonrenewable resource is used up, it may take millions of years to replenish it.

Resources are said to be *depleted* when a large fraction of the resource has been used up. **Figure 12** shows a mine where copper, a nonrenewable resource, is removed from the Earth's crust. Some renewable resources can also be depleted. For example, if trees are harvested faster than they can grow naturally in an area, deforestation will result.

Pollution One effect of the Industrial Revolution is that societies began to produce wastes faster than the wastes could be disposed of. These wastes accumulate in the environment and cause pollution. **Pollution** is an undesired change in air, water, or soil that adversely affects the health, survival, or activities of humans or

BRAIN FOOD

Tracking the Ivory-Billed Woodpecker

The ivory-billed woodpecker, one of the largest species of woodpeckers, lived for centuries in the forests of Cuba and the southern United States. As a result of logging and habitat loss, the bird had not been seen in the United States since the 1950s, and was considered to be extinct. However, in April of 1999 a college student on a spring break hunting trip in southeastern Louisiana provided a reliable report of an ivory-billed woodpecker sighting. Several groups of researchers have searched this area of Louisiana since the report, but unfortunately, the status of the ivory-billed woodpecker remains unresolved. Encourage students to find out about the current status of the ivory-billed woodpecker hunt.

other organisms. Much of the pollution that troubles us today is produced by human activities. Air pollution in Mexico City, as shown in **Figure 13**, is dangerously high, mostly because of car exhaust.

There are two main types of pollutants. *Biodegradable pollutants* are pollutants that can be broken down by natural processes. They include materials such as human sewage or a stack of newspapers. Degradable pollutants are a problem only when they accumulate faster than they can be broken down. Pollutants that cannot be broken down by natural processes, such as mercury, lead, and some types of plastic, are called *nondegradable pollutants*. Because nondegradable pollutants do not break down easily, they can build up to dangerous levels in the environment.

Loss of Biodiversity The term **biodiversity** refers to the number and variety of species that live in an area. Earth has been home to hundreds of millions of species. Yet only a fraction of those species are alive today—the others are extinct. Extinction is a natural process, and several large-scale extinctions, or *mass extinctions*, have occurred throughout Earth's history. For example, at the end of the Permian period, 250 million years ago, as much as 95 percent of all species became extinct. So why should we be concerned about the modern extinction of an individual species such as the Tasmanian tiger shown in **Figure 14**?

The organisms that share the world with us can be considered natural resources. We depend on other organisms for food, for the oxygen we breathe, and for many other things. A species that is extinct is gone forever, so a species can be considered a nonrenewable resource. We have only limited information about how modern extinction rates compare with those of other periods in Earth's history. But many scientists think that if current rates of extinction continue, it may cause problems for human populations in the future. Many people also argue that all species have potential economic, ecological, scientific, aesthetic, and recreational value, so it is important to preserve them.

Figure 13 ▶ The problem of air pollution in Mexico City is compounded because the city is located in a valley that traps air pollutants.

Figure 14 ▶ The Tasmanian tiger may be the only mammal to become extinct in the past 200 years on the island of Tasmania. During the same period of time, on nearby Australia, as much as 50 percent of all mammals became extinct.

SECTION 1 Review

1. **Explain** how hunter-gatherers affected the environment in which they lived.

2. **Describe** the major environmental effects of the agricultural revolution and the Industrial Revolution.

3. **Explain** how environmental problems can be local, regional, or global. Give one example of each.

4. **Identify** an example of a natural source of pollution.

CRITICAL THINKING

5. **Analyzing Relationships** How did the Industrial Revolution affect human population growth?

6. **Making Inferences** Fossil fuels are said to be nonrenewable resources, yet they are produced by the Earth over millions of years. By what time frame are they considered nonrenewable? Write a paragraph that explains your answer. **WRITING SKILLS**

15

SECTION 2
The Environment and Society

Objectives

▶ Describe "The Tragedy of the Commons."
▶ Explain the law of supply and demand.
▶ List three differences between developed and developing countries.
▶ Explain what sustainability is, and describe why it is a goal of environmental science.

Key Terms

law of supply and demand
ecological footprint
sustainability

When we think about environmental problems and how to solve them, we have to consider human societies, how they act, and why they do what they do. One way to think about society and the environment is to consider how a society uses common resources. A neighborhood park, for example, is a common resource that people share. On a larger scale, the open ocean is not owned by any nation, yet people from many countries use the ocean for fishing and for transporting goods. How do we decide how to share common resources? In 1968, ecologist Garrett Hardin published an essay titled "The Tragedy of the Commons," which addressed these questions and became the theoretical backbone of the environmental movement.

"The Tragedy of the Commons"

In his essay, Hardin argued that the main difficulty in solving environmental problems is the conflict between the short-term interests of individuals and the long-term welfare of society. To illustrate his point, Hardin used the example of the *commons*, as shown in **Figure 15**. Commons were areas of land that belonged to a whole village. Anyone could graze cows or sheep on the commons. It was in the best short-term interest of an individual to put as many animals as possible on the commons. Individuals thought, If I don't use this resource, someone else will. And anyway, the harm my animals cause is too little to matter.

However, if too many animals grazed on the commons, the animals destroyed the grass. Then everyone suffered because no one could raise animals on the commons. Commons were eventually replaced by closed fields owned by individuals. Owners were careful not to put too many animals on their land, because overgrazing meant that fewer animals could be raised the next year. The point of Hardin's essay is that someone or some group has to

Figure 15 ▶ Hardin observed that when land was held in common (left), individuals tended to graze as many animals as possible. Overgrazing led to the destruction of the land resources. When commons were replaced by enclosed fields owned by individuals (right), people tended to graze only the number of animals that the land could support.

16

take responsibility for maintaining a resource. If no one takes that responsibility, the resource can be overused and become depleted.

Earth's natural resources are our modern commons. Hardin thought that people would continue to deplete natural resources by acting in their own self-interest to the point of society's collapse. But Hardin did not consider the social nature of humans. Humans live in groups and depend on one another. In societies, we can solve environmental problems by planning, organizing, considering the scientific evidence, and proposing a solution. The solution may override the interests of individuals in the short term, but it improves the environment for everyone in the long term.

Economics and the Environment

In addition to social pressures, economic forces influence how we use resources. Many of the topics you will explore later in this book are affected by economic considerations.

Supply and Demand One basic rule of economics is the **law of supply and demand,** which states that the greater the demand for a limited supply of something, the more that thing is worth. One example of this rule is shown in **Figure 16,** which illustrates the relationship between the production of oil and the price of oil over 20 years. Many environmental solutions have to take the relationship between supply and demand into account. For example, if the supply of oil decreases, we have three choices: pay the higher price, use less oil, or find new sources of energy.

Costs and Benefits The cost of environmental solutions can be high. To determine how much to spend to control air pollution, a community may perform a cost-benefit analysis. A *cost-benefit analysis* balances the cost of the action against the benefits one expects from it. The results of a cost-benefit analysis often depend on who is doing the analysis. To an industry, the cost of pollution control may outweigh the benefits, but to a nearby community, the benefits may be worth the high price. The cost of environmental regulations is often passed on to the consumer or the taxpayer. The consumer then has a choice—pay for the more expensive product that meets environmental regulations or seek out a cheaper product that may not have the same environmental safeguards.

Risk Assessment One of the costs of any action is the risk of an undesirable outcome. Cost-benefit analysis involves *risk assessment,* which is one tool that helps us create cost-effective ways to protect our health and the environment. To come up with an effective solution to an environmental problem, the public must perceive the risk accurately. This does not always happen. In one study, people were asked to assess the risk from various technologies. The public generally ranked nuclear power as the riskiest technology on the list, whereas experts ranked it 20th—less risky than riding a bicycle.

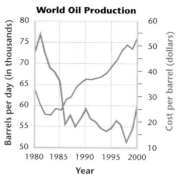

World Oil Production

Source: U.S. Department of Energy.

Figure 16 ▶ In general, when the production of oil declines, the price of a barrel of oil increases.

MATHPRACTICE

Market Equilibrium
In economics, the point where supply and demand are in balance is known as *market equilibrium*. In Figure 16, market equilibrium for oil was reached in 1986. What was the cost of a barrel of oil in that year? How many barrels of oil were produced in that year? By how much did the cost of a barrel of oil decline from 1981 to 1986?

17

MISCONCEPTION ALERT

"The Tragedy of the Commons"
Students may assume that because "The Tragedy of the Commons" is a parable that it is a very simple theory that does not relate to real experiences. Explain to students that the opposite is true. Even though "The Tragedy of the Commons" is simple, the underlying concept applies to many problems in environmental science.

SKILL BUILDER ———— GENERAL

Writing Ask students to write their own parables about an environmental issue. The issue that they choose could be either local or global. Remind students that the parables should be simple and straightforward.

Teach

Interpreting Statistics ———— GENERAL

Oil Prices After reviewing **Figure 16,** students may think that the amount of oil production is the only factor that determines the cost of a barrel of oil. Ask them to think about other factors that influence the cost of oil. The cost of oil depends on many other factors such as world demand, time of year, political influence, and artificial market manipulation. Also ask students to think about what factors could be included in the cost of oil but are not. (Sample answers: The cost of cleaning up oil spills, greenhouse gas emissions, or health-care problems.) These costs are called *externalities*.

READING SKILL BUILDER ———— BASIC

K-W-L To use the K-W-L strategy, ask students to make a chart in their *EcoLog.* The chart should have three columns, and each column should be labeled "What you **K**now," "What you **W**ant to know," and "What you **L**earned." Then ask students the following questions to answer in the first column: "What is meant by *supply and demand*? What is meant by *cost-benefit analysis*? What is *risk assessment*?" Tell students that these are ways that the environment and resources are evaluated by economists. Then ask students to fill in the second column. After students read the Economics and the Environment subsection, have them revisit their chart and fill in the third column.

MATHPRACTICE

Answers

$26 per barrel; $60,000 \times 365 = 21,900,000$; $22

Using the Figure —— GENERAL

Fresh Vegetables Figure 17 shows two different vegetable markets. One is in a developing nation while the other is in a developed nation. Ask students the following questions, "What do you notice about the variety of vegetables in each photograph? What do you notice about the packaging in each market? Which market do you think used more energy to get the vegetables to the market? Which market do you think has the freshest vegetables?" **LS Visual**

Homework —— GENERAL

Examining Food Packaging Ask students to look at the food they have at home. Ask students to look at how the food is packaged. Ask students how much of the food is in cans, how much is in bags, and how much is in boxes. Ask students, "How much of the packaging contributes to the waste we generate? What types of resources are used to make the packaging for food? How does the packaging help the food?"

Geofact

Minerals of South Africa The resources a country has are a result of geologic processes. South Africa, for example, has some of the most productive mineral deposits in the world. In fact, the country is nearly self-sufficient in the mineral resources that are important to modern industry. South Africa is the world's largest producer of gold, platinum, and chromium.

Figure 17 ▶ Developed and developing nations have different consumption patterns and different environmental problems. Both of these photos show food markets. What do you think the environmental problems of each consumption pattern are?

Developed and Developing Countries

The decisions and actions of all people in the world affect our environment. But the unequal distribution of wealth and resources around the world influences the environmental problems that a society faces and the choices it can make. The United Nations generally classifies countries as either developed or developing. *Developed countries* have higher average incomes, slower population growth, diverse industrial economies, and stronger social support systems. They include the United States, Canada, Japan, and the countries of Western Europe. *Developing countries* have lower average incomes, simple and agriculture-based economies, and rapid population growth. In between are middle-income countries, such as Mexico, Brazil, and Malaysia.

Population and Consumption

Almost all environmental problems can be traced back to two root causes. First, the human population in some areas is growing too quickly for the local environment to support. Second, people are using up, wasting, or polluting many natural resources faster than they can be renewed, replaced, or cleaned up.

Local Population Pressures When the population in an area grows rapidly, there may not be enough natural resources for everyone in the area to live a healthy, productive life. Often, as people struggle for survival in severely overpopulated regions, forests are stripped bare, topsoil is exhausted, and animals are driven to extinction. Malnutrition, starvation, and disease can be constant threats. Even though there are millions of people starving in developing countries, the human population tends to grow most rapidly in these countries. Food production, education, and job creation cannot keep pace with population growth, so each person gets fewer resources as time goes by. Of the 4.5 billion people in developing countries, fewer than half have access to enough food, safe drinking water, and proper sanitation.

INCLUSION Strategies

• Gifted and Talented

Have students draw a timeline of the past 200 years on a piece of unlined paper. They should indicate on the timeline when five species of animals became extinct. For each of the extinct species, students should include habitats, last sightings, and possible reasons for extinction. Lower-level students should be able to label dates the species became extinct and draw pictures or locate and print pictures of the species on the Internet. They should staple or glue the pictures on the timeline. Compile the timelines to share with the class.

Table 3 ▼

Indicators of Development for the United States, Japan, Mexico, and Indonesia					
	Measurement	U.S.	Japan	Mexico	Indonesia
Health	life expectancy in years	77	81	71.5	68
Population growth	per year	0.8%	0.2%	1.7%	1.8%
Wealth	gross national product per person	$29,240	$32,350	$3,840	$640
Living space	people per square mile	78	829	133	319
Energy use	per person per year (millions of Btu)	351	168	59	18
Pollution	carbon dioxide from fossil fuels per person per year (tons)	20.4	9.3	3.5	2.2
Waste	garbage produced per person per year (kg)	720	400	300	43

Consumption Trends For many people in the wealthier part of the world, life is better than ever before. Pollution controls improve every year, and many environmental problems are being addressed. In addition, the population has stabilized or is growing slowly. But to support this quality of life, developed nations are using much more of Earth's resources than developing nations are. Developed nations use about 75 percent of the world's resources, even though they make up only about 20 percent of the world's population. This rate of consumption creates more waste and pollution per person than in developing countries, as shown in **Table 3**.

Ecological Footprints One way to express the differences in consumption between nations is as an ecological footprint, as shown in **Figure 18**. An **ecological footprint** shows the productive area of Earth needed to support one person in a particular country. It estimates the land used for crops, grazing, forest products, and housing. It also includes the ocean area used to harvest seafood and the forest area needed to absorb the air pollution caused by fossil fuels.

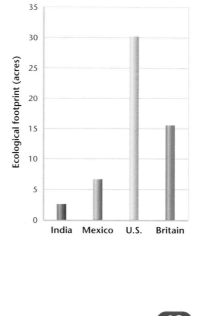

Figure 18 ▶ An ecological footprint is a calculation of the amount of land and resources needed to support one person from a particular country. The ecological footprint of a person in a developed country is, on average, four times as large as the footprint of a person in a developing country.

Ecological footprint (acres) — India, Mexico, U.S., Britain

19

Using the Figure — GENERAL

Developed Vs. Developing
Table 3 shows indicators for development from four different countries. Ask students to compare the life expectancy between the countries and ask what they notice. Ask students, "Why do you think people in developed countries live longer?" Next ask students to compare the other characteristics. Ask students, "Does the living space tell you anything about the conditions that people live in?" Note that Japan has a high standard of living but also has the highest population density. Finally ask, "What does the amount of waste generated per person tell you about the differences between developed and developing nations?" **LS Logical**

Using the Figure — GENERAL

Different Footprints Figure 18 shows the ecological footprint for four countries. Ask students, "Which country has the largest ecological footprint? Why do you think the United States has the largest ecological footprint? Which regions of the United States have the largest ecological footprints? Which regions have the smallest ecological footprints?" **LS Logical**

BRAIN FOOD

Reducing Your Ecological Footprint
You can do many things to reduce your ecological footprint. You can eat locally grown foods, eat less meat, carpool, ride public transportation, ride a bike, and use lights and appliances that conserve energy. Ask students to make a list of five things that they could do to reduce their ecological footprint.

◀ Internet Activity — GENERAL

Ecological Footprints The Internet has a number of different calculators for determining ecological footprints. Ask students to find at least three of these calculators and determine their ecological footprint. Ask students how the results from the different calculators compare. Ask students why they think these differences occur.

Teach, continued

Discussion —— BASIC

Environmental Problems in Your Neighborhood Ask students to imagine what they would think if they heard that a major employer in their area was found to be dumping poisonous chemicals, which showed up in the local water supply. Ask students, "How would you recommend solving the problem?" Then tell students that the company has stated that if it is forced to clean up its operations, it will simply close its plant and open one in another country that has less-strict environmental controls. Ask students, "Does this information change your opinion about what should be done?" Point out that knowing the best way to solve environmental problems is difficult because all possible actions often have both positive and negative consequences.

Group Activity —— GENERAL

Town Council Meeting Select five students to represent the town council. Then, select a current environmental topic for your area. Poll remaining students in the class to see if there are opposing viewpoints in the class. If not, assign students to one side or the other for the topic. Have students research their position on the topic and prepare a 3 minute address to the town council. The town council members should take notes about each address. After all the students have expressed their opinions to the council, have the council vote on the issue. Then allow each member of the council to state why he or she voted for a certain issue. Co-op Learning

FIELD ACTIVITY

Critical Thinking and the News
Find a news article or watch a news broadcast about a current environmental issue. In your *Ecolog*, write down your initial reaction including your thoughts, feelings, and questions.

Now, look or think again, and answer the following questions:
• Did the report present different sides of the issue? Describe the sides.
• Did the report seem to favor one side over the other? How could you tell?
• Did the report use images, sounds, or words that made you feel a certain way?
• Did the report provide any facts that helped you form an opinion? Try to list the facts.
• Were sources of the facts provided? Did the sources seem reliable?
• Were the opinions of any expert scientists presented? Who were the scientists?
• Is there any information that was *not* provided that might be important? Give examples.
• When you think about the issue more, does your opinion change?

Figure 19 ▶ Anyone can express an opinion on environmental issues at state and local public hearings.

☑ **internet** connect

www.scilinks.org
Topic: Solving Environmental Problems
SciLinks code: HE4013

SC*LINKS* Maintained by the National Science Teachers Association

20

Environmental Science in Context

As you have learned, environmental problems are complex. Simple solutions are rare, and they sometimes cause more damage than the original problem did. To complicate matters, in recent years, the environment has become a battleground for larger issues that affect human societies. For example, how do you balance the rights of individuals and property owners with the needs of society as a whole? Or, when economic or political refugees emigrate—legally or illegally—what can be done about the devastation they may cause to the local environment? How do human rights relate to the environment?

Critical Thinking and the Environment People on any side of an environmental issue may feel passionately about their cause, and they can distort information and mislead people about the issues. Research done by environmental scientists is often used to make political points or is misrepresented to support controversial viewpoints. In addition to the scientific data, the economic dimensions of an environmental issue can be oversimplified. To further complicate things, the media often sensationalizes environmental issues. So, as you make your own decisions about the environment, it is essential that you use your critical-thinking skills.

Learning to think critically about what you see in newspapers, on TV, and on the Internet will help you make informed decisions. As you explore environmental science further, you should remember a few things. First, be prepared to listen to many viewpoints. People have many different reasons for the opinions they form. Try to understand what those reasons are before reacting to their ideas. If you want your ideas to be heard, it is important that you listen to the opinions of others, as shown in **Figure 19**. Also, identify your own bias. How does it affect the way you interpret the issue?

Student Opportunities

Set Up A Recycling Program There are many things that students can do to help work towards a sustainable world. Recycling is one easy way to make a difference. Glass, plastic, paper, and aluminum are easy to recycle. Most communities have a recycling center that will accept these materials. The center often has information on what can and cannot be recycled. Students could see that containers for recyclable materials are placed on campus, or if the containers are already there, students could start a campaign to encourage other students to use the containers. Recycling saves energy and resources and reduces the amount of trash that goes to landfills.

Second, investigate the source of the information you encounter. Ask yourself whether the authors have reason for bias. Also, question the conclusions that are drawn from data. Ask yourself if the data support the claims that are made. Be especially critical of information posted on the Internet—flashy graphics and persuasive text might be hiding a biased agenda. Finally, gather all the information you can before drawing a conclusion.

A Sustainable World

Despite the differing points of view on the environment, most people support a key goal of environmental science: achieving sustainability. **Sustainability** is the condition in which human needs are met in such a way that a human population can survive indefinitely. A sustainable world is not an unchanging world—technology advances and human civilizations continue to be productive. But at the present time we live in a world that is far from sustainable. The standard of living in developed countries is high because those countries are using resources faster than they can be replaced.

The problems described in this chapter are not insurmountable. Achieving a sustainable world requires everyone's participation. If all parts of society—individual citizens, industry, and government—cooperate, we can move toward sustainability. For example, you read about how Seattle's Lake Washington is cleaner and healthier now than it was 30 years ago. Another example is the bald eagle, which was once on the brink of extinction. Today bald eagles are making a comeback, because of the efforts to preserve their habitat and to reduce pollution from the pesticide DDT.

Nevertheless, our environmental problems are significant and require careful attention and action. The 21st century will be a crucial time in human history, a time when we must find solutions that allow people on all parts of our planet to live in a clean, healthy environment and have the resources they need for a good life.

Figure 20 ▶ These high school students are taking action to improve their environment. They are cleaning up trash that is clogging an urban creek.

Connection to Astronomy

Another Earth? If the environment on Earth changed drastically, would we have anywhere to go? There are no other planets in our solar system with an adequate range of temperatures, a breathable atmosphere, or the resources needed to sustain humans with our present technology. There may be other planets like Earth in the universe, but the closest planets we know of are in other solar systems that are light-years away.

SECTION 2 Review

1. **Describe** three differences between developing and developed nations using the examples in Table 3. Would you classify Mexico as a developing nation? Explain your answer.

2. **Explain** why critical thinking is an important skill in environmental science.

3. **Explain** the law of supply and demand, and give an example of how it relates to the environment.

CRITICAL THINKING

4. **Applying Ideas** The law of supply and demand is a simplification of economic patterns. What other factors might affect the cost of a barrel of oil?

5. **Evaluating Ideas** Write a description of "The Tragedy of the Commons." Do you think that Hardin's essay is an accurate description of the relationship between individuals, society, and the environment? **WRITING SKILLS**

Answers to Section Review

1. Answers may vary. Mexico has characteristics of developing and developed countries.

2. Answers may vary. Critical thinking is a skill that enables one to evaluate ideas and information.

3. The law of supply and demand states that the greater the demand for a limited supply of something is, the more the supply is worth. Oil is one example. When demand is high the price usually increases.

4. Answers may vary. There are many other factors that might affect the price of oil. The cost of shipping and transporting the oil may increase. The cost of finding and drilling wells may increase the price. Students may also note that the majority of oil producing nations are part of a trade organization called OPEC. OPEC can control the price and production of oil.

5. Answers may vary. "The Tragedy of the Commons" states that when land held in common is used for grazing, people tend to overgraze the land. When land is divided up and privately owned, people tend not to overgraze the land. To some extent this is accurate. People tend to care more for an area when it is under their direct control.

Alternative Assessment ——— GENERAL

Conduct a Survey Suggest that students work in pairs to develop and conduct a survey to determine the attitudes of family members, classmates, neighbors, teachers, and friends toward specific environmental issues. Encourage students to include their survey results in their *EcoLog.*

Chapter Resource File

- **Chapter Test** GENERAL
- **Chapter Test** ADVANCED
- **Concept Review** GENERAL
- **Critical Thinking** ADVANCED
- **Test Item Listing** GENERAL
- **Math/Graphing Lab** GENERAL
- **Research Lab** ADVANCED
- **Consumer Lab** GENERAL
- **Long-Term Project** BASIC

1 Understanding Our Environment

Key Terms

environmental science, 5
ecology, 6
agriculture, 10
natural resource, 14
pollution, 14
biodiversity, 15

Main Ideas

▶ Environmental science is an interdisciplinary study of human interactions with the living and nonliving world. One important foundation of environmental science is the science of ecology.

▶ Environmental change has occurred throughout Earth's history.

▶ Hunter-gatherer societies cleared grassland by setting fires and may have contributed to the extinction of some large mammals.

▶ The agricultural revolution caused human population growth, habitat loss, soil erosion, and the domestication of plants and animals.

▶ The Industrial Revolution caused rapid human population growth and the increased use of fossil fuels. Most modern environmental problems began during the Industrial Revolution.

▶ The major environmental problems we face today are resource depletion, pollution, and loss of biodiversity.

2 The Environment and Society

law of supply and demand, 17
ecological footprint, 19
sustainability, 21

▶ "The Tragedy of the Commons" was an influential essay that described the relationship between the short-term interests of the individual and the long-term interests of society.

▶ The law of supply and demand states that when the demand for a product increases while the supply remains fixed, the cost of the product will increase.

▶ Environmental problems in developed countries tend to be related to consumption. In developing nations, the major environmental problems are related to population growth.

▶ Describing how sustainability can be achieved is a primary goal of environmental science.

Using Key Terms

Use each of the following terms in a separate sentence.

1. *agriculture*
2. *natural resource*
3. *pollution*
4. *ecological footprint*
5. *sustainability*

Use the correct key term to complete each of the following sentences.

6. The_____ Revolution was characterized by a shift from human and animal power to fossil fuels.

7. Resources that can theoretically last forever are called _____ resources.

8. _____ is a term that describes the number and variety of species that live in an area.

✔ STUDY TIP

Root Words As you study it may be helpful to learn the meaning of important root words. You can find these roots in most dictionaries. For example, *hydro-* means "water." Once you learn the meaning of this root, you can learn the meanings of words such as *hydrothermal*, *hydrologist*, *hydropower*, and *hydrophobic*.

Understanding Key Ideas

9. An important effect that hunter-gatherer societies may have had on the environment was
   a. soil erosion.
   b. extinction.
   c. air pollution.
   d. All of the above

10. An important effect of the agricultural revolution was
    a. soil erosion.
    b. habitat destruction.
    c. plant and animal domestication.
    d. All of the above

11. Which of the following does *not* describe an effect of the Industrial Revolution?
    a. Fossil fuels became important energy sources.
    b. The amount of land and labor needed to produce food increased.
    c. Artificial substances replaced some animal and plant products.
    d. Machines replaced human muscle and animal power.

12. Pollutants that are not broken down by natural processes are
    a. nonrenewable.
    b. nondegradable.
    c. biodegradable.
    d. Both (a) and (c)

13. All of the following are renewable resources *except*
    a. energy from the sun.
    b. minerals.
    c. crops.
    d. fresh water.

14. In his essay, "The Tragedy of the Commons," one factor that Garrett Hardin failed to consider was
    a. the destruction of natural resources.
    b. human self-interest.
    c. the social nature of humans.
    d. None of the above

15. The term used to describe the productive area of Earth needed to support the lifestyle of one person in a particular country is called
    a. supply and demand.
    b. the ecological footprint.
    c. the consumption crisis.
    d. sustainability.

ANSWERS

Using Key Terms

1. Sample answer: Agriculture is the practice of growing, breeding and caring for plants and animals that are used for food, clothing, housing, transportation, and other purposes.

2. Sample answer: Natural resources are either renewable or nonrenewable.

3. Sample answer: Toxic chemicals released into the environment cause pollution.

4. Sample answer: A developed nation has an ecological footprint that is about four times larger than a developing nation.

5. Sample answer: Sustainability is a goal of environmental science.

6. Industrial
7. renewable
8. Biodiversity

Understanding Key Ideas

9. b
10. d
11. b
12. b
13. b
14. c
15. b

23

Assignment Guide	
Section	**Questions**
1	1–3, 6–13, 16, 17, 21, 25–29, 31, 35
2	4, 5, 14, 15, 18–20, 22–24, 32–34, 36, 37

Short Answer

16. Sample answer: Geology might be involved in environmental science because scientists might use geology to find out how different rock types might affect pollution flow. Chemistry might be involved in environmental science because scientists might use chemistry to identify how different chemicals interact in the environment.

17. Answers may vary. Biodegradable pollutants can cause environmental problems if they accumulate in the environment and they do not degrade quickly enough.

18. Sample answer: The Earth's resources are shared by everyone, if societies do not use resources responsibly, the resources can become depleted or polluted.

19. Answers may vary.

20. Answers may vary. Students might suggest looking at the sources for the information, the references provided, the presentation of data, or the credentials of the author.

21. Sample answer: A species can be a natural resource because a species is part of the environment.

Interpreting Graphics

22. Developing countries have a much higher rate of population growth than developed countries. Developed countries use about twice as much energy as developing countries, even though the populations of developed countries are much smaller.

23. Developing countries will probably consume a higher percentage of commercial energy over time, because population growth is increasing in developing countries.

Short Answer

16. Give an example of how environmental science might involve geology and chemistry.

17. Can biodegradable pollutants cause environmental problems? Explain your answer.

18. In what ways are today's environmental resources like the commons described in the essay "The Tragedy of the Commons"?

19. How could environmental concerns conflict with your desire to improve your standard of living?

20. If you were evaluating the claims made on a Web site that discusses environmental issues what types of information would you look for?

21. Can species be considered natural resources? Explain your answer.

Interpreting Graphics

The graphs below show the difference in energy consumption and population size in developed and developing countries. Use the graphs to answer questions 22–24.

Commercial Energy Consumption

Developing countries 32%

Developed countries 68%

Population Size

(Line graph: Population (in billions) on y-axis from 0 to 6; Year on x-axis from 1900 to 2000. Developing countries line rises steeply; Developed countries line rises slowly.)

22. Describe the differences in energy consumption and population growth between developed and developing countries.

23. Do you think that the percentage of commercial energy consumed by developing countries will increase or decrease? Explain your answer.

24. Why is information on energy consumption represented in a pie graph, while population size is shown in a line graph?

Concept Mapping

25. Use the following terms to create a concept map: *geology, biology, ecology, environmental science, chemistry, geography,* and *social sciences*.

Critical Thinking

26. Analyzing Ideas Are humans part of the environment? Explain your answer.

27. Drawing Conclusions Why do you think that fossil fuels were not widely used until the Industrial Revolution? Write a paragraph that describes your thoughts. **WRITING SKILLS**

28. Evaluating Assumptions Once the sun exhausts its fuel and burns itself out, it cannot be replaced. So why is the sun considered a renewable resource?

29. Evaluating Assumptions Read the description of the Industrial Revolution. Were all the effects of the Industrial Revolution negative? Explain your answer. **READING SKILLS**

Cross-Disciplinary Connection

30. Demographics Obtain the 1985 and 2000 census reports for your town or city. Look for changes in demographic characteristics, such as population size, income, and age. Make a bar graph that compares some of the characteristics you chose. How does your city or town compare with national trends? What might be some of the environmental implications of these trends?

Portfolio Project

31. Make a Diagram Many resources can be traced to energy from the sun. For example, plants living in swamps millions of years ago used energy from the sun to grow. Over time, some of these plants became coal deposits. When we burn coal today, we are using energy that radiated from the sun millions of years ago. Choose a resource, and create a diagram that traces the resource back to energy from the sun.

24

24. The pie graph represents part of a total amount, while the chart of population size shows change over time.

Concept Mapping

25. Answers to the concept mapping questions are on pp. 667–672.

Critical Thinking

26. Sample answer: Humans are part of the environment because they are animals that use resources.

27. Sample answer: Fossil fuels were not widely used before the Industrial Revolution because

there was not an efficient way to mine or transport the fuels and there were not many machines that could use fossil fuels.

28. Answers may vary. The sun is usually considered a renewable resource because it is much older than a human lifetime.

29. Sample answer: Many of the effects of the Industrial Revolution were positive. Machines made many processes much more efficient and reduced the need for human labor and animal power to do many tasks. The Industrial Revolution also improved the standard of living.

Use the table below to answer questions 32–34.

	U.S.	Japan	Indonesia
People per square mile	78	829	319
Garbage produced per person per year	720 kg	400 kg	43 kg

32. **Analyzing Data** Make a bar graph that compares the garbage produced per person per year in each country.

33. **Making Calculations** Calculate how much garbage is produced each year per square mile of each country listed in the table.

34. **Evaluating Data** Use the information in the table to evaluate the validity of the following statement: In countries where population density is high, more garbage is produced per person.

 **WRITING SKILLS**

35. **Communicating Main Ideas** Briefly describe the relationship between humans and the environment through history.

36. **Writing Persuasively** Write a persuasive essay explaining the importance of science in a debate about an environmental issue.

37. **Outlining Topics** Write a one-page outline that describes population and consumption in the developing and developed world.

STANDARDIZED TEST PREP

For extra practice with questions formatted to represent the standardized test you may be asked to take at the end of your school year, turn to the sample test for this chapter in the Appendix.

 READING SKILLS

Read the passage below, and then answer the questions that follow.

Think about what you did this morning. From the moment you got up, you were making decisions and acting in ways that affect the environment. The clothes you are wearing, for example, might be made of cotton. Several years ago the fibers of cotton in your shirt might have sprouted as seedlings in Egypt or Arizona. The cotton seedlings were probably irrigated with water diverted from a nearby river or lake. Chemicals such as pesticides, herbicides, and fertilizers helped the seedlings grow into plants. Furthermore, the metal in the machines that harvested the cotton was mined from the Earth's crust. In addition, the vehicles that brought the shirt to the store where you bought it were powered by fossil fuels. Fossil fuels came from the bodies of tiny sea creatures that lived millions of years ago. All of these connections can make environmental science a complex and interesting field.

1. According to the passage, which of the following conclusions is true?
   a. Decisions we make in everyday life do not affect our environment.
   b. Cotton comes from minerals in the Earth's crust.
   c. Many different things in the environment are connected and interrelated.
   d. There is no connection between the resources needed to grow a field of cotton and a cotton shirt.

2. Which of the following statements best describes the meaning of the term *irrigation*?
   a. Irrigation is a connection between living things in the environment.
   b. Irrigation is the artificial process by which water is supplied to plants.
   c. Irrigation is the process of diverting water from a stream or lake.
   d. Irrigation is the process by which cotton seedlings grow into plants.

25

30. Answers may vary. The change in demographics will probably show change in population and income levels. The change may also show an increase in age. Increased population affects the resources used, and increased income probably translates to a larger ecological footprint.

Portfolio Project

31. Sample answer: Trees grow because they use the energy from the sun. Trees are cut down to burn as fuelwood. When wood is burned, the energy is released.

Math Skills

32. **Garbage Produced per Person per Year**

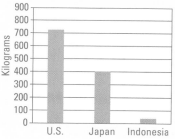

33. U.S. 56,160 kg/mi^2; Japan 331,600 kg/mi^2; Indonesia 13,717 kg/mi^2

34. The statement is false because the population density is high in Japan and Indonesia, yet the garbage produced per person is lower than the United States.

Writing Skills

35. Answers may vary.
36. Answers may vary.
37. Answers may vary.

Reading Skills

1. c
2. c

WHAT'S IN AN ECOSYSTEM?

Teacher's Notes

Time Required

one or two 45-minute class periods

Lab Ratings

EASY —————————→ HARD

TEACHER PREPARATION ▲
STUDENT SETUP ▲▲▲
CONCEPT LEVEL ▲▲
CLEANUP ▲▲

Skills Acquired

- Measuring
- Collecting Data
- Communicating
- Predicting

The Scientific Method

In this lab, students will:
- Make Observations
- Draw Conclusions
- Communicate the Results

Materials

The materials listed are enough for a group of 3 to 4 students.

Objectives

▶ **Survey** an area of land and determine the land's physical features and the types of organisms that live there.

▶ **USING SCIENTIFIC METHODS** **Identify** possible relationships between the organisms that live in the area of land you surveyed.

Materials

hand lens
markers or felt-tip pens of several different colors
notebook
pen or pencil
poster board
stakes, (4)
string, about 50 m
tape measure or metric ruler

optional materials: field guides to insects or plants

▶ **Marking a Site** Use stakes and string to mark a site that you will observe in detail.

What's in an Ecosystem?

How well do you know the environment around your home or school? You may walk through it every day without noticing most of the living things it contains or thinking about how they survive. Ecologists, on the other hand, observe organisms and seek to understand how ecosystems work. In this lab, you will play the role of an ecologist by closely observing part of your environment.

Procedure

1. Use a tape measure or meter stick to measure a 10 m × 10 m site to study. Place one stake at each corner of the site. Loop the string around each stake, and run the string from one stake to the next to form boundaries for the site.

2. Survey the site, and then prepare a site map of the physical features of the area on the poster board. For example, show the location of streams, sidewalks, trails, or large rocks, and indicate the direction of any noticeable slope.

3. Create a set of symbols to represent the organisms at your site. For example, you might use green triangles to represent trees, blue circles to represent insects, or brown squares to represent animal burrows or nests. At the bottom or side of the poster board, make a key for your symbols.

4. Draw your symbols on the map to show the location and relative abundance of each type of organism. If there is not enough space on your map to indicate the specific kinds of plants and animals you observed, record them in your notebook.

5. In your notebook, record any observations of organisms in their environment. For example, note insects feeding on plants or seeking shelter under rocks. Also describe the physical characteristics of your study area. Consider the following characteristics:

   a. **Sunlight Exposure** How much of the area is exposed to sunlight?

   b. **Soil** Is the soil mostly sand, silt, clay, or organic matter?

   c. **Rain** When was the last rain recorded for this area? How much rain was received?

   d. **Maintenance** Is the area maintained? If so, interview the person who maintains it and find out how often the site is watered, fertilized, treated with pesticides, and mowed.

   e. **Water Drainage** Is the area well drained, or does it have pools of water?

   f. **Vegetation Cover** How much of the soil is covered with vegetation? How much of the soil is exposed?

6. After completing these observations, identify a 2 m × 2 m area that you would like to study in more detail. Stake out this area, and wrap the string around the stakes.

26

Tips and Tricks

Before you perform this lab, obtain as many field guides about the natural history of your region as possible. Local libraries and used book stores should have these guides. The Internet should also offer information. Encourage the groups to study the field guides before performing the lab, so that they do not go into the field "blind." Award points to groups that use the field guides to correctly identify plants and animals in their plots.

Encourage groups to survey terrain that has a variety of features. Wooded, hilly areas are good spots for this lab. If you have additional time for this activity, have groups switch plots until each group has surveyed every plot. If you save the results of this lab from year to year, students can compare the changes over time.

Pick a site that is fairly undisturbed. Tie 10 m sections of string to the stakes ahead of time because setting up the plot can be time-consuming. Use a hammer to secure the stakes into the ground. Before beginning the activity, discuss with students what organisms they might expect to find at their site, how those organisms might interact, and what abiotic factors might come into play.

7. Use your hand lens to inspect the area, and record the insects you see. Be careful not to disturb the soil or the organisms. Then record the types of insects and plants you see.

8. Collect a small sample of soil, and observe it with your hand lens. Record a description of the soil and the organisms that live in it.

Analysis

1. Organizing Data Return to the classroom, and display your site map. Use your site map, your classmates' site maps, and your notes, to answer the following questions. Write your answers in your notebook.

2. Analyzing Data Write one paragraph that describes the 10 m × 10 m site you studied.

3. Analyzing Data Describe the 2 m × 2 m site you studied. Is this site characteristic of the larger site?

Conclusions

4. Interpreting Conclusions What are the differences between the areas that your classmates studied? Do different plants and animals live in different areas?

5. Making Predictions As the seasons change, the types of organisms that live in the area you studied may also change. Predict how your area might change in a different season or if a fire or flood occurred. If possible, return to the site at different times throughout the year and record your observations.

▶ **Site Maps** Your site map should be as detailed as possible, and it should include a legend.

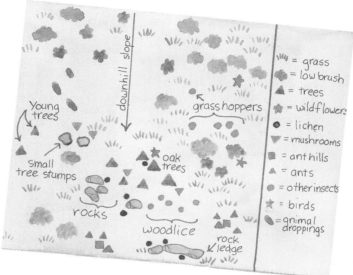

Extension

1. Asking Questions Based on what you have learned, think of a question that explores how the components of the area you observed interact with each other. For example, you might want to consider the influence of humans on the site; study a particular predator/prey relationship; or explore the effects of physical features, such as water or sunlight, on the growth or behavior of organisms. Write a description of how you would investigate this topic. **WRITING SKILLS**

27

CLASSROOM TESTED & APPROVED

Jim Watson
Dalton High School
Dalton, Georgia

CHICKEN OF THE TREES

Background

The green iguana (*Iguana iguana*) is found throughout Central America. In some parts of its habitat, it has been hunted to extinction. The green iguana has been used for centuries as a food source. It is called *chicken of the trees* or *gallina de palo* in Spanish. The name comes from the fact that iguanas have a white meat that tastes like chicken. The demand for iguanas and iguana eggs for food has caused them to nearly become extinct.

Releasing the iguanas into the wild is a key part of Dr. Werner's program. Attempting to raise the iguanas to a marketable size in captivity would be a costly process because iguanas constantly need food and require large habitats. By releasing the iguanas into the wild, much of the expense is reduced. However, to increase the productivity of the iguanas, feeding stations are set up in the jungle.

By supplying extra food for the iguanas—broken rice, meat, bone, fish meal, papayas, mangos, bananas, avocados, and a variety of leaves and flowers—many more iguanas can grow to maturity than could be naturally supported by the habitat. In fact, scientists estimate that 6 to 10 times more iguanas can be raised by supplied food.

CHICKEN OF THE TREES

In the stillness of predawn, the air warms over the Carara Biological Preserve in Costa Rica. Several thousand eggs in sun-heated incubators just below the surface of the Earth stir in response. Within these eggs are tiny iguanas—lizards that will eventually emerge, grow to a length of 1.5 m to 2.0 m (5 ft to 6.5 ft), and weigh up to 6 kg (13 lb).

What's going on here? Well, these giant lizards are being raised so that they can be released into the rain forest. It's part of a project led by German-born scientist Dr. Dagmar Werner. Her goal is to help restore an iguana population that has been severely reduced in the past several decades.

The lizard has suffered from the effects of hunting, pollution, and habitat destruction by people who clear the rain forest for farming. Prime iguana habitat is at the edge of the forest—where a combination of open areas, scrub, and trees occur. Historically, these areas are the type of habitat that humans

▶ These are iguanas at the Carara Biological Preserve in Costa Rica.

most often destroy when converting forestland to farmland. People cut down the forest at its edges—which just happens to be prime habitat for iguanas and other plants and animals.

Back at the Iguana Ranch

Dr. Werner's "iguana ranch" preserve has artificial nests where females can lay their eggs in a predator-free environment. After they hatch, the young lizards are placed in a temperature- and humidity-controlled incubator and given a special diet. As a result, the iguanas grow faster and stronger and are better protected from predators than their noncaptive counterparts.

In the first five years of her project, more than 80,000 iguanas were released into the wild. Ordinarily, less than 2 percent of all iguanas survive to adulthood in the wild, but Dr. Werner's iguanas have a 77 percent survival rate. Dr. Werner knows this because after she releases the iguanas into the rain forest, the lizards are tracked and monitored to determine whether they have successfully adapted to life in the wild.

Passing It On

Since the 1980s, Dr. Werner has improved the iguanas' chances of survival by breeding them and releasing thousands of young iguanas into the wild. But Dr. Werner soon realized that this effort was not enough, so she began training other people to do the same.

Because she knew there was no time to lose, Dr. Werner took an immediate and drastic approach to solving the problem. She combined her captive-breeding program at the preserve with an education program that shows farmers that there is more than one way to make a profit from the rain forest. Instead

28

▶ Dr. Werner and an associate discuss how the iguana can be farmed.

of raising cattle (and cutting down rain forest to do so), she encourages local farmers to raise iguanas, which can be released into the wild or sold for food. Known as the "chicken of the trees," this lizard has been a favored source of meat among native rain-forest inhabitants for thousands of years.

Not only do farmers profit from the sale of iguana meat, they also produce iguana leather and other handicrafts from the lizard.

Fundación Pro Iguana Verde

With Dr. Werner's methods, farmers can release many iguanas into the wild and earn a good living. But convincing farmers to use her methods hasn't been easy. According to Dr. Werner, "Many locals have never thought of wild animals as creatures that must be protected in order to survive. That's why so many go extinct." To get her message across, Dr. Werner has established the Fundación Pro Iguana Verde (the Green Iguana Foundation). This organization sponsors festivals and education seminars in local communities. These activities promote the traditional appeal of the iguana, increase civic pride in the animal, and heighten awareness about the iguana's economic importance.

By demonstrating that the needs of all concerned parties can be met when attempting to save an endangered species, Dr. Werner has revolutionized the concepts of species preservation and economic development. This hard-working scientist has hit upon a solution that may encourage farmers throughout Central America to "have their lizards and eat them too."

▶ Dr. Werner has established an innovative way to raise the number of iguanas living in the wild.

What Do You Think?

How does Dr. Werner's project protect iguanas and help local farmers too? Why do you think that she trains farmers to raise and value iguanas—what could her larger goal be? Can you think of a similar project that would be suitable for your area?

Sustainable Development

Dr. Werner's iguana ranch is just one example of a sustainable development project run by a non-governmental organization (NGO). One goal that most sustainable development projects share is encouraging local support and involvement in the project. If a project is supported by the local population the project is more likely to match the needs of the population and to be economically sustainable. Students may be interested in finding out about other NGOs that are working on environmental issues. One notable organization is Ashoka, based in Arlington, Virginia. Ashoka locates and supports individuals called *social entrepreneurs*. Like business entrepreneurs, social entrepreneurs have developed innovative, pattern breaking ideas that would produce maximum results for a minimum of investment. Social entrepreneurs, however, focus on areas such as healthcare, the environment, and human rights. Worldwide, Ashoka supports nearly 1,000 social entrepreneurs. Students may be interested in finding out more about Ashoka or another NGO that offers internships or other career opportunities.

29

Answers to What Do You Think?

Dr. Werner's project helps the iguanas because it takes hunting pressure off the iguana population. Her program helps the local farmers because it increases their income and prevents habitat destruction. Dr. Werner's larger goals are for the protection of the environment. She is working toward sustainable use of the forest resources.

Tools of Environmental Science
Chapter Planning Guide

PACING	CLASSROOM RESOURCES	LABS, DEMONSTRATIONS, AND ACTIVITIES
BLOCKS 1, 2 & 3 · 135 min pp. 30–37		
Chapter Opener		
Section 1 Scientific Methods	OSP **Lesson Plan** * CRF **Active Reading** * **BASIC** TT Bellringer * TT Rainfall and Tree Ring Width * TT John Snow's Cholera Spot Map * CD **Interactive Exploration 7** Shut Your Trap!	SE **QuickLab** Hypothesizing and Predicting, p. 32 TE **Activity** Park it Right Here, p. 33 **BASIC** TE **Group Activity** The Metric Game, p. 34 **ADVANCED** TE **Internet Activity** Dendrochronology Tutorials, p. 35 **ADVANCED** SE **Skills Practice Lab** Scientific Investigations, pp. 54–55 ◆ CRF **Datasheets for In-Text Labs** * CRF **Design Your Own Lab** * ◆ **GENERAL** CRF **CBL™ Probeware Lab** * ◆ **GENERAL**
BLOCKS 4 & 5 · 90 min pp. 38–44		
Section 2 Statistics and Models	OSP **Lesson Plan** * CRF **Active Reading** * **BASIC** TT Bellringer * TT Size Distribution of Dwarf Wedge Mussels * TT Conceptual Model of Mercury Contamination *	TE **Group Activity** That's MY Birthday, p. 38 **GENERAL** TE **Activity** M&M® Samples and Population, p. 40 ◆ **GENERAL** TE **Group Activity** Teaching To the Class, p. 41 **GENERAL** TE **Group Activity** Modeling Lava Fractures, p. 42 ◆ **GENERAL** SE **Field Activity** Conceptual Model, p. 43 TE **Group Activity** Making Models, p. 43 **BASIC** CRF **Observation Lab** * **BASIC**
BLOCKS 6 & 7 · 90 min pp. 45–49		
Section 3 Making Informed Decisions	OSP **Lesson Plan** * CRF **Active Reading** * **BASIC** TT Bellringer * CD **Interactive Tutor** How Science Works: Informed Decisions	TE **Internet Activity** Acronym Glossary, p. 46 **GENERAL** CRF **Consumer Lab** * ◆ **ADVANCED** CRF **Long-Term Project** * ◆ **GENERAL**

BLOCKS 8 & 9 · 90 min

Chapter Review and Assessment Resources

- SE **Chapter Review** pp. 51–53
- SE **Standardized Test Prep** pp. 638–639
- CRF **Study Guide** * ■ **GENERAL**
- CRF **Chapter Test** * ■ **GENERAL**
- CRF **Chapter Test** * **ADVANCED**
- CRF **Concept Review** * ■ **GENERAL**
- CRF **Critical Thinking** * **ADVANCED**
- OSP **Test Generator**
- CRF **Test Item Listing** *

Online and Technology Resources

 Holt Online Learning

Visit **go.hrw.com** for access to Holt Online Learning or enter the keyword **HE6 Home** for a variety of free online resources.

 **One-Stop Planner® CD-ROM**

This CD-ROM package includes
- Lab Materials QuickList Software
- Holt Calendar Planner
- Customizable Lesson Plans
- Printable Worksheets
- ExamView® Test Generator
- Interactive Teacher Edition
- Holt PuzzlePro® Resources
- Holt PowerPoint® Resources

 Holt Environmental Science Interactive Tutor CD-ROM

This CD-ROM consists of interactive activities that give students a fun way to extend their knowledge of environmental science concepts.

KEY

TE	Teacher Edition	**CRF**	Chapter Resource File	*	Also on One-Stop Planner
SE	Student Edition	**TT**	Teaching Transparency	▪	Also Available in Spanish
OSP	One-Stop Planner	**CD**	Interactive CD-ROM	◆	Requires Advance Prep

ENRICHMENT AND SKILLS PRACTICE	SECTION REVIEW AND ASSESSMENT	CORRELATIONS
SE Pre-Reading Activity, p. 30 **TE** Using the Figure, p. 30 `GENERAL`		**National Science Education Standards**
SE Case Study The Experimental Method In Action, p. 33 **TE** Skill Builder Writing, p. 33 `BASIC` **TE** Using the Figure Drought in Jamestown, p. 35 `GENERAL` **TE** Inclusion Strategies, p. 36	**SE** Section Review, p. 37 **TE** Reteaching, p. 37 `BASIC` **TE** Quiz, p. 37 `GENERAL` **TE** Alternative Assessment, p. 37 `GENERAL` **CRF** Quiz * ▪ `GENERAL`	ST 2c SPSP 6d
TE Reading Skill Builder Paired Summarizing, p. 39 `BASIC` **SE** Graphic Organizer Venn Diagram, p. 39 `GENERAL` **TE** Skill Builder Math, p. 39 `BASIC` **SE** MathPractice Probability, p. 40 **TE** Skill Builder Math, p. 41 `BASIC` **TE** Reading Skill Builder Paired Reading, p. 42 `BASIC` **TE** Using the Figure Mercury Sources, p. 43 `GENERAL`	**TE** Homework, p. 38 `GENERAL` **TE** Reteaching, p. 39 `BASIC` **TE** Homework, p. 42 `BASIC` **SE** Section Review, p. 44 **TE** Reteaching, p. 44 `BASIC` **TE** Quiz, p. 44 `GENERAL` **TE** Alternative Assessment, p. 44 `GENERAL` **CRF** Quiz * ▪ `GENERAL`	SPSP 6d
TE Skill Builder Writing, p. 45 `GENERAL` **SE** Case Study Saving the Everglades, pp. 46–47 **TE** Inclusion Strategies, p. 47 **TE** Interpreting Statistics What Happened to the Warblers? p. 48 `GENERAL` **TE** Student Opportunities Understanding Local Issues, p. 48 **TE** Career Architect, p. 48 **SE** Maps in Action A Topographic Map of Keene, New Hampshire, p. 56 **SE** Society & the Environment Bats and Bridges, p. 57 **CRF** Map Skills Maps as Models * `GENERAL`	**TE** Reteaching, p. 47 `BASIC` **SE** Section Review, p. 49 **TE** Reteaching, p. 49 `BASIC` **TE** Quiz, p. 49 `GENERAL` **TE** Alternative Assessment, p. 49 `ADVANCED` **CRF** Quiz * ▪ `GENERAL`	SPSP 6d

 Guided Reading Audio CDs

These CDs are designed to help auditory learners and reluctant readers. (Audio CDs are also available in Spanish.)

 www.scilinks.org

Maintained by the **National Science Teachers Association**

TOPIC: Experimenting in Science
SciLinks code: HE4040

TOPIC: Statistics in Science
SciLinks code: HE4105

TOPIC: Using Models
SciLinks code: HE4119

TOPIC: Environmental Decision Making
SciLinks code: HE4035

 CNN Videos

Each video segment is accompanied by a Critical Thinking Worksheet.

Earth Science Connections

Segment 5 Earthquake Seekers

Chapter Enrichment

This Chapter Enrichment provides relevant and interesting information to expand and enhance your classroom instruction of the chapter material.

1 Scientific Methods

Serendipity: Whale Bones, Bacteria, and Laundry Detergents

Chance and luck often play a role in scientific discoveries. Recently, a group of scientists made a chance discovery while looking for a rare species of mussel around hydrothermal vents on the ocean floor. The scientists discovered whale bones that were coated by slimy mats of bacteria. In the dark, cold water the bacterial mats digest oils leaking out of whale carcasses. This may sound disgusting, but the discovery may be great news for the environment. Researchers are looking at ways to incorporate the bacterial enzymes into laundry detergents. If the researchers succeed, they could create a natural method of removing grease stains using a clothes washer's cold water cycle.

▶ Jane Goodall with a young chimpanzee

Cold Fusion: The Result No One Could Repeat

On March 23, 1989, two chemists from the University of Utah announced that they had produced a "cold fusion" nuclear reaction in a jar of water. Nuclear fusion, the combining of atomic nuclei, was thought possible only at temperatures of more than one million degrees Fahrenheit. Fusion at relatively low temperatures offered the possibility of a cheap and limitless energy source. Other scientists, companies, and government agencies rushed to try to reproduce the results in hopes of discovering an inexpensive and reliable source of power. But in the months following the announcement, no one was able to reproduce the original results and the scientists had to retract their discovery. This incident underscored the importance of having scientific work peer reviewed before publishing results.

BRAIN FOOD

The longest running ecological experiment in the world is the Park Grass Experiment in Rothamsted, UK. The experiment began in 1856 to study the effects of different types of fertilizer on the yield of a hay field. Today, the experiment continues to explore the gene flow and evolution of various plant species.

Getting A Handle On Epidemiology

Five years before the cholera outbreak described in this chapter, John Snow wrote and distributed a pamphlet which hypothesized that cholera is caused by a pathogen that is found in vomit and feces. But little attention was paid to Snow's hypothesis until 1854, when he pointed out that certain water districts contaminated with sewage suffered a high density of cholera cases. He advised public officials to remove the handle of the Broad Street Pump, thereby averting further illness. According to the Center for Disease Control, the pump handle endures today as the symbol of epidemiology.

A Recipe for Mice

Jan Baptista van Helmont, a prominent Flemish scientist, once published this recipe for creating mice: "Place a dirty shirt or some rags in an open pot or barrel containing a few grains of wheat or some wheat bran, and in 21 days, mice will appear. There will be adult males and females present, and they will be capable of mating and reproducing more mice." This brilliant Flemish scientist was not kidding. The theory of spontaneous generation was accepted by scientists for thousands of years until it was finally challenged—using the scientific method—by Francesco Redi, Lazzaro Spallanzani, and Louis Pasteur, over the course of three centuries.

SECTION

2 Statistics and Models

Scatter Plots and Correlation

How would you use a graph to determine if there is a relationship, or *correlation,* between two environmental variables, such as the weight of young robins and their exposure to a pesticide? One way would be to use a graph called a *scatter plot.* A scatter plot is a graph that uses individual points to show the relationship, if any, between two variables. The variables comprise the *x* and *y* axes. Scatter plots differ from other types of graphs in that they are comprised of individual points "scattered" on the background of the graph. The pattern of the scattered points indicates the relationship between the variables. Once scientists have graphed their data, they usually use statistical tests to pinpoint the relationship between the variables.

▶ Satellite image, San Francisco Bay area

SECTION

3 Making Informed Decisions

Aldo Leopold and the Land Ethic

Aldo Leopold was a wildlife biologist, forester, professor, and author. He is also one of the founders of the conservation movement in the United States. In his book *A Sand County Almanac,* published in 1949, Leopold described the idea of a "land ethic." To Leopold, a land ethic is based on the premise that the community to which humans belong extends to the land—that we are morally obligated to consider the health of the land in making our decisions. Leopold argued that the people are part of the land, rather than that the people owned the land. *A Sand County Almanac* inspired many Americans to rethink their relationship with the land and the environment.

The Precautionary Principle

In making decisions, politicians often ask for ironclad certainty or proof, especially when the action in question affects the environment. But scientists do not always know beyond a doubt that a certain action will have a certain environmental impact. Furthermore, scientists are trained to state their opinions in terms of probabilities rather than certainties. So, many scientists use the precautionary principle put forth in 1992 by the Intergovernmental Agreement on the Environment. The precautionary principle states the following: "Where there are threats of serious or irreversible environmental damage, lack of full scientific certainty should not be used as a reason for postponing measures to prevent environmental degradation." In other words, even if you are not 100 percent sure of a negative outcome you should still act to prevent it.

▶ A roseate spoonbill in the Everglades

Tools of Environmental Science

Overview

Tell students that this chapter will *not* focus on the physical tools of environmental science, such as microscopes or computers. This chapter discusses tools of the mind: the mental and conceptual tools that scientists use to explore and understand the environment. This chapter also explains the value of making informed, thoughtful decisions about the environment.

Using the Figure — GENERAL

Scientists have mounted cameras on the heads of Weddell seals to gather information about their habits. The resulting film has yielded something unexpected: observations of the Antarctic silverfish and the Antarctic toothfish, two ecologically important species. Ask students to consider how these videos might help scientists. (These videos might help scientists learn about the hunting habits of seals. They might also help scientists better understand feeding relationships.) **LS** Logical

PRE-READING ACTIVITY

Have pairs of students use their FoldNotes to study key terms from the chapter. Instruct one student to use the FoldNote to provide the key term, and have the other student give the definition. Have the student who provides the key term correct the other student's definition.

VIDEO SELECT

For information about videos related to this chapter, go to **go.hrw.com** and type in the keyword **HE4 TOOV.**

1 **Scientific Methods**
2 **Statistics and Models**
3 **Making Informed Decisions**

PRE-READING ACTIVITY

FOLDNOTES **Key-Term Fold**
Before you read this chapter, create the **FoldNote** entitled "Key-Term Fold" described in the Reading and Study Skills section of the Appendix. Write a key term from the chapter on each tab of the key-term fold. Under each tab, write the definition of the key term.

This photograph shows a researcher filming a Weddell seal in Antarctica. Although scientists often use sophisticated tools in their work, their most important tools are those they carry with them—their senses and their habits of mind.

30

Chapter Correlations   *National Science Education Standards*

ST 2c Creativity, imagination, and a good knowledge base are all required in the work of science and engineering. **(Section 1)**

SPSP 6d Individuals and society must decide on proposals involving new research and the introduction of new technologies into society. Decisions involve assessment of alternatives, risks, costs, and benefits and consideration of who benefits and who suffers, who pays and gains, and what the risks are and who bears them. Students should understand the appropriateness and value of basic questions—"What can happen?"—"What are the odds?"—and "How do scientists and engineers know what will happen?" **(Section 1, Section 2, Section 3)**

Scientific Methods

The word *science* comes from the Latin verb *scire*, meaning "to know." Indeed, science is full of amazing facts and ideas about how nature works. But science is not just something you know; it is also something you do. This chapter explores how science is done and examines the tools scientists use.

The Experimental Method
You have probably heard the phrase, "Today scientists discovered…" How do scientists make these discoveries? Scientists make most of their discoveries using the *experimental method*. This method consists of a series of steps that scientists worldwide use to identify and answer questions. The first step is observing.

Observing Science usually begins with observation. Someone notices, or observes, something and begins to ask questions. An **observation** is a piece of information we gather using our senses— our sight, hearing, smell, and touch. To extend their senses, scientists often use tools such as rulers, microscopes, and even satellites. For example, a ruler provides our eyes with a standard way to compare the lengths of different objects. The scientists in **Figure 1** are observing the tail length of a tranquilized wolf with the help of a tape measure. Observations can take many forms, including descriptions, drawings, photographs, and measurements.

Students at Keene High School in New Hampshire have observed that dwarf wedge mussels are disappearing from the Ashuelot River, which is located near their school. The students have also observed that the river is polluted. These observations prompted the students to take the next step in the experimental method—forming hypotheses.

Objectives
► List and describe the steps of the experimental method.
► Describe why a good hypothesis is not simply a guess.
► Describe the two essential parts of a good experiment.
► Describe how scientists study subjects in which experiments are not possible.
► Explain the importance of curiosity and imagination in science.

Key Terms
   observation
   hypothesis
   prediction
   experiment
   variable
   experimental group
   control group
   data
   correlation

Figure 1 ▶ These scientists are measuring the tail of a tranquilized wolf. What questions could these observations help the scientists answer?

31

Focus

Overview
Before beginning this section, review with your students the Objectives in the Student Edition. The section focuses on the experimental method, but it also explains the value of the correlation method for use when experiments are impossible or unethical. Students learn about scientific habits of mind, including curiosity, skepticism, intellectual honesty, and imagination.

💿 Bellringer
Ask students, "Why might the scientists be measuring the wolf's tail in **Figure 1**? Once they make the measurement, how might they use it?" (The scientists might be measuring the tail to record characteristics of the members of a wolf pack. The scientists could use this information to study how the pack changes over time.) **LS Logical**

Motivate

Identifying Preconceptions — BASIC
Scientific Methods An experiment may grow out of an observation, but a scientist does not necessarily know beforehand what he or she is going to observe. For example, if a scientist goes to a stream to make observations about the population trends of a type of frog, she might discover that another frog species is missing from a lake and decide to investigate that species instead. Ask students, "Where would the scientist go from there?" (Much like a reporter, a scientist observes and follows observations to seek out a story. The scientist might try to find out if the frogs have disappeared from other bodies of water in the surrounding area. The scientist will look for explanations that he or she can test.) **LS Logical**

Chapter Resource File
• Lesson Plan
• Active Reading BASIC
• Section Quiz GENERAL

🖳 Transparencies

TT Bellringer

Skills Acquired

- Predicting
- Interpreting

Answers

1. Answers may vary. Students should observe that fine-grained soils are more likely to wash away.

2. Answers may vary.

MISCONCEPTION ALERT

Hypotheses Vs. Predictions
Many people confuse hypotheses with predictions. Explain that a hypothesis is a general statement that offers an explanation of a problem that has been observed. Hypotheses can generally be supported or contradicted by experimentation. Point out that a prediction is based on a hypothesis. A prediction is meant to explain what will happen in a specific situation, such as during an experiment, if the hypothesis turns out to be right. Reinforce this distinction by having students form hypotheses and then form predictions based on their hypotheses.

QuickLAB

Hypothesizing and Predicting

Procedure

1. Place a **baking tray** on a table, and place a **thin book** under one end of the tray.
2. Place **potting soil**, **sand**, and **schoolyard dirt** in three piles at the high end of the baking tray.
3. Use a **toothpick** to poke several holes in a **paper cup**.
4. Write down a hypothesis to explain why soil gets washed away when it rains.
5. Based on your hypothesis, predict which of the three soils will wash away most easily.
6. Pour **water** into the cup, and slowly sprinkle water on the piles.

Analysis

1. What happened to the different soils?
2. Revise your hypothesis, if necessary, based on your experiment.

Hypothesizing and Predicting Observations give us answers to questions, but observations almost always lead to more questions. To answer a specific question, a scientist may form a hypothesis. A **hypothesis** (hie PAHTH uh sis) is a testable explanation for an observation. A hypothesis is more than a guess. A good hypothesis should make logical sense and follow from what you already know about the situation.

The Keene High School students observed two trends: that the number of dwarf wedge mussels on the Ashuelot River is declining over time and that the number of dwarf wedge mussels decreases at sites downstream from the first study site. These trends are illustrated in **Figure 2**. Students tested the water in three places and found that the farther downstream they went, the more phosphate the water contained. Phosphates are chemicals used in many fertilizers.

Armed with their observations, the students might make the following hypothesis: *phosphate fertilizer from a golf course is washing into the river and killing dwarf wedge mussels*. To test their hypothesis, the students make a **prediction**, a logical statement about what will happen if the hypothesis is correct. The students might make the following prediction: *mussels will die when exposed to high levels of phosphate in their water*.

It is important that the students' hypothesis—high levels of phosphate are killing the mussels—can be disproved. If students successfully raised mussels in water that has high phosphate levels, their hypothesis would be incorrect. Every time a hypothesis is disproved, the number of possible explanations for an observation is reduced. By eliminating possible explanations a scientist can zero in on the best explanation with more confidence.

Figure 2 ▶ The diagram below illustrates the trends observed by the students at Keene High School.

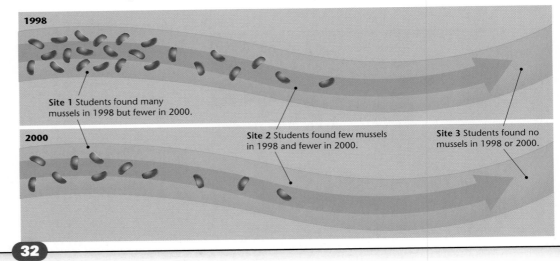

1998

Site 1 Students found many mussels in 1998 but fewer in 2000.

2000

Site 2 Students found few mussels in 1998 and fewer in 2000.

Site 3 Students found no mussels in 1998 or 2000.

32

MISCONCEPTION ALERT

Laws and Theories Many students believe that if a theory is accepted by enough people for a long enough period of time, the theory will "grow up" to be a law. Emphasize that laws and theories are both accepted, but that they serve different functions. A law, such as the law of gravity, is a concise statement of fact that is accepted as true and universal. Theories, such as the theory of evolution by natural selection, are statements that are products of many scientific observations and may encompass numerous hypotheses or laws. Like a law, a theory is accepted as true, but theories are much more complex. A law can be compared to a rubber ball. When dropped under constant conditions, the ball will always bounce exactly as predicted. Bouncing is the only action the ball performs. On the other hand, a theory can be compared to the operation of a car. A car has many components, all performing different tasks and working in unison. A part of the car may be improved, such as the brakes or air bags, but the general function of the car remains constant.

Experimenting The questions that arise from observations often cannot be answered by making more observations. In this situation scientists usually perform one or more experiments. An **experiment** is a procedure designed to test a hypothesis under controlled conditions.

Experiments should be designed to pinpoint cause-and-effect relationships. For this reason, good experiments have two essential characteristics: a single variable is tested, and a control is used. The **variable** (VER ee uh buhl) is the factor of interest, which, in our example, would be the level of phosphate in the water. To test for one variable, scientists usually study two groups or situations at a time. The variable being studied is the only difference between the groups. The group that receives the experimental treatment is called the **experimental group.** In our example, the experimental group would be those mussels that receive phosphate in their water. The group that does not receive the experimental treatment is called the **control group.** In our example, the control group would be those mussels that do not have phosphate added to their water. If the mussels in the control group thrive while most of those in the experimental group die, the experiment's results support the hypothesis that phosphates from fertilizer are killing the mussels.

internet connect

www.scilinks.org
Topic: Experimenting in Science
SciLinks code: HE4040

SCILINKS Maintained by the National Science Teachers Association

The Experimental Method In Action at Keene High School

▶ **Keene High School** students are conducting an experiment to study the effect of phosphate levels on the growth rates of freshwater mussels.

Keene High School students collected mussels (nonendangered relatives of the dwarf wedge mussel) and placed equal numbers of them in two aquariums. They ensured that the conditions in the aquariums were identical—same water temperature, food, hours of light, and so on. The students added a measured amount of phosphate to the aquarium of the experimental group. They added nothing to the aquarium of the control group.

A key to the success of an experiment is changing only one variable and having a control group. What would happen if the aquarium in which most of the mussels died had phosphate in the water and was also warmer? The students would not know if the phosphate or the higher temperature killed the mussels.

Another key to experimenting in science is *replication*, or recreating the experimental conditions to make sure the results are consistent. In this case, using six aquariums—three control and three experimental— would help ensure that the results are not simply due to chance.

CRITICAL THINKING

1. Applying Ideas Why did the students ensure that the conditions in both aquariums were identical?

2. Evaluating Hypothesis How would you change the hypothesis if mussels died in both aquariums?

33

The Metric Game Point out that most of the measurements in this book use SI units. Understanding the metric system is an essential component of scientific literacy. Divide students into teams and have each team send a representative to the board. Select an item in the room (such as a book, the board, or water in a glass) and provide a metric unit with which to measure the item. Have each team estimate the measurement of the item. After measuring the item, the team with the closest estimate is awarded one point. Allow students to take turns measuring the items so that students become familiar with the meter stick, balance, graduated cylinder, and other measuring tools. Repeat this exercise throughout the year. **LS Visual**

MISCONCEPTION
///ALERT

Information on the Internet
The Internet is an increasingly important means not only for disseminating scientific results and studies but also for linking relevant studies together. The Internet also offers a lot of misinformation and bias. As a class, brainstorm a list of environmental science keywords. Then have students use the keywords to find examples of both unbiased and biased information on the Internet.

Figure 3 ▶ This scientist is analyzing his data with the help of a computer.

Table 1 ▼

Pollutant Concentrations		
Site	Nitrates	Phosphates
1	0.3	0.02
2	0.3	0.06
3	0.1	0.07

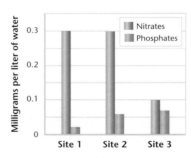

Figure 4 ▶ The table (top) presents data on the amount of phosphates and nitrates found at Sites 1, 2, and 3 on the Ashuelot River in 2000. The bar graph (bottom) displays this information in graphical form.

34

Organizing and Analyzing Data Keeping careful and accurate records is extremely important in science. A scientist cannot rely on experimental results that are based on sloppy observations or incomplete records. The information that a scientist gathers during an experiment, which is often in numeric form, is called **data.**

Organizing data into tables and graphic illustrations helps scientists analyze the data and explain the data clearly to others. The scientist in **Figure 3** is analyzing data on pesticides in food. Graphs are often used by scientists to display relationships or trends in the data. For this reason, graphs are especially useful for illustrating conclusions drawn from an experiment.

One common type of graph is called a *bar graph*. Bar graphs are useful for comparing the data for several things in one graph. The bar graph in **Figure 4** displays the information contained in the table above it. Graphing the information makes the trends easy to see. The graph shows that phosphates decrease downstream and that nitrates increase downstream.

Drawing Conclusions Scientists determine the results of their experiment by analyzing their data and comparing the outcome of their experiment with their prediction. Ideally, this comparison provides the scientist with an obvious conclusion. But often the conclusion is not obvious. For example, in the mussel experiment, what if three mussels died in the control tank and five died in the experimental tank? The students could not be certain that phosphate is killing the mussels. Scientists often use mathematical tools to help them determine whether such differences are meaningful or are just a coincidence. Scientists also repeat their experiments.

Repeating Experiments Although the results from a single experiment may seem conclusive, scientists look for a large amount of supporting evidence before they accept a hypothesis. The more often an experiment can be repeated with the same results, in different places and by different people, the more sure scientists become about the reliability of their conclusions.

Communicating Results Scientists publish their results to share what they have learned with other scientists. When scientists think their results are important, they usually publish their findings as a scientific article. A scientific article includes the question the scientist explored, reasons why the question is important, background information, a precise description of how the work was done, the data that were collected, and the scientist's interpretation of the data.

"There are two possible outcomes: If the result confirms the hypothesis, then you've made a measurement. If the result is contrary to the hypothesis, then you've made a discovery."

—Enrico Fermi

Ask students to explain the following quote in relation to scientific methods and the formation of a hypothesis.
LS Verbal

Never Cry Wolf Students may not be familiar with naturalist Farley Mowat's novel *Never Cry Wolf*. In this book, the Canadian government sent Mowat out to investigate wolves. Their hypothesis was that the growing wolf population threatened people and caribou. Mowat's data did not confirm the hypothesis. In fact, the wolves were helping to maintain a healthy caribou population. Ask students to read *Never Cry Wolf*, or watch the movie, and identify different aspects of the scientific method in Mowat's research. **LS Logical**

The Correlation Method

Whenever possible, scientists study questions by using experiments. But many questions cannot be studied experimentally. The question "What was Earth's climate like 60 million years ago?" cannot be studied by performing an experiment because the scientists are 60 million years too late. "Does smoking cause lung cancer in humans?" cannot be studied experimentally because doing experiments that injure people would be unethical.

When using experiments to answer questions is impossible or unethical, scientists test predictions by examining **correlations,** or reliable associations between two or more events. For example, scientists know that the relative width of a ring on a tree trunk is a good indicator of the amount of rainfall the tree received in a given year. Trees produce wide rings in rainy years and narrow rings in dry years. Scientists have used this knowledge to investigate why the first European settlers at Roanake Island, Virginia (often called the Lost Colony) disappeared and why most of the first settlers at Jamestown, Virginia, died. As shown in **Figure 5,** the rings of older trees on the Virginia coast indicate that the Lost Colony and the Jamestown Colony were founded during two of the worst droughts the coast had experienced in centuries. The scientists concluded that the settlers may have been the victims of unfortunate timing.

Although correlation studies are useful, correlations do not necessarily prove cause-and-effect relationships between two variables. For example, the correlation between increasing phosphate levels and a declining mussel population on the Ashuelot River does not prove that phosphates harm mussels. Scientists become more sure about their conclusions if they find the same correlation in different places and as they eliminate possible explanations.

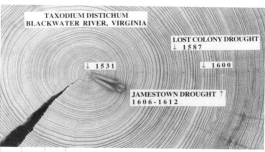

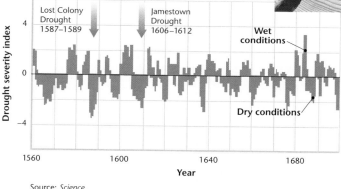

Figure 5 ▶ This cross section of a baldcypress from southeastern Virginia (above) shows a record of rainfall beginning in 1531. The graph translates the relative tree ring width into what is called a *drought index,* which allowed scientists to compare rainfall between different years.

Source: *Science.*

BRAIN FOOD

Phenology in Alaska In an attempt to study climate change, scientists have utilized the records from a yearly betting pool in Nenana, Alaska. For most of the past century, area residents bet yearly on when the ice of the Nenana River would crack, signaling the start of the spring thaw. Participants pay a few dollars to enter, and the winners split the pot. The wagering has resulted in a nearly 100-year record of the date the river has broken up each year. A Stanford University scientist who researches *phenology,* the study of the timing of natural events, used the records as data for a correlation study. In 2001 the ice breakup occurred, on average, 5.5 days sooner than it did in 1917. This indicates that the climate around Nenana has warmed over the course of 80 years.

Jane Goodall Jane Goodall is famous for studying chimpanzees in the Gombe Forest in Tanzania. Before she began her research in 1960, people knew little about chimps and thought they were dangerous. She would spend days at a time patiently observing the chimps in their habitat. She initially had little money for her studies but was able to survive on little food and drink. She made the most of the resources she had, going on to be one of the most successful woman scientists in history. To learn more about the Gombe chimps and the researchers studying them today, students can visit the Jane Goodall Institute's Center for Primate Studies on the Internet.

Scientific Methods Point out that the title of this section is plural. Emphasize that there is no single scientific method. Scientists approach problems from a variety of viewpoints. They conduct their research using available tools, data, time, and people. Research often leads scientists to develop new tools and techniques, but the basic methods remain unchanged.

Transparencies
TT John Snow's Cholera Spot Map

36

INCLUSION *Strategies* — **GENERAL**

• *Visually Impaired*   • *Developmentally*
• *Learning Disabled*   *Delayed*

Have students observe people, objects, systems and the environment of the classroom for 8–10 minutes. Then have them label four note cards with the following: "See," "Hear," "Smell," and "Touch." Students can draw or write about their observations for each of the senses. Observations should indicate time, date, and room conditions. This exercise can also be repeated, and students can compare and discuss the results.

Connection to Biology

Discovering Penicillin Alexander Fleming discovered penicillin by accident. Someone left a window open near his dishes of bacteria, and the dishes were infected with spores of fungi. Instead of throwing the dishes away, Fleming looked at them closely and saw that the bacteria had died on the side of a dish where a colony of green *Penicillium* mold had started to grow. If he had not been a careful observer, penicillin might not have been discovered. You may find *Penicillium* yourself on moldy bread.

Scientific Habits of Mind

Scientists actually approach questions in many different ways. But good scientists tend to share several key habits of mind, or ways of approaching and thinking about things.

Curiosity Good scientists are endlessly curious. Jane Goodall, pictured in **Figure 6**, is an inspiring example. She studied a chimpanzee troop in Africa for years. She observed the troop so closely that she came to know the personalities and behavior of each member of the troop and greatly contributed to our knowledge of that species.

The Habit of Skepticism Good scientists also tend to be skeptical, which means that they don't believe everything they are told. For example, 19th century doctors were taught that men and women breathe differently—men use the diaphragm (the sheet of muscle below the rib cage) to expand their chests, whereas women raise their ribs near the top of their chest. Finally, a female doctor found that women seemed to breathe differently because their clothes were so tight that their ribs could not move far enough to pull air into their lungs.

Openness to New Ideas As the example above shows, skepticism can go hand in hand with being open to new ideas. Good scientists keep an open mind about how the world works.

Figure 6 ▶ Jane Goodall is famous for her close observations of chimpanzees—observations fueled in part by her endless curiosity.

Notable Quotes — **GENERAL**

"The whole of science is nothing more than a refinement of everyday thinking."

—Albert Einstein

Discuss the quotation with students. Ask, "Are there tools of environmental science that are useful to people in everyday situations?" If necessary, prompt them to consider how a mechanic uses the experimental method to diagnose a problem with a car. **LS Logical**

Intellectual Honesty A scientist may become convinced that a hypothesis is correct even before it has been fully tested. But when an experiment is repeated, the results may be different from those obtained the first time. A good scientist is willing to recognize that the new results may be accurate, even though that means admitting that his or her hypothesis might be wrong.

Imagination and Creativity Good scientists are not only open to new ideas but able to conceive of new ideas themselves. The ability to see patterns where others do not or imagine things that others cannot allows a good scientist to expand the boundaries of what we know.

An example of an imaginative and creative scientist is John Snow, shown in **Figure 7.** Snow was a physician in London during a cholera epidemic in 1854. Cholera, a potentially fatal disease, is caused by a bacterium found in water that is polluted with human waste. Few people had indoor plumbing in 1854. Most people got their water from public pumps; each pump had its own well. In an attempt to locate the polluted water source, Snow made a map showing where the homes of everyone who died of cholera were located. The map also showed the public water pumps. In an early example of a correlation study, he found that more deaths occurred around a pump in Broad Street than around other pumps in the area. London authorities ended the cholera epidemic by taking the handle off the Broad Street pump so that it could no longer be used. Using observation, imagination, and creativity, Snow solved an environmental problem and saved lives.

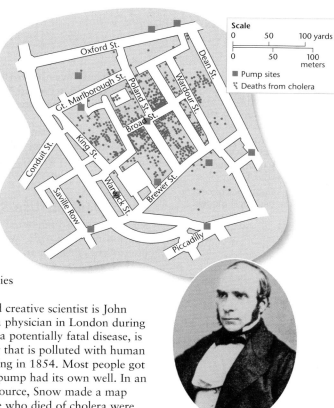

Scale
| 0 | 50 | 100 yards |
| 0 | 50 | 100 meters |

■ Pump sites
⚲ Deaths from cholera

Figure 7 ▶ John Snow (bottom) created his famous spot map (top), which enabled him to see a pattern no one had noticed before.

SECTION 1 Review

1. **Describe** the steps of the experimental method.

2. **Name** and explain the importance of three scientific habits of mind.

3. **Explain** why a hypothesis is not just a guess.

4. **Explain** how scientists try to answer questions that cannot be tested with experiments.

CRITICAL THINKING

5. **Analyzing Methods** Read the description of experiments. Describe the two essential parts of a good experiment, and explain their importance. **READING SKILLS**

6. **Analyzing Relationships** How can a scientist be both skeptical and open to new ideas at the same time? Write a one-page story that describes such a situation. **WRITING SKILLS**

37

Close

Reteaching ——— BASIC

Scientific Methods Remind students that scientists seek to answer questions about the natural world and how it works, but not all questions can be answered using scientific methods. Ask students to give examples of questions that people might have. Then ask students which questions scientists could answer and which ones they could not. **LS Logical**

Quiz ——— GENERAL

1. Why should the results of an experiment be repeatable? (If the results of an experiment cannot be replicated, scientists cannot trust the reliability of their conclusions.)

2. Why are experiments designed to have as few variables as possible? (Limiting variables allows scientists to pinpoint causes and effects.)

3. Why is the correlation method a useful accompaniment to the experimental method? (Drawing conclusions based upon correlations allows scientists to study subjects when experiments are impossible or unethical.)

Alternative Assessment ——— GENERAL

Identifying Scientific Methods Provide students with a copy of an article about an environmental issue from a magazine such as *Discover*, *National Geographic*, or *Scientific American*. Have students highlight and label parts of the article that demonstrate the scientific method.

Answers to Section Review

1. First, make observations or measurements to gather information. Second, use these observations to form a hypothesis, and make a prediction based on the hypothesis. Next, conduct an experiment to test the hypothesis under controlled conditions. Organize and analyze information, or data, that is gathered from the experiment. Use the data to form conclusions about the original hypothesis. Repeat the experiment and share the results.

2. Sample answers: Curiosity leads scientists to ask new questions. Skepticism leads scientists to question explanations they doubt. Openness to new ideas prevents scientists

from limiting their thinking. Intellectual honesty helps ensure accurate conclusions. Imagination and creativity help scientists conceive new ideas and explanations.

3. A hypothesis considers information gathered by observation, while a guess might not. A hypothesis is also a testable explanation, while a guess may not be testable.

4. Scientists study correlations, reliable associations between two events, to answer questions that cannot be investigated with experiments. The more correlations that exist between variables, the more sure scientists can be of their conclusions.

5. The two essential parts of a good experiment are testing only one variable and using a control. It is important to test only one variable so you can be sure that this variable is the cause of any changes that occur. It is important to use a control so that you have something to compare with the experimental treatment.

6. Answers may vary. A skeptical scientist does not believe everything he or she is told. In being skeptical about established ideas, a scientist is open to new ones.

Overview

Before beginning this section, review with your students the Objectives in the Student Edition. This section discusses statistics and explains how scientists apply statistics to data. Students will also learn about the importance of physical, graphical, conceptual, and mathematical models in science.

 Bellringer

Bring in a few news clippings that include statistics. Discuss the statistics and how they are used with the class. Point out, in particular, advertisements that include statistics to promote a product. Discuss with the class the difference between responsible uses of statistics and misleading uses of statistics.
LS Verbal/Visual

Motivate

Group Activity —— GENERAL

That's MY Birthday Write the 12 months of the year on the board. Ask everyone in the room to tell you his or her birthday. If more than one student shares a birthday, call attention to this fact. Ask students if they are surprised. Start a discussion about the probability of this occurrence. Ask interested students to research why this might be the case. (Probability is not as simple as it seems. In a group of 23 people, there is a 50 percent chance that two will share a birthday. In a group of 40, the chance is nearly 90 percent. Have students research in the library or on the Internet to find the probability of a shared birthday.) **LS** Logical

Objectives

▶ Explain how scientists use statistics.
▶ Explain why the size of a statistical sample is important.
▶ Describe three types of models commonly used by scientists.
▶ Explain the relationship between probability and risk.
▶ Explain the importance of conceptual and mathematical models.

Key Terms

statistics
mean
distribution
probability
sample
risk
model
conceptual model
mathematical model

Environmental science provides a lot of data that need to be organized and interpreted before they are useful. **Statistics** is the collection and classification of data that are in the form of numbers. People commonly use the term statistics to describe numbers, such as the batting record of a baseball player. Sportswriters also use the methods of statistics to translate a player's batting record over many games into a batting average, which allows people to easily compare the batting records of different players.

How Scientists Use Statistics

Scientists are also interested in comparing things, but scientists use statistics for a wide range of purposes. Scientists rely on and use statistics to summarize, characterize, analyze, and compare data. Statistics is actually a branch of mathematics that provides scientists with important tools for analyzing and understanding their data.

Consider the experiment in which students studied mussels to see if the mussels were harmed by fertilizer in their water. Students collected data on mussel length and phosphate levels during this experiment. Some mussels in the control group grew more than some mussels in the experimental group, yet some grew less. How could the students turn this data into meaningful numbers?

Statistics Works with Populations Scientists use statistics to describe statistical populations. A *statistical population* is a group of similar things that a scientist is interested in learning about. For example, the dwarf wedge mussels shown in **Figure 8** are part of the population of all dwarf wedge mussels on the Ashuelot River.

Figure 8 ▶ Students found these dwarf wedge mussel shells in a muskrat den. These mussels are part of the statistical population of all dwarf wedge mussels on the Ashuelot River.

Chapter Resource File

• **Lesson Plan**
• **Active Reading** BASIC
• **Section Quiz** GENERAL

Transparencies

TT Bellringer
TT Size Distribution of Dwarf Wedge Mussels

Homework —— GENERAL

Critiquing Advertorials Ask students to search newspapers and magazines for advertorials. These features often resemble magazine or newspaper content but are created by advertisers. Ask students to annotate an advertorial with their own critique of the statements and statistics provided in the text. A good method to prepare a critique of the advertorial is to photocopy and reduce the feature so there is a margin of white space for students to work in. **LS** Logical

What Is the Average? Although statistical populations are composed of similar individuals, these individuals often have different characteristics. For example, in the population of students in your classroom, each student has a different height, weight, and so on.

As part of their experiments, the Keene High School students measured the lengths of dwarf wedge mussels in a population, as shown in **Figure 8**. By adding the lengths of the mussels and then dividing by the number of mussels, students calculated the average length of the mussels, which in statistical terms is called the *mean*. A **mean** is the number obtained by adding up the data for a given characteristic and dividing this sum by the number of individuals. For scientists, the mean provides a single numerical measure for a given aspect of a population. Scientists can easily compare different populations by comparing their means. The mean length of the mussels represented in **Figure 9** is about 30 mm.

The Distribution The bar graph in **Figure 9** shows the lengths of dwarf wedge mussels in a population. The pattern that the bars create when viewed as a whole is called the *distribution*. A **distribution** is the relative arrangement of the members of a statistical population. In **Figure 9**, the lengths of the individuals are arranged between 15 and 50 mm.

The overall shape of the bars, which rise to form a hump in the middle of the graph, is also part of the distribution. The line connecting the tops of the bars in **Figure 9** forms the shape of a bell. The graphs of many characteristics of populations, such as the heights of people, form bell-shaped curves. A bell-shaped curve indicates a *normal distribution*. In a normal distribution, the data are grouped symmetrically around the mean.

internet connect

www.scilinks.org
Topic: Statistics in Science
SciLinks code: HE4105

SciLINKS. Maintained by the National Science Teachers Association

Graphic Organizer **Venn Diagram** Create the Graphic Organizer entitled "Venn Diagram" described in the Appendix. Label the circles with "Statistics" and "Models." Then, fill in the diagram with the characteristics that each way of interpreting the data shares with the other.

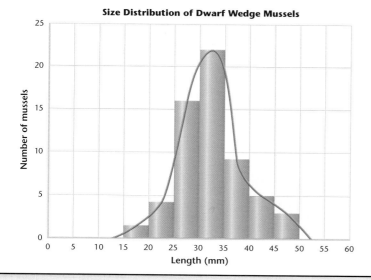

Size Distribution of Dwarf Wedge Mussels

Number of mussels (y-axis: 0, 5, 10, 15, 20, 25)
Length (mm) (x-axis: 0, 5, 10, 15, 20, 25, 30, 35, 40, 45, 50, 55, 60)

Figure 9 ▶ This bar graph shows the distribution of lengths in a population of dwarf wedge mussels. The location of each bar on the *x*-axis indicates length. The height of each bar represents the number of mussels as shown on the *y*-axis. For example, the second bar indicates that there are four mussels between 20 and 25 mm long.

Reteaching ——— BASIC

Distribution Students may have trouble with the concept of distributions. Show them a number of graphs throughout the book and explain the arrangement of data. You may want to discuss some general grading strategies with students to further explain what a bell-shaped curve is. **LS** Visual

English Language Learners

READING SKILL BUILDER ——— BASIC

Paired Summarizing Have students pair with a partner. Each student should read and summarize the meaning of the terms *distribution* and *mean*. Ask each pair to share an example of a distribution and a mean. **LS** Interpersonal

SKILL BUILDER ——— BASIC

Math Skills Ask students to create a set of data that has a mean of 12 and that includes at least five numbers. Then, ask students to create a line graph that shows the distribution of this data. Ask them if the data appears to be in a bell-shaped curve. (Accept any set of data that has a mean of 12.) **LS** Logical

Graphic Organizer GENERAL

Venn Diagram
You may want to have students work in groups to create this Venn Diagram. Have one student draw the map and fill in information provided by other students from the group.

LANGUAGE ARTS ——— CONNECTION — ADVANCED

"Rosencrantz and Guildenstern Are Dead"
In Tom Stoppard's play, minor characters from Shakespeare's "Hamlet" become the protagonists. Ask students to read the first act of the play or watch the movie of the same name. In the play, a character is tossing a coin with some improbable results. Despite continuous tossing, the coin always comes up heads. Ask students to interpret this scene. Ask, "What does Tom Stoppard mean to say about the world that the characters inhabit?" (Students may think that Stoppard was trying to present a world in which the laws of logic and reason are suspended. In truth, Rosencrantz and Guildenstern inhabit a world that is determined by mathematical concepts of probability. Mathematicians define *probability* as the long-run pattern that predictably emerges from a series of random outcomes. If 100 coin flips were made, the results can be surprisingly uneven. Coin tosses only yield about a .5 probability after one thousand flips. The characters may function with free will, but they are beholden to a fate—or a probability—that they cannot perceive.) **LS** Logical

MATHPRACTICE
Answers

$$\frac{4}{20} = \frac{20}{100} = 0.2$$

There is a probability of 0.2 that the next pine tree will have pine cones.

Activity ————— GENERAL

M&M® Samples and Population
The distribution of colors in a typical bag of M&M's® is available on the Internet. Download this information for the following activity. Randomly draw five M&M's® from a bag and ask students to compare the color distribution of these five to the distribution given. Draw 10, 20, and finally, 30 candies, having students consider the distribution each time. Ask students how this exercise demonstrates the importance of sample size in science. As an extension, students can graph the results of larger samples.

MISCONCEPTION
///ALERT\\\

Probability and Possibility
There is a difference between *probability* and *possibility*. When something is possible, it can occur. Mathematical probability refers to the likelihood of possible events. For example, it is possible that a tornado might sweep up a winning lottery ticket and drop it gently at someone's feet on his or her birthday, but it is not probable.

MATHPRACTICE

Probability Probability is often determined by observing ratios or patterns. For example, imagine that you count 20 pine trees in a forest and notice that four of those trees have pine cones. What is the probability that the next pine tree you come across will have pine cones?

Figure 10 ▶ Most people are familiar with statistics regarding the weather, such as the chance, or probability, of a thunderstorm.

40

What Is the Probability? The chance that something will happen is called **probability.** For example, if you toss a penny, what is the probability that it will come up heads? Most people would say "half and half," and they would be right. The chance of a tossed penny coming up heads is ½, which can also be expressed as 0.5 or 50%. In fact, probability is usually expressed as a number between 0 and 1 and written as a decimal rather than as a fraction. Suppose the penny comes up heads 7 out of 10 times. Does this result prove that the probability of a penny coming up heads is 0.7? No, it does not. So what is the problem?

The problem is that the *sample size*—the number of objects or events sampled—is too small to yield an accurate result. In statistics, a **sample** is the group of individuals or events selected to represent the population. If you toss a penny 10 times, your sample size is 10. If you continue tossing 1,000 times, you are almost certain to get about 50% heads and 50% tails. In this example, the sample is the number of coin tosses you make, while the population is the total number of coin tosses possible. Scientists try to make sure that the samples they take are large enough to give an accurate estimate for the whole population.

Statistics in Everyday Life
You have probably heard, "There is a 50 percent chance of rain today." **Figure 10** shows an example of a natural event that we often associate with probability—a thunderstorm. You encounter statistics often and use them more than you may think. People are constantly trying to determine the chance of something happening. A guess or gut instinct is probably just an unconscious sense of probability.

Understanding the News The news contains statistics every day, even if they are not obvious. For example, a reporter may say, "A study shows that forest fires increased air pollution in the city last year." We could ask many statistical questions about this news item. We might first ask what the average amount of air pollution in the city is. We could gather data on air pollution levels over the past 20 years and graph them. Then we could calculate the mean, and ask ourselves how different last year's data were from the average. We might graph the data and look at the distribution. Do this year's pollution levels seem unusually high compared to levels in other years? Recognizing and paying attention to statistics will make you a better consumer of information, including information about the environment.

HISTORY ———
CONNECTION — GENERAL

Pascal's Triangle In 1653, French mathematician Blaise Pascal created "Pascal's Triangle," an arrangement of numbers which describes simple probabilities. For example, if there are five slippers in a closet, and you want to figure out how many different ways you can choose two of them to wear, Pascal's Triangle can help determine how many different ways you can choose two objects from a set of five. Pascal's Triangle can be easily downloaded from the Internet and is discussed in many upper-level math texts. Bring in a number of objects such as hats, slippers, bracelets, or ties to clearly demonstrate how Pascal's Triangle is useful in the study of probability. **LS Visual/Logical**

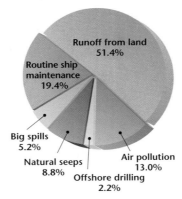

Runoff from land
51.4%

Routine ship maintenance
19.4%

Big spills
5.2%

Natural seeps
8.8%

Offshore drilling
2.2%

Air pollution
13.0%

Thinking About Risk In scientific terms, **risk** is the probability of an unwanted outcome. For example, if you have a 1 in 4 chance of failing a class, the risk is ¼, or 0.25. **Figure 11** shows a well-publicized environmental problem—oil spills. But as you can see in the pie graph, the risk of pollution from large oil spills is much smaller than the risk of oil pollution from everyday sources.

The most important risk we consider is the risk of death. Most people overestimate the risk of dying from sensational causes, such as plane crashes, and underestimate the risk from common causes, such as smoking. Likewise, most citizens overestimate the risk of sensational environmental problems and underestimate the risk of ordinary ones, as shown in **Table 2.**

Table 2 ▼

Perceptions of Risk by Experts and Ordinary Citizens		
	High risk	**Low risk**
Experts	ozone depletion; global climate change	oil spills; radioactive materials; water pollution
Citizens	ozone depletion; radioactive waste; oil spills	global climate change; water pollution

Source: U.S. Environmental Protection Agency.

Connection to Law

Oil Tankers The Oil Pollution Act of 1990 was a direct response to the *Exxon Valdez* oil spill. The controversial bill had been debated for 14 years; it passed swiftly in the aftermath of the disaster. Under the law, all oil tankers operating in United States waters must be protected with double hulls by 2015.

41

REAL-LIFE CONNECTION ── GENERAL

Federal Emergency Management Agency (FEMA) After a disaster occurs, such as a hurricane, flood, drought, or terrorist attack, FEMA helps a community recover. Another responsibility of FEMA is *mitigation*, the effort to lessen the impact of a disaster before it occurs. With mitigation in mind, FEMA provides the public with information on how to protect houses and public buildings from flooding, earthquakes, and wind. This information is based upon risks that have been determined by experts. The probability of a flood is something that can be calculated by analyzing the geography and the historic weather and climactic patterns of a region. Mitigation is not limited to FEMA. Architects and structural engineers also consider how to engineer buildings and other structures to withstand the forces of earthquakes, floods, and tornadoes.

Discussion — GENERAL

Giant's Causeway Giant's Causeway in Northern Ireland is said to have been built by the mythical giant Finn McCool. According to legend, McCool built the plateau, composed of about 40,000 columns, to woo a lady giant. Geologists know that when lava cools it forms regular polygonal columns like the ones at Giant's Causeway. But scientists did not know why, until physicists Alberto G. Rojo and Eduardo A. Jagla started to "play" with computer models. "If you want to fracture a material with the least energy, hexagons are the way to do it," explains Rojo. The scientists realized that they could model their results using simple materials. With cornstarch, water, and tape, they duplicated the fracturing of lava. Ask students to categorize the two types of models that Rojo and Jagla employed. (The first model was a computer model, which is conceptual and mathematical. The second model is a physical model.)

Group Activity — GENERAL

Modeling Lava Fractures To demonstrate how lava behaves as it cools, mix cornstarch with water in equal parts by volume. Spread the mixture onto a flat surface to a depth of one-half of an inch. When the mixture dries, place a wide swath of clear tape across the top and pull up. The mix of polygons that appear on the plate mimic those of Giant's Causeway, as well as those in the Palisades in New Jersey and in Devil's Postpile, in California. **LS Kinesthetic/Visual**

Figure 12 ▶ This plastic model of a DNA molecule is an example of a physical model.

Geofact

Fossil-Fuel Deposits Fossil fuels, such as coal and oil, are often buried deep underground in particular rock formations. We find kinds of fossil fuels by drilling for rocks that indicate the presence of fossil fuels and then we make models of where the coal or oil is likely to be found.

Figure 13 ▶ This map of the Denver, Colorado, area is an example of a graphical model.

Models

You are probably already familiar with models. Museums have models of ships, dinosaurs, and atoms. Architects build models of buildings. Even crash-test dummies are models. **Models** are representations of objects or systems. Although people usually think of models as things they can touch, scientists use several different types of models to help them learn about our environment.

Physical Models All of the models mentioned above are physical models. *Physical models* are three-dimensional models you can touch. Their most important feature is that they closely resemble the object or system they represent, although they may be larger or smaller.

One of the most famous physical models was used to discover the structure of DNA. The two scientists that built the structural model of DNA knew information about the size, shape, and bonding qualities of the subunits of DNA. With this knowledge, the scientists created model pieces that resembled the subunits and the bonds between them. These pieces helped them figure out the potential structures of DNA. Discovering the structure of DNA furthered other research that helped scientists understand how DNA replicates in a living cell. **Figure 12** shows a modern model of a DNA molecule. The most useful models teach scientists something new and help to further other discoveries.

Graphical Models Maps and charts are the most common examples of *graphical models*. Showing someone a road map is easier than telling him or her how to get somewhere. An example of a graphical model is the map of the Denver, Colorado, area in **Figure 13.** Scientists use graphical models to show things such as the positions of the stars, the amount of forest cover in a given area, and the depth of water in a river or along a coast.

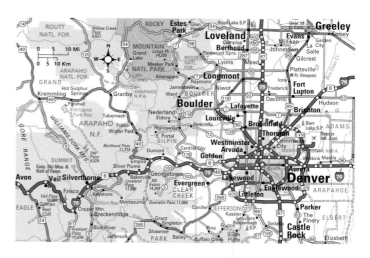

READING SKILL BUILDER — BASIC

Paired Reading Group students in pairs. Have each student read the "Models" section silently while taking note of any passages he or she finds confusing. After students finish reading, ask one student to summarize the section and the second student to add any ideas that were omitted. Both readers should help each other with any parts the other person did not understand. Have the students prepare a list of questions (with answers) to ask the class. Students may still have unanswered questions. Ask the class if it can provide the answers. **LS Verbal/Interpersonal**

Homework — BASIC

Identifying Models Have students use the weather section of the newspaper to identify uses of graphical, mathematical, and conceptual models. For example, a mathematical model may be used to predict the likelihood of rain on a certain day. A graphical model, such as a map, may be used to show the locations of pressure fronts or storms. A conceptual model may be used to explain long-term trends in the weather. **LS Logical**

Conceptual Models A **conceptual model** is a verbal or graphical explanation for how a system works or is organized. A flow-chart diagram is an example of a type of conceptual model. A flow-chart uses boxes linked by arrows to illustrate what a system contains and how those contents are organized.

Consider the following example. Suppose that a scientist is trying to understand how mercury, a poisonous metal, moves through the environment to reach people after the mercury is released from burning coal. The scientist would use his or her understanding of mercury in the environment to build a conceptual model, as shown in **Figure 14.** Scientists often create such diagrams to help them understand how a system fits together—what components the system contains, how the components are arranged, and how they affect one another.

Conceptual models are not always diagrams. They can also be verbal descriptions or even drawings of how something works or is put together. For example, the famous model of an atom as a large ball being circled by several smaller balls is a conceptual model of the structure of an atom. As this example shows, an actual model can be more than one type. An atomic model made of plastic balls is both a conceptual model as well as a physical model.

FIELD ACTIVITY

Conceptual Model Accompany your class outdoors. Observe your surroundings, and write down observations about what you see. In your *Ecolog,* draw a conceptual model of something you observe. Your model should be of a system with components that interact, such as a small community of organisms.

internet connect

www.scilinks.org
Topic: Using Models
SciLinks code: HE4119

SCiLINKS Maintained by the National Science Teachers Association

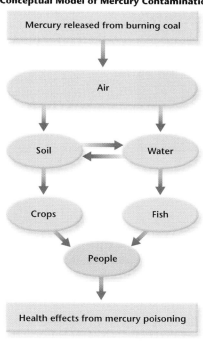

Conceptual Model of Mercury Contamination

Mercury released from burning coal

↓

Air

↓   ↓

Soil ⇄ Water

↓   ↓

Crops   Fish

↓   ↓

People

↓

Health effects from mercury poisoning

Figure 14 ▶ This conceptual model shows how mercury released from burning coal could end up reaching people, where it could cause poisoning.

43

Models Reinforce students' comprehension of models by dividing the class into groups of three. Ask each group to have a representative explain to you what a physical, graphical, conceptual, and mathematical model is. Be sure to ask for an example of each.

Quiz ———————— GENERAL

1. Why do murders make front-page news, but lung cancer deaths do not? (Answers may vary. Students should incorporate the idea that rare and sensational events receive more attention than daily risks.)

2. Provide an example of a conceptual model. (Answers may vary.)

Alternative
Assessment ———— GENERAL

Risks Ask students to identify different risks in their lives—things they are scared of or things they may live with that do not bother them. They should also include dangers that seem very remote as well as environmental issues. Ask students to rank the risks and then do research to determine which are likely to be high risks for them and which are likely to be low risks. Ask them to share their results with you or with the class.
LS Intrapersonal

Mathematical Models A **mathematical model** is one or more equations that represents the way a system or process works. You can represent many common situations using math models.

For example, suppose that the grapes in a fruit basket at home are getting moldy. You notice that every day the mold covers two more grapes. Here is a mathematical model for the number of moldy grapes on Tuesday:

$$M_{Tue} = M_{Mon} + 2, \text{ where } M = \text{number of moldy grapes}$$

Mathematical models are especially useful in cases with many variables, such as the many things that affect the weather.

Because mathematical models use numbers and equations, people may think the models are always right. But weather models, for example, sometimes predict rain on dry days. In fact, people are the ones who interpret data and write the equations. If the data or the equations are wrong, the model will not be realistic and so will provide incorrect information. Like all models, mathematical models are only as good as the data that went into building them.

People may think of mathematical models as being confined to blackboards and paper, but scientists can use the models to create amazing, useful images. Look at the image of the San Francisco Bay Area in **Figure 15**. This is a "false color" digital satellite image. The satellite measures energy reflected from the Earth's surface. Scientists use mathematical models to relate the amount of energy reflected from objects to the objects' physical condition.

Figure 15 ▶ This is a satellite image of the San Francisco Bay Area. Scientists use mathematical models to understand the terrain from the way objects on the surface reflect light. In this image, healthy vegetation is red.

SECTION 2 Review

1. Explain why sample size is important in determining probability.

2. Explain what "the mean number of weeds in three plots of land" means.

3. Describe three types of models used by scientists.

CRITICAL THINKING

4. Analyzing Relationships Explain the relationship between probability and risk.

5. Applying Ideas Write a paragraph that uses examples to show how scientists use statistics. **WRITING SKILLS**

6. Evaluating Ideas Why are conceptual and mathematical models especially powerful?

Answers to Section Review

1. A statistical sample needs to be large enough to reflect a population. If a sample is not large enough it can easily misrepresent probability.

2. The statement means that someone has determined how many weeds are in each of three plots of land, added these numbers together, and divided by three. In this case, the "mean" represents the average number of weeds in the three plots.

3. Answers may vary. Students should include three of the following: Physical models are three dimensional and closely resemble the object sor system they represent. Graphical models, which include maps and charts, illustrate data such as positions or amounts graphically. Conceptual models show how something works or how it is organized. Mathematical models use one or more equations to represent how a system or process works.

4. Probability is the chance that an event will occur. Risk is the probability of an unwanted outcome.

5. Answers may vary.

6. Answers may vary. Students may say that conceptual and mathematical models represent ideas and relationships clearly and precisely.

Making Informed Decisions

Scientific research is an essential first step in solving environmental problems. However, many other factors must also be considered. How will the proposed solution affect people's lives? How much will it cost? Is the solution ethical? Questions like these require an examination of **values,** which are principles or standards we consider important. What values should be considered when making decisions that affect the environment? **Table 3** lists some values that often affect environmental decisions. You might think of others as well.

An Environmental Decision-Making Model

Forming an opinion about an environmental issue is often difficult and may even seem overwhelming. It helps to have a systematic way of analyzing the issues and deciding what is important. One way to guide yourself through this process is by using a decision-making model. A **decision-making model** is a conceptual model that provides a systematic process for making decisions.

Figure 16 shows one possible decision-making model. The first step of the model is to gather information. In addition to watching news reports and reading newspapers, magazines, and books about environmental issues, you should listen to well-informed people on all sides of an issue. Then consider which values apply to the issue. Explore the consequences of each option. Finally, evaluate all of the information and make a decision.

Table 3 ▼

Values That Affect Environmental Decision Making	
Value	**Definition**
Aesthetic	what is beautiful or pleasing
Economic	the gain or loss of money or jobs
Environmental	the protection of natural resources
Educational	the accumulation and sharing of knowledge
Ethical/moral	what is right or wrong
Health	the maintenance of human health
Recreational	human leisure activities
Scientific	understanding of the natural world
Social/cultural	the maintenance of human communities and their values and traditions

Objectives

▶ Describe three values that people consider when making decisions about the environment.

▶ Describe the four steps in a simple environmental decision-making model.

▶ Compare the short-term and long-term consequences of two decisions regarding a hypothetical environmental issue.

Key Terms

value
decision-making model

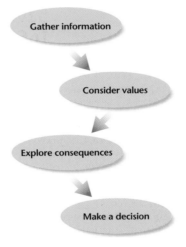

Figure 16 ▶ The diagram above shows a simple decision-making model.

45

SKILL BUILDER — GENERAL

Writing Ask students to imagine that they live in a town with a coal-fired power plant nearby. Many people are employed by the power plant and at a nearby coal mine. A neighboring county, which is generally more affluent, is building a wind farm to generate electricity. The wind farm will be operational in a few years. The press has raised the question of whether the existing plant should be closed. Ask students to write a sentence for each value in **Table 3,** explaining how it relates to this situation. **LS Logical**

Focus

Before beginning this section, review with your students the Objectives in the Student Edition. This section introduces the idea that environmental decisions involve different values that are often competing. Difficult decisions can be managed systematically by using a decision-making model.

🔊 Bellringer

Ask students to write down a problem in their life that presented a difficult decision. Ask them to consider how they usually approach decisions. Ask, "Is it by flipping a coin or by talking to your friends? How do you weigh what is important to you?" **LS Intrapersonal**

Motivate

Discussion — GENERAL

Alien Invasion Present students with the following scenario: A highly-evolved alien race has invaded Earth and is quickly descending upon your home town. You are forced to evacuate immediately, perhaps leaving your home behind forever. You do not know where you will go or what you will do next. There is only time to take 10 items with you. Food is supplied on the evacuation ship. Ask students what items they would decide to take. Discuss the pros and cons of each choice and come up with a common list for the class. Ask students what values influenced their decision. (Answers may vary. Ethical, educational, environmental, health, and social values may influence students' choices.) **LS Verbal**

Discussion ———

Emotions and Decision-Making

Point out to students that it is important to distinguish between emotional arguments and factual arguments while making decisions. People often confuse the two. Have students come up with a list of emotional statements and a list of factual statements and explore the differences between the two. You might also stress that emotional arguments have value and should be considered. **LS** Verbal

REAL-LIFE ———
CONNECTION — **GENERAL**

NIMBY NIMBY, which stands for **N**ot **I**n **M**y **B**ackyard, refers to the reaction many people have towards something they consider unpleasant being located near their homes. NIMBY may apply to things that are dangerous, unsightly, noisy or inconvenient. Have students give examples of what might elicit a NIMBY response. (Examples may include the proposed building of nuclear power plants, power lines, landfills, highways, airports, or cellular phone antennas.)

Figure 17 ▶ The map (above) shows the proposed nature preserve, which would be home to warblers like the one pictured (right).

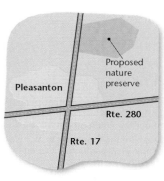

A Hypothetical Situation

Consider the following hypothetical example. In the town of Pleasanton, in Valley County, biologists from the local college have been studying the golden-cheeked warbler, shown in **Figure 17.** The warblers have already disappeared from most areas around the state, and the warbler population is declining in Valley County. The biologists warn county officials that if the officials do not take action, the state fish and wildlife service may list the bird as an endangered species.

Pleasanton is growing rapidly, and much of the new development is occurring outside the city limits. This development is destroying warbler habitat. Valley County already has strict environmental controls on building, but these controls do not prevent the clearing of land.

Several groups join together to propose that the county buy several hundred acres of land where the birds are known to

CASE STUDY

Saving the Everglades: Making Informed Decisions

The Florida Everglades is an enormous, shallow freshwater marsh. The water in the Everglades slowly flows from Lake Okeechobee to Florida Bay. Much of the marsh is filled with sawgrass, mangroves, and other water-loving plants. It is also home to wildlife, from 40 species of fish to panthers, alligators, and wading birds such as herons and roseate spoonbills.

In the 1880s, marshlands were considered wastelands. Developers began to drain the Everglades. They replaced marsh with houses and sugarcane fields. Between 1940 and 1971, the Army Corps of Engineers built dikes, canals, and pumping stations that drained even more water. The Corps also straightened the Kissimmee River, which runs into Lake Okeechobee.

Scientists have shown that what remains of the Everglades is dying. Its islands and mangrove swamps are vanishing, its water is polluted with fertilizer from farms, and its wading-bird colonies are much smaller than before. These effects have economic consequences. Because much of the Everglades' water has been diverted from Florida Bay into the Atlantic Ocean, the towns of southeast Florida are running out of fresh water and much of the marine life in Florida Bay has died.

In the 1990s, a commission reported that the destruction of the Everglades had jeopardized the state's tourism industry, farming, and the economic future of south Florida. The solution was obvious—undo the water diversion dikes and dams and restore water to the Everglades.

▶ **The roseate spoonbill** is a colorful resident of the Everglades.

46

🡒 Internet Activity ——— **GENERAL**

Acronym Glossary Acronyms are popular in environmental science. Have students research environmental science acronyms on the Internet so that they can create an acronym glossary. Knowledge of these acronyms will help students better understand articles that relate to environmental issues. The following acronyms will help students get started:

NGO—**N**on **G**overnmental **O**rganization— refers to organizations that do not receive government funds. **NIMEY**—**N**ot **I**n **M**y **E**lection **Y**ear—refers to a politician's decision to forestall something unpleasant for the constituency during an election year. **GOOMBY**— **G**et **O**ut **O**f **M**y **B**ackyard—refers to an individual's desire to remove an environmental hazard from a neighborhood or community. **BANANA**—**B**uild **A**bsolutely **N**othing **A**nywhere **N**ear **A**nything—refers to a person who is opposed to any form of real estate development. **LULU** stands for **L**ocally **U**ndesirable **L**and **U**se. Other acronyms include **EPA, USGS, FWS, NRDC, IPCC, WWF** and **WMO.**

breed and save that land as a nature preserve. The groups also propose limiting development on land surrounding the preserve. The group obtains enough signatures on a petition to put the issue to a vote, and the public begins to discuss the proposal.

Some people who own property within the proposed preserve oppose the plan. These property owners have an economic interest in this discussion. They believe that they will lose money if they are forced to sell their land to the county instead of developing it.

Other landowners support the plan. They fear that without the preserve the warbler may be placed on the state's endangered species list. If the bird is listed as endangered, the state will impose a plan to protect the bird that will require even stricter limits on land development. People who have land near the proposed preserve think their land will become more valuable. Many residents of Pleasanton look forward to hiking and camping in the proposed preserve. Other residents do not like the idea of more government regulations on how private property can be used.

Ecofact

The Everglades Scientists have identified more than 400 endangered species of plants and animals that live in the Florida Everglades.

internet connect

www.scilinks.org
Topic: Environmental Decision Making
SciLinks code: HE4035

SCiLINKS Maintained by the National Science Teachers Association

▶ **The Everglades** can be thought of as a shallow, slow-moving river that empties into Florida Bay.

In 2000, the $7.8 billion Everglades Restoration Plan was signed into law. The plan was put together by groups that had been fighting over the Everglades for decades: environmentalists, politicians, farmers, tourism advocates, and developers. Over the course of 5 years, members from the groups met and crafted a plan. At first people were afraid to break up into committees for fear that other people would make deals behind their backs. The director instituted social gatherings, and the members got to know and trust each other.

In the end, no one was completely satisfied, but all agreed that they would be better off with the plan than without it. Already Florida has restored 7 mi of the Kissimmee River to its original path. Native plants are absorbing some of the pollution that has killed an estimated $200 million worth of fish and wildfowl. The Everglades Restoration Plan is not perfect, but the process of creating and approving it shows how science and thoughtful negotiation can help solve complex environmental problems.

CRITICAL THINKING

1. Analyzing Processes Explain why it was so difficult for people to agree on how to restore the Everglades.

2. Analyzing Relationships If your county decided to build a landfill, do you think the decision-making process would resemble the Everglades example?

47

INCLUSION Strategies

• *Visually Impaired*   • *Developmentally Delayed*
• *Learning Disabled*

Ask students to locate Key Terms in the chapter by using the Chapter Highlights and page numbers. Students should copy the sentence where the term appears in the text. Students can also write an original sentence using the term. If the student has difficulty with writing, the student can read or compose the sentences aloud into a tape recorder. Students may use the tape or written activity as a study guide for independent or small group work.

Reteaching ——— BASIC

Revisiting Einstein After discussing decision-making models, ask students to think again about the following quote from Albert Einstein: "The whole of science is nothing more than a refinement of everyday thinking." Ask students to review their comments in their *EcoLog,* and ask them if they have any further insights. How is the decision-making model similar to the scientific method? **LS Logical**

CASE STUDY

Saving the Everglades Much of south Florida would be uninhabitable if it were not for the Central and Southern Florida Project (CS&F Project), which was initiated in 1948. The U.S. Army Corps of Engineers constructed one of the largest water management projects in the world to drain portions of the Everglades in an attempt to mitigate flooding, conserve water, preserve fish and wildlife, and make land available for development. Unfortunately, the project had adverse effects on south Florida ecosystems. In 1992 the Central & South Florida Project began to be re-evaluated, leading to the comprehensive restoration efforts in place today. These efforts must consider how to protect the ecosystem while ensuring that Florida's water needs continue to be met.

Answers to Critical Thinking

1. Answers may vary but students should focus on the idea that people are motivated by different values and interests. Even when different groups could agree that action had to be taken, they argued on the scope and methods of restoration.

2. Answers may vary, but students should recognize that democratic community decisions occur when people with different values come to an agreement and establish trust.

Interpreting Statistics ——— GENERAL

What Happened to the Warblers?
Ask students to analyze the data in **Figure 18.** Have them create a table or bar graph that shows the population of warblers every two years. Have students brainstorm ways that biologists could have collected this data. Begin a discussion about whether the data in such a graph could be misleading. Encourage skepticism and imagination in the class. (Biologists could have collected this data by surveying a sample of the entire county area for warblers. The data could be misleading if the scientists did not sample a representative area. For example, the warblers might prefer to nest in certain trees or in specific areas. If the scientists sampled the land area randomly, the chances are that their results would misrepresent the size of the warbler population.)
LS Visual

Student Opportunities

Understanding Local Issues
Making environmental decisions involves a variety of people who have different roles in the community. Have students pick a local environmental issue and identify some of the people involved with it. Then encourage them to choose one of these people and interview them. Have students summarize the interviews and bring the summaries to class. Then have students discuss the issue using information from the people involved. **LS** Verbal

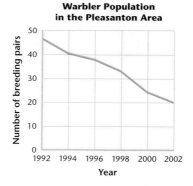

Warbler Population in the Pleasanton Area

Figure 18 ▶ The population of golden-cheeked warblers in the Pleasanton area has declined in recent years.

How to Use the Decision-Making Model
The hypothetical situation in Pleasanton can be used to illustrate how to use the decision-making model. Michael Price is a voter in Valley County who will vote on whether the county should create a nature preserve to protect the golden-cheeked warbler. The steps Michael took to make his decision about the proposal are outlined below.

Gather Information Michael studied the warbler issue thoroughly by watching local news reports, reading the newspaper, learning more about golden-cheeked warblers from various Web sites, and attending forums where the issues were discussed. An example of scientific information that Michael considered includes the graph of warbler population decline in **Figure 18.** Several of the arguments on both sides made sense to him.

Consider Values Michael made a table similar to **Table 4** to clarify his thoughts. The values listed are environmental, economic, and recreational. Someone else might have thought other values were more important to consider.

Should Valley County Set Aside a Nature Preserve?			
	Environmental	**Economic**	**Recreational**
Positive short-term consequences	Habitat destruction in the nature preserve area is slowed or stopped.	Landowners whose property was bought by the county receive a payment for their land. Property outside the preserve area can be developed with fewer restrictions.	Parts of the preserve are made available immediately for hiking and picnicking.
Negative short-term consequences	Environmental controls are made less strict in parts of the county outside the preserve area.	Property owners inside the preserve area do not make as much money as if they had developed their land. Taxpayers must pay higher taxes to buy preserve land.	Michael could not think of any negative short-term consequences.
Positive long-term consequences	The population of warblers increases, and the bird does not become endangered. Other species of organisms are also protected. An entire habitat is preserved.	Property near the preserve increases in value because it is near a natural area. Businesses move to Valley County because of its beauty and recreational opportunities, which results in job growth. The warbler is not listed as endangered, which avoids stricter controls on land use.	Large areas of the preserve are available for hiking and picnicking. Landowners near the preserve may develop campgrounds with bike trails, swimming, and fishing available on land adjacent to the preserve.
Negative long-term consequences	Other habitat outside the preserve may be damaged by overdevelopment.	Taxpayers must continue to pay for maintaining the preserve. Taxpayers lose the tax revenue that this land would have provided if it was developed.	State officials might restrict some recreational activities on private land within the preserve.

48

Career

Architect Architect Samuel Mockbee asked, "Does the architect have a role in addressing political or economic inequities, or transportation issues, or environmental issues? Because I think we do." In 1993, Mockbee and Dennis K. Ruth started a program called the "Rural Studio" at the University of Alabama. Students work with poor, rural residents of Hale County, Alabama, to improve their quality of life by designing affordable housing and other buildings using materials that would otherwise end up in landfills. They work with a community to address its needs from within. Mockbee decided to use his talent to combine a social commitment to the poor with an environmental focus. Have students find out more about the Rural Studio and the connections between architecture and environmental science. Students can also look for information about "Green Architecture" programs on the Internet.

Explore Consequences Michael decided that in the short term the positive and negative consequences listed in his table were almost equally balanced. He saw that some people would suffer financially from the plan, but others would benefit. Taxpayers would have to pay for the preserve, but all the residents would have access to land that was previously off-limits because it was privately owned. Some parts of the county would have more protection from development, and some would have less.

It was the long-term consequences of the plan that allowed Michael to make his decision. Michael realized that environmental values were an important factor in his decision. The idea of a bird becoming extinct distressed him. Also, protecting warbler habitat now would be less costly than protecting it later under a state-imposed plan.

Michael considered that there were long-term benefits to add to the analysis as well. He had read that property values were rising more rapidly in counties where land was preserved for recreation. He found that people would pay more to live in counties that have open spaces. Michael had found that Valley County contained very little preserved land. He thought that creating the preserve would bring the county long-term economic benefits. He also highly valued the aesthetic and recreational benefits a preserve would offer, such as the walking trail in **Figure 19.**

Make A Decision Michael chose to vote in favor of the nature preserve. Other people who looked at the same table of pros and cons might have voted differently. If you were a voter in Valley County, how would you have voted?

As you learn about issues affecting the environment, both in this course and in the future, use this decision-making model as a starting point to making your decisions. Make sure to consider your values, weigh pros and cons, and keep in mind both the short-term and long-term consequences of your decision.

Figure 19 ▶ Land set aside for a nature preserve can benefit people as well as wildlife.

SECTION 3 Review

1. **Explain** the importance of each of the four steps in a simple decision-making model.

2. **List** and define three possible values to consider when making environmental decisions.

3. **Describe** in a short paragraph examples of two situations in which environmental values come into conflict with other values. **WRITING SKILLS**

CRITICAL THINKING

4. **Making Decisions** Pick one of the situations you described in question 3. Make a decision-making table that shows the positive and negative consequences of either of two possible decisions.

5. **Analyzing Ideas** Suggest how to make the decision-making model presented here more powerful.

Answers to Section Review

1. Gathering information is important to fully understand a problem. It is necessary to consider values so that a decision is made based on what is important to the decision-maker. Exploring the consequences is important because it is necessary to consider the long-term impact of a decision. It is important to make a decision so that something can be done and the decision maker's values can be implemented.

2. Answers may vary, but may include the following: Environmental values are values based on how important nature and the environment are to you. Economic values are values based on monetary costs and benefits. Recreational values are values based on the importance of leisure and having fun.

3. Answers may vary. Accept any reasonable, thoughtful answer.

4. The tables may vary but should show an understanding of the decision-making model and a thoughtful exploration of an issue.

5. Answers may vary. Students may offer improvements such as quantifying the positive and negative consequences to better assess the situation.

Close

Reteaching ——— BASIC

Applying the Decision-Making Model Have students identify an environmental issue that affects your area. Ask students to research the issue and then prepare an environmental decision-making table similar to **Table 4.** Review and discuss the tables in class. Then have students write a paragraph that describes their decision-making process and the decision they made. Ask students to use the decision-making model throughout the school year. **LS Intrapersonal**

Quiz ——— GENERAL

1. What are four values that are very important to you when you make everyday decisions? Explain why. (Answers may vary.)

2. Why is it important to identify the different values that influence decision making? (Answers may vary. Students should recognize that different values influence decisions and these values can conflict with each other.)

Alternative Assessment ——— ADVANCED

Land Use One of the most controversial topics in urban development is land use, the way that a city will be organized in order to be a functional place. Ask students to choose a city in the United States or abroad that interests them and analyze the infrastructure, including public transportation, schools, public land, shopping districts, and living areas. Ask students to include their findings in a report that focuses on the environmental characteristics of the city they studied. Students should supplement their report with a map illustrating land use in the city they chose. **LS Verbal/Visual**

Alternative Assessment — GENERAL

Models in Our Lives Ask students to identify five models they use in their daily lives. Have them describe how they use each model and identify whether it is a physical, mathematical, conceptual, or graphical model. Students should include examples of five models they use in their **Portfolio**.

Decision-Making Have students think of an environmental issue that might also be an emotional issue for them. Have them create a decision-making table that includes risks and use the table to help make a decision. Ask students to make persuasive use of statistics from the newspaper, Internet, or books to support their reasoning.
LS Intrapersonal

CHAPTER 2 Highlights

1 Scientific Methods

Key Terms

observation, 31
hypothesis, 32
prediction, 32
experiment, 33
variable, 33
experimental group, 33
control group, 33
data, 34
correlation, 35

Main Ideas

▶ Science is a process by which we learn about the world around us. Science progresses mainly by the experimental method.

▶ The experimental method involves making observations, forming a hypothesis, performing an experiment, interpreting data, and communicating results.

▶ In cases in which experiments are impossible, scientists look for correlations between different phenomena.

▶ Good scientists are curious, creative, honest, skeptical, and open to new ideas.

2 Statistics and Models

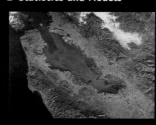

Key Terms

statistics, 38
mean, 39
distribution, 39
probability, 40
sample, 40
risk, 41
model, 42
conceptual model, 43
mathematical model, 44

Main Ideas

▶ Scientists use statistics to classify, organize, and interpret data.

▶ Measures such as means and probabilities are different ways of describing populations and events.

▶ Statistics provides a powerful tool for evaluating information about the environment.

▶ Scientists use models, including conceptual and mathematical models, to understand the systems they study.

3 Making Informed Decisions

Key Terms

value, 45
decision-making model, 45

Main Ideas

▶ Making environmental decisions involves gathering information, considering values, and exploring consequences.

▶ Decisions about the environment should be made thoughtfully. Using a decision-making model will provide you with a systematic process for making knowledgeable decisions.

▶ Making a table that lists positive and negative short- and long-term consequences will help you recognize and weigh your values about an environmental decision.

50

Chapter Resource File

- **Chapter Test** GENERAL
- **Chapter Test** ADVANCED
- **Concept Review** GENERAL
- **Critical Thinking** ADVANCED
- **Test Item Listing**
- **Observation Lab** BASIC
- **Design Your Own Lab** GENERAL
- **CBL™ Probeware Lab** GENERAL
- **Consumer Lab** GENERAL
- **Long-Term Project** GENERAL

CHAPTER 2 Review

Using Key Terms

Use each of the following terms in a separate sentence.

1. *experiment*
2. *correlation*
3. *model*
4. *distribution*
5. *values*

For each pair of terms, explain how the meanings of the terms differ.

6. *hypothesis* and *prediction*
7. *risk* and *probability*
8. *distribution* and *population*
9. *sample* and *population*

✔ **STUDY TIP**

Imagining Examples To understand how key terms apply to actual examples, work with a partner and take turns describing an environmental problem and explaining how the key terms relate to the problem.

Understanding Key Ideas

10. Scientists form _____ hypotheses to answer questions.
    a. accurate
    b. short
    c. mathematical
    d. testable

11. Risk is the _____ of a negative outcome.
    a. sample
    b. statistic
    c. probability
    d. event

12. If the results of your experiment do not support your hypothesis, you should
    a. publish your results anyway.
    b. consider the results abnormal and continue working.
    c. find a way to rationalize your results.
    d. try another method.

13. In a population, characteristics such as size will often be clustered around the
    a. sample.
    b. mean.
    c. distribution.
    d. collection.

14. Models used by scientists include
    a. conceptual models.
    b. variable models.
    c. physical models.
    d. Both (a) and (c)

15. Reading scientific reports is an example of
    a. assessing risk.
    b. considering values.
    c. gathering information.
    d. exploring consequences.

16. A conceptual model represents a way of thinking about
    a. relationships.
    b. variables.
    c. data.
    d. positions.

17. In an experiment, the experimental treatment differs from the control treatment only in the _____ being studied.
    a. experiment
    b. variable
    c. hypothesis
    d. data

18. To fully understand a complex environmental issue, you may need to consider
    a. economics.
    b. values.
    c. scientific information.
    d. All of the above

19. Scientists _____ experiments to make sure the results are meaningful.
    a. perform
    b. repeat
    c. conclude
    d. communicate

51

Assignment Guide

Section	Questions
1	1, 2, 6, 10, 12, 17, 19–21, 28, 33, 35–37
2	3, 4, 7–9, 11, 13, 14, 16, 22, 23, 29–32
3	5, 15, 18, 24, 34, 38

ANSWERS

Using Key Terms

1. Sample answer: The scientist conducted an experiment to verify her hypothesis.
2. Sample answer: Storm clouds and rain have a strong correlation.
3. Sample answer: The doctor showed his patient a model to demonstrate how the heart works.
4. Sample answer: To see if there are more elderly people than young people, you could look at a distribution of ages.
5. Sample answer: My mother always considers her values before making a decision.
6. A hypothesis is a testable explanation for an observation, while a prediction is an educated guess of what will happen when the hypothesis is tested.
7. A risk is the chance that an unwanted event will occur. A probability is the likelihood that something will happen.
8. A distribution is the arrangement of the members of a population in relation to a characteristic. A population is a group of similar things that is being studied.
9. A population is a group of similar things that is being studied, while a sample is a smaller group studied to represent the population.

Understanding Key Ideas

10. d
11. c
12. a
13. b
14. d
15. c
16. a
17. b
18. d
19. b

Short Answer

20. When an observation warrants further examination, a good scientist knows what questions to investigate to fully understand the observation.

21. In an experiment, the control group provides a standard to compare with the experimental treatment. The control group provides a set of data unaffected by the variable or with the variable at a baseline value.

22. Statistics help people quantify and analyze different kinds of information, including information about the environment.

23. Environmental scientists use mathematical models to express quantifiable relationships in the most precise form possible.

24. Making a table can help in a decision-making situation by organizing all the positive and negative consequences of a decision for comparison.

Interpreting Graphics

25. The density of the alligator population fell drastically—from almost 22 per kilometer of shoreline to about 3 per kilometer of shoreline.

26. The alligator concentration increased. In 1994 there was less than one alligator per kilometer. In 1998 there were almost 5 alligators per kilometer.

27. The alligator population was more than five times greater in 1986.

Concept Mapping

28. Answers to the concept mapping questions are on pp. 667–672

Short Answer

20. Explain the statement, "A good scientist is one who asks the right questions."

21. Explain the role of a control group in a scientific experiment.

22. How are statistics helpful for evaluating information about the environment?

23. Explain why environmental scientists use mathematical models.

24. How does making a table help you evaluate the values and concerns you have when making a decision?

Interpreting Graphics

The graph below shows the change in size of a shoreline alligator population over time. Use the graph to answer questions 25–27.

25. What happened to the density of alligators between 1986 and 1988?

26. What happened to the trend in the alligator concentration between 1994 and 1998?

27. How many times greater was the alligator population in 1986 than it was in 2000?

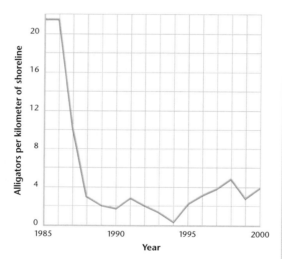

Concept Mapping

28. Use the following terms to create a concept map: *control group, experiment, experimental group, prediction, data, observations, conclusions,* and *hypothesis.*

Critical Thinking

29. Drawing Conclusions What does a scientist mean by the statement, "There is an 80 percent probability that a tornado will hit this area within the next 10 years"?

30. Making Inferences How does a map of Denver allow you to navigate around the city?

31. Evaluating Assumptions Are complicated models always more accurate? Write a paragraph that uses examples to explain your answer. **WRITING SKILLS**

32. Interpreting Statistics Explain what the following statement proves: "We sampled pet owners and found that three out of five surveyed own dogs and two out of five surveyed own cats."

Cross-Disciplinary Connection

33. Language Arts The word *serendipity,* which means "luck in finding something accidentally," came from a Persian fairy tale called *The Three Princes of Serendip.* In the story, each of the princes discovers something by accident. Research and write a short report on a serendipitous discovery about the environment. **WRITING SKILLS**

Portfolio Project

34. Make a Poster Choose an environmental issue in your area. You can choose a real-life problem that you have heard about on the news, such as improving the sewage system or building a new landfill, or you can choose a project that you think should be considered. Research the issue at your school or local library. Prepare a poster listing the groups of people likely to be involved in the decision and the factors that may be taken into consideration, including economic, social, and environmental factors.

Critical Thinking

29. The scientist means that there is an 8 out of 10 chance that a tornado will strike the area over the next 10 years.

30. Answers may vary, but students should recognize that a map is a graphical model of the city that illustrates the relative size and location of streets and landmarks.

31. No, complicated models are not necessarily more accurate. Adding detail to a model is only justified if that detail truly reflects relevant aspects of the system in question.

32. The statement proves very little because it does not state the sample size in relation to the population size. It only proves that 60% of those surveyed own dogs and 40% own cats.

Cross-Disciplinary Connection

33. Answers may vary.

Portfolio Project

34. Answers may vary.

This table shows the results of an experiment that tested the hypothesis that butterflies are attracted to some substances but not to others. Twenty-four trays containing four substances were placed in random order on a sandbank to see if butterflies landed on the trays. The number of butterflies that landed on each type of tray and stayed for more than 5 min during a 2 h period was recorded in the table. Use the data in the table below to answer questions 35–36.

Butterfly Feeding Preferences				
	Sugar solution	Nitrogen solution	Water	Salt solution
Number of butterflies attracted	5	87	7	403

35. Evaluating Data Do the results in the table show that butterflies are attracted to salt solution but not any other substance? Why or why not? What other data would you like to see to help you evaluate the results of this experiment?

36. Analyzing Data Are there any controls shown in this table?

WRITING SKILLS

37. Communicating Main Ideas How is the experimental method an important scientific tool?

38. Writing Persuasively Write a letter to the editor of your local paper outlining your opinion on a local environmental issue.

STANDARDIZED TEST PREP

For extra practice with questions formatted to represent the standardized test you may be asked to take at the end of your school year, turn to the sample test for this chapter in the Appendix.

READING SKILLS

Read the passage below, and then answer the questions that follow.

Jane and Jim observed a group of male butterflies by the roadside. Jane said that this behavior was called puddling and that the butterflies were counting each other to see if there was room to set up a territory in the area. Jim said he did not think butterflies could count each other and suggested the butterflies were feeding on nitrogen in the sand. Jane agreed that the butterflies appeared to be feeding, but she said that they may not be feeding on nitrogen, because female butterflies need more nitrogen than males.

Jim and Jane decided to perform some experiments on the butterflies. They put out trays full of sand in an area where butterflies had been seen. Two trays contained only sand. Two contained sand and water, two contained sand and a salt solution, and two contained sand and a solution containing nitrogen. Butterflies came to all the trays, but they stayed for more than 1 min only at the trays that contained the salt solution.

1. Which of the following statements is a useful hypothesis that can be tested?
   a. Male butterflies mate with female butterflies.
   b. Salt is a compound and nitrogen is an element.
   c. Butterflies are never seen in groups except on sandy surfaces.
   d. Butterflies are attracted to salt.

2. Which of the following conclusions is supported by the observations Jane and Jim made?
   a. Male butterflies can count each other.
   b. The butterflies were probably feeding on nitrogen in the sand.
   c. The butterflies were probably feeding on salts in the sand.
   d. Female butterflies need less nitrogen than male butterflies.

Math Skills

35. No, butterflies do not appear to be attracted only to salt solution, because 87 stayed for more than 5 minutes at the nitrogen solution. How the butterflies behaved afterwards and whether any died might be interesting data. The experiment should be repeated before the results are evaluated. Data on combinations of ingredients might also be helpful.

36. Yes, the tray with water functions as the control. The only difference between this tray and the others is in the substances that are dissolved in the water.

Writing Skills

37. Answers may vary. The experimental method is an important scientific tool because it provides a method for using observations to provide explanations for phenomena in the natural world. The experimental method allows people to study the impact of a variable as objectively as possible.

38. Answers may vary.

Reading Skills

1. d
2. c

SCIENTIFIC INVESTIGATIONS

Teacher's Notes

Time Required

one 45-minute class period

Lab Ratings

EASY ———→ HARD

TEACHER PREPARATION
STUDENT SETUP
CONCEPT LEVEL
CLEANUP

Skills Acquired

- Predicting
- Experimenting
- Measuring
- Collecting Data
- Classifying
- Organizing and Analyzing Data
- Communicating

The Scientific Method

In this lab, students will:
- Make Observations
- Ask Questions
- Test the Hypothesis
- Analyze the Results
- Draw Conclusions
- Communicate the Results

Materials

The materials listed are for a group of 3 to 4 students. For each working group, you will need two ice cubes.

Objectives

▶ USING SCIENTIFIC METHODS **Formulate** a hypothesis about the relationship between temperature and fermentation by yeast.

▶ USING SCIENTIFIC METHODS **Test** your hypothesis.

▶ **Analyze** your data.

▶ **Explain** whether your data support or refute your hypothesis.

Materials

beakers, 100 mL (3)
beakers, 400 mL (3)
clock
delivery tubes, rubber or plastic (3)
graph paper
ice cubes
solution of yeast, corn syrup, and water
stoppers, no. 2, one-hole (3)
test tubes, 20 mm × 200 mm (3)
thermometer

▶ **Step 3** Carbon dioxide bubbles will be released from the delivery tube.

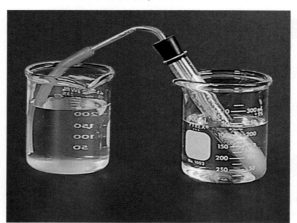

Scientific Investigations

A scientist considers all the factors that might be responsible for what he or she observes. Factors that can vary and that can be measured are called *variables*. The variable that you experimentally manipulate is the *independent variable*. The variable that you think will respond to this manipulation is the *dependent variable*.

You can practice the scientific method as it relates to everyday observations, such as the observation that bread dough rises when it is baked. According to a bread recipe, you dissolve a package of yeast in warm water and add flour, corn syrup, salt, and oil. Yeast is a microorganism that plays an important role in making bread. Yeast obtains energy by converting sugar to alcohol and carbon dioxide gas in a process called *fermentation*. The carbon dioxide forms bubbles, which make the bread dough rise. But what role, if any, does temperature play in this process? In this investigation, you will work as part of a team to try to answer these questions. Together, you will form a hypothesis and conduct an experiment that tests your hypothesis.

Procedure

1. Restate the question relating temperature to fermentation in yeast as a hypothesis.

2. Set up three test tubes containing yeast, water, and corn syrup stoppered with a gas-delivery tube. Label the test tubes "A", "B", and "C". Place each test tube in a water bath of different temperature. Place tube A in a water bath cooled by a few ice cubes, place tube B in room-temperature water, and place tube C in a warm water bath.

3. Allow the apparatus to sit for 5 min. Then place the open end of the delivery tube under water and begin to collect data on gas production. For the next 10 min, count the number of gas bubbles released from each tube, and record your data in the table on the next page.

4. Prepare a graph of data by placing time on the *x*-axis and the total number of gas bubbles released on the *y*-axis. Plot three curves on the same graph, and label each with the temperature you recorded for each test tube. Compare your graph with that of three other teams before handing in your report.

Tips and Tricks

This lab could be a 2-day activity; the first day could focus on hypotheses writing practice, and the second day could be used to complete the lab. Large empty jars may be substituted for 400 mL beakers. To prepare the yeast liquid, add 1 tbsp dry active yeast and 200 mL corn syrup to 1 L warm water. Combine the ingredients 15–20 minutes before class. Note that sugar will be provided in the corn syrup. To assure consistency, you will want to make sure that each test tube contains the same level of yeast solution. Test tubes should be filled about halfway. Using water temperatures around 4°C–10°C, 20°C–25°C, and 40°C–45°C will provide good results.

Carbon Dioxide Bubbles Released by Yeast										
Time (min)	1	2	3	4	5	6	7	8	9	10
Tube A:_____										
Tube B:_____										
Tube C:_____										

DO NOT WRITE IN THIS BOOK

Analysis

1. **Classifying Data** Which set of conditions is most similar to the conditions for the bread dough in the recipe? Why were two other conditions used in this experiment?

2. **Classifying Data** What was the independent variable in this experiment? Explain your answer.

3. **Classifying Data** What was the dependent variable in this experiment? Explain your answer.

Conclusions

4. **Drawing Conclusions** Write a conclusion for this experiment. Describe how the independent and dependent variables are related. Tell how the data supports your conclusion.

5. **Evaluating Results** What does temperature have to do with making bread dough rise?

6. **Evaluating Methods** Why did you compare your results with those of other teams before writing your conclusions?

7. **Applying Conclusions** Science is not just something you know but also something you do. Explain this statement in light of what you have learned in this investigation.

Extension

1. **Designing Experiments** Formulate a new hypothesis about the effect of different types of sugar on carbon dioxide production by yeast. Test your new hypothesis under controlled conditions. Did your results support your hypothesis? Research the types of sugar you used, and write a short explanation for your findings.

▶ **Recording Data** Count the number of bubbles produced under each experimental condition and record the data in a table.

55

Answers to Analysis

1. The solution sitting in the warm bath is most similar to the recipe conditions. The other conditions were used to test the role of temperature in fermentation.

2. The independent variable was the temperature of the liquid bath, because the temperature is the only quality that varies between the three beakers of yeast liquid.

3. The dependent variable in this experiment was the number of gas bubbles produced. The number of carbon dioxide bubbles is a measure of the rate at which the yeast are fermenting the sugar solution.

Answers to Conclusions

4. Answers may vary, but students will probably find higher water temperatures correlate positively with the rate of fermentation.

5. Yeast becomes active in warm conditions. Active yeast ferments sugars in bread dough, creating alcohol and carbon dioxide bubbles. These bubbles make the dough rise.

6. You should compare results with the results of other teams because this practice is similar to repeating the experiment.

7. Answers may vary, but after performing this experiment students should have confidence in their abilities to use scientific methods to actively make discoveries.

Answers to Extension

1. Answers may vary.

TEACHER TESTED & APPROVED

Catherine Cummings
Currituck County
School System
Currituck, North Carolina

Chapter Resource File

• Datasheets for In-Text Labs
• Lab Notes and Answers

A TOPOGRAPHIC MAP OF KEENE, NEW HAMPSHIRE

Discussion ——— GENERAL

Contour Intervals Reproduce a mountainous portion of a contour map on the board. Discuss with students where the steepest slopes are (where the lines are closest together.) Then change the contour interval by erasing every other contour line. Point out to students that the contour interval is now twice as large. Discuss with students the advantages and the disadvantages of a map with a larger contour interval. (The map may seem easier to read, but detail is lost.)

LS Visual

English Language Learners

Group Activity ——— BASIC

Contour Mapping Give each group a small bucket, a piece of clear, stiff plastic to cover the bucket, a ruler, a felt-tip pen, and some clay. Have groups create a landform out of the clay and place it in the bucket. Then have groups place the plastic over the bucket and mark the edges for alignment purposes. Groups should trace the outline of the bottom of the "hill" onto the plastic and then pour 1 cm of water into the bucket. After replacing the plastic sheet and aligning the corner marks, have students trace the waterline on the plastic. Students should repeat this until they reach the top of the hill.

LS Visual/Kinesthetic

Transparencies
TT Topographic Map of Keene, New Hampshire

A TOPOGRAPHIC MAP OF KEENE, NEW HAMPSHIRE

MAP SKILLS

Topographic maps use contour lines to indicate areas that share a common elevation. Where the lines are close together, the terrain is steep. Where the lines are far apart, the landscape is flat. In this map, the Ashuelot River flows downhill from Site 1 to Site 3. Use the map to answer the questions below.

1. **Using a Key** Use the scale at the top of the map to calculate the distance between Sites 1 and 2 and between Sites 2 and 3.

2. **Understanding Topography** Are the hills to the east and west of the town of Keene more likely to

drain into the river around Site 3 or Site 2? Explain your answer.

3. **Identifying Trends** Which site is more likely to be polluted? Explain your answer.

4. **Analyzing Data** Trace the sections of the Ashuelot River between each site to determine the length of stream between each site.

5. **Interpreting Landforms** A flood plain is an area that is periodically flooded when a river overflows its banks. Interpret the contour lines to locate the flood plain of the Ashuelot River.

56

Answers to Map Skills

1. The distance between Site 1 and Site 2 is approximately 3.5 km. The distance between Site 2 and Site 3 is approximately 3.1 km.

2. The hills are more likely to drain to Site 3 because many of the tributaries of the Ashuelot River flow into the river downstream from Site 2.

3. Site 3 is the most likely to be polluted because it is downstream from the other sites. In addition, Site 3 is located in downtown Keene.

4. Accept any measurement between 8 km and 9 km.

5. Answers may vary. The flood plain of the Ashuelot River is widest south of Site 2. A large swamp begins in this area.

BATS AND BRIDGES

A large colony of Mexican free-tailed bats lives under the Congress Avenue Bridge in Austin, Texas. These bats eat millions of insects a night, so they are welcome neighbors. Communities around the country and around the world have learned of the bats and have asked Austin for help in building bat-friendly bridges. But all that the people of Austin knew was that the bats appeared after the Congress Avenue Bridge was rebuilt in the 1980s. What attracted the bats? The people of Austin had to do a little research.

A Crevice Will Do

In the wild, bats spend the day sleeping in groups in caves or in crevices under the flaking bark of old trees. They come back to the same place every day to roost. Deep crevices in tree bark are rare now that many of our old forests have been cut down, and many bats are in danger of extinction.

In the 1990s, the Texas Department of Transportation and Bat Conservation International, a non-profit organization located in Austin, set out to discover what made a bridge attractive to bats. They collected data on 600 bridges, including some that had bat colonies and some that did not. They answered the following questions: Where was the bridge located? What was it made of? How was it constructed? Was it over water or land? What was the temperature under the bridge? How was the land around the bridge used?

Some Bridges are Better

Statistical analysis of the data revealed a number of differences between bridges occupied by bats and bridges unoccupied by bats. Which differences were important to the bats and which were not? The researchers returned to the Congress Avenue Bridge in Austin to find out. Crevices under the bridge appeared to be crucial, and the crevices had to be the right size. Free-tailed bats appeared to prefer crevices 1 to 3 cm wide and about 30 cm deep in hidden corners of the bridge, and they prefer bridges made of concrete, not steel.

The scientists looked again at their data on bridges. They discovered that 62 percent of bridges in central and southern Texas that had appropriate crevices were occupied by bats. Now, the Texas Department of Transportation is adding bat houses to existing bridges that do not have crevices. These houses are known as Texas Bat-Abodes, and they can make any bridge bat friendly.

Bat Conservation International is collecting data on bats and bridges everywhere. Different bat species may have different preferences. A Texas Bat-Abode might not attract bats to a bridge in Minnesota or Maine. If we can figure out what features attract bats to bridges, we can incorporate these features into new bridges and make more bridges into bat-friendly abodes.

▶ **Mexican free-tailed bats** leave their roost under the Congress Avenue Bridge in Austin, Texas, to hunt for insects.

What Do You Think?

Many bridges in the United States could provide roosting places for bats. Do you think communities should try to establish colonies of bats under local bridges? How should communities make this decision, and what information would they need to make it wisely?

BATS AND BRIDGES

Background

People have discussed the idea of attracting bats to artificial roosts since at least 1900, when Dr. Charles Campbell designed and installed bat boxes in San Antonio, Texas. Dr. Campbell installed the bat homes in an attempt to reduce the population of malarial mosquitoes in the area. The importance of preserving and creating bat habitat has gained popularity since the 1980s, largely due to the efforts of Bat Conservation International.

◤ Internet Activity — GENERAL

Bats on the Web There are numerous Internet sites that offer information on bats. Have students find a bat Web site and research the topic of their choice. Students may learn how to build a bat house, research how sonar works, or investigate the ecological role that certain bat species play. Have students share their findings with the class.

Answers to What Do You Think?

Answers may vary according to student's opinions, but should take into consideration personal values, community values, and the decision-making models.

CHAPTER 3

The Dynamic Earth
Chapter Planning Guide

PACING	CLASSROOM RESOURCES	LABS, DEMONSTRATIONS, AND ACTIVITIES
BLOCKS 1 & 2 · 90 min pp. 58–66 **Chapter Opener**		
Section 1 The Geosphere	**OSP** Lesson Plan * **CRF** Active Reading * **BASIC** **TT** Bellringer * **TT** Earth's Layers * **TT** Tectonic Plates * **CD** Interactive Tutor Geosphere	**TE** Activity Tectonic Jigsaw Puzzle, p. 59 **BASIC** **TE** Internet Activity Eratosthenes Experiment, p. 60 **GENERAL** **TE** Group Activity Plate Movement, p. 62 **GENERAL** **TE** Demonstration Pressure and Temperature, p. 62 **BASIC** **TE** Group Activity Analyzing Seismograms, p. 64 **BASIC** **TE** Group Activity Locating Volcanoes, p. 65 ◆ **BASIC** **CRF** Math/Graphing Lab * **ADVANCED** **CRF** Consumer Lab * ◆ **GENERAL** **CRF** Long-Term Project * ◆ **GENERAL**
BLOCKS 3 & 4 · 90 min pp. 67–72 **Section 2** The Atmosphere	**OSP** Lesson Plan * **CRF** Active Reading * **BASIC** **TT** Bellringer * **TT** Earth's Atmosphere * **TT** Energy in Earth's Atmosphere * **CD** Interactive Tutor Hydrosphere	**TE** Activity How Heavy is a Cloud? p. 67 **GENERAL** **TE** Group Activity Mapping the Aurora, p. 69 **BASIC** **SE** QuickLab The Heat Is On! p. 70 **TE** Demostration Understanding Diffraction, p. 70 **ADVANCED** **TE** Activity Weather Journal, p. 71 **BASIC** **TE** Group Activity Modeling Convection Currents, p. 71 **BASIC** **SE** Field Activity Exploring the Greenhouse Effect, p. 72 **CRF** CBL™ Probeware Lab * ◆ **GENERAL** **CRF** Modeling Lab * **BASIC**
BLOCKS 5, 6, 7 & 8 · 180 min pp. 73–81 **Section 3** The Hydrosphere and Biosphere	**OSP** Lesson Plan * **CRF** Active Reading * **BASIC** **TT** Bellringer * **TT** Surface Currents of the World * **CD** Interactive Tutor Atmosphere **CD** Interactive Tutor Biosphere **CD** Interactive Tutor The Water Cycle	**TE** Demonstration The Coriolis Effect, p. 73 **ADVANCED** **TE** Activity Tracking Icebergs, p. 75 **GENERAL** **SE** QuickLab Make a Hydrothermal Vent, p. 78 **TE** Group Activity Local Aquifers, p. 79 **BASIC** **TE** Internet Activity Researching the Biosphere, p. 80 **GENERAL** **SE** Exploration Lab Beaches, pp. 86–87 ◆ **CRF** Datasheets for In-Text Labs *

BLOCKS 9 & 10 · 90 min

Chapter Review and Assessment Resources

- **SE** Chapter Review pp. 83–85
- **SE** Standardized Test Prep pp. 640–641
- **CRF** Study Guide * ■ **GENERAL**
- **CRF** Chapter Test * ■ **GENERAL**
- **CRF** Chapter Test * **ADVANCED**
- **CRF** Concept Review * ■ **GENERAL**
- **CRF** Critical Thinking * **ADVANCED**
- **OSP** Test Generator
- **CRF** Test Item Listing *

Online and Technology Resources

Holt Online Learning

Visit **go.hrw.com** for access to Holt Online Learning or enter the keyword **HE6 Home** for a variety of free online resources.

One-Stop Planner® CD-ROM

This CD-ROM package includes
- Lab Materials QuickList Software
- Holt Calendar Planner
- Customizable Lesson Plans
- Printable Worksheets
- ExamView® Test Generator
- Interactive Teacher Edition
- Holt PuzzlePro® Resources
- Holt PowerPoint® Resources

Holt Environmental Science Interactive Tutor CD-ROM

This CD-ROM consists of interactive activities that give students a fun way to extend their knowledge of environmental science concepts.

KEY

TE	Teacher Edition	CRF	Chapter Resource File	*	Also on One-Stop Planner
SE	Student Edition	TT	Teaching Transparency	■	Also Available in Spanish
OSP	One-Stop Planner	CD	Interactive CD-ROM	◆	Requires Advance Prep

ENRICHMENT AND SKILLS PRACTICE	SECTION REVIEW AND ASSESSMENT	CORRELATIONS
SE **Pre-Reading Activity,** p. 58 TE **Using the Figure,** p. 58 `GENERAL`		**National Science Education Standards**
TE **Inclusion Strategies,** p. 60 TE **Skill Builder** Math, p. 61 `BASIC` TE **Skill Builder** Vocabulary, p. 62 `BASIC` SE **Graphic Organizer** Comparison Table, p. 65 `GENERAL`	SE **Section Review,** p. 66 TE **Reteaching,** p. 66 `BASIC` TE **Quiz,** p. 66 `GENERAL` TE **Alternative Assessment,** p. 66 `GENERAL` CRF **Quiz** * ■ `GENERAL`	ES 1a ES 2a ES 3c
TE **Career** Atmospheric Sciences, p. 70	SE **Section Review,** p. 72 TE **Reteaching,** p. 72 `BASIC` TE **Quiz,** p. 72 `GENERAL` TE **Alternative Assessment,** p. 72 `GENERAL` CRF **Quiz** * ■ `GENERAL`	ES 1a ES 2a ES 3c ES 1c ES 1d
TE **Inclusion Strategies,** p. 73 SE **Case Study** Hydrothermal Vents, pp. 74–75 SE **MathPractice** The Influence of the Gulf Stream, p. 77 TE **Student Opportunities** Internships at Sea, p. 77 TE **Skill Builder** Vocabulary, p. 80 `BASIC` SE **Maps in Action** Earthquake Hazard Map of the Contiguous United States, p. 88 SE **Science & Technology** Tracking Ocean Currents with Toy Ducks, p. 89 CRF **Map Skills** Flowing Downhill * `GENERAL`	TE **Homework,** p. 74 `GENERAL` SE **Section Review,** p. 81 TE **Reteaching,** p. 81 `BASIC` TE **Quiz,** p. 81 `GENERAL` TE **Alternative Assessment,** p. 81 `GENERAL` CRF **Quiz** * ■ `GENERAL`	ES 1a ES 2a ES 3c ES 1c ES 1d LS 4a

Guided Reading Audio CDs

These CDs are designed to help auditory learners and reluctant readers. (Audio CDs are also available in Spanish.)

SCiLINKS.
NSTA
www.scilinks.org

Maintained by the **National Science Teachers Association**

TOPIC: Composition of the Earth
SciLinks code: HE4016

TOPIC: Layers of the Atmosphere
SciLinks code: HE4061

TOPIC: The Biosphere
SciLinks code: HE4127

CNN Videos

Each video segment is accompanied by a Critical Thinking Worksheet.

Earth Science Connections

Segment 4 Seattle Quake

Segment 7 Earth's Core

Segment 16 Beach Erosion Tools

Science, Technology & Society

Segment 22 Solar Storms

CHAPTER
3

Chapter Enrichment

This Chapter Enrichment provides relevant and interesting information to expand and enhance your classroom instruction of the chapter material.

The Geosphere

▶ **Earth's lithosphere is divided into tectonic plates**

What Is a Tectonic Plate?

A tectonic plate is a block of lithosphere that consists of the crust and the rigid, upper part of the mantle. The largest plates are the Pacific, the Antarctic, and the Eurasian plates, which measure millions of square kilometers in area. Smaller plates measure a little more than 100,000 square kilometers in area. One of the smallest plates is the Juan de Fuca plate, which is located in the Pacific Ocean off the northwest coast of North America. The thicknesses of tectonic plates also vary depending on whether plates are composed largely of oceanic or continental lithosphere. Continental lithosphere is thicker than oceanic lithosphere because continental lithosphere is older and colder than oceanic lithosphere.

Tectonic Plate Motion

Tectonic plates either converge (push together), diverge (pull apart), or slip past one another. At divergent plate boundaries, such as the Mid-Atlantic Ridge, Earth's crust is pulled apart and thinned. At convergent plate boundaries, where two plates are colliding, forces can cause one of two things to happen. If both plates are of equal density, plates push together, forcing the crust upward. This process creates mountain ranges, such as the Himalayas. When one plate is denser than the other, one plate is subducted. Subduction is the process by which a denser plate slides under a plate that is less dense, forcing the denser plate downward into the

mantle. When two plates slip past one another, massive movement along faults can cause violent seismic activity. The San Andreas Fault, where the North American Plate and the Pacific Plate are slipping past one another, is the site of much seismic activity.

The Core

Earth's core, which is composed of iron, nickel, and about 10 percent of an unknown light element, is divided into a solid inner core and a liquid outer core. The inner core is extremely dense—more than 13 times denser than water. The reason the inner core is solid, whereas the outer core is liquid, is because of the extreme pressure that exists in the inner core. This pressure is millions of times higher than the atmospheric pressure at Earth's surface. Temperatures are probably greater than 4,000°C. Together, the diameter of the inner and outer core span approximately 7,000 km and account for one-third of Earth's total mass. The core is the source of Earth's magnetic field. Movements of convection cells in the liquid outer core maintain that magnetism.

The Atmosphere

Air Pressure

Gravity pulls air toward Earth's surface. This invisible force is what gives air molecules weight. In turn, the weight of air molecules exerts a force on Earth that is called *air pressure,* or atmospheric pressure. Air near Earth's surface is denser and exerts more pressure. At higher altitudes, there are fewer air molecules pressing down from above. This causes air pressure and density to decrease.

Convection

The interaction of cold and warm air masses creates wind in much the same way as cold and warm water masses in the ocean create currents. Less dense warm air rises, and more dense cool air sinks. The process is called *convection,* and it affects both liquids (water) and gases (air). This pattern of uneven heating across the globe accounts for Earth's major wind belts.

▶ **The aurora borealis**

The average amount of energy Earth absorbs from the sun is roughly equal to the amount it loses to space. This energy balance exists on a global level but does not hold true for different latitudes. For example, the polar regions lose more energy to space than they absorb, and the tropics absorb more energy than they lose. Global winds and the global ocean help maintain the energy balance by circulating heat.

Composition of Earth's Early Atmosphere

It has taken 4.6 billion years for our atmosphere to evolve into its present composition of 78 percent nitrogen, 21 percent oxygen, and minor percentages of other gases. During Earth's early history, there was no oxygen in the atmosphere. The only gases in the atmosphere were emitted from volcanoes. Ancient volcanoes released massive quantities of gases from deep in Earth's interior and pumped them into the atmosphere. Volcanic eruptions produce mostly water vapor and carbon dioxide, along with sulfur dioxide, chlorine, and nitrogen. The large amount of carbon dioxide released into the atmosphere probably kept the early Earth much warmer than it is today.

BRAIN FOOD

A column of air with a base of one square centimeter that extended from the surface of the ocean to the top of the atmosphere would weigh about 1 kilogram!

SECTION 3 The Hydrosphere and Biosphere

Ocean Currents

Ocean currents help animals and plants colonize new islands. The Galápagos Islands are 1,100 km west of South America. Of the reptiles on the islands, 20 of the 22 species are found nowhere else in the world. Scientists speculate that the ancestors of these reptiles found their way from South America to the islands on natural rafts carried by the Humboldt and South Equatorial Currents (the same currents that carried Thor Heyerdahl to Polynesia). Reptiles such as lizards, snakes, and iguanas sometimes ride these rafts until they wash up on the shores of distant islands. Have interested students find out more about the unique plants and animals of the Galápagos Islands.

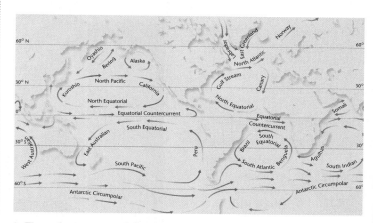

▶ **The surface currents of the global ocean**

Currents and the Atacama Desert

The Atacama Desert in northern Chile is the driest place on Earth and receives rainfall only two to four times a century. Surprisingly, the ocean is the cause of this dryness. The Humboldt Current brings cold water up from the Antarctic, releasing cold air at the surface. This cold air forces warm air upward, producing a thermal inversion that creates fog but little rain over the desert. The inversion also makes the Atacama cold, even though it is not far from the equator. There is little vegetation in the Atacama.

Overview

Explain to students that Earth is an integrated system consisting of four interacting components—the geosphere, the atmosphere, the hydrosphere, and the biosphere. One example of this complex interaction is the effect that the movement of winds has on currents in the global ocean. Wind is caused in part by the unequal heating of the atmosphere by the sun. When winds blow over the surface of the global ocean, surface currents are produced. Surface currents moderate the climates of the land areas they flow past and affect the distribution of nutrients and life in the global ocean.

Using the Figure — GENERAL

In this image of Earth, taken from space, some aspects of the changing nature of our planet are evident. For example, cloud formations composed of water vapor are visible as they move through Earth's lower atmosphere. Satellite imaging can be used to document many global phenomena. Have students download satellite images from the NASA Web site and discuss these images in class. **LS** Visual

PRE-READING ACTIVITY

Have students exchange FoldNotes and check each other's notes on The Geosphere, Atmosphere, and Hydrosphere. Encourage students to use a colored pen to add additional notes about The Geosphere, Atmosphere, and Hydrosphere.

VIDEO SELECT

For information about videos related to this chapter, go to **go.hrw.com** and type in the keyword **HE4 EARV.**

The Dynamic Earth

1 **The Geosphere**

2 **The Atmosphere**

3 **The Hydrosphere and Biosphere**

PRE-READING ACTIVITY

FOLDNOTES

Pyramid Before you read this chapter, create the **FoldNote** entitled "Pyramid" described in the Reading and Study Skills section of the Appendix. Label the sides of the pyramid with "Geosphere," "Atmosphere," "Hydrosphere," and "Biosphere." As you read the chapter, write information you learn about each category under the appropriate flap.

Landmasses are moving slowly across our planet's surface. The atmosphere is a swirling mix of gases and vapor. Our planet, which may appear placid from space, is not stable and unchanging.

58

Chapter Correlations — *National Science Education Standards*

ES 1a Earth systems have internal and external sources of energy, both of which create heat. The sun is the major external source of energy. Two primary sources of internal energy are the decay of radioactive isotopes and the gravitational energy from the earth's original formation. **(Section 1, Section 2, and Section 3)**

ES 2a The earth is a system containing essentially a fixed amount of each stable chemical atom or element. Each element can exist in several different chemical reservoirs. Each element on earth moves among reservoirs in the solid earth, oceans, atmosphere, and organisms as part of geochemical cycles. **(Section 1, Section 2, and Section 3)**

ES 3c Interactions among the solid earth, the oceans, the atmosphere, and organisms have resulted in the ongoing evolution of the earth system. We can observe some changes such as earthquakes and volcanic eruptions on a human time scale, but many processes such as mountain building and plate movements take place over hundreds of millions of years. **(Section 1,**

Section 2, and Section 3)

ES 1c Heating of earth's surface and atmosphere by the sun drives convection within the atmosphere and oceans, producing winds and ocean currents. **(Section 2 and Section 3)**

ES 1d Global climate is determined by energy transfer from the sun at and near the earth's surface. This energy transfer is influenced by dynamic processes, such as cloud cover and the earth's rotation, and static conditions, such as the position of mountain ranges and oceans. **(Section 2 and Section 3)**

LS 4a The atoms and molecules on the earth cycle among the living and nonliving components of the biosphere. **(Section 3)**

The Geosphere

Molten rock from Earth's interior flows over the surface of the planet, and violent eruptions blow the tops off of volcanoes. Hurricanes batter beaches and change coastlines. Earthquakes shake the ground and topple buildings and freeway overpasses. None of this activity is caused by people. Instead, it is the result of the dynamic state of planet Earth. What are the conditions that allow us to survive on a constantly changing planet?

The Earth as a System

The Earth is an integrated system that consists of rock, air, water, and living things that all interact with each other. Scientists divide this system into four parts. As shown in **Figure 1,** the four parts are the geosphere (rock), the atmosphere (air), the hydrosphere (water), and the biosphere (living things).

The solid part of the Earth that consists of all rock, and the soils and sediments on Earth's surface, is the **geosphere.** Most of the geosphere is located in Earth's interior. At the equator, the average distance through the center of the Earth to the other side is 12,756 km. The atmosphere is the mixture of gases that makes up the air we breathe. Nearly all of these gases are found in the first 30 km above the Earth's surface. The hydrosphere makes up all of the water on or near the Earth's surface. Much of this water is in the oceans, which cover nearly three-quarters of the globe. Water is also found in the atmosphere, on land, and in the soil. The biosphere is made up of parts of the geosphere, the atmosphere, and the hydrosphere. The biosphere is the part of the Earth where life exists. It is a thin layer at Earth's surface that extends from about 9 km above the Earth's surface down to the bottom of the ocean.

Objectives

▶ Describe the composition and structure of the Earth.

▶ Describe the Earth's tectonic plates.

▶ Explain the main cause of earthquakes and their effects.

▶ Identify the relationship between volcanic eruptions and climate change.

▶ Describe how wind and water alter the Earth's surface.

Key Terms

geosphere
crust
mantle
core
lithosphere
asthenosphere
tectonic plate
erosion

Figure 1 ▶ The Earth is an integrated system that consists of the geosphere, the atmosphere, the hydrosphere, and the biosphere (inset).

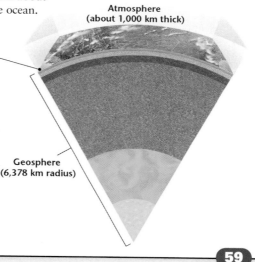

Atmosphere
(about 1,000 km thick)

Hydrosphere 29 km

Biosphere
20 km

9 km

11 km

Geosphere
(6,378 km radius)

59

Chapter Resource File

• Lesson Plan
• Active Reading BASIC
• Section Quiz GENERAL

Transparencies

TT Bellringer

SECTION 1

Focus

Overview

Before beginning this section, review with your students the Objectives listed in the Student Edition. In this section, students learn that Earth's interior is divided into layers based on both composition and physical characteristics. The section discusses plate tectonics and the effects of tectonic plate motion, and concludes with a discussion of erosion and the alteration of Earth's surface by wind and water.

Bellringer

Using **Figure 1,** have students reduce the proportions of the geosphere, the atmosphere, and the biosphere to sizes that are easier to visualize. For instance, have students reduce the geosphere to 6 m, the atmosphere to 1 m, and the biosphere to .02 m. Using the classroom or hall, compare the relative proportions of each of these spheres of the Earth. Note how a small area of the Earth system—the biosphere—can support life.
LS Visual

Motivate

Activity ———— BASIC

Tectonic Jigsaw Puzzle Ask students to trace a world map that shows the outlines of the continents on a sheet of paper. Tell students to cut out the shapes of the continents, and ask them how they would arrange the pieces to make a well-fitting whole. Have them compare the coastline of West Africa with the eastern coastline of South America. Then show them a map of Pangaea. Explain that the land was once joined together to form one giant supercontinent. Ask them to hypothesize what could have caused such a giant landmass to break apart. **LS** Kinesthetic

Teach

GEOPHYSICS
CONNECTION — GENERAL

The World's Deepest Hole The deepest hole in the world is a drill hole located on Russia's Kola Peninsula. It is about 12 km deep. The Russians drilled the hole between 1970 and 1989, and penetrated halfway through the crust of the Baltic continental shield. The oldest rocks they uncovered are 2.7 billion years old, representing over half of Earth's 4.6-billion-year history. Have students hypothesize what some of the obstacles might be when drilling a hole to this depth. Have them research the answers to their questions. **LS Kinesthetic**

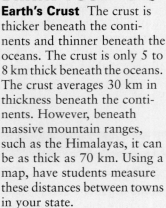

BRAIN FOOD

Earth's Crust The crust is thicker beneath the continents and thinner beneath the oceans. The crust is only 5 to 8 km thick beneath the oceans. The crust averages 30 km in thickness beneath the continents. However, beneath massive mountain ranges, such as the Himalayas, it can be as thick as 70 km. Using a map, have students measure these distances between towns in your state.

Transparencies
TT Earth's Layers

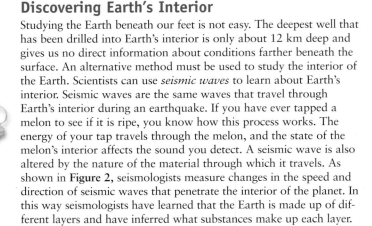

internet connect
www.scilinks.org
Topic: Composition of the Earth
SciLinks code: HE4016

Discovering Earth's Interior

Studying the Earth beneath our feet is not easy. The deepest well that has been drilled into Earth's interior is only about 12 km deep and gives us no direct information about conditions farther beneath the surface. An alternative method must be used to study the interior of the Earth. Scientists can use *seismic waves* to learn about Earth's interior. Seismic waves are the same waves that travel through Earth's interior during an earthquake. If you have ever tapped a melon to see if it is ripe, you know how this process works. The energy of your tap travels through the melon, and the state of the melon's interior affects the sound you detect. A seismic wave is also altered by the nature of the material through which it travels. As shown in **Figure 2**, seismologists measure changes in the speed and direction of seismic waves that penetrate the interior of the planet. In this way seismologists have learned that the Earth is made up of different layers and have inferred what substances make up each layer.

The Composition of the Earth Scientists divide the Earth into three layers—the crust, the mantle, and the core—based on the composition of each layer. These layers are made up of progressively denser materials toward the center of the Earth. **Figure 3** shows a cross section of the Earth. Earth's thin outer layer, the **crust,** is made almost entirely of light elements. It makes up less than 1 percent of the planet's mass. The crust is Earth's thinnest

Figure 2 ▶ Seismologists have measured changes in the speed and direction of seismic waves that travel through Earth's interior. Through this process, they have learned that the Earth is made up of different layers.

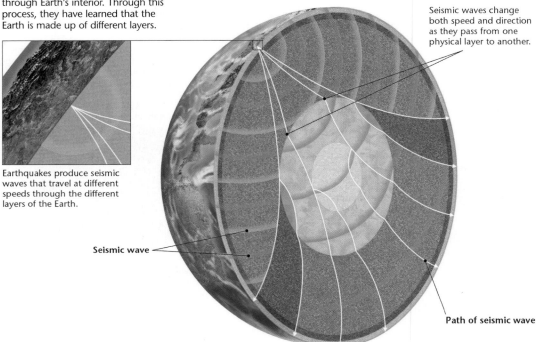

Earthquakes produce seismic waves that travel at different speeds through the different layers of the Earth.

Seismic waves change both speed and direction as they pass from one physical layer to another.

Seismic wave

Path of seismic wave

60

Internet Activity — GENERAL

Eratosthenes Experiment Have students look up a project called "Eratosthenes Experiment: A Worldwide Science and Math Experiment" on the Internet. This one-day project (which occurs twice a year during the equinox) allows students from around the globe to calculate Earth's circumference by measuring the shadow of the sun and entering the data into a simple equation. They can share data with students around the world online. **LS Logical**

INCLUSION Strategies

• *Learning Disabled*
• *Attention Deficit Disorder*
• *English Language Learners*

Ask students to copy all key terms that appear in bold type throughout the chapter in a Vocabulary Notebook. The students can copy the sentence in which the key term appears to help them understand the word in the context of the chapter. The notebooks can be used for independent study or in small groups as a study guide or discussion prompt.

layer. It is 5 km to 8 km thick beneath the oceans and is 20 km to 70 km thick beneath the continents. The **mantle,** which is the layer beneath the crust, makes up 64 percent of the mass of the Earth. The mantle is approximately 2,900 km thick and is made of rocks of medium density. Earth's innermost layer is the **core.** The core is composed of the densest elements. It has a radius of approximately 3,400 km.

The Structure of the Earth The Earth can be divided into five layers based on the physical properties of each layer. Earth's outer layer is the **lithosphere.** It is a cool, rigid layer that is 15 km to 300 km thick that includes the crust and uppermost part of the mantle. It is divided into huge pieces called *tectonic plates.* The **asthenosphere** is the layer beneath the lithosphere. The asthenosphere is a plastic, solid layer of the mantle made of rock that flows very slowly and allows tectonic plates to move on top of it. Beneath the asthenosphere is the mesosphere, the lower part of the mantle.

The Earth's outer core is a dense liquid layer. At the center of the Earth is the dense, solid inner core, which is made up mostly of the metals iron and nickel. The temperature of the inner core is estimated to be between 4,000°C to 5,000°C. Even though the inner core is so hot, it is solid because it is under enormous pressure. Earth's outer and inner core together make up about one-third of Earth's mass.

Geofact

Pangaea Two hundred and forty-five million years ago almost all the land on Earth was joined into one supercontinent known as *Pangaea,* which is Greek for "all earth." Pangaea was surrounded by a world ocean called *Panthalassa,* which means "all sea." By the time the dinosaurs became extinct 65 million years ago, Pangaea had separated into all the present continents with positions close to the present positions.

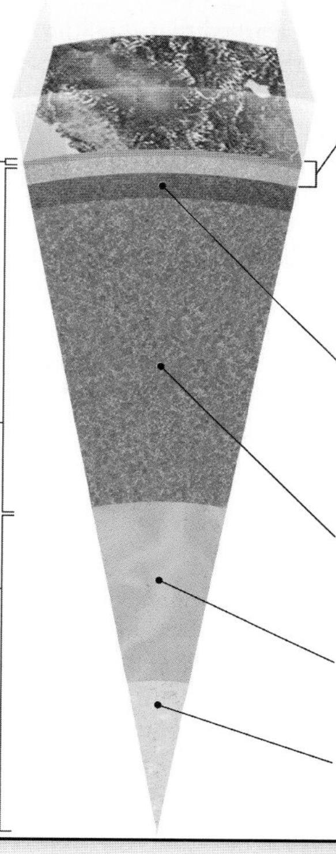

Figure 3 ▶ Earth's Layers
Scientists divide the Earth into different layers based on composition and physical properties.

Crust 5–70 km thick; the solid, brittle, outermost layer of the Earth; continental crust is thick and made of lightweight materials, whereas oceanic crust is thin and made of denser materials

Mantle 2,900 km thick; the layer of the Earth between the crust and the core; made of dense, iron-rich minerals

Core 3,428 km radius; a sphere of hot, dense nickel and iron at the center of the Earth

Lithosphere 15–300 km thick; the cool, rigid, outermost layer of the Earth; consists of the crust and the rigid, uppermost part of the mantle; divided into huge pieces called tectonic plates, which move around on top of the asthenosphere and can have both continental and oceanic crust

Asthenosphere 250 km thick; the solid, plastic layer of the mantle between the mesosphere and the lithosphere; made of mantle rock that flows very slowly, which allows tectonic plates to move on top of it

Mesosphere 2,550 km thick; the "middle sphere"; the lower layer of the mantle between the asthenosphere and the outer core

Outer Core 2,200 km thick; the outer shell of Earth's core; made of liquid nickel and iron

Inner Core 1,228 km radius; a sphere of solid nickel and iron at the center of the Earth

61

BIOLOGY CONNECTION **GENERAL**

Intraterrestrials Some scientists speculate that life on Earth may have first evolved deep in the crust and then migrated to the surface as environmental conditions became more favorable. It is not known how deep in Earth's interior life-forms exist. However, this subsurface community of microbes comprises a large portion of Earth's biomass; some scientists have hypothesized that the biomass below the surface may far exceed the biomass above the surface. To enable them to survive under such intense heat and pressure, these extremophiles (a term that describes microbes that live in harsh environments such as intense heat or cold) may have developed unique enzymes that could be of use to medical science and in biotechnological and industrial applications. One potential application may involve using such enzymes for the bio-cleanup of contaminated groundwater.

MISCONCEPTION ALERT

Catastrophism Until the late 1700s, most Europeans believed that all physical changes on the planet occurred through a rapid series of catastrophes. When the Scottish geologist James Hutton proposed the principle of uniformitarianism in 1785, it marked a revolution in thinking. This theory states that "the present is the key to the past." The same gradual geologic processes (such as erosion) that shape Earth's surface now are what shaped them in the past. This theory includes catastrophes such as earthquakes and volcanic eruptions.

CHEMISTRY CONNECTION **ADVANCED**

Earth's Mantle The mantle is different in chemical composition from the crust and the core. Most of the mantle is composed of peridotite, a rock of medium density that contains the mineral olivine and minerals in the pyroxene and garnet groups. The upper mantle, which contains the asthenosphere, may be partially molten in places. Deeper in the mantle, pressure causes minerals to become completely solid. Geologists think convection currents in the mantle may drive tectonic plate movements.

SKILL BUILDER **GENERAL**

Math The Atlantic Ocean is spreading at a rate of 1 to 2 cm per year, and the eastern Pacific seafloor is spreading between 3 to 8 cm per year. Using the average rate of spreading for the Atlantic Ocean, have students calculate how many years it would take the seafloor of the Atlantic Ocean to spread one kilometer (66,667 years). Have students use the average rate of spreading for the Pacific to calculate how long it would take the seafloor of the Pacific Ocean to spread one kilometer. (18,182 years) **LS Logical**

Plate Movement Challenge students to model the types of movement at each type of plate boundary. Have them work in pairs with materials they have chosen. Possible materials include sheets of foam padding, cardboard, modeling clay, or phone books. As students demonstrate plate movements, they should be prepared to explain the density of each plate and the differences between the forces at work. **LS** Visual

Demonstration ── BASIC

Pressure and Temperature If the pressure on a material is increased quickly, the temperature also increases. Bring two deflated bicycle inner tubes. Leave one tube flat, and pump air into the other. Have students feel the two tubes and compare their temperatures. They should note that as the inflated tube became pressurized, its temperature also increased. **LS** Kinesthetic

SKILL BUILDER ── BASIC

Vocabulary The word *Pangaea* is derived from Greek and means "all Earth." Discuss the meaning of other words that have the "pan" root, such as pandemonium and panorama. **LS** Verbal | English Language Learners

Transparencies

TT Tectonic Plates

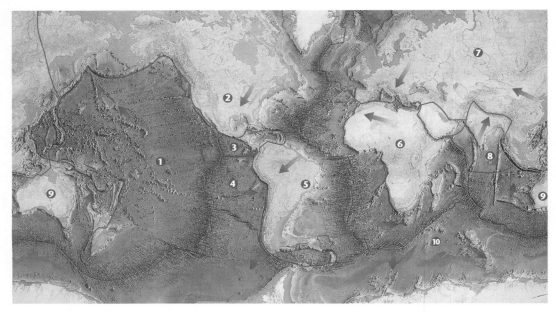

Figure 4 ▶ Earth's lithosphere is divided into pieces called *tectonic plates*. The tectonic plates are moving in different directions and at different speeds.

Major Tectonic Plates
1. Pacific plate
2. North American plate
3. Cocos plate
4. Nazca plate
5. South American plate
6. African plate
7. Eurasian plate
8. Indian plate
9. Australian plate
10. Antarctic plate

Plate Tectonics

You learned that the lithosphere—the rigid, outermost layer of the Earth—is divided into pieces called **tectonic plates.** These plates glide across the underlying asthenosphere in much the same way as a chunk of ice drifts across a pond. The continents are located on tectonic plates and move around with them. The major tectonic plates include the Pacific, North American, South American, African, Eurasian, and Antarctic plates. **Figure 4** illustrates the major tectonic plates and their direction of motion.

Plate Boundaries Much of the geologic activity at the surface of the Earth takes place at the boundaries between tectonic plates. Plates may separate from one another, collide with one another, or slip past one another. Enormous forces are generated at tectonic plate boundaries, where the crust is pulled apart, is squeezed together, or is constantly slipping. The forces produced at the boundaries of tectonic plates can cause mountains to form, earthquakes to shake the crust, and volcanoes to erupt.

Plate Tectonics and Mountain Building Tectonic plates are continually moving around the Earth's surface. When tectonic plates collide, slip by one another, or pull apart, enormous forces cause rock to break and buckle. Where plates collide, the crust becomes thicker and eventually forms mountain ranges. The Himalaya Mountains, as shown in **Figure 5**, began to form when the tectonic plate containing Asia and the tectonic plate containing India began to collide 50 million years ago.

HISTORY
── CONNECTION ── GENERAL

Alfred Wegener German meteorologist Alfred Wegener (1880–1930) developed the theory of continental drift. Relying on evidence from various scientific fields, Wegener proposed that all the continents were once joined into one landmass, which he named Pangaea. Like many revolutionary scientific theories, continental drift was scorned at its inception. Wegener's ideas were not given full credence until decades later when oceanographic research found evidence for the process of seafloor spreading.

GEOLOGY
── CONNECTION ── ADVANCED

The Iron Catastrophe As the mass of the newly formed Earth increased, its gravitational force increased. Earth's interior then started to melt. The densest of Earth's abundant elements is iron. As iron melted, iron droplets sank toward Earth's interior, where they coalesced. As the process continued, iron began melting and and sinking in the interior with increased frequency and speed. Because of this catastrophic rapidity, the core formation event was named "the iron catastrophe."

Earthquakes

A *fault* is a break in the Earth's crust along which blocks of the crust slide relative to one another. When rocks that are under stress suddenly break along a fault, a series of ground vibrations is set off. These vibrations of the Earth's crust caused by slippage along a fault are known as *earthquakes*. Earthquakes are occurring all of the time, but many are so small that we cannot feel them. Other earthquakes are enormous movements of the Earth's crust that cause widespread damage.

The Richter scale is used by scientists to quantify the amount of energy released by an earthquake. The measure of the energy released by an earthquake is called *magnitude*. The smallest magnitude that can be felt is approximately 2.0, and the largest magnitude that has ever been recorded is 9.5. Each increase of magnitude by one whole number indicates the release of 31.7 times more energy than the whole number below it. For example, an earthquake of magnitude 6.0 releases 31.7 times the energy of an earthquake of magnitude 5.0. Earthquakes that cause widespread damage have magnitudes of 7.0 and greater.

Where Do Earthquakes Occur? Areas of the world where earthquakes occur are shown on the map in **Figure 6.** As you can see from the map, the majority of earthquakes take place at or near tectonic plate boundaries because of the enormous stresses that are generated when tectonic plates separate, collide, or slip past each other. Over the past 15 million to 20 million years, large numbers of earthquakes have occurred along the San Andreas fault, which runs almost the entire length of California. The San Andreas fault is the line where parts of the North American plate and the Pacific plate are slipping past one another.

Figure 5 ▶ The Himalaya Mountains are still growing today because the tectonic plate containing Asia and the tectonic plate containing India continue to collide.

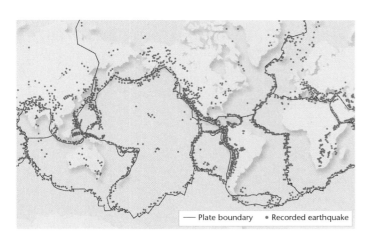

— Plate boundary · Recorded earthquake

Figure 6 ▶ The largest and most active earthquake zones lie along tectonic plate boundaries.

63

Group Activity — BASIC

Analyzing Seismograms Have students go to the U.S. Geological Survey Seismogram Display on the Internet to analyze real-time views of seismograms. Have them note what the different levels of seismic activity mean and compare them to a seismograph of a major earthquake. **LS Visual**

MISCONCEPTION ///ALERT\\\

Useful Volcanoes Volcanic eruptions are often thought of as being destructive to life and property. In fact, volcanic eruptions are beneficial to Earth's environment in a number of ways. Volcanoes are instrumental in creating new crust and providing opportunities for life. Each year, thousands of square kilometers of lava flow onto the Earth's surface and onto the ocean floor to form new crust. In as little as a couple years, volcanic rock weathers to form fertile soils, such as those found in California, Hawaii, and southern Italy. Most metallic elements, such as gold, silver, copper, lead, and zinc, are mined from volcanic rock. Geothermal energy is also a byproduct of active volcanoes. For example, most homes around Reykjavík, Iceland, are heated by water piped from nearby volcanic hot springs.

Connection to Biology

Can Animals Predict Earthquakes? Can animals that live close to the site of an earthquake detect changes in their physical environment prior to an earthquake? Documentation of unusual animal behavior prior to earthquakes can be found as far back as 1784. Examples of this odd behavior include zoo animals refusing to enter shelters at night, snakes and small mammals abandoning their burrows, and wild birds leaving their usual habitats. These behaviors reportedly happened within a few days, hours, or minutes of earthquakes.

Figure 7 ▶ The Ring of Fire
Tectonic plate boundaries are places where volcanoes usually form. The Ring of Fire contains nearly 75 percent of the world's active volcanoes that are on land. A large number of people live on or near the Ring of Fire.

Earthquake Hazard Despite much study, scientists cannot predict when earthquakes will take place. However, information about where they are most likely to occur can help people prepare for them. An area's earthquake-hazard level is determined by past and present seismic activity. The Maps in Action activity located at the end of this chapter shows earthquake-hazard levels for the contiguous United States.

Earthquakes are not restricted to high-risk areas. In 1886, an earthquake shook Charleston, South Carolina, which is considered to be in a medium-risk area. Because the soil beneath the city is sandy, this earthquake caused extensive damage. During shaking from a strong earthquake, sand acts like a liquid and causes buildings to sink. In areas that are prone to earthquakes, it is worth the extra investment to build bridges and buildings that are at least partially earthquake resistant. Earthquake-resistant buildings are slightly flexible so that they can sway with the ground motion.

Volcanoes

A *volcano* is a mountain built from magma—melted rock—that rises from the Earth's interior to the surface. Volcanoes are often located near tectonic plate boundaries where plates are either colliding or separating from one another. Volcanoes may occur on land or under the sea, where they may eventually break the ocean surface as islands. As **Figure 7** shows, the majority of the world's active volcanoes on land are located along tectonic plate boundaries that surround the Pacific Ocean.

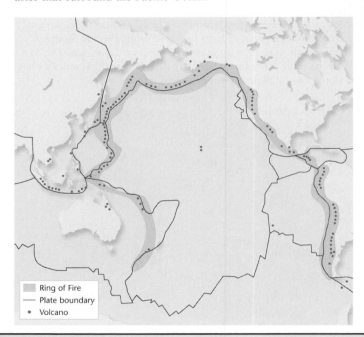

Ring of Fire
— Plate boundary
• Volcano

64

HISTORY — CONNECTION — GENERAL

The Youngest Mountain in the World
One morning, a farmer in Michoacan, Mexico, was astonished to discover a volcano had appeared in his cornfield. Paracutín was born on February 20, 1943, as volcanic ash spewed forth from the budding volcano. By the end of the first day, the volcano was 7.5 m high. It was 50 m high on the second day and roughly 140 m high by the end of the first week. The volcano formed most of its cone and reached an elevation of 336 m in its first year. Eruptions continued for nearly 10 years. The mountain now rises 410 m high. The locals called the volcano "El Monstre" in recognition of the widespread damage it caused. Have students find pictures of Paracutín on the Internet. **LS Logical**

Local Effects of Volcanic Eruptions Volcanic eruptions can be devastating to local economies and can cause great human loss. Clouds of hot ash, dust, and gases can flow down the slope of a volcano at speeds of up to 200 km/hr and sear everything in their path. During an eruption, volcanic ash can mix with water and produce a mudflow. In 1985, Nevado del Ruiz in Colombia erupted, melting ice at the volcano's summit. A mudflow raced downhill and engulfed the town of Armero. In addition, ash that falls to the ground can cause buildings to collapse under its weight, bury crops, and damage the engines of vehicles. Volcanic ash may also cause breathing difficulties.

Global Effects of Volcanic Eruptions Major volcanic eruptions, such as the eruption of Mount St. Helens shown in **Figure 8**, can change Earth's climate for several years. In large eruptions, clouds of volcanic ash and sulfur-rich gases may reach the upper atmosphere. As the ash and gases spread across the planet, they can reduce the amount of sunlight that reaches the Earth's surface. This reduction in sunlight can cause a drop in the average global surface temperature. In the 1991 eruption of Mount Pinatubo in the Philippines, large clouds of ash and gases entered the Earth's atmosphere. The amount of sunlight that reached the Earth's surface was estimated to have decreased by 2 to 4 percent. As a result, the average global temperature dropped by several tenths of a degree Celsius over a period of several years.

Figure 8 ► On May 18, 1980, Mount St. Helens in Washington State erupted. Sixty-three people lost their lives, and 400 km² of forest were destroyed in an eruption that blew away the top 410 m of the volcano.

Graphic Organizer **Comparison Table**

Create the **Graphic Organizer** entitled "Comparison Table" described in the Appendix. Label the columns with "Local Effects" and "Global Effects." Label the rows with "Volcanic Eruptions" and "Earthquakes." Then, fill in the table with details about the characteristics and the effects of each type of natural disaster.

65

Tectonic Processes Reinforce the idea of tectonic processes by having students build a small model of two tectonic plates with clay. Have them model divergent plates or convergent plates that result in either subduction (one plate being subsumed under the other) or uplift (mountain-building), and give an example of which plates are currently undergoing these processes. **LS** Kinesthetic

Quiz — GENERAL

1. What are the four components of the Earth system? (The geo-sphere, the hydropshere, the atmosphere, and the biosphere.)

2. What is the importance of the asthenosphere? (It is the layer of the mantle that is viscous enough to allow Earth's tectonic plates to move around on it.)

3. Where do earthquakes and vol-canic eruptions most often occur? (They occur primarily at plate boundaries.)

Alternative Assessment — GENERAL

The Big Event Have students write a short story in which they are scientists (either seismologists or vulcanologists). The *big event* is about to occur (either an earth-quake or a volcanic eruption), and they are following its progress. Have students tell the story of what hap-pened and how the phenomenon progressed, describing events in as much scientific detail as possible. **LS** Logical   English Language Learners

Figure 9 ▶ Over long periods of time, erosion can produce spectacular landforms on Earth's surface.

Erosion

Forces at the boundaries of tectonic plates bring rock to the surface of the Earth. At the Earth's surface, rocks are altered by other forces. The Earth's surface is continually battered by wind and scoured by running water, which moves rocks around and changes their appearance. The removal and transport of surface material is called **erosion.** Erosion wears down rocks and makes them smoother as time passes. The older a mountain range is, the longer the forces of erosion have acted on it. This information helped geologists learn that the round-topped Appalachian Mountains in the eastern United States are older than the jagged Rocky Mountains in the west.

Water Erosion Erosion by both rivers and oceans can produce dramatic changes on Earth's surface. Waves from ocean storms can erode coastlines to give rise to a variety of spectacular landforms. Over time, rivers can carve deep gorges into the landscape, as shown in **Figure 9.**

Wind Erosion Like moving water, wind can also change the landscape of our planet. In places where plants grow, their roots hold soil in place. But in places where there are few plants, wind can blow soil away very quickly. Beaches and deserts, which have loose, sandy soil, are examples of places where few plants grow. Soft rocks, such as sandstone, erode more easily than hard rocks, such as granite, do. In parts of the world, spectac-ular rock formations are sometimes seen where pinnacles of hard rock stand alone because the softer rock around them has eroded by wind and/or water.

SECTION 1 Review

1. **Name** and describe the physical and compositional layers into which scientists divide the Earth.

2. **Explain** the main cause of earthquakes and their effects.

3. **Describe** the effects that a large-scale volcanic erup-tion can have on the global climate.

4. **Describe** how wind and water alter the Earth's surface.

CRITICAL THINKING

5. **Analyzing Processes** How might the surface of the Earth be different if it were not divided into tec-tonic plates?

6. **Compare and Contrast** Read about the effects of erosion on mountains on this page. From what you have read, describe the physical features you would associate with a young mountain range and an old mountain range. **READING SKILLS**

66

Answers to Section Review

1. The three compositional layers are the crust, the mantle, and the core. The five physical lay-ers are the lithosphere, the asthenosphere, the mesosphere, the outer core, and the inner core.

2. When rocks that are under stress suddenly break along a fault, a series of ground vibra-tions is set off. Many earthquakes are so small that we cannot feel them; others cause wide-spread damage.

3. Large-scale volcanic eruptions can cause a drop in the average global surface temperature.

4. Earth's surface is continually battered by wind and scoured by running water. Wind can blow

soil away very quickly and erode soft rock. Waves can erode coastlines, and rivers can carve deep gorges.

5. Answers may vary but might include the idea that Earth's surface would be largely flat, the positions of the continents and oceans would not change, there would be fewer islands, and there would be fewer earthquakes.

6. A young mountain range would have tall, jagged peaks and deep valleys. An old moun-tain range would have rounded peaks and shallow valleys.

The Atmosphere

Earth is surrounded by a mixture of gases known as the **atmosphere.** Nitrogen, oxygen, carbon dioxide, and other gases are all parts of this mixture. Earth's atmosphere changes constantly as these gases are added and removed. For example, animals remove oxygen when they breathe in and add carbon dioxide when they breathe out. Plants take in carbon dioxide and add oxygen to the atmosphere when they produce food. Gases can be added to and removed from the atmosphere in ways other than through living organisms. A volcanic eruption adds gases. A vehicle both adds and removes gases.

The atmosphere also insulates Earth's surface. This insulation slows the rate at which the Earth's surface loses heat. The atmosphere keeps Earth at temperatures at which living things can survive.

Composition of the Atmosphere

Figure 10 shows the percentages of gases that make up Earth's atmosphere. Nitrogen makes up 78 percent of the Earth's atmosphere. It enters the atmosphere when volcanoes erupt and when dead plants and animals decay. Oxygen, the second most abundant gas in Earth's atmosphere, is primarily produced by plants. Gases including argon, carbon dioxide, methane, and water vapor make up the rest of the atmosphere.

In addition to gases, the atmosphere contains many types of tiny, solid particles, or atmospheric dust. Atmospheric dust is mainly soil but includes salt, ash from fires, volcanic ash, particulate matter from combustion, skin, hair, bits of clothing, pollen, bacteria and viruses, and tiny, liquid droplets called *aerosols.*

Objectives

▶ Describe the composition of the Earth's atmosphere.
▶ Describe the layers of the Earth's atmosphere.
▶ Explain three mechanisms of heat transfer in Earth's atmosphere.
▶ Explain the greenhouse effect.

Key Terms

atmosphere
troposphere
stratosphere
ozone
radiation
conduction
convection
greenhouse effect

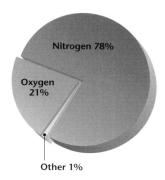

Figure 10 ▶ Ninety-nine percent of the air we breathe is made up of nitrogen and oxygen.

Figure 11 ▶ This sunrise scene that was taken from space captures the tropopause, the transitional zone that separates the troposphere (yellow layer) from the stratosphere (white layer). The tropopause is the illuminated brown layer.

67

Chapter Resource File

• Lesson Plan
• Active Reading BASIC
• Section Quiz GENERAL

Transparencies

TT Bellringer

METEOROLOGY
CONNECTION ── GENERAL

The Tropopause The tropopause, shown in **Figure 11,** is the boundary between the troposphere and stratosphere. It is not found at a uniform altitude everywhere in the atmosphere. At higher latitudes, the tropopause is lower than it is near the equator. It is highest over the warm waters of the western Pacific Ocean near the equator. Ask students why this might be. (Cold conditions create a lower tropopause because the troposphere is denser under those conditions.) **LS** Logical

Focus

Overview

Before beginning this section, review with your students the Objectives listed in the Student Edition. This section explores the composition and the layers of atmosphere. Energy in the atmosphere and the greenhouse effect are also discussed.

🔊 Bellringer

Ask students, "Why do planes fly so high, often 10 km above the ground?" (Most commercial planes fly at about 10,000 m because the air at that altitude is less than half as dense as air 1,500 m above the ground. Planes experience less drag and require less fuel when they move through thinner, less dense air. The air is also less turbulent above cloud level.) **LS** Logical

Motivate

Activity ── GENERAL

How Heavy is a Cloud? Ask students to estimate the weight of the average cumulus cloud. Then ask them to perform a calculation that will allow them to determine the answer. Tell students that if the average cumulus cloud is 1,000 m long, 1,000 m wide, and 1,000 m high, it has a volume of 1,000,000,000 m³, or 1 km³. Then have students determine the weight of the cloud by multiplying that volume by the weight of water in 1 m³ of a cumulus cloud (0.5 g). (1,000,000,000 m³ × 0.5 g = 500,000,000 g, or 500,000 kg, or 500 metric tons.) **LS** Logical

BIOLOGY CONNECTION ADVANCED

Oxygen and Climbing Mount Everest At 29,035 feet above sea level, Mount Everest is the highest point in the world. Airplane cabins at this altitude must be pressurized to compensate for the lack of oxygen in the upper atmosphere. Mount Everest offers the same challenges to mountain climbers. As climbers gain altitude, each breath contains less oxygen. Without oxygen, dizziness, numbness, and death ensue. Most climbers use oxygen tanks in order to reach the summit of Mount Everest. A handful of people have successfully climbed the mountain without oxygen. However, without supplemental oxygen, bodily and brain functions are so impaired that climbers have to recover after every step. Above 20,000 feet hair and nails stop growing, and there is not enough oxygen to heal cuts. The area above 25,000 feet is called the "Death Zone," because the body is expending more energy at the expense of body maintenance than it replaces—in other words, dying.

REAL-LIFE CONNECTION BASIC

Turbulence Airplanes often encounter turbulence. Turbulence is caused when two air streams that are moving at different speeds converge. This causes a patch of disturbed, choppy air.

Transparencies

TT Earth's Atmosphere

internet connect

www.scilinks.org
Topic: Layers of the Atmosphere
SciLinks code: HE4061

SciLINKS. Maintained by the National Science Teachers Association

Geofact

The Mesosphere In geology, the term *mesosphere*, which means "middle sphere," refers to the 2,550 km thick compositional layer of the Earth that lies below the asthenosphere. The mesosphere is also the name of the atmospheric layer that extends from 50 to 80 km above Earth's surface.

Figure 12 ▶ The layers of the atmosphere are defined by changes in temperature and pressure. The red line indicates temperature, and the green line indicates pressure in pascals.

Air Pressure Earth's atmosphere is pulled toward Earth's surface by gravity. As a result of the pull of gravity, the atmosphere is denser near Earth's surface. Almost the entire mass of Earth's atmospheric gases is located within 30 km of our planet's surface. Fewer gas molecules are found at altitudes above 30 km; therefore, less pressure at these altitudes pushes downward on atmospheric gases. The air also becomes less dense as elevation increases, so breathing at higher elevations is more difficult.

Layers of the Atmosphere

The atmosphere is divided into four layers based on temperature changes that occur at different distances above the Earth's surface. **Figure 12** shows the four layers of Earth's atmosphere.

The Troposphere The atmospheric layer nearest Earth's surface is the troposphere. The **troposphere** extends to 18 km above Earth's surface. Almost all of the weather occurs in this layer, as shown in **Figure 13**. The troposphere is Earth's densest atmospheric layer. Temperature decreases as altitude increases in the troposphere.

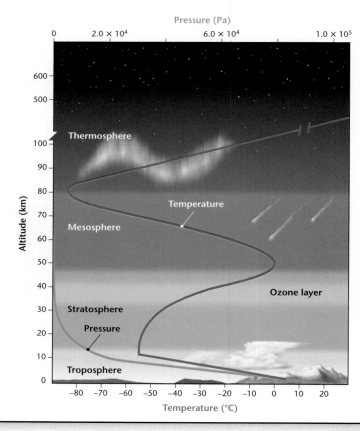

Wind Speed Students may be confused between the terms *tropical storm* and *hurricane*. The difference between a hurricane and a tropical storm is wind speed. To qualify as a tropical storm, a storm must originate over a warm, tropical ocean. Wind speeds must be at least 63 km/h but no greater than 119 km/h. If wind speed is greater than 119 km/h, the storm is a hurricane. Also, a tropical storm doesn't have an eye like a hurricane.

BRAIN FOOD

Hurricanes and Tropical Storms There are few hurricanes in the North Pacific Ocean and North Atlantic Ocean. The cold waters in these parts of the global ocean account for the absence of hurricanes in the northeastern and northwestern parts of the United States. Hurricanes and tropical storms need hot, humid air to survive, so they lose energy quickly once they travel over land or cool water.

The Stratosphere Above the troposphere is the stratosphere. The **stratosphere** extends from 18 km to an altitude of about 50 km. Temperatures rise as altitude increases in the stratosphere. This change happens because ozone in the stratosphere absorbs the sun's ultraviolet (UV) energy and warms the air. **Ozone, O_3,** is a molecule that is made up of three oxygen atoms. Almost all the ozone in the atmosphere is concentrated in the ozone layer in the stratosphere. Because ozone absorbs UV radiation, it reduces the amount of UV radiation that reaches the Earth. UV radiation that reaches Earth can damage living cells.

The Mesosphere The layer above the stratosphere is the *mesosphere*. This layer extends to an altitude of about 80 km. The mesosphere is the coldest layer of the atmosphere, and its temperatures have been measured as low as −93°C.

The Thermosphere The atmospheric layer located farthest from Earth's surface is the *thermosphere*. In the thermosphere, nitrogen and oxygen absorb solar radiation, which results in temperatures that have been measured above 2,000°C. Even though air temperatures in this layer are very high, the thermosphere would not feel hot to us. Air particles that strike one another transfer heat. The air in the thermosphere is so thin that air particles rarely collide, so little heat is transferred.

Nitrogen and oxygen atoms in the lower region of the thermosphere (about 80 km to 550 km above Earth's surface) absorb harmful solar radiation, such as X rays and gamma rays. This absorption causes atoms to become electrically charged. Electrically charged atoms are called *ions*. The lower thermosphere is called the *ionosphere*. Sometimes ions radiate energy as light. These lights often glow in spectacular colors in the night skies near the Earth's North and South Poles, as shown in **Figure 14.**

Figure 13 ▶ Scientists on board a research plane from the National Oceanic and Atmospheric Administration (NOAA) are making measurements of temperature, humidity, barometric pressure, and wind speed as they fly over the eye of a hurricane.

Figure 14 ▶ The *aurora borealis,* or Northern Lights, can be seen in the skies around Earth's North Pole.

69

QuickLAB

Skills Acquired
- Constructing Models
- Experimenting
- Interpreting

Answers
1. radiation

Demonstration ——— ADVANCED

Understanding Diffraction Most natural light contains many colors. Have each student build a spectroscope to analyze the light spectrum.

1. Cut two circles of poster board that are slightly larger than the openings on a toilet paper roll. Make a small hole in one of the circles of poster board big enough for a piece of diffraction grating (about 2 cm). Tape the grating to the circle, and tape the circle over the end of the tube.

2. Cut a narrow slot about 2 cm long across the other circle.

3. Peering through the roll with the grating close to your eye, look toward a light source. Rotate the circle with the slot until it produces the widest spectrum.

4. Have students draw color sketches of the spectrums emitted by the different sources. Compare them and analyze the main gases present in each form of light. **LS** **Visual/Kinesthetic**

QuickLAB

The Heat Is On!

Procedure

1. Fill two **250 mL beakers** with **water**. Use a **thermometer** to record the initial temperature of the water in both beakers. The temperature of the water should be the same for both beakers.

2. Wrap one beaker with **white paper**, and wrap one with **black paper**. Secure the paper with a piece of **tape**.

3. Place a **150 W floodlight** 50 cm away from the beakers, and turn the light on.

4. Record the temperature of the water in both beakers at 1 min, 5 min, and 10 min.

Analysis

1. By what mechanism is energy being transferred to the beakers? Explain your answer.

Energy in the Atmosphere

As shown in **Figure 15,** energy from the sun is transferred in Earth's atmosphere by three mechanisms: radiation, convection, and conduction. **Radiation** is the transfer of energy across space and in the atmosphere. When you stand before a fire or a bed of coals, the heat you feel has reached you by radiation. **Conduction** is the flow of heat from a warmer object to a colder object when the objects are placed in direct physical contact. **Convection** is the transfer of heat by air currents. Hot air rises and cold air sinks. Thus, if you hold your hand above a hot iron, you will feel the heat because a current of hot air rises up to your hand.

Heating of the Atmosphere Solar energy reaches the Earth as electromagnetic radiation, which includes visible light, infrared radiation, and ultraviolet light. The sun releases a vast amount of radiation, but our planet only receives about two-billionths of this energy. This seemingly small amount of radiation contains a tremendous amount of energy, however. As shown in **Figure 15,** about half of the solar energy that enters the atmosphere passes through the atmosphere and reaches Earth's surface. The rest of the energy is absorbed or reflected in the atmosphere by clouds, gases, and dust, or it is reflected by the Earth's surface. On a sunny day, rocks may become too hot to touch. If the Earth's

Figure 15 ▶ Thermal Radiation
Three important mechanisms responsible for transferring heat in the atmosphere are ❶ radiation, ❷ conduction, and ❸ convection.

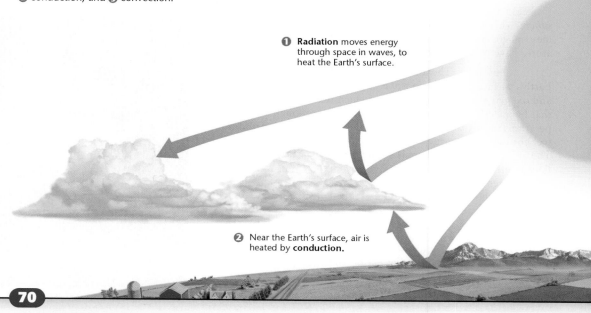

❶ **Radiation** moves energy through space in waves, to heat the Earth's surface.

❷ Near the Earth's surface, air is heated by **conduction.**

Transparencies

TT Energy in Earth's Atmosphere

Career

Atmospheric Sciences Ask students how scientists study phenomena such as hurricanes, tornadoes, and lightning. Ask them to research a career dealing with the atmospheric sciences. Have them report back to the class what a person with that job does and what courses they had to study to perform that job. If possible, contact a meteorologist in your local area to come speak to the class about his or her field of study. **LS** **Interpersonal**

surface continually absorbed energy, the Earth would get hotter and hotter. The Earth does not continue to get warmer, because the oceans and the land radiate the energy they have absorbed back into the atmosphere.

You may have noticed that dark-colored objects become much hotter in the sun than light-colored objects. Dark-colored objects absorb more solar radiation than light-colored objects, so dark-colored objects have more energy to release as heat. This is one reason the temperature in cities is higher than the temperature in the surrounding countryside.

The Movement of Energy in the Atmosphere Air that is constantly moving upward, downward, or sideways causes Earth's weather. In the troposphere, currents of less dense air, warmed by the Earth's surface, rise into the atmosphere and currents of denser cold air sink toward the ground. As a current of air rises into the atmosphere, it begins to cool. Eventually, the air current becomes more dense than the air around it and sinks instead of continuing to rise. So, the air current moves back toward Earth's surface until it is heated by the Earth and becomes less dense. Then, the air current begins to rise again. The continual process of warm air rising and cool air sinking moves air in a circular motion, called a *convection current*. A convection current is shown in **Figure 15.**

Activity ——— **BASIC**

Weather Journal Have students keep a weather journal. Students will use the weather report from their local newspaper over a one-week period to construct an overview of local weather conditions. **LS Logical**

Group Activity ——— **BASIC**

Modeling Convection Currents Have students model convection currents in beakers. Use four beakers of the same size that do not have pour spouts. Fill two with warm water and two with cold water. Color the warm water yellow with food coloring, and put blue dye in the cold water. Have students create two scenarios. In one scenario, the water from the warm-water beaker will be poured over the cold-water beaker. In the other, the water from the cold-water beaker will be poured over the warm-water beaker. Start by placing a thin, plastic object, such as a playing card or a laminated driver's license, over the mouth of the warm-water beaker. Holding it carefully in position, place the warm-water beaker on the cold-water beaker so that their mouths touch. Carefully slip the card from its place while holding the beaker securely. Observe what happens in the two beakers. Repeat this experiment by placing the cold-water beaker on top of the warm-water beaker. Ask students to explain what happened. (Cold water is denser than warm water. When cold water is poured into warm, less-dense water, the warm water rises to the top and the cold water sinks. This movement illustrates a convection current. The same thing happens with air. Convection currents cause thunderstorms to form when cold and warm air masses collide. On the other hand, warm water over cold is stable and remains layered, like the stratosphere.) **LS Visual**

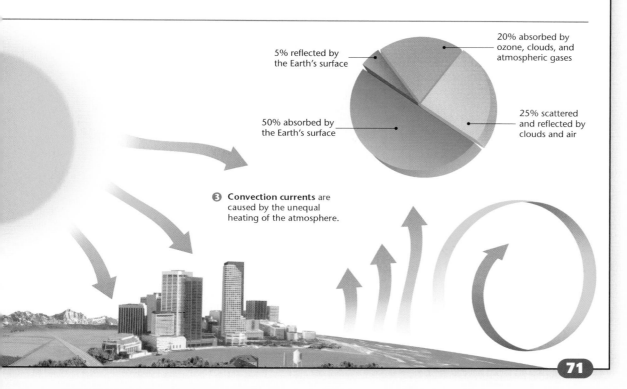

5% reflected by the Earth's surface

20% absorbed by ozone, clouds, and atmospheric gases

50% absorbed by the Earth's surface

25% scattered and reflected by clouds and air

❸ **Convection currents** are caused by the unequal heating of the atmosphere.

71

MISCONCEPTION ALERT

Wind Many students think that wind is not made of anything. Point out that wind is moving air that is composed of matter. Air behaves like a fluid and flows due to gravity and differences in pressure. Air is pulled toward Earth's surface by gravity. The amount of force with which it is drawn downward is referred to as the air pressure. Wind is simply the result of horizontal variations in air pressure. If there is a great difference in the air pressure between two places at the same altitude, this will cause strong wind. Calm days are the result of relatively uniform air pressures.

Close

Reteaching ─── BASIC

The Atmosphere Divide the class into pairs. Have each student write five questions that can be answered by reading the section. Have students use the questions to quiz each other on the section. **LS** Interpersonal

Quiz ─── GENERAL

1. What are the four layers of Earth's atmosphere? (troposphere, stratosphere, mesosphere, and thermosphere)

2. What causes the movement of energy in the atmosphere? (It is caused by differential heating by solar radiation.)

3. What happens when gases trap heat near the Earth? (It is part of Earth's natural process called the *greenhouse effect*, which allows the Earth's surface to remain warm enough to support life.)

Alternative Assessment ─── GENERAL

Birthing a Weather Event Have students "birth" a weather event such as a hurricane, tornado, or thunderstorm. Have them create the conditions necessary for that event to occur and then track it as it progresses. In their *EcoLog*, have students describe the consequences of the event on local human populations and the environment. **LS** Logical

FIELD ACTIVITY

Exploring the Greenhouse Effect Some of your classmates and teachers probably drive to school. Given what you know about the reflection and absorption of heat, go to the parking lot on a sunny day and hypothesize which cars will have the hottest interiors. Base your hypothesis on such variables as the color of car interiors and whether the windows are tinted or untinted. Record your observations in your *Ecolog*.

Figure 16 ▶ The gases in the atmosphere act like a layer of glass. Both glass and the gases in the atmosphere allow solar energy to pass through. But glass and some of the gases in the atmosphere absorb heat and stop the heat from escaping to space.

The Greenhouse Effect

The gases in Earth's atmosphere act like the glass in the car shown in **Figure 16**. Sunlight that penetrates Earth's atmosphere heats the surface of the Earth. The Earth's surface radiates heat back to the atmosphere, where some of the heat escapes into space. The remainder of the heat is absorbed by greenhouse gases, which warms the air. Heat is then radiated back toward the surface of the Earth. This process, in which gases trap heat near the Earth, is known as the **greenhouse effect.** Without the greenhouse effect, the Earth would be too cold for life to exist.

The gases in our atmosphere that trap and radiate heat are called *greenhouse gases*. None of the greenhouse gases have a high concentration in Earth's atmosphere. The most abundant greenhouse gases are water vapor, carbon dioxide, methane, and nitrous oxide. The quantities of carbon dioxide and methane in the atmosphere vary considerably as a result of natural and industrial processes, and the amount of water varies because of natural processes.

❶ Sunlight passes through the glass into the car.

❷ The interior absorbs radiant energy, changing it into heat.

❸ The glass in the car stops most of the heat from escaping, increasing the temperature inside the car.

SECTION 2 Review

1. **Describe** the composition of Earth's atmosphere.

2. **Describe** a characteristic of each layer of the atmosphere.

3. **Explain** the three mechanisms of heat transfer in Earth's atmosphere.

4. **Describe** the role of greenhouse gases in Earth's atmosphere.

CRITICAL THINKING

5. **Analyzing Processes** Read about the density of Earth's atmosphere under the heading "Air Pressure." Write a paragraph that explains why Earth's atmosphere becomes less dense with increasing altitude above Earth. **WRITING SKILLS**

6. **Analyzing Processes** How does human activity change some greenhouse-gas levels?

72

Answers to Section Review

1. Earth's atmosphere is composed mainly of nitrogen and oxygen and small quantities of other gases.

2. Answers may vary. Sample Answers: The troposphere is the warmest layer of the atmosphere. Ozone is concentrated in a layer in the stratosphere, where it absorbs ultraviolet rays, protecting organisms from genetic damage. The mesosphere absorbs little heat from the sun, and the air is thin and cold. Nitrogen and oxygen in the thermosphere absorb high-energy solar radiation such as X-rays and gamma rays.

3. The three mechanisms of heat transfer in the atmosphere are radiation, conduction, and convection.

4. Greenhouse gases absorb heat and radiate it back to the surface of the Earth.

5. Fewer gas molecules are found with increasing altitude above Earth's surface. Therefore, less pressure at these altitudes pushes downward on atmospheric gases. As a result, the air is less dense as altitude increases.

6. Answers may vary. One answer is that the burning of fossil fuels increases the amount of carbon dioxide in the atmosphere.

The Hydrosphere and Biosphere

Life on Earth is restricted to a very narrow layer around the Earth's surface. In this layer, called the *biosphere,* everything that organisms need to survive can be found. One of the requirements of living things is liquid water.

The Hydrosphere and Water Cycle

The hydrosphere includes all of the water on or near the Earth's surface. The hydrosphere includes water in the oceans, lakes, rivers, wetlands, polar ice caps, soil, rock layers beneath Earth's surface, and clouds.

The continuous movement of water into the air, onto land, and then back to water sources is known as the **water cycle,** which is shown in **Figure 17. Evaporation** is the process by which liquid water is heated by the sun and then rises into the atmosphere as water vapor. Water continually evaporates from Earth's oceans, lakes, streams, and soil, but the majority of the water evaporates from the oceans. In the process of **condensation,** water vapor forms water droplets on dust particles. These water droplets form clouds, in which the droplets collide, stick together, and create larger, heavier droplets. These larger droplets fall from clouds as rain in the process called **precipitation.** Precipitation may also take the form of snow, sleet, or hail.

Objectives

▶ Name the three major processes in the water cycle.

▶ Describe the properties of ocean water.

▶ Describe the two types of ocean currents.

▶ Explain how the ocean regulates Earth's temperature.

▶ Discuss the factors that confine life to the biosphere.

▶ Explain the difference between open and closed systems.

Key Terms

water cycle
evaporation
condensation
precipitation
salinity
fresh water
biosphere
closed system
open system

② **CONDENSATION** ❶ **EVAPORATION**

❸ **PRECIPITATION**

Figure 17 ▶ The major processes of the water cycle include ❶ evaporation, ② condensation, and ❸ precipitation.

73

Chapter Resource File

- **Lesson Plan**
- **Active Reading** BASIC
- **Section Quiz** GENERAL

Transparencies

TT Bellringer

INCLUSION Strategies

- *Developmentally Delayed*
- *Learning Disabled*
- *Attention Deficit Disorder*

Ask students to use **Figure 17** to make a colorful poster or bulletin board depicting the major processes of the water cycle. The students can label the parts of the cycle with large, clear print. Students can demonstrate an understanding of the concept by giving an oral presentation of their work. The displays can be used during the course of the chapter.

Focus

Overview

Before beginning this section, review with your students the Objectives listed in the Student Edition. This section explores the hydrosphere and the biosphere.

Bellringer

Ask students to estimate the percentage of Earth's water that is salt water and the percentage that is fresh water. (97 percent is salt water; 3 percent is fresh water.) Have students design a way to represent these percentages. **LS Logical**

Motivate

Demonstration ── ADVANCED

The Coriolis Effect The Coriolis effect affects the direction in which currents and winds flow on Earth. The materials needed for this demonstration are a circular piece of cardboard and a pencil, pen, or marker.

1. Pin or nail the cardboard to a surface in such a way that it can rotate freely.

2. Rotate the cardboard smoothly with one hand. With the other hand, draw a straight line from the center of the cardboard toward a particular fixed direction.

3. Repeat the procedure by rotating the cardboard in the opposite direction.

4. If the cardboard serves as a model for the Coriolis effect, what conclusions can you make about the direction of apparent deflection in the Northern Hemisphere? in the Southern Hemisphere? Is the Coriolis effect in the Southern Hemisphere a mirror image of the Coriolis effect in the Northern Hemisphere? **LS Visual/Kinesthetic**

Discussion — BASIC

Weather Control Ask students if humans should attempt to control weather processes. Also ask if it is a good thing to strive for "convenient" weather and what the potential local and global effects of manipulating the weather might be. **LS** Verbal

REAL-LIFE — CONNECTION — GENERAL

Cloud Seeding Through a process called *cloud seeding,* attempts have been made to modify the weather. The process involves injecting particles of silver iodide into clouds. This causes ice crystals to attach to the silver iodide particles. The ice crystals grow larger and eventually become heavy enough to fall as precipitation. Another method involves injecting water droplets into the base of a cloud. Updrafts carry the droplets up through the cloud. The droplets grow in size as they merge with other water droplets in the cloud. Eventually, they become heavy enough to fall. The technique has only been able to enhance precipitation by 15 percent. Many attempts have actually caused precipitation to diminish. For example, overseeding clouds can produce so many small ice particles that there are no longer enough water droplets and vapor to grow and merge. Cloud seeding has also been attempted to reduce the destructiveness of hail, to reduce hurricane winds, and to disperse fog.

Connection to Geology

Submarine Volcanoes Geologists estimate that approximately 80 percent of the volcanic activity on Earth takes place on the ocean floor. Most of this activity occurs as magma slowly flows onto the ocean floor where tectonic plates pull away from each other. But enormous undersea volcanoes are also common. Off the coast of Hawaii, a submarine volcano called the *Loihi Seamount* rises 5,185 m from the ocean floor. Loihi is just 915 m below the ocean's surface, and in several thousand years, this volcano may become the next Hawaiian Island.

Earth's Oceans

We talk about the Atlantic Ocean, the Pacific Ocean, the Arctic Ocean, and the Indian Ocean. However, if you look at **Figure 18,** you see that these oceans are all joined. This single, large, interconnected body of water is called the *world ocean.* Its waters cover a little over 70 percent of the Earth's surface. As we will see, the world ocean plays many important roles in regulating our planet's environment.

The largest ocean on Earth is the Pacific Ocean. It covers a surface area of approximately 165,640,000 km^2 and has an average depth of 4,280 m. The deepest point on the ocean floor is in the Pacific Ocean. The point is called the Challenger Deep and is located east of the Philippine Islands at the bottom of the Mariana Trench. The Challenger Deep is 11,033 m below sea level, which is deeper than Mount Everest is tall. Oceanographers often divide the Pacific Ocean into the North Pacific and South Pacific based on the direction of surface current flow in each half of the Pacific Ocean. Surface currents in the Pacific move in a clockwise direction north

CASE STUDY — Hydrothermal Vents

The light from your tiny research submarine illuminates the desert-like barrenness of the deep-ocean floor. Suddenly, the light catches something totally unexpected—an underwater oasis teeming with sea creatures that no human has laid eyes on before. At the center of this community is a tall chimney-like structure from which a column of black water is rising.

This scene is much like the one which John Corliss and John Edmond witnessed when they discovered the first deep-sea hydrothermal vents and the odd community of creatures that inhabit them. Corliss's and Edmond's discovery was made during a dive in the submarine *Alvin* in early 1977. The dive site was in the eastern Pacific Ocean near the Galápagos Islands. Since

the original dive, many more hydrothermal vents have been located on the ocean bottom.

Hydrothermal vents are openings in the ocean floor where super-hot, mineral-rich waters stream into the ocean. Hydrothermal vents form where tectonic plates are separating and where deep fractures are opening in the Earth's crust. Water seeps down into some of these fractures to a depth where it is heated by molten rock and enriched with minerals. The water returns to the ocean floor through other fractures and then pours into the ocean. Water often streams through structures called *chimneys.* Chimneys form when the minerals in the vent water—mostly iron and sulfur—precipitate as the

► Superheated, mineral-rich water streams through a chimney at a hydrothermal vent on the floor of the Pacific Ocean.

water cools from above 100°C to less than 50°C. The tallest chimney reported to date is 49 m. Vent

74

Homework — GENERAL

Threatened Coral Reefs The organisms that make up coral reefs live in a delicate balance with their environment. Water pollution has already harmed a significant number of reefs. Coral reefs are also susceptible to larger environmental changes. Warmer sea temperature is believed to be causing the widespread bleaching of reefs. Corals live within a narrow temperature range—between 26°C to 27°C (78.8°F to 80.6°F). They become stressed at temperatures above 29°C (84.2°F). Corals

aren't the only marine animals that are affected by temperature changes. An increase or decrease of only a few Celsius degrees can kill certain fish, mollusks, and marine plants. Have each student choose a major world coral reef and create a posterboard display that includes information about the reef, the organisms that live around it, and the environmental threats the reef faces. Possible reefs include the Florida reef tract in the Florida Keys and the Great Barrier Reef in Australia. **LS** Logical

of the equator, whereas surface currents flow in a counterclockwise direction south of the equator.

The second largest ocean on Earth is the Atlantic Ocean. It covers a surface area of 81,630,000 km², which is about half the area of the Pacific Ocean. Like the Pacific Ocean, the Atlantic Ocean can be divided into a north and south half based on the directions of surface current flow north and south of the equator.

The Indian Ocean covers a surface area of 73,420,000 km² and is the third-largest ocean on Earth. It has an average depth of 3,890 m.

The smallest ocean is the Arctic Ocean, which covers 14,350,000 km². The Arctic Ocean is unique because much of its surface is covered by floating ice. This ice, which is called *pack ice*, forms when either waves or wind drive together frozen seawater, known as sea ice, into a large mass.

Figure 18 ▶ The Pacific, Atlantic, Indian, and Arctic Oceans are interconnected into a single body of water, the world ocean, which covers 70 percent of Earth's surface.

▶ Over 300 species of organisms have been found in hydrothermal vent communities, including species of tube worms that may grow to a length of 3 m.

water can reach temperatures as high as 400°C.

The pressure at the ocean bottom is tremendous. No sunlight penetrates these depths, and hydrothermal vents spew minerals into their surroundings. Still, at least 300 species of organisms—all new to scientists—live near hydrothermal vents. These organisms include tube worms, giant clams, mussels, shrimp, crabs, sea anemones, and octopuses.

How is life at hydrothermal vents possible? Bacteria that live in vent communities can use hydrogen sulfide escaping from the vents as an energy source. Some animals that live in vent communities consume these bacteria to obtain their energy. Other animals have bacteria living inside their bodies that supply them with energy.

CRITICAL THINKING

1. Applying Processes Some scientists have suggested that life may have originated in or near hydrothermal vents because vent organisms are able to obtain their energy from chemicals in the absence of sunlight. Does this suggestion seem realistic?

2. Making Predictions How might the creatures that live in hydrothermal vent communities be of benefit to humankind in the future?

75

The Dead Sea Located in hot desert lands between Israel and Jordan, the surface of the 50-mile-long Dead Sea is 1,340 feet below sea level—the lowest elevation on Earth. The Dead Sea is a deep inland water body that filled a deep fissure left by plate separation. The Dead Sea is fed by rivers that flow into it. With no outlet, all inflow makes its way back into the water cycle through evaporation. The Dead Sea is almost seven times saltier than the ocean. Along its bottom are solid deposits of sodium chloride and other minerals. The extreme salinity of the water makes it exceptionally dense. A human body will easily float upon the surface of the water. Swallowing a cupful of Dead Sea water can kill a person. Nothing lives in the Dead Sea except for microbes that have adapted to the extreme salinity. Microbes that live, grow, and multiply in saline environments, such as the Dead Sea, are called *halophiles*. Have students research halophiles and report their findings to the class.
LS **Logical**

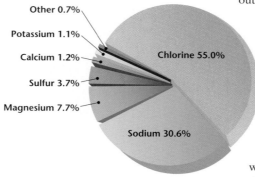

Figure 19 ▸ This pie graph shows the percentages by weight of dissolved solids found in ocean water. Sodium and chlorine, the two elements that form salt, are the most important dissolved solids in ocean water.

Other 0.7%
Potassium 1.1%
Calcium 1.2%
Sulfur 3.7%
Magnesium 7.7%
Chlorine 55.0%
Sodium 30.6%

Ocean Water The difference between ocean water and fresh water is that ocean water contains more salts. These salts have dissolved out of rocks on land and have been carried down rivers into the ocean over millions of years. Underwater volcanic eruptions also add salts to the ocean.

Most of the salt in the ocean is sodium chloride, which is made up of the elements sodium and chlorine. **Figure 19** shows the concentration of these and other elements in ocean water. The **salinity** of ocean water is the concentration of all the dissolved salts it contains. The average salinity of ocean water is 3.5 percent by weight. The salinity of ocean water is lower in places that get a lot of rain or in places where fresh water flows into the sea. Salinity is higher where water evaporates rapidly and leaves the salts behind.

Temperature Zones **Figure 20** shows the temperature zones of the ocean. The surface of the ocean is warmed by the sun. In contrast, the depths of the ocean, where sunlight never reaches, are very cold and have a temperature only slightly above freezing. Surface waters are stirred up by waves and currents, so the warm surface zone may be as much as 350 m deep. Below the surface zone is the thermocline, which is a layer about 300 to 700 m deep where the temperature falls rapidly with depth. If you have ever gone swimming in a deep lake in the summer, you have probably encountered a shallow thermocline. Sun warms the surface of the lake to a comfortable temperature, but if you

Figure 20 ▸ Water in the ocean can be divided into three zones based on temperature.

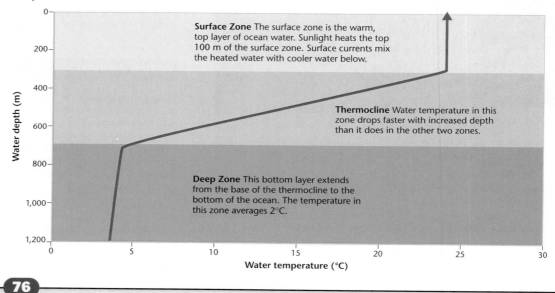

Surface Zone The surface zone is the warm, top layer of ocean water. Sunlight heats the top 100 m of the surface zone. Surface currents mix the heated water with cooler water below.

Thermocline Water temperature in this zone drops faster with increased depth than it does in the other two zones.

Deep Zone This bottom layer extends from the base of the thermocline to the bottom of the ocean. The temperature in this zone averages 2°C.

BRAIN FOOD 🧠

Phytoplankton Marine and freshwater plankton drift along with the tides, currents, and winds. Plankton with the ability to photosynthesize are called *phytoplankton*. Because phytoplankton depend upon specific conditions for their growth, they are a good indicator of environmental change. They are the first link in the complex aquatic food web. If phytoplankton die off, other aquatic organisms suffer. Furthermore, 90 percent of all photosynthesis and oxygen release occurs in the oceans.

drop your feet, they fall into cold water that may be only slightly above freezing. The boundary between the warm and cold water is the thermocline.

A Global Temperature Regulator One of the most important functions of the world ocean is to absorb and store energy from sunlight. This capacity of the ocean to absorb and store energy from sunlight regulates temperatures in Earth's atmosphere.

The world ocean absorbs over half the solar radiation that reaches the planet's surface. The ocean both absorbs and releases heat more slowly than land does. As a consequence, the temperature of the atmosphere changes much more slowly than it would if there were no ocean on Earth. If the ocean did not regulate atmospheric and surface temperatures, the temperature would be too extreme for life on Earth to exist.

Local temperatures in different areas of the planet are also regulated by the world ocean. Currents that circulate warm water cause the land areas they flow past to have a more moderate climate. For example, the British Isles are warmed by the waters of the Gulf Stream, which moves warm waters from lower latitudes toward higher latitudes, as in **Figure 21.**

MATHPRACTICE

The Influence of the Gulf Stream The temperature of the British Isles is moderated by the Gulf Stream. Falmouth, England, and Winnipeg, Canada, are located at approximately 50° north latitude. Falmouth, which is located in extreme southwest England near the Atlantic Ocean, has average high temperatures of 18°C in June, 19°C in July, and 19°C in August. Winnipeg, which is located in the interior of North America, has average high temperatures of 22°C in June, 25°C in July, and 23°C in August. What is the difference in average high temperatures in degrees Celsius between Falmouth and Winnipeg?

MATHPRACTICE

Answers

18°C + 19°C + 19°C = 56°C ÷ 3 = 18.67°C
22°C + 25°C + 23°C = 70°C ÷ 3 = 23.33°C
23.33°C − 18.67°C = 4.66°C difference between average high temperatures in Falmouth and Winnipeg during summer

OCEANOGRAPHY
► CONNECTION ─── GENERAL

The Gulf Stream Due to its moderating effect on the climate of western Europe, the Gulf Stream is one of the most important currents in the global ocean. In western Scotland, the warming effects of the Gulf Stream allow semitropical plants and palm trees to grow despite being at the same latitude as Siberia and Hudson's Bay. The Gulf Stream flows northeast from the Straits of Florida to the Grand Banks near Newfoundland. The Gulf Stream is fed by the warm North and South Equatorial Currents, which join together (along with water from the Gulf of Mexico) to create the important warm current. The surface temperature of the Gulf Stream is 25°C (77°F). The Gulf Stream flows at about 3 mph and is identifiable by its bright blue hue and high salinity. It measures about 50 miles wide near its point of origin and widens to about 300 miles off of New York. When the Gulf Stream reaches the Grand Banks, it mixes with the cold Labrador Current and then flows across the ocean, eventually branching off into several tongues that flow toward Europe. To obtain a five-day Gulf Stream weather update, as well as a color map of the Gulf Stream, have students log on to the Marine Weather Web site. **LS Logical**

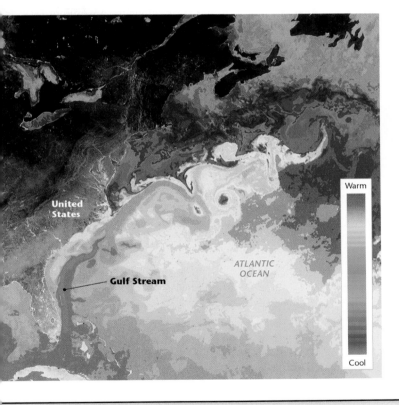

United States

Gulf Stream

ATLANTIC OCEAN

Warm

Cool

Figure 21 ► In this infrared satellite image, the Gulf Stream is moving warm water (shown in red, orange, and yellow) from lower latitudes into higher latitudes. The British Isles are warmed by the waters of the Gulf Stream.

77

Student Opportunities

Internships at Sea Many oceanographic institutes offer opportunities for interested students to learn more about marine environments and oceanographic research. As a class, research these opportunities on the Internet. Have teams of students compile files that contain information on internships, summer programs, and other opportunities that are available to high school students. Each team can focus on one or two of the following institutes:

Scripps, Wood's Hole, Acadia, Florida Marine Research Institute, Rosensteil School, and Lamont-Doherty. Students might also be interested in opportunities offered by the Jason Project. Even if students can't spend the summer in a research vessel, you will find classroom resources that will enrich student learning. Save the files in your classroom as a resource for future students.

Teach, continued

QuickLAB

Skills Acquired
- Experimenting
- Identifying and Recognizing Patterns

Answers
1. There should be very little mixing of hot and cold water.

OCEANOGRAPHY
CONNECTION — **ADVANCED**

Antarctic Bottom Water

Antarctic Bottom Water is cold, dense water that sinks slowly to the ocean depths and carries oxygen with it. It is the densest water mass in the global ocean. The process of Antarctic Bottom Water formation begins when Antarctic ice cools the air. The cool air sinks and gains speed as it flows down the slopes of ice sheets. This wind hits sea ice and pushes it away from shore. Cold winds chill exposed water and freeze it, which squeezes the salt content out. The water below becomes denser with the additional salt and sinks. This cold bottom water begins to slowly circulate around the globe, creeping along at 2 to 10 cm/sec (0.8 to 4 in/sec) or less. It reaches places as far away as the North Atlantic. In fact, Antarctica is the source of 75 percent of the world's bottom water.

Transparencies
TT Surface Currents of the World

QuickLAB

Make a Hydrothermal Vent

Procedure
1. Fill a **large glass container** or **aquarium** with very **cold water.**
2. Tie one end of a **piece of string** around the neck of a **small bottle.**
3. Fill the small bottle with **hot water**, and add a few drops of **food coloring.**
4. Keep the small bottle upright while you lower it into the glass container until it rests flat on the bottom.

Analysis
1. Did the food coloring indicate that the hot water and cold water mixed?

Ocean Currents Streamlike movements of water that occur at or near the surface of the ocean are called surface currents. Surface currents are wind driven and result from global wind patterns. **Figure 22** shows the major surface currents of the world ocean. Surface currents may be warm-water currents or cold-water currents. Currents of warm water and currents of cold water do not readily mix with one another. Therefore, a warm-water current like the Gulf Stream can flow for hundreds of kilometers through cold water without mixing and losing its heat.

Surface currents can influence the climates of land areas they flow past. As we have seen, the Gulf Stream moderates the climate in the British Isles. The Scilly Isles in England are as far north as Newfoundland in northeast Canada. However, palm trees grow on the Scilly Isles, where it never freezes, whereas Newfoundland has long winters of frost and snow.

Deep currents are streamlike movements of water that flow very slowly along the ocean floor. Deep currents form when the cold, dense water from the poles sinks below warmer, less dense ocean water and flows toward the equator. The densest and coldest ocean water is located off the coast of Antarctica. This cold water sinks to the bottom of the ocean and flows very slowly northward to produce a deep current called the Antarctic Bottom Water. The Antarctic Bottom Water creeps along the ocean floor for thousands of kilometers and reaches a northernmost point of approximately 40° north latitude. It takes several hundred years for water in this deep current to make this trip northward.

Figure 22 ▶ The oceans' surface currents circulate in different directions in each hemisphere.

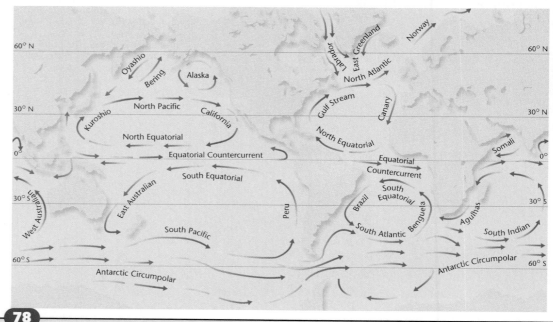

78

LITERATURE
CONNECTION — **GENERAL**

Endurance: Shackleton's Incredible Voyage
In 1914, British explorer Ernest Shackleton attempted to reach the South Pole in his ship *Endurance.* Just one day's sail from Antarctica, *Endurance* became trapped in sea ice, and became frozen fast for a period of 10 months. Pressure from the ice eventually crushed the ship, and Shackleton and his men were forced to spend a period of five months camping on ice floes. At the end of this five-month period, Shackleton made a daring attempt to seek help and organize a team to rescue the men that he was forced to leave behind by crossing 800 miles of open ocean to South Georgia Island. After crossing the mountains of South Georgia, Shackleton was able to reach the island's whaling station. There, he assembled the rescue team that would return to Antarctica to save all of the men he left behind. This heroic epic is documented in the book *Endurance: Shackleton's Incredible Voyage* by Alfred Lansing, which is available in both print version and as an audio book.

Fresh Water

Most of the water on Earth is salt water in the ocean. A little more than 3 percent of all the water on Earth is **fresh water.** Most of the fresh water is locked up in icecaps and glaciers that are so large they are hard to imagine. For instance, the ice sheet that covers Antarctica is as large as the United States and is up to 3 km thick. The rest of Earth's fresh water is found in lakes, rivers, wetlands, the soil, rock layers below the surface, and in the atmosphere.

River Systems A river system is a network of streams that drains an area of land. A river system contains all of the land drained by a river, including the main river and all its tributaries. As shown in **Figure 23,** *tributaries* are smaller streams or rivers that flow into larger ones. Some river systems are enormous. For example, most of the precipitation that falls between the Rocky Mountains in the west and the Appalachian Mountains in the east eventually drains into the Mississippi River. The Mississippi River system covers about 40 percent of the contiguous United States.

Groundwater

Rain and melting snow sink into the ground and run off the land. Some of this water ends up in streams and rivers, but most of it trickles down through the ground and collects as *groundwater.* Groundwater fulfills the human need for fresh drinking water and supplies water for many agricultural and industrial uses. But groundwater accounts for less than 1 percent of all the water on Earth.

Aquifers A rock layer that stores and allows the flow of groundwater is called an *aquifer.* The surface of the land where water enters an aquifer is called a *recharge zone.* **Figure 24** shows the location of aquifers in the contiguous United States.

Figure 23 ▶ This photo shows a network of tributaries flowing into a river in the wetlands of southern Louisiana.

☐ Aquifers

Figure 24 ▶ Much of the United States is underlain by aquifers. The brown areas are rocks that contain relatively little stored water.

79

GEOGRAPHY — CONNECTION GENERAL

The Middle East and Water The Middle East is one of the most water-scarce regions in the world. Eleven of the fourteen countries in the region already experience a scarcity of water. Places like Saudi Arabia, Kuwait, Bahrain, and the United Arab Emirates have so little fresh water available that they operate costly desalination plants. But more water resources are still needed to keep up with the growing population of the Middle East. One expert quipped that the Middle East "ran out of water" in 1972. Since then, more water has been withdrawn from aquifers and rivers than has been replenished. Jordan and Yemen annually extract 30 percent more water from aquifers than has been replaced. Saudi Arabia represents one of the most unsustainable water-use cases in the world. Extremely arid, Saudi Arabia must mine fossil groundwater—which cannot be replenished—to meet its water needs.

Group Activity — BASIC

Local Aquifers Have students draw a map of their state and draw in where aquifers exist. The locations of aquifers can be found in USGS Internet resources. Have students identify how big each aquifer is and how many people rely on it. **LS** Visual

ECONOMICS — ● CONNECTION GENERAL

Water Banking One method people in arid areas have begun to use to "harvest" more water to meet their needs is water banking. Popular in areas like southern Australia and the southwestern United States, water banking involves storing river water, urban storm water, and reclaimed water in underground aquifers for future use. For example, the state of Nevada currently receives credits for unused water from the Colorado River that is stored in Arizona's groundwater basin. If this procedure becomes more widespread, states can pay the cost of storing water in other states that have such water banks. Have students research and model different forms of water banking. **LS** Logical

Vocabulary *Ecology* is derived from the Greek *oikos,* which means "house." The word *ecology,* then, means "house study." What are some other words that have the *eco*– root? (ecosystem, economics)

English Language Learners

Internet Activity — GENERAL

Researching the Biosphere
The SeaWiFS Web site offers a variety of resources and tools to investigate the ocean and human impact on the oceanic environment. Ask students to browse the SeaWiFs site. Have them report on some aspect of the site that interests them. **LS** Verbal

BIOLOGY
CONNECTION — GENERAL

Biomes The land areas of the biosphere can be grouped into a few areas called *biomes*. Biomes are determined by temperature, precipitation, latitude, and altitude. Some land or terrestrial biomes of the world include tropical rain forests, grasslands, deserts, and tundra. The name biome generally describes the type of vegetation that grows there. The type of vegetation that grows in a biome also determines the other organisms that can live there. Have students research and report on their favorite biomes. **LS** Logical

---

internet connect
www.scilinks.org
Topic: **The Biosphere**
SciLinks code: **HE4127**

SCiLINKS. Maintained by the National Science Teachers Association

The Biosphere

The Earth is often compared to an apple, and the biosphere is compared to the apple's skin. This comparison illustrates how small the layer of the Earth that can support life is in relation to the size of the planet. Scientists define the **biosphere** as the narrow layer around Earth's surface in which life can exist. The biosphere is made up of the uppermost part of the geosphere, most of the hydrosphere, and the lower part of the atmosphere. The biosphere extends about 11 km into the ocean and about 9 km into the atmosphere, where insects, the spores of bacteria, and pollen grains have been discovered.

Life exists on Earth because of several important factors. Life requires liquid water, temperatures between 10°C and 40°C, and a source of energy. The materials that organisms require must continually be cycled. Gravity allows a planet to maintain an atmosphere and to cycle materials. Suitable combinations of the things that organisms need to survive are found only in the biosphere.

The biosphere is located near Earth's surface because most of the sunlight is available near the surface. Plants on land and in the oceans are shown in **Figure 25**. Plants need sunlight to produce their food, and almost every other organism gets its food from plants and algae. Most of these algae float at the surface of the ocean. These tiny, free-floating, marine algae are known as phytoplankton. Except for bacteria that live at hydrothermal vents, most of the organisms that live deep in the ocean feed on dead plants and animals that drift down from the surface.

Figure 25 ▶ This illustration of the biosphere shows the concentration of plant life on land and in the oceans. The colors represent different concentrations of plant life in different regions.

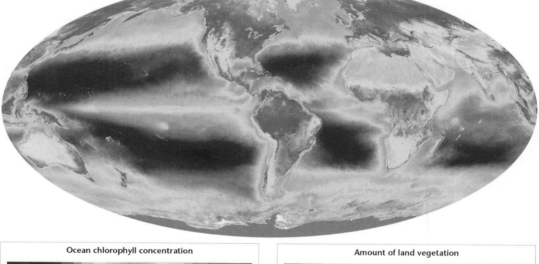

Ocean chlorophyll concentration	
Low	High

Amount of land vegetation	
Low	High

BRAIN FOOD

Rain Forests Rain forests are essential in maintaining global climate by absorbing carbon dioxide, which has been connected with global warming. Tropical rain forests absorb more carbon dioxide than any other land ecosystem on Earth.

Energy Flow in the Biosphere

While energy is constantly added to the biosphere from the sun, matter is not. The energy used by organisms must be obtained in the biosphere and must be constantly supplied for life to continue. When an organism dies, its body is broken down and the nutrients in it become available for use by other organisms. This flow of energy allows life on Earth to continue to exist.

In a **closed system,** energy enters the environment, but matter does not. Today, the Earth is mostly a closed system with respect to matter but is still an open system for energy. Energy enters the biosphere in the form of sunlight, which plants use to make their food. When an animal eats a plant, the energy stored in the plant is transferred to the animal. Animals in turn eat other animals. At each stage in the food chain, some of the energy is lost to the environment as heat, which is eventually lost to space.

In an **open system,** both matter and energy are exchanged between a system and the surrounding environment. Matter was added to the early Earth through the collisions of comets and meteorites with our planet. Now, however, little matter reaches our planet in this way.

Figure 26 ▶ The Eden Project is an attempt to model the biosphere. In this project, plants from all over the world will live in a closed system. The Eden Project is housed within a series of domes that were constructed in an old clay pit in England.

SECTION 3 Review

1. **Name** and describe each of the three major processes in the water cycle.

2. **Describe** the properties of ocean water.

3. **Describe** the two types of ocean currents.

4. **Name** two factors that confine living things to the biosphere.

CRITICAL THINKING

5. **Analyzing Processes** Read about the ocean's role in regulating temperature under the heading "A Global Temperature Regulator." How might Earth's climate change if the land area on Earth were greater than the area of the world ocean? **READING SKILLS**

6. **Analyzing Relationships** Why is the human body considered an open system?

Answers to Section Review

1. The three major processes in the water cycle are evaporation—liquid water is converted to water vapor, condensation—water vapor forms droplets on dust particles, and precipitation—large droplets fall from clouds.

2. Salinity and temperature are two important properties of ocean water.

3. The two types of ocean currents are surface currents, which are wind driven and may be hot or cold, and deep currents, which are cold and flow slowly along the ocean floor.

4. Answers may vary. A moderate temperature and an energy source (such as sunlight) are two physical factors that constrain living things to the biosphere.

5. Answers may vary. Sample answer: Earth's seasonal temperatures might change more quickly.

6. Answers may vary. Answers may include that humans obtain energy from plant and animal life and release energy and matter during life processes and when they die.

Close

Reteaching — BASIC

Water and Temperature As a class, review the roles temperature plays in the hydrosphere. Then have students form pairs and work together to come up with as many examples of heating and cooling as possible. As a class, compile a list of all of the examples, and evaluate each pair's findings. **LS Logical**

Quiz — GENERAL

1. Where are most of the hottest and coldest places on Earth located? (Most are located far inland, away from the ocean. The ocean has a moderating effect on temperature.)

2. Why is ocean water salty? (Water dissolves salts out of rocks on land and washes them into the sea. Evaporation concentrates the salts.)

3. How do cold ocean currents and warm currents interact? (They interact through convection. Warm currents rise, and cold currents sink. Because of differences in density, warm currents and cold currents do not easily mix.)

4. Where is the biosphere located and why? (It is located near Earth's surface, because that is where the sunlight and liquid water are.)

Alternative Assessment — GENERAL

Global Navigation Ask students to imagine they are planning a voyage around the world. They can choose any route they wish, but they must sail with the currents. Have them map out their selected route, showing the names of the currents, their point of origin, and their point of destination. **LS Visual**

Alternative Assessment — GENERAL

Connections in the Earth System

Create a chart that has the four parts of the Earth system—the geosphere, the atmosphere, the hydrosphere, and the biosphere—listed separately and circled in each of the four corners of the chart. Have students use lines to connect different parts of the Earth system to each other. Students should create as many connections between parts of the Earth system as they can recall from reading the chapter. One example of a connection would be the winds of the atmosphere creating the surface currents of the ocean in the hydrosphere. Another example would be photosynthesizing plants in the biosphere creating the oxygen in the atmosphere. **LS** Visual

Chapter Resource File

- **Chapter Test** GENERAL
- **Chapter Test** ADVANCED
- **Concept Review** GENERAL
- **Critical Thinking** ADVANCED
- **Test Item Listing** GENERAL
- **Math/Graphing Lab** ADVANCED
- **Modeling Lab** BASIC
- **CBL™ Probeware Lab** GENERAL
- **Consumer Lab** GENERAL
- **Long-Term Project** GENERAL

CHAPTER 3 Highlights

1 The Geosphere

Key Terms

geosphere, 59
crust, 60
mantle, 61
core, 61
lithosphere, 61
asthenosphere, 61
tectonic plate, 62
erosion, 66

Main Ideas

▶ The solid part of the Earth that consists of all rock, and the soils and sediments on Earth's surface, is the geosphere.

▶ Earth's interior is divided into layers based on composition and structure.

▶ Earth's surface is broken into pieces called *tectonic plates,* which collide, separate, or slip past one another.

▶ Earthquakes, volcanic eruptions, and mountain-building are all events that occur at the boundaries of tectonic plates.

▶ Earth's surface features are continually altered by the action of water and wind.

2 The Atmosphere

Key Terms

atmosphere, 67
troposphere, 68
stratosphere, 69
ozone, 69
radiation, 70
conduction, 70
convection, 70
greenhouse effect, 72

Main Ideas

▶ The mixture of gases that surrounds the Earth is called the *atmosphere*.

▶ The atmosphere is composed almost entirely of nitrogen and oxygen.

▶ Earth's atmosphere is divided into four layers based on changes in temperature that take place at different altitudes.

▶ Heat is transferred in the atmosphere by radiation, conduction, and convection.

▶ Some of the gases in Earth's atmosphere slow the escape of heat from Earth's surface in what is known as the greenhouse effect.

3 The Hydrosphere and Biosphere

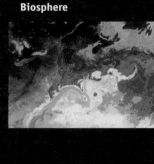

Key Terms

water cycle, 73
evaporation, 73
condensation, 73
precipitation, 73
salinity, 76
fresh water, 79
biosphere, 80
closed system, 81
open system, 81

Main Ideas

▶ The hydrosphere includes all of the water at or near Earth's surface.

▶ Water in the ocean can be divided into three zones—the surface zone, the thermocline, and the deep zone—based on temperature.

▶ The ocean absorbs and stores energy from sunlight, regulating temperatures in the atmosphere.

▶ Surface currents in the ocean affect the climate of the land they flow near.

▶ The biosphere is the narrow layer at the surface of the Earth where life can exist.

▶ The Earth is largely a closed system with respect to matter but an open system with respect to energy.

Using Key Terms

Use each of the following terms in a separate sentence.

1. *tectonic plate*
2. *erosion*
3. *radiation*
4. *ozone*
5. *salinity*

For each pair of terms, explain how the meanings of the terms differ.

6. *lithosphere* and *asthenosphere*
7. *conduction* and *convection*
8. *crust* and *mantle*
9. *closed system* and *open system*

✔ **STUDY TIP**

The Importance of Nouns Most multiple-choice questions center around the definitions of nouns. When you study, pay attention to the definitions of nouns that appear to be important in the text. These nouns will often be boldfaced key terms or italicized secondary terms.

Understanding Key Ideas

10. The thin layer at Earth's surface where life exists is called the
    a. geosphere.
    b. atmosphere.
    c. hydrosphere.
    d. biosphere.

11. The thin layer of the Earth upon which tectonic plates move around is called the
    a. mantle.
    b. asthenosphere.
    c. lithosphere.
    d. outer core.

12. Seventy-eight percent of Earth's atmosphere is made up of
    a. oxygen.
    b. hydrogen.
    c. nitrogen.
    d. carbon dioxide.

13. The ozone layer is located in the
    a. stratosphere.
    b. mesosphere.
    c. thermosphere.
    d. troposphere.

14. Convection is defined as the
    a. transfer of energy across space.
    b. direct transfer of energy.
    c. trapping of heat near the Earth by gases.
    d. transfer of heat by currents.

15. Which of the following gases is *not* a greenhouse gas?
    a. water vapor
    b. nitrogen
    c. methane
    d. carbon dioxide

16. Liquid water turns into gaseous water vapor in a process called
    a. precipitation.
    b. convection.
    c. evaporation.
    d. condensation.

17. Currents at the surface of the ocean are moved mostly by
    a. heat.
    b. wind.
    c. salinity.
    d. the mixing of warm and cold water.

18. Which of the following statements about the biosphere is *not* true?
    a. The biosphere is a system closed to matter.
    b. Energy enters the biosphere in the form of sunlight.
    c. Nutrients in the biosphere must be continuously recycled.
    d. Matter is constantly added to the biosphere.

83

Assignment Guide

Section	Questions
1	1, 2, 6, 8, 11, 19, 26, 27, 31, 36
2	3, 4, 7, 12–16, 20, 21, 28, 34, 36
3	5, 9, 10, 17, 18, 22–25, 29, 30, 32, 33, 35, 36

ANSWERS

Using Key Terms

1. Sample answer: Earth's lithosphere is broken up into segments called tectonic plates.
2. Sample answer: Wind and water alter the landscape by a process called erosion.
3. Sample answer: The transfer of energy across space and in the atmosphere is called radiation.
4. Sample answer: Ozone is a molecule with three oxygen atoms that absorbs UV light in the stratosphere.
5. Sample answer: Salinity is the total quantity of dissolved solids in sea water.
6. The lithosphere refers to the crust and part of the upper mantle. The lithosphere is divided into tectonic plates. The asthenosphere is a layer at the top of the mantle that is solid yet plastic and flows slowly enough to allow plates to move on top of it.
7. Conduction is the direct transfer of energy, and convection is the transfer of heat by currents.
8. The crust is the outermost layer of the Earth that is made up of lower-density minerals. Beneath the crust is the mantle, which is composed of denser, iron-rich minerals.
9. In an open system, matter and energy enter and leave the system. In a closed system, energy enters the system but matter does not.

Understanding Key Ideas

10. d
11. b
12. c
13. a
14. d
15. b
16. c
17. b
18. d

Short Answer

19. As seismic waves travel through Earth's interior, they are altered by the nature of the material through which they travel. By measuring changes in the speed and direction of the waves as they penetrate the interior, scientists can infer what types of materials each layer is made of.

20. Gravity pulls gas molecules toward Earth's surface. Because of gravity, Earth's lower atmosphere is denser than its upper atmosphere.

21. Convection currents transport heat in the atmosphere through the interaction between warm and cool currents. As warm air rises into the atmosphere, it cools and eventually becomes cool enough to sink. As the cool air moves back toward Earth's surface, it is heated and rises once again.

22. Land near the ocean changes temperature less rapidly than land farther from the ocean because of the moderating effect of the ocean. Whereas land loses heat rapidly after the sun goes down, the ocean does not.

23. Life on Earth is confined to a narrow layer near Earth's surface because living things require a moderate temperature, liquid water, and a source of energy.

Short Answer

19. How do seismic waves give scientists information about Earth's interior?

20. Explain the effect of gravity on Earth's atmosphere.

21. Explain how convection currents transport heat in the atmosphere.

22. Why does land that is near the ocean change temperature less rapidly than land that is located farther inland?

23. Why is life on Earth confined to such a narrow layer near the Earth's surface?

Interpreting Graphics

The map below shows the different amounts of chlorophyll in the ocean. Chlorophyll is the pigment that makes plants and algae green. Chlorophyll identifies the presence of marine algae. The red and orange colors on the map show the highest amounts of chlorophyll, the blue and purple colors on the map show the smallest amounts of chlorophyll. Use the map to answer questions 24–25.

24. Is there a greater concentration of marine algae at location A or at location B?

25. What conclusion can you reach about conditions in the parts of the ocean where marine algae may prefer to live?

Chlorophyll Content

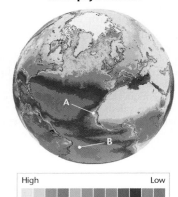

| High | | | | | | | | | | | Low |

Concept Mapping

26. Use the following terms to create a concept map: *geosphere, crust, mantle, core, lithosphere, asthenosphere,* and *tectonic plate.*

Critical Thinking

27. Making Predictions The eruption of Mount Pinatubo in 1991 reduced global temperature by several tenths of a Celsius degree for several years. Write a paragraph predicting what might happen to Earth's climate if several large-scale eruptions took place at the same time? **WRITING SKILLS**

28. Analyzing Processes Read about the heating of Earth's surface and the absorption of incoming solar radiation under the heading "Heating of the Atmosphere." How might the Earth be different if the Earth's surface absorbed greater or lesser percentages of radiation? **READING SKILLS**

29. Analyzing Processes Surface currents are deflected by continental landmasses. How might the pattern of Earth's surface currents change if the Earth had no landmasses? Where on the world ocean might the majority of warm surface currents be located? Where would the cold surface currents be located?

Cross-Disciplinary Connection

30. History Scientists believe that some human migration between distant landmasses may have taken place on rafts powered only by the wind and ocean currents. Look at Figure 22, which shows the Earth's surface currents. Hypothesize potential migratory routes these early seafarers may have followed.

Portfolio Project

31. Plotting Seismic Activity Most earthquakes take place near tectonic plate boundaries. Using the encyclopedia, the Internet, or another source, find at least 20 locations where major earthquakes took place during the 20th century. Plot these locations on a map of the world that shows Earth's tectonic plates. Did the majority of earthquakes occur at or near tectonic plate boundaries?

Interpreting Graphics

24. There is more marine algae at location A.

25. Marine algae live where there is sunlight, nutrients, and warm water. Regions that have heavy "vertical mixing" in the upper layer of the ocean tend to be areas with the greatest algae populations. Antarctica is such a region, because the chilled surface water there sinks and is replaced by upwelling water in which there are abundant nutrients.

Concept Mapping

26. Concept mapping answers are on pp. 667–672.

27. The larger and more violent an eruption is, the more material is ejected into the stratosphere, which blocks sunlight and lowers average global temperature. Simultaneous eruptions could potentially block out so much sunlight that Earth's temperature could grow cold enough to kill off some animal and plant species, particularly those species that are adapted to a narrow temperature range. Although humans can survive in a wide range of temperatures, their food supply could be compromised by crop failures and by the collapse of parts of the food chain.

MATH SKILLS

Use the graph below to answer questions 32–33.

32. Analyzing Data Rearrange the oceans in order of highest depth-to-area ratio to lowest depth-to-area ratio.

33. Making Calculations On the graph, you are given the average depths of the four oceans. From this data, calculate the average depth of the world ocean.

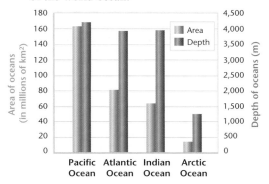

WRITING SKILLS

34. Communicating Main Ideas Describe the three important ways in which the movement of energy takes place in Earth's atmosphere.

35. Writing Persuasively Write a persuasive essay that explains why the Earth today should be regarded as a closed system for matter rather than an open system.

36. Outlining Topics Write a one-page outline that describes some of the important interactions that take place in the Earth system.

STANDARDIZED TEST PREP

For extra practice with questions formatted to represent the standardized test you may be asked to take at the end of your school year, turn to the sample test for this chapter in the Appendix.

READING SKILLS

Read the passage below, and then answer the questions that follow.

Researchers at Ohio State University have developed a video camera that photographs the foamy bubbles left when a wave breaks on a beach. Software analyzes images from the camera and uses the movement of the foam to calculate the speed and direction of currents along the shore. How do we know that the software produces an accurate picture of the currents? To test this process, researchers set up a video camera on the beach at Duck, North Carolina, where dozens of underwater sensors already measure currents directly. A comparison of the currents detected by the video camera and by the sensors showed a close match. The Ohio State University researchers believe data from the video camera would be even more accurate if the camera were directly above the breaking waves. The researchers' next step will be to mount the camera on a blimp suspended over a beach in Monterey, California.

1. According to the passage, which of the following conclusions is true?
   a. The video camera uses wave speed to calculate the direction and speed of currents.
   b. Underwater sensors are less accurate at measuring currents than video cameras.
   c. Video cameras do not measure currents directly.
   d. Underwater sensors detect the movement of foam.

2. What is the importance of foam in measuring currents?
   a. Foam can be measured directly by both video cameras and underwater sensors.
   b. The movement of foam can be used to calculate the direction and speed of a current.
   c. Foam from breaking waves can be detected by placing a video camera at any point on a beach.
   d. both (b) and (c)

36. Answers may vary. Students might discuss, for example, how the movement of winds in the atmosphere affects currents in the oceans (hydrosphere), which in turn affects aquatic life and the movement of nutrients.

Reading Skills

1. c
2. b

28. If greater percentages of radiation were absorbed by Earth's surface, the surface would be considerably hotter. Currently, Earth's surface deflects a portion of the solar radiation that reaches its surface. If less radiation was absorbed, Earth's surface would grow cooler. Even small temperature changes (either hotter or cooler) could severely compromise the viability of the biosphere.

29. If Earth had no landmasses, surface currents would move with the wind direction and the Coriolis effect. Warm surface currents would originate near the equator, and cold currents would originate near the poles.

Cross-Disciplinary Connection

30. Answers may vary.

Portfolio Project

31. Answers may vary.

Math Skills

32. In order of highest depth-to-area ratio to the lowest: Arctic, Indian, Atlantic, Pacific.

33. 4,250 (Pacific) + 4,000 (Atlantic) + 4,000 (Indian) + 1,250 (Arctic) = 13,500 ÷ 4 = 3,375 meters on average.

Writing Skills

34. Answers may vary but should include the movement of energy by convection, conduction, and radiation.

35. Answers may vary. Students should note that in open systems matter and energy enter and leave the system. In a closed system, energy enters the system but matter does not. The atoms that make up Earth, which is mostly a closed system, have been the same for billions of years. Students may cite asteroids and meteors as rare examples of how matter enters the Earth system today.

BEACHES

Teacher's Notes

Time Required

two 45-minute class periods

Lab Ratings

EASY —————→ HARD

TEACHER PREPARATION
STUDENT SETUP
CONCEPT LEVEL
CLEANUP

Skills Acquired

- Constructing Models
- Experimenting
- Identifying and Recognizing Patterns
- Interpreting

The Scientific Method

In this lab, students will:
- Make Observations
- Ask Questions
- Analyze the Results
- Draw Conclusions

Materials

The materials listed on the page are enough for groups of two to three students. After the students have made their plaster blocks in the prelab activity, tell them the plaster must be completely dry before they peel away the milk cartons. You may wish to have your first-period class set up the stream tables, and then use the same water throughout the day with your other classes. Have the last class of the day clean up the materials.

Objectives

▶ **Examine** models that show how the forces generated by wave action build, shape, and erode beaches.

▶ **USING SCIENTIFIC METHODS**
Hypothesize ways in which beaches can be preserved from the erosive forces of wave activity.

Materials

metric ruler
milk cartons, empty, small (2)
pebbles
plaster of Paris
plastic container (large) or long wooden box lined with plastic
rocks, small
sand, 5 to 10 lb
wooden block, large

▶ **Step 2** Use a wooden block to generate waves at the end of the container opposite the beach.

Beaches

Almost one-fourth of all of the structures that have been built within 150 m of the U.S. coastline, including the Great Lakes, will be lost to beach erosion over the next 60 years, according to a June 2000 report released by the Federal Emergency Management Agency (FEMA). The supply of sand for most beaches has been cut off by dams built on rivers and streams that would otherwise carry sand to the sea. Waves generated by storms also erode beaches. Longshore currents, which are generated by waves that break at an angle to a shoreline, transport sediment continuously and change the shape of a shoreline.

You will now observe a series of models. These models will help you understand how beaches can be both washed away and protected from the effects of waves and longshore currents.

Procedure

1. One day before you begin the investigation, make two plaster blocks. Mix a small amount of water with plaster of Paris until the mixture is smooth. Add five or six small rocks to the mixture for added weight. Pour the plaster mixture into the milk cartons. Let the plaster harden overnight. Carefully peel the milk cartons away from the plaster.

2. Prepare a wooden box lined with plastic or other similar large, shallow container. Make a beach by placing a mixture of sand and small pebbles at one end of the container. The beach should occupy about one-fourth the length of the container. See step 2. In the area in front of the sand, add water to a depth of 2 to 3 cm. Use the large wooden block to generate several waves by moving the block up and down in the water at the end of the container opposite the beach. Continue this wave action until about half the beach has moved. Record your observations.

3. Remove the water, and rebuild the beach. In some places, breakwaters have been built offshore in an attempt to protect beaches from washing away. Build a breakwater by placing two plaster blocks across the middle of the container. Using the metric ruler, leave a 4 cm space between the blocks. See step 3. Use a wooden block to generate waves. Describe the results.

4. Drain the water, and make a new beach along one side of the container for about half its length. See step 4. Using the wooden block, generate a series of waves from the same end of the container as the end of the beach. Record your observations.

5. Rebuild the beach along the same side of the container. A jetty or dike can be built out into the ocean to intercept and break up a longshore current. Make a jetty by placing one of the small plaster blocks in the sand. See step 5. As you did in the previous steps, use the wooden block to generate waves. Describe the results.

6. Remove the wet sand, and put it in a container. Dispose of the water. (Note: Follow your teacher's instructions for disposal of the sand and water. Never pour water containing sand into a sink.)

Analysis

1. **Describing Events** In step 2 of the procedure, what happened to the beach when water was first poured into the container? What happened to the particles of fine sand? Predict what would happen to the beach if it had no source of additional sand.

2. **Analyzing Results** In step 3 of the procedure, did the breakwater help protect the beach from washing away?

3. **Describing Events** What happened to the beach that you made in step 4 of the procedure? What happened to the shape of the waves along the beach?

4. **Analyzing Results** What effect did the jetty have on the beach that you made in step 5 of the procedure?

Conclusions

5. **Drawing Conclusions** What can be done to preserve a beach area from being washed away as a result of wave action and longshore currents?

6. **Drawing Conclusions** What can be done to preserve a beach area that has been changed as a result of excessive use by people?

Extension

1. **Building Models** Make a beach that would be in danger of being washed away by a longshore current. Based on what you have learned, build a model in which the beach would be preserved by a breakwater or jetties. Explain how your model illustrates ways in which longshore currents can be intercepted and broken up.

▶ **Step 3** Build a breakwater by placing two plastic blocks across the middle of the container.

▶ **Step 4** Make a beach lengthwise along one side of the container. The length of the beach should equal one-half the length of the container.

▶ **Step 5** Place one of the small plaster blocks in the sand to make a jetty.

87

Chapter Resource File
- Datasheets for In-Text Labs
- Lab Notes and Answers

CLASSROOM TESTED & APPROVED

Don Kanner
Lane Technical
High School
Chicago, Illinois

Tips and Tricks

Tell students to pour the water slowly into the stream tables to prevent alteration of the shoreline. Emphasize that the waves they generate in the stream tables should be a consistent size. If stream tables are not available, large shallow pans can be used.

Answers to Analysis

1. Fine sand and some pebbles were washed down into the water. The fine sand moved into deeper water. Without additional sand, it will turn into a pebble beach.

2. Yes, only a small part of the waves got through, and they disappeared quickly. The movement of the sand was reduced to practically nothing.

3. The sand on the beach moved into the water as the sand was moved by the current. The waves were refracted toward the shore.

4. Sand built up behind the jetty. Some sand remained in front of the jetty.

Answers to Conclusions

5. Beaches can be preserved with rock pilings and other jetties built out into the water. This would deter the longshore movement of sand. Dredging can replace beach sand lost by wave action and longshore currents.

6. Walkways can be built over dunes, protective vegetation can be planted, and houses can be built somewhere other than directly on the dunes.

Answers to Extension

1. Answers may vary. Students' models should include breakwaters or jetties, and students should explain how they protect the beach from currents. Students may also propose beach restoration. Beach restoration involves dredging sand offshore and transporting it to beaches.

EARTHQUAKE HAZARD MAP OF THE CONTIGUOUS UNITED STATES

Group Activity — **BASIC**

Earthquake Hazard Have students research major earthquakes that have occurred in the United States over the past several centuries. Have them pinpoint the epicenters of these earthquakes on the Earthquake Hazard Map of the Contiguous United States. Can students see a correlation between the epicenters of large earthquakes and the earthquake hazard levels as shown on the map? **LS** Visual

GEOLOGY —
● **CONNECTION** ▶ **GENERAL**

The San Andreas Fault The San Andreas fault is the boundary between the Pacific plate and the North American plate. The fault stretches approximately 1,100 km from Cape Mendocino in Northern California to the Salton Sea in Southern California. The Pacific side of the San Andreas fault is moving in a northwesterly direction at an average rate of 3.4 cm per year. This means that Los Angeles is moving toward San Francisco. Ask students, "If Los Angeles is located 550 km by air from San Francisco, how many years will it take for residents of Los Angeles to shake hands with residents in San Francisco?" (550 km × 100,000 = 55,000,000 cm ÷ 3.4 cm = 16,476,170 years) **LS** Logical

---

🖥️ **Transparencies**

TT Earthquake Hazard Map of the Contiguous United States

MAPS in action

EARTHQUAKE HAZARD MAP OF THE CONTIGUOUS UNITED STATES

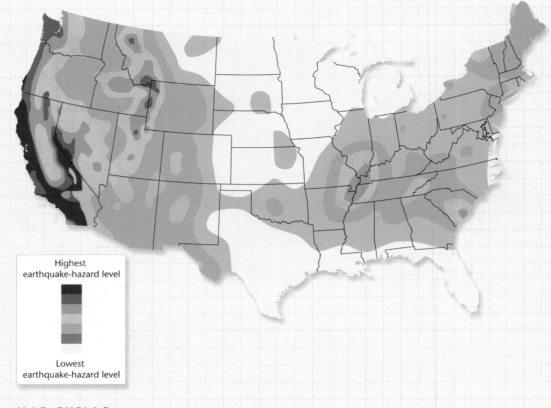

Highest
earthquake-hazard level

Lowest
earthquake-hazard level

MAP SKILLS

Use the earthquake-hazard map of the contiguous United States to answer the questions below.

1. **Using a Key** Which area of the contiguous United States has a very high earthquake-hazard level?

2. **Using a Key** Determine which areas of the contiguous United States have very low earthquake-hazard levels.

3. **Analyzing Relationships** In which areas of the contiguous United States would scientists most likely set up earthquake-sensing devices?

4. **Inferring Relationships** Most earthquakes take place near tectonic plate boundaries. Based on the hazard levels, where do you think a boundary between two tectonic plates is located in the United States?

5. **Forming a Hypothesis** The New Madrid earthquake zone passes through southeastern Missouri and western Tennessee and has experienced some of the most widely felt earthquakes in U.S. history. Yet this earthquake zone lies far from any tectonic plate boundary. Propose a hypothesis that would explain these earthquakes.

88

Answers to Map Skills

1. The west coast of the United States has the highest earthquake-hazard level.

2. Most of the rest of the United States has low earthquake-hazard levels. Exceptions are the New Madrid earthquake zone and the area around Charleston, South Carolina.

3. Scientists would most likely set up earthquake-sensing devices where there are the most earthquakes, primarily along the west coast of the United States.

4. A tectonic plate boundary is located along the west coast of California, where the North American plate and Pacific plate form a

boundary and along the coast of Oregon and Washington, where the Juan de Fuca plate is subducting under North America.

5. Although the New Madrid earthquake zone is located far from a tectonic plate boundary, an older active fault system may underlie this area.

SCIENCE & TECHNOLOGY

TRACKING OCEAN CURRENTS WITH TOY DUCKS

Scientists usually study ocean currents by releasing labeled drift bottles from various points and recording where they are found. However, only about 2 percent of drift bottles are recovered, so this type of research takes a long time. A large toy spill is helping scientists track surface currents in the Pacific Ocean.

Toys Ahoy!

In 1993, thousands of bathtub toys were found on Alaskan beaches. When oceanographers heard about this, they placed advertisements in newspapers up and down the Alaskan coast asking people who found the toys to call them. They discovered that in 1992 a container ship that was traveling northwest of Hawaii ran into a storm. Several containers were washed overboard and burst open. One of these held 29,000 plastic toys. Ten months later, the toys—blue turtles, yellow ducks, red beavers, and green frogs—began washing up near Sitka, Alaska. In the following years, toys began to be found farther north, in the Bering Sea. The map below shows where the containers went overboard and where the toys were found.

The Data in the Deep Blue Sea

Obviously, the toys had traveled east from where they were spilled. But what did this reveal about the currents in the North Pacific? The answer is not as obvious as it might seem. First, floating objects are moved by wind as well as by currents. The floating toys stuck up about 4 cm above the water, which may have caused them to be moved by the wind as well as by currents. The toys started out in cardboard and plastic packages. Did the packages make them sink when they were first released? To find the answer, scientists obtained some of the packaged toys from the manufacturer in China and dropped them in buckets of sea water. The glue in the packaging dissolved within a day and released the toys. So it was obvious that the toys had floated most of the way to where they were found.

Experiments showed how fast they moved under the influence of wind without any current. The toys had floated past a weather station where many drift bottles had been released and also past the place where 61,000 shoes had fallen off a ship in 1990. About two percent of the shoes were recovered in Alaska. Comparing data from the toys and the shoes with other data from as far back as 1946, the researchers concluded that the current across the northeast Pacific Ocean moves little from year to year. But the data showed that in 1990 and 1992 the current was unusually far north.

Data that help us understand ocean currents and many other natural processes come not just from scientific experiments. Data sometimes come from the most unusual sources.

▶ This map is a computer simulation that shows the possible trajectory of the toys and their estimated locations on certain dates as they floated across the Pacific Ocean from the point of the spill to recovery points in Alaska.

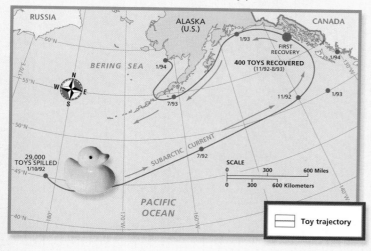

What Do You Think?

Take a look at Figure 22. If the toys continue to be carried by surface currents, where might they be found in the future? How might the height of the toys above the water's surface have influenced the speed at which they traveled?

⬛ **Internet** Activity — GENERAL

Tracking Cargo Spills Surface currents move in large, slow circles called *gyres*. In 1990, a Korean ship carrying a load of athletic shoes was bound for the United States. During a storm in the North Pacific Ocean, more than 60,000 shoes were washed overboard. Six months later, the shoes began to wash up on North American shores from Oregon to British Columbia. This accident turned out to be a big bonus to oceanographers. They created a computer model to predict where the gyres would carry the shoes next. True to the model, the shoes began to wash up in Hawaii three years later. With the help of dedicated beachcombers, this accident has enabled oceanographers to create a detailed map of surface currents in the Pacific Ocean. Have students look at a map of global ocean currents and identify the currents that carried the shoes. Students may also enjoy visiting Web sites that track the movements of other cargo spills.

89

Answers to What Do You Think?

If the toys keep being pushed along by surface currents, they might be carried through the Bering Sea and into the Arctic Ocean. The toys stuck up about 4 cm above the ocean's surface. This allowed them to be carried along by the wind as well as by the current, increasing their travel speed.

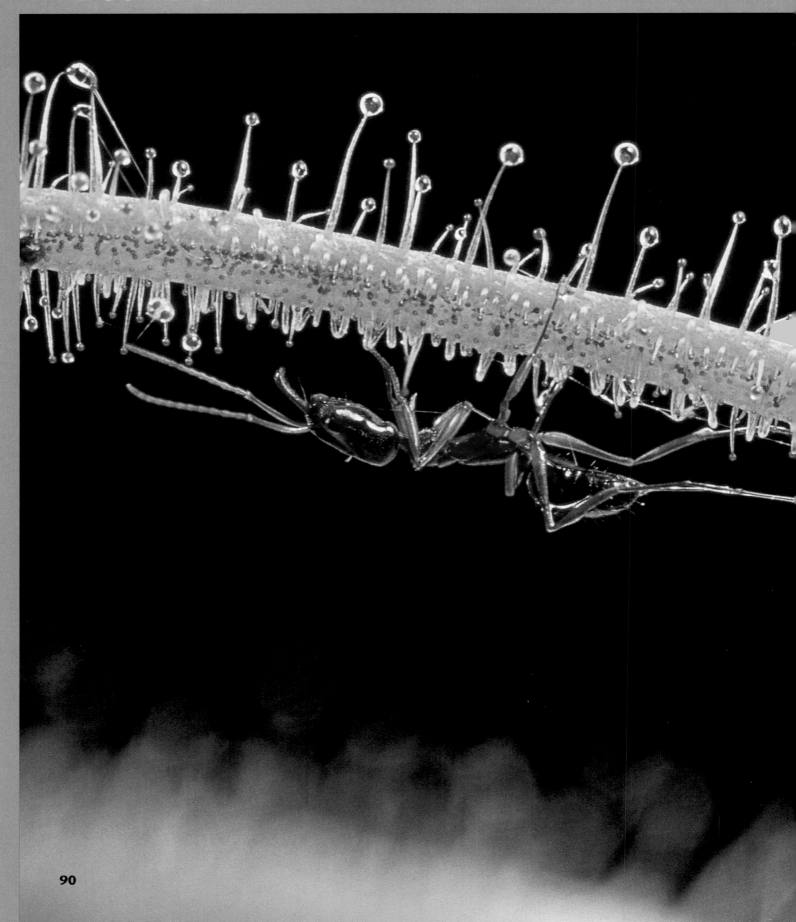

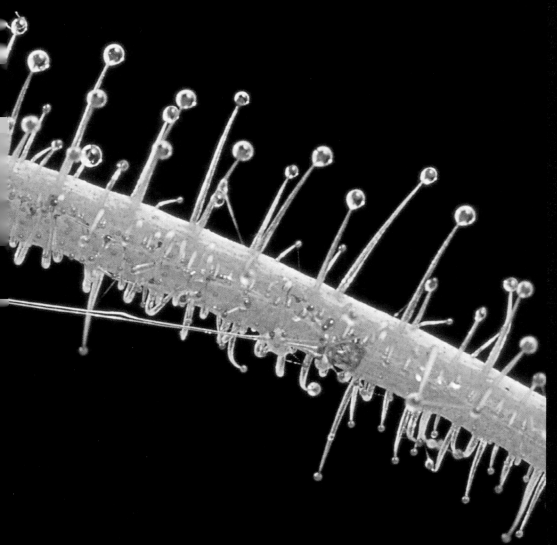

CHAPTER 4

The Organization of Life

CHAPTER 5

How Ecosystems Work

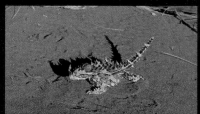

CHAPTER 6

Biomes

CHAPTER 7

Aquatic Ecosystems

This Australian plant called the *fork-leafed sundew* gets the nutrients that it needs to survive by dissolving insects that get stuck on its sticky tips.

The Organization of Life
Chapter Planning Guide

PACING	CLASSROOM RESOURCES	LABS, DEMONSTRATIONS, AND ACTIVITIES
BLOCKS 1, 2, 3 & 4 · 180 min pp. 92–96 **Chapter Opener**		
Section 1 Ecosystems: Everything Is Connected	**OSP** Lesson Plan * **CRF** Active Reading * BASIC **TT** Bellringer * **TT** Levels of Ecological Organization * **CD** Interactive Tutor Ecosystem Structure	**TE** Activity Connections Web, p. 93 GENERAL **TE** Demonstration Identifying Ecosystem Components, p. 94 ◆ BASIC **TE** Activity Ecosystem Connections, p. 94 ◆ GENERAL **TE** Group Activity Golf Course Impacts, p. 95 ◆ ADVANCED **TE** Internet Activity Self-sustaining Colonies, p. 95 GENERAL **SE** Inquiry Lab How Do Brine Shrimp Select a Habitat? pp. 112–113 ◆ **CRF** Datasheets for In-Text Labs * **CRF** CBL™ Probeware Lab * ◆ ADVANCED
BLOCKS 5 & 6 · 90 min pp. 97–101 **Section 2** Evolution	**OSP** Lesson Plan * **CRF** Active Reading * BASIC **TT** Bellringer * **TT** The Evolution of Thicker Fur in a Deer Population * **TT** The Evolution of Pesticide Resistance * **CD** Interactive Tutor Evolution	**TE** Group Activity Natural Variety, p. 97 GENERAL **TE** Group Activity Adaptations Worldwide, p. 98 GENERAL **TE** Demonstration Dog Breeds, p. 100 ◆ BASIC **TE** Internet Activity Gardens and Artificial Selection, p. 100 GENERAL **SE** Field Activity Artificial Selection, p. 101 **CRF** Math/Graphing Lab * ◆ GENERAL **CRF** Observation Lab * ◆ BASIC
BLOCKS 7 & 8 · 90 min pp. 102–107 **Section 3** The Diversity of Living Things	**OSP** Lesson Plan * **CRF** Active Reading * BASIC **TT** Bellringer * **TT** Classification of Living Things *	**TE** Group Activity Mushroom Walk, p. 103 ◆ GENERAL **TE** Activity Pond Protists, p. 104 ◆ GENERAL **SE** QuickLab Pollen and Flower Diversity, p. 105 **TE** Group Activity Angiosperms in the Classroom, p. 105 GENERAL **TE** Group Activity Pollinator Game, p. 106 GENERAL **CRF** Consumer Lab * ◆ BASIC **CRF** Long-Term Project * ◆ GENERAL

BLOCKS 9 & 10 · 90 min

Chapter Review and Assessment Resources

- **SE** Chapter Review pp. 109–111
- **SE** Standardized Test Prep pp. 642–643
- **CRF** Study Guide * ■ GENERAL
- **CRF** Chapter Test * ■ GENERAL
- **CRF** Chapter Test * ADVANCED
- **CRF** Concept Review * ■ GENERAL
- **CRF** Critical Thinking * ADVANCED
- **OSP** Test Generator
- **CRF** Test Item Listing *

Online and Technology Resources

go.hrw.com Holt Online Learning

Visit **go.hrw.com** for access to Holt Online Learning or enter the keyword **HE6 Home** for a variety of free online resources.

One-Stop Planner® CD-ROM

This CD-ROM package includes
- Lab Materials QuickList Software
- Holt Calendar Planner
- Customizable Lesson Plans
- Printable Worksheets
- ExamView® Test Generator
- Interactive Teacher Edition
- Holt PuzzlePro® Resources
- Holt PowerPoint® Resources

Holt Environmental Science Interactive Tutor CD-ROM

This CD-ROM consists of interactive activities that give students a fun way to extend their knowledge of environmental science concepts.

KEY

TE	Teacher Edition	**CRF**	Chapter Resource File	*	Also on One-Stop Planner
SE	Student Edition	**TT**	Teaching Transparency	■	Also Available in Spanish
OSP	One-Stop Planner	**CD**	Interactive CD-ROM	◆	Requires Advance Prep

ENRICHMENT AND SKILLS PRACTICE	SECTION REVIEW AND ASSESSMENT	CORRELATIONS
SE Pre-Reading Activity, p. 92 **TE** Using the Figure, p. 92 GENERAL		**National Science Education Standards**
TE Using the Figure Connecting Ecosystem Levels, p. 95 GENERAL	**SE** Section Review, p. 96 **TE** Reteaching, p. 96 BASIC **TE** Quiz, p. 96 GENERAL **TE** Alternative Assessment, p. 96 GENERAL **CRF** Quiz * ■ GENERAL	LS 4c
SE Case Study Darwin's Finches, pp. 98–99 **TE** Skill Builder Writing, p. 99 ADVANCED **SE** MathPractice Plumper Pumpkins, p. 100	**TE** Reteaching, p. 98 BASIC **SE** Section Review, p. 101 **TE** Reteaching, p. 101 BASIC **TE** Quiz, p. 101 GENERAL **TE** Alternative Assessment, p. 101 GENERAL **CRF** Quiz * ■ GENERAL	LS 3a LS 3b
SE Graphic Organizer Spider Map, p. 103 GENERAL **TE** Skill Builder Vocabulary, p. 105 GENERAL **SE** MathPractice Insect Survival, p. 106 **TE** Skill Builder Writing, p. 106 GENERAL **TE** Inclusion Strategies, p. 106 **SE** Making a Difference Butterfly Ecologist, pp. 114–115 **CRF** Map Skills Park Habitat * GENERAL	**TE** Homework, p. 103 GENERAL **SE** Section Review, p. 107 **TE** Reteaching, p. 107 BASIC **TE** Quiz, p. 107 GENERAL **TE** Alternative Assessment, p. 107 GENERAL **CRF** Quiz * ■ GENERAL	LS 3b LS 3e

 Guided Reading Audio CDs

These CDs are designed to help auditory learners and reluctant readers. (Audio CDs are also available in Spanish.)

 NSTA
www.scilinks.org

Maintained by the **National Science Teachers Association**

TOPIC: Ecosystems
SciLinks code: HE4027

TOPIC: Evolution
SciLinks code: HE4039

TOPIC: Invertebrates
SciLinks code: HE4057

 CNN Videos

Each video segment is accompanied by a Critical Thinking Worksheet.

Earth Science Connections

Segment 17 Birds from Dinosaurs?

Science, Technology & Society

Segment 1 Tapping into Yellowstone's Hot Springs

CHAPTER
4

Chapter Enrichment

This Chapter Enrichment provides relevant and interesting information to expand and enhance your classroom instruction of the chapter material.

SECTION 1 Ecosystems: Everything is Connected

▶ **A vacant lot can be an ecosystem**

A Leaf Cycle

When a leaf drops from a tree into a stream, it becomes a food source in a complex aquatic ecosystem. Initially, decomposers, such as bacteria and fungi, colonize the leaf and start to break it down. Stream insects may start to eat the bacteria and fungi on the leaf. Insects that comprise a functional feeding group called *shredders* (which include mayflies, caddisflies, and larval beetles) will start to physically break down the leaf. The action of the stream also contributes to breaking the leaf apart. When the leaf is broken into fine pieces, *collectors*, such as caddisflies, may catch it and eat it. Algae and river plants may also use dissolved nutrients from the leaf. All of the nutrients and energy from the leaf move through the system, as predator insects eat insects that used the leaf for food. When the predator insects emerge from the stream, birds may eat them. While nesting in trees, these birds excrete droppings that provide nutrients that can be used by the tree to form new leaves.

BRAIN FOOD

Male crickets chirp to attract mates and defend territories. You can calculate the temperature outside by counting cricket chirps. Just count the number of chirps heard in 15 seconds and add 40. That number will equal the approximate temperature in degrees Fahrenheit.

SECTION 2 Evolution

Sexual Selection

Through the process of sexual selection, traits may evolve that do not seem adaptive to the environment. Male peacocks' long, decorative tail feathers are an example of a sexually selected trait. To avoid predators, it would be more adaptive for male peacocks to have a shorter, less flamboyant tail. However, the tail is used to attract mates. An attractive male can mate with more females and pass on more of his genes to successive generations. Males may also develop traits that allow them to fight other males to win mates. The fiddler crab possesses such a trait. It has an enlarged claw, which is used in combat with other males.

BRAIN FOOD

Bacterial populations may evolve traits that help them survive treatment with antibiotics. But they also may resist death by forming a biofilm community. In a biofilm community, bacteria produce a slimy layer of sugars and proteins that forms a barrier against the outside environment. Antibiotics have difficulty penetrating a biofilm, so the encased bacteria can reproduce freely. Bacteria-filled biofilms can cause chronic infections when they form on medical devices implanted in the body.

Social Darwinism

The philosopher Herbert Spencer, a contemporary of Charles Darwin, came up with the phrase "survival of the fittest." Spencer started the idea of Social Darwinism,

▶ **A chameleon snatching its prey**

in which he traced the "evolution" of society from the simple to the complex. The concept of Social Darwinism has been simplified and misused to argue that certain races or sexes are superior or inferior to others. It has also been used as an excuse to ignore the poor, because helping them would allow their "inferior" genes to survive and thrive. Social Darwinism was the basis for the Nazi eugenics movement that sought to "weed out inferior individuals." Millions lost their lives due, in part, to this tragically misguided notion that certain groups of humans are better than others. This movement shows the danger of uncritically applying biological theories to social situations to advance the welfare of certain groups.

SECTION

3 The Diversity of Living Things

Discovering New Mammals

It is exciting when scientists find new species of plants or insects that have gone unnoticed, but it is rare when scientists find a new mammal. Between 1992 and 1997, scientists in Vietnam discovered three new species of deerlike mammals. The most recent was discovered in August 1997. This newly discovered mammal, the Truong Son muntjac, is about one-third of a meter (14 in.) tall, weighs 15.5 kg (34 lb), has a black coat and very short antlers, and barks like a dog. These mammals stay well hidden in the thick Vietnamese forest. The two other mammal species discovered recently are the Vu Quang ox and the giant muntjac.

BRAIN FOOD
The short-haired domestic cat is probably descended from a small, African wildcat called the *Caffre cat*. The ancient Egyptians tamed these cats around 2500 BCE. Long-haired domestic cats may have descended from an Asian wildcat. The tabby markings common in domestic cats today probably descended from these ancient ancestors.

BRAIN FOOD
Nematodes are one of the most diverse group of multicellular animals on Earth. There are an estimated 500,000 to 1,000,000 species of nematodes. These animals are commonly called "roundworms," but some can look like dragons, pinecones, or hedgehogs. Their heads also vary, looking like anything from a catfish head to a suction cup. Most are small, but some can reach 8 meters (about 9 yards) in length. Nematodes live in every ecosystem on Earth, and almost every plant or animal has some kind of nematode parasite.

Archaebacteria in Ecosystems

The archaebacteria kingdom consists of organisms that can live in extremely hot, salty, or completely anaerobic conditions.

Thermophiles ("heat lovers") occur in hot, acidic environments, such as deep sea thermal vents. These bacteria oxidize sulfur to grow.

Halophiles ("salt lovers") can grow in saltwater concentrations of up to 32 percent. Halophiles are found in the Great Salt Lake and the Dead Sea, as well as in foods preserved by salting.

Anaerobic bacteria can function without oxygen. They form methane by the reduction of carbon dioxide. These bacteria can be found in the digestive system of cows.

► Snow leopards live in the mountains of South Asia

Overview

Tell students that this chapter introduces the concept of ecosystems. In ecosystems, the biotic (living) and abiotic (non-living) components interact to form an interconnected system. Species adapt to their environment through the process of evolution by natural selection. The six-kingdom system of organization helps scientists to classify organisms and study their differences.

Using the Figure — GENERAL

In a coral reef ecosystem, reef-building coral combine with algae to produce a colony that gathers energy from the sun, and creates shelter for many organisms. Ask students to identify ways that this ecosystem is organized and to identify some of the possible interactions between organisms in the photo. (sponges grow on coral, animals hide in coral, water provides nutrients) Animals that use coral reefs include sponges, sea worms, crustaceans, mollusks, sea urchins, jellyfish, turtles, sea anemones, and many varieties of fish. Unfortunately, coral reefs are very sensitive to changing environmental conditions, and pollution, fishing, and boating can damage coral reef ecosystems.
LS Visual

PRE-READING ACTIVITY

Encourage students to use their FoldNote as a study guide to quiz themselves for a test on the chapter material. Students may want to create Layered Book FoldNotes for different topics within the chapter.

VIDEO SELECT

For information about videos related to this chapter, go to **go.hrw.com** and type in the keyword **HE4 EVOV**.

The Organization of Life

1 **Ecosystems: Everything Is Connected**

2 **Evolution**

3 **The Diversity of Living Things**

PRE-READING ACTIVITY

FOLDNOTES

Layered Book
Before you read this chapter, create the **FoldNote** entitled "Layered Book" described in the Reading and Study Skills section of the Appendix. Label the tabs of the layered book with "Ecosystem," "Population," "Community," and "Habitat." As you read the chapter, write information you learn about each category under the appropriate flap.

A coral reef is an ecosystem that contains a wide variety of species. How many different species can you find in this photograph?

92

Chapter Correlations | *National Science Education Standards*

LS 4c Organisms both cooperate and compete in ecosystems. The interrelationships and interdependencies of these organisms may generate ecosystems that are stable for hundreds or thousands of years. **(Section 1)**

LS 3a Species evolve over time. Evolution is the consequence of the interactions of (1) the potential for a species to increase its numbers, (2) the genetic variability of offspring due to mutation and recombination of genes, (3) a finite supply of the resources required for life, and (4) the ensuing selection by the environment of those offspring better able to survive and leave offspring. **(Section 2)**

LS 3b The great diversity of organisms is the result of more than 3.5 billion years of evolution that has filled every available niche with life forms. **(Section 2 and Section 3)**

LS 3e Biological classifications are based on how organisms are related. Organisms are classified into a hierarchy of groups and subgroups based on

similarities which reflect their evolutionary relationships. Species is the most fundamental unit of classification. **(Section 3)**

Ecosystems: Everything Is Connected

You may have heard the concept that in nature everything is connected. What does this mean? Consider the following example. In 1995, scientists interested in controlling gypsy moths, which kill oak trees, performed an experiment. The scientists removed most mice, which eat young gypsy moths, from selected plots of oak forest. The number of young gypsy moth eggs and young increased dramatically. The scientists then added acorns to the plots. Mice eat acorns. The number of mice soon increased, and the number of gypsy moths declined as the mice ate them as well.

This result showed that large acorn crops can suppress gypsy moth outbreaks. Interestingly, the acorns also attracted deer, which carried ticks. Young ticks soon infested the mice. Wild mice carry the organism that causes Lyme disease. Ticks can pick up the organism when they bite mice. Then the ticks can bite and infect humans. This example shows that in nature, things that we would never think were connected—mice, acorns, ticks, and humans—can be linked to each other in a complex web.

Defining an Ecosystem

The mice, deer, moths, oak trees, and ticks in the previous example are all part of the same ecosystem. An **ecosystem** (EE koh SIS tuhm) is all of the organisms living in an area together with their physical environment. An oak forest is an ecosystem. The coral reef on the opposite page is an ecosystem. Even a vacant lot, as shown in **Figure 1**, is an ecosystem.

Objectives

▶ Distinguish between the biotic and abiotic factors in an ecosystem.
▶ Describe how a population differs from a species.
▶ Explain how habitats are important for organisms.

Key Terms

ecosystem
biotic factor
abiotic factor
organism
species
population
community
habitat

📧 internet connect

www.scilinks.org
Topic: Ecosystems
SciLinks code: HE4027

SCiLINKS. Maintained by the National Science Teachers Association

Figure 1 ▶ This vacant lot is actually a small ecosystem. It includes various organisms, including plants and insects, as well as soil, air, and sunlight.

93

REAL-LIFE CONNECTION — GENERAL

Complex Systems Use the following example to illustrate the complexity and interconnectedness of the components within an ecosystem. In San Diego, California, a marsh habitat, home to two endangered bird species, was destroyed to build a freeway. To get a permit to build the freeway, the city had to agree to "rebuild" the ecosystem for the birds. After five years and $500,000, scientists and officials found that replacing an ecosystem is something that we do not know much about.

For example, when the endangered birds were released into their re-created ecosystem, they would not nest because the marsh grass was not tall enough. The grass was shorter because a tiny predator beetle that fed on marsh-grass-eating insects was not present in the new ecosystem. Without the beetle to control the insect population, the marsh grass could not grow to its full height. This was only one of many problems that the city faced in its ecosystem re-creation project.

SECTION 1

Focus

Overview

Before beginning this section, review with your students the Objectives in the Student Edition. This section will introduce students to the different components that make up an ecosystem and how populations and communities are structured into ecosystems.

🔊 Bellringer

Ask students to think about all the things they need for survival. (Answers may vary, but should include at least food, water, oxygen, and shelter.) Ask them to think about the kinds of ecosystems that might produce these necessities.

Motivate

Activity ——————— GENERAL

Connections Web Draw a "connections web" on the board. Start with a common animal, such as a blue jay, a raccoon, or a wasp. Write its name on the board and circle it. Then have students name interactions that this animal has with other plants or animals in its environment. For instance, a blue jay eats different insects, so you could draw a line connecting the bird to an insect. Extend the connections and include abiotic factors (the insect feeds on a plant, which uses sunlight). Continue until the web becomes complex. Introduce the idea of an ecosystem, emphasizing that your web represents only a fraction of the interactions in a natural ecosystem at any given time. **LS** Visual

Chapter Resource File

• Lesson Plan
• Active Reading **BASIC**
• Section Quiz **GENERAL**

🖥 Transparencies

TT Bellringer

Demonstration — BASIC

Identifying Ecosystem Components Bring in some magazines that have photos of natural systems. Pick out three pictures that represent different ecosystems. Show each picture to the class, and have them brainstorm all the components of each ecosystem. Write the components of each system in its own section on the board. After the class is finished naming the components, compare the systems. Identify the components that are common to all systems. Discuss why some of the components are needed in all ecosystems. **LS** Visual

Activity — GENERAL

Ecosystem Connections In this exercise, students will learn how forest and stream ecosystems are linked. Bring in some pre-assembled wooden frames and some window screen. Have students staple the screen to the frames to form a frame that will catch and hold leaves. Take these materials out to a stream area with trees next to it. (Make sure the area is somewhat isolated, so the materials are not disturbed.) Have students tie the screens to vegetation that is close to the stream edge. Tell them to try to keep the screens flat. Also have them attach a note to the screens identifying them as a science project. Take students out to the stream every week to collect any material that has been caught in the screens. Have students bring the material back to the classroom to identify and weigh it. **LS** Kinesthetic

Figure 2 ▶ Like all ecosystems, this desert in France contains certain basic components. What components can you identify?

The Living Soil Soil, which is part of nearly all ecosystems on land, is formed in part by living organisms, which break down dead leaves and organisms. Small, plantlike organisms even help break down rocks!

Figure 3 ▶ This caribou is a biotic factor in a cold, northern ecosystem in Denali National Park, Alaska.

94

Ecosystems Are Connected People often think of ecosystems as isolated from each other, but ecosystems do not have clear boundaries. Things move from one ecosystem into another. Soil washes from a mountain into a lake, birds migrate from Michigan to Mexico, and pollen blows from a forest into a field.

The Components of an Ecosystem

In order to survive, ecosystems need five basic components. These are energy, mineral nutrients, water, oxygen, and living organisms. As shown in **Figure 2**, plants and rock are two of the most obvious components of most land ecosystems. The energy in most ecosystems comes from the sun.

To appreciate how all of the things in an ecosystem are connected, think about how a car works. The engine alone is made up of hundreds of parts that all work together. If even one part breaks, the car might not run. Likewise, if one part of an ecosystem is destroyed or changes, the entire system may be affected.

Biotic and Abiotic Factors An ecosystem is made up of both living and nonliving things. **Biotic factors** are the living and once living parts of an ecosystem, including all of the plants and animals. Biotic factors include dead organisms, dead parts of organisms, such as leaves, and the organisms' waste products. The biotic parts of an ecosystem interact with each other in various ways. They also interact with the **abiotic** (ay bie AHT ik) **factors,** the nonliving parts of the ecosystem. Abiotic factors include air, water, rocks, sand, light, and temperature. **Figure 3** shows several biotic and abiotic factors in an Alaskan ecosystem.

Scientists organize living things into various levels. **Figure 4** shows how an ecosystem fits into the organization of living things. The illustration shows the different levels of ecological organization, from the individual organism to the biosphere.

BRAIN FOOD

Diverse Temperate Forests The soil of a temperate forest has a biological diversity that rivals the soil found in tropical rain forests. In fact, invertebrates in forest soils may be the most important factor in determining the long-term productivity of a forest. Soil arthropods include beetles, centipedes, pseudoscorpions, springtails, and mites. Have students collect several 20 cm^3 soil samples and use a microscope to examine the samples for soil arthropods.

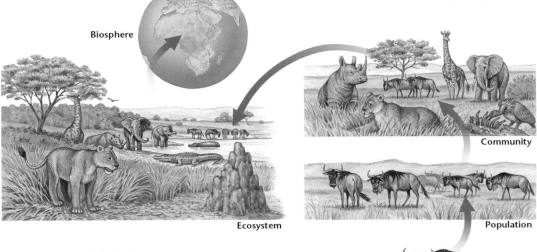

Biosphere

Community

Population

Organism

Figure 4 ▶ An individual organism is part of a population, a community, an ecosystem, and the biosphere.

Ecosystem

Organisms An **organism** is an individual living thing. You are an organism, as is an ant crawling across the floor, an ivy plant on the windowsill, and a bacterium in your intestines.

A **species** is a group of organisms that are closely related and that can mate to produce fertile offspring. All humans, for example, are members of the species *Homo sapiens*, while all black widow spiders are members of the species *Latrodectus mactans*. Every organism is a member of a species.

Populations Members of a species may not all live in the same place. Field mice in Maine and field mice in Florida will never interact even though they are members of the same species. An organism lives as part of a population. A **population** is all the members of the same species that live in the same place at the same time. For example, all the field mice in a corn field make up one population of field mice.

An important characteristic of a population is that its members usually breed with one another rather than with members of other populations. The bison in **Figure 5** (right) will usually mate with another member of the same herd, just as the wildflowers (left) will usually be pollinated by other flowers in the same field.

Figure 5 ▶ The two populations shown here are a population of purple-flowered musk thistle (left) and a herd of bison (right).

95

Reteaching — BASIC

Finding Interactions Have students consider the following organisms: honeybees, sunflowers, earthworms, red-winged blackbirds, and moles. Ask students to brainstorm and draw a possible ecosystem using these organisms. Ask them to label at least three interactions. (Possible interactions include: honeybees pollinate sunflowers, which produce seeds for blackbirds; earthworms process soil for sunflowers; moles eat earthworms; sunflowers use water, air, and soil.) **LS** Logical

Quiz — GENERAL

1. Think of your favorite animal. What are the components of its habitat that allow it to survive? (Answers may vary.)

2. How could soil and a mouse interact in an ecosystem? (The mouse could dig a burrow in the soil. The soil sustains plants and fungi that provide food for the mouse. When the mouse dies, it provides food for the microbes in the soil.)

Alternative Assessment — GENERAL

Organism Biographies Ask students to pick their favorite organism and write a "biography" for that organism that includes three chapters: one on habitat, a second on the survival of that organism's population within the community, and a third on abiotic characteristics that are crucial to the survival of the organism. Invite students to include their "biography" in their **Portfolio.** **LS** Verbal

Communities An organism does not live alone and neither does a population. Every population is part of a **community,** a group of various species that live in the same place and interact with each other. A pond community, for example, includes all of the populations of plants, fish, and insects that live in and around the pond. All of the living things in an ecosystem belong to one or more communities.

The most obvious difference between communities is the types of species they have. Land communities are often dominated by a few species of plants. In turn, these plants determine what other organisms live in that community. For example, the most obvious feature of a Colorado forest might be its ponderosa pine trees. This pine community will have animals, such as squirrels, that live in and feed on these trees.

Figure 6 ▶ Salamanders, such as this red backed salamander, live in habitats that are moist and shaded.

Habitat

The squirrel discussed above lives in a pine forest. All organisms live in particular places. The place an organism lives is called its **habitat.** A howler monkey's habitat is the rain forest, a cactus's habitat is a desert, and a waterlily's habitat is a pond. The salamander shown in **Figure 6** is in its natural habitat, the damp forest floor. An organism's habitat may be thought of as its "address."

Every habitat has specific characteristics that the organisms that live there need to survive. A coral reef contains sea water, coral, sunlight, and a wide variety of other organisms. If any of these factors change, then the habitat changes.

Organisms tend to be very well suited to their natural habitats. Indeed, animals and plants usually cannot survive for long periods of time away from their natural habitat. For example, a fish that lives in the crevices of a coral reef will die if the coral reef is destroyed.

SECTION 1 Review

1. **List** the abiotic and biotic components you see in the northern ecosystem in Figure 3.

2. **Describe** a population not mentioned in this section.

3. **Describe** which factors of an ecosystem are not part of a community.

4. **Explain** the difference between a population and a species.

CRITICAL THINKING

5. **Analyzing Relationships** Write your own definition of the term *community,* using the terms *biotic factors* and *abiotic factors.* **WRITING SKILLS**

6. **Understanding Concepts** Why might a scientist say that an animal is becoming rare because of habitat destruction?

96

Organisms tend to be well suited to where they live and what they do. **Figure 7** shows a chameleon (kuh MEEL ee uhn) capturing an insect. Insects are not easy to catch, so how does the chameleon do it? Chameleons can change the color and pattern of their skin, and then blend into their backgrounds. Their eyes are raised on little, mobile turrets that enable the lizards to look around without moving. An insect is unlikely to notice such an animal sitting motionless on a branch. When the insect moves within range, the chameleon shoots out an amazingly long tongue to grab the insect, while the chameleon's big hind feet hold it securely to the branch.

Evolution by Natural Selection

How do organisms become so well suited to their environments? In 1859, English naturalist Charles Darwin proposed an answer. Darwin observed that organisms in a population differ slightly from each other in form, function, and behavior. Some of these differences are *hereditary* (huh RED i TER ee)—that is, passed from parent to offspring. Darwin proposed that the environment exerts a strong influence over which individuals survive to produce offspring. Some individuals, because of certain traits, are more likely to survive and reproduce than other individuals. Darwin used the term **natural selection** to describe the unequal survival and reproduction that results from the presence or absence of particular traits.

Darwin proposed that over many generations natural selection causes the characteristics of populations to change. A change in the genetic characteristics of a population from one generation to the next is known as **evolution.**

Objectives

► Explain the process of evolution by natural selection.
► Explain the concept of adaptation.
► Describe the steps by which a population of insects becomes resistant to a pesticide.

Key Terms

natural selection
evolution
adaptation
artificial selection
resistance

Connection to Geology

Darwin and Fossils In the 1800s, fossil hunting was a popular hobby. The many fossils that people found started arguments about where fossils came from. Darwin's theory of evolution proposed that fossils are the remains of extinct species from which modern species evolved. When his book on the theory of evolution was first published in 1859, it became an immediate bestseller.

internet connect

www.scilinks.org
Topic: Evolution
SciLinks code: HE4039

SCiLINKS. Maintained by the National Science Teachers Association

Figure 7 ► A chameleon catches an unsuspecting insect that has strayed within range of the lizard's long, fast-moving tongue.

97

Chapter Resource File

• Lesson Plan
• Active Reading BASIC
• Section Quiz GENERAL

Transparencies

TT Bellringer

PALEONTOLOGY
CONNECTION ──── ADVANCED

Evolution in the Fossil Record Have students do research on either the fossil *Archaeopteryx* (thought to be the first feathered organism and the precursor to birds) or the ancient precursors of the horse. Both organisms are good examples of evolution represented in the fossil record. Encourage students to include their research in their **Portfolio.** LS Intrapersonal

SECTION 2

Focus

Overview

Before beginning this section, review with your students the Objectives in the Student Edition. In this section, students learn how organisms become adapted to their environments through the process of evolution by natural selection.

Bellringer

Have students look at **Figure 7** and write down the characteristics they think help the chameleon when it hunts. Ask them to compare their thoughts to the information in the first paragraph of the section.

Motivate

Group Activity ──── GENERAL

Natural Variety Members of a population naturally vary from one another. Have pairs measure each other's shoe size in centimeters. On the board, create a table to record the data for each student. Then, have students construct a bar graph using the results. Explain to students that Darwin, in his theory of evolution used examples of natural variation within a population to show how individuals with characteristics that are better suited to their environment have a better chance of survival. These individuals can produce more surviving offspring, and the more suitable trait becomes more common. Ask students if they think a bigger or smaller shoe size could be a more suitable trait in a human population. Ask them to think of characteristics that might have been advantageous to the survival of early humans. (Sample answers: problem-solving capabilities, the ability to cooperate with other humans, or the ability to recognize enemies)
LS Interpersonal

How Evolution Occurs Ask students, "If a squirrel starts to nest under your desk, and this allows it to survive and produce more offspring than other squirrels in the area, has the squirrel evolved?" (No, it has just changed its behavior; an individual cannot evolve.)
LS Logical

Group Activity — GENERAL

Adaptations Worldwide Assign student groups to the following areas: African Desert, South American Tropical Rainforest, North American Prairie, Australian Outback, Lake Baikal in Siberia, Alaskan Tundra. Have groups brainstorm adaptations that they think would suit plants and animals that live in their assigned region. Initially, they do not have to think of actual species, just the types of organisms and their adaptations. Discuss the adaptations in class. Then, ask the groups to do some research to find representative plants and animals from their region. Have each group create a poster with their guesses about adaptations on the top, and pictures of the real organisms with descriptions of their adaptations below the guesses. **LS Interpersonal**
Co-op Learning

Transparencies

TT The Evolution of Thicker Fur in a Deer Population

Nature Selects Darwin thought that nature selects for certain traits, such as sharper claws or lighter feathers, because organisms with these traits are more likely to survive and reproduce. Over time, the population includes a greater and greater proportion of organisms with the beneficial trait. As the populations of a given species change, so does the species. **Table 1** summarizes Darwin's

Table 1 ▼

Evolution by Natural Selection	
1. Organisms produce more offspring than can survive.	In nature, organisms have the ability to produce more offspring than can survive to become adults.
2. The environment is hostile and contains limited resources.	The environment contains things and situations that kill organisms, and the resources needed to live, such as food and water, are limited.
3. Organisms differ in the traits they have.	The organisms in a population may differ in size, coloration, resistance to disease, and so on. Much of this variation is inherited.
4. Some inherited traits provide organisms with an advantage.	Some inherited traits give organisms an advantage in coping with environmental challenges. These organisms are more likely to survive longer and produce more offspring; they are "naturally selected for."
5. Each generation contains proportionately more organisms with advantageous traits.	Because organisms with more advantageous traits have more offspring, each generation contains a greater proportion of offspring with these traits than the previous generation did.

CASE STUDY

Darwin's Finches

▶ Notice the beaks in the two species of Darwin's finches. What do you think the finches eat?

Before Charles Darwin formulated his theory of evolution, he sailed around the coast of South America. The plants and animals he saw had a great effect on his thinking about how modern organisms had originated. He was surprised by the organisms he saw on islands because they were often unusual species found nowhere else.

He was particularly impressed by the organisms in the Galápagos Islands, an isolated group of volcanic islands in the Pacific Ocean west of Ecuador. The islands contain 13 unique species of birds, which have become known as Darwin's finches. All the species look gener-

ally similar, but each species has a specialized bill adapted to eating a different type of food. Some species have large, parrotlike bills adapted to cracking big seeds, some species have slim bills that are used to sip nectar from flowers, and some species have even become insect eaters. Darwin speculated that all the Galápagos finches had evolved from a single species of seed-eating finch that found its way to the islands from the South American mainland. Populations of the finches became established on the various islands, and those finches were able to eat what they found on their island in order to survive.

Princeton University biologists Peter and Rosemary Grant have spent 25 years studying Darwin's finches on Daphne Major, one of the Galápagos Islands. Here, one

98

MISCONCEPTION ALERT

Survival of the Fittest Ask students what the term "survival of the fittest" means to them. They may say that when two organisms fight each other to survive, the "fittest" one will win the contest. This is not the correct meaning of the term. The "fittest" individual, the one that survives to pass its genes on, is the one most adapted to its current or changing environment. This individual may belong to a species of birds with small bills that can only crack small seeds. If smaller seeds become less available, and a few individuals have slightly larger bills that allow them to crack bigger seeds, those individuals are more likely to survive and pass their genes on to future generations. This example involves no fighting. Instead, it involves the spread of advantageous characteristics through a population.

theory of evolution by natural selection. An example of evolution is shown in **Figure 8**, in which a population of deer became isolated in a cold area. Many died, but some had genes for thicker, warmer fur. These deer were more likely to survive, and their young with thick fur were also more likely to survive to reproduce. The deer's thick fur is an **adaptation,** an inherited trait that increases an organism's chance of survival and reproduction in a certain environment.

Figure 8 ▶ These steps show the evolution of thicker fur in a population of deer.

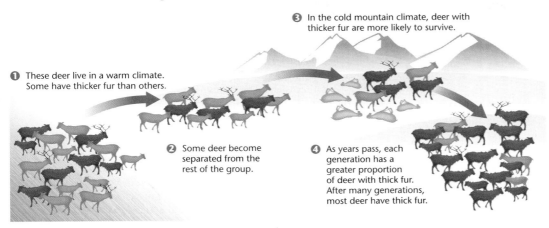

❸ In the cold mountain climate, deer with thicker fur are more likely to survive.

❶ These deer live in a warm climate. Some have thicker fur than others.

❷ Some deer become separated from the rest of the group.

❹ As years pass, each generation has a greater proportion of deer with thick fur. After many generations, most deer have thick fur.

species, the medium ground finch, has a short, stubby beak and eats seeds as well as a few insects. The ground finches have few predators. The Grants found that the main factor that determined whether a finch lived or died was how much food was available. During a long drought in 1977, many plants died and the small seeds that the finches

eat became scarce. Only finches that had large beaks survived. Large beaks allowed them to eat larger seeds from the larger plants that had survived the drought.

The finches that survived the drought passed their genes for large beaks to their offspring. Two years later, the Grants found that the beaks of medium ground finches on Daphne Major were nearly 4 percent larger, on average, than they were before the drought. The Grants had observed evolution occurring in birds over a short period of time, something that had seldom been seen before.

The Galápagos Islands are well suited for research on evolution because the islands are strongly influenced by El Niño and La Niña weather patterns.

These weather patterns produce alternating periods of very wet and dry weather in a relatively short period of time. The weather determines which plants live and which plants die. Then, this effect exerts selective pressure on the animals that depend on particular plants for food or for places to reproduce.

CRITICAL THINKING

1. Making Inferences What is the shortest period in which a population of Darwin's finches can evolve?

2. Analyzing Relationships Would you expect that the finches that evolved bigger beaks in this study might one day evolve smaller beaks?

99

Cultural Awareness

Crop Origins in the New World Teosinte, a wild grass, was domesticated in Mexico more than 7,000 years ago, and eventually became our modern corn. Other important crops that were probably first cultivated in Mexico include cotton, tomatoes, chili peppers, tobacco, cacao, pineapple, squash, and avocadoes. Potatoes, tomatoes, and peanuts may have originated in Peru and were brought to Mexico by human travelers.

Demonstration —— BASIC

Dog Breeds Ask students why they think dogs were bred to appear so different. (Dogs were bred to do different jobs, requiring different sizes and personalities.) Bring in a book of dog breeds from the local library, and show a few examples of retrieving, hunting, working, and companion dogs. Ask students to describe the characteristics of each dog type that make it suitable for its work. (Sample answer: Some small terriers were bred to catch rats, so they are small, fast, and tenacious.) If students have a dog at home, have them research the kind of work for which their dog was bred and compare this work with their dog's characteristics. **LS** Visual

Internet Activity —— GENERAL

Gardens and Artificial Selection
Bring in some flower and vegetable seed catalogs. Have students each pick a flower or vegetable and research the plant from which it originated. Ask them to print a picture of the original plant, if it is available, and to compare the differences between the artificially-selected plant and the original. Ask students what characteristic was selected for in each case. (It is usually the size of the fruit or flower.) As an extension, grow a few heirloom plants for students to observe during the school year. **LS** Visual   English Language Learners

Figure 9 ▶ This Hawaiian honeycreeper is using its curved beak to sip nectar from a lobelia flower.

MATH**PRACTICE**

Plumper Pumpkins
Each year a farmer saves and plants only the seeds from his largest pumpkins. If he starts with pumpkins that average 5 kg and each year grows pumpkins 3 percent more massive, on average, than those he grew the year before, what will be the average mass of his pumpkins after 10 years?

Coevolution Organisms evolve adaptations to other organisms as well as to their physical environment. The process of two species evolving in response to long-term interactions with each other is called *coevolution* (koh EV uh LOO shuhn). One example is shown in **Figure 9**. The honeycreeper's beak is long and curved, which lets it reach the nectar at the base of the long, curved flower. The flower has evolved structures that ensure that the bird gets pollen from the flower on its head as it sips nectar. When the bird moves to another flower, some of the pollen will rub off. In this way, the bird helps lobelia plants reproduce. The honeycreeper's adaptation is a long, curved beak. The plant has two adaptations. One is sweet nectar, which attracts the birds. The other is a flower structure that forces pollen onto a bird's head when the bird sips nectar.

Evolution by Artificial Selection

Many populations of plants and animals do not live in the wild but instead are cared for by humans. People control how these plants and animals reproduce and therefore how they evolve. The wolf and the Chihuahua in **Figure 10** are closely related. Over thousands of years, humans bred the ancestors of today's wolves to produce the variety of dog breeds we now have. The selective breeding of organisms by humans for specific characteristics is called **artificial selection.**

The fruits, grains, and vegetables we eat were also produced by artificial selection. Humans saved the seeds from the largest, sweetest fruits and most nutritious grains. By selecting for these traits, farmers directed the evolution of crop plants. As a result, crops produce fruits, grains, and roots that are larger, sweeter, and often more nutritious than their wild relatives do. Native Americans cultivated the ancestor of today's corn from a grasslike plant in the mountains of Mexico. Modern corn is very different from the wild plant that was its ancestor.

Figure 10 ▶ As a result of artificial selection, the Chihuahua on the right looks very different from its wolf ancestor on the left.

100

BRAIN FOOD

Wasp and Fig Coevolution Figs and tiny wasps called *fig wasps* have coevolved a unique relationship. Fig flowers are completely enclosed in a structure called a *syconium.* Fig wasp eggs hatch in a syconium, the wasp eats the ovule on which the egg hatches, and the wasp matures to mate there. The male fig wasp lives and dies within the structure, but the female wasp chews its way out. A female, covered with pollen after mating, enters another syconium through a pore in the structure. Once inside, the female spreads pollen on some of the flowers, lays eggs, and dies, thereby continuing the cycle.

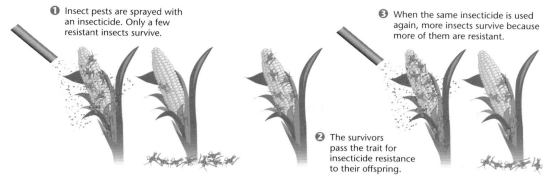

❶ Insect pests are sprayed with an insecticide. Only a few resistant insects survive.

❸ When the same insecticide is used again, more insects survive because more of them are resistant.

❷ The survivors pass the trait for insecticide resistance to their offspring.

Figure 11 ▶ The evolution of resistance to a pesticide starts when the pesticide is sprayed on the corn. Most of the insects are killed, but a few resistant ones survive. After each spraying, the insect population contains a larger proportion of resistant organisms.

Evolution of Resistance

Sometimes humans cause populations of organisms to evolve unwanted adaptations. You may have heard about insect pests that are resistant to pesticides and about bacteria that are resistant to antibiotics. What is resistance, and what does it have to do with evolution?

Resistance is the ability of one or more organisms to tolerate a particular chemical designed to kill it. An organism may be resistant to a chemical when it contains a gene that allows it to break the chemical down into harmless substances. By trying to control pests and bacteria with chemicals, humans promote the evolution of resistant populations.

Pesticide Resistance Consider the evolution of pesticide resistance among corn pests, as shown in **Figure 11.** A pesticide is sprayed on corn to kill grasshoppers. Most of the grasshoppers die, but a few survive. The survivors happen to have a gene that protects them from the pesticide. The surviving insects pass on the gene to their offspring. Each time the corn is sprayed, the insect population changes to include more and more resistant members. After many sprayings, the entire population may be resistant, making the pesticide useless. The faster an organism reproduces, the faster its populations can evolve.

FIELD ACTIVITY

Artificial Selection Look around your school grounds and the area around your home for possible examples of artificial selection. Observe and report on any examples you can find.

Dogs are one example of artificial selction mentioned in this chapter, but you will probably find many more plant examples. Record your observations in your *EcoLog.*

SECTION 2 Review

1. **Explain** what an adaptation is, and provide three examples.

2. **Explain** the process of evolution by natural selection.

3. **Describe** one way in which artificial selection can benefit humans.

4. **Explain** how a population of insects could become resistant to a pesticide.

CRITICAL THINKING

5. **Understanding Concepts** Read the description of evolution by natural selection in this section and describe the role that the environment plays in the theory. READING SKILLS

6. **Identifying Relationships** A population of rabbits evolves thicker fur in response to a colder climate. Is this an example of coevolution? Explain your answer.

101

Answers to Section Review

1. An adaptation is an inherited trait that increases an organism's chances of survival and reproduction, such as thick fur, sharp claws, or a sticky tongue.

2. Organisms within populations differ in their traits. These differences make some organisms more likely to survive and reproduce in their environment than others in their population are. The genetic characteristics of populations change over time in response to these likelihoods, which is a process called *evolution.*

3. Answers may vary. Sample answer: Artificially selecting the most nutritious rice can help nourish more people.

4. When a pesticide is sprayed on insects, many of them die. But the insects that survive have traits that allow them to resist the pesticide. Those resistant insects produce offspring that are resistant, so the population becomes resistant.

5. If the environment contains limited resources, organisms that have certain characteristics or traits may be more likely to survive and reproduce in the environment than other organisms in the same environment.

6. No. This is an adaptation to the physical environment, so it is natural selection. Coevolution involves two organisms that have evolved mutually beneficial traits.

Overview

Before beginning this section, review with your students the Objectives in the Student Edition. This section describes the diversity of living organisms and the way that scientists classify organisms.

Bellringer

Ask students, "Do you know how a scientist would classify you?" (Kingdom: Animalia; Phylum: Chordata; Class: Mammalia; Order: Primates; Family: Hominidae; Genus: *Homo*; Species: *Homo sapiens*.)

Motivate

Identifying Preconceptions —— BASIC

Extreme Bacteria Ask students to list places where bacteria grow. Then ask if they think that bacteria grow in hot water. They may say no, because common knowledge dictates that washing your hands in hot water kills germs. Explain that Archaebacteria live in extreme conditions, such as in scalding hot springs. **LS** Intrapersonal

Chapter Resource File

- **Lesson Plan**
- **Active Reading** BASIC
- **Section Quiz** GENERAL

Transparencies

TT Bellringer

TT Classification of Living Things

SECTION 3
The Diversity of Living Things

Objectives

▶ Name the six kingdoms of organisms and identify two characteristics of each.
▶ Explain the importance of bacteria and fungi in the environment.
▶ Describe the importance of protists in the ocean environment.
▶ Describe how angiosperms and animals depend on each other.
▶ Explain why insects are such successful animals.

Key Terms

bacteria
fungus
protist
gymnosperm
angiosperm
invertebrate
vertebrate

Life on Earth is incredibly diverse. Take a walk in a park, and you will see trees, birds, insects, and maybe fish in a stream. All of these organisms are living, but they are all very different from one another. How do scientists organize this variety into categories they can understand?

Most scientists classify organisms into six *kingdoms*, as described in **Table 2**, based on different characteristics. Members of the six kingdoms get their food in different ways and are made up of different types of *cells*, the smallest unit of biological organization. The cells of animals, plants, fungi, and protists contain a *nucleus* (NOO klee uhs), which consists of a membrane that surrounds a cell's genetic material. A characteristic shared by bacteria, fungi, and plants is the *cell wall*, a structure that surrounds a cell and provides it with rigid support.

Bacteria

Bacteria are microscopic, single-celled organisms that usually have cell walls and reproduce by dividing in half. Bacteria also lack nuclei, unlike all other organisms. Scientists have found two main kinds of bacteria, archaebacteria (AHR kee bak TIR ee uh) and eubacteria (YOO bak TIR ee uh). Most bacteria, including the kinds that cause disease and those found in garden soil, are eubacteria. Bacteria live in every habitat on Earth, from hot springs to the bodies of animals.

Table 2 ▼

The Kingdoms of Life		
Kingdom	**Characteristics**	**Examples**
Archaebacteria	single celled; lack cell nuclei; reproduce by dividing in half; found in harsh environments	methanogens (live in swamps and produce methane gas) and extreme thermophiles (live in hot springs)
Eubacteria	single celled; lack cell nuclei; reproduce by dividing in half; incredibly common	proteobacteria (common in soils and in animal intestines) and cyanobacteria (also called *blue-green algae*)
Fungi	absorb their food through their body surface; have cell walls; most live on land	yeasts, mushrooms, molds, mildews, and rusts
Protists	most single celled but some have many cells; most live in water	diatoms, dinoflagellates (red tide), amoeba, trypanosomes, paramecia, and *Euglena*
Plants	many cells; make their own food by photosynthesis; have cell walls	ferns, mosses, trees, herbs, and grasses
Animals	many cells; no cell walls; ingest their food; live on land and in water	corals, sponges, worms, insects, fish, reptiles, birds, and mammals

---

MISCONCEPTION ALERT

Classification of Life This textbook classifies organisms into six kingdoms. Other scientists use three superkingdoms: Bacteria, Archaea, and Eucarya. Still others use a five-kingdom system. Who is right? All of them are, in a way. Scientists use the best data they have to organize species in a way that reflects how the species relate to each other. But each classification system focuses on different characteristics. It is difficult to determine how all organisms evolved, so it is difficult to find the "right" way to classify them. Alternate systems enliven the debate and inspire scientists to find the closest underlying relationships. Ask students to give examples of how things may be classified differently but still logically. (One example is the way grade levels are organized. One school district may divide its schools into Primary (1–6), Jr. High (7–8) and High School (9–12). Other districts may have a Middle School that encompasses grades 6–9. Neither way is right or wrong, and each way reflects logical relationships between the grades.) **LS** Logical

Bacteria and the Environment Bacteria play many important roles in the environment. Some kinds of bacteria break down the remains and wastes of other organisms and return nutrients to the soil. Others recycle mineral nutrients, such as nitrogen and phosphorous. For example, certain kinds of bacteria play a very important role by converting nitrogen in the air into a form that plants can use. Nitrogen is important because it is a main component of proteins and genetic material.

Bacteria also allow many organisms, including humans, to extract certain nutrients from their food. The bacteria in **Figure 12** are *Escherichia coli*, or *E. coli*, a bacterium found in the intestines of humans and other animals. Here, *E. coli* helps digest food and release vitamins that humans need.

Fungi

A **fungus** (plural, *fungi*) is an organism whose cells have nuclei, cell walls, and no chlorophyll (the pigment that makes plants green). Cell walls act like miniature skeletons that allow fungi, such as the mushrooms in **Figure 13**, to stand upright. A mushroom is the reproductive structure of a fungus. The rest of the fungus is an underground network of fibers. These fibers absorb food from decaying organisms in the soil.

Indeed, all fungi absorb their food from their surroundings. Fungi get their food by releasing chemicals that help break down organic matter, and then absorbing the nutrients. The bodies of most fungi are a huge network of threads that grow through the soil, dead wood, or other material on which the fungi are feeding. Like bacteria, fungi play an important role in the environment by breaking down the bodies and body parts of dead organisms.

Like bacteria, some fungi cause diseases, such as athlete's foot. Other fungi add flavor to food. The fungus in blue cheese, shown in **Figure 13**, gives the cheese its strong flavor. And fungi called *yeasts* produce the gas that makes bread rise.

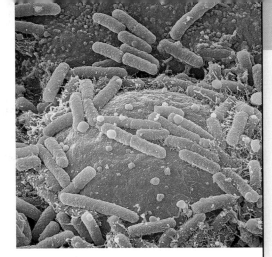

Figure 12 ▶ The long, orange objects in the image above are *E. coli* bacteria as they appear under a microscope.

Graphic Organizer **Spider Map**
Create the **Graphic Organizer** entitled "Spider Map" described in the Appendix. Label the circle "Kingdoms." Create a leg for each kingdom. Then, fill in the map with details about the organisms in each kingdom.

Figure 13 ▶ A mushroom (left) is the reproductive structure of a fungus that lives in the soil. The cheese (above) gets its taste and its blue color from a fungus.

103

Teach

Group Activity ── GENERAL

Mushroom Walk Take your class on a walk in the woods to find the fungal structures commonly known as mushrooms. Have students study the fungi they find. CAUTION: Warn students not to eat any of the fungi, as some may be poisonous. Use a mushroom identification guide, and work with students to make spore prints and identify the types of fungi. Have students wear gloves while making spore prints. Also have students identify where each fungus fits into its woodland ecosystem. Ask, "Does it decompose dead wood or attack other fungi?" Have students draw the different types of fungi and include those drawings in their **Portfolio.**
LS Kinesthetic/Visual

BRAIN FOOD 🧠

Humongous Fungus The world's largest individual organism is a fungus! One particular honey mushroom, *Armillaria ostoyae*, has been growing for about 2,400 years and now covers 2,200 acres in the Malheur National Forest in eastern Oregon. That makes it 3.5 miles across and as big as 1,665 football fields. This giant fungus was determined to be one organism by a Forest Service scientist, Catherine Parks, who tested its DNA at various locations. Unfortunately, this fungus kills trees as it spreads, so it's not that popular with the Forest Service. They do realize, however, that this particular fungus has played a role in its ecosystem for thousands of years.

Homework ── GENERAL

Ruminant Animals Bring in a diagram of a cow's digestive system. Show students how the stomachs are arranged, and tell them that each stomach contains microbes that help the cow process its food. Ask students to research how the microbes help the cow derive nutrients from grass and hay. (Microbes that can break down cellulose, which is found in the tough cell walls of plants, ferment grass and hay. This process releases fatty acids from the plant matter that can be absorbed and used by the cow.)
LS Intrapersonal

Graphic Organizer GENERAL

Spider Map
You may want to have students work in groups to create this Spider Map. Have one student draw the map and fill in the information provided by other students from the group.

Activity ──────── GENERAL

Pond Protists Gather 2 or 3 plastic lab bottles with lids, some microscopes, glass slides, cover slips, medicine droppers, and a book on protist identification. Then visit a local pond where it is easy to get a water sample. Have students fill the bottles with pond water. Back in class have students prepare slides using drops of pond water. Have students look for protists and other organisms under the microscope. Ask students to draw what they see and to include those drawings in their **Portfolio.**
LS Kinesthetic | English Language Learners

Teaching Tip ──────── GENERAL

Pond Creatures The following protists are likely to be found in pond water: *diatoms:* one-celled organisms with geometric shapes, they are made of silica and are usually yellow-brown in color; *amoebas:* one-celled organisms that "ooze" or travel by cytoplasmic movement; *ciliates:* single-celled protozoans with hair-like structures (cilia) around their entire cell margin; *volvox:* colonial green algae that look like spiky golfballs; *euglenas:* single-celled protozoans with a tail called a *flagellum; dinoflagellates:* tiny algae with two flagella, one extending like a tail and one running in a groove around the center of the organism; *other types of green algae:* these can be various shapes and sizes.

Figure 14 ▶ Kelp (left) are huge protists with many cells that live attached to the ocean floor. The microscopic diatoms (right) are protists that live in the plankton.

Connection to Physics

Cell Size Every cell must exchange substances with its environment across its surface. The larger the cell, the smaller its surface is compared with its volume. So the larger the cell, the more slowly substances move from outside the cell to its interior. This relationship limits most cells to microscopic sizes.

Figure 15 ▶ Lower plants, such as these mosses and ferns, live in damp places.

Protists

Most people have some idea what bacteria and fungi are, but few could define a protist. **Protists** are a diverse group of organisms that belong to the kingdom Protista. Some, such as amoebas, are animallike. Others, such as the kelp in **Figure 14**, are plantlike. Still others are more like fungi. Most protists are one-celled microscopic organisms. This group includes amoebas and *diatoms* (DIE uh TAHMS). Diatoms, shown in **Figure 14**, float on the ocean surface. The most infamous protist is *Plasmodium*, the one-celled organism that causes the disease malaria.

From an environmental standpoint, the most important protists are probably algae. Algae are plantlike protists that can make their own food using the sun's energy. Green pond "scum" and seaweed are examples of algae. They range in size from the giant kelp to the one-celled *phytoplankton*, which are the initial source of food in most ocean and freshwater ecosystems.

Plants

Plants are many-celled organisms that make their own food using the sun's energy and have cell walls. Most plants live on land, where the resources a plant needs are separated between the air and the soil. Sunlight, oxygen, and carbon dioxide are in the air, and minerals and water are in the soil. Plants have roots that tap resources underground and leaves that intercept light and gases in the air. Leaves and roots are connected by *vascular tissue*, a system of tubes that carries water and food. Vascular tissue has thick cell walls, so a wheat plant or a tree is like a building supported by its plumbing.

Lower Plants The first land plants had no vascular tissue, and they also had swimming sperm. As a result, these early plants could not grow very large and had to live in damp places. Their descendants alive today are small plants such as mosses. Ferns and club mosses were the first vascular plants. Some of the first ferns were as large as small trees, and some of these tree ferns live in the tropics and in New Zealand today. Some examples of lower plants are shown in **Figure 15**.

LANGUAGE ARTS ──────
● CONNECTION ── BASIC

Horton Hears a Who Check out the Dr. Seuss book *Horton Hears a Who* from the local library. Many of your students may have read this book as small children. Tell students that, even though this is a children's book, it has many ideas that relate to the ecological concepts in this chapter. Ask for some volunteers to read the story out loud. After students have read the story, ask them to discuss how it relates to species diversity and ecosystems. **LS Verbal** | English Language Learners

ARTS ──────
● CONNECTION ── BASIC

Diatoms—Jewels of the Pond Diatoms are truly beautiful organisms. Since they are made of silica, they look like small glass boxes. If you find a good assortment of diatoms in a pond water sample, have students sketch what they see looking through a microscope. If you can't find a good array, have students find photos of diatoms on the Internet to sketch. Some universities have Web sites with Scanning Electron Microscope (SEM) galleries that include images of diatoms. Ask students to include the sketches in their **Portfolio.**
LS Visual | English Language Learners

Gymnosperms Pine trees and other evergreens are common examples of gymnosperms (JIM noh SPUHRMZ). **Gymnosperms** are woody plants whose seeds are not enclosed in fruits. Gymnosperms such as pine trees are also called *conifers* because they bear cones, as shown in **Figure 16.**

Gymnosperms have several adaptations that allow them to live in drier conditions than lower plants can. Gymnosperms produce *pollen,* which protects and moves sperm between plants. These plants also produce *seeds,* which protect developing plants from drying out. And a conifer's needle-like leaves lose little water. Much of our lumber and paper comes from gymnosperms.

Angiosperms Most land plants today are **angiosperms** (AN jee oh SPUHRMZ), flowering plants that produce seeds in fruit. All of the plants in **Figure 17** are angiosperms. The flower is the reproductive structure of the plant. Some angiosperms, such as grasses, have small flowers that produce pollen that is carried by the wind. Other angiosperms have large flowers that attract insects or birds to carry their pollen to other plants. Many flowering plants depend on animals to disperse their seeds and carry their pollen. For example, a bird that eats a fruit will drop the seeds elsewhere, where they may grow into new plants.

Most land animals are dependent on flowering plants. Most of the food we eat, such as wheat, rice, beans, oranges, and lettuce, comes from flowering plants. Building materials and fibers, such as oak and cotton, also come from flowering plants.

Figure 16 ▶ This gymnosperm has male and female reproductive structures called *cones.*

Figure 17 ▶ This meadow contains a wide array of angiosperms, including grasses, trees, and wildflowers.

105

QuickLAB

Pollen and Flower Diversity

Procedure
1. Use a **cotton swab** to collect pollen from a common **flowering plant.**
2. Tap the cotton swab on a **microscope slide** and cover the slide with a **cover slip.**
3. Examine the slide under a **microscope,** and draw the pollen grains in your *EcoLog.*
4. Repeat this exercise with a **grass plant in bloom.**

Analysis
1. Based on the structure of the flower and the pollen grains, explain which plant is pollinated by insects and which is pollinated by wind.

Teach, *continued*

MATHPRACTICE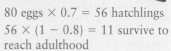

Answers

80 eggs × 0.7 = 56 hatchlings

56 × (1 − 0.8) = 11 survive to
reach adulthood

Group Activity ———— GENERAL

Pollinator Game Organize the
class into two groups, and give each
group a pile of blank index cards.
Have students conduct research to
find names of specific plants that
have insect pollinators. Ask the
groups to write the name of each
plant on one side of each card and
that of the pollinator on the other.
Tell the groups not to share information. After the groups have each
created six pollination pairs, have
each group take turns trying to
identify the pollinator associated
with the plants the other group has
researched. The group that identifies the most correct pairs wins
the game. (You may want to give
flowers as a prize.) **LS Interpersonal**

SKILL BUILDER — GENERAL

Writing Ask students to write an
essay about their favorite animal.
Have them include information
about where the animal lives, what
it eats, if it is a vertebrate or invertebrate, how it reproduces, if it is
solitary or social, and information
on its other traits. Ask students
to include information about how
the animal may have adapted to its
current environment. **LS Verbal**

MATHPRACTICE

Insect Survival Most
invertebrates produce large
numbers of offspring. Most of
these offspring die before reaching
adulthood. Suppose an insect lays
80 eggs on a plant. If 70 percent
of the eggs hatch and 80 percent
of those that hatch die before
reaching adulthood, how many
insects will reach adulthood?

---

internet connect

www.scilinks.org
Topic: Invertebrates
SciLinks code: HE4057

SCI**LINKS** Maintained by the
National Science
Teachers Association

Animals

Animals cannot make their own food like plants can. They have
to take in food from their environment. In addition, animal cells
have no cell walls, so animals' bodies are soft and flexible. Some
animals have evolved hard skeletons against which their muscles
can pull to move the body. As a result, animals are much more
mobile than plants and all animals move around in their environments during at least one stage in their lives.

Invertebrates Animals that lack backbones are **invertebrates**
(in VUHR tuh brits). Many invertebrates live attached to hard
surfaces in the ocean and filter their food out of the water. These
organisms move around only when they are larvae. At this early
stage of life, they are part of the ocean's plankton. Filter feeders
include corals, various worms, and mollusks such as clams and
oysters. **Figure 18** shows a variety of invertebrates. Other invertebrates, including squid in the ocean and insects on land, move
around actively in search of food.

More insects exist on Earth than any other type of animal.
They are successful for several reasons. Insects have a waterproof
external skeleton, they move quickly, and they reproduce quickly.
Also, most insects can fly. Their small size allows them to live on
little food and to hide from enemies in small spaces, such as a
seed or in the hair of a mammal.

Many insects and plants have evolved together and depend on
each other to survive. Insects carry pollen from male parts of flowers
to fertilize a plant's egg, which develops into a fruit. Without
insect pollinators, we would not have tomatoes, cucumbers, apples,
and many other crops. Insects are also valuable because they eat
other insects that we consider to be pests. But, humans and insects
are often enemies. Bloodsucking insects transmit human diseases,
such as malaria, sleeping sickness, and West Nile virus. Insects
probably do more damage indirectly, however, by eating our crops.

Figure 18 ▶ Examples of invertebrates include a banana slug (left),
a leaf-footed bug (middle), and
a cuttlefish (right).

106

INCLUSION Strategies

- *Learning Disabled*
- *Developmentally Delayed*
- *Attention Deficit Disorder*

Ask students to create a chart to compare invertebrates and vertebrates. Ask students to
fold a piece of lined paper in half vertically.
Have students label one half "Invertebrates"
and the other half "Vertebrates." Students
can fill in the characteristics of these animals
using the textbook as a reference. Have students list at least three examples for each
type of animal. The charts can be displayed
or presented to groups of students.

Figure 19 ► Examples of vertebrates include the toco toucan (left), the blue-spotted stingray (middle), and the snow leopard (right).

Vertebrates Animals that have backbones are called **vertebrates.** Members of three vertebrate groups are shown in **Figure 19.** The first vertebrates were fish, but today most vertebrates live on land. Amphibians, which include toads, frogs, and salamanders, are partially aquatic. Nearly all amphibians must return to water to lay their eggs.

The first land vertebrates were the reptiles, which today include turtles, lizards, snakes, and crocodiles. These animals were successful because they have an almost waterproof egg, which allows the egg to hatch on land, away from predators in the water.

Birds are warm-blooded vertebrates with feathers. Bird eggs have hard shells. Adult birds keep their eggs and young warm until they develop insulating layers of fat and feathers. *Mammals* are warm-blooded vertebrates that have fur and feed their young milk. The ability to maintain a high body temperature allows birds and mammals to live in cold areas, where other animals cannot survive.

Eco fact

Conserving Water Arthropods and vertebrates are the only two groups of animals that have adaptations that prevent dehydration so effectively that some of them can move about freely on land on a dry, sunny day.

SECTION 3 Review

1. **Describe** how animals and angiosperms depend on each other. Write a short paragraph to explain your answer. **WRITING SKILLS**

2. **Describe** the importance of protists in the ocean.

3. **Name** the six kingdoms of life, and give two characteristics of each.

4. **Explain** the importance of bacteria and fungi in the environment.

CRITICAL THINKING

5. **Analyzing Relationships** Explain how the large number and wide distribution of angiosperm species is related to the success of insects.

6. **Understanding Concepts** Write a short paragraph that compares the reproductive structures of gymnosperms and angiosperms. **WRITING SKILLS**

107

Alternative Assessment ——— GENERAL

Ecosystem Alterations Ask students to think of three organisms from each of the six kingdoms of life. Have them connect all the organisms so that they operate as an ecosystem. Students may represent this ecosystem as a drawing or as words arranged in a web-like chart. Ask students to eliminate two species and to rearrange the ecosystem to represent that loss. Finally, have students triple the population of one of the species. Ask them to document how the ecosystem changes. **LS** Visual

Clarifying Natural Selection

Lamarck was a famous scientist who suggested (incorrectly) that the neck of the giraffe became longer because it was stretching to reach high leaves. Ask students to discuss why Lamarck was wrong. (If the parents lengthened their necks slightly by reaching for food, the offspring would still start with the same length neck as their parents. Necks would not lengthen over generations without a genetic basis.) Have students use Darwin's theory of evolution by natural selection to devise a possible explanation for the giraffe's long neck. (Giraffe ancestors with slightly longer necks were able to reach more food and were able to survive and reproduce better than their counterparts with shorter necks. Generations of longer-necked animals survived and reproduced more than others did, causing the evolution of long-necked giraffes.) **LS** Logical

CHAPTER 4 Highlights

1 Ecosystems: Everything Is Connected

Key Terms

ecosystem, 93
biotic factor, 94
abiotic factor, 94
organism, 95
species, 95
population, 95
community, 96
habitat, 96

Main Ideas

▶ Ecosystems are composed of many interconnected parts that often interact in complex ways.

▶ An ecosystem is the community of all the different organisms living in an area and their physical environment.

▶ An ecosystem contains biotic (living) and abiotic (nonliving) components.

▶ Organisms live as populations of one species in communities with other species. Each species has its own habitat, or type of place that it lives.

2 Evolution

natural selection, 97
evolution, 97
adaptation, 99
artificial selection, 100
resistance, 101

▶ The naturalist Charles Darwin used the term natural selection to describe the unequal survival and reproduction that results from the presence or absence of particular traits.

▶ Darwin proposed that natural selection is responsible for evolution—a change in the genetic characteristics of a population from one generation to the next.

▶ By selecting which domesticated animals and plants breed, humans cause evolution by artificial selection.

▶ We have unintentionally selected for pests that are resistant to pesticides and for bacteria that are resistant to antibiotics.

3 The Diversity of Living Things

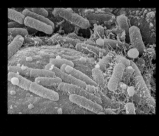

bacteria, 102
fungus, 103
protist, 104
gymnosperm, 105
angiosperm, 105
invertebrate, 106
vertebrate, 107

▶ Organisms can be divided into six kingdoms, which are distinguished by the types of cells they possess and how they obtain their food.

▶ Bacteria and fungi play the important environmental roles of breaking down dead organisms and recycling nutrients.

▶ Gymnosperms are evergreen plants, many of which bear cones, while angiosperms produce flowers and bear seeds in fruit.

▶ Insects, invertebrates that are the most successful animals on Earth, affect humans in both positive and negative ways.

▶ Vertebrates, or animals with backbones, include fish, amphibians, reptiles, birds, and mammals.

108

INCLUSION Strategies

• *Learning Disabled*
• *Developmentally Delayed*

Using **Figure 4** as a model, ask students to design and draw an ecosystem from a different environment such as a forest, ocean, or desert. Have students label the organism, population, community, and ecosystem. Students can use pictures from magazines or their original drawings to show their understanding of the ecosystem.

Chapter Resource File

• **Chapter Test** GENERAL
• **Chapter Test** ADVANCED
• **Concept Review** GENERAL
• **Critical Thinking** ADVANCED
• **Test Item Listing**
• **Math/Graphing Lab** GENERAL
• **Observation Lab** BASIC
• **CBL™ Probeware Lab** ADVANCED
• **Consumer Lab** BASIC
• **Long-Term Project** BASIC

Using Key Terms

Use each of the following terms in a separate sentence.

1. *adaptation*
2. *invertebrate*
3. *abiotic factor*
4. *habitat*
5. *species*

For each pair of terms, explain how the meanings of the terms differ.

6. *community* and *population*
7. *evolution* and *natural selection*
8. *gymnosperm* and *angiosperm*
9. *bacteria* and *protists*

✔ STUDY TIP

Make an Outline After reading each section, summarize the main ideas into a short outline, leaving space between each entry. Then write the key terms under the subsection in which they are introduced, followed by a short definition for each.

Understanding Key Ideas

10. Which of the following pairs of organisms belong to the same population?
    a. a dog and a cat
    b. a marigold and a geranium
    c. a human mother and her child
    d. a spider and a cockroach

11. Which of these phrases does *not* describe part of the process of evolution by natural selection?
    a. the environment contains limited resources
    b. organisms produce more offspring than will survive to reproduce
    c. communities include populations of several species
    d. organisms in a population differ in their traits

12. Which of the following components of an ecosystem are *not* abiotic factors?
    a. wind
    b. small rocks
    c. sunlight
    d. tree branches

13. Some snakes produce a powerful poison that paralyzes their prey. This poison is an example of
    a. coevolution.
    b. an adaptation.
    c. a reptile.
    d. a biotic factor.

14. Angiosperms called roses come in a variety of shapes and colors as a result of
    a. natural selection.
    b. coevolution.
    c. different ecosystems.
    d. artificial selection.

15. Single-celled organisms that live in swamps and produce methane gas are
    a. protists.
    b. archaebacteria.
    c. fungi.
    d. eubacteria.

16. Which of the following statements about protists is *not* true?
    a. Most of them live in water.
    b. Some of them cause diseases in humans.
    c. They contain genetic material.
    d. Their cells have no nucleus.

17. Which of the following statements about plants is *not* true?
    a. They make their food from carbon dioxide and water through photosynthesis.
    b. Land plants have cell walls that help hold their stems upright.
    c. They have adaptations that help prevent water loss.
    d. Plants absorb food through their roots.

109

ANSWERS

Using Key Terms

1. Sample answer: Thick fur is an adaptation that helps deer survive in cold weather.

2. Sample answer: An invertebrate does not have a backbone.

3. Sample answer: Water, rocks, light, temperature, and air are all abiotic factors in an ecosystem.

4. Sample answer: A habitat is the place where an organism lives.

5. Sample answer: A species is a group of organisms that are closely related and can breed to produce fertile offspring.

6. A community is a group of various species that live in the same place and interact with each other. A population is an interbreeding group of individuals of the same species in a given area.

7. Evolution is the change in the genetic characteristics of a population from one generation to the next. Natural selection is a process that drives evolution in nature.

8. A gymnosperm is a plant whose seeds are not enclosed in fruit. An angiosperm is a plant that produces flowers and seeds enclosed in fruit.

9. Bacteria are single-celled organisms that lack nuclei. Protists can be multi-celled, and their cells contain nuclei.

Understanding Key Ideas

10. c
11. c
12. d
13. b
14. d
15. b
16. d
17. d

Assignment Guide	
Section	**Questions**
1	3–6, 10, 12, 18, 19, 25, 31
2	1, 7, 11, 13, 14, 20, 21, 23, 24, 26, 28–30, 34, 35
3	2, 8, 9, 15–17, 22, 27, 31–33

CHAPTER 4 Review

Short Answer

18. Energy, mineral nutrients, water, oxygen, and living organisms are all necessary to maintain an ecosystem.

19. Biotic factors are the living or once-living factors, and abiotic factors are the non-living factors.

20. An adaptation is an inherited trait that increases an organism's chance of survival. Evolution is a change in the genetic characteristics of a population from one generation to the next. Populations evolve adaptations.

21. Step 1: The pesticide kills most of the insects, but some are not killed; Step 2: The surviving insects pass the trait for resistance to their offspring; Step 3: Each time the pesticide is used, the resistant insects become a larger portion of the population, which is evolution.

22. Archaebacteria are single-celled, lack nuclei, and live in extreme environments. Eubacteria are single-celled, lack nuclei, and live in everyday environments. Fungi absorb food through their surfaces and have cell walls. Protists are mostly single-celled, and most live in water. Plants are multi-celled, produce food through photosynthesis, and have cell walls. Animals are multi-celled and ingest their food.

Interpreting Graphics

23. The population dropped in size after each spraying.

24. Yes. The population still alive after each spraying steadily increased during the summer indicating that a larger and larger proportion of the population is resistant.

Short Answer

18. List the five components that an ecosystem must contain to survive indefinitely.

19. What is the difference between biotic and abiotic factors in an ecosystem?

20. What is the difference between adaptation and evolution?

21. Describe the three steps by which a population of insects becomes resistant to a pesticide.

22. List the six kingdoms of organisms and the characteristics of each kingdom.

Interpreting Graphics

Below is a graph that shows the number of aphids on a rose bush during one summer. The roses were sprayed with a pesticide three times, as shown. Use the graph below to answer questions 23 and 24.

23. What evidence is there that the pesticide killed aphids?

24. Aphids have a generation time of about 10 days. Is there any evidence that the aphids evolved resistance to the pesticide during the summer? Explain your answer.

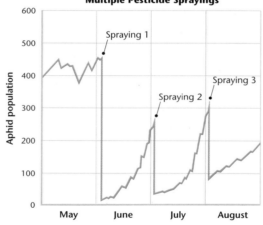

Aphid Population Changes Over Multiple Pesticide Sprayings

Concept Mapping

25. Use the following terms to create a concept map: *ecosystem, abiotic factor, biotic factor, population, species, community,* and *habitat.*

Critical Thinking

26. **Analyzing Ideas** Can a person evolve? Read the description of evolution in this chapter and explain why or why not. **READING SKILLS**

27. **Making Inferences** A scientist applies a strong fungicide, a chemical that kills fungi, to an area of forest soil every week during October and November. How might this area look different from the surrounding ground at the end of the experiment?

28. **Drawing Conclusions** In what building in your community do you think bacteria are evolving resistance to antibiotics most rapidly? Explain your answer.

29. **Evaluating Assumptions** Many people assume that the human population is no longer evolving. Do you think these people are right? Explain your answer.

Cross-Disciplinary Connection

30. **Geography** Find out how the isolation of populations on islands has affected their evolution. Research a well-known example, such as the animals and plants of Madagascar, the Galápagos Islands, and the Hawaiian Islands. Write a short report on your findings. **WRITING SKILLS**

Portfolio Project

31. **Study an Ecosystem** Observe an ecosystem near you, such as a pond or a field. Identify biotic and abiotic factors and as many populations of organisms as you can. Do not try to identify the organisms precisely. Just list them, for example, as spiders, ants, grass, not as a specific type. Make a poster showing the different populations. Put the organisms into columns to show which of the kingdoms they belong to.

110

Concept Mapping

25. Answers to the concept mapping questions are on pages pp. 667–672.

Critical Thinking

26. No, individual people cannot evolve. Individuals with different traits can reproduce, and the traits that are best suited to the environment may become more common in the population. In this way, the genetic makeup of any population can gradually change.

27. Answers may vary. Leaf litter may accumulate because fungi are not there to help break it down. Plants that have fungal associations may not grow there.

28. Answers may vary. Many antibiotics are used in a hospital, so hospitals could be an area where bacteria evolve antibiotic resistance.

29. Answers may vary. Natural selection may not be taking place in most modern human populations because we are protected from most environmental factors.

Cross-Disciplinary Connection

30. Answers may vary.

MATH SKILLS

Use the graph below to answer questions 32–33.

32. Analyzing Data The graph below shows the mass of different types of organisms found in a meadow. How much greater is the mass of the plants than that of the animals?

33. Analyzing Data What is the ratio of the mass of the bacteria to the mass of the fungi?

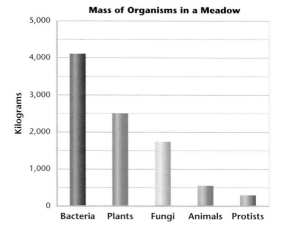

Mass of Organisms in a Meadow

WRITING SKILLS

34. Communicating Main Ideas Why is evolution considered to be such an important idea in biology?

35. Outlining Topics Outline the essential steps in the evolution of pesticide resistance in insects.

STANDARDIZED TEST PREP

For extra practice with questions formatted to represent the standardized test you may be asked to take at the end of your school year, turn to the sample test for this chapter in the Appendix.

READING SKILLS

Read the passage below, and then answer the questions that follow.

Some Central American acacia trees, called *ant acacias,* have a mutually beneficial relationship with ants that live on them. The trees have several structures that benefit the ants. The trees have hollow thorns in which the ants live, glands that produce sugary nectar, and swollen leaf tips, which the ants remove and feed to their larvae.

The ants reduce the damage that other organisms do to the tree. They remove dust, fungus spores, and spider webs. They destroy seedlings of other plants that sprout under the tree, so that the tree can obtain water and nutrients without competition from other plants. The ants sting animals that try to eat the tree.

Proof that the ants are valuable to the acacia tree comes from studies in which the ants are removed. Fungi invade the tree, it is eaten by herbivores, and it grows more slowly. When ants are removed from the tree, it usually dies in a few months.

1. According to the passage, which of the following statements is not true?
   a. Ants and ant acacias have evolved a relationship beneficial to both of them.
   b. The ants prevent fungi from growing on the acacia.
   c. The tree would benefit from not having ants.
   d. The ants benefit from living on the tree.

2. What is the advantage to an acacia of not having other plants grow nearby?
   a. Ants cannot crawl onto the acacia from the other plants.
   b. The acacia keeps more ants for itself.
   c. This reduces competition for water and nutrients.
   d. This reduces competition for fungi.

Portfolio Project
31. Answers may vary.

Math Skills
32. Total plant mass is 2500 kg and total animal mass is approximately 500 kg. Therefore, the mass of the plants is about 5 times the mass of the animals.

33. $4{,}100 \div 1{,}700 = 2.4$; The ratio of bacteria to fungi is about 2.4 to 1.

Writing Skills
34. Answers may vary. Evolution provides a simple mechanism (natural selection) that explains why organisms are well suited to their environments, why populations change over time, and why and how new species develop.

35. Answers may vary. Insects are treated with a pesticide; most die, but some are not harmed; over generations, insects that can survive the pesticide become a dominant part of the population; the pesticide eventually becomes ineffective because the insect population has evolved to be resistant to it.

Reading Skills
1. c
2. c

How Do Brine Shrimp Select a Habitat?

Teacher's Notes

Time Required
two 45-minute class periods

Lab Ratings

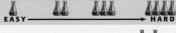

EASY ——————→ HARD

TEACHER PREPARATION 🧪🧪
STUDENT SETUP 🧪🧪🧪
CONCEPT LEVEL 🧪🧪
CLEANUP 🧪🧪

Skills Acquired
• Designing Experiments
• Experimenting
• Communicating

Scientific Methods
In this lab, students will:
• Make Observations
• Ask Questions
• Test the Hypothesis
• Analyze the Results
• Draw Conclusions
• Communicate the Results

Materials
Materials listed are enough for groups of 3 to 4 students. After the lab, return the brine shrimp to an aquarium or a pet supply store.

Objectives

▶ **USING SCIENTIFIC METHODS** **Observe** the behavior of brine shrimp.

▶ **USING SCIENTIFIC METHODS** **Identify** a variable, and design an experiment to test the effect of the variable on habitat selection by brine shrimp.

Materials

aluminum foil
brine shrimp culture
corks sized to fit tubing
Detain™ or methyl cellulose
fluorescent lamp or grow light
funnel
graduated cylinder or beaker
hot-water bag
ice bag
metric ruler
Petri dish
pipet
plastic tubing, 40cm × 1cm, clear, flexible
screen, pieces
screw clamps
tape
test-tube rack
test tubes with stoppers

▶ **Making a Test Chamber** Use a screw clamp to divide one section of tubing from another.

How Do Brine Shrimp Select a Habitat?

Different organisms are adapted for life in different habitats. For example, brine shrimp are small crustaceans that live in saltwater lakes. Organisms select habitats that provide the conditions, such as a specific temperature range and amount of light, to which they are best adapted. In this investigation, you will explore habitat selection by brine shrimp and determine which environmental conditions they prefer.

Procedure

Establish a Control Group

1. To make a test chamber and to establish a control group, divide a piece of plastic tubing into four sections by making a mark at 10 cm, 20 cm, and 30 cm from one end. Label the sections "1", "2", "3", and "4".

2. Place a cork in one end of the tubing. Then transfer 50 mL of brine shrimp culture to the tubing. Place a cork in the other end of the tubing. Set the tube aside, and let the brine shrimp move about the tube for 30 min.

3. After 30 min, divide the tubing into four sections by placing a screw clamp at each mark on the tubing. While someone in your group holds the corks firmly in place, tighten the middle clamp at 20 cm and then tighten the other two clamps.

4. Remove the cork from the end of section 1 and pour the contents of section 1 into a test tube labeled "1." Repeat this step for the other sections by loosening the screw clamps and pouring the contents of each section into their corresponding test tubes.

5. To get an accurate count for the number of brine shrimp in each test tube, place a stopper on test tube 1, and invert the tube gently to distribute the shrimp. Use a pipet to transfer a 1 mL sample of the culture to a Petri dish. Add a few drops of Detain™ to the sample so that the brine shrimp move slower. Count and record the number of brine shrimp in the Petri dish.

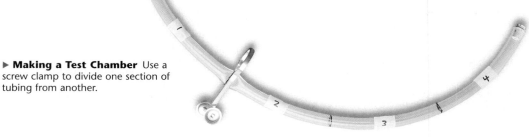

112

Tips and Tricks

Students are asked to design the conditions for the brine shrimp themselves. Possible variables for this experiment include temperature and light. To test temperature, students can place a section of tubing on a hot water bag, place a section on an ice bag, and leave two sections at room temperature. (Caution: Do not use water with a temperature higher than 70°C.) Each of the room temperature sections should be right next to the cold or hot section so that shrimp don't have to migrate through a habitat they do not like to get to one they

prefer. To test light shine a grow light on one section and cover another section with aluminum foil. Again leave two sections in natural light and put those on either side of the extremes. Be sure that students include a control in their experiment. Students should make sure other factors besides the one they are testing are not different across the test tube. If students do not keep all other factors the same, they will not be able to tell which environmental factors the brine shrimp selected.

6. Empty the Petri dish, and take two more 1 mL samples of brine shrimp from test tube 1. Calculate the average of the three samples recorded for test tube 1.

7. Repeat steps 5 and 6 for each of the remaining test tubes to count the number of brine shrimp in each section of tubing.

Ask a Question

8. Write a question you would like to explore about brine shrimp habitat selection. For example, you can explore how temperature or light affects brine shrimp. To explore the question, design an experiment that uses the materials listed for this lab.

9. Write a procedure and a list of safety precautions for your group's experiment. Have your teacher approve your procedure and precautions before you begin the experiment.

10. Set up and conduct your group's experiment.

Analysis

1. **Constructing Graphs** Make a bar graph of your data. Plot the environmental variable on the *x*-axis and the number of brine shrimp on the *y*-axis.

2. **Evaluating Results** How did the brine shrimp react to changes in the environment?

3. **Evaluating Methods** Why did you have to have a control in your experiment?

4. **Evaluating Methods** Why did you record the average of three samples to count the number of brine shrimp in each test tube in step 6?

Conclusions

5. **Drawing Conclusions** What can you conclude from your results about the types of habitat that brine shrimp prefer?

Extension

1. **Formulating Hypotheses** Now that you have observed brine shrimp, write a hypothesis about how brine shrimp select a habitat that could be explored with another experiment, other than the one you performed in this lab. Formulate a prediction based on your hypothesis.

2. **Evaluating Hypotheses** Conduct an experiment to test your prediction. Write a short explanation of your results. Did your results support your prediction? Explain your answer.

▶ **Brine Shrimp** These crustaceans have specific habitat preferences.

TEACHER TESTED & APPROVED

Tammie Niffeneger
Port Washington
High School
Port Washington, Wisconsin

Answers to Analysis

1. Answers may vary.

2. Answers may vary. Have students compare results so they can understand how brine shrimp reacted to different variables.

3. A control is there to show what happens to the organisms during "normal," non-experimental circumstances. It shows how much of a difference there is between treated and untreated organisms.

4. If you sample several times as opposed to once, it is more likely that any differences you see will be due to the factor you are testing, not because of a random sampling error or natural variation in the brine shrimp population.

Answers to Conclusions

5. Answers may vary depending on students' results.

Answers to Extension

1. Answers may vary. If one of the factors appears to influence habitat selection, students could test that factor in conjunction with another factor. This kind of experiment could help determine if environmental factors interact with each other to change behavior.

2. Answers may vary.

Chapter Resource File

- Datasheets for In-Text Labs
- Lab Notes and Answers

BUTTERFLY ECOLOGIST

Making a difference

BUTTERFLY ECOLOGIST

Discussion — GENERAL

Annual Life Cycle of a Monarch

There are four generations of monarchs in the United States each year. First generation butterflies emerge from eggs laid by females that overwintered in Mexico. Four days after they emerge, the first generation migrates north, laying eggs as it migrates. Eggs from this generation are only found in the southern United States. The second generation is then spread throughout the United States. When this second generation emerges, it migrates farther north to avoid hot temperatures. Females lay eggs as they migrate. Individuals of the third generation that emerge during the summer then lay eggs. Individuals in the third generation that emerge during early fall, as well as all of the fourth generation, migrate to Mexico. This generation of monarchs does not reproduce right away. They slip into a stage called *diapause*, which is a stage of delayed maturity. In March, generations three and four move into the southern United States to lay eggs, and the first generation of the next year emerges.

Imagine millions of butterflies swirling through the air like autumn leaves, clinging in tightly packed masses to tree trunks and branches, and covering low-lying forest vegetation like a luxurious, moving carpet. According to Alfonso Alonso-Mejía, this is quite a sight to see.

Every winter Alfonso climbs up to the few remote sites in central Mexico where about 150 million monarch butterflies spend the winter. He is researching the monarchs because he wants to help preserve their habitat and the butterflies themselves. His work helped him earn a Ph.D. in ecology from the University of Florida.

Monarchs are famous for their long-distance migration. The butterflies that eventually find their way to Mexico come from as far away as the northeastern United States and southern Canada. Some of them travel an amazing 3,200 km before reaching central Mexico.

Wintering Habitat at Risk

Unfortunately, the habitat that the monarchs travel long distances to reach is increasingly threatened by logging and other human activities. Only 9 to 11 of the monarchs' wintering sites remain (monarchs colonize more sites in some years than in others). Five of those sites are set aside as sanctuaries for the butterflies, but even these sanctuaries are endangered by people who cut down fir trees for firewood or for commercial purposes.

Alfonso's work is helping Mexican conservationists better understand and protect monarch butterflies. Especially important is Alfonso's discovery that monarchs depend on bushlike vegetation, called *understory vegetation,* that grows beneath the fir trees.

Keeping Warm

Alfonso's research showed that when the temperature falls below freezing, as it often does in the mountains where the monarchs winter, understory vegetation can mean the difference between life and death for some monarchs. These conditions are life threatening to monarchs because low temperatures (–1°C to 4°C, or 30°F to 40°F) limit their movement. In fact, the butterflies are not even able to fly at such low temperatures. At extremely cold temperatures (–7°C to –1°C, or 20°F to 30°F), monarchs resting on the forest floor are in danger of freezing to death. But if the forest has understory vegetation, the monarchs can slowly climb the vegetation until they are at least 10 cm above the ground. This tiny difference in elevation can provide a microclimate that is warm enough to ensure the monarchs' survival.

▶ **Butterfly Man** Alfonso examines a monarch as part of his efforts to understand its ecology.

114

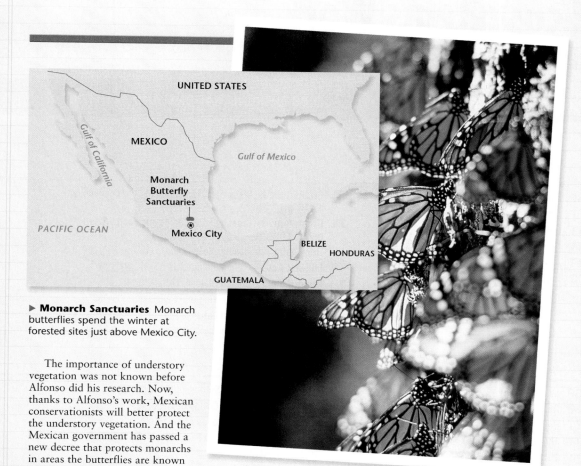

Monarch Sanctuaries Monarch butterflies spend the winter at forested sites just above Mexico City.

The importance of understory vegetation was not known before Alfonso did his research. Now, thanks to Alfonso's work, Mexican conservationists will better protect the understory vegetation. And the Mexican government has passed a new decree that protects monarchs in areas the butterflies are known to use.

The Need for Conservation

Although the monarchs continue to enjoy the forests where they overwinter, those forests are still threatened. There is little forest left in this area, and the need for wood increases each year. Alfonso hopes his efforts will help protect the monarch both now and in the future.

Now that he has completed his Ph.D., Alfonso is devoting himself to preserving monarchs and other organisms. He works as director for conservation and development for the Smithsonian Institutions

A Sea of Orange At their over-wintering sites in Mexico, millions of monarchs cover trees and bushes in a fluttering carpet of orange and black.

Monitoring and Assessment of Biodiversity (MAB) program.

Information...

If you are interested in learning more about monarchs, including their spectacular migration, visit the Website for Monarch Watch. Monarch Watch is an organization based at the University of Kansas that is dedicated to educating people about the monarch and promoting its conservation.

What Do You Think?

As a migrating species, monarchs spend part of their lives in the United States and part in Mexico. Should the U.S. and Mexico cooperate in their efforts to understand and manage the monarch? Should nations set up panels to manage other migrating species, such as many songbirds?

115

Answers to What Do You Think?

Animals that migrate between two countries need the habitat that exists in both countries to survive. If countries do not cooperate to understand and protect the animals, one of the countries may destroy critical habitat. Even if necessary habitat is ample in one country, a loss of habitat in the other country may cause the animal's population to decline or disappear. Therefore, cooperation between governments is essential when managing threatened or endangered migratory species.

How Ecosystems Work
Chapter Planning Guide

PACING	CLASSROOM RESOURCES	LABS, DEMONSTRATIONS, AND ACTIVITIES
BLOCKS 1, 2 & 3 · 135 min pp. 116–123 **Chapter Opener**		
Section 1 Energy Flow in Ecosystems	**OSP** Lesson Plan * **CRF** Active Reading * BASIC **TT** Bellringer * **TT** A Food Chain * **TT** A Food Web * **CD** Interactive Exploration 3 Sea Sick	**TE** **Group Activity** Classroom Hydrothermal Vent Community, p. 118 GENERAL **TE** **Group Activity** Creating Food Chains and Food Webs, p. 119 GENERAL **TE** **Demonstration** Cellular Respiration, p. 120 ◆ GENERAL **TE** **Internet Activity** Biomagnification, p. 121 GENERAL **TE** **Group Activity** Human Diets, p. 122 GENERAL **SE** **Exploration Lab** Dissecting Owl Pellets, pp. 138–139 ◆ **CRF** **Datasheets for In-Text Labs** * **CRF** **CBL™ Probeware Lab** * ◆ GENERAL
BLOCKS 4 & 5 · 90 min pp. 124–128 **Section 2** The Cycling of Materials	**OSP** Lesson Plan * **CRF** Active Reading * BASIC **TT** Bellringer * **TT** The Carbon Cycle * **TT** The Nitrogen Cycle * **TT** The Phosphorous Cycle * **CD** Interactive Tutor The Carbon, Nitrogen, and Phosphorus Cycles	**TE** **Activity** Carbon Cycle Stories, p. 124 GENERAL **SE** **QuickLab** Make Every Breath Count, p. 125 **TE** **Activity** Observing Nitrogen Fixing Bacteria, p. 126 GENERAL **TE** **Internet Activity** Nitrogen Fertilizers, p. 127 ADVANCED **CRF** **Observation Lab** * ◆ BASIC
BLOCKS 6 & 7 · 90 min pp. 129–133 **Section 3** How Ecosystems Change	**OSP** Lesson Plan * **CRF** Active Reading * BASIC **TT** Bellringer * **TT** Secondary Succession: Old-Field Succession * **CD** Interactive Tutor Ecological Succession	**SE** **Field Activity** Investigating Succession, p. 132 **TE** **Activity** Musical Succession, p. 132 BASIC **CRF** **Consumer Lab** * ◆ GENERAL **CRF** **Long-Term Project** * ADVANCED

BLOCKS 8 & 9 · 90 min

Chapter Review and Assessment Resources

- **SE** Chapter Review pp. 135–137
- **SE** Standardized Test Prep pp. 644–645
- **CRF** Study Guide * ■ GENERAL
- **CRF** Chapter Test * ■ GENERAL
- **CRF** Chapter Test * ADVANCED
- **CRF** Concept Review * ■ GENERAL
- **CRF** Critical Thinking * ADVANCED
- **OSP** Test Generator
- **CRF** Test Item Listing *

Online and Technology Resources

 Holt Online Learning

Visit **go.hrw.com** for access to Holt Online Learning or enter the keyword **HE6 Home** for a variety of free online resources.

 **One-Stop Planner® CD-ROM**

This CD-ROM package includes
- Lab Materials QuickList Software
- Holt Calendar Planner
- Customizable Lesson Plans
- Printable Worksheets
- ExamView® Test Generator
- Interactive Teacher Edition
- Holt PuzzlePro® Resources
- Holt PowerPoint® Resources

 Holt Environmental Science Interactive Tutor CD-ROM

This CD-ROM consists of interactive activities that give students a fun way to extend their knowledge of environmental science concepts.

KEY

TE	Teacher Edition	**CRF**	Chapter Resource File	*	Also on One-Stop Planner
SE	Student Edition	**TT**	Teaching Transparency	■	Also Available in Spanish
OSP	One-Stop Planner	**CD**	Interactive CD-ROM	◆	Requires Advance Prep

ENRICHMENT AND SKILLS PRACTICE	SECTION REVIEW AND ASSESSMENT	CORRELATIONS
SE Pre-Reading Activity, p. 116 **TE** Using the Figure, p. 116 `GENERAL`		**National Science Education Standards**
TE Using the Figure Life Depends on the Sun, p. 118 `BASIC` **TE** Skill Builder Vocabulary, p. 118 `BASIC` **SE** MathPractice A Meal Fit for a Grizzly Bear, p. 119 **TE** Skill Builder Math, p. 119 `GENERAL` **TE** Skill Builder Vocabulary, p. 119 `BASIC` **TE** Inclusion Strategies, p. 119 **SE** Case Study DDT in an Aquatic Food Chain, pp. 120–121 **TE** Career Toxicologist, p. 120 **TE** Inclusion Strategies, p. 122 **TE** Skill Builder Math, p. 122 `GENERAL`	**SE** Section Review, p. 123 **TE** Reteaching, p. 123 `BASIC` **TE** Quiz, p. 123 `GENERAL` **TE** Alternative Assessment, p. 123 `GENERAL` **CRF** Quiz * ■ `GENERAL`	LS 1e LS 4b LS 5f
TE Using the Figure The Nitrogen Cycle, p. 126 `ADVANCED`	**SE** Section Review, p. 128 **TE** Reteaching, p. 128 `BASIC` **TE** Quiz, p. 128 `GENERAL` **TE** Alternative Assessment, p. 128 `GENERAL` **CRF** Quiz * ■ `GENERAL`	ES 2a ES 2b LS 5f SPSP 4b
SE Graphic Organizer Chain-of-Events Chart, p. 129 `GENERAL` **SE** Case Study Communities Maintained by Fire, pp. 130–131 **TE** Skill Builder Writing, p. 130 `ADVANCED` **TE** Skill Builder Writing, p. 131 `GENERAL` **SE** Maps in Action Doppler Radar Tracking of Bats and Insects in Central Texas, p. 140 **SE** Society & the Environment Eating the Bait, p. 141 **CRF** Map Skills Global Warming * `GENERAL`	**SE** Section Review, p. 133 **TE** Reteaching, p. 133 `BASIC` **TE** Quiz, p. 133 `GENERAL` **TE** Alternative Assessment, p. 133 `GENERAL` **CRF** Quiz * ■ `GENERAL`	

 Guided Reading Audio CDs

These CDs are designed to help auditory learners and reluctant readers. (Audio CDs are also available in Spanish.)

 SCILINKS
NSTA
www.scilinks.org

Maintained by the **National Science Teachers Association**

Topic: Food Chains, Food Webs, and Trophic Levels
SciLinks Code: HE4043

Topic: Nitrogen Cycle
SciLinks Code: HE4073

Topic: Ecological Succession
SciLinks Code: HE4024

 CNN Videos

Each video segment is accompanied by a Critical Thinking Worksheet.

Science, Technology & Society
Segment 21 Taking Earth's Pulse

CHAPTER
5

Chapter Enrichment

This Chapter Enrichment provides relevant and interesting information to expand and enhance your classroom instruction of the chapter material.

SECTION 1 Energy Flow in Ecosystems

Organisms of a Hydrothermal Vent Community

Hydrothermal vents begin to form when parts of the ocean's crust move apart, creating cracks. Seawater is heated by magma under the cracks and flows out of the crust at a different location. Hydrothermal vents can be incredibly diverse and support life forms that do not derive energy from the sun. Some of the organisms that thrive in the hydrothermal vent community include:

- Bacteria that are able to thrive in a super-heated environment

- Tube worms and bacteria that live together in a symbiotic relationship (tubeworms provide a substrate for the bacteria, the bacteria provide food for the tubeworms)

- Pompeii Worms, the most heat-hardy animals on Earth (they can survive temperatures of 80°C)

- Spider crabs, crustaceans that look similar to the Lunar Lander

- Giant clams

- Deep sea octopuses

▶ **Tube worms in a hydrothermal vent community**

The Discovery of Photosynthesis

It was not until 1727 that scientists suspected that plants received energy from anywhere but the soil. That year, the Reverend Stephen Hales, an English plant physiologist, suggested that light was involved in plant growth. The first scientist to show that plants take in carbon dioxide from the atmosphere and release oxygen was Swiss naturalist Jean Senebier in 1782. By 1804, another Swiss scientist, N. T. de Saussure, made the first quantitative measurement of photosynthesis, when he found that plants gained more weight during photosynthesis than could be accounted for by just taking in carbon dioxide. Saussure discovered that water accounted for the extra weight and was therefore involved in the process. In 1864, German botanist Julius Sachs discovered that in addition to oxygen, starch was a product of photosynthesis.

SECTION 2 The Cycling of Materials

Nitrogen Fixation by Bacteria

Bacteria fix atmospheric nitrogen for legumes, but how do the bacteria end up in the roots of legumes in the first place? The answer is an interesting example of symbiosis. *Rhizobia* are the bacteria that can colonize legumes. Even though these bacteria can live in the soil, they can only fix nitrogen when they are inside root nodules of a legume. Rhizobia colonization occurs in the following way:

1. Legume root hairs secrete a substance that attracts rhizobia.

2. When rhizobia get close enough to the roots, the root hair curls around the bacteria.

3. An enzyme is secreted (it is unclear which organism secretes it) that softens the root hair cell wall.

4. An infection thread from the rhizobia grows into the root cortex, causing the root to form nodules, which are round tissue growths with bacteria inside.

5. Rhizobia within the nodules become irregularly shaped bacteroids (enlarged bacterial cells) and can begin fixing atmospheric nitrogen.

▶ **Nitrogen-fixing bacteria live inside the root nodules of legumes**

Nodules are full of a special type of hemoglobin (leghemoglobin), which provides the right amount of oxygen to the bacteroids. Leghemoglobin contains the same substance as animal blood hemoglobin, so when a root nodule is cut open, it will be red inside.

Carbon Sinks that Fight Global Warming

For many years, researchers thought that new plantation forests were more capable of trapping carbon dioxide than old, wild forests. In 2000, that information was called into question. A study conducted by researchers at the Max Planck Institute for Biogeochemistry in Germany, found that old forests lock away huge amounts of carbon in trunks, branches, roots, and the soil around the roots. When an old growth forest is cut, it can no longer absorb and store more carbon. It would take hundreds of years for young, growing forests to lock away the same amount of carbon. This study disputes the perception that old forests release as much carbon dioxide as they capture, due to increased levels of decaying wood. Instead, carbon binds to decayed leaves, twigs, and roots in these forests, and can remain sequestered in the soil for thousands of years. This study emphasizes that old forests must be preserved, and regulations should focus on limiting carbon dioxide emissions to decrease global warming.

BRAIN FOOD

Many other plants besides legumes have symbiotic associations with microbes. Trees such as pine, oak, and birch, form symbiotic relationships with many species of fungi. The fungi and roots combine to form ectomycorrhizae, which means "outside fungus root." Fungi wrap around the roots, forming a sheath that can make up 40% of the dry weight of the root system. Fungal hyphae (like growing fingers) penetrate the areas between the living root cells to form a sheath. The sheath of fungi is thought to make the trees more resistant to cold and dry conditions, because the sheath of fungi facilitates the transfer of organic carbon to the roots.

Acid Precipitation: Causes and Effects

Acid precipitation is caused by an atmospheric reaction between water, oxygen, and nitrogen oxides or sulfur dioxide. This reaction creates a mild solution of nitric acid or sulfuric acid, which falls from the sky. About 2/3 of all sulfur dioxide and 1/4 of all nitrogen oxide is produced by the burning of fossil fuels for electrical power generation. Acid precipitation was first noticed when fish in remote wilderness areas began to die mysteriously. Acid precipitation was later found to damage forests and soils, besides acidifying and damaging lakes. Acid precipitation has also caused the rapid decay of some historic stone buildings and monuments.

SECTION

3 How Ecosystems Change

▶ Lodgepole pine trees growing after a fire

Succession After a Fire

Some typical species that establish themselves in an area after a fire may be fireweed, grasses, wild roses, pea vine, willow, and other shrubs. Then, fast-growing trees such as poplar and birch may invade the area. These trees start to shade out species in the understory, and may increase moisture in the area by trapping water in their roots, and reducing evaporation from wind. Slower growing species, such as pines, may then invade the area. In some cases, deciduous trees such as oak, maple, and basswood will form the climax community. A climax community may remain in an area for many decades, if disturbances are rare.

More About Lichens

A lichen is an organism made up of two types of microorganisms: algae or cyanobacteria and fungi. There are three types of lichens: crustose, foliose, and fruticose. The crustose form is the type that adheres closely to rocks and to the bark of trees. Lichens can survive on rock because they are able to withstand dry conditions and extreme temperatures. Lichens break down rocks by producing organic acids and by holding water. The water may freeze and thaw in cracks of the rock to produce bigger cracks, which further breaks down the rock. Some lichens can fix nitrogen, because their algal component is cyanobacteria or blue-green algae. These lichens can be important sources of nitrogen in nutrient-poor communities, such as the tundra. Although lichens are adapted to withstand extreme conditions, they are vulnerable to air pollution. Sulfur dioxide in particular, inhibits photosynthesis in lichens, which causes the fungi to take over. This breaks the mutualistic relationship and destroys the lichen.

Overview

Tell students that in this chapter, they will learn how the flow of energy, cycling of materials, and ecological succession combine to affect how ecosystems work. Organisms need energy to stay alive. Some organisms, such as plants, can directly convert usable energy from the sun. The cycling of materials such as carbon, nitrogen, and phosphorus is essential to keep nutrients balanced in ecosystems. Human activities can affect these cycles. Through ecological succession, ecosystems can change over time.

Using the Figure — GENERAL

Frogs and dragonflies are both consumers that are high in an aquatic food chain. Ask the students to discuss how energy would continue to be transferred in this food chain. (Sample answer: a heron might gain energy by eating the green frog. When the heron dies, microbes might gain energy by breaking down the heron.) **LS** Logical

PRE-READING ACTIVITY

Encourage students to use their FoldNote as a study guide to quiz themselves for a test on the chapter material. Students may want to create Double-Door FoldNotes for different topics within the chapter.

For information about videos related to this chapter, go to **go.hrw.com** and type in the keyword **HE4 ECOV**.

How Ecosystems Work

1 Energy Flow in Ecosystems
2 The Cycling of Materials
3 How Ecosystems Change

PRE-READING ACTIVITY

FOLDNOTES **Double-Door Fold**
Before you read this chapter, create the **FoldNote** entitled "Double-Door Fold" described in the Reading and Study Skills section of the Appendix. Write "Energy flow in ecosystems" on one flap of the double door and "Movement of materials in ecosystems" on the other flap. As you read the chapter, compare the two topics, and write characteristics of each on the inside of the appropriate flap.

This green frog gets the energy it needs to survive by eating other organisms, such as dragonflies.

116

Chapter Correlations   *National Science Education Standards*

LS 1e Plant cells contain chloroplasts, the site of photosynthesis. Plants and many microorganisms use solar energy to combine molecules of carbon dioxide and water into complex, energy rich organic compounds and release oxygen to the environment. This process of photosynthesis provides a vital connection between the sun and the energy needs of living systems. **(Section 1)**

LS 4b Energy flows through ecosystems in one direction, from photosynthetic organisms to herbivores to carnivores and decomposers. **(Section 1)**

LS 5f As matter and energy flows through different levels of organization of living systems—cells, organs, organisms, communities—and between living systems and the physical environment, chemical elements are recombined in different ways. Each recombination results in storage and dissipation of energy into the environment as heat. Matter and energy are conserved in each change. **(Section 1 and Section 2)**

ES 2a The earth is a system containing essentially a fixed amount of each stable chemical atom or element. Each element can exist in several different chemical reservoirs. Each element on earth moves among reservoirs in the solid earth, oceans, atmosphere, and organisms as part of geochemical cycles. **(Section 2)**

ES 2b Movement of matter between reservoirs is driven by the earth's internal and external sources of energy. These movements are often accompanied by a change in the physical and chemical properties of the matter. Carbon, for example, occurs in carbonate rocks such as limestone, in the atmosphere as carbon dioxide gas, in water as dissolved carbon dioxide, and in all organisms as complex molecules that control the chemistry of life. **(Section 2)**

SPSP 4b Materials from human societies affect both physical and chemical cycles of the earth. **(Section 2)**

SECTION 1
Energy Flow in Ecosystems

Just as a car cannot run without fuel, an organism cannot survive without a constant supply of energy. Where does an organism's energy come from? The answer to that question depends on the organism, but the ultimate source of energy for almost all organisms is the sun.

Life Depends on the Sun

Energy from the sun enters an ecosystem when a plant uses sunlight to make sugar molecules by a process called **photosynthesis.** During photosynthesis, plants, algae, and some bacteria capture solar energy. Solar energy drives a series of chemical reactions that require carbon dioxide and water, as shown in **Figure 1.** The result of photosynthesis is the production of sugar molecules known as *carbohydrates.* Carbohydrates are energy-rich molecules which organisms use to carry out daily activities. As organisms consume food and use energy from carbohydrates, the energy travels from one organism to another. Plants, such as the sunflowers in **Figure 2,** produce carbohydrates in their leaves. When an animal eats a plant, some energy is transferred from the plant to the animal. Organisms use this energy to move, grow, and reproduce.

Objectives

► Describe how energy is transferred from the sun to producers and then to consumers.
► Describe one way in which consumers depend on producers.
► List two types of consumers.
► Explain how energy transfer in a food web is more complex than energy transfer in a food chain.
► Explain why an energy pyramid is a representation of trophic levels.

Key Terms

photosynthesis
producer
consumer
decomposer
cellular respiration
food chain
food web
trophic level

Figure 1 ► During photosynthesis, plants use carbon dioxide, water, and solar energy to make carbohydrates and oxygen.

Figure 2 ► The cells in the leaves of these sunflowers contain a green chemical called *chlorophyll.* Chlorophyll helps plants trap energy from the sun to produce energy-rich carbohydrates.

Chapter Resource File

- Lesson Plan
- **Active Reading** BASIC
- **Section Quiz** GENERAL

Transparencies

TT Bellringer

Overview

Before beginning this section, review with your students the Objectives in the Student Edition. This section discusses the principle that sunlight is the ultimate energy source for nearly all living things. The section describes energy transfer in ecosystems. Students will also learn about the roles of producers, consumers, and decomposers in food chains and food webs.

Bellringer

Ask students to write in their *EcoLog* three plants or animals and the animals that eat them. Also ask them to write down any plants they know of that eat animals (Venus flytrap and pitcher plant are two examples). Tell them to be creative, and to think about animals and plants on different continents.

Motivate

Identifying Preconceptions ——

Energy: The Missing Ingredient
Write the following ingredients on the board: 1/2 bathtub full of oxygen, 50 glasses of water, 1/2 cup of sugar, 1/2 cup of calcium, 1/10 thimbleful of salt, a small pinch of assorted elements such as phosphorus, potassium, nitrogen, sulfur, magnesium, and iron, and a mystery ingredient.

Ask students to write down what these ingredients will make, and what they think the mystery ingredient is. Tell students, "Believe it or not, these are the basic ingredients that you are made of and energy is the missing ingredient." Then ask, "Where does that energy come from?" (the food we eat) Ask, "Where does the energy in food come from?" (ultimately, the sun)
LS Logical

Using the Figure — BASIC

Life Depends on the Sun Use the photographs in **Figure 3** to help students visualize how life depends on the sun. Ask, "In this series of photographs, how does the coyote depend on the sun?" (The coyote depends on the sun because it gains the energy it needs to survive by eating the rabbit. The rabbit gains its energy from the clover, which in turn gains energy from the sun.) **LS** Visual

Group Activity — GENERAL

Classroom Hydrothermal Vent Community Bring in some pieces of poster board and photos of organisms associated with deep ocean hydrothermal vent communities. Ask students to draw individual pictures of creatures that depend on the bacteria associated with hydrothermal vents. Ask students to sketch bacterial colonies and the vents themselves. Tell students to clearly label each creature with its common and scientific name, if possible. Ask students to cut out their drawings and arrange them into communities on the pieces of poster board. After each community is constructed, have students explain how all the creatures interact with each other. If students are unsure of the role of some of the organisms, ask them to research these roles before the next class period. **LS** Visual/Kinesthetic

SKILL BUILDER — BASIC

Vocabulary The Greek suffix *–troph* means "feeder." Therefore, *trophic* levels are "feeding" levels.

Figure 3 ▶ Transfer of Energy
Almost all organisms depend on the sun for energy. Plants like the clover shown above get energy from the sun. Animals like the rabbit and coyote get their energy by eating other organisms.

Figure 4 ▶ The tube worms (above) depend on bacteria that live inside them to survive. The bacteria (right) use energy from hydrogen sulfide to make their own food.

118

From Producers to Consumers When a rabbit eats a clover plant, the rabbit gets energy from the carbohydrates the clover plant made through photosynthesis. If a coyote eats the rabbit, some of the energy is transferred from the rabbit to the coyote. In the example shown in **Figure 3**, the clover is the producer. A **producer** is an organism that makes its own food. Producers are also called *autotrophs*, self-feeders. Both the rabbit and the coyote are **consumers,** organisms that get their energy by eating other organisms. Consumers are also called *heterotrophs*, other-feeders. In the example shown in **Figure 3**, the clover, rabbit, and coyote get their energy from the sun. Some producers get energy directly from the sun by absorbing it through their leaves. Consumers get energy indirectly from the sun by eating producers or other consumers.

An Exception to the Rule: Deep-Ocean Ecosystems In 1977, scientists discovered areas on the bottom of the ocean off the coast of Ecuador that were teeming with life, even though sunlight did not reach the bottom of the ocean. The scientists found large communities of worms, clams, crabs, mussels, and barnacles living near thermal vents in the ocean floor. These deep-ocean communities exist in total darkness, where photosynthesis cannot occur. So where do these organisms get their energy? Bacteria, such as those pictured in **Figure 4**, live in some of these organisms and use hydrogen sulfide to make their own food. Hydrogen sulfide is present in the hot water that escapes from the cracks in the ocean floor. Therefore, the bacteria are producers that can make food without sunlight. These bacteria are eaten by the other underwater organisms and thus support a thriving ecosystem.

MISCONCEPTION ALERT

Thermal Vents Are Diverse With Life
Students may think that hydrothermal vents are hot, hostile places that support only bacteria and a few other unique organisms. However, more than 300 species are associated with deep ocean thermal vents. Bacteria colonize the vents, and serve as the food base for other organisms such as mussels, shrimp, and crabs. Alternately, students may have heard that the hydrothermal vents produce very hot water, and they may think that the organisms living around the vents are living there to escape the frigid ocean conditions. Even though the water around the vents can reach up to 400 °C (752 °F), temperatures can drop to just above freezing just outside the vents. Most of the organisms that feed on the vent bacteria live in the cold zone.

What Eats What

Table 1 below classifies organisms by the source of their energy. Consumers that eat only producers are called *herbivores*, or plant eaters. Rabbits are herbivores and so are cows, sheep, deer, grasshoppers, and many other animals. Consumers, such as lions and hawks, that eat only other consumers are called *carnivores*, or flesh eaters. You already know that humans are consumers, but what kind of consumers are we? Because most humans eat both plants and animals, we are called *omnivores*, or eaters of all. Bears, pigs, and cockroaches are other examples of omnivores. Some consumers get their food by breaking down dead organisms and are called **decomposers.** Bacteria and fungi are examples of decomposers. The decomposers allow the nutrients in the rotting material to return to the soil, water, and air.

Table 1 ▼

What Eats What in an Ecosystem		
	Energy source	**Examples**
Producer	makes its own food through photosynthesis or chemical sources	grasses, ferns, cactuses, flowering plants, trees, algae, and some bacteria
Consumer	gets energy by eating producers or other consumers	mice, starfish, elephants, turtles, humans, and ants

Types of Consumers in an Ecosystem		
	Energy source	**Examples**
Herbivore	producers	cows, sheep, deer, and grasshoppers
Carnivore	other consumers	lions, hawks, snakes, spiders, sharks, alligators, and whales
Omnivore	both producers and consumers	bears, pigs, gorillas, rats, raccoons, cockroaches, some insects, and humans
Decomposer	breaks down dead organisms in an ecosystem and returns nutrients to soil, water, and air	fungi and bacteria

Figure 5 ▶ Bears, such as the grizzly bear below, are omnivores. Grizzly bears eat other consumers, such as salmon, but they also eat various plants.

 119

MATHPRACTICE

A Meal Fit for a Grizzly Bear Grizzly bears are omnivores that can eat up to 15 percent of their body weight per day when eating salmon and up to 33 percent of their body weight when eating fruits and other vegetation. How many pounds of salmon can a 200 lb grizzly bear eat in one day? How many pounds of fruits and other vegetation can the same bear eat in one day?

MATHPRACTICE

Answers

15% of 200 lbs = 30 lbs; 33% of 200 lbs = 66 lbs

Group Activity ── GENERAL

Creating Food Chains and Food Webs Pass out 1 or 2 blank index cards to each student. Ask students to write an organism on each card (tell them to use large letters, so that everyone can see the name of each organism at the back of the classroom). Divide the middle of the board into the following sections (from top to bottom): Carnivore, Herbivore, and Producer. Write Decomposer to the right and Omnivore to the left of Herbivore. Have students stick their cards onto the correct section of the board with tape. Then ask a couple of students to stay at the board and, using suggestions from the class, have them arrange the cards into specific food chains/food webs (drawing lines between the connected elements). Many of the food webs will get messy, but that should help students to understand that ecological interactions in the real world can be complicated. If none of the students contribute any decomposers to the board, have them brainstorm some possible organisms that break down animals or plants, and link them to the food chains/food webs.
LS Kinesthetic/Visual

SKILL BUILDER ── GENERAL

Math Have students figure out how many calories there are in 30 lbs of salmon if each pound of salmon = approx. 654 cal (30 lbs × 654 cal = 19620 cal). Have students calculate how many calories there are in 66 lbs of blueberries if each pound of blueberries = approx. 261 cal (66 lbs × 261 cal = 17226 cal). **LS** Logical

Transparencies
TT A Food Chain
TT A Food Web

SKILL BUILDER ── BASIC

Vocabulary To help students learn some of the new vocabulary in this section, direct their attention to **Table 1.** Read the table out loud, discussing each portion. Then use magazine photos and other visual aids to provide more examples of each type of organism. Check student comprehension by holding up various photos and asking students whether the organism shown is a producer or a consumer. If it is a consumer, ask them which type of consumer it is. (herbivore, carnivore, omnivore, or decomposer)
English Language Learners
LS Visual

INCLUSION Strategies

- *Learning Disabled*
- *Developmentally Delayed*
- *Attention Deficit Disorder*

Ask students to draw a picture of an imaginary ecosystem. The picture should include organisms that are producers and consumers as defined in **Table 1.** The types of consumers and producers should be labeled clearly. The student may present their ecosystem to the class or small group to show their understanding of the concept.

Demonstration —— GENERAL

Cellular Respiration Obtain the following materials: 50 mL Erlenmeyer flask, 2.5 mL of yeast, 40 mL of apple cider, and a latex balloon.

Mix the yeast and cider in the flask and place the balloon securely on top of the flask. Yeast will respire aerobically and anaerobically, producing CO_2, which will inflate the balloon in about 2–3 days. During these days, encourage students to research how respiration works in yeast. Have students record their research and daily observations in their *EcoLog.* LS Visual/Kinesthetic

LANGUAGE ARTS —
● CONNECTION — ADVANCED

Silent Spring The book *Silent Spring* was written to educate the public on the potential harm to the environment from the overuse of chemical pesticides, such as DDT. The author, Rachel Carson, was a former U.S. Fish and Wildlife Service employee who researched her findings carefully before publishing them. When *Silent Spring* was published, chemical companies responded with a campaign touting the benefits of pesticides and attacking Carson. This actually helped to publicize the book. Her book was responsible for changes to pesticide regulations in the U.S. Urge students to check out this important book at their local library.

Connection to Chemistry

Chemical Equations Chemical reactions are represented by chemical equations. A chemical equation is a shorthand description of a chemical reaction using chemical formulas and symbols. The starting materials in a reaction are called *reactants*, and the substances formed from a reaction are called *products*. The number of atoms of each element in the reactants equals the number of atoms of those elements in the products to make a balanced equation.

Figure 6 ▶ Through cellular respiration, cells use glucose and oxygen to produce carbon dioxide, water, and energy.

Cellular Respiration: Burning the Fuel

So far, you have learned how organisms get energy. But how do they use the energy they get? To understand the process, use yourself as an example. Suppose you have just eaten a large meal. The food you ate contains a lot of energy. Your body gets the energy out of the food by using the oxygen you breathe to break down the food. By breaking down the food, your body obtains the energy stored in the food.

The process of breaking down food to yield energy is called **cellular respiration,** which occurs inside the cells of most organisms. This process is different from *respiration*, which is another name for breathing. During cellular respiration, cells absorb oxygen and use it to release energy from food. As you can see in **Figure 6**, the chemical equation for cellular respiration is essentially the reverse of the equation for photosynthesis. During cellular respiration, sugar and oxygen combine to yield carbon dioxide, water, and, most importantly, energy.

$$C_6H_{12}O_6 + 6O_2 \longrightarrow 6CO_2 + 6H_2O + \text{energy}$$

CASE STUDY

DDT in an Aquatic Food Chain

In the 1950s and 1960s, something strange was happening in the estuaries near Long Island Sound, near New York and Connecticut. Birds of prey, such as ospreys and eagles, that fed on fish in the estuaries had high concentrations of the pesticide DDT in their bodies. But when the water in the estuary was tested, it had low concentrations of DDT.

What accounted for the high levels of DDT in the birds? Poisons that dissolve in fat, such as DDT, can become more concentrated as they move up a food chain in a process called *biological magnification*. When the pesticide enters the water, algae and bacteria take in the poison. When fish eat the algae and bacteria, the poison dissolves into the fat of the fish rather than diffusing back into the water. Each time a bird feeds

on a fish, the bird accumulates more DDT in its fatty tissues. In some estuaries on Long Island Sound, DDT

concentrations in fatty tissues of organisms were magnified almost 10 million times from the bottom to the

▶ A high concentration of DDT decreases the thickness and the strength of eggshells of many birds of prey.

120

Career

Toxicologist Toxicologists study adverse effects of chemicals on living organisms and ecosystems. They look at relationships between chemicals and disease, and environmental risks associated with various chemicals. A toxicologist can be trained as a biologist or chemist. Toxicologists typically work in government, universities, chemical industries, or consulting firms. Toxicologists are concerned with the short and long-term safety of chemicals. Many participate in setting guidelines for the regulation of certain chemicals. If your local university or college has a toxicology program, contact the department to invite a toxicology professor or graduate student to speak with the class.

You use a part of the energy you obtained through cellular respiration to carry out your daily activities. Every time you walk, breathe, read a book, think, or play a sport, you use energy. The energy you obtain is also used to make more body tissues and to fight diseases so that you grow and stay healthy. Excess energy you obtain is stored as fat or sugar. All living things use cellular respiration to get the energy they need from food molecules. Even organisms that make their own food through photosynthesis use cellular respiration to obtain energy from the carbohydrates they produce.

Energy Transfer

Each time one organism eats another organism, a transfer of energy occurs. We can trace the transfer of energy as it travels through an ecosystem by studying food chains, food webs, and trophic levels. Food chains, food webs, and trophic levels can tell us how energy is transferred as well as how much energy is transferred between organisms in an ecosystem. Studying the paths of energy between organisms can also tell us which organisms in an ecosystem depend on other organisms to survive.

▶ Poisons such as DDT have the greatest affect on organisms at the top of food chains. For example, the osprey shown here would have a greater concentration of DDT in its body than the perch it's about to eat.

top of the food chain. Large concentrations of DDT may kill an organism, weaken its immune system, cause deformities, or impair its ability to reproduce. DDT can also weaken the shells of bird eggs. When eggs break too soon, bird embryos die. Therefore, the effects of these chemicals cause a tremendous drop in the population of carnivorous bird species.

The U.S. government recognized DDT as an environmental contaminant and in 1972 banned its sale except in emergencies. The aquatic food chains immediately started to recover, and the populations of ospreys and eagles started to grow.

Food chains are still not free of DDT. DDT is still legal in some countries, where it is used in large quantities to eliminate mosquitoes that carry the disease malaria. As a result, migratory birds may be exposed to DDT while wintering in locations outside the United States.

CRITICAL THINKING

1. Analyzing Processes DDT does not dissolve readily in water. If it did, how would the accumulation of the pesticide in organisms be affected?

2. Evaluating Information Even though DDT is harmful to the environment, why is it still used in some countries?

Group Activity —— GENERAL

Human Diets Divide the class into Group A and Group B. Tell students in Group A that their assignment is to determine the cost per ounce of rice, dried beans, and rolled oats. Tell students in Group B that their assignment is to research and determine the cost per ounce of ground beef, chicken, and pork chops. Then, ask each group to determine the number of calories per ounce and the cost of one calorie for each type of food. Have Group A and Group B compare their results. (Costs per calorie are lower for grains and other plant products than for meats.) Ask, "What is the significance of your calculations for humans?" (Humans could have more food for more people and spend less money by eating foods lower on the food chain.) **LS Logical** Co-op Learning

● INCLUSION Strategies

• *Learning Disabled*
• *Developmentally Delayed*
• *Attention Deficit Disorder*

Ask students to draw a food chain from a different ecosystem than the one shown in **Figure 7.** The food chain should include producers and consumers. Students can cut and paste each organism onto a note card. On the back of the card, have students indicate the name of the organism and whether it is a producer or consumer in the food chain. Have students exchange note cards with each other and reorder them as a study activity.

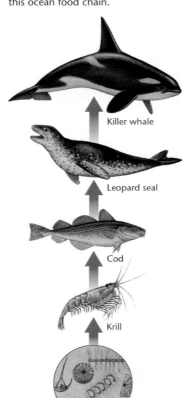

Figure 7 ▶ Energy is transferred from one organism to another in a food chain, such as the one shown below. Algae are the producers in this ocean food chain.

Killer whale

Leopard seal

Cod

Krill

Algae

Figure 8 ▶ This food web shows how the largest organisms, such as a killer whale, depend on the smallest organisms, such as algae, in an ocean ecosystem.

Food Chains and Food Webs A **food chain** is a sequence in which energy is transferred from one organism to the next as each organism eats another organism. **Figure 7** shows a typical food chain in an ocean ecosystem. Algae are eaten by krill, which are eaten by cod. The cod are eaten by leopard seals, which are eaten by killer whales.

Energy flow in an ecosystem is much more complicated than energy flow in a simple food chain. Ecosystems almost always contain many more species than a single food chain contains. In addition, most organisms, including humans, eat more than one kind of food. So a food web, such as the one shown in **Figure 8,** includes more organisms and multiple food chains linked together. A **food web** shows many feeding relationships that are possible in an ecosystem. Notice that the food chain is just one strand in the larger food web.

Trophic Levels Each step in the transfer of energy through a food chain or food web in an ecosystem is known as a **trophic level.** In **Figure 8,** the algae are in the bottom trophic level, the krill are in the next level, and so on. Each time energy is transferred from one organism to another, some of the energy is lost as heat and less energy is available to organisms at the next trophic level. Some of this energy is lost during cellular respiration. Organisms use much of the remaining energy to carry out the functions of living, such as producing new cells, regulating body temperature, and moving

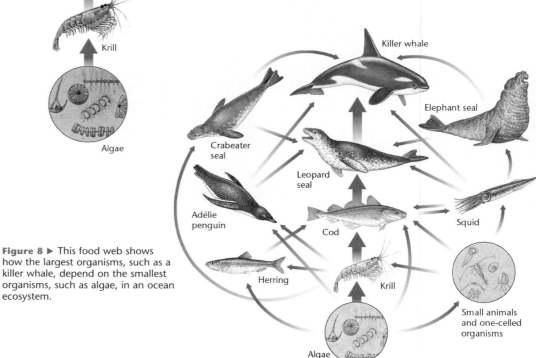

Killer whale

Elephant seal

Crabeater seal

Leopard seal

Adélie penguin

Cod

Squid

Herring

Krill

Small animals and one-celled organisms

Algae

Math Skills Write the following energy pyramid on the board, and ask students to figure out the percentage of energy loss at each level (from studies done at Cayuga Lake, New York):

1. Humans: 1.2 calories
2. Trout: ~6 calories
3. Smelt (a small fish): 30 calories
4. Small aquatic animals: 150 calories
5. Algae: 1000 calories

(Energy loss from levels 5–4: 85%; 4–3: 80%; 3–2: 80%; 2–1: 80%) **LS Logical**

ANATOMY
● CONNECTION —— GENERAL

Chew on This Ask students to name a typical mammalian carnivore (any wild cat), a herbivore (rabbit or deer), and an omnivore (humans are best for this exercise). Have them describe the shape of each organism's teeth (molars, canines, incisors). Bring pictures of skulls, or actual skulls in as a resource. Ask students to describe how the shape of teeth relates to feeding groups. (Cats have sharp canines for grabbing and piercing flesh. Rabbits have incisors for cutting plant material, and flat molars for grinding, and deer have only molars. Humans have a mix of grinding and sharp teeth to eat all types of food.) **LS Visual**

around. About 90 percent of the energy at each trophic level is used in these ways. The remaining 10 percent of the energy becomes part of the organism's body and is stored in its molecules. This 10 percent that is stored is all that is available to the next trophic level when one organism consumes another organism.

Energy Pyramids One way to visualize the loss of energy from one trophic level to the next trophic level is to draw an energy pyramid like the one shown in **Figure 9**. Each layer in the energy pyramid represents one trophic level. Producers form the base of the pyramid, the lowest trophic level, which contains the most energy. Herbivores contain less energy and make up the second level. Carnivores that feed on herbivores form the next level, and carnivores that feed on other carnivores make up the top level. Organisms in the upper trophic levels store less energy than both herbivores and producers. A pyramid is a good way to illustrate trophic levels because the pyramid becomes smaller toward the top, where less energy is available.

How Energy Loss Affects an Ecosystem The decreased amount of energy at each trophic level affects the organization of an ecosystem. First, because so much energy is lost at each level, there are fewer organisms at the higher trophic levels. For example, zebras and other herbivores outnumber lions on the African savanna by about 1,000 to 1. In this example, there simply are not enough herbivores to support more carnivores.

Second, the loss of energy from trophic level to trophic level limits the number of trophic levels in an ecosystem. Ecosystems rarely have more than four or five trophic levels because the ecosystem does not have enough energy left to support higher levels. For example, a lion typically needs up to 250 km² of land to hunt for food. Therefore, an animal that feeds on lions would have to expend a lot of energy to harvest the small amount of energy available at the top trophic level. The organisms that do feed on organisms at the top trophic level are usually small, such as parasitic worms and fleas that require a very small amount of energy.

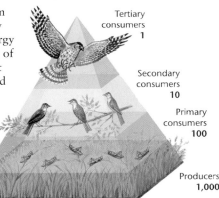

Figure 9 ▶ This energy pyramid shows how energy is lost from one trophic level to the next. The grass at the bottom level stores 1,000 times more energy than the hawk at the top level.

Tertiary consumers **1**

Secondary consumers **10**

Primary consumers **100**

Producers **1,000**

internet connect

www.scilinks.org
Topic: Food Chains, Food Webs, and Trophic Levels
SciLinks code: HE4043

SciLINKS. Maintained by the National Science Teachers Association

SECTION 1 Review

1. **Describe** how energy is transferred from one organism to another.

2. **Describe** the role that producers play in an ecosystem.

3. **Explain** the difference between an herbivore and an omnivore.

4. **Compare** energy transfer in a food chain to energy transfer in a food web.

CRITICAL THINKING

5. **Interpreting Graphics** Look at Figure 8. What feeding relationships does the crabeater seal have?

6. **Inferring Relationships** Read the paragraph under the heading, "Trophic Levels" in this section. Could more people be supported by 20 acres of land if they ate only plants instead of both plants and animals? Explain your answer. **READING SKILLS**

Overview

Before beginning this section, review with your students the Objectives in the Student Edition. This section describes the carbon, nitrogen, and phosphorus cycles. The section also describes how human activities affect these cycles.

 **Bellringer**

Ask students to list 3 products that they recycle. Ask, "Where do the products come from? Where will the products go after they are recycled?" Discuss their answers in relation to natural cycles.
LS Intrapersonal

Motivate

Activity —————— GENERAL

Carbon Cycle Stories Tell students that, on average, it takes a carbon atom about 300 years to complete one cycle. Based on that information, have each student write a short story describing who or what might have used the same carbon atom that he or she is using now.

Chapter Resource File
• **Lesson Plan**
• **Active Reading** BASIC
• **Section Quiz** GENERAL

Transparencies
TT Bellringer
TT The Carbon Cycle

SECTION 2
The Cycling of Materials

Objectives

▶ Describe the short-term and long-term process of the carbon cycle.
▶ Identify one way that humans are affecting the carbon cycle.
▶ List the three stages of the nitrogen cycle.
▶ Describe the role that nitrogen-fixing bacteria play in the nitrogen cycle.
▶ Explain how the excess use of fertilizer can affect the nitrogen and phosphorus cycles.

Key Terms

carbon cycle
nitrogen-fixing bacteria
nitrogen cycle
phosphorus cycle

What will happen to the next ballpoint pen you buy? You will probably use it until its ink supply runs out and then throw it away. The plastic and steel the pen is made of will probably never be reused. By contrast, materials in ecosystems are constantly reused. In this section, you will read about three cycles by which materials are reused—the carbon cycle, the nitrogen cycle, and the phosphorus cycle.

The Carbon Cycle

Carbon is an essential component of proteins, fats, and carbohydrates, which make up all organisms. The **carbon cycle** is a process by which carbon is cycled between the atmosphere, land, water, and organisms. As shown in **Figure 10,** carbon enters a short-term cycle in an ecosystem when producers, such as plants, convert carbon dioxide in the atmosphere into carbohydrates during photosynthesis. When consumers eat producers, the consumers obtain carbon from the carbohydrates. As the consumers break down the food during cellular respiration, some of the carbon is released back into the atmosphere as carbon dioxide. Organisms that make their own food through photosynthesis also release carbon dioxide during cellular respiration.

Some carbon enters a long-term cycle. For example, carbon may be converted into *carbonates*, which make up the hard parts of bones and shells. Bones and shells do not break down easily.

Figure 10 ▶ The Carbon Cycle
Carbon exists in air, water, soil, and living organisms. What role do plants play in the carbon cycle?

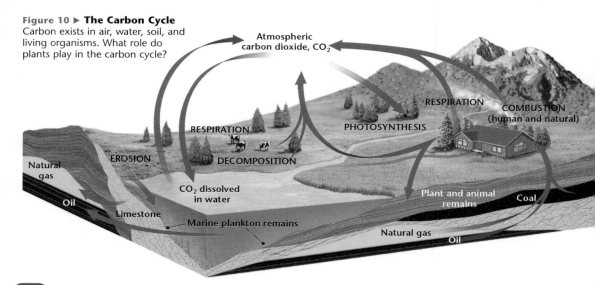

BRAIN FOOD

Biogeochemical Cycles The cycling of materials is accomplished in huge cycles called *biogeochemical cycles*. Biogeochemical means "life-earth-chemical." Different elements require different lengths of time to go from the living part of an ecosystem, such as an organism, to the nonliving part, such as soil, water, or atmosphere, and back to the body of a living organism again. The amount of matter processed through biogeochemical cycles is enormous. For example, about 90 million metric tons of nitrogen are cycled per year—90 percent of it is cycled by bacteria and 10 percent of it is cycled by human-produced fertilizers. Although animals are usually thought of as returning CO_2 to the atmosphere, both plants and decomposers return more CO_2 than animals do—plants return 42 percent, decomposers return 46 percent, and animals return 12 percent.

Over millions of years, carbonate deposits produce huge formations of limestone rocks. Limestone is one of the largest *carbon sinks*, or carbon reservoirs, on Earth.

Some carbohydrates in organisms are converted into fats, oils, and other molecules that store energy. The carbon in these molecules may be released into the soil or air after an organism dies. When these molecules are released they can form deposits of coal, oil, and natural gas underground known as *fossil fuels*. Fossil fuels are essentially stored carbon left over from bodies of plants and animals that died millions of years ago.

How Humans Affect the Carbon Cycle When we burn fossil fuels, we release carbon into the atmosphere. The carbon returns to the atmosphere as carbon dioxide. Cars, factories, and power plants rely on these fossil fuels to operate. In the year 2000, vehicles, such as the truck in **Figure 11**, were the source of one-third of all carbon dioxide emitted in the United States. All together, about 6 billion metric tons of carbon a year are released into the atmosphere as carbon dioxide. Natural burning of wood or forest fires combined with the burning of fossil fuels make up this 6 billion metric tons. About half of this carbon dioxide remains in the atmosphere, so over a period of years, the amount of carbon dioxide in the atmosphere has steadily increased.

Increased levels of carbon dioxide may contribute to global warming, which is an overall increase in the temperature of the Earth. What happens to the carbon dioxide that is not absorbed by the atmosphere? Scientists estimate that over a billion metric tons of carbon dioxide dissolves into the ocean, which is a carbon sink. Plants probably absorb the remaining carbon dioxide.

Figure 11 ▶ This truck releases carbon into the atmosphere when it burns fuel to operate.

Make Every Breath Count

Procedure
1. Pour **100 mL** of **water** from a **graduated cylinder** into a **250 mL beaker**. Add several drops of **bromthymol blue** to the beaker of water. Make sure you add enough to make the solution a dark blue color.
2. Exhale through a **straw** into the solution until the solution turns yellow. (CAUTION: Be sure not to inhale or ingest the solution.)
3. Pour the yellow solution into a large **test tube** that contains a **sprig** of *Elodea*.
4. **Stopper** the test tube, and place it in a sunny location.
5. Observe the solution in the test tube after 15 minutes.

Analysis
1. What do you think happened to the carbon dioxide that you exhaled into the solution? What effect do plants, such as the *Elodea*, have on the carbon cycle?

Connection to ▶ Biology

The Rise of Carbon Dioxide
In the past 150 years, more than 350 billion tons of carbon have been released into the air in the form of carbon dioxide. The concentration of carbon dioxide today has increased 30 percent since preindustrial times. If the present amount of carbon dioxide emission continues, this concentration will double by 2080. Many scientists speculate that as a result, Earth's temperature may rise by 3°C.

QuickLAB

Skills Acquired:
• Predicting
• Interpreting

Teacher's Notes: Brom thymol will stain clothes, so you may want to have students wear aprons for this activity.

Answers
1. The carbon dioxide that was exhaled into the solution was used by the sprig of *Elodea*, which caused the color of the solution to turn from yellow and back to blue. Plants, such as the *Elodea* convert atmospheric carbon dioxide into carbohydrates during photosynthesis. Plants also release carbon dioxide into the atmosphere during cellular respiration.

EARTH SCIENCE — CONNECTION ADVANCED

Carbon Sequestration The increased use of fossil fuels has caused the amount of carbon dioxide in the atmosphere to also increase. In order to reduce the amount of atmospheric CO_2, some scientists are researching carbon sequestration. Carbon sequestration involves capturing atmospheric CO_2 and storing it in the terrestrial biosphere, underground, or in the oceans. The U.S. Department of Energy is currently focusing on ways to enhance carbon sequestration in the terrestrial biosphere by increasing the storage of carbon in biomass and soils. They are also focusing on enhancing the net oceanic uptake of CO_2 from the atmosphere by fertilizing phytoplankton as well as by injecting liquid CO_2 to ocean depths greater than 1000 m. Scientists are also investigating methods of pumping carbon dioxide into abandoned mines and petroleum wells.

Using the Figure — ADVANCED

The Nitrogen Cycle Refer students to **Figure 12** and ask, "If only bacteria can use nitrogen from the atmosphere, how do plants and animals take part in the nitrogen cycle?"(Nitrogen-fixing bacteria convert atmospheric nitrogen into compounds plants can use. Plants absorb these compounds and use them to make proteins. Herbivores obtain nitrogen by eating plants. Carnivores get nitrogen by eating herbivores and other carnivores. Nitrogen is released in animal wastes and when decomposers break down dead plant and animal matter.) **LS Logical**

Activity — GENERAL

Observing Nitrogen Fixing Bacteria Obtain samples of fresh soybeans, alfalfa, or clover with their root nodules intact. Have students squash a nodule and then swab the material on a microscope slide. Have students stain the sample with methylene blue. Then have students apply a cover slip and examine the nitrogen-fixing bacteria that live within the nodules. Suggest that they draw what they see. Next, have students dry the plant samples and mount them on poster board along with their drawings of the nitrogen-fixing bacteria. **Caution:** Methylene blue will stain skin and clothes. **LS Visual**

Transparencies
TT The Nitrogen Cycle

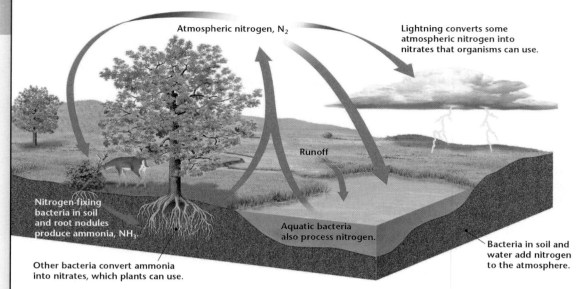

Atmospheric nitrogen, N₂

Lightning converts some atmospheric nitrogen into nitrates that organisms can use.

Runoff

Nitrogen-fixing bacteria in soil and root nodules produce ammonia, NH₃.

Aquatic bacteria also process nitrogen.

Bacteria in soil and water add nitrogen to the atmosphere.

Other bacteria convert ammonia into nitrates, which plants can use.

Figure 12 ▶ The Nitrogen Cycle
Nitrogen could not be cycled in the atmosphere without nitrogen-fixing bacteria. What role do animals play in the nitrogen cycle?

Figure 13 ▶ The swellings on the roots of this soybean plant are called *nodules*. Nitrogen-fixing bacteria, as shown magnified at right, live inside the nodules of plants.

The Nitrogen Cycle

All organisms need nitrogen to build *proteins,* which are used to build new cells. Nitrogen makes up 78 percent of the gases in the atmosphere. However, most organisms cannot use atmospheric nitrogen. It must be altered, or fixed, before organisms can use it. The only organisms that can fix atmospheric nitrogen into chemical compounds are a few species of bacteria known as nitrogen-fixing bacteria. All other organisms depend upon these bacteria to supply nitrogen. Nitrogen-fixing bacteria are a crucial part of the nitrogen cycle, a process in which nitrogen is cycled between the atmosphere, bacteria, and other organisms. As shown in **Figure 12**, bacteria take nitrogen gas from the air and transform it into molecules that living things can use.

Nitrogen-fixing bacteria, shown in **Figure 13**, live within nodules on the roots of plants called *legumes*. Legumes include beans, peas, and clover. The bacteria use sugars provided by the legumes to produce nitrogen-containing compounds such as nitrates. The excess nitrogen fixed by the bacteria is released into the soil. In addition, some nitrogen-fixing bacteria live in the soil rather than inside the roots of legumes. Plants that do not have nitrogen-fixing bacteria in their roots get nitrogen from the soil. Animals get nitrogen by eating plants or other animals, both of which are sources of usable nitrogen.

Decomposers and the Nitrogen Cycle In the nitrogen cycle, nitrogen moves between the atmosphere and living things. After nitrogen cycles from the atmosphere to living things, nitrogen is again returned to the atmosphere with the help of bacteria. These decomposers are essential to the nitrogen cycle because they break down wastes, such as urine, dung, leaves, and other decaying

126

ECOLOGY
● CONNECTION — ADVANCED

How Humans Affect the Nitrogen Cycle
The article *Human Alteration of the Global Nitrogen Cycle: Causes and Consequences* was written by a distinguished group of ecologists, which included Peter M. Vitousek, Gene E. Likens, David W. Schindler, and G. David Tilman. According to their findings, humans have doubled the natural rate of nitrogen entering the land-based nitrogen cycle, through the use of chemical fertilizers and fossil fuels. The increase in nitrogen has

increased water and atmospheric pollution, has led to the acidification of lakes, streams and soils, and has contributed to greenhouse gases. Some plants adapted to soils low in nitrogen have been negatively affected or replaced by nitrogen-loving plants. This has affected the animals that consume those plants. Encourage students to read this article, and write about it. The article can be found online, or in the journal *Ecological Applications* (Volume 7, August 1997).

plants and animals and return the nitrogen that these wastes and dead organisms contain to the soil. If decomposers did not exist, much of the nitrogen in ecosystems would be stored in wastes, corpses, and other parts of organisms. After decomposers return the nitrogen to the soil, bacteria transform a small amount of the nitrogen into nitrogen gas, which then returns to the atmosphere and completes the nitrogen cycle. So once nitrogen enters an ecosystem, most of it stays within the ecosystem, cycles between organisms and the soil, and is constantly reused.

The Phosphorus Cycle

Phosphorus is an element that is a part of many molecules that make up the cells of living organisms. For example, phosphorus is an essential material needed to form bones and teeth in animals. Plants get the phosphorus they need from soil and water, while animals get their phosphorus by eating plants or other animals that have eaten plants. The **phosphorus cycle** is the movement of phosphorus from the environment to organisms and then back to the environment. This cycle is slow and does not normally occur in the atmosphere because phosphorus rarely occurs as a gas.

As shown in **Figure 14**, phosphorus may enter soil and water in a few ways. When rocks erode, small amounts of phosphorus dissolve as phosphate in soil and water. Plants absorb phosphates in the soil through their roots. In addition, phosphorus is added to soil and water when excess phosphorus is excreted in waste from organisms and when organisms die and decompose. Some phosphorus also washes off the land and eventually ends up in the ocean. Many phosphate salts are not soluble in water, so they sink to the bottom of the ocean and accumulate as sediment.

internet connect

www.scilinks.org
Topic: Nitrogen Cycle
SciLinks code: HE4073

SCILINKS. Maintained by the National Science Teachers Association

Geofact

Minerals in Your Mouth
Phosphorus is the 11th most abundant element in the Earth's crust and occurs naturally as phosphate in the mineral apatite. Apatite can exist in igneous, metamorphic, and sedimentary rocks as well as in your teeth and bones.

Figure 14 ▶ The Phosphorus Cycle Phosphorus moves from phosphate deposits in rock to the land, then to living organisms, and finally to the ocean.

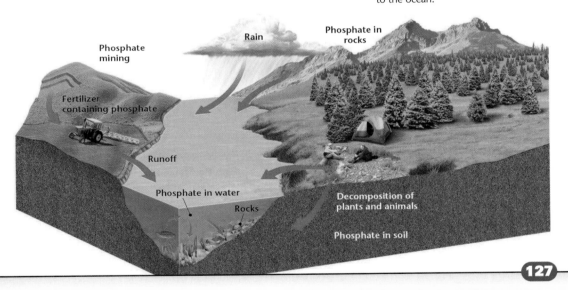

Phosphate in rocks

Rain

Phosphate mining

Fertilizer containing phosphate

Runoff

Phosphate in water

Rocks

Decomposition of plants and animals

Phosphate in soil

127

Discussion — ADVANCED
Nutrient Cycling and Ecosystem Structure Ask groups to discuss the following question: How does the structure of an ecosystem influence nutrient cycling? (Structured interrelationships such as those that exist among soil, water, nutrients, producers, consumers, and decomposers allow for the transfer of nutrients from one thing to another in a chain of events. If one link in the chain is missing, the transfer cannot be completed.) Have each group report their answers to the whole class, and discuss their answers.
LS Logical/Interpersonal

Internet Activity — ADVANCED
Nitrogen Fertilizers Modern agriculture relies on the use of nitrogen-rich fertilizers. Ask students to find out how these fertilizers are produced and what impact they have on the environment. Why would some environmentalists say that nitrogen fertilizers harm the environment twice? (First, energy is needed to produce nitrogen-rich fertilizers. The energy needed to produce fertilizers usually comes from fossil fuels, which add carbon to the atmosphere. Second, when excess nitrogen compounds run off fields and into bodies of water, they can cause an overgrowth of algae and aquatic plants. When aquatic plants die and decay, bacteria use oxygen to decompose the plant remains. When there is an abundance of plant remains, decomposers can deplete the water of oxygen which other aquatic organisms need to survive.) Have students record their findings in their *EcoLog*. **LS** Logical

Notable Quotes — GENERAL

"In nature there are neither rewards nor punishments—there are consequences."
—R. G. Ingersoll

Ask students, "What are some of the consequences of humans affecting the carbon, nitrogen, and phosphorus cycles?" (Sample answers: global climate change, destruction of ecological systems, algal blooms, fish kills, acid rain, and soil erosion)

Transparencies

TT The Phosphorus Cycle

Reteaching ━━━━ BASIC

Comparing the Cycles Have students study the carbon, nitrogen, and phosphorus cycles shown in **Figures 10, 12,** and **14.** Ask them to list the ways that the three cycles are similar and different. (The cycles are similar in that all move materials between different components of the biosphere, have no beginning or end, and conserve the materials in the cycle. The cycles differ in that the phosphorus cycle only remotely involves living organisms.)
LS Logical/Visual

Quiz ━━━━ GENERAL

1. What are the major sources of phosphorus? (Weathered rocks, the soil, and decaying organisms are all sources of phosphorus.)

2. How are the following elements used in organisms: carbon, nitrogen, and phosphorus? (Carbon is a component of proteins, fats and carbohydrates. Nitrogen builds proteins. Phosphorus builds bones, teeth, and is essential to many molecules.)

Alternative Assessment ━━━━ GENERAL

Connecting to the Carbon Cycle Ask students to write an essay for their **Portfolio,** describing the role they play in the carbon cycle.
LS Verbal

Figure 15 ▶ More than 30 percent of fertilizer may flow with runoff from farmland into nearby waterways. Large amounts of fertilizer in water can cause an excessive growth of algae (below).

Fertilizers and the Nitrogen and Phosphorus Cycles People often apply fertilizers to stimulate and maximize plant growth. Fertilizers contain both nitrogen and phosphorus. The more nitrogen and phosphorus that is available to a plant, the faster and bigger the plant tends to grow. However, if excessive amounts of fertilizer are used, the fertilizer can enter terrestrial and aquatic ecosystems through runoff. Excess nitrogen and phosphorus in an aquatic ecosystem or nearby waterway can cause rapid and over-abundant growth of algae, which results in an *algal bloom.* An algal bloom, as shown in **Figure 15,** is a dense, visible patch of algae that occurs near the surface of water. Algal blooms, along with other plants and the bacteria that break down dead algae, can deplete an aquatic ecosystem of important nutrients such as oxygen. Fish and other aquatic organisms need oxygen to survive.

Acid Precipitation We affect the nitrogen cycle every time we burn coal, wood, or oil. When we burn fuel, a large amount of nitric oxide is released into the atmosphere. Nitric oxide is a harmful gas, and when it is released into the air, it can combine with oxygen and water vapor to form nitric acid. Nitric acid can dissolve in rain and snow, which contributes to acid precipitation.

SECTION 2 Review

1. Describe the two processes of the carbon cycle.

2. Describe how the burning of fossil fuels affects the carbon cycle.

3. Explain how the excess use of fertilizer affects the nitrogen cycle and the phosphorus cycle.

4. Explain why the phosphorus cycle occurs more slowly than both the carbon cycle and the nitrogen cycle.

CRITICAL THINKING

5. Making Comparisons Write a short paragraph that describes the importance of bacteria in the carbon, nitrogen, and phosphorus cycles. What role does bacteria play in each cycle? **WRITING SKILLS**

6. Applying Ideas What is one way that a person can help reduce the level of carbon dioxide in the atmosphere? Can you think of more than one way?

128

Answers to Section Review

1. Producers convert carbon dioxide to carbohydrates during photosynthesis. Carbon is then passed on to consumers when they eat producers and other consumers. Producers and consumers release carbon dioxide back into the atmosphere during cellular respiration.

2. Fossil fuels contain carbon from plants and animals that died millions of years ago. When fossil fuels are burned, they release carbon into the atmosphere.

3. Excess use of fertilizer containing nitrogen and phosphorus can enter aquatic ecosystems in runoff, causing algal blooms which deplete them of oxygen.

4. Phosphorus is rarely cycled in the atmosphere, because it rarely occurs as a gas. Gases, such as atmospheric carbon and nitrogen, are easily exchangeable.

5. Some bacteria take in carbon dioxide during photosynthesis and release carbon dioxide during cellular respiration in the carbon cycle. Bacteria take in and fix nitrogen for other organisms to use in the nitrogen cycle. Decomposing bacteria also release carbon, nitrogen and phosphorus back into the environment.

6. Answers may vary.

How Ecosystems Change

Ecosystems are constantly changing. A forest hundreds of years old may have been a shallow lake a thousand years ago. A dead tree falls to the ground and lets sunlight reach the forest floor. The sunlight causes some seeds to germinate, and soon wildflowers and shrubs cover the forest floor. Mosses, shrubs, and small trees cover the concrete of a demolished city building. These are all examples of environmental change that scientists define as ecological succession.

Ecological Succession

Ecological succession is a gradual process of change and replacement of the types of species in a community. In nature, the process of ecological succession may take hundreds or thousands of years. Each new community that arises often makes it harder for the previous community to survive. For example, the younger beech trees in **Figure 16** will have a hard time competing with the older beech trees for sun. However, if a shade-loving species of tree began to grow in the forest, the new species might replace the smaller beech trees.

Primary succession is a type of succession that occurs on a surface where no ecosystem existed before. Primary succession can occur on rocks, cliffs, and sand dunes. **Secondary succession,** the more common type of succession, occurs on a surface where an ecosystem has previously existed. Secondary succession occurs in ecosystems that have been disturbed or disrupted by humans, animals, or by natural processes such as storms, floods, earthquakes, and volcanoes.

Objectives

▶ List two examples of ecological succession.
▶ Explain how a pioneer species contributes to ecological succession.
▶ Explain what happens during old-field succession.
▶ Describe how lichens contribute to primary succession.

Key Terms

ecological succession
primary succession
secondary succession
pioneer species
climax community

Graphic Organizer **Chain-of-Events Chart**
Create the **Graphic Organizer** entitled "Chain-of-Events Chart" described in the Appendix. Then, fill in the chart with details about each step of ecological succession.

Figure 16 ▶ Taller beech trees compete with shorter, young beech trees for sun and make it hard for the younger trees to survive.

129

Focus

Overview

Before beginning this section, review with your students the Objectives in the Student Edition. This section introduces the concept of ecological succession. The section distinguishes between secondary and primary succession and explains the importance of pioneer species.

Bellringer

Ask students to consider the following question: "Is your school experiencing ecological succession?" Then walk outside with the students, and look for evidence of moss and lichens on the outside of the school building. Look for cracks in the parking lot where grass or weeds are breaking through. Talk about what will happen if this vegetation continues to grow and is not disturbed. (It would probably take a long time, possibly hundreds of years, for lichens or moss to break down the school building, but they would gradually break it down if they were not disturbed.)

Motivate

Discussion ———— GENERAL

Change Over Time Ask the class to imagine that on graduation day their school is bulldozed and a tall fence is placed around the bare school grounds. Then ask them to picture returning to the school grounds in 1 month, 1 year, 5 years, and 100 years. Call on individual students to describe what they think the school grounds would be like at these time intervals. (Living things would return to the area, but the types of living things would change over the years. Weeds might be the first to appear, then brush and small trees.) Explain that this process is called *ecological succession*.
LS Verbal

Chapter Resource File

• Lesson Plan
• Active Reading **BASIC**
• Section Quiz **GENERAL**

Transparencies

TT Bellringer

Graphic Organizer **GENERAL**

Chain of Events Chart

You may want to use this Graphic Organizer to assess students' prior knowledge before beginning a discussion of each step of ecological succession. You may also choose to use a similar activity as a quiz to assess students' knowledge of ecological succession or to assess students' knowledge after students have read the description of ecological succession.

Discussion — GENERAL

Yellowstone Fire System Twelve years after the Yellowstone fire, researchers Millspaugh, Whitlock, and Bartlein wrote: "The historic trend toward infrequent severe fires, such as those in 1988, will be short-lived and in all likelihood replaced by a regime of many small fires . . . The forests of central Yellowstone will change not so much in composition as in stand-age distribution. Thus the disturbance regime will serve to perpetuate lodgepole pine where it now grows and allow its expansion to higher elevations." Ask students to discuss this quote, and what the researchers are implying. (Now that fires are allowed to burn in national parks, natural fire succession will be restored. Smaller fires will keep the fuel load down, and the lodgepole pine ecosystem will expand and remain viable.)
LS Logical

Debate — GENERAL

Fires in National Parks Have students divide into two groups and ask them to research and debate the pros and cons of the National Park Service's fire policy. You might want to get the discussion moving by asking some questions such as: "Would controlled burns be a good idea for maintaining ecosystems? Since national parks attract many visitors, would fires affect tourism and hurt the economic future of the park system? Is it dangerous to let fires burn in areas that humans use? Would smaller frequent fires help to maintain each park's ecosystem? Should people be allowed to build near the boundaries of fire-maintained parks?" **LS** Verbal

Figure 17 ▶ When Mount St. Helens erupted in 1980, much of the forest around the volcano was destroyed.

Secondary Succession In 1980, the volcano Mount St. Helens erupted in Washington State. The eruption at Mount St. Helens has been described as one of the worst volcanic disasters because more than 44,460 acres of forest were burned and flattened by the force of hot ash and other volcanic debris, as shown in **Figure 17.** After the eruption, plants began to colonize the volcanic debris. Such plants are called **pioneer species**—the first organisms to colonize any newly available area and begin the process of ecological succession. Over time, pioneer species will make the new area habitable for other species. If you visited Mount St. Helens today, you would find that the forest is in the process of secondary succession. **Figure 18** shows how after 12 years, plants and flowers had covered most of the lava and new trees and shrubs had started to grow. If these organisms at Mount St. Helens continue to grow, over time they will eventually form a climax community. A **climax community** is

 CASE STUDY

Communities Maintained by Fire

Fires set by lightning or human activities occasionally sweep through large areas. Burned areas undergo secondary succession. In the forests of the Rocky Mountains, for example, burned areas are rapidly colonized by fireweed, which clothes the slopes with purple flowers. In some places, fire determines the nature of the climax community. In the United States, ecological communities that are maintained by fire include the chaparral of California, the temperate grassland of the Midwest, and many southern and western pine forests.

Plants native to these communities are adapted to living with fire. A wildfire that is not unusually hot may not harm fire-adapted pine trees, but it can kill deciduous trees—those trees that lose their leaves in winter. Seeds of

▶ Fireweed is one type of plant that colonizes land after the land has been burned by fire.

some species, such as longleaf pine trees, will not germinate until exposed to temperatures of several hundred degrees. When a fire sweeps through a forest, the fire kills plants on the ground and stimulates the pine seeds to germinate.

Longleaf pines have a strange growth pattern. When they are young, they have long needles that reach down to the ground, and the trees remain only approximately a half of a meter high for many years, while they store nutrients. If a fire occurs, it sweeps through the tops of the tall trees that survived the last fire and the young longleaf pines near the ground may escape the fire. Then, the young pines use their stored food to grow very rapidly. A young pine can grow as much as 2 m/y. Soon the young pines are tall enough so that a fire near the ground would not harm them.

If regular fires are prevented in a fire-adapted community, deciduous trees may invade the area. These trees form a thick barrier near the ground. In addition, their dead leaves and

130

BRAIN FOOD

Secondary Succession Some animals play a part in preparing an ecosystem for secondary succession. For example, pocket gophers in old abandoned farm fields cycle fresh soil from underground, and deposit the soil in mounds outside of their burrows. These mounds attract seeds from pioneer plant species that are floating through the air. Eventually, taller grasses from the field shade out the pioneer species, and these grasses take over the mounds and the field.

SKILL BUILDER — ADVANCED

Writing Ask students if the following statement is true or false: "An ecosystem has historical aspects; the present is related to the past, and the future is related to the present." (True. The Earth's cycles and the predictable patterns of change, such as those evident in succession, connect an ecosystem's past to its present.) Suggest that students think about the statement and write an essay outlining their thoughts in their **Portfolio. LS** Verbal

a final and stable community. Even though a climax community continues to change in small ways, this type of community may remain the same through time if it is not disturbed.

Fire and Secondary Succession Natural fires caused by lightning are a necessary part of secondary succession in some communities, as discussed in the Case Study below. Some species of trees, such as the Jack pine, can release their seeds only after they have been exposed to the intense heat of a fire. Minor forest fires remove accumulations of brush and deadwood that would otherwise contribute to major fires that burn out of control. Some animal species also depend on occasional fires because they feed on the vegetation that sprouts after a fire has cleared the land. Therefore, foresters sometimes allow natural fires to burn unless the fires are a threat to human life or property.

Figure 18 ▶ The photo above was taken 12 years after the eruption of Mount St. Helens and shows evidence of secondary succession.

▶ These young lodgepole pine trees have started growing after a devastating forest fire.

▶ This firefighter is helping maintain a controlled fire in South Dakota. Some fires are set on purpose by fire officials to bring nutrients to soil from burned vegetation.

branches pile up on the ground and form extra fuel for fires. When a fire does occur, it is hotter and more severe than usual. The fire destroys not only the deciduous trees but also the pines. It may end up as a devastating wildfire.

Although it may seem odd, frequent burning is essential to preserve many plant communities and the animals that depend on them. This is the reason the U.S. Park

Service adopted the policy of letting fires in national parks burn if they do not endanger human life or property.

This policy caused a public outcry when fires burned Yellowstone National Park in 1988, because people did not understand the ecology of fire-adapted communities. The fires later became an opportunity for visitors to learn about the changes in an ecosystem after a fire.

131

Activity — **BASIC**

Musical Succession Organize students into four groups, and give one of the following objects to each student in each group: a crumpled sheet of paper; a pencil; a hard-cover book; and a plastic jar filled with beans. Tell students they are going to make a succession band. Ask group 1 to start crumpling their paper, as if playing an instrument. After a minute or so, ask group 2 to tap their pencils quietly. Allow this to go on for a few minutes. Then ask group 3 to start slamming the covers of their books softly at first, but then with more force (but ask them not to damage the books!). Groups 1 and 2 should be told to fade out. Have Group 4 come in, shaking their jars quietly and slowly, and then more quickly and loudly. Have students shaking the jars try to drown out the book slammers. Ask the paper crumplers and pencil tappers to try to drown out the fourth group (they should not be able to). Bring students to a halt. Ask students how this "band" represents succession. (Each group of instruments succeeded the next. The sound from groups 3 and 4 could drown out groups 1 and 2. The last group finally formed a climax community that was not subject to competition from the first three groups.) **LS** Auditory/Kinesthetic

Transparencies

TT Secondary Succession: Old-Field Succession

Year 1: annual plants | Year 2: perennial plants and grasses | Year 3–10: shrubs | About year 20: young pine forest | After about 150 years: mature oak forest

Figure 19 ▶ The illustration above shows what an abandoned farm area might look like during old-field succession. Why do you think young oak trees begin to appear around year 20?

FIELD ACTIVITY

Investigating Succession Explore two or three blocks in your neighborhood, and find evidence of succession. Make notes in your *EcoLog* about the location and the evidence of succession that you observe. Pay attention to sidewalks, curbs, streets, vacant lots, and buildings, as well as parks, gardens, fields, and other open areas. Create a map from your data that identifies where succession is taking place in your neighborhood.

132

Old-field Succession Another example of secondary succession is *old-field succession*, which occurs when farmland is abandoned. When a farmer stops cultivating a field, grasses and weeds quickly grow and cover the abandoned land. The pioneer grasses and weeds grow rapidly and produce many seeds to cover large areas.

Then over time, taller plants, such as perennial grasses, grow in the area. These plants shade the ground, which keeps light from the shorter pioneer plants. The long roots of the taller plants also absorb most of the water in the soil and deprive the pioneer plants of adequate water to survive. The pioneer plants soon die from lack of sunlight and water. As succession continues, the taller plants are deprived of light and water by growing trees. Finally, slower-growing trees, such as oaks, hickories, beeches, and maples, take over the area and block out the sunlight to the smaller trees. As shown in **Figure 19**, after about a century, the land can return to the climax community that existed before the farmers cleared it to plant crops.

Primary Succession On new islands created by volcanic eruptions, in areas exposed when a glacier retreats, or on any other surface that has not previously supported life, primary succession can occur. Primary succession is much slower than secondary succession because primary succession begins where there is no soil. It can take several hundred to several thousand years to produce fertile soil naturally. Imagine that a glacier melts and exposes an area of bare rock. The first pioneer species to colonize the bare rock will probably be bacteria and lichens, which can live without soil. Lichens, as shown in **Figure 20**, are important early pioneers in primary

BRAIN FOOD

Mammals and Succession This chapter focuses on plant communities and how they change over time. But how do the animal communities respond to that change? Have students think about the mammals that live in an old field. (Mice, voles and shrews live in most old fields. Other mammals might include ground squirrels, gophers, and badgers.) Ask, "What happens when taller grass takes over the field?" (There will be more cover, so more of the same animals might move into the area, or reproduction might increase.) "If trees start moving in, how will that change the community?" (The grass might be shaded out, so voles would be reduced; shrews that eat voles might move elsewhere; other animals that live mainly in open areas would move out, such as ground squirrels, gophers, and badgers; species that use trees and open areas would start to colonize (some types of mice); the forest creatures would follow (deer, chipmunks, forest mice and voles.) Encourage students to further investigate the links between mammal communities and succession. **LS** Logical

succession. They are the colorful, flaky patches that you see on trees and rocks. A lichen is a producer that is actually composed of two different species, a fungus and an alga. The alga photosynthesizes, while the fungus absorbs nutrients from rocks and holds water. Together, they begin to break down the rock.

As the growth of the lichen breaks down the rock, water may freeze and thaw in cracks, which breaks up the rock further. Soil slowly accumulates as dust particles in the air are trapped in cracks in the rock. Dead remains of lichens and bacteria also accumulate in the cracks. Mosses may later grow larger and break up the rock even more. When the mosses die, they decay and add material and nutrients to the growing pile of soil. Thus, fertile soil forms from the broken rock, decayed organisms, water, and air. Primary succession can also be seen in any city street as shown in **Figure 20.** Mosses, lichens, and weeds can establish themselves in cracks in a sidewalk or building. Fungi and mosses can also invade a roof that needs repair. Even New York City would eventually turn into a cement-filled woodland if it were not constantly cleaned and maintained.

internet connect

www.scilinks.org
Topic: Ecological
Succession
SciLinks code: HE4024

SCI LINKS. Maintained by the National Science Teachers Association

Figure 20 ▶ Lichens (left) are colonizing a boulder in Wyoming. Over a long period of time, lichens can break down rock into soil. Plants that grow through cracks in city sidewalks (below) can also be described as pioneers of primary succession.

SECTION 3 Review

1. **Compare** primary and secondary succession.

2. **Describe** what role a pioneer species plays during the process of ecological succession.

3. **Explain** why putting out forest fires may be damaging in the long run.

4. **Describe** the role lichens play in primary succession. Write a short paragraph to explain your answer. **WRITING SKILLS**

CRITICAL THINKING

5. **Analyzing Processes** Over a period of 1,000 years, a lake becomes a maple forest. Is this process primary or secondary succession? Explain your answer.

6. **Analyzing Relationships** How are lichens similar to the pioneer species that colonize abandoned farm areas? How are they different?

133

Close

Reteaching — BASIC
Succession Quizzes Have each student come up with five quiz questions based on material in this section. Have students pair off and exchange quizzes. Each student should answer a quiz and grade his or her partner's quiz. Discuss students answers with the class.
LS Verbal

Quiz — GENERAL
1. What are some of the ways an ecological community changes into a different community through secondary succession? (Answers may vary. Pioneer species can move in and colonize an area such as in an old field; taller perennial grasses can compete with shorter grasses for sun and water and eventually take over; disturbances such as a fire or a volcanic eruption can occur.)

2. Would a newly-formed volcanic island be a site of primary succession or secondary succession? (primary succession)

Alternative Assessment — GENERAL
Succession Animation Give students 50 index cards. On the cards, have them illustrate succession in an ecosystem of their choice. The cards should show gradual changes in the same area over time. Have them clip the index cards together, and animate the scene by flipping through the cards rapidly using their thumb. If students videotape their animations, they can share them with the class. **LS Kinesthetic/Visual**

Answers to Section Review

1. Primary succession occurs on a surface where no ecosystem previously existed. Secondary succession occurs on a surface where an ecosystem previously existed.

2. Pioneer species are the first organisms to colonize any newly available area and begin the process of ecological succession.

3. Some communities rely on fire to reproduce or eliminate invading species. Putting out forest fires in these communities will cause them to be taken over by other species and change into a different community.

4. Lichens can live without soil, so they colonize bare rock and begin to break it down through physical and chemical processes.

5. secondary succession; primary succession only occurs in areas where there was no previous ecosystem

6. Lichens and the first old-field succession plants are pioneers. Lichens, however, do not need soil to colonize an area.

CHAPTER 5 Highlights

Alternative Assessment —— GENERAL

Schoolyard Succession Have students observe the process of succession during the school year. Early in the school year, get permission to mark off a section of the school grounds that is approximately 4 ft. × 4 ft. Post "Do Not Mow" signs in the marked-off area. Have students monitor the area by keeping a regular journal of the changes that they observe over time. Their journal can be written or recorded on videotape. Students may also wish to collect and preserve a few specimens of the plants and insects that occupy the area. Be sure students carefully label their notes and collections with a complete date and a detailed description of their findings. The project could be continued from one school year to the next, with current students building on the documentation of previous students. **LS Kinesthetic/Visual**

Succession Collage Bring in some posterboard, glue, and copies of National Geographic, or other magazines with pictures of natural scenes and organisms. Organize students into small groups, and assign a type or stage of ecological succession to each group. Have students find pictures of landscapes and organisms that represent their type of community. Ask them to cut out the pictures and to arrange and glue them into ecosystems on the posterboard. Instruct students to label each component of their system, and to provide a rationale as to why it was included in their collage. **LS Visual**

1 Energy Flow in Ecosystems

Key Terms

photosynthesis, 117
producer, 118
consumer, 118
decomposer, 119
cellular respiration, 120
food chain, 122
food web, 122
trophic level, 122

Main Ideas

▶ The majority of the Earth's organisms depend on the sun for energy. Producers harness the sun's energy directly through photosynthesis, while consumers use the sun's energy indirectly by eating producers or other consumers.

▶ The paths of energy transfer can be followed through food chains, food webs, and trophic levels.

▶ Only about 10 percent of the energy that an organism consumes is stored and transferred when that organism is eaten.

2 The Cycling of Materials

Key Terms

carbon cycle, 124
nitrogen-fixing bacteria, 126
nitrogen cycle, 126
phosphorus cycle, 127

Main Ideas

▶ Materials in ecosystems are recycled and reused by natural processes.

▶ Carbon, nitrogen, and phosphorus are essential for life, and each of them follows a recognizable cycle.

▶ Humans can affect the cycling of materials in an ecosystem through activities such as burning fossil fuels and applying fertilizer to soil.

3 How Ecosystems Change

Key Terms

ecological succession, 129
primary succession, 129
secondary succession, 129
pioneer species, 130
climax community, 130

Main Ideas

▶ Organisms in an environment sometimes follow a pattern of change over time known as ecological succession.

▶ Secondary succession occurs on a surface where an ecosystem has previously existed. Primary succession occurs on a surface where no ecosystem existed before.

▶ Climax communities are made up of organisms that take over an ecosystem and remain until the ecosystem is disturbed again.

134

Chapter Resource File

- **Chapter Test** GENERAL
- **Chapter Test** ADVANCED
- **Concept Review** GENERAL
- **Critical Thinking** ADVANCED
- **Test Item Listing**
- **Observation Lab** BASIC
- **CBL™ Probeware Lab** GENERAL
- **Consumer Lab** GENERAL
- **Long-Term Project** ADVANCED

CHAPTER 5 Review

Using Key Terms

Use each of the following terms in a separate sentence.

1. *photosynthesis*
2. *trophic level*
3. *carbon cycle*
4. *nitrogen-fixing bacteria*
5. *decomposers*

For each pair of terms, explain how the meanings of the terms differ.

6. *producer* and *consumer*
7. *primary succession* and *secondary succession*
8. *nitrogen cycle* and *phosphorus cycle*
9. *food chain* and *food web*

✔ STUDY TIP

Taking Multiple-Choice Tests When you take multiple-choice tests, be sure to read all of the choices before you pick the correct answer. Be patient, and eliminate choices that are obviously incorrect.

Understanding Key Ideas

10. Which of the following statements is *not* true of consumers?
    a. They get energy indirectly from the sun.
    b. They are also called *heterotrophs*.
    c. They make their own food.
    d. They sometimes eat other consumers.

11. Which of the following is correctly arranged from the lowest trophic level to the highest trophic level?
    a. bacteria, frog, eagle, mushroom
    b. algae, deer, wolf, hawk
    c. grass, mouse, snake, eagle
    d. grass, bass, minnow, snake

12. Communities of bacteria have been found living thousands of feet underwater. Which of the following statements is a proper conclusion to draw about these bacteria?
    a. Somehow they are conducting photosynthesis.
    b. They are living on borrowed time.

c. They were somehow introduced by human activities.
d. They use an energy source other than sunlight.

13. Which of the following pairs of organisms probably belong to the same trophic level?
    a. humans and bears
    b. bears and deer
    c. humans and cows
    d. both (a) and (c)

14. The energy lost between trophic levels
    a. can be captured only by parasitic organisms.
    b. cools the surrounding environment.
    c. is used in the course of normal living.
    d. evaporates in the atmosphere.

15. From producer to secondary consumer, about what percentage of energy is lost?
    a. 10 percent
    b. 90 percent
    c. 99 percent
    d. 100 percent

16. Which of the following statements about the nitrogen cycle is *not* true?
    a. Animals get nitrogen by eating plants or other animals.
    b. Plants generate nitrogen in their roots.
    c. Nitrogen moves back and forth between the atmosphere and living things.
    d. Decomposers break down waste to yield ammonia.

17. Which of the following are most likely to be the pioneer organisms on an area of bare rock?
    a. saplings
    b. shrubs
    c. lichens
    d. perennial grasses

18. Excessive use of fertilizer that contains nitrogen and phosphorus
    a. affects the carbon cycle.
    b. may cause algal blooms in waterways.
    c. causes soil erosion.
    d. contributes to primary succession.

135

Assignment Guide

Section	Questions
1	1, 2, 5–6, 8–15, 19–21, 24–28, 31–32
2	3–4, 16, 18, 22–23, 29, 34–37
3	7, 17, 30, 33

ANSWERS

Using Key Terms

1. Sample answer: Photosynthesis uses carbon dioxide, water, and energy to produce sugars and oxygen.
2. Sample answer: In an aquatic ecosystem, algae are in the bottom trophic level.
3. Sample answer: Organisms add carbon to the carbon cycle during cellular respiration.
4. Sample answer: Nitrogen-fixing bacteria live inside the root nodules of legumes.
5. Sample answer: Decomposers break down dead organic matter.
6. A producer is an organism that makes its own food. A consumer is an organism that gets its energy by eating other organisms.
7. Primary succession is a type of succession that occurs on a surface where no ecosystem existed before. Secondary succession occurs on a surface where an ecosystem has previously existed.
8. The nitrogen cycle is the movement of nitrogen between the atmosphere, bacteria, and other organisms. The phosphorus cycle is the movement of phosphorus from the environment to organisms and back.
9. A food chain is a sequence in which energy is transferred from one organism to another. A food web is made up of many food chains and shows the possible feeding relationships in an ecosystem.

Understanding Key Ideas

10. c
11. c
12. d
13. a
14. c
15. c
16. b
17. c
18. b

Short Answer

19. Answers may vary. Cellular respiration provides the carbon dioxide that fuels photosynthesis; photosynthesis provides the oxygen that fuels cellular respiration.

20. Answers may vary. A large amount of energy is lost at each level, so there has to be a limit on the number of trophic levels. For example, in an African savanna ecosystem, animals that are larger than lions and that eat lions do not exist, because there is not enough energy to support a trophic level higher than lions on the African savanna.

21. Answers may vary. If decomposers did not exist, dead organic matter would not be broken down. Also, carbon, nitrogen, and other nutrients essential to life would stay in an unusable form and nutrients would not be cycled in ecosystems.

22. Answers may vary. Phosphorus rarely occurs as a gas. And phosphorus takes a long period of time to weather from rock. Phosphorus in water gets trapped in an unusable form called phosphate, which sinks to the bottom of the ocean and lakes. Much of the phosphorus on Earth is locked in the form of phosphate-bearing rock.

23. Answers may vary. Carbon dioxide is the gaseous form of carbon in the carbon cycle. Carbon dioxide becomes dispersed in the atmosphere, and is readily available for use by plants.

Short Answer

19. Explain the relationship between cellular respiration and photosynthesis.
20. Why is the number of trophic levels that can exist limited?
21. Why are decomposers an essential part of an ecosystem?
22. Write a short paragraph that explains why the phosphorus cycle occurs slower than the carbon and nitrogen cycles. **WRITING SKILLS**
23. Describe the role of carbon dioxide in the carbon cycle.

Interpreting Graphics

Use the diagram to answer questions 24–26.

24. How many organisms depend on the squid as a source of food?
25. If the population of Adélie penguins decreased drastically in this ecosystem, what effect would the decreased number of penguins have on elephant seals?
26. What role do algae play in this food web?

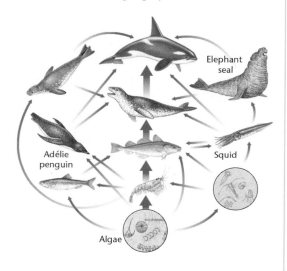

Elephant seal

Adélie penguin

Squid

Algae

Concept Mapping

27. Use the following terms to create a concept map: *algae, humans, solar energy, carnivores, consumers, producers, directly, herbivores, indirectly,* and *omnivores.*

Critical Thinking

28. **Comparing Processes** How are producers and decomposers opposites of each other?
29. **Inferring Relationships** Abandoned fields in the southwestern part of the United States are often taken over by mesquite trees, which can grow in nutrient-poor soil. If the land is later cleared of mesquite, the soil is often found to be enriched with nitrogen and is more suitable for crops. What might be the reason for this phenomenon?
30. **Understanding Concepts** Read the description under the head, "What Eats What" in this chapter, and explain why decomposers are considered to be consumers. **READING SKILLS**
31. **Drawing Conclusions** Suppose that a plague eliminates all the primary consumers in an ecosystem. What will most likely happen to organisms in other trophic levels in this ecosystem?

Cross-Disciplinary Connection

32. **Mathematics** If a lake contains 600,000 kg of plankton and the top consumers are a population of 40 pike, which each weigh an average of 15 kg, how many trophic levels does the lake contain? Make a graph or pyramid that illustrates the trophic levels.

Portfolio Project

33. **Research Local Succession** Do a special project on succession. Find areas in your community that have been cleared of vegetation and left unattended at different times in the past. Ideally, you should find several areas that were cleared at different times, including recently and decades ago. Photograph each area, and arrange the pictures to show how succession takes place in your geographic region.

Interpreting Graphics

24. four
25. The elephant seal population might decrease because the killer whale would have to hunt more of the seals to make up for the loss of penguins as prey.
26. Algae are the base food for the entire food web.

Concept Mapping

27. Answers to the concept mapping questions are on pp. 667–672.

Critical Thinking

28. Producers transform chemicals into a form usable by consumers; decomposers transform dead organisms back to chemicals that producers can use.
29. Mesquite is a legume and has nitrogen-fixing bacteria in its roots that enrich the soil.
30. Decomposers are considered to be consumers because they get their food by breaking down dead organisms. Therefore, they depend on other organisms to gain energy.

Use the data in the table below to answer questions 34–35.

Percentage of Fertilizer Use per Year	
Region of the World	Percentage
North America	17
Asia	52
Africa	3
Europe	18
Latin America and the Caribbean	8
Oceania	2

34. Making Calculations If 137.25 million metric tons of fertilizer is used worldwide per year, how many million metric tons does Asia use?

35. Graphing Data Make a bar graph that compares the percentage of fertilizer use in different regions worldwide per year.

 WRITING SKILLS

36. Communicating Main Ideas Describe the importance of the carbon, nitrogen, and phosphorus cycles to humans.

37. Writing from Research Research information on how countries regulate carbon dioxide emissions. Write an essay that describes the laws regulating carbon dioxide emissions and the solutions some countries have devised to decrease the amount of carbon dioxide emitted.

 STANDARDIZED TEST PREP

For extra practice with questions formatted to represent the standardized test you may be asked to take at the end of your school year, turn to the sample test for this chapter in the Appendix.

 READING SKILLS

Read the passage below, and then answer the questions that follow.

The Peruvian economy and many sea birds depend on normal atmospheric conditions. But sometimes, usually in December, the normal east-to-west winds do not form over the Pacific Ocean. Instead, winds push warm water eastward toward the coast of South America. When these conditions occur, the warm surface water cuts off the upwelling of nutrients. This event is called El Niño, which means "the child," because it happens near Christmas.

Because all convection cells are linked in the atmosphere, the effects of El Niño extend beyond Peru. Under a strong El Niño, northeastern Australia can suffer summer drought, which leads to reduced grain production there. The southeastern United States gets higher rainfall in El Niño years, which boosts agriculture and decreases forest fires.

1. According to the passage, a possible cause of reduced grain production in Australia is
   a. a rate of convection that is higher than the average rate.
   b. an amount of rainfall that is higher than the average amount.
   c. a reduced fish population.
   d. a summer drought.

2. According to the passage, which of the following statements is true?
   a. The effects of El Niño do not extend beyond Peru.
   b. During El Niño years, the U.S. agricultural industry suffers.
   c. El Niño is caused by winds that push warm water eastward toward South America.
   d. Australia's agricultural industry benefits the most from strong winds during El Niño.

31. Producers may flourish, because they are not being eaten, but secondary consumers would dwindle and die or move out of the area. Primary consumers from a different area may be attracted to the area of high primary production, and their predators may follow. This type of an event would probably change the ecosystem significantly.

Cross-Disciplinary Connection

32. The total mass for the pike would be 600 kg. Since each trophic level loses 90% of its energy, this would be a system with 4 trophic levels (1st level = 600,000 kg, 2nd level = 60,000 kg, 3rd level = 6000 kg, 4th level (pike) = 600 kg).

Portfolio Project

33. Answers may vary.

Math Skills

34. $137.25 \times 0.52\% = 71.37$ million metric tons

35. The graph should be arranged from lowest to highest use. The countries should be on the x-axis and percentage should be on the y-axis.

Writing Skills

36. Answers may vary. Be sure that students link the cycles to the foods they eat and the waste materials they excrete.

37. Students may want to review the Kyoto Protocol on the Web. The Kyoto Protocol is an international agreement to reduce CO_2 emissions.

Reading Skills

1. d
2. c

137

DISSECTING OWL PELLETS

Teacher's Notes

Time Required
one 45 minute class period

Lab Ratings

EASY ———————————————→ HARD

TEACHER PREPARATION 🧪🧪
STUDENT SETUP 🧪🧪🧪
CONCEPT LEVEL 🧪🧪
CLEANUP 🧪🧪🧪

Skills Acquired
• Constructing Models
• Classifying
• Inferring
• Organizing and Analyzing Data

The Scientific Method
In this lab, students will:
• Make Observations
• Ask Questions
• Analyze the Results
• Draw Conclusions

Materials
The materials listed are enough for a group of 4 to 5 students. You may want to provide bleach to disinfect the skulls. If skulls are disinfected, students can take them home. This would make the lab more memorable for students.

Objectives

▶ **Examine** the remains of an owl's diet.
▶ USING SCIENTIFIC METHODS **Construct** a food chain based on your observations.

Materials

disposable gloves
dissecting needle
dissecting pan
egg cartons
forceps
owl pellet(s)
piece of white paper
small animal identification field guide that includes skull illustrations

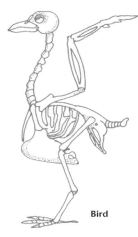

Bird

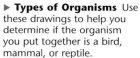
▶ **Types of Organisms** Use these drawings to help you determine if the organism you put together is a bird, mammal, or reptile.

Reptile

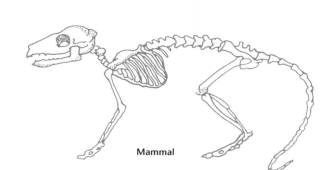
Mammal

138

Dissecting Owl Pellets

Owls are not known as finicky eaters. They prey on almost any animal that they can swallow whole. Like many other birds, owls have an interesting adaptation—a special structure called a gizzard. The gizzard acts as a filter and prevents the indigestible parts of their prey, such as fur, feathers, and bones—from passing into their intestines. These indigestible parts are passed to a storage pouch, where they accumulate. A few hours after consuming a meal, the owl coughs up the accumulated indigestible material, which has been compressed into a pellet. By examining such a pellet, you can tell what the owl ate. In addition, by examining the remains of the owl's prey found in the pellet, you can get a good idea of what the prey ate. Using this information, you can construct a food chain of the owl and its prey.

Procedure

1. Work in groups of three or four. Place an owl pellet in the dissecting pan, and remove it from its aluminum-foil casing.
2. Examine the owl pellet. Using the dissecting needle and forceps, carefully break apart the owl pellet. Separate the fur or feathers from the bones. Be careful not to damage the small bones. Place the bones onto a piece of white paper.
3. Identify the major components of the pellet.
4. If the pellet contains remains from more than one organism, determine as best as you can how many different animals and species are present.
5. Attempt to group the remains by type of organism. Count the number of skulls to find out how many prey were in the pellet. Decide which bones belong with which skulls. Then try to assemble complete skeletons. Sample skeletal diagrams are shown below.

Safety Cautions

Because students will be dealing with material associated with a bird, there is a risk of exposure to Salmonella bacteria. Be sure that students use the gloves provided, and you may also want to provide masks for students. Also require that students wash their hands thoroughly after the exercise.

6. Closely examine the skulls of each prey. Compare the skulls to the diagrams of skulls on this page. What purpose do the teeth or bills seem to have—tearing flesh, chewing plant parts, or grinding seeds? If you are able to identify the prey, find out their typical food sources.

7. On a separate piece of paper, construct a simple food chain based on your findings.

8. Compare your findings with those of other groups of students.

Analysis

1. **Examining Data** How many skeletons were you able to make from your pellet? What kinds of animals did you identify in the owl pellet?

2. **Organizing Data** Compare your findings with those of your classmates by using the following questions:

   a. What animals were represented most often in the pellets?

   b. What common traits do these animals have?

   c. How many animals found in the pellets were herbivores? How many were carnivores?

Conclusions

3. **Interpreting Information** What biological relationships were you able to determine from your examination of the owl pellets?

4. **Evaluating Data** Of the animals you found in your pellet, how many different trophic levels are represented?

5. **Drawing Conclusions** Most owls hunt at night and sleep during the day. From that information, what can you infer about their prey?

Extension

1. **Research and Communications** Research information on an owl species and the types of organisms found in its habitat. Make a poster of a food web, including the owl species. Be sure to include producers, consumers, and decomposers.

▶ **Identify the Prey** Use these drawings to identify the owl's prey.

Shrew

Vole

Mouse

Mole

Frog

Snake

Rat

Rabbit

Answers to Analysis

1. Students are likely to find the remains of a variety of small animals such as mice, lizards, shrews, voles, young squirrels, or rabbits.

2. Answers may vary.

Answers to Conclusions

3. Answers may vary. In many cases, owls feed on different types of animals that are either primary or secondary consumers.

4. Answers may vary. There may be one or two trophic levels represented.

5. Answers may vary. Most of the animals that owls eat are active at night but not always. If students find a ground squirrel or chipmunk skull, the owl was probably active at dawn or dusk, or even during the day.

Answers to Extension

1. Owl pellets are a great way to learn about the diet of owls, and to learn about which animals are represented in a local ecosystem.

Chapter Resource File

- **Datasheets for In-Text Labs**
- **Lab Notes and Answers**

139

Richard P. Filson
Edison High School
Stockton, California

DOPPLER RADAR TRACKING OF BATS AND INSECTS IN CENTRAL TEXAS

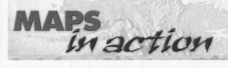
DOPPLER RADAR TRACKING OF BATS AND INSECTS IN CENTRAL TEXAS

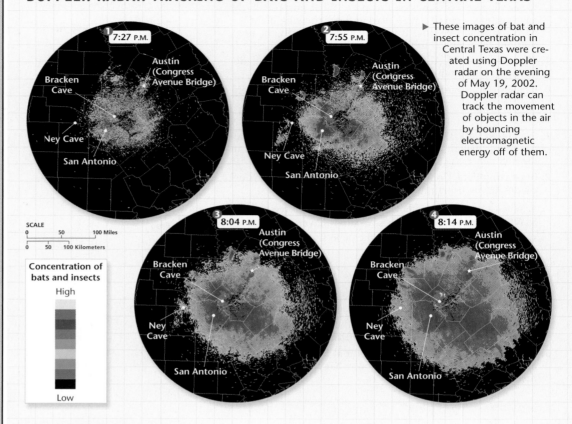

▶ These images of bat and insect concentration in Central Texas were created using Doppler radar on the evening of May 19, 2002. Doppler radar can track the movement of objects in the air by bouncing electromagnetic energy off of them.

Concentration of bats and insects
High
Low

MAP SKILLS

Use the Doppler radar images of bats and insects in Central Texas to answer the questions below.

1. **Analyzing Data** At what time was the bat and insect concentration the lowest? At what time was the bat and insect concentration the highest?

2. **Using a Key** Use the concentration key to determine which area of Central Texas has the highest concentration of bats and insects at 8:14 P.M.

3. **Analyzing Data** Approximately how many kilometers wide is the concentration of bats and insects at 7:27 P.M.? at 8:14 P.M.?

4. **Inferring Relationships** Bracken Cave is home to 20 million bats that eat millions of pounds of insects nightly. Approximately how far is Bracken Cave from the city of San Antonio? If the bat population in the cave drastically decreased, what effect would this decrease have on the people living in San Antonio and Central Texas?

5. **Identifying Trends** These Doppler radar images of bats and insects were taken in the beginning of the summer season. How might these four images look in the month of December?

140

Internet Activity

Scientists can use Doppler weather radar to track the movement and concentration of bats and insects in areas such as Central Texas. The four Doppler images show the concentrations of Mexican free-tailed bats and the insects they eat, on the evening of May 19, 2002. Bats migrate to Central Texas during the warmer months to feed on billions of insects. Bracken Cave is home to a record number 20 million bats, which is the world's largest colony. Have students find out more about radar entomology and radar ornithology by researching on the Internet. The Radar Ornithology Lab at Clemson University offers a wealth of online information.

Transparencies

TT Doppler Radar Tracking of Bats and Insects in Central Texas

Answers to Map Skills

1. 7:27 P.M.; 8:14 P.M.

2. Answers may vary. Bracken Cave would be the most accurate area showing the highest concentration.

3. Answers may vary. At 7:27 P.M., it was approximately 200 km. At 8:14 P.M., it was approximately 350 km.

4. Answers may vary. Bracken Cave is approximately 50 km from San Antonio. If the bat population drastically decreased, the insect population in San Antonio would probably increase.

5. In December, these four images would probably look very different due to the colder weather. The concentration of bats and insects would probably be much lower.

SOCIETY & the Environment

EATING THE BAIT

Most of the food we eat comes from agriculture and farming, but we also rely heavily on the fishing industry to provide us with fresh fish. Because of a high demand for fish, however, many fish species have become overharvested. Many organisms depend on these *fisheries*, places where fish are caught, to survive. The swordfish and cod fisheries of the North Atlantic and the salmon fishery off the northwestern coast of the United States are examples of fisheries that have become depleted. These fisheries now contain so few fish that harvesting these fish is not economical.

▶ **Overfishing** of organisms from higher trophic levels has forced the commercial fishing industry to harvest organisms in lower trophic levels in order to fulfill the demand for fresh fish.

Fishing Down the Food Chain

Fish such as cod, tuna, and snapper are top carnivores in ocean food chains and food webs. As these fish have disappeared, species from lower trophic levels have begun to appear in fish markets. Fish that were once swept back into the sea when they were caught in nets by accident are now being kept and sold. Organisms from lower trophic levels such as mullet, squid, mackerel, and herring, which were typically used as bait to catch larger fish, now appear on restaurant menus.

According to data from the United Nations on worldwide fish harvests, the overall trophic level at which most fish are caught has declined since the 1950s. Overfishing of organisms in lower trophic levels disrupts food chains and food webs. If the food webs of ocean ecosystems collapse, the commercial fishing industry will also collapse. For example, in the North Atlantic cod fisheries, the cod began to disappear, so the fishermen concentrated on the cods' prey, which is shrimp. Cod are higher trophic level organisms, while shrimp are in the lower trophic levels and feed on algae and detritus. If the shrimp and the cod become overfished, the other organisms that depend on both the shrimp and cod to survive are affected.

Creating Sustainable Fisheries

One aim of environmental science is to determine how fisheries can be managed so that they are sustainable or capable of supplying the same number of fish to be harvested each year. However, few, if any, countries manage their

▶ A squid is an example of an organism from a lower trophic level that was used for bait but is now sold in restaurants.

fisheries in this way. Almost all countries permit unsustainable, large harvests. One solution to overfishing is to establish "no-take" zones. These are areas of the sea where no fishing is permitted. Studies have shown that fish populations grow rapidly in "no-take" zones. When a population grows in a "no-take zone," the higher trophic level organisms leave the zone and become available to fishermen. "No-take" zones help populations recover and allow food chains and food webs to remain intact.

What Do You Think?

The next time you go to a fish market or seafood restaurant, take note of the different types of species for sale. Write down the names of the species, and try to assign each species to a trophic level. How many of the species for sale belong to lower trophic levels? How many belong to higher trophic levels? How do prices differ between the species for sale?

141

Answers to What Do You Think?

Answers may vary, depending on your area. You might record the types of fish seen at the nearest grocery store, so that students know which trophic level is most represented in your area.

EATING THE BAIT

Background

Greater demand and more efficient methods of catching fish have led to the depletion of predator fish populations, and subsequently, the depletion or collapse of those fisheries. Cod and salmon fisheries are examples of fisheries that have collapsed in North America. As humans move down the trophic level to "eat the bait," we compete with the depleted populations of predator fish, so their numbers do not rise. We also deplete the "bait" trophic level, which as a result depletes two trophic levels. Other animals that consume either predator or bait fish will also be negatively affected. Fish populations have been shown to grow quickly in "No-take" zones, so these zones might promote local recovery of populations.

REAL-LIFE CONNECTION | GENERAL

Overfishing: A Global Problem
Many scientists are afraid that we may overfish many species in the ocean to the point of extinction. Some overfished areas of the ocean have been compared to an African savanna where termites are at the top of the food chain due to poaching. Have students search for articles on overfishing and factory trawling on the Internet. Ask them to find out which fish are suffering population declines, and how the fish are used by humans. Many news outlets have covered this story in recent years, and students will get an idea of the severity of this problem. After students have done some research, lead a discussion designed to pinpoint problems and potential solutions. Then ask students to make a seafood guide that is designed to help people avoid eating species that have been overfished.

PACING	CLASSROOM RESOURCES	LABS, DEMONSTRATIONS, AND ACTIVITIES
BLOCKS 1 & 2 · 90 min pp. 142–145 **Chapter Opener**		
Section 1 What Is a Biome?	**OSP** Lesson Plan * **CRF** Active Reading * BASIC **TT** Bellringer * **TT** Biomes of the World * **TT** Temperature Vs. Precipitation * **TT** Latitude Vs. Altitude * **CD** Interactive Tutor Biosphere	**CRF** Design Your Own Lab * ◆ BASIC **CRF** Math/Graphing Lab * GENERAL **CRF** CBL™ Probeware Lab * ◆ ADVANCED
BLOCKS 3 & 4 · 90 min pp. 146–154 **Section 2** Forest Biomes	**OSP** Lesson Plan * **CRF** Active Reading * BASIC **TT** Bellringer * **TT** Biome Climatograms: A *	**TE** Demonstration Exotic Fruit, p. 146 GENERAL **TE** Group Activity Rain Forest Collage, p. 147 ◆ BASIC **TE** Demonstration Visual Precipitation, p. 147 GENERAL **TE** Group Activity Light in the Forest, p. 148 BASIC **TE** Activity We're All Thumbs! p. 149 BASIC **TE** Activity Ecotourism, p. 150 GENERAL **TE** Group Activity Pacific Northwest Food Web, p. 151 ADVANCED **TE** Activity Forest Soil Field Trip, p. 152 ADVANCED **TE** Activity Touring the Forest, p. 153 ADVANCED **TE** Group Activity Biome Components, p. 153 GENERAL **CRF** Consumer Lab * BASIC
BLOCKS 5, 6 & 7 · 135 min pp. 155–163 **Section 3** Grassland, Desert, and Tundra Biomes	**OSP** Lesson Plan * **CRF** Active Reading * BASIC **TT** Bellringer * **TT** Biome Climatograms: B *	**TE** Activity Grass Roots, p. 157 ◆ GENERAL **TE** Group Activity Plant a Prairie! p. 157 GENERAL **TE** Activity Native Grasses, p. 158 GENERAL **SE** QuickLab Sponging It Up, p. 158 **TE** Demonstration Plants of the Chaparral, p. 159 BASIC **SE** Field Activity Miniature Desert, p. 161 **TE** Activity Animal Adaptations to Heat, p. 161 ADVANCED **TE** Internet Activity Xeriscaping, p. 161 GENERAL **TE** Activity Tundra Creatures, p. 162 GENERAL **TE** Group Activity Arctic Science, p. 162 ADVANCED **SE** Exploration Lab Identify Your Local Biome, pp. 168–169 ◆ **CRF** Datasheets for In-Text Labs * **CRF** Long-Term Project * ◆ GENERAL

BLOCKS 8 & 9 · 90 min

Chapter Review and Assessment Resources

SE Chapter Review pp. 165–167
SE Standardized Test Prep pp. 646–647
CRF Study Guide * ■ GENERAL
CRF Chapter Test * ■ GENERAL
CRF Chapter Test * ADVANCED
CRF Concept Review * ■ GENERAL
CRF Critical Thinking * ADVANCED
OSP Test Generator
CRF Test Item Listing *

Online and Technology Resources

 Holt Online Learning

Visit **go.hrw.com** for access to Holt Online Learning or enter the keyword **HE6 Home** for a variety of free online resources.

 One-Stop Planner® CD-ROM

This CD-ROM package includes
• Lab Materials QuickList Software
• Holt Calendar Planner
• Customizable Lesson Plans
• Printable Worksheets
• ExamView® Test Generator
• Interactive Teacher Edition
• Holt PuzzlePro® Resources
• Holt PowerPoint® Resources

 Holt Environmental Science Interactive Tutor CD-ROM

This CD-ROM consists of interactive activities that give students a fun way to extend their knowledge of environmental science concepts.

KEY

TE	Teacher Edition	**CRF**	Chapter Resource File	*	Also on One-Stop Planner
SE	Student Edition	**TT**	Teaching Transparency	■	Also Available in Spanish
OSP	One-Stop Planner	**CD**	Interactive CD-ROM	◆	Requires Advance Prep

ENRICHMENT AND SKILLS PRACTICE	SECTION REVIEW AND ASSESSMENT	CORRELATIONS
SE Pre-Reading Activity, p. 142 **TE** Using the Figure, p. 142 **GENERAL**		**National Science Education Standards**
TE Using the Figure Climatograms, p. 144 **GENERAL** **TE** Career Meteorologist, p. 144 **TE** Reading Skill Builder Paired Summarizing, p. 144 **BASIC**	**TE** Homework, p. 144 **GENERAL** **SE** Section Review, p. 145 **TE** Reteaching, p. 145 **BASIC** **TE** Quiz, p. 145 **GENERAL** **TE** Alternative Assessment, p. 145 **ADVANCED** **CRF** Quiz * ■ **GENERAL**	
TE Using the Figure Climatograms, p. 147 **GENERAL** **SE** Case Study Deforestation, Climate, and Floods, pp. 150–151 **TE** Inclusion Strategies, p. 151 **TE** Reading Skill Builder Concept Mapping, p. 153 **BASIC**	**SE** Section Review, p. 154 **TE** Reteaching, p. 154 **BASIC** **TE** Quiz, p. 154 **GENERAL** **TE** Alternative Assessment, p. 154 **GENERAL** **CRF** Quiz * ■ **GENERAL**	LS 3a LS 4c LS 4d LS 6b
TE Inclusion Strategies, p. 159 **TE** Reading Skill Builder Reading Organizer, p. 159 **BASIC** **SE** Graphic Organizer Venn Diagram, p. 162 **GENERAL** **TE** Student Opportunities Toolik Field Station, p. 162 **SE** MathPractice U.S. Oil Production, p. 163 **SE** Points of View The Future of the Arctic National Wildlife Refuge, pp. 170–171 **CRF** Map Skills African Biomes * **GENERAL**	**TE** Homework, p. 157 **ADVANCED** **SE** Section Review, p. 163 **TE** Reteaching, p. 163 **BASIC** **TE** Quiz, p. 163 **GENERAL** **TE** Alternative Assessment, p. 163 **GENERAL** **CRF** Quiz * ■ **GENERAL**	LS 3a LS 4c LS 4d LS 6b

Guided Reading Audio CDs

These CDs are designed to help auditory learners and reluctant readers. (Audio CDs are also available in Spanish.)

NSTA

www.scilinks.org

Maintained by the **National Science Teachers Association**

TOPIC: Biomes
SciLinks code: HE4007

TOPIC: Threats to Rain Forests
SciLinks code: HE4112

TOPIC: Temperate Deciduous Forests
SciLinks code: HE4110

 CNN Videos

Each video segment is accompanied by a Critical Thinking Worksheet.

Earth Science Connections

Segment 12 Egypt's Pyramids

Chapter Enrichment

This Chapter Enrichment provides relevant and interesting information to expand and enhance your classroom instruction of the chapter material.

1 What is a Biome?

Adapting to the Taiga

The species that live in each biome are different. But, you may notice that some species in one biome may look and act similar to species in another biome. This occurs because similar niches become available in each biome. To fill those niches, animals develop adaptations that cause them to look and behave in ways similar to other animals in other biomes. For example, boreal forests, or taiga, in Canada, Russia, and Scandinavia all have species that fit the following niches:

- large omnivores, such as brown bears and polar bears
- large predators, such as wolves, lynxes, and tigers
- medium-sized predators, such as foxes, martens, and weasels
- medium-sized, hoofed herbivores, such as deer, elk, and reindeer
- medium-sized seed eaters, such as flying squirrels, tree squirrels, and ground squirrels
- small herbivores, such as hares and pikas
- small rodents that eat vegetation, such as voles, mice, and lemmings
- small insect eaters, such as shrews, moles, and bats

Taiga

In areas that receive heavy snow, such as the taiga of Northern Europe and Northern Russia, a large accumulation of snow can break trees and limit the development of forests. Snow and ice weighing as much as 3,000 kg can accumulate on trees in these regions. The abrasive action of high winds and snow can also strip the foliage and bark from the trees. This type of abrasion can lead to extreme water loss and eventually the death of the tree.

► Plants of the tundra biome

BRAIN FOOD

Desert biomes are characterized by a lack of water. They can be divided into four types: hot and dry, semiarid, coastal, and cold. Hot and dry deserts have very short periods of rain, and include the Mojave Desert in California and the Sahara Desert in Africa. Semiarid deserts are cooler during the day than hot and dry deserts, and warmer at night. An example of a semiarid desert is the sagebrush of Utah. Coastal deserts, such as those in Chile, are dry but not very hot. Coastal deserts are warm in the summer and cool in the winter. Cold deserts have short, moist summers and long, cold winters. Some cold deserts are located in Antarctica and Greenland.

2 Forest Biomes

People of the Forest

The Republic of Cameroon in western Africa is home to the Baka Pygmy tribes, which inhabit the country's tropical forests. These tribes were probably the first inhabitants of Cameroon, and they still rely on forest resources and hunting to survive. Because of this reliance, they have clashed with rangers that are employed to protect dwindling species, such as the elephant. Pygmies have also seen their land converted into other uses, such as developing it to build sawmills. Indigenous people, such as the Pygmies, have little political power to help them protect their way of life, which is tied to the resources of the forest.

Acclimation to Cold

Trees and shrubs that are adapted to colder climates survive by acclimating to cold weather. During the acclimation process, water that is inside a plant but outside the cells, is the first to form ice crystals. These crystals do not break a plant's cell walls, so they do not cause injury or death to the plant. The amount of energy in the ice outside the cell is lower compared to the amount of energy in the water inside the cell. This difference in energy forces water to be drawn out of the cell to freeze. As the temperature decreases, water continues to move

▶ A temperate deciduous forest during winter

out of the cell, but the water within the cell does not freeze. The movement of water out of the cell's cytoplasm lowers the cell's freezing point because it increases the number of solutes in the cell's cytoplasm. Unless the temperature gets so cold that it causes the cell's cytoplasmic water to freeze, the tree or shrub will survive the winter without injury. However, if the temperature drops too quickly injury or death may occur. Without an acclimation period, water within a plant's cells will freeze and break the cell walls.

BRAIN FOOD

Since 1989, the Rain Forest Alliance has sponsored a wood certification program called SmartWood. This program was created to monitor and reward sustainable forestry practices. To be certified as SmartWood, forests must be managed to maintain watershed integrity and wildlife habitat. All harvesting must be sustainable. Finally, all forestry activities must provide social and economic benefits to local communities. SmartWood certification was started in order to improve rain forest forestry, but has expanded to include all forest types. SmartWood products help to fill the demand for earth-friendly products, while working to improve the lives and environment of the people who grow and harvest the wood.

SECTION 3 Grassland, Desert, and Tundra Biomes

Prairie Gardens

Landscaping with native prairie plants is becoming a popular hobby for many people in areas that were once temperate grassland. Native plant societies and initiatives such as Illinois' Corridors for Tomorrow, which plants native prairie species along roadsides, have reintroduced people to beautiful native wildflowers that attract butterflies and birds. Besides providing food and shelter to important native insects and birds, prairie plants save money and time because once the plants are established, they need very little maintenance. Prairie plants also help the environment because they do not require fertilizers or pesticides to grow and flourish.

Special Stomates

Stomates are the openings in a plant's leaves or stems that allow carbon dioxide to enter the plant so that it can perform photosynthesis. Stomates also cause the plant to lose water when they are open during the process of transpiration. To minimize loss of water in the desert, some desert plants have developed fewer, smaller stomates. These stomates may only open during the night. The special stomates may also be located deeper in the plant's tissues, as opposed to on the surface of the leaf or stem.

Insulation in the Desert

It may not seem to make sense, but many desert mammals have more fur on their bodies to beat the heat. Thicker fur can insulate animals from the heat of the sun. Some hoofed mammals, such as camels, have thicker fur on the top of their bodies and very little fur on their bellies and legs. This insulates them from the heat of the sun above, while radiating excess heat from the body parts that have less fur.

Overview

Tell students that the purpose of this chapter is to help them understand the different terrestrial biomes found throughout the world. Biomes are described by their vegetation, temperature, and precipitation. The terrestrial biomes of the world include tropical rain forest, temperate forest, taiga, temperate grassland, desert, tundra, chaparral, and savanna. Threats to habitats in each biome are also described.

Using the Figure — GENERAL

Animals such as this thorny devil have adapted to the desert's high temperatures and low precipitation. Ask students to identify adaptations of the thorny devil in the photograph. (Sample answers: the colors of the thorny devil's skin help it blend in with the desert environment, it has pointy spikes on its rough skin to help it ward off predators. And its thick skin helps retain water in the hot, dry climate.) Ask students to think of other desert animals and explain how each has adapted to living in a hot, dry place. (Sample answer: Armadillos have thick skin and are nocturnal.)
LS Visual

PRE-READING ACTIVITY

Encourage students to use their FoldNote as a study guide to quiz themselves for a test on the chapter material. Students may want to create Four-Corner Fold FoldNotes for different topics within the chapter.

VIDEO SELECT

For information about videos related to this chapter, go to **go.hrw.com** and type in the keyword **HE4 TERV**.

Biomes

142

1 **What Is a Biome?**
2 **Forest Biomes**
3 **Grassland, Desert, and Tundra Biomes**

PRE-READING ACTIVITY

FOLDNOTES **Four-Corner Fold**

Before you read this chapter, create the **FoldNote** entitled "Four-Corner Fold" described in the Reading and Study Skills section of the Appendix. Label each flap of the four-corner fold with "Forest Biomes," "Grassland Biomes," "Desert Biomes," and "Tundra Biomes." As you read the chapter, define each biome, and write characteristics of each biome on the appropriate fold.

This thorny devil lives in the desert of Australia. The grooves in its rough skin help it collect water to drink. Water from rain or condensation lands on its back and runs along the tiny grooves to its mouth.

Chapter Correlations   *National Science Education Standards*

LS 3a Species evolve over time. Evolution is the consequence of the interactions of (1) the potential for a species to increase its numbers, (2) the genetic variability of offspring due to mutation and recombination of genes, (3) a finite supply of the resources required for life, and (4) the ensuing selection by the environment of those offspring better able to survive and leave offspring. **(Section 2 and Section 3)**

LS 4c Organisms both cooperate and compete in ecosystems. The interrelationships and interdependencies of these organisms may generate ecosystems that are stable for hundreds or thousands of years. **(Section 2 and Section 3)**

LS 4d Living organisms have the capacity to produce populations of infinite size, but environments and resources are finite. This fundamental tension has profound effects on the interactions between organisms. **(Section 2 and Section 3)**

LS 6b Organisms have behavioral responses to internal changes and to

external stimuli. Responses to external stimuli can result from interactions with the organism's own species and others, as well as environmental changes; these responses either can be innate or learned. The broad patterns of behavior exhibited by animals have evolved to ensure reproductive success. Animals often live in unpredictable environments, and so their behavior must be flexible enough to deal with uncertainty and change. Plants also respond to stimuli. **(Section 2 and Section 3)**

What Is a Biome?

Earth is covered by many types of ecosystems. Ecologists group these ecosystems into larger areas known as biomes. A **biome** is a large region characterized by a specific type of climate and certain types of plants and animal communities. Each biome is made up of many individual ecosystems. The map in **Figure 1** shows the locations of the world's major land, or terrestrial, biomes. In this chapter, you will take a tour through these terrestrial biomes— from lush rain forests to scorching deserts and the frozen tundra. When you read about each biome, notice the adaptations that organisms have to their very different environments.

Biomes and Vegetation

Biomes are described by their vegetation because plants that grow in an area determine the other organisms that can live there. For example, shrubs called *rhododendrons* grow in northern temperate forests because they cannot survive high temperatures. However, mahogany trees grow in tropical rain forests because they cannot survive cold, dry weather. Organisms that depend on mahogany trees will live where mahogany trees grow.

Plants in a particular biome have characteristics, specialized structures, or adaptations that allow the plants to survive in that biome. These adaptations include size, shape, and color. For example, plants that grow in the tundra tend to be short because they cannot obtain enough water to grow larger. They also have a short summer growing season, while desert plants, such as cactuses, do not have leaves. Instead, cactuses have specialized structures to conserve and retain water.

Objectives

▶ **Describe** how plants determine the name of a biome.

▶ **Explain** how temperature and precipitation determine which plants grow in an area.

▶ **Explain** how latitude and altitude affect which plants grow in an area.

Key Terms

biome
climate
latitude
altitude

Figure 1 ▶ The ecosystems of the world can be grouped into regions called *biomes*. These biomes shown below are named for the vegetation that grows there.

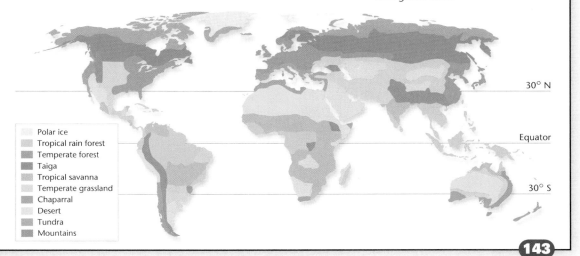

- Polar ice
- Tropical rain forest
- Temperate forest
- Taiga
- Tropical savanna
- Temperate grassland
- Chaparral
- Desert
- Tundra
- Mountains

30° N

Equator

30° S

143

MISCONCEPTION ///ALERT\\\

Biome Vs. Ecosystem Students may have difficulty distinguishing between the terms *biome* and *ecosystem*. This is not an easy distinction—in fact, the terms are not used consistently by scientists. A biome and an ecosystem are defined by abiotic and biotic factors, and by the types of interactions among the organisms that live in each. To help students understand the difference, discuss the following hierarchy: the biosphere is divided into biomes, which are divided into smaller ecosystems.

Chapter Resource File

- **Lesson Plan**
- **Active Reading** BASIC
- **Section Quiz** GENERAL

Transparencies

TT Bellringer
TT Biomes of the World
TT Temperature Vs. Precipitation

Focus

Overview

Before beginning this section, review with your students the Objectives in the Student Edition. In this section, students are introduced to biomes. This section also explains how climate and biomes vary with latitude and altitude.

Bellringer

Have small groups of students look at a world map. Ask students to find the latitudes of their city or town, and then find a large city in Finland, Argentina, Vermont, and Cameroon. Have them compare these latitudes to the map in **Figure 1** in order to find the biomes associated with these cities. Ask students to think about why the biomes might be different in each of these locations. (Finland has taiga, Argentina has grasslands, Vermont has temperate deciduous forest, and Cameroon has tropical rain forests. Different biomes have developed at different latitudes because of the different climatic conditions.) **LS** Visual

English Language Learners

Motivate

Identifying Preconceptions — BASIC

Biome Mix-up Perform this exercise before students begin to read this section. Write the names of the biomes of the world on the board. Mix the names up, so they are not in order by latitude. Now ask students to tell you whether the temperature of each biome is hot, moderate, or cold. Then ask students to tell you whether each biome is wet, moderate, or dry. Write their guesses next to each biome. Then have students look at **Figure 3,** to find out how accurate they were. **LS** Logical

Teach

Climatograms Ask students to trace or draw **Figure 3** in their *EcoLog*. Students do not need to draw all of the vegetation accurately; they should just sketch a few key plants. Have students leave space near the name of each biome. Then have students find the climatograms (the graphs that give monthly averages of the temperature and precipitation for each biome) throughout the rest of the chapter. On their sketch, ask them to record the range of temperature and the approximate annual precipitation for each biome from the climatograms. (You might want to review how to read a climatogram with students before they do this.) Have them discuss whether or not the climatograms match the general trends indicated in **Figure 3.** (they should match approximately) Ask students to figure out where chaparral fits into the figure. (It should fit between temperate grassland and desert.) **LS** Visual

Homework — GENERAL

Plants and Animals Ask students to research two of their favorite plants or animals to find out which biome each organism lives in. Have students write a short paragraph about each organism that details its maximum and minimum temperature and precipitation needs, its typical biome, and some of the adaptations that allow it to survive in that biome. Have them include a picture of the organism above the descriptive paragraph. Encourage students to add these paragraphs to their **Portfolio.** **LS** Intrapersonal

Figure 2 ▶ Plants in the tundra biome, such as those shown above, are usually short because the soil is frozen most of the year, which prevents the plants from obtaining much water.

Figure 3 ▶ Temperature and precipitation help determine the type of vegetation in an ecosystem. As temperature and precipitation decrease, the climate of an area becomes drier and vegetation becomes sparser.

Biomes and Climate

Biomes are defined by their plant life, but what factors determine which plants can grow in a certain area? The main factor is climate. **Climate** refers to the weather conditions, such as temperature, precipitation, humidity, and winds, in an area over a long period of time. Temperature and precipitation are the two most important factors that determine a region's climate.

Temperature and Precipitation The climate of a biome is determined by average temperature and precipitation. Most organisms are adapted to live within a particular range of temperatures and will not survive at temperatures too far above or below their range.

Precipitation also limits the organisms that are found in a biome. All organisms need water, and the larger an organism is, the more water it needs. For example, biomes that do not receive enough rainfall to support large trees support communities dominated by small trees, shrubs, and grasses. In biomes where rainfall is not frequent, the vegetation is made up of mostly cactuses and desert shrubs. The plants in **Figure 2** grow close to the ground in the tundra because there is not enough water to support larger plants and trees. In extreme cases, lack of rainfall results in no plants, no matter what the temperature is. As shown in **Figure 3**, the higher the temperature and precipitation are, the taller and denser the vegetation is. Notice how much more vegetation exists in a hot, wet tropical rain forest than in a dry desert.

144

Career

Meteorologist A meteorologist is a scientist who studies atmospheric phenomena, such as pressure fronts and humidity, often in order to predict weather. Meteorologists use sophisticated machinery, such as Doppler radar, to read cloud patterns and pressure systems within the layers of the atmosphere. They also predict severe weather, such as tornadoes and hurricanes, in order to save lives and minimize property damage. Contact a local news station to see if you can set up a tour of the weather prediction facility, or have the meteorologist visit the class to give a presentation on weather and climate in your area.

READING SKILL BUILDER — BASIC

Paired Summarizing Have students form pairs. Then ask students to quiz each other on the temperature and precipitation traits of each biome by using **Figure 3.** Have each student within the pair switch off and work together until both of them can name all the traits without looking at the figure. **LS** Auditory

Mountains (ice and snow)

Tundra (herbs, lichens, and mosses)

Taiga (coniferous forests)

Temperate deciduous forests

Tropical rain forests

Altitude

Tropical rain forests

Temperate deciduous forests

Taiga

Tundra

Polar ice

Latitude

Latitude and Altitude Biomes, climate, and vegetation vary with latitude and altitude. **Latitude** is the distance north or south of the equator and is measured in degrees. **Altitude** is the height of an object above sea level. Climate varies with latitude and altitude. For example, climate gets colder as latitude and altitude increase. So, climate also gets colder as you move farther up a mountain.

Figure 4 shows that as latitude and altitude increase, biomes and vegetation change. For example, the trees of tropical rain forests usually grow closer to the equator, while the mosses and lichens of the tundra usually grow closer to the poles. The land located in the temperate region of the world, between about 30° and 60° north latitude and 30° and 60° south latitude, is where most of the food in the world is grown. This region includes biomes such as temperate forests and grasslands, which usually have moderate temperatures and fertile soil that is ideal for agriculture.

Figure 4 ▶ Latitude and altitude affect climate and vegetation in a biome.

internet connect

www.scilinks.org
Topic: Biomes
SciLinks code: HE4007

*SCi*LINKS. Maintained by the National Science Teachers Association

SECTION 1 Review

1. **Describe** how plants determine the name of a biome.

2. **Explain** how temperature affects which plants grow in an area.

3. **Explain** how precipitation affects which plants grow in an area.

4. **Define** *latitude* and *altitude*. How is latitude different from altitude? How do these factors affect the organisms that live in a biome?

CRITICAL THINKING

5. **Making Inferences** The equator passes through the country of Ecuador. But the climate in Ecuador can range from hot and humid to cool and dry. Write a short paragraph that explains what might cause this range in climate. **WRITING SKILLS**

6. **Analyzing Relationships** Look at Figure 1, and locate the equator and 30° north latitude. Which biomes are located between these two lines?

145

Answers to Section Review

1. Scientists name biomes after their vegetation because the plants that grow in an area determine what other organisms can live there.

2. Plants are adapted to a particular range of temperature and can usually survive only in a climate with that particular range.

3. Plants are adapted to a particular level of precipitation. In general, the larger a plant is, the more water it needs.

4. Latitude is the distance north or south of the equator. Altitude is the height of an object above sea level. Both latitude and altitude determine the temperature and precipitation

of a biome. Therefore, if an organism lives in a biome that is close to the equator and at a low altitude, the organism must be able to survive in a very warm, moist environment.

5. Sample answer: Part of the Andes Mountains is located in Ecuador. The resulting wide range of altitudes in Ecuador creates the wide range of climates.

6. Desert, chaparral, tropical savanna, temperate grassland, temperate forest, and tropical rain forest are the biomes that are located between the equator and 30° north latitude.

Overview

Before beginning this section, review with your students the Objectives in the Student Edition. This section describes the tropical rain forest, temperate rain forest, temperate deciduous forest, and taiga.

🔊 Bellringer

Light a wooden match in front of students, and allow it to burn. Then explain that in the time it took the match to burn (45 seconds), approximately 97 acres of rain forest were destroyed by slash-and-burn techniques. This amount is roughly the size of one football field every second.

Motivate

Demonstration —— GENERAL

Exotic Fruit Many rain forest fruits, such as carambola ("star fruit"), are considered to be exotic throughout much of the United States. However, most of these products are quite common in more tropical areas of the world. Bring in some samples of rain forest fruits and nuts (which are usually available at specialty grocery stores) to the classroom. Find out in advance where each product comes from (the plant and its country), and share this information with students. Allow students to feast on the food as you discuss its origins. Be sure that students with allergies to nuts do not eat them. **LS** Visual

---

Forest Biomes

Objectives

▶ List three characteristics of tropical rain forests.
▶ Name and describe the main layers of a tropical rain forest.
▶ Describe one plant in a temperate deciduous forest and an adaptation that helps the plant survive.
▶ Describe one adaptation that may help an animal survive in the taiga.
▶ Name two threats to the world's forest biomes.

Key Terms

tropical rain forest
emergent layer
canopy
epiphyte
understory
temperate rain forest
temperate deciduous forest
taiga

The air is hot and heavy with humidity. You walk through the shade of the tropical rain forest, step carefully over tangles of roots and vines, and brush past enormous leaves. Life is all around you, but you see little vegetation on the forest floor. Birds call, and monkeys chatter from above.

Tropical Rain Forests

Of all the biomes in the world, forest biomes are the most widespread and the most diverse. The large trees of forests need a lot of water, so forests exist where temperatures are mild to hot and where rainfall is plentiful. Tropical, temperate, and coniferous forests are the three main forest biomes of the world.

Tropical rain forests are located in a belt around the Earth near the equator, as shown in **Figure 5.** They help regulate world climate and play vital roles in the nitrogen, oxygen, and carbon cycles. Tropical rain forests are always humid and warm and get about 200 to 450 cm of rain a year. Because they are near the equator, tropical rain forests get strong sunlight year-round and maintain a relatively constant temperature year-round. This climate is ideal for a wide variety of plants and animals, as shown in **Figure 6.** The warm, wet conditions also nourish more species of plants than any other biome does. While one hectare (10,000 m²) of temperate forest usually contains a few species of trees, the same area of tropical rain forest may contain more than 100 species.

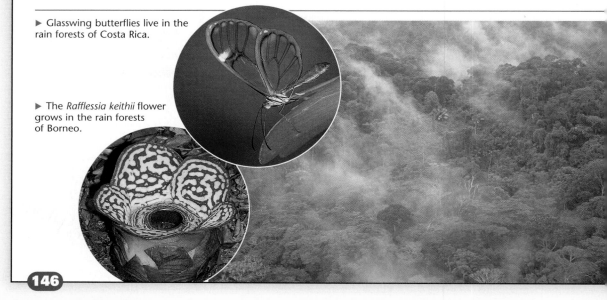

▶ Glasswing butterflies live in the rain forests of Costa Rica.

▶ The *Rafflessia keithii* flower grows in the rain forests of Borneo.

146

A Dark and Tangled Jungle Ever notice how Hollywood creates tropical rain forests to appear as an impenetrable tangle of plants? Check out some old Tarzan movies, or current movies with a jungle theme, from your local library. Watch some scenes that take place in the "jungle." Then, compare what a rain forest really looks like by renting a video about the ecology of the rain forest. Rain forests are not covered in dense jungle undergrowth. Tall trees dominate the rain forest and the rain forest floor is typically not covered with plants.

Chapter Resource File

Lesson Plan
Active Reading BASIC
Section Quiz GENERAL

Transparencies

TT Bellringer

Limon,
Costa Rica

Tropical
rain forest

30° N

Equator

30° S

Nutrients in Tropical Rain Forests You might think that the diverse plant life grows on rich soil, but it does not. Most nutrients are within plants, not within soil. Rapid decay of plants and animals occurs with the help of decomposers, organisms that break down dead organisms. Decomposers on the rain-forest floor break down dead organisms and return nutrients to the soil, but plants quickly absorb the nutrients. Some trees in the tropical rain forest support fungi that feed on dead organic matter on the rain-forest floor. In this relationship, fungi transfer the nutrients from the dead organic matter directly to the tree. Nutrients from dead organic matter are removed so efficiently that runoff from rain forests is often as pure as distilled water. Most tropical soils that are cleared of plants for agriculture lack nutrients and cannot support crops for more than a few years. Many of the trees form above-ground roots called *buttresses* or *braces* that grow sideways from the trees and provide extra support to the tree in the thin soil.

Tropical Rain Forest
(Limon, Costa Rica)

Months

Figure 5 ▶ The world's tropical rain forests have heavy, year-round rainfall and fairly constant, warm temperatures.

▶ The rain forests that blanket the Andes Mountains of Ecuador are always humid and warm.

Figure 6 ▶ Species of Tropical Rain Forests

▶ These mountain gorillas live in the rain forests of Rwanda.

▶ Scarlet macaws live in the trees of rain forests of Peru.

 147

BRAIN FOOD

Limiting Factors Every organism in an ecosystem needs certain resources, both biotic and abiotic. When a resource is in such short supply that it limits the growth of a population, it is called a *limiting factor*. Abiotic factors such as temperature, water, light, and nutrients in soil are usually the most limiting. As you proceed through the chapter, have students identify the limiting factors for each biome.

Teach

Using the Figure ── GENERAL

Climatograms **Figure 5** is the first of many climatograms that students will encounter in this chapter. This graph has two *y*-axes: one for precipitation and one for temperature. The blue bars correspond with the precipitation axis, and the red line corresponds with the temperature. Make sure students are reading the values correctly by asking them questions about the figure.
LS Logical

Group Activity ───── BASIC

Rain Forest Collage Bring in magazines with good photos of tropical animals and plants (such as *National Geographic*). Organize the class into six groups and assign each group the task of finding photos of rain forest mammals, birds, reptiles, amphibians, insects, trees, and epiphytes. Have the groups cut these photos out and assemble them as an ecosystem on one piece of poster board. Have the groups label each organism with its common and scientific name. If there is space, also have them describe each species' role in the ecosystem. Display the poster in your classroom.
LS Interpersonal

Co-op Learning   English Language Learners

Demonstration ──── GENERAL

Visual Precipitation Have students add up the total annual precipitation represented in **Figure 5** for the Costa Rican rain forest. (approx. 338 cm, or 3.38 m per year) Provide them with some string, and have them measure out a piece that is as long as the yearly precipitation. Have them do the same for the total annual precipitation represented in **Figure 25** for the desert in Egypt (and for the precipitation from other biomes, if they are interested). Hang the strings on a wall in the classroom. **LS** Visual   English Language Learners

Transparencies

TT Biome Climatograms: A

Group Activity —— BASIC

Light in the Forest Have students simulate the way light is filtered through the forest by having them put together tubes of layered paper. Divide students into four groups, and give each group a 12-inch piece of cardboard tube. Have each group cut their tube into a different number of pieces (2, 3, 4 and 5 pieces). Then, have them tape a thin piece of writing paper to 1, 2, 3 or 4 of their pieces of tube. Ask students to tape the tubes back together. Then, shine a flashlight over the top of each of the tubes. Compare the amount of light that gets through 1, 2, 3 or 4 sheets of paper. With hundreds of leaves for light to travel through, it is easy to see why the rain forest floor is dark. **LS** Visual/Kinesthetic

Co-op Learning   English Language Learners

BOTANY ——
CONNECTION —— GENERAL

Familiar Epiphytes Bromeliads and orchids are two families of mostly-epiphytic plants common in rain forest canopies. They are also popular plants in homes and greenhouses. Bromeliad species that students may have seen include pineapples, Spanish moss, and ball moss. Bring in examples of these types of plants. Have students observe the adaptations these plants have for "living on air," such as water-catching mechanisms and dangling roots. Tell students that some bromeliads are carnivorous—they trap and digest insects to get nutrients.

Emergent trees

Upper canopy

Lower canopy

Understory

Bright light

Filtered light

Dense shade

Figure 7 ▶ The plants in tropical rain forests form distinct layers. The plants in each layer are adapted to a particular level of light. The taller trees absorb the most light, while the plants near the forest floor are adapted to growing in the shade.

Connection to Chemistry

Medicines from Plants Many of the medicines we use come from plants native to tropical rain forests. Chemists extract and test chemicals found in plants to determine if the chemicals can cure or fight diseases. Rosy periwinkle, a plant that grows in the tropical rain forests of Madagascar, is the source of two medicines, vinblastine and vincristine. Vinblastine is used to treat Hodgkin's disease, a type of cancer, and vincristine is used to treat childhood leukemia.

Layers of the Rain Forest In tropical rain forests, different types of plants grow in different layers, as shown in **Figure 7.** The four main layers above the forest floor are the emergent layer, the upper canopy, the lower canopy, and the understory. The top layer is called the **emergent layer.** This layer consists of the tallest trees, which reach heights of 60 to 70 m. The trunks of trees this tall can measure up to 5 m around. Trees in the emergent layer grow and emerge into direct sunlight. Animals such as eagles, bats, monkeys, and snakes live in the emergent layer.

The next layer, considered the primary layer of the rain forest, is called the **canopy.** Trees in the canopy can grow more than 30 m tall. The tall trees form a dense layer that absorbs up to 95 percent of the sunlight. The canopy can be split into an upper canopy and a lower canopy. The lower canopy receives less light than the upper canopy does. Plants called **epiphytes,** such as the orchid in **Figure 8,** use the entire surface of a tree as a place to live. Epiphytes grow on tall trees for support and grow high in the canopy, where their leaves can reach the sunlight needed for photosynthesis. Growing on tall trees also allows them to absorb the water and nutrients that run down the tree after it rains. Most animals that live in the rain forest live in the canopy because they depend on the abundant flowers and fruits that grow there.

Below the canopy, very little light reaches the next layer, called the **understory.** Trees and shrubs adapted to shade grow in the understory. Most plants in the understory do not grow more than 3.5 m tall. Herbs with large, flat leaves that grow on the forest floor capture the small amount of sunlight that penetrates the understory. These plants must be able to grow in the darker spots. When fallen trees create an opening in the canopy, tree seedlings that are adapted to grow quickly compete with other seedlings on the forest floor for sunlight.

CHEMISTRY ——
CONNECTION —— ADVANCED

Tropical Medicinals Many rain forest plants and animals have yielded chemicals used to create powerful new drugs. Have each student research one of these medicinal plants or animals (besides the rosy periwinkle). Ask them to identify the active ingredient in each of these organisms, the drug that it helped to create, and what the drug helps to cure. Have students find the chemical structure of the active ingredient, if possible. Encourage students to write up their findings to include in their **Portfolio. LS** Intrapersonal

Notable Quotes —— BASIC

"There is more information of a higher order and complexity stored in a few square kilometers of forest than there is in all the libraries of mankind."

— Eugene Odum, Ecologist.

Have students discuss the types of information one could gather from a forest. (relationships among species, genetic information, evolutionary history, chemical formulas, artistic forms, perfumes, different tastes)

Species Diversity in Rain Forests The tropical rain forest is the biome with the greatest amount of species diversity. The diversity of rain-forest vegetation has led to the evolution of a diverse community of animals. Most rain-forest animals are specialists that use specific resources in particular ways to avoid competition. Some rain-forest animals have amazing adaptations for capturing prey, and other animals have adaptations that they use to escape predators. For example, the collared anteater in **Figure 8** uses its long tongue to reach insects in small cracks and holes where other animals cannot reach. The wreathed hornbill (shown below) uses its strong, curved beak to crack open nutshells. Insects, such as the Costa Rican mantis in **Figure 8**, use camouflage to avoid predators and may be shaped like leaves or twigs.

Ecofact

A Little Land, A Lot of Species
Tropical rain forests cover less than 7 percent of Earth's land surface but contain at least 50 percent of all the plant and animal species in the world.

Figure 8 ▶ Examples of plant and animal adaptations in the tropical rain forest include ❶ the long tongue of a collared anteater, ❷ the strong, curved beak of a wreathed hornbill, ❸ the shape of a Costa Rican mantis, and ❹ an orchid attached to a tall tree.

149

Activity ——— GENERAL

Ecotourism Before you cover this chapter with students, order some brochures from various rain forest ecotourism companies, or find brochures and advertisements on the Internet. Be sure to get brochures from multiple countries. Pass these brochures out to students, and have them outline the kinds of activities included in each trip, and the kinds of organisms they would see in each forest. Have students compare their outlines, and discuss the diversity of species presented in the brochures. Discuss why certain kinds of organisms are valued by humans and others are not. This exercise should also help students to see the efforts by some countries to gain economically from their ecological treasures.
LS Interpersonal

Cultural Awareness GENERAL

Threatened Cultures Point out to students that the destruction of many ecosystems threatens the livelihood of groups of indigenous people. Have students research some of these groups, such as the Yanomami of Brazil and Venezuela. The organization *Cultural Survival* offers current information about these groups. Ask students to find out how these groups have adapted to the ecosystem where they live, and what are the current threats to their culture and way of life. Have students write a short report and share their findings with the class. Encourage students to include their report in their **Portfolio.**
LS Interpersonal

www.scilinks.org
Topic: Threats to
Rain Forests
SciLinks code: HE4112

SCLINKS Maintained by the National Science Teachers Association

Threats to Rain Forests Tropical rain forests once covered about 20 percent of Earth's surface. Today, they cover only about 7 percent. Every minute of every day, 100 acres of tropical rain forest are cleared for logging operations, agriculture, or oil exploration. *Habitat destruction* occurs when land inhabited by an organism is destroyed or altered. If the habitat that an organism depends on is destroyed, the organism is at risk of disappearing.

Animals and plants are not the only organisms that live in rain forests. An estimated 50 million native peoples live in tropical rain forests. These native peoples are also threatened by habitat destruction. Because they obtain nearly everything they need from the forest, the loss of their habitat could be devastating. This loss of habitat may force them to leave their homes and move into cities. This drastic change of lifestyle may also cause the native peoples to lose their culture and traditions along the way.

Plants and animals that live in rain forests are also threatened by trading. Many plant species found only in tropical rain forests are valuable and marketable to industries. Animals are threatened by exotic-pet trading. Some exotic-pet traders illegally trap animals, such as parrots, and sell them in pet stores at high prices.

CASE STUDY

Deforestation, Climate, and Floods

A plant absorbs water from the soil through its roots and transports the water to its stems and leaves. Water then evaporates from pores in plant leaves into the atmosphere through a process called *transpiration*. A large tree may transpire as much as five tons of water on a hot day. Water absorbs heat when it evaporates. Therefore, the temperature is much cooler under a tree on a hot day than under a wood or brick shelter. Trees that provide shade around homes keep homes much cooler in the summer.

When rain falls on a forest, much of the rain is absorbed by plant roots and transpired into the air as water vapor. Water vapor forms rain clouds. Much of this water will fall as rain somewhere downwind from the forest. Because of the role trees play in

transpiration, *deforestation,* the clearing of trees, can change the climate. If a forest is cut down or replaced by smaller plants, much of the rainfall is not absorbed by plants. Instead, the rain runs off the soil and causes flooding as well as soil erosion. So, the climate downwind from the forest becomes drier.

Deforestation led to the disastrous flooding of the Yangtze River in China in 1998. More than 2,000 people died in the floods, and at least 13 million people had to leave their homes. When the Yangtze River flooded, the water poured into a flood plain where over 400 million people lived. It is estimated that 85 percent of the forest in the Yangtze River basin has been cut down. The millions of tons of water that these trees once absorbed now

▶ A man makes his way past flooded buildings in his street on a makeshift raft after the Yangtze River flooded in July 1998. Water of the Yangtze River reached record-high levels.

flows freely down the river and spreads across fields and into towns during the seasonal monsoon rains.

Deforestation has also caused terrible floods in places such as Bangladesh. The Ganges River rises high in the Himalaya Mountains and flows through Bangladesh. Deforestation of the Himalaya

150

REAL-LIFE CONNECTION ——— GENERAL

Costa Rica Some countries with tropical rain forests, such as Costa Rica, have created large preserves to protect their unique ecological areas. To generate income for the country, Costa Rica has developed ecotourism in these preserves. While the Costa Rican program has been a success and a model for other countries, it is also a special case. Costa Rica is a fairly prosperous and educated country, largely free of the poverty and civil unrest that plague some Central and South American countries. For other countries, setting aside large amounts of

land as ecological reserves may not be politically or economically realistic. Ask students, "Why would a poverty-stricken country be less likely to follow the Costa Rican example?" (An increase in population and poverty would force people to use whatever resources were available.) Ask students, "What can we as residents in a developed nation, do to help less developed nations follow the Costa Rican example?" (Answers may vary. Suggestions include offering grants, increasing ecotourism, or buying the land from the country.) **LS** Logical

Temperate Forests

Temperate rain forest occurs in North America, Australia, and New Zealand. Temperate rain forests have large amounts of precipitation, high humidity, and moderate temperatures. The Pacific Northwest shown in **Figure 9**, houses North America's only temperate rain forest, where tree branches are draped with mosses and tree trunks are covered in lichens. The forest floor is blanketed with lush ferns. Evergreen trees that are 90 m tall, such as the Sitka spruce and the Douglas fir, dominate the forest. Other large trees, such as western hemlock, Pacific silver fir, and redwood, can also be found in temperate rain forests.

Even though the temperate rain forest of the Pacific Northwest is located north of most other rain forests, it still maintains a moderate temperature year-round. The temperate rain forest also rarely freezes because the nearby Pacific Ocean waters keep temperatures mild by blowing cool ocean wind over the forest. As this ocean wind meets the coastal Olympic Mountains, a large amount of rainfall is produced. This rainfall keeps the temperate rain forest cool and moist.

Figure 9 ▶ The only temperate rain forest in North America is located in the Pacific Northwest, as shown above in Olympic National Park in Washington State.

▶ Deforestation reduces the amount of water that is absorbed by plants after it rains. The more trees that are cleared from a forest, the more likely a flood will occur in that area.

Mountains left few trees to stop the water flowing down the mountain. So, most of the water flows into the river when it rains. Heavy rains have eroded and carried away so much soil from the slopes of the mountains that the soil has formed a new

island in the Bay of Bengal, which is off the coast of Bangladesh.

People are beginning to understand the connection between deforestation and floods. People held protests in northern Italy in 2000 after floods covered a town that had

never been flooded before. The townspeople claimed that authorities had permitted developers to cover the hills with homes. These developers cut down most of the trees and covered much of the land with asphalt. After heavy rains, the water was no longer absorbed by trees and soil, so the water flowed down the hills and flooded the town.

CRITICAL THINKING

1. Identifying Relationships How might deforestation in China and other countries affect the overall climate of the Earth?

2. Analyzing a Viewpoint Imagine that you are a city council member and must vote on whether to clear a forest so that a mall can be built. List the pros and cons of each viewpoint. After reviewing your list, how would you vote? Explain your answer.

151

Pacific Northwest Food Web In the past, the northern spotted owl has received a great deal of attention in a battle between loggers and environmentalists over the fate of old-growth forests in the Pacific Northwest. Students may not realize that the spotted owl is part of a complicated food web that also contains martens, flying squirrels, red-backed voles, Douglas firs, pseudoscorpions, termites, and black and yellow mycorrhizal fungi. Have students form teams, and have each team investigate the ecological role of one of these organisms. Then have students work together to make a bulletin board display. At the center of the display, place an illustration or photograph of a tree from an old-growth forest. Groups should add a picture of their organism to the display along with a written description of its niche and interaction with the tree.

LS **Logical** Co-op Learning

INCLUSION Strategies

- *Attention Deficit Disorder*
- *Learning Disabled*
- *Developmentally Delayed*

To help students understand the concept of habitat destruction, ask them to write a story from the perspective of an animal living on land that has been destroyed or altered. The story should include how the animal will adapt or move as a result of the habitat destruction. The story can be word processed or dictated into a tape recorder. Drawings of the land before and after the destruction can be added.

CASE STUDY

Deforestation, Climate, and Floods Have students research the major floods in history. Have students research how and why these floods occurred and the amount of damage that each flood caused. Have them share their research. Discuss with students the importance of preserving areas to prevent flooding.

Answers to Critical Thinking

1. Deforestation may affect the overall climate of the Earth because deforestation releases

carbon into the atmosphere and removes the trees that would normally take up carbon. Carbon dioxide can contribute to global warming.

2. Answers may vary. Students should determine whether forests are plentiful locally and whether the removal of the forest would cause other problems, such as flooding, erosion, or species loss.

Activity — ADVANCED

Forest Soil Field Trip Forests throughout the world can have strikingly different soils. Each type of forest tends to have its own characteristic soil type. For example, the soil in tropical rain forests is acidic, with few nutrients. If possible, have students investigate the soil in two forested areas around your town or city. Each forest should feature a different dominant tree species. In each area have students dig a hole and measure a cross section of the soil. They should also draw a picture that illustrates the layers they see, and use their drawings to compare the two different areas. Encourage students to note any small organisms that they uncover (without disturbing the organisms). While students are out in the woods, also have them measure the thickness of the leaf litter layer at each site. This layer serves as a protection for animals that use the forest floor, and it is an important nutrient source for forest plants. After students are finished taking soil and litter measurements, have them use a soil test kit to analyze the soil for minerals and pH levels. Suggest that students include their field notes, drawings, and soil-test results in their **Portfolio.**
LS Kinesthetic

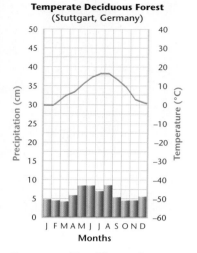

Figure 10 ▶ The difference between summer and winter temperatures in temperate deciduous forests is extreme.

internet connect

www.scilinks.org
Topic: Temperate Deciduous Forests
SciLinks code: HE4110

SCiLINKS. Maintained by the National Science Teachers Association

Figure 11 ▶ The change of seasons in a temperate deciduous forest in Michigan is shown below.

Temperate Deciduous Forests

If you walk through a North American deciduous forest in the fall, you will immerse yourself in color. Leaves in every shade of orange, red, and yellow crackle beneath your feet. Most birds have flown south. The forest is quieter than it was in the summer. You see mostly chipmunks and squirrels gathering and storing the food they will need during the long, cold winter.

In **temperate deciduous forests,** trees drop their broad, flat leaves each fall. These forests once dominated vast regions of the Earth, including parts of North America, Europe, and Asia. Today, temperate deciduous forests are generally located between 30° and 50° north latitude, as shown in **Figure 10.** The range of temperatures in a temperate deciduous forest can be extreme, and the growing season lasts for only four to six months. Summer temperatures can soar to 35°C. Winter temperatures often fall below freezing, so little water is available for plants. Temperatures vary due to a change of seasons, as shown in **Figure 11.** Temperate deciduous forests are moist. They receive 75 to 125 cm of precipitation annually. The rain and snow help decompose dead organic matter, such as fallen leaves, which in turn contributes to the rich, deep soils of temperate deciduous forests.

Plants of Temperate Deciduous Forests Like the plants of tropical rain forests, the plants in deciduous forests grow in layers. Tall trees, such as maple, oak, and birch, dominate the forest canopy. Small trees and shrubs cover the understory. Because the forest floor in a deciduous forest gets more light than that of a rain forest does, more plants such as ferns, herbs, and mosses grow in a deciduous forest.

152

HISTORY — CONNECTION — GENERAL

The Father of Forests John Muir (1838–1914) was an American naturalist, explorer, and writer. His conservation efforts included helping persuade Congress to establish both Yosemite and Sequoia National Parks in 1890 and establishing the Sierra Club in 1892. He traveled to many parts of the world and is known for his work explaining Yosemite's glacial origins, as well as for his discovery of a glacier in Muir Woods, a redwood forest near San Francisco which was named after him in 1908. His books include *The Mountains of California* (1894), *Our National Parks* (1901), and *The Yosemite* (1912). Have students research and report on John Muir, or another American naturalist, such as Aldo Leopold or John James Audobon.

Temperate-forest plants are adapted to survive seasonal changes. In the fall, most deciduous trees begin to shed their leaves. In the winter, moisture in the soil changes to ice, which causes the remaining leaves to fall to the ground. Also, herb seeds, bulbs, and rhizomes, which are underground stems, become dormant in the ground and are insulated by the soil, leaf litter, and snow. In the spring, when sunlight increases and temperatures rise, trees grow new leaves, seeds germinate, and rhizomes and roots grow new shoots and stems.

Animals of Temperate Deciduous Forests The animals of temperate deciduous forests are adapted to use the forest plants for food and shelter. Squirrels eat the nuts, seeds, and fruits in the treetops. Bears feast on the leaves and berries of forest plants. Grasshoppers, such as the one shown in **Figure 12,** eat almost all types of vegetation found throughout the forest, while deer and other herbivores nibble leaves from trees and shrubs.

Many birds nest in the relative safety of the canopy. Most of these birds are migratory. Because the birds cannot survive the harsh winters, each fall they fly south for warmer weather and for more available food. Each spring, they return north to nest and feed. Animals that do not migrate use various strategies for surviving the winter. For example, mammals and insects reduce their activity so that they do not need as much food for energy.

Taiga

The **taiga** is the northern coniferous forest that stretches in a broad band across the Northern Hemisphere just below the Arctic Circle. As shown in **Figure 13,** winters are long (6 to 10 months) and have average temperatures that are below freezing and that often fall to −20°C. In the taiga, the forest floor is dark and has little vegetation. Many trees seem like straight, dead shafts of bark and wood—until you look up and see their green tops. The growing season in the taiga may be as short as 50 days depending on latitude. Plant growth is most abundant during the summer months because of nearly constant daylight and larger amounts of precipitation.

Figure 12 ▶ Grasshoppers, woodpeckers, and deer are among many animals that live in the temperate deciduous forest.

Figure 13 ▶ The taiga has long, cold winters and small amounts of precipitation, as shown in the climatogram below.

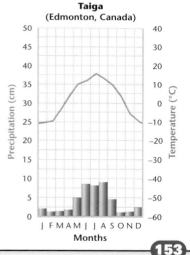

153

READING SKILL BUILDER ————— BASIC

Concept Mapping Have students create a concept map using the animals of the temperate deciduous forest mentioned on this page. Ask them to map the animals by what they eat, where they might live, and whether or not they leave in winter. After they are finished with their maps, ask students to pair up to discuss how they created the connections on their maps. Also have students include their map in their **Portfolio. LS Logical**

Activity ————— ADVANCED

Touring the Forest Have students develop a travel brochure that advertises camping vacations in one of the four forest ecosystems presented in this section. In addition to making a list of the supplies that campers will need, students should provide information about the climate (including any important seasonal variations, such as a rainy season or temperature variations), a description of the plants and animals that campers might encounter (including any that might be dangerous or endangered), and recommendations on how campers can lessen the environmental impact of their activities. The brochures should include photos or illustrations of the forest and its inhabitants, as well as a map showing the location of the forest described. Have students also map any areas that are off-limits to camping, due to sensitive species or dangerous trails. Encourage students to include their brochure in their **Portfolio. LS Visual**

Group Activity ————— GENERAL

Biome Components Organize students into four groups, and provide each group with a large sheet of poster board or butcher paper. Each sheet should be labeled with a type of forest biome discussed in this section. Ask students to illustrate in light colors the vegetation that grows in the listed forest biome. Then have the groups switch posters and use darker colors to add the appropriate animals for the biome. Encourage students to also include decomposers, such as insects, fungi, and bacteria. Have students switch illustrations once again, this time describing the interrelationships between the organisms shown. Finally, students can switch posters again to label and describe some of the organisms' adaptations. Display these posters in your room. **LS Visual** Co-op Learning

Close

Reteaching ─── BASIC

News Flashes Some students may have trouble distinguishing between the different forest biomes. Ask volunteers to prepare a three to five minute news report featuring the characteristics of, the location of, and the threats facing the different forest biomes. Then have students present their news flashes to the class. Ask all students to take notes and include them in their **Portfolio.** LS Verbal/Interpersonal

Quiz ─── GENERAL

1. Do animals of the tropical rain forest generally migrate? Explain your answer. (No. The climate is stable year-round, so a reliable food source is available year-round.)

2. How do small mammals survive winter in the taiga? (They may burrow to stay warm.)

3. Most rain forest animals are food generalists, true or false? (False. Because the plants provide a stable, consistently available food source, animals in the tropics are mainly specialists.)

Alternative Assessment ─── GENERAL

Detailing a Biome Organize the class into groups of five or six students. Give each group a detailed picture of a forest biome, and ask them to describe the characteristics of the biome. (Answers should include a description of where the biome is located, what types of animal and plant species live there, and what special adaptations help those organisms to survive in the biome.) LS Visual/Interpersonal

Plants of the Taiga A *conifer* is a tree that has seeds that develop in cones. Most conifers do not shed their needle-shaped leaves, which help them survive harsh winters. The leaves' narrow shape and waxy coating retain water for the tree when the moisture in the ground is frozen. As shown in **Figure 14**, a conifer's pointed shape also helps the tree shed snow to the ground so that it does not become weighed down.

Conifer needles contain substances that make the soil acidic when the needles fall to the ground. Most plants cannot grow in acidic soil, which is one reason the forest floor of the taiga has few plants. In addition, soil forms slowly in the taiga because the climate and acidity of the fallen leaves slow decomposition.

Animals of the Taiga The taiga has many lakes and swamps that in summer attract birds that feed on insects, fish, or other aquatic organisms. Many birds migrate south to avoid winter in the taiga. Some year-round residents, such as shrews and rodents, may burrow underground during the winter, because the deep snow cover insulates the ground. Moose and snowshoe hares eat any vegetation they can find. As shown in **Figure 15**, some animals, such as snowshoe hares, have adapted to avoid predation by lynxes, wolves, and foxes by shedding their brown summer fur and growing white fur that camouflages them in the winter snow.

Figure 14 ▶ The taiga has cold winter temperatures, a small amount of annual precipitation, and coniferous trees. The seeds of conifers are protected inside tough cones like the one above. Also, the narrow shape and waxy coating of conifer needles help the tree retain water.

Figure 15 ▶ In the taiga, a snowshoe hare's fur changes color according to the seasons to help camouflage the animal from predators.

SECTION 2 Review

1. **List** three characteristics of tropical rain forests.

2. **Name** the main layers of a tropical rain forest. What kinds of plants grow in each layer?

3. **Describe** two ways in which tropical rain forests of the world are being threatened.

4. **Describe** how a plant survives the change of seasons in a temperate deciduous forest. Write a short paragraph to explain your answer. **WRITING SKILLS**

CRITICAL THINKING

5. **Evaluating Information** Which would be better suited for agricultural development: the soil of a tropical rain forest or the soil of a temperate deciduous forest? Explain your answer.

6. **Identifying Relationships** How does a snowshoe hare avoid predation by other animals during the winter in a taiga biome? How might this affect the animal that depends on the snowshoe hare for food?

154

Answers to Section Review

1. Tropical rain forests are typically humid and warm, receive 200 to 450 cm of rain per year, and have nutrient-poor soil.

2. The main layers of a tropical rain forest are the emergent layer, the upper and lower canopies, and the understory. The emergent layer consists of the tallest trees. The upper and lower canopies consist of tall trees and epiphytes. Trees and shrubs that are adapted to shade, grow in the understory.

3. Answers may vary, but should include two of the following: logging, farming, ranching, oil exploration, and trading.

4. Sample answer: In order to survive the change of seasons in the temperate deciduous forest, a plant loses its leaves during the fall and winter and remains dormant until spring.

5. temperate deciduous forest; Organic matter decays slowly in the temperate forest, and forms a deep, rich soil that would be better suited for agriculture.

6. A snowshoe hare avoids predation by shedding its brown summer fur and growing white fur that camouflages it in the winter snow. Predators will have a difficult time finding a snowshoe hare during the winter.

Grassland, Desert, and Tundra Biomes

In climates that have less rainfall, forest biomes are replaced by savanna, grassland, and chaparral biomes. As less rain falls in these biomes, they change into desert and tundra biomes. As precipitation decreases in an area, the diversity of the species in the area also decreases. But while the number of different species is often smaller in areas that have less precipitation, the number of individuals of each species present may be very large.

Savannas

Parts of Africa, western India, northern Australia, and some parts of South America are covered by grassland called *savanna*. **Savannas** are located in tropical and subtropical areas near the equator and between tropical rain forest and desert biomes. Because savannas are full of grasses, scattered trees, and shrubs, savannas contain a large variety of grazing animals and the predators that hunt them. As shown in **Figure 16**, savannas receive little precipitation throughout the year. Savannas have a wet season and a dry season. Many animals of the savanna are active only during the wet season. Grass fires sweep across the savanna during the dry season and help restore nutrients to the soil.

Plants of the Savanna Because most of the rain falls during the wet season, plants must be able to survive prolonged periods without water. Therefore, some trees and grasses have large horizontal root systems by which they obtain water during the dry season. These root systems also enable plants to quickly grow again after a fire. The coarse savanna grasses have vertical leaves that expose less of their surface area to the hot sun to further help the grasses conserve water. Some trees of the savanna also lose their leaves during the dry season to conserve water. Trees and shrubs often have thorns or sharp leaves that keep hungry herbivores away.

Objectives

▶ Describe the difference between tropical and temperate grasslands.
▶ Describe the climate in a chaparral biome.
▶ Describe two desert animals and the adaptations that help them survive.
▶ Describe one threat to the tundra biome.

Key Terms

savanna
temperate grassland
chaparral
desert
tundra
permafrost

Figure 16 ▶ Savannas have periods of heavy rainfall followed by periods of drought.

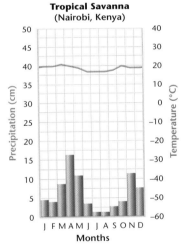

Tropical Savanna
(Nairobi, Kenya)

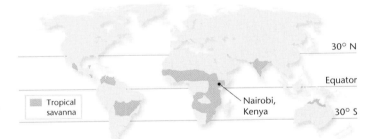

Tropical savanna

Nairobi, Kenya

30° N

Equator

30° S

155

Focus

Overview

Before beginning this section, review with your students the Objectives in the Student Edition. This section introduces students to the characteristics of the savanna, grassland, desert, chaparral, and tundra biomes. Students will also learn how the adaptations of plants and animals in each biome help them to survive.

🔊 Bellringer

Before students begin to read this section, ask them to list five characteristics of grasslands. You can guide students by asking them about types of grasslands, temperature, precipitation, seasons, soil, plants, and animals.

Motivate

Identifying Preconceptions — GENERAL

Ask students to write down as many examples of African animals as they can think of in their *EcoLog*. Ask one student to write down all of the names the class could think of on the board. Circle the animals that are found on the African savanna—the vast grasslands of Africa. Most of the animals will probably be from the savanna. Have students discuss how these animals are adapted to survive in the savanna biome. Also remind students that the savanna is not the same as the "jungle" (rain forest). **LS Logical**

REAL-LIFE CONNECTION — GENERAL

The Serengeti The tropical savanna is unparalleled in its diversity of large mammals. The major predators in a tropical savanna are lions, leopards, cheetahs, hyenas, and wild dogs. The prey in the savanna are just as varied. Unfortunately, the savannas are threatened by overgrazing, overhunting, and farming activities. However, some ranchers are using native species instead of cattle on the land, which can result in less degradation and more profit for the farmer. Studies have shown that when herds of native antelopes are raised instead of cattle, ranchers receive a much higher meat yield. Also, antelopes need much less water and do not degrade the environment as much as cattle do. Many farmers in areas such as Central Texas are farming large African game animals for sport hunting or food. Students can find the Web sites of these ranches on the Internet. Ask students to identify the similarities between the African savanna and Central Texas.

Chapter Resource File

• **Lesson Plan**
• **Active Reading** BASIC
• **Section Quiz** GENERAL

Transparencies

TT Bellringer
TT Biome Climatograms: B

Discussion ── GENERAL

Introducing Grazers When cattle or other nonnative species are introduced to grasslands, the result is often degradation of the environment. Ask students to discuss the following question: "Why would grazers native to a grassland biome have less of an impact on the area than newly-introduced grazers?" (When a community has remained stable for a long period of time, the animals and plants that comprise it are generally well adapted to each other's presence. For example, a native grass will be adapted to survive the amount of grazing that is normal for the region's native herbivores. When a new species is introduced, different consumption patterns might be too severe for a native grass to withstand, and it may die out. The loss of the grass species could then affect the entire ecosystem.) **LS** Verbal

Cultural Awareness ── GENERAL

Prairie Fires Students may be interested to know that the Native Americans who inhabited the plains of North America discovered a method for avoiding prairie fires. First, they would start another small fire near their community. Then, the small fire would burn out a safety zone in a short time. Finally, when a large fire bore down upon the community, it would pass by the already burned area. This method of avoidance was later used by European settlers on the plains.

Figure 17 ▶ Herbivores of the savanna reduce their competition for food by feeding on vegetation located at different heights. Elephants feed on tree leaves, while impala graze on grasses.

Geofact

Deep Soil Gravel or sand becomes fertile soil when decomposers slowly break down organic matter such as dead leaves. Decomposers work most effectively in hot, wet weather. As a result, the world's deepest soil is in grasslands. In grassland biomes, winters are cold and summers are dry, which causes leaves to break down slowly. So, organic matter builds up over time. Some North American prairies had more than 2 m of topsoil when the first farmers arrived.

Animals of the Savanna Grazing herbivores such as the elephants shown in **Figure 17**, have adopted a migratory way of life. They follow the rains to areas of newly sprouted grass and watering holes. Some predators follow and stalk the migratory animals for food. Many savanna animals give birth only during the rainy season, when food is most abundant and the young are more likely to survive. Also, some species of herbivores reduce competition for food by eating vegetation at different heights than other species do. For example, small gazelles graze on grasses, black rhinos browse on shrubs, and giraffes feed on tree leaves.

Temperate Grasslands

A **temperate grassland** is a biome that is dominated by grasses and that has very few trees. Most temperate grasslands have hot summers and cold winters. The amount of rainfall that a temperate grassland receives is moderate compared to the amount a forest receives. On average, a temperate grassland can receive 50 to 88 cm of precipitation per year, as shown in **Figure 18**. Although temperate grasslands may seem harsh and dry, they have the most fertile soil of any biome. So, many grassland biomes have been replaced with crops such as corn, soybeans, and wheat. Few natural temperate grasslands remain because many have been replaced by farms and grazing areas.

Figure 18 ▶ Temperate grasslands are characterized by small amounts of rainfall, periodic droughts, and high temperatures in the summer.

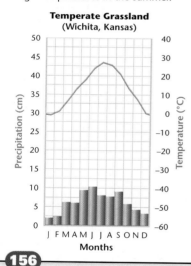

Temperate Grassland (Wichita, Kansas)

Wichita, Kansas

■ Temperate grassland

30° N

Equator

30° S

156

Temperate grasslands are located on the interiors of continents where too little rain falls for trees to grow. Grassland biomes include the prairies of North America, the steppes of Russia and Ukraine, and the pampas of South America, as shown in **Figure 19.** Mountains often play a crucial role in maintaining grasslands. For example, in North America, rain clouds from the west are blocked by the Rocky Mountains, so the shortgrass prairie east of the mountains receives only about 25 cm of rain a year. Rainfall increases as you move eastward, so taller grasses and some shrubs can grow in areas where more rain falls. Heavy precipitation is rare in the grasslands, so sizzling temperatures in the summer make the grasslands susceptible to fires, which are common in grassland biomes.

Plants of Temperate Grasslands Prairie grasses and wildflowers are perennials, plants that survive from year to year. The root systems of prairie grasses form dense layers that survive drought and fire as well as hold the soil in place. The amount of rainfall in an area determines the types of plants that will grow in that area. **Figure 20** shows how root depth and grass height vary depending on the amount of rainfall. Few trees survive on the grasslands because of the lack of rainfall, fire, and the constant winds.

Figure 19 ▶ Temperate grasslands can be named according to the vegetation that grows there. Steppes (left), have shorter grasses and are located in Europe and Asia. Pampas (right), are made up of clusters of feathery grasses and are located in South America.

Connection to History

The State of Bison More than 60 million bison once roamed the temperate grasslands of North America. But these large grass-eating mammals were almost brought to extinction by the late 1800s because of hunting by western settlers. By 1889, fewer than 1,100 bison remained in North America! The first bill to save the bison was introduced by Congress in 1874. In 1903, President Theodore Roosevelt started the National Wildlife Refuge System to provide protected areas for bison and other animals. Today, North America has more than 200,000 bison.

Shortgrass prairie (about 25 cm rain per year)	Mixed or middlegrass prairie (about 50 cm rain per year)	Tallgrass prairie (up to 88 cm rain per year)

Figure 20 ▶ The height of grassland plants and the depth of their roots depend on the amount of rainfall that the grasslands receive.

157

Homework ——— ADVANCED

Bison Today Have students research and write a report on the status of bison in North America today. Ask them to find out where the bison are, how and if they are being managed, where the largest herds are, if they are conflicting with cattle and other animals on the grasslands, and how bison products are used by humans. If you are in an area where bison are ranched, ask to interview a rancher. LS **Verbal**

LANGUAGE ARTS ———
CONNECTION ADVANCED

PrairyErth Students who are interested in the American prairie may enjoy the book *PrairyErth (A Deep Map): An Epic History of the Tallgrass Prairie Country* by William Least Heat-Moon. In this unique work, the author explores the plants, animals, and people that molded the history and ecology of a tallgrass prairie remnant in central Kansas. Encourage students who read this book to write a book report for their **Portfolio,** linking information from this chapter to details in the book. LS **Intrapersonal**

Activity ——— GENERAL

Grass Roots Gather some trowels and plastic bags, and take students out to a weedy grassland on or near the school grounds. Have students dig out clumps of grass, being careful to try to dig out all of the root system as well. Since grass roots spread horizontally and vertically quite extensively, students will find that it is difficult to get the whole system out without breaking some of the roots. Ask students to try to shake as much soil off of the roots as possible. Then have them put the grass and root systems into plastic bags, and bring them back to the classroom for further study. Carefully rinse each specimen, and ask students to measure both above ground and below ground components of the grass. Students will find that the below ground component, with all its branching, is much longer. Ask them why the root system is so large. (Large root systems help grasses to survive drought, fire, and grazing in a grassland ecosystem, by allowing them to store water and nutrients below ground. Shoots also form along spreading roots, which helps the grass to cover more area and reproduce asexually.) LS **Kinesthetic**

Group Activity ——— GENERAL

Plant a Prairie! If your climate is appropriate, get permission to plant a prairie garden on your school grounds. Many sources of native prairie seeds exist, and seed suppliers can help your students learn how to prepare the ground and tend the garden properly. With proper maintenance, which includes periodic burning, this garden will thrive with a wide variety of grass and flower species. Your pocket prairie will also attract butterflies and other interesting insects, and can be a great educational resource. LS **Kinesthetic**

Figure 21 ▶ Prairie dogs, such as those shown here, live in temperate grasslands. Prairie dogs live in colonies and burrow in the ground to build mounds, holes, and tunnels.

Activity ——————— GENERAL

Native Grasses Have students collect a variety of samples of grasses native to their area and display pressed specimens on a poster board. Students could also write a short report to accompany their display. The report should indicate the major characteristics of each type of grass represented, such as height, abundance, and habitat. Suggest that students include their display in their **Portfolio.** **LS** Visual

Quick**LAB**

Skills Acquired:

- Experimenting
- Identifying and Recognizing Patterns

Teacher's Notes: If you cannot find a sunny place outside, place the sponges under a heat lamp in the classroom.

Answers

1. The sponge that was not covered in plastic wrap lost the most mass because the water evaporated more quickly from that sponge.

2. The plastic wrap is similar to the adaptations of chaparral plants because they have thick, leathery leaves that help the plant retain water.

Quick**LAB**

Sponging It Up

Procedure

1. Completely saturate **two small sponges** with **water** and allow the excess water to drain off.
2. Measure each sponge's mass by using an **electric balance.** Record the mass.
3. Using **plastic wrap**, completely cover one of the sponges.
4. Place the sponges outside in a sunny place for 10 to 15 minutes.
5. Measure each sponge's mass after removing it from outside. Record the mass.

Analysis

1. Which sponge lost the most mass? Why?
2. How was the covering you created for the sponge similar to the adaptations of the plants in the chaparral biome?

Figure 22 ▶ Temperate woodlands are usually too dry to support a forest, but they receive sufficient precipitation to support vegetation that grows in bunches, such as the piñon and juniper trees shown here.

Animals of Temperate Grasslands Grazing animals, such as pronghorn antelope and bison, have large, flat back teeth for chewing the coarse prairie grasses. Other grassland animals, such as badgers, prairie dogs, and owls, live protected in underground burrows as shown in **Figure 21.** The burrows shield the animals from fire and weather and protect them from predators on the open grasslands.

Threats to Temperate Grasslands Farming and overgrazing have changed the grasslands. Grain crops cannot hold the soil in place as well as native grasses can because the roots of crops are shallow, so soil erosion eventually occurs. Erosion is also caused by overgrazing. When grasses are constantly eaten and trampled, the grasses cannot regenerate or hold the soil. This constant use can change fruitful grasslands into less productive, desertlike biomes.

Chaparral

Plants that have leathery leaves are commonly found in temperate woodland biomes. Temperate woodland biomes have fairly dry climates but receive enough rainfall to support more plants than a desert does. Temperate woodlands consist of scattered tree communities made up of coniferous trees such as piñon pines and junipers, as shown in **Figure 22.**

Chaparral is a type of temperate woodland biome that is dominated by more broad-leafed evergreen shrubs than by evergreen trees. Look at the famous white letters that spell Hollywood

HISTORY ——————
●─ CONNECTION ── GENERAL ─

The American Dust Bowl In the 1930s, extreme weather conditions and poor farming practices created a "Dust Bowl" in the formerly rich agricultural areas of the Great Plains. Beginning in 1930, a drought early in the year prevented planting seeds and continued for four years. Wind blowing across abandoned fields kicked up incredible dust storms across the region. Heat followed the drought of the first four years, killing people as well as causing more dust storms. On a Sunday in 1935, an epic dust storm roared across the land at 60 mph. This day was

called "Black Sunday" by those who witnessed the storm. Drought and dirt storms continued through the decade, blowing away the precious, nutrient-rich topsoil that had been formed over centuries by the decay of the prairie grasses. Fortunately, the United States learned from the Dust Bowl years, and began a soil conservation service to spread information about soil management practices. Have students search the Internet for information about the Dust Bowl and soil conservation methods.

across green and brown California hills in **Figure 23**. Now imagine the scrub-covered settings common in old westerns. Both of these landscapes are part of the chaparral biome. As shown in **Figure 24**, chaparral is located in the middle latitudes, about 30° north and south of the equator. Chaparral is located primarily in coastal areas that have Mediterranean climates. Chaparral biomes typically have warm, dry summers and mild, wet winters.

Plants of the Chaparral Most chaparral plants are low-lying, evergreen shrubs and small trees that tend to grow in dense patches. Common chaparral plants include chamise, manzanita, scrub oak, olive trees, and herbs, such as sage and bay. These plants have small, leathery leaves that retain water. The leaves also contain oils that promote burning, which is an advantage because natural fires destroy trees that might compete with chaparral plants for light and space. Chaparral plants are so well adapted to fire that they can resprout from small bits of surviving plant tissue. The flammable oils give plants such as sage their characteristic taste and smell.

Animals of the Chaparral A common adaptation of chaparral animals is camouflage, shape or coloring that allows an animal to blend into its environment. Animals such as quail, lizards, chipmunks, and mule deer have a brownish gray coloring that lets them move through the brush without being noticed.

Threats to the Chaparral Worldwide, the greatest threat to chaparral is human development. Because chaparral biomes get a lot of sun, are near the oceans, and a have a mild climate year-round, humans tend to develop land for commercial and residential use.

Figure 23 ▶ The chaparral biome in the Hollywood hills is home to plants such as the manzanita, which is shown above.

Figure 24 ▶ Chaparral biomes are located in areas that have Mediterranean climates.

Santa Barbara, California

Chaparral

30° N

Equator

30° S

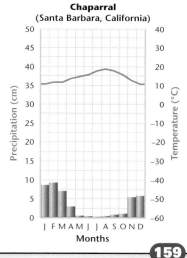

Chaparral
(Santa Barbara, California)

Precipitation (cm)

Temperature (°C)

J F M A M J J A S O N D
Months

159

Demonstration ——— BASIC
Plants of the Chaparral Bring in some fresh leaves from plants that can be found in the Mediterranean chaparral region, such as rosemary, sage, lavender, or bay laurel. Pass the leaves around, and invite students to crush them and smell the essential oils in the leaves. Ask students, "Why do these plants have oils in their leaves?" (The oils allow fires to start easily in the chaparral, and fire keeps trees and other competitors out of chaparral areas.) Have students describe the leaves on these plants. (They are small or thin and leathery.) Ask students, "Why are the leaves small with tough skin?" (To prevent water loss in the dry conditions of the chaparral biome.) If you live near chaparral, bring in samples of leaves from native plants in your area and compare them to leaves of the plants of the Mediterranean region.
LS Kinesthetic

REAL-LIFE ——
● CONNECTION

Chaparral Fire Control Fire is a natural and necessary force in the chaparral. However, large human populations live near areas of chaparral, and work to eliminate fires to protect their homes and businesses. Suggest that students research some of the consequences of human attempts to control chaparral wildfires. Also, ask them to find out whether government land managers in areas of chaparral practice prescribed burning. Have them record their findings in their *EcoLog.* **LS Interpersonal**

READING SKILL BUILDER ——— BASIC

Reading Organizer Have students look at the climatograms in the chapter to determine if each biome's seasonal temperature and rainfall are stable or fluctuating. Then have students match the biomes that have similar seasonal temperature or precipitation. For example, the chaparral and the tropical rain forest would be similar because their temperatures remain relatively stable throughout the year. However, chaparral has a rainy season in the winter months, so chaparral would be more similar to savanna and desert biomes with respect to precipitation. **LS Logical**

INCLUSION Strategies

• Attention Deficit Disorder
• Learning Disabled
• Developmentally Delayed

Ask students to write the name of one of the major forest, grassland, desert, and tundra biomes on a note card. On the back of each note card, students can write descriptive phrases, pictures, or words about the biome. The note cards can be used for individual or small group review sessions.

MISCONCEPTION ALERT

Oregon's Rain Shadow Many students may think that Oregon is a state that is completely dominated by lush forests. In fact, some of the state has a shrub-steppe ecosystem that can receive as little as 25 cm of rainfall per year. This situation is due to an effect called *rain shadow*. In western Oregon, the Cascade Range of the Rocky Mountains (which runs from the northern to the southern edges of the state) creates a barrier that blocks ocean moisture from moving to the eastern side of the mountains. Because of this effect, ecosystems on the western side of the mountains are dominated by old-growth forest, while ecosystems on the eastern side are dominated by sagebrush. Have students research the types of plants and animals found in Oregon's shrub-steppe ecosystem. Also have them look at a world map to identify other areas that may have arid ecosystems caused by the rain shadow effect. Have students research those areas to determine if they were correct.

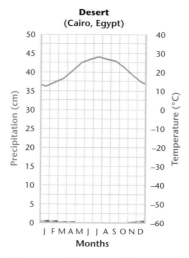

Figure 25 ▶ Deserts are the driest places on Earth. They typically receive less than 25 cm of precipitation a year.

Deserts

When some people think of a desert, they think of the hot sand that surrounds the Egyptian pyramids. Other people picture the Sonoran Desert and its mighty saguaro cactuses, or the magnificent rock formations of Monument Valley in Arizona and Utah. Many kinds of deserts are located throughout the world, but one characteristic that they share is that they are the driest places on Earth.

Deserts are areas that receive less than 25 cm of precipitation a year and have little or no vegetation. Deserts, as shown in **Figure 25**, also have extreme temperatures. Hot deserts, such as Arizona's Sonoran Desert and the Sahara in Africa, are closer to the equator than are cold deserts, such as the Gobi Desert in China and the Great Basin of the western United States. Deserts are often located in areas near large mountain ranges because mountains can block the passage of moisture-filled clouds, which limits precipitation.

▶ The flapnecked chameleon lives in the deserts of Botswana.

▶ The plants of the Sonoran Desert in Arizona collect and retain water after rain showers to survive the hot desert temperatures.

160

Cultural Awareness — GENERAL

People of the Desert Just as plants and animals are adapted to desert life, people have also found a way to survive there. Have students research groups of people who live in the world's deserts, such as the San of the Kalahari, the Tuareg of the Sahara, or the Native Americans of the Southwest. Have students report their findings to the class and include their report in their **Portfolio.** **LS** Interpersonal

ARTS CONNECTION

Home, Home on the Biome Ask students to find a song or poem that was inspired by one of the biomes in this chapter. Have students try to find at least one selection for each type of biome. Have students list, highlight, or pantomime the unique aspects of the biomes that are mentioned within the selections. **LS** Auditory

Plants of the Desert All desert plants have adaptations for obtaining and conserving water, which allows the plants to live in dry, desert conditions. Plants called *succulents*, such as cactuses, have thick, fleshy stems and leaves that store water. Their leaves also have a waxy coating that prevents water loss. Sharp spines on cactuses keep thirsty animals from devouring the plant's juicy flesh. Rainfall rarely penetrates deeply into the soil, so many plants' roots spread out just under the surface of the soil to absorb as much rain as possible.

Instead of living in dry conditions, some desert plants are adapted to survive for long periods of time without water. When conditions are too dry, some plants die and drop seeds that stay dormant in the soil until the next rainfall. Then, new plants quickly germinate, grow, and bloom before the soil becomes dry again. Some desert plants have adapted so that they can survive even if their water content drops to as low as 30 percent of their mass. Water levels below 50 to 75 percent are fatal for most plants.

Animals of the Desert Reptiles, such as Gila monsters and rattlesnakes, have thick, scaly skin that prevents water loss. Amphibians, such as the spadefoot toad, survive scorching desert summers by *estivating*—burying themselves in the ground and sleeping through the dry season. Some animals, such as the elf owl shown in **Figure 26**, nest in cactuses to avoid predators. Desert insects and spiders are covered with body armor that helps them retain water. In addition, most desert animals are nocturnal, which means they are active mainly at night or at dusk, when the air is cooler.

FIELD ACTIVITY

Miniature Desert Create a miniature desert by growing a small cactus garden. Purchase two or three small cactus plants, or take several cuttings from a large cactus. To take cuttings, carefully break off the shoots growing at the base of the parent cactus. Place the plants in rocky or sandy soil similar to the soil in a desert. Keep the cactuses in bright sunlight, and do not water them frequently. Record your observations of your cactus garden in your *Ecolog*.

Figure 26 ▶ Desert plants survive harsh conditions by growing deep roots to reach groundwater and by having specialized structures that limit the loss of water. Desert animals bury themselves underground or burrow in cactuses to avoid extreme temperatures and predators.

▶ Elf owls burrow in cactuses to avoid hot temperatures during the day.

▶ This sidewinder has a unique way of moving so that only small portions of its body are in contact with the hot sands at any one time.

161

Internet Activity ——— **GENERAL**

Xeriscaping As urban and suburban desert areas become more populous and water in those areas becomes more scarce, some people have begun to landscape their yards using native plants instead of typical grass. This method of landscaping, called *xeriscaping*, can be an interesting way to learn about native ecosystems while conserving water resources. Have students use the Internet to research the species of plants used to xeriscape areas in a desert region. Ask groups to create a landscape plan for a home in an urban desert area. **LS Interpersonal** Co-op Learning

REAL-LIFE CONNECTION — **BASIC**

Waxy Resistance To help students understand how wax helps a plant retain water, bring in some hand lotion that contains beeswax, and have students put some on their hands. Explain that the waxy barrier keeps the skin on their hands from absorbing or excreting water. Then, have them dunk their hands in water, to observe how water beads on their skin.

Debate ——— **GENERAL**

ATVs in the Desert Encourage students to debate the use of all-terrain vehicles in a desert ecosystem. First, have students brainstorm or research the possible negative effects of driving all-terrain vehicles in desert ecosystems. (Effects could include crushed vegetation, the destruction of bird eggs laid on the ground, pollution from leaking motor oil and discarded trash.) Then have them find or think of reasons why these vehicles should be restricted to designated trails in the desert. (If some areas were designated as trails, this would keep all-terrain vehicles out of the more sensitive areas of deserts.) Finally, ask students to debate whether all-terrain vehicles should be used at all in the desert. **LS Verbal**

Activity ——— **ADVANCED**

Animal Adaptations to Heat Animals that live in extremely hot and dry conditions have developed many adaptations that allow them to retain water and stay cool. Some of these adaptations are listed in the Student Edition, but many more exist. Have students research the cavernous sinus system, which keeps mammal brains cool even in intense heat. In this system of heat exchange, blood from the veins is cooled in the nose by the evaporation of water in the nasal passages. This cooler blood then moves through the veins up into the cavernous sinus in the head. In the cavernous sinus, veins and arteries come into contact with each other, and the warmer arterial blood is cooled before it reaches the animal's brain. Have students sketch a diagram of the cavernous sinus system to include in their **Portfolio**. **LS Visual**

Tundra Creatures Review with students the characteristics of the tundra. (extreme cold, low precipitation, limited winter sunlight, permafrost, large numbers of insects in the summer, small plants with shallow roots) Based on these conditions, have students invent a hypothetical organism that could thrive in the Arctic tundra. Have them write an essay about their creature and its adaptations. (Students should describe the creature's appearance, size, color, skin covering, defense mechanisms, diet, and reproductive strategy.) Encourage students to draw a sketch to accompany their essay and share the new creatures with other class members. Have students add their work to their **Portfolio.** **LS** Visual

Arctic Science Experiments to determine how global warming will affect arctic areas are being coordinated by the Arctic System Science (ARCSS) Data Coordination Center, through the International Tundra Experiment (ITEX) project. Have students locate information from ARCSS about predicted tundra thaw depths, vegetation changes, snow melt, methane flux from warming tundra, and other possible effects of global warming. Have them create a short news announcement that explains this information, and encourage them to share this with the rest of the school.
LS Interpersonal | Co-op Learning

Graph*ic*

(Organizer) GENERAL

Venn Diagram
You may want to have students work in groups to create this Venn Diagram. Have one student draw the map and fill in information provided by other students from the group.

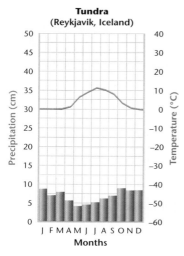

Tundra
(Reykjavik, Iceland)

Figure 27 ▶ The precipitation that the tundra biome receives remains frozen much of the year.

Graph*ic*

(Organizer) **Venn Diagram**
Create the **Graphic Organizer** entitled "Venn Diagram" described in the Appendix. Label the circles with "Tundra" and "Desert." Then, fill in the diagram with characteristics that each biome shares with the other.

Figure 28 ▶ During its brief summer, the Alaskan tundra is covered by flowering plants and lichens.

Tundra

Tundra is a biome that is dominated by grasses, lichens, and herbs and that is located primarily north of the Arctic Circle, as shown in **Figure 27.** The tundra soil supports mostly tough grasses and shrubs. Summers are short in the tundra, so only the top few centimeters of soil thaw. Plants flower in the summer, as shown in **Figure 28.** Underneath the topsoil is a layer of soil called **permafrost,** which is permanently frozen soil. The tundra becomes dotted with bogs and swamps when the top layer of soil thaws. In summer, these wet areas are ideal breeding grounds for huge numbers of swarming insects, such as mosquitoes and blackflies, and for the many birds that feed on the insects.

Vegetation of the Tundra Mosses and lichens, which can grow without soil, cover vast areas of rocks in the tundra. The soil is thin, so plants have wide, shallow roots to help anchor them against the icy winds. Most flowering plants of the tundra, such as campion and gentian, are short. Growing close to the ground keeps the plants out of the wind and helps them absorb heat from the sunlit soil. Woody plants and perennials such as willow and juniper have evolved dwarf forms and grow flat or grow along the ground.

Student Opportunities

Toolik Field Station One of the designated long-term ecological research (LTER) sites in North America is located above the treeline in arctic Alaska. The Toolik Field Station is managed by the Institute of Arctic Biology at the University of Alaska, Fairbanks. At the station, researchers focus on ecological projects related to the tundra environment. Sample projects have examined freeze avoidance in hibernating ground squirrels, changes in tundra plants and soil due to artificial warming (as a model for global warming), fish community structure and vertical migration of zooplankton in arctic lakes, changes in rivers due to artificial fertilization, and the hydrology of tundra wetlands. Students who are interested in studying biology as undergraduates may want to explore research opportunities at the Toolik Station. Encourage those who are interested to find the Toolik Web site on the Internet, and to contact the University of Alaska for more information.

Figure 29 ▶ Many migratory animals, such as geese (left) and caribou (below), return to the tundra each year to breed.

Animals of the Tundra Millions of migratory birds fly to the tundra to breed in the summer. Food is abundant in the form of plants, mollusks, worms, and especially insects. Caribou, as shown in **Figure 29**, migrate throughout the tundra in search of food and water. Hunters such as wolves roam the tundra and prey on caribou, deer, moose, and smaller animals, such as lemmings, mice, and rabbits. These rodents burrow underground during the winter but they are still active. Many animals that live in the tundra year-round, such as arctic foxes, lose their brown fur and grow white fur that camouflages them with the winter snow. These animals are also extremely well insulated.

Threats to the Tundra The tundra is one of the most fragile biomes on the planet. The food chains are relatively simple, so they are easily disrupted. Because conditions are so extreme, the land is easily damaged and slow to recover. Until recently, these areas were undisturbed by humans. But oil has been located in some tundra regions, such as Prudhoe Bay in northern Alaska. Oil exploration, extraction, and transport has disrupted the habitats of the plants and animals in many parts of the tundra. Pollution caused by spills or leaks of oil and other toxic materials may also poison the food and water sources of the organisms that live in the tundra.

MATHPRACTICE

U.S. Oil Production On average, the United States produces an estimated 8.1 million barrels of oil per day. How many millions of barrels of oil does the United States produce in 1 year? If all of the oil-producing countries of the world produce an estimated 74.13 million barrels of oil per day, what percentage of worldwide oil does the United States produce?

SECTION 3 Review

1. **Describe** two desert animals and the adaptations that help them survive.

2. **Describe** how tropical grasslands differ from temperate grasslands.

3. **Compare** the plants that live in deserts with the plants that live in the tundra biome.

4. **Describe** one threat to the tundra biome.

CRITICAL THINKING

5. **Making Inferences** Former grasslands are among the most productive farming regions. Read the description of temperate grasslands in this section and explain why this statement is true. **READING SKILLS**

6. **Analyzing Relationships** Explain why elephants and caribou, which live in very different biomes, both migrate.

163

Answers to Section Review

1. Sample Answer: Elf owls burrow in cactuses to avoid hot daytime temperatures. Sidewinders move so that only small areas of their bodies contact the hot sand at a given time.

2. Tropical grasslands are located in tropical and subtropical areas near the equator, and they are full of grasses, scattered trees, and shrubs. Temperate grasslands are located farther from the equator and are dominated by grasses. They have extremely fertile soil.

3. Plants that grow in deserts and plants native to tundra are adapted to low moisture and extreme temperatures. Both desert and tundra

plants grow and flower quickly in the short growing season. Desert plants have special structures that allow them to trap and store water. Tundra plants grow low to the ground to stay out of the cold and dry wind.

4. Oil exploration is a threat to the tundra.

5. Moderate to warm temperatures, consistent rain during the growing season, and fertile soils make for productive farming regions.

6. Large herbivores migrate due to the cycles of rain and drought in both savannas and the tundra. Migratory herbivores follow the rains to find newly sprouted grass and water.

MATHPRACTICE
Answers
8.1 million × 365 = 2.96 billion barrels of oil per year; (8.1 million ÷ 74.13 million) × 100 = 10.9% of oil worldwide

Close

Reteaching ———— BASIC

Biome Collage Ask students to cut out pictures from science or nature magazines and use them to create a colorful collage of a savanna, temperate grassland, chaparral, desert, or tundra ecosystem. Hang the collages in the classroom, and suggest that students include their collage in their **Portfolio.** Ⓛ Visual

Quiz ———— GENERAL

1. Name two adaptations that animals might have to desert conditions. (Sample answers: Animals may have thick skin or body armor to prevent water loss, they may estivate during the summer, or they may be nocturnal.)

2. How is the tundra similar to the desert? (Water is relatively scarce in both biomes.)

3. In which biome do plants tend to have deep root systems? (Tallgrass prairie, which is a type of temperate grassland.)

Alternative Assessment ———— GENERAL

Similar Adaptations Ask students to list similar plant and animal adaptations for desert and tundra biomes. (Some adaptations include: avoiding extreme heat or cold by burrowing and sleeping; a thick outer covering of skin or fur to keep heat in or out; shallow roots that take advantage of a small amount of rain; a thick cuticle that minimizes water loss) Have students describe why animals would evolve similar strategies to different situations. Ⓛ Logical/Verbal

CHAPTER 6 Highlights

Alternative Assessment — GENERAL

Biome Expert Chat Organize the class into four groups, and assign two different biomes to each group. Tell students that they have been asked to appear on a major talk show, and they must prepare answers to the questions below. Then, have students in each group tailor the questions to their biomes. Give students 2–3 days, and then stage mock interviews in front of the rest of the class where one person in each group is the TV host and the others are the "authorities." Some students may want to videotape their interview.

Interview Questions

1. What can you tell our viewers about the climate conditions of the _____ (first biome)? (Answers may vary, but students should include information about temperature and precipitation.)

2. After visiting the _____ (first biome), spending time in the _____ (second biome) must have been quite a shock. Is this true? Why or why not? (Answers may vary depending on the pair of biomes. Students should point out differences in plant and animal life as well as differences in climate.)

3. What do the biomes have in common? (Answers may vary.)

4. What was the most unusual plant or animal that you encountered in either of the two biomes, and how was it adapted to survive in its environment? (Answers may vary.) **LS** Verbal

1 What Is a Biome?

2 Forest Biomes

3 Grassland, Desert, and Tundra Biomes

Key Terms

biome, 143
climate, 144
latitude, 145
altitude, 145

tropical rain forest, 146
emergent layer, 148
canopy, 148
epiphyte, 148
understory, 148
temperate rain forest, 151
temperate deciduous forest, 152
taiga, 153

savanna, 155
temperate grassland, 156
chaparral, 158
desert, 160
tundra, 162
permafrost, 162

Main Ideas

▶ Scientists classify the ecosystems of the world into large areas called *biomes*.

▶ Biomes are described by their plant life because the plants that grow in an area determine what other organisms live there.

▶ Temperature, precipitation, latitude, and altitude are factors that affect climate, which determines the types of the plants that can grow in an area.

▶ The major forest biomes include tropical rain forests, temperate rain forests, temperate deciduous forests, and taiga.

▶ Tropical rain forests receive heavy rains and high temperatures throughout the year. They receive about 200 to 450 cm of rainfall a year. They are the most diverse of all biomes.

▶ Temperate deciduous forests experience seasonal variations in temperature and precipitation. They receive 75 to 125 cm of precipitation a year.

▶ Forest biomes are threatened by deforestation through logging, ranching, and farming.

▶ Savannas are located north and south of tropical rain forests and have distinct wet seasons. Savannas receive 90 to 150 cm of precipitation a year.

▶ Temperate grasslands get too little rainfall to support trees. Grasslands are dominated mostly by different types of grasses and flowering plants. Shortgrass prairies receive about 25 cm of precipitation a year.

▶ Deserts are the driest biomes on Earth. Deserts receive less than 25 cm of precipitation a year.

▶ Plants and animals found in each biome adapt to the environment in which they live.

164

SKILL BUILDER — GENERAL

Writing Have each student research and write an essay about the daily activities of a rain forest insect. What does it eat, when does it forage, what are its predators, and how does it avoid them? Encourage students to add this essay to their **Portfolio. LS** Verbal

Chapter Resource File

- **Chapter Test** GENERAL
- **Chapter Test** ADVANCED
- **Concept Review** GENERAL
- **Critical Thinking** ADVANCED
- **Test Item Listing**
- **Design Your Own Lab** BASIC
- **Math/Graphing Lab** GENERAL
- **CBL™ Probeware Lab** ADVANCED
- **Consumer Lab** BASIC
- **Long-Term Project** GENERAL

Using Key Terms

Use each of the following terms in a separate sentence.

1. *biome*
2. *climate*
3. *epiphyte*
4. *tundra*
5. *permafrost*

For each pair of terms, explain how the meanings of the terms differ.

6. *understory* and *canopy*
7. *latitude* and *altitude*
8. *chaparral* and *desert*
9. *tropical rain forest* and *temperate deciduous forest*

✔ STUDY TIP

Concept Maps Remembering words and understanding concepts are easier when information is organized in a way that you recognize. For example, you can use key terms and key concepts to create a concept map that links them together in a pattern you will understand and remember.

Understanding Key Ideas

10. Approximately what percentage of the Earth's species do tropical rain forests contain?
    a. 7 percent
    b. 20 percent
    c. 40 percent
    d. 50 percent

11. Animal species of the tropical rain forest
    a. compete more for available resources than species native to other biomes do.
    b. have adaptations to use resources and avoid competition.
    c. have adaptations to cope with extreme variations in climate.
    d. are never camouflaged.

12. Migration of animals in the savanna is mostly a response to
    a. predation.
    b. altitude.
    c. rainfall.
    d. temperature.

13. Spadefoot toads survive the dry conditions of the desert by
    a. migrating to seasonal watering holes.
    b. finding underground springs.
    c. burying themselves in the ground.
    d. drinking cactus juice.

14. The tundra is most suitable to a vertebrate that
    a. requires nesting sites in tall trees.
    b. is coldblooded.
    c. has a green outer skin for camouflage.
    d. can migrate hundreds of kilometers each summer.

15. A biome that has a large amount of rainfall, high temperature, and poor soil is a
    a. temperate woodland.
    b. temperate rain forest.
    c. tropical rain forest.
    d. savanna.

16. The two main factors that determine where organisms live are
    a. soil type and precipitation.
    b. temperature and precipitation.
    c. altitude and precipitation.
    d. temperature and latitude.

17. Which of the following biomes contains large trees?
    a. savanna
    b. temperate rain forest
    c. chaparral
    d. desert

18. The most common types of plants in the taiga biome are
    a. deciduous trees.
    b. short shrubs.
    c. coniferous trees.
    d. grasses.

165

Assignment Guide	
Section	Questions
1	1, 2, 7, 16, 21, 24–26, 32–33
2	3, 6, 9–11, 15, 17–19, 23, 29, 30
3	4, 5, 8, 12–14, 20, 22, 27, 28, 31

ANSWERS

Using Key Terms

1. Sample answer: A biome is a large region characterized by a specific type of climate and certain types of plant and animal communities.

2. Sample answer: Climate refers to weather conditions in an area over a long period of time.

3. Sample answer: Epiphytes grow on tall trees in the canopy for support.

4. Sample answer: Tundra is a biome dominated by grasses, lichens, and herbs and is primarily located north of the Arctic Circle.

5. Sample answer: Permafrost is a layer of soil beneath topsoil that is permanently frozen.

6. The canopy is the main layer of trees in a forest. The understory is below the canopy and extends to shorter plants and shrubs on the ground.

7. Latitude is the distance from the equator. Altitude is the height above sea level.

8. Chaparral has moderately warm temperatures all year and is dry. The desert is drier and is hotter in the summer.

9. Tropical rain forests are located near the equator and maintain a relatively constant warm temperature year-round. Temperate deciduous forests are generally located between 30° and 50° north latitude and have extreme ranges of temperatures.

Understanding Key Ideas

10. d
11. b
12. c
13. c
14. d
15. c
16. b
17. b
18. c

Short Answer

19. The rain forest floor lacks vegetation because the soil is poor in nutrients and receives very little sunlight.

20. Large root systems hold soil in place and prevent erosion.

21. Biomes vary in climate, and the climate gets colder with increasing altitude. So, different biomes are located in different altitudes across a mountain.

22. Permafrost in the tundra can preserve organisms such as a mammoth because the cold temperatures and frozen ground prevent and slow bacterial activity. Therefore, a mammoth would not decompose quickly if buried in the tundra.

23. A decrease in the number of trees will cause less carbon to be absorbed and stored. This will lead to an increase in carbon dioxide, a greenhouse gas that can cause warmer temperatures. A decrease in the number of trees also leads to flooding because the trees are no longer there to absorb and store water.

Interpreting Graphics

24. Cold temperatures and limited precipitation exclude most trees from tundra.

25. The amount of vegetation decreases.

26. Temperate grasslands have moderate temperature and precipitation.

Concept Mapping

27. Answers to the concept mapping questions are on pp. 667–672.

Short Answer

19. Unlike the jungles you see in movies, the floor of an undisturbed tropical rain forest usually lacks much vegetation. Explain why it lacks vegetation.

20. What is the relationship between root systems and erosion in a grassland ecosystem?

21. How might a mountain affect where particular types of biomes are located?

22. Well-preserved mammoths have been found buried in the tundra. Explain why the tundra preserves animal remains well.

23. How does deforestation contribute to a change in climate and increase the chance of floods in a biome?

Interpreting Graphics

Use the diagram below to answer questions 24–26.

24. Why are tall trees found in taiga biomes but not in tundra biomes?

25. As moisture decreases, what happens to the amount of vegetation in an area?

26. What does the diagram tell you about the temperature of and precipitation in temperate grasslands?

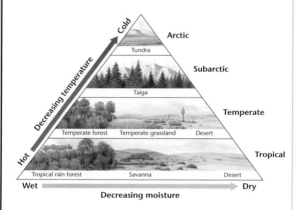

Critical Thinking

28. Similar environments led to the evolution of similar organisms with similar characteristics and strategies, even though the species live in different locations.

29. Answers may vary. Two factors that are responsible for the biodiversity of this biome are constant warm temperatures and large amounts of annual rainfall. Deforestation caused by logging and agricultural operations contributes to the decline of tropical rain forests.

30. The squirrels are not adapted to the rain forest, so they may not be able to eat or

---

Concept Mapping

27. Use the following terms to create a concept map: *threats to an ecosystem, erosion, overgrazing, logging, grasslands, rain forests, tundra, deserts, oil extraction,* and *irrigation.*

Critical Thinking

28. **Comparing Processes** American prairies and Asian steppes contain different plant species but are dominated by grasses. Write a short paragraph that explains why the two grasslands contain different species but the same types of plants. **WRITING SKILLS**

29. **Classifying Information** Read the description of tropical rain forests in this chapter, and list two factors that are responsible for the biodiversity of this biome. Describe two reasons for the decline of tropical rain forests, and discuss actions that some countries have taken to protect tropical rain forests. **READING SKILLS**

30. **Analyzing Relationships** If you took a population of squirrels from the southeastern United States and introduced them into a Central American rain forest, they would probably not survive. Why do you think the squirrels would not survive even though they are naturally adapted to life in a forest?

31. **Making Inferences** How might prairie fires set from natural and human causes have affected the evolution of fire resistance in prairie grasses?

Cross-Disciplinary Connection

32. **Geography** Use a world map to find locations of the various biomes. Then, make a poster that contains photos or illustrations of plants and animals native to each biome.

Portfolio Project

33. **Food Webs in Your Community** Do a special project on the ecosystems in your community. Use field guides of your area to find out what plants and animals live in your community. Then, draw a food web that shows how organisms in each ecosystem could be related.

---

compete for the food that is available. The displaced squirrels are also not adapted to constant high temperatures, high humidity, and high precipitation.

31. Frequent fires may have increased the evolutionary pressure for prairie grasses to evolve fire resistance.

Cross-Disciplinary Connection

32. Answers may vary.

Portfolio Project

33. Answers may vary.

MATH SKILLS

Use the table below to answer quesions 34–35.

Amount of Tropical Rainforest		
Country	Amount of tropical rain forest (km²)	Amount of annual deforestation (km²/y)
A	1,800,000	50,000
B	55,000	3,300
C	22,000	6,000
D	530,000	12,000
E	80,000	700

34. Making Calculations What percentage of tropical rain forest is being destroyed each year in country A? in country D?

35. Interpreting Statistics According to the table, which country's tropical rain forest will be completely destroyed first? Which country's rain forest will be completely destroyed last?

WRITING SKILLS

36. Communicating Main Ideas Describe the importance of conserving the biomes of the world. What can you do to help conserve the world's biomes?

37. Writing From Research Choose one biome and research the threats that exist against it. Write a short essay that describes the threats and any actions that are being taken to help save the biome.

STANDARDIZED TEST PREP

For extra practice with questions formatted to represent the standardized test you may be asked to take at the end of your school year, turn to the sample test for this chapter in the Appendix.

READING SKILLS

Read the passage below, and then answer the questions that follow.

The Tropics and other regions of high biodiversity include some of the economically poorest countries on Earth. These countries are trying to use their natural resources to build their economies and to raise the standard of living for their citizens. Several conservation strategies offer ways for developing countries to benefit economically from preserving their biodiversity.

For example, in a *debt-for-nature swap*, richer countries or private conservation organizations will sometimes pay some of the debts of a developing country. In exchange, the developing country agrees to take steps to protect its biodiversity, such as setting up a preserve or launching an education program for its citizens. Another idea to help local people make money from intact ecosystems is to set up a national park to attract tourists. People who want to see the ecosystem and its unique organisms will pay money for nature guides, food, and lodging. This idea is called *ecotourism*.

1. The main objective of both *debt-for-nature swap* and *ecotourism* is
   a. economic gain.
   b. education of citizens.
   c. preservation of biodiversity.
   d. Both (a) and (c)

2. According to the passage, which of the following statements is true?
   a. Regions of high biodiversity are not worth saving.
   b. Intact ecosystems are those ecosystems that are most developed.
   c. Debt-for-nature swap is an example of international compromise.
   d. Launching education programs for citizens does not help protect ecosystems.

Math Skills

34. 50,000 ÷ 180,000 × 100% = 2.8% in country A; 12,000 ÷ 530,000 × 100% = 2.3% in country D

35. The tropical rain forest in country C will be completely destroyed first. The tropical rain forest in country E will be completely destroyed last.

Writing Skills

36. Each biome represents a wealth of adaptations and biological variety. Losing any biome would rob the world of that evolutionary history and diversity. Biomes can be conserved in part by the development of sustainable forestry and agriculture and by setting aside areas for preservation. Support of international conservation and human rights organizations can also help to preserve biomes.

37. Answers may vary.

Reading Skills

1. d

2. c

IDENTIFY YOUR LOCAL BIOME

Teacher's Notes

Time Required
one 45-minute class period

Lab Ratings

EASY ——————→ HARD

TEACHER PREPARATION
STUDENT SETUP
CONCEPT LEVEL
CLEANUP

Skills Acquired
- Classifying
- Collecting Data
- Inferring
- Interpreting
- Organizing and Analyzing Data

The Scientific Method
In this lab, students will:
- Make Observations
- Ask Questions
- Analyze the Results
- Draw Conclusions

Materials
The materials on the student page are enough for a group of four students.

Objectives
▶ **Collect** information from international, national, and local resources about the biome in which you live.
▶ **USING SCIENTIFIC METHODS** **Perform** field observations to identify the name of the biome in which you live.

Materials
binoculars (optional)
field guide to local flora and fauna
globe or atlas
graph paper (optional)
notebook
pencil or pen
ruler

▶ **Climatograms** The temperature and precipitation for Austin, Texas is shown in this climatogram.

Identify Your Local Biome
In what biome do you live? Do you live in a temperate deciduous forest, a desert, or a temperate grassland, such as a prairie or savanna? In this lab, you will explore certain characteristics of the biome in which you live. With the information you gather, you will be able to identify which biome you live in.

Procedure
1. Use a globe or atlas to determine the latitude at which you live. Record this information.
2. Consider the topography of the place where you live. Study the contour lines on a map or surface variations on a globe. What clues do you find that might help identify your biome? For example, is your area located in near a mountain or an ocean? Record your findings.
3. Prepare a climatogram of your local area. A climatogram is a graph that shows average monthly values for two factors: temperature and precipitation. Temperature is expressed in degrees Celsius and is plotted as a smooth curve. Precipitation values are given in centimeters and are plotted as a histogram.

   To make a climatogram of your area, obtain monthly averages for one year of precipitation and temperature from your local TV or radio weather station. Make a data table, and record these values. Next, draw the vertical and horizontal axes of your climatogram in your notebook or on graph paper. Then, show the temperature scale along the vertical axis on the right side of the graph and the precipitation scale along the vertical axis on the left side of the graph. Show months of the year along the horizontal axis. Finally, plot your data.

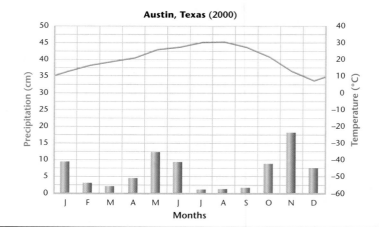

Tips and Tricks
If you are near a university or college, contact the biology department, and ask if a professor or student would be willing to help your students identify the local flora and fauna. Climatograms will probably be influenced by human habitations and roads. These structures can elevate ambient temperature. Therefore, students' climatograms may not match any in the chapter. Have students identify organisms living in natural vegetation as well as those living in human habitation. Have them record the differences in their lab write-up.

4. Go outside to observe the plants growing in your area. Bring a field guide, and respond to the following items in your notebook.

a. Sketch or describe as many plants that are common in the area as you can. Use your field guide to identify each of these species.

b. Describe three or more adaptations of each plant to the local climate.

c. Which of the plants that you observed are native to your area? Which have been introduced by humans? Which of the introduced plants can survive on their own in local conditions? Which of the introduced plants require extensive human care to remain alive?

d. Look for evidence that animals have left behind—footprints, nests, dens or burrows, hair or feathers, scratches, or urine markings. Sketch or describe as many different animal species as possible. Identify each species by using your field guide.

e. Describe three or more adaptations that each animal has developed in order to survive in local climatic conditions.

Analysis

1. Analyzing Data Compare your local climatogram to the biome climatograms shown in this chapter. Which biome has a climatogram most similar to your climatogram?

2. Analyzing Results Consider your latitude, topographical findings, and observations of local plants and animals. Combine this information with your climatogram, and determine which biome best matches the area in which you live.

Conclusions

3. Evaluating Results Does your local climatogram match any of the seven major terrestrial climatograms shown in the chapter? Explain how any differences between your local biome and the biome in the chapter that your local biome most clearly matches might influence the adaptations of local animals and plants.

4. Applying Conclusions Organisms create features of the biome in which they live. What features of your biome are created by the organisms that live there?

Extension

1. Classifying Information Name three adaptations you observed in the plants that grow naturally in your area. Name at least three adaptations you observed in local animals. Explain in detail how each of these adaptations meets the conditions of your biome.

▶ **Biomes** These two cities are located in two different biomes. Stamford, Vermont (top) is located in a temperate deciduous forest, and Tucson, Arizona (bottom) is located in a desert.

Answers to Analysis
1. Answers may vary.
2. Answers may vary.

Answers to Conclusions
3. Answers may vary.
4. Answers may vary but may include dams, dens, burrows, or other features.

Answers to Extension
1. Answers may vay.

Chapter Resource File

• Datasheets for In-Text Labs
• Lab Notes and Answers

169

CLASSROOM TESTED & APPROVED

Lowell S. Bailey
Bedford - N. Lawrence
High School
Bedford, Indiana

THE FUTURE OF THE ARCTIC NATIONAL WILDLIFE REFUGE

Background

In 1998, the U.S. Geological Survey estimated that there is a mean of 7.7 billion barrels of technically recoverable oil in the coastal zone area of the Arctic National Wildlife Refuge. Proponents of drilling have cited 6 to 16 billion barrels of recoverable oil. The Department of the Interior estimates that there are 10.4 billion barrels of oil, which would only run the state of Texas for 9 years.

MISCONCEPTION /// **ALERT** \\\

Recreational Opportunities in ANWR Students may think that the Arctic National Wildlife Refuge is off limits to all non-indigenous human activity. Recreational opportunities for visitors do exist. The U.S. Forest Service has created a brochure about the refuge, with information about hiking, camping, and river exploration. No specific trail maps exist, but areas are open for those interested in a true wilderness experience.

POINTS of view

THE FUTURE OF THE ARCTIC NATIONAL WILDLIFE REFUGE

During the 1970s, Congress passed the Alaska National Interests Land Conservation Act. The legislation gave Congress the responsibility for determining how Alaska's lands will be used. Included in this responsibility is the fate of the Arctic National Wildlife Refuge (ANWR). Oil company geologists believe that oil reserves are under several areas of the northern Alaska coast, including the refuge. Debate has raged about whether Congress should maintain ANWR as a wildlife refuge or open it to oil exploration. Advocates of the refuge feel that the environmental cost of oil exploration would be too high. But those who favor oil exploration in the refuge believe the oil reserves must be tapped to meet U.S. oil needs and to maintain economic security.

▶ A small group of muskoxen huddles on the arctic tundra of the Arctic National Wildlife Refuge.

Protect the Refuge

Conservationists and ecologists are concerned about the impact that oil exploration would have on animals that live in the refuge. Oil exploration would occur within the 1.5 million acre coastal plain of the

▶ Migratory birds, such as the Canada geese below, nest and raise their young on the refuge's tundra.

ANWR. This area includes the breeding ground and grazing area for one of the last great herds of caribou in North America. Biologists think that forcing the herd into other areas of the refuge would deprive the caribou of their main food source and would expose calves to increased predation.

In addition, migratory birds from all over the world travel to the refuge's tundra to nest and raise their young during the short arctic summers. Scientists believe that oil exploration would disrupt the nesting and feeding of these birds so much that the birds would be unable to finish rearing their young before the first freeze of early September.

The ANWR is also a habitat for more than 7,000 native peoples. Some of these people depend on the caribou of the ANWR for food, clothing, and tools. Oil exploration and drilling could displace the caribou population, which would drastically affect the culture and way of life for these people.

Opponents of oil exploration in the refuge also point to the environmental damage that has already been done in nearby Prudhoe Bay. When oil was found there, oil companies joined forces to extract the oil and built a pipeline across the state to reach tankers on the southern Alaska coast. Advocates of the refuge say that the use of this pipeline has exposed the fragile tundra ecosystem to toxic chemicals and destroyed natural habitats. They fear the same fate for the refuge if oil exploration is permitted there.

Advocates of protecting the refuge also point out that no one knows how much oil is available in the refuge. They also point out that even if all of the oil that could possibly be in the refuge were extracted, the oil would supply the United States for only nine months.

Conservationists contend that the development and use of renewable energy resources, such as wind and fuel cells, could reduce the dependence on oil. Laws that require stricter energy conservation measures could also reduce the need for oil in the United States.

170

REAL-LIFE — **CONNECTION** — **GENERAL**

Environmental Damage in an Alaskan Wildlife Refuge The Kenai National Wildlife Refuge in Alaska was established by President Franklin Delano Roosevelt in 1941 to protect moose and other wildlife. After oil was discovered in the refuge, drilling occurred, which has caused extensive environmental damage. More than 350 spills have released 270,000 gallons of oil into the area, causing groundwater contamination. Because of toxic spills, wildlife have been exposed to oil, benzene, PCBs, xylene, and heavy metals.

▶ The pipeline shown above carries oil across the entire state of Alaska.

Open the Refuge

Advocates of oil exploration in the ANWR believe that the current U.S. demand for oil cannot be met by energy conservation alone. Advocates insist that the United States must utilize every domestic source of oil available, including the ANWR. The advocates also point out that the United States depends too much on oil from other countries that control its price and availability. A significant amount of our oil is imported from the Middle East, a politically unstable area. If those countries restrict sales of oil to the United States, our economy could be seriously affected. Those who favor exploration think economic security should take priority over environmental concerns.

Advocates of oil exploration in the refuge also stress that much of the oil in the Prudhoe Bay area has already been extracted and that oil production will soon begin to decline. The industrial complex that is already in place for the production of oil in Prudhoe Bay could be used for the production of oil from the nearby refuge. New construction in the area would be limited.

Because of the decline of oil in Prudhoe Bay, advocates support oil drilling in the ANWR for economic reasons. The oil industry supports two-thirds of Alaska's economy and employs 1 percent of the population. If the refuge were open for drilling, oil companies would profit and more jobs might be available for the people of Alaska.

Government studies indicate a 19 to 46 percent chance of finding oil in the refuge, which is a percentage that the oil industry believes justifies exploration. People who favor exploration also suggest that oil companies can now extract oil with less environmental damage than was caused in Prudhoe Bay.

People who oppose the protection of the wildlife refuge believe the economic benefits of oil exploration in this area outweigh any remaining risks of environmental damage.

▶ Exploration for oil would occur in the northern coastal plain of the Arctic National Wildlife Refuge.

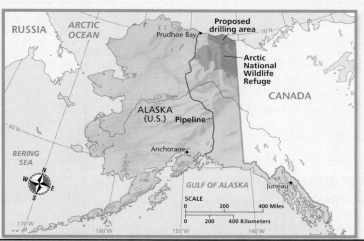

What Do You Think?

Oil exploration in ANWR could have a negative impact on the animals and people living in and around the refuge, but oil in the refuge may help the U.S. meet its future energy needs. Without knowing what the consequences or benefits will be, do you think that the U.S. should permanently protect the ANWR or open it for oil exploration? Explain.

171

Answers to What Do You Think?
Answers may vary.

Discussion ——— GENERAL

Outlining the Points of View
Have students work in groups to create an outline of the main points of view expressed in this feature. Ask them to analyze whether each point is adequately answered by the opponent's side. (For example, those people in favor of opening the refuge respond to the concern over the caribou population by stating that the caribou population actually increased in the Prudhoe Bay area.) Suggest that students identify the type of value being invoked by each response (such as environmental, economic, or aesthetic). Then for each point, have students decide which point of view is the most persuasive. Finally, based on their analysis, student groups should discuss the debate and decide which point of view is the most convincing. Encourage each group member to write a short statement explaining which point of view is closest to his or her own point of view about the situation. **LS Verbal**

Cultural Awareness

Indigenous Cultures and ANWR When polled about the Arctic National Wildlife Refuge, 78 percent of the people in the Inupiat village of Kaktovik (within the refuge) agreed that the refuge should be open to drilling. They are afraid of losing jobs and their associated benefits, such as health care and educational opportunities. Nearly half of those polled, however, worried that exposure to Western culture would lead to increased alcohol and drug abuse in their village.

Aquatic Ecosystems
Chapter Planning Guide

PACING	CLASSROOM RESOURCES	LABS, DEMONSTRATIONS, AND ACTIVITIES
BLOCKS 1, 2 & 3 · 135 min pp. 172–178 **Chapter Opener**		
Section 1 Freshwater Ecosystems	**OSP** Lesson Plan * **CRF** Active Reading * **BASIC** **TT** Bellringer * **TT** Life Zones in a Lake * **TT** Freshwater Wetlands in the United States * **CD** Interactive Tutor Aquatic Environments **CD** Interactive Exploration 4 Flood Bank	**TE** Demonstration Observing Aquariums, p. 173 **BASIC** **TE** Internet Activity Wetland Plants, p. 176 **BASIC** **TE** Activity Wetlands in Your Neighborhood, p. 177 **BASIC** **SE** Skills Practice Lab Eutrophication: Too Much of a Good Thing? pp. 190–191 ◆ **CRF** Datasheets for In-Text Labs * **CRF** CBL™ Probeware Lab * ◆ **ADVANCED**
BLOCKS 4 & 5 · 90 min pp. 179–185 **Section 2** Marine Ecosystems	**OSP** Lesson Plan * **CRF** Active Reading * **BASIC** **TT** Bellringer * **TT** The Formation of Estuaries * **TT** Coral Reefs of the World * **TT** Life Zones in the Ocean *	**SE** QuickLab Estuaries, p. 180 **TE** Demonstration Salinity Change, p. 180 **BASIC** **TE** Activity Save the Chesapeake Bay, p. 181 **BASIC** **TE** Internet Activity Tides and Estuaries, p. 181 **ADVANCED** **TE** Demonstration Barrier Islands, p. 182 **BASIC** **TE** Group Activity Coral Reefs, p. 183 **BASIC** **SE** Field Activity Make a Miniature Aquatic Ecosystem, p. 184 **TE** Group Activity Marine Ecosystems, p. 184 **ADVANCED** **CRF** Field Activity * ◆ **GENERAL** **CRF** Long-Term Project * ◆ **GENERAL** **CRF** Observation Lab * ◆ **BASIC** **CRF** Consumer Lab * **GENERAL**

BLOCKS 6 & 7 · 90 min

Chapter Review and Assessment Resources

SE Chapter Review pp. 187–189
SE Standardized Test Prep pp. 648–649
CRF Study Guide * ■ **GENERAL**
CRF Chapter Test * ■ **GENERAL**
CRF Chapter Test * **ADVANCED**
CRF Concept Review * ■ **GENERAL**
CRF Critical Thinking * **ADVANCED**
OSP Test Generator
CRF Test Item Listing *

Online and Technology Resources

Visit **go.hrw.com** for access to Holt Online Learning or enter the keyword **HE6 Home** for a variety of free online resources.

This CD-ROM package includes
• Lab Materials QuickList Software
• Holt Calendar Planner
• Customizable Lesson Plans
• Printable Worksheets
• ExamView® Test Generator
• Interactive Teacher Edition
• Holt PuzzlePro® Resources
• Holt PowerPoint® Resources

Holt Environmental Science Interactive Tutor CD-ROM

This CD-ROM consists of interactive activities that give students a fun way to extend their knowledge of environmental science concepts.

KEY

TE	Teacher Edition	**CRF**	Chapter Resource File	*****	Also on One-Stop Planner
SE	Student Edition	**TT**	Teaching Transparency	■	Also Available in Spanish
OSP	One-Stop Planner	**CD**	Interactive CD-ROM	◆	Requires Advance Prep

ENRICHMENT AND SKILLS PRACTICE	SECTION REVIEW AND ASSESSMENT	CORRELATIONS
SE Pre-Reading Activity, p. 172 **TE** Using the Figure, p. 172 (GENERAL)		**National Science Education Standards**
TE Inclusion Strategies, p. 174 **TE** Using the Figure Identifying Wetlands, p. 175 (GENERAL) **TE** Skill Builder Writing, p. 175 (ADVANCED) **TE** Reading Skill Builder Reading Organizer, p. 175 (BASIC) **TE** Student Opportunities Wetland Nurseries, p. 176 **SE** MathPractice Wetland Conversion, p. 177	**TE** Homework, p. 176 (ADVANCED) **TE** Homework, p. 177 (ADVANCED) **SE** Section Review, p. 178 **TE** Reteaching, p. 178 (BASIC) **TE** Quiz, p. 178 (GENERAL) **TE** Alternative Assessment, p. 178 (BASIC) **CRF** Quiz * ■ (GENERAL)	LS 4e SPSP 3c SPSP 4a SPSP 5b SPSP 6e
SE Case Study Restoration of the Chesapeake Bay, pp.180–181 **TE** Reading Skill Builder Brainstorming, p. 181 (GENERAL) **TE** Reading Skill Builder Comparing and Contrasting, p. 182 (BASIC) **SE** Graphic Organizer Cause-and-Effect Map, p. 183 (GENERAL) **TE** Inclusion Strategies, p. 183 **TE** Career Oceanographer, p. 184 **SE** Maps in Action Wetlands in the United States, 1780s Vs. 1980s, p. 192 **SE** Science & Technology Creating Artificial Reefs, p. 193 **CRF** Map Skills Aquatic Diversity * (GENERAL)	**TE** Homework, p. 184 (ADVANCED) **SE** Section Review, p. 185 **TE** Reteaching, p. 185 (BASIC) **TE** Quiz, p. 185 (GENERAL) **TE** Alternative Assessment, p. 185 (GENERAL) **CRF** Quiz * ■ (GENERAL)	LS 4e SPSP 3c SPSP 4a SPSP 5b SPSP 6e

Guided Reading Audio CDs

These CDs are designed to help auditory learners and reluctant readers. (Audio CDs are also available in Spanish.)

www.scilinks.org

Maintained by the **National Science Teachers Association**

TOPIC: Lakes and Ponds
SciLinks code: HE4058

TOPIC: Estuaries
SciLinks code: HE4037

 CNN Videos

Each video segment is accompanied by a Critical Thinking Worksheet.

Science, Technology & Society

Segment 17 Battling Over the Oregon Inlet

Earth Science Connections

Segment 21 Salty Water

CHAPTER

7

Chapter Enrichment

This Chapter Enrichment provides relevant and interesting information to expand and enhance your classroom instruction of the chapter material.

SECTION

1 Freshwater Ecosystems

Lakes and Ponds

The five most common types of lakes are glacial lakes, barrier lakes, crater lakes, tectonic lakes, and artificial lakes. Glacial lakes are formed by the action of glaciers. Glaciers scour the land surface and leave depressions that fill with water. Barrier lakes are formed when landslides or glacial tills block streams or rivers and hold back water. Crater lakes form in the craters of volcanoes. Tectonic lakes fill the bottoms of rift valleys that formed when the Earth's crust was pulled apart by tectonic movement. Artificial lakes are formed when animals, such as humans dig depressions in the land or build dams.

Seasonal Changes in Lakes

Most ponds and lakes do not remain static over a year's time. Seasonal changes in temperature, wind, and stream flow cause the movement of water, organisms, and nutrients within a lake. However, during summer and winter months, most lakes become highly *stratified* by temperature. At the deepest layers, cold water is trapped and light levels are low. In these conditions, most of the oxygen that is present is used up, and the main organisms that remain are anaerobic microbes and bottom-feeding animals adapted to living in dark areas such as, clams and catfish. *Ecologists* use the oxygen content of deep water in summer as an indicator of the degree of eutrophication of lakes (oxygen is used up fastest at the bottom of high-nutrient lakes). In spring and fall, weather changes often cause *turnover* in lakes. Wind and/or temperature changes at the surface will cause mixing with the lower layers. After this turnover, the nutrients, oxygen, and many of the organisms will again be mixed throughout the lake.

Lake Baikal

Lake Baikal in Russia is the oldest lake in the world. It is a tectonic lake that is 650 km long, 80 km wide, and up to 1.6 km deep. The lake holds 23,000 km³ of water. This is 20% of all the liquid fresh water on Earth's surface. Lake Baikal is home to 1,200 species of animals and 1,000 species of plants. About 80% of these species are found only in Lake Baikal. The shores of Lake Baikal are also home to 1 million people, and 3.8 million people live in the lake's watershed. Pollution is becoming a serious problem in spite of Lake Baikal's immense size. One industry contributing to the pollution is the pulp industry. At Baikalsk, a pulp mill was started in 1954. Since it began operation, it has been dumping wastes produced from bleaching cellulose. The bleach kills the plants and animals that live in a 200 square km area where the wastes are dumped. During and since the 1990s, many efforts have been made to stop pollution of the lake and to clean up existing problems.

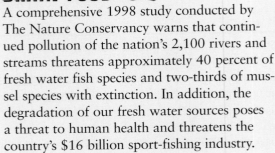

BRAIN FOOD

A comprehensive 1998 study conducted by The Nature Conservancy warns that continued pollution of the nation's 2,100 rivers and streams threatens approximately 40 percent of fresh water fish species and two-thirds of mussel species with extinction. In addition, the degradation of our fresh water sources poses a threat to human health and threatens the country's $16 billion sport-fishing industry.

▶ River headwaters (top) tend to flow faster and have more oxygen than sections downstream (bottom)

SECTION

2 Marine Ecosystems

Salinity and Composition of Sea Water

Salinity is usually measured in parts per thousand (ppt). Sea water contains, on average, 35 ppt of chemical elements. This means that for every 1,000 g of sea water, there are 35 g of dissolved salts. The total volume of sea water in the oceans is 1.5×10^9 km^3 of water, containing 5×10^{22} g of dissolved chemical elements. Sea water contains about 70 different chemical elements. The most common chemical elements found in sea water are chlorine, sodium, sulfate, magnesium, calcium, potassium, bicarbonate, and bromine.

Upwelling

Along some coastlines, surface currents carry water away from the shore. These flow patterns cause upwelling, the rise of colder deep water that contains many nutrients required by phytoplankton. Where the nutrients reach the surface waters, phytoplankton flourish. Phytoplankton are food for many species of commercial fish. Upwelling is common near the equator and in many coastal zones, usually on the western margins of continents. Human populations in these coastal areas depend on the fishing industry as a major source of income and food. Many species of commercial fish harvested off the western coasts of North and South America are part of a food chain supported by upwelling.

Blue Whales

Blue whales have a feeding structure called a *baleen* that they use to strain small organisms from sea water. The whales' diet is made up of small, shrimplike creatures called *krill*. Blue whales swallow as much as 64,600 L of water at one time and then force the water out through the baleen. Then they swallow the organisms that remain. Blue whales can consume as much as 3,600 kg of krill in a day! During the summer, blue whales migrate to the North and South Poles, where krill are abundant because of long daylight hours. Blue whales then return to warmer waters to give birth to their young. Blue whales have a lifespan of about 90 years.

Horseshoe Crabs

Horseshoe crabs have changed little in the past 250 million years. Their success results from their tolerance of wide temperature and salinity ranges, their hard, protective shell, and their ability to live for up to a year

▶ **Horseshoe crabs in Delaware Bay**

without eating. There are four extant species of horseshoe crabs. Three species are found in Asia and the fourth species is found from Nova Scotia to the Yucatán Peninsula. Though often overlooked, these crabs play a vital role in coastal ecosystems. Horseshoe crabs lay millions of eggs during huge annual congregations to places such as Delaware Bay. The eggs are an essential food source for migrating birds. Without the nutrients from the horseshoe crab eggs, some bird species would be unable to complete their yearly migration. Horseshoe crabs are also important for medical research. Their blue blood is copper based and contains a substance that forms clots around bacteria. Blood is collected from horseshoe crabs and is used to test the purity of medicines.

BRAIN FOOD

Sound waves travel for hundreds of kilometers in the oceans' deep-water channels. Whales and other marine animals take advantage of these properties for long-range communication and to search for food. Whales communicate with clicks, whistles, squeaks, and songs that convey information. Scientists aren't sure what the songs mean, but they do know that whales can communicate at distances as great as 1,600 km!

Overview

Tell students that this chapter discusses freshwater and marine ecosystems. Freshwater ecosystems are made up of lakes, rivers, and wetlands. Marine ecosystems include estuaries, coral reefs, and oceans. Aquatic ecosystems perform many environmental functions and support many plant and animal species. Unfortunately, pollution, development, and overuse threaten many of these ecosystems.

Using the Figure — GENERAL

Manatees live in both fresh water and salt water. In North America, most manatees are found in the estuaries, bays, and coastal ecosystems of states such as Florida. Manatees are often called sea cows because they lazily graze on aquatic vegetation. They are 3 to 4 m in length and usually weigh 360 to 545 kg. Unfortunately, manatees are sometimes hit by boats because they live in shallow water and cannot swim fast enough to avoid boats.

PRE-READING ACTIVITY

You may want to use this FoldNote in a classroom discussion to review material from the chapter. On the board, write each category from the Two-Panel Flip Chart. Then, ask students to provide information for each category. Under the appropriate category on the board, write the information that students provide.

VIDEO SELECT

For information about videos related to this chapter, go to **go.hrw.com** and type in the keyword **HE4 AQUV**.

Aquatic Ecosystems

1 Freshwater Ecosystems
2 Marine Ecosystems

PRE-READING ACTIVITY

FOLDNOTES

Two-Panel Flip Chart
Before you read this chapter, create the **FoldNote** entitled "Two-Panel Flip Chart" described in the Reading and Study Skills section of the Appendix. Label the flaps of the two-panel flip chart with "Freshwater Ecosystems" and "Marine Ecosystems." As you read the chapter, write information you learn about each category under the appropriate flap.

Manatees live in both freshwater and saltwater ecosystems. Manatees are herbivores and will eat at least 27 kg (60 lb) of aquatic plants per day.

172

Chapter Correlations   *National Science Education Standards*

LS 4e Human beings live within the world's ecosystems. Increasingly, humans modify ecosystems as a result of population growth, technology, and consumption. Human destruction of habitats through direct harvesting, pollution, atmospheric changes, and other factors is threatening current global stability, and if not addressed, ecosystems will be irreversibly affected. **(Section 1 and Section 2)**

SPSP 3c Humans use many natural systems as resources. Natural systems have the capacity to reuse waste, but that capacity is limited. Natural systems can change to an extent that exceeds the limits of organisms to adapt naturally or humans to adapt technologically. **(Section 1 and Section 2)**

SPSP 4a Natural ecosystems provide an array of basic processes that affect humans. Those processes include maintenance of the quality of the atmosphere, generation of soils, control of the hydrologic cycle, disposal of wastes, and recycling of nutrients. Humans are changing many of these basic processes, and the changes may be detrimental to humans. **(Section 1 and**

Section 2)

SPSP 5b Human activities can enhance potential for hazards. Acquisition of resources, urban growth, and waste disposal can accelerate rates of natural change. **(Section 1 and Section 2)**

SPSP 6e Humans have a major effect on other species. For example, the influence of humans on other organisms occurs through land use—which decreases space available to other species—and pollution—which changes the chemical composition of air, soil, and water. **(Section 1 and Section 2)**

Freshwater Ecosystems

The types of organisms in an aquatic ecosystem are mainly determined by the water's *salinity*—the amount of dissolved salts the water contains. As a result, aquatic ecosystems are divided into freshwater ecosystems and marine ecosystems.

Freshwater ecosystems include the sluggish waters of lakes and ponds, such as the lake shown in **Figure 1,** and the moving waters of rivers and streams. They also include areas where land, known as a **wetland,** is periodically underwater. Marine ecosystems include the diverse coastal areas of marshes, swamps, and coral reefs as well as the deep, vast oceans.

Characteristics of Aquatic Ecosystems

Factors such as temperature, sunlight, oxygen, and nutrients determine which organisms live in which areas of the water. For instance, sunlight reaches only a certain distance below the surface of the water, so most photosynthetic organisms live on or near the surface.

Aquatic ecosystems contain several types of organisms that are grouped by their location and by their adaptations. Three groups of aquatic organisms include plankton, nekton, and benthos. **Plankton** are the organisms that float near the surface of the water. Two types of plankton are microscopic plants called *phytoplankton,* and microscopic animals called *zooplankton.* Phytoplankton produce most of the food for an aquatic ecosystem. **Nekton** are free-swimming organisms, such as fish, turtles, and whales. **Benthos** are bottom-dwelling organisms, such as mussels, worms, and barnacles. Many benthic organisms live attached to hard surfaces. Decomposers, organisms that break down dead organisms, are also a type of aquatic organism.

Objectives

▶ Describe the factors that determine where an organism lives in an aquatic ecosystem.

▶ Describe the littoral zone and the benthic zone that make up a lake or pond.

▶ Describe two environmental functions of wetlands.

▶ Describe one threat against river ecosystems.

Key Terms

wetland
plankton
nekton
benthos
littoral zone
benthic zone
eutrophication

Figure 1 ▶ Lake Louise in Alberta, Canada, is an example of a freshwater ecosystem.

173

CHEMISTRY
CONNECTION — GENERAL

Water Molecules The molecular structure of water gives it some very unusual properties. When water freezes, its molecular structure expands. Thus, solid water (ice) is less dense than liquid water. (The opposite is true for almost every other substance.) If water did not have this unique property, little aquatic life would exist outside the Tropics because in the winter, ice would sink and lakes and ponds would freeze solid, killing aquatic organisms.

Focus

Overview

Before beginning this section, review with your students the Objectives listed in the Student Edition. This section discusses the characteristics of freshwater ecosystems. Students also learn about how wetlands are an important kind of freshwater ecosystem.

Bellringer

Ask students to describe in their *EcoLog* a river and a lake that they have seen or visited. Tell students that they should describe how the two are similar and how they are different. Ask students to reread their descriptions after completing this section. **LS** Logical

Motivate

Demonstration —— BASIC

Observing Aquariums If you have an aquarium in the classroom, ask students to observe it. If you do not have an aquarium, show the class a photograph of one. Ask students to describe the different plants and animals in the aquarium. (The animals and plants observed depend on the aquarium.) Ask, "What requirements do the fish in the aquarium need to survive?" (The fish in the aquarium need food, oxygen, and the correct temperature of water.) Ask students, "What would happen if a fish from the ocean were added to the aquarium?" (Assuming the aquarium is a freshwater aquarium, the fish would die.)

Chapter Resource File

• Lesson Plan
• Active Reading BASIC
• Section Quiz GENERAL

Transparencies

TT Bellringer

Discussion — GENERAL

Beavers With the exception of humans, beavers may do more to reshape their landscape than any other mammal. Beavers build dams that can create ponds and divert streams to create areas of wetland. Because the water in a pond is standing, it seeps into the ground and increases the soil moisture in the area. Higher soil moisture allows different communities of plants to thrive. The ponds and associated wetlands provide homes and food for dozens of different species that otherwise could not survive in the area. Also, the dams trap sediments that wash off the land during floods. As a beaver pond slowly fills with sediments, the pond is converted into a meadow. Scientists estimate that 200 million beavers once lived in the continental U.S. In the 18th century, the beaver population was nearly decimated by hunting and trapping. The beaver dams fell into disrepair, and the ponds and wetlands that the dams had created were replaced with flowing streams. The loss of beavers changed many areas and affected the plants and animals that lived there. **Verbal**

Figure 2 ▶ Amphibians, such as this bull frog, live in or near lakes and ponds.

internet connect

www.scilinks.org
Topic: Lakes and Ponds
SciLinks code: HE4058

SCi LINKS. Maintained by the National Science Teachers Association

Figure 3 ▶ A pond or lake ecosystem is structured according to how much light is available. Tiny plants called *phytoplankton* and tiny animals called *zooplankton* live in open water, where more sunlight is available.

Lakes and Ponds

Lakes, ponds, wetlands, rivers, and streams make up the various types of freshwater ecosystems. Lakes, ponds, and wetlands can form naturally where groundwater reaches the Earth's surface. Beavers can also create ponds by damming up streams. Humans intentionally create artificial lakes by damming flowing rivers and streams to use them for power, irrigation, water storage, and recreation.

Life in a Lake Lakes and ponds can be structured into horizontal and vertical zones. In the nutrient-rich **littoral zone** near the shore, aquatic life is diverse and abundant. Plants, such as cattails and reeds, are rooted in the mud underwater, and their upper leaves and stems emerge above the water. Plants that have floating leaves, such as pond lilies, are rooted here also. Farther out from the shore, in the open water, plants, algae, and some bacteria capture solar energy to make their own food during *photosynthesis*. As shown in **Figure 3**, the types of organisms present in a pond or lake ecosystem depend on the amount of sunlight available.

Some bodies of fresh water have areas so deep that there is too little light for photosynthesis. Bacteria live in the deep areas of the fresh water to decompose dead plants and animals that drift down from the land and water above. Fish adapted to cooler, darker water also live there. Eventually, dead and decaying organisms reach the **benthic zone,** the bottom of a pond or lake, which is inhabited by decomposers, insect larvae, and clams.

Animals that live in lakes and ponds have adaptations that help them obtain what they need to survive. Water beetles use the hairs under their bodies to trap surface air so that they can breathe during their dives for food. Whiskers help catfish sense food as they swim over dark lake bottoms. In regions where lakes partially freeze in winter, amphibians burrow into the littoral mud to avoid freezing temperatures.

Sunlight

LITTORAL ZONE

Decomposers

Phytoplankton and zooplankton

BENTHIC ZONE

174

Transparencies

TT Life Zones in a Lake

MISCONCEPTION ALERT

Freshwater Plankton Many people think that plankton live only in the ocean. Plankton is a broad term that describes microscopic plants—phytoplankton—and microscopic animals—zooplankton. Some zooplankton are actually large enough to be seen with the unaided eye. Many times, zooplankton are larvae of aquatic mollusks or crustaceans.

INCLUSION *Strategies*

• *Learning Disabled*
• *Attention Deficit Disorder*

Have students develop a poster that shows the characteristics of plankton, nekton, and benthos. The poster should include examples of each type of organism and information about where in the ocean each organism is found. Students can provide photographs of each aquatic organism from the Internet.

How Nutrients Affect Lakes **Eutrophication** is an increase in the amount of nutrients in an aquatic ecosystem. A lake that has a large amount of plant growth due to nutrients, as shown in **Figure 4,** is known as a *eutrophic lake*. As the amount of plants and algae grows, the number of bacteria feeding on the decaying organisms also grows. These bacteria use the oxygen dissolved in the lake's waters. Eventually, the reduced amount of oxygen kills oxygen-loving organisms. Lakes naturally become eutrophic over a long period of time. However, eutrophication can be accelerated by runoff. Runoff is precipitation, such as rain, that can carry sewage, fertilizers, or animal wastes from land into bodies of water.

Freshwater Wetlands

Freshwater wetlands are areas of land that are covered with fresh water for at least part of the year. The two main types of freshwater wetlands are marshes and swamps. *Marshes* contain nonwoody plants, such as cattails, while *swamps* are dominated by woody plants, such as trees and shrubs.

Wetlands perform several important environmental functions, as shown in **Table 1.** Wetlands act as filters or sponges because they absorb and remove pollutants from the water that flows through them. Therefore, wetlands improve the water quality of lakes, rivers, and reservoirs downstream. Wetlands also control flooding by absorbing extra water when rivers overflow, which protects farms and urban and residential areas from damage. Many of the freshwater game fish caught in the United States each year use the wetlands for feeding and spawning. In addition, these areas provide a home for native and migratory wildlife, including the blue herons shown in **Figure 5.** Wetland vegetation also traps carbon that would otherwise be released as carbon dioxide, which may be linked to rising atmospheric temperatures. Some wetlands are used to produce many commercially important products, such as cranberries.

Figure 4 ▶ A eutrophic lake, like the one above, contains large amounts of plants as a result of high levels of nutrients.

Figure 5 ▶ Wetlands provide habitat for many plants and animals, including the blue herons shown below.

Table 1 ▼

Environmental Functions of Wetlands
• trapping and filtering sediments, nutrients, and pollutants, which keep these materials from entering lakes, reservoirs, and oceans
• reducing the likelihood of a flood, protecting agriculture, roads, buildings, and human health and safety
• buffering shorelines against erosion
• providing spawning grounds and habitat for commercially important fish and shellfish
• providing habitat for rare, threatened, and endangered plants and animals
• providing recreational areas for activities such as fishing, bird-watching, hiking, canoeing, photography, and painting

175

Freshwater wetlands

Teach, *continued*

Homework ── ADVANCED

The Everglades Nearly 90 years ago, an enthusiastic scientist with the United States Department of Forestry received a small packet of seeds in the mail from a colleague in Australia. The seeds were of the melaleuca tree. The scientist scattered the seeds across the Florida Everglades, hoping the water-loving trees would dry up the "mucky wasteland" to make way for agricultural and residential development. Since then, the melaleuca has spread across thousands of acres of South Florida wetlands. The tree has no natural consumers in South Florida and has driven out many native plants and animals. Scientists are currently trying to eradicate the tree by experimenting with another import from Australia—a melaleuca-eating insect. Ask students to find out more about the problems with melaleuca trees and the eradication program. Students should prepare a brief report on their findings. **LS Verbal**

🧭 Internet Activity ── BASIC

Wetland Plants Have students search the Internet to find five common wetland plants that grow in their area. Then have students locate and print out a picture of each plant. Students can start their search at their state's natural resources or conservation department.

Transparencies

TT Freshwater Wetlands in the United States

Connection to **History**

The Florida Everglades Because of the work of many writers, conservationists, and naturalists, former U.S. President Truman dedicated the Everglades National Park in 1947. The park was established to protect the wildlife and habitat of the Florida Everglades. The Florida Everglades is one of only three sites on Earth declared an International Biosphere Reserve, a World Heritage Site, and a Wetland of International Importance. The other two sites are located in Tunisia and Bulgaria.

Figure 7 ▶ A marsh is a type of wetland that contains nonwoody plants.

Marshes As shown in **Figure 6,** most freshwater wetlands are located in the southeastern United States. The Florida Everglades is the largest freshwater wetland in the United States. Freshwater marshes tend to occur on low, flat lands and have little water movement. In shallow waters, plants such as reeds, rushes, and cattails root themselves in the rich bottom sediments. As shown in **Figure 7,** the leaves of these and other plants stick out above the surface of the water year-round.

The benthic zones of marshes are nutrient rich and contain plants, numerous types of decomposers, and scavengers. Waterfowl, such as grebes and ducks, have flat beaks adapted for sifting through the water for fish and insects. Water birds, such as herons, have spearlike beaks that they use to grasp small fish and to probe for frogs buried in the mud. Marshes also attract many migratory birds from temperate and tropical habitats.

There are several kinds of marshes, each of which is characterized by its salinity. Brackish marshes have slightly salty water, while salt marshes contain saltier water. In each marsh type, organisms are adapted to live within the ecosystem's range of salinity.

176

Student Opportunities

Wetland Nurseries Tampa BayWatch is a nonprofit organization that monitors, restores, and protects the marine and wetland environments in the Tampa area. Over 2,000 students at area high schools help the restoration efforts by constructing, maintaining, and transplanting wetland plants. These nurseries help raise plants that are used in restoration projects. Encourage interested students to find out about wetland restoration efforts in their area. Students could also find out if they can help by establishing wetland nurseries in other areas.

Figure 8 ▶ The American alligator is a common reptile that lives in marshes and swamps.

Swamps Swamps occur on flat, poorly drained land, often near streams. Swamps are dominated by woody shrubs or water-loving trees, depending on the latitude and climate in which the swamps are located. Mangrove swamps occur in warm climates near the ocean, so their water is salty. Freshwater swamps are the ideal habitat for many amphibians, such as frogs and salamanders, because of the continuously moist environment. Swamps also attract birds, such as wood ducks, that nest in hollow trees near or over the water. Reptiles, such as the American alligator in **Figure 8**, are the predators of swamps and will eat almost any organism that crosses their path.

Human Impact on Wetlands Wetlands were previously considered to be wastelands that provide breeding grounds for insects. Therefore, many have been drained, filled, and cleared for farms or residential and commercial development, as shown in **Figure 9**. For example, the Florida Everglades once covered 8 million acres of south Florida but it now covers less than 2 million acres. The important role of wetlands as purifiers of wastewater and in flood prevention is now recognized. Wetlands are vitally important as habitats for wildlife. Law and the federal government protect many wetlands, and most states now prohibit the destruction of certain wetlands.

MATHPRACTICE

Wetland Conversion
From 1982 to 1992, approximately 1.6 million acres of wetlands on nonfederal lands in the United States were converted for other uses. Fifty-seven percent of the wetlands were converted into land for development. Twenty percent of the wetlands were converted into land for agriculture. How many acres of land were converted into land for development? How many acres of land were converted into land for agriculture?

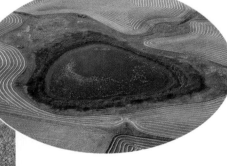

Figure 9 ▶ The wetland above has been drained for agricultural purposes. Wetlands such as the one above typically serve as breeding areas for ducks. The oil rig on the left is located in a marsh off the coast of Louisiana.

177

MATHPRACTICE

Answers
1. 57% × 1,600,000 acres = 912,000 acres
2. 20% × 1,600,000 acres = 320,000 acres

Debate ——————— GENERAL

The Florida Everglades The Florida Everglades is the largest wetlands area in North America. In the 1800s, the first canals were dug to begin draining the Everglades to make way for development. This trend has continued until recent years. Many plans have been devised to save and protect the Everglades. The Comprehensive Everglades Restoration Plan is a 20-year project that will cost an estimated $7.8 billion to complete. Divide the class into two groups. Ask one group to research the advantages of the project and the other group to research the disadvantages of the project. Organize a debate and have the two groups present their findings to a mock hearing of the Florida State Legislature. **LS** Verbal

Activity ——————— BASIC

Wetlands in Your Neighborhood
Ask students to find and observe at least one wetland near their school or home (if students cannot find a wetland to observe, have them perform this activity on an aquatic ecosystem of their choice). Have students write a brief physical description of the wetland and explain why it qualifies as a wetland. Have students answer questions about the wetland such as, "What types of plants and animals live in the wetland? Is the wetland threatened or in danger of being drained or developed?" Have students present their descriptions to the class. **LS** Interpersonal

Homework ——————— ADVANCED

Gone Fishin' Tell students to imagine that they are going fishing in a small freshwater pond located in their home state. Have them do research and then write a short paragraph about the types of fish they are likely to catch and the type of bait they will need. Students may begin their research with the department that oversees natural resources in their state. Suggest that interested students follow up their research with an actual fishing trip. They could then compare the results of the trip with their research findings. Suggest that students include a description of their findings in their **Portfolio.**

Reteaching ——— BASIC

Concept Mapping Have students create a concept map using the following terms: *freshwater wetlands, lakes, littoral zone, freshwater ecosystems, eutrophication, traps and filters pollutants, runoff, benthic zone, buffers shorelines, phytoplankton, zooplankton,* and *decomposers.* Remind students that they must supply words that show clear connections between the terms used in their maps.

Quiz ——— GENERAL

1. How does a lake become eutrophic? (A nutrient-rich lake supports a lot of plants and algae. The lake's bacterial population grows as bacteria break down dead plants and algae and use up oxygen that other organisms need to survive.)

2. What are phytoplankton and what role do they play in an aquatic ecosystem? (Phytoplankton are microscopic plants near the surface of the water that produce most of the food for an aquatic ecosystem.)

Alternative Assessment ——— BASIC

A Fish Tale Have students imagine that they are a fish journeying through three freshwater ecosystems: a pond, a lake, and a river. Ask students to describe what they see, including details about the characteristics of each body of water and the organisms encountered along the way. Encourage students to include their description in their **Portfolio.**

Figure 10 ▶ The water flow of a river slows and the habitat changes as narrow headwaters give way to wide channels downstream.

Rivers

Many rivers originate from snow melt in mountains. At its headwaters, a river is usually cold and full of oxygen and runs swiftly through a shallow riverbed. As a river flows down a mountain, a river may broaden, become warmer, wider, and slower, and decrease in oxygen. **Figure 10** compares the water flow of two sections of two different rivers. A river changes with the land and the climate through which it flows. Runoff, for example, may wash nutrients and sediment from the surrounding land into a river. These materials affect the growth and health of the organisms in the river.

Life in a River Near the churning headwaters, mosses anchor themselves to rocks by using rootlike structures called *rhizoids.* Plankton do not live in the headwaters because the current is too strong for them to float. However, trout and minnows are adapted to the cold, oxygen-rich headwaters. Trout are powerful swimmers and have streamlined bodies that present little resistance to the strong current. Farther downstream, plankton can float in the warmer, calmer waters. Other plants, such as the crowfoot, set roots down in the river's rich sediment. The leaves of some plants, such as the arrowhead, will vary in shape according to the strength of a river's current. Fish such as catfish and carp also live in the calmer waters.

Rivers in Danger Industries use river water in manufacturing processes and as receptacles for waste. For many years, people have used rivers to dispose of their sewage and garbage. These practices have polluted rivers with toxins, which have killed river organisms and made river fish inedible. Today, runoff from the land puts pesticides and other poisons into rivers and coats riverbeds with toxic sediments. Dams also alter the ecosystems in and around a river.

SECTION 1 Review

1. **List** two factors that determine where an organism lives in an aquatic ecosystem.

2. **Compare** the littoral zone of a lake with the benthic zone of a lake.

3. **List** two environmental functions that wetlands provide. How do these functions affect you?

4. **Describe** one threat against river ecosystems.

CRITICAL THINKING

5. **Identifying Relationships** A piece of garbage that is thrown into a stream may end up in a river or an ocean. What effects might one piece of garbage have on an aquatic ecosystem? What effects might 100 pieces of garbage have on an aquatic ecosystem?

6. **Analyzing Processes** Write a short paragraph that explains how fertilizing your yard and applying pesticides can affect the health of a river ecosystem.
WRITING SKILLS

178

Answers to Section Review

1. Answers may vary. Sample answer: temperature and nutrients

2. The littoral zone exists in shallow, sunlit water, while the benthic zone occupies the dark bottom of a lake or pond.

3. Wetlands trap and filter pollutants, which improves water quality downstream. Wetlands also absorb excess water and reduce the likelihood of a flood.

4. Answers may vary. Sample answer: Sewage dumping threatens river ecosystems.

5. Answers may vary. In general, more garbage will have more of a harmful effect on the ecosystem.

6. Sample answer: Rain can wash fertilizers and pesticides off my yard and into a river. Fertilizers can cause an increase in plant growth, and pesticides can kill plants and animals in the river.

SECTION 2
Marine Ecosystems

Marine ecosystems of the world are made up of a wide variety of plant and animal communities. Marine ecosystems are located mainly in coastal areas and in the open ocean. Organisms that live in coastal areas adapt to changes in water level and salinity. Organisms that live in the open ocean adapt to changes in temperature and the amount of sunlight and nutrients available.

Coastal Wetlands

Coastal land areas that are covered by salt water for all or part of the time are known as *coastal wetlands*. Coastal wetlands provide habitat and nesting areas for many fish and wildlife. Coastal wetlands also absorb excess rain, which protects areas from flooding, they filter out pollutants and sediments, and they provide recreational areas for boating, fishing, and hunting.

Estuaries Many coastal wetlands form in estuaries. An **estuary** is an area in which fresh water from a river mixes with salt water from the ocean. As the two bodies of water meet, currents form and cause mineral-rich mud and other nutrients to fall to the bottom. **Figure 11** illustrates how the waters mix in such a way that the estuary becomes a nutrient trap. These nutrients then become available to producers, and in some shallow areas, marsh grass will grow in the mud. Estuaries are very productive ecosystems because they constantly receive fresh nutrients from the river and from the ocean. The surrounding land, such as the mainland or a peninsula, protects estuaries from the harsh force of ocean waves.

Objectives

▶ Explain why an estuary is a very productive ecosystem.
▶ Compare salt marshes and mangrove swamps.
▶ Describe two threats to coral reefs.
▶ Describe two threats to ocean organisms.

Key Terms

estuary
salt marsh
mangrove swamp
barrier island
coral reef

🎵 **internet** connect

www.scilinks.org
Topic: Estuaries
SciLinks code: HE4037

SCi*LINKS* Maintained by the
National Science
Teachers Association

Figure 11 ▶ The mixing of fresh water and salt water at the mouth of a river creates a nutrient-rich estuary.

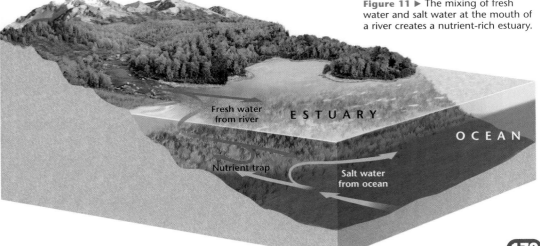

Fresh water from river

ESTUARY

OCEAN

Nutrient trap

Salt water from ocean

179

Chapter Resource File

• Lesson Plan
• Active Reading **BASIC**
• Section Quiz **GENERAL**

Transparencies

TT Bellringer
TT The Formation of Estuaries

SECTION 2

Focus

Overview

Before beginning this section, review with your students the Objectives in the Student Edition. This section describes marine ecosystems, such as estuaries, coral reefs, and open oceans. Students also learn about threats to these ecosystems.

🔊 Bellringer

Provide specimens or photographs of common organisms that live in various marine ecosystems, such as coral, shells, sponges, and starfish. Show them to the class one at a time, and ask students which marine ecosystem each organism is from. Then ask students what part or zone of the marine ecosystem the organism is adapted to. (For example, a piece of coral would be found near the surface in shallow ocean waters.)
LS Visual

Motivate

Identifying
Preconceptions — **GENERAL**

Singing Whales At the beginning of class, have students close their eyes and sit quietly. Play a brief audiotape of a dolphin or whale song. Afterward, ask students, "What is the largest animal on Earth?" (the blue whale) Inform students that blue whales can weigh over 110 metric tons (over 242,000 lb), which is about 15 times the weight of the largest African elephant, and are larger than any dinosaur. Ask students, "Why can't animals of that size live on land?" (They would not be able to support their weight.) "Can you think of other organisms that might form part of the blue whale's marine community?" (Answers may vary. Blue whales live at various depths in the open ocean. They may share their community with other whales, fish, dolphins, and seals.) **LS** Auditory

Chapter 7 • Aquatic Ecosystems   **179**

QuickLAB

Skills Acquired:
- Experimenting
- Interpreting

Answers

1. The salt water will sink below the fresh water and will not mix completely.

Demonstration ── BASIC

Salinity Change Students may be unfamiliar with salinity changes as the tides come in and out in estuaries. To show students how salinity changes, use a tall jar or large graduated cylinder and a hydrometer. Hydrometers may be found at aquarium stores, where they are used to measure salinity in a marine aquarium. Fill the jar or cylinder three-quarters full of tap water, and place the hydrometer in the jar. Record the reading. In a separate container, dissolve as much table salt as possible in hot tap water. Have students watch the hydrometer as you slowly add the salt-saturated water. Ask students to describe how the hydrometer changed as the salt solution was added. **LS Kinesthetic**

QuickLAB

Estuaries

Procedure

1. Place a few drops of **red food coloring** in a **test tube** filled with **water**.
2. In a separate **test tube**, add **salt water** and a few drops of **yellow food coloring**.
3. Gently place some of the fresh water solution on top of the salt water solution.

Analysis

1. How do fresh water and salt water interact in an estuary?

Plants and Animals of Estuaries For a week each spring, horseshoe crabs, shown in **Figure 12**, crawl out of the ocean and onto the beaches of Delaware Bay. In the shallow areas along the shore, the crabs mate and lay billions of eggs. Many migrating shorebirds depend on these eggs for food.

Estuaries support many marine organisms because estuaries receive plenty of light for photosynthesis and plenty of nutrients for plants and animals. Rivers supply nutrients that have been washed from the land, and because the water is shallow, sunlight can reach all the way to the bottom of the estuary. The light and nutrients support large populations of rooted plants as well as plankton. The plankton in turn provide food for larger animals, such as fish. Dolphins, manatees, seals, and other mammals often feed on fish and plants in estuaries. Oysters, barnacles, and clams live anchored to marsh grass or rocks and feed by filtering plankton out of the water. Organisms that live in estuaries are able to tolerate variations in salinity because the salt content of the water varies as fresh water and salt water mix when tides go in and out.

CASE STUDY

Restoration of the Chesapeake Bay

The Chesapeake Bay is the largest estuary in the United States. The bay produces large amounts of seafood each year, supports many species of wildlife, and provides recreation for millions of people.

However, the ecosystems of the bay are threatened by several environmental problems. For example, pollution builds up because the tide flushes pollutants out of the bay very slowly. Pollution builds up because only a very narrow opening joins the bay and the ocean. By 1980, the Chesapeake Bay was severely polluted with toxic industrial chemicals. Pesticides as well as excess nutrients ran into the bay from housing developments, farms, and wastewater (including sewage). Marsh grasses and plankton were dying, and fish, oysters, and crabs were disappearing. Birds of prey, such as bald eagles, had almost

vanished. Therefore, fishers, environmentalists, and residents became alarmed and launched campaigns to save the bay.

Restoring Chesapeake Bay habitats and water quality is not easy. Maryland and Virginia, the main bordering states of the bay, have different environmental laws. Also, the bay's watershed covers parts of four other states. Interested groups would have to work together if they were to restore the bay. The Chesapeake Bay Program was set up as a partnership between the Environmental Protection Agency, the District of Columbia, Maryland, Pennsylvania, Virginia, and citizen advisory groups. Goals included reducing chemical pollution, removing dams that prevented fish from migrating, and reforesting river banks to reduce soil erosion.

▶ The Chesapeake Bay forms where the Potomac, Rappahannock, and other rivers meet the Atlantic Ocean.

Remarkable progress has been made in the last 20 years. About half of the wastewater flowing into

180

GEOGRAPHY ──
● CONNECTION ── GENERAL

Coastal Settlements Throughout history, a disproportionate amount of human settlement has occurred in coastal areas. There are several reasons. First, having access to a body of water facilitates travel and trade with other areas. In addition, the fish and other organisms in oceans represent a valuable and readily available resource. Also, the

climate in coastal areas is usually moderated by the nearby water body. Historically, the ocean and its organisms have figured prominently in human culture. For instance, the ocean is a powerful symbol in many religions, and cowrie shells were perhaps the first form of established currency.

Estuaries provide protected harbors, access to the ocean, and connection to a river. As a result, many of the world's major ports are built on estuaries. Of the 10 largest urban areas in the world, 6 were built on estuaries. These 6 cities are Tokyo, New York, Shanghai, Buenos Aires, Rio de Janeiro, and Bombay.

Threats to Estuaries Estuaries that exist in populated areas were often used as places to dump waste. Estuaries that are filled with waste can be developed and used as building sites. This practice occurred extensively in California, which now has plans to restore some of its estuary wetlands. The pollutants that damage estuaries are the same ones that pollute lakes, rivers, and the oceans: sewage, industrial waste containing toxic chemicals, and agricultural runoff of soil containing pesticides and fertilizers. Most of these pollutants break down over time, but estuaries cannot cope with the amounts produced by dense human populations.

Figure 12 ▶ Horseshoe crabs go to the Delaware Bay, an estuary between New Jersey and Delaware, to lay their eggs.

▶ This great egret lives in one of the estuaries that borders the Chesapeake Bay.

the bay is now biologically treated to remove pollutants and excess nutrients. Bald eagles are back, and industry has reduced the chemical pollutants released into the bay by nearly 70 percent. Planting trees has restored forested buffers to about 60 percent of the river banks, and populations of fish, such as striped bass, are increasing.

However, the number of people in the bay area is increasing and the number of miles these people drive

each year has increased even faster. In the last 30 years, miles traveled by vehicles increased four times as fast as the population. This has led to runoff from streets and lawns and pollution from vehicle exhaust, all of which harm the bay. The oyster harvest has decreased and the forested part of the bay's watershed is still decreasing.

You can help save your local watershed in the following ways: by reducing the number of miles you

drive, trying to conserve electricity and water, planting native vegetation, using only a small amount of fertilizer or water on your lawn or garden, and properly disposing of hazardous wastes such as motor oil, antifreeze, and cleaning fluids. You can help by picking up trash that others leave behind. You can also join a citizens group to help preserve estuaries.

CRITICAL THINKING

1. Predicting Consequences If the Chesapeake Bay Program had never been founded, what might have happened to the Chesapeake Bay? Explain how one organism may have been affected.

2. Identifying Relationships How may the use of less fertilizer on plants and lawns help the Chesapeake Bay and other estuaries?

181

Internet Activity ——— **ADVANCED**

Tides and Estuaries Have students select an estuary in the United States. Ask students to use the Internet to find the tide charts for a location on the estuary and a nearby spot on the coast. Tide charts may be found at the Web sites of newspapers for cities near the estuary. Ask students to write a brief description of how the tide in the estuary differs from the tide at a nearby point on the coast. Ask students to compare the tide levels on the coast and in the estuary. How do the tides affect the water in the estuary? **LS Intrapersonal**

READING SKILL BUILDER ——— GENERAL

Brainstorming After reading the subsection on plants and animals of estuaries, have students brainstorm the adaptations that plants and animals would need to live in estuaries. You may need to prompt students into thinking about temperature changes, seasonal changes, and migratory predators. **LS Verbal**

Activity ——— **BASIC**

Save the Chesapeake Bay Ask students to make posters to increase public awareness about the problems in the Chesapeake Bay. Tell students to select one aspect of the problems in the bay and have them develop a poster that explains the issue and offers suggestions for solutions. Encourage students to illustrate their posters. Display the posters around the classroom or in the hallway. **LS Interpersonal**

CASE STUDY

The Chesapeake Bay The Chesapeake Bay is the largest estuary in the United States. It is 320 km long and varies from 6.4 km wide at its narrowest point to 48 km wide at its widest point. The watershed for the bay covers parts of six states and more than 164,000 km². The watershed is five times larger than Maryland and one-fourth the size of Texas. In 2000, approximately 16 million people lived in the watershed. More than 2,700 species of plants and animals live in the Chesapeake Bay. Today, the biggest threat to the bay is nutrients in runoff from crops and lawns. Tell students that cleaning up the bay requires changes in the entire watershed and involves not only the federal government but also governments from six states.

Answers to Critical Thinking

1. If the Chesapeake Bay Program was not founded, the Chesapeake Bay may not have been protected or restored. Without this protection, pollution could have increased and could have killed fish living in the bay. Fewer fish could have affected dolphins, which depend on the fish for food.

2. Using less fertilizer on plants and lawns could decrease the amount of fertilizer in runoff and could reduce nutrient pollution in the Chesapeake Bay.

Debate ──────── GENERAL

Disappearing Barrier Islands
Divide the class into two groups. Assign one group to be for enhancement projects that rebuild and protect barrier islands and the other group to be against such programs. Have the groups research the issues and then debate the different viewpoints. **LS Verbal**

Demonstration ──────── BASIC

Barrier Islands Place a large, flat, clear glass pan on an overhead projector or on a tabletop. Pour about 2 cm of water into the pan. Tell students that the water represents the ocean and that one edge of the pan represents the coastline. To make waves, tap with a pen on the surface of the water opposite the coastline. Students should be able to see the waves. Ask students, "What happens when waves reach the coastline?" (The waves slap against the side of the pan.) Next, pour a line of sand about 5 cm from the edge called the shoreline. The sands should just barely reach the surface. Repeat the wave-making process. Ask students, "What happens to the waves now that there is a barrier island?" (The waves usually stop at the barrier island.) "What effect would the barrier islands have on coastal erosion?" (The barrier islands would reduce coastal erosion.) **LS Kinesthetic**

Figure 13 ▶ Mangrove swamps are found along warm, tropical coasts and are dominated by salt-tolerant mangrove trees.

Ecofact

Mangrove Swamps Mangroves cover 180 billion square meters of tropical coastlines around the world. The largest single mangrove swamp is 5.7 billion square meters, located in the Sundarbans of Bangladesh. This single mangrove swamp provides habitat for the Bengal tiger and helps supply approximately 300,000 people with food, fuel, building materials, and medicines.

Figure 14 ▶ This barrier island is located off the coast of Long Island, New York. Barrier islands are separated from the mainland and help protect the shore of the mainland from erosion.

182

Salt Marshes Marsh grasses dominate much of the shoreline of the Gulf of Mexico and the Atlantic Coast of the United States. These **salt marshes** develop in estuaries where rivers deposit their load of mineral-rich mud. Here, thousands of acres of salt marsh support a community of clams, fish, and aquatic birds. The marsh also acts as a nursery in which many species of shrimps, crabs, and fishes find protection when they are small. As they grow to maturity and migrate to the sea, they are eaten by larger fish or caught by commercial fisheries. Salt marshes, like other wetlands, absorb pollutants and protect inland areas.

Mangrove Swamps Swamps located along coastal areas of tropical and subtropical zones are called **mangrove swamps.** Plants called mangrove trees dominate mangrove swamps. Mangrove trees, such as those shown in **Figure 13,** grow partly submerged in the warm, shallow, and protected salt water of mangrove swamps. The swamps help protect the coastline from erosion and reduce the damage from storms. They provide the breeding and feeding grounds for about 2,000 animal species. Like salt marshes, mangrove swamps have been filled with waste and destroyed in many parts of the world.

Rocky and Sandy Shores Rocky shores have many more plants and animals than sandy shores do. The rocks provide anchorage for seaweed and the many animals that live on them, such as sea anemones, mussels, and sponges. Sandy shores dry out when the tide goes out, and many of the tiny organisms that live between the sand grains eat the plankton that are stranded on the sand. These organisms are the main food for a number of shorebirds. **Barrier islands,** such as the one in **Figure 14,** typically run parallel to the shore. These long, thin islands help protect the mainland and the coastal wetlands.

HISTORY
──── **CONNECTION** ──── GENERAL

North Carolina's Outer Banks Some of the best-known barrier islands in the United States are the Outer Banks of North Carolina. The islands were the site of England's "lost colony" and were used by the pirate Blackbeard. Nearby dunes hosted the Wright Brother's historic flight. Also, the scenic Outer Banks contains the country's first national seashore, Cape Hatteras.

READING SKILL BUILDER ──────── BASIC

Comparing and Contrasting Divide the class into small groups. Have each group read about salt marshes and mangrove swamps. Then, ask each group to compare salt marshes with freshwater marshes and mangrove swamps with freshwater swamps. Ask each group to develop a chart describing the similarities and differences.

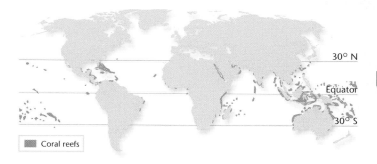

Coral reefs

Figure 15 ▶ Coral reefs are found in warm, shallow waters, where there is enough light for photosynthesis. Coral reefs support a great diversity of species.

Graphic
Organizer **Cause-and-Effect Map**

Create the **Graphic Organizer** entitled "Cause-and-Effect Map" described in the Appendix. Label the effect with "Disappearing Coral Reefs." Then, fill in the map with causes of disappearing coral reefs and details about the causes and effects.

Coral Reefs

Coral reefs are limestone ridges built by tiny coral animals called *coral polyps*. Coral polyps secrete skeletons of limestone (calcium carbonate), which slowly accumulate and form coral reefs. Thousands of species of plants and animals live in the cracks and crevices of coral reefs, which makes coral reefs among the most diverse ecosystems on Earth.

Corals live only in clear and warm salt water where there is enough light for photosynthesis, so coral reefs are found in shallow, tropical seas, as shown in **Figure 15**. Only the outer layer of a reef contains living corals, which build their rock homes with the help of photosynthetic algae. Corals, such as those shown in **Figure 16**, are predators that never chase their prey. They use stinging tentacles to capture small animals, such as zooplankton, that float or swim close to the reef. Because of their convoluted shape, reefs provide habitats for a magnificent variety of tropical fish, and for snails, clams, and sponges.

Disappearing Coral Reefs Coral reefs are productive but fragile ecosystems. An estimated 27 percent of the coral reefs in the world are in danger of destruction from human activities. If the water surrounding a reef is too hot or too cold or if fresh water drains into the water surrounding a reef, the corals may die. If the water is too muddy, polluted, or too high in nutrients, the algae that live within the corals will either die or grow out of control. If the algae grows out of control, it may kill the corals.

Oil spills, sewage, pesticide, and silt runoff have been linked to coral-reef destruction. Furthermore, overfishing can devastate fish populations and upset the balance of a reef's ecosystem. Because coral reefs grow slowly, a reef may not be able to repair itself after chunks of coral are destroyed by careless divers, fisheries, shipwrecks, ships dropping anchor, or people breaking off pieces of it for decorative items or building materials.

Figure 16 ▶ Coral reefs (bottom) are limestone ridges built by tiny coral animals. Coral animals have coral tentacles (top) that emerge from protective structures to capture food.

183

Group Activity —— ADVANCED

Marine Ecosystems Divide the class into three groups. Assign each group a portion of this section to teach to the class. Group one will teach about the estuary ecosystem. Group two will teach about coral reefs. Group three will teach about the ocean. Tell each group they must create a lesson plan for their topic that includes objectives, the actual lesson, a handout (daily assignment), and a visual aid. Give students two days to prepare. Then have them present their lesson to the class. Suggest that students include their lesson plan in their **Portfolio.** **Interpersonal**

Transparencies

TT Life Zones in the Ocean

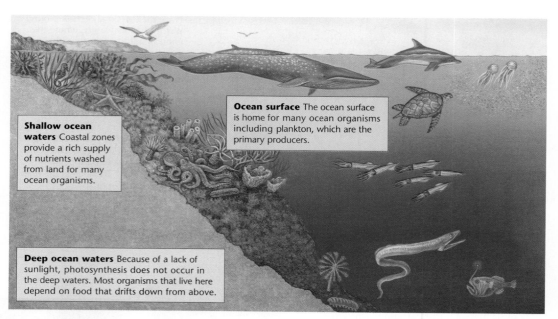

Shallow ocean waters Coastal zones provide a rich supply of nutrients washed from land for many ocean organisms.

Ocean surface The ocean surface is home for many ocean organisms including plankton, which are the primary producers.

Deep ocean waters Because of a lack of sunlight, photosynthesis does not occur in the deep waters. Most organisms that live here depend on food that drifts down from above.

Figure 17 ▶ The amount of sunlight available determines which organisms can live in each layer of the ocean.

FIELD ACTIVITY

Make a Miniature Aquatic Ecosystem Make your own aquarium by collecting organisms from an aquatic ecosystem near your home or school. Be sure to collect some water from the aquatic ecosystem. Bring your collection back to school and set up an aquarium. If necessary, research the Internet to find out the special care that your ecosystem may require. Examine a few drops of your collected water under the microscope. Be sure to look for algae or other forms of life. Record and draw your observations in your *EcoLog*. Observe and record the changes you see in your aquarium over the next 3 weeks. What conditions are needed to keep your miniature ecosystem healthy?

184

Oceans

Because water absorbs light, sunlight that is usable by plants for photosynthesis penetrates only about 100 m (330 ft) into the ocean before all of the sunlight is absorbed. As a result, much of the ocean's life is concentrated in the shallow, coastal waters. Here, sunlight penetrates to the bottom and rivers wash nutrients from the land. Seaweed and algae grow anchored to rocks, and phytoplankton drift on the surface. Invertebrates and fish that feed on these plants are also concentrated near the shore.

Plants and Animals of Oceans In the open ocean, phytoplankton grow only in areas where there is enough light and nutrients. As a result, the open ocean is one of the least productive of all ecosystems. Phytoplankton have buoyancy devices, such as whip-like flagella, that prevent them from sinking into deep water, which is too dark for photosynthesis. The sea's smallest herbivores are the zooplankton, which live near the surface with the phytoplankton they eat. The zooplankton include jellyfish, tiny shrimp, and the larvae of fish and bottom-dwelling animals, such as oysters and lobsters. Fish feed on the plankton as do marine mammals such as whales.

The depths of the ocean are perpetually dark, so most food at the ocean floor consists of dead organisms that fall from the surface. Decomposers, filter feeders, and the organisms that eat them live in the deep areas of the ocean. **Figure 17** illustrates the types of organisms that may be found in the layers of the ocean at various depths, depending on available sunlight.

Career

Oceanographer Oceanographers study the chemical and physical properties of seawater and the way ocean water moves (e.g., currents, tides, and waves). Oceanographers also study the sea floor and how it is changing because of plate tectonic forces. Oceanographers use many tools to study the ocean, such as satellites, radio buoys, ships, and submersibles. Oceanographers take classes in marine biology, geology, ecology, chemistry, physics, and geography.

Homework —— ADVANCED

Oceans of the World Have students research the major oceans of the world and create a magazine-style article that compares the oceans. Ask students to include descriptive examples of living organisms unique to each ocean. Students could include their article in their **Portfolio.** **Verbal**

Threats to the Oceans Although oceans are huge, they are steadily becoming more polluted. Most ocean pollution arises from activities on land. For example, plant nutrients washing off the land as runoff from fertilized fields may cause algal blooms, some of which are poisonous. Industrial waste and sewage discharged into rivers is the biggest source of coastal pollution in the United States.

Overfishing and certain fishing methods are also destroying some fish populations. Immense trawl nets can entangle organisms that are larger than the net holes. Marine mammals such as dolphins, which must breathe air, can drown in the nets. Although it is against the law, some ships discard fishing lines into the ocean, where they can strangle and kill fish and seals. The sea lion in **Figure 18** was strangled by a net off the coast of California.

Figure 18 ▶ This sea lion was strangled by a fishing net.

Arctic and Antarctic Ecosystems The arctic ecosystems at the North and South Poles depend on marine ecosystems because nearly all the food comes from the ocean.

The Arctic Ocean is rich in nutrients from the surrounding landmasses, and it supports large populations of plankton which feed a rich diversity of fish in the open water and under the ice. The fish are food for ocean birds, whales, and seals. Beluga whales, shown in **Figure 19**, feed on nearly 100 different arctic organisms. Fish and seals also provide food for polar bears and people on land.

The Antarctic is the only continent never colonized by humans. It is governed by an international commission and is used mainly for research. Even during the summer, only a few plants grow at the rocky edges of the continent. As in the Arctic, plankton form the basis of the Antarctic food web. The plankton nourish large numbers of fish, whales, and birds such as penguins, which cannot fly because their wings have evolved for swimming.

Figure 19 ▶ Beluga whales inhabit the Arctic Ocean.

SECTION 2 Review

1. **Explain** why estuaries are very productive ecosystems. Why are estuaries vulnerable to the effects of pollution?

2. **Compare** salt marshes with mangrove swamps.

3. **Describe** two factors that can damage coral reefs.

4. **List** two ways in which animals of the oceans are threatened.

CRITICAL THINKING

5. **Predicting Consequences** Suppose the sea level were suddenly to rise by 100 m. What would happen to the world's coral reefs? Explain.

6. **Analyzing Processes** Read the description of estuaries in this section and explain why cities are often built on estuaries. How would building a city on an estuary affect the plants and animals living in an estuary? **READING SKILLS**

185

Answers to Section Review

1. Estuaries are very productive because they constantly receive nutrients from rivers and from the ocean. Estuaries are vulnerable to pollution because ocean, lake, and river pollutants enter estuaries and because dense human settlements surround most estuaries.

2. Salt marshes are dominated by marsh grasses and develop in estuaries, while mangrove swamps are dominated by mangrove trees and develop in tropical and subtropical areas.

3. Water that is too hot or cold can damage reefs by killing corals. Muddy water can kill the algae that live within corals.

4. Overfishing is destroying some fish populations, and trawl nets can entangle and drown marine mammals, such as sea lions.

5. A 100 m rise in sea level would kill off most of the world's coral reef ecosystems, because the algae in corals need water shallow enough to allow sufficient light through for photosynthesis.

6. Cities are often built on estuaries because estuaries provide protected harbors, access to the ocean, and connection to a river. Building a city on an estuary would reduce populations of many animals through fishing and pollution.

Close

Reteaching ——— BASIC

Marine Ecosystem Types As a class, review the types of marine ecosystems covered in this section. Have students get into small groups to quiz each other on the main characteristics of each marine ecosystem. Then, have students draw diagrams of each marine ecosystem and label each component and/or layer. **English Language Learners**

Quiz ——— GENERAL

1. Why are estuaries a mineral-rich environment? (Estuary waters mix to trap mineral-rich mud washed from rivers.)

2. How are salt marshes and mangrove swamps different? (Marsh grasses dominate salt marshes, while mangrove trees dominate mangrove swamps. Salt marshes are most commonly found on the Atlantic coast and along the Gulf of Mexico. Mangrove swamps are found in tropical and subtropical zones.)

3. What are the primary producers in the open ocean? (plankton)

4. Name two threats to the ocean. (pollution and overfishing)

Alternative Assessment ——— GENERAL

Underwater Tour Have students write and perform a skit in which 8 to 10 students operate a special submarine that takes a small group of guests (the rest of the class) on an underwater tour through the different depths of the ocean, from the surface to the benthic zone. Then have each student in the crew describe characteristics of the ocean at each level and provide facts about the organisms encountered. Have the guests devise questions for the crew members.

Alternative Assessment — GENERAL

Nutrient Flow Have students make a poster that describes the flow of nutrients from the top of a mountain to the open ocean. The poster should illustrate how nutrients pass through all the different ecosystems on their way from the mountaintop to the open ocean. Encourage students to find photographs or make drawings of each of the ecosystems that the nutrients pass through. **LS** Visual

Researching Estuaries Have students research a major estuary (besides the Chesapeake Bay) and create a report highlighting its historical significance, the seafood harvested from it, its current environmental health, and efforts to preserve or restore the ecosystem.

Chapter Resource File

- **Chapter Test** GENERAL
- **Chapter Test** ADVANCED
- **Concept Review** GENERAL
- **Critical Thinking** ADVANCED
- **Test Item Listing**
- **Field Activity** GENERAL
- **Observation Lab** BASIC
- **CBL™ Probeware Lab** ADVANCED
- **Consumer Lab** GENERAL
- **Long-Term Project** GENERAL

1 Freshwater Ecosystems

Key Terms

wetland, 173
plankton, 173
nekton, 173
benthos, 173
littoral zone, 174
benthic zone, 174
eutrophication, 175

Main Ideas

▶ Aquatic ecosystems can be classified as freshwater ecosystems or marine ecosystems. The plants and animals in aquatic ecosystems are adapted to specific environmental conditions.

▶ Freshwater ecosystems include lakes, ponds, freshwater wetlands, rivers, and streams. The types of freshwater ecosystems are classified by the depth of the water, the speed of the water flow, and the availability of minerals, sunlight, and oxygen.

▶ Freshwater wetlands serve many functions within ecosystems. They trap and filter sediments and pollutants; reduce the likelihood of a flood; and buffer shorelines against erosion.

2 Marine Ecosystems

estuary, 179
salt marsh, 182
mangrove swamp, 182
barrier island, 182
coral reef, 183

▶ Marine ecosystems are identified by the presence of salt water and include coastal wetlands, coral reefs, oceans, and polar ecosystems.

▶ Estuaries are among the most productive of ecosystems because they constantly receive fresh nutrients from a river and from an ocean. Estuaries provide habitat for a multitude of plants and animals.

▶ Coral reefs are susceptible to destruction because they must remain at tropical temperatures and they must receive a large amount of sunlight. Coral reefs provide habitat for approximately one-fourth of all marine organisms.

▶ Almost every person has an impact on aquatic ecosystems. Through understanding how we affect aquatic ecosystems, we can reduce the negative effects we have on them.

CHAPTER 7 Review

Using Key Terms

Use each of the following terms in a separate sentence.

1. *wetland*
2. *mangrove swamp*
3. *estuary*
4. *eutrophication*
5. *benthos*

For each pair of terms, explain how the meanings of the terms differ.

6. *littoral zone* and *benthic zone*
7. *plankton* and *nekton*
8. *salt marsh* and *barrier island*
9. *wetland* and *coral reef*

✔ **STUDY TIP**

Graph Skills Taking the following steps when reading a graph will help you correctly interpret the information. Be sure to read the title so that you understand what the graph represents. If the graph has axes, read the titles of both the *x*- and the *y*-axis. Examine the range of values on both the *x*- and the *y*-axis. Finally, examine the data on the graph, reading them from left to right, and put into words what you think the graph represents.

Understanding Key Ideas

10. Wetlands are important to fisheries in the United States because
    a. wetlands are the easiest place to catch fish.
    b. wetlands are the breeding grounds for insects that are eaten by fish.
    c. wetlands provide the most desirable species of fishes.
    d. many of the fish caught each year use wetlands for feeding and spawning.

11. Animals that live in estuaries
    a. tend to produce few offspring.
    b. are usually found in unpolluted environments.
    c. must be adapted to varying levels of salinity.
    d. are adapted to cold-water conditions.

12. Bacteria can kill organisms in eutrophic lakes by
    a. feeding on decaying plants and animals.
    b. reducing oxygen dissolved in the water.
    c. Both (a) and (b)
    d. Neither (a) nor (b)

13. Arctic ecosystems are considered marine ecosystems because
    a. arctic ecosystems contain an enormous amount of frozen sea water.
    b. arctic ecosystems are inhabited by few organisms.
    c. sunlight is limited.
    d. phytoplankton form the basis of arctic food webs.

14. Which of the following statements does *not* describe a function of wetlands?
    a. Wetlands buffer shorelines against erosion.
    b. Wetlands provide spawning grounds for commercially important fish and shellfish.
    c. Wetlands filter pollutants.
    d. Wetlands make good hazardous waste dumpsites.

15. Tiny animals, called *coral polyps*, that excrete limestone create
    a. barrier islands.
    b. coral reefs.
    c. swamps.
    d. salt marshes.

16. Mangrove trees grow
    a. along riverbanks.
    b. in freshwater wetlands.
    c. in tropical areas and in subtropical areas.
    d. in the benthic zones of lakes.

17. The Florida Everglades
    a. is the largest freshwater marsh in the United States.
    b. protects threatened and endangered wildlife.
    c. provides habitat for migratory birds.
    d. All of the above

18. Which of the following actions is an example of how humans affect wetlands?
    a. draining a wetland to create farmland
    b. clearing a wetland to build a housing development
    c. using a wetland as a landfill
    d. all of the above

187

Assignment Guide	
Section	Questions
1	1, 4–7, 9, 10, 12, 14, 17, 18, 20–22, 27, 28, 31, 33–35
2	2, 3, 8, 9, 11, 13, 15, 16, 18, 19, 21, 23–27, 29, 30, 32, 33, 36, 37

ANSWERS

Using Key Terms

1. Sample answer: A wetland is an area where land is periodically underwater.
2. Sample answer: Mangrove trees grow in mangrove swamps.
3. Sample answer: An estuary is the area where a river meets the ocean.
4. Sample answer: Excess nutrients in a lake can cause an explosion of plant algae and bacteria called eutrophication.
5. Sample answer: Mussels, clams, and barnacles are bottom-dwelling animals called benthos.
6. In a pond, the littoral zone is the shallow margin, while the benthic zone is the deep, open-water region.
7. Plankton are tiny organisms that float near the water's surface. Nekton are free-swimming organisms, such as fish and turtles.
8. A salt marsh is an ecosystem that develops in an estuary and is dominated by marsh grasses. A barrier island is a long, thin island that protects the shore from erosion.
9. A wetland is an area of land covered by water for at least part of the year, while coral reefs are limestone ridges built by corals.

Understanding Key Ideas

10. d
11. c
12. c
13. d
14. d
15. b
16. c
17. d
18. d

CHAPTER **7** Review

Short Answer

19. Answers may vary. Estuaries can be described as the "best of both worlds" because they lie at the interface of rivers and oceans and thus receive nutrients from both.

20. Answers may vary. Plankton are very small plants or animals that live suspended in the water. Nekton are free-swimming organisms, such as fish. Benthos are bottom-dwelling organisms.

21. Answers may vary. Wetlands trap and filter sediments and pollutants, absorb potential floodwater, and buffer the shorelines against erosion.

22. In a eutrophic lake, high nutrient levels foster plant and algae growth, which supports bacteria that feed on decaying plants and algae. Rising bacterial populations use up the oxygen in the water, which kills oxygen-loving organisms.

23. Mangrove swamps are dominated by mangrove trees.

Interpreting Graphics

24. 255,000 km² × 0.27 = 68,931 km²

25. Coral reefs at a low risk of being destroyed would be clustered in remote areas away from human populations.

26. Coral reefs at a high risk of being destroyed would be found in heavily populated or polluted areas.

Short Answer

19. How does the phrase "best of both worlds" relate to an estuary?

20. Explain the difference between the types of organisms that make up these classes: plankton, nekton, and benthos.

21. List three functions of wetlands.

22. Describe what happens when a lake is considered to be eutrophic.

23. What type of vegetation dominates mangrove swamps?

Interpreting Graphics

The pie graph below shows the percentage of coral reefs at risk in the world. Use the pie graph to answer questions 24–26.

24. If there is a total of 255,300 km² of coral reefs in the world, how many square kilometers of coral reefs are at a high risk of being destroyed?

25. Where would you expect to find coral reefs that are at a low risk of being destroyed?

26. Where would you expect to find coral reefs that are at a high risk of being destroyed?

The World's Coral Reefs at Risk

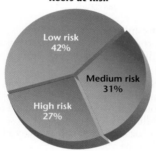

Low risk 42%

Medium risk 31%

High risk 27%

Source: World Resources Institute.

Concept Mapping

27. Use the following terms to create a concept map: *lakes, estuaries, aquatic ecosystems, coral reefs, freshwater wetlands, freshwater ecosystems, rivers, oceans, marshes, marine ecosystems, swamps, coastal ecosystems,* and *mangrove swamps.*

Critical Thinking

28. **Analyzing Relationships** Write a short paragraph that explains the relationship between the speed of a river and the oxygen content of a river. **WRITING SKILLS**

29. **Determining Cause and Effect** Explain what may happen if the use of fertilizer on farms and lawns around an estuary is not controlled.

30. **Making Comparisons** Read the paragraph under the heading "Threats to Estuaries" in this chapter. How do these threats compare to those described under the heading, "Threats to the Oceans?" **READING SKILLS**

31. **Analyzing Relationships** Explain why planting trees along a riverbank might benefit a river ecosystem.

Cross-Disciplinary Connection

32. **Demography** Six out of 10 of the largest urban areas were built on estuaries. Three of these cities are Tokyo, New York, and Rio de Janeiro. Research the populations of each of these cities, and predict what may happen if population numbers continue to increase.

Portfolio Project

33. **Research a Local Aquatic Ecosystem** Observe an aquatic ecosystem near your school or home. This ecosystem can be as simple as a pond or stream or as complex as a lake or estuary. Observe the types of plants and animals in the aquatic ecosystem. Record any interactions among these organisms that you observe. When you have recorded all of your data and observations, write a one-page report on the aquatic ecosystem.

Concept Mapping

27. Answers to the concept mapping questions are on pp. 667–672.

Critical Thinking

28. Answers may vary. In general, fast flowing rivers will have a high oxygen content and slow flowing rivers will have a low oxygen content.

29. Runoff from the fertilized areas will increase the amount of nutrients in the estuary, which will create eutrophic conditions that can "choke" the estuary.

30. Answers may vary. Estuaries located in populated areas are threatened by development. Many cities are built on estuaries because estuaries provide access to a river and to an ocean. Estuaries are also threatened by pollution. Oceans are threatened by pollution and overfishing. Certain fishing methods such as discarded fishing lines can threaten ocean organisms.

31. Planting trees along a riverbank will decrease erosion and flooding, and the trees will absorb some of the nutrients from the water.

MATH SKILLS

Use the graph below to answer questions 34–35.

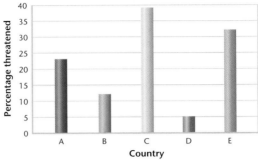

Threatened Freshwater Fish Species

34. **Analyzing Data** The graph below illustrates the percentage of freshwater fish species that are threatened in specific countries. What percentage of freshwater fish species are threatened in country B? in country D?

35. **Evaluating Data** If the number of freshwater fish species in country C totals 599 different species, how many of these species are threatened?

WRITING SKILLS

36. **Communicating Main Ideas** What effect does overfishing have on estuaries? What effect does overfishing have on oceans?

37. **Writing from Research** Research endangered marine mammals of ocean and polar ecosystems. Write a one-page report on the factors that have caused these mammals to become endangered.

 STANDARDIZED TEST PREP

For extra practice with questions formatted to represent the standardized test you may be asked to take at the end of your school year, turn to the sample test for this chapter in the Appendix.

 READING SKILLS

Read the passage below, and then answer the questions that follow.

In the United States during the last 200 years, over 99 percent of native prairies have been replaced with farmland or urban development and most of the old-growth forests have been cut. Loss of so many of these habitats has resulted in losses of biodiversity.

A new discipline, called *conservation biology*, seeks to identify and maintain natural areas. In areas where human influence is greater, such as agricultural areas, former strip mines, and drained wetlands, biologists may have to reverse major changes and replace missing ecosystem components. For example, returning a strip-mined area to grassland may require contouring the land surface, introducing bacteria to the soil, planting grass and shrub seedlings, and even using periodic fires to manage the growth of vegetation. Restoring an area to its natural state is called *restoration ecology*.

1. Which of the following phrases describes a likely task of a restoration ecologist?
   a. raising funds needed to create a national park
   b. returning missing ecosystem components to a drained wetland
   c. educating citizens about the need to protect a local habitat
   d. both (a) and (b)

2. According to the passage, which of the following statements is true?
   a. Former strip mines tend to have a high level of biodiversity.
   b. A conservation biologist would most likely oppose the development of areas around the Grand Canyon.
   c. Periodic fires in some ecosystems do not help manage excess growth of vegetation.
   d. Most prairie ecosystems located in the United States have been preserved.

189

Cross-Disciplinary Connection

32. Answers may vary. As population increases, pollution increases, which can have a negative effect on the estuaries.

Portfolio Project

33. Answers may vary depending on the type of aquatic ecosystem that is studied and the time of year.

Math Skills

34. about 12%; 5%
35. 599×0.39 = about 234 species

Writing Skills

36. Answers may vary. Overfishing removes fish populations from some estuaries. Marine mammals such as dolphins and seals feed on fish, so overfishing of estuaries drives these animals to look elsewhere for food and probably reduces their populations, too. Overfishing is destroying some fish populations in oceans as well. Destroying fish populations affects entire ecosystems because different fish eat and are eaten by various other organisms at different stages in their lives.

37. Answers may vary.

Reading Skills

1. b
2. b

CHAPTER 7
Skills Practice Lab: OBSERVATION

EUTROPHICATION: TOO MUCH OF A GOOD THING?

Teacher's Notes

Time Required
one 45-minute class period to perform procedure; 3 weeks to complete observations

Lab Ratings

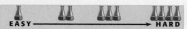

EASY ————————▶ HARD

TEACHER PREPARATION 🧪🧪
STUDENT SETUP 🧪🧪
CONCEPT LEVEL 🧪🧪🧪
CLEANUP 🧪🧪

Skills Acquired
- Predicting
- Designing Experiments
- Experimenting
- Collecting Data
- Interpreting
- Organizing and Analyzing Data

The Scientific Method
In this lab, students will:
- Make Observations
- Ask Questions
- Test the Hypothesis
- Analyze the Results
- Draw Conclusions

Materials
The materials listed are enough for a group of two students.

Objectives

▶ **USING SCIENTIFIC METHODS** **Observe** the effects of nitrates and phosphates on an aquatic ecosystem.
▶ **Compare** the growth of organisms in different levels of nutrients.
▶ **Predict** possible effects nitrates and phosphates would have on an aquatic ecosystem in your area.

Materials

- distilled water
- eyedropper
- fertilizer, household use
- fluorescent lamp
- graduated cylinder
- guide to pond life identification
- jars, 1 qt (3)
- microscope
- microscope slides with coverslips
- plastic wrap
- pond water that contains viable organisms
- stirring rod
- wax pencil

▶ **Step 5** Observe a drop of pond water under the microscope.

190

Eutrophication: Too Much of a Good Thing?

Plants depend on nutrients such as phosphates and nitrates to survive. However, when people release large amounts of these nutrients into rivers and lakes, *artificial eutrophication* can occur. In artificial eutrophication, nutrients cause algae and plant life to grow rapidly and then die off and decay. When microorganisms decompose the algae and plant matter, they use up oxygen in the water, which causes the death of fish and other animals that depend on oxygen for survival. Eutrophication is commonly caused by phosphates, which are often found in detergents, and by nitrates, which are found in animal wastes and fertilizers. In this lab, you will observe artificial eutrophication in an aquatic ecosystem.

Procedure

1. Working with your team, use a wax pencil to label one jar "Control," a second jar "Fertilizer," and a third jar "Excess fertilizer."

2. Put 750 mL of distilled water in each of the three jars. Read the label on the fertilizer container to determine the recommended dilution of fertilizer for watering plants. To the "Fertilizer" jar, add the amount of fertilizer recommended for a quart of water. To the "Excess fertilizer" jar, add 10 times this amount of fertilizer. Stir the contents of each jar thoroughly to dissolve the fertilizer.

3. Obtain a sample of pond water. Stir it gently but thoroughly to ensure that the organisms in it are evenly distributed. Measure 100 mL of pond water into each of the three jars.

4. Cover each jar loosely with plastic wrap. Place all three jars about 20 cm from a fluorescent lamp. (Do not place the jars in direct sunlight, as this may cause them to heat up too much.)

5. Observe a drop of pond water from your sample, under the microscope. On a sheet of paper, draw at least four different organisms that you see. Determine whether the organisms are algae (usually green) or consumers (usually able to move). Describe the total number and type of organisms that you see.

Safety Cautions
You can substitute plastic materials for glass materials in this activity. Broken glassware should be disposed of properly. As a class, review the safety cautions on the container of household fertilizer. Make sure that students follow all cautions when using and disposing of the fertilizer. Students should wash their hands immediately after handling the pond water.

6. Based on what you have learned about eutrophication, make a prediction about how the pond organisms will grow in each of the three jars.

7. Observe the jars when you first set them up and at least once every three days for the next 3 weeks. Make a data table to record the date, color, odor, and any other observations you make for each jar.

8. When life-forms begin to be visible in the jars (probably after a week), use an eyedropper to remove a sample of organisms from each jar and observe the sample under the microscope. Record your observations.

9. At the end of your 3-week observation period, again remove a sample from each jar and observe it under the microscope. Draw at least four of the most abundant organisms that you see, and describe how the number and type of organisms have changed.

▶ **Step 7** Record your observations of the jars every 3 days for 3 weeks.

Analysis

1. **Describing Events** After three weeks, which jar shows the most abundant growth of algae? What may have caused this growth?

2. **Analyzing Data** Did you observe any effects on organisms other than algae in the jar that had the most abundant algae growth? Explain.

Conclusions

3. **Applying Conclusions** Did your observations match your predictions? Explain.

4. **Drawing Conclusions** How can artificial eutrophication be prevented in natural water bodies?

Extension

1. **Designing Experiments** Modify the experiment by using household dishwashing detergent instead of household fertilizer. Are the results different?

2. **Research and Communications** Research the watersheds that are located close to your area. How might activities such as farming and building affect watersheds?

191

Chapter Resource File

- Datasheets for In-Text Labs
- Lab Notes and Answers

TEACHER
TESTED & APPROVED

Denise Sandefur
Nucla High School
Nucla, Colorado

Tips and Tricks

If quart jars are not available, pint jars could be easily substituted by using one-half of the amount of distilled water, pond water, and fertilizer.

If students obtain their own pond water, be sure that they obtain the water from a healthy pond. The location and appearance of the pond should be considered. For example, if the plants around the pond seem wilted or damaged, herbicides may have been sprayed recently and could in turn affect the outcome of the lab.

Count estimation and extrapolation are additional skills that could be acquired during this lab. For example, have students take a percentage (25%) of the field of view under the microscope, count the numbers of organisms found within that percentage, and multiply the count by four. This calculation will give an estimate of the count of each sample and a better assessment of the effects of the regular samples and the fertilized samples. These calculations can be displayed on graphs.

Answers to Analysis

1. Answers may vary. The jar with excess fertilizer should show the most growth. The extra fertilizer should stimulate additional plant and algae growth.

2. Answers may vary. Other organisms may increase in abundance as their food supply (algae) increases. However, if algae have begun to die and decay, the decay process may deplete oxygen in the jar, causing oxygen-dependent life-forms to die as well.

Answers to Conclusions

3. Answers may vary.

4. Artificial eutrophication can be prevented by controlling the application of fertilizers in areas where runoff is common.

Answers to Extension

1. Answers may vary. The samples with dishwashing detergent and excess dishwashing detergent should show the same amount of growth as the samples with fertilizer and excess fertilizer did.

2. Answers may vary.

WETLANDS IN THE UNITED STATES, 1780s Vs. 1980s

Background

Many factors have contributed to the loss of wetlands between 1780 and 1980. The population of the United States has increased dramatically, and people have moved throughout the country. The pressures from increased population and development have affected wetlands. People tend to settle near wetlands because wetlands are a water and food source. Other factors have also increased the destruction of wetlands. For example, technological improvements have increased the use of land for agriculture, which has increased erosion. And the building of dams to impound water and control floods has reduced wetlands as well. In general, the greatest loss of wetlands has occurred in the eastern United States. This region has the highest population and has had a large population for the longest time.

Transparencies

TT Wetlands in the United States, 1780s Vs. 1980s

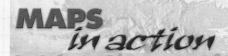

WETLANDS IN THE UNITED STATES, 1780s Vs. 1980s

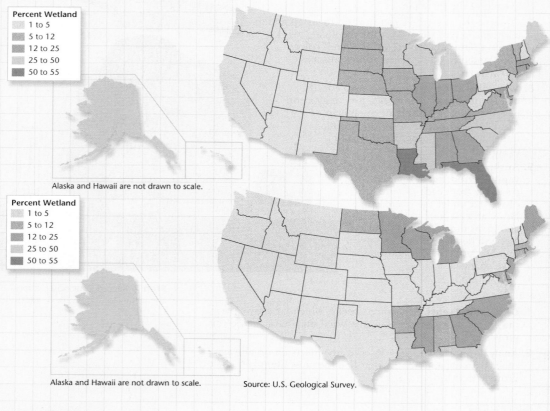

Alaska and Hawaii are not drawn to scale.

Percent Wetland
- 1 to 5
- 5 to 12
- 12 to 25
- 25 to 50
- 50 to 55

Alaska and Hawaii are not drawn to scale.

Source: U.S. Geological Survey.

MAP SKILLS

Use the maps of wetland loss in the United States to answer the questions below.

1. **Using a Key** Use the key to determine how many states had a decrease in wetland distribution from 5 to 12 percent to 1 to 5 percent.

2. **Analyzing Data** Is there any state on the map of wetland distribution in the 1980s that has the same percentage of wetland distribution as it did in the 1780s? If so, how many?

3. **Analyzing Data** Which states have had the greatest decrease in wetland distribution since the 1780s?

4. **Making Inferences** What might have caused Florida's and Louisiana's wetlands to decrease in distribution?

5. **Using a Key** Use the key to determine how many states had a decrease in wetland distribution from 25 to 50 percent to 12 to 25 percent.

6. **Identifying Trends** If these trends of wetland loss continue, what might a map of wetland distribution of the United States look like circa 2040?

192

Answers to Map Skills

1. 11 states

2. yes; 18 states have the same percentage.

3. Arkansas, Illinois, Indiana, and Ohio

4. Development may have taken place and destroyed wetlands in Florida and Louisiana.

5. 9 states

6. Answers may vary. Students should suggest that many more states will have lower percentages of wetlands.

CREATING ARTIFICIAL REEFS

Hundreds of years ago, people found that the fishing is often good over a sunken ship. The fishing is especially good if the wreck is in a protected area, where it will not be broken up by heavy surf or covered with silt from a river.

The reason fishing is often good under these conditions is because many marine organisms, such as seaweed, corals, and oysters live only where they can attach to a hard surface in clear water. So, the rocky shores of New England and of the West Coast support many more species than sandy areas do.

The Formation of a Reef

Organisms that attach to a hard bottom attract other species and eventually form a reef community. When seaweed grows on a rock, snails and crustaceans that eat the seaweed will join the community,

as will sea anemones, which feed on the eggs, larvae, and waste produced by other organisms. Corals may settle on the bottom and add their hard skeletons to the reef. Then, fish arrive to feed on the reef organisms and to reproduce in the cracks and crevices of the reef that protect them from predators. If the reef is in a relatively calm area, it will become a diverse community that fishermen and scuba divers can enjoy.

In recent years, many communities have created artificial reefs by sinking various objects in the ocean. Barges full of broken up concrete are a common choice. The San Diego Oceans Foundation sank a decommissioned Canadian warship to create an artificial reef off Mission Beach, California. The warship was sunk to attract tourists who are recreational divers. A nonprofit group is working to bring the decommissioned USAF's *General Hoyt S. Vandenburg* to Key West to be sunk as a diving reef that will enhance the Florida Keys National Marine Sanctuary. A reef founda-

▶ The gun turret (below) and the aircraft (left) are examples of objects used to create artificial reefs.

tion in Georgia produces objects called reef balls. Reef balls are hollow, concrete balls that have holes in them. They have legs that "stick" out to prevent them from rolling around on the ocean floor. The balls come in various sizes and can be towed behind even a small boat to where they will be sunk. Reef balls are used to provide additional hard surfaces for restoring damaged coral reefs as well as to create new fishing and diving reefs.

Safe Artificial Reefs

Sinking artificial reefs is not problem free. Obviously, the reef must be deep enough so that it does not interfere with the movement of ships. But we cannot be sure what depth of water the ships of the future will need or that shipping channels will be in the same places. In addition, if the object is to attract divers, any parts that might trap or entangle a diver must be removed. Also, the reef must not pollute the water or fall apart and leave debris drifting in the sea. In the case of ships that are used to create artificial reefs, stripping out all the hazardous materials, such as plastics, insulation, and oil, is necessary.

What Do You Think?

What are the benefits to creating artificial reefs? How are most artificial reefs created? Are there any disadvantages to creating artificial reefs? Explain why creating artificial reefs may be helpful in aquatic ecosystems other than the ocean. Research the Internet to find out if there are any artificial reefs in your community.

193

Answers to What Do You Think?

Answers may vary. Artificial reefs benefit organisms by giving them a place to live, feed, and breed. Artificial reefs may be beneficial in any aquatic ecosystem because they can replace lost habitat that is needed by the organisms living there.

Background

Artificial reefs are usually constructed of materials that are environmentally friendly. The most ideal materials are concrete and steel. It is important to use materials that do not break down and cause pollution. Old ships and planes make ideal artificial reefs because of the materials from which they are constructed. Before the plane or ship is sunk to make a reef, all fuel and oil are removed so that they do not leak out and pollute the area. Other structures that are good for artificial reefs are retired oil rigs and pieces of old concrete foundations.

Artificial reefs can be created in many types of aquatic ecosystems. Some cities have programs to create artificial reefs in lakes. Sometimes the reefs are built of concrete scrap, rocks, or even old Christmas trees. These artificial reefs provide valuable habitat for game fish and make excellent fishing spots.

MISCONCEPTION ALERT

Artificial Reefs The term *artificial reef* is somewhat misleading. The artificial reefs act just like natural outcroppings of rock. They are quickly colonized by sponges, corals, and other invertebrates. Fish are attracted to the artificial reefs as a place for protection, for feeding, and for breeding. In a short time, an artificial reef is transformed into a living reef. The entire surface of the reef is covered with living organisms.

POPULATIONS

CHAPTER 8

Understanding Populations

CHAPTER 9

The Human Population

CHAPTER 10

Biodiversity

This school of young striped cat-fish near the coast of Australia gathers into a huge, writhing ball to defend against predators. Forming a ball makes the fish look like one large organism, and the fish's stripes may make it hard for a predator to see individual fish.

Understanding Populations
Chapter Planning Guide

PACING	CLASSROOM RESOURCES	LABS, DEMONSTRATIONS, AND ACTIVITIES
BLOCKS 1, 2 & 3 · 135 min pp. 196–202 **Chapter Opener**		
Section 1 How Populations Change in Size	**OSP** Lesson Plan * **CRF** Active Reading * BASIC **TT** Bellringer * **TT** Population Change & Exponential Growth * **TT** Population Changes and Carrying Capacity * **CD** Interactive Tutor Describing Populations **CD** Interactive Tutor How Populations Change in Size	**TE** Activity Two Types of Growth, p. 197 GENERAL **SE** QuickLab Population Growth, p. 198 **TE** Demonstration Classroom Density, p. 198 BASIC **TE** Internet Activity Island Carrying Capacities, p. 200 GENERAL **SE** Skills Practice Lab Studying Population Growth, pp. 214–215 ◆ **CRF** Datasheets for In-Text Labs * **CRF** CBL™ Probeware Lab * ◆ ADVANCED **CRF** Consumer Lab * ◆ GENERAL **CRF** Modeling Lab * GENERAL
BLOCKS 4 & 5 · 90 min pp. 203–209 **Section 2** How Species Interact with Each Other	**OSP** Lesson Plan * **CRF** Active Reading * BASIC **TT** Bellringer * **TT** Types of Species Interactions * **TT** Niche Restriction Due to Competition * **CD** Interactive Tutor Resource Limitation and Carrying Capacity **CD** Interactive Tutor Predator-Prey Relationships **CD** Interactive Exploration 2 What's Bugging You?	**TE** Activity Constructing a Personal Niche Map, p. 203 BASIC **SE** Field Activity Observing Competition, p. 204 **TE** Group Activity Species Interactions Skit, p. 204 GENERAL **TE** Internet Activity Restricted and Potential Niches, p. 205 GENERAL **TE** Demonstration Predator Adaptations, p. 206 GENERAL **TE** Group Activity Evolving Relationships, p. 208 GENERAL **CRF** Long-Term Project * ◆ GENERAL

BLOCKS 6 & 7 · 90 min

Chapter Review and Assessment Resources

- **SE** Chapter Review pp. 211–213
- **SE** Standardized Test Prep pp. 650–651
- **CRF** Study Guide * ■ GENERAL
- **CRF** Chapter Test * ■ GENERAL
- **CRF** Chapter Test * ADVANCED
- **CRF** Concept Review * ■ GENERAL
- **CRF** Critical Thinking * ADVANCED
- **OSP** Test Generator
- **CRF** Test Item Listing *

Online and Technology Resources

 Holt Online Learning

Visit **go.hrw.com** for access to Holt Online Learning or enter the keyword **HE6 Home** for a variety of free online resources.

 **One-Stop Planner® CD-ROM**

This CD-ROM package includes
- Lab Materials QuickList Software
- Holt Calendar Planner
- Customizable Lesson Plans
- Printable Worksheets
- ExamView® Test Generator
- Interactive Teacher Edition
- Holt PuzzlePro® Resources
- Holt PowerPoint® Resources

 Holt Environmental Science Interactive Tutor CD-ROM

This CD-ROM consists of interactive activities that give students a fun way to extend their knowledge of environmental science concepts.

ENRICHMENT AND SKILLS PRACTICE	SECTION REVIEW AND ASSESSMENT	CORRELATIONS
SE Pre-Reading Activity, p. 196 **TE** Using the Figure, p. 196 `GENERAL`		**National Science Education Standards**
TE Using the Figure Exponential Growth, p. 199 `GENERAL` **TE** Using the Figure What Limits Population Growth? p. 200 `BASIC` **TE** Inclusion Strategies, p. 200 **SE** MathPractice Growth Rate, p. 201 **TE** Skill Builder Understanding Equations, p. 201 `ADVANCED`	**TE** Homework, p. 197 `GENERAL` **SE** Section Review, p. 202 **TE** Reteaching, p. 202 `BASIC` **TE** Quiz, p. 202 `GENERAL` **TE** Alternative Assessment, p. 202 `GENERAL` **CRF** Quiz * ■ `GENERAL`	LS 4c LS 4d SPSP 2a SPSP 2c
TE Using the Figure Indirect Interactions, p. 204 `GENERAL` **TE** Reading Skill Builder Brainstorming, p. 204 `BASIC` **SE** Graphic Organizer Spider Map, p. 205 `GENERAL` **SE** Case Study Predator-Prey Adaptations, pp. 206–207 **TE** Using the Figure Specialists Vs. Generalists, p. 207 `GENERAL` **TE** Skill Builder Graphing, p. 207 `BASIC` **TE** Skill Builder Writing, p. 208 `GENERAL` **TE** Skill Builder Vocabulary, p. 209 `GENERAL` **SE** Points of View Where Should the Wolves Roam? pp. 216–217 **CRF** Map Skills Tracking Cichlids * `GENERAL`	**TE** Homework, p. 204 `GENERAL` **TE** Homework, p. 206 `GENERAL` **SE** Section Review, p. 209 **TE** Reteaching, p. 209 `BASIC` **TE** Quiz, p. 209 `GENERAL` **TE** Alternative Assessment, p. 209 `GENERAL` **CRF** Quiz * ■ `GENERAL`	LS 4c LS 4d

Guided Reading Audio CDs

These CDs are designed to help auditory learners and reluctant readers. (Audio CDs are also available in Spanish.)

SCiLINKS.
NSTA

www.scilinks.org

Maintained by the **National Science Teachers Association**

TOPIC: Populations and Communities
SciLinks code: HE4085

TOPIC: Coevolution
SciLinks code: HE4014

 CNN Videos

Each video segment is accompanied by a Critical Thinking Worksheet.

Science, Technology & Society

Segment 25 Salmon Sound Barriers

Chapter Enrichment

This Chapter Enrichment provides relevant and interesting information to expand and enhance your classroom instruction of the chapter material.

SECTION 1 How Populations Change in Size

Community Ecology

The field known as community ecology studies the interactions of populations of organisms within ecosystems. The field is relatively new to science, and developed such fundamental concepts as niche, mutualism, and coevolution. It is important to keep in mind that many of these basic definitions and concepts are convenient generalizations, while the realities of interactions within ecosystems are quite complex. Most ecologists admit that our study of the interactions within ecosystems has only scratched the surface of their complexity.

Turtle Reproduction

Loggerhead Sea Turtles are an example of an animal that puts its reproductive energy into producing many offspring but provides no maternal care. This strategy allows the organism to fill the habitat with so many young that some are bound to survive. Female turtles mate during migration between nesting and feeding grounds. On the nesting grounds, one female may lay up to 190 eggs in one nest. Females also nest multiple times during one season. During their first two years of life, the baby turtles are particularly vulnerable to predation.

▶ **Just-hatched baby sea turtles heading to the ocean**

Limiting Competition

Allelopathy is a strategy that plants may use to compete for limited resources, such as soil nutrients. With allelopathy, a plant produces a chemical that prevents any other seedlings from germinating in the area around the plant. One example of an allelopathic relationship is the lady fern and the spruce tree. Lady ferns produce a phenolic compound that kills spruce seedlings during germination. Black walnut trees, sunflowers, junipers, and sagebrushes also exhibit allelopathy. The term allelopathy means "causing another suffering".

SECTION 2 How Species Interact with Each Other

Seasonal Competition

During the breeding season, most birds compete fiercely for mates, territory, and food. When breeding is finished, and as summer fades to autumn and winter, non-migratory birds in northern climates sometimes form multi-species feeding groups that cooperate to find food. Black-capped Chickadee pairs, for example, will often join with unrelated juveniles, Tufted Titmice or White-breasted Nuthatches to forage. This behavior may help the birds to search for scarce food more effectively or to avoid predation.

Barnacles

Barnacles are crustaceans that glue themselves to marine surfaces. Many barnacle species build calcium-carbonate plates that surround and protect them. They also build a shell that can be opened or closed, to protect the barnacle from predators and from drying out. From inside the shell, barnacles are able to collect food and release gametes. When young larvae mature, they float through the water, settle on marine surfaces and build their new homes. Barnacles, and many of the organisms who share their intertidal-zone habitat, are subject to a great deal of competition and predation (by snails, crabs, birds, etc). Intertidal zones are well-studied habitats with clearly-delineated sub-zones.

Two Types of Mimicry

Some insect species that taste good to predators may mimic others that taste terrible or are harmful to eat. This kind of mimicry is called *Batesian mimicry,* after the British scientist H. W. Bates, who introduced the concept in 1861. One example of Batesian mimicry is a grasshopper in Borneo that mimics a tiger beetle. Predators avoid the tiger beetle because it is aggressive. Because the grasshopper looks just like the beetle, predators then tend to avoid it as well. It was originally thought that the edible Viceroy butterfly was a Batesian mimic of the unpalatable Monarch butterfly. However, the Viceroy has turned out to be unpalatable as well. The Viceroy is a case of *Mullerian mimicry*—when several species benefit by having similar markings because predators learn to avoid all similarly-marked species.

▶ A tropical frog with warning coloration

Sneaky Mussels

Many freshwater mussels have a slightly parasitic stage in their life history. A larval form called a *glochidia* is released from the adult and must attach to a fish's gills and then find a new home. Some mussels mimic fish prey items to lure fish, so that the fish will ingest the glochidia. For example, a female mussel may wave a tissue resembling a minnow or insect to attract a fish. When the fish approaches, the female releases the glochidia close to the fish's mouth.

Ant Partnerships

There are numerous examples of mutualisms between an ant species and another insect, or a plant or fungus. In Ecuador, *Azteca* ants protect *Cecropia* plants from herbivorous beetles. In West Africa, ants in the genus *Petalomyrmex* colonize and protect shoots of the plant *Leonardoxa africana* from grazing herbivores. Other ant species are "shepherds" of aphids or butterfly larvae, or "farmers" of fungi.

▶ Ants in a mutualistic relationship with a tree

Bacteria within Bacteria

One hypothesis for the origin of many of the organelles in eukaryotic cells is that they began as symbiotic bacteria. Carol von Dohlen and her colleagues at Utah State have discovered modern examples of bacteria that reside within host bacteria. Von Dohlen and her colleagues knew from previous studies that the citrus mealybug digests its food with the help of bacteria living in its gut. There was chemical evidence that two different species of proteobacteria were involved. But using electron microscopes, Von Dohlen's team could only see one type of proteobacteria, inside strange "symbiotic spheres." Then microscope supervisor William McManus suggested that the spheres could be the second form of bacteria. He was correct! This finding, reported in the journal *Nature,* was the first documented case of one bacterium living inside another.

Overview

Tell students that this chapter introduces basic concepts that ecologists use in studying ecosystems. The chapter explores the properties of populations, how populations increase or decrease in response to their environment, and how populations of different species interact. Many of these concepts are still being developed as scientists realize the true complexity of ecosystems.

Using the Figure — GENERAL

Tell students that the orcas in the photograph (also called "killer whales") hunt and eat sea lions. Ask students, "Would a change in the numbers of sea lions have an effect on the orcas?" (Yes, probably.) Ask students, "Would it make a difference if the sea lions were the only food source for the orcas? (Yes. If the orcas could eat other food, they would be less affected by the numbers of sea lions.) Ask students to think of other reasons that the numbers of orcas or sea lions might change. (Answers may vary.)
LS Logical

PRE-READING ACTIVITY

You may want to use this FoldNote in a classroom discussion to review material from the chapter. On the board, write each category from the Three-Panel Flip Chart. Then, ask students to provide information for each category. Under the appropriate category on the board, write the information that students provide.

VIDEO SELECT

For information about videos related to this chapter, go to **go.hrw.com** and type in the keyword **HE4 POCV**.

Understanding Populations

1 How Populations Change in Size

2 How Species Interact with Each Other

PRE-READING ACTIVITY

FOLDNOTES

Three-Panel Flip Chart
Before you read this chapter, create the **FoldNote** entitled "Three-Panel Flip Chart" described in the Reading and Study Skills section of the Appendix. Label the flaps of the chart with "Population," "Niche," and "Species Interaction." As you read the chapter, write information you learn about each category under the appropriate panel.

Orcas (also called killer whales) hunt sea lions. The interaction between these groups affects the number of individuals and the behavior of individuals in each group.

196

Chapter Correlations   *National Science Education Standards*

LS 4c Organisms both cooperate and compete in ecosystems. The interrelationships and interdependencies of these organisms may generate ecosystems that are stable for hundreds or thousands of years. **(Sections 1 and 2)**

LS 4d Living organisms have the capacity to produce populations of infinite size, but environments and resources are finite. This fundamental tension has profound effects on the interactions between organisms. **(Sections 1 and 2)**

SPSP 2a Populations grow or decline through the combined effects of births and deaths, and through emigration and immigration. Populations can increase through linear or exponential growth, with effects on resource use and environmental pollution. **(Section 1)**

SPSP 2c Populations can reach limits to growth. Carrying capacity is the maximum number of individuals that can be supported in a given environment. The limitation is not the availability of space, but the number of people in relation to resources and the capacity of earth systems to support human beings.

Changes in technology can cause significant changes, either positive or negative, in carrying capacity. **(Section 1)**

How Populations Change in Size

Biologist Charles Darwin once calculated that a single pair of elephants could theoretically produce 19 million descendants within 750 years. Darwin made the point that the actual number of elephants is limited by their environment.

One way to study the relationship of elephants with their environment is at the level of populations. Such a study would include tracking the number of elephants in an area and observing the animals' interactions with their environment.

What Is a Population?

A **population** is all the members of a species living in the same place at the same time. The bass in an Iowa lake make up one population. **Figure 1** shows other examples of populations. A population is a reproductive group because organisms usually breed with members of their own population. For example, daisies in an Ohio field will breed with each other and not with daisies in a Maryland population. The word *population* refers to the group in general and also to the size of the population—the number of individuals it contains.

Objectives

▶ **Describe** the three main properties of a population.
▶ **Describe** exponential population growth.
▶ **Describe** how the reproductive behavior of individuals can affect the growth rate of their population.
▶ **Explain** how population sizes in nature are regulated.

Key Terms

population
density
dispersion
growth rate
reproductive potential
exponential growth
carrying capacity

Figure 1 ▶ The palm trees on an island (left) and a school of fish (below) are examples of populations.

Chapter Resource File

• Lesson Plan
• Active Reading BASIC
• Section Quiz GENERAL

Transparencies

TT Bellringer

Homework ———————— GENERAL

Favorite Populations Ask students to research information about populations of their favorite plant or animal. Have them answer the following questions: "Where in the world can you find populations of this organism? What kinds of resources are limiting to its growth? How are the individuals dispersed within their habitat? How do the organisms find each other to mate?" Have the students record their findings in their *EcoLog.* LS Intrapersonal

Focus

Overview

Before beginning this section, review with your students the Objectives listed in the Student Edition. This section introduces the general characteristics of populations, explores how populations can grow at different rates, and explains why there are natural limits to population growth.

Bellringer

Have students write down the definition of a population in their *EcoLog.* (A population is all members of the same species that live in the same place at the same time and breed with each other.) Have them record examples of populations in their neighborhood. (Humans, squirrels, trees, grass, weeds, mice, cats, microbes, etc.) Ask them to draw or describe where these populations fit in an individual-to-ecosystem hierarchy. LS Logical | English Language Learners |

Motivate

Activity ———— GENERAL

Two Types of Growth Present the following scenario to students: "You have just been offered a job that will last one month. You have two salary options. You can either receive $10 a week with a $5 per week raise every week, or you can receive one penny for your first day on the job, and then double the previous day's pay for each of the remaining 30 days." Ask students to determine which salary option they would prefer. Provide students with calculators to calculate their salaries. (The "double penny" option yields a much higher salary. The weekly payment option would yield $70 for the month, while the "double penny" option would yield over $10 million.) Tell students that populations may increase in size in either of these two ways, and that this chapter will describe situations in which these two types of growth may occur. LS Logical

Classroom Density To strengthen students' concept of population density, ask students to mark or rope off a corner of the room that is 2 meters on each side. Have 12 students stand within the area, and ask the class to calculate the density of that population of students (12 students/4 m² = 3 students/m²). Now ask those 12 students to double their area (4 m × 4 m = 16 m²). Ask the class to calculate the new density (12 students/16 m² = 0.75 students/m²). Ask, "Which population has the higher density and why?" (The first one, because there were more students per unit of space.)

LS Kinesthetic/ Logical | English Language Learners

MISCONCEPTION ///ALERT \\\

Populations Are Difficult to Contain The strict ecological definition of a population is tricky to apply to organisms in natural ecosystems. In the strictest sense, a population is only those members of a species that actually are interbreeding. Also, populations are in constant flux as individuals reproduce, die, or migrate in or out of a given area. Thus, the term *population* may commonly be used, for convenience, to refer to the number of members of a species within a defined area at a given time (whether or not they interbreed). A common example of this use is when referring to "the human population of the United States."

Figure 2 ▶ Populations may have very different sizes, densities, and dispersions. Flamingos (right) are usually found in huge, dense flocks, while most snakes (above) are solitary and dispersed randomly.

QuickLAB

Population Growth

Procedure

1. Model the change in size of a population by applying the equation at right.
2. Start with **100 g (3.5 oz)** of **dry beans**. Count out five beans to represent the starting population of a species.
3. Assume that each year 20 percent of the beans each have two offspring. Also assume that 20 percent of the beans die each year.
4. Calculate the number of beans to add or subtract for 1 y. Round your calculations to whole numbers. Add to or remove beans from your population as appropriate.
5. Continue modeling your population changes over the course of 10 y. Record each change.

Analysis

1. Make a graph of your data. Describe the changes in your population.

198

QuickLAB

Skills Acquired:

- Collecting Data
- Organizing and Analyzing Data

Answers

6. Sample results:

Year	1	2	3	4	5	6	7	8	9	10
Starting	5	6	7	9	11	13	15	18	22	27
Births	2	2	3	4	4	5	6	8	9	11
Deaths	1	1	1	2	2	3	3	4	4	5
Ending	6	7	9	11	13	15	18	22	27	33

Properties of Populations

Populations may be described in terms of size, density, or dispersion, as shown in **Figure 2.** A population's **density** is the number of individuals per unit area or volume, such as the number of bass per cubic meter of water in a lake. A population's **dispersion** is the relative distribution or arrangement of its individuals within a given amount of space. A population's dispersion may be *even, clumped,* or *random.* Size, density, dispersion, and other properties can be used to describe populations and to predict changes within them.

How Does a Population Grow?

A population gains individuals with each new offspring or birth and loses them with each death. The resulting population change over time can be represented by the equation below. A change in the size of a population over a given period of time is that population's **growth rate.** The growth rate is the *birth rate* minus the *death rate.*

Over time, the growth rates of populations change because birth rates and death rates increase or decrease. Growth rates can be positive, negative, or zero. For a population's growth rate to be zero, the average number of births must equal the average number of deaths. A population would remain the same size if each pair of adults produced exactly two offspring, and each of those offspring survived to reproduce. If the adults in a population are not replaced by new births, the growth rate will be negative and the population will shrink.

Transparencies

TT Population Change & Exponential Growth

How Fast Can a Population Grow?

A female sea turtle may lay 2,000 eggs in her lifetime in nests she digs in the sand. **Figure 3** shows newly hatched sea turtles leaving their nest for the ocean. If all of them survived, the turtle population would grow rapidly. But they do not all survive. Populations usually stay about the same size from year to year because various factors kill many individuals before they can reproduce. These factors control the sizes of populations. In the long run, the factors also determine how the population evolves.

Reproductive Potential A species' *biotic potential* is the fastest rate at which its populations can grow. This rate is limited by the maximum number of offspring that each member of the population can produce, which is called its **reproductive potential.** Some species have much higher reproductive potentials than others. Darwin calculated that it could take 750 years for a pair of elephants to produce 19 million descendants. In contrast, a bacterium can produce 19 million descendants in a few days or weeks.

Reproductive potential increases when individuals produce more offspring at a time, reproduce more often, and reproduce earlier in life. Reproducing earlier in life has the greatest effect on reproductive potential. Reproducing early shortens the *generation time,* the average time it takes a member of the population to reach the age when it reproduces.

Small organisms, such as bacteria and insects, have short generation times. These organisms can reproduce when they are only a few hours or a few days old. As a result, their populations can grow quickly. In contrast, large organisms, such as elephants and humans, become sexually mature after a number of years. The human generation time is about 20 years, so humans have a much lower reproductive potential than insects.

Exponential Growth Populations sometimes undergo **exponential growth,** which means they grow faster and faster. For example, if a pair of dogs gives birth to 6 puppies, there will be 6 dogs in one generation. If each pair of dogs in that generation has 6 puppies, there will be 18 dogs in the next generation. The following generation will contain 54 dogs, and so on. If the number of dogs is plotted versus time on a graph, the graph will have the shape shown in **Figure 4.** In exponential growth, a larger number of individuals is added to the population in each succeeding time period.

Exponential growth occurs in nature only when populations have plenty of food and space, and have no competition or predators. For example, populations of European dandelions and starlings imported into the United States underwent exponential growth. Similar population explosions occur when bacteria or molds grow on a new source of food.

Figure 3 ▶ Most organisms have a reproductive potential that far exceeds the number of their offspring that will survive. Very few of these baby sea turtles will survive long enough to breed.

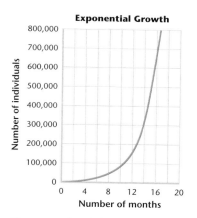

Exponential Growth

Number of individuals (vertical axis): 0, 100,000, 200,000, 300,000, 400,000, 500,000, 600,000, 700,000, 800,000
Number of months (horizontal axis): 0, 4, 8, 12, 16, 20

Figure 4 ▶ Population growth is graphed by plotting population size over a period of time. Exponential population growth will look like the curve shown here.

199

BIOLOGY CONNECTION — GENERAL

Coyote Reproduction Coyotes are known to have reproductive rates that vary in response to food availability, or in response to pressures such as being hunted. Coyotes may alter the size or frequency of their litters of pups. This phenomenon thwarts many attempts to exterminate coyotes.

Discussion — ADVANCED

Reproductive Strategies Ask students to consider the following statement: "Animals have different strategies for maximizing reproductive success." The sea turtles in **Figure 3** use a lot of energy to produce many offspring at a time, but put little or no energy into offspring care. Songbirds, on the other hand, usually use a small amount of energy to produce 2–3 eggs, but then spend a great deal of energy ensuring that those few offspring survive. Ask students to give examples of how other animals maximize their reproductive success. Have students discuss how the evolution of reproductive strategies in those animals relates to their parental behaviors. (Animals that do not protect their broods gain evolutionary success by producing many offspring; animals that produce fewer offspring gain success by spending more energy per offspring.) **LS** Verbal

Using the Figure — GENERAL

Exponential Growth Use **Figure 4** to reinforce the meaning of exponential population growth. Ask students to describe the curve and explain how it shows exponential growth. (The curve rises more and more steeply, meaning the population increases by greater amounts during each time period.) Ask, "What would the graph look like if it showed linear (or arithmetic) growth?" (It would show a straight line, increasing by the same amount during each time period.) Ask: "How does exponential growth relate to reproductive potential?" (Most organisms have the potential to reproduce "multiples" of themselves, thus creating exponential growth rates. Exponential growth is a mathematical description of nearly-unlimited population growth.) **LS** Visual/Logical

Using the Figure — BASIC

What Limits Population Growth?
Have students study **Figure 5** and explain how the environment affects population growth. (As the population uses up limited resources, its numbers fluctuate and stabilize around the carrying capacity.) Then ask students to explain what caused the initial population crash. (There were more organisms than the resources could support. The population exceeded the carrying capacity.)
LS Visual

INCLUSION Strategies

• *Learning Disabled*
• *Attention Deficit Disorder*

On an index card, have each student draw two penny-sized circles. On a second card, the student should double the number of circles. Continue doubling the number of circles on each of the next cards until no more complete circles can fit on a card. Tell students the circles represent members of a population and the white space on the cards represent the available resources. Students should be able to discuss what problems would occur when the white space (resources) is depleted and how it would affect the population.

Transparencies

 TT Population Changes and Carrying Capacity

📶 **internet** connect

www.scilinks.org
Topic: Populations and Communities
SciLinks code: HE4085

SC*LINKS* Maintained by the National Science Teachers Association

 Ecofact

Carrying Capacity of Islands
Islands are good places to study carrying capacity because islands have clear boundaries. The Pribilof Islands near Alaska were the site of a well-studied population explosion and crash. In 1911, 25 reindeer were introduced on one of the islands. By 1938, the herd had grown to 2,000 animals. The reindeer ate mostly lichens, which grow back very slowly. By 1950, there were only 8 reindeer alive on the island.

Figure 5 ▶ An example of carrying capacity is shown by the dashed yellow line in the graph (right). This line seems to be a limit on the size of the example population (blue line). When rabbits were introduced into Australia (below), their population quickly exceeded the carrying capacity of the area.

What Limits Population Growth?

Because natural conditions are neither ideal nor constant, populations cannot grow forever and rarely grow at their reproductive potential. Eventually, resources are used up or the environment changes, and deaths increase or births decrease. Under the forces of natural selection in a given environment, only some members of any population will survive and reproduce. Thus, the properties of a population may change over time.

Carrying Capacity The blue line in **Figure 5** represents a population that seems to be limited to a particular size. This theoretical limit, shown by the dashed yellow line, is called *carrying capacity*. The **carrying capacity** of an ecosystem for a particular species is the maximum population that the ecosystem can support indefinitely. A population may increase beyond this number, but it cannot stay at this increased size. Because ecosystems change, carrying capacity is difficult to predict or calculate exactly. However, carrying capacity may be estimated by looking at average population sizes or by observing a population crash after a certain size has been exceeded.

The history of rabbits in Australia demonstrates both exponential growth and carrying capacity. Originally, there were no rabbits in the native ecosystems of Australia. When rabbits were introduced there in 1859, their numbers increased rapidly because they had plenty of vegetation to eat, no competition, and no predators. But eventually, disease and starvation caused the rabbit population to crash. Over time, the vegetation recovered, and the rabbit population increased again. The population continues to increase and decrease, but less dramatically.

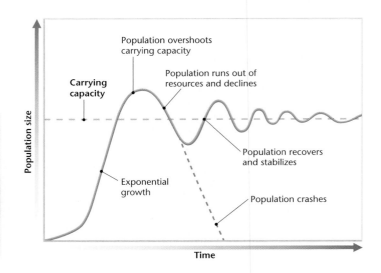

Island Carrying Capacities Have students research a specific population of organisms that was introduced to or invaded an island. (Suggest Australia, Guam, Hawaii, or Madagascar). Ask students to write an essay for their **Portfolio** that focuses on population trends of that animal over time, and the way resources control those trends. Ask them to use past data to predict future population trends. **LS Verbal**

MISCONCEPTION ///ALERT\\\

Population Change Includes Migration
Another possible element in the equation for population change is migration. Populations can increase by births and when individuals move into a population (immigration). Populations can decrease by deaths and when individuals move out of a population (emigration).

Resource Limits A species reaches its carrying capacity when it consumes a particular natural resource at the same rate at which the ecosystem produces the resource. That natural resource is then called a *limiting resource* for the species in that area. For example, plant growth is limited by supplies of water, sunlight, and mineral nutrients. The supply of the most severely limited resources determines the carrying capacity of an environment for a particular species at a particular time.

Competition Within a Population The members of a population use the same resources in the same ways, so they will eventually compete with one another as the population approaches its carrying capacity. An example is the fate of mealworm larvae in a sack of flour. Adults of this type of beetle will find a sack of flour, lay their eggs in the sack, and leave. Most of the first larvae to hatch will have plenty of flour to eat and will grow to adulthood. However, the sack has a limited amount of food, and mealworms from eggs that were laid later may not have enough food to survive to adulthood.

Instead of competing directly for a limiting resource, members of a species may compete indirectly for social dominance or for a territory. A *territory* is an area defended by one or more individuals against other individuals. The territory is of value not only for the space but also for the shelter, food, or breeding sites it contains. Competition within a population is part of the pressure of natural selection. Many organisms expend a large amount of time and energy competing with members of the same species for mates, food, or homes for their families. Some examples of competition within species are shown in **Figure 6.**

Figure 6 ▶ Members of a population often compete with each other. These plants (above) are growing over each other as they compete for light. These wolves (left) are competing for food and for social dominance.

MATHPRACTICE

Growth Rate A growth rate is a change in a population's size over a specific period of time.

$$\text{growth rate} = \frac{\text{change in population}}{\text{time}}$$

Imagine a starting population of 100 individuals. If there were 10 births and 5 deaths in a given year, what was the population's growth rate for the year? In the next year, if there were 20 births and 10 deaths, what would the new growth rate be? If births increased by 10 and deaths increased by 5 for each of the next 5 years, how would you describe the growth of this population?

201

MATHPRACTICE

Answers

With an original population size of 100, and 10 births and 5 deaths, the change in population would be +5, so the population would increase to 105. The next year, with 20 births and 10 deaths, the change in population would be +10, and the new population size would be 115. In each of the next five years, the rate of population increase would accelerate.

Calculations:

Year	1	2	3	4	5	6
Starting pop.	100	105	115	130	150	175
Births	10	20	30	40	50	60
Deaths	5	10	15	20	25	30
Ending pop.	**105**	**115**	**130**	**150**	**175**	**205**
Growth rate	+5	+10	+15	+20	+25	+30
% change	5.0	9.5	13.0	15.4	16.7	17.1

Reteaching ——— BASIC

Carrying Capacity and Energy Flow Ask students to think about energy flow in ecosystems. Have them draw an energy pyramid for grass, gazelles and lions. (Grass forms a large base level, with gazelles in the middle, and lions as a small top level.) Ask students, "How does the loss of energy at each trophic level affect the carrying capacity of these populations?" (The carrying capacity is lower at the top of the pyramid. A lion population needs a larger number of gazelle prey, which need to feed on a very large amount of grass.) **LS** Visual

Quiz ——— GENERAL

1. In what ways are disease and predation density-dependent? (In a dense population, increased physical contact and waste products mean that disease could spread more easily. Dense prey populations make it easier for predators to find prey.)

2. What are some examples of resources that could determine carrying capacity in an ecosystem? (water, sunlight, amount of soil, or any item that regenerates at a fixed rate and is consumed by all members of a species)

Alternative Assessment ——— GENERAL

Carrying Capacity Comics
Have students design a comic strip to show what might happen when a species exceeds its carrying capacity. Suggest that they include their comic strip in their **Portfolio. LS** Visual/Kinesthetic

Figure 7 ▶ The way a disease spreads through a population is affected by the population's density. These pine trees have been infected by a disease carried by the southern pine beetle. This disease has spread rapidly through U.S. timber forests.

Connection to ▶ History

Density and Disease The black plague of 14th-century Europe was spread in a density-dependent pattern. About one-third of Europe's population died from the highly contagious disease. Most of the deaths occurred in the crowded towns of the time, and fewer deaths occurred in the countryside.

Figure 8 ▶ Weather events usually affect every individual in a similar way, so such events are considered density-independent regulation.

Two Types of Population Regulation

Population size can be limited in ways that may or may not depend on the density of the population. Causes of death in a population may be *density dependent* or *density independent*.

When a cause of death in a population is *density dependent*, deaths occur more quickly in a crowded population than in a sparse population. This type of regulation happens when individuals of a population are densely packed together, such as when a population is growing rapidly. Limited resources, predation, and disease result in higher rates of death in dense populations than in sparse populations. The pine trees in **Figure 7** are infected with a disease that is spreading in a density-dependent pattern. Many of the same kind of pine tree are growing close to each other, so a disease-carrying beetle easily spreads the disease from one tree to another.

When a cause of death is *density independent*, a certain proportion of a population may die regardless of the population's density. This type of regulation affects all populations in a general or uniform way. Severe weather and natural disasters are often density-independent causes of death. The winter storm shown in **Figure 8** froze crops and fruiting trees regardless of the density of plants in the area.

SECTION 1 Review

1. Compare two populations in terms of size, density, and dispersion. Choose any populations you know of.

2. Describe exponential population growth.

3. Describe three methods by which the reproductive behavior of individuals can affect the growth rate of a population.

4. Explain how population sizes in nature are regulated.

CRITICAL THINKING

5. Making Predictions How accurately do you think the size of a population can be predicted? What information might be needed to make this prediction?

6. Compare and Contrast Read the description of the populations of rabbits in Australia and reindeer in the Pribilof Islands. List the similarities and differences between these two histories. **READING SKILLS**

202

Answers to Section Review

1. Answers may vary. Students should mention size, density, and dispersion.

2. Exponential growth will increase by a multiplicative factor (may double or triple with each generation), while linear growth merely adds the same number to each generation.

3. Populations can have increased growth rates if individuals reproduce earlier in life, reproduce more often, or produce more offspring at a time.

4. Interactions with the environment will change the characteristics of a population over time. The population may change in size, density, dispersion, or niche.

5. Answers may vary. To predict future population size, one would need to know all the environmental factors that might act on the population, including predators, prey, abiotic factors, climate, etc. One might only be able to accurately predict population size if all of these factors were stable.

6. Answers may vary. Students should describe how each population exceeded its resources, then crashed.

SECTION 2
How Species Interact with Each Other

What's the difference between lions in a zoo and lions in the wild? In the wild, lions are part of a community and a food web. In the African savanna, lions hunt zebras, fight with hyenas, and are fed upon by fleas and ticks. Interactions like these were part of the evolution of the lions that you see in zoos. Any species is best understood by looking at all of the relationships the species has within its native communities.

An Organism's Niche

The unique role of a species within an ecosystem is its niche (NICH). A niche includes the species' physical home, the environmental factors necessary for the species' survival, and all of the species' interactions with other organisms. A niche is different from a habitat. An organism's *habitat* is a location. However, a niche is an organism's pattern of use of its habitat.

A niche can also be thought of as the functional role, or job, of a particular species in an ecosystem. For example, American bison occupied the niche of large grazing herbivores on American grasslands. Kangaroos occupy a similar niche on Australian grasslands. Herbivores often interact with carnivores, such as lions, if they both exist in the same habitat. Some parts of a lion's niche are shown in **Figure 9**.

Objectives

▶ Explain the difference between niche and habitat.
▶ Give examples of parts of a niche.
▶ Describe the five major types of interactions between species.
▶ Explain the difference between parasitism and predation.
▶ Explain how symbiotic relationships may evolve.

Key Terms

niche
competition
predation
parasitism
mutualism
commensalism
symbiosis

Figure 9 ▶ Parts of a lion's niche are shown here. Can you think of other parts?

MISCONCEPTION
///ALERT\\\

Niche Versus Habitat The concept of the niche was introduced in 1917, and originally meant an organism's physical habitat. Niche was later defined as an interaction between an organism and other species. The definition that is most accepted by ecologists today describes a niche as a tolerance range for a set of resources used by an organism. For example, an organism can tolerate a range of temperatures or types of food. The modern definition makes the concept of niche multidimensional, and allows for the variety that we see in nature.

GEOPHYSICS
● CONNECTION — ADVANCED

Evolution and Empty Niches Around 5.5 million years ago, most of the Mediterranean Sea dried up, and all gobies (a family of fish) that were living there disappeared. When the Straits of Gibraltar opened, the Atlantic Ocean flooded into the empty Mediterranean basin, and gobies from the Atlantic were able to move in. With no competition from other gobies, they were able to "fill" a wide variety of "empty" ecological niches. As a result, many unique species of gobies evolved within the Mediterranean Sea.

SECTION 2

Focus

Overview

Before beginning this section, review with your students the Objectives listed in the Student Edition. This section explores the many ways in which species interact with each other and with their habitat.

🔊 Bellringer

Ask students to write a short paragraph describing a carnivore, such as a fox, through the eyes of its intended meal, such as a mouse.
LS Logical

Motivate

Activity ———— BASIC

Constructing a Personal Niche Map Invite students to think about where they fit in their community. Where do they gather their resources? What kinds of resources do they prefer? Who do they compete or cooperate with? Where do they live, and where do they hang out? Do they spend certain parts of the day in certain places? Tell students that all organisms use their habitats in a unique way, and this defines their niche. Ask students to draw a "niche map" that shows how they use their habitat. Ask volunteers to share their niche maps with the class, and compare niches (no one will have exactly the same niche). Have students add their niche map to their **Portfolio.**
LS Intrapersonal/ Visual | English Language Learners

Chapter Resource File

• **Lesson Plan**
• **Active Reading** BASIC
• **Section Quiz** GENERAL

Transparencies

TT Bellringer

Chapter 8 • Understanding Populations   203

Using the Figure — GENERAL

Indirect Interactions Have students look at the table within **Figure 10** and note that four of the five types of interactions are shown in the picture. Have students brainstorm other ways that the organisms in the scene might interact, either directly or indirectly. (Sample answers: All of the organisms probably compete for water; the fox and coyote compete for den space; the yucca competes with other plants for space and nutrients; parasites on the rat may spread to its predators; the rat and wren both may have other predators, may eat and spread seeds of plants, and may prey on the moths; they all provide food for microbes.) **LS Visual**

Group Activity — GENERAL

Species Interactions Skit Divide students into five groups. Assign one of the different types of species interaction from **Figure 10** to each group, but tell students not to let the other groups know which interaction they have. Ask each group to plan a skit that illustrates their assigned interaction. Tell them to be creative—the skit does not have to reflect a real habitat, it just needs to explain the concept. Have each group perform their skit for the class, and ask the audience to guess the interaction. **LS Interpersonal/ Kinesthetic**

| Co-op Learning | English Language Learners |

Transparencies

TT Types of Species Interactions

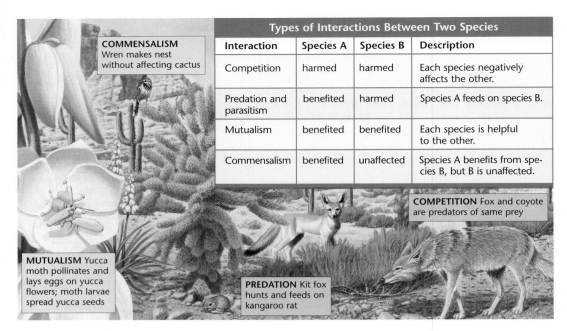

Types of Interactions Between Two Species			
Interaction	Species A	Species B	Description
Competition	harmed	harmed	Each species negatively affects the other.
Predation and parasitism	benefited	harmed	Species A feeds on species B.
Mutualism	benefited	benefited	Each species is helpful to the other.
Commensalism	benefited	unaffected	Species A benefits from species B, but B is unaffected.

COMMENSALISM Wren makes nest without affecting cactus

COMPETITION Fox and coyote are predators of same prey

MUTUALISM Yucca moth pollinates and lays eggs on yucca flowers; moth larvae spread yucca seeds

PREDATION Kit fox hunts and feeds on kangaroo rat

Figure 10 ▶ Species Interactions

FIELD ACTIVITY

Observing Competition You can study competition between bird species at home or at school. Build a bird feeder using a plastic milk jug, a metal pie pan, or some other inexpensive material. Fill the feeder with unsalted bread crumbs, sunflower seeds, or commercial birdseed.

Observe the birds that visit the feeder. Sit quietly in the same spot, and make observations at the same time each day for several days in a row.

In your *EcoLog,* keep a record that includes data about the kinds of birds that use the feeder, the kinds of seeds that the birds prefer, the factors that affect how much the birds eat, and the kinds of birds that are better competitors for the birdseed.

204

Ways in Which Species Interact

Interactions between species are categorized at the level where one population interacts with another. The five major types of species interactions, summarized in **Figure 10,** are competition, predation, parasitism, mutualism, and commensalism. These categories are based on whether each species causes benefit or harm to the other species in a given relationship. Keep in mind that the benefit or harm is in terms of total effects over time. Also note that other types of interaction are possible. Many interactions between species are indirect, and some interactions do not fit a category clearly. Other types of interactions seem possible but are rarely found. Therefore, many interactions are neither categorized nor well studied.

Competition

For most organisms, competition is part of daily life. Seed-eating birds compete with each other for seed at a bird feeder, and weeds compete for space in a sidewalk crack. **Competition** is a relationship in which different individuals or populations attempt to use the same limited resource. Each individual has less access to the resource and so is harmed by the competition.

Competition can occur both within and between species. We have learned that members of the same species must compete with each other because they require the same resources—they occupy the same niche. When members of different species compete, we say that their niches *overlap,* which means that each species uses some of the same resources in a habitat.

Homework — GENERAL

Artistic Interactions Ask students to create a portrait, a comic strip, watercolor painting, or other piece of art depicting one of the five types of species relationships. Have students include their work in their **Portfolio. LS Visual**

READING SKILL BUILDER — BASIC

Brainstorming Have students look at **Figure 10** carefully, and copy the table in their *EcoLog.* Ask them to suggest and then write or draw additional examples of each type of interaction in their tables. **LS Visual** | English Language Learners |

Indirect Competition Species can compete even if they never come into direct contact with each other. Suppose that one insect species feeds on a certain plant during the day and that another species feeds on the same plant during the night. Because they use the same food source, the two species are indirect competitors. Similarly, two plant species that flower at the same time may compete for the same pollinators even if the plants do not compete in any other way. Humans rarely interact with the insects that eat our food crops, but those insects are still competing with us for food.

Adaptations to Competition When two species with similar niches are placed together in the same ecosystem, we might expect one species to be more successful than the other species. The better-adapted species would be able to use more of the niche. But in the course of evolution, adaptations that decrease competition will also be advantageous for species whose niches overlap.

One way competition can be reduced between species is by dividing up the niche in time or space. *Niche restriction* is when each species uses less of the niche than they are capable of using. Niche restriction is observed in closely related species that use the same resources within a habitat. For example, two similar barnacle species compete for space in the intertidal zone of rocky shorelines. One of the species, *Chthamalus stellatus*, is found only in the upper level of the zone when the other species is present. But when the other species is removed from the area, *C. stellatus* is found at deeper levels, as shown in **Figure 11**. In the presence of competition, the actual niche used by a species may be smaller than the potential niche. Ecologists have observed various other ways of dividing up a niche among groups of similar species.

Graphic Organizer Spider Map
Create the **Graphic Organizer** entitled "Spider Map" described in the Appendix. Label the circle "Species Interactions." Create a leg for each type of species interaction. Then, fill in the map with details about each type of species interaction.

Graphic Organizer GENERAL

Spider Map
You may want to have students work in groups to create this Spider Map. Have one student draw the map and fill in the information provided by other students from the group.

Internet Activity — GENERAL

Restricted and Potential Niches
In a study done by Robert MacArthur in 1958, a pattern of division, similar to **Figure 11,** was found in warblers that feed on insects in coniferous trees (spruce, fir, pine). MacArthur observed that each species of warbler foraged for insects in a different part of the tree, thus reducing competition among the different species. Ask students to search the Internet for this example of niche partitioning. Have them draw a representative profile of a conifer. Have them indicate, with different colors, the parts of the tree that each warbler uses when the other warblers are present. Also, ask students to try to find out what each warbler's potential niche is (the parts of the tree each warbler might use if the other species are absent). If they can find this information, have them draw a dashed line around those parts of the tree. Students could also research and diagram other examples of niche restriction and potential niches.
LS Intrapersonal/Visual

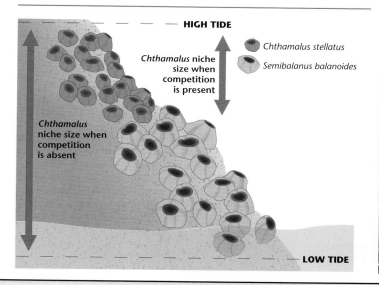

HIGH TIDE

Chthamalus niche size when competition is present

Chthamalus stellatus
Semibalanus balanoides

Chthamalus niche size when competition is absent

LOW TIDE

Figure 11 ▶ The barnacle species *Chthamalus stellatus* uses less of its potential niche when competing for space with a similar barnacle species, *Semibalanus balanoides.*

205

Transparencies

TT Niche Restriction Due to Competition

Demonstration —— GENERAL

Predator Adaptations Ask students to think about the various special adaptations that a predator might have. Start with the head, and ask students to name all the possible features of a successful predator. Draw, or have volunteers draw, each feature on the board as it is suggested. Allow the drawing to get as wild as possible. Have the volunteers label each feature as they draw. (Features may include: a long, sticky tongue; sharp claws, teeth, or bill; keen eyesight or night vision; keen sense of smell; big ears that can move independently; long, muscular legs; long, broad wings; an ejectable stomach or digestive fluids; stealthy silent feathers or paws; camouflage; venom.)
LS Interpersonal/ Visual

English Language Learners

Homework —— GENERAL

Predatory Strategies Have pairs of students research a specific predator-prey relationship of their choice. Encourage them to consider a variety of types of organisms. Have students produce a poster or report describing the effects of this relationship on each of the species. Ask students to include information related to concepts in this chapter, such as the size, dispersal, or habitat of each population, as well as the special adaptations of each species. Encourage students to add this report to their **Portfolio.**
LS Intrapersonal

Figure 12 ▶ This predatory bird had to outrun its prey. Many organisms are adapted to avoid predation.

Predation

An organism that feeds on another organism is called a *predator*, and the organism that is fed upon is the *prey*. This kind of interaction is called **predation.** Examples of predation include snakes eating mice, bats eating insects, or whales consuming krill. **Figure 12** shows a predatory bird with its captured prey.

Predation is not as simple to understand as it seems. We may think of predators as meat-eating animals, but there can be less obvious kinds of predators. In complex food webs, a predator may also be the prey of another species. Most organisms have evolved some mechanisms to avoid or defend against predators.

Some predators eat only specific types of prey. For example, the Canadian lynx feeds mostly on snowshoe hares during the winter. In this kind of

CASE STUDY

Predator-Prey Adaptations

Most organisms are vulnerable to predation, so there is strong selective pressure for adaptations that serve as defenses against predators.

Many animals are *camouflaged*—disguised so that they are hard to see even when they are in view. Visual camouflage is very obvious to us, because vision is the dominant sense in humans. Many predators also have keen vision. An animal's camouflage usually disguises its recognizable features. The eyes are the most recognizable part of the animal, and hundreds of species have black stripes across their eyes for disguise. Dark bands of color, such as those on many snakes, may also break up the apparent bulk of the animal's body.

Some predators do not chase their prey but wait for the prey to come near enough to be caught. Praying mantises and frogs are examples of these types of predators. Such predators are usually camouflaged so that the prey does not notice them waiting to attack.

Animals, and more often plants, may contain toxic chemicals that harm or deter predators. Many animals that have chemical defenses have a striking coloration. This *warning coloration* alerts potential predators to stay away and protects the prey species from damage. Patterns with black stripes and red, orange, or yellow are common in many species of bees, wasps, skunks, snakes, and poisonous frogs.

Warning coloration works well against predators that can learn and that have good vision.

▶ Patterns of black and red, orange, or yellow are common warning signs.

During the course of evolution, members of several well-protected species have come to resemble each other. For example, both bees and wasps often have black and yellow stripes. This is an example of *mimicry* of one species by another. The advantage of mimicry is that the more individual organisms that have the same pattern, the less chance any one individual has of being killed. Also, predators learn to avoid all animals that have similar warning patterns.

REAL-LIFE
● CONNECTION —— GENERAL

Biomimicry The design principle of *biomimicry* uses nature as a model in order to come up with designs and processes that will help solve human problems. It is based on the premise that millions of years of evolution have produced useful innovations that should be copied. Some examples of products that have been inspired by nature include Velcro (barbs on seed pods) and a hay-baling machine called a Haying Mantis™. Have students research biomimicry or brainstorm their own ideas, and summarize their favorite ideas in their **Portfolio.** **LS** Intrapersonal

BRAIN FOOD

Seductive Sirens In the normal firefly (or "lightning bug") courtship ritual, a male advertises for a mate by flashing a certain pattern and watching for a female to flash that pattern back. However, there are some species of fireflies in which a predatory female (so-called "Femme Fatale") will mimic the flash pattern of another species, and eat the male of the other species when he approaches.

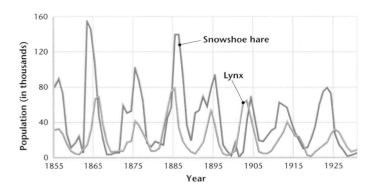

Figure 13 ▶ Populations of predators depend on populations of prey, so changes in one of these populations may be linked to changes in the other. This graph shows population estimates over time for Canadian lynx and their favorite food, snowshoe hares.

close relationship, the sizes of each population tend to increase and decrease in linked patterns, as shown in **Figure 13.** However, many predators will feed on whichever type of prey is easiest to capture.

▶ Both predators and prey may exhibit adaptations such as camouflage or mimicry. The spider that looks like an ant (left) is a predator of insects. The protective quills of this porcupine (right) are a simple but effective way to repel predators.

Occasionally, a harmless species is a mimic of a species that has chemical protection. You have probably tried to get away from insects that you thought were wasps or bees. In fact, some of them were probably flies. Several species of harmless insects have evolved to mimic wasps and bees. On the other hand, sometimes a predator may look like another, less threatening species. Some species of spiders may be mistaken for ants or other types of insects.

A simple defense against predation is some type of *protective covering.* The quills of a porcupine, the spines of a cactus, and the shell of a turtle are all examples of protective covering.

CRITICAL THINKING

1. Making Comparisons For each of these types of adaptations, give an additional example that you have seen or heard of.

2. Determining Cause and Effect Write a paragraph to explain how one of these adaptations might have evolved. **WRITING SKILLS**

Using the Figure — GENERAL
Specialists Vs. Generalists Tell students that some predators, such as the lynx, are *specialists* and eat only a certain type of prey. Other predators, such as wolves and lions, are *generalists,* feeding on the prey that is easiest to find and capture. Specialized predators are closely tied to the population of their prey. This is not true of generalized predators. Ask students how **Figure 13** demonstrates this. (Each increase or decrease in the hare population seems to be matched by an similar change in the lynx population.)

CASE STUDY

Predator-Prey Adaptations
Have a predator-prey costume party. Have students create costumes to represent an organism of their own invention, complete with adaptations. Ask students to make their costumes at home, or provide craft materials. Serve food with wild colors and designs. During the party, have each student role-play their organism. Students could wear a nametag listing their species name and special adaptation(s), and introduce themselves to the other partygoers. "Predators" could hunt for "prey," with success or failure dependent on the adaptations listed on their nametags. Give prizes for the most successful predators and prey, or for the best costume.
LS Kinesthetic/ Interpersonal   English Language Learners

Answers to Critical Thinking
1. Answers may vary. Students should give examples of camouflage, warning coloration, mimicry, and protective covering.
2. Answers may vary.

SKILL BUILDER ———— BASIC

Graphing Simulate the effect of prey population density on the success of a predator. Blindfold a volunteer "predator." Loosely tape 25 poker chips or beans to a large piece of cardboard. Have another student keep time, and allow the predator 30 seconds to pick up as many "prey" as possible. (Adjust the time limit if appropriate.) Conduct five more trials, increasing the amount of prey by 15 each time. Have students record the predator's success in a table, plot the data on a graph, and write an explanation of the results.   English Language **LS** Visual/Logical   Learners

BRAIN FOOD
Feature Recognition Experiments have shown that most animals recognize other animals by three main features: the eyes, the silhouette, and the body's bulk. Most forms of camouflage disguise one of these features. Have students design an experiment to test their own abilities to recognize other animals.

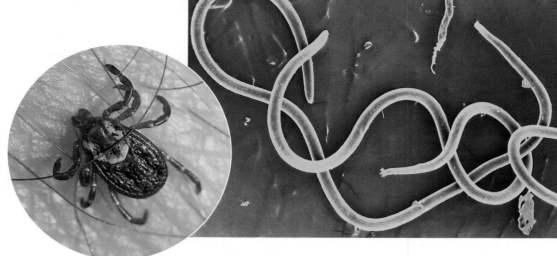

Teach, *continued*

Group Activity — GENERAL

Evolving Relationships Describe some examples of mutualisms to students, or have them research examples. Divide the class into groups, and have each group choose one example. Have the groups hypothesize as to how a parasitism or commensalism might have evolved into this mutualism. Ask each group to present their ideas to the class. **LS** Verbal/Interpersonal

|SKILL BUILDER — GENERAL

Writing Have students write a humorous advertisement for a newspaper or a commercial for television about a species seeking a new relationship with another species. The written ads can be combined in a newspaper format, or students can act out their television ads for the class. Encourage students to include their work in their **Portfolio. LS** Intrapersonal

BOTANY
CONNECTION — GENERAL

Interactions With Plants Plants may also act as predators, prey, parasites, commensals or mutualists. Some examples are: Venus Fly-Trap (predatory), any plant that is eaten whole by herbivores (prey), mistletoe (parasite), bromeliads (commensal epiphytes, like Spanish moss), and legumes (mutual with bacteria in root nodules). Have students think of examples of mimicry or protective covering that they have seen in plants.

Figure 14 ▶ Parasites such as ticks (left) and intestinal worms (right) could be harmful to you. You probably try to avoid these parasites, almost as if they were predators. In what ways are parasites like predators?

Figure 15 ▶ These acacia trees in Central America have a mutualistic relationship with these ants. The trees provide food and shelter to the ants, and the ants defend the tree.

Parasitism

An organism that lives in or on another organism and feeds on the other organism is a *parasite*. The organism the parasite takes its nourishment from is known as the *host*. The relationship between the parasite and its host is called **parasitism.** Examples of parasites are ticks, fleas, tapeworms, heartworms, bloodsucking leeches, and mistletoe.

The photos of parasites in **Figure 14** may make you feel uneasy, because parasites are somewhat like predators. The differences between a parasite and a predator are that a parasite spends some of its life in or on the host, and that parasites do not usually kill their hosts. In fact, the parasite has an evolutionary advantage if it allows its host to live longer. However, the host is often weakened or exposed to disease by the parasite.

Mutualism

Many species depend on another species for survival. In some cases, neither organism can survive alone. A close relationship between two species in which each species provides a benefit to the other is called **mutualism.** Certain species of bacteria in your intestines form a mutualistic relationship with you. These bacteria help break down food that you could not otherwise digest or produce vitamins that your body cannot make. In return, you give the bacteria a warm, food-rich habitat.

Another case of mutualism happens in the ant acacia trees of Central America, shown in **Figure 15.** Most acacia trees have spines that protect them against plant-eating animals, but the ant acacias have an additional protection—an ant species that lives only on these trees. The trees provide these ants shelter within hollow thorns as well as food sources in sugary nectar glands and nutrient-rich leaf tips. In turn, the ants defend the tree against herbivores and many other threats.

208

MISCONCEPTION ///ALERT\\\

Herbivores are Like Both Predators and Parasites An herbivore is in a way like a parasite because it feeds on a plant without killing it (usually). However, herbivores do not always live in or on plants, and in that way are not like parasites. Herbivores could be considered predators because they "hunt" for and sometimes kill certain plants. As with all inter-species relationships, a given relationship may not fit any definition perfectly.

LANGUAGE ARTS
CONNECTION — GENERAL

Parasite Rex Have students read passages from Carl Zimmer's *Parasite Rex: Inside the Bizarre World of Nature's Most Dangerous Creatures.* The book offers a guided tour to the hidden, fascinating world of parasites, and shows how parasites can control the fate of entire ecosystems and even steer the course of evolution. After reading, ask students if their understanding of the significance of parasites in ecosystems has changed.

Commensalism

A relationship in which one species benefits and the other species is neither harmed nor helped is called **commensalism.** An example is the relationship between sharks and a type of fish called remoras, which are shown in **Figure 16.** Remoras attach themselves to sharks and feed on scraps of food left over from the shark's meals. Another example of commensalism is when birds nest in trees, but only if the birds do not cause any harm to the tree. Even a seemingly harmless activity might have an effect on another species.

Symbiosis and Coevolution

A relationship in which two organisms live in close association is called **symbiosis.** Many types of species interactions are considered symbiotic in some cases. Symbiosis is most often used to describe a relationship in which at least one species benefits.

Over time, species in close relationships may *coevolve.* These species may evolve adaptations that reduce the harm or improve the benefit of the relationship. Recall that harm and benefit are measured in total effects over time. For example, coevolution can be seen in the relationships of flowering plants and their pollinators. Many types of flowers seem to match the feeding habits of certain species of insects or other animals that spread pollen.

Connection to Biology

An Ecosystem in Your Body
Our health is affected by our relationships with microorganisms in our digestive system, skin, blood, and other parts of our body. For example, live-culture yogurt is considered a healthy food because the kinds of bacteria it contains are beneficial to us. The bacteria assist our digestion of dairy products and also compete with other microorganisms, such as yeast, that might cause infections.

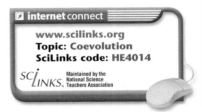

💻 **internet** connect
www.scilinks.org
Topic: Coevolution
SciLinks code: HE4014
SCiLINKS Maintained by the National Science Teachers Association

SECTION 2 Review

1. **List** as many parts as you can of the niche of an organism of your choice.

2. **Give examples** of species that have the same habitat but not the same niche that a lion has.

3. **Describe** the five types of species interactions.

CRITICAL THINKING

4. **Making Comparisons** Read the definition of parasites and predators, and then explain how parasites differ from predators. **READING SKILLS**

5. **Analyzing Relationships** Choose an example of mutualism, and then describe the long process by which the relationship could have developed.

Answers to Section Review

1. Answers may vary. Students should mention habitat traits, food sources, and other species with which the organism interacts, and consider interactions with abiotic factors and with members of each biological kingdom.

2. Answers may vary. Any prey organism or any plant would qualify. Predators that do not eat the same food would qualify. Even predators that eat the same food, but also eat different foods, would qualify. Examples: zebras, grass, anteaters, hyenas.

3. Competition harms both organisms, predation and parasitism harms one and benefits one, mutualism benefits both, commensalism benefits one and has no effect on the other.

4. Unlike predators, a parasite spends some of its life in or on the host, and generally does not kill the host.

5. Answers may vary. A mutualistic relationship may evolve from a close relationship that did not originally benefit both parties. As two species interact closely over time, either organism may evolve methods that take better advantage of the relationship with the other. However, organisms cannot evolve a trait on purpose.

Alternative Assessment ——— GENERAL

Species Management Plan Ask students to devise a management plan for a species found in a zoo, aquarium, or wildlife refuge. In preparing their plan, students could answer questions such as, "How could you ensure the population has all it needs? How should the population size be controlled? Which other species should be allowed in the same area? What restrictions or regulations should be placed on the use of the area?"
LS Verbal

Species Interactions Ask students to collaborate on a collage of photographs and words from magazines and newspapers that illustrate and describe each of the five interactions between organisms. Provide magazines, newspapers, poster board, scissors, and glue. When it is finished, display the collage in the classroom.
LS Visual

Co-op Learning	English Language Learners

Competition Debate Divide students into two groups. Then ask each group to consider the following statement, "Humans must compete in order to survive." Have one group defend the statement and have the other argue against it. Have students in each group brainstorm ideas before they begin to formally argue. Ask them to describe specific situations in history to support their arguments.
LS Verbal/Interpersonal

---

1 How Populations Change in Size

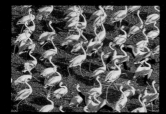

Key Terms

population, 197
density, 198
dispersion, 198
growth rate, 198
reproductive potential, 199
exponential growth, 199
carrying capacity, 200

Main Ideas

▶ Each population has specific properties, including size, density, and pattern of dispersion.

▶ Each population has a characteristic reproductive potential. This is the fastest possible growth rate of the population.

▶ When a population has few limits to its growth, it may have an exponential growth rate. Usually, population growth is limited by factors such as disease and competition.

▶ Carrying capacity is the maximum population a habitat can support over time.

▶ A population that grows rapidly may be subject to density-dependent regulation.

2 How Species Interact with Each Other

niche, 203
competition, 204
predation, 206
parasitism, 208
mutualism, 208
commensalism, 209
symbiosis, 209

▶ The niche of an organism is its pattern of use of its habitat and its interactions with other organisms.

▶ Interactions between species are categorized based on the relative benefit or harm that one species causes the other. The categories are competition, predation, parasitism, mutualism, and commensalism.

▶ Competition between species occurs when their niches overlap. The competition may be direct or indirect.

▶ Pairs of species that have close relationships often evolve adaptations that either increase the benefit of or reduce the harm from the relationship.

210

INCLUSION Strategies ——— GENERAL

• *Attention Deficit Disorder*
• *Learning Disabled*

Have each student develop a poster about predation. The poster should include a definition, three examples of a predator-prey relationship, and some mechanisms that organisms have evolved which allow them to avoid predators. Lower level students can simply draw pictures of predators and their prey. The students may present their work in small groups or display the posters in the classroom.

Chapter Resource File

• **Chapter Test** GENERAL
• **Chapter Test** ADVANCED
• **Concept Review** GENERAL
• **Critical Thinking** ADVANCED
• **Test Item Listing**
• **Modeling Lab** GENERAL
• **CBL™ Probeware Lab** ADVANCED
• **Consumer Lab** GENERAL
• **Long-Term Project** GENERAL

Using Key Terms

Use each of the following terms in a separate sentence.

1. *reproductive potential*
2. *carrying capacity*
3. *competition*
4. *symbiosis*

For each pair of terms, explain how the meanings of the terms differ.

5. *niche* and *habitat*
6. *predator* and *prey*
7. *predation* and *parasitism*
8. *mutualism* and *commensalism*

> ✔ **STUDY TIP**
>
> **Review with a Partner** To review the main ideas of the text, try summarizing with a partner. Take turns reading a passage, and then try to summarize aloud what you have read. Try not to look back at the text. Then, discuss and review the text with your partner to check your understanding.

Understanding Key Ideas

9. In which of the following pairs do both organisms belong to the same population?
   a. a rose and a carnation
   b. a zebra and a horse
   c. two residents of New York City
   d. two similar species of monkeys

10. A population of some species is most likely to grow exponentially
    a. if the species is already very common in the area.
    b. when the species moves into a new area of suitable habitat.
    c. when it uses the same habitat as a similar species.
    d. if the population size is already large.

11. A population will most likely deplete the resources of its environment if the population
    a. grows beyond carrying capacity.
    b. must share resources with many other species.
    c. moves frequently from one habitat to another.
    d. has a low reproductive potential.

12. The growth rate of a population of geese will probably increase within a year if
    a. more birds die than are hatched.
    b. several females begin laying eggs at younger ages than their mothers did.
    c. most females lay two eggs instead of three during a nesting season.
    d. some birds get lost during migration.

13. Which of the following is an example of competition between species?
    a. two species of insects feeding on the same rare plant
    b. a bobcat hunting a mouse
    c. a lichen, which is an alga and a fungus living as a single organism
    d. a tick living on a dog

14. Which of the following statements about parasitism is true?
    a. The presence of a parasite does not affect the host.
    b. Parasitism is a cooperative relationship between two species.
    c. Parasites always kill their hosts.
    d. Parasitism is similar to predation.

15. Ants and acacia trees have a mutualistic relationship because
    a. they are both adapted to a humid climate.
    b. they are part of the same ecosystem.
    c. they benefit each other.
    d. the ants eat parts of the acacia tree.

16. Which of the following is an example of coevolution?
    a. flowers that can be pollinated by only one species of insect
    b. rabbits that invade a new habitat
    c. wolves that compete with each other for territory
    d. bacteria that suddenly mutate in a lab

211

Assignment Guide

Section	Questions
1	1, 2, 9–12, 19–23, 26–28, 30–33
2	3–8, 13–18, 24, 25, 33, 34

ANSWERS

Using Key Terms

1. Sample answer: Reproductive potential is the fastest possible rate at which a population can reproduce.

2. Sample answer: The carrying capacity of an ecosystem for a given species is determined by a limiting resource.

3. Sample answer: Organisms are subject to competition from members of their own species, and often from other species as well.

4. Sample answer: Any close relationship between organisms can be said to be a symbiosis.

5. A habitat is the environment in which an organism lives; a niche is the pattern of use of its environment.

6. Predator and prey are the two parties in a predation relationship. The predator hunts and eats the prey.

7. In predation, one organism eats another or part of another. In parasitism, one organism lives in or on another, and usually does not kill its host.

8. In mutualism, both organisms benefit. In commensalism, one benefits and one experiences no positive or negative effect.

Understanding Key Ideas

9. c.
10. b.
11. a.
12. b.
13. a.
14. d.
15. c.
16. a.

Short Answer

17. This is parasitism. The organism benefits to the detriment of the cow, and it lives in the cow during part of its life.

18. Some species may use the same resource, but at different times of the day. Individuals of an earlier generation may deplete a resource so that those in a later generation cannot use it to survive.

19. A few individual snail kites may have had slightly hooked bills that allowed them to eat more snails than other individuals. These birds were able to produce more hook-billed offspring, and gradually the population evolved to include only hook-billed individuals.

20. The snails would die off, and then the snail kite population would starve and die off.

Interpreting Graphics

21. The population seemed to increase in a linear fashion until the mid-1930s, when it increased exponentially and then crashed in the late-1930s.

22. The food resource was probably depleted by 1937, which caused the population crash.

23. Answers may vary. It's difficult to estimate the carrying capacity, because the population did not recover to a stable level after crashing. It looks like the level of 250 individuals was steady over time in the early-1920s.

Concept Mapping

24. Answers to the concept mapping questions are on pp. 667–672.

Short Answer

17. A tapeworm lives in the intestines of a cow and feeds by absorbing food that the cow is digesting. What kind of relationship is this? Explain your answer.

18. Explain how two species can compete for the same resource even if they never come in contact with each other.

19. Snail kites are predatory birds that feed only on snails. The kites use their hooked, needle-like beaks to pull snails from their shells. Explain how these specialized beaks might have evolved in these birds.

20. What would happen to the population of snail kites mentioned in question 19 if the snails' habitat was destroyed? Explain your answer.

Interpreting Graphics

The graph below shows the population of some reindeer that were introduced to an Alaskan island in 1910. Use the graph to answer questions 21–23.

21. Describe this population's changes over time.

22. What might have happened in 1937?

23. How would you estimate this island's carrying capacity for reindeer? Explain your answer.

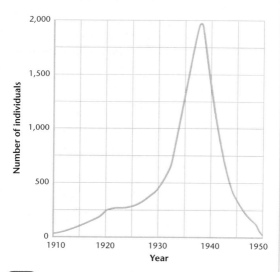

Concept Mapping

24. Use the following terms to create a concept map: *symbiosis, predation, predator, prey, parasitism, parasite, host, mutualism,* and *commensalism.*

Critical Thinking

25. Analyzing Relationships Read the explanations of competition and predation. If one species becomes extinct, and then soon after, another species becomes extinct, was their relationship most likely competition or predation? Explain your answer. **READING SKILLS**

26. Evaluating Hypotheses Scientists do not all agree on the specific carrying capacity of Earth for humans. Why might this carrying capacity be difficult to determine?

27. Evaluating Conclusions A scientist finds no evidence that any of the species in a particular community are competing and concludes that competition never played a role in the development of this community. Could this conclusion be valid? Write a paragraph to explain your answer. **WRITING SKILLS**

Cross-Disciplinary Connection

28. Health Viruses are the cause of many infectious diseases, such as common colds, flu, and chickenpox. Viruses can be passed from one person to another in many different ways. Under what conditions do you think viral diseases will spread most rapidly between humans? What can be done to slow the spread of these viruses?

Portfolio Project

29. Create a Niche Map Create a visual representation of the niche of an organism of your choice. Research the organism's habitat, behaviors, and interactions with other species. If possible, observe the organism (without disturbing it) for a day or more. Create a piece of art to show all of the interactions that this organism has with its environment.

Critical Thinking

25. Answers may vary. The relationship cannot be known from the information given. It could have been competition—if both populations were using one common food source, then both would decline if their food source was lost. However, it could have been predation—if the first species to die off was the prey of the second, the predator might have had no other food source.

26. Answers may vary. Humans have often altered their use of habitats (changed their niche) and have been clever about exploiting resources in new ways so that resources in a

particular area are no longer limiting. However, these activities remove resources from other areas and species, and drastically alter ecosystems. Humans may be able to stretch the Earth's carrying capacity to the sum of its ecosystems.

27. Answers may vary. It would be difficult to say that competition never played a role unless you knew the complete history of all organisms in the ecosystem. Moreover, competition—at least within a species—is inevitable in natural ecosystems.

MATH SKILLS

Use the equation below to answer questions 30–31.

30. Extending an Equation The equation gives the change in a population over a given amount of time (for example, an increase of 100 individuals in one year). Use the two parts on the right side of the equation to write an inequality that would be true if the population were increasing. Rewrite the inequality for a decreasing population.

31. Analyzing an Equation Suppose you are studying the small town of Hill City, which had a population of 100 people in the first year of your study. One year later, 10 people have died, and only 9 mothers have given birth. Yet the population has increased to 101. How could this increase happen?

 WRITING SKILLS

32. Communicating Main Ideas Why do population sizes not grow indefinitely?

33. Creative Writing Write a science fiction story about life without competition.

34. Writing from Research Find information in encyclopedias or natural history references about different kinds of mutualism. Summarize the similarities and differences between the various relationships. Focus on the ways in which each species benefits from the other species.

STANDARDIZED TEST PREP

For extra practice with questions formatted to represent the standardized test you may be asked to take at the end of your school year, turn to the sample test for this chapter in the Appendix.

 READING SKILLS

Read the passage below, and then answer the questions that follow.

Excerpt from Charles Darwin, On the Origin of Species, *1859.*

I should premise that I use the term struggle for existence in a large and metaphorical sense, including dependence of one being on another, and including (which is more important) not only the life of the individual, but success in leaving progeny. Two canine animals in a time of dearth, may truly be said to struggle with each other which shall get food and live. But a plant on the edge of the desert is said to struggle for life against the drought, though more properly it should be said to be dependent on the moisture. A plant which annually produces a thousand seeds, of which on average only one comes to maturity, may more truly be said to struggle with the plants of the same and other kinds which already clothe the ground . . . In these several senses, which pass into each other, I use for convenience sake the general term of struggle for existence.

1. Which of the following statements best describes the author's main purpose in this passage?
   a. to describe the process of reproduction
   b. to persuade the reader that all animals struggle for existence
   c. to explain the meaning of the author's use of the phrase *struggle for existence*
   d. to argue that life in the desert depends on moisture

2. Which of the following statements most closely matches what the author means by the phrase *struggle for existence*?
   a. whenever plants or animals interact in nature
   b. whenever plants or animals compete to survive and to produce offspring
   c. when plants produce many more seeds than are likely to grow
   d. when animals compete for food during difficult times

Cross-Disciplinary Connection

28. Viruses will be spread quickly in dense populations. The spread of viral diseases can be slowed by limiting contact or transfer of infected materials between individuals (hygiene, physical separation).

Portfolio Project

29. Answers may vary. Be sure the students include all possible interactions between the organism, its habitat, and other species.

Math Skills

30. For a population to increase, births must be greater than deaths. To decrease, deaths must be greater than births.

31. Answers may vary. Two people may have moved to town, or perhaps a mother has given birth to twins. Also, the original equation left out the possibility that individuals might move in or out of a population (immigration and emigration).

Writing Skills

32. Answers may vary. Limited resources in an area limit the size of any population, so it cannot increase indefinitely.

33. Answers may vary.

34. Answers may vary.

Reading Skills

1. c
2. b

213

STUDYING POPULATION GROWTH

Teacher's Notes

Time Required
two 45-minute class periods

Lab Ratings

EASY ———————→ HARD

TEACHER PREPARATION 🝊🝊🝊
STUDENT SETUP 🝊🝊
CONCEPT LEVEL 🝊
CLEANUP 🝊🝊

Skills Acquired
- Predicting
- Organizing and Analyzing Data

The Scientific Method
In this lab, students will:
- Make Observations
- Draw Conclusions

Safety Cautions
Caution students to treat all micro-organisms as potential pathogens, and tell them that methylene blue is harmful if swallowed and can cause skin or eye irritation. Call the Poison Control center if a student ingests any of this substance. Remind students to wear gloves and goggles, to keep their hands away from their faces, and to wash their hands after this lab. Review sterile techniques.

The yeast/methylene blue mixtures should be disposed of as hazardous waste. Microscope slides and cover slips should be disposed of as contaminated glassware.

Objectives
▶ **USING SCIENTIFIC METHODS** **Observe, record, and graph** the growth and decline of a population of yeast cells in an experimental environment.

▶ **USING SCIENTIFIC METHODS** **Predict** the carrying capacity of an environment for a population.

▶ **Infer** the limiting resource of an environment.

Materials
compound microscope
methylene blue solution, 1%
micrometer, stage type or
  eyepiece disc for microscope
microscope slide, with coverslip
  (5)
pipet, 1 mL (5)
test tube (5)
yeast culture, in an Erlenmeyer
  flask (5)

▶ **Budding Yeast** This live, budding yeast cell is magnified 3,025 times by a scanning electron microscope.

Studying Population Growth
You have learned that a population will keep growing until limiting factors slow or stop this growth. How do you know when a population has reached its carrying capacity? In this lab, you will observe the changes in a population of yeast cells. The cells will grow in a container and have limited food over several days.

Procedure
1. Your teacher will prepare several cultures of household baker's yeast (fungi of the genus *Saccharomyces*) in flasks. Each yeast culture will have grown for a different period of time in the same type of environment. Each flask will have been prepared with 500mL lukewarm water, 1 g of active dry yeast, and 20 g of sugar. The sugar is the only food source for the yeast.

2. Make two data tables like the one shown below. One table will contain your observations of living yeast cells, and the other table will be for observations of dead yeast cells.

Cell Counts						
Time (h)	1	2	3	4	Average	Class Average
0						
12						
24						
36						
48						

DO NOT WRITE IN THIS BOOK

3. Take a sample of yeast culture from the first flask (Time 0). Swirl the flask gently to mix the yeast cells evenly, and then immediately use the pipet to transfer 1 mL of yeast culture to a test tube. Add two drops of methylene blue solution to the test tube. The methylene blue will stain the dead yeast cells a deep blue but will not stain the living cells.

4. Make a wet mount by placing a small drop of your mixture of yeast and methylene blue on a microscope slide. Cover the slide with a coverslip.

5. Observe the mounted slide under the low power of a compound microscope. (Note: Adjust the light so that you can clearly see both stained and unstained cells.) After focusing, switch the microscope to high power (400× or 1,000×).

6. Count the live (unstained) yeast cells and the dead (stained) cells that you see through the microscope. Use the micrometer ruler, or ask your teacher for the best counting method. Record the numbers of live cells and dead cells in your data tables.

Materials
The materials listed on the student page are enough for a group of 3 to 4 students. A stage micrometer, available from laboratory suppliers, is a microscopic scale printed on a microscope slide. Similar micrometer discs that fit into the microscope eyepiece are also available.

To prepare the yeast cultures, obtain 3 standard supermarket packages (7 g) of active dry yeast (not fast-rising or instant yeast).

Activate each of 5 batches of yeast culture at 12-hour intervals prior to the time students will start their observations (0, 12, 24, 36, and 48 hours). To activate the yeast, dissolve 20 g of sugar in 500 mL of warm water (approximately 32°C or 90°F) in a flask, then add 1 g (~1/4 tsp) yeast and swirl thoroughly. Keep the flasks covered in a warm, dark area until beginning the lab.

7. Move the microscope slide slightly, and then make another count of the number of living and dead cells that you can see. Repeat this step until you have made four counts of the cells on the slide.

8. Calculate and record the average number of live cells per observation. Record this number in your data table. Do the same calculation for the dead cells.

9. Predict how many live and dead cells you expect to count in the samples from the other flasks. Record your prediction.

10. Repeat steps 3–8 for each of the flasks to obtain data that represents the growth of a population over a 48-hour period.

11. Clean up your work area, and store all lab equipment appropriately. Ask your teacher how to dispose of the yeast samples and any extra or spilled chemicals.

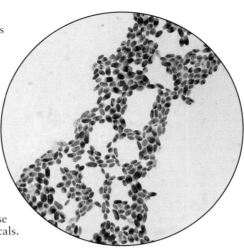

▶ **Stained Yeast Cells** These yeast cells have been stained with methylene blue and magnified 1,000 times with a microscope. Methylene blue gives a deep blue color to dead yeast cells but not to live yeast cells.

Analysis

1. **Analyzing Data** Share your data with the rest of the class. Calculate and record the class averages for each set of observations.

2. **Constructing Graphs** Graph the changes in the average numbers of live yeast cells and dead yeast cells over time. Plot the average number of cells per observation on the *y*-axis and the time (in hours from start of culture) on the *x*-axis.

3. **Describing Events** Describe the general population changes you observed in the yeast cultures over time.

Conclusions

4. **Evaluating Methods** Why were several counts taken and then averaged for each time period?

5. **Evaluating Results** Were your predictions of the yeast cell counts close to the actual average counts? How close were your predictions relative to the variation among all the samples?

6. **Applying Conclusions** Did the yeast cell populations appear to reach a certain carrying capacity? What was the limiting resource in the experimental environment of the flasks?

Extension

1. **Designing Experiments** Form a hypothesis about another factor that might limit the yeast's population growth, and explain how you would test this hypothesis.

Linda Culp
Thorndale High School
Thorndale, Texas

Tips and Tricks

Tell students that they are observing yeast cultures at several stages of growth, to simulate the observation of a single population over time. Point out the generation of gas bubbles by the actively-growing cultures. Demonstrate for students how a culture is prepared (Time 0), how to properly swirl the flask of yeast culture (failure to evenly distribute cells will alter counts), how to use a pipet the transfer a sample into a test tube, how to place a small drop of yeast-stain mixture onto a slide, and how add the cover slip, and how to focus the microscope.

Tell students that the methylene blue only stains the dead yeast cells because living cells actively keep out the stain. Tell students to include the buds on actively-dividing yeast cells in their cell counts. You may want to experiment with and demonstrate a method of counting the cells that works best for the microscopes and materials you have. Variation among counts taken from the same flask should be less significant than differences between flasks at different stages.

Answers to Analysis

1. Answers may vary. Student counts should be similar to the class average.

2. Students should create a double line graph to compare how the density of live and dead cells changes over time.

3. Answers may vary. Yeast populations should increase exponentially, stabilize, and then decline.

Answers to Conclusions

4. Counting several random samples allows for variation in yeast dispersion within the flask or within the microscope slide.

5. Answers may vary.

6. Answers may vary. A maximum population size generally indicates the carrying capacity. The limiting resources are food (sugar) and habitat (pH-neutral water).

Answers to Extension

1. Answers may vary.

WHERE SHOULD THE WOLVES ROAM?

Background

Research done in northeastern Minnesota by world-renowned wolf expert L. David Mech sought to understand how wolves and humans compete for the White-tailed Deer resource. Mech and colleagues found that buck harvest by human hunters was reduced by large wolf populations only in the poorest-quality habitats. In better habitats, where deer were more prevalent, wolves had no significant impact on buck harvest.

>POINTS >of view

WHERE SHOULD THE WOLVES ROAM?

The gray wolf was exterminated from much of the northwestern United States by the 1920s. Ranchers and federal agents killed the animal to protect livestock. The Rocky Mountain gray wolf was listed as an endangered species in 1973. Then in the 1980s, the U.S. Fish and Wildlife Service began a plan to restore wolf populations in the United States. The agency decided to reintroduce wolves into certain areas. Biologists looked for areas where wolves could have large habitats and enough food. Three areas were chosen, as shown in the figure below.

Between 1995 and 1996, 64 wolves were released in Yellowstone National Park and 34 wolves were released in central Idaho. The original goal was to have breeding populations of at least 100 wolves in each location by 2002. This goal has now become reality.

The wolf reintroduction efforts remain controversial. Some people would prefer that the wolves become extinct. On the other side, some people think that the government has not done enough to protect wolf populations. Read the following points of view, and then analyze the issue for yourself.

Wolves Should Not Be Reintroduced

Some opponents of the reintroduction plan argue that wolves are not truly endangered. Biologists estimate that hundreds of wolves live in Minnesota, and thousands live in Alaska and Canada. Because there are large numbers of wolves in the wild, some people feel that wolves should not receive special treatment as endangered species.

Many hunters also oppose the plan to reintroduce wolves. Both hunters and wolves hunt for large game animals such as deer, elk, or moose. Hunters believe the wolves might create too much competition for the game animals. Some studies suggest that populations of game animals will decrease if hunted by both humans and wolves.

▶ **A wild wolf** is a now rare sight in most of the United States. Efforts to reintroduce and protect wolf populations are controversial.

Hunters point out that hunting is an important part of the economies of the western states. Also, licensed hunting has become part of the way large parks and wildlife preserves are managed. Hunting is sometimes allowed by park and game managers to control wildlife populations. Hunting fees also help fund wildlife management efforts, such as habitat improvement and biological studies.

Ranchers are among the people who most strongly oppose wolf reintroduction. Ranchers worry that wolves will kill their livestock. Ranchers argue that they cannot afford to lose their livestock. Even though there is a program to pay ranchers for lost livestock, the program will last

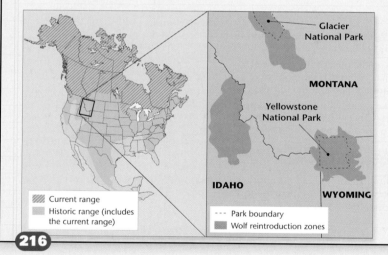

Current range

Historic range (includes the current range)

Glacier National Park

MONTANA

Yellowstone National Park

IDAHO

WYOMING

- - - Park boundary

Wolf reintroduction zones

▶ **The breeding range of the gray wolf** (far left) has been lost in most of the United States. The U.S. government has reintroduced wolves into parts of Montana, Wyoming, and Idaho (left).

216

📄 Internet Activity ——— GENERAL

Wolf Population Status The United States Fish and Wildlife Service, and many others who are interested in wolves, post information on their websites about the status of wolf populations, and plans for recovery and management of wolves in different areas. Have students research wolf populations in an area that they have visited or that is nearest to them. Students may also be interested in finding out about wolf conservation around the world. Countries such as Belarus, Finland, Spain, Italy and Poland have wolf populations that are increasing. Encourage students to research the ways that some of these countries are dealing with the challenges of human/wolf interactions and livestock protection. Students can find out about the work done by organizations such as The U.K. Wolf Conservation Trust, Wolf Haven International, The Ethiopian Wolf Conservation Programme, or The Wolf Trust.

only as long as wolves are classified as an endangered species. When there are many wolves again in the target areas, the wolf will no longer have endangered status. Ranchers point out that the payment program will disappear when it would be needed most.

Other groups that oppose wolf reintroduction include groups that use public lands for activities such as logging or mining. These groups worry that protection for the wolves may lead to the prevention of other uses of land in the target areas.

Wolves Should Be Reintroduced

A basic argument in favor of wolf reintroduction is that the federal government must uphold the law by trying to restore wolf populations in the United States. But supporters of wild wolves give other reasons too.

Many environmentalists and scientists believe that the reintroduction plan could restore a balance to the Yellowstone ecosystem. Predation by wolves would keep the herds of elk, moose, and deer from growing too dense and overgrazing the land.

The argument that wolves help control wild herds is like the argument in favor of hunting. Some wolf supporters even say that licensed hunting of wolves should be allowed. In this way, hunters might support the reintroduction plans, and populations of both wolves and game animals could be managed.

In response to ranchers' concerns that wolves will attack their livestock, biologists say that this is not likely to be a problem. There

▶ **The International Wolf Center in Minnesota** tries to educate the public about wolves. The center's "Wolves and Humans" exhibit is shown here.

is evidence that most wolves prefer to hunt wild animals rather than domestic animals. Wolves rarely attack livestock when large herds of wild game are nearby. In fact, from 1995 to 1997, fewer than five wolf attacks on livestock were reported in the United States.

Still, some supporters of reintroduction have tried to address the concerns of ranchers. One group raised money to pay ranchers for livestock killed by wolves. Other groups conduct studies and educational programs or talk with local landowners. Most wolf supporters are trying to create reintroduction plans that will work for both humans and wolves.

In response to fears that the wolves pose a danger to humans, supporters say this is also unlikely. There have been no verified attacks on humans by healthy wolves in North America. Wolf experts insist that wolves are shy

animals that prefer to stay away from people.

Most wolf supporters admit that there are only a few places where wolves may live without causing problems. Supporters of the plan believe that the target areas are places where wolves can carry out a natural role without causing problems for humans.

What Do You Think?

Like many plans to protect endangered species, the plan to reintroduce wolves causes some people to weigh their own interests against the needs of a single species. Do you feel that the decision is a simple one? Can you think of other ways to look at this issue? Explain your answers.

Depredation and Animal Husbandry In areas of dense wolf populations, ranchers may worry about wolf predation on their livestock (called depredation). A study was conducted to compare ranches that had experienced high depredation with ranches that had little depredation. The only real differences found were that ranches with high depredation rates had larger herds of cattle that were farther away from ranch buildings. It was unclear whether different animal husbandry practices played a role. Various animals have been used to guard sheep from wolves. Shepherd dogs have been bred for this purpose, but another animal that serves well is the donkey. Donkeys will stay with the flock and bond with the sheep. They will loudly bray when confronted by a predator, and will chase and even kill a canine predator that threatens the herd. Students may want to find out other alternatives for defending livestock from wolves.

217

Answers to What Do You Think?

Answers will vary. Other considerations may include: the value of wolves for eco-tourism; the fact that coyotes have taken over wolves' niches in many places; the idea that national parks are supposed to be preserves of our natural heritage, which includes all the species therein; a romantic or iconic image of wolves; that idea that human life and property is paramount to any other species; the possible influence of childhood fairy tales.

PACING	CLASSROOM RESOURCES	LABS, DEMONSTRATIONS, AND ACTIVITIES
BLOCKS 1 & 2 · 90 min pp. 218–224 **Chapter Opener**		
Section 1 Studying Human Populations	**OSP** Lesson Plan * **CRF** Active Reading * BASIC **TT** Bellringer * **TT** Human Population Over Time * **TT** Age-Structure Diagrams * **TT** Total Fertility Rate & Population Growth in the U.S. * **TT** The Demographic Transition *	**TE** Activity Growth Rates, p. 219 GENERAL **TE** Activity Modeling Infant Mortality, p. 222 ◆ GENERAL **TE** Internet Activity Demographic Data, p. 223 ADVANCED **CRF** Math/Graphing Lab * GENERAL **CRF** Math/Graphing Lab * ◆ GENERAL **CRF** Long-Term Project * ◆ BASIC
BLOCKS 3, 4 & 5 · 135 min pp. 225–231 **Section 2** Changing Population Trends	**OSP** Lesson Plan * **CRF** Active Reading * BASIC **TT** Bellringer * **TT** Worldwide Trends in Fertility & Population Growth *	**TE** Group Activity Suburban, Metropolitan, Rural, or ? p. 226 GENERAL **SE** Field Activity Does Your Local Area Have Population Pressures? p. 227 **TE** Group Activity UN Population Goals, p. 230 ADVANCED **SE** QuickLab Estimating Fertility Rates, p. 231 ◆ **SE** Skills Practice Lab How Will Our Population Grow? pp. 236–237 **CRF** Datasheets for In-Text Labs * **CRF** Consumer Lab * ◆ ADVANCED **CRF** Modeling Lab * ◆ GENERAL

BLOCKS 6 & 7 · 90 min

Chapter Review and Assessment Resources

- **SE** Chapter Review pp. 233–235
- **SE** Standardized Test Prep pp. 652–653
- **CRF** Study Guide * ■ GENERAL
- **CRF** Chapter Test * ■ GENERAL
- **CRF** Chapter Test * ADVANCED
- **CRF** Concept Review * ■ GENERAL
- **CRF** Critical Thinking * ADVANCED
- **OSP** Test Generator
- **CRF** Test Item Listing *

Online and Technology Resources

 Holt Online Learning

Visit go.hrw.com for access to Holt Online Learning or enter the keyword **HE6 Home** for a variety of free online resources.

 **One-Stop Planner® CD-ROM**

This CD-ROM package includes
- Lab Materials QuickList Software
- Holt Calendar Planner
- Customizable Lesson Plans
- Printable Worksheets
- ExamView® Test Generator
- Interactive Teacher Edition
- Holt PuzzlePro® Resources
- Holt PowerPoint® Resources

 Holt Environmental Science Interactive Tutor CD-ROM

This CD-ROM consists of interactive activities that give students a fun way to extend their knowledge of environmental science concepts.

KEY

TE	Teacher Edition	**CRF**	Chapter Resource File	*	Also on One-Stop Planner
SE	Student Edition	**TT**	Teaching Transparency	■	Also Available in Spanish
OSP	One-Stop Planner	**CD**	Interactive CD-ROM	◆	Requires Advance Prep

ENRICHMENT AND SKILLS PRACTICE	SECTION REVIEW AND ASSESSMENT	CORRELATIONS
SE Pre-Reading Activity, p. 218 **TE** Using the Figure, p. 218 `GENERAL`		**National Science Education Standards**
TE Skill Builder Math, p. 219 `ADVANCED` **TE** Using the Figure Age-Structure Diagrams, p. 220 `BASIC` **TE** Skill Builder Graphing, p. 220 `GENERAL` **TE** Skill Builder Vocabulary, p. 220 `GENERAL` **SE** MathPractice Extending the Equation for Population Change, p. 221 **TE** Interpreting Statistics Historical Correlation, p. 221 `GENERAL` **TE** Career Statistician, p. 221 **TE** Inclusion Strategies, p. 222 **TE** Using the Figure The Demographic Transition, p. 223 `BASIC` **SE** Maps in Action Fertility Rates and Female Literacy in Africa, p. 238	**SE** Section Review, p. 224 **TE** Reteaching, p. 224 `BASIC` **TE** Quiz, p. 224 `GENERAL` **TE** Alternative Assessment, p. 224 `BASIC` **CRF** Quiz * ■ `GENERAL`	SPSP 2a SPSP 2b
SE Graphic Organizer Spider Map, p. 226 `GENERAL` **TE** Reading Skill Builder Reading Organizer, p. 227 `BASIC` **SE** Case Study Thailand's Population Challenges, pp. 228–229 **TE** Inclusion Strategies, p. 229 **TE** Interpreting Statistics Declining Fertility Rates, p. 230 `GENERAL` **SE** Society & the Environment Lost Populations: What Happened? p. 239 **CRF** Map Skills Population Density * `GENERAL`	**TE** Homework, p. 229 `BASIC` **SE** Section Review, p. 231 **TE** Reteaching, p. 231 `BASIC` **TE** Quiz, p. 231 `GENERAL` **TE** Alternative Assessment, p. 231 `ADVANCED` **CRF** Quiz * ■ `GENERAL`	SPSP 2b SPSP 3a SPSP 3b

Guided Reading Audio CDs

These CDs are designed to help auditory learners and reluctant readers. (Audio CDs are also available in Spanish.)

www.scilinks.org

Maintained by the **National Science Teachers Association**

TOPIC: Demographic Transition
SciLinks code: HE4018

TOPIC: Human Demographics
SciLinks code: HE4056

TOPIC: Developed and Developing Countries
SciLinks code: HE4021

CNN Videos

Each video segment is accompanied by a Critical Thinking Worksheet.

Science, Technology & Society

Segment 15 Fingerprinting *E. coli*

CHAPTER	Chapter Enrichment
9	

This Chapter Enrichment provides relevant and interesting information to
expand and enhance your classroom instruction of the chapter material.

SECTION 1 — Studying Human Populations

The Bubonic Plague

The Bubonic Plague or Black Death was an epidemic that
started in Asia and spread to Europe and North Africa
in the 1300s. The plague is caused by a bacterium that
can be transferred from rats to humans by fleas. Effects
include swelling of the lymph nodes, fever, chills, coughing, breathing difficulty, and, if untreated, death. The role
of rats and fleas in spreading the disease was not understood until the late 1800s. Between 1347 and 1350, as
many as 20 million people in Europe died from the disease. The plague heavily affected urban areas. At that
time, urban areas were crowded and unsanitary. Piles of
garbage were breeding grounds for rats. Epidemics of the
plague spread across Europe until the 1800s, but none
were as severe as the initial outbreak. The plague still
occurs around the world today, but the threat of another
widespread epidemic is unlikely.

The Industrial Revolution

Death rates and life expectancy changed in a variety of
ways during the Industrial Revolution in Europe and the
United States. The movement of people into cities for
employment in factories sometimes led to overcrowding
and epidemics of disease. Scientific study of these epidemics led to improvements in sanitation and nutrition.
Furthermore, scientists began to understand that bacteria
were the cause of many diseases. This knowledge led to
new methods of prevention and treatment. Infant mortality rates declined dramatically. Industrial machinery
enabled the production of more food, clothes, and other
materials. New forms of transportation made food, medicines and other goods available to more people. Thus,
in the last 300 years, advances in health care, transportation, and agriculture contributed to rapid human
population growth.

The Ratio between Males and Females

The theoretical ratio between human males and females
should be 1:1. However, the reality is not quite that
simple. The conception rate for human males is about
125 to every 100 females. However, a larger proportion
of males is lost through stillbirths and miscarriages.
Thus, the live birth ratio between males and females is
105:100. This means that human populations will have
slightly more males than females at birth. However,
these ratios may even out in the total population because
males generally have lower life expectancies and steeper
survivorship curves.

► Women learning
to read in Bolivia

Carrying Capacity

Carrying capacity relates to the amount of resources present in an ecosystem to support a maximum population.
The carrying capacity of Earth for humans is difficult to
calculate. This makes it difficult to predict future population growth because no one knows when the carrying
capacity will be reached or exceeded. The carrying capacity for humans is difficult to determine because humans
are adept at exploiting new resources when the need
arises. This subject is hotly debated among scientists,
sociologists, and in political circles.

Less Developed Nations and Population

The less developed countries currently make up the
greatest portion of world population growth. Most of
the more-developed countries have a zero or negative
rate of growth. Worldwide, 96 percent of the world's
population growth takes place in the less developed
countries. According to data collected in 1998, 99 percent of all births and 77 percent of all deaths took place
in less developed countries. It is predicted that in more-
developed countries, ongoing population growth will
mostly be the result of immigration from less developed
countries.

▶ **A city street in Bangkok, Thailand**

SECTION

2 Changing Population Trends

Water Supply Contamination

Water supplies can be contaminated in a number of different ways. Problems with water supplies are perpetuated when people lack sufficient antibiotics, healthcare, and sewage treatment facilities. This scenario is common in developing countries. The most common form of

BRAIN FOOD

The five most populated cities in the world:

1. Seoul, South Korea   10.0 million
2. Bombay, India   9.9 million
3. Karachi, Pakistan   9.8 million
4. São Paulo, Brazil   9.8 million
5. Shanghai, China   9.2 million

The five most populated metropolitan areas:

1. Tokyo, Japan   34.5 million
2. New York, New York, USA   21.4 million
3. Seoul, South Korea   20.3 million
4. Mexico City, Mexico   19.3 million
5. Bombay, India   19.0 million

The five most populated countries in 2001:

1. China   1,273 million
2. India   1,033 million
3. United States   285 million
4. Indonesia   206 million
5. Brazil   172 million

water contamination is from sewage pollution. If untreated or poorly treated sewage enters waterways, it has the potential to transmit microbes. Some of the most common waterborne diseases are cholera, giardiasis, and amebiasis. Each of these diseases can cause severe diarrhea, with a risk of dehydration and even death. Cholera is caused by the bacterium *Vibrio cholerae*; giardiasis is caused by the protozoan *Giardia lamblia*; and amebiasis is caused by the protozoan *Entamoeba histolytica*. Each of these microorganisms have a life stage that passes to a new host through food or water that has been contaminated with feces from an infected host. All of these diseases can be treated with strong antibiotics.

The Nile River and Fertile Land

The Nile River is the second longest river in the world. The Nile River valley has been continuously settled for more than 5,000 years. One reason for the continuous settlement of the area is the fertile soil and abundant water in an otherwise arid area. For the past 10,000 years, annual floods have deposited new layers of silt through the Nile River valley. This silt provided a fresh input of nutrients each year which made agriculture sustainable. The annual floods on the river were not predictable but they did occur every year. Today, many dams have been built on the Nile River. These dams control the floodwaters and make it possible for people to live closer to the river. However, controlling the floodwaters has also stopped silt from being deposited on the land. As a result, the land in the Nile River valley is losing its ability to support the same level of agricultural production. The Egyptian government is looking for ways to move housing and agriculture into previously unused parts of the desert.

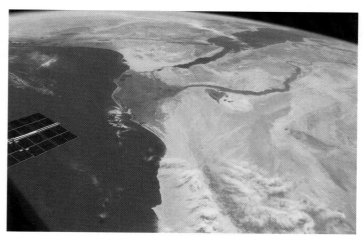

▶ **Aerial view of the Nile river and its delta in Egypt**

Overview

Tell students that this chapter presents concepts of human demography. The chapter builds on basic concepts of population ecology, extending into the more complicated human context. Predicting and managing human population growth has grown more challenging in recent centuries. However, some patterns can be seen in the economic and political development of different countries.

Using the Figure — GENERAL

The photograph shows a public beach in China. Ask students to estimate the density of people in the photo (perhaps 1 or 2 persons per m²). Discuss the difference between population size and population density. (Size is the total number of individuals in a specified group; density is the number of per given unit of area.) Discuss student reactions to the population density shown in the photograph.

PRE-READING ACTIVITY

You may want students to work in pairs to create this FoldNote.
Have one student fill in the first two columns, and have the other student fill in the last two columns.

VIDEO SELECT

For information about videos related to this chapter, go to **go.hrw.com** and type in the keyword **HE4 POPV**.

The Human Population

1 **Studying Human Populations**

2 **Changing Population Trends**

PRE-READING ACTIVITY

FOLDNOTES

Table Fold
Before you read this chapter, create the **FoldNote** entitled "Table Fold" described in the Reading and Study Skills section of the Appendix. Label the columns of the table fold with "Changes in Human Population," and "Effects of Population Change." Label the rows with "Before 1700," "From 1700 to 2000," and "From 2000 to 2050 and Beyond." As you read the chapter, write examples of each topic under the appropriate column.

China has one of the largest populations in the world, with more than 1 billion people. However, China's population is projected to stop growing by the year 2050, mostly because Chinese families are having fewer children.

218

Chapter Correlations *National Science Education Standards*

SPSP 2a Populations grow or decline through the combined effects of births and deaths, and through emigration and immigration. Populations can increase through linear or exponential growth, with effects on resource use and environmental pollution. **(Section 1)**

SPSP 2b Various factors influence birth rates and fertility rates, such as average levels of affluence and education, importance of children in the labor force, education and employment of women, infant mortality rates, costs of raising children, availability and reliability of birth control methods, and religious beliefs and cultural norms that influence personal decisions about family size. **(Section 1 and Section 2)**

SPSP 3a Human populations use resources in the environment in order to maintain and improve their existence. Natural resources have been and will continue to be used to maintain human populations. **(Section 2)**

SPSP 3b The earth does not have infinite resources; increasing human consumption places severe stress on the natural processes that renew some resources, and it depletes those resources that cannot be renewed. **(Section 2)**

Studying Human Populations

The human population of Earth grew faster in the 20th century than it ever has before. However, this rapid growth has led to environmental problems around the globe. Thus, we must try to understand and predict changes in human populations.

Demography is the study of populations, but most often refers to the study of human populations. Demographers study the historical size and makeup of the populations of countries to make comparisons and predictions. Demographers also study properties that affect population growth, such as economics and social structure. Countries with similar population trends are often grouped into two general categories. *Developed countries* have higher average incomes, slower population growth, diverse industrial economies, and stronger social support systems. *Developing countries* have lower average incomes, simple and agriculture-based economies, and rapid population growth.

The Human Population Over Time

After growing slowly for thousands of years, the human population grew rapidly in the 1800s, as shown in **Figure 1**. The human population underwent *exponential growth,* meaning that population growth rates increased during each decade. These increases were mostly due to increases in food production and improvements in hygiene that came with the industrial and scientific revolutions. However, it is unlikely that the Earth can sustain this growth for much longer.

Objectives

▶ Describe how the size and growth rate of the human population has changed in the last 200 years.

▶ Define four properties that scientists use to predict population sizes.

▶ Make predictions about population trends based on age structure.

▶ Describe the four stages of the demographic transition.

▶ Explain why different countries may be at different stages of the demographic transition.

Key Terms

demography
age structure
survivorship
fertility rate
migration
life expectancy
demographic transition

Figure 1 ▶ After growing slowly for thousands of years, the human population grew rapidly in the 1800s. What caused this change?

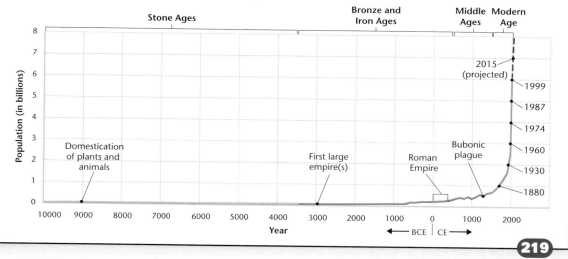

World Population Over Time

SECTION 1

Focus

Overview

Before beginning this section, review with your students the Objectives in the Student Edition. This section analyzes the factors that affect human population growth and decline.

⏱ Bellringer

Have students describe the changes in the human population in **Figure 1.** Have them hypothesize why the population has recently grown so rapidly.

Motivate

Activity ———— GENERAL

Growth Rates Demographers use per-capita annual growth rates as a simple measure of population trends. For example, a birth rate of 18 per 1,000 people and a death rate of 8 per 1,000 people gives a net population change of +10 per 1,000. This is expressed as a 1% annual growth rate per capita. Annual growth rates can be calculated in the same way one calculates compound interest. Have students multiply 1 by 1.01 on a calculator, and count the number of times they press the "=" key until they get to 2.0 (72). This demonstrates the effect of a constant 1% rate of compound interest (or annual growth) over a period of 72 years. Ask students to try the calculation with 1.02, and explain the results (2% growth over 35 years). ⓛⓢ **Kinesthetic**

Chapter Resource File

• **Lesson Plan**
• **Active Reading** BASIC
• **Section Quiz** GENERAL

Transparencies

TT Bellringer
TT Human Population over Time

SKILL BUILDER ———— ADVANCED

Math The time it takes in years for a population to double in size is called *doubling time,* and is directly related to the growth rate of a population. Tell students that, at a constant per-capita annual growth rate of 1%, a human population will double roughly every 72 years. Ask students how often a population will double, at a constant 2% growth rate (roughly every 36 years). Tell students they can also calculate doubling time using a banker's "shortcut," by dividing the annual percentage growth rate into 72. For example, at 2% growth, the doubling time is 72 ÷ 2 = 36 years. Ask students to calculate the doubling time for growth rates of 0.5% (about 144 years) and 1.7% (about 42 years). Have students look at **Figure 2** and estimate the growth rates during the 1800s (between 0.5 and 1%) and during the 1900s (between 1 and 2%). Discuss how this shows that a population with a positive per-capita growth rate will eventually double, even at a low growth rate. ⓛⓢ **Logical**

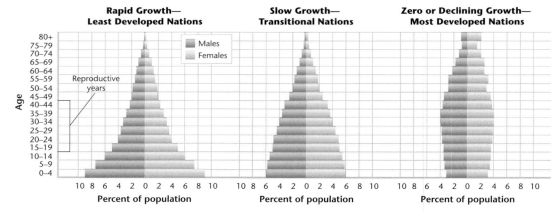

Rapid Growth—Least Developed Nations **Slow Growth—Transitional Nations** **Zero or Declining Growth—Most Developed Nations**

Males
Females

Reproductive years

Age

80+
75–79
70–74
65–69
60–64
55–59
50–54
45–49
40–44
35–39
30–34
25–29
20–24
15–19
10–14
5–9
0–4

10 8 6 4 2 0 2 4 6 8 10   10 8 6 4 2 0 2 4 6 8 10   10 8 6 4 2 0 2 4 6 8 10
Percent of population   Percent of population   Percent of population

Figure 2 ▶ Age-Structure Diagrams These graphs allow demographers to compare the distribution of ages and sexes in a population. Each graph shows a typical shape for a population with a particular rate of growth. Note that people between 15 and 44 years of age are most likely to produce children.

Forecasting Population Size

Will your community need more schools in the next 20 years, or will it need more retirement communities? Will people move in and create demand for more roads and utility services? Demographers look at many properties of populations to predict such changes. Population predictions are often inaccurate, however, because human behavior can change suddenly.

Age Structure Demographers can make many predictions based on age structure—the distribution of ages in a specific population at a certain time. For example, if a population has more young people than older people, the population size will likely increase as the young people grow up and have children. Age structure can be graphed in a *population pyramid*, a type of double-sided bar graph like those shown in **Figure 2.** The figure shows typical age structures for countries that have different rates of growth. Countries that have high rates of growth usually have more young people than older people. In contrast, countries that have slow growth or no growth usually have an even distribution of ages in the population. When parents are having fewer children, the population will have fewer young people.

Survivorship Another way to predict population trends is to study survivorship. Survivorship is the percentage of members of a group that are likely to survive to any given age. To predict survivorship, a demographer studies a group of people born at the same time and notes when each member of the group dies. The results plotted on a graph might look like one of the types of *survivorship curves* in **Figure 3.** Wealthy developed countries such as Japan and Germany currently have a Type I survivorship curve because most people live to be very old. Type II populations have a similar death rate at all ages. Type III survivorship is the pattern in very poor human populations in which many children die. Both Type I and Type III survivorship may result in populations that remain the same size or grow slowly.

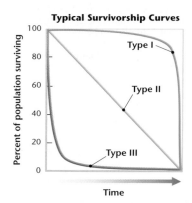

Typical Survivorship Curves

Percent of population surviving

100
80
60
40
20
0

Type I
Type II
Type III

Time

Figure 3 ▶ Survivorship curves show how much of the population survives to a given age. A Type I curve is seen in populations where most members survive to be very old. A Type III curve is seen in populations where many children die.

220

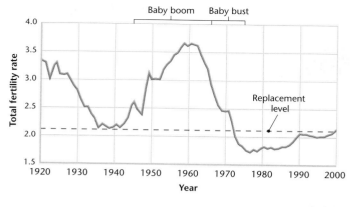

Figure 4 ▶ The total fertility rate in the United States went through many changes from 1900 to 2000. The *baby boom* was a period of high fertility rates, and the *baby bust* was a period of decreasing fertility.

Fertility Rates The number of babies born each year per 1,000 women in a population is called the **fertility rate.** Demographers also calculate the *total fertility rate,* or the average number of children a woman gives birth to in her lifetime.

A graph of historical total fertility rates for the United States is shown in **Figure 4.** In 1972, the total fertility rate dropped below replacement level for the first time in U.S. history. *Replacement level* is the average number of children each parent must have in order to "replace" themselves in the population. This number is about 2.1, or slightly more than 2, because not all children born will survive and reproduce.

Total fertility rates in the United States remained below replacement level for most of the 1990s. However, the population continued to grow, as shown in **Figure 5.** One reason for this growth was that the children of the baby boom grew up and had children.

Migration Another reason the population continued to grow was that immigration increased. The movement of individuals between areas is called **migration.** Movement into an area is *immigration* and movement out of an area is *emigration.* Migration between and within countries is a significant part of population change. The populations of many developed countries might be decreasing if not for immigration.

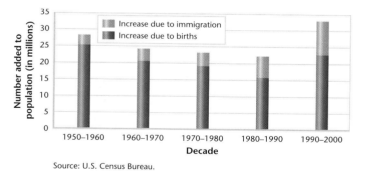

Source: U.S. Census Bureau.

MATHPRACTICE

Extending the Equation for Population Change

The following equation is a simple way to calculate the change in a population over a period of time:

$$\text{change in population} = (\text{births} - \text{deaths})$$

However, this equation does not account for changes due to migration. Rewrite the equation to include *immigration* and *emigration.*

Next, create an example word problem that would require the use of this new equation. Trade problems with a classmate, and try to solve the classmate's new word problem.

Figure 5 ▶ The population of the United States has continued to grow in the last century because of births as well as immigration.

221

Interpreting Statistics ——— GENERAL

Historical Correlation Have students consult U.S. history references, and locate the following events or periods within **Figure 4:** the Great Depression, World War II, and periods of economic growth. Have students find out how old most "baby boomers" are now. Have students compare immigration rates during each of these decades. **LS** Logical

MISCONCEPTION /// ALERT \\\

Fertility and Growth Rates
Students may think that a population should stop growing when total fertility rates are declining or low (such as happened in the US during the "baby bust" period). Emphasize to students that there is a delay of approximately twenty years (one generation) before a group of kids grows up and begins having their own children (and thus contributing to the fertility rate). After the baby boom, a large "bulge" in the population grew up and eventually had their own kids. Even though the baby boomers had lower fertility rates than their parents, they still contributed to population growth because there were so many baby boomers having at least one child.

MATHPRACTICE

Answers

Sample equation:
(change in population) =
(births + immigration) −
(deaths + emigration)
Word problems may vary.

🖥 Transparencies

TT Total Fertility Rate & Population Growth in the U.S.

Career

Statistician Statisticians are people who are specially trained in collecting and analyzing large amounts of data. Statisticians are employed by the Census Bureau, by insurance companies, and by other businesses that rely on data analysis. Statisticians receive training in math and statistical analysis. They also need to take courses in a variety of fields (biology, population dynamics, economics, and geography). Statisticians do a lot of "number crunching," which means analyzing data to look for trends, patterns, and correlations. Modern statisticians use sophisticated mathematical modeling software to analyze large sets of data and look for complex or subtle relationships between numbers.

Activity ———————— GENERAL

Modeling Infant Mortality Have students use beads, beans, calculators, spreadsheet software, or pencil-and-paper to come up with a way to model the growth of a population. (Example: Each pair of "parent" beans may "give birth" to a certain number of "child" beans in each "generation.") Next, have students devise a way to model infant mortality and changing life expectancies. (Example: Use different piles of different colors to represent pre-reproductive children, reproductive-age adults, and post-reproductive elderly in each generation. Vary the "death rate" of the groups and the length of time an individual spends at each stage.) Have students test whether reducing child mortality increases the growth rate of their population (it should). **LS Logical/Kinesthetic**

HEALTH ———————
● CONNECTION ▶ GENERAL

Is AIDS a Contagious Epidemic?
HIV/AIDS is not contagious in the same way as tuberculosis, flu, or measles. The transmissibility of HIV is very low, but an infected person remains contagious for years. The important factors in the spread of sexually-transmitted diseases (STDs) are the number and type of sexual contacts an infected person has. Contagious diseases are more likely to become epidemics in crowded populations. One likely cause for the increased spread of AIDS was increased urbanization in sub-Saharan Africa since the 1950s.

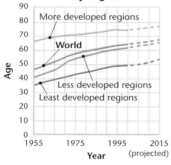

Figure 6 ▶ Today, people usually live longer because of improvements in healthcare, nutrition, and sanitation.

Figure 7 ▶ Since 1900, average life expectancy has increased worldwide (red line), although it remains lower in less developed countries (blue and purple lines).

Average Life Expectancy by Region

- More developed regions
- World
- Less developed regions
- Least developed regions

Age / Year (1955, 1975, 1995, 2015 (projected))

Source: UN Population Division.

222

Declining Death Rates

The dramatic increase in Earth's human population in the last 200 years has happened because death rates have declined more rapidly than birth rates. Death rates have declined mainly because more people now have access to adequate food, clean water, and safe sewage disposal. The discovery of vaccines in the 20th century also contributed to declining death rates, especially among infants and children. These factors are shown in **Figure 6.**

Life Expectancy The average number of years a person is likely to live is that person's **life expectancy.** Life expectancy is most affected by *infant mortality*, the death rate of infants less than a year old. In 1900, worldwide life expectancy was about 40 years and the infant mortality rate was very high. By 2000, the rate of infant mortality was less than one-third of the rate in 1900. The graph in **Figure 7** shows that average life expectancy has increased to more than 67 years worldwide. For people in many developed countries, life expectancy is almost 80 years.

Expensive medical care is not needed to prevent infant deaths. The infant mortality rate differs greatly among countries that have the same average income. Instead, infant health is more affected by the parents' access to education, food, fuel, and clean water. Even in poor areas, many people now know that babies simply need to be fed well and kept clean and warm. If these basic needs are met, most children will have a good chance of surviving.

Meanwhile, new threats to life expectancy arise as populations become denser. Contagious diseases such as AIDS and tuberculosis are a growing concern in a world where such diseases can spread quickly. Life expectancy in sub-Saharan Africa has been reduced in recent decades due to epidemics of AIDS.

BRAIN FOOD 🧠

Life Expectancy on the Rise Life expectancy has increased dramatically over the past century in more-developed countries. Life expectancy in Britain went from 48 for males in 1901 to 75 in 2000 and from 49 to 80 for females in the same time period. In the United States life expectancy has reached over 74 for males and over 80 for females.

● INCLUSION Strategies

• Learning Disabled

Have students create posters on the effects of the bubonic plague on the populations of Europe. Students may work alone or in groups to make a series of posters which: describe the effects of the plague; show the geographic distribution of the plague; list or graph estimated population sizes and deaths; and explain how the plague epidemic was eventually controlled.

The Demographic Transition

In most developed countries, populations have stopped growing. How can populations quadruple in a single century, then stop growing or even shrink in the next century? The **demographic transition** is a model that describes how these changes can occur. The model is based on observations of the history of many developed countries. The theory behind the demographic transition is that industrial development causes economic and social progress that then affects population growth rates. The graph in **Figure 8** compares expected trends in birth rates, death rates, and population sizes during each of the four stages of the transition.

Stages of the Transition In the first stage of the demographic transition, a society is in a preindustrial condition. The birth rate and the death rate are both at high levels and the population size is stable. Most of the world was in this condition until about 1700, when the scientific and industrial revolutions began.

In the second stage, a population explosion occurs. Death rates decline as hygiene, nutrition, and education improve. But birth rates remain high, so the population grows very fast. During this stage, the population could double in less than 30 years.

In the third stage of the demographic transition, population growth slows because the birth rate decreases. As the birth rate becomes close to the death rate, the population size stabilizes. However, the population is much larger than before the demographic transition. In most countries that have passed through the transition, the population quadrupled during the 20th century.

In the fourth stage, the birth rate drops below replacement level, so the size of the population begins to decrease. It has taken from one to three generations for the demographic transition to occur in most developed countries.

internet connect

www.scilinks.org
Topic: Demographic **Transition**
SciLinks code: HE4018

SCiLINKS. Maintained by the National Science Teachers Association

Ecofact

The Population Clock The Earth's human population is increasing by 2.4 people each second, which accounts for births and deaths. This means Earth gains over 200,000 people per day, or over 75 million per year.

Figure 8 ▶ The four stages of the demographic transition are shown here. Note the relative changes in birth rates, death rates, and population size. Do you think that all countries will fit this pattern?

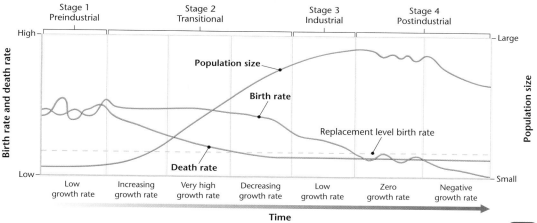

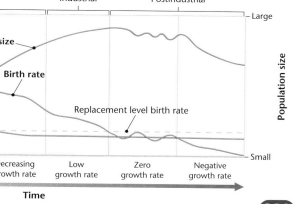

223

Close

Reteaching ─────── BASIC

Key Terms, Graphs, and Equations Review the key terms in this section with students. Next, review graphs and equations that use these key terms. Have students look for relationships between concepts that appear in graphs and headings. Have students reproduce the graphs and equations in their *EcoLog.*

English Language Learners

Quiz ─────── GENERAL

1. What factors have led to an increased life expectancy? (improvements in healthcare, nutrition, education, and sanitation)

2. Why does a decline in birth rate take several generations to have an effect on a population? (The decline in birth rate takes several generations to affect a population because the population is still producing replacements. After the children reach child bearing years, the number of new births can decrease. This process takes time.)

Alternative Assessment ─────── BASIC

Age Structure Supply students with enough demographic data to create an age structure diagram for their local region. This information can be found in almanacs, county records, reference books, or on the Internet. Ask students to explain what the diagram shows about the population.
LS Intrapersonal

English Language Learners

Connection to Biology

Female Influence Females have the primary influence over reproductive rates in most species of animals, because they invest more energy in reproduction than males do. Females usually produce and lay eggs, carry the fetus, give birth, and care for the young offspring. The time and resources a female must invest in each successful offspring is usually greater than the energy a male must invest.

internet connect

www.scilinks.org
Topic: Human Demographics
SciLinks code: HE4056

SCI LINKS. Maintained by the National Science Teachers Association

Women and Fertility The factors most clearly related to a decline in birth rates are increasing education and economic independence for women. In the demographic transition model, the lower death rate of the second stage is usually the result of increased levels of education. Educated women find that they do not need to bear as many children to ensure that some will survive. Also, the women may learn family planning techniques. They are able to contribute to their family's increasing prosperity while spending less energy bearing and caring for children. Some countries that want to reduce birth rates have placed a priority on the education of females, as shown in **Figure 9.**

Large families are valuable in communities in which children work or take care of older family members. But as countries modernize, parents are more likely to work away from home. If parents must pay for child care, children may become a financial burden rather than an asset. The elderly will not need the support of their children if pensions are available. All of these reasons contribute to lower birth rates. Today, the total fertility rate in developed countries is about 1.6 children per woman, while in developing countries, the rate is about 3.1 children per woman.

Figure 9 ▶ These women in Bolivia are learning to read. Many countries include the education of women in development efforts.

SECTION 1 Review

1. **Describe** how the size and growth rate of the human population has changed in the last 200 years.

2. **Define** four properties that scientists use to predict population sizes.

3. **Explain** what we can predict about a population's likely growth rates based on its current age structure.

4. **Describe** the four stages of the demographic transition.

CRITICAL THINKING

5. **Analyzing Relationships** Read the description of life expectancy in this section. Explain why the oldest people in a population may be much older than the average life expectancy. **READING SKILLS**

6. **Evaluating Theories** Do you think that all countries will follow the pattern of the demographic transition? Explain your answer.

224

Answers to Section Review

1. In the past 200 years, the population has increased from about 1 billion to over 6 billion and the growth rate has increased. This change was made possible by improvements in healthcare, nutrition, and sanitation.

2. age structure, survivorship, fertility rates, and migration

3. If a population has more young people than older people, the population size will probably increase as the young people grow up and have children.

4. In stage 1 the birth and death rates are high so the population grows slowly or not at all. In stage 2 birth rates remain high but death rates decline so the population increases. In stage 3 the birth rate and death rate are about equal, so the population stabilizes. In stage 4 the birth rate declines while the death rate stays low, so the population decreases.

5. Life expectancy is affected by infant mortality. If infant mortality is high, it brings down life expectancy. Many people live past life expectancy because the number is an average of all life spans.

6. Answers may vary. Some countries do not have the resources to lower death rates and keep birth rates high.

Changing Population Trends

Some countries have followed the model of the demographic transition—they have reached large and stable population sizes and have increased life expectancies. But throughout history, and currently in many parts of the world, populations that have high rates of growth create environmental problems. A rapidly growing population uses resources at an increased rate and can overwhelm the infrastructure of a community. **Infrastructure** is the basic facilities and services that support a community, such as public water supplies, sewer lines, power plants, roads, subways, schools, and hospitals. The symptoms of overwhelming population growth include suburban sprawl, overcrowded schools, polluted rivers, barren land, and inadequate housing, as shown in **Figure 10.** You may have seen some of these problems in your community.

Problems of Rapid Growth

People cannot live without sources of clean water, burnable fuel, and land that can be farmed to produce food. A rapidly growing population can use resources faster than the environment can renew them, unless resources come from elsewhere. Standards of living decline when wood is removed from local forests faster than it can grow back, or when wastes overwhelm local water sources. Vegetation, water, and land are the resources most critically affected by rapid growth.

Objectives

► Describe three problems caused by rapid human population growth.

► Compare population growth problems in more-developed countries and less developed countries.

► Analyze strategies countries may use to reduce their population growth.

► Describe worldwide population projections into the next century.

Key Terms

infrastructure
arable land
urbanization
least developed countries

Land Area per Person If each person alive on Earth in the year 2000 was given an equal portion of existing surface land, each person would get about 7.3 acres (0.025 km², or about four football fields). In the year 2050, each person might get 4.8 acres of land (0.016 km², or about two and a half football fields).

Figure 10 ► Rapid population growth can put pressure on water sources, land, and materials used for fuel or shelter. The makeshift housing shown here is one consequence of unmanaged growth.

225

Population Size, Growth Rate, and Density Students may confuse the concepts of population size, population growth rates, and population density. Remind students that large populations—as well as small ones—may have a high growth rate, a low growth rate, or a zero growth rate. The growth rate is not necessarily dependent on the size of the population, unless the population has reached the carrying capacity of the environment. Population density is measured over a defined area, so that one can compare the population size and densities of different countries, or look at local distribution (such as urban or rural.) Most of the problems discussed in this section are due to rapid local population growth that is not alleviated by emigration.

Focus

Overview

Before beginning this section, review with your students the Objectives in the Student Edition. This section describes some of the problems associated with rapid human population growth, and compares population trends in more-developed and less developed regions.

Bellringer

Ask students to think about what would happen if the population in their city or town doubled but the number of schools, roads, stores, and houses remained the same. Have students describe the changes they think would take place. **LS Intrapersonal**

Motivate

Identifying Preconceptions ── BASIC

U.S. Population Growth Most of the world's developed countries have low population growth rates. However, the United States has a high growth rate. On the board, write "U.S. Population" and then list the following numbers: 1950: 150 million; 1998: 270 million; 2050: 430 million. Ask students how fast the U.S. population is increasing (close to 2.5 million per year). Ask students to discuss possible results of this growth (answers may vary). **LS Logical**

Chapter Resource File

• Lesson Plan
• Active Reading BASIC
• Section Quiz GENERAL

Transparencies

TT Bellringer

Teach

Group Activity ——— GENERAL

Suburban, Metropolitan, Rural, or ? Tell students that in the U.S., cities are usually spread out into metropolitan areas, with people who live in the suburban areas commuting large distances to work. In Europe, the cities are more compact and do not usually have a large suburban population. In many countries, the majority of people live on farms or ranches, or in small villages. Have students work in groups, and assign one kind of settlement pattern to each group. Instruct groups to create a list of the pros and cons of each pattern of settlement. Have each group present their findings to the class.
LS Interpersonal Co-op Learning

Debate ——— GENERAL

Limiting Population Growth
Some people think that unless humans find new sources of energy (or can emigrate to another habitable planet), we will one day exceed the carrying capacity of Earth. Some people argue that population growth should be controlled to save the resources of Earth. Others argue that limiting population growth is a fundamental violation of personal rights. Furthermore, advances in technology and international cooperation, combined with natural demographic patterns may address the problems associated with future population growth. Ask students to chose a side and form a debate in class. Have students research their positions before beginning the debate. **LS** Verbal

Graphic Organizer ——— GENERAL

Spider Map
You may want to have students work in groups to create this Spider Map. Have one student draw the map and fill in the information provided by other students from the group.

Figure 11 ▶ These women in Myanmar are gathering firewood for cooking and boiling water. Gathering fuel is part of daily survival in many developing countries.

Graphic Organizer — Spider Map
Create the **Graphic Organizer** entitled "Spider Map" described in the Appendix. Label the circle "Problems of Rapid Population Growth." Create a leg for each problem of rapid population growth. Then, fill in the map with details about each problem of rapid population growth.

Figure 12 ▶ This woman is washing clothes in the Rio Grande on the U.S.-Mexico border. In areas that have no sewage or water treatment systems, people may use the same water supply for drinking, bathing, washing, and sewage disposal.

226

A Shortage of Fuelwood In many of the poorest countries, wood is the main fuel source. When populations are stable, people use fallen tree limbs for fuel, which does not harm the trees, as shown in **Figure 11.** When populations grow rapidly, deadwood does not accumulate fast enough to provide enough fuel. People begin to cut down living trees, which reduces the amount of wood available in each new year. Parts of Africa, Asia, and India have been cleared of vegetation by people collecting fuelwood.

A supply of fuel ensures that a person can boil water and cook food. In many parts of the world, water taken directly from wells or public supplies is not safe to drink because it may carry waterborne parasites or other diseases. The water can be sterilized by boiling it, but fuel is needed to do so. Also, food is often unsafe or harder to digest unless it is cooked. Without enough fuelwood, many people suffer from disease and malnutrition.

Unsafe Water In places that lack infrastructure, the local water supply may be used not only for drinking and washing but also for sewage disposal. As a result, the water supply becomes a breeding ground for organisms that cause diseases such as dysentery, typhoid, and cholera.

Many cities have populations that are doubling every 15 years, and water systems cannot be expanded fast enough to keep up with this growth. In the year 2001, over 1 billion people worldwide lacked safe drinking water and more than 3 million died of diseases that were spread through water. The Rio Grande, shown in **Figure 12,** is one example of an unsafe water source used by many people.

Cultural Awareness ——— GENERAL

The Chipko Movement in India In India, a man named Bahuguna realized that many poor, rural people in his country depended on fuelwood for survival, and that trees were being over-harvested. Bahuguna started a movement called Chipko, which in Hindi means "to hug" (as in "to hug a tree"). Chipko supporters join hands to surround, or collectively "hug," a tree in order to protect it from being cut down by loggers. The larger goals of the Chipko movement are to teach everyone the value of trees and the necessity of replanting them. Bahuguna visits thousands of rural areas in India each year to teach people that for every tree cut down, at least one must be planted to replace it. Many trees have been planted through this program, and some deforested areas are being reforested. Usually, the trees planted are rapidly growing trees that can be used for fuelwood, so that other areas of native forest are protected. Have students find out about other indigenous movements to protect natural resources such as the rubber tappers in Brazil.

Lima, Peru, is another example of an area with unsafe water. More than half of the population of Lima is housed in shantytowns that have no plumbing. The bacteria that cause cholera thrived in Lima's unmanaged water sources in 1991. In that year, Lima's population suffered the first epidemic of cholera that had occurred in the Western Hemisphere in 75 years.

Impacts on Land Every person needs space to live in, and people prefer to live where they have easy access to resources and a comfortable lifestyle. Growing populations may have a shortage of **arable land,** which is land that can be used to grow crops. Growing populations also make trade-offs between competing uses for land such as agriculture, housing, or natural habitats.

For example, Egypt has a population of more than 69 million that is growing at 2 percent per year. For food and exportable products, Egypt depends on farming within the narrow Nile River valley, shown in **Figure 13.** Most of the country is desert, and less than 4 percent of Egypt's land is arable. However, the fertile Nile River valley is also where the jobs are located, and where most Egyptians live. Egyptians continue to build housing on what was once farmland, which reduces Egypt's available arable land.

Much of the world's population is undergoing **urbanization,** which means that more people are living in cities than in rural areas. In the United States, many cities are becoming large metropolitan areas. People often find work in the cities but move into suburban areas around the cities. This *suburban sprawl* leads to traffic jams, inadequate infrastructure, and the reduction of land for farms, ranches, and wildlife habitat. Meanwhile, housing within cities becomes more costly, more dense, and in shorter supply.

FIELD ACTIVITY

Does Your Local Area Have Population Pressures? Take an informal survey of your community's population trends. Gather information by taking a walk, reading the local newspaper, or by contacting your local government or chamber of commerce offices.

Try to answer the following questions:

• Is your local population growing or shrinking? How much change is due to migration?

• What growth-related problems are citizens and government planners aware of?

• What solutions are being proposed or debated?

Record your results in your *EcoLog.*

Figure 13 ► Egypt's population is mostly crowded into the narrow Nile River valley (green areas in center of photo at right). The United States has more arable land, but suburban sprawl (left) creates many problems.

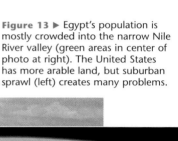

READING SKILL BUILDER — BASIC

Reading Organizer Encourage students to make a graphic reading organizer of the problems associated with rapid population growth. Students should draw a concept map of the problems and issues based on the headings in this section. After they read the section, encourage students to look over their concept map and fill in any additional details that might be missing.

LS Logical   English Language Learners

ENGINEERING CONNECTION — ADVANCED

Biomass Fuels Fuelwood and other fuels derived directly from plants or animals are *biomass fuels.* Biomass fuels are sometimes called "hidden" sources of energy, because they are not traded on a large scale and they are often not tracked by governments. In regions that are agriculturally productive but where the wood supply is limited dung is used as fertilizer, as a source of cooking fuel, and as a basic building material. Crop residues (such as grain straw) are also used when fuelwood is scarce. However, crop residues and dung can be inefficient fuels that are a significant source of air pollution in developing countries. Gathering dung can also involve a great deal of labor. In some countries, people spend entire days gathering dung for fuel.

ECONOMICS CONNECTION — GENERAL

The Economic Roles of Women It is generally agreed that women's contribution to most forms of productive and subsistence agriculture is poorly estimated in official statistics. For example, a 1994 labor force survey conducted in an Asian country reported that 92% of women over 10 years of age were "inactive." The survey also claimed that only 0.5% of the female population participated in agriculture. In a conversation with a researcher who questioned these figures, the government statistician responsible for this survey said: "They expect me to count women who collect fodder, fuel, and water. That's just about every woman in the country. They must be crazy if they think I'm going to do that!" The contribution of women is underreported in many economic contexts. Have students interview their mothers and grandmothers about the work done by women in their families.

LANGUAGE ARTS
CONNECTION **ADVANCED**

Influential Ideas on Population Growth Several books have had a major impact on popular and scientific thinking about human population growth. Thomas Malthus' *Essay on the Principle of Population* was published in 1798. Charles Darwin's theory of natural selection was influenced by Malthus' observation that plants, animals, and humans are capable of producing far more offspring than can possibly survive. Malthus concluded that unless family size was regulated, worldwide poverty and famine would become inevitable. Malthus' views were not popular among social reformers of his time. Malthus did not adequately consider the environmental and social factors that influence population growth. Jonathan Swift's *A Modest Proposal* considered population issues from a humorous perspective. In later decades, Paul Ehrlich generated controversy with his 1968 book *The Population Bomb*, and its 1990 sequel *The Population Explosion*. Ehrlich's ideas continue to be heavily debated. Have students read excerpts from Ehrlich and from other sources that disagree with his ideas. Select a moderator and organize a class debate on these issues. **LS Verbal/Logical**

Connection to Law

International Development
The United Nations (UN) has an important role in understanding and assisting the development of nations. The UN holds conferences, publishes research, creates treaties, manages international programs, and dispenses funds.

The UN also creates formal designations, such as *least developed countries*. Demographers, foreign aid programs, and international treaties may use these designations.

A Demographically Diverse World

As you have seen, demographers may categorize countries as either developed or developing. However, demographers may prefer the terms *more developed* and *less developed* to describe countries or regions, because the reality of development is complex and politically sensitive.

Not every country in the world is progressing through each stage of the demographic transition according to the model. Some countries now have modern industries, but incomes remain low. A few countries have achieved stable and educated populations with little industrialization. Some countries seem to remain in the second stage of the model. These countries have rapid population growth, but are unable to make enough educational and economic gains to reduce the birth rate and move into the third stage.

In recent years, the international community has begun to focus on the **least developed countries.** These countries show few signs of development and in some cases have increasing death rates, while birth rates remain high. Least developed countries are officially identified by the United Nations. These countries may be given priority for foreign aid and development programs to address their population and environmental problems.

Thailand's Population Challenges

Population growth is a major concern for many developing countries. But the options are limited for a country that has a poor economy and growing demands for limited resources. Thailand is one country that has effectively and purposefully slowed its population growth.

Around 1970, Thailand's population was growing at a rate of more than 3 percent per year, and the average Thai family had 6.4 children. The country had increasing environmental problems, such as air pollution in major cities and unsafe water supplies. Thailand's emissions of carbon dioxide from burning fossil fuels almost doubled between 1990 and 1997. In Thailand's capital, Bangkok, one-ninth of residents

▶ **Bangkok, Thailand** is one of the most crowded and polluted cities in the world. However, population growth is slowing in Thailand, and some environmental problems are starting to be solved.

228

Notable Quotes ——— GENERAL

"A baby born in the United States will damage the planet 20 to 100 times more in a lifetime than a baby born into a poor family in an LDC (less developed country). Each rich person in the United States does 1,000 times more damage than a poor person in an LDC."

—Dr. Paul Ehrlich

Have students debate whether Ehrlich's statement is true. Discuss whether the following estimates support Ehrlich's statement: Worldwide, the United States produces 21 percent of the goods and services, uses one-third of the processed mineral resources and one-fourth of the nonrenewable energy, and produces one-third of the pollution and trash.

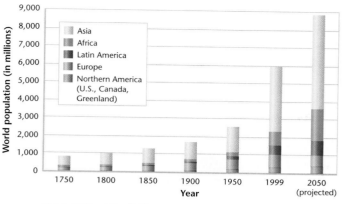

Source: UN Population Division.

Figure 14 ▶ Different regions of the world are growing at different rates. Which regions will contribute the most growth?

Growth rates for different parts of the world are shown in **Figure 14.** Populations are relatively stable in Europe, the United States, Canada, Russia, South Korea, Thailand, China, Japan, Australia, and New Zealand. In contrast, populations are still growing rapidly in less developed regions. Most of the world's population is now within Asia.

☑ **internet** connect

www.scilinks.org
Topic: Developed and Developing Countries
SciLinks code: HE4021

SCiLINKS® Maintained by the National Science Teachers Association

Thailand's Population Strategies

- improved healthcare for mothers and children

- openness of the people, government, and community leaders to changing social traditions

- cooperation of private and nonprofit organizations with the government

- increases in women's rights and ability to earn income

- economic incentives such as building loans for families who participated in the family planning programs

- creative family-planning programs promoted by popular government leaders

- high literacy rates of women (80 percent in 1980 and 94 percent in 2000)

have respiratory problems, and many people die of waterborne diseases each year.

In 1971, Thailand's government adopted a policy to reduce Thailand's population growth. The policy included increased education for women, greater access to healthcare and contraceptives, and economic incentives to parents who have fewer children. Fifteen years later, the country's population growth rate had been cut to about 1.6 percent. By 2000, the growth rate had fallen to 1.1 percent and the age structure was more evenly distributed. These changes also reflected a decline in the infant mortality rate.

How did Thailand make such major changes with limited resources? Demographers believe the changes are due to the combination of strategies shown in the table at left.

CRITICAL THINKING

1. Applying Ideas For what reasons could Thailand be described as a developing country in the 1970s? In what ways was it able to change?

2. Expressing Viewpoints Do you approve of all of the strategies that the government of Thailand employed in order to reduce their population growth? Do the goals justify the strategies they used? Write a persuasive paragraph to defend your opinion. **WRITING SKILLS**

229

Homework —— BASIC

Immigration Issues Have students research the ways that immigration of workers and refugees (from war, famine, or disaster) has caused social disruptions in developed countries. Students may search books, newspapers, magazines, websites, or television news programs and documentaries to collect information. Tell students to find out how some residents of developed countries react to immigrants in their area or about the effects that large numbers of immigrants have on the economies and social service expenditures of developed nations. Students may create written reports or produce their own documentary to present the issues to the class. **LS** Interpersonal

CASE STUDY

Thailand's Population Challenges Have students locate Thailand on a map. After students have read the Case Study, ask them the following questions: "Why did Thailand focus its attention on women? Would these strategies be effective in the U.S.? What about other countries?" Then ask students to find current information on population issues in a country near Thailand. **LS** Verbal

Answers to Critical Thinking

1. In 1970, Thailand had a high birth rate and its population was increasing.

2. Answers may vary.

⬤ **INCLUSION Strategies**

• *Gifted and Talented*

Ask students to write an opinion paper about China's campaign to reduce birth rates. The paper should include information about the current population of China and China's strategies for reducing birth rates. Students can take a position in favor of or against population control in China.

🌐 **Cultural Awareness** GENERAL

City of Many Names To the Thai people the city of Bangkok is known as Krueng Thep or City of Angels. The full name of Bangkok is Krungthepmahanakhon Amornrattanakosin Mahintharayutthaya Mahadilokphop Noppharat Ratchathaniburirom Udomratchaniwetmahasathan Amonphiman Awatansathit Sakkathattiyawitsanukamprasit, the longest name of any capital.

📠 **Transparencies**

TT Worldwide Trends in Fertility & Population Growth

Figure 15 ▶ China has implemented a long campaign to reduce birth rates. Their strategies have included economic rewards to promote single-child families and advertising such as the billboard shown here.

Group Activity —— ADVANCED

UN Population Goals Have students work in groups to discuss the ICPD Goals in **Table 1.** Ask students to determine how each goal relates to the other goals, and suggest actions to reach the goals. Have each group report which goal they believe should be the highest priority. **LS** Interpersonal

Interpreting Statistics —— GENERAL

Declining Fertility Rates Ask students to look at **Figure 16** and describe the trends in each region and worldwide. Ask: "At what points did the fertility trends change?" (Wherever the lines abruptly change slope. In the more developed regions, around 1965 and 1980; in the less-developed regions, around 1970.) If the world trend line continues on its present path into the future, when would the population stabilize?" (At about 2050, at fertility rates of about 2.1.) "What will happen if the line in the more-developed regions continues to slope downward?" (Those regions will have shrinking populations, unless they continue to have immigration.) "Why do some of the lines become more level while others become steeper?" (Because fertility behavior may change rapidly or slowly.) "Do these trends match the model of the demographic transition?" (not completely; In some cases, less-developed regions are beginning to reduce birth rates by choice, without first being fully industrialized or having a proportional increase in population size.) **LS** Logical/Visual

Table 1 ▼

ICPD Goals for 2015

- Provide universal access to a full range of safe and reliable family-planning methods and related reproductive health services.

- Reduce infant mortality rates to below 35 infant deaths per 1,000 live births and mortality rates of children under five years old to below 45 deaths per 1,000 live births.

- Close the gap in maternal mortality between developing and developed countries. Achieve a maternal mortality rate below 60 deaths per 100,000 live births.

- Increase life expectancy at birth to more than 75 years. In countries with the highest mortality, increase life expectancy at birth to more than 70 years.

- Achieve universal access to and completion of primary education. Ensure the widest and earliest possible access by girls and women to secondary and higher levels of education.

Source: UN Population Fund.

Figure 16 ▶ Worldwide Trends in Fertility Most countries' fertility rates are dropping toward replacement level.

230

Managing Development and Population Growth

Humans throughout history have witnessed the negative effects of population growth. Today, less developed countries face the likelihood that continued population growth will prevent them from imitating the development of the world's economic leaders. Some governments have tried to move forward in the demographic transition by directly reducing birth rates. Countries such as China, Thailand, and India have created campaigns to reduce the fertility rates of their citizens. These campaigns might include public advertising, as shown in **Figure 15,** or family planning programs, economic incentives, or legal punishments.

In 1994, the United Nations held the International Conference on Population and Development (ICPD). This conference involved debates about the relationships between population, development, and environment. **Table 1** shows the main goals that resulted from the conference. Many countries favor stabilizing population growth through investments in development, especially through improvements in women's status. In fact, worldwide fertility rates are dropping, as shown in **Figure 16.**

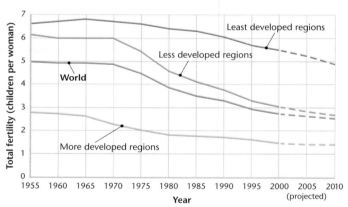

Source: UN Population Division.

Replacement-Level Fertility Fertility is said to be at "replacement level" when couples have just enough children to replace the adults in a population. Replacement-level is higher than two children because not all females survive through their childbearing years and because there are slightly more males than females born. In the more-developed countries, where infant and child mortality are low, replacement-level is about 2.1 children per couple. In countries with higher mortality rates, more children are needed to "replace" the population.

Growth Is Slowing

The human population of the world is now more than 6 billion and is still increasing. The worldwide population growth rate peaked at about 87 million people per year between 1985 and 1990. In contrast, the population grew by only 81 million people per year from 1990 to 1995.

Fertility rates have declined since about 1970 in both more-developed and less developed regions. However, rates are still much higher in less developed regions. Demographers predict that this trend will continue and that worldwide population growth will be slower in this century than in the last century. If current trends continue, most countries will have replacement level fertility rates by 2050. If so, world population growth would eventually stop.

Projections to 2050 United Nations projections of world population growth to 2050 are shown in **Figure 17**. The medium-growth line assumes that worldwide fertility rates will decline to replacement level by 2050. The high- and low-growth lines would result from higher or lower fertility rates. Most demographers predict the medium growth rate and a world population of 9 billion in 2050.

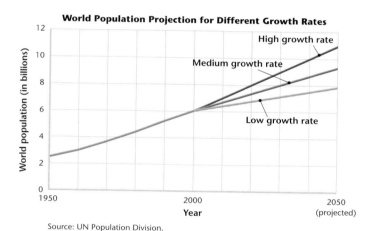

World Population Projection for Different Growth Rates

High growth rate
Medium growth rate
Low growth rate

Source: UN Population Division.

Figure 17 ▶ Current fertility trends will result in a world population of about 9 billion in 2050 (middle line). Economic or political changes could lead to higher or lower numbers.

QuickLAB

Estimating Fertility Rates

Procedure

1. Your goal is to estimate the average fertility rate of the mothers of students in your school. Design and conduct a quick survey of other students in the school.

2. Create one or two survey questions that will collect the needed information. Be sure that the questions are sensitive to personal differences and are not judgmental.

3. Devise a method to make the survey anonymous. You might simply pass out a questionnaire to another class.

4. Get your teacher's approval for your survey questions and method, and then conduct your survey.

Analysis

1. Analyze your results, and prepare a short summary of your findings.

QuickLAB

Skills Acquired:

• Collecting Data

Teacher's Notes: Approve the student questionnaires before they begin collecting data.

Answers

1. Answers may vary. Encourage students to present their findings in a graph or table.

Close

Reteaching ——— BASIC

Demographic Methods Ask students to look at **Figure 17,** and to notice that the chart shows data projected into the year 2050. Ask: "On what do demographers base these predictions?" (trends in growth rates and fetility rates) "What factors could change the projected figures?" (changes in economic trends, environmental disasters, war, changes in women's status, government policies, etc.)

Quiz ——— GENERAL

1. Why is fuel such as wood important to people? (it is needed for heating, cooking, and boiling water)

2. What are some of the problems of suburban sprawl? (traffic problems, inadequate infrastructure, and the loss of land for agriculture and wildlife)

Alternative Assessment ——— ADVANCED

Population Perspectives Have students prepare a brief report summarizing opposing points of view regarding the problems of population growth. Tell students to compare how people interpret the same data, how environmental science concepts are applied, and contradicting predictions of future population growth.

SECTION 2 Review

1. **Describe** three problems caused by rapid human population growth.

2. **Compare** population growth in more-developed countries to population growth in less developed countries.

3. **Describe** worldwide population projections for the next 50 years.

CRITICAL THINKING

4. **Analyzing a Viewpoint** Write a comparison of the pros and cons of the strategies nations have used to reduce population growth. **WRITING SKILLS**

5. **Analyzing Relationships** Do you think that simply changing birth rates will cause a nation to undergo further development?

231

Answers to Section Review

1. Sample answer: the shortage of fuelwood, poor water quality, and impacts on land

2. Sample answer: Less developed countries typically have higher birth rates and lower life expectancies than developed countries.

3. Sample answer: The worldwide population is expected to continue to increase for the next 50 years and reach between 8 and 11 billion. The exact amount of increase is uncertain.

4. Sample answer: Efforts to limit population growth have worked in some places. Reducing population growth can reduce the impact on the environment and improve the quality of life. However, government control of fertility takes away personal freedoms.

5. Sample answer: There may be some cases where birth rates could impact people's economic status directly. An individual's economic status seems to have more of an influence on birth rates. Many other factors affect both the birth rates and the economic status of a nation.

Alternative Assessment —— GENERAL

Population Strategies Have students research a country that is trying to control its population growth. Have students make a poster that illustrates the ways the country is controlling population growth.

Debate It is well known that as standards of living increase, human population growth slows. People in developed countries use about 83 percent of the world's resources, have few children, and enjoy a relatively high standard of living. People in developing countries use less than 2 percent of the world's resources, have large families, and often have a poor standard of living. Ask students to debate the following question: "Should people in developed countries conserve resources so that more resources can be used to raise the standard of living (and lower the birth rate) in developing countries?" Remind students that this may result in a lower standard of living for people in developed countries. Encourage students to cite information from this and other chapters in this book to support their arguments. Have one student act as moderator and list the arguments in favor of each point of view. **LS Verbal**
Co-op Learning

1 Studying Human Populations

Key Terms

demography, 219
age structure, 220
survivorship, 220
fertility rate, 221
migration, 221
life expectancy, 222
demographic transition, 223

Main Ideas

▶ Human population growth has accelerated in the last few centuries. The main reasons for this growth were improvements in hygiene and increases in food production, which accompanied the industrial and scientific revolutions.

▶ Demographers try to predict population trends using data such as age structure, survivorship, fertility rates, migration, and life expectancy.

▶ In the demographic transition model, countries progress through four stages of change in birth rates, death rates, and population size.

2 Changing Population Trends

infrastructure, 225
arable land, 227
urbanization, 227
least developed countries, 228

▶ When a growing population uses resources faster than they can be renewed, the resources most critically affected are fuelwood, water, and arable land.

▶ In this century, countries may be labeled more developed or less developed. Not all countries are going through the demographic transition in the same way that the more-developed countries did.

▶ Some countries attempt to reduce birth rates directly through public advertising, family planning programs, economic incentives, or legal punishments for their citizens.

232

Homework —— BASIC

Local Infrastructure Have students prepare a brief report on how your local community plans for infrastructure management and population growth. Remind students to consider utility services (water, sewage, electricity, garbage), schools, hospitals, traffic, housing development, and other uses of land and services. Have students add their report to their **Portfolio. LS Intrapersonal**

Chapter Resource File

• Chapter Test **GENERAL**
• Chapter Test **ADVANCED**
• Concept Review **BASIC**
• Critical Thinking **ADVANCED**
• Test Item Listing
• Math/Graphing Lab **GENERAL**
• Math/Graphing Lab **GENERAL**
• Modeling Lab **GENERAL**
• Consumer Lab **BASIC**
• Long-Term Project **BASIC**

Using Key Terms

Use each of the following terms in a separate sentence.

1. *demography*
2. *demographic transition*
3. *infrastructure*
4. *least developed countries*

For each pair of terms, explain how the meanings of the terms differ.

5. *age structure* and *survivorship*
6. *infant mortality* and *life expectancy*
7. *death rate* and *fertility rate*
8. *urbanization* and *migration*

✔ STUDY TIP

Quantitative Terms Look for key terms in the graphs in this chapter. In your *EcoLog*, copy the graphs and write brief descriptions of how key terms may relate to the graphs and to other key terms. For example, copy Figure 3, and write "High infant mortality results in low life expectancy and Type III survivorship."

Understanding Key Ideas

9. Age structure data include all of the following *except*
   a. the number of members of a population who are between 5 and 11 years old.
   b. the ratio of males to females in a population.
   c. the amount of population change due to immigration or emigration.
   d. the ratio of older people to younger people in a population.

10. Human population growth accelerated in recent centuries mostly because of
    a. the bubonic plague.
    b. better hygiene and food.
    c. the discovery of electricity.
    d. improved efficiency of fuel use.

11. Which countries have Type I survivorship?
    a. the most developed countries
    b. the least developed countries
    c. countries in the second stage of the demographic transition
    d. countries in the first stage of the demographic transition

12. The demographic transition is a(n)
    a. untested hypothesis.
    b. natural law.
    c. model based on observed patterns.
    d. international law.

13. A country in the second stage of the demographic transition may have all of the following *except*
    a. increasing agricultural production.
    b. improving healthcare and education.
    c. decreasing population size.
    d. decreasing death rates.

14. Which of the following resources is likely to be impacted the most by a rapidly growing population?
    a. clothing
    b. food
    c. housing
    d. water

15. Which of the following diseases is often spread through unsafe public water sources?
    a. dysentery
    b. flu
    c. chickenpox
    d. AIDS

16. Which of the following uses of wood is the most important for basic human needs?
    a. heating the home
    b. boiling water
    c. making tools
    d. building shelter

17. In this century, the world population is likely to
    a. remain the same.
    b. continue to grow exponentially.
    c. decline rapidly because fertility rates are already below replacement level.
    d. stabilize after fertility rates fall below replacement level.

233

Assignment Guide

Section	Questions
1	1, 2, 5–13, 18–21, 27, 30–33, 35–37
2	3, 4, 14–17, 22, 23, 27–29, 31, 38

ANSWERS

Using Key Terms

1. Sample answer: Demography is the study of populations, particularly human populations.

2. Sample answer: The demographic transition is a model of the changes in a region's population as it industrializes.

3. Sample answer: A rapid population increase can put stress on the infrastructure of a city.

4. Sample answer: Least developed countries are identified as having high birth and death rates and low growth rates.

5. Age structure is the distribution of ages and sexes within a population at a particular time. Survivorship is the percentage of members of a group that are likely to survive to a given age.

6. Infant mortality is the death rate of infants less than a year old. Life expectancy is the average number of years that a person is likely to live.

7. Fertility rate is the rate at which women are giving birth in a population. Death rate is the rate at which members of a population are dying.

8. Urbanization is the movement of people into cities and urban areas. Migration is any movement of people from one area to another.

Understanding Key Ideas

9. c
10. b
11. a
12. c
13. c
14. d
15. a
16. b
17. d

CHAPTER 9 Review

Short Answer

18. Life expectancy has increased because of better nutrition, healthcare, clean water, and safe sewage disposal.

19. The age structure of a population helps predict future population growth because it shows the number of people who are in child bearing years and the number who will be entering child bearing years.

20. Sample answer: A more educated population usually has a lower fertility rate because they often choose to have fewer children.

21. Birth rate, death rate, and overall population size are the factors that change during demographic transition.

22. The key resources affected by rapidly growing populations are fuelwood, water, and land.

23. Europe, northern America, Russia, South Korea, Thailand, China, Japan, Australia, and New Zealand are considered to be more developed. Southeast Asia, the Middle East, Central and South America, Africa, and India are generally less developed.

Interpreting Graphics

24. Africa and Latin America have only recently exceeded Europe and northern America in population size.

25. Europe is projected to decline in population size.

Short Answer

18. What are the main reasons that life expectancy has increased worldwide?

19. How does the age structure of a population help predict future population growth?

20. What is the relationship between education and fertility rates in a human population?

21. Which properties of a population change during the demographic transition?

22. Which key resources are impacted the most by rapidly growing populations?

23. Which regions of the world are generally more developed? less developed?

Interpreting Graphics

The graph below shows each region's contribution to world population growth. Use the graph to answer questions 24–26.

24. Which region(s) are projected to increase in population size?

25. Which region(s) are projected to decline in population size?

26. Can you assume that all the countries within each region have the same growth patterns? Explain your answer.

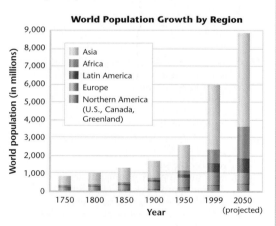

World Population Growth by Region

Concept Mapping

27. Use the following terms to create a concept map: *rapid human population growth, demographic transition, survivorship, fertility rate, fuelwood, water,* and *land.*

Critical Thinking

28. **Analyzing Predictions** Why are human population trends difficult to predict? Describe an example of an event that would change most demographic predictions.

29. **Analyzing Methods** In what ways does the study of human populations differ from the study of wildlife ecology?

30. **Identifying Relationships** What other factors, besides those already mentioned, might have an effect on fertility rates in a given population?

31. **Evaluating Theories** Write an evaluation of the demographic transition as a theory of how populations will develop. How useful is the demographic transition model in predicting the future? What assumptions are made by the theory? What criticisms could be made of the theory? **WRITING SKILLS**

Cross-Disciplinary Connection

32. **Careers** Demographers are employed by many kinds of organizations including governments, health organizations, and insurance companies. How can their skills be useful to each of these organizations?

33. **Social Studies** Find out the demographic history, for the last 100 years, of a developing country of your choice. Explain how closely this country's pattern of development follows the demographic transition model.

Portfolio Project

34. **Research Demographic Trends** Look up population statistics for your local city, county, or state. Read and take notes about recent demographic trends and predictions for the next few decades. Make a summary of your findings. **READING SKILLS**

26. Answers may vary. Individual countries in any region may vary from the general trends. Some countries within a region are more developed than most of their neighbors. For example, Japan is wealthier and has a lower fertility rate than most other Asian countries.

Concept Mapping

27. Answers to the concept mapping questions are on pp. 667–672.

Critical Thinking

28. Answers may vary. There are many factors that can affect population changes. Diseases, famines, economic changes, war, and the availability of resources each could alter demographic trends.

29. Answers may vary. Human populations are different from wildlife in that they are widespread throughout the world, migrate more often and more freely over greater areas, and can manipulate their environment and exploit resources using tools and technology.

30. Answers may vary. Fertility rates might also be affected by culture, laws, personal choice, or the availability of healthcare.

MATH SKILLS

Use the graph below to answer questions 35–37.

35. Analyzing Data At which times did the fertility rate change most drastically in the United States?

36. Graphing Data Sketch a copy of the graph below. Smooth the bumps to give an idea of general trends.

37. Drawing Conclusions On your new graph, draw a second line to show the changes in population size that you would predict to result from the given fertility rates over time.

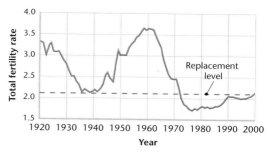

WRITING SKILLS

38. Writing Persuasively Write an opinion article for a newspaper or magazine. Argue either for or against a policy related to immigration or family planning.

39. Writing Using Research Look up recent census data from your city, county, or state. Write a paragraph that describes the major demographic trends of the last few years.

STANDARDIZED TEST PREP

For extra practice with questions formatted to represent the standardized test you may be asked to take at the end of your school year, turn to the sample test for this chapter in the Appendix.

READING SKILLS

Read the passage below, and then answer the questions that follow.

Excerpt from UN Population Fund, The State of World Population 2001, 2001.

Worldwide, women have primary responsibility for rearing children and ensuring sufficient resources to meet their needs. In the rural areas of developing countries, women are also the main managers of essential household resources like clean water, fuel for cooking and heating, and fodder for domestic animals.

Women make up more than half of the world's agricultural workforce. They grow crops for the home and market and often produce most staple crops. In the world's poorest countries, women head almost a quarter of rural households.

However, although women have the primary responsibility for managing resources, they usually do not have control. National law or local customs often deny women the right to secure title or inherit land, which means they have no collateral to raise credit and improve their conditions.

Women often lack rights in other aspects of their lives, reinforcing gender inequalities. High fertility and large families are still a feature of rural life, though the rationale has long since passed. In part, this reflects women's lack of choice in the matter.

1. Which of the following are *not* cited in the passage as major responsibilities of women?
   a. management of household resources
   b. agricultural work
   c. government leadership
   d. rearing children

2. The passage implies that improving women's rights would lead to
   a. the ability of women to earn more money.
   b. increased availability of fuel.
   c. poorer rural households.
   d. larger families.

235

31. Answers may vary. The model is useful in predicting the future because many different populations have demonstrated similar patterns of development. Critics could point out that the model is oversimplified. Countries may have different demographic trends.

Cross-Disciplinary Connection

32. Demographers study data such as birth rate, death rate, life expectancy, and migration. They can help governments plan for future development, private medical-aid organizations determine where assistance is needed, and insurance companies determine rates for premiums.

33. Answers may vary depending on the country that the student chooses.

Portfolio Project

34. Answers may vary.

Math Skills

35. The greatest changes were during the approximate periods of 1925–1935, 1941–1950, and 1962–1975.

36. Student graphs should look similar to the example graph but should have smooth curves.

37. Student graphs should show that the population keeps growing, but population growth accelerates during periods of high fertility and for about 20 years after. Population growth should slow down about 20 years after fertility declines. Population growth should be negative for about 20 years past periods in which the fertility rate is below the replacement level. Note that U.S. population growth has not followed these projections exactly, because of immigration and other variables.

Writing Skills

38. Answers may vary.

39. Answers may vary.

Reading Skills

1. c

2. a

HOW WILL OUR POPULATION GROW?

Teacher's Notes

Time Required

two 45-minute class periods

Lab Ratings

EASY ──────────▶ HARD

TEACHER PREPARATION 🧪
STUDENT SETUP 🧪🧪
CONCEPT LEVEL 🧪🧪🧪
CLEANUP 🧪

Skills Acquired

- Identifying and Recognizing Patterns
- Interpreting
- Organizing and Analyzing Data

The Scientific Method

In this lab, students will:
- Test the Hypothesis
- Analyze Results
- Draw Conclusions

Materials

The materials are enough for a group of 3 to 4 students.

Chapter Resource File

- **Datasheets for In-Text Labs**
- **Lab Notes and Answers**

Objectives

▶ **USING SCIENTIFIC METHODS** **Predict** which variable has a greater effect on population growth rates.
▶ **Calculate** changes for a given population over a 50-year period.
▶ **Graph** the resulting population's age structure by creating a population pyramid.
▶ **Compare** the effects of fertility variables on population growth rates.

Materials

calculator or computer
colored pencils or markers
graph paper
notebook
pen or pencil
ruler

▶ **Age Structure** You will make an age-structure diagram, such as this graph of the U.S. population in 2000.

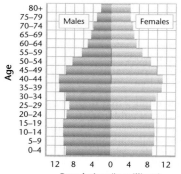

United States, 2000

Population (in millions)

Age

How Will Our Population Grow?

If you were a demographer, you might be asked to determine how a population is likely to change in the future. You have learned that the rate of population growth is affected by both the number of children per family and the age at which people have children. But which factor has a greater effect? To explore this question, you will use age-structure diagrams—also called population pyramids—such as the one shown below.

Procedure

1. In this lab you will calculate future population trends for an imaginary city. To compare how fertility variables may affect population growth, each group of students will test the effects of different assumptions. Assume the following about the population of this city:

Assumptions About the Population

- Half the population is male and half is female.
- Every woman will have all of her children during a given five-year period of her life.
- Everyone who is born will live to the age of 85 and then die.
- No one will move into or out of the city.

---

2. Your teacher will divide the class into four groups. Each group will project population growth using the following assumptions:

Assumptions About the Women in the Population

Group	Each woman gives birth to	While in the age range of
A	5 children	15–19
B	5 children	25–29
C	2 children	15–19
D	2 children	25–29

3. Predict which of the four groups will have the greatest population growth in 50 years. Write down the order you would predict for the relative size of the groups from largest population to smallest population.

Tips and Tricks

Demonstrate the method of "moving up" age groups to fill in each new column in the tables, and calculating new totals and births for each period of time. Use the sample answers in the table as a starting point. Test student understanding of the procedure by asking questions such as: "If there are 5,000 twenty- to twenty-four-year-olds in 2010 and each woman in that age group has three children, how many zero- to four-year-olds will there be in 2015?" (7,500) Students may need extra help understanding age-structure diagrams. Ask students questions about sample diagrams to check

understanding. Emphasize the division between left and right for gender, and explain that each diagram represents a cross-section of the population at a single point in time (for example, one census year).

To complete this lab, students must understand and accept the assumptions given in step 2 of the procedure. Note that these assumptions are simplified so that the models will work. Check that students are able to identify the variables being tested in each different scenario (number of children per woman and age of reproduction.)

4. The table at right shows the population of our imaginary city for the year 2000. Use the data in the table to make an age-structure diagram (population pyramid) for the city. Use the example diagram at left to help you.

5. Make a table similar to the one shown at right. Add columns for the years 2005, 2010, and for every fifth year until the year 2050.

6. Calculate the number of 0- to 4-year-olds in the year 2005. To do this, first determine how many women will have children between 2000 and 2005. Remember that half of the population in each age group is female, and that members of the population will reproduce at specific ages. Multiply the number of child-bearing women by the number of children that each woman will have. For example, Group A will have 12,500 new births by 2005.

7. Fill in the entire column for the year 2005. Determine the number of people in each age group by "shifting" each group from 2000. For example, the number of 5- to 9-year-olds in 2005 will equal the number of 0- to 4-year-olds in 2000.

8. Calculate the total population for each five-year period.

9. Repeat the process described in steps 3–8 for each column, to complete the table through the year 2050.

Analysis

1. **Constructing Graphs** Plot the growth of the population on a line graph. You may want to use a computer to graph the results.

2. **Constructing Graphs** Make a population pyramid for the population in 2050.

Conclusions

3. **Evaluating Data** Compare your graphs with the graphs of the other three groups. Were your predictions correct?

4. **Drawing Conclusions** Which variable had a greater effect on population growth—the number of children each woman had or the age at which each woman had children?

5. **Interpreting Information** Did any of the groups show no growth in the population? Explain these results.

Extension

1. From the age-structure diagram on the previous page, what would you predict to happen to the U.S. population in the next 20 years? in the next 50 years? What parts of the age structure are most important to these predictions?

Population in Each Age Group, 2000–2050			
Age	2000	2005	2010
80+	100		
75–79	500		
70–74	600		
65–69	700		
60–64	800		
55–59	900		
50–54	1,000		
45–49	1,250		
40–44	1,500		
35–39	2,000		
30–34	2,500		
25–29	3,000		
20–24	4,000		
15–19	5,000		
10–14	6,500		
5–9	8,000	10,000	
0–4	10,000	12,500	
Total	48,350		
Females that give birth		2,500	
New births		12,500	

▶ **Sample Population Data** Use this table as an example to calculate the age structure for each generation of your imaginary population. Add columns for five-year periods up to 2050. Examples of some of Group A's results are shown in red.

237

Answers to Extension

1. If current fertility trends continue, and without considering immigration, the U.S. age structure will begin to even out as current generations age. The "bulge" of middle-aged people will move into older brackets. The lower levels will slowly get smaller relative to the upper levels, making the pyramid more "top-heavy." Within 20 years, the age-structure will be somewhat evenly distributed. Within 50 years, the average person will be much older.

Jennifer Seelig-Fritz
North Springs High School
Atlanta, Georgia

Jennifer Seelig-Fritz
North Springs High School
Atlanta, Georgia

Answers to Analysis

1. Sample graphs:

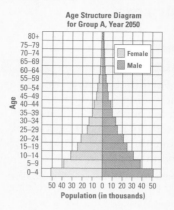

Age Structure Diagram for Group A, Year 2050

The shape of age structure diagrams for groups A–D will vary.

2. Sample graphs:

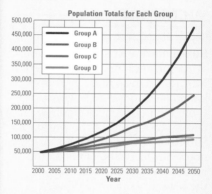

Population Totals for Each Group

Answers to Conclusions

3. All graphs should show population growth. Group A should have the greatest population growth. Students should note whether or not their prediction was correct.

4. The number of children per woman had the greatest effect on population growth. The age of reproduction was also important but had a lesser effect.

5. No. All groups showed some growth because the population started out "bottom-heavy" and people were still having some children. Groups C and D would eventually show zero growth.

FERTILITY RATES AND FEMALE LITERACY IN AFRICA

Interpreting Statistics — GENERAL

The Fertility-Literacy Link Some of the countries on the map may seem to contradict the hypothesis that fertility rates and female literacy rates are linked. However, for the past decade, fertility rates in Africa have been declining while literacy rates have been increasing in most African countries. Some African countries that seem to be exceptions are those that have recently made efforts to increase education or decrease fertility.

SOCIOLOGY
● CONNECTION — GENERAL

Sub-Saharan Africa The region of Africa south of Tunisia, Algeria, Morocco, Libya, and Egypt is known as *Sub-Saharan Africa*. From the perspective of more-developed regions of the world, Sub-Saharan Africa may seem notorious for wars, famines, and natural disasters. However, the region has many unique characteristics and challenges. Despite some recent development gains, Sub-Saharan Africa contains many of the world's poorest countries and a growing share of the world's poorest people. African countries generally have declining exports, simple economies, little influence in the global economy, and a steady loss of skills and capital to other regions. Average African income per capita in 2000 was lower than it was at the end of the 1960s.

MAPS in action

FERTILITY RATES AND FEMALE LITERACY IN AFRICA

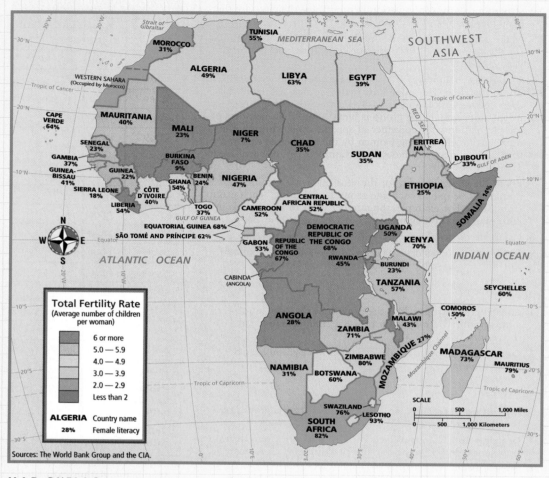

Sources: The World Bank Group and the CIA.

MAP SKILLS

Use the map of Africa to answer the questions below.

1. **Describing Locations** Which regions of Africa have the highest female literacy (percentage of females who can read and write)? the lowest female literacy? Which regions have the highest fertility rates? the lowest fertility rates?

2. **Analyzing Data** Choose 20 countries and make a graph comparing the total fertility rates and female literacy of each country.

3. **Comparing Data** Worldwide, the average total fertility rate is about 2.8 children per woman, and the average female literacy is 74 percent. How does Africa compare with the rest of the world in both aspects?

Answers to Map Skills

1. Answers may vary. The highest female literacy rates occur in the southern nations of South Africa, Lesotho, Swaziland, Zimbabwe, Madagascar, and Mauritius. Many other nations throughout Africa have fairly low female literacy rates. The highest fertility rates are in central Africa, and the lowest are in the northern and southern coastal countries.

2. Answers may vary. Check students' data against the map.

3. Answers may vary. Africa has generally higher fertility rates and lower female literacy than most other regions of the world.

Transparencies

TT Female Literacy and Fertility in Africa

SOCIETY & the Environment

LOST POPULATIONS: WHAT HAPPENED?

At various points in human history, entire populations have disappeared and left mysterious remains such as the Egyptian pyramids and the Anasazi pueblos in the southwestern United States. Why did these people and their civilizations disappear? Archeologists often find evidence that environmental destruction was one of the reasons the populations disappeared.

Easter Island

On Easter Island in the Pacific Ocean, the first European visitors were amazed to find huge stone heads that were miles from the quarries where the heads had been made. It seemed impossible that the islanders could have moved the heads. There were no horses, oxen, or carts on the island and there were also no trees, which could have been used as rollers to move the heads. The islanders were using grass and reeds to make fires because the island was barren grassland. The island had no tree or shrub that was more than 3 m tall.

A Changed Environment

Researchers have now shown that Easter Island was very different when it was first colonized by Polynesians around 400 CE. In the oldest garbage heaps on the island, archaeologists have found that one-third of the bones came from dolphins. To hunt dolphins, the islanders must have had strong canoes made of wood from tall trees. Pollen grains, which are used to identify plants, show that the island was once covered by a forest that contained many species found nowhere else in the world.

But by 1600 CE, trees were rare and the Easter Island palm tree was extinct. The palm seeds were probably eaten by rats that the Polynesians had brought to the island. With the destruction of the forest, every species of native land bird also became extinct, and the human population crashed.

The people of Easter Island destroyed their environment by overusing its natural resources and introducing new species such as chickens and rats. The people were reduced from a complex civilization to a primitive lifestyle. Easter Island is a small-scale example of what ecologists worry could happen to Earth's entire human population.

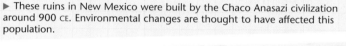

▶ These ruins in New Mexico were built by the Chaco Anasazi civilization around 900 CE. Environmental changes are thought to have affected this population.

▶ These large stone figures found on Easter Island were made by a civilization that has disappeared.

What Do You Think?

Industrialized countries have started to invest in environmental improvements, such as replanting forests that have been destroyed and protecting endangered species. Do you think this makes these countries safe from the kind of environmental disasters that destroyed the Easter Island civilization?

239

ARCHEOLOGY
CONNECTION — GENERAL

Ancient Polynesian Islanders A team of researchers is excavating thousands of archeological sites on the islands of Hawaii and Maui, searching for clues to cultural and environmental sustainability. The multidisciplinary team from five universities is led by archeology professor Patrick Kirch, an expert on the Pacific Islands. The team's goal is to fully understand the interactions between humans and the environment in the region prior to 1800. During that time, the island populations increased dramatically, societies grew more complex, farming-based economies developed, and local environments were dramatically changed. "Although the project is focused on the Hawaiian Islands, the issues addressed are global," said Kirch. "Many of the cultural and natural co-evolutionary processes that happened in Hawaii over the millennium prior to European contact have also happened elsewhere and are taking place today on a global environmental scale."

Society & the Environment

LOST POPULATIONS: WHAT HAPPENED?

HISTORY
CONNECTION — ADVANCED

Other Lost Civilizations There are many examples of large, sophisticated populations about which we know little. Ruins of such civilizations have been found in many countries including China, India, Egypt, Persia, Greece, Rome, Japan, England, Ireland, North and South America, Australia, and many island nations. Have students choose a region to research and ask them to prepare a brief presentation on the lost civilizations of their chosen region. Students should include any theories to explain the origins and demise of these civilizations.

SKILL BUILDER — GENERAL

Writing Ask students to write a response to this prompt: "Imagine that you were one of the people who lived on Easter Island or in another lost civilization. Write some entries you might have made into a daily journal." English Language Learners

LS Verbal

Answers to What Do You Think?
Answers may vary.

Biodiversity
Chapter Planning Guide

PACING	CLASSROOM RESOURCES	LABS, DEMONSTRATIONS, AND ACTIVITIES
BLOCKS 1 & 2 • 90 min pp. 240–244 **Chapter Opener**		
Section 1 What Is Biodiversity?	**OSP** Lesson Plan* **CRF** Active Reading* BASIC **TT** Bellringer* **TT** Known and Estimated Number of Species on Earth* **TT** Sea Otters as an Example of a Keystone Species* **TT** A Genetic Bottleneck*	**TE** Activity Keystone Metaphor, p. 242 BASIC **TE** Internet Activity A Wild Pharmacy, p. 243 GENERAL **CRF** Field Activity* ◆ BASIC
BLOCKS 3 & 4 • 90 min pp. 245–251 **Section 2** Biodiversity at Risk	**OSP** Lesson Plan* **CRF** Active Reading* BASIC **TT** Bellringer* **TT** Biodiversity and Extinction Over Geologic Time* **TT** Species Known to Be Threatened or Extinct Worldwide* **TT** Global Biodiversity Hotspots* **CD** Interactive Exploration 1 Something's Fishy	**TE** Demonstration Exotic Species, p. 245 BASIC **TE** Group Activity Public Perceptions, p. 246 GENERAL **TE** Group Activity Exotic Alert, p. 247 GENERAL **TE** Activity Species Bulletin, p. 247 GENERAL **TE** Activity Biodiversity Hotspot Fair, p. 250 BASIC **CRF** Research Lab* ◆ GENERAL
BLOCKS 5, 6, 7 & 8 • 180 min pp. 252–257 **Section 3** The Future of Biodiversity	**OSP** Lesson Plan* **CRF** Active Reading* BASIC **TT** Bellringer*	**TE** Group Activity Fun Park USA, p. 252 GENERAL **TE** Activity Humans and Wildlife, p. 253 GENERAL **SE** QuickLab Design a Wildlife Preserve, p. 254 **TE** Internet Activity Species Survival Plans, p. 255 GENERAL **SE** Field Activity Simple Biodiversity Assessment, p. 257 **SE** Exploration Lab Differences in Diversity, pp. 262–263 ◆ **CRF** Datasheets for In-Text Labs* **CRF** Field Activity* ◆ ADVANCED **CRF** Consumer Lab* ◆ GENERAL **CRF** Long-Term Project* ◆ GENERAL

BLOCKS 9 & 10 • 90 min

Chapter Review and Assessment Resources

- **SE** Chapter Review pp. 259–261
- **SE** Standardized Test Prep pp. 654–655
- **CRF** Study Guide* ■ GENERAL
- **CRF** Chapter Test* ■ GENERAL
- **CRF** Chapter Test* ADVANCED
- **CRF** Concept Review* ■ GENERAL
- **CRF** Critical Thinking* ADVANCED
- **OSP** Test Generator
- **CRF** Test Item Listing*

Online and Technology Resources

Holt Online Learning

Visit **go.hrw.com** for access to Holt Online Learning or enter the keyword **HE6 Home** for a variety of free online resources.

One-Stop Planner® CD-ROM

This CD-ROM package includes
- Lab Materials QuickList Software
- Holt Calendar Planner
- Customizable Lesson Plans
- Printable Worksheets
- ExamView® Test Generator
- Interactive Teacher Edition
- Holt PuzzlePro® Resources
- Holt PowerPoint® Resources

Holt Environmental Science Interactive Tutor CD-ROM

This CD-ROM consists of interactive activities that give students a fun way to extend their knowledge of environmental science concepts.

KEY

TE Teacher Edition	**CRF** Chapter Resource File	* Also on One-Stop Planner
SE Student Edition	**TT** Teaching Transparency	■ Also Available in Spanish
OSP One-Stop Planner	**CD** Interactive CD-ROM	◆ Requires Advance Prep

ENRICHMENT AND SKILLS PRACTICE	SECTION REVIEW AND ASSESSMENT	CORRELATIONS
SE Pre-Reading Activity, p. 240 **TE** Using the Figure, p. 240 (GENERAL)		**National Science Education Standards**
	SE Section Review, p. 244 **TE** Reteaching, p. 244 (BASIC) **TE** Quiz, p. 244 (GENERAL) **TE** Alternative Assessment, p. 244 (GENERAL) **CRF** Quiz * ■ (GENERAL)	LS 3b LS 4e HNS 1c SPSP 3a
TE Skill Builder Vocabulary, p. 245 (ADVANCED) **TE** Reading Skill Builder Reading Organizer, p. 246 (BASIC) **TE** Skill Builder Writing, p. 246 (GENERAL) **TE** Student Opportunities Controlling Invasive Species, p. 247 **SE** MathPractice Estimating Species Loss, p. 248 **SE** Case Study A Genetic Gold Rush, pp. 248–249 **TE** Using the Figure Biodiversity Hotspots, p. 250 (GENERAL) **TE** Inclusion Strategies, p. 250	**TE** Homework, p. 245 (GENERAL) **TE** Homework, p. 247 (GENERAL) **TE** Homework, p. 249 (GENERAL) **SE** Section Review, p. 251 **TE** Reteaching, p. 251 (BASIC) **TE** Quiz, p. 251 (GENERAL) **TE** Alternative Assessment, p. 251 (GENERAL) **CRF** Quiz * ■ (GENERAL)	LS 3b LS 4e SPSP 6e
SE Graphic Organizer Spider Map, p. 253 (GENERAL) **TE** Student Opportunities Biodiversity Field Trip, p. 253 **TE** Using the Figure Coffee and Habitat, p. 254 (GENERAL) **SE** MathPractice Measuring Risk, p. 256 **SE** Making a Difference Dr. E. O. Wilson, pp. 264–265 **CRF** Map Skills Animal Ranges * (GENERAL)	**SE** Section Review, p. 257 **TE** Reteaching, p. 257 (BASIC) **TE** Quiz, p. 257 (GENERAL) **TE** Alternative Assessment, p. 257 (GENERAL) **CRF** Quiz * ■ (GENERAL)	HNS 1c SPSP 3a

 Guided Reading Audio CDs

These CDs are designed to help auditory learners and reluctant readers. (Audio CDs are also available in Spanish.)

www.scilinks.org

Maintained by the **National Science Teachers Association**

TOPIC: Biodiversity
SciLinks code: HE4005

TOPIC: Medicines from Plants
SciLinks code: HE4065

TOPIC: Endangered Species
SciLinks code: HE4031

TOPIC: Preserving Ecosystems
SciLinks code: HE4088

 CNN Videos

Each video segment is accompanied by a Critical Thinking Worksheet.

Science, Technology & Society
Segment 8 Bioengineered Plants

CHAPTER

10

Chapter Enrichment

This Chapter Enrichment provides relevant and interesting information to expand and enhance your classroom instruction of the chapter material.

SECTION

1 What is Biodiversity?

Keystone Species

The keystone species concept has been fundamental to ecology and conservation biology since its introduction by zoologist Robert T. Paine in 1969. According to Paine, a keystone species is one whose importance to its community or ecosystem is greater than would be expected from its relative abundance or total biomass. The "top predator" in a food pyramid often fits this role, although keystone species could exist at any level. Well-studied examples of keystone species include the sea otter in sea kelp-bed communities, the predatory starfish *Pisaster* in rocky intertidal zones, elephants in the African grasslands, and the extinct dodo bird of Mauritius. The keystone concept has provided a model for understanding ecological communities and has influenced efforts to conserve species and habitats. When keystone species are identified and protected, entire ecosystems may be preserved. However, the keystone role of a species may not be known until it has been extirpated (locally wiped out) and its ecosystem changes.

▶ A sea otter eating a sea urchin

The Dodo Tree

In 1973, a scientist was studying the endangered *tambalacoque* trees on Mauritius, an island off the coast of Africa. Only 13 of these trees were still growing, and all were nearing the end of their 300 year lifespan. The scientist knew that the native dodo birds went extinct about 300 years before. Accounts of early settlers and fossil remains showed that dodos used to eat the fruit of

the tambalacoque. The scientist hypothesized that the tree's seeds sprouted only after passing through the dodo's gizzard, which would grind away the seed's tough outer layer. He fed the large seeds to turkeys, which also have gizzards. Some of the seeds passed through and sprouted—the first new tambalacoque trees in 300 years!

BRAIN FOOD

In 2002, scientists from Denmark and Germany announced the discovery of a new order of insects, which they named *Mantophasmatodea*. The carnivorous insects, found in the mountains of Namibia, resemble something between crickets and juvenile praying mantids. The discovery made headlines because scientists had not found a new insect order in 87 years. A British scientist commented that if this discovery had been among mammals, it would be "like discovering bats."

SECTION

2 Biodiversity at Risk

Volcanic Hot Spots and Biodiversity

By superimposing maps of volcanos and earthquakes on maps showing biodiversity, researchers have found that volcanic hotspots are also biodiversity hotspots. Scientists are not entirely sure why volcanic hotspots are so diverse. One reason is probably that mountains form in these areas, creating differences in altitude and corresponding habitats. Another reason is that nutrient-rich volcanic deposits promote abundant vegetative growth.

United States Biodiversity

Biodiversity hotspots are mostly found in the tropics. To learn about biodiversity in places that are usually not considered hotspots (such as urban parks), a group of scientists have held "bio-blitzes" at an urban park in Connecticut. The purpose of the blitzes is to identify as many species as possible within twenty-four hours. In 2001, scientists and volunteers participating in the blitz identified 2,519 species in the park. Organisms included birds, mammals, and many invertebrates. Projects like the bio-blitz help to highlight the idea that everywhere, the world is teeming with life.

▶ The endangered Dragon Tree, a huge yucca endemic to the Canary Islands

Traditional Ecological Knowledge

Robin Kimmerer, a plant ecologist in New York, thinks that ecology curricula should include traditional ecological knowledge (TEK). TEK is ecological knowledge held by people in traditional, natural resource-based societies. Kimmerer notes that TEK is usually gathered through long-term observations over the course of many generations. Thus, TEK complements scientific studies, which tend to be short term. Kimmerer also notes that studying TEK in native languages supports the link between conservation of biodiversity and conservation of cultural diversity.

SECTION

3 The Future of Biodiversity

Botanical Gardens and Species Survival

The 1,600 botanical gardens around the world not only display beautiful and interesting plants, they also serve as plant research and conservation centers. Researchers cultivate endangered plants at many botanical gardens. Growing these species helps horticulturalists understand the plants' needs so people can successfully reintroduce

the plants into wild habitats. Professionals that maintain botanical gardens communicate with each other through organizations such as Botanic Gardens Conservation International (BGCI), located in the United Kingdom. BGCI members have created a Botanic Gardens Conservation Strategy document, which outlines a plan to save plant species around the world. The BGCI Web site produces manuals for raising rare plants, information from botanical gardens around the world, and the most current conservation information.

Island Biogeography

In the 1960s, two young ecologists, Robert MacArthur and Edward O. Wilson, proposed the theory of island biogeography. According to the theory, the smaller an island is and the further it is from the mainland, the fewer species it will contain, and vice versa. Many parks and preserves on land function almost like ecological islands in a sea of human development. Island biogeography has helped scientists plan the size and location of parks to maximize the biodiversity they can conserve.

Cloning Endangered Species

Cloning, a controversial procedure that creates an embryo out of adult tissue, may become another tool in the effort to preserve endangered species. In 2001, Italian scientists successfully cloned a European mouflon, an endangered Mediterranean sheep. The animal was cloned by first removing the nucleus from an unfertilized domestic sheep's egg. The nucleus from a mouflon body cell was then inserted into the unfertilized egg. After the egg divided, the embryo was implanted into a domestic sheep. Habitat preservation is the best way to save species. But cloning could reintroduce genes from dead or infertile animals as a means of increasing the genetic diversity of the population.

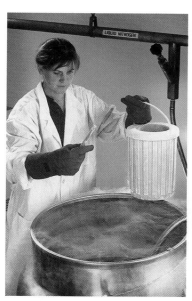

▶ Germ plasm storage in liquid nitrogen

Overview

Tell students that this chapter will introduce the concept of biodiversity. The diverse life on Earth provides humans with a variety of crops, medicines, and recreational experiences. Currently, human activities threaten many species with extinction, but people around the world are working to protect biodiversity.

Using the Figure — GENERAL

The photo shows native plants in spring bloom in Llano County, Texas. Have students describe the diversity of plants in the photo (noting number, color, shapes, flower types). Have interested students locate this area on a map. Texas is home to 5,500 plant species, 425 of which occur only in Texas. **LS** Visual

PRE-READING ACTIVITY

You may want to collect students' FoldNotes to determine their prior knowledge. You may also want to modify your lesson plan to answer questions that students list in the "Want" column of their table.

Biodiversity

240

1 **What Is Biodiversity?**
2 **Biodiversity at Risk**
3 **The Future of Biodiversity**

PRE-READING ACTIVITY

FOLDNOTES | **Tri-Fold**
Before you read this chapter, create the **FoldNote** entitled "Tri-Fold" described in the Reading and Study Skills section of the Appendix. Write what you know about biodiversity in the column labeled "Know." Then, write what you want to know in the column labeled "Want." As you read the chapter, write what you learn about biodiversity in the column labeled "Learn."

How many species are in this photo? Scientists know that this region of Central Texas is home to an unusual number of unique species. However, many more species remain unknown to science, both in faraway jungles and in our own backyards.

For information about videos related to this chapter, go to **go.hrw.com** and type in the keyword **HE4 BIOV**.

Chapter Correlations *National Science Education Standards*

LS 3b The great diversity of organisms is the result of more than 3.5 billion years of evolution that has filled every available niche with life forms. **(Section 1 and Section 2)**

LS 4e Human beings live within the world's ecosystems. Increasingly, humans modify ecosystems as a result of population growth, technology, and consumption. Human destruction of habitats through direct harvesting, pollution, atmospheric changes, and other factors is threatening current global stability, and if not addressed, ecosystems will be irreversibly affected. **(Section 1 and Section 2)**

SPSP 6e Humans have a major effect on other species. For example, the influence of humans on other organisms occurs through land use—which decreases space available to other species—and pollution—which changes the chemical composition of air, soil, and water. **(Section 2)**

HNS 1c Scientists are influenced by societal, cultural, and personal beliefs

and ways of viewing the world. Science is not separate from society but rather science is a part of society. **(Section 1 and Section 3)**

SPSP 3a Human populations use resources in the environment in order to maintain and improve their existence. Natural resources have been and will continue to be used to maintain human populations. **(Section 1 and Section 3)**

What Is Biodiversity?

Every day, somewhere on Earth, a unique species of organism becomes *extinct* as the last member of that species dies—often because of human actions. Scientists are not sure how many species are becoming extinct or even how many species there are on Earth. How much extinction is natural? Can we—or should we—prevent extinctions? The study of biodiversity helps us think about these questions, but does not give us all the answers.

A World Rich in Biodiversity

The term **biodiversity,** short for "biological diversity," usually refers to the number and variety of different species in a given area. Certain areas of the planet, such as tropical rain forests, contain an extraordinary variety of species. The complex relationships between so many species are hard to study, but humans may need to understand and preserve biodiversity for our own survival.

Unknown Diversity The study of biodiversity starts with the unfinished task of cataloging all the species that exist on Earth. As shown in **Figure 1,** the number of species known to science is about 1.7 million, most of which are insects. However, the actual number of species on Earth is unknown. Most scientists agree that we have not studied Earth's species adequately, but they accept an estimate of greater than 10 million for the total number of species. New species are considered *known* when they are collected and described scientifically. Unknown species exist in remote wildernesses, deep in the oceans, and even in cities. Some types of species are harder to study and receive less attention than large, familiar species. For example, less is known about insects and fungi than is known about trees and mammals.

Objectives

▶ Describe the diversity of species types on Earth, relating the difference between known numbers and estimated numbers.

▶ List and describe three levels of biodiversity.

▶ Explain four ways in which biodiversity is important to ecosystems and humans.

▶ Analyze the potential value of a single species.

Key Terms

biodiversity
gene
keystone species
ecotourism

Figure 1 ▶ Number of Species on Earth About 1.7 million species on Earth are known to science. Many more species are *estimated* to exist, especially species of smaller organisms. Scientists continue to revise these estimates.

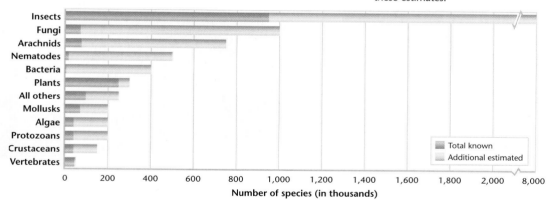

Source: World Conservation Monitoring Center.

241

How Many Species on Earth?

Determining the total number of species on Earth is neither simple nor precise. Estimates of the total number of species range from 8 million to 50 million. Students may wonder how such a disparity could exist. Explain that there is no reasonable way to count all of the species on Earth. Instead, scientists develop systems for making educated guesses. The following are two such systems: 1) By thoroughly sampling organisms in a relatively unstudied region, scientists determine how many of those organisms are of unknown species. That number is then used to make estimates of the proportion of unclassified species in all areas. 2) By working with the premise that many more large species have been identified than small ones, scientists find a ratio between the number of large and small organisms in a well-studied area. They compare this ratio to the total number of large species on Earth to extrapolate the number of as-yet-unidentified small species.

Overview

Before beginning this section, review with your students the Objectives in the Student Edition. This section will describe scientists' current knowledge of biodiversity. It will also highlight the benefits of biodiversity, including healthy ecosystems, medicines, and foods.

🔊 Bellringer

Remind students that a species is a group of organisms that are closely related and can mate to produce fertile offspring. Ask students to think about the number of species they see or interact with on an average day. Have them list as many as they can think of in their *EcoLog.* **LS** Logical

Motivate

Discussion ——— GENERAL

Unnecessary Species Have student groups work together to answer the following question: "If you could get rid of one species, what would it be?" (Students will probably name species that can harm humans, such as mosquitoes, bees, or poison ivy.) Have each group share their answer and have the other groups offer reasons that the given species should be saved. (Sample answer: Bees pollinate flowers and make honey.) Discuss examples of when we should dedicate money, time, and resources to protecting a species, and who should make such decisions. **LS** Interpersonal

Chapter Resource File

• **Lesson Plan**
• **Active Reading** BASIC
• **Section Quiz** GENERAL

Transparencies

TT Bellringer

TT Known and Estimated Number of Species on Earth

Teach

Activity — BASIC

Keystone Metaphor Explain that a keystone species is one so valuable to an ecosystem that the ecosystem cannot function properly if that species dies out. Tell students that ecology borrowed the word *keystone* from architecture. In masonry (the profession of building with stone and brick), a keystone is the central stone of an arch. Until the keystone is in place, a true arch will not support itself. The architectural formation is entirely dependent on the keystone, just as an ecosystem is dependent on its keystone species. Display for students a drawing or photograph of an arch with an obvious keystone. Students can also build a simple arch with a keystone. Have students illustrate the concept of a keystone species with a drawing, a short story, a poem, or a song. Have students share their work with the class and include it in their **Portfolio.**

 Visual/ Kinesthetic

English Language Learners

Transparencies

TT Sea Otters as an Example of a Keystone Species

Figure 2 ▶ Scientists continue to find and describe new species. Specimens may be stored in collections such as this one, with a small tag that says where and when the specimen was found.

Figure 3 ▶ The sea otters of North America are an example of a keystone species, upon which a whole ecosystem depends.

Levels of Diversity Biodiversity can be studied and described at three levels. *Species diversity* refers to all the differences between populations of species, as well as between different species. This kind of diversity has received the most attention and is most often what is meant by *biodiversity*. *Ecosystem diversity* refers to the variety of habitats, communities, and ecological processes within and between ecosystems. *Genetic diversity* refers to all the different *genes* contained within all members of a population. A **gene** is a piece of DNA that codes for a specific trait that can be inherited by an organism's offspring.

Benefits of Biodiversity

Biodiversity can affect the stability of ecosystems and the sustainability of populations. In addition, there are many ways that humans clearly use and benefit from the variety of life-forms on Earth. Biodiversity may be more important than we realize.

Species Are Connected to Ecosystems We depend on healthy ecosystems to ensure a healthy biosphere that has balanced cycles of energy and nutrients. Species are part of these cycles. When scientists study any species closely, they find that it plays an important role in an ecosystem. Every species is probably either dependent on or depended upon by at least one other species in ways that are not always obvious. When one species disappears from an ecosystem, a strand in a food web is removed. How many threads can be pulled from the web before it collapses? We often do not know the answer until it is too late.

Some species are so clearly critical to the functioning of an ecosystem that they are called **keystone species.** One example of a keystone species is the sea otter. **Figure 3** shows how the loss of sea otter populations led to the loss of the kelp beds along the U.S. Pacific coast and how the recovery of otter populations led to the recovery of the kelp communities.

❶ In the 1800s, sea otters were hunted for their fur. They disappeared from the Pacific coast of the U.S.

❷ Sea urchins, with no more predators, multiplied and ate all of the kelp. The kelp beds began to disappear from the area.

❸ In 1937, a small group of surviving otters was discovered. With protection and scientific efforts, the otter populations grew.

❹ The otters once again preyed on the sea urchins. The kelp beds regenerated.

242

MISCONCEPTION ALERT

Keystone Vs. Flagship Species
Students may think that a keystone species is the best-known species in an ecosystem, but this is not necessarily true. A keystone species is simply one whose effect on its ecosystem is greater than you would assume given its mass or abundance. A flagship species, on the other hand, is a species that is well-known and popular. Organizations often use flagship species, such as the panda bear, to attract support for conservation.

BRAIN FOOD

Populations Need Diversity Genetic diversity helps populations survive in their environment. If a disease strikes a natural population of ox, genetic variation might enable some individuals to resist the disease. And the fact that not every animal carried the disease would reduce the number of animals that could spread the disease. If the same disease struck a herd of cloned ox that did not have genetic resistance to the disease, all of the ox would become sick.

Species and Population Survival The level of genetic diversity within populations is a critical factor in species survival. Genetic variation increases the chances that some members of a population may survive environmental pressures or changes. Small and isolated populations are less likely to survive such pressures. When a population shrinks, its genetic diversity decreases as though it is passing through a *bottleneck*, represented in **Figure 4**. Even if such a population is able to increase again, there will be inbreeding within a smaller variety of genes. Then, members of the population may become more likely to inherit genetic diseases.

Medical, Industrial, and Agricultural Uses People throughout history have used the variety of organisms on Earth for food, clothing, shelter, and medicine. About one quarter of the drugs prescribed in the United States are derived from plants. Almost all antibiotics are derived from chemicals found in fungi. **Table 1** lists some plants from which medicines are derived.

For some industries, undiscovered and poorly studied species represent a source of potential products. New chemicals and industrial materials may be developed from chemicals discovered in all kinds of species. The scientific community continues to find new uses for biological material and genetic diversity, from combating diseases to understanding the origins of life.

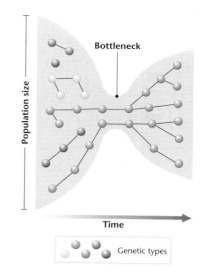

Figure 4 ▶ When a population is reduced to a few members, this creates a *bottleneck* of reduced genetic variation. Even if the population grows again, its chances of long-term survival are lower.

Table 1 ▼

Common Medicines Derived from Plants		
Medicine	Origin	Use
Neostigmine	calabar bean (Africa)	treatment of glaucoma and basis for synthetic insecticides
Turbocurarine	curare vine (South America)	surgical muscle relaxant; treatment of muscle disorders; and poison for arrow tips
Vincristine, vinblastine	rosy periwinkle (Madagascar)	treatment of pediatric leukemia and Hodgkin's disease
Bromelain	pineapple (South America)	treatment to control tissue inflammation
Taxol	Pacific yew (North America)	anticancer agent
Novacaine, cocaine	coca plant (South America)	local anesthetic and basis for many other anesthetics
Cortisone	wild yam (Central America)	hormone used in many drugs
L-dopa (levodopa)	velvet bean (tropical Asia)	treatment of Parkinson's disease
Reserpine	Indian snakeroot (Malaysia)	treatment to reduce high blood pressure

🖳 internet connect

www.scilinks.org
Topic: Biodiversity
SciLinks code: HE4005
Topic: Medicines from Plants
SciLinks code: HE4065

SCI*LINKS*. Maintained by the National Science Teachers Association

243

Discussion ────── **ADVANCED**

Making a Case for Conservation
To try to make a case for preserving species and ecosystems, some scientists have tried to calculate the monetary value of ecosystem services or of plants that could potentially produce new medicines. This is the basis for the argument that saving biodiversity results in economic gain. However, some scientists argue that it is impossible to put a price tag on biodiversity and that this approach ignores the aesthetic and ethical reasons for preserving species. Have students research both arguments. Then ask students to argue the case for preserving biodiversity, using the best points from both perspectives.
LS Verbal/Interpersonal

ANIMAL BIOLOGY
● **CONNECTION** ──── **GENERAL**

Chimp Medicines People aren't the only animals to use wild plants as medicines. In 1987, a Tanzanian chimpanzee that was suffering from diarrhea was observed chewing on the pith of *Vernonia amygdalina*, a small tree. Researchers found that, after eating pith from that tree, the chimp recovered in twenty-four hours. When they examined dung excreted twenty hours after the chimp ate the plant, they found fewer intestinal parasites. Later research confirmed that a compound in the pith had antiparasitic and antibacterial properties. Observations such as this have led to the creation of the science of *zoopharmacognosy*, which seeks to learn about medicinal plants by observing other animals. Encourage students to research this science to find out what other plants have been used as medicines by various animals (one example is grass eaten by dogs and cats). **LS** Intrapersonal

🖥 **Transparencies**
TT A Genetic Bottleneck

LANGUAGE ARTS
● **CONNECTION** ── **GENERAL**

Nature Writing Introduce students to *Walden* by Henry David Thoreau. In this book, Thoreau escapes society and lives in a cabin on Walden Pond for a year. His memoir of this time is a poetic description of wildlife around the pond and the changing of the seasons. Read excerpts out loud in class, and encourage students to read further. Then have students keep a nature journal for a few weeks. Ask them to visit the same natural area at least five times for at least an hour each time. Have them write about wildlife and plant life they observe. Encourage them to add sketches to the journal.

🖳 Internet Activity ───── **GENERAL**

A Wild Pharmacy Have students search the Internet for other examples of medicines that were derived from wild plants. Ask the class to create a table similar to **Table 1,** and have them compile a comprehensive list of medicines, their wild origins, and their uses. Post the table outside the door of your classroom so others in the school can learn about the natural origins of common medicines. **Caution:** discourage students from experimenting with wild plants by emphasizing that few people know the safe dosage and preparation for such medicines. **LS** Intrapersonal

Figure 5 ▶ A produce market in Bolivia shows a diversity of native foods. Food crops that originated in the American tropics include corn, tomatoes, squash, and many types of beans and peppers.

Close

Reteaching ——— BASIC

Concept Mapping Have students create a concept map using the main ideas in this section. Tell them to use headings and key terms as well as italicized terms within the map. LS Logical

Quiz ——— GENERAL

1. What might happen within an ecosystem when a keystone species is driven to extinction? (The balance of populations of other organisms might be upset, thus affecting the entire ecosystem.)

2. True or false: Humans don't need to preserve wild plants because we can breed new crops. Explain. (False. New crops are generally bred by combining genetic material from other plant populations.)

Alternative Assessment ——— GENERAL

Why Should We Care? Divide the class into groups, and assign each group one of the subheadings in this section. Have students plan and organize a "Why Should We Care?" fair. Their assignment is to develop a creative means to convince the school that their topic is important. For example, students could pass out flyers, show a video, perform a skit, make a mural or a series of posters, or even create a T-shirt design. Each group should have a slogan or logo, and their presentation should be supported with facts and data. LS Interpersonal Co-op Learning

Table 2 ▼

Origins of Some Foods
North America, Central America, and South America
• corn (maize), tomato, bean (pinto, green, and lima), peanut, potato, sweet potato, avocado, pumpkin, pineapple, cocoa, vanilla, and pepper (green, red, and chile)
Northeastern Africa, Central Asia, and Near East
• wheat (several types), sesame, chickpea, fig, lentil, carrot, pea, okra, date, walnut, coffee, cow, goat, pig, and sheep
India, East Asia, and Pacific Islands
• soybean, rice, banana, coconut, lemon, lime, orange, cucumber, eggplant, turnip, tea, black pepper, and chicken

Humans benefit from biodiversity every time they eat. Most of the crops produced around the world originated from a few areas of high biodiversity. Some examples of crop origins are shown in **Figure 5** and **Table 2.** Most new crop varieties are *hybrids*, crops developed by combining genetic material from other populations. History has shown that depending on too few plants for food is risky. For example, famines have resulted when an important crop was wiped out by disease. But some crops have been saved from diseases by being crossbred with wild plant relatives. In the future, new crop varieties may come from species not yet discovered.

Ethics, Aesthetics, and Recreation Some people believe that we should preserve biodiversity for ethical reasons. They believe that species and ecosystems have a right to exist whether or not they have any other value. To people of some cultures and religions, each organism on Earth is a gift with a higher purpose.

People also value biodiversity for aesthetic or personal enjoyment—keeping pets, camping, picking wildflowers, or watching wildlife. Some regions earn the majority of their income from **ecotourism,** a form of tourism that supports the conservation and sustainable development of ecologically unique areas.

SECTION 1 Review

1. **Describe** the general diversity of species on Earth in terms of relative numbers and types of organisms. Compare known numbers to estimates.

2. **Describe** the three levels of biodiversity. Which level is most commonly meant by *biodiversity?*

3. **Explain** how biodiversity is important to ecosystems, and give examples of how it is important to humans.

CRITICAL THINKING

4. **Analyzing a Viewpoint** Is it possible to put a price on a single species? Explain your answer.

5. **Predicting Consequences** What is your favorite type of organism? If this organism were to go extinct, how would you feel? What would you be willing to do to try to save it from extinction? Write a short essay describing your reaction. **WRITING SKILLS**

244

Answers to Section Review

1. A high diversity of species is known to exist among insects and land plants; fewer species are *known* to exist among vertebrates, bacteria, algae, arachnids, protists, nematodes, mollusks and crustaceans. However, much greater diversity is *estimated* to exist among insects, fungi, bacteria, algae, and arachnids. Vertebrates are the best-studied organisms but are relatively less diverse than other groups.

2. The three levels of biodiversity are species diversity, ecosystem diversity, and genetic diversity. Species diversity is most commonly meant by the term *biodiversity.*

3. An ecosystem with its biodiversity intact will be more stable and have more sustainable populations. Stable, functioning ecosystems provide many benefits to humans, including clean air and water, food, fuel, fiber, medicines, and recreational opportunities.

4. Answers may vary. Accept any well-reasoned, thoughtful response.

5. Answers may vary.

Biodiversity at Risk

The last of the dinosaurs died about 65 million years ago, when a series of changes in the Earth's climate and ecosystems caused the extinction of about half the species on Earth. The extinction of many species in a relatively short period of time is called a *mass extinction*. Earth has experienced several mass extinctions, as shown in **Figure 6**, each probably caused by a global change in climate. It takes millions of years for biodiversity to rebound after a mass extinction.

Current Extinctions

Scientists are warning that we are in the midst of another mass extinction. The rate of extinction is estimated to have increased by a multiple of 50 since 1800. Between 1800 and 2100, up to 25 percent of all species on Earth may have become extinct. The current mass extinction is different from those of the past because humans are the primary cause of the extinctions.

Species Prone to Extinction Cockroaches and rats are not likely to become extinct because they have large populations that adapt easily to many habitats. But species with small populations in limited areas can easily become extinct. Species that are especially at risk of extinction include those that migrate, those that need large or special habitats, and those that are exploited by humans.

An **endangered species** is a species that is likely to become extinct if protective measures are not taken immediately. A **threatened species** is a species that has a declining population and that is likely to become endangered if it is not protected. Additional categories of risk exist for certain legal and biological purposes.

Objectives

▶ Define and give examples of *endangered* and *threatened species*.

▶ Describe several ways that species are being threatened with extinction globally.

▶ Explain which types of threats are having the largest impact on biodiversity.

▶ List areas of the world that have high levels of biodiversity and many threats to species.

▶ Compare the amount of biodiversity in the United States to that of the rest of the world.

Key Terms

endangered species
threatened species
exotic species
poaching
endemic species

Figure 6 ▶ Biodiversity has generally increased over time, as indicated here by the numbers of families of marine animals. The past five mass extinctions were probably caused by global climate changes.

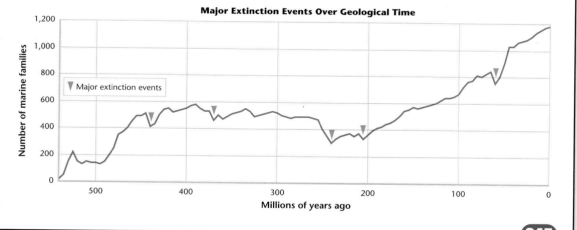

Major Extinction Events Over Geological Time

▽ Major extinction events

Number of marine families (y-axis: 0, 200, 400, 600, 800, 1,000, 1,200)

Millions of years ago (x-axis: 500, 400, 300, 200, 100, 0)

245

SKILL BUILDER ——— **ADVANCED**

Vocabulary *Extirpation* is the complete removal of a particular type of organism from an area, usually a specified geographic area. *Extermination* means driving out or completely destroying a type of organism.

Homework ——— **GENERAL**

Documenting Extinction Ask students to research the history of a plant or animal that has become extinct. Have them analyze the factors that contributed to the species' demise and create a video, multimedia presentation, or poster to present their findings. Encourage students to evaluate which types of efforts to save species were more or less effective, and which might help to save other threatened or endangered species. **LS Intrapersonal**

SECTION 2

Focus

Overview

Before beginning this section, review with your students the Objectives in the Student Edition. In this section, students will learn about the various causes of species extinctions, where most species are being threatened, and which factors are causing the most extinctions.

🔊 Bellringer

Ask students to write the names of all the endangered species that they can think of in their *EcoLog*. Have them try to think of any that are local. Have them write down some of the reasons why they think the species is endangered. **LS Intrapersonal**

Motivate

Demonstration ——— BASIC

Exotic Species Bring in pictures of various exotic organisms that are sold as pets or for décor. As you show each picture, ask students if they know anyone who has the organism at home. Tell students that many exotic species are collected directly from the wild in tropical rain forests or other foreign habitats. Although trade laws now protect many species, some are still both legally and illegally captured and sold. Consumers should ask for documentation of the origin and legality of an exotic species before buying it. **LS Visual**

Chapter Resource File

• Lesson Plan
• Active Reading BASIC
• Section Quiz GENERAL

Transparencies

TT Bellringer
TT Biodiversity and Extinction Over Geologic Time

Table 3 ▼

Species Known to Be Threatened or Extinct Worldwide			
Type of species	Number threatened (all categories of risk)	Number extinct (since ~1800)	Percent of species that may be threatened
Mammals	1,130	87	26
Birds	1,183	131	12
Reptiles	296	22	3.3
Amphibians	146	5	3.1
Fishes	751	92	3.7
Insects	555	73	0.054
Other crustaceans	555	73	1.03
Mollusks and worms	944	303	1.3
Plants	30,827	400	0.054

Source: UN Environment Programme.

Teach

Group Activity — GENERAL

Public Perceptions Organize students into groups of four. Ask each group to design a survey to assess whether or not the general public feels that the growing rate of species extinction is a serious problem. Ask students to include at least five questions, and have them make some of the questions open-ended. Have the groups go to public places in their community to ask people their questions. Some groups may want to videotape the responses (have the students make sure the subjects don't mind being taped, and that taping is allowed in the area where they are asking the questions). Then have each group share the public's responses with the rest of the class. Have the class compile the responses into a press release for the school or local newspaper.

 Verbal/Interpersonal Co-op Learning

 — BASIC

Reading Organizer Have students read through each section and outline the major ideas. Then ask them to write a summary statement explaining why the chapter organization makes sense. If they feel that a different organization would be better, have them create a different outline and explain why it makes more sense to them.

Interpersonal English Language Learners

Transparencies

TT Species Known to Be Threatened or Extinct Worldwide

Figure 7 ▶ Fewer than 80 Florida panthers (right) remain in the wild. Almost all of the habitat (below) of this cougar subspecies has been destroyed or fragmented by commercial and housing development.

Range at about 1500 CE
Current range

Source: Florida Fish and Wildlife Conservation Commission.

How Do Humans Cause Extinctions?

In the past 2 centuries, human population growth has accelerated and so has the rate of extinctions. The numbers of worldwide species known to be threatened, endangered, or recently extinct are listed in **Table 3**. The major human causes of extinction today are the destruction of habitats, the introduction of nonnative species, pollution, and the overharvesting of species.

Habitat Destruction and Fragmentation As human populations grow, we use more land to build homes and harvest resources. In the process, we destroy and fragment the habitats of other species. It is estimated that habitat loss causes almost 75 percent of the extinctions now occurring.

Due to habitat loss, the Florida panther is one of the most endangered animals in North America. The panther and its historical range are shown in **Figure 7**. Two hundred years ago, cougars—

246

SKILL BUILDER — GENERAL

Writing Have students research and write a report on the Florida panther for their **Portfolio.** Several resources are available to learn more about the near extinction of the Florida panther. A detailed account can be found in *The Florida Panther: Life and Death of a Vanishing Carnivore*, by David Maehr, Island Press, Washington, D.C., 1997.

Intrapersonal/Verbal

REAL-LIFE
● **CONNECTION** — ADVANCED

Local Landscapes Obtain aerial photographs of your local area at a college map library, state or county land department, or on the Internet. Try to get a current photo and photos at 5–10 year intervals for the last 30–40 years. Have students put transparencies over the photos and outline areas that are still covered with native vegetation. Then have students compare the drawings. Discuss how habitat fragmentation caused by human settlement has changed over time and whether habitat loss increased dramatically at any point. **Visual**

a species that includes panthers and mountain lions—ranged from Alaska to South America. Cougars require expansive ranges of forest habitat and large amounts of prey. Today, much of the cougars' habitat has been destroyed or broken up by roads, canals, and fences. In 2001, fewer than 80 Florida panthers made up the only remaining wild cougar population east of the Mississippi River.

Invasive Exotic Species An **exotic species** is a species that is not native to a particular region. Even such familiar organisms as cats and rats are considered to be exotic species when they are brought to regions where they never lived before. Exotic species can threaten native species that have no natural defenses against them. The invasive fire ants in **Figure 8** threaten livestock, people, and native species throughout the southeastern United States.

Harvesting, Hunting, and Poaching Excessive hunting and harvesting of species can also lead to extinction. In the United States in the 1800s and 1900s, 2 billion passenger pigeons were hunted to extinction and the bison was hunted nearly to extinction. Thousands of rare species worldwide are harvested and sold for use as pets, houseplants, wood, food, or herbal medicine.

Many countries now have laws to regulate hunting, fishing, harvesting, and trade of wildlife. However, these activities continue illegally, a crime known as **poaching.** In poor countries especially, local species are an obvious source of food, medicine, or income. Moreover, not all threatened species are legally protected.

Pollution Pesticides, cleaning agents, drugs, and other chemicals used by humans are making their way into food webs around the globe. The long-term effects of chemicals may not be clear until after many years of use. The bald eagle is a well-known example of a species that was endangered because of a pesticide known as DDT. Although DDT is now illegal to use in the United States, it is still manufactured here and used around the world.

Connection to **Ecology**

Extinction and Global Change Scientists have worried for some time that environmental pollutants might cause drastic changes in our atmosphere and biosphere. However, it is difficult to draw a direct link from global changes to specific extinctions.

In recent decades, scientists have observed a worldwide decline in amphibian species. Unlike most cases of habitat loss or overhunting, there are no clear causes for these extinctions. But there is growing evidence to indicate two probable causes: the pollution of water sources with hormone-like chemicals and increased UV radiation exposure due to the thinning of the Earth's ozone layer.

Figure 8 ▶ Mounds made by imported fire ants cover many fields in the southeastern United States. As with many invasive exotic species, these ants had no natural predators and little competition from native species when they were first brought into the country by accident.

247

Group Activity ——— **GENERAL**

Exotic Alert Organize the class into groups of three or four students, and give each group the name of an exotic organism that has negatively affected the natural environment of your city, state, or region. Also provide students with a brief summary of the damage that the organism has caused since its arrival in the area. Have students use the information to create a TV news segment, commercial, mock horror movie, folklore legend, or song to show how the exotic species has upset the natural equilibrium of an ecosystem and to educate people about the species. **LS** Kinesthetic   Co-op Learning

Activity ——— **GENERAL**

Species Bulletin Tell students to suppose that they are public affairs agents for the U.S. Fish and Wildlife Service. Their assignment is to develop a campaign to educate the general public about the laws surrounding the capture, trade, or use of threatened species and endangered species. Encourage students to present their public service announcement in the form of a flyer, poster, TV or radio commercial, or newspaper advertisement. Suggest that students include their announcement in their **Portfolio.** **LS** Interpersonal

HISTORY ———
CONNECTION ——— **GENERAL**

Buffalo Legacy The American buffalo (also called bison) was nearly hunted to extinction in the 1800s. When Europeans first came to North America, at least 60 million buffalo roamed the continent's vast prairies and forests. Early travelers across the Great Plains said there were so many buffalo that they covered the land like a dark sea. By 1906 only about 300 buffalo remained. Many of the buffalo had been killed just for their tongues (considered a delicacy) and hides. The rest of the animal was left to rot. Today, laws protect the buffalo, and their population has grown to more than 200,000.

Student Opportunities

Controlling Invasive Species Suggest that students contact a local conservation organization or outdoor education group to inquire about local opportunities to combat invasive species. Invasive plants are a problem in many areas. Some communities organize volunteers to find and remove invasive plants. These efforts avoid the use of chemical pesticides and educate people about the ecology of their area. **LS** Kinesthetic   English Language Learners

Homework ——— **GENERAL**

The Impact of Lawn Chemicals Ask students to find out if their parents use chemicals to control weeds and pests in their lawn. Have students research the effects of lawn chemicals on wildlife. Also have them research strategies for a pesticide-free lawn. **LS** Intrapersonal

Local Endemics Have students research endemic species in their community or state. Information about endemics should be available from the state's department of natural resources, a state university extension service, or local wildlife groups. Invite a local plant or wildlife expert to talk about local endemic species and their special needs. **LS** Intrapersonal

MATHPRACTICE

Answers

(1 million spp ÷ 5 million spp) ÷ (0.005 spp/y) = 40 y

Note to Teacher:

The answer of 40 years is a rough estimate. A more accurate calculation would compound the rate of loss annually to obtain the answer of 44.5 years.

LAW
● CONNECTION ─ GENERAL

Patenting Plants Students may be suprised to learn that a plant can be patented. Most plant-related patent laws in the U.S. were created in 1930 to cover new crop varieties. To be patented, a plant must be a "new variety that it is distinct from existing forms." Uncultivated plants cannot be patented. Patenting of genetic material or whole organisms has come under debate, especially when companies want exclusive rights to plant products. Have students research this topic, and then write questions to ask a patent lawyer via email or a classroom visit. **LS** Verbal

MATHPRACTICE

Estimating Species Loss
The annual loss of tropical forest habitat is estimated at about 1.8 percent per year. Some scientists estimate that this habitat loss results in a loss of about 0.5 percent of species per year. Given a low estimate of only 5 million species on Earth, how many years would it take for 1 million species to be lost, if current rates of habitat loss continue?

Areas of Critical Biodiversity

Certain areas of the world contain a greater diversity of species than other areas do. An important feature of such areas is that they have a large portion of **endemic species,** meaning species that are native to and found only within a limited area. Ecologists often use the numbers of endemic species of plants as an indicator of overall biodiversity, because plants form the basis of ecosystems on land. Ecologists increasingly point out the importance of biodiversity in oceans, though marine ecosystems are also complex and poorly understood.

Tropical Rain Forests The remaining tropical rain forests cover less than 7 percent of the Earth's land surface. Yet biologists estimate that over half of the world's species live in these forests. Most of these species have never been described. Unknown numbers of species are disappearing as tropical forests are cleared for farming or cattle grazing. Meanwhile, tropical forests are among the few places where some native people maintain traditional lifestyles and an intimate knowledge of their forest homes. The case study below explains the increasing value of such knowledge in the global marketplace.

CASE STUDY

A Genetic Gold Rush in the Rain Forests

How much is a species worth? To some people, there is money to be made in centers of biodiversity such as rain forests. Thus, the Amazonian rain forests are witnessing an increase in foreign visitors—not just tourists, but scientists searching for genes, glory, or enlightenment into the mysteries of these quickly disappearing treasures.

To biologists, the prospect of discovering new species may be a chance at fame. The first scientist to collect and describe a species often gets to choose a name for that species. For other scientists, researching the unknown inner workings of the rain forests is an adventure similar to the adventures of explorers charting new lands.

But like the quests of early European explorers of the Americas, some reasons to venture into the wilderness may be economic. The *biotechnology* industry is based on the application of biological science to create new products such as drugs. This industry depends on Earth's variety of organisms—especially their genetic material—for research and development.

In fact, the Brazilian government has taken notice of the increased international interest in the Amazon's amazing biological assets. The government has claimed the right to tax or patent any genetic material harvested from within its borders.

Other researchers are more interested in another special feature of the Amazon—native peoples. Some Amazonian natives, such as the Yanomamö, are still living a lifestyle of intimate connection to their forest

▶ This botanist is researching the uses of rain-forest plants and other species with the help of local people.

MISCONCEPTION ALERT

Estimating Species Loss Estimating current rates of species extinction is as imprecise as estimating the numbers of species on Earth. Most biologists agree that humans are currently causing extinctions at about 0.1 percent/year, a rate 1,000 times greater than the estimated prehuman rate of 0.0001 percent/ year. However, scientists have used various methods to estimate current extinction rates. One method is to extrapolate from the extinction rates of well-studied species. This method could cause underestimation because the last members of well-known species are often kept alive through extreme effort. Another method is to look at the rate at which threatened species digress in status from less endangered to extinct. A third method is to estimate the numbers of species per area in specific kinds of habitat and then extrapolate extinctions due to known rates of habitat loss. This method is complicated because certain areas of the world hold high concentrations of species while suffering greater destruction of habitat.

Coral Reefs and Coastal Ecosystems Like rain forests, coral reefs occupy a small fraction of the marine environment yet contain the majority of the biodiversity there. Reefs provide millions of people with food, tourism revenue, coastal protection, and sources of new chemicals. One study in 1998 estimated the value of these services to be $375 billion per year. But reefs are poorly studied and not as well protected by laws as terrestrial areas are. Nearly 60 percent of Earth's coral reefs are threatened by human activities, such as development along waterways, overfishing, and pollution. Similar threats affect coastal ecosystems, such as swamps, marshes, shores, and kelp beds. Coastal areas are travel routes for many migrating species as well as links to ecosystems on land.

Islands When an island rises from the sea, it is colonized by a limited number of species from the mainland. These colonizing species may then evolve into several new species. Thus, islands often hold a very distinct but limited set of species. For example, the Hawaiian Islands have 28 species of an endemic family of birds called *honeycreepers*. However, honeycreepers and many other island species are endangered because of invasive exotic species.

Geofact

The World's Largest Reef The Great Barrier Reef of Australia is the largest and probably the oldest reef system in the world. It stretches for 2,000 km (1,250 mi) and consists of 3,400 individual reefs.

▶ The Yanomamö are among the few native peoples of the tropical rain forests who still live traditional lifestyles and use their knowledge of the forests to meet all of their needs.

for their use of the skin excretions of poison dart frogs for hunting.

Often, researchers originally learned of a useful species from a local shaman, or medicine man. Biochemistry researchers have been amazed by the complex combinations of new chemicals they have discovered in many rain-forest species. Some of these chemicals are already being used in research and medicine.

home, in much the same way as they have for thousands of years.

An important value of such native peoples is their vast knowledge of the variety of species in the ecosystems where they live. Their knowledge includes more than just being able to recognize or name species. For example, the Yanomamö make use of thousands of plants, fungi, and animals for food, drugs, weapons, and art. Amazonian natives such as the Yanomamö are probably best known

CRITICAL THINKING

1. Expressing Viewpoints To whom do you think the genetic material of the rain forests should belong? What are some ways this benefit of biodiversity might be shared with the whole world?

Threats to Honeycreepers
According to the National Wildlife Federation, Hawaii's honeycreepers are the "world's most imperiled group of birds." Many invasive exotic species in Hawaii have affected honeycreepers and their habitats. Introduced cattle destroyed honeycreeper habitat by overeating the vegetation. Feral pigs continue to destroy habitat. Introduced mosquitoes carry avian malaria. Rats eat eggs and young birds. Introduced birds have out-competed the honeycreepers for habitat. Surprisingly, there is some hope for remaining honeycreepers. Scientists have found that although 60% of non-native birds tested in Oahu have avian malaria, none of the native *ámakihi* honeycreepers were infected. The reason for this is not known yet but rekindles optimism for the survival of this colorful group of birds.

Homework ——— **GENERAL**

The Biodiversity Treaty When the Biodiversity Treaty was signed in 1993, it was still uncertain how the international community would raise the funds and allocate the money for protecting the biodiversity of developing nations. Have students research what leadership steps have been taken by the United States since the treaty was signed. Have them summarize their findings in a "news report" or a critical review, and suggest that they add their report or review to their **Portfolio.** **LS** Intrapersonal

---

CASE STUDY

A Genetic Gold Rush The Brazilian government is working on laws to help protect its citizens from increasing "biopiracy" in the Amazon region. For example, the Kaxinawa Indians of southwestern Brazil claim that they were tricked into sharing their knowledge of 300 native species by a nonprofit charity that gave them baseball hats and aspirin as payment. Brazil's challenge is to allow economic use of rain forest products while also receiving appropriate compensation for rain forest inhabitants.

Answers to Critical Thinking

1. Answers may vary. Students may suggest that the genetic material belongs to rain forest residents or to their country. They may also suggest that genetic material from those plants should not be patented, so that researchers around the world would be able to study the material and produce their own medicines.

Teach, continued

Using the Figure — GENERAL

Biodiversity Hotspots Ask each student to pick one of the hotspots identified in **Figure 9.** Have students research the major factors threatening organisms in each specific hotspot. Enlarge the hotspot map, and have each student record the threats next to each hotspot. Post the figure in the classroom. **LS** Intrapersonal/Visual

Activity — BASIC

Biodiversity Hotspot Fair
Devote a day to exploring the amazing organisms and landscapes of the 25 biodiversity hotspots in **Figure 9.** Have different groups of students choose one hotspot, then prepare a display about it. Ask each group to highlight the region's rare species and threats to those species, as well as aspects of the region's ecosystems and interesting cultural information. As a motivation, have a competition to see which group can list or show the most species from their region. **LS** Visual

| Co-op Learning | English Language Learners |

REAL-LIFE CONNECTION — GENERAL

Ecotourism Request some travel brochures for areas in some of the biodiversity hotspots shown in **Figure 9.** Have students learn about travel opportunities and ecotourism in each hotspot. Invite someone from the community who has taken a trip to one of the biodiversity hotspots to give a talk or slide show about their experience. **LS** Visual

internet connect
www.scilinks.org
Topic: Endangered Species
SciLinks code: HE4031
Maintained by the National Science Teachers Association

Biodiversity Hotspots The most threatened areas of high species diversity on Earth have been labeled *biodiversity hotspots.* Twenty-five of these areas, shown in **Figure 9,** have been identified by international conservationists. The hotspot label was developed by ecologists in the late 1980s to identify areas that have high numbers of endemic species but that are also threatened by human activities. Most of these hotspots have lost at least 70 percent of their original natural vegetation. The hotspots include mostly tropical rainforests, coastal areas, and islands. In Madagascar, for example, only 18 percent of the original forests remain. More than 80 percent of Madagascar's 10,000 flowering plant species are endemic, as are 91 percent of its 300 reptile species. All 33 species of lemur, which make up a tenth of the world's primate species, are found only in Madagascar.

Figure 9 ▶ Conservationists have identified these 25 *biodiversity hotspots* (green). Examples of endangered species from some areas are shown.

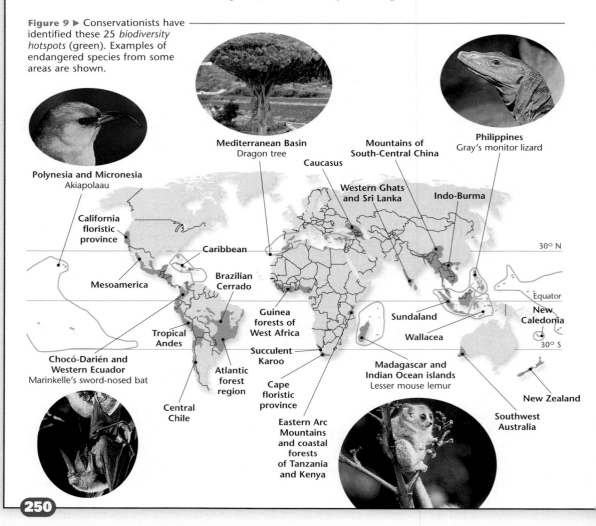

Polynesia and Micronesia — Akiapolaau
Mediterranean Basin — Dragon tree
Mountains of South-Central China
Philippines — Gray's monitor lizard
Caucasus
California floristic province
Western Ghats and Sri Lanka
Indo-Burma
Caribbean
Mesoamerica
Brazilian Cerrado
30° N
Tropical Andes
Guinea forests of West Africa
Sundaland
New Caledonia
Equator
Chocó-Darién and Western Ecuador — Marinkelle's sword-nosed bat
Succulent Karoo
Wallacea
30° S
Atlantic forest region
Madagascar and Indian Ocean islands — Lesser mouse lemur
Central Chile
Cape floristic province
New Zealand
Eastern Arc Mountains and coastal forests of Tanzania and Kenya
Southwest Australia

250

GEOLOGY CONNECTION — GENERAL

Wallace's Line The great biodiversity in the region of Indonesia results in part from its location between previously-isolated continents. On the western side, most species are related to Asian populations, whereas those on the eastern side are more often related to Australian populations. English naturalist and biogeographer Alfred Wallace noticed this apparent split and drew an imaginary line through the region. Today, biogeographers refer to Wallace's line and refer to the region as Wallacea, in honor of Alfred Wallace. **LS** Visual/Verbal

INCLUSION Strategies

• *Learning Disabled*
• *Attention Deficit Disorder*

Have students create an Extinction Notebook. Students should begin by titling four sheets of paper with the ways that humans cause extinctions (listed in this section). On each page, students should describe the extinction pressure and give examples of animals affected. Students can then draw pictures of the animals, their habitat, or the human influence on their habitat. Student work can be assessed with an oral or tape recorded presentation of the findings.

Biodiversity in the United States You may notice that three of the biodiversity hotspots in **Figure 9** are partly within U.S. borders. The United States includes a wide variety of unique ecosystems, including the Florida Everglades, the California coastal region, Hawaii, the Midwestern prairies, and the forests of the Pacific Northwest. The United States holds unusually high numbers of species of freshwater fishes, mussels, snails, and crayfish. Species diversity in the United States is also high among groups of land plants such as pine trees and sunflowers. Some examples of the many unique native species are shown in **Figure 10.**

The California Floristic Province, a biodiversity hotspot, is home to 3,488 native plant species. Of these species, 2,124 are endemic and 565 are threatened or endangered. The threats to this area include the use of land for agriculture and housing, dam construction, overuse of water, destructive recreation, and mining—all stemming from local human population growth.

Figure 10 ▶ Examples of endemic species of the United States include ❶ the cecropia moth, (declining populations), ❷ the tulip poplar tree (limited distribution), ❸ the crayfish *Cambarus mongalensis* (limited distribution), ❹ the desert pupfish (endangered), and ❺ the northern spotted owl (threatened).

SECTION 2 Review

1. **Describe** four ways that species are being threatened with extinction globally.

2. **Define** and give examples of *endangered species* and *threatened species.*

3. **List** areas of the Earth that have high levels of biodiversity and many threats to species.

4. **Compare** the amount of biodiversity in the United States to that of the rest of the world.

CRITICAL THINKING

5. **Interpreting Graphics** The biodiversity hot spots shown in Figure 9 share several characteristics besides a great number of species. Look at the map, and name as many shared characteristics as you can.

6. **Expressing Opinions** Which of the various threats to biodiversity do you think will be most difficult to stop? Which are hardest to justify? Write a paragraph to explain your opinion. **WRITING SKILLS**

251

Answers to Section Review

1. Threats include habitat destruction, introduction of exotics, poaching, and pollution.

2. Endangered species are species likely to become extinct if protective measures are not taken soon. Examples are the Florida panther and the Gray's monitor lizard. Threatened species are species with declining populations that are likely to become endangered if not protected. Students may note local threatened species.

3. Threatened areas of high biodiversity include the Caribbean, the Tropical Andes, Central Chile, Madagascar, New Zealand, the Phillippines, Southwest Australia, and the Mediteranean Basin.

4. Some areas of the United States, such as Hawaii, California, and Florida, have unusually high biodiversity. High biodiversity has evolved in U.S. watersheds and among gymnosperms, composite flowers, and freshwater fish. Biodiversity in the rest of the United States is not as high as in many tropical regions.

5. Hotspots are frequently in coastal zones or islands in the tropics.

6. Answers may vary. All of the threats have been difficult to stop.

Close

Reteaching — BASIC

Habitat Loss Ask students "Why is habitat loss a problem? Why don't the animals just move to some nearby area?" Tell the students that you are going to simulate habitat removal by "clearing" desks around the room. Eliminate a desk as "habitat" by placing a marker on it (such as a piece of colored paper). Tell students that the person seated at that desk may no longer use it, and that they will be eliminated unless they are seated at the end of the game. Tell them that no one can share seats, nor can anyone force another to leave their seat. Once there are no spare seats, students whose desks have been cleared will be permanently eliminated. At the end of the game, explain that competition for resources will only allow a certain number of organisms to exist in a habitat. Organisms that already live in a habitat have an advantage, because they already know how to use and defend the resources that are there. New organisms that cannot compete with those residents may die. **LS Kinesthetic**

Quiz — GENERAL

1. Which type of threat is having the greatest impact on biodiversity? (habitat loss)

2. How might pollutants threaten species? (Because they get into food webs, pollutants such as pesticides may threaten species that are sensitive to those chemicals.)

Alternative Assessment — GENERAL

Speciesopoly Have students work in groups to create board games that dramatize threats to species around the world. Have them trade the games and play them. **LS Interpersonal** Co-op Learning

Transparencies

TT Global Biodiversity Hotspots

Focus

Overview

Before beginning this section, review with your students the Objectives in the Student Edition. This section discusses the techniques being used around the world to protect species and ecosystems.

 **Bellringer**

Have students brainstorm techniques that could be used to save threatened species. After reading the section, have them look back at the list, to see if they thought of things that were not discussed.
LS Logical

Motivate

Group Activity —— GENERAL

Fun Park USA Tell students that a corporation called Fun Park USA plans to build a huge amusement park in a wild area close to your town. Ask five volunteers to take the following roles: (1) a Fun Park marketing executive; (2) a lawyer who is fighting for Fun Park's right to use the land as it desires; (3) a scientist who says that the park would have an impact on biodiversity; (4) a city representative who has tried to get a local species classified as endangered; and (5) a citizen who likes amusement parks and wildlife. Tell the actors to present the scenario as a short skit in which they demonstrate the conflict and work to resolve it.
LS Kinesthetic/Interpersonal

The Future of Biodiversity

Objectives

▶ List and describe four types of efforts to save individual species.
▶ Explain the advantages of protecting entire ecosystems rather than individual species.
▶ Describe the main provisions of the Endangered Species Act.
▶ Discuss ways in which efforts to protect endangered species can lead to controversy.
▶ Describe three examples of worldwide cooperative efforts to prevent extinctions.

Key Terms

germ plasm
Endangered Species Act
habitat conservation plan
Biodiversity Treaty

Slowing the loss of species is possible, but to do so we must develop new approaches to conservation and sensitivity to human needs around the globe. In this section, you will read about efforts to save individual species and to protect entire ecosystems.

Saving Species One at a Time

When a species is clearly on the verge of extinction, concerned people sometimes make extraordinary efforts to save the last few individuals. These people hope that a stable population may be restored someday. Methods to preserve individual species often involve keeping and breeding the species in captivity.

Captive-Breeding Programs Sometimes, wildlife experts may attempt to restore the population of a species through *captive-breeding* programs. These programs involve breeding species in captivity, with the hope of reintroducing populations to their natural habitats. One example of a captive-breeding program involves the California condor, shown in **Figure 11**.

Condors are scavengers. They typically soar over vast areas in search of dead animals to eat. Habitat loss, poaching, and lead poisoning brought the species near extinction. In 1986, the nine remaining wild California condors were captured by wildlife experts to protect the birds and to begin a breeding program. As of 2002, 58 condors had been returned to the wild and 102 were living in captivity. The question remains whether the restored populations will ever reproduce in the wild.

Preserving Genetic Material One way to save the essence of a species is by preserving its genetic material. **Germ plasm** is any form of genetic material, such as that contained within the reproductive, or germ, cells

Figure 11 ▶ The California condor (above) nearly became extinct in the 1980s. A captive-breeding program (right) is returning some condors to the wild.

 252

Chapter Resource File

• **Lesson Plan**
• **Active Reading** BASIC
• **Section Quiz** GENERAL

Transparencies

TT Bellringer

BRAIN FOOD 🧠

Przewalski's Horse Przewalski's horse is the only wild horse species that has survived since the Ice Age. It is a short, stocky animal with a golden brown coat, a black mane, and a long, black tail. In 1879, Russian explorer Nikolai Przewalski discovered the horse in the Gobi Desert. Although Przewalski's horse went extinct in the wild, western collectors and zoos bred them in captivity. Recently, conservationists have reintroduced three small herds into their native habitat.

of plants and animals. Germ-plasm banks store germ plasm for future use in research or species-recovery efforts. Material may be stored as seeds, sperm, eggs, or pure DNA. Germ plasm is usually stored in special controlled environments, such as that shown in **Figure 12,** to keep the genetic material intact for many years. Farmers and gardeners also preserve germ plasm when they save and share seeds.

Zoos, Aquariums, Parks, and Gardens The original idea of zoos was to put exotic animals on display. However, in some cases, zoos now house the few remaining members of a species and are perhaps the species' last hope for survival. Zoos, wildlife parks, aquariums, and botanical gardens are living museums of the world's biodiversity. Botanical gardens, such as the one shown in **Figure 13,** house about 90,000 species of plants worldwide. Even so, these kinds of facilities rarely have enough resources or knowledge to preserve more than a fraction of the world's rare and threatened species.

More Study Needed Ultimately, saving a few individuals does little to preserve a species. Captive species may not reproduce or survive again in the wild. Also, small populations are vulnerable to infectious diseases and genetic disorders caused by inbreeding. Conservationists hope that these strategies are a last resort to save species.

Figure 12 ▶ This scientist is handling samples of genetic material that are preserved in controlled conditions. The samples may be able to reproduce organisms many years from now.

Figure 13 ▶ This botanical garden is contained within a clear dome in Queen Elizabeth Park in Vancouver, Canada. The dome houses over 500 species of plants from all over the world as well as over 100 species of tropical birds.

Graphic

(Organizer) **Spider Map**
Create the **Graphic Organizer** entitled "Spider Map" described in the Appendix. Label the circle "Conservation Efforts." Create a leg for each kind of conservation effort. Then, fill in the map with details about each of the ways people today are trying to preserve diversity.

(253)

BOTANY
● **CONNECTION** **GENERAL**

Heirloom Crops "Heirloom" crops are varieties of common garden vegetables (such as tomatoes and beans) that are open-pollinated and from which seeds or cuttings are saved and replanted over many generations. Heirloom varieties have unique properties that result from breeding by gardeners; they are often well-adapted for a particular region and may have natural pest resistance. Growers of heirloom plants are both preserving biodiversity and conducting genetic engineering "in the field." In contrast, the leading commercial crop varieties have low genetic diversity and are often sterile hybrids for which new seed must be purchased each year. Have students compare descriptions of fruits, vegetables, and flowers in different seed or plant catalogs and discuss the differences in class. **LS** Visual

Using the Figure — GENERAL

Coffee and Habitat Encourage students to read more about shade-grown coffee in **Figure 14** and how it preserves or restores habitat for birds. Also have them research some of the disadvantages of shade-grown coffee. Have them discuss how coffee farmers or coffee sellers could overcome these disadvantages. **LS Intrapersonal**

QuickLAB

Skills Acquired:
• Constructing Models
• Inferring
• Organizing and Analyzing Data
• Communicating

Teacher's Notes: After students complete the Quicklab, invite a local official to visit your classroom to discuss the students' ideas. Have them ask the official about ways to protect the land and about efforts to preserve local biodiversity.

Answers

1. Answers may vary. Students should create plans that anticipate human encroachment but maximize protection for existing habitat and allow connections and buffer zones around habitat preserves.

Figure 14 ▶ Another conservation strategy is to promote more creative and sustainable land uses. This coffee crop is grown in the shade of native tropical trees instead of on cleared land. This practice is restoring habitat for many migrating songbirds.

internet connect

www.scilinks.org
**Topic: Preserving
Ecosystems**
SciLinks code: HE4088

SCILINKS Maintained by the
National Science
Teachers Association

QuickLAB

Design a Wildlife Preserve

Procedure
1. Imagine you have enough money and political support to set aside some land in your community to be habitat for local wildlife. Your goal is to decide which areas to preserve.
2. Find out which species in your area would need this protection the most, where they currently exist, and what their habitat needs are.
3. Use a **colored pencil** to draw some proposed preserve areas on a copy of a **local map.**

Analysis
1. Explain why you chose the areas you did. Can you connect or improve any existing areas of habitat? How could you reduce various threats to the species?

Preserving Habitats and Ecosystems

The most effective way to save species is to protect their habitats. But setting aside small plots of land for a single population is usually not enough. A species confined to a small area could be wiped out by a single natural disaster. Some species require a large range to find adequate food, find a suitable mate, and rear their young. Therefore, protecting the habitats of endangered and threatened species often means preserving or managing large areas.

Conservation Strategies Most conservationists now give priority to protecting entire ecosystems rather than individual species. By protecting entire ecosystems, we may be able to save most of the species in an ecosystem instead of only the ones that have been identified as endangered. The general public has begun to understand that Earth's biosphere depends on all its connected ecosystems in ways we may not yet fully realize or be able to replace.

To protect biodiversity worldwide, conservationists focus on the hotspots described in the previous section. However, they also support additional strategies. One strategy is to identify areas of native habitat that can be preserved, restored, and linked into large networks. Another promising strategy is to promote products that have been harvested with sustainable practices, such as the shade-grown coffee shown in **Figure 14**.

More Study Needed Conservationists emphasize the urgent need for more serious study of the workings of species and ecosystems. Only in recent decades has there been research into such basic questions as, How large does a protected preserve have to be to maintain a certain number of species? How much fragmentation can a particular ecosystem tolerate? The answers may be years or decades away, but decisions affecting biodiversity continue to be made based on available information.

Cultural Awareness GENERAL

Culture and Wildlife Cultural views have helped many indigenous societies to preserve the biodiversity of their region. For example, the Kuna tribe of Panama has strict rules involving the harvest of lobsters, turtles, and game animals. These rules tend to discourage overkill and waste. The Wet'suwet'en and Gitksan peoples of western Canada protect their staple food, salmon, with religious customs. They believe that salmon spirits offer their bodies to humans for food but will punish humans who waste or misuse this gift

or disrupt salmon habitat. Ask students to discuss how cultural views about wildlife have affected the survival or demise of plants or animals in your state, the nation, or other countries. (Examples: Cultural views have protected organisms that are revered, such as the bald eagle or a state flower or bird, hurt species that are considered undesirable, such as cougars, or wolves, and hurt some species that are considered abundant, such as whales, American bison, and many game birds.) **LS Interpersonal**

Legal Protections for Species

Many nations have laws and regulations designed to prevent the extinction of species, and those in the United States are among the strongest. Even so, there is controversy about how to enforce such laws and about how effective they are.

U.S. Laws In 1973, the U.S. Congress passed the **Endangered Species Act** and has amended it several times since. This law, summarized in **Table 4,** is designed to protect plant and animal species in danger of extinction. Under the first provision, the U.S. Fish and Wildlife Service (USFWS) must compile a list of all endangered and threatened species in the United States. As of 2002, 983 species of plants and animals were listed as endangered or threatened. Dozens more are considered for the list each year. The second main provision of the act protects listed species from human harm. Anyone who harms, buys, or sells any part of these species is subject to a fine. The third provision prevents the federal government from carrying out any project that jeopardizes a listed species.

Recovery and Habitat Conservation Plans Under the fourth main provision of the Endangered Species Act, the USFWS must prepare a *species recovery plan* for each listed species. These plans often propose to protect or restore habitat for each species. However, attempts to restrict human uses of land can be controversial. Real-estate developers may be prohibited from building on their own land because it contains critical habitat for a species. People may lose income when land uses are restricted and may object when their interests are placed below those of another species.

Although battles between developers and environmentalists are widely publicized, in most cases compromises are eventually worked out. One form of compromise is a **habitat conservation plan**—a plan that attempts to protect one or more species across large areas of land through trade-offs or cooperative agreements. The region of California shown in **Figure 15** is part of a habitat conservation plan.

Table 4 ▼

Major Provisions of the Endangered Species Act
• The U.S. Fish and Wildlife Service (USFWS) must compile a list of all endangered and threatened species.
• Endangered and threatened animal species may not be caught or killed. Endangered and threatened plants on federal land may not be uprooted. No part of an endangered and threatened species may be sold or traded.
• The federal government may not carry out any project that jeopardizes endangered species.
• The U.S. Fish and Wildlife Service must prepare a species recovery plan for each endangered and threatened species.

Figure 15 ► This region of San Diego, California, is home to several endangered species. A habitat conservation plan attempts to protect these species by managing a large group of lands in the area.

255

CULTURAL
CONNECTION — GENERAL

Women and Biodiversity Conservation Several women's organizations in central Asia are working to reverse the legacy of severe environmental degradation left by the former Soviet Union. Women's groups in Kazakstan and Kyrgyzstan provide environmental education for children, teaching them about their connection to the environment. An organization called Ekolog has been formed to protect biodiversity in the region. Elena Mukhina of Uzbekistan is a biologist with Ekolog who has focused on the protection of biodiversity in the local zapovednik (nature preserve) system. As an English speaker, Mukhina communicates information from the Internet about conservation techniques to zapovednik workers. Encourage students to research other efforts by women around the world to preserve biodiversity and stop environmental damage.

LS Intrapersonal

MATHPRACTICE

Answers

1,400 / 20,500 = 0.068 or 6.8 percent. If there are an estimated 10 million species on Earth, 680,000 of them are at some risk.

Figure 16 ▶ Scenes like this one of elephant tusk poaching were common before the worldwide ban on the sale of ivory as part of CITES.

MATHPRACTICE

Measuring Risk There are many ways to categorize a species' degree of risk of extinction. The IUCN and the Nature Conservancy have multiple ranks for species of concern, ranging from "presumed extinct" to "secure." According to one study of 20,500 species in the United States, 1,400 of those species were at some risk. Calculate this number of species at risk as a percentage. Use this percentage to estimate how many species may be at risk around the world.

256

International Cooperation

At the global level, the International Union for the Conservation of Nature and Natural Resources (IUCN) facilitates efforts to protect species and habitats. This organization is a collaboration of almost 200 government agencies and over 700 private conservation organizations. The IUCN publishes *Red Lists* of species in danger of extinction around the world. The IUCN also advises governments on ways to manage their natural resources, and works with groups like the World Wildlife Fund to sponsor conservation projects. The projects range from attempting to stop poaching in Uganda to preserving the habitat of sea turtles on South American beaches.

International Trade and Poaching One product of the IUCN has been an international treaty called *CITES* (the Convention on International Trade in Endangered Species). The CITES treaty was the first effective effort to stop the slaughter of African elephants. Elephants were being killed by poachers who would sell the ivory tusks. Efforts during the 1970s and 1980s to limit the sale of ivory did little to stop the poaching. Then in 1989, the members of CITES proposed a total worldwide ban on all sales, imports, and exports of ivory, hoping to put a stop to scenes like those in **Figure 16**.

Some people worried that making ivory illegal might increase the rate of poaching instead of decrease it. They argued that illegal ivory, like illegal drugs, might sell for a higher price. But after the ban was enacted, the price of ivory dropped, and elephant poaching declined dramatically.

The Biodiversity Treaty One of the most ambitious efforts to tackle environmental issues on a worldwide scale was the United Nations Conference on Environment and Development, also known as the first *Earth Summit*. More than 100 world leaders and 30,000 other participants met in 1992 in Rio de Janeiro, Brazil.

BRAIN FOOD

Advances in Captive Breeding In August 2000, the Cincinnati Zoo was the birthplace of a special ocelot kitten named *Sihil*, a Mayan word meaning "to be born again." The kitten was born through a procedure called *embryo transfer*. Ocelots have been on the U.S. endangered species list since 1972. Most of the approximately 120 ocelots in North American zoos are *generic*, meaning that they are of unknown ancestry. Recent changes to the Ocelot Species Survival Plan recommended that a unique subspecies, the Brazilian ocelot, be imported and bred in U.S. facilities. Instead of removing ocelots from the wild, researchers proposed to collect live embryos from pregnant Brazilian ocelots, then freeze and transport the embryos to U.S. zoos. The U.S. zoos would then implant the embryos into generic ocelot mothers, resulting in purebred Brazilian kittens. Sihil's birth showed that the technique would work.

An important result of the Earth Summit was an international agreement called the **Biodiversity Treaty.** The treaty's goal is to preserve biodiversity and ensure the sustainable and fair use of genetic resources in all countries. However, the treaty took many years to be adopted into law by the U.S. government. Some political groups objected to the Treaty, especially to the suggestion that economic and trade agreements should take into account any impacts on biodiversity that might result from the agreements. The international community will thus continue to have debates like those that have surrounded the Endangered Species Act in the United States.

Private Conservation Efforts Many private organizations work to protect species worldwide, often more effectively than government agencies. The World Wildlife Fund encourages the sustainable use of resources and supports wildlife protection. The Nature Conservancy has helped purchase millions of hectares of habitat preserves in 29 countries. Conservation International helps identify biodiversity hotspots and develop ecosystem conservation projects in partnership with other organizations and local people. Greenpeace International organizes direct and sometimes confrontational actions, such as the one shown in **Figure 17,** to counter environmental threats.

Balancing Human Needs

Attempts to protect species often come into conflict with the interests of the world's human inhabitants. Sometimes, an endangered species represents a source of food or income. In other cases, a given species may not seem valuable to those who do not understand the species' role in an ecosystem. Many conservationists feel that an important part of protecting species is making the value of biodiversity understood by more people.

Figure 17 ▶ These Greenpeace activists are blocking the path of a Japanese whaling ship. Do you think this is an effective way to protect species?

FIELD ACTIVITY

Simple Biodiversity Assessment Discover the diversity of weeds and other plants in a small area. Yards, gardens, and vacant lots are good places to conduct such a study. Mark off a 0.5 m² section. Use a field guide to identify every plant species that you can. At least identify how many different kinds of plants there are. You may want to sketch or photograph some of the plants. Then count the number of each kind of plant you identified. Record your results in your *EcoLog.*

SECTION 3 Review

1. **Describe** four types of efforts to save individual species.

2. **Explain** the advantages of protecting entire ecosystems rather than individual species.

3. **Describe** the main provisions of the Endangered Species Act.

4. **Give** examples of worldwide cooperative efforts to prevent extinctions.

CRITICAL THINKING

5. **Analyzing Methods** Read the headings in this section. Which type of effort to preserve species do you think is most worthwhile? READING SKILLS

6. **Comparing Viewpoints** Discuss ways in which efforts to protect species can lead to controversy.

7. **Inferring Relationships** Why was a complete ban of ivory sales more effective than a limited ban?

257

CHAPTER 10 Highlights

Alternative Assessment ── GENERAL

Teaching Biodiversity Have students plan a lesson about biodiversity for a nearby elementary or middle-school class. Ask students to each write up a lesson plan. Have them use materials they have created or used for this chapter as visual aids. **LS** Intrapersonal/ Interpersonal

Survival Plans Tell students to imagine that they are wildlife biologists who have studied a rare species for five years. They have submitted paperwork showing that there is a dire need to classify the species as an endangered species, but have just learned that the U.S. Fish and Wildlife Service has determined that protecting this species is not as high a priority as protecting others. Have groups of students choose the species that they are trying to protect, then brainstorm and research other laws or programs (besides the Endangered Species Act) that might help protect the species. Ask each group to present a plan of action to the class. **LS** Interpersonal   Co-op Learning

Biodiversity and Aesthetics Have students compile a multimedia collection or artistic presentation that demonstrates the aesthetic value of species diversity to humans. Suggest that they consider games, songs, décor, art, and clothing throughout human history that have been inspired by other species. **LS** Auditory/Visual

1 What Is Biodiversity?

Key Terms

biodiversity, 241
gene, 242
keystone species, 242
ecotourism, 244

Main Ideas

▶ Biodiversity usually refers to the number and variety of different species in a given area, but it can also describe genetic variation within populations or variation across ecosystems.

▶ The study of biodiversity starts with the unfinished task of identifying and cataloging all species on Earth. Although scientists disagree about the probable number of species on Earth, they do agree that we need to study biodiversity more thoroughly.

▶ Humanity benefits from biodiversity in several ways and perhaps in some unknown ways.

2 Biodiversity at Risk

Key Terms

endangered species, 245
threatened species, 245
exotic species, 247
poaching, 247
endemic species, 248

Main Ideas

▶ Many scientists are now concerned that loss of biodiversity is the most challenging environmental issue we face.

▶ The most common cause of extinction today is the destruction of habitats by humans. Unregulated hunting and the introduction of nonnative species also contribute to extinctions.

▶ Certain areas of the world contain a greater diversity of species than other areas. An important feature of such areas is that they have a large portion of endemic species.

▶ The United States has a very important role in preserving biodiversity.

3 The Future of Biodiversity

Key Terms

germ plasm, 252
Endangered Species Act, 255
habitat conservation plan, 255
Biodiversity Treaty, 257

Main Ideas

▶ Most major conservation efforts now concentrate on protecting entire ecosystems rather than individual species.

▶ The Endangered Species Act establishes protections for endangered and threatened species in the United States. The act has generated some controversy and has been amended several times.

▶ International cooperation has led to increased recognition and protection of biodiversity worldwide.

▶ The desire to protect biodiversity often conflicts with other human interests.

258

Chapter Resource File

- Chapter Test GENERAL
- Chapter Test ADVANCED
- Concept Review GENERAL
- Critical Thinking ADVANCED
- Test Item Listing
- Field Activity BASIC
- Field Activity ADVANCED
- Research Lab GENERAL
- Consumer Lab GENERAL
- Long-Term Project GENERAL

Using Key Terms

Use each of the following terms in a separate sentence.

1. *keystone species*
2. *ecotourism*

For each pair of terms, explain how the meanings of the terms differ.

3. *hunting* and *poaching*
4. *endemic species* and *exotic species*
5. *endangered species* and *threatened species*
6. *gene* and *germ plasm*
7. *CITES* and *Biodiversity Treaty*

✔ **STUDY TIP**

Use a Map As you review the chapter, refer to an atlas, to the maps in the Appendix, or to previous chapters about biomes to compare information. Draw your own map or make a list of the locations of some of the interesting species and ecosystems that you learn about.

Understanding Key Ideas

8. The term *biodiversity* refers to
   a. the variety of species on Earth.
   b. the extinction of the dinosaurs.
   c. habitat destruction, invasive exotic species, and poaching.
   d. the fact that 40 percent of prescription drugs come from living things.

9. Most of the living species known to science
   a. are large mammals.
   b. live in deserts.
   c. live in the richer countries of the world.
   d. are insects.

10. Some species are so important to the functioning of an ecosystem that they are called
    a. threatened species.
    b. keystone species.
    c. endangered species.
    d. extinct species.

11. A mass extinction is
    a. a rapid increase in biodiversity.
    b. the introduction of exotic species.
    c. the extinction of many species in a short period of time.
    d. a benefit to the environment.

12. When sea otters disappeared from the Pacific coast of North America,
    a. the area became overrun with kelp.
    b. the number of fish in the kelp beds increased.
    c. the number of sea urchins in the kelp beds increased.
    d. the area became overrun with brown seaweed.

13. Which of the following statements about the Endangered Species Act is *not* true?
    a. Parts of an endangered animal, such as feathers or fur, may be traded or sold but only if the animal is not killed.
    b. A species is considered endangered if it is expected to become extinct in the near future.
    c. The federal government cannot carry out a project that may jeopardize an endangered plant.
    d. A recovery plan is prepared for all animals that are listed as endangered.

14. Because of efforts by the Convention on International Trade in Endangered Species (CITES),
    a. the poaching of elephants increased.
    b. the cost of ivory worldwide increased.
    c. the international trade of ivory was banned worldwide.
    d. a captive-breeding program for elephants was established.

15. Emphasizing the preservation of entire ecosystems will
    a. cause the economic needs of farmers to suffer in order to save a single species.
    b. decrease biodiversity, especially in tropical rain forests, coral reefs, and islands.
    c. throw the food webs of many ecosystems out of balance.
    d. save many unknown species from extinction.

259

Assignment Guide

Section	Questions
1	1, 2, 8–10, 12, 20–23, 31
2	3–5, 11, 12, 16, 17, 25, 27–29, 32
3	6, 7, 13–15, 18, 19, 24, 26, 30, 33, 34

ANSWERS

Using Key Terms

1. Sample answer: A keystone species is critical to the functioning of its ecosystem.
2. Sample answer: Ecotourism allows a country to make money from its biodiversity without harvesting its plants or animals.
3. Hunting is the legal harvesting of game animals. Poaching is illegal harvesting.
4. Endemic species are species that evolved and exist within a limited geographic region. Exotic species are species that have been introduced to a new region.
5. Endangered species are considered in danger of extinction unless protective measures are taken immediately. Threatened species are species with declining populations that are likely to become endangered if they are not protected.
6. Genes are pieces of DNA that code for particular traits that can be inherited. Germ plasm is any form of genetic material, such as that within seeds and sperm.
7. CITES is an international treaty governing trade in endangered species. The Biodiversity Treaty is an international agreement that calls for the preservation of biodiversity and the sustainable and fair use of genetic resources.

Understanding Key Ideas

8. a
9. d
10. b
11. c
12. c
13. a
14. c
15. d

Short Answer

16. Hunting was a major cause of extinctions in the U.S. in the 1800s and 1900s.

17. Exotic species are species that are not native to a particular region. Exotic species can become new predators, parasites, or competitors of native species, thus endangering them.

18. An ecosystem approach protects most or all of the species in an ecosystem, not only the ones that are known to be endangered. Also, the Earth's biosphere depends on many connected ecosystems.

19. Answers may vary. Needs include housing and food. Interests include recreation and higher standards of living.

Interpreting Graphics

20. No. These numbers relate only to known species. There are millions of unknown species.

21. Small invertebrates (mollusks and worms, insects and crustaceans) are probably underrepresented.

22. Worldwide, most of the species that have been identified as endangered or threatened are larger vertebrates. In the U.S., a greater proportion of invertebrates and fishes has been identified.

23. Fishes, mollusks, worms, insects, crustaceans, and amphibians might need further research world-wide.

Concept Mapping

24. Answers to the concept mapping questions are on pages 667–672.

Short Answer

16. When was hunting a major cause of extinctions in the United States?

17. What are exotic species, and how do they endanger other species?

18. Why do biologists favor using an ecosystem approach to preserve biodiversity?

19. Describe three ways that preserving biodiversity can come into conflict with human interests.

Interpreting Graphics

The graph below shows the numbers of various types of species that are officially listed as endangered or threatened in the United States and internationally. Use the graph to answer questions 20–23.

20. Do these numbers necessarily reflect *all* species that may be in danger? Explain your answer.

21. Which types of species might be underrepresented here?

22. Compare the United States and world listings. What trends do you see in the types of species listed?

23. Given this information, which types of species might need further research worldwide?

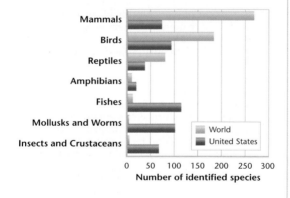

Number of identified species

Critical Thinking

25. Extinctions in the past may have been more related to climate change and local natural disasters that were not tied to human activity. Extinctions today are more directly tied to human activity, such as habitat destruction.

26. Answers may vary. The facilities probably could not restore many populations. Successful populations require the natural conditions of their native habitat, which are difficult to recreate. Also, many endangered animal populations have not been identified.

27. The loss of huge tracts of rain forests might cause many extinctions. These species might have provided new sources of food and medicine that could have benefited people. Rain forest loss could also negatively affect the biosphere because the water and nutrient cycles would be altered.

Concept Mapping

24. Use the following terms to create a concept map: *biodiversity, species, gene, ecosystem, habitat loss, poaching, exotic species, germ plasm, captive breeding programs,* and *habitat preservation.*

Critical Thinking

25. **Comparing Processes** Read the passage in this chapter that describes current extinctions. How are the extinctions that are occurring currently different from most extinctions in the past? **READING SKILLS**

26. **Analyzing Methods** With unlimited funding, could zoos and captive-breeding programs restore most endangered animal populations? Explain your answers.

27. **Determining Cause and Effect** How might the loss of huge tracts of tropical rain forests have an effect on other parts of the world?

Cross-Disciplinary Connection

28. **Literature** Try to remember or find some children's stories that include wild animals that are currently endangered, threatened, or extinct. Write a description of how these animals are portrayed in the stories. Also compare the animals in the stories to what you know about the real animals. **WRITING SKILLS**

29. **Geography** Obtain a list of the plants and animals that are endangered in your state. Find out where these species live, and mark the locations on a map of your state. Research the effects of habitat loss on species in your county or in surrounding areas.

Portfolio Project

30. **Endangered Species Outreach** Create a special project about one endangered species of your choice. Consider using a poster, an oral presentation, or a video to inform your classmates about your chosen species or to persuade them of the importance of saving the species.

MATH SKILLS

Use the table below to answer questions 31–32.

31. Analyzing Data Which of the types of species in the table below are most accurately described? What do the numbers indicate about how well various species are studied?

32. Applying Quantities Which of the types of species may represent the greatest unknown loss of biodiversity? Which type of species is probably least important for further research into biodiversity?

Estimates of Knowledge of Earth's Species

Type of species	Number of species described	Described species as % of total	Number threatened or extinct	Accuracy of estimates
Bacteria	4,000	0.40	(unknown)	very poor
Vertebrates	52,000	94.55	3,843	good
Crustaceans	40,000	26.67	628	moderate
Plants	270,000	84.38	31,277	good

WRITING SKILLS

33. Writing Persuasively Write a letter to the editor of a publication or to an elected representative in which you express your opinion regarding protections of endangered species that might affect your local area.

34. Outlining Topics Outline the major strategies for protecting biodiversity that have been described in this chapter. List pros and cons of each strategy.

STANDARDIZED TEST PREP

For extra practice with questions formatted to represent the standardized test you may be asked to take at the end of your school year, turn to the sample test for this chapter in the Appendix.

READING SKILLS

Read the passage below, and then answer the questions that follow.

Excerpt from M. Reaka-Kudla, D. Wilson, and E. Wilson, eds., Biodiversity II, 1996.

Aside from the academic tradition of biodiversity, another powerful influence, related to biodiversity, brought our culture to its current level of technological development: the exploration of the New World. From the thirteenth to the nineteenth centuries, technological developments in navigation allowed European voyagers to embark on an unprecedented exploration of the globe. These expeditions revolutionized knowledge of the geography, human culture, and biology of the world at the time. This ultimately led to a reevaluation of human society's place in the world and an understanding of the evolution of all living things. But the exploration also allowed the acquisition of untold wealth in living and non-living natural resources, which was brought back from the New World and invested in the culture of western Europe.

1. What do the authors probably mean by the term *influence*?
   a. a force of cultural change
   b. a new type of scientific discovery
   c. a source of geographic information
   d. a form of navigation

2. Which of the following are not mentioned by the authors as factors in our current level of technological development?
   a. geographical information
   b. knowledge of a variety of species
   c. new forms of government
   d. evolutionary theory

3. Which of the following did the authors most likely discuss in the paragraph just *before* this passage?
   a. natural resources of the New World
   b. religious beliefs of native peoples
   c. academic tradition of European biology
   d. history of European expeditions

261

Cross-Disciplinary Connection

28. Answers may vary.

29. Answers may vary. Students may find this information on Web sites for their state department of natural resources, university extension service, or county land department.

Portfolio Project

30. Answers may vary.

Math Skills

31. Vertebrates and plants are most accurately described. This indicates that vertebrates and plants are better studied than crustaceans and bacteria are.

32. Bacteria may represent the greatest unknown loss of biodiversity since the level of accuracy of estimates is very poor. Vertebrates are probably least important for future research because nearly 95% of all species have been described.

Writing Skills

33. Answers may vary. Students should emphasize the reasons to protect the species and ways to minimize economic loss from protective measures.

34. Answers may vary. Students should include pros and cons of the following three strategies: (1) saving species one at a time; (2) preserving habitats and ecosystems; and (3) legal protections for species.

Reading Skills

1. a

2. c

3. c

DIFFERENCES IN BIODIVERSITY

Teacher's Notes

Time Required
two 45-minute class periods

Lab Ratings

EASY ———————→ HARD

TEACHER PREPARATION ▲
STUDENT SETUP ▲▲
CONCEPT LEVEL ▲▲▲
CLEANUP ▲▲

Skills Acquired
• Measuring
• Collecting Data
• Classifying
• Inferring
• Organizing and Analyzing Data

The Scientific Method
In this lab, students will:
• Make Observations
• Analyze the Results
• Draw Conclusions
• Communicate the Results

Materials
Materials listed are enough for groups of 3 or 4 students. Provide plant and insect identification guides. Local species "checklists" may be available from parks and wildlife departments or groups. You might also provide cups or bags for soil samples and allow students to examine these with microscopes.

Objectives

▶ **USING SCIENTIFIC METHODS** **Observe and measure** differences in species diversity between two locations.

▶ **USING SCIENTIFIC METHODS** **Graph and analyze** data collected to reflect differences in species diversity.

▶ **Evaluate** the possible reasons for observed differences in biodiversity.

▶ **USING SCIENTIFIC METHODS** **Infer** other human activities that may influence local biodiversity.

Materials
graph paper
hand lens
meterstick or tape measure
pen or pencil
string or chalk line

optional materials: local-area field guides for plants, animals, and soil organisms; shovel or trowel

▶ **Step 2** Measure and mark off sample areas for your observation and counts of species diversity.

262

Differences In Diversity

Biodiversity is most obvious and dramatic in tropical rain forests and coral reefs, but you do not have to travel that far to observe differences in species diversity or to see the effects that humans can have on biodiversity.

Recall that biodiversity is most often defined as the number of different species that are present in a given area. This measure can be estimated by making a sample count of species within a representative area. It is often easiest and most effective to collect or observe small organisms, such as insects and soil dwellers, or stationary organisms, such as plants and trees. In this activity, you will investigate the differences in species diversity in two areas that are close to each other, but that are affected differently by humans. You may work in teams or groups.

Procedure

1. Choose two sites for your analysis. Site 1 should be an area that has been greatly affected by humans, such as your school building and the surrounding sidewalks, parking area, or groomed lawns. Site 2 should be an area within view of site 1 but that is less affected by humans, such as a wooded area or a vacant lot overgrown with weeds. If directed by your teacher, you may choose more than two sites. Also ask your teacher about your sample square size.

2. At each site, measure a 5 m × 5 m square area using the meterstick or tape measure. You might use the edge of a building as a side of your square, or you might use trees as the corners. Mark the measurement of the area with string or a chalk line, as shown in the photograph.

3. Observe each site carefully, and record a detailed description of each site. Include as many features as possible, such as location, soil condition, ways the area is used, amount of sun or rain exposure, and other factors that might affect the organisms that exist there.

4. For each site, create a table like the table below.

Species Counts Per Site		
Species type	Site number ___	Site number ___
Animals		
Plants	DO NOT WRITE	IN THIS BOOK
Fungi and other soil organisms		

Safety Caution
Before beginning this activity, select and visit the sites your students may use, to check for safe conditions. If the activity is conducted as a field trip, a ratio of 1 adult per 10 students is recommended. Review with students the proper precautions to take when working with live organisms. Teach students to watch for, identify, and avoid any hazardous local species, such as stinging insects or irritating plants. Also watch for injury hazards such as traffic, trash, water, obstacles, or precipices. Have students use a "buddy system," where each student is paired with and is responsible for another student.

Tips and Tricks
Before the investigation, encourage students to make thorough, descriptive field notes of the study sites. You may want to determine the best sample plot size for students to use. Smaller plot sizes may be appropriate for areas that are densely vegetated. Also encourage some way to randomize the exact location of sample squares, to avoid bias. If time is limited, have groups share their counts. Encourage students to cautiously dig in the soil and look under rocks but to restore anything they disturb. Remind students that finding and counting the variety of different species is more important than counting the number of individual organisms.

5. Using your hand lens, find as many different species as possible within the site. Record each new species by placing a slash or tick mark in the column for each different species identified in each general category. You do not need to identify every organism by scientific name, but using field guides may help you have an idea of what you are finding. You may also make more specific categories (such as birds, insects, grasses, and trees) if you are able. Be careful not to disturb the area unnecessarily.

6. Repeat steps 2–5 for each site. If directed by your teacher, compare your data with those of other groups.

7. After you have made and recorded all of your observations, put away your materials and restore anything you disturbed at the sites.

Analysis

1. **Constructing Graphs** Create a bar graph of the number of species counted at each site. As directed by your teacher, you may combine all species counts into one total per site or graph each category of organisms separately.

2. **Analyzing Results** Based on your observations of the organisms found at the sites, which area reflected a higher level of biodiversity?

3. **Interpreting Results** What factors may have contributed to the differences in biodiversity at the sites?

Conclusions

4. **Drawing Conclusions** What can you conclude about the effect of human activities on biodiversity?

5. **Applying Conclusions** What other human activities, besides those you observed directly, could have affected the biodiversity present at your sites?

6. **Evaluating Methods** Do you feel that the method used in this lab was an effective way to identify biodiversity in an area? Why or why not? How could it have been improved?

Extension

1. **Research and Communications** If you were able to use local field guides, what can you generalize about the organisms that you were able to identify? Pay attention to aspects such as how easily recognized each organism is, how common it is in your local area, where it is found outside of your area, or what other unique facts are known about the biology or habitat needs of the organism.

▶ **Step 5** Observe and record how many different types of organisms you find within each sample area.

Answers to Analysis

1. Answers may vary. Student data may vary widely. You may want to have students create pie charts using the combined class data for each type of area.

2. Answers may vary. The site less affected by human activity may have greater biodiversity. If the human-affected site has more biodiversity, however, many of the species may be non-native. Check that students have thoroughly searched for a variety of organisms.

3. Answers may vary. Factors may include soil quality, presence of pesticides, human disturbances such as lawn mowing and the presence of diverse plants that provide habitat for other organisms.

Answers to Conclusions

4. Answers may vary. Human activities in urban areas often decrease biodiversity.

5. Human activities such as the application of lawn chemicals, watering, and driving or walking on the area may have affected its biodiversity.

6. Answers may vary. To improve the lab, students could be encouraged to identify species instead of just counting organisms, vary sample plot sizes, take more samples, and more thoroughly sample all of the species in an area over time.

Answers to Extension

1. Answers may vary.

Dr. E. O. Wilson: Champion of Biodiversity

Background

Dr. Wilson became an important ant biologist at an early age. At 13, he was the first person to discover red imported fire ants in the United States—in a vacant lot near his home in Mobile, Alabama. Years later, as a college student, Dr. Wilson was instrumental in producing research that decoded the fire ant's chemical communication system. Dr. Wilson refers to ants as "the little creatures that run the Earth." He has also said that "if human beings were not so impressed by size alone, they would consider an ant more wonderful than a rhinoceros."

Wilson has discovered hundreds of new species of organisms and written 20 books on various subjects. His recent books include *The Future of Life,* and *Consilience: The Unity of Knowledge.* Wilson has received over seventy awards and twenty-five honorary degrees. Time Magazine called him "an ecological Paul Revere" and listed him as one of America's 25 most influential people.

Making a difference

DR. E. O. WILSON: CHAMPION OF BIODIVERSITY

Dr. Edward Osborne Wilson deserves some of the credit for the fact that this book includes a chapter called "Biodiversity." A few decades ago, the word *biodiversity* was used by few scientists and was found in few dictionaries. Dr. Wilson has helped make the concept and value of biodiversity widely recognized, through his extensive research, publishing, organizing, and social advocacy.

Since his early career as a pioneer in the fields of entomology and sociobiology, Dr. Wilson has gained recognition for many additional accomplishments. He has written two Pulitzer Prize-winning nonfiction books, and has received the National Medal of Science and dozens of other scientific awards and honors. Wilson is widely recognized as one of the most influential scientists and citizens of our time.

It All Started with Bugs

Even before his scientific career, Wilson developed a fascination with insects and the natural world. He always had high expectations of himself but made the best of circumstances. Although his parents were divorced and his father's government career required frequent moves, Wilson found companionship in the woods of the southern United States or the museums of Washington, D.C. After injuries damaged his vision and hearing, Wilson focused his scientific skills on the smaller forms of life.

By the time he earned his master's degree at the University of Alabama at the age of 20, Wilson was well known as a promising *entomologist*—an expert on the insect world. His specialty is the study of ants and their complex social behaviors. So it makes sense that Wilson next went to study at Harvard University, home to the world's largest ant collection. While at Harvard, he earned his Ph.D., conducted field research around the world, collected more than 100 previously undescribed species, and wrote several books on insect physiology and social organization. He eventually became curator of the Museum of Entomology at Harvard.

Clearly, Wilson has a passion for insects. "There is a very special pleasure in looking in a microscope and saying I am the first person to see a species that may be millions of years old," he says. Some of Wilson's research has focused on the social behavior of ants. Among other important scientific findings, Wilson was the first to demonstrate that ant behavior and communication is based mostly on chemical signals.

From Insects to Humans

In 1971, Wilson published *The Insect Societies,* which surveyed the evolution of social organization among wasps, ants, bees, and termites. Wilson began to extend his attempts to understand the relationship of biology and social behavior to other animals, including humans. In 1975, Wilson published a controversial book exploring these new ideas, called *Sociobiology.* Now an accepted branch of science, sociobiology is the study of the biological basis of social behavior in animals, including humans.

During Wilson's studies of the behavior of ants and other social insects, he became interested in the

▶ Dr. Wilson with one of his favorite subjects—ants.

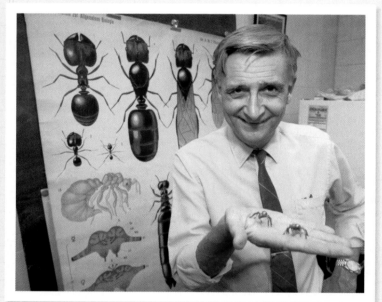

Notable Quotes   GENERAL

"One planet, one experiment."

—E.O. Wilson

Have students discuss the quote in the context of biodiversity. Ask them to discuss the many ways that humans might be "experimenting" with the planet, and what the consequences might be. **LS Verbal**

Student Opportunities

Small But Important In a speech to National Park Service personnel, Dr. Wilson called for more biological inventories in the National Parks to include overlooked organisms such as fungi, insects and microbes. According to Wilson, monitoring and protecting small but important species may protect larger species that rely on them. Encourage students to call managers or naturalists in their favorite National Park to find out what kind of research they are currently conducting there. Have students ask whether there are opportunities for internships in the park.

insects' role in the ecosystems where he studied them. Some of his research involved camping for months at a time in a remote wilderness such as the Amazon basin, carefully studying the activities of certain species. His writings include amazing tales of watching huge colonies of "driver" ants swarm out over an area, capturing and killing a great many other species in their path.

If you have ever played the popular computer game *SimAnt*™, Dr. Wilson again deserves credit for providing the inspiration. In 1990, Wilson received his second Pulitzer Prize for co-authoring *The Ants*, an enormous encyclopedia of the ant world. In addition to describing 8,800 known species of ants, the book details the great variations among ant species in terms of anatomy, biochemistry, complex social behaviors, and especially their critical role in many ecosystems. Wilson reminds us that ants "are some of the most abundant and diverse of the Earth's 1.4 million species. They're among the little creatures that run the earth. If ants and other small animals were to disappear, the Earth would rot. Fish, reptiles, birds—and humans— would crash to extinction."

Onward to Biodiversity

As with many great scientists, each thing Dr. Wilson studies leads him to new questions and new ideas. During his research in remote lands, Wilson spent time reflecting and writing on the nature of ecosystems, the importance of biodiversity, and the role of humans in relation to these. In 1992, he put many of these ideas into another popular book called *The Diversity of Life*. This book combined Wilson's engaging writing style and personal expertise with the latest ecological research.

▶ Dr. Wilson (center) speaks to politicians and the public about the need to conserve our planet's biodiversity.

The book showed both how such incredible biodiversity has evolved on the Earth and how this asset is being lost because of current human activities. The book clearly explained for the general public many of the problems and potential solutions regarding biodiversity that we have studied in this chapter.

Urgent Work

Despite his fame, Wilson is a soft-spoken fellow who would prefer to live a quiet life with his research and with his family in their home in the woods of Massachusetts. But the urgent problem of species loss makes Wilson willing to face the public. "Humanity is entering a bottleneck of overpopulation and environmental degradation unique in history. We need to carry every species through the bottleneck . . . Along with culture itself, they will be the most precious gift we can give future generations."

In 1986, Wilson served as one of the leaders of the first National Forum on Biodiversity, and then as editor of *Biodiversity,* the resulting collection of reports. Wilson continues to engage in public and

private meetings with scientists and policy makers around the globe, urging them to support conservation efforts based on sound science.

Dr. Wilson recently began promoting the need for a global biodiversity survey. This project would involve an international scientific effort on par with the Human Genome Project. Wilson states that "to describe and classify all of the species of the world deserves to be one of the great scientific goals of the new century."

What Do You Think?

Do you find insects interesting? Could you imagine yourself as an entomologist? Do you think that Dr. Wilson made a goal early in his life to be an internationally famous conservationist? What has led him to take on this role?

265

ECONOMICS CONNECTION

Buying Endangered Ecosystems
Dr. Wilson has advocated a simple strategy for preserving ecosystems: buying them. At a recent conference, Wilson and other scientists estimated that four billion dollars would buy the remaining tropical forest and twenty-four billion could buy 2.4 million square kilometers of biodiversity hotspots. By purchasing these areas, Wilson estimates that we could preserve about seventy percent of all the world's biodiversity. Money could also be spent to buy logging rights on important ecosystems, and trees could be conserved instead of being cut. Dr. Wilson once compared the destruction of forests to "burning a Renaissance painting to cook a meal."

Student Opportunities — ADVANCED

Finding All Species Have students research efforts to catalog all of the species on Earth. One nonprofit organization, the All Species Foundation, is dedicated to discovering and classifying every species in the world within the next few decades. The foundation intends to train students all over the world in the methods of taxonomy and to "make taxonomy cool." The United Nations and several major conservation groups are supporting similar efforts. These efforts are taking advantage of new technologies such as global positioning systems, remote-controlled equipment, sophisticated databases, and the Internet.

CHAPTER 11
Water

CHAPTER 12
Air

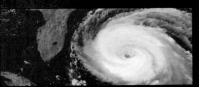

CHAPTER 13
Atmosphere and Climate Change

CHAPTER 14
Land

CHAPTER 15
Food and Agriculture

For thousands of years, humans have altered the environment to grow food. These rice paddies in China are built to trap water from the monsoon rains.

PACING	CLASSROOM RESOURCES	LABS, DEMONSTRATIONS, AND ACTIVITIES
BLOCKS 1 & 2 • 90 min pp. 268–275 **Chapter Opener**		
Section 1 Water Resources	**OSP Lesson Plan** * **CRF Active Reading** * **BASIC** **TT** Bellringer * **TT** Watersheds of the World * **TT** Groundwater and the Water Table * **CD Interactive Tutor** The Water Cycle	**TE Demonstration** How Much Water Is There? p. 269 ◆ **BASIC** **TE Group Activity** Why Can't We Drink Salt Water? p. 271 ◆ **BASIC** **TE Group Activity** Modeling an Aquifer, p. 274 **GENERAL**
BLOCKS 3 & 4 • 90 min pp. 276–283 **Section 2** Water Use and Management	**OSP Lesson Plan** * **CRF Active Reading** * **BASIC** **TT** Bellringer * **TT** Drinking-Water Treatment * **CD Interactive Exploration 4** Flood Bank	**TE Activity** Modeling Water Management Projects, p. 279 **ADVANCED** **CRF Modeling Lab** * ◆ **GENERAL** **CRF CBL™ Probeware Lab** * ◆ **GENERAL** **CRF Consumer Lab** * ◆ **GENERAL**
BLOCKS 5, 6 & 7 • 135 min pp. 284–293 **Section 3** Water Pollution	**OSP Lesson Plan** * **CRF Active Reading** * **BASIC** **TT** Bellringer * **TT** The Wastewater Treatment Process * **TT** How Pollutants Enter Groundwater * **TT** Major North American Oil Spills *	**SE Field Activity** Identifying Sources of Pollution, p. 285 **SE QuickLab** Measuring Dissolved Oxygen, p. 288 **TE Group Activity** Mapping Pollution, p. 288 **ADVANCED** **TE Activity** Artificial Eutrophication in a Fishbowl, p. 289 ◆ **GENERAL** **TE Internet Activity** Researching Phosphates, p. 289 **GENERAL** **TE Group Activity** Demonstrating Biomagnification, p. 292 **BASIC** **TE Demonstration** Deadly Six Packs, p. 292 **GENERAL** **SE Field Activity** Coastal Cleanups, p. 293 **SE Exploration Lab** Groundwater Filters, pp. 298–299 ◆ **CRF Datasheets for In-Text Labs** * **CRF Design Your Own Lab** * ◆ **GENERAL** **CRF Long-Term Project** * **ADVANCED**

BLOCKS 8 & 9 • 90 min

Chapter Review and Assessment Resources

- **SE Chapter Review** pp. 295–297
- **SE Standardized Test Prep** pp. 656–657
- **CRF Study Guide** * ■ **GENERAL**
- **CRF Chapter Test** * ■ **GENERAL**
- **CRF Chapter Test** * **ADVANCED**
- **CRF Concept Review** * ■ **GENERAL**
- **CRF Critical Thinking** * **ADVANCED**
- **OSP Test Generator**
- **CRF Test Item Listing** *

Online and Technology Resources

Holt Online Learning

Visit **go.hrw.com** for access to Holt Online Learning or enter the keyword **HE6 Home** for a variety of free online resources.

One-Stop Planner® CD-ROM

This CD-ROM package includes
- Lab Materials QuickList Software
- Holt Calendar Planner
- Customizable Lesson Plans
- Printable Worksheets
- ExamView® Test Generator
- Interactive Teacher Edition
- Holt PuzzlePro® Resources
- Holt PowerPoint® Resources

Holt Environmental Science Interactive Tutor CD-ROM

This CD-ROM consists of interactive activities that give students a fun way to extend their knowledge of environmental science concepts.

ENRICHMENT AND SKILLS PRACTICE	SECTION REVIEW AND ASSESSMENT	CORRELATIONS
SE Pre-Reading Activity, p. 268 **TE** Using the Figure, p. 268 `GENERAL`		National Science Education Standards
TE Interpreting Statistics Proportions, p. 270 `BASIC` **TE** Using the Figure Watersheds of the World, p. 271 `GENERAL` **SE** Case Study The Ogallala Aquifer, pp. 272–273 **TE** Inclusion Strategies, p. 272	**SE** Section Review, p. 275 **TE** Reteaching, p. 275 `BASIC` **TE** Quiz, p. 275 `GENERAL` **TE** Alternative Assessment, p. 275 `GENERAL` **CRF** Quiz * ■ `GENERAL`	
TE Interpreting Statistics Bar Graphs, p. 277 `GENERAL` **TE** Skill Builder Graphing, p. 277 `BASIC` **SE** Graphic Organizer Comparison Table, p. 278 `GENERAL` **TE** Skill Builder Math, p. 280 `GENERAL` **SE** MathPractice Israeli Agriculture, p. 281 **TE** Reading Skill Builder Brainstorming, p. 282 `GENERAL` **TE** Skill Builder Vocabulary, p. 282 `GENERAL` **TE** Inclusion Stategies, p. 282	**TE** Homework, p. 277 `ADVANCED` **TE** Homework, p. 279 `GENERAL` **TE** Homework, p. 282 `ADVANCED` **SE** Section Review, p. 283 **TE** Reteaching, p. 283 `BASIC` **TE** Quiz, p. 283 `GENERAL` **TE** Alternative Assessment, p. 283 `GENERAL` **CRF** Quiz * ■ `GENERAL`	SPSP 3a SPSP 3b
TE Student Opportunities Volunteering, p. 288 **TE** Career Water Management, p. 288 **SE** MathPractice Parts per Million, p. 290 **TE** Using the Figure Sources of Groundwater Pollution, p. 290 `GENERAL` **TE** Using the Figure Oil Spills p. 291 `ADVANCED` **TE** Student Opportunities Beach Sweeps, p. 291 **TE** Reading Skill Builder Paired Summarizing, p. 292 `GENERAL` **SE** Points of View The Three Gorges Dam, pp. 300–301 **CRF** Map Skills Canada's Water * `GENERAL`	**TE** Reteaching, p. 285 `BASIC` **TE** Homework, p. 287 `ADVANCED` **TE** Homework, p. 290 `ADVANCED` **TE** Homework, p. 292 `ADVANCED` **SE** Section Review, p. 293 **TE** Reteaching, p. 293 `BASIC` **TE** Quiz, p. 293 `GENERAL` **TE** Alternative Assessment, p. 293 `GENERAL` **CRF** Quiz * ■ `GENERAL`	SPSP 4b SPSP 6e

Guided Reading Audio CDs

These CDs are designed to help auditory learners and reluctant readers. (Audio CDs are also available in Spanish.)

www.scilinks.org

Maintained by the **National Science Teachers Association**

TOPIC: Watersheds
SciLinks code: HE4124

TOPIC: Groundwater
SciLinks code: HE4052

TOPIC: Aquifers
SciLinks code: HE4004

TOPIC: Water Conservation
SciLinks code: HE4121

TOPIC: Water Pollution
SciLinks code: HE4122

CNN Videos

Each video segment is accompanied by a Critical Thinking Worksheet.

Science, Technology & Society
Segment 19 Chemistry of Dry Cleaning

Chapter Enrichment

This Chapter Enrichment provides relevant and interesting information to expand and enhance your classroom instruction of the chapter material.

1 Water Resources

Towing Icebergs

Towing icebergs was first demonstrated in 1971 in Newfoundland. It is now a common practice in the management of icebergs for the offshore oil industry. Often large icebergs are towed to slightly alter their path. To tow an iceberg, a ship travels around the iceberg while laying out a floating tow line. The iceberg is lassoed, and then tow tension is applied carefully to avoid rolling the iceberg or pulling the line over the top.

▶ A composite photo of an iceberg

For years scientists have tried to develop ways to harness icebergs and tow them to areas of the world that experience water shortages, such as parts of Africa, the Middle East, or even California. Because glacial ice is formed from falling snow, which is condensed water vapor from the atmosphere, water from icebergs is very pure fresh water, and there are not likely to be many pollutants in it. People have towed icebergs short distances to maneuver them out of the way of ships, but how do you get an iceberg from the Arctic or the Antarctic to the equator without it melting? One idea is to "wrap" the iceberg in something like a large plastic bag to help prevent some melting, and contain the meltwater that is released. While the idea is still being investigated, it is currently considered too expensive.

BRAIN FOOD

Tiny worms, less than 1 mm long, live in temperate glacial ice. The worms survive by eating icebound pollen grains, fern spores, and red algae that is found in the glacial ice. The town of Cordova, Alaska, celebrates the ice worm every year with an Iceworm Festival.

Finding Groundwater

Surface water is easy to find—you can see springs, lakes, and rivers. But how do scientists know where to find water that is hundreds of feet underground? Springs are a good indication that groundwater is present. But most

of the time geologists, hydrologists, and geophysicists study and interpret conductivity maps, which show the Electrical Earth Resistivity (EER) of the ground at different depths beneath the surface. EER maps are made by passing an electrical current through the ground using a battery and four electrodes.

This process records the different amounts of resistivity from the underlying rock and soil. The presence of groundwater increases the electrical conductivity of rock and soil formations. So, by mapping the electrical conductivity of underground materials, scientists can determine the location of groundwater. EER maps are also useful for locating deposits of oil, gas, and minerals.

BRAIN FOOD

The longest river in the world is the Nile, which is about 4,180 mi long. The longest river in North America is the Missouri River, which is about 2,300 mi long. The oldest river in the United States is the New River in West Virginia, which is approximately 300 million years old. The Rio Grande is the youngest river in the United States. It is less than 2 million years old.

2 Water Use and Management

Sewer Systems

Cities use either combined or separated sewer systems. Combined systems, found in older cities, carry both storm water and municipal wastewater to wastewater treatment plants. During heavy rains, combined systems can overload and send untreated water, which contains raw sewage, directly into rivers or bays. Separated systems treat only municipal wastewater. Though they do not overload during storms, they send all storm water directly into local bodies of water.

▶ A Roman aqueduct in Spain

Bioremediation

In recent years scientists have learned how to use living organisms to clean up water pollution. Some naturally occurring bacteria will feed on harmful organic chemicals, such as petroleum compounds or dioxin compounds. By selectively breeding these bacteria (which are harmless to humans), scientists have developed strains of bacteria that voraciously consume and render harmless a number of common pollutants. Such bacteria have been used successfully to help clean up oil and chemical spills at sea. But perhaps the most promising application of these bacteria is in the remediation of polluted groundwater. Polluted aquifers are difficult to remediate because they are underground. Bioremediation gives environmental workers the tools they need to address this critical need. The aquifer to be remediated is "seeded" with a colony of the bacteria and a supply of nutrients. The bacteria multiply exponentially as they encounter and consume the pollutant. When the pollutant is exhausted, the bacteria die.

BRAIN FOOD

Water hyacinths and some anaerobic bacteria excel at removing toxins from an ecosystem. In fact, both organisms have been employed in commercial waste processing facilities.

Who Owns the Oceans?

Part of the problem of ocean pollution is uncertainty about who has jurisdiction over the oceans. In the past, international law has permitted nations to exercise complete control over their territorial waters, which extend 4.8 km from the coast. The rest of the world's oceans were high seas and were open to everyone. In the 20th century, some countries claimed territorial waters to 20 km, and sometimes to 320 km, from their coasts.

In an attempt to clarify the situation, the Third United Nations Conference on the Law of the Sea met between the years 1973 and 1982. The conference resulted in the Law of the Sea Treaty. The Law of the Sea Treaty states that the laws of a coastal nation extend to 22 km from its coastline. This area is called a nation's *territorial sea*. The area that extends 370 km from land is called a nation's *exclusive economic zone*. A nation has control over economic activity, environmental preservation, and research in this area. The rest of the world's oceans are designated as communal property to be controlled by the International Seabed Authority. The final agreement was signed by 134 countries. However, some of the most powerful developed nations, including the United States, did not sign the treaty. Several of these nations objected to the treaty's restrictions on seabed mineral mining.

► A fish kill from thermal pollution in Brazil

Low Frequency Active Sonar

The U.S. Navy developed Low Frequency Active Sonar (LFA) to detect submarines. LFA functions like a searchlight by scanning the ocean with intense sound across vast distances. The noise LFA would produce in the oceans would be billions of times more intense than the levels known to disturb large whales. These active sonar systems can kill marine animals, even those as large as whales and dolphins. The acoustic trauma causes hemorrhaging in and around the ears. Environmental groups have urged the Navy to evaluate the impact that using this sonar will have on the marine environment.

As part of an experiment, a healthy Navy diver was exposed briefly to LFA sonar at 160 decibels (a fraction of the intensity it normally operates at). After 12 min, the diver experienced dizziness and drowsiness. The diver was hospitalized and then suffered memory dysfunction and seizures. Two years later he was still being treated with antiseizure and antidepressant medications. Whales rely on their sensitive hearing to send signals to each other, follow migratory routes, and find food. Noise that interferes with this ability could threaten their ability to survive.

Overview

Tell students that the purpose of this chapter is to help them understand the importance of water in our environment. Although there is an abundance of water on Earth, only a small percentage is fresh water suitable for human use. Pollution and misuse has caused fresh water to become one of our most threatened resources. Many of the world's rivers, lakes, oceans, and aquifers are contaminated with industrial chemicals, agricultural runoff, and sewage. Damage to a water ecosystem can be immediate and widespread, but attempts to clean up water pollution may take a very long time to succeed.

Using the Figure — GENERAL

The photograph shown is a composite of several icebergs from different locations. Ask students to estimate the percentage of ice that is underwater in an average iceberg. (about 90 percent) Ask students, "What would be the obstacles to towing icebergs to places that experience water shortages such as Africa, the Middle East, or California?"
LS Visual

PRE-READING ACTIVITY

FOLDNOTES Encourage students to use their FoldNote as a study guide to quiz themselves for a test on the chapter material. Students may want to create Layered Book FoldNotes for different topics within the chapter.

VIDEO SELECT

For information about videos related to this chapter, go to **go.hrw.com** and type in the keyword **HE4 WATV**.

Water

1 Water Resources
2 Water Use and Management
3 Water Pollution

PRE-READING ACTIVITY

FOLDNOTES **Layered Book**

Before you read this chapter, create the **FoldNote** entitled "Layered Book" described in the Reading and Study Skills section of the Appendix. Label the tabs of the layered book with "Water Resources," "Water Use," "Water Management," and "Water Pollution." As you read the chapter, write information you learn about each category under the appropriate flap.

This composite photograph shows what an iceberg might look like if you could see the entire iceberg. Did you know that some countries are considering towing icebergs to their coasts and melting the ice to provide drinking water?

268

Chapter Correlations   *National Science Education Standards*

SPSP 3a Human populations use resources in the environment in order to maintain and improve their existence. Natural resources have been and will continue to be used to maintain human populations. **(Section 2)**

SPSP 3b The earth does not have infinite resources; increasing human consumption places severe stress on the natural processes that renew some resources, and it depletes those resources that cannot be renewed. **(Section 2)**

SPSP 4b Materials from human societies affect both physical and chemical cycles of the earth. **(Section 3)**

SPSP 6e Humans have a major effect on other species. For example, the influence of humans on other organisms occurs through land use—which decreases space available to other species—and pollution—which changes the chemical composition of air, soil, and water. **(Section 3)**

Water Resources

The next time you drink a glass of water, think about where the water came from. Did you know that some of the water in your glass may have been part of a rainstorm that pounded the Earth long before life existed? Or that water may have been part of a dinosaur that lived millions of years ago. Some of the water we drink today has been around since water formed on Earth billions of years ago. Water is essential to life on Earth. Humans can survive for more than a month without food, but we can live for only a few days without water.

Two kinds of water are found on Earth. Fresh water—the water that people can drink—contains little salt. Salt water—the water in oceans—contains a higher concentration of dissolved salts. Most human uses for water, such as drinking and agriculture, require fresh water.

The Water Cycle

The Earth is often called "the Water Planet" because it has an abundance of water in all forms: solid, liquid, and gas. Water is a renewable resource because it is circulated in the water cycle, as shown in **Figure 1.** In the water cycle, water molecules travel between the Earth's surface and the atmosphere. Water evaporates at the Earth's surface and leaves behind salts and other impurities. Water vapor, which is a gas, rises into the air. As water vapor rises through the atmosphere, the gas cools and condenses into drops of liquid water that form clouds. Eventually the water in clouds falls back to Earth and replenishes the Earth's fresh water. The oceans are an important part of the water cycle because almost all of Earth's water is in the oceans.

Objectives

▶ Describe the distribution of Earth's water resources.

▶ Explain why fresh water is one of Earth's limited resources.

▶ Describe the distribution of Earth's surface water.

▶ Describe the relationship between groundwater and surface water in a watershed.

Key Terms

surface water
river system
watershed
groundwater
aquifer
porosity
permeability
recharge zone

Figure 1 ▶ The water cycle is the continuous movement of water between Earth and its atmosphere.

CONDENSATION

PRECIPITATION

EVAPORATION

269

MISCONCEPTION ALERT

The Water Cycle Students may not know that plants and soil are also part of the water cycle. Plants cycle water through the water cycle as they draw liquid water from the ground and *transpire*, or release water vapor through leaf pores. You can demonstrate this for students by placing a seedling or small plant in a clear plastic cup and sealing the cup with plastic wrap. After a few hours in the sun, water from the soil and the plant will condense on the plastic wrap.

Chapter Resource File

• **Lesson Plan**
• **Active Reading** BASIC
• **Section Quiz** GENERAL

Transparencies

TT Bellringer

Focus

Overview

Before beginning this section, review with your students the Objectives listed in the Student Edition. This section reviews the water cycle and then explores the distribution of fresh water on Earth's surface and underground.

Bellringer

Ask students to think about where water comes from. Then have them answer the following question in their *EcoLog:* "Is there more or less water on Earth than there was 1 billion years ago?" (Water on Earth came primarily from the outgassing of rocks in the mantle. As the rocks cooled and changed state, they released tremendous amounts of water. Impacts from comets and other bodies during Earth's formation also brought water. The amount of water on Earth has not changed significantly since Earth formed.)
LS Logical

Motivate

Demonstration — BASIC

How Much Water Is There?
Begin with 4 L of water. Add a few drops of food coloring so that students can see the water volume easily. Explain that this volume of water represents all of the water on Earth. Then pour about 120 mL from the container into a graduated cylinder. This volume represents the fresh water available in the world. Remove 92 mL of water to represent the water trapped in glaciers and ice caps. Remove 20 mL of water to represent ground water. The remaining 8 mL represents all of the surface water on Earth.
LS Visual

English Language Learners

Interpreting Statistics — BASIC

Proportions The pie graph in **Figure 2** shows two different sets of statistics on global water distribution. As in any pie graph, the percentages given for all of the pieces of the pie add up to exactly 100 percent. Have students check this graph by adding up the percentages given for each part of the pie. Fresh water and saltwater should equal 100 percent, and the three parts that make up fresh water should add up to 100 percent. Then have students calculate the percentage of all water that is composed of groundwater. (0.22 × 0.03 = 0.0066, or 0.66 percent)
LS Visual/Logical

Discussion — BASIC

The Hydrosphere Remind students about what they learned about the water cycle and the hydrosphere. Explain that water is a renewable resource, which means that when it is used it is replenished by nature. However, if the water is polluted or made unusable at a rate faster than nature can replenish it, then it becomes nonrenewable.

Transparencies
TT Watersheds of the World

Global Water Distribution

To understand why fresh water is such a limited resource you have to understand how little fresh water is found on Earth. Although 71 percent of the Earth's surface is covered with water, nearly 97 percent of Earth's water is salt water in oceans and seas. **Figure 2** illustrates this relationship. Of the fresh water on Earth, about 77 percent is frozen in glaciers and polar icecaps. Only a small percentage of the water on Earth is liquid fresh water that humans can use. The fresh water we use comes mainly from lakes and rivers and from a relatively narrow zone beneath the Earth's surface.

SALT WATER 97%

FRESH WATER 3%

Icecaps and glaciers 77%

Other fresh water 1%

Groundwater 22%

Figure 2 ▶ This pie graph shows the distribution of water on Earth. What percentage of the Earth's fresh water is in lakes and rivers?

Surface Water

Surface water is fresh water on Earth's land surface. Surface water is found in lakes, rivers, streams, and wetlands. Although surface water makes up a small fraction of the fresh water on Earth, the distribution of surface water has played a vital role in the development of human societies. Throughout history, people have built cities, towns, and farms near reliable sources of surface water. Some of the oldest cities in the world were built near rivers. Today, most large cities depend on surface water for their water supplies. Rivers, lakes, and streams provide drinking water, water to grow crops, food such as fish and shellfish, power for industry, and a means of transportation by boat.

Figure 3 ▶ Watersheds of the World
This map shows the Earth's major watersheds. The highlighted area of the satellite image below shows that the Mississippi River watershed covers almost half of the United States.

North America
1. Yukon
2. Mackenzie
3. Columbia
4. Colorado
5. Rio Grande
6. Mississippi

South America
7. Orinoco
8. Amazon
9. Paraná

GEOGRAPHY CONNECTION — GENERAL

The Mississippi River Watershed The Mississippi River system drains water from about 30 states. In fact, its watershed covers about 40 percent of the land area of the contiguous United States. The Mississippi River and its tributaries also supply fresh water to thousands of communities across the central United States. Have students examine **Figure 3** and locate the major tributaries of the Mississippi River. Point out that the Continental Divide is indicated with a dotted yellow line. Have students determine whether their community is in the Mississippi watershed. Ask students what tributary their community is near. In their **EcoLog**, have them trace the path of water from their community to the ocean, if possible. Have a class discussion about how waste from a city can ultimately end up in the Gulf of Mexico.
LS Visual

River Systems Have you ever wondered where all the water in a river comes from? Streams form as water from falling rain and melting snow drains from mountains, hills, plateaus, and plains. As streams flow downhill, they combine with other streams and form rivers. The more streams that run into a river, the larger the river becomes. As streams and rivers move across the land, they form a flowing network of water called a **river system.** If a river system is viewed from above, it can look like the roots of a tree that are feeding into a trunk. The Mississippi, the Amazon, and the Nile are enormous river systems because they collect the water that flows from vast areas of land. The Amazon River system is the largest river system in the world—it drains an area of land that is nearly the size of Europe.

Watersheds The area of land that is drained by a river is known as a **watershed.** The watershed of the Mississippi River is shown in the satellite image in **Figure 3.** Pollution anywhere in a watershed may end up polluting a river. The amount of water that enters a watershed varies throughout the year. Rapidly melting snow as well as spring and summer rains can dramatically increase the amount of water in a watershed. Other times of the year, the river system that drains a watershed may be reduced to a trickle. Communities that depend on rivers for water can be severely affected by these changes to the river system.

Connection to **Biology**

Amazon River Dolphins The Amazon River dolphin is one of the world's few freshwater dolphin species. The dolphins are almost completely blind, but they can easily navigate through the silty waters of the Amazon by using sonar.

internet connect

www.scilinks.org
Topic: Watersheds
SciLinks code: HE4124

SCiLINKS Maintained by the National Science Teachers Association

Using the Figure — **GENERAL**
Watersheds of the World Have student groups choose a landmass shown in **Figure 3** and investigate its watersheds in greater detail. Student groups can prepare a posterboard map of their landmass that shows river systems, tributaries, major cities, topographical details, and cultural details. Have the class piece the posterboards together to create a world map. Use the world map for discussion as you teach this chapter. When you teach about dams, for example, students can find the major dams on the rivers they researched and mark the dams on the map.
LS Visual/Interpersonal

Group Activity — **BASIC**
Why Can't We Drink Salt Water? Cut a potato into two equal pieces. Determine and record the mass of each piece. Then prepare two beakers as follows. In beaker 1, dissolve 10 g of salt into 250 mL of tap water. Tell students that the solution in the beaker equals the approximate concentration of sea water. Place 250 mL of tap water into beaker 2 (the control). Place one potato piece into beaker 1, and place the other piece into beaker 2. Let the beakers sit overnight. The following day, have students observe the texture of each potato piece. Blot the pieces dry, and determine the mass of each piece again. Compare the results to the original masses, and discuss the results with students. (The potato in the salt water shriveled up, while the other potato appeared unchanged. The salt has caused the potato's cells to lose water and shrink.) Ask students, "Why can't people drink salt water?" (One reason is that their cells will lose water.) **LS Kinesthetic**

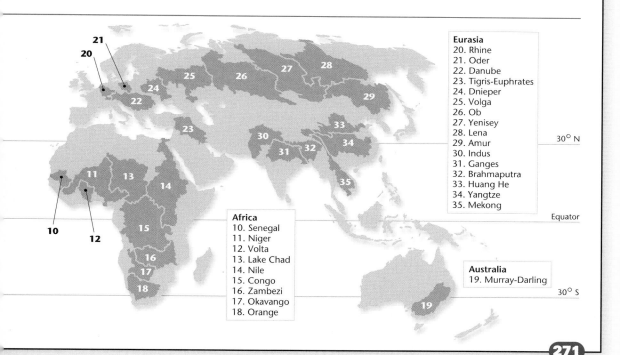

Eurasia
20. Rhine
21. Oder
22. Danube
23. Tigris-Euphrates
24. Dnieper
25. Volga
26. Ob
27. Yenisey
28. Lena
29. Amur
30. Indus
31. Ganges
32. Brahmaputra
33. Huang He
34. Yangtze
35. Mekong

Africa
10. Senegal
11. Niger
12. Volta
13. Lake Chad
14. Nile
15. Congo
16. Zambezi
17. Okavango
18. Orange

Australia
19. Murray-Darling

30° N

Equator

30° S

271

LANGUAGE ARTS
CONNECTION — **GENERAL**

Mark Twain Mark Twain (born Samuel Clemens) is one of the best known chroniclers of early life along the Mississippi River. He grew up in the port town of Hannibal, Missouri and later was a steamboat pilot on the Mississippi River, so he had an intimate knowledge of river culture. Portraits of characters and life along the river fill the pages of novels such as *Huckleberry Finn* and *Tom Sawyer.* In fact, his pen name—Mark Twain— is taken from river lingo and means "two fathoms deep." Have students read passages from Twain's novels in class. Ask students to hypothesize why the Mississippi River has played such an important role in the history of the United States. **LS Verbal**

Discussion ———— GENERAL

Where Is the Water? To emphasize that the limited amount of available fresh water is not very evenly distributed, use a detailed physical map of the world to highlight the areas where water is abundant and the areas where water is scarce. Point to Canada, and tell students that in this country hundreds of thousands of lakes (including the Great Lakes) scattered across thinly populated areas make up about 20 percent of the world's surface freshwater supply. Next point to Lake Baikal, in Siberia (near the Mongolian border). Explain that this lake contains another 20 percent of the world's surface freshwater supply, but also explain that the lake serves only a small population of people. Then point to the Amazon River. Tell students that the river contains another 20 percent of the world's surface freshwater supply, but again tell them it supports only a small number of the world's people. Then point to China and India, countries that have combined human populations of more than 2 billion. Explain that there are several large river systems in these countries and that most are heavily polluted. Yet the rivers supply water to very large populations. Explain that southwestern Asia has a fast-growing population in a largely arid climate. Encourage discussion about the world's unequal water distribution and the problems this distribution creates. **LS** Verbal/Visual

Geofact

How Much Groundwater Is There on Earth? There are about 50 million cubic kilometers of groundwater on Earth. That means there is about 20 times more water underground than in all of the rivers and lakes on Earth!

Groundwater

Most of the fresh water that is available for human use cannot be seen—it exists underground. When it rains, some of the water that falls onto the land flows into lakes and streams. But much of the water percolates through the soil and down into the rocks beneath. Water stored beneath the Earth's surface in sediment and rock formations is called **groundwater.**

As water travels beneath the Earth's surface, it eventually reaches a level where the rocks and soil are saturated with water. This level is known as the *water table.* In wet regions, the water table may be at the Earth's surface and a spring of fresh water may flow out onto the ground. But in deserts, the water table may be hundreds of meters beneath the Earth's surface. The water table is actually not as level as its name implies. The water table has peaks and valleys that match the shape of the land above it. Just as surface water flows downhill, groundwater tends to flow slowly from the peaks of the water table to the valleys.

Aquifers An underground formation that contains groundwater is called an **aquifer.** The water table forms the upper boundary of

CASE STUDY

The Ogallala Aquifer: An Underground Treasure

Anyone who has eaten food produced in the United States has probably enjoyed the benefits of the Ogallala Aquifer, one of the largest known aquifers in the world. This enormous underground water system formed from glaciers that melted at the end of the last Ice Age, 12,000 years ago. Today, the Ogallala Aquifer supplies about one-third of all the groundwater used in the United States.

People began to use the Ogallala Aquifer extensively for irrigation in the 1940s. With help from this ancient water source, farmers turned the Great Plains into one of the most productive farming regions in the world. For many years, farmers seemed to enjoy a limitless supply of fresh water. But in recent years, the

Ogallala Aquifer has started to show its limits. Water is being withdrawn from the aquifer 10 to 40 times faster than it is being replaced. In some places, the water table has dropped more than 30 m (100 ft) since pumping began.

Humans are not the only living things that depend on the Ogallala Aquifer. In some areas, the aquifer flows onto the surface and creates wetlands, which are a vital habitat for many organisms, especially birds. These wetlands are often the first habitats to disappear when the water table falls.

Many people are working together to try to conserve the Ogallala Aquifer. For example, some farmers have begun to limit irrigation during bird migrations in order to allow surface-water levels

▶ **The Ogallala Aquifer** holds about 4 quadrillion liters of water—enough to cover the United States to a depth of 0.5 m (1.5 ft).

BRAIN FOOD

Lake Baikal Lake Baikal, located in Siberia near the Mongolian border, is the deepest, biggest, and one of the oldest lakes in the world. It is almost 1 mi deep, and measures 395 mi by 50 mi. It makes up 20 percent of the world's surface freshwater. If the lake was drained, it would take all the water of the five Great Lakes of the United States to fill it back up. Legend says that Genghis Khan, the fierce Mongolian warrior, was born on Olkhon Island, in Lake Baikal.

INCLUSION Strategies

- *Learning Disabled*
- *Attention Deficit Disorder*

Have students design a poster of the water cycle. Have students include definitions for evaporation, condensation, and precipitation and include examples of different forms of precipitation. Also have the students describe why oceans are important in the water cycle. Students can use their posters as study guides by presenting the information to a group or to a partner one or more times.

an aquifer. Most aquifers consist of materials such as rock, sand, and gravel that have a lot of spaces where water can accumulate. These aquifers hold water in much the same way that a sponge holds water. Groundwater can also dissolve rock formations, such as those made of limestone, and fill vast caves with water, which creates underground lakes. Aquifers are an important water source for many cities and for agriculture.

Porosity and Permeability How can a rock formation hold millions of gallons of water? Although most rocks appear solid, many kinds of rocks contain small holes, or pore spaces. **Porosity** is the amount of space between the particles that make up a rock. Water in an aquifer is stored in the pore spaces and flows from one pore space to another. The more porous a rock is, the more water it can hold. The ability of rock or soil to allow water to flow through it is called **permeability.** Materials such as gravel that allow the flow of water are *permeable*. Materials such as clay or granite that stop the flow of water are *impermeable*. The most productive aquifers usually form in permeable materials, such as sandstone, limestone, or layers of sand and gravel.

internet connect
www.scilinks.org
Topic: Groundwater
SciLinks code: HE4052
Topic: Aquifers
SciLinks code: HE4004

SCiLINKS Maintained by the National Science Teachers Association

▶ **Sandhill cranes** are among the many kinds of birds that rely on water from the Ogallala Aquifer.

CRITICAL THINKING

1. Applying Ideas Most of the water in the Ogallala Aquifer came from glaciers that melted thousands of years ago. What is the aquifer's primary water source today?

2. Expressing Viewpoints Do you think residents of the Great Plains are the only people who have an interest in conserving the Ogallala Aquifer? Write an editorial that expresses your viewpoint.
WRITING SKILLS

to rise. Other farmers have adopted water-saving irrigation systems and are planting crops such as wheat or grain sorghum, which require less water than corn or cotton.

Many farmers and other residents of the Great Plains recognize the value of the Ogallala Aquifer and are fighting to preserve it. They are pressuring politicians to replace policies that encourage wasting water with policies that promote water conservation. These efforts may help save this underground treasure.

273

Group Activity —— GENERAL

Modeling an Aquifer Have student groups build a model of an aquifer in an aquarium using clay, gravel, and sand. Have students add food coloring to a watering can filled with water to simulate recharge. Then have them pour the water over the recharge zone. As an extension, have them build structures on the recharge zone. Ask, "How do the structures affect percolation?" Have them withdraw water from the aquifer using straws to simulate wells.

LS Kinesthetic/Interpersonal

MISCONCEPTION ///ALERT\\\

Withdrawal and Consumption
To describe human use of available water resources, hydrologists use the terms *consumption* and *withdrawal*. These terms can be confusing. All water that is used by humans is withdrawn. Water that has been withdrawn and is no longer available for reuse—because it has evaporated or it has been used by people, animals, or plants—has been consumed. For example, people might pump a certain amount of water from a well to water a lawn. When the lawn is watered, much of the water percolates through the soil and recharges the groundwater. However, some of the water is used by the grass, and the remainder evaporates. This water has been consumed.

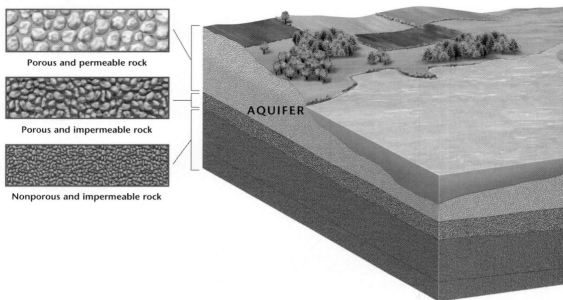

Figure 4 ▶ Groundwater and the Water Table
Aquifers are underground formations that hold water. Impermeable rock can be porous or nonporous, but only permeable rock allows water to pass through it.

Porous and permeable rock

Porous and impermeable rock

Nonporous and impermeable rock

AQUIFER

The Recharge Zone To reach an aquifer, surface water must travel down through permeable layers of soil and rock. Water cannot reach an aquifer from places where the aquifer is covered by impermeable materials. Notice the permeable layers above the aquifer in **Figure 4.** The area of the Earth's surface where water percolates down into the aquifer is called the **recharge zone.** Recharge zones are environmentally sensitive areas because any pollution in the recharge zone can also enter the aquifer.

The size of an aquifer's recharge zone is affected by the permeability of the surface above the aquifer. Structures such as buildings and parking lots can act as impermeable layers to reduce the amount of water entering an aquifer. Communities should carefully manage recharge zones, because surface water can take a very long time to refill an aquifer. In fact, aquifers can take tens of thousands of years to recharge.

Wells If you go nearly anywhere on Earth and dig a hole deep enough, you will eventually find water. A hole that is dug or drilled to reach groundwater is called a well. For thousands of years, humans have dug wells to reach groundwater. We dig wells because groundwater may be a more reliable source of water

ASTRONOMY ——
● CONNECTION —— GENERAL

Is There an Aquifer on Mars? Like Earth, Mars also has a dynamic geologic history. Massive volcanoes, canyons, lava fields, plateaus, and flood channels exist on Mars. Scientists now estimate that Mars has an expansive aquifer, about the size of the United States. This aquifer may have been a source for the water that filled ancient lakes and oceans on northern Mars, and the water that carved Mars' vast flood channels. Have students research and report on the current findings regarding water on other planets and moons in our solar system.

Transparencies
TT Groundwater and the Water Table

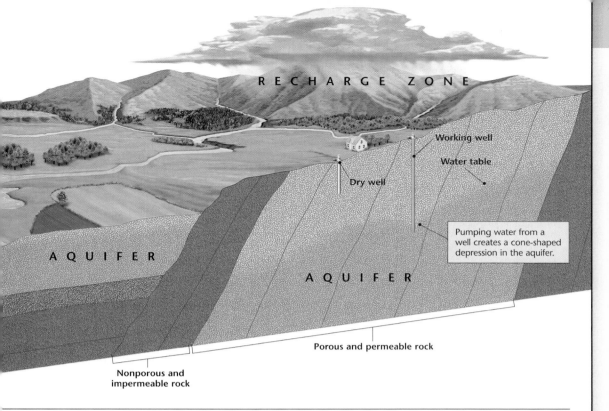

RECHARGE ZONE

Working well

Water table

Dry well

AQUIFER

AQUIFER

Pumping water from a well creates a cone-shaped depression in the aquifer.

Porous and permeable rock

Nonporous and impermeable rock

than surface water and because water is filtered and purified as it travels underground. The height of the water table changes seasonally, so wells are drilled to extend below the water table. However, if the water table falls below the bottom of the well during a drought, the well will dry up. In addition, if groundwater is removed faster than it is recharged, the water table may fall below the bottom of a well. To continue supplying water, the well must be drilled deeper.

SECTION 1 Review

1. **Describe** the distribution of water on Earth. Where is most of the fresh water located?

2. **Explain** why fresh water is considered a limited resource.

3. **Explain** why pollution in a watershed poses a potential threat to the river system that flows through it.

4. **Describe** how water travels through rock.

CRITICAL THINKING

5. **Making Comparisons** Read the description of aquifers. How are aquifers like water-filled sponges?
   READING SKILLS

6. **Analyzing Relationships** Describe the relationship between groundwater and surface water in a watershed. What human activities in a recharge zone can affect the groundwater?

275

Answers to Section Review

1. Most of the water on Earth is salt water found in the oceans. The largest percentage of fresh water is found in icecaps and glaciers. Fresh water is also found on Earth's surface in rivers and lakes and below Earth's surface in aquifers.

2. Fresh water is considered a limited resource because it makes up a very small percentage of water on Earth. The percentage of fresh water available for human use is even smaller.

3. Any pollution in a watershed can enter a river system when rainfall carries the pollution across Earth's surface and into rivers.

4. Some types of rock are permeable and porous. Water can flow through the rock from one pore space to another.

5. Most aquifers are made of permeable rocks or sediments. This type of formation is similar to a water-filled sponge.

6. Surface water in a watershed can percolate through the ground and recharge an aquifer. Human activities in the recharge zone can limit the amount of water that reaches the aquifer. For example, construction on the recharge zone can limit the percolation of water from the surface. Wells drilled in the recharge zone will also affect

the water table. Human activities can also pollute an aquifer. For example, pollutants in the recharge zone can be carried into an aquifer as groundwater percolates into the aquifer.

Close

Reteaching — BASIC

The Water Cycle Organize the class into pairs. Ask one student in each pair to draw a diagram that illustrates the water cycle, such as the diagram in **Figure 1.** Ask the other student in the pair to add explanatory labels that describe what happens at each stage of the water cycle.
English Language Learners

Quiz — GENERAL

1. What factors affect the level of the water table? (the amount of water in the ground, the surface features, the type of rock, and the amount of water that is withdrawn)

2. What is a watershed? (the area of land that is drained by a river)

3. What are the percentages of the fresh water found in icecaps and glaciers and in the ground? (77 percent and 22 percent)

4. What are the states of matter that water in the water cycle goes through? (liquid, solid, and gas)

Alternative Assessment — GENERAL

Water Distribution Have students design their own way to demonstrate the percentages of water found in different areas on Earth. They may use a pie chart, graph, illustration, or some other device.
LS Visual
English Language Learners

Overview

Before beginning this section, review with your students the Objectives listed in the Student Edition. This section describes global and domestic patterns of water use. The section then discusses water management projects and concludes with a discussion of water conservation.

Bellringer

Ask students to imagine they are camping in the desert. Ask them to think of ways to get water if none was nearby. (Sample answer: capturing water by laying out leaves overnight to collect dew or condensation or by digging up roots of succulent plants such as cactus, which store water in their roots) **LS Logical**

Motivate

Identifying Preconceptions — GENERAL

Water Use Ask students to list all the ways that they use water in their daily lives. Students will probably mention that water is used for drinking, for bathing, and for cleaning things. Then point out that it can take about 400 gallons of water to produce a cotton shirt. Also, 49 gallons of water are needed to produce a glass of milk, and it can take more than 2,000 gallons of water to produce a single serving of steak. As a class, discuss the different ways that water is used to produce each of those items.
LS Verbal

SECTION 2
Water Use and Management

Objectives

▶ Identify patterns of global water use.

▶ Explain how water is treated so that it can be used for drinking.

▶ Identify how water is used in homes, in industry, and in agriculture.

▶ Describe how dams and water diversion projects are used to manage freshwater resources.

▶ Identify five ways that water can be conserved.

Key Terms

potable
pathogen
irrigation
dam
reservoir
desalination

You may have heard the expression "We all live downstream." When a water supply is polluted or overused, everyone living downstream can be affected. The number of people who rely on the Earth's limited freshwater reserves is increasing every day. In fact, a shortage of clean, fresh water is one of the world's most pressing environmental problems. According to the World Health Organization, more than 1 billion people lack access to a clean, reliable source of fresh water.

Global Water Use

To understand the factors that affect the world's supply of fresh water, we must first explore how people use water. **Figure 5** shows the three major uses for water—residential use, agricultural use, and industrial use.

Most of the fresh water used worldwide is used to irrigate crops. Patterns of water use are not the same everywhere, however. The availability of fresh water, population sizes, and economic conditions affect how people use water. In Asia, agriculture accounts for more than 80 percent of water use, whereas it accounts for only 38 percent of water use in Europe. Industry accounts for about 19 percent of the water used in the world. The highest percentage of industrial water use occurs in Europe and North America. Globally, about 8 percent of water is used by households for activities such as drinking and washing.

Residential Water Use

There are striking differences in residential water use throughout the world. For example, the average person in the United States uses about 300 L (80 gal) of water every day. But in India, the

❶ First Filtration The source water supply is filtered to remove large organisms and trash.

❷ Coagulation Alum is rapidly mixed into the water and forms sticky globs called *flocs*. Bacteria and other impurities cling to the flocs, which settle to the bottom of a tank.

❸ Second Filtration Layers of sand, gravel, and hard coal filter the remaining impurities.

276

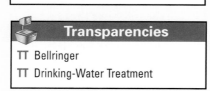

Chapter Resource File

• Lesson Plan

• Active Reading **BASIC**

• Section Quiz **GENERAL**

Transparencies

TT Bellringer

TT Drinking-Water Treatment

REAL-LIFE
● CONNECTION — GENERAL

Where Does Your Water Come From? Have students find out where their tap water comes from. Ask them, "Does your community's water come mainly from surface water or groundwater? How many people can current water sources support?" If possible, plan a trip to your local water treatment plant or contact the plant administrators to see if they have a classroom resource kit.

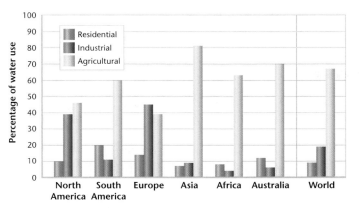

Source: World Resources Institute.

Figure 5 ▶ Europe is the only continent that uses more water for industry than for agriculture.

average person uses only 41 L of water every day. In the United States, only about half of residential water use is for activities inside the home, such as drinking, cooking, washing, and toilet flushing. The remainder of the water used residentially is used outside the home for activities such as watering lawns and washing cars. **Table 1** shows how the average person in the United States uses water.

Water Treatment Most water must be treated to make it **potable,** or safe to drink. Water treatment removes elements such as mercury, arsenic, and lead, which are poisonous to humans even in low concentrations. These elements are found in polluted water, but they can also occur naturally in groundwater. Water treatment also removes **pathogens,** which are organisms that cause illness or disease. Bacteria, viruses, protozoa, and parasitic worms are common pathogens. Pathogens are found in water contaminated by sewage or animal feces. There are several methods of treating water to make it potable. **Figure 6** shows a common drinking water treatment method that includes both physical and chemical treatment.

Table 1 ▼

Daily Water Use in the United States (per Person)	
Use	Water (L)
Lawn watering and pools	95
Toilet flushing	90
Bathing	70
Brushing teeth*	10
Cleaning (inside and outside)	20
Cooking and drinking	10
Other	5

*with water running
Source: U.S. Environmental Protection Agency.

Figure 6 ▶ Drinking-Water Treatment

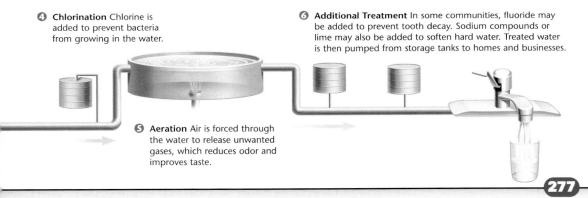

❹ **Chlorination** Chlorine is added to prevent bacteria from growing in the water.

❻ **Additional Treatment** In some communities, fluoride may be added to prevent tooth decay. Sodium compounds or lime may also be added to soften hard water. Treated water is then pumped from storage tanks to homes and businesses.

❺ **Aeration** Air is forced through the water to release unwanted gases, which reduces odor and improves taste.

277

Teach

Interpreting Statistics —————— GENERAL

Bar Graphs The bar graph in **Figure 5** shows how water is used in different regions of the world. Students may think that Australia and Asia use a greater volume of water for agriculture than other regions of the world. Point out that the graph shows percentages of water used, not the amounts of water used. For each region, the three bars do not equal 100 percent because of water lost to evaporation. Have students explain why the percentage of water used for agricultural purposes by the world is less than the percentage used by Asia and Australia. (The percentages shown for the world are an average of each region of the world.) **LS** **Logical/Visual**

SKILL BUILDER —— BASIC

Graphing Have students make a bar graph of the information shown in **Table 1.** As an extension, students could make a pie graph showing the relative percentage of each type of water use.

Homework —————— ADVANCED

World Water Resources Have each student in the class choose a country and investigate its water resources. Ask students, "What percentage of water is used by households, by industry, and by agriculture? What are the country's water resources? What are the major threats to the country's water supply?" In class, have students share the answers to these questions and share any other information they find.

Computer Manufacturing and Water Use Buying a computer is a major expense. Yet many people do not know the effect that computer manufacturing has on the environment. Computer chips go through an elaborate cleaning process that involves a large volume of water and 500 to 1,000 different chemicals, such as arsenic. The production process generates toxic wastes, and can contaminate groundwater and soil. The process consumes vast amounts of electricity and water. A chip manufacturing plant can use millions of gallons of water every day.

Wastewater from the process is chemically treated and released into the city's water treatment centers. As a result, Silicon Valley, the heart of computer-chip manufacturing in the United States, has 23 Superfund sites created by the high-tech industry. The disposal of computers presents additional problems. Deteriorating hard drives and monitors can release toxic metals, such as lead, cadmium, and mercury.

Some European manufacturers are including the cost of recycling a computer in its price. Ask students to debate whether the cost of clean-up should be included in the price of a computer.
LS Interpersonal

Comparison Table
You may want to use this Graphic Organizer in a game to review material before the test. Divide the class into two teams. Ask students questions about material from the Comparison Table. Give points to each team that provides correct answers.

Figure 7 ▶ Water is a very important industrial resource. These nuclear power plant cooling towers release the steam produced from water used to cool a nuclear reactor.

Industrial water use (world) 19%

Other water use (world) 81%

Graphic
Organizer   Comparison Table

Create the **Graphic Organizer** entitled "Comparison Table" described in the Appendix. Label the columns with "Residential Water Use," "Industrial Water Use," and "Agricultural Water Use." Label the rows with "Characteristics" and "Water Conservation." Then, fill in the table with details about the characteristics and the ways water can be conserved in each type of water use.

278

Industrial Water Use

Industry accounts for 19 percent of water used in the world. Water is used to manufacture goods, to dispose of waste, and to generate power. The amount of water needed to manufacture everyday items can be astounding. For instance, nearly 1,000 L of water are needed to produce 1 kg of aluminum, and almost 500,000 L of water are needed to manufacture a car. Vast amounts of water are required to produce computer chips and semiconductors.

Most of the water that is used in industry is used to cool power plants. **Figure 7** shows water being released as steam from nuclear power plant cooling towers. Power-plant cooling systems usually pump water from a surface water source such as a river or lake, carry the water through pipes in a cooling tower, and then pump the water back into the source. The water that is returned is usually warmer than the source, but it is generally clean and can be used again.

Agricultural Water Use

Did you know that it can take nearly 300 L (80 gal) of water to produce one ear of corn? That's as much water as an average person in the United States uses in a day! Agriculture accounts for 67 percent of the water used in the world. Plants require a lot of water to grow, and as much as 80 percent of the water used in agriculture evaporates and never reaches plant roots.

Irrigation Fertile soil is sometimes found in areas of the world that do not have abundant rainfall. In regions where rainfall is inadequate, extra water can be supplied by irrigation. **Irrigation** is a method of providing plants with water from sources other than direct precipitation. The earliest form of irrigation probably involved flooding fields with water from a nearby river.

ARCHEOLOGY ——
● CONNECTION — GENERAL

Tales of the Teeth Trace elements from local bedrock that enter the water supply can be used to identify where a person was born. Adult tooth enamel forms during childhood and is highly resistant to chemical change, even after death. So even if a person relocates as an adult, trace elements in the teeth will tell where the person is from. For example, lead is a trace element that is used to identify the teeth of ancient people. Every place in the world has a specific lead isotope signature.

When trace amounts of lead make their way into the local water, and human diets, the lead leaves traces in the teeth. Scientists extract the lead from the enamel of ancient human teeth and analyze it using mass spectrometry. Using this method, British archeologists were recently able to pinpoint the birthplace of a Roman princess who lived 1,600 years ago! Encourage students to find out about another archeological discovery that was made using this method.

Many different irrigation techniques are used today. For example, some crops, such as cotton, are irrigated by shallow, water-filled ditches, as shown in **Figure 8**. In the United States, high-pressure overhead sprinklers are the most common form of irrigation. This method of irrigation is inefficient because nearly half the water evaporates and never reaches the plant roots. Irrigation systems that use water more efficiently are becoming more common.

Other water use (world) 33%

Agricultural water use (world) 67%

Figure 8 ▶ High-pressure overhead sprinklers (left) are inefficient because a lot of water is lost to evaporation. Water-filled ditches (above) irrigate cotton seedlings.

Water Management Projects

For thousands of years, humans have altered streams and rivers to make them more useful. Nearly two thousand years ago, the Romans built aqueducts, huge canals that brought water from the mountains to the dry areas of France and Spain. One such aqueduct is shown in **Figure 9**. Some of these aqueducts are still used today. Engineering skills have improved since the time of the Romans, and water projects have become more complex.

People often prefer to live in areas where the natural distribution of surface water is inadequate. Water management projects, such as dams and water diversion canals, are designed to meet these needs. Water management projects can have various goals, such as bringing in water to make a dry area habitable, creating a reservoir for recreation or drinking water, or generating electric power. Water management projects have changed the American Southwest and have proved that if water can be piped in, people can live and grow crops in desert areas.

Figure 9 ▶ This Spanish aqueduct was built almost two thousand years ago by the Romans.

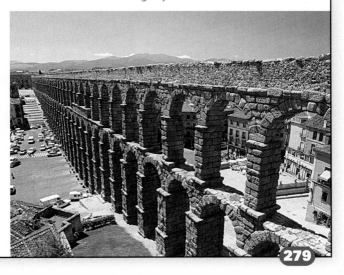

279

HISTORY ——
● **CONNECTION** ── **GENERAL**

Roman Aqueducts Water was so important to the Romans for maintaining and expanding their empire that they spent as much time building aqueducts throughout their vast empire as they did building grand monuments, such as the Coliseum in Rome. In the ancient city of Rome there were 11 major aqueducts built between 312 BCE and 226 CE, the longest of which was 94 km long. When the population was well over one million

inhabitants, scientists have calculated that the aqueducts brought enough water to supply each person with more than a cubic meter of water per day. The aqueducts were constructed to run on a shallow gradient from their source in the hills surrounding Rome into the central cistern in the city. This type of construction sometimes required great feats of engineering to maintain that constant gradient over rolling terrain and through steep valleys.

Activity —— **ADVANCED**
Modeling Water Management Projects Have students build a working model of an aqueduct, a water diversion project, or a dam. Students could also design a model of the different irrigation methods discussed in this section. For example, to model ditch irrigation, have students place soil in a container and build a main channel with several smaller irrigation channels intersecting the main water supply. The students should notice the effectiveness of dispersing large amounts of water over large distances by using this method. Ask students to identify the potential water conservation issues presented by each irrigation method.
LS Kinesthetic

Cultural Awareness **GENERAL**
The Taukachi-Konkan Much of the west coast of South America has a hyperarid climate. Years often pass without any rainfall. Yet in this harsh place a group of people called the Taukachi-Konkan flourished almost 4,000 years ago. This culture was able to exist because its people had mastered irrigation. An elaborate irrigation system captured the water of nearby mountain streams and distributed it among a network of terraced fields where crops were grown. The Taukachi-Konkan culture existed for several hundred years.

Homework —— **GENERAL**

Cadillac Desert Have students read passages from *Cadillac Desert*, by Marc Reisner, or watch the four-part video series. Discuss with the class the issues introduced by Reisner, such as the history of water projects in the American West and their political and environmental consequences.

Discussion ———— BASIC

Kinetic and Potential Energy

Water flowing downstream can be compared to a rock tumbling down a mountainside. Water's potential energy is constantly being converted into kinetic energy. The reservoir of a hydroelectric dam has tremendous potential energy; when the gates of the dam are open, that potential energy is converted into kinetic energy that spins a turbine. The turbine then generates electrical energy. Ask students to think of other devices that use the potential energy of water. (Examples include a flush toilet and a water tank.)

|SKILL| BUILDER —— GENERAL

Math A series of hydroelectric dams on the Columbia River, in the Pacific Northwest, has severely reduced the salmon population. As salmon migrate they must pass through the turbines in each dam. Surprisingly, only about 15 percent of the fish die in passing through a given dam. However, because the fish must pass through eight dams on the Columbia River, there is a large cumulative loss. Ask students how many fish would survive out of an initial population of 1,000. (272)

Water Diversion Projects To supply dry regions with water, all or part of a river can be diverted into canals that carry water across great distances. **Figure 10** shows a canal that diverts the Owens River in California to provide drinking water for Los Angeles. Another river, the Colorado River, is diverted to provide water for several states. The Colorado River begins as a glacial stream in the Rocky Mountains and quickly grows larger as other streams feed into it. As the river flows south, however, it is divided to meet the needs of seven western states. So much of the Colorado River's water is diverted for irrigation and drinking water in states such as Arizona, Utah, and California that the river often runs dry before it reaches Mexico and flows into the Gulf of California. In fact, the Colorado River reaches the Gulf only in the wettest years.

Dams and Reservoirs A **dam** is a structure built across a river to control the river's flow. When a river is dammed, an artificial lake, or **reservoir,** is formed behind the dam. Water from a reservoir can be used for flood control, drinking water, irrigation, recreation, and industry. Dams are also used to generate electrical energy. Hydroelectric dams use the power of flowing water to turn a turbine that generates electrical energy. About 20 percent of the world's electrical energy is generated by hydroelectric dams, such as the one shown in **Figure 11.**

Although dams provide many benefits, interrupting a river's flow can also have far-reaching consequences. When the land behind a dam is flooded, people are often displaced and entire ecosystems can be destroyed. It is estimated that 50 million people around the world have been displaced by dam projects. Dams also affect

Figure 10 ▶ This canal carries water more than 300 km across mountains and deserts to supply drinking water to Los Angeles, California.

Figure 11 ▶ Dams, such as this one in Zimbabwe, are built to manage freshwater resources.

LANGUAGE ARTS ———
● CONNECTION —— GENERAL

Water, Water, Everywhere Samuel Taylor Coleridge wrote "The Rime of the Ancient Mariner" in 1798. The following is an excerpt from the poem:

> Water, water, everywhere,
> And all the boards did shrink
> Water, water, everywhere,
> Nor any drop to drink . . .
> And every tongue, through utter drought,

> Was withered at the root;
> We could not speak, no more than if
> We had been choked with soot.

Suggest that students write a short summary explaining what this passage implies about water as a resource. As an extension, have students read the entire poem in class.
LS Verbal

the land below them. As a river enters a reservoir, it slows down and deposits some of the sediment it carries. This fertile sediment builds up behind a dam instead of enriching the land farther down the river. As a result, the farmland below a dam may become less productive. Dam failure can be another problem—if a dam bursts, people living along the river below the dam can be killed. In the United States, the era of large dam construction is probably over. But in developing countries, such as Brazil, India, and China, the construction of large dams continues.

Water Conservation

As water sources become depleted, water becomes more expensive. This is because wells must be dug deeper, water must be piped greater distances, and polluted water must be cleaned up before it can be used. Water conservation is one way that we can help ensure that everyone will have enough water at a reasonable price.

Water Conservation in Agriculture Most of the water loss in agriculture comes from evaporation, seepage, and runoff, so technologies that reduce these problems go a long way toward conserving water. *Drip irrigation systems* offer a promising step toward conservation. Shown in **Figure 12**, drip irrigation systems deliver small amounts of water directly to plant roots by using perforated tubing. Water is released to plants as needed and at a controlled rate. These systems are sometimes managed by computer programs that coordinate watering times by using satellite data. Using precise information, a well-designed drip irrigation system loses very little water to evaporation, seepage, or runoff.

Water Conservation in Industry As water resources have become more expensive, many industries have developed water conservation plans. In industry today, the most widely used water conservation practices involve the recycling of cooling water and wastewater. Instead of discharging used water into a nearby river, businesses often recycle water and use it again. Thus, the production of 1 kg of paper now consumes less than 30 percent of the water it required 50 years ago. Small businesses are also helping conserve water. Denver, Colorado, was one of the first cities to realize the value of conserving water in business. In an innovative program, the city pays small businesses to introduce water conservation measures. The program not only saves money for the city and for businesses but also makes more water available for agricultural and residential use.

MATHPRACTICE

Israeli Agriculture
From 1950 to 1980, Israel reduced the amount of water loss in agriculture from 83 percent to 5 percent, mainly by switching from overhead sprinklers to water-saving methods such as drip irrigation. If a small farm uses 10,000 L of water a day for overhead sprinkler irrigation, how much water would be saved in one year by using a drip irrigation system that consumes 75 percent less water?

Figure 12 ▶ Drip irrigation systems use perforated tubing to deliver water directly to plant roots.

281

MATHPRACTICE
Answers
10,000 L/day × 365 days = 3,650,000 L
3,650,000 L × 0.75 = 2,737,500 L of water saved in a year

Discussion ——— GENERAL

When Water Is Scarcer Than Oil
In the desert countries of northern Africa and the Middle East, water is a scarce and valuable resource. Throughout this area of more than 6.5 million square kilometers (2.5 million square miles), no permanent streams flow except for the Nile, which derives its water from the moist Tropics far to the south. More than 1.6 million square kilometers (1 million square miles) of this region typically have no rainfall for years at a time. However, much of this desert region has large aquifers. The water that these aquifers contain dates to much wetter times thousands of years ago. Occasionally, the water table reaches the surface to form an oasis. Wells scattered sparsely across the desert supply the rest of the water used throughout the region. Tell students that in some regions of Saudi Arabia and Kuwait, wells drilled for water more often strike oil, and it is considered to be inconvenient when this occurs. Some political theorists have stated that if another major war is fought in the Middle East, it may be fought over water resources rather than oil. Discuss this claim with students. **LS** Verbal

Notable Quotes ——— GENERAL

"All is born of water, all is sustained by water."
—Johann Wolfgang von Goethe
Goethe was a German poet and dramatist. Discuss the meaning of this quote with students.

Brainstorming Before students read this page, have them brainstorm different ways to conserve water at home. After students have read this page, have them compare their list with what they read.

English Language Learners

SKILL BUILDER — GENERAL

Vocabulary The word *xeriscape* comes from the Greek root "xeri-" which involves "dry." Xeriscaping is the process of using native plants that conserve water.

English Language Learners

Homework — ADVANCED

Water Conservation Plans
Have students record the amount of water they use when brushing their teeth, when showering or bathing, when washing dishes, and when doing other activities. To accomplish this task, have students record the amount of water used in one minute by each device and then multiply that number times the number of minutes they use the device each day. Then ask students to devise a plan to reduce the amount of water they use by one-third. The plan should be reasonable and should not involve great sacrifice. After students have completed this activity, ask, "If everyone in the United States followed your plan, how much water could be saved?" Suggest that students include their plan and notes in their *EcoLog.*

Figure 13 ▶ This xeriscaped yard in Arizona features plants that are native to the state. What kinds of plants are native to your region?

Table 2 ▼

What You Can Do to Conserve Water
• Take shorter showers, and avoid taking baths unless you keep the water level low.
• Install a low-flow shower head in your shower.
• Install inexpensive, low-flow aerators in your water faucets at home.
• Purchase a modern, low-flow toilet, install a water-saving device in your toilet, or simply place a water-filled bottle inside your toilet tank to reduce the water used for each flush.
• Do not let the water run while you are brushing your teeth.
• Fill up the sink basin rather than letting the water run when you are shaving, washing your hands or face, or washing dishes.
• Wash only full loads in your dishwasher and washing machine.
• Water your lawn sparingly.

Water Conservation at Home Although households use much less water than agriculture or industry, a few changes to residential water use will make a significant contribution to water conservation. People can conserve water by changing a few everyday habits and by using only the water that they need. Some of these conservation methods are shown in **Table 2.**

Water-saving technology, such as low-flow toilets and shower heads, can also help reduce household water use. These devices are required in some new buildings. Many cities will also pay residents to install water-saving equipment in older buildings.

About one-third of the water used by the average household in the United States is used for landscaping. To conserve water, many people water their lawns at night to reduce the amount of water lost to evaporation. Another way people save water used outside their home is a technique called *xeriscaping* (ZIR i SKAY ping). Xeriscaping involves designing a landscape that requires minimal water use. **Figure 13** shows one example of xeriscaping in Arizona.

Can one person make a difference? When you multiply one by the millions of people who are trying to conserve water—in industry, on farms, and at home—you can make a big difference.

Solutions for the Future

In some places, conservation alone is not enough to prevent water shortages, and as populations grow, other sources of fresh water need to be developed. Two possible solutions are desalination and transporting fresh water.

282

REAL-LIFE CONNECTION — GENERAL

Local Water Use Have students research and discuss local water conservation issues. Ask students, "What is the main source of water for your area? Does another community or industry depend on that water supply? Or, if no current water problem exists, what possible situations could occur to affect the local water supply?" (Sample answers: pollution, biological pathogens, and extended drought)

INCLUSION Strategies

• *Learning Disabled*
• *Developmentally Delayed*
• *Attention Deficit Disorder*

Ask students to make a list of their personal water usage for a 24-hour period. For each usage on the list, students can identify at least one way they could conserve water the next time they use it. Also, ask students to identify major water-conservation ideas that could be implemented in their school, community, or home.

Desalination Some coastal communities rely on the oceans to provide fresh water. **Desalination** (DEE SAL uh NAY shuhn) is the process of removing salt from salt water. Some countries in drier parts of the world, such as the Middle East, have built desalination plants to provide fresh water. Most desalination plants heat salt water and collect the fresh water that evaporates. **Figure 14** shows one such plant in Kuwait. Because desalination consumes a lot of energy, the process is too expensive for many nations to consider.

Transporting Water In some areas of the world where freshwater resources are not adequate, water can be transported from other regions. For example, the increasing number of tourists visiting some Greek islands in the Mediterranean Sea have taxed the islands' freshwater supply. As a result, ships travel regularly from the mainland towing enormous plastic bags full of fresh water. The ships anchor in port, and fresh water is then pumped onto the islands. This solution is also being considered in the United States, where almost half of the available fresh water is in Alaska. Scientists are exploring the possibility of filling huge bags with water from Alaskan rivers and then towing the bags down the coast to California, where fresh water is often in short supply.

Because 76 percent of the Earth's fresh water is frozen in icecaps, icebergs are another potential freshwater source. For years, people have considered towing icebergs to communities that lack fresh water. But an efficient way to tow icebergs is yet to be discovered.

internet connect
www.scilinks.org
Topic: Water Conservation
SciLinks code: HE4121
SCiLINKS. Maintained by the National Science Teachers Association

Figure 14 ▶ Most desalination plants, such as this one in Kuwait, use evaporation to separate ocean water from the salt it contains.

SECTION 2 Review

1. **Describe** the patterns of global water use for each continent shown in the bar graph in Figure 5.

2. **Describe** the drinking water treatment process in your own words.

3. **Describe** the benefits and costs of dams and water diversion projects.

4. **List** some things you can do to help conserve the world's water supply. Give at least three examples.

CRITICAL THINKING

5. **Making Comparisons** Write a description of the evaporative method of desalination using terms from the water cycle. **WRITING SKILLS**

6. **Identifying Alternatives** Describe three ways that communities can increase their freshwater resources.

283

Answers to Section Review

1. Answers may vary. Most of the water that is used in the world is used for agriculture. South America, Africa, Australia, and Asia have the highest percentage of agricultural water use. North America and Europe have the highest percentage of industrial water use.

2. Answers may vary. The drinking-water treatment process begins with the filtration of source water to remove large particles. Then, alum is added to the water to remove bacteria and other impurities. The water is then passed through layers of sand, gravel, and hard coal to filter out the remaining impurities. Next, drinking water

is chlorinated and aerated. Finally, drinking water is treated with compounds such as sodium, fluoride, or lime and is stored in tanks.

3. Answers may vary. Dams and water diversion projects help us manage our water resources. These projects can help control flooding, provide water for agriculture and drinking, and generate electricity. Water management projects can also disrupt habitats, displace people, and reduce the fertility of agricultural land.

4. Answers may vary. Ways to conserve water include using less water to accomplish daily

tasks, installing water-saving technology, and xeriscaping.

5. Most desalination processes use energy to heat a body of water. When the water reaches a specific temperature, some of it evaporates and becomes water vapor. Salts and other impurities are left behind. The evaporated water cools and becomes pure, liquid water that is collected for use.

6. Answers may vary. Communities can increase freshwater supplies by developing water management projects, by obtaining fresh water from desalination, and by conserving water.

Overview

Before beginning this section, review with your students the Objectives listed in the Student Edition. This section explores the effects of water pollution and discusses the major laws designed to improve water quality in the United States.

 **Bellringer**

Ask students to define the term *water pollution*. Is a cup of coffee polluted water? Is a muddy stream polluted? **LS** Logical

Motivate

Discussion —————— GENERAL

Can I Drink the Water? Write the following categories on the board: "the tap, a spring or well, a mountain stream, a nearby creek or lake, and a drainage ditch after a heavy rain." Ask students whether they would consider drinking water from each source. For each source, record the student responses. You will probably notice that students became less willing to trust the quality of the water as you proceeded through the categories. Point out that many people instinctively recognize the likelihood that many of our freshwater sources are polluted. If any students answer "no" to the first category, tap water, ask them to explain. (Many people avoid tap water because they fear that it contains high concentrations of chemicals or pollutants.) **LS** Verbal

Objectives

▶ **Compare** point-source pollution and nonpoint-source pollution.
▶ **Classify** water pollutants by five types.
▶ **Explain** why groundwater pollution is difficult to clean.
▶ **Describe** the major sources of ocean pollution, and explain the effects of pollution on ecosystems.
▶ **Describe** six major laws designed to improve water quality in the United States.

Key Terms

water pollution
point-source pollution
nonpoint-source pollution
wastewater
artificial eutrophication
thermal pollution
biomagnification

Figure 15 ▶ Point-source pollution comes from a single, easily identifiable source. In this photo, the waste from an iron mine is being stored in a pond.

Table 3 ▼

Sources of Point Pollution

- leaking septic-tank systems
- leaking storage lagoons for polluted waste
- unlined landfills
- leaking underground storage tanks that contain chemicals or fuels such as gasoline
- polluted water from abandoned and active mines
- water discharged by industries
- public and industrial waste-water treatment plants

You might think that you can tell if a body of water is polluted by the way that the water looks or smells, but sometimes you can't. There are many different forms of water pollution. **Water pollution** is the introduction of chemical, physical, or biological agents into water that degrade water quality and adversely affect the organisms that depend on the water. Almost all of the ways that we use water contribute to water pollution. However, the two underlying causes of water pollution are industrialization and rapid human population growth.

In the last 30 years, developed countries have made great strides in cleaning up many polluted water supplies. Despite this progress, some water is still dangerously polluted in the United States and in other countries. In developing parts of the world, water pollution is a big problem. Industry is usually not the major cause of water pollution in developing countries. Often, the only water available for drinking in these countries is polluted with sewage and agricultural runoff, which can spread waterborne diseases. To prevent water pollution, people must understand where pollutants come from. As you will learn, water pollution comes from two types of sources: point and nonpoint sources.

Chapter Resource File

- **Lesson Plan**
- **Active Reading** BASIC
- **Section Quiz** GENERAL

 Transparencies

TT Bellringer

Point-Source Pollution

When you think of water pollution, you probably think of a single source, such as a factory, a wastewater treatment plant, or a leaking oil tanker. These are all examples of **point-source pollution,** which is pollution discharged from a single source. **Table 3** lists some additional examples of point-source pollution. Point-source pollution can often be identified and traced to a source. But even when the source of pollution is known, enforcing cleanup is sometimes difficult.

Nonpoint-Source Pollution

Nonpoint-source pollution comes from many different sources that are often difficult to identify. For example, a river can be polluted by runoff from any of the land in its watershed. If a farm, a road, or any other land surface in a watershed is polluted, runoff from a rainstorm can carry the pollution into a nearby river, stream, or lake. **Figure 16** shows common sources of nonpoint pollutants. **Table 4** lists some additional causes of nonpoint pollution.

Because nonpoint pollutants can enter bodies of water in many different ways, they are extremely difficult to regulate and control. The accumulation of small amounts of water pollution from many sources is a major pollution problem—96 percent of the polluted bodies of water in the United States were contaminated by nonpoint sources. Controlling nonpoint-source pollution depends to a great extent on public awareness of the effects of activities such as spraying lawn chemicals and using storm drains to dispose of used motor oil.

FIELD ACTIVITY

Identifying Sources of Pollution Walk around your neighborhood, and record potential sources of nonpoint pollution. See Table 4 for examples. Count the number of potential sources of nonpoint pollution, and suggest ways to reduce each source of pollution in your *EcoLog*.

Table 4 ▼

Nonpoint Sources of Pollution
• chemicals added to road surfaces (salt and other de-icing agents)
• water runoff from city and suburban streets that may contain oil, gasoline, animal feces, and litter
• pesticides, herbicides, and fertilizer from residential lawns, golf courses, and farmland
• feces and agricultural chemicals from livestock feedlots
• precipitation containing air pollutants
• soil runoff from farms and construction sites
• oil and gasoline from personal watercraft

Figure 16 ▶ Sources of Nonpoint Pollution Examples of nonpoint-source pollution include ❶ livestock polluting water holes that can flow into streams and reservoirs, ❷ oil on a street, which can wash into storm sewers and then drain into waterways, and ❸ thousands of watercraft, which can leak gasoline and oil.

285

Teach

Discussion ———— GENERAL

Household Chemicals Household chemicals are a major source of water pollutants. Homeowners frequently dump chemicals such as paints, solvents, motor oil, and household cleaners down the drain or onto the ground—often in violation of state or local laws. State and local agencies often provide guidelines for the disposal of common household chemicals that are hazardous. As a class, search on the Internet to find out what state and local laws govern the disposal of waste in your community. Then create a guide for local homeowners.
LS Visual

Reteaching ———— BASIC

Point and Nonpoint Pollution To ensure that students grasp the difference between point and nonpoint pollution, first review the definitions of the terms and then provide an example of each type of pollution. Quiz students by giving an example of each type of pollution and by then selecting a student to identify the example as either point or nonpoint pollution. The following are some examples you may wish to use: parking-lot runoff (nonpoint), untreated sewage (point), runoff from feedlots (nonpoint), a restaurant's drainage pipe emptying runoff into a river (point), and the litter in storm runoff (nonpoint).
English Language Learners

REAL-LIFE ——— CONNECTION — GENERAL

Two Cycle Engines What do personal watercraft and snowmobiles have in common? Most have two-cycle engines. Each year, two-cycle engines leak millions of gallons of unburned fuel into ecosystems. An estimated 20 to 25 percent of fuel in two-cycle engines fails to combust, and it flushes into the water as raw fuel vapor emissions. Noise pollution from these vehicles also affects aquatic environments. Have students find out about personal watercraft and water pollution. Encourage them to research newer designs that produce less pollution than two-cycle engines. Students may also be interested in learning about the controversy involving snowmobile use in the U.S. park system.

Table 5 ▼

Pollutant Types and Sources		
Type of pollutant	Agent	Major sources
Pathogens	disease-causing organisms, such as bacteria, viruses, protozoa, and parasitic worms	mostly nonpoint sources; sewage or animal feces, livestock feedlots, and poultry farms; sewage from overburdened wastewater treatment plants
Organic matter	animal and plant matter remains, feces, food waste, and debris from food-processing plants	mostly nonpoint sources
Organic chemicals	pesticides, fertilizers, plastics, detergents, gasoline and oil, and other materials made from petroleum	mostly nonpoint sources; farms, lawns, golf courses, roads, wastewater, unlined landfills, and leaking underground storage tanks
Inorganic chemicals	acids, bases, salts, and industrial chemicals	point sources and nonpoint sources; industrial waste, road surfaces, wastewater, and polluted precipitation
Heavy metals	lead, mercury, cadmium, and arsenic	point sources and nonpoint sources; industrial discharge, unlined landfills, some household chemicals, and mining processes; heavy metals also occur naturally in some groundwater
Physical agents	heat and suspended solids	point sources and nonpoint sources; heat from industrial processes and suspended solids from soil erosion

Teach, continued

Discussion —— GENERAL

Pollutant Sources Share the following information with your students:

Common Sources of Point Pollution in the United States

- 23 million septic-tank systems
- 190,000 storage lagoons for polluted waste
- 9,000 municipal landfills
- about 2 million underground storage tanks containing pollutants such as gasoline
- thousands of public and industrial wastewater treatment plants

Common Sources of Nonpoint Pollution in the United States

- highway construction and maintenance, including eroding soil and toxic chemicals
- storm-water runoff including oil, gasoline, dog feces, and litter from city and suburban streets
- pesticides from cropland
- 50 million tons of fertilizer applied to crops, lawns and golf courses every year
- 10 million tons of dry salt applied to highways for snow and ice control every year

Brainstorm ways to reduce each of these pollutant sources.

Principal Water Pollutants

There are many different kinds of water pollutants. **Table 5** lists some common types of pollutants and some of the possible sources of each pollutant.

Wastewater

Do you know where water goes after it flows down the drain in a sink? The water usually flows through a series of sewage pipes that carry it—and all the other wastewater in your community—to a wastewater treatment plant. **Wastewater** is water that contains waste from homes or industry. At a wastewater treatment plant, water is filtered and treated to make the water clean enough to return to a river or lake.

Treating Wastewater A typical residential wastewater treatment process is illustrated in **Figure 17.** Most of the wastewater from homes contains biodegradable material that can be broken down by living organisms. For example, wastewater from toilets and kitchen sinks contains animal and plant wastes, paper, and soap, all of which

286

HEALTH
● CONNECTION — GENERAL

Pollution and the Poor Studies show that low-income populations are more affected by the environmental pollution than upper-income populations. Waste dumps, toxic incinerators, chemical plants, and other industries are often located in low-income communities because the local population is usually not as politically organized and has fewer resources to oppose the industries. In addition, a much higher percentage of low-income individuals are employed in industries in which the workers handle toxic chemicals. Have student groups research in periodicals or on the Internet to find out about a low-income community that is affected by industrial pollution. Suggest that students find out about Superfund sites in East St. Louis, or read excerpts from *A Civil Action.* This issue is also explored in the movie *Erin Brockovich.*

are biodegradable. But wastewater treatment plants may not remove all of the harmful substances in water. Some household and industrial wastewater and some storm-water runoff contains toxic substances that cannot be removed by the standard treatment.

Sewage Sludge If you look again at **Figure 17**, you will see that one of the products of wastewater treatment is *sewage sludge*, the solid material that remains after treatment. When sludge contains dangerous concentrations of toxic chemicals, it must be disposed of as hazardous waste. The sludge is often incinerated, and then the ash is buried in a secure landfill. Sludge can be an expensive burden to towns and cities because the volume of sludge that has to be disposed of every year is enormous.

The problem of sludge disposal has prompted many communities to look for new uses for this waste. If the toxicity of sludge can be reduced to safe levels, sludge can be used as a fertilizer. In another process, sludge is combined with clay to make bricks that can be used in buildings. In the future, industries will probably find other creative ways to use sludge.

Connection to **History**

***Cryptosporidium* Outbreak** In 1993, a pathogen called *Cryptosporidium parvum* contaminated the municipal water supply of Milwaukee, Wisconsin. The waterborne parasite caused more than 100 deaths, and 400,000 people experienced a flulike illness. *Cryptosporidium* is found in animal feces, but the parasite usually occurs in low levels in water supplies. The outbreak in Milwaukee was probably caused by an unusual combination of heavy rainfall and agricultural runoff that overburdened the city's water treatment plants.

MISCONCEPTION ALERT

Waterborne Disease in North America Most people in the United States think of waterborne diseases as a problem only in developing countries. But despite elaborate wastewater purification systems, outbreaks still occur. In 1993, the same year in which the *Cryptosporidium* outbreak affected 400,000 people in Milwaukee, a similar outbreak occurred in Round Rock, Texas. In 1993 and 1994, an epidemic of another, deadlier, waterborne illness—cholera—swept through Latin America as far north as northern Mexico. The epidemic raised fears that the United States could eventually be affected.

Homework ——— **ADVANCED**

Water Filters People who are camping often use water filters to remove contaminants from water. A common concern is giardiasis, which is caused by microscopic waterborne parasites. Giardia causes diarrhea, nausea, and abdominal cramps, and can be severely debilitating. Water can be purified by using hand-held filters that pump water through porous ceramic or carbon filters, which act as a screen for microscopic parasites and debris. Alternatives to water filters include boiling water for at least five minutes and using iodine tablets. Have students research and report on the effectiveness of different water filters for home and for camping.

Figure 17 ▶ **Wastewater Treatment Process**

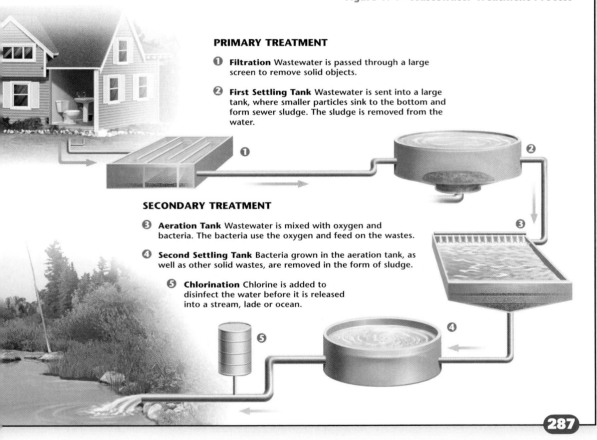

PRIMARY TREATMENT

❶ **Filtration** Wastewater is passed through a large screen to remove solid objects.

❷ **First Settling Tank** Wastewater is sent into a large tank, where smaller particles sink to the bottom and form sewer sludge. The sludge is removed from the water.

SECONDARY TREATMENT

❸ **Aeration Tank** Wastewater is mixed with oxygen and bacteria. The bacteria use the oxygen and feed on the wastes.

❹ **Second Settling Tank** Bacteria grown in the aeration tank, as well as other solid wastes, are removed in the form of sludge.

❺ **Chlorination** Chlorine is added to disinfect the water before it is released into a stream, lade or ocean.

287

Transparencies

TT The Wastewater Treatment Process

CHEMISTRY CONNECTION **GENERAL**

Dioxins Dioxin refers to a group of chemicals that contain chlorine, carbon, oxygen, and hydrogen. Dioxins are a byproduct of chemical processes such as paper bleaching that use chlorine and compounds composed of hydrogen and carbon. Dioxins may cause liver and nerve damage, genetic, reproductive, and immune system problems, and many other problems. According to the EPA, dioxins are some of the most toxic chemicals known to science, and they are the most dangerous environmental pollutants in North America.

QuickLAB

Skills Acquired:
• Experimenting
• Measuring
• Interpreting

Answers
1. The sample that was shaken had the highest dissolved-oxygen level, and the sample that was boiled had the lowest dissolved-oxygen level.
2. Rapids and waterfalls (modeled by the shaken sample) tend to increase the level of dissolved oxygen in a stream. Thermal pollution (modeled by the boiled water) lowers dissolved-oxygen levels.

Group Activity —— ADVANCED
Mapping Pollution Tell students to imagine that pollutants have entered a waterway near your school. Using a map or series of maps (starting with a large-scale local map and working toward smaller scales), have students trace the path that these pollutants would take on their way to the sea. Ask them to make a rough estimate of the time required for the pollutant to reach the sea. (Sample answer: (Assuming that the gross distance via the stream is 1,600 km [1,000 mi], adding about 25 percent for meanders, and using an average flow speed of around 3.2 km/h [2 mi/h], one possible answer is 625 hours, or a little over 26 days.) **LS** Logical

QuickLAB

Measuring Dissolved Oxygen

Procedure
1. Start with **three water samples**, each in a **plastic jar** that is ¾ full. Two water samples should be tap water from a faucet without an aerator. One sample should be water that has been boiled and allowed to cool.
2. Using a **dissolved-oxygen test kit**, test the boiled water and one other water sample.
3. Tighten the lid on the third sample, and then vigorously shake the sample for one minute. Unscrew the lid, and then recap the jar.
4. Repeat step 3 twice. Then, uncap the jar quickly, and test the sample.

Analysis
1. Which sample had the highest dissolved oxygen level? Which sample had the lowest level?
2. What effects do rapids and waterfalls have on the levels of dissolved oxygen in a stream? What effect does thermal pollution have?

Figure 18 ▶ In an effort to limit artificial eutrophication, some states have either banned phosphate detergents or limited the amount of phosphates in detergents.

Artificial Eutrophication

Most nutrients in water come from organic matter, such as leaves and animal waste, that is broken down into mineral nutrients by decomposers such as bacteria and fungi. Nutrients are an essential part of any aquatic ecosystem, but an overabundance of nutrients can disrupt an ecosystem. When lakes and slow-moving streams contain an abundance of nutrients, they are eutrophic (yoo TROH fik).

Eutrophication is a natural process. When organic matter builds up in a body of water, it will begin to decay and decompose. The process of decomposition uses up oxygen. As oxygen levels decrease, the types of organisms that live in the water change over time. For example, as a body of water becomes eutrophic, plants take root in the nutrient-rich sediment at the bottom. As more plants grow, the shallow waters begin to fill in. Eventually, the body of water becomes a swamp or marsh.

The natural process of eutrophication is accelerated when inorganic plant nutrients, such as phosphorus and nitrogen, enter the water from sewage and fertilizer runoff. Eutrophication caused by humans is called **artificial eutrophication.** Fertilizer from farms, lawns, and gardens is the largest source of nutrients that cause artificial eutrophication. Phosphates in some laundry and dishwashing detergents are another major cause of eutrophication. Phosphorus is a plant nutrient that can cause the excessive growth of algae. In bodies of water polluted by phosphorus, algae can form large floating mats, called *algal blooms,* as shown in **Figure 18.** As the algae die and decompose, most of the dissolved oxygen is used and fish and other organisms suffocate in the oxygen-depleted water.

Student Opportunities

Volunteering There are many different opportunities for students to help reduce water pollution. If students are interested in such activities, suggest that they check with the local Chamber of Commerce or other local organizations for opportunities. Students can find volunteer opportunities in different parts of the country and information about environmental careers at the Web site for the Environmental Careers Organization.

Career

Water Management Give students some ideas for possible careers relating to water issues such as hydrology, conservation management, or stream ecology. Explain that a hydrologist is someone who studies the properties, distribution, and effects of water on Earth's surface, the soil, the underlying rocks, and the atmosphere. A stream ecologist is someone who studies the health of organisms and ecological processes in streams, particularly the detrimental impacts of human activity on a stream. Invite a professional in one of these areas to come to class to talk about his or her field of work.

Figure 19 ▶ Fish kills, such as this one in Brazil, can result from thermal pollution.

Thermal Pollution

If you look at **Figure 19**, you might assume that a toxic chemical caused the massive fish kill in the photo. But the fish were not killed by a chemical spill—they died because of thermal pollution. When the temperature of a body of water, such as a lake or stream, increases, **thermal pollution** can result. Thermal pollution can occur when power plants and other industries use water in their cooling systems and then discharge the warm water into a lake or river.

Thermal pollution can cause large fish kills if the discharged water is too warm for the fish to survive. But most thermal pollution is more subtle. If the temperature of a body of water rises even a few degrees, the amount of oxygen the water can hold decreases significantly. As oxygen levels drop, aquatic organisms may suffocate and die. If the flow of warm water into a lake or stream is constant, it may cause the total disruption of an aquatic ecosystem.

Groundwater Pollution

Pollutants usually enter groundwater when polluted surface water percolates down from the Earth's surface. Any pollution of the surface water in an area can affect the groundwater. Pesticides, herbicides, chemical fertilizers, and petroleum products are common groundwater pollutants. Leaking underground storage tanks are another major source of groundwater pollution. It is estimated that there are millions of underground storage tanks in the United States. Most of the tanks—located beneath gas stations, farms, and homes—hold petroleum products, such as gasoline and heating fuel. As underground storage tanks age, they may develop leaks, which allow pollutants to seep into the groundwater.

Connection to Chemistry

Dissolved Oxygen One of the most important measures of the health of a body of water is the amount of dissolved oxygen in the water. Gaseous oxygen enters water by diffusion from the surrounding air, as a byproduct of photosynthesis, and as a result of the rapid movement (aeration) of water. The amount of oxygen that water can hold is determined by the water's temperature, pressure, and salinity. Slow-moving waters tend to have low levels of dissolved oxygen, while rapidly flowing streams have higher levels. Artificial eutrophication and thermal pollution also reduce levels of dissolved oxygen. When dissolved oxygen levels remain below 1 to 2 mg/L for several hours, fish and other organisms suffocate, and massive fish kills can result.

289

Activity — GENERAL

Artificial Eutrophication in a Fishbowl Collect samples of pond water in two small, plastic fishbowls. Fishbowl A will be the control and fishbowl B will be the experimental factor. Ask students to observe both water samples. Add 2 mg of phosphate or nitrate fertilizer to fishbowl B. Place the bowls in the sun. Have students observe the samples over several days. The water in fishbowl B should turn pea-green and brackish, and algae levels will increase. The water in fishbowl A should remain clear. If microscopes are available, have students observe both water samples and record their observations in their *EcoLog*.

Ask students to describe the process that occurred in fishbowl B. (artificial eutrophication) Ask students where the algae came from. (The algae were present in the pond water and the nutrients promoted algae growth.) Ask students to evaluate whether or not this activity was an accurate model of artificial eutrophication. English Language Learners

LS Visual

MISCONCEPTION // ALERT

What Is Beach Tar? Along the Gulf of Mexico, people sometimes encounter blobs of sticky tar on the beach. Offshore drilling activity is often blamed for the tar. In fact, drilling activity contributes very little to the problem. In the past, the major source of tar was oil tankers. Until recently, it was common for oil tankers to flush their holds using sea water. This practice has been banned. Since the ban, there has been a substantial decline in the amount of beach tar. Natural seepage from underwater oil deposits is also a source of beach tar.

⬆ Internet Activity — GENERAL

Researching Phosphates Before the advent of phosphate-free detergents, detergents were a primary source of phosphate pollution in rivers and lakes. High levels of nitrates and phosphates lead to artificial eutrophication. Phosphates were initially added to detergents to make water softer and to help remove dirt. Have students research on the Internet to find out if any detergents currently on the market in the United States contain phosphates. As an extension, encourage students to research the contamination of water by phosphates in the developing world.

Teach, continued

Using the Figure — GENERAL

Sources of Groundwater Pollution **Figure 20** shows many different sources of groundwater pollution. At the bottom of the image is a petroleum storage tank. Underground storage tanks for gasoline, found at almost every filling station and on many farms, can present major groundwater contamination problems. Older metal tanks and abandoned tanks are especially problematic. Most modern storage tanks are made of fiberglass or some other durable, noncorroding synthetic material and they are protected by concrete barriers. In many cases, state or local statutes have required the replacement of metal tanks with noncorroding, double-containment tanks. Have students find the location of underground storage tanks in your community. Maps are available on the Internet. Search by using your county name and the keywords "underground storage tanks" and "maps." Discuss the uses for and problems associated with the tanks. **LS Visual**

Transparencies
TT How Pollutants Enter Groundwater

MATHPRACTICE

Parts per Million Water contamination is often measured in parts per million (ppm). If the concentration of a pollutant is 5 ppm, there are 5 parts of the pollutant in 1 million parts of water. If the concentration of gasoline is 3 ppm in 650,000 L of water, how many liters of gasoline are in the water?

The location of aging underground storage tanks is not always known, so the tanks often cannot be repaired or replaced until after they have leaked enough pollutants to be located. Modern underground storage tanks are contained in concrete and have many features to prevent leaks. Other sources of groundwater pollution include septic tanks, unlined landfills, and industrial wastewater lagoons, as shown in **Figure 20.**

Cleaning Up Groundwater Pollution Groundwater pollution is one of the most challenging environmental problems that the world faces. Even if groundwater pollution could be stopped tomorrow, some groundwater would remain polluted for generations to come. As you have learned, groundwater recharges very slowly. The process for some aquifers to recycle water and purge contaminants can take hundreds or thousands of years. Groundwater is also difficult to decontaminate because the water is dispersed throughout large areas of rock and sand. Pollution can cling to the materials that make up an aquifer, so even if all of the water in an aquifer were pumped out and replaced with clean water, the groundwater could still become polluted.

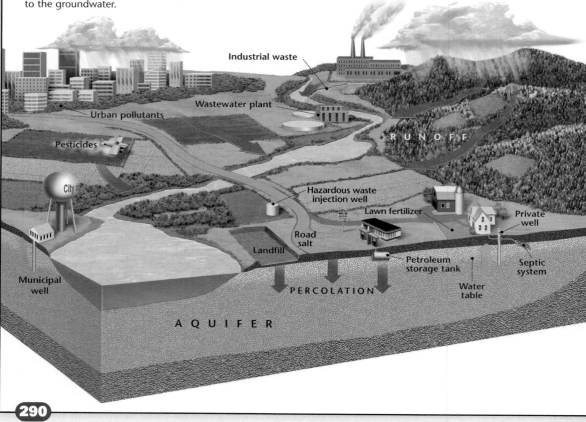

Figure 20 ▶ This diagram shows some of the major sources of groundwater pollution. Runoff and percolation transport contaminants to the groundwater.

Homework — ADVANCED

Consumer Focus: Bottled Water According to the USDA, the use of bottled water in the United States has increased 380 percent since 1980. Many people now rely exclusively on bottled water to meet their drinking water needs. But there are no federal standards for drinking water purity. Have students form groups to investigate the bottled water that is available in your community. Each group should have a specific task. One group should investigate the source of each type of bottled water. The group should find out whether the source is a spring, a well, a reservoir, or a drinking water treatment plant. A second group should investigate the quality of the drinking water. Have these students use microscopes or hand lenses to observe particulates and organisms in the water. This group can also test for total solids or phosphates. The third group should research in periodicals and on the Internet to find current information about the quality of bottled water. Have the class share the data with the school. **LS Interpersonal** Co-op Learning

Figure 21 ▶ **Major North American Oil Spills**

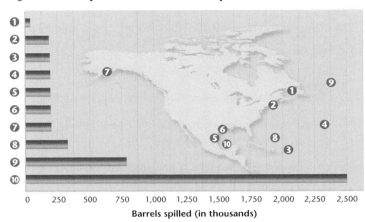

❶ *Kurdistan* Gulf of St. Lawrence, Canada, 1979

❷ *Argo Merchant* Nantucket, MA, 1976

❸ Storage Tank Benuelan, Puerto Rico, 1978

❹ *Athenian Venture* Atlantic Ocean, 1988

❺ *Unnamed Tanker* Tuxpan, Mexico, 1996

❻ *Burmah Agate* Galveston Bay, TX, 1979

❼ *Exxon Valdez* Prince William Sound, AK, 1989

❽ *Epic Colocotronis* Caribbean Sea, 1975

❾ *Odyssey* North Atlantic Ocean, 1988

❿ Exploratory Well Bay of Campeche, 1979

Ocean Pollution

Although oceans are the largest bodies of water on Earth, they are still vulnerable to pollution. Pollutants are often dumped directly into the oceans. For example, ships can legally dump wastewater and garbage overboard in some parts of the ocean. But at least 85 percent of ocean pollution—including pollutants such as oil, toxic wastes, and medical wastes—comes from activities on land. If polluted runoff enters rivers, for example, the rivers may carry the polluted water to the ocean. Most activities that pollute oceans occur near the coasts, where much of the world's human population lives. As you might imagine, sensitive coastal ecosystems, such as coral reefs, estuaries, and coastal marshes, are the most affected by pollution.

Oil Spills Ocean water is also polluted by accidental oil spills. Disasters such as the 1989 *Exxon Valdez* oil spill in Prince William Sound, Alaska, make front-page news around the world. In 2001, a fuel-oil spill off the coast of the Galápagos Islands captured public attention. Each year, approximately 37 million gallons of oil from tanker accidents are spilled into the oceans. **Figure 21** shows some of the major oil spills that occurred off the coast of North America in the last 30 years.

Such oil spills have dramatic effects, but they are responsible for only about 5 percent of oil pollution in the oceans. Most of the oil that pollutes the oceans comes from cities and towns. Every year, as many as 200 million to 300 million gallons of oil enter the ocean from nonpoint sources on land. That's almost 10 times the amount of oil spilled by tankers. In fact, in one year, the road runoff from a coastal city of 5 million people could contain as much oil as a tanker spill does. Limiting these nonpoint sources of oil pollution would go a long way toward keeping the oceans clean.

Ecofact

Cruise Ship Discharges In one year, ships dump almost 7 billion kilograms of trash into the ocean. About 75 percent of all ship waste comes from cruise ships. According to most international law, cruise ships are allowed to dump non-plastic waste—including untreated sewage—into the ocean. Increasing public pressure has begun to cause the cruise-ship industry to change this practice, however.

▣ internet connect

www.scilinks.org
Topic: Water Pollution
SciLinks code: HE4122

SCiLINKS. Maintained by the National Science Teachers Association

---

Using the Figure — **ADVANCED**

Oil Spills Some of the major oil spills that have occurred off the coast of North America are shown in **Figure 21.** Have students form groups to research the major oil spills that have occurred off the coast of each continent. As a class, create a world map of major oil spills that have occurred in the past 30 years. Students can add to the map by showing major oil transportation routes, offshore drilling sites, and areas that are environmentally sensitive. Co-op Learning

Discussion — **GENERAL**

Pollution From Ships The dumping of waste by ships and offshore operations remains a serious problem. Ask students to suggest ways of reducing waste dumping in the open ocean. (Students may suggest requiring ships to have certified waste-handling systems, an inventory system in which waste items must be accounted for, a system of unannounced spot inspections of ships and their waste-handling systems, a designated waste officer who is responsible for reporting to regulatory authorities, or severe penalties for violators.) **LS** Verbal

Transparencies
TT Major North American Oil Spills

Student Opportunities — **GENERAL**

Beach Sweeps Students who live in coastal communities may have heard about or participated in beach sweeps. These are periodic events during which beaches are cleaned of litter and debris. Ask students to consider the following question: "The beaches of the Gulf of Mexico are easily polluted for a number of reasons; what might the reasons be?" Have students examine a map to answer the question. (The Gulf of Mexico is ringed by highly populated areas; many rivers drain into it; and it has a very large volume of ships. Some students may also mention oil-drilling activity. Any floatable object that falls or is dropped into the Gulf stands a good chance of winding up on a beach.) Remind students that recent agreements ban or restrict the disposal of non degradable waste by ships at sea. However, there is currently no way to effectively enforce these bans and restrictions.

Group Activity —— BASIC

Demonstrating Biomagnification
Divide the class into the trophic levels of a food chain. Ten students can serve as the back row (plankton). Give the back row two paper clips each. The next row (shrimp) will include five students. The shrimp receive four paper clips each from the plankton. In the next row (small fish), three students get the paper clips from the shrimp (two fish will get seven paper clips). In the next row (big fish), two students receive the paper clips from the small fish (one big fish will eat two small fish). The teacher (osprey) receives all the paper clips.

Co-op Learning   English Language Learners

Demonstration —— GENERAL

Deadly Six Packs Provide each student with a plastic loop cut from a six-pack of drinks. Have students slide the ring over their tightly closed fingers (excluding the thumb). The band should be just slightly below the first knuckles, where the fingers join the hand. Tell students that this simulates what happens when these rings get caught around the head of a bird, turtle, seal, or fish. Ask students to remove the loop without using the thumb or the other hand. Their struggle simulates the ordeal that marine animals face when they become entangled in debris. Stimulate discussion by asking, "Will the ring break down in water? What places on an animal's body could the ring become attached?" **LS Kinesthetic**

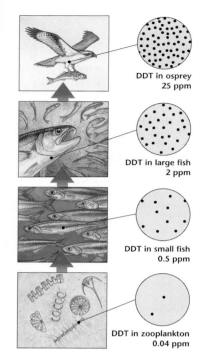

Figure 22 ▶ The accumulation of pollutants at successive levels of the food chain is called biomagnification.

DDT in osprey 25 ppm

DDT in large fish 2 ppm

DDT in small fish 0.5 ppm

DDT in zooplankton 0.04 ppm

Figure 23 ▶ The Cuyahoga River was so polluted with petroleum and petroleum byproducts that it caught on fire and burned in 1969.

Water Pollution and Ecosystems

Water pollution can cause immediate damage to an ecosystem. For example, toxic chemicals spilled directly into a river can kill nearly all living things for miles downstream. But the effects of water pollution can be even more far reaching. Many pollutants accumulate in the environment because they do not decompose quickly. As the pollutant levels increase, they can threaten an entire ecosystem.

Consider a river ecosystem. Soil tainted with pesticides washes into the river and settles to the river bottom. Some of the pesticides enter the bodies of tiny, bottom-dwelling organisms, such as insect larvae and crustaceans. A hundred of these organisms are eaten by one small fish. A hundred of these small fish are eaten by one big fish. A predatory bird, such as an eagle, eats 10 big fish. Each organism stores the pesticide in its tissues, so at each step along the food chain, the amount of the pesticide passed on to the next organism increases. This accumulation of pollutants at successive levels of the food chain is called **biomagnification.** Biomagnification, which is illustrated in **Figure 22,** has alarming consequences for organisms at the top of the food chain. Biomagnification is one reason why many U.S. states limit the amount of fish that people can eat from certain bodies of water.

Cleaning Up Water Pollution

In 1969, the Cuyahoga River in Cleveland, Ohio, was so polluted that the river caught on fire and burned for several days, as shown in **Figure 23.** This shocking event was a major factor in the passage of the Clean Water Act of 1972. The stated purpose of the act was to "restore and maintain the chemical, physical, and biological integrity of the nation's waters." The goal of the act was to make all surface water clean enough for fishing and swimming by 1983. This goal was not achieved; however, much progress has been made since the act was passed. The percentage of lakes and rivers that are fit for swimming and fishing has increased by about 30 percent, and many states have passed stricter water-quality standards of their own. Many toxic metals are now removed from wastewater before the water is discharged.

The Clean Water Act opened the door for other water-quality legislation, some of which is described in **Table 6.** For example, the Marine Protection, Research, and Sanctuaries Act of 1972 strengthened the laws against ocean dumping.

The Oil Pollution Act of 1990 requires all oil tankers traveling in U.S. waters to

READING SKILL BUILDER —— GENERAL

Paired Summarizing Have students work with a partner and take turns reading and summarizing the laws described in **Table 6.** Each pair should come up with a written description of the laws in their own words. Student pairs should also offer examples of situations in which each law could be applied.

Homework —— ADVANCED

Art Project Tell students to imagine that the United States Postal Service has decided to issue a series of stamps that will encourage people to honor and respect the world's oceans. The Postal Service has commissioned your class to provide the designs. Students may design a single stamp or a whole series, and they may communicate the design either in writing or with sketches. Suggest that students include their designs in their **Portfolio.** **LS Visual/Intrapersonal**

Table 6 ▼

Federal Laws Designed to Improve Water Quality in the United States
1972 Clean Water Act (CWA) The CWA set a national goal of making all natural surface water fit for fishing and swimming by 1983 and banned pollutant discharge into surface water after 1985. The act also required that metals be removed from wastewater.
1972 Marine Protection, Research, and Sanctuaries Act, amended 1988 This act empowered the EPA to control the dumping of sewage wastes and toxic chemicals in U.S. waters.
1975 Safe Drinking Water Act (SDWA), amended 1996 This act introduced programs to protect groundwater and surface water from pollution. The act emphasized sound science and risk-based standards for water quality. The act also empowered communities in the protection of source water, strengthened public right-to-know laws, and provided water system infrastructure assistance.
1980 Comprehensive Environmental Response Compensation and Liability Act (CERCLA) This act is also known as the Superfund Act. The act makes owners, operators, and customers of hazardous waste sites responsible for the cleanup of the sites. The act has reduced the pollution of groundwater by toxic substances leached from hazardous waste dumps.
1987 Water Quality Act This act was written to support state and local efforts to clean polluted runoff. It also established loan funds to pay for new wastewater treatment plants and created programs to protect major estuaries.
1990 Oil Pollution Act This act attempts to protect U.S. waterways from oil pollution by requiring that oil tankers in U.S. waters be double-hulled by 2015.

FIELD ACTIVITY

Coastal Cleanups You can be a part of a coastal cleanup. Every September, people from all over the world set aside one day to help clean up debris from beaches. You can join this international effort by writing to The Center for Marine Conservation.

If you do participate in a coastal cleanup, keep a record of the types of trash you find in your *EcoLog*.

have double hulls by 2015 as an added protection against oil spills. Legislation has improved water quality in the United States, but the cooperation of individuals, businesses, and the government will be essential to maintaining a clean water supply in the future.

SECTION 3 Review

1. **Explain** why point-source pollution is easier to control than nonpoint-source pollution.

2. **List** the major types of water pollutants. Suggest ways to reduce the levels of each type of pollutant in a water supply.

3. **Describe** the unique problems of cleaning up groundwater pollution.

4. **Describe** the source of most ocean pollution. Is it point-source pollution or nonpoint-source pollution?

CRITICAL THINKING

5. **Interpreting Graphics** Read the description of biomagnification. Draw a diagram that shows the biomagnification of a pollutant in an ecosystem. **READING SKILLS**

6. **Applying Ideas** What can individuals do to decrease ocean pollution? Write and illustrate a guide that gives at least three examples. **WRITING SKILLS**

293

Close

Reteaching ——— BASIC

Pollutant Types As a class, review the major pollutant types and sources. Then have students form pairs and work together to come up with as many examples of each pollutant type and source as possible. As a class, compile a list of all of the examples, and evaluate each pair's findings.

Quiz ——— GENERAL

1. Is acid precipitation an example of point-source or nonpoint-source pollution? Explain why. (Acid precipitation is nonpoint pollution because it is caused by the accumulation of pollutants in the atmosphere from several different sources.)

2. Describe three laws designed to protect domestic water quality. (Answers may vary.)

Alternative Assessment ——— GENERAL

Controlling Nonpoint Pollution Tell students to imagine that they have been appointed to a blue-ribbon committee that is studying the problem of nonpoint pollution. Each member of the committee has been requested to provide at least five recommendations for controlling the problem of nonpoint pollution.

Answers to Section Review

1. Point-source pollution is relatively easy to control because the source of the pollutants is known. Nonpoint-source pollution comes from many different sources, so the sources are more difficult to identify.

2. The major types of water pollutants are pathogens, organic matter, organic chemicals, inorganic chemicals, heavy metals, and physical agents. Suggestions may vary. Pathogen levels can be reduced by improving sanitation and treating water. The levels of organic matter can be reduced by preventing the matter from entering a water supply and by using filtration. The levels of organic and inorganic

chemicals can be reduced by using less of them and by ensuring that they do not enter a water supply. The levels of heavy metals can be reduced by ensuring that industrial wastes are disposed of properly or recycled. Two ways to reduce pollution from physical agents include allowing water to cool to an acceptable level before discharging it and maintaining ground cover to reduce soil erosion.

3. Answers may vary. Groundwater pollution is difficult to clean up because groundwater recharges very slowly and because pollutants cling to the materials that make up an aquifer.

4. Most ocean pollution comes from nonpoint sources on land.

5. Answers may vary. Students should indicate that the concentration of a pollutant increases in successive levels of the food chain.

6. Answers may vary. Individuals can reduce the amount of pollution on land, particularly near coastal areas. Individuals could also support legislation intended to reduce ocean pollution or volunteer in a coastal cleanup.

Alternative Assessment — GENERAL

Profiling Local Water Resources

Have students create a three-part folder that profiles the water resources of your community. Students should use the information in each section to describe your community's water resources. When possible, have students create figures to illustrate their points. For example, students could create a map that shows the water resources of your county. Students can also create a diagram to show drinking water treatment in your community. Students can describe how your community uses water, and what potential sources of pollution threaten the water supply in your area. Students should also describe how wastewater is treated.

Chapter Resource File

- **Chapter Test** GENERAL
- **Chapter Test** ADVANCED
- **Concept Review** GENERAL
- **Critical Thinking** ADVANCED
- **Test Item Listing**
- **Design Your Own Lab** GENERAL
- **Modeling Lab** GENERAL
- **CBL Probeware™ Lab** GENERAL
- **Consumer Lab** BASIC
- **Long-Term Project** ADVANCED

CHAPTER 11 Highlights

1 Water Resources

Key Terms

surface water, 270
river system, 271
watershed, 271
groundwater, 272
aquifer, 272
porosity, 273
permeability, 273
recharge zone, 274

Main Ideas

▶ Only a small fraction of Earth's water supply is fresh water. The two main sources of fresh water are surface water and groundwater.

▶ River systems drain the land that makes up a watershed. The amount of water in a river system can vary in different seasons and from year to year.

▶ Groundwater accumulates in underground formations called *aquifers*. Surface water enters an aquifer through the aquifer's recharge zone.

▶ If the water in an aquifer is pumped out faster than it is replenished, the water table drops, which can affect humans and animals that depend on the groundwater.

2 Water Use and Management

Key Terms

potable, 277
pathogen, 277
irrigation, 278
dam, 280
reservoir, 280
desalination, 283

Main Ideas

▶ There are three main types of water use: residential, industrial, and agricultural. Worldwide, most water use is agricultural.

▶ Dams and water diversion projects are built to manage surface-water resources. Damming and diverting rivers can have environmental and social consequences.

▶ Water conservation is necessary to maintain an adequate supply of fresh water. Desalination and transporting water are options to supplement local water supplies.

3 Water Pollution

Key Terms

water pollution, 284
point-source pollution, 285
nonpoint-source pollution, 285
wastewater, 286
artificial eutrophication, 288
thermal pollution, 289
biomagnification, 292

Main Ideas

▶ Water can become polluted by chemical, physical, or biological agents. Most water pollution in the United States is caused by nonpoint-source pollutants.

▶ Groundwater pollution is difficult to clean up because aquifers recharge slowly and because pollutants cling to the materials that make up an aquifer.

▶ Ocean pollution is mainly caused by coastal, nonpoint-source pollutants.

▶ Government legislation, such as the Clean Water Act of 1972, has succeeded in reducing surface-water pollution. Future challenges include reducing nonpoint-source pollution and groundwater pollution.

294

Using Key Terms

Use each of the following terms in a separate sentence.

1. *aquifer*
2. *recharge zone*
3. *irrigation*
4. *wastewater*
5. *biomagnification*

For each pair of terms, explain how the meanings of the terms differ.

6. *surface water* and *groundwater*
7. *porosity* and *permeability*
8. *watershed* and *river system*
9. *point-source pollution* and *nonpoint-source pollution*

✓ **STUDY TIP**

Root Words To practice vocabulary, write the key terms and definitions on a piece of paper and fold the paper lengthwise so that the definitions are covered. First, see how many definitions you already know. Then, write the definitions you do not know on another piece of paper, and practice until you know all of the terms.

Understanding Key Ideas

10. Which of the following processes is *not* a part of the water cycle?
    a. evaporation
    b. condensation
    c. biomagnification
    d. precipitation

11. Most of the fresh water on Earth is
    a. located underground in aquifers.
    b. frozen in the polar icecaps.
    c. located in rivers, lakes, streams, and wetlands.
    d. found in Earth's atmosphere.

12. Which of the following processes is *not* used in a conventional method of water treatment?
    a. filtration
    b. coagulation
    c. aeration
    d. percolation

13. Which of the following is *not* an example of point-source pollution?
    a. oil that is escaping from a damaged tanker
    b. heavy metals that are leaching out of an underground mine
    c. water runoff from residential lawns
    d. untreated sewage that is accidentally released from a wastewater treatment plant

14. Which of the following pollutants causes artificial eutrophication?
    a. heavy metals from unlined landfills
    b. inorganic plant nutrients from wastewater and fertilizer runoff
    c. toxic chemicals from factories
    d. radioactive waste from nuclear power plants

15. Pumping large amounts of water from an aquifer may cause the
    a. water table to rise.
    b. recharge zone to shrink.
    c. wells in an area to run dry.
    d. percolation of groundwater to stop.

16. Oil pollution in the ocean is mostly caused by
    a. major oil spills, such as the 1989 *Exxon Valdez* oil spill.
    b. the cumulative effect of small oil spills and leaks on land.
    c. decomposed plastic materials.
    d. intentional dumping of excess oil.

17. Thermal pollution has a harmful effect on aquatic environments because
    a. water has been circulated around power-plant generators.
    b. it increases the number of disease-causing organisms in aquatic environments.
    c. it reduces the amount of dissolved oxygen in aquatic environments.
    d. it decreases the nutrient levels in aquatic environments.

295

Assignment Guide

Section	Questions
1	1, 2, 6–8, 10, 11, 15, 18, 26, 30, 32
2	3, 12, 19, 20, 28, 29, 31, 32, 36
3	4, 5, 9, 13, 14, 16, 17, 21, 22, 27, 32, 35

ANSWERS

Using Key Terms

1. Sample answer: An aquifer is an underground formation that contains a large volume of groundwater.

2. Sample answer: The recharge zone is the area of land above an aquifer through which surface water percolates.

3. Sample answer: Irrigation is a method of supplying water to plants from sources other than direct precipitation.

4. Sample answer: Wastewater is water that contains wastes from homes or industry.

5. Sample answer: Biomagnification is the process by which pollutants become concentrated at successive levels of the food chain.

6. Surface water includes all above-ground sources, such as lakes, ponds, rivers, oceans, and streams. Groundwater is water in underground sources, such as aquifers.

7. Porosity is the amount of space between the particles that make up a rock. Permeability is a measure of the ability of rock or soil to allow water to flow through it.

8. A watershed is the region of land drained by a river system. A river system is a flowing network of rivers and streams that drain a watershed.

9. Point-source pollution comes from a single, identifiable source. Nonpoint-source pollution comes from many sources.

Understanding Key Ideas

10. c
11. b
12. d
13. c
14. b
15. c
16. b
17. c

CHAPTER **11** Review

Short Answer

18. The construction of buildings and parking lots can cover the land surface with an impermeable layer. This layer can limit the amount of water that reaches an aquifer.

19. Overhead sprinklers tend to spray a lot of water into the air, which causes a large percentage of the water to be lost to evaporation. Drip irrigation is a more efficient method.

20. Answers may vary. Dams provide electricity and drinking water, and they reduce flooding. Dams can also disrupt entire ecosystems, deprive land of fertile sediments, and displace people.

21. The process of eutrophication occurs as nutrient levels build up in a body of water and the population of decomposers increases greatly. As the decomposers increase, they consume more and more of the dissolved oxygen in the water. Decreasing oxygen levels can kill other organisms that live in the water. Fertilizer from farms, lawns, and gardens, as well as phosphate detergents released from households will accelerate eutrophication.

22. In primary treatment, wastewater is filtered and passed to a settling tank. In secondary treatment, wastewater is filtered again, and then bacteria and oxygen are added to the water. In the final stage, chlorine is added to disinfect the water.

Short Answer

18. What effect can buildings and parking lots have on an aquifer's recharge zone?

19. Why is the use of overhead sprinklers for irrigation inefficient? What is a more efficient method of irrigation?

20. List three advantages and three disadvantages of dams.

21. What is the process of eutrophication, and how do human activities accelerate it?

22. Describe the steps that are involved in the primary and secondary treatment of wastewater.

Interpreting Graphics

The graph below shows the annual flow, or discharge, of the Yakima River in Washington. Use the graph to answer questions 23–25.

23. In which months is the river's discharge highest? What might explain these discharge rates?

24. What might cause the peaks in river discharge between November and March?

25. How might the data be different if the hydrograph readings were taken below a large dam on the Yakima River?

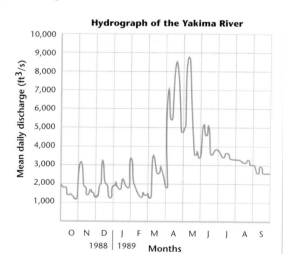

Hydrograph of the Yakima River

Mean daily discharge (ft³/s) vs. Months

O N D | J F M A M J J A S
1988 | 1989

Concept Mapping

26. Use the following terms to create a concept map: *Earth's surface, rivers, underground, fresh water, water table, 3 percent,* and *icecaps.*

Critical Thinking

27. **Making Comparisons** Read the description of artificial eutrophication in this chapter. Do you think artificial eutrophication is more disturbing to the stability of a water ecosystem than natural eutrophication is? **READING SKILLS**

28. **Analyzing Relationships** Water resources are often shared by several countries. A river, for example, might flow through five countries before it reaches an ocean. When water resources are shared, how should countries determine water rights and environmental responsibility?

29. **Making Inferences** Explain why it takes 36 gallons of water to produce a single serving of rice, but it takes more than 2,000 gallons of water to produce a single serving of steak. What do you think the water is used for in each case?

30. **Making Inferences** Why is there so little fresh water in the world? Do you think that there would have been more fresh water at a different time in Earth's history?

Cross-Disciplinary Connection

31. **Social Studies** Find out how freshwater resources affected the development of one culture in history. Use at least five key terms from this chapter to write a two-paragraph description of how the availability of fresh water affected the culture you chose. **WRITING SKILLS**

Portfolio Project

32. **Investigation** Find out about the source of the tap water in your home. Where does the tap water come from, and where does your wastewater go? Does the water complete a cycle? Make a poster to illustrate your findings. You may want to work with several classmates and visit the sites you discover.

Interpreting Graphics

23. The river's discharge is highest in April and May. April and May could be the rainy season.

24. Answers may vary. Spring snowmelt and high amounts of rainfall might affect the river's discharge.

25. Answers may vary. The dam would probably regulate the flow of the river so that there would be less variation in discharge.

Concept Mapping

26. Answers to the concept-mapping questions are on pp. 467–472.

Critical Thinking

27. Answers may vary. Natural eutrophication is a slow, stable process that allows for gradual changes in the types of plants and animals that live in a body of water. Because artificial eutrophication is an accelerated process, the body of water and the organisms that live in it do not have time to adapt to the new conditions. Thus, artificial eutrophication can be more disturbing to the stability of an ecosystem.

28. Answers may vary.

29. Answers may vary.

30. Answers may vary.

MATH SKILLS

The graph below illustrates the pumping rates for a set of wells that provide water to a small community. Use the graph to answer question 33.

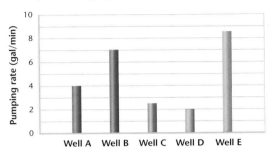

33. Analyzing Data How many gallons does Well B pump per day? What is the average pump rate for all of the wells? In one hour, how many more gallons of water will Well A pump than Well C?

34. Making Calculations If placing a container of water in your toilet tank reduces the amount of water per flush by 2 L, how much water would be saved each day if this were done in 80 million toilets? (Assume that each toilet is flushed five times per day.) Convert your answer into gallons (1 L = .26 Gal).

 WRITING SKILLS

35. Communicating Main Ideas Why is water pollution a serious problem?

36. Writing Persuasively Write a letter to a senator in which you voice your support or criticism of a hypothetical water diversion project.

 STANDARDIZED TEST PREP

For extra practice with questions formatted to represent the standardized test you may be asked to take at the end of your school year, turn to the sample test for this chapter in the Appendix.

 READING SKILLS

Read the passage below, and then answer the questions that follow.

Water use is measured in two ways: by withdrawal and by consumption. Withdrawal is the removal and transfer of water from its source to a point of use, such as a home, business, or industry. Most of the water that is withdrawn is eventually returned to its source. For example, much of the water used in industries and in homes is treated and returned to the river or lake it came from. When water is withdrawn and is not returned to its source, the water is consumed. For example, when a potted plant is watered, almost all of the water eventually enters the atmosphere by *evapotranspiration* through the leaves of the plant. The evaporated water was consumed because it was not directly returned to its source.

1. According to the passage, which of the following statements is true?
  a. Water that is consumed was never withdrawn.
  b. Water that is withdrawn cannot be consumed.
  c. A fraction of the water withdrawn is usually consumed.
  d. All of the water withdrawn is consumed.

2. Which phrase best describes the meaning of the term *evapotranspiration*?
  a. the absorption of water by plant leaves
  b. the process by which potted plants transpire their leaves by evaporation
  c. the process by which the atmosphere maintains water levels in plant leaves
  d. the process by which water evaporates from plant leaves

3. Which of the following statements is an example of consumption?
  a. A river is diverted to irrigate crops.
  b. A power plant takes in cool water from a lake and returns the water to the lake.
  c. A dam forms a reservoir on a river.
  d. An aquifer is recharged by surface water.

Cross-Disciplinary Connection

31. Answers may vary. Students could research the Maya, the Romans, the early Mesopotamian civilizations, the ancient Egyptians, or the Inca.

Portfolio Project

32. Answers may vary.

Math Skills

33. In 1 min pump B will provide 7 gal. In 1 h, pump B will provide 420 gal. In 1 day, the well will provide 10,080 gal. The average pump rate for all the wells is 4.8 gal/min. In 1 h, well A will provide 90 more gallons of water.

34. 2 L × 5 flushes/day = 10 L/day
10 L × 80,000,000 = 800,000,000 L
800,000,000 L × .26 = 208,000,000 gal.

Writing Skills

35. Answers may vary. Everyone depends on water resources, so if they become polluted, everyone suffers.

36. Answers may vary.

Reading Skills

  1. c
  2. d
  3. a

297

GROUNDWATER FILTERS

Teacher's Notes

Time Required
one 45-minute class period

Lab Ratings

EASY ——————→ HARD

TEACHER PREPARATION 🧪🧪
STUDENT SETUP 🧪
CONCEPT LEVEL 🧪🧪
CLEANUP 🧪🧪

Skills Acquired
- Predicting
- Constructing Models
- Measuring
- Communicating

The Scientific Method
In this lab, students will:
- Make Observations
- Analyze the Results
- Draw Conclusions
- Communicate Results

Materials
The materials provided are enough for groups of four students. You will need to poke holes in the soda bottle caps beforehand. To prepare the glucose solution, add one glucose tablet to 200 mL of water. (Tablets and test paper are available at most pharmacies.) Large empty jars may be substituted for beakers. To prepare other contaminant solutions, use 5 mL of cooking oil, vinegar, or liquid laundry detergent per 500 mL of water.

Objectives

▶ **Construct** a model of the Earth's natural groundwater filtering system.

▶ **USING SCIENTIFIC METHODS** **Test** the ability of your groundwater filters to filter contaminants out of different solutions.

Materials

beakers, 750 mL (5)
glucose solution
glucose test paper
graduated cylinder
gravel
metric ruler
soda bottles, 2 to 3L (4)
red food coloring
sand
soil
stirring rod
wax pencil

optional contaminants:
cooking oil, detergent, fertilizer, vinegar, soda

optional filter materials:
alum, charcoal

▶ **Filter Apparatus** Your ground filtration models should be layered as shown here.

Groundwater Filters

As surface water travels downward through rock and soil, the water is filtered and purified. As a result, the water in aquifers is generally cleaner than surface water. In this investigation, you will work in small teams to explore how layers of the Earth act as a filter for groundwater. You will make models of the Earth's natural filtration system and test them to see how well they filter various substances.

Procedure

1. Label four beakers as follows: "Contaminant: glucose," "Contaminant: soil," "Contaminant: food coloring," and "Water (control)."

2. Fill these beakers two-thirds full with clean tap water. Then add to each beaker the contaminant listed on its label. (The table on the next page shows how much of each contaminant you should use.) Stir each mixture thoroughly.

3. Copy the data table into your notebook. Carefully observe each beaker, and record your observations. Use some of the glucose test paper to test the glucose level in the glucose beaker.

4. Make four separate filtration systems similar to the one shown below. Your teacher will provide you with bottle caps that have holes poked through them. Fasten each cap to a bottle. Cut the bottom off of each soda bottle, and fill each bottle with layers of gravel, sand, and soil. Consider using the optional filter materials, such as alum or charcoal, but be sure to make each model identical to the next.

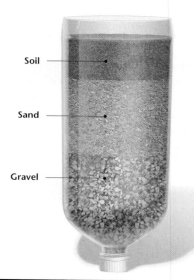

Soil

Sand

Gravel

Ammonia can also be used in the same concentration. Use 3 tablespoons of powdered laundry detergent or 2 tablespoons of high-nitrogen-content fertilizer per 500 mL of water (mix the night before if possible). To detect the contaminants, use pH paper (turns red in vinegar and blue in fertilizer and ammonia). You can detect detergent by covering the container and gently shaking it to produce suds. Cooking oil should be visible in the filtrate.

Safety Cautions
Contact your local Environmental Protection Agency, Department of Waste Management, or Health Department to find out how to properly dispose of contaminants. Do **NOT** allow students to pour contaminated water down the drain or into the garbage.

Observations of Substances in Surface Water		
Contaminant	Before filtration	After filtration
Glucose (15 mL)		
Soil (15 mL)	DO NOT WRITE IN THIS BOOK	
Food coloring (15 drops)		
Water (control)		

5. You are now going to pour each mixture through a filtration system. But first predict how well the filters will clean each water sample. Write your predictions in your notebook.

6. Stir a contaminant mixture in its beaker, and immediately pour the mixture through a filtration system into a clean beaker. Observe the resulting "groundwater," and record your observations in the table you created. CAUTION: Do not taste any of the substances you are testing.

7. Repeat this procedure for each mixture. Clean and relabel the contaminant beakers as you go along.

Analysis

1. **Analyzing Results** Test the glucose-water mixture for the presence of glucose. Can you see the glucose?

2. **Analyzing Results** Was the soil removed from the water by filtering? Was the food coloring removed? How do you know?

Conclusions

3. **Drawing Conclusions** How accurate were your initial predictions?

4. **Drawing Conclusions** What conclusions can you draw about the filtration model and the materials you used?

Extension

1. **Making Predictions** Choose a substance from the materials list that has not been tested. Predict what will happen if you mix this substance in the water supply.

2. **Evaluating Results** Now test your prediction. Use the filter that was the control in the earlier experiment. How did your results compare with your prediction?

3. **Analyzing Results** Compare your results with the results of other teams. What precautions do you recommend for keeping groundwater clean?

▶ **Step 6** Pour each sample of contaminated surface water through a filter.

299

Chapter Resource File
• Datasheets for In-Text Labs
• Lab Notes and Answers

CLASSROOM TESTED & APPROVED

Arthur Goldsmith
Hallandale High School
Hallandale, Florida

Tips and Tricks

You may consider making the cooking oil a required substance in this lab. If you do, you can demonstrate how 1 qt of motor oil can contaminate hundreds of thousands of gallons of water.

Answers to Analysis

1. Glucose is present. It can't be seen, but it is detectable with the test paper.

2. Most of the soil was removed; most of it is no longer visible. Food coloring was not removed; it is still visible.

Answers to Conclusions

3. Answers may vary.

4. Answers may vary. Students should note that food coloring and glucose (substances dissolved in the water) pass through the filter. They may conclude from this observation that any hazardous chemicals that dissolve in water pose a threat to our groundwater.

Answers to Extension

1. Answers may vary. Students should note that substances that dissolve in water will probably not be filtered.

2. Answers may vary.

3. Answers may vary. Water is an excellent solvent, so harmful substances, such as fertilizers, insecticides, and hazardous wastes, can potentially be carried by water through the ground. Precautions for keeping groundwater clean include recycling motor oil instead of dumping it, minimizing the use of fertilizers and pesticides, and protecting groundwater recharge zones from pollution and hazardous materials.

THE THREE GORGES DAM

HISTORY
CONNECTION — GENERAL

Dams in the United States

Settlement of the American West followed the taming of its rivers. Damming and controlling rivers created usable farmland, allowed settlement in flood plains, and powered industry. Beginning in the 1930s, massive hydroelectric dams were built across many western rivers. Following this example, many countries have begun to dam their rivers to provide water for irrigation and drinking, to control flooding, and to produce electricity. While dams provide many benefits, large dam projects face increasing criticism for numerous social, environmental, economic, and safety concerns. Around the world, as many as 50 million people have been displaced by dams. Many have also suffered from the downstream effects of dams: farmland deprived of flood sediments and water for irrigation becomes less productive, fisheries become less productive, and epidemics of waterborne disease often follow large dam projects. Although hydropower is hailed as a cheap, nonpolluting source of energy, the true cost and impact of hydropower is often great. Ask students, "Should projects such as China's Three Gorges Dam be stopped? After America has benefited so much from dam projects, can we ask developing countries not to exploit their water resources?" Encourage students to find out more and debate these issues in class.

POINTS of view

THE THREE GORGES DAM

China's Yangtze River is the third-longest river in the world after the Nile and the Amazon. The Yangtze River flows through the Three Gorges region of central China, which is famous for its natural beauty and historical sites. For thousands of years, the area's sheer cliffs have inspired paintings and poems. This idyllic region seems like the sort of place that would be protected as a park or reserve. But in fact, it is the construction site for the Three Gorges Dam—the largest hydroelectric dam project in the world. When the Yangtze River is dammed, it will rise to form a reservoir that is 595 km (370 mi) long—as long as Lake Superior. In other words, the reservoir will be about as long as the distance between Los Angeles and San Francisco!

Benefits of the Dam

The dam has several purposes. It will control the water level of the Yangtze River to prevent flooding.

About 1 million people died in the last century from flooding along the river. The damage caused by a severe flood in 1998 is estimated to cost as much as the entire dam project. The dam will also provide millions of people with hydroelectric power. China now burns air-polluting coal to meet 75 percent of the country's energy needs. Engineers project that when the dam is completed, its turbines will provide enough electrical energy to power a city that is 10 times the size of Los Angeles, California. When the Yangtze's flow is controlled, the river will be deep enough for large ships to navigate on it, so the dam will also increase trade in a relatively poor region of China.

Some Disadvantages

The project has many drawbacks, however. The reservoir behind the dam will flood an enormous area. Almost 2 million people living in

▶ **The Three Gorges Dam** is named for the beautiful canyons it will flood. When completed, the dam may meet 20 percent of China's energy needs with hydroelectric power.

the affected areas must be relocated—there are 13 cities and hundreds of villages in the area of the proposed reservoir. As the reservoir's waters rise, they will also destroy fragile ecosystems and valuable archeological sites.

Opponents of the project claim that the dam will increase pollution levels in the Yangtze River. Most of the cities and factories along the river dump untreated wastes directly into the water. Some people think the reservoir will become the world's largest sewer when 1 billion tons of sewage flow into the reservoir every year.

Long-Term Concerns

People have also raised long-term concerns about the project. The dam is being built over a fault line. Scientists question whether the dam would be able to withstand earthquakes that may occur along the fault. If the dam burst, towns and

▶ The reservoir that will form behind the Three Gorges Dam is shown in yellow.

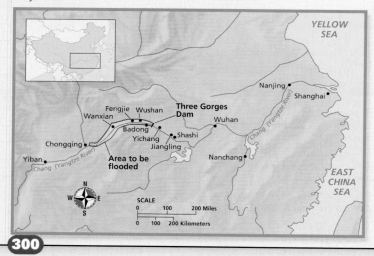

🖅 Internet Activity — GENERAL

Researching the Three Gorges Dam Have students write down five questions that they would like to answer about the Three Gorges Dam. Have students research on the Internet to answer their questions and to find out the current status of the Three Gorges Dam.

cities downstream would be flooded. Another concern is that the dam may quickly fill with sediment. The Yangtze picks up enormous amounts of yellowish soil and sediment as it flows across China. When the river is slowed by the dam, much of the silt will be deposited in the new reservoir. As sediment builds up behind the dam, the deposited sediment will reduce the size of the reservoir—limiting the flood-prevention capacity of the dam. In addition, productive farming regions below the dam will be deprived of the fertile sediment that is deposited every year when the river floods.

The enormous reservoir may also cause disease among the local population. The potential heath risks include an increase in encephalitis and malaria. The most deadly disease spread by the Three Gorges Dam could be a parasitic disease called schistosomiasis.

Hidden Costs?

Supporters of the dam claim that the project will cost $25 billion, while opponents claim that the costs will be closer to $75 billion. The true cost of the dam may never be known because corruption and inefficiency have plagued the project from the start. Controversy over the dam has prompted the U.S. government and the World Bank to withhold money for the project. Public opposition to the project has been silenced since the Tiananmen Square crackdown. But with help from private investment companies from the United States, the Chinese government is continuing with the project, and the dam is slowly being built. The world's third-longest river will soon swell in the middle, and China will change along with it.

▶ When the dam waters rise, these ancient temples will be flooded.

▶ Engineers discuss plans at the dam construction site. More than 20,000 people are working at the construction site.

What Do You Think?

Hundreds of dams in the western United States provide electrical energy, drinking water, and water for crops, but the dams also flooded scenic canyons and destroyed ecosystems. Now that the environmental consequences of large dams are known, do you think that China should reconsider the Three Gorges Dam project?

301

Dam Projects Currently, 600,000 mi of rivers in the United States lie behind an estimated 60,000 to 80,000 dams. One-third of the dams in North America are considered multipurpose dams and are designed to do a combination of some or all of the following: control flooding, provide hydroelectric power, provide water for irrigation, provide recreation, and contribute to the water supply. In the United States, one of the biggest debates over a dam is about the existence of Glen Canyon Dam in northern Arizona, which was constructed in 1963 and simultaneously created the Lake Powell Reservoir. Creating the dam flooded miles of the Colorado River, destroyed habitats, and submerged hidden canyons and archeological sites. The controversy over the dam influenced the birth of the modern environmental movement. Encourage students to debate the pros and cons of a proposed large dam project such as the Sardas Sarovar Dam, in India, Bujagali Falls, in Uganda, dams on the Biobío River, in Chile, or the Three Gorges, in China. The International Rivers Network provides online information about dam projects around the world. **LS** Interpersonal

Answers to What Do You Think?
Answers may vary.

ECONOMICS ──
● CONNECTION ─── ADVANCED ⟩

Financing Three Gorges The China Development Bank, or CDB, is the main organization responsible for financing the Three Gorges Dam project. The World Bank, the U.S. Export-Import Bank, and the Asian Development Bank all refused to finance the project. Because of the difficulty in attempting to sell bonds specifically for Three Gorges, the CDB issued general obligation bonds. These bonds raise money for the CDB but do not specifically say that the money will go towards the Three Gorges project. China International Capital Corporation, or CICC, serves as the advisor for raising funds. As of 2002, CICC is 35 percent owned by Morgan Stanley, a financial services company based in the United States. Morgan Stanley, as well as Citigroup and Merrill Lynch, was involved in underwriting the CDB bonds. This means that the companies did not purchase the bonds, but rather agreed to insure the bonds so that they could be sold. Morgan Stanley says it is committed to protecting the environment, but some environmental groups still hold the company responsible for supporting the Three Gorges Dam project.

CHAPTER 12

Air
Chapter Planning Guide

PACING	CLASSROOM RESOURCES	LABS, DEMONSTRATIONS, AND ACTIVITIES
BLOCKS 1 & 2 · 90 min pp. 302–308 **Chapter Opener**		
Section 1 What Causes Air Pollution?	**OSP** Lesson Plan * **CRF** Active Reading * BASIC **TT** Bellringer * **TT** The Formation of Smog * **TT** Temperature Inversion *	**TE** Group Activity Collecting Particulate Matter, p. 304 GENERAL **TE** Group Activity Reducing Auto Emissions, p. 305 GENERAL **TE** Group Activity Air Pollution and the Individual, p. 306 GENERAL **TE** Demonstration Temperature Inversion, p. 307 ◆ BASIC **CRF** Field Activity * ◆ GENERAL **CRF** Consumer Lab * BASIC **CRF** Long-Term Project * ◆ GENERAL
BLOCKS 3 & 4 · 90 min pp. 309–313 **Section 2** Air, Noise, and Light Pollution	**OSP** Lesson Plan * **CRF** Active Reading * BASIC **TT** Bellringer * **TT** Indoor Air Pollutants *	**TE** Group Activity Search the School, p. 311 GENERAL **TE** Group Activity Asbestos Contamination, p. 312 GENERAL **SE** Field Activity Light Pollution, p. 313
BLOCKS 5, 6 & 7 · 135 min pp. 314–317 **Section 3** Acid Precipitation	**OSP** Lesson Plan * **CRF** Active Reading * BASIC **TT** Bellringer * **TT** The pH Scale * **TT** How Acid Precipitation Forms * **TT** A Global Look at Acid Precipitation * **CD** Interactive Exploration 1 Something's Fishy **CD** Interactive Exploration 5 Moose Malady	**SE** QuickLab Neutralizing Acid Precipitation, p. 316 **SE** Exploration Lab The Acid Test, pp. 322–323 ◆ **CRF** Datasheets for In-Text Labs * **CRF** Observation Lab * ◆ ADVANCED **CRF** CBL™ Probeware Lab * ◆ GENERAL

BLOCKS 8 & 9 · 90 min

Chapter Review and Assessment Resources

- **SE** Chapter Review pp. 319–321
- **SE** Standardized Test Prep pp. 658–659
- **CRF** Study Guide * ■ GENERAL
- **CRF** Chapter Test * ■ GENERAL
- **CRF** Chapter Test * ADVANCED
- **CRF** Concept Review * ■ GENERAL
- **CRF** Critical Thinking * ADVANCED
- **OSP** Test Generator
- **CRF** Test Item Listing *

Online and Technology Resources

 Holt Online Learning

Visit **go.hrw.com** for access to Holt Online Learning or enter the keyword **HE6 Home** for a variety of free online resources.

 **One-Stop Planner® CD-ROM**

This CD-ROM package includes
- Lab Materials QuickList Software
- Holt Calendar Planner
- Customizable Lesson Plans
- Printable Worksheets
- ExamView® Test Generator
- Interactive Teacher Edition
- Holt PuzzlePro® Resources
- Holt PowerPoint® Resources

 Holt Environmental Science Interactive Tutor CD-ROM

This CD-ROM consists of interactive activities that give students a fun way to extend their knowledge of environmental science concepts.

KEY

TE	Teacher Edition	**CRF**	Chapter Resource File	*	Also on One-Stop Planner
SE	Student Edition	**TT**	Teaching Transparency	■	Also Available in Spanish
OSP	One-Stop Planner	**CD**	Interactive CD-ROM	◆	Requires Advance Prep

ENRICHMENT AND SKILLS PRACTICE	SECTION REVIEW AND ASSESSMENT	CORRELATIONS
SE Pre-Reading Activity, p. 302 **TE** Using the Figure, p. 302 GENERAL		**National Science Education Standards**
TE Inclusion Strategies, p. 303 **TE** Skill Builder Paired Summarizing, p. 304 GENERAL **SE** MathPractice Utility Incentives for Zero-emission Vehicles, p. 306 **TE** Student Opportunities The Yellow Bikes Program, p. 306	**TE** Homework, p. 306 BASIC **SE** Section Review, p. 308 **TE** Reteaching, p. 308 BASIC **TE** Quiz, p. 308 GENERAL **TE** Alternative Assessment, p. 308 GENERAL **CRF** Quiz * ■ GENERAL	PS 3b SPSP 4a
SE Case Study The Health Effects of Ground-Level Ozone, pp. 310–311 **TE** Using the Figure Indoor Air Pollution, p. 311 GENERAL	**TE** Homework, p. 310 GENERAL **TE** Homework, p. 312 GENERAL **SE** Section Review, p. 313 **TE** Reteaching, p. 313 BASIC **TE** Quiz, p. 313 GENERAL **TE** Alternative Assessment, p. 313 GENERAL **CRF** Quiz * ■ GENERAL	SPSP 4a
SE Graphic Organizer Chain-of-Events Chart, p. 315 GENERAL **TE** Inclusion Strategies, p. 315 **TE** Skill Builder Terms and Concepts, p. 316 GENERAL **TE** Skill Builder Writing, p. 316 GENERAL **SE** Maps in Action Light Sources, p. 324 **SE** Society & the Environment The Donora, Pennsylvania, Killer Smog, p. 325 **CRF** Map Skills Pollution Levels * GENERAL	**SE** Section Review, p. 317 **TE** Reteaching, p. 317 BASIC **TE** Quiz, p. 317 GENERAL **TE** Alternative Assessment, p. 317 GENERAL **CRF** Quiz * ■ GENERAL	SPSP 4a SPSP 4b SPSP 6e

Guided Reading Audio CDs

These CDs are designed to help auditory learners and reluctant readers. (Audio CDs are also available in Spanish.)

www.scilinks.org

Maintained by the **National Science Teachers Association**

TOPIC: Air Pollution
SciLinks code: HE4003

TOPIC: Respiratory Disorders
SciLinks code: HE4094

TOPIC: Acid Rain
SciLinks code: HE4001

CNN Videos

Each video segment is accompanied by a Critical Thinking Worksheet.

Earth Science Connections

Segment 9 Radon Risk

Segment 23 Acid Rain

CHAPTER
12

Chapter Enrichment

This Chapter Enrichment provides relevant and interesting information to expand and enhance your classroom instruction of the chapter material.

1 What Causes Air Pollution?

▶ Industry and transportation are the two main sources of urban air pollution

Moving Mountains to Fight Air Pollution

Lanzhou, China, has a terrible claim to fame—the air is actually gritty. Two million people live in this city in northwestern China, where smoke from coal and exhaust from cars combine with dust particles blowing off the region's dry, crumbling mountains. Because the city is located in the bottom of a valley, there is little chance for air to circulate. Nonsmokers who live in Lanzhou suffer the effects of smoking about a pack of cigarettes a day just from breathing the air.

To help air circulation and to clean the air, the citizens of Lanzhou decided to move a mountain. Using explosives, 100 feet were removed from the top of a mountain in the late 1990s. However, the dust produced by the blast made air quality even worse. The environmental protection bureau of Lanzhou is investigating converting to unleaded gasoline, planting trees, and the use of cleaner forms of energy—natural gas, wind, solar, and hydroelectric power.

Solar Cars

Australia receives a lot of solar energy, so it makes a perfect setting for the World Solar Challenge. In this 3,000 km race across the country, teams of students and researchers enter cars that use solar power instead of gasoline. Automobile manufacturers often sponsor

the designs of the cars. The first race, held in 1987, boasted 23 entrants. The most recent race had 38 participants. The winning car, a Dutch model, averaged 98.81 km/h (57.04 mph). It finished the race in 32 hours and 9 minutes. The race is all part of an effort to increase environmental awareness. It is also a chance for car manufacturers to put their names on some pretty fantastic designs!

Ethanol-Powered Cars in Brazil

During the oil shortage of the 1970s, Brazil, the world's largest producer of sugar, developed a line of cars that burned sugar cane-based hydrous ethanol, or alcohol, instead of gasoline. Alcohol is a clean-burning fuel. But when a shortage of alcohol occurred in the late 1980s, some Brazilians tried in desperation to get the cars to run on homemade "moonshine." Their attempts failed, and many of the country's cars have sat dormant since then. However, recent treaties that demand a reduction in greenhouse gases, in conjunction with a bumper crop of sugar cane, have led to a new wave of technological innovation. Some cars can run on gasoline, alcohol, or a blend of the two.

2 Air, Noise, and Light Pollution

Air Pollution in Mexico City

In Mexico City, many factors combine to cause the city's notoriously poor air quality. First of all, in 2000, Mexico City was the world's second most populous city, with more than 18 million inhabitants (the exact number is

▶ A police officer directs traffic in Bangkok, Thailand

unknown). Its roads are traveled by millions of vehicles each day, many of them older, poorly maintained models that spew pollutants into the air. Because of the city's high elevation (2,309 m or 7,575 ft), even modern cars run less efficiently and pollute more than they would at sea level. Second, the city is the site of heavy industry, which pumps large quantities of pollutants into the air. Third, Mexico City's pollution problem is exacerbated by topography. The city is located in a deep valley, surrounded by high mountains that effectively trap pollutants. These mountains may be difficult to see—as smog has reduced visibility that used to be 16 km in the 1930s to less than 4 km today.

The Indoor Air Quality in U.S. Schools

According to the U.S. Environmental Protection Agency, 20 percent of the U.S. population, or approximately 55 million Americans, are students in the nation's 115,000 elementary schools, middle schools, and high schools. Half of these schools suffer from problems related to poor indoor air quality. Because of the amount of time students spend in schools and the susceptibility of children to air pollutants, students are at greatest risk from poor indoor air quality at these facilities.

BRAIN FOOD

The World Bank estimates that the cost of China's pollution is $54 billion or as much as 8 percent of its gross domestic product, taking into consideration the cost of workers' sick days and the required health care from diseases caused by pollution. When analysts look at the situation in this way, this amount of money nearly cancels out the considerable economic progress China has made over the last several years.

Shhh! Quiet Construction Underway

The United States enacted the Noise Control Act in 1972. The act declared that it was U.S. policy to "promote an environment for all Americans free from noise that jeopardizes their health or welfare." It was the first attempt to coordinate efforts for research and activities on noise control, to set noise standards, and to provide information about noise pollution to the public. Communities set their own noise standards as well. But perception of noise in the U.S. is still different from that in other

countries. Some European countries have such comprehensive (and strongly enforced) noise laws that construction equipment has been designed to be "quiet."

Nature's Alarm

Physiologically, noise is often regarded as stressful, but it can be stressful to a far greater degree than one might think. Even noises we can't hear can affect us. For example, humans typically can hear frequencies from 20 Hz to 20,000 Hz. Large animals, such as whales, elephants, rhinos, and tigers make noises that register below 20 hertz. But a tiger's roar can stun human trainers, even those who are experienced with a tiger's roar. The low frequency and the intensity of this sound can travel long distances.

SECTION

3 Acid Precipitation

The Rain in Spain . . . May Be Cleaner than the Rain in Sweden

Due to the wind flow direction over Europe, as much as 56 percent of acid rain in Sweden begins as air pollution in other countries. Sulfur dioxide released into the air largely from industry in Germany and the United Kingdom has resulted in Swedish lakes becoming increasingly acidic over the past 70 years. Fish cannot live in lakes with a pH below 5.5, and 5,000 Swedish lakes now have a pH below 5.0. Ten thousand other lakes have a pH level below 6.0. Since the 1930s, the United Kingdom has been making a concerted effort to reduce air pollution, and the air in that country has indeed become cleaner. However, one of the cornerstones of the project—building taller smokestacks— means that more pollutants are transported to other countries by winds.

▶ These fish were killed by a rapid reduction in pH

Overview

Tell students that the purpose of this chapter is to help them understand the different kinds of pollutants that affect air quality. Students will learn ways in which air pollution is produced and ways in which it can be reduced. Air pollution has short-term and long-term health effects, and it is often difficult to determine the source of some pollutants. The health effects of noise pollution are also discussed.

Using the Figure — GENERAL

Nagano, Japan, hosted the 1998 Winter Olympics, which were intended to be the first "environmentally sensitive" Olympics. Despite these claims, the Nagano Olympics had many environmental problems. Air pollution from trash-burning outside Olympic lodges and automobile emissions combined with air pollution already present in the city. There were at least 44 athletes at the games who suffered from asthma. Ask students to brainstorm a list of ways that air pollution could affect Olympic athletes. Ask, "Should environmental concerns such as air pollution be an important factor in the Olympic Committee's choice to host Olympic games?" **LS Visual**

PRE-READING ACTIVITY

You may want to assign this FoldNote activity as homework.
Collect the FoldNotes to check students' understanding of the material.

VIDEO SELECT

For information about videos related to this chapter, go to **go.hrw.com** and type in the keyword **HE4 AIRV**.

Air

302

PRE-READING ACTIVITY

FOLDNOTES

Booklet
Before you read this chapter, create the **FoldNote** entitled "Booklet" described in the Reading and Study Skills section of the Appendix. Label each page of the booklet with a main idea from the chapter. As you read the chapter, write what you learn about each main idea on the appropriate page of the booklet.

Nagano, Japan, is just one of many cities around the world that suffers from unhealthy levels of air pollution.

Chapter Correlations *National Science Education Standards*

PS 3b Chemical reactions may release or consume energy. Some reactions such as the burning of fossil fuels release large amounts of energy by losing heat and by emitting light. Light can initiate many chemical reactions such as photosynthesis and the evolution of urban smog. **(Section 1)**

SPSP 4a Natural ecosystems provide an array of basic processes that affect humans. Those processes include maintenance of the quality of the atmosphere, generation of soils, control of the hydrologic cycle, disposal of wastes, and recycling of nutrients. Humans are changing many of these basic processes, and the changes may be detrimental to humans. **(Section 1, Section 2, and Section 3)**

SPSP 4b Materials from human societies affect both physical and chemical cycles of the earth. **(Section 3)**

SPSP 6e Humans have a major effect on other species. For example, the influence of humans on other organisms occurs through land use—which

decreases space available to other species—and pollution—which changes the chemical composition of air, soil, and water. **(Section 3)**

What Causes Air Pollution?

In Mexico City, children rarely use the color blue when they make paintings of the sky. This metropolitan area of 20 million people is known as the most dangerous city in the world for children because of its very polluted air. When pollution levels are high, students are banned from playing outdoors until the emergency passes.

Clean air consists mostly of nitrogen and oxygen gas, as well as very small amounts of argon, carbon dioxide, and water vapor. When harmful substances build up in the air to unhealthy levels, the result is **air pollution.** Substances that pollute the air can be in the form of solids, liquids, or gases.

Most air pollution is the result of human activities, but pollutants can also come from natural sources. A volcano, for example, can spew clouds of particles and sulfur dioxide, SO_2, into the atmosphere. Natural pollutants also include dust, pollen, and spores.

Primary and Secondary Pollutants

A pollutant that is put directly into the air by human activity is called a **primary pollutant.** An example of a primary pollutant is soot from smoke. **Figure 1** shows some sources of primary air pollutants. **Secondary pollutants** form when a primary pollutant comes into contact with other primary pollutants or with naturally occurring substances such as water vapor and a chemical reaction takes place. An example of a secondary pollutant is ground-level ozone. Ground-level ozone forms when the emissions from cars, trucks, and natural sources react with the ultraviolet rays of the sun and then mix with the oxygen in the atmosphere.

Objectives

▶ Name five primary air pollutants, and give sources for each.

▶ Name the two major sources of air pollution in urban areas.

▶ Describe the way in which smog forms.

▶ Explain the way in which a thermal inversion traps air pollution.

Key Terms

air pollution
primary pollutant
secondary pollutant
smog
temperature inversion

Figure 1 ▶ Each day in the United States, hundreds of thousands of tons of polluting emissions that result from human activity enter the air.

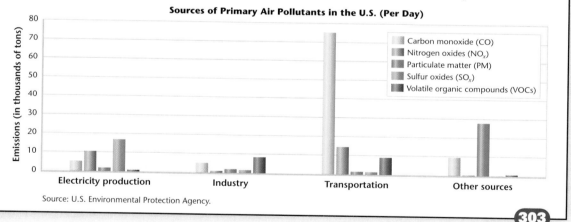

Sources of Primary Air Pollutants in the U.S. (Per Day)

- Carbon monoxide (CO)
- Nitrogen oxides (NO_x)
- Particulate matter (PM)
- Sulfur oxides (SO_x)
- Volatile organic compounds (VOCs)

Source: U.S. Environmental Protection Agency.

303

Chapter Resource File

- Lesson Plan
- Active Reading **BASIC**
- Section Quiz **GENERAL**

Transparencies

TT Bellringer

INCLUSION Strategies

- *Attention Deficit Disorder*
- *Learning Disabled*

Have students use a word processor, pencil and paper, or posterboard to list the sources of primary air pollutants in **Figure 1.** Have students describe each type of pollutant and explain how it is created. Have students estimate how much pollution is created by each source of primary air pollution.

SECTION 1

Focus

Overview

Before beginning this section, review with your students the Objectives listed in the Student Edition. This section defines primary air pollutants and lists their sources. Pollution from automobiles and industry is discussed, along with smog and temperature inversions.

🔊 Bellringer

Write the following quotation on the board. "I thought I saw a blue jay this morning. But the smog was so bad that it turned out to be a cardinal holding its breath." —Michael J. Cohen. Tell students that sometimes humor gets people's attention about a serious issue more readily than any other method. Divide the class into small groups, and tell them to use humor, as Cohen did, to complete the following sentence: "You know the air is polluted when. . . ." **LS** Verbal

Motivate

Discussion ———— BASIC

Comparing Cigarette Smoke and Pollution Poll the class, and ask how many students smoke. Ask everyone, "Are you concerned about the effects of smoking?" Ask students to list some of the possible effects of smoking. Point out that some people breathe in air that is so polluted it is equivalent to smoking a pack of cigarettes or more a day. Ask, "If cigarette manufacturers are being held accountable for the health effects of the use of their products, should industries be accountable for the effects of the pollutants they discharge into the air?" **LS** Interpersonal

Table 1 ▼

Teach

Group Activity —— GENERAL

Collecting Particulate Matter
Check the air quality near your school by collecting particulate matter. Remove the protective backing from an 8 1/2 × 11 in sheet of clear contact paper, and place it over a sheet of graph paper. Pin the papers to a piece of cardboard with the sticky side up. Place the cardboard somewhere where it will be undisturbed for one day. Students can collect the contact paper and can use the grids on the graph paper and a magnifying glass to count the number of particles they collected. Students should also note particle sizes. As an extension, try this experiment at different times of the year and in different locations in your community.
LS Visual/Kinesthetic

MISCONCEPTION ALERT

Invisible Pollutants Many people believe that polluted air must be visibly smoky or must be brown or black in color. Stress that some of the most dangerous air pollutants are those that can't be seen with the naked eye. Challenge students to use the Internet to find out about the various pollutants monitored by the Environmental Protection Agency and other organizations that monitor air quality. Have students compile their results in a table that lists the acceptable amounts allowed in the air, the levels in your community, and the health problems associated with each pollutant.

Primary Air Pollutants			
Pollutant	Description	Primary Sources	Effects
Carbon monoxide (CO)	CO is an odorless, colorless, poisonous gas. It is produced by the incomplete burning of fossil fuels.	Sources of CO are cars, trucks, buses, small engines, and some industrial processes.	CO interferes with the blood's ability to carry oxygen, slowing reflexes and causing drowsiness. In high concentrations, CO can cause death.
Nitrogen oxides (NO_x)	When combustion (burning) temperatures exceed 538°C, nitrogen and oxygen combine to form nitrogen oxides.	NO_x comes from burning fuels in vehicles, power plants, and industrial boilers.	NO_x can make the body vulnerable to respiratory infections, lung diseases, and cancer. NO_x contributes to the brownish haze seen over cities and to acid precipitation.
Sulfur dioxide (SO_2)	SO_2 is produced by chemical interactions between sulfur and oxygen.	SO_2 comes mostly from burning fossil fuels.	SO_2 contributes to acid precipitation as sulfuric acid. Secondary pollutants that result from reactions with SO_2 can harm plant life and irritate the respiratory systems of humans.
Volatile organic compounds (VOCs)	VOCs are organic chemicals that vaporize readily and form toxic fumes.	VOCs come from burning fuels. Vehicles are a major source of VOCs.	VOCs contribute to smog formation and can cause serious health problems, such as cancer. They may also harm plants.
Particulate matter (particulates or PM)	Particulates are tiny particles of liquid or solid matter.	Most particulates come from construction, agriculture, forestry, and fires. Vehicles and industrial processes also contribute particulates.	Particulates can form clouds that reduce visibility and cause a variety of respiratory problems. Particulates have also been linked to cancer. They may also corrode metals and erode buildings and sculptures.

internet connect
www.scilinks.org
Topic: Air Pollution
SciLinks code: HE4003
SCiLINKS. Maintained by the National Science Teachers Association

Sources of Primary Air Pollutants As shown in **Table 1** above, household products, power plants, and motor vehicles are sources of primary air pollutants such as carbon monoxide, nitrogen oxide, sulfur dioxide, and chemicals called *volatile organic compounds* (VOCs). Carbon monoxide gas is an important component of the exhaust from vehicles. Vehicles are also a major source of nitrogen oxide emissions. Coal-burning power plants are another source of nitrogen oxide. Sulfur dioxide gases are formed when coal and oil, which contain sulfur, are burned. Power plants, refineries, and metal smelters contribute much of the sulfur dioxide emissions to the air. Vehicles and gas station spillage make up most of the human-made emissions of volatile organic compounds. VOCs are also found in many household products.

Particulate matter can also pollute the air and is usually divided into fine and coarse particles. Fine particles enter the air from fuel burned by vehicles and coal-burning power plants. Sources of coarse particles are cement plants, mining operations, incinerators, wood-burning fireplaces, fields, and roads.

SKILL BUILDER —— GENERAL

Paired Summarizing Have students pair with a partner and take turns reading and summarizing the primary pollutants identified in **Table 1.** Each pair should come up with a description of each kind of pollutant in their own words. Student pairs should also offer two examples of each category of primary pollutant. **LS** Interpersonal

The History of Air Pollution

Air pollution is not a new phenomenon. Whenever something burns, pollutants enter the air. Two thousand years ago, Seneca, a Roman philosopher and writer, complained about the foul air in Rome. In 1273, England's King Edward I ordered that burning a particularly dirty kind of coal called sea-coal was illegal. One man was even hanged for disobeying this medieval "clean air act."

The world air-quality problem is much worse today because modern industrial societies burn large amounts of fossil fuels. As shown in **Figure 2**, most air pollution in urban areas comes from motor vehicles and industry.

Motor Vehicle Emissions

Almost one-third of our air pollution comes from gasoline burned by vehicles. According to the U.S. Department of Transportation, Americans drove their vehicles over 2.6 trillion miles in 1998. Over 90 percent of that mileage was driven by passenger vehicles. The rest was driven by trucks and buses.

Controlling Vehicle Emissions The Clean Air Act, passed in 1970 and strengthened in 1990, gives the Environmental Protection Agency (EPA) the authority to regulate vehicle emissions in the United States. The EPA required the gradual elimination of lead in gasoline, and as a result, lead pollution has been reduced by more than 90 percent in the United States. In addition, catalytic converters, which are required in automobiles, clean exhaust gases of pollutants before the pollutants are able to exit the tailpipe. The EPA estimates that cars and trucks today burn fuel 35 percent more efficiently and with 95 percent fewer emissions of pollutants, excluding carbon dioxide, than they did 30 years ago.

Geofact

Sea-Coal In 12th-century London, wood was becoming too scarce and too expensive to use as a fuel source. Large deposits of coal called *sea-coal* that are found off the northeast coast of England provided a plentiful alternative. However, this soft coal did not burn efficiently. The sea-coal produced mostly smoke and not much heat. The smoke from the coal emanated from London homes and factories and combined with fog to produce smog.

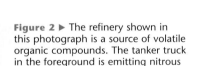

Connection to Law

Off with His Head! Around 1300 CE, King Edward II of England forbade the burning of coal while Parliament was in session. "Be it known to all within the sound of my voice," King Edward II said, "whosoever shall be found burning coal shall suffer the loss of his head."

Figure 2 ▶ The refinery shown in this photograph is a source of volatile organic compounds. The tanker truck in the foreground is emitting nitrous oxide into the atmosphere.

305

Group Activity ── GENERAL

Air Pollution and the Individual

Divide the class into small groups, and have members of each group work together to discuss the following questions: "How do I contribute to air pollution? At what point would the level of air pollutants in the air be high enough for me to take action?"

LS Interpersonal

Homework ── BASIC

Evaluating Car Use Tell students to observe their transportation habits for a week. Have them make a list of every place they regularly travel by car. Ask students, "What is the farthest distance that you regularly travel by car? What is the closest distance?" Then tell students to make a list of places that they regularly walk to. Ask students, "What is the farthest distance you walk? What is the shortest distance you walk? Are there any trips that might make more sense to make on foot? How many hours traveling by car might you save weekly if you walked instead?" Suggest that students include their responses in their *EcoLog.*

MATH**PRACTICE**

Utility Incentives for Zero-emission Vehicles

The Los Angeles Department of Water and Power provides discounts of $0.025 per kilowatt hour (kWh) for electricity used to recharge electric vehicles. If the energy charge per kWh is $0.02949 and you use 150 kWh hours of electricity per month to recharge your vehicle, how much money would you save on your electric bill each month? each year? How much would you save if you had three electric cars?

California Zero-Emission Vehicle Program In 1990, the California Air Resources Board established the zero-emission vehicle (ZEV) program. *Zero-emission vehicles* are vehicles that have no tailpipe emissions, no emissions from gasoline, and no emission-control systems that deteriorate over time. **Figure 3** illustrates the catalytic converter emission-control system that is in use today as well as the ways an automobile contributes to air pollution.

By the year 2016, 16 percent of all vehicles sold in California are required to be zero-emission vehicles. This requirement includes sports utility vehicles (SUVs), trucks, small vans, and automobiles. At present, ZEVs such as electric vehicles are for sale in California, and vehicles with advanced batteries are being demonstrated by the major automakers. Vehicles powered by hydrogen fuel are being developed and will qualify as zero-emission vehicles. Partial zero-emissions vehicles are also included in the program. These vehicles include hybrid-electric cars and cars powered by methanol fuel cells. Zero-emission vehicle programs have also been adopted by Maine, Massachusetts, New York, and Vermont.

Figure 3 ▶ The catalyst material in a catalytic converter (top) causes a chemical reaction that changes exhaust emissions to less harmful substances. The bottom illustration shows a car's contribution to air pollution.

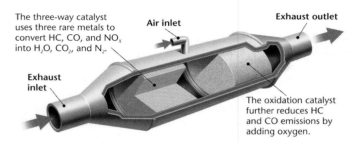

The three-way catalyst uses three rare metals to convert HC, CO, and NO_x into H_2O, CO_2, and N_2.

Air inlet

Exhaust outlet

Exhaust inlet

The oxidation catalyst further reduces HC and CO emissions by adding oxygen.

Interior
▶ Car seats may be covered in plastic that contains a volatile organic compound called *vinyl chloride.*

Body and Frame
▶ Steel smelters send thousands of metric tons of sulfur dioxide into the air each year.

▶ Many auto factories in Mexico, Eastern Europe, and some Asian countries lack pollution-control devices.

Fuel Tank
▶ When filling the car with gasoline, VOCs escape into the atmosphere.

Exhaust
▶ Car exhaust is a major source of nitrogen oxides, carbon monoxide, and hydrocarbons.

▶ In developing countries, car exhaust may contain over a thousand poisonous substances.

▶ Each car releases 4.5 metric tons (5 tons) of carbon dioxide every year.

Student Opportunities

The Yellow Bikes Program In an effort to make cities more livable—and fun—several urban communities have begun a "Yellow Bikes" program. Members of the community donate hundreds of old bikes, which are repaired to an acceptable condition and are distributed around a city. Portland, Oregon, Austin, Texas, and Minneapolis-St. Paul, Minnesota, each developed their own version of this program, which originated in Holland.

The bikes are left unlocked in public spaces for community use. The idea is that a resident uses the bike for a trip (or two, at most) to avoid using a car—or even walking. The programs were designed to be fun, athletic, and community-oriented. However, these programs have not always been a success. Have students investigate the possibility of instituting a "Yellow Bikes" program in their area.

LS Logical

Industrial Air Pollution

Many industries and power plants that generate our electricity must burn fuel to get the energy they need. They usually burn fossil fuels. Burning fossil fuels releases huge quantities of sulfur dioxide and nitrogen oxide into the air. Power plants that produce electricity emit at least two-thirds of all sulfur dioxide and more than one-third of all nitrogen oxides that pollute the air.

Some industries also produce VOCs, which are chemical compounds that form toxic fumes. As shown in **Figure 4,** some of the chemicals used in dry cleaning are sources of VOCs. Oil refineries, chemical manufacturing plants, furniture refinishers, and automobile repair shops also contribute to the VOCs in the air. When people use some of the products that contain VOCs, more VOCs are added to the air.

Regulating Air Pollution From Industry The Clean Air Act requires many industries to use scrubbers or other pollution-control devices. Scrubbers remove some of the more harmful substances that would otherwise pollute the air. A *scrubber,* as shown in **Figure 5,** is a machine that moves gases through a spray of water that dissolves many pollutants. Ammonia is an example of a pollutant gas that can be removed from the air by a scrubber.

Electrostatic precipitators are machines used in cement factories and coal-burning power plants to remove dust particles from smokestacks. In an electrostatic precipitator, gas containing dust particles is blown through a chamber containing an electrical current. An electrical charge is transferred to the dust particles, which causes them to stick to one another and the sides of the chamber. The clean gas is released from the chamber, and the concentrated dust particles can then be collected and removed. Electrostatic precipitators remove 22 million metric tons (20 million tons) of ash generated by coal-burning power plants from the air each year in the United States.

Figure 4 ▶ In 1996, the federal government established standards to reduce emissions of VOC-producing chemicals used in dry cleaning.

Air Pollution's Impact on Birds Scientists in Finland have documented the effects of harmful emissions from a copper smelter in Finland on two species of birds that live nearby. The two species of birds respond differently to the pollutants containing heavy metals and acidic substances. One species appears to suffer directly from the toxic effects of the pollutants. The other species suffers because the amount of insect food for its nestlings has been reduced. When heavy metal emissions from the smelter decreased, a rapid improvement in breeding success and decrease in the heavy metal found in the bones of nestlings was observed.

Figure 5 ▶ Scrubbers work by spraying gases with water, which removes many pollutants.

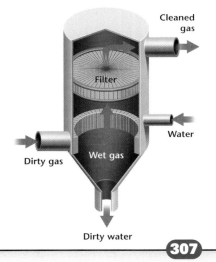

307

Pollution Sources Write the following pollution sources on the board: Industrial and Commercial, Transportation, and Noncommercial. Have students list as many causes of air pollution as possible, and help them place the pollutants in the appropriate categories. Finally, discuss the types of pollutants that each source usually emits and possible remedies for those pollutants.

Quiz ——— GENERAL

1. What does VOC stand for? (volatile organic compound)

2. What are the components of smog? (sunlight, air, automobile exaust, and ozone are the main cause of smog.)

Alternative Assessment ——— GENERAL

Smog and Ozone Alerts Ask students to prepare a script for a simulated TV news report on a hypothetical smog or ozone alert in your area. Students' scripts should be creative yet plausible and coherent. Ask students to describe the source of the pollutants and any aggravating factors. They should also include advice about how people can minimize their exposure during the alert and about what individuals can do to help alleviate the problem. LS Verbal

Transparencies

TT The Formation of Smog

TT Temperature Inversion

Figure 6 ▶ The diagram below shows how smog is formed. Large cities with dry, sunny climates and millions of automobiles often suffer from smog.

❷ Ozone reacts with automobile exhaust to form smog.

Smog

Ozone

Automobile exhaust

❶ Automobile exhaust reacts with air and sunlight to form ozone.

Figure 7 ▶ Normal air circulation is shown at left, whereas a temperature inversion, in which pollutants are trapped near the Earth's surface, is shown at right.

Smog When air pollution hangs over urban areas and reduces visibility, it is called **smog.** As you can see in **Figure 6,** smog results from chemical reactions that involve sunlight, air, automobile exhaust, and ozone. Pollutants released by vehicles and industries are the main causes of smog. Los Angeles, California, Denver, Colorado, and Phoenix, Arizona, are examples of cities that have smog.

Temperature Inversions The circulation of air in the atmosphere usually keeps air pollution from reaching dangerous levels. During the day, the sun heats the surface of the Earth and the air near the Earth. The warm air rises through the cooler air above and carries pollutants away from the ground and into the atmosphere.

Sometimes, however, pollution is trapped near the Earth's surface by a temperature inversion. Usually, air temperatures decrease with height, but in an area with a **temperature inversion,** the air above is warmer than the air below. **Figure 7** shows how a temperature inversion traps pollutants near the Earth's surface. The warmer air above keeps the cooler air at the surface from moving upward. So, pollutants are trapped below with the cooler air. If a city is located in a valley, the city has a greater chance of experiencing temperature inversions. Los Angeles, which is surrounded on three sides by mountains, often has temperature inversions that trap smog in the city.

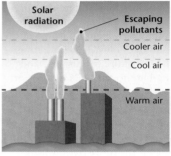

Normal situation

Solar radiation — Escaping pollutants

Cooler air

Cool air

Warm air

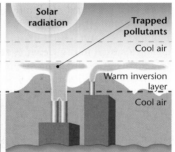

Temperature inversion

Solar radiation — Trapped pollutants

Cool air

Warm inversion layer

Cool air

SECTION 1 Review

1. Name five primary air pollutants, and give important sources for each.

2. Name the two major sources of air pollution in urban areas.

3. Describe the way in which smog forms.

4. Define the term *temperature inversion.* Explain how temperature inversion traps pollutants near Earth's surface.

CRITICAL THINKING

5. Making Decisions Read the passage on the California Zero-Emission Vehicle Program. Should automobile makers be made to adhere to quotas of zero-emission vehicles set by states, even if the quota causes automakers to lose revenue? READING SKILLS

6. Analyzing Relationships Can you think of any other possible type of pollution-control device that could be used to remove particulates from smokestacks in a manner similar to an electrostatic precipitator?

308

Answers to Section Review

1. Answers will vary, but should reflect the material in Table 1.

2. The two major sources of air pollution in urban areas are motor vehicles and industry.

3. Automobile exhaust reacts with air and sunlight to form ozone, and ozone reacts with automobile exhaust to form smog.

4. A temperature inversion is an atmospheric situation in which cool air is trapped close to the Earth by a layer of warmer air above. This

situation can trap pollutants, which are generally carried away from the surface of the Earth by the rising of warm air. When the air below is cooler, pollutants are unable to rise.

5. Answers may vary. Students should weigh both short-term and long-term economic and environmental consequences.

6. Answers may vary. Students may describe pollution-control devices that use centrifugal force to collect pollutants.

Air, Noise, and Light Pollution

Air pollution can cause serious health problems. People who are very young or very old and people who have heart or lung problems can be most affected by air pollutants. Decades of research have shown convincing evidence linking air pollution to disease. But because pollution adds to the effects of existing diseases, no death certificates list the cause of death as air pollution. Instead, diseases such as emphysema, heart disease, and lung cancer are cited as causes of death. The American Lung Association has estimated that Americans pay tens of billions of dollars a year in health costs to treat respiratory diseases caused by air pollution.

Short-Term Effects of Air Pollution on Health

Many of the effects of air pollution on people's health are short-term and are reversible if their exposure to air pollution decreases. The short-term effects of air pollution on people's health include headache; nausea; irritation to the eyes, nose, and throat; tightness in the chest; coughing; and upper respiratory infections, such as bronchitis and pneumonia. Pollution can also make the condition of individuals who suffer from asthma and emphysema worse.

Long-Term Health Effects of Air Pollution

Long-term effects on health that have been linked to air pollution include emphysema, lung cancer, and heart disease. Long-term exposure to air pollution may worsen medical conditions suffered by older people and may damage the lungs of children.

Objectives

▶ Describe three possible short-term effects and long-term effects of air pollution on human health.
▶ Explain what causes indoor air pollution and how it can be prevented.
▶ Describe three human health problems caused by noise pollution.
▶ Describe solutions to energy waste caused by light pollution.

Key Terms

sick-building syndrome
asbestos
decibel (dB)

▣ internet connect
www.scilinks.org
Topic: Respiratory Disorders
SciLinks code: HE4094
SCiLINKS. Maintained by the National Science Teachers Association

Figure 8 ▶ This police officer wears a smog mask as he directs traffic in Bangkok, Thailand.

309

Chapter Resource File

• Lesson Plan
• Active Reading BASIC
• Section Quiz GENERAL

Transparencies

TT Bellringer

HEALTH
━━ ● ● CONNECTION ━ GENERAL

Respiratory Diseases Have students find out about respiratory diseases, such as asthma, that can be aggravated by air pollution. Have students compile their findings into tables that list the disease, its symptoms, how it is treated, the age groups mostly commonly afflicted, and the relationship between the disease and air pollutants. LS Logical

---

Focus

Overview

Before beginning this section, review with your students the Objectives listed in the Student Edition. This section discusses the effects of air pollution on human health. Indoor air pollution is evaluated as a major source of health problems. Noise and light pollution are also discussed.

🔊 Bellringer

Tell students that human lungs have about 300 million tiny air pockets called *alveoli*. Altogether, the surface area in the lungs of humans is equivalent to the area of a tennis court. Ask students to think about why the surface area of human lungs is so large, and have them describe the effects of pollutants on the human lungs. (Lungs have such a large surface area because they are the site at which individual gas molecules are exchanged between the human blood stream and the air. Pollutants can damage the tiny alveoli.)

Motivate

Discussion ━━━ GENERAL

Air Pollution Bring a filter mask to class. Have each student make a list of three situations in which one might wear such a mask. For example, surgeons wear filter masks to prevent the transfer of disease-causing microbes, and sandblasters wear masks to avoid inhaling dust and paint chips. Tell students that people living in areas with heavily polluted air wear filter masks to protect themselves from pollutants in the air they breathe.

Discussion —————— GENERAL

Indoor Air Pollution Explain that the air inside buildings may be polluted by a variety of sources. Ask students to list possible sources of indoor air pollution. If students have difficulty coming up with examples, tell them that air pollution is often invisible. Chalk dust, cooking oils, carpets, insulation, tobacco smoke, paints, glues, copier machines, space heaters, gas appliances, and fireplaces are just a few of the sources of indoor pollution.

Homework —————— GENERAL

Radon Gas Radon is a naturally occurring gas that results from the decay of uranium, particularly in igneous rocks such as granite. Have students research the air pollution and health problems associated with radon. Students may be able to obtain a radon map for your county or state that will help them assess the potential for significant radon concentrations in your community. Encourage them to write a short informative essay based on their findings.
LS Interpersonal

Connection to **Chemistry**

Formaldehyde Formaldehyde is a colorless gas that has a strong odor. It is a very common industrial and commercial chemical that is used to make building materials and household products. In the home, significant amounts of formaldehyde are found in adhesives in plywood, particle board, furniture, and carpet. Other sources of formaldehyde may be foam insulation, gas stoves, tobacco smoke, and dry-cleaned clothing. The health effects of formaldehyde may include eye irritation, burning sensations in the throat, nausea, and difficulty breathing.

Indoor Air Pollution

The quality of air inside a home or a building is sometimes worse than the quality of the air outside. Plastics and other industrial chemicals are major sources of pollution. These compounds can be found in carpets, building materials, paints, and furniture, particularly when these items are new. **Figure 9** shows examples of some indoor air pollutants.

Buildings that have very poor air quality have a condition called **sick-building syndrome.** Sick-building syndrome is most common in hot places where buildings are tightly sealed to keep out the heat. In Florida, for example, a new, tightly sealed county courthouse had to be abandoned. Half of the people who worked there developed allergic reactions to fungi that were growing in the air-conditioning ducts, ceiling tiles, carpets, and furniture.

Identifying and removing the sources of indoor air pollution is the most effective way to maintain good indoor air quality. Ventilation, or mixing outdoor air with indoor air, is also necessary for good air quality. When activities such as renovation and painting, which cause indoor air pollution, are undertaken, ventilation should be increased.

CASE STUDY

The Health Effects of Ground-Level Ozone

You have learned that the ozone layer in the stratosphere shields the Earth from the harmful effects of ultraviolet radiation from the sun. At the surface of the Earth, however, ozone is a human-made air pollutant that at certain concentrations may affect human health.

Ozone forms from the reaction of volatile organic compounds (VOCs) and nitrogen oxides (NO_x) in the presence of heat and sunlight. High concentrations of ozone form in the atmosphere on sunny days that have high temperatures in the late spring, summer, and early fall. The sources of VOCs and NO_x emissions are largely motor vehicles, power plants, gasoline vapors, and chemical solvents. Most ozone pollution forms in urban

▶ Children who engage in vigorous outdoor activities where pollutant concentrations are often high may have a greater risk of developing asthma or other respiratory illnesses.

and suburban areas. However, pollutants may be transported hundreds of kilometers from their source.

As ozone concentrations in the atmosphere increase, greater numbers of people may experience harmful health effects of ozone on the lungs. Some of the short-term effects of ozone on health include irritation of the respiratory system, a reduction in lung function, the aggravation of asthma, and inflammation to the lining of the lungs. Scientists believe that ozone may

have other damaging effects on human health. Lung diseases such as bronchitis and emphysema may be aggravated by ozone. Scientists

310

Notable Quotes —————— GENERAL

"Ill air slays sooner than the sword."
— Ratis Raving, c. 1450

"Fresh air keeps the doctor poor."
— Danish proverb

Write these quotes on the board, and ask students to explain them. Ask students to think up their own humorous or pithy sayings about air quality.

REAL-LIFE CONNECTION —— GENERAL

OSHA The Occupational Safety and Health Administration (OSHA) enforces regulations concerning a number of airborne substances in the workplace. If possible, invite an OSHA representative to speak to the class about such regulations. Or ask an office manager to speak to the class about steps that must be taken to meet OSHA regulations and to control indoor air pollution in the workplace.

Bleach, sodium hydroxide, and hydrochloric acid from household cleaners

Nitrogen oxides from unvented gas stove, wood stove, or kerosene heater

Fungi and bacteria from dirty heating and air conditioning ducts

Tetrachloro-ethylene from dry-cleaning fluid

Carbon monoxide from faulty furnace and car left running

Methylene chloride from paint strippers and thinners

Paradichloro-benzene from moth-ball crystals and air fresheners

Radon-222 from uranium-containing rocks under the house

Tobacco smoke from cigarettes and pipes

Formaldehyde from furniture, carpeting, particleboard, and foam insulation

Gasoline from car and lawn mower

Figure 9 ▶ Some indoor air pollutants and their sources are shown here.

▶ A therapist performs a lung-function test on a patient by using a machine that measures various aspects of lung function.

believe that permanent lung injury may result from repeated short-term exposure to ozone pollution. Children who are regularly exposed to high concentrations of ozone may have reduced lung function as adults. Exposure to ozone may also accelerate the natural decline in lung function that is part of the aging process.

Those who are most at risk from ozone include children, adults who exercise or work outdoors, older people, and people who suffer from respiratory diseases. In addition, there are some healthy individuals who have unusually high susceptibility to ozone.

CRITICAL THINKING

1. Making Decisions Write a brief paragraph explaining whether or not lung-function tests should be mandatory for children who live in urban areas where high concentrations of ozone are frequent. **WRITING SKILLS**

2. Making Decisions If lung-function tests become mandatory, who will pay for these tests, and who will provide the equipment? Would these tests be performed at school, in a doctor's office, or at a hospital?

311

CASE STUDY

Ozone Action Days Several cities in the United States have established Ozone Action Day programs. These are days when meteor-ologists have predicted conditions that will likely cause high ozone levels. Hotlines and Web sites provide daily information on ozone levels. In larger cities, the Environmental Protection Agency has required the levels of pollutants, including ozone levels, be reported in terms of the Air Quality Index, or AQI. A reading over 100 is considered unhealthy for children and people with respiratory prob-lems. When the reading exceeds 150, even

healthy people are encouraged to limit out-door activities. You can use the following tips to reduce ozone levels:

- Limit driving and use public transportation.
- Refuel vehicles after 7:00 P.M.
- Don't mow the lawn until late evening and postpone the use of volatile household products.

Answers to Critical Thinking

1. Answers may vary.
2. Answers may vary.

Asbestos Contamination Divide the class into three groups, and have each group research one of the following topics: (1) Is asbestos a problem in your area? If so, where and why? (2) What are the effects of asbestos on human health? (3) How can asbestos contamination be dealt with? When all groups have finished compiling their data, have them write them up and have an "asbestos-awareness day" during which each group shares its findings with the class. Suggest that students include their findings in their *EcoLog*.
LS Interpersonal Co-op Learning

REAL-LIFE ———
● CONNECTION ——— GENERAL

Our Noisy Oceans Scientists are warning that underwater noise pollution is disrupting the ability of marine mammals to navigate and communicate. Whales and dolphins rely on super-sensitive hearing and underwater sounds to communicate over large distances. In the ocean, sound waves travel faster than they do on land and they lose less energy over distance. Ships, military sonar, oil exploration and drilling, motor boats, and jet skis are all sources of disruptive noise. For example, the noise made by a tanker is louder than the sounds a blue whale uses to communicate. Loudness is only one factor that affects whales and dolphins. Pitch and frequency of noises may also disturb communication and navigation.

Figure 10 ▶
Asbestos (right) forms in long, thin fibers. The worker above is removing debris from a structure that was built with asbestos.

Table 2 ▼

Intensity of Common Noises	
Noise	**Intensity (dB)**
Rocket engine	180
Jet engine	140
Rock-and-roll concert	120
Car horn	110
Chainsaw	100
Lawnmower	90
Doorbell	80
Conversation	60
Whisper	30
Faintest sound heard by the human ear	0

Radon Gas Radon gas is colorless, tasteless, and odorless. It is also radioactive. *Radon* is one of the elements produced by the decay of uranium, a radioactive element that occurs naturally in the Earth's crust. Radon can seep through cracks and holes in foundations into homes, offices, and schools, where it adheres to dust particles. When people inhale the dust, radon enters their lungs. In the lungs, radon can destroy the genetic material in cells that line the air passages. Such damage can lead to cancer, especially among people who smoke. Radon is the second-leading cause of lung cancer in the United States.

Asbestos Several minerals that form in long, thin fibers and that are valued for their strength and resistance to heat are called **asbestos.** Asbestos is primarily used as an insulator and as a fire retardant, and it was used extensively in building materials. The U.S. government banned the use of most asbestos products in the early 1970s. Exposure to asbestos in the air is dangerous. Asbestos fibers that are inhaled can cut and scar the lungs, which causes the disease asbestosis. Victims of the disease have more and more difficulty breathing and may eventually die of heart failure. Schools in the United States have taken this threat seriously. Billions of dollars have been spent to remove asbestos from school buildings. **Figure 10** shows asbestos fibers and asbestos removal from a building.

Noise Pollution

A sound of any kind is called a noise. However, some noises are unnecessary and can cause noise pollution. Noise is a pollutant that affects human health and the quality of human life. Airplanes, construction equipment, city traffic, factories, home appliances, and lawnmowers are some of the examples of things that make unnecessary sounds that commonly travel through the air. Health problems that can be caused by noise pollution include loss of hearing, high blood pressure, and stress. Noise can also cause loss of sleep, which may lead to decreased productivity at work and in the classroom.

The intensity of sound is measured in units called **decibels (dB).** The quietest sound that a human ear can hear is represented by 0 db. For each increase in decibel intensity, the decibel level is 10 times higher than the previous level. For example, 20 dB is 10 times the intensity of 10 dB, 30 dB is 100 times the intensity of 10 dB, and 40 dB is 1,000 times the intensity of 10 dB. **Table 2** shows the intensity of some common sounds. A sound of 120 dB is at the threshold of pain. Permanent deafness may come as a result of continuous exposure to sounds over 120 dB.

Homework ——————— GENERAL

Noise Pollution in Your Community Ask students how noise pollution affects their lives and how it makes them feel. Have them brainstorm a list of possible sources of noise pollution in your community. Have groups use sound level meters and tape recorders to map and record the noise levels in different areas of your community. Students should map the average peak sound levels 10 times in one day using the sound level meters.

Students can create an audio journal of common sounds using the tape recorder. Combine their results to make a class map of the loudest and quietest areas of your community and an audio field guide to local noises. Remind students that noise levels greater than 120 dB can permanently damage their ears.
LS Auditory/Interpersonal

English Language Learners

Light Pollution

Unlike air or water pollution, light pollution does not present a direct hazard to human health. However, light pollution does negatively affect our environment. The use of inefficient lighting in urban areas is diminishing our view of the night sky. In urban areas, the sky is often much brighter than the natural sky.

A more important environmental concern of inefficient lighting is energy waste. For example, energy is wasted when light is directed upward into the night sky and lost to space, as shown in **Figure 11**. Examples of inefficient lighting are billboards and other signs that are lit from below, the lighting of building exteriors, and poor-quality street lights. One solution to energy waste includes shielding light so it is directed downward. Using time controls so that light is used only when needed and using low-pressure sodium sources—the most energy-efficient source of light—wherever possible are two other solutions.

Figure 11 ▶ This view of Seattle shows how lighting in urban areas can cause skyglow, which is an effect of light that can dramatically reduce our view of the night sky.

FIELD ACTIVITY

Light Pollution At night, in your neighborhood or from your front porch, note any efficient or inefficient uses of light that you see, and write down your observations in your *EcoLog*.

SECTION 2 Review

1. **Describe** the long-term effects and the short-term effects of air pollution on health.

2. **Describe** two ways in which indoor air pollution can be prevented.

3. **Describe** some of the human health problems caused by noise pollution.

4. **Describe** several solutions to the energy waste associated with light pollution.

CRITICAL THINKING

5. **Making Comparisons** Read the descriptions of noise and light pollution in this section. Explain ways in which noise pollution and light pollution are similar and ways they are different. `READING SKILLS`

6. **Analyzing Relationships** Molds can grow in new, tightly sealed buildings where the humidity is high and the ventilation is poor. Explain how you would control the growth of mold in this type of environment.

313

Answers to Section Review

1. Short-term effects include nausea, irritation of mucous membranes, and respiratory trouble. Long-term effects include emphysema, lung cancer, and heart disease.

2. Identifying and removing sources of indoor air pollution and improving ventilation (the mixing of outdoor with indoor air) help to control indoor air pollution.

3. Human health problems caused by noise pollution include loss of hearing, high blood pressure, stress, and loss of sleep.

4. Directing light downward, using time controls to use light only when needed, and using low-pressure sodium sources are solutions to energy waste related to light pollution.

5. Noise pollution and light pollution are similar in that they both lower quality of life. But noise pollution can cause human health problems, including deafness.

6. Sample answers: One could ensure that there is air flow in the building and dehumidify the building.

Overview

Before beginning this section, review with your students the Objectives listed in the Student Edition. This section explores the causes and effects of acid precipitation. International efforts to control acid precipitation are also discussed.

 Bellringer

Explain to students that regions directly downwind from major producers of particulate pollutants often have greater rainfall than the areas that are upwind. Ask students to consider why. (The particles act as condensation nuclei for water vapor. If enough condensation occurs, rain, sleet, or snow will fall.)
 **Logical**

Motivate

Discussion ——————— BASIC

Acid Precipitation and Cars
The paint on a car is beginning to fade after a few short years. Tell students that the cause of the fading is acid precipitation. Ask students if they are familiar with acid precipitation and some of its sources. Tell students that nitrogen oxides (NO_x) released by automobile emissions are one source of acid precipitation. The NO_x combines with water in the atmosphere and forms nitric acid, and this nitric acid falls to Earth's surface as acid precipitation. Make students aware of the irony that pollutants introduced into the atmosphere by automobiles can also damage automobiles.

SECTION 3
Acid Precipitation

Objectives

▶ Explain the causes of acid precipitation.
▶ Explain how acid precipitation affects plants, soils, and aquatic ecosystems.
▶ Describe three ways that acid precipitation affects humans.
▶ Describe ways that countries are working together to solve the problem of acid precipitation.

Key Terms

acid precipitation
pH
acid shock

Imagine that you are hiking through the forests of the Adirondack Mountains in New York. You come to a lake and sit down to rest. You are amazed at how clear the water is; it is so clear that you can see the bottom of the lake. But after a few minutes you feel uneasy. Something is wrong. What is it? Suddenly, you realize that the lake has no fish.

What Causes Acid Precipitation?

This lake and thousands of lakes throughout the world are victims of acid precipitation, which is also known as acid rain. **Acid precipitation** is precipitation such as rain, sleet, or snow that contains a high concentration of acids. When fossil fuels are burned, they release oxides of sulfur and nitrogen. When the oxides combine with water in the atmosphere, they form sulfuric acid and nitric acid, which fall as acid precipitation. This acidic water flows over and through the ground, and into lakes, rivers, and streams. Acid precipitation can kill living things, and can result in the decline or loss of some local animal and plant populations.

A **pH** (power of hydrogen) number is a measure of how acidic or basic a substance is. A pH scale is shown in **Figure 12.** As you can see from the scale, the lower the pH number is, the more acidic a substance is; the higher a pH number is, the more basic a substance is. Each whole number on the pH scale indicates a tenfold change in acidity.

Figure 12 ▶ The pH scale measures how basic or how acidic a substance is. Below are the pH measurements of some common substances.

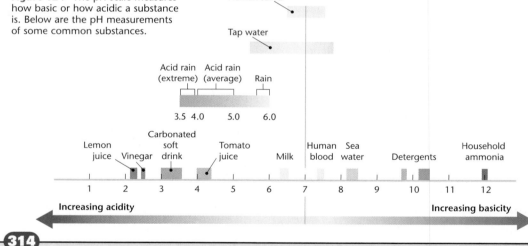

Chapter Resource File

• **Lesson Plan**
• **Active Reading** BASIC
• **Section Quiz** GENERAL

Transparencies

TT Bellringer
TT The pH Scale
TT How Acid Precipitation Forms

CHEMISTRY ——————
● CONNECTION —— ADVANCED

Sources of Acid Precipitation Power plants are some of the largest contributors of pollutants that cause acid precipitation. Sulfur dioxide, SO_2, is a noxious gas with a characteristic pungent smell; it is toxic to both plants and animals. SO_2 is produced mainly by power plants that burn high-sulfur coal (coal that contains iron sulfides) to generate electricity. A large power plant may burn 9,070 metric tons (10,000 tons) of coal per day. If the coal contains 2 percent sulfur, about 182 metric tons (200 tons) of SO_2 are produced every day.

Pure water has a pH of 7.0. Normal precipitation is slightly acidic, because atmospheric carbon dioxide dissolves into the precipitation and forms carbonic acid. Normal precipitation has a pH of about 5.6. Precipitation is considered acid precipitation if it has a pH of less than 5.0. **Figure 13** shows how acid precipitation forms.

The pH of precipitation varies between different geographic areas. For example, Eastern Europe and parts of Scandinavia have precipitation with a pH of 4.3 to 4.5, whereas the remainder of Europe has precipitation with pH values of 4.5 to 5.1. The pH of precipitation in the eastern United States and Canada ranges from 4.2 to 4.8. The most acidic precipitation in North America occurs around Lake Erie and Lake Ontario. It has a pH of 4.2.

How Acid Precipitation Affects Soils and Plants

Plant communities have adapted over long periods of time to the acidity of the soil in which they grow. Acid precipitation can cause a drop in the pH of soil and water. This increase in the concentration of acid is called *acidification*. Acidification changes the balance of a soil's chemistry in several ways. When the acidity of soil increases, some nutrients are dissolved and washed away by rainwater. Increased acidity causes aluminum and other toxic metals to be released and possibly absorbed by the roots of plants. Aluminum also causes root damage. Sulfur dioxide in water vapor clogs the openings on the surfaces of plants. **Figure 14** shows the harmful effects of acid precipitation on trees.

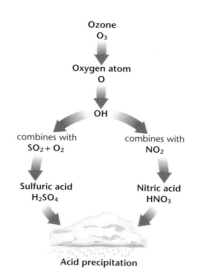

Figure 13 ▶ Sulfur oxides and nitrogen oxides combine with water in the atmosphere to form sulfuric and nitric acids. Rainfall that contains these acids is called *acid precipitation*.

Graphic Organizer

Chain-of-Events Chart

Create the **Graphic Organizer** entitled "Chain-of-Events Chart" described in the Appendix. Then, fill in the chart with details about each step of the formation of acid precipitation.

Figure 14 ▶ The trees in this forest in Poland show the dramatic effect that acid precipitation can have on plants. Damage to more than 16 million acres in nine European countries has been linked to acid precipitation.

315

Graphic Organizer GENERAL

Chain of Events Chart
You may want to use this Graphic Organizer to assess students' prior knowledge before beginning a discussion of each step of the formation of acid precipitation. You may also choose to use a similar activity as a quiz to assess students' knowledge of formation of acid precipitation or to assess students' knowledge after students have read the description of formation of acid precipitation.

METEOROLOGY
CONNECTION — **ADVANCED**

The Adirondack Mountains Ask students to think about why the Adirondack Mountains have been greatly affected by acid precipitation. Display a map that shows the topography of the region. Point out that because of their topography, the Adirondacks consistently trigger rainfall and snowfall on a large scale. Furthermore, since prevailing winds blow from west to east, the Adirondacks are the first mountain range encountered by winds from the upper Midwest industrial belt. Consequently, the Adirondacks receive the full, undiluted load of these pollutants. **LS Visual**

MISCONCEPTION ALERT

The pH Scale Students are often confused by the pH scale. The term *pH* is French and translates as "power of hydrogen." It refers to the concentration of hydronium ions in a solution. A decrease of one number on the pH scale represents an increase in the concentration of hydronium ions by a power of 10. Thus an acid with a pH of 2 is 100 times as concentrated as an acid with a pH of 4. Remind students that acidic solutions have a pH less than 7 and that basic solutions have a pH greater than 7.

INCLUSION Strategies

• Gifted and Talented
Have students research the effects of acid precipitation on structures such as statues, tombstones, and buildings. Direct their research by mentioning that the famous Egyptian obelisk that now resides in New York City's Central Park has suffered more erosion in the last 100 years than in the previous 3,000. Encourage students to document their findings with pictures whenever possible. Also suggest that students find out about any actions that are being taken to protect these structures. Students may also wish to research whether anything is being done by modern artists and architects to guard against damage by acid precipitation. Suggest that students share their findings with the class in the form of a newspaper article or an academic report, with supporting photographs or drawings. Encourage students to include their article or report in their **Portfolio.**

Terms and Concepts Prepare index cards that list terms, places, objects, and concepts mentioned in this section. For example, you could list acid precipitation, fossil fuels, acid shock, fish, amphibians, Adirondack Mountains, and the Canada-U.S. Air Quality Agreement. Make one card for each student. Once each student has a card, call on a student to read his or her card aloud. Then let the other students who think their card is related to the initial one explain the relationship. Or you could call on a second student and have the class explain whether the two cards are related and, if so, how. Record student responses in the form of a concept map on the board.
LS Verbal/Logical

QuickLAB

Skills Acquired:
- Constructing Models
- Experimenting

Answers

1. The mixture will become less acidic; as the acidic vinegar dissolves the basic limestone, the solution will become more basic.

Figure 15 ▶ Fish are vulnerable to acid shock, a sudden influx of acidic water into a lake or stream that causes a rapid change in pH.

QuickLAB

Neutralizing Acid Precipitation

Procedure

1. Pour 1/2 Tbsp of **vinegar** into one cup of **distilled water**, and stir the mixture well. Check the pH of the mixture by using **pH paper**. The pH should be about 4.

2. Crush one stick of **blackboard chalk** into a powder. Pour the powder into the vinegar and water mixture. Check the pH of the mixture.

Analysis

1. Did the vinegar and water mixture become more or less acidic after the powdered chalk was poured in?

Acid Precipitation and Aquatic Ecosystems

Aquatic animals are adapted to live in an environment with a particular pH range. If acid precipitation falls on a lake and changes the water's pH, acid can kill aquatic plants, fish, and other aquatic animals. The change in pH is not the only thing that kills fish. Acid precipitation causes aluminum to leach out of the soil surrounding a lake. The aluminum accumulates in the gills of fish and interferes with oxygen and salt exchange. As a result, fish are slowly suffocated.

The effects of acid precipitation are worst in the spring, when acidic snow that accumulated in the winter melts and rushes into lakes and other bodies of water. This sudden influx of acidic water that causes a rapid change in the water's pH is called **acid shock.** This phenomenon causes large numbers of fish in a population to die, as shown in **Figure 15.** Acid shock also affects the reproduction of fish and amphibians. They produce fewer eggs, and these eggs often do not hatch. The offspring that do survive often have birth defects and cannot reproduce.

To counteract the effects of acid precipitation on aquatic ecosystems, some states in the United States and some countries spray powdered limestone (calcium carbonate) on acidified lakes in the spring to help restore the natural pH of the lakes. Because lime has a pH that is basic, the lime raises the pH of the water. Unfortunately, enough lime cannot be spread to offset all acid damage to lakes.

Acid Precipitation and Humans

Acid precipitation can affect humans in a variety of ways. Toxic metals such as aluminum and mercury can be released into the environment when soil acidity increases. These toxic metals can find their way into crops, water, and fish. The toxins then poison the human body.

Acid precipitation can lead to other human health problems. Research has indicated that there may be a correlation between large amounts of acid precipitation received by a community and an increase in respiratory problems in the community's children.

The standard of living of some people is affected by acid precipitation. Decreases in numbers of fish caused by the acidification of lakes and streams can influence the livelihood of commercial fisherman and people involved in the sport-fishing industry. Forestry is also affected when trees are damaged by acid precipitation.

Acid precipitation can dissolve the calcium carbonate in common building materials, such as concrete and limestone. Some of the world's most important and historic monuments, including those made of marble, are being affected by acid precipitation. For example, sulfur dioxide has caused black crusts to form on the carbonate stones of historic Greek monuments.

BIOLOGY CONNECTION — ADVANCED

Acid Precipitation in North America Ask students, "What areas in North America are losing species due to acid precipitation?" Discuss the decline of Atlantic salmon in Nova Scotia. Atlantic salmon have difficulty reproducing in bodies of water with a pH of 5.5. Because the pH scale works logarithmically, then a pH of 5.0 is toxic. Fourteen rivers in Nova Scotia have a pH of less than 4.7, and Atlantic salmon do not live in these rivers. Another 20 rivers have pH levels of 4.7 to 5.0, and salmon live only in a few tributaries of these rivers.
LS Logical

SKILL BUILDER — GENERAL

Writing Ask students to write a paragraph on whether they think it would be better to prevent the release of acid-producing pollutants through expensive pollution-control equipment or to find ways of cleaning up the environmental effects of acid precipitation after it has occurred. (Most students will probably choose the first option because it is more expensive to clean up pollution than to prevent it and because cleaning up after the fact cannot bring back organisms that have died from the direct or indirect effects of acid precipitation.) Suggest that students include their paragraph in their *EcoLog.*

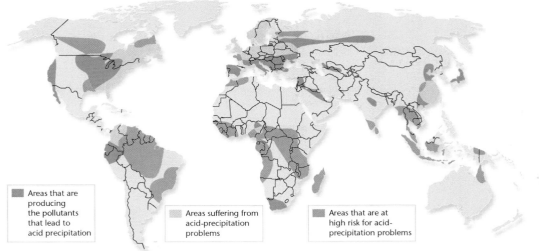

Areas that are
producing
the pollutants
that lead to
acid precipitation

Areas suffering from
acid-precipitation
problems

Areas that are at
high risk for acid-
precipitation problems

Figure 16 ▶ Acid precipitation is a global problem.

International Conflict and Cooperation

One problem in controlling acid precipitation is that pollutants may be released in one geographical area and fall to the ground hundreds of kilometers away. For example, almost half of the acid precipitation that falls in southeastern Canada results from pollution produced in Ohio, Indiana, Pennsylvania, Illinois, Missouri, West Virginia, and Tennessee. **Figure 16** shows areas of the world that produce pollutants and areas which are then affected by acid precipitation.

Because acid precipitation falls downwind, the problem of solving acid precipitation has been difficult, especially on the international level. In the spirit of cooperation, Canada and the United States signed the Canada–U.S. Air Quality Agreement in 1991. Both countries agreed to reduce acidic emissions that flowed across the Canada–U.S. boundary. More international agreements such as this may be necessary to control the acid-precipitation problem.

internet connect

www.scilinks.org
Topic: Acid Rain
SciLinks code: HE4001

SCiLINKS Maintained by the National Science Teachers Association

SECTION 3 Review

1. **Explain** how acid precipitation forms.

2. **Describe** the harmful effects that acid precipitation can have on plants, soils, and aquatic ecosystems.

3. **Describe** three ways in which acid precipitation can affect humans.

4. **Describe** a way in which countries are working together to solve the problem of acid precipitation.

CRITICAL THINKING

5. **Inferring Relationships** In addition to negatively affecting forestry and the fishing industry, how might acid precipitation affect local economies?

6. **Analyzing Viewpoints** Write a short essay in which you discuss whether or not a country that releases significant amounts of pollutants into the air that fall as acid precipitation in another country should be expected to pay some of the costs of cleanup. **WRITING SKILLS**

317

Answers to Section Review

1. When fossil fuels are burned, they release oxides of sulfur and nitrogen. When the oxides combine with water in the atmosphere, they form sulfuric and nitric acid, which fall as acid precipitation.

2. Answers may vary. Acid precipitation can cause a decrease in the pH of soil and water, which can harm plants and aquatic ecosystems.

3. Acid precipitation can release toxic metals into the environment. These metals may enter the

human body. Acid precipitation can lead to human health problems. Acid precipitation can decrease the human standard of living by affecting fishing and forestry.

4. Sample answer: Countries have enacted international agreements to control acid precipitation.

5. Answers may vary. Sample answer: Acid precipitation may affect farm crops locally.

6. Answers may vary.

Close

Reteaching — BASIC

Subsection Review Have students recap the main points of each subsection. Ask students to draw diagrams illustrating these main points. Quiz students orally to ensure their understanding. *English Language Learners*

Quiz — GENERAL

1. What does pH stand for? (power of hydrogen)

2. Describe one way in which countries are working together to solve the problem of acid precipitation. (Canada and the United States signed the Canada–U.S. Air Quality Agreement in 1991 and agreed to reduce acidic emissions that crossed the Canada–U.S. boundary.)

3. When soil acidity increases, what are some toxins that can be released? (aluminum and mercury)

4. What is an indirect effect that pollution can have on birds? (Pollution can reduce the amount of suitable food for nestlings.)

Alternative Assessment — GENERAL

The Effects of Acid Precipitation in Forests Tell students that acid precipitation has killed 30 percent of the trees in a hypothetical forest ecosystem. Ask students to predict the indirect effects of that precipitation on the ecosystem. Students can respond with a short paragraph or with a concept map.

Transparencies

TT A Global Look at Acid Precipitation

CHAPTER **12** **Highlights**

Alternative Assessment — GENERAL

Mapping Air Quality in the U.S.

Have students create maps to illustrate the industrial centers of the United States. Ask each student to choose an industry to focus on, such as car production, mining, or smelting. Alternatively, they could choose the most populated areas, or urban areas without public transportation. They can use color codes to illustrate the areas that are most likely to suffer from poor air quality. When possible, have students create charts or graphs to back up their work. Ask them to be prepared to describe to the class the meanings of the different symbols or colors they chose.

LS Visual/Intrapersonal

Chapter Resource File

- **Chapter Test** GENERAL
- **Chapter Test** ADVANCED
- **Concept Review** GENERAL
- **Critical Thinking** ADVANCED
- **Test Item Listing**
- **Field Activity** GENERAL
- **Observation Lab** ADVANCED
- **CBL™ Probeware Lab** GENERAL
- **Consumer Lab** BASIC
- **Long-Term Project** GENERAL

---

1 What Causes Air Pollution?

Key Terms

air pollution, 303
primary pollutant, 303
secondary pollutant, 303
smog, 308
temperature inversion, 308

Main Ideas

► Primary pollutants are pollutants put directly in the air by human activity.

► Secondary pollutants are formed when a primary pollutant comes into contact with other primary pollutants or with naturally occurring substances and a chemical reaction takes place.

► Most air pollution comes from vehicles and industry.

► Air pollution that hangs over cities and reduces visibility is called *smog*.

► Pollution can be trapped near the surface of the Earth by a condition known as temperature inversion.

---

2 Air, Noise, and Light Pollution

Key Terms

sick-building syndrome, 310
asbestos, 312
decibel (dB), 312

Main Ideas

► Air pollution may have both long- and short-term effects on human health.

► The air indoors may be more polluted than the air outside. Plastics, cleaning chemicals, and building materials are major sources of indoor air pollution.

► Noise is a pollutant that affects human health and the quality of life.

► Inefficient lighting diminishes our view of the night sky and wastes energy.

---

3 Acid Precipitation

Key Terms

acid precipitation, 314
pH, 314
acid shock, 316

Main Ideas

► Acid precipitation is precipitation such as rain, sleet, or snow that contains a high concentration of acids.

► Acid shock occurs when a sudden influx of acidic water enters a lake or stream and causes a rapid change in pH that harms aquatic life.

► Pollutants released in one geographical area may fall to the ground hundreds of kilometers away as acid precipitation—sometimes in another country.

318

Using Key Terms

Use each of the following terms in a sentence.

1. *air pollution*
2. *smog*
3. *temperature inversion*
4. *sick-building syndrome*
5. *pH*

For each pair of terms, explain how the meanings of the terms differ.

6. *primary pollutant* and *secondary pollutant*
7. *asbestos* and *radon*
8. *pH* and *acid precipitation*
9. *acidification* and *acid shock*

✓ **STUDY TIP**

Predicting Exam Questions Before you take a test, do you ever attempt to predict what the questions will be? For example, of the 10 multiple-choice questions that appear on this page, how many would you have predicted to be asked in a review of this chapter? Before your next test, predict and answer possible exam questions.

Understanding Key Ideas

10. Which of the following air pollutants is *not* a primary pollutant?
    a. particulate matter
    b. ozone
    c. sulfur dioxide
    d. volatile organic compounds

11. A device used to clean exhaust gases before they exit an automobile's tailpipe is called a(n)
    a. electrostatic precipitator.
    b. catalytic converter.
    c. scrubber.
    d. None of the above

12. The majority of sulfur dioxide produced by industry comes from
    a. oil refineries.
    b. dry cleaners.
    c. chemical plants.
    d. coal-burning power plants.

13. Which of the following substances is *not* involved in the chemical reaction that produces smog?
    a. sunlight
    b. particulate matter
    c. automotive exhaust
    d. ozone

14. Which of the following respiratory diseases is considered a long-term effect of air pollution on human health?
    a. emphysema
    b. bronchitis
    c. pneumonia
    d. all of the above

15. Which of the following substances is a colorless, tasteless, and odorless radioactive gas?
    a. asbestos
    b. carbon monoxide
    c. radon
    d. ozone

16. A sound measuring 40 dB has how many times the intensity of a sound that measures 10 dB?
    a. 4 times
    b. 30 times
    c. 400 times
    d. 1,000 times

17. Which of the following choices is *not* an effective solution to the energy waste related to inefficient lighting?
    a. using low-pressure sodium lighting sources
    b. pointing lights on billboards and street signs upward
    c. placing light sources on time controls
    d. shielding light to direct it downward

18. Which of the following numbers on the pH scale would indicate that a substance is acidic?
    a. 5.0
    b. 7.0
    c. 9.0
    d. none of the above

19. Normal precipitation has a pH of
    a. 7.0.
    b. 5.6.
    c. 5.1.
    d. 4.5.

319

Assignment Guide

Section	Questions
1	1–3, 6, 10–13, 20, 27, 28, 33–36
2	4, 7, 14–17, 21, 22, 29, 31, 32, 36
3	5, 8, 9, 18, 19, 23–26, 30, 36

ANSWERS

Using Key Terms

1. Sample answer: Air pollution results when harmful substances build up to unhealthy levels in the air.

2. Sample answer: Smog results from chemical reactions that involve sunlight, air, automobile exhaust, and ozone.

3. Sample answer: A temperature inversion traps pollutants near Earth's surface.

4. Sample answer: Sick-building syndrome is a condition that occurs in buildings that have poor air quality.

5. Sample answer: pH is a measure of how acidic or basic a substance is.

6. Primary pollutants are released into the air by human activity. Secondary pollutants occur when primary pollutants come into contact with other primary pollutants or with naturally occurring substances.

7. Radon is a colorless, tasteless, and odorless radioactive gas. Asbestos is any of several minerals that form in long, thin fibers and are valued for their strength and heat resistance.

8. pH is a measure of the acidity or basicity of a substance. Acid precipitation is precipitation that contains a high concentration of acids.

9. The increase in the concentration of acid is acidification, whereas acid shock is the sudden influx of acid causing a sudden change in the pH of water.

Understanding Key Ideas

10. b
11. b
12. d
13. b
14. a
15. c
16. d
17. b
18. a
19. b

CHAPTER 12 Review

Short Answer

20. A zero-emission vehicle is a vehicle with no tailpipe emissions, no emissions from gasoline, and no emission-control systems that deteriorate over time. They include electric vehicles, vehicles with advanced batteries, and vehicles powered by hydrogen fuel.

21. Answers may vary. Some indoor air pollutants include fungi and bacteria, asbestos, nitrogen oxides, carbon monoxide, tobacco smoke, formaldehyde, and gasoline.

22. Radon can lead to the destruction of genetic material in the cells that line the air passages of the lungs, and this can potentially lead to lung cancer.

23. Powdered lime is a chemical that is used to counteract the effects of acid precipitation on aquatic ecosystems.

24. In short, the nations that produce the pollutants that cause acid precipitation aren't always the nations that are ultimately affected by acid precipitation.

Interpreting Graphics

25. Western New York and Pennsylvania, and central Maryland have the most acid precipitation, according to the map.

26. The areas with the highest pH are far from the industrial centers.

Concept Mapping

27. Answers to the concept mapping questions are on pp. 667–672.

Short Answer

20. Define the term *zero-emission vehicle*. What types of vehicles qualify as zero-emission vehicles?

21. Identify five indoor air pollutants and examples of sources of each pollutant.

22. Explain the health hazards that radon gas poses for humans.

23. Identify a chemical that is used to counteract the effects of acid precipitation on aquatic ecosystems.

24. Explain why acid precipitation is a source of international conflict and why international cooperation is necessary to resolve the problem.

Interpreting Graphics

The map below shows the pH of precipitation that has been measured at field stations in the northeastern United States. Use the map and legend to answer questions 25–26.

25. Which area(s) of the northeastern United States have the most-acid precipitation?

26. Are the areas that have the highest pH located close to or far from major cities?

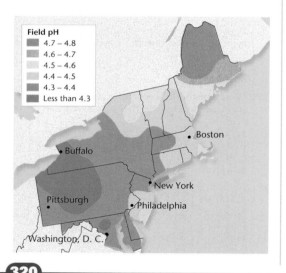

Field pH
- 4.7 – 4.8
- 4.6 – 4.7
- 4.5 – 4.6
- 4.4 – 4.5
- 4.3 – 4.4
- Less than 4.3

Boston
Buffalo
New York
Pittsburgh
Philadelphia
Washington, D. C.

320

Critical Thinking

28. Answers may vary. Students should include a thoughtful examination of economic values and environmental values.

29. Answers may vary. Students should include quality of life concerns as well as economic ones.

30. Students should infer that wind in this part of North America blows predominantly north to northeast.

Concept Mapping

27. Use the following terms to create a concept map: *air pollution, primary pollutant, volatile organic compound, scrubber, secondary pollutant, smog,* and *temperature inversion.*

Critical Thinking

28. **Making Decisions** Five states now have zero-emission vehicle programs in place that will help decrease some primary pollutants. What would be the advantages or disadvantages of a federal program that required automobile makers to produce a set number of ZEVs nationwide?

29. **Making Decisions** In some cities, noise-pollution laws, such as restrictions placed on the use of leaf blowers, have been put in place. Do you think the benefits of noise reduction outweigh the costs of enforcing the law?

30. **Inferring Relationships** As you read under the head "International Conflict and Cooperation," about half of the acid precipitation that falls in southeastern Canada is produced by pollutants from the United States. How do the acid pollutants get from their sources to southeastern Canada? **READING SKILLS**

Cross-Disciplinary Connection

31. **Health** Asbestos, lead paint, tobacco, and many other products have been linked to adverse effects on human health. Research one such case that has been brought into the courts. Describe the allegations and the outcome of the trial and write a paragraph that explains whether you agree or disagree with the decision. **WRITING SKILLS**

Portfolio Project

32. **Make a Poster** Create a poster similar to the diagram that appears in Figure 9. This diagram may be of your home, your garage, a portion of your school, or a particular classroom in your school. Use the diagram to identify and label potential sources of indoor air pollutants. Photographs may be used to document these sources.

Cross-Disciplinary Connection

31. Answers may vary.

Portfolio Project

32. Answers may vary.

MATH SKILLS

Use the graph below to answer questions 33 and 34.

33. Analyzing Data The graph below shows the change in air-pollution emissions in the United States between 1970 and 1997. Excluding NO_x, which emissions category experienced the greatest decrease over this period of time?

34. Interpreting Graphics Why is lead, Pb, shown separately from the other air pollutants?

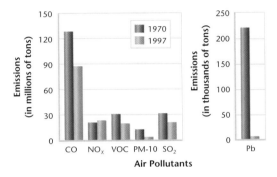

 WRITING SKILLS

35. Outlining Topics Outline the major sources of air pollution in the United States. Include information about pollution sources and pollution types.

36. Writing Persuasively Write a letter to a legislator that expresses your concern about a particular aspect of air, noise, or light pollution that is important to you.

STANDARDIZED TEST PREP

For extra practice with questions formatted to represent the standardized test you may be asked to take at the end of your school year, turn to the sample test for this chapter in the Appendix.

 READING SKILLS

Read the passage below, and then answer the questions that follow.

Lichens are unique organisms that consist of a fungus and microscopic alga that live together and function as a single organism. The alga is the photosynthetic partner, whereas the fungus absorbs water and minerals and anchors the plant. Lichens form crusts or leafy growths on rocks, trees, and bare ground. Lichens do not have roots. Instead, they absorb the nutrients they need directly from rain. Lichens grow very slowly and can live for centuries. Species of lichens have adapted to almost every environment in the world.

Lichens are sensitive to air pollution, particularly sulfur dioxide. When lichens are exposed to high levels of sulfur dioxide, they absorb the sulfur that is contained in rain. The sulfur destroys chlorophyll and inhibits photosynthesis. So, lichens are good indicators of air pollution. Lichens usually disappear from areas where sulfur dioxide levels are high. Where the air is free of pollutants, a greater number of lichens will usually be present. In areas where sulfur dioxide pollution is decreasing, lichens will slowly return and colonize the area.

1. Which of the following statements about lichens is true?
   a. Lichens are present when sulfur dioxide levels are high.
   b. Lichens absorb nutrients through their root systems.
   c. Lichens photosynthesize.
   d. Lichens grow only where the climate is moderate.

2. Where would you be most likely to see the greatest number of lichens?
   a. in areas where sulfur dioxide levels are high
   b. in areas where sulfur dioxide levels are low
   c. in areas where sulfur dioxide levels are decreasing
   d. in areas where sulfur dioxide is absent

Math Skills

33. CO

34. Lead is shown separately from the other air pollutants because lead emissions are measured in thousands of tons rather than millions of tons.

Writing Skills

35. Outlines should include the five categories of primary pollutants and the concepts of secondary pollutants (including ozone) and indoor air pollution.

36. Answers will vary.

Reading Skills

   1. c
   2. d

THE ACID TEST

Teacher's Notes

Time Required
one 45-minute class period

Lab Ratings

EASY ————————→ HARD

TEACHER PREPARATION 🧪🧪
STUDENT SETUP 🧪🧪🧪
CONCEPT LEVEL 🧪🧪
CLEANUP 🧪🧪🧪

Skills Acquired
- Predicting
- Constructing Models
- Experimenting
- Collecting Data
- Organizing and Analyzing Data
- Communicating

The Scientific Method
In this lab, students will
- Make Observations
- Ask Questions
- Analyze the Results
- Draw Conclusions
- Communicate the Results

Materials
The materials listed on the student page are enough for groups of two or three students.

Objectives

▶ **Perform** a chemical reaction that produces sulfur dioxide, a component of acid precipitation.

▶ **USING SCIENTIFIC METHODS**
Hypothesize what the effects of acids that contain sulfur on plants will be.

Materials

beaker, 50 mL
clear plastic bags, large (2)
houseplants of the same type, potted (2)
sodium nitrite (2 g)
sulfuric acid, 1 M (2 mL)
twist tie or tape

▶ **Step 1** Place a plant and a beaker that contains sodium nitrite into a plastic bag. Do not seal the bag.

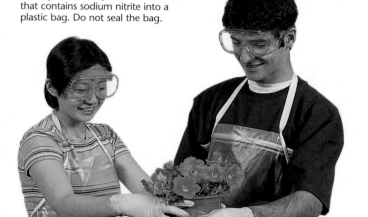

322

The Acid Test

Acid precipitation is one of the effects of air pollution. When pollutants that contain nitrogen or sulfur react with water vapor in clouds, dilute acid forms. These acids fall to Earth as acid precipitation.

Often, acid precipitation does not occur in the same place where the pollutants are released. The acid precipitation usually falls some distance downwind—sometimes hundreds of kilometers away. Thus, the sites where pollutants that cause acid precipitation are released may not suffer the effects of acid precipitation.

Coal-burning power plants are one source of air pollution. These power plants release sulfur dioxide into the air. Sulfur dioxide reacts with the water vapor in air to produce acid that contains sulfur. This acid later falls to Earth as acid precipitation.

In this investigation, you will create a chemical reaction that produces sulfur dioxide. The same acids that result from coal-burning power plants will form. You will see the effects of acid precipitation on living things—in this case, plants.

Procedure

1. Place 2 g of sodium nitrite in a beaker. Place a plant and the beaker inside a plastic bag. Do not seal the bag. CAUTION: Steps 2–4 should be carried out *only* under a fume hood or outdoors.

2. Carefully add 2 mL of a 1 M solution of sulfuric acid to the beaker. Immediately seal the bag tightly, and secure the bag with a twist tie or tape. CAUTION: Because this reaction produces sulfur dioxide, a toxic gas, the bag should have no leaks. If a leak occurs, move away from the bag until the reaction is complete and the gas has dissipated.

3. Seal the same type of plant in an identical bag that does not contain sodium nitrite or sulfuric acid.

Safety Cautions

Before attempting this activity, become familiar with the material safety data sheet for sulfur dioxide. When working with caustic or poisonous chemicals, use extreme caution. Allow only your most mature students to handle these materials. Alternatively, you may wish to handle the chemicals yourself or perform the procedure as a demonstration. For this activity, it is essential that a functioning fume hood is used to safely remove the sulfur

dioxide gas. A functioning eyewash station should be immediately accessible. Test the bags beforehand for possible leaks. Use only bags that are free of leaks. Sulfur dioxide is very poisonous. Keep students at least 5 m from the simulation for the duration of the reaction. Pour all leftover acid solution into a safe container, and neutralize it to pH 7 by adding a dilute base, drop by drop, before pouring it down the drain.

Day	Control Plant	Experimental Plant
1		
2	DO NOT WRITE IN THIS BOOK	
3		

4. After 10 minutes, cut both bags open. Stay at least 5 m from the bags as the sulfur dioxide gas dissipates. Keep the plants and bags under the fume hood.

5. Predict the effects of the experiment on each plant over the next few days. Record your predictions.

6. Observe both plants over the next three days. Record your observations below.

Analysis

1. **Examining Data** How closely did your predictions about the effects of the experiment on each plant match your observations?

2. **Explaining Events** What does this experiment suggest about the effects of acid precipitation on plants?

Conclusions

3. **Drawing Conclusions** In what ways is this a realistic model of acid precipitation?

4. **Drawing Conclusions** In what ways is this experiment *not* a realistic simulation of acid precipitation?

Extension

1. **Analyzing Models** Would you expect to see similar effects occur as rapidly, more rapidly, or less rapidly in the environment? Explain your answer.

2. **Building Models** Acid precipitation is damaging to plants because the sulfur dioxide contained in the water vapor clogs the openings on the surfaces of plants and interferes with photosynthesis. What kind of a safe model would demonstrate the damaging effects of acid precipitation in the form of water vapor on plant photosynthesis? Would this model be a realistic simulation of acid precipitation?

323

Chapter Resource File
• Datasheets for In-Text Labs
• Lab Notes and Answers

CLASSROOM TESTED & APPROVED

Michael Lubich
Maple Town High School
Greensboro, Pennsylvania

Answers to Analysis

1. Answers may vary.

2. The acid precipitation damages the vegetation.

Answers to Conclusion

3. All of the basic ingredients are the same: sulfur dioxide, water, and living plants.

4. The acidity is much greater in the simulation than in real life; the length of exposure is much shorter; little or no precipitation actually forms; the simulation takes place in a sealed environment; in the simulation, the effects of exposure to acid occur more rapidly.

Answers to Extension

1. Effects would occur less rapidly in real life because the acid would be much less concentrated.

2. Answers may vary.

LIGHT SOURCES

Demonstration — GENERAL

Light Pollution Light pollution affects the quality of life, and it is a serious problem for astronomers. Demonstrate how ambient light affects the number of visible stars by using a slide projector, a piece of aluminum foil, and a flashlight. Poke small holes in the foil, and in a dark room, project light through the foil. Tell students that they might see this number of stars on a dark night. Ask students to count the stars. Then shine the flashlight on the screen, and ask students to count the stars again. Discuss with students some natural and artificial sources of ambient light. **LS Visual**

Student Opportunities

Reducing Light Pollution There are many different ways to reduce light pollution in your community. Interested students can contact organizations such as the International Dark Sky Association to find out what they can do. Have groups of students develop informational brochures that they can use to educate the public about the problems of light pollution. The brochures should list simple solutions to address this problem. **LS Visual**

Transparencies
TT Light Sources

LIGHT SOURCES

MAP SKILLS

This map of what the Earth looks like from space at night shows light sources that are human in origin. The map is a composite image made from hundreds of images taken by orbiting satellites. Use the map of light sources on Earth to answer the questions below.

1. **Inferring Relationships** Using the brightness of the light sources on the map as a key, can you estimate the locations of some of the most densely populated areas on Earth? Where are some of these areas?

2. **Inferring Relationships** Some climatic conditions on Earth, such as extreme cold, heat, wetness, or a thin atmosphere, make parts of our planet less

habitable than others. Examples of areas on our planet that do not support large populations include deserts, high mountains, polar regions, and tropical rain forests. From the map, can you identify regions of the Earth where climatic conditions may not be able to support large human populations? What are some of these places?

3. **Finding Locations** Many large cities are seaports that are located along the coastlines of the world's oceans. From the map, can you pick out light sources along coastlines that might indicate the sites of large ports? Identify some of these cities by name.

4. **Inferring Relationships** From the differences in the density of the light sources, can you pick out any international borders?

324

Answers to Map Skills

1. Some of the most densely populated places on Earth would be Western Europe, Japan, the eastern half of the United States, the West Coast of the United States, southeastern Canada, and central Mexico.

2. Areas of the Earth that are not able to support large human populations are the polar regions, the deserts of north Africa, central Asia, and western Australia, and the rain forests of South America.

3. Some easily recognizable port cities would be Rio de Janeiro, Brazil; Buenos Aires, Argentina; Cape Town, South Africa; Sydney, Australia; and Hong Kong, China, among others.

4. The international border between Canada and the United States is fairly easy to make out using the differences in density of light sources.

SOCIETY & the Environment

THE DONORA, PENNSYLVANIA, KILLER SMOG

For the residents of the small Monongahela Valley town of Donora, Pennsylvania, living with the smoke that billowed from the local zinc smelter was an everyday occurrence—until October 26, 1948. On that night, a temperature inversion and an absence of wind began to trap a deadly mixture of sulfur dioxide, carbon monoxide, and metal dust that would hang in the valley air for five days. Over that period of time, 20 residents lost their lives and 7,000 other residents—about half of the town's population—suffered some form of respiratory problems.

The Weekend of the Killer Smog

By Saturday afternoon, October 29, 1948, the yellowish smog had become so thick that spectators in the stands at a local high-school football game could not see the players on the field. Only the whistles of the referees could be heard. By nightfall, driving was unsafe. This proved to be catastrophic because doctors recommended that any residents who suffered from respiratory ailments be evacuated from town. In an attempt to alleviate the suffering of people who were struggling to breathe, several local firemen carried oxygen tanks through the streets to different homes. Because of the low visibility, the firemen had to feel their way along buildings and fences. Because the supply of oxygen was limited, only a few breaths of oxygen could be given to each person. Eleven people died that night. A makeshift morgue was set up in the local community center.

Even as the killer smog choked the valley, the zinc smelter continued production throughout the night. The smelter continued sending more gases and dust into the air over Donora. The smelter was shut down only when the magnitude of the problem became apparent—6:00 A.M. on Sunday, October 30, 1948.

Later that day, a drizzling rain began to fall and washed the pollutants from the sky. By the time the rain fell, 20 people ages 52 to 85, who suffered from respiratory ailments, were dead. Thousands of other people were at home in bed or were filling the corridors and examining rooms of the two area hospitals. People who were less affected by the smog suffered from nausea and vomiting, headaches, and abdominal cramps. Some victims were choking or coughing up blood. The zinc smelter resumed operation on Monday morning, October 31.

The Aftermath

The smog of Donora was one of the United States' most serious environmental disasters. Shortly after the incident, investigations were undertaken by the Pennsylvania Department of Health, the U.S. Public Health Service, and other agencies. This was the first time an organized attempt was made to document the effects of air pollution on health in the United States. The knowledge that air pollution could be linked directly to the deaths of individuals resulted in legislation at the local, regional, state, and federal levels. These laws were set to limit emissions of sulfur dioxide, carbon monoxide, particulate matter, and other pollutants. The greatest legacy of the Donora tragedy was passage of the Clean Air Act of 1970.

▶ This historical photo from the *Pittsburgh Gazette* captures the town of Donora, Pennsylvania, as it is enveloped in smog at noon on Saturday, October 28, 1948.

What Do You Think?

Who do you think should be held responsible for the Donora, Pennsylvania, disaster? Explain your answer. Given what you know about the regulation of industrial pollutants under the Clean Air Act, do you think another incident such as the Donora killer smog could happen in the United States today?

Background

As tragic an event as the Donora killer smog was, London suffered an even worse air-quality disaster. In 1952, London abandoned its electric tramcars and introduced a system of diesel buses. Suddenly, there was a new source of air pollution. The city was already plagued by smoke from coal fires burning in homes and the output from local factories. When a temperature inversion trapped a mass of warm air over the city, the pollutants couldn't rise and dissipate. The sulfurous and photochemical smog that resulted was trapped for four days in 1952. With reduced oxygen and a large amount of particulate matter suspended in the air, the hospitals filled with people. The death rate at the time, normally 135 people per week, skyrocketed to over 500. Between 2,000 and 4,000 people died of "the killer fog." As a result England passed its own Clean Air Act in 1956.

325

Answers to What Do You Think?

Answers may vary. Students may consider the smelter responsible or perhaps the state or federal government. Under the Clean Air Act, a similar incident is certainly much less likely to occur, largely because of the mandatory installation of pollutant-control systems in the smokestacks of industrial plants that produce air pollutants. Our understanding of the conditions that cause temperature inversion and the ability of meteorologists to predict days when air quality will be unhealthful also make the possibility of a similar incident highly unlikely.

Atmosphere and Climate Change
Chapter Planning Guide

PACING	CLASSROOM RESOURCES	LABS, DEMONSTRATIONS, AND ACTIVITIES
BLOCKS 1, 2 & 3 • 135 min pp. 326–334 **Chapter Opener**		
Section 1 Climate and Climate Change	**OSP** Lesson Plan * **CRF** Active Reading * BASIC **TT** Bellringer * **CD** Interactive Tutor Latitude and Climate	**TE** Demonstration Latitude and Climate, p. 328 BASIC **TE** Demonstration Convection Currents, p. 329 GENERAL **SE** QuickLab Investigating Prevailing Winds, p. 330 **TE** Demonstration Air Currents and the Spinning Earth, p. 330 BASIC **TE** Group Activity Rain Shadows in Satellite Imagery, p. 333 GENERAL **SE** Inquiry Lab Build a Model of Global Air Movement, pp. 350–351 ◆ **CRF** Datasheets for In-Text Labs *
BLOCKS 4 & 5 • 90 min pp. 335–338 **Section 2** The Ozone Shield	**OSP** Lesson Plan * **CRF** Active Reading * BASIC **TT** Bellringer * **TT** CFCs and Ozone Depletion * **TT** Ozone Depletion and UV Radiation * **CD** Interactive Tutor The Ozone Layer	**TE** Activity Modeling Ozone Reactions, p. 336 GENERAL **TE** Activity Testing Sunscreens, p. 337 ◆ GENERAL **CRF** Math/Graphing Lab * GENERAL **CRF** Consumer Lab * ◆ GENERAL
BLOCKS 6 & 7 • 90 min pp. 339–345 **Section 3** Global Warming	**OSP** Lesson Plan * **CRF** Active Reading * BASIC **TT** Bellringer * **TT** How the Greenhouse Effect Works * **TT** Increases in Carbon Dioxide, 1958–2000 * **TT** The Global Temperature Record * **TT** Total World Emissions of Carbon Dioxide * **CD** Interactive Tutor Atmosphere	**TE** Demonstration Decreasing Land Surface, p. 339 GENERAL **SE** Field Activity Carbon Dioxide, p. 340 **TE** Activity Testing a Climate Model, p. 340 ADVANCED **TE** Internet Activity Researching Alternative-Fuel Vehicles, p. 340 GENERAL **TE** Internet Activity Global Warming and Climate Change, p. 341 ADVANCED **TE** Group Activity How Much Does a Car Really Cost? p. 343 ADVANCED **CRF** Modeling Lab * ◆ ADVANCED **CRF** Long-Term Project * ◆ BASIC **CRF** CBL™ Probeware Lab * ◆ GENERAL

BLOCKS 8 & 9 • 90 min

Chapter Review and Assessment Resources

SE Chapter Review pp. 347–349
SE Standardized Test Prep pp. 660–661
CRF Study Guide * ■ GENERAL
CRF Chapter Test * ■ GENERAL
CRF Chapter Test * ADVANCED
CRF Concept Review * ■ GENERAL
CRF Critical Thinking * ADVANCED
OSP Test Generator
CRF Test Item Listing *

Online and Technology Resources

Holt Online Learning

Visit go.hrw.com for access to Holt Online Learning or enter the keyword **HE6 Home** for a variety of free online resources.

One-Stop Planner® CD-ROM

This CD-ROM package includes
• Lab Materials QuickList Software
• Holt Calendar Planner
• Customizable Lesson Plans
• Printable Worksheets
• ExamView® Test Generator
• Interactive Teacher Edition
• Holt PuzzlePro® Resources
• Holt PowerPoint® Resources

Holt Environmental Science Interactive Tutor CD-ROM

This CD-ROM consists of interactive activities that give students a fun way to extend their knowledge of environmental science concepts.

KEY

TE	Teacher Edition	**CRF**	Chapter Resource File	*	Also on One-Stop Planner
SE	Student Edition	**TT**	Teaching Transparency	■	Also Available in Spanish
OSP	One-Stop Planner	**CD**	Interactive CD-ROM	◆	Requires Advance Prep

ENRICHMENT AND SKILLS PRACTICE	SECTION REVIEW AND ASSESSMENT	CORRELATIONS
SE Pre-Reading Activity, p. 326 **TE** Using the Figure, p. 326 `GENERAL`		**National Science Education Standards**
TE Inclusion Strategies, p. 327 **SE** Case Study Ice Cores, pp. 330–331 **TE** Skill Builder Graphing, p. 332 `BASIC` **TE** Skill Builder Math, p. 332 `GENERAL` **SE** Math Practice Precipitation Extremes on Earth, p. 334	**TE** Homework, p. 331 `GENERAL` **SE** Section Review, p. 334 **TE** Reteaching, p. 334 `BASIC` **TE** Quiz, p. 334 `GENERAL` **TE** Alternative Assessment, p. 334 `GENERAL` **CRF** Quiz * ■ `GENERAL`	ES 1c ES 1d SPSP 4a SPSP 4b SPSP 5d
TE Inclusion Strategies, p. 337	**SE** Section Review, p. 338 **TE** Reteaching, p. 338 `BASIC` **TE** Quiz, p. 338 `GENERAL` **TE** Alternative Assessment, p. 338 `GENERAL` **CRF** Quiz * ■ `GENERAL`	ES 1d SPSP 4a SPSP 4b SPSP 5d
SE Graphic Organizer, Spider Map, p. 343 `GENERAL` **TE** Skill Builder Math, p. 344 `GENERAL` **SE** Making a Difference, Ozone Scientist, pp. 352–353 **CRF** Map Skills Different Winds * `GENERAL`	**SE** Section Review, p. 345 **TE** Reteaching, p. 345 `BASIC` **TE** Quiz, p. 345 `GENERAL` **TE** Alternative Assessment, p. 345 `GENERAL` **CRF** Quiz * ■ `GENERAL`	SPSP 4a SPSP 4b SPSP 5d

Guided Reading Audio CDs

These CDs are designed to help auditory learners and reluctant readers. (Audio CDs are also available in Spanish.)

www.scilinks.org

Maintained by the **National Science Teachers Association**

TOPIC: Winds
SciLinks code: HE4125

TOPIC: Ozone Shield
SciLinks code: HE4080

TOPIC: Global Warming
SciLinks code: HE4049

CNN Videos

Each video segment is accompanied by a Critical Thinking Worksheet.

Earth Science Connections

Segment 25 Hurricane Double Whammy

Segment 15 Antarctic Meltdown

Segment 26 Hottest Years

Chapter Enrichment

This Chapter Enrichment provides relevant and interesting information to expand and enhance your classroom instruction of the chapter material.

1 Climate and Climate Change

Oceanic Currents and Climate

San Francisco, California, and Wichita, Kansas, are located at about the same latitude. However, during the month of January, the average high temperature in Wichita is lower than the average low temperature in San Francisco. On the other hand, during the month of July, the average low temperature in Wichita is about equal to the average high temperature in San Francisco. This unusual effect occurs because San Francisco's climate is moderated by the cool water of the California Current, which flows southward along the California coastline. Wichita, which is located in the middle of the continent, is far from any such moderating influences and thus has a greater range of temperatures.

Cutting-Edge Climate Research

One method of learning about past climate changes is through the examination of fossil pollen, called *palynology*. The accumulation of pollen grains and spores creates a record of the past vegetation in an area. The plant assemblage of an area can be reconstructed from the accumulated pollen. The reconstructed plant assemblage can tell palynologists whether the climate was warmer, colder, wetter, or drier than it is today. Scientists are also investigating climate change by measuring the isotope records in fossils and by studying insect assemblages (changes in groups of local or regional insects may indicate climatic change over time). Data about historic climate change can also be inferred from living organisms such as lichens.

The Early Atmosphere

According to current scientific theories, if the early atmosphere had contained abundant amounts of oxygen, life on Earth might never have evolved. In the presence of oxygen, the amino acid building blocks of life could not have been synthesized by abiotic processes, such as lightning or volcanism. Just after Earth formed, there was no atmosphere

▶ **The sun at solar maximum**

or oceans. There was, however, a lot of volcanic activity as the planet began the long process of cooling. Outgassing, the release of gases from Earth's interior by volcanic activity, is believed to have created the early atmosphere. Huge amounts of water vapor may have been pumped into the atmosphere by constant volcanic activity, resulting in torrential downpours. At first, Earth's surface may have been so hot that the water turned to steam as soon as it reached the ground. However, after millions of years, the surface cooled enough for oceans to form in topographically low places. Later, photosynthetic organisms appeared and began to contribute more oxygen.

2 The Ozone Shield

BRAIN FOOD

UV radiation has the power to both form and break down complex organic molecules. For this reason, unfiltered UV rays may have been essential to the emergence of life on Earth. UV radiation can also cause genetic mutations. Currently, most organisms on Earth depend on the ozone layer to protect them from the effects of UV radiation.

Misunderstandings About Ozone

There are several misconceptions about problems related to the ozone hole (a thinning region of the Earth's stratospheric ozone layer) over Antarctica. One misconception is that the observed variations in the size of the ozone hole are natural, seasonal phenomena. Although the ozone concentrations over Antarctica fluctuate naturally with the seasons, the addition of CFCs to the atmosphere has caused unprecedented year-round degradation of the ozone layer. Another misconception is that ozone-damaging chlorine comes mostly from natural sources, such as volcanoes and ocean salts. Most chlorine from natural sources is water-soluble and is rained out in the troposphere before it can reach the ozone layer in the stratosphere. On the other hand, chlorine in CFCs is insoluble in water and not very reactive. Therefore, CFCs can migrate into the stratosphere, where UV radiation breaks down the molecules.

This releases volatile chlorine atoms that react with ozone molecules. A third misconception is that current CFC levels (only a few parts per trillion) are insignificant. Ozone concentration in the ozone layer is less than 10 parts per million, and one chlorine atom can destroy up to 100,000 ozone molecules. Another

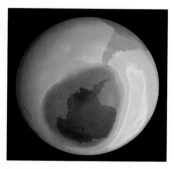

▶ **Satellite image of the ozone hole over the Antarctic in 2000**

misconception is that CFC replacements such as HCFCs (hydrochlorofluorocarbons) or HFCs (hydrofluorocarbons) are "environmentally safe." HCFCs still have the potential to destroy ozone, although the hydrogen in HCFCs makes them more reactive in the troposphere and less likely to reach the ozone layer. Also, both HCFC and HFC gases trap heat and thus contribute to global warming.

SECTION

3 Global Warming

Rising Sea Levels

Current trends of increasing global temperatures will melt ice in glaciers and polar regions and expand seawater volume, which will impact islands and coastlines around the world. Potential consequences of rising sea levels in coastal urban centers include accelerated beach erosion, increased frequency and severity of flooding, and the inundation of coastal wetlands. Tunnels, bridge entrances, airport runways, and beachfront property are all vulnerable to flooding. Certain activities may exacerbate the effects of rising sea levels. Building on soft sediment compresses the sediment and lowers the land surface. The construction of flood-control devices on rivers disrupts the natural replenishment of sediment by flooding; so as coastline erodes, it is not replaced.

Viking Migrations and Climate Change

During the Medieval Warm Period (MWP, approximately 800 CE to 1200 CE), the Vikings expanded their settlements into Greenland under the leadership of Eric the Red. Ironically, Eric named the island Greenland in order to attract more settlers. At the time, the land could barely support the communities that settled there. As the Little Ice Age began and temperatures fell, the hardships became unendurable. Grains would no longer grow, and livestock died of starvation, leaving the settlers with almost no food supply. Viking remains from the early 1400s indicate that the last Viking inhabitants of Greenland were significantly shorter (possibly by as much as seven inches) than the original settlers. The last Viking settlers suffered from malnutrition and overexposure to cold. Because of the increased hardships, the Greenland Vikings eventually died out. Those Vikings who survived this period were able to move south to Iceland, where grains could grow and livestock could be raised successfully.

▶ **An iceberg splits off from the Ross Ice Shelf in Antarctica**

BRAIN FOOD

The Larsen B ice shelf is a large floating ice mass that is located on the eastern side of the Antarctic Peninsula. In a 35-day period between January 31 and March 7, 2001, approximately 3,250 km² of the Larsen B ice shelf broke apart, shattering into thousands of icebergs. The area lost in the event was greater than the area of the state of Rhode Island (2,717 km²), and the volume of ice released was estimated to be 720 billion tons—enough ice to fill 290 trillion five-pound bags. This is the largest ice shelf loss to take place on the Antarctic Peninsula during the last three decades. These retreats are linked to climate warming in the region. There has been an 0.5°C increase each decade since the late 1940s.

Overview

Tell students that the purpose of this chapter is to describe Earth's climate and explore ways that human activities may be causing climate change. Two serious issues are explored: the hole in the stratospheric ozone layer and global warming, the rise in average global temperature.

Using the Figure — GENERAL

This photograph of Hurricane Fran was taken by satellite in 1996. Tell students to notice which way the storm clouds are rotating (counterclockwise, typical in the Northern Hemisphere). Explain that the circular opening in the cloud layer at the center of the storm is a relatively calm spot known as the eye. Have students look at a map of ocean currents. Explain that hurricanes form in the trade wind belt as the result of colliding air masses of different temperatures. Typically, the areas on the northeastern sides of these storms receive the most rainfall because rotation pulls moist air from over the water onto dry land.

LS Visual | English Language Learners

VIDEO SELECT

For information about videos related to this chapter, go to **go.hrw.com** and type in the keyword **HE4 ATMV**.

Atmosphere and Climate Change

1 Climate and Climate Change
2 The Ozone Shield
3 Global Warming

PRE-READING ACTIVITY

FOLDNOTES **Tri-Fold**
Before you read this chapter, create the **FoldNote** entitled "Tri-Fold" described in the Reading and Study Skills section of the Appendix. Write what you know about climate change in the column labeled "Know." Then, write what you want to know in the column labeled "Want." As you read the chapter, write what you learn about climate change in the column labeled "Learn."

The climate on Earth can be very extreme. This satellite image of Hurricane Fran was taken before it struck the coastline of North Carolina in early September 1996.

326

Chapter Correlations   *National Science Education Standards*

ES 1c Heating of earth's surface and atmosphere by the sun drives convection within the atmosphere and oceans, producing winds and ocean currents. **(Section 1)**

ES 1d Global climate is determined by energy transfer from the sun at and near the earth's surface. This energy transfer is influenced by dynamic processes, such as cloud cover and the earth's rotation, and static conditions, such as the position of mountain ranges and oceans. **(Section 1 and Section 2)**

SPSP 4a Natural ecosystems provide an array of basic processes that affect humans. Those processes include maintenance of the quality of the atmosphere, generation of soils, control of the hydrologic cycle, disposal of wastes, and recycling of nutrients. Humans are changing many of these basic processes, and the changes may be detrimental to humans. **(Section 1, Section 2, and Section 3)**

SPSP 4b Materials from human societies affect both physical and chemical cycles of the earth. **(Section 1, Section 2, and Section 3)**

SPSP 5d Natural and human-induced hazards present the need for humans to assess potential danger and risk. Many changes in the environment designed by humans bring benefits to society, as well as cause risks. Students should understand the costs and trade-offs of various hazards—ranging from those with minor risk to a few people to major catastrophes with major risk to many people. The scale of events and the accuracy with which scientists and engineers can (and cannot) predict events are important considerations. **(Section 1, Section 2, and Section 3)**

Climate and Climate Change

Weather is the state of the atmosphere at a particular place at a particular moment. **Climate** is the long-term prevailing weather conditions at a particular place based upon records taken. To understand the difference between weather and climate, consider Seattle, Washington, and Phoenix, Arizona. These two cities may have the same weather on a particular day. For example, it may be raining, warm, or windy in both places. But their climates are quite different. Seattle's climate is cool and moist, whereas Phoenix's climate is hot and dry.

What Factors Determine Climate?

Climate is determined by a variety of factors. These factors include latitude, atmospheric circulation patterns, oceanic circulation patterns, the local geography of an area, solar activity, and volcanic activity. The most important of these factors is distance from the equator. For example, the two locations shown in **Figure 1** have different climates mostly because they are at different distances from the equator.

Objectives

▶ Explain the difference between weather and climate.
▶ Identify four factors that determine climate.
▶ Explain why different parts of the Earth have different climates.
▶ Explain what causes the seasons.

Key Terms

climate
latitude
El Niño
La Niña

Figure 1 ▶ At left is Trunk Bay on the island of St. John in the U.S. Virgin Islands, which is located near the equator. Below is a photograph of the Antarctic Peninsula.

327

INCLUSION Strategies

• *Gifted and Talented*

Have students determine the latitude of their town or city by using a map. Have students research monthly meteorological data for their hometown in a reference book or on the Internet. Then, have students identify one city 5° north of their hometown and one city 5° south of their hometown. Have students develop a chart that compares the climate in their hometown with the climates in the other two cities.

SECTION 1

Focus

Overview

Before beginning this section, review with your students the Objectives listed in the Student Edition. This section explains the difference between weather and climate and then explores the factors influencing climate.

Bellringer

Ask students to think of a place on Earth where they would like to live or visit some day. Have them write a description of the climate at their chosen location and list various factors that might influence the climate there. Have students write their description in their *EcoLog*. Ask them to think about the reasons they chose those particular factors as important contributors to the climate. **LS Logical**

Motivate

Discussion ———— GENERAL

Ideal Climates Ask students to share their responses to the Bellringer activity. Write student responses on the board. After several students have volunteered their information, ask them to compare the latitudes of the various locations, the climates, and their postulated climate influences. Note the similarities and differences on the board, and then link these comparisons to the factors described in this section. **LS Verbal**

Chapter Resource File

• **Lesson Plan**
• **Active Reading** BASIC
• **Section Quiz** GENERAL

Transparencies

TT Bellringer

Teach

Demonstration — BASIC

Latitude and Climate Using a globe or world map, indicate the following three continental locations and relate the following information about their climates. Pictures of the three areas might be helpful. Fairbanks, Alaska (64°50' north latitude), has pleasant, mild summers but extremely cold winters; temperatures there may vary by about 40°C (72°F) over the course of a year. On the other hand, in Memphis, Tennessee (35°20' north latitude), summers are warm to hot, winters are moderate, and the average annual temperature varies only by about 28°C (50°F). Mexico City, Mexico (19°24' north latitude), has an even narrower annual temperature range of about 6°C (10°F) and experiences relatively slight seasonal temperature differences. Have students find the average annual maximum and minimum temperatures for the area in which they live. Have them calculate the average annual temperature range.

LS Visual

Figure 2 ▶ At the equator, sunlight hits the Earth vertically. The sunlight is concentrated on a smaller surface area at the equator. Away from the equator, sunlight hits the Earth at an oblique angle and spreads over a larger surface area.

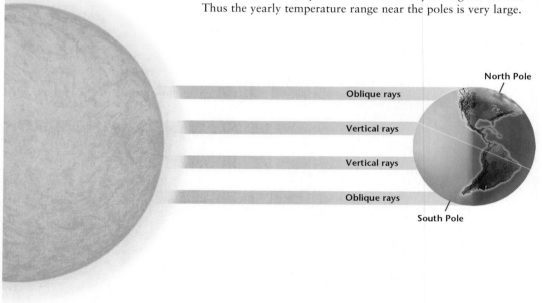

North Pole

Oblique rays

Vertical rays

Vertical rays

Oblique rays

South Pole

Latitude

The distance from the equator measured in degrees north or south of the equator is called **latitude.** The equator is located at 0° latitude. The most northerly latitude is the North Pole, at 90° north, whereas the most southerly latitude is the South Pole, at 90° south.

Low Latitudes Latitude strongly influences climate because the amount of solar energy an area of Earth receives depends on its latitude. More solar energy falls on areas near the equator than on areas closer to the poles, as shown in **Figure 2.** The incoming solar energy is concentrated on a small surface area at the equator.

In regions near the equator, night and day are both about 12 hours long throughout the year. In addition, temperatures are high year-round, and there are no summers or winters.

High Latitudes In regions closer to the poles, the sun is lower in the sky. This reduces the amount of energy arriving at the surface. In the northern and southern latitudes, sunlight hits the Earth at an oblique angle and spreads over a larger surface area than it does at the equator. Yearly average temperatures near the poles are therefore lower than they are at the equator. The hours of daylight also vary. At 45° north and south latitude, there is as much as 16 hours of daylight each day during the summer and as little as 8 hours of sunlight each day during the winter. Near the poles, the sun sets for only a few hours each day during the summer and rises for only a few hours each day during the winter. Thus the yearly temperature range near the poles is very large.

328

GEOLOGY
CONNECTION — GENERAL

Endangered Aquifers Climate change is likely to adversely affect important water sources. In Texas, the Edwards and Ogallala Aquifers are two of the main sources of fresh water. This water is used not only by cities such as San Antonio, Austin, and Amarillo but also by farms and ranches. Scientists think that if the average temperature becomes warmer, there will be an increase in water demand from cities and farms, an increase in evapotranspiration, and a decrease in aquifer recharge. If more water is removed from an aquifer than is replaced, the aquifer may become depleted.

Atmospheric Circulation

Three important properties of air illustrate how air circulation affects climate. First, cold air sinks because it is denser than warm air. As cold air sinks, it compresses and warms. Second, warm air rises. It expands and cools at it rises. Third, warm air can hold more water vapor than cold air can. Therefore, when warm air cools, the water vapor it contains may condense into liquid water to form rain, snow, or fog.

Solar energy heats the ground, which warms the air above it. This warm air rises, and cooler air moves in to replace it. This movement of air within the atmosphere is called *wind*. Because the Earth rotates, and because different latitudes receive different amounts of solar energy, the pattern of global atmospheric circulation shown in **Figure 3** results. This circulation pattern determines Earth's precipitation pattern. For example, the intense solar energy striking the Earth's surface at the equator causes the surface as well as the air above the equator to become very warm. This warm air can hold large amounts of water that evaporate from the equatorial oceans and land. As this warm air rises, however, it cools, which reduces some of its ability to hold water. Thus, areas near the equator receive large amounts of rain.

internet connect

www.scilinks.org
Topic: Winds
SciLinks code: HE4125

SCI*LINKS* Maintained by the
National Science
Teachers Association

Ecofact

Deserts Air that is warmed at the equator rises and flows northward and southward to 30° north and south latitude, where it sinks. The sinking air is compressed and its temperature increases. As the temperature of the air increases, the air is able to hold a larger quantity of water vapor. Evaporation from the land surface is so great beneath these sinking warm air masses that little water returns to Earth in the form of precipitation. Thus, most of the Earth's deserts lie at 30° north and south latitude.

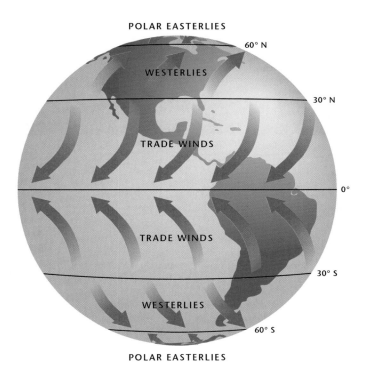

POLAR EASTERLIES

60° N

WESTERLIES

30° N

TRADE WINDS

0°

TRADE WINDS

30° S

WESTERLIES

60° S

POLAR EASTERLIES

Figure 3 ▶ Three belts of prevailing winds occur in each hemisphere.

Demonstration ——— **GENERAL**

Convection Currents Materials: a clear container with straight sides such as a casserole dish; very cold water; an immersion heater; food coloring in a dropper bottle. Tell students that you will demonstrate how air and water currents are created by differences in temperature. Unplug the overhead projector. Fill the container with very cold water and carefully place it on the projector surface. Plug in the projector and make sure that it is projecting clearly. Place the immersion heater in the water and turn it on. Tell students that as the water around the heating element is warmed, the warm water will rise and the surrounding cool water will flow in behind it. Tell the students that warm and cold water each bend light at different angles, so the differences are shown on the projection surface. Add a few drops of the food coloring, and ask students to observe the flow indicated by its movement. Ask the students, "What does this show us about current formation?" (Hot air and water rise, cold air and water sink, and this causes circulation, or currents in the fluids.) Then ask, "What is the main source of energy to heat air and water on Earth?" (sunlight)

LS Interpersonal/Visual

BRAIN FOOD

Chaos and Structure Climatologists say that the atmosphere is both chaotic and structured. How can this be? The answer is a function of scale. When viewed as a whole, the atmosphere appears to be highly structured, but when observed up close, it appears chaotic. For example, the troposphere is a discrete structural element in relation to the other atmospheric layers, but if you look more closely, you see that it is composed of currents of air that are undergoing constant change.

METEOROLOGY ———
● **CONNECTION** ——— **GENERAL**

Tornadoes in the United States Tornadoes occur almost exclusively in the United States and southern Canada. Cold, dry air from the north and warm, moist air from the south often collide on the flat region of the Great Plains. When this happens, a low pressure area is formed and air masses spiral around it, creating funnel clouds.

Skills Acquired:

• Constructing Models
• Experimenting
• Identifying and Recognizing Patterns
• Interpreting

Answers

1. The water simulates the direction of the prevailing winds because both the cardboard and the Earth spin around an axis.

Demonstration ── BASIC

Air Currents and the Spinning Earth Lightly dust a globe with flour. Have a student fill an eyedropper with water and hold it over the North Pole. Ask a student to spin the globe counterclockwise while the student with the dropper drips water onto the globe. Let the globe come to rest, and then ask students to describe the tracks made by the water. (They arc down and toward the west, opposite the direction of the globe's spin.) Tell students that this pattern simulates the Coriolis effect, an apparent deflection in the path of an object due to the rotation of the Earth. For example, if you traveled by air due south from the North Pole toward the equator, the line traced on the Earth's surface would curve to the west because the Earth rotates counterclockwise—from west to east. **LS Visual/Kinesthetic**

QuickLAB

Investigating Prevailing Winds

Procedure

1. Cut a 20 cm diameter disk out of **cardboard**.
2. Insert a **pencil** through the center of the disk. Place the tip of the eraser on a table so that the cardboard is tilted slightly.
3. Place a few drops of **water** near the center of the cardboard, and spin it on the pencil tip. What happens?

Analysis

1. How is the motion of the water related to the prevailing winds?

Global Circulation Patterns Cool air normally sinks, but cool air over the equator cannot descend because hot air is rising below the cool air. This cool air rises and is forced away from the equator toward the North and South Poles. At about 30° north latitude and 30° south latitude, air begins to accumulate in the upper atmosphere. Some of this air sinks back to the Earth's surface and becomes warmer as it descends. The warm, dry air moves across the surface and causes water to evaporate from the land below, creating dry conditions.

Air descending at 30° north latitude and 30° south latitude either moves toward the equator or flows toward the poles. Air moving toward the poles warms while it is near Earth's surface. At about 60° north latitude and 60° south latitude, this air collides with cold air traveling from the poles. The warm air rises. When this rising air reaches the top of the troposphere, a small part of this air returns back to the circulation pattern between 60° and 30° north latitude and 60° and 30° south latitude. However, most of this uplifted air is forced toward the poles. Cold, dry air descends at the poles, which are essentially very cold deserts.

CASE STUDY

Ice Cores: Reconstructing Past Climates

Imagine having at your fingertips a record of Earth's climate that extends back several thousand years. Today, ice cores are providing scientists an indirect glimpse of Earth's climate history. These ice cores have been drilled out of ice sheets thousands of meters thick in Canada, Greenland, and Antarctica.

How do scientists reconstruct the climate history of our planet from ice cores? As snow falls to Earth, the snow carries substances that are in the air at the time. If snow falls in a cold climate where it does not melt, the snow turns to ice because of the weight of the snow above it. The substances contained in snow, such as soot, dust, volcanic ash, and chemical compounds, are buried year after

Above present normal temperature
Below present normal temperature

+3°C
0°C
–3°C

12,000 10,000 8,000 6,000 4,000 2,000 0

Years before present

Source: National Glaciological Program.

▶ With the help of ice cores, scientists are beginning to reconstruct Earth's climate history over hundreds of thousands of years.

year, one layer on top of another. Air between snowflakes and grains becomes trapped in bubbles when the snow is compacted. These bubbles of air can provide information about the composition of the atmosphere over time.

How do scientists date ice cores? Scientists have learned to recognize that differences exist

between snow layers that are deposited in the winter and in the summer. Knowing these differences allows scientists to count and place dates with the annual layers of ice.

Scientists can discover important events in Earth's climate history by studying ice cores. For example, volcanoes produce large quantities of dust, so a history of volcanic

GEOLOGY
● CONNECTION ── ADVANCED

Climate History and Speleothems In addition to studying ice cores and coral reefs, some geologists are currently evaluating the usefulness of speleothems (stalagmites, stalactites, and flowstone) as indicators of past climatic change. Speleothems are formed by the slow deposition of calcium carbonate in successive growth layers, so they contain a history of past climates. The age of each layer is

determined by radiometric dating. Then, by measuring the ratio of isotopes of oxygen in a given layer of a speleothem, climate data can be inferred. These measurements tell scientists what the climate was like at the time the growth layer was deposited. A series of samples from an entire stalagmite can potentially tell scientists how the climate in that region fluctuated over thousands of years.

Prevailing Winds Winds that blow predominantly in one direction throughout the year are called *prevailing winds*. Because of the rotation of the Earth, these winds do not blow directly northward or southward. Instead these winds are deflected to the right in the Northern Hemisphere and to the left in the Southern Hemisphere.

Belts of prevailing winds are produced in both hemispheres between 30° north and south latitude and the equator. These belts of wind are called the *trade winds*. The trade winds blow from the northeast in the Northern Hemisphere and from the southeast in the Southern Hemisphere.

Prevailing winds known as the westerlies are produced between 30° and 60° north latitude and 30° and 60° south latitude. In the Northern Hemisphere, these westerlies are southwest winds, and in the Southern Hemisphere, these westerlies are northwest winds, as shown in **Figure 4**. The polar easterlies blow from the poles to 60° north and south latitude.

Figure 4 ▶ The red areas indicate fires around Sydney, Australia, the smoke from which shows the direction the wind is blowing.

Homework ——— **GENERAL**

Annual Precipitation Ask students to compare the annual precipitation amounts for two cities in the United States over several decades. For each city, have students record the annual precipitation amounts for each decade and the overall average annual precipitation. Suggest that they record these data in a table. If you wish to provide this information for the students, the data may be found via several Internet sources, including the TNRIS Research and Distribution Center or NNDC Climate Data Online.
LS Logical

CASE STUDY

Answers to Critical Thinking

1. Information about past carbon dioxide concentrations helps scientists understand whether present carbon dioxide concentrations are unusual in Earth's history.

2. Answers may vary. Students may say that ice cores should contain evidence of Earth's volcanic history, because the dust and ash produced by volcanoes would be preserved in the ice.

▶ Whether scientists work on ice cores in the field or in the laboratory, all ice cores must be handled in such a way that the cores do not become contaminated by atmospheric pollutants.

CRITICAL THINKING

1. Expressing Viewpoints How might information about past carbon dioxide concentrations on Earth contribute to scientists' understanding of present carbon dioxide concentrations?

2. Applying Ideas What information, besides what is mentioned in this Case Study, might scientists learn about Earth's climatic history from ice cores?

activity is preserved in ice cores. Most important, a record of concentration of carbon dioxide, an important greenhouse gas, has been preserved in air bubbles trapped in the ice. Some scientists who study ice cores have come to believe that rapid, global climate change may be more the norm than the exception. Evidence of increases in global temperature of several Celsius degrees over several decades has been discovered in ice cores from thousands of years ago.

331

MISCONCEPTION ALERT

Drain Spin It is a common misconception that because of the Coriolis force, water spins counterclockwise as it drains from sinks in the Northern Hemisphere and spins in the opposite direction in the Southern Hemisphere. This is not true. The Coriolis force does influence long-lasting vortices, such as hurricanes and large mid-latitude storms. Thus, the air flowing around a hurricane generally moves counterclockwise in the Northern Hemisphere and clockwise in the Southern Hemisphere. But with everyday small-scale rotations such as a draining sink, the Earth's rotation has a negligible effect. Have students design a simple experiment to disprove this common misconception.

Graphing Have students choose a city along the west coast of the United States and draw a line graph of its annual precipitation over the span of four decades (for example, 1962–2002). Ask students to note the five years that have the highest annual precipitation. Then ask them to infer which years may have been El Niño years. (El Niño increases precipitation in the Eastern Pacific region, so their inferences should correspond with the following documented El Niño years: 1951, 1953, 1957–58, 1965, 1972–73, 1976, 1982–83, 1986–87, 1991–92, 1994, and 1997.) Ask students how they would determine which years were La Niña years. (Precipitation should be lower.) **LS** Visual

Math Point out to students that scientists are often interested in determining how far measurements deviate from the average. Ask students to calculate the deviation in the annual rainfall amounts during El Niño years in the city they chose in the activity above. (average annual precipitation − lowest annual precipitation = deviation from the norm for the low end; highest annual precipitation − average annual precipitation = deviation from the norm for the high end) Note that the deviation from the norm is expressed as "− (low end deviation) to + (high end deviation)." **LS** Logical

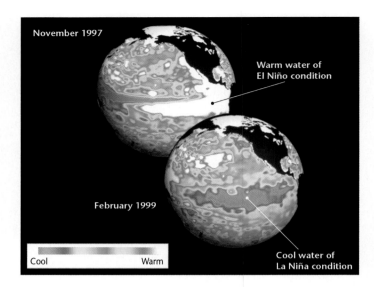

Figure 5 ▶ The El Niño-Southern Oscillation (ENSO) is a periodic change in the location of warm and cold water masses in the Pacific Ocean. The phase of ENSO in which the eastern Pacific surface water is warm is called *El Niño,* and the phase where it is cool is called *La Niña.*

November 1997

Warm water of El Niño condition

February 1999

Cool   Warm

Cool water of La Niña condition

Oceanic Circulation Patterns

Ocean currents have a great effect on climate because water holds large amounts of heat. The movement of surface ocean currents is caused mostly by winds and the rotation of the Earth. These surface currents redistribute warm and cool masses of water around the planet. Some surface currents warm or cool coastal areas year-round.

Surface currents affect the climate in many parts of the world. Here, we will only discuss surface currents that change their pattern of circulation over time.

El Niño—Southern Oscillation **El Niño** (el NEEN yoh) is the name given to the short-term (generally 6- to 18-month period), periodic change in the location of warm and cold water masses in the Pacific Ocean. During an El Niño, winds in the western Pacific Ocean, which are usually weak, strengthen and push warm water eastward. Rainfall follows this warm water eastward and produces increased rainfall in the southern half of the United States and in equatorial South America. El Niño causes drought in Indonesia and Australia. During **La Niña** (lah NEEN yah), on the other hand, the water in the eastern Pacific Ocean is cooler than usual. El Niño and La Niña are opposite phases of the *El Niño–Southern Oscillation* (ENSO) cycle. El Niño is the warm phase of the cycle, and La Niña is the cold phase, as illustrated in **Figure 5.**

Pacific Decadal Oscillation The *Pacific Decadal Oscillation* (PDO), shown in **Figure 6,** is a long-term, 20- to 30-year change in the location of warm and cold water masses in the Pacific Ocean. PDO influences the climate in the northern Pacific Ocean and North America. It affects ocean surface temperatures, air temperatures, and precipitation patterns.

Figure 6 ▶ This satellite image shows the cool phase of the Pacific Decadal Oscillation. During this phase, cooler water (purple and blue) can be seen in the eastern Pacific Ocean. During the warm phase, the situation is reversed.

332

GEOLOGY CONNECTION — GENERAL

Future Climate Change Due to Plate Tectonics Tectonic activity will continue to rearrange the Earth's continents in the future. Europe and North America will continue to spread apart, allowing greater circulation between the Arctic and Atlantic Oceans. At the same time, Antarctica will move away from the South Pole. Ask students to imagine that they have been transported 50 million years into the future. How is the Earth different in terms of the events just described? Have students create a story or comic about

the Earth of the distant future. (Sample answer: Earth will be much warmer. The sea level will also be higher because, unlike today, ocean currents will reach both polar regions, warming them. Antarctica, having moved away from the South Pole, will no longer be an ice-bound continent. Additionally, the continents will be rearranged, and some of today's prominent geographic features—such as the Rocky Mountains—will be significantly eroded.) Suggest that students include their story in their **Portfolio. LS** Intrapersonal

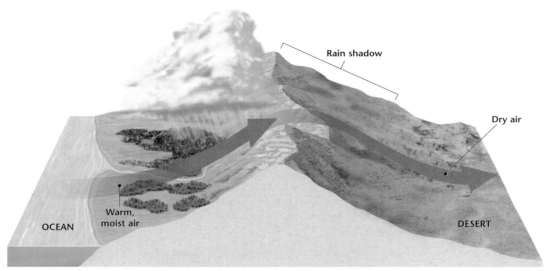

Rain shadow

Dry air

Warm, moist air

OCEAN

DESERT

Topography

Mount Kilimanjaro, a 5,896 m extinct volcano in Tanzania, is located about 3° south of the equator, but snow covers its peak year-round. Kilimanjaro illustrates the important effect that height above sea level (elevation) has on climate. Temperatures fall by about 6°C (about 11°F) for every 1,000 m increase in elevation.

Mountains and mountain ranges also influence the distribution of precipitation. For example, consider the Sierra Nevada mountains of California. Warm air from the Pacific Ocean blows east, hits the mountains, and rises. As the air rises, it cools, which causes it to rain on the western side of the mountains. By the time the air reaches the eastern side of the mountains, it is dry. This effect is known as a rain shadow, as shown in **Figure 7**.

Other Influences on Earth's Climate

Both the sun and volcanic eruptions influence Earth's climate. At a *solar maximum*, shown in **Figure 8**, the sun emits an increased amount of ultraviolet (UV) radiation. UV radiation produces more ozone. This increase in ozone warms the stratosphere. The increased solar radiation can also warm the lower atmosphere and surface of the Earth a little.

In large-scale volcanic eruptions, sulfur dioxide gas can reach the upper atmosphere. The sulfur dioxide gas, which can remain in the atmosphere for up to three years, reacts with smaller amounts of water vapor and dust in the stratosphere. This reaction forms a bright layer of haze that reflects enough sunlight to cause the global temperature to decrease.

Figure 7 ▶ Moist ocean air moves up the coastal side of a mountain range. The air cools and releases its moisture as rain or snow. Air then becomes drier as it crosses the range. When the dry air descends on the inland side of the mountains, the air warms and draws up moisture from the surface.

Figure 8 ▶ The sun has an 11-year cycle in which it goes from a maximum of activity to a minimum.

333

MATH PRACTICE

Answer

Cherrapunji receives 10,649.2 mm (10.6 m) more precipitation.

Close

Reteaching ——— BASIC

Modeling the Seasons Circle a light source with a globe, tilting the axis to model the revolution of Earth around the sun. Rotate the globe counterclockwise as you revolve to create day and night. Then ask volunteers to model the seasons, equinoxes, and solstices for each hemisphere. **LS** Kinesthetic

Quiz ——— GENERAL

1. How does latitude affect climate? (The closer a location is to the equator, the warmer the climate, the smaller the variations in temperature extremes, and the smaller the seasonal variations.)

2. How do ice cores provide information on climate change? (Gases contained in air bubbles and other substances trapped within ice cores are chemical "snapshots" of the time the ice or snow was deposited.)

Alternative Assessment ——— GENERAL

No More Seasons! Have students imagine they are on an advisory board to eliminate Earth's seasonal changes. Have students describe how they would accomplish this. (Answers may vary. One plan might be to eliminate Earth's axial tilt.)

MATHPRACTICE

Precipitation Extremes on Earth Cherrapunji, India, which is located in eastern India near the border of Bangladesh, is the wettest spot on Earth. Cherrapunji has an annual average precipitation of 1,065 cm. Arica, Chile, is located in extreme northern Chile near the Peruvian border. Arica is the driest spot on Earth and has an annual average precipitation of 0.8 mm. What is the difference in millimeters between the annual average precipitation in Cherrapunji and the annual average precipitation in Arica?

Figure 9 ▶ Because of the Earth's tilt, the angle at which the sun's rays strike the Earth changes as the Earth orbits the sun. This change in angle accounts for seasonal climate differences around the world. The seasons for the Northern Hemisphere are shown here.

Seasonal Changes in Climate

You know that temperature and precipitation change with the seasons. But do you know what causes the seasons? As shown in **Figure 9,** the seasons result from the tilt of Earth's axis (about 23.5° relative to the plane of its orbit). Because of this tilt, the angle at which the sun's rays strike the Earth changes as the Earth moves around the sun.

During summer in the Northern Hemisphere, the Northern Hemisphere tilts toward the sun and receives direct sunlight. The number of hours of daylight is greatest in the summer. Therefore, the amount of time available for the sun to heat the Earth becomes greater. During summer in the Northern Hemisphere, the Southern Hemisphere tilts away from the sun and receives less direct sunlight. During summer in the Southern Hemisphere, the situation is reversed. The Southern Hemisphere is tilted toward the sun, whereas the Northern Hemisphere is tilted away.

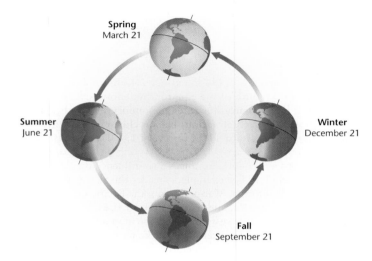

Spring March 21

Summer June 21

Winter December 21

Fall September 21

SECTION 1 Review

1. **Explain** the difference between weather and climate.

2. **Name** four factors that determine climate.

3. **Explain** why different parts of the Earth have different climates.

4. **Explain** what causes the seasons.

CRITICAL THINKING

5. **Making Comparisons** At the equator, there are no summers or winters, only wet and dry seasons. Based on what you have learned about atmospheric circulation patterns, why do you think there are no seasons? Write a paragraph that explains your answer. **WRITING SKILLS**

6. **Analyzing Processes** If the Earth were not tilted in its orbit, how would the climates and seasons be affected at the equator and between 30° north and south latitude?

Answers to Section Review

1. Weather is the state of the atmosphere at a moment in time and can occur at any location. Climate is the overall pattern of weather at a location recorded over a period of years.

2. Latitude, atmospheric circulation patterns, oceanic circulation patterns, local geography, solar activity, and volcanic activity are all factors. Latitude is the most important factor.

3. Because different amounts of solar energy fall on different parts of the Earth, different parts of the Earth have different climates. Sunlight is concentrated on a smaller surface area at the equator and is spread over a larger surface area away from the equator.

4. The seasons are caused by the tilt of the Earth on its axis as the Earth revolves around the sun.

5. There are no seasons at the equator because the sun's rays strike the Earth at almost a 90° angle all year round.

6. If the Earth were not tilted in its orbit, the climate and seasons at the equator would probably be largely unaffected. Between the equator and 30° north and south latitude, there would be no seasonal variation and the climate would be different.

The Ozone Shield

The **ozone layer** is an area in the stratosphere where ozone is highly concentrated. *Ozone* is a molecule made of three oxygen atoms. The ozone layer absorbs most of the ultraviolet (UV) light from the sun. Ultraviolet light is harmful to organisms because it can damage the genetic material in living cells. By shielding the Earth's surface from most of the sun's ultraviolet light, the ozone in the stratosphere acts like a sunscreen for the Earth's inhabitants.

Chemicals That Cause Ozone Depletion

During the 1970s, scientists began to worry that a class of human-made chemicals called **chlorofluorocarbons (CFCs)** might be damaging the ozone layer. For many years CFCs were thought to be miracle chemicals. They are nonpoisonous and nonflammable, and they do not corrode metals. CFCs quickly became popular as coolants in refrigerators and air conditioners. They were also used as a gassy "fizz" in making plastic foams and were used as a propellant in spray cans of everyday products such as deodorants, insecticides, and paint.

At the Earth's surface, CFCs are chemically stable. So, they do not combine with other chemicals or break down into other substances. But CFC molecules break apart high in the stratosphere, where UV radiation, a powerful energy source, is absorbed. Once CFC molecules break apart, parts of the CFC molecules destroy protective ozone.

Over a period of 10 to 20 years, CFC molecules released at the Earth's surface make their way into the stratosphere. **Figure 10** shows how the CFCs destroy ozone in the stratosphere. Each CFC molecule contains from one to four chlorine atoms, and scientists have estimated that a single chlorine atom in the CFC structure can destroy 100,000 ozone molecules.

Objectives

▶ Explain how the ozone layer shields the Earth from much of the sun's harmful radiation.
▶ Explain how chlorofluorocarbons damage the ozone layer.
▶ Explain the process by which the ozone hole forms.
▶ Describe the damaging effects of ultraviolet radiation.
▶ Explain why the threat to the ozone layer is still continuing today.

Key Terms

ozone layer
chlorofluorocarbons (CFCs)
ozone hole
polar stratospheric clouds

Figure 10 ▶ The CFC molecule in this illustration contains a single chlorine atom. This chlorine atom continues to enter the cycle and repeatedly destroys ozone molecules.

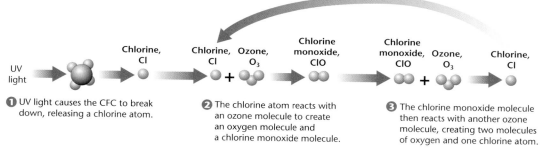

UV light

Chlorine, Cl

Chlorine, Cl + Ozone, O₃

Chlorine monoxide, ClO

Chlorine monoxide, ClO + Ozone, O₃

Chlorine, Cl

❶ UV light causes the CFC to break down, releasing a chlorine atom.

❷ The chlorine atom reacts with an ozone molecule to create an oxygen molecule and a chlorine monoxide molecule.

❸ The chlorine monoxide molecule then reacts with another ozone molecule, creating two molecules of oxygen and one chlorine atom.

335

MISCONCEPTION ///ALERT\\\

Rates of Ozone Depletion Students may encounter a variety of estimates of the current rates of ozone depletion as well as a variety of estimates of ozone damage that a given chemical can do. The ozone layer has always been dynamic, with ozone molecules naturally being formed and destroyed as sunlight and chemicals interact in the atmosphere. Putting additional ozone-destroying compounds such as CFCs into the atmosphere causes ozone to be destroyed at a rate faster than the (otherwise natural) rate of ozone destruction and faster than ozone is naturally created. The actual rate of ozone formation or destruction varies with: latitude; amount of sunlight; stratospheric cloud formations; overall concentrations of chemicals such as CFCs; and other factors. Thus, the statement that "a single chlorine atom can destroy up to 100,000 ozone molecules" is true under the most extreme combination of these conditions.

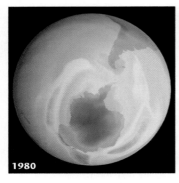
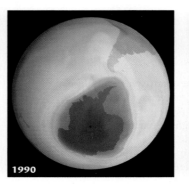
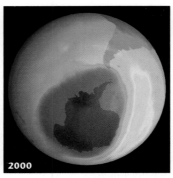

1980 · **1990** · **2000**

Figure 11 ▶ These satellite images show the growth of the ozone hole, which appears purple, over the past two decades.

Activity —————— GENERAL

Modeling Ozone Reactions Tell students that ozone is constantly being created and destroyed by natural processes. In the stratosphere, some diatomic oxygen, O_2, is dissociated by high-energy solar radiation into two molecules of monatomic oxygen, O. O reacts with O_2 to form ozone, O_3. During the day, however, ozone has a short lifetime because it is rapidly degraded by solar radiation into O and O_2. But because new ozone is constantly being formed, the net level of ozone does not change much. Have students construct models of O_2 molecules with plastic-foam balls and straws. Ask them to demonstrate the O_2 being broken apart by solar radiation and then to recombine one of the O atoms with another O_2 molecule to form O_3. To take this activity one step further, have students demonstrate the effect of CFC molecules on ozone molecules as shown in **Figure 10.** For each group of students, provide the following materials: 4 small balls for oxygen and 1 large ball with 4 small balls for the CFC molecule (the carbon center plus chlorine atoms). For an environmentally friendly model, substitute limes for the oxygen, lemons for the chlorine atoms, and an orange for the center of the CFC molecule. Use toothpicks to put the molecule together. **LS Kinesthetic/Logical**

Connection to **Meteorology**

Polar Stratospheric Clouds
Because the stratosphere is extremely dry, clouds normally do not form in this layer of the atmosphere. However, during polar winters, temperatures become low enough to cause condensation and cloud formation. These clouds, which occur at altitudes of about 21,000 m, are known as polar stratospheric clouds, or PSCs. Because of their iridescence, PSCs are called mother-of-pearl or nacreous clouds. Outside of the poles, the stratosphere is too warm for these clouds to form. Because these clouds are required for the breakdown of CFCs, ozone holes are confined to the Antarctic and Arctic regions.

336

The Ozone Hole

In 1985, an article in the scientific journal *Nature* reported the results of studies by scientists working in Antarctica. The studies revealed that the ozone layer above the South Pole had thinned by 50 to 98 percent. This was the first news of the **ozone hole,** a thinning of stratospheric ozone that occurs over the poles during the spring.

After the results from the studies from Halley Bay were published, NASA scientists reviewed data that had been sent to Earth by the *Nimbus 7* weather satellite since the satellite's launch in 1978. They were able to see the first signs of ozone thinning in the data from 1979. Although the concentration of ozone fluctuates during the year, the data showed a growing ozone hole, as shown in **Figure 11.** Ozone levels over the Arctic have decreased as well. In fact, March 1997 ozone levels over part of Canada were 45 percent below normal.

How Does the Ozone Hole Form? During the dark polar winter, strong circulating winds over Antarctica, called the *polar vortex*, isolate cold air from surrounding warmer air. The air within the vortex grows extremely cold. When temperatures fall below about −80°C, high-altitude clouds made of water and nitric acid, called **polar stratospheric clouds,** begin to form.

On the surfaces of polar stratospheric clouds, the products of CFCs are converted to molecular chlorine. When sunlight returns to the South Pole in spring, molecular chlorine is split into two chlorine atoms by ultraviolet radiation. The chlorine atoms rapidly destroy ozone. The destruction of ozone causes a thin spot, or ozone hole, which lasts for several months. Some scientists estimate that as much as 70 percent of the ozone layer can be destroyed during this period.

Because ozone is also being produced as air pollution, you may wonder why this ozone does not repair the ozone hole in the stratosphere. The answer is that ozone is very chemically reactive. Ozone produced by pollution breaks down or combines with other substances in the troposphere long before it can reach the stratosphere to replace ozone that is being destroyed.

PHYSICS ————
● **CONNECTION** —— ADVANCED

Ultraviolet Radiation The ultraviolet portion of the electromagnetic spectrum is divided into ultraviolet A (UV-A), ultraviolet B (UV-B), and ultraviolet C (UV-C). UV-A is not absorbed by stratospheric ozone but is generally not damaging to living things. Both UV-B and UV-C are absorbed by stratospheric ozone in the ozone layer above the Earth and are capable of causing damage to organisms. However, UV-C is almost completely absorbed by ozone, and little UV-C penetrates to Earth's surface. Thus, even a thinning ozone layer can still absorb most

UV-C. On the other hand, UV-B is only partially absorbed by stratospheric ozone and can therefore cause damage to organisms. Sunburn is evidence that a person has been exposed to too much UV-B too quickly. A person's potential to develop skin cancer is related to their exposure to UV-B radiation. The depletion of stratospheric ozone could cause the amount of UV-B radiation reaching Earth's surface to increase significantly. This, in turn, could lead to an increase in the damaging effects of UV-B radiation to the DNA of humans and other animals.

Effects of Ozone Thinning on Humans As the amount of ozone in the stratosphere decreases, more ultraviolet light is able to pass through the atmosphere and reach Earth's surface, as shown in **Figure 12.** UV light is dangerous to living things because it damages DNA. DNA is the genetic material that contains the information that determines inherited characteristics. Exposure to UV light makes the body more susceptible to skin cancer, and may cause certain other damaging effects to the human body.

Effects of Ozone Thinning on Animals and Plants High levels of UV light can kill single-celled organisms called *phytoplankton* that live near the surface of the ocean. The loss of phytoplankton could disrupt ocean food chains and reduce fish harvests. In addition, a reduction in the number of phytoplankton would cause an increase in the amount of carbon dioxide in the atmosphere.

Some scientists believe that increased UV light could be especially damaging for amphibians, such as toads and salamanders. Amphibians lay eggs that lack shells in the shallow water of ponds and streams. UV light at natural levels kills many eggs of some species by damaging unprotected DNA. Higher UV levels might kill more eggs and put amphibian populations at risk. Ecologists often use the health of amphibian populations as an indicator of environmental change due to the environmental sensitivity of these creatures.

UV light can damage plants by interfering with photosynthesis. This damage can result in lower crop yields. The damaging effects of UV light are summarized in **Table 1**.

Table 1 ▼

Damaging Effects of UV Light	
Humans	• increased incidence of skin cancer • premature aging of the skin • increased incidence of cataracts • weakened immune response
Amphibians	• death of eggs • genetic mutations among survivors • reduction of populations
Marine Life	• death of phytoplankton in surface water • disruption of food chain • reduction in the number of photosynthesizers
Land Plants	• interference with photosynthesis • reduced crop yields

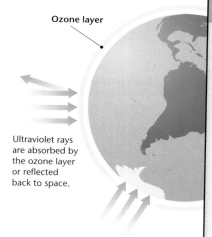

Ozone layer

Ultraviolet rays are absorbed by the ozone layer or reflected back to space.

Ultraviolet rays penetrate to the Earth's surface through the ozone hole.

Figure 12 ▶ Depletion of the ozone layer allows more ultraviolet (UV) radiation to reach the surface of the Earth.

INCLUSION Strategies

• *Attention Deficit Disorder*
• *Learning Disabled*

Have students develop a set of study cards about the ozone layer. On separate index cards, have students write a definition of *ozone*, describe the composition of ozone, describe why ozone is important in our lives, and list the potential effects of ozone depletion. The completed cards may be used for studying or may be glued to a posterboard for a class presentation.

Activity ———— GENERAL
Testing Sunscreens Materials: sunprint paper; SPF 8, SPF 15, and SPF 30 sunscreens; sunscreen testing sheet; cellophane tape; microscope slides (3); a pan of water; and paper towels. Each student should be given a sunscreen-testing sheet with three marked rectangles, one for each sunscreen. Students should frame each of the rectangles on their testing sheet with cellophane tape. There should be three stations set up, each with a microscope slide and a bottle of sunscreen. Students should move to all three stations. Using the long edge of the microscope slide, they should smear sunscreen inside the borders of the rectangles on the sunscreen-testing sheets. Each layer should be the thickness of the cellophane tape. When the three sunscreen samples have been collected, have the students tape the sunscreen testing sheet to the blue side of the sunprint paper. Students should take the testing sheet outside and place it in the sun. Leave the testing sheet in place until the blue color turns almost white (two to three minutes on a sunny day, and six to 10 minutes on a cloudy day). To develop the sunprint, students should remove the tape that is adhering the sunprint paper to the testing sheet and rinse the exposed sunprint paper in the pan of water for about one minute. Then, the sunprint should be placed on a paper towel to dry. Using tissues, students should clean the sunscreens off the testing sheets. Ask students to observe the results and enter their observations into their *EcoLog.* **LS** Visual/Kinesthetic

Transparencies

TT Ozone Depletion and UV Radiation

CFCs and Ozone Check that students understand how the damaging effects of CFCs on the ozone layer could be increasing even though CFC use is decreasing. Explain that the CFCs causing trouble today may have entered the atmosphere before CFC use was banned in many countries. Ask students to predict future changes in the levels of CFCs in the stratosphere. (There will be a lag between the reduction of CFC production at Earth's surface and the decrease of CFCs in the stratosphere. Therefore, CFC levels may continue to increase for a few more years, level off, and then decrease.)

LS Logical | English Language Learners

Quiz ——— GENERAL

1. What is the ozone layer, and why is it important? (It is a layer in the stratosphere where ozone is highly concentrated. It absorbs UV light from the sun that can damage genetic material in living cells.)

2. What chemicals are currently causing the majority of ozone depletion? (chlorofluorocarbons)

Alternative Assessment ——— GENERAL

Life Without the Ozone Layer Tell students to imagine that they live in a world in which the ozone layer does not exist. Have them write a diary entry describing a typical day in their life. (Answers should demonstrate that students understand the protective role of the ozone layer.) **LS** Intrapersonal

Protecting the Ozone Layer

In 1987, a group of nations met in Canada and agreed to take action against ozone depletion. Under an agreement called the Montreal Protocol, these nations agreed to sharply limit their production of CFCs. A second conference on the problem was held in Copenhagen, Denmark, in 1992. Developed countries agreed to eliminate most CFCs by 1995. The United States pledged to ban all substances that pose a significant danger to the ozone layer by 2000.

After developed countries banned most uses of CFCs, chemical companies developed CFC replacements. Aerosol cans no longer use CFCs as propellants, and air conditioners are becoming CFC free. Because many countries were involved and decided to control CFCs, many people consider ozone protection an international environmental success story. **Figure 13** illustrates the decline in world CFC production since the 1987 Montreal Protocol.

The battle to protect the ozone layer is not over. CFC molecules remain active in the stratosphere for 60 to 120 years. CFCs released 30 years ago are still destroying ozone today, so it will be many years before the ozone layer completely recovers.

Internet connect
www.scilinks.org
Topic: Ozone Shield
SciLinks code: HE4080
SciLINKS Maintained by the National Science Teachers Association

Figure 13 ▶ Chlorofluorocarbon production has declined greatly since developed countries agreed to ban CFCs in 1987.

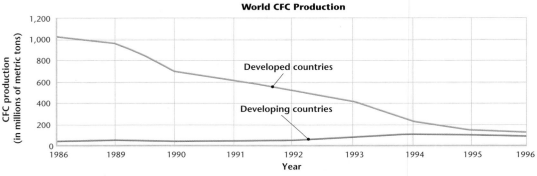

World CFC Production

Source: UN Environment Programme.

SECTION 2 Review

1. Describe the process by which chlorofluorocarbons destroy ozone molecules in the stratosphere.

2. Describe the process by which the ozone hole forms over Antarctica in spring.

3. List five harmful effects that UV radiation could have on plants or animals as a result of ozone thinning.

4. Explain why it will take years for the ozone layer to recover even though the use of CFCs has declined significantly. Write a paragraph that explains your answer. **WRITING SKILLS**

338

CRITICAL THINKING

5. Making Decisions If the ozone layer gets significantly thinner during your lifetime, what changes might you need to make in your lifestyle?

6. Analyzing Relationships CFCs were thought to be miracle chemicals when they were first introduced. What kinds of tests could be performed on any future miracle chemical to make sure serious environmental problems do not result from its use?

Answers to Section Review

1. UV light breaks down CFCs, releasing a chlorine atom (Cl), which in turn reacts with ozone (O_3) to create chlorine monoxide (ClO) and oxygen (O_2). The ClO reacts again with O_3, creating O_2 molecules and Cl atoms.

2. Products of CFCs are converted to molecular chlorine on the surface of polar stratospheric clouds. During the polar spring, sunlight splits the molecular chlorine into two chlorine atoms, which break apart ozone molecules.

3. The effects of excessive UV light include skin cancer, cataracts, and a weakened immune response in humans; genetic mutations and the

death of eggs in amphibians; a decrease in marine photosynthesis; interference with the food chain; and a decrease in photosynthesis and reduced crop yields among land plants.

4. CFC molecules can be active for 60 to 120 years. Thus, CFCs released years ago will be damaging the ozone layer for years to come.

5. Answers may vary but should include reducing exposure to UV light.

6. Answers may vary. Substances should be tested for their potential effects on the environment.

Global Warming

Have you ever gotten into a car that has been sitting in the sun for a while with all its windows closed? Even if the day is cool, the air inside the car is much warmer than the air outside. On a hot summer day, opening the door to the car can seem like opening the door of an oven.

The reason heat builds up inside the car is that the sun's energy streams into the car through the clear glass windows in the form of sunlight. The carpets and upholstery in the car absorb the light and change it into heat energy. Heat energy does not pass through glass as easily as light energy does. Sunlight continues to stream into the car through the glass, but heat cannot get out. The heat continues to build up and is trapped inside the car. A greenhouse works the same way. By building a house of glass, gardeners trap the sun's energy and grow delicate plants in the warm air inside the greenhouse even when there is snow on the ground outside.

The Greenhouse Effect

The Earth is similar to a greenhouse. The Earth's atmosphere acts like the glass in a greenhouse. As shown in **Figure 14**, sunlight streams through the atmosphere and heats the Earth. As this heat radiates up from Earth's surface, some of it escapes into space. The rest of the heat is absorbed by gases in the troposphere and warms the air. This process of heat absorption is called the *greenhouse effect*.

Not every gas in our atmosphere absorbs heat in this way. The gases that do absorb and radiate heat are called **greenhouse gases.** The major greenhouse gases are water vapor, carbon dioxide, chlorofluorocarbons, methane, and nitrous oxide. Of these, water vapor and carbon dioxide account for most of the absorption of heat that occurs in the atmosphere.

Objectives

▶ Explain why Earth's atmosphere is like the glass in a greenhouse.

▶ Explain why carbon dioxide in the atmosphere appears to be increasing.

▶ Explain why many scientists think that the Earth's climate may be becoming increasingly warmer.

▶ Describe what a warmer Earth might be like.

Key Terms

greenhouse gases
global warming
Kyoto Protocol

Figure 14 ▶ How the Greenhouse Effect Works

❷ Energy from the sun is absorbed by Earth's surface and then radiated into the atmosphere as heat, some of which escapes into space.

❸ Greenhouse gases also absorb some of the sun's energy and radiate it back toward the lower atmosphere and Earth's surface.

❶ Solar radiation passes through the atmosphere and warms Earth's surface.

339

Chapter Resource File

• Lesson Plan
• Active Reading BASIC
• Section Quiz GENERAL

Transparencies

TT Bellringer
TT How the Greenhouse Effect Works

MISCONCEPTION ///ALERT\\\

A Good Greenhouse Effect Point out that the greenhouse effect is a natural process in Earth's atmosphere. It is estimated that without the natural greenhouse effect, Earth's average temperature would be about −16°C (4°F)! Ask students to consider how the Earth might be different without an atmosphere that traps heat. (Sample answer: The planet would probably be too cold for current life-forms to exist because all water, including the oceans, would be frozen.)

SECTION **3**

Focus

Overview

Before beginning the section, review with your students the Objectives listed in the Student Edition. This section describes how trace amounts of gases in the atmosphere trap heat similar to the way a greenhouse does. Increasing levels of these gases, especially carbon dioxide, appear to be causing the Earth to warm.

◉ Bellringer

Ask students if they think the average seasonal temperatures have changed during the past few years. Ask how they think recent temperatures may compare with temperatures from thousands or millions of years ago. Ask students how climate change may affect ecosystems and human populations. Have students record their responses in their *EcoLog*.
LS Logical   English Language Learners

Motivate

Demonstration ——— GENERAL

Decreasing Land Surface Find a topographic map of a low-lying coastal region of the United States, such as Florida. Locate the 150 ft contour line, and make sure all students can see it. Explain that if the polar icecaps were to melt entirely, everything below the 150 ft contour line could be underwater. Ask, "What might make the polar icecaps melt?" (an increase in temperature at the poles) Then ask if any students have heard of global warming—an increase in average global temperature. Tell students that if global warming is occurring, a worst-case scenario suggests that water levels could rise to the level you indicated on the map. Ask students to think about the profound effects of such an event on humans and other organisms.
LS Kinesthetic

Teach

Discussion —— ADVANCED

More Climactic Influences? Ask students to think of variables (other than those in the text) that could potentially affect the global climate and that are probably not included in current climate models. (Examples include: asteroid impacts, mass extinctions of plants or animals, and nuclear war.) Make a list of student responses. **LS** Verbal

Activity —— ADVANCED

Testing a Climate Model Have students use a computer simulation program to manipulate the levels of atmospheric gases, such as oxygen and carbon dioxide, and observe the simulation's response. Ask them to suggest other variables to manipulate, and have them make predictions about the resulting effects. **LS** Logical

Internet Activity —— GENERAL

Researching Alternative-Fuel Vehicles Ask students to research an alternative-fuel vehicle currently available to the public. Have them compare the efficiency and emission statistics of that car with those of a comparably sized fossil fuel-burning vehicle. Ask them to evaluate both vehicles and summarize their findings in a brief report or display. **LS** Intrapersonal/Logical

FIELD ACTIVITY

Carbon Dioxide Create a question dealing with carbon dioxide or carbon dioxide levels in the atmosphere. Investigate the FAQ section of the Carbon Dioxide Information Analysis Center's Web site to see if your question has already been answered. If not, click on "Ask Us a Question," and e-mail your question to the center. Report your findings to the class.

Measuring Carbon Dioxide in the Atmosphere In 1958, a geochemist named Charles David Keeling installed an instrument at the top of a tall tower on the volcano Mauna Loa in Hawaii. Keeling wanted to precisely measure the amount of carbon dioxide in the air, far away from forests and cities. In a forest, carbon dioxide levels rise and fall with the daily rhythms of photosynthesis. Near cities, carbon dioxide from traffic and industrial pollution raises the local concentration of the gas. The winds that blow steadily over Mauna Loa have come thousands of miles across the Pacific Ocean, far from most forests and human activities, swirling and mixing as they traveled. Keeling reasoned that at Mauna Loa, the average carbon dioxide levels for the entire Earth could be measured.

Keeling's first measurement, in March of 1958, was 314 parts per million of carbon dioxide in the air, or 0.0314 percent. The next month the levels rose slightly. By summer the levels were falling, but in the winter they rose again. During the summer, growing plants use more carbon dioxide for photosynthesis than they release in respiration. This causes carbon dioxide levels in the air to decrease in the summer. In the winter, dying grasses and fallen leaves decay and release the carbon that was stored in them during the summer. So, carbon dioxide levels rise.

Rising Carbon Dioxide Levels After only a few years of measuring carbon dioxide, it became obvious that the levels were changing in ways other than just the seasonal fluctuations. Each year, the high carbon dioxide levels of winter were higher, and each year, the summer levels did not fall as low. **Figure 15** shows the carbon dioxide measured from 1958 to 2000. By 2000, the average level of carbon dioxide was about 368 parts per million. Thus, in 42 years, carbon dioxide has gone from 314 to 368 parts per million, an increase of 54 parts per million or 17 percent. This increase may be due to the burning of fossil fuels.

Figure 15 ▶ The graph shows that the average yearly concentration of carbon dioxide in the atmosphere has increased since 1958.

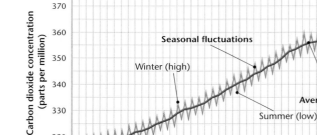

340

Transparencies

TT Increases in Atmospheric Carbon Dioxide, 1958–2000

Greenhouse Gases and the Earth's Temperature Many scientists think that because greenhouse gases trap heat near the Earth's surface, more greenhouse gases in the atmosphere will result in an increase in global temperature. A comparison of carbon dioxide in the atmosphere and average global temperatures for the past 400,000 years supports that view.

Today, we are releasing more carbon dioxide than any other greenhouse gas into the atmosphere. Millions of tons of carbon dioxide are released into the atmosphere each year from power plants that burn coal or oil and from cars that burn gasoline. Millions of trees are burned in tropical rain forests to clear the land for farming. Thus, the amount of carbon dioxide in the atmosphere increases. We are also releasing other greenhouse gases, such as CFCs, methane, and nitrous oxide, in significant amounts. **Table 2** shows the sources of some major greenhouse gases.

How Certain Is Global Warming?

Many scientists think that the increasing greenhouse gases in our atmosphere result in increasing the average temperature on Earth. The result, they believe, will be a warmer Earth. This predicted increase in global temperature is known as **global warming.** As is shown by the graph in **Figure 16,** Earth's average global temperature increased during the 20th century. Many scientists project that the warming trend that began in the 20th century will continue throughout the 21st century. However, not all scientists agree that the observed global warming is due to greenhouse gases. Some scientists believe that the warming is part of natural climatic variability. They point out that widespread fluctuations in temperature have occurred throughout geologic time.

Table 2 ▼

Major Greenhouse Gases and Their Sources
Carbon dioxide, CO_2: burning fossil fuels and deforestation
Chlorofluorocarbons (CFCs): refrigerants, aerosols, foams, propellants, and solvents
Methane, CH_4: animal waste, biomass burning, fossil fuels, landfills, livestock, rice paddies, sewage, and wetlands
Nitrous Oxide, N_2O: biomass burning, deforestation, burning of fossil fuels, and microbial activity on fertilizers in the soil
Water vapor, H_2O: evaporation, plant respiration

Figure 16 ▶ As shown by the graph, the surface of the Earth warmed during the 20th century.

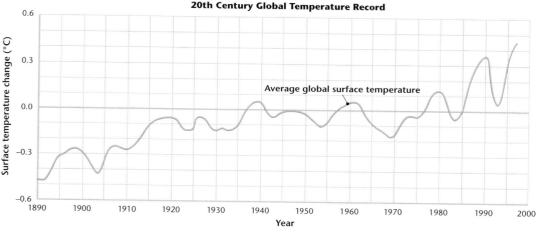

20th Century Global Temperature Record

Average global surface temperature

Source: National Center for Atmospheric Research.

341

ASTRONOMY
CONNECTION **GENERAL**

Since the *Voyager* spacecraft passed Neptune's moon Triton in 1989, scientists have noticed an interesting trend in Triton's atmosphere. Images from the Hubble Space Telescope taken in 1998 indicate that Triton is going through a rapid period of global warming. As Triton warms, frozen nitrogen on its surface melts and contributes nitrogen gas to its thin atmosphere. This process has happened so rapidly that the atmospheric pressure of Triton has doubled in less than 10 years! Scientists hope to use the global warming trends on Triton to understand warming patterns on Earth. Because Triton is a much simpler world than Earth—Triton has a thinner atmosphere, a surface of frozen nitrogen, and no oceans—it is a good place to study environmental change.

✔ **Internet Activity** —**ADVANCED**

Global Warming and Climate Change Many nonprofit groups, such as The Environmental Literacy Council, as well as government agencies, such as NASA and NOAA, have Web sites with information and activities to help students understand the issues of global warming and climate change. Have students visit several Web sites on the topic and make notes of the major points presented. Ask students to answer the following questions based on their research: "Is global warming perceived as a problem by many people in different parts of the world? What evidence exists for global warming? What kinds of actions are scientists and environmental advocates asking people, industries, and governments to take?" Have students work in pairs or groups and present a summary of their findings to the class. **LS Verbal**

BIOLOGY
CONNECTION **GENERAL**

Methane Producers Ruminants are animals such as cattle, sheep, goats, pigs, and horses that possess a four-chambered stomach. In the chamber called the *rumen,* bacteria break down food and generate methane as a byproduct. Methane is a greenhouse gas, and methane emissions from livestock have become a significant source of methane in Earth's atmosphere. It is estimated that livestock account for 15 percent of annual methane emissions—or 65 million to 85 million tons of the annual global production of 400 million to 600 million tons of methane. Termites are another source of methane. Methane is produced in the digestive tracts of termites when cellulose is broken down by symbiotic microorganisms. Termites produce about 15 million tons of methane a year.

Transparencies

TT The Global Temperature Record

The Permian-Triassic Extinctions Several mass extinctions have occurred in Earth's history. One such extinction occurred at the Permo-Triassic boundary, which marks the transition from the Paleozoic Era to the Mesozoic Era approximately 248 million years ago. It is estimated that as many as 95 percent of all species perished during this catastrophic extinction. There are several theories as to the cause of the extinction, including meteorite impacts, supernova explosions, volcanism, sea-level rise combined with continental drift, or increased concentrations of CO_2 combined with ocean anoxia. However, scientists think that many scenarios may have worked in concert. One factor that most explanations have in common is an increase in the greenhouse effect, which may have caused global climate change.

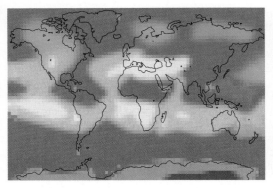 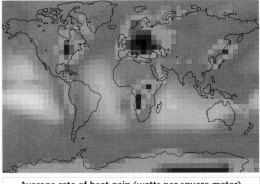

Average rate of heat gain (watts per square meter)

−1   0   1   2   3

Figure 17 ▶ These maps were developed from computer models. The map on the left shows the effect of greenhouse gases on the Earth before sulfur pollution was added. The map on the right shows how the addition of the sulfur pollution variable shows a cooling effect.

Connection to Biology

Ocean Warming Commercial fishing in the northern Atlantic Ocean depends heavily on a fish called a *cod*. In recent years, the numbers of cod in the North Atlantic have greatly decreased. In 2001, English scientists embarked on a research project to find out if the decline in the numbers of cod was linked to the changing global climate. They sailed the ocean waters between Greenland and Iceland collecting samples of zooplankton. The scientists found that zooplankton levels have drastically decreased since 1963, the date of the last survey. The scientists believe that slowly warming ocean-water temperatures have in some way affected zooplankton in the North Atlantic Ocean and have in turn impacted the animals such as the cod that rely on the zooplankton as food.

Modeling Global Warming Scientists are currently unable to make accurate predictions about the rate of global warming because climatic patterns are complex and too many variables must be taken into account to be solved even using today's fastest computers. Predictions about climate change are based on computer models, such as the models shown in **Figure 17**. Scientists write equations representing the atmosphere and oceans. Scientists also enter data about carbon dioxide levels, prevailing winds, and many other variables. The computer models predict how phenomena such as temperature, rainfall patterns, and sea level will be affected. Computer modeling is complicated by the Earth's feedback processes that sometimes make it necessary to use different equations under changing simulated environments. These feedback processes are related to clouds, water vapor, ice, circulation changes within the oceans, changes in ocean chemistry, and changes in vegetation. Computer models of Earth's climate are becoming more reliable as more data are available, as additional factors are taken into consideration, and as faster computers are built.

The Consequences of a Warmer Earth

The Earth's climate has changed dramatically in the past. Many of those changes, however, occurred over thousands or millions of years. Scientists are not sure how quickly the Earth will warm or how severe the effects of global warming might be. Different computer models give different answers to these questions.

The impacts of global warming could include a number of potentially serious environmental problems. These problems range from the disruption of global weather patterns and global rise in sea level to adverse impacts on human health, agriculture, and animal and plant populations. Other impacts on the environment that could not be predicted by computer models might also arise.

342

BRAIN FOOD

General Circulation Models One of the most common climate model types is the General Circulation Model (GCM). GCMs are very complex, three-dimensional models that represent the effects of influences such as greenhouse gases, ocean and land temperatures, rainfall, and the cryosphere (the frozen regions of Earth's surface) on the climate system. The Ames Mars Atmosphere Modeling Group employs GCMs to study the atmosphere of Mars. Huge storms occur on Mars, swirling a large amount of dust into the air. Scientists computed the effect on Martian surface temperatures as a result of this dust. They then applied similar GCMs to an Earth atmosphere filled with the particles that could be released by a nuclear war and discovered that Earth's climate would be significantly cooled by such an event, leading to a nuclear winter.

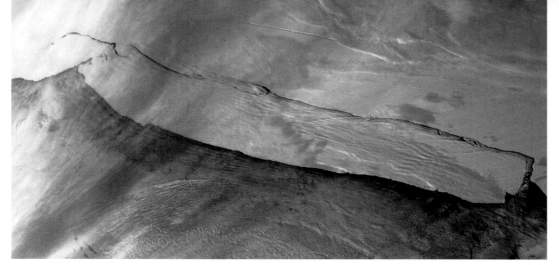

Figure 18 ▶ This is a satellite map of a 11,000 km² iceberg—the size of Connecticut!—that split off from the Ross Ice Shelf in Antarctica in March of 2000. Many scientists believe that scenarios like this would become more common if the poles grow warmer.

Melting Ice and Rising Sea Levels If the global temperature increased, the amount of ice and snow at the poles would decrease. The melting of ice and snow at the poles would cause sea levels around the world to rise. The rise in sea level might affect coastal areas in a number of ways. Coastal wetlands and other low-lying areas might be flooded. Enormous numbers of people who live near coastlines could lose their homes and sources of income. Beaches could be extensively eroded. The salinity of bays and estuaries might increase, which could adversely affect marine fisheries. Also, coastal freshwater aquifers could become too salty to be used as sources of fresh water.

Global Weather Patterns If the Earth warms up significantly, the surface of the oceans will absorb more heat, which may make hurricanes and typhoons more common. Some scientists are concerned that global warming will also cause a change in ocean current patterns, such as shutting off the Gulf Stream. Such a change could significantly affect the world's weather. For instance, some regions might have more rainfall than normal, whereas other regions might have less. Severe flooding could occur in some regions at the same time that droughts devastate other regions.

Human Health Problems Warmer average global temperatures could pose threats to human health. Greater numbers of heat-related deaths could occur as a result of global warming. Both very young people and very old people would have the greatest risk of heat exhaustion during periods of high temperatures. Concentrations of ground-level ozone could increase as air temperatures rise. Consequently, respiratory illnesses could increase, especially in urban areas. Furthermore, global warming could cause insectborne diseases to spread. Warmer temperatures might enable mosquitoes, which carry diseases such as malaria and encephalitis, to greatly increase in number.

Graphic Organizer **Spider Map** Create the **Graphic Organizer** entitled "Spider Map" described in the Appendix. Label the circle "Consequences of Global Warming." Create a leg for each consequence of global warming. Then, fill in the map with details about each consequence of global warming.

Graphic Organizer GENERAL

Spider Map
You may want to have students work in groups to create this Spider Map. Have one student draw the map and fill in the information provided by other students from the group.

BRAIN FOOD 🧠

Volcanic Eruptions and Disease The particulates and aerosols released during a volcanic eruption can have long-term effects on climate. These climate changes affect the biosphere in many ways, including decreasing temperatures. The resulting cooler temperatures can significantly reduce crop yields, leading to severe famine. Severe famine, in turn, may lead to lower resistance to disease in both animals and humans, which could potentially trigger plague pandemics.

Group Activity ADVANCED
How Much Does a Car Really Cost? Owning a car has costs beyond the original purchase price, when one considers the cost of owning, operating, and maintaining a car. Ask groups of students to select a vehicle and research what it would cost them over a period of one year. Provide them with the research materials (technical data sheets from car dealerships and new vehicle buyer's guides), as well as the annual mileage and fuel price. Have them list the make, model, engine size (in liters), retail cost, fuel efficiency, annual maintenance and insurance costs, annual registration fees, and other annual fees (such as parking). For the operating costs, have them calculate the annual fuel consumption (annual mileage/fuel efficiency), annual fuel cost (fuel price × fuel consumption), and total annual operating costs (fuel cost + maintenance + insurance + registration fees + other fees). Then have the students calculate the yearly cost of ownership (down payment + 12 monthly payments + insurance and repair costs). Finally, provide them with the CO_2 tailpipe emission factor (2.36 kg/L) and the upstream emission factor (0.65 kg/L). Next, ask them to calculate the annual CO_2 emissions for exhaust (annual fuel consumption × tailpipe emission factor) and upstream (annual fuel consumption × upstream emission factor) and then to calculate the total annual emissions. Students not only will be surprised by the real cost of having a car but also will realize the amount of CO_2 emitted by vehicles. Note that these calculations still do not include the cost of optional insurance. **LS** Interpersonal/Logical
Co-op Learning

Math Have students examine **Figure 22,** and ask them to calculate the percentage difference in projected emissions between 1995 and 2035 for a given country. (Example: for China, 17% − 11% = 6%) Then, have students convert the percentages to tons. (Example: for China, 11% × 6.46 billion tons = .71 billion tons for 1995; 17% × 11.71 billion tons = 1.99 billion tons for 2035) Next, instruct the students to calculate the difference in tons of emissions between 1995 and 2035. (Example: for China, 1.99 billion tons − .71 billion tons = 1.28 billion tons) Finally, ask them to calculate the percentage increase in projected emissions based on the difference in tons of emissions. (Example: the total ton increase between 1995 and 2035, 11.71 billion tons − 6.46 billion tons = 5.25 billion tons; the percentage increase in emissions for China [1.28 billion tons/5.25 billion tons] × 100 = 24.4%) **LS** Logical

METEOROLOGY
CONNECTION — GENERAL

Glacial Retreat Scientists have noted that glaciers and snowfields in the Himalayas are retreating at an accelerating rate. As a result, glacial lakes have been forming and rapidly filling on the surfaces of glaciers for the past few decades. At present, more than 40 lakes in Nepal and Bhutan could soon overflow, endangering tens of thousands of people in these two countries. The cause of the glacial melt can be traced to a regional temperature increase of 1°C since the 1970s.

Figure 19 ▶ If climate change caused extreme weather to become more frequent, global agriculture would become severely impacted.

internet connect

www.scilinks.org
Topic: Global Warming
SciLinks code: HE4049

SCI*LINKS* Maintained by the National Science Teachers Association

Figure 20 ▶ In spite of its name, the crabeater seal actually feeds on zooplankton. This seal is a resident of Antarctica.

Agriculture Agriculture would be most severely impacted by global warming if extreme weather events, such as droughts, became more frequent. The effects of drought are shown in **Figure 19.** Higher temperatures could result in decreased crop yields. The demand for irrigation could increase, which would further deplete aquifers that have already been overused.

Effects on Plants and Animals Climate change could alter the range of plant species and could change the composition of plant communities. A warmer climate could cause trees to colonize northward into cooler areas. Forests could shrink in area in the southern part of their range and lose diversity.

Global warming may cause a shift in the geographical range of some animals. For example, birds that live in the Northern Hemisphere may not have to migrate as far south during the winter. Warming in the surface waters of the ocean might cause a reduction of zooplankton, tiny, shrimplike animals, that many marine animals depend on for food. The crabeater seal, shown in **Figure 20,** would be just one of the animals affected by a reduction in zooplankton. Warming in tropical waters may kill the microscopic algae that nourish corals, thus destroying coral reefs.

Recent Findings

The Intergovernmental Panel on Climate Change (IPCC) is a network of approximately 2,500 of the world's leading climatologists from 70 countries. In 2001, the IPCC issued its Third Assessment Report (TAR). TAR describes what is currently known about the global climate system and provides future estimates about the state of the global climate system. Some of the findings of the IPCC included that the average global surface temperature increased by 0.6°C during the 20th century, that snow cover and ice extent have decreased, and that the average global sea level has risen. The IPCC has also reported that concentrations of atmospheric greenhouse gases have continued to increase as a result of human activities. It has also predicted that human influences will continue to change the composition of the Earth's atmosphere and continue to warm the Earth throughout the 21st century.

344

BRAIN FOOD
Warmer Climate Equals Worse Storms?
Some scientists think that as the global climate warms, the tropopause will rise and the height of the troposphere will increase. This would allow thunderclouds to grow even taller and therefore would increase the severity of storms. Scientists point to massive deposits of rock layers in the Cretaceous period, a time in Earth's history when average global temperatures were very warm as support of their theory. The deposits seem to have been generated by erosion from storms.

Reducing the Risk

In 1997, representatives from 160 countries met and set timetables for reducing emissions of greenhouse gases. These timetables will go into effect when a treaty called the *Kyoto Protocol* is ratified by 55 percent of the attending nations. The **Kyoto Protocol** requires developed countries to decrease emissions of carbon dioxide and other greenhouse gases by an average of 5 percent below their 1990 levels by 2012. In March of 2001, the United States decided not to ratify the Kyoto Protocol. Most developed nations are going ahead with the treaty.

The need to slow global warming has been recognized by the global community. Some nations and organizations have engaged in reforestation projects to reduce carbon dioxide, such as the project shown in **Figure 21.** However, the attempt to slow global warming is made difficult by the economic, political, and social factors faced by different countries. Conflict has already arisen between developed and developing countries over future CO_2 emissions, the projections of which are shown in **Figure 22.**

Figure 21 ▶ Because plants take in carbon dioxide during photosynthesis, reforestation projects such as this project in Haiti may help to offset a portion of global carbon dioxide emissions.

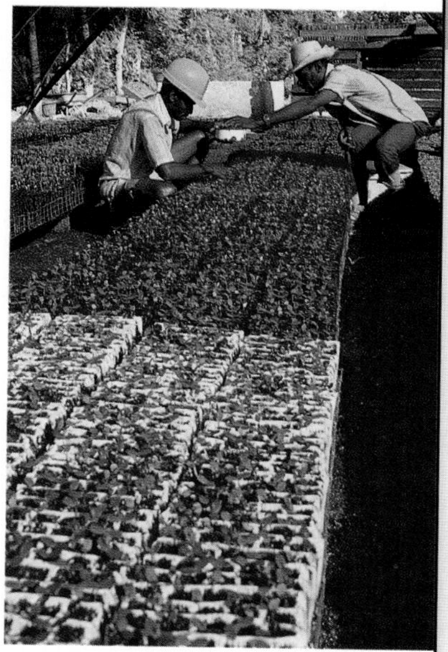

Total World Emissions of CO_2

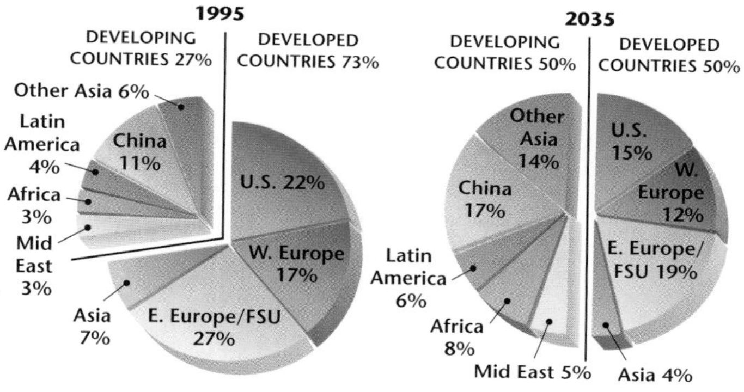

1995
DEVELOPING COUNTRIES 27%
DEVELOPED COUNTRIES 73%
Other Asia 6%
Latin America 4%
China 11%
Africa 3%
Mid East 3%
U.S. 22%
W. Europe 17%
Asia 7%
E. Europe/FSU 27%
2035
DEVELOPING COUNTRIES 50%
DEVELOPED COUNTRIES 50%
Other Asia 14%
China 17%
Latin America 6%
Africa 8%
Mid East 5%
U.S. 15%
W. Europe 12%
E. Europe/ FSU 19%
Asia 4%

Source: U.S. Environmental Protection Agency.

Figure 22 ▶ Developing countries are projected to make up half of all CO_2 emissions by 2035. In 1995, total carbon released as CO_2 was 6.46 billion tons (5.86 billion metric tons). In 2035, total carbon emissions are projected to be 11.71 billion tons (10.62 billion metric tons).

SECTION 3 Review

1. **Explain** why Earth's atmosphere is like the glass in a greenhouse.
2. **Explain** why carbon dioxide in the atmosphere appears to be increasing.
3. **Explain** why many scientists believe Earth's climate may be getting increasingly warmer.
4. **Name** some of the possible consequences of a warmer Earth.

CRITICAL THINKING

5. **Making Predictions** Read the text under the heading "Modeling Global Warming." What difficulties do scientists face when they attempt to construct models that accurately predict the rate of global warming? READING SKILLS
6. **Analyzing Relationships** How could environmental problems in developing countries that result from global climate change affect the economies of developed countries, such as the United States?

345

Answers to Section Review

1. Solar radiation warms the Earth's surface and atmosphere, and greenhouse gases absorb the heat and radiate it back toward Earth's surface. A greenhouse lets in solar radiation to heat the air, and the glass prevents the warm air from escaping.
2. Measurements made by C. D. Keeling at the top of Mauna Loa showed that carbon dioxide in the atmosphere has been increasing by a few parts per million each year—from 315 ppm in 1958 to 363 ppm in 2000.
3. Most atmospheric scientists think that more greenhouse gases released into the atmosphere will result in an increase in global temperature.
4. Answers may vary but might include the disruption of global weather patterns, a global rise in sea level, and adverse impacts on human health, agriculture, and animal and plant populations.
5. Answers may vary but should include the complexity of climate models and Earth's feedback processes and the lack of computers fast enough to analyze the many variables involved in global warming.
6. Answers may vary but should include potential environmental and economic impacts.

Close

Reteaching — BASIC

Outlining Have students create an outline to recap their understanding of each subsection. Ask students to draw diagrams illustrating the main points. Quiz students orally, as necessary, to ensure their understanding. English Language Learners

Quiz — GENERAL

1. What is global warming? (Global warming is the predicted increase in the average temperature on Earth. Global warming may occur naturally, but it is exacerbated by greenhouse gases.)
2. What greenhouse gas increased significantly after the Industrial Revolution? (carbon dioxide)
3. Is global warming definitely happening? (Yes, the average global temperature has increased 0.6°C in the last 100 years. Although the Earth is probably undergoing a natural warming trend, the increase of greenhouse gases is contributing to the warming trend.)

Alternative Assessment — GENERAL

CO_2 Reduction Tell students to imagine they are employed by the EPA to design a public-service advertising campaign to encourage the reduction of carbon dioxide. Students' efforts should be persuasive, creative, factual, and informative. Have students include their work in their **Portfolio.** LS Visual

Transparencies

TT Total World Emissions of Carbon Dioxide

CHAPTER 13 Highlights

Alternative Assessment ——— GENERAL

Debate Tell students there is no wrong answer to the following question and that they should feel free to answer honestly. Then ask, "Which of you are convinced that humans are causing the Earth's climate to undergo a long-term warming trend?" Have the students who raise their hands form a group and work together to provide evidence to support their view. (sample evidence: warmer-than-average temperatures over the last few years, less snowfall, retreating glaciers, and rising sea levels) Ask students who are not convinced to form another group. They should explain why and what kind of evidence they will need to be persuaded. (possible explanation: variations in global climate are natural and frequent; sample evidence: meteorological, oceanographic, and atmospheric data from the past and records of volcanic events) Once students have discussed their opposing points of view, ask them to assume that global warming is happening—regardless of the cause. Then have them work as a class to list possible pros and cons associated with the phenomenon. (negatives: longer, hotter, drier summers; lower capacity to grow some foodstuffs; more frequent violent storms; disrupted ecosystems; increased insect problems; and flooded coastal areas; positives: longer growing seasons; possible rainfall increase; milder winters)

LS Verbal

English Language Learners

1 Climate and Climate Change

Key Terms

climate, 327
latitude, 328
El Niño, 332
La Niña, 332

Main Ideas

▶ Climate represents the long-term prevailing weather conditions at a particular place based on records taken.

▶ Factors that determine climate include latitude, atmospheric and oceanic circulation patterns, local geography, and solar and volcanic activity. Latitude is the most important determining factor of climate.

▶ The angle at which the sun's rays strike the Earth changes as the Earth moves around the sun. This change in angle is what causes the seasons to change.

2 The Ozone Shield

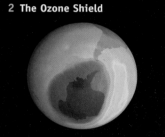

ozone layer, 335
chlorofluorocarbons (CFCs), 335
ozone hole, 336
polar stratospheric clouds, 336

▶ The ozone layer in Earth's stratosphere absorbs most of the ultraviolet (UV) light from the sun.

▶ Chlorofluorocarbons are human-made chemicals that destroy ozone molecules and damage the ozone layer.

▶ Ozone levels measured over the polar regions have been decreasing over the past several decades.

▶ The thinning of the ozone layer may increase the harmful effects of ultraviolet light that reaches Earth's surface.

3 Global Warming

greenhouse gases, 339
global warming, 341
Kyoto Protocol, 345

▶ Gases that absorb and radiate heat in the atmosphere are called *greenhouse gases*. The important greenhouse gases are water vapor, carbon dioxide, CFCs, methane, and nitrous oxide.

▶ The predicted increase in global temperature that occurs as a result of increasing greenhouse gases in the atmosphere is called *global warming*.

▶ Because climate patterns are complex, scientists use computer models to attempt to predict the rate of global warming.

▶ Global warming could produce a number of potentially serious environmental problems.

▶ In 1997, representatives from 160 countries ratified the Kyoto Protocol, which set timetables for reducing emissions of greenhouse gases.

346

Chapter Resource File

- **Chapter Test** GENERAL
- **Chapter Test** ADVANCED
- **Concept Review** GENERAL
- **Critical Thinking** ADVANCED
- **Test Item Listing**
- **Math/Graphing Lab** GENERAL
- **Modeling Lab** ADVANCED
- **CBL™ Probeware Lab** GENERAL
- **Consumer Lab** GENERAL
- **Long-Term Project** BASIC

Using Key Terms

Use each of the following terms in a separate sentence.

1. *latitude*
2. *El Niño*
3. *chlorofluorocarbons*
4. *polar stratospheric clouds*
5. *Kyoto Protocol*

For each pair of terms, explain how the meanings of the terms differ.

6. *weather* and *climate*
7. *El Niño* and *La Niña*
8. *ozone layer* and *ozone hole*
9. *greenhouse gases* and *global warming*

✓ STUDY TIP

Qualifiers When taking a test that contains multiple-choice, true/false, fill-in-the-blank, or matching questions, locate qualifiers in the sentences. Qualifiers are words, such as adjectives or adverbs, or groups of words that modify or limit the meaning of another word or group of words. *Never, always, all, some, none, greatest,* and *least* are examples of qualifiers. These words are the keys to the meaning of sentences.

Understanding Key Ideas

10. The belt of prevailing winds that is produced between 30° and 60° north latitude and 30° and 60° south latitude is called the
    a. doldrums.
    b. westerlies.
    c. polar easterlies.
    d. trade winds.

11. Which of the following statements about El Niño is true?
    a. El Niño is the cold phase of the El Niño–Southern Oscillation cycle.
    b. El Niño is a long-term change in the location of warm and cold water masses in the Pacific Ocean.
    c. El Niño produces storms in the northern Pacific Ocean.

    d. El Niño produces winds in the western Pacific Ocean that push warm water eastward.

12. Polar stratospheric clouds convert the products of CFCs into
    a. carbon dioxide.
    b. hydrochloric acid.
    c. nitric acid.
    d. molecular chlorine.

13. Which of the following is *not* an adverse effect of high levels of ultraviolet light?
    a. disruption of photosynthesis
    b. disruption of ocean food chains
    c. premature aging of the skin
    d. increased amount of carbon dioxide in the atmosphere

14. In which season (in the Northern Hemisphere) does carbon dioxide in the atmosphere decrease as a result of natural processes?
    a. fall
    b. winter
    c. summer
    d. spring

15. Which of the following gases is a greenhouse gas?
    a. carbon dioxide
    b. water vapor
    c. methane
    d. all of the above

16. Which of the following substances is *not* a source of methane?
    a. fossil fuels
    b. sewage
    c. fertilizer
    d. rice

17. The average global temperature increased by how many Celsius degrees during the 20th century?
    a. 0.4°C
    b. 0.6°C
    c. 0.8°C
    d. 1.0°C

18. Which of the following countries decided not to ratify the Kyoto Protocol?
    a. Russia
    b. United States
    c. Canada
    d. Australia

347

Assignment Guide

Section	Questions
1	1, 2, 6, 7, 10, 11, 19, 20, 26, 27
2	3, 4, 8, 12, 13, 21, 22, 28, 30, 32, 34
3	5, 9, 14–18, 23–25, 29, 31, 33, 35, 36

ANSWERS

Using Key Terms

1. Sample answer: Latitude is the distance north or south of the equator measured in degrees up to 90°.

2. Sample answer: El Niño is caused when winds in the western Pacific Ocean strengthen and push warm water eastward.

3. Sample answer: Chlorofluorocarbons, also known as CFCs, are atmospheric pollutants that damage Earth's protective ozone layer.

4. Sample answer: Polar stratospheric clouds develop over the Antarctic, collecting chlorine molecules on their surface until the spring, when UV light breaks apart the molecular chlorine, which then reacts with ozone molecules.

5. Sample answer: The Kyoto Protocol is a treaty designed by 160 countries to reduce worldwide emissions of greenhouse gases.

6. Weather is a daily event that can occur at any location, whereas climate is the overall pattern of weather at a location as recorded over a span of years.

7. Sample answer: El Niño and La Niña are part of the El Niño–Southern Oscillation Cycle (ENSO). El Niño is the phase of ENSO in which water in the eastern Pacific Ocean is warm, and La Niña is the phase when it is cool.

8. Sample answer: The ozone layer is a protective layer of ozone in the stratosphere that shields the Earth and its inhabitants from the harmful effects of ultraviolet radiation. The ozone hole is a thinner region of the ozone layer above Antarctica that is natural in origin but is further depleted by CFCs.

9. Sample answer: Greenhouse gases, which include carbon dioxide, methane, nitrous oxide, chlorofluorocarbons, and water vapor, accumulate in the atmosphere and trap heat. The resulting increase in temperature contributes to increased global warming.

Understanding Key Ideas

10. b
11. d
12. d
13. d
14. c
15. d
16. c
17. b
18. b

Chapter 13 • Atmosphere and Climate Change **347**

Short Answer

19. The three important properties that influence how air circulation affects global climate are that warm air rises, cool air sinks, and warm air can hold more water vapor than cold air can.

20. Sample answer: Local geography affects precipitation when mountains produce a rain-shadow effect.

21. Chlorofluorocarbons seemed like miracle chemicals because they are useful in a variety of applications and they are nonflammable, nonpoisonous, and noncorrosive to metals.

22. Stratospheric ozone protection is thought of as a success story because most countries replaced CFCs with less harmful chemicals, thus significantly decreasing the amount of CFCs and therefore potential damage to the ozone layer.

23. In order to design computer models of global warming, scientists must create mathematical models to represent the atmosphere and oceans, and then enter data about variables such as carbon dioxide concentrations and prevailing winds. The computer models will predict the results, including temperature and rainfall patterns.

24. Rising sea levels can cause flooding of swamps and marshlands, beach erosion, flooding of coastal cities, increased salinity in coastal waters, and the intrusion of salt water into freshwater aquifers.

Short Answer

19. Name three properties of air that are important for understanding how air circulation affects global climate.

20. Explain how local geography can influence the local pattern of precipitation.

21. Describe the properties chlorofluorocarbons possess that made them seem like miracle chemicals when they were discovered.

22. Explain why stratospheric ozone protection has been considered an environmental success story.

23. Explain the general process scientists use to make computer models of global warming.

24. Describe some of the environmental problems that rising sea level might cause.

25. Describe what is currently known about the state of the climate system as reported in the 3rd Assessment Report of the Intergovernmental Panel on Climate Change.

Interpreting Graphics

The graph below shows the average monthly temperature of two locations that are at the same latitude but are in different parts of the United States. Use the graph to answer questions 26–27.

26. Which location has the smallest temperature range between summer and winter?

27. What factors could cause the difference in climate between the two locations?

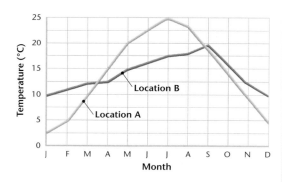

Concept Mapping

28. Use the following terms to create a concept map: *ozone layer, ultraviolet (UV) light, chlorofluorocarbons, polar vortex, polar stratospheric clouds,* and *ozone hole.*

Critical Thinking

29. Making Predictions Over a long period of time, how might living things adapt to increased carbon dioxide levels and global warming? Do you think most species will adapt, or are many species likely to go extinct? Write a short essay that explains your answers. WRITING SKILLS

30. Analyzing Relationships Read about the harmful effects that ultraviolet light can have on humans as a result of ozone thinning under the head "Effects of Ozone Thinning on Humans." However, ultraviolet light serves an extremely important function that benefits life on Earth. Can you recall what that function is and how it helps make life on Earth possible? READING SKILLS

Cross-Disciplinary Connection

31. Economics Insurance companies set some of their rates by estimating the number of destructive natural events, such as hurricanes and floods, that will occur in the next 20 years. Explain why insurance companies would be interested in knowing scientists' predictions about global warming for the next two decades.

Portfolio Project

32. Designing a Pamphlet Design a pamphlet that documents the harmful effects of ultraviolet light on living things. Table 1 can be used as a source of information. You might also collect information by checking out the Web sites of the American Cancer Society and the Environmental Protection Agency. Distribute the pamphlet to your classmates, and include it in your portfolio.

348

25. Findings of the Intergovernmental Panel on Climate Change in its Third Assessment Report include that the average global surface temperature increased by 0.6°C during the 20th century, that concentrations of atmospheric greenhouse gases have continued to increase because of human activities, and that most of the warming that took place during the 20th century was the result of human activity.

Interpreting Graphics

26. Location B

27. Answers may vary. Sample answers: latitude, atmospheric and oceanic circulation patterns, and local geography

Concept Mapping

28. Answers to the concept mapping questions are on pp. 667–672.

MATH SKILLS

33. Making Calculations In 1958, the carbon dioxide level measured in Earth's atmosphere was approximately 315 parts per million. In 2000, the carbon dioxide level in the atmosphere had increased to approximately 368 ppm. What was the average annual increase in carbon dioxide in the atmosphere between 1958 and 2000 measured in parts per million?

WRITING SKILLS

34. Communicating Main Ideas Imagine that you are a scientist who is studying the effects of chlorofluorocarbons on stratospheric ozone. Follow the path of a chlorine atom from the time it is released into the atmosphere from a CFC source through the time it has destroyed ozone molecules. Summarize your findings in a brief essay.

35. Writing Persuasively Imagine you are a scientist who has been studying the subject of global warming. You have been asked by the President of the United States to write a recommendation for his environmental policy on the subject. The President has asked you to provide important facts that can be used to promote the proposed policies. Summarize your recommendations in a brief letter.

36. Writing Persuasively You are the mayor of a low-lying coastal town. Write a plan of expansion for your town. The plan should takes global warming into account. Report your plan of expansion in front of the class.

STANDARDIZED TEST PREP

For extra practice with questions formatted to represent the standardized test you may be asked to take at the end of your school year, turn to the sample test for this chapter in the Appendix.

READING SKILLS

Read the passage below, and then answer the questions that follow.

During photosynthesis, a plant takes in carbon dioxide from the air. Some of the carbon in the carbon dioxide becomes part of the plant's body. That carbon is not returned to the air until the leaves fall or the plant dies and decays.

Some plants, however, never completely decay. Instead, they are covered by sediments. After millions of years of being buried, the plants become coal, oil, or natural gas, which are fossil fuels. When fossil fuels are burned, they release the stored carbon as carbon dioxide. Millions of tons of carbon dioxide are released into the atmosphere each year from power plants that burn coal or oil and from cars that burn gasoline.

The burning of living plants also releases carbon dioxide. This process increases the carbon dioxide in the air in two ways. First, a burning plant gives off carbon dioxide. Second, when a living plant is burned, there is one less plant to remove carbon dioxide from the air by photosynthesis. As millions of trees are burned in tropical rain forests to clear the land for farming, the amount of carbon dioxide in the atmosphere increases.

1. According to the above paragraph, plants give off carbon dioxide
   a. when they are buried under sediments.
   b. when they die and decay.
   c. when they are burned.
   d. Both (b) and (c)

2. According to the paragraph above, which of the following is a process that does *not* add carbon dioxide to the atmosphere?
   a. burning gasoline in cars
   b. photosynthesis
   c. burning trees in tropical rain forests
   d. burning coal in power plants

Critical Thinking

29. Answers may vary. However, the response should note that if the change is slow enough, organisms may adapt; if the change is abrupt, then some species may become extinct.

30. Ultraviolet radiation produces ozone molecules in the stratosphere.

Cross-Disciplinary Connection

31. Answers may vary. Global warming could change atmospheric circulation, thereby altering current weather patterns. Changes in the number and severity of destructive natural events might cause insurance companies to change their rates.

Portfolio Project

32. Answers may vary.

Math Skills

33. 363 ppm – 315 ppm = 48 ppm
48 ppm/39 y = 1.2 ppm/y

Writing Skills

34. Answers may vary.

35. Answers may vary.

36. Answers may vary.

Reading Skills

1. d
2. b

BUILD A MODEL OF GLOBAL AIR MOVEMENT

Teacher's Notes

Time Required
one 45-minute class period

Lab Ratings

EASY ———————→ HARD

TEACHER PREPARATION 🔥🔥
STUDENT SETUP 🔥🔥🔥
CONCEPT LEVEL 🔥🔥
CLEANUP 🔥🔥

Skills Acquired
• Constructing Models
• Organizing and Analyzing Data

The Scientific Method
In this lab, students will:
• Make Observations
• Analyze the Results
• Draw Conclusions

Materials and Equipment
The materials listed are enough for groups of two to four students.

Objectives

▶ **Examine** a model that shows how the movement of air creates a system of wind currents on Earth.

▶ **USING SCIENTIFIC METHODS**
Hypothesize why the closed system of an aquarium is like the Earth and its atmosphere.

Materials

aquarium, 15 gal, glass, with cover
goose-neck lamp, adjustable, with a 100 W incandescent bulb
ice cubes, large (24)
incense stick
masking tape
matches
thermometers, outdoor (2)

▶ **Step 1** Attach a thermometer to each end of the aquarium, making certain that the thermometers can be read from the outside of the aquarium.

Build a Model of Global Air Movement

Warm air rises and cools, and cold air sinks and warms. This is true whether we are observing the temperature and air circulation in a room or around the globe. On Earth, this movement of air creates a system of wind currents that you will demonstrate by building a model. You will build a closed system in which ice represents the polar regions and a lamp represents the equator. You will follow the movement of the air over these regions by watching a trail of smoke as it traces the path of air. (Remember that in the global circulation pattern, warm air moving toward the poles collides with cold air that is traveling from the poles. This collision, which takes place at about 60° north latitude and 60° south latitude, causes the warm air to rise.)

Procedure

1. Stack the ice cubes on the bottom of the aquarium against one end of it. Place the lamp outside the other end of the aquarium, and direct the bulb at the bottom half of that end. Use masking tape to attach one thermometer to each end inside of the aquarium. Make sure the thermometers can be read from the outside of the aquarium. Place the cover on the aquarium.

2. Wait 5 minutes. Then read and record the temperature at each end of the aquarium.

3. Light the end of the incense stick so that it produces a steady plume of smoke.

Tips and Tricks
Any closed aquarium may be used as long as it is twice as long as it is deep—for example, 12 in × 12 in × 24 in. Very small tanks are not recommended. The cover can be any material that will prevent room air currents from influencing air flow in the tank. Any lamp that can direct heat in a single direction may be used. Infrared lightbulbs will create the most dramatic effect. The number of ice cubes may need to be adjusted. The incense stick should be about 12 inches long. The thermometers must be hung inside each end of the aquarium.

Chapter Resource File
• **Datasheets for In-Text Labs**
• **Lab Notes and Answers**

4. Lift the aquarium cover very slightly so that you can insert the incense stick. Hold it steadily in place over the ice about 5 cm from the cover, and observe the smoke.

5. Observe the movement of the smoke. How does the smoke behave? Draw a diagram of the aquarium. Use arrows to indicate the movement of the smoke in the aquarium.

6. Remove several ice cubes and record your observations.

7. Remove all of the ice cubes and record your observations.

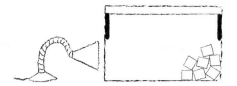

▶ **Diagramming Smoke Flow**
Make a simple diagram of the aquarium showing the position of the light source and the ice cubes. Draw arrows to indicate the movement of the smoke in the aquarium.

Analysis

1. Explaining Events Did the air movement pattern change after some ice was removed? Explain your answer.

2. Explaining Events Did the air movement pattern change after all of the ice was removed? Explain your answer.

Conclusions

3. Drawing Conclusions Why is the difference between temperatures at the two ends of the aquarium an important factor in the flow of heat through the aquarium?

4. Making Predictions Predict how air movement patterns will change if polar ice begins to thaw because of global warming.

Extension

1. Analyzing Models A closed system is a collection of elements that nothing can escape from or enter. Your aquarium is an example of a closed system. Convection is the movement of warm air relative to cooler air. Discuss your observations of convection in the closed system of the aquarium. How can you apply this information to the movement of air over the Earth?

2. Analyzing Models How is the Earth and its atmosphere like the closed system of your aquarium? What factors that affect air movement, climate, and weather exist on Earth but not in your model?

351

Linnaea Smith
Bastrop High School
Bastrop, Texas

2. Earth is nearly a closed system, but it is much larger and has many more factors that influence air movement. The temperature difference between the ends of the tank simulates the difference between the polar regions and the equator. The aquarium system is too small to simulate the deserts and temperate zones that exist on Earth. Also, ocean currents and landmasses were not simulated in the model. Earth's rotation on its axis causes a Coriolis force to act on air masses, further altering their paths.

Answers to Analysis

1. Air descends more slowly because the area over the ice is warmer when some ice is removed. Also, some smoke took a different path and does not descend at all.

2. When all of the ice was removed, the air movement came to a stop. Without the ice, the lamp heated the air until there was no temperature difference inside the aquarium.

Answers to Conclusions

3. Without a temperature differential, the air will not move. Air moves by convection. The smaller the temperature differential, the slower the air movement.

4. If the temperature gradient between the equator and the poles decreased as a result of polar ice melting, the global air circulation would decrease. The winds would possibly decrease, and circulation patterns would change.

Answers to Extension

1. Just as air warms and rises over the warm side of the tank, air warms and rises over the equator. In the tank, cool air continually moves in behind the rising mass of air, warming and rising and finally cooling and falling as it moves away from the heat source. Similarly, warm air from the equatorial region is pushed out toward the polar regions, where it condenses and falls. Cool air falling over polar regions will push the mass of air below it toward the equator, where it will warm and rise, creating a circular air flow. The air is constantly being pushed by the air behind it. In this way, air circulation patterns are established over Earth.

OZONE SCIENTIST

Background

Susan Solomon is a leading researcher of the effect of pollution on the atmosphere. Inspired as a child by the undersea adventures of Jacques Cousteau, she chose to study the chemistry of the real world. She received a bachelor's degree in chemistry from Illinois Institute of Technology and both a master's and doctoral degree from the University of California at Berkeley, where she specialized in atmospheric chemistry. While working at the National Oceanic and Atmospheric Administration in Boulder, Colorado, she led an expedition to Antarctica in 1986 to study a phenomenon that was soon to be called the ozone hole. She is the recipient of numerous awards, including the Department of Commerce Gold Medal for Exceptional Service, the J. B. MacElwane award from the American Geophysical Union, and the prestigious 1999 National Medal of Science. She has published a book titled *The Coldest March* in which she argues that terrible weather, not poor preparation, was the largest factor in the death of British polar explorer Robert Falcon Scott in 1913 after reaching the South Pole.

Making a difference

OZONE SCIENTIST

Susan Solomon will not soon forget crawling across the roof of an Antarctic field station in windchill temperatures of −62°C (−80°F), moving heavy equipment, and adjusting mirrors while the winds howled and whipped about her. Sounds like an adventure, right? It sure was! But it is just part of what Solomon has done to establish herself as one of the world's leading authorities on ozone destruction.

Q: Is it true that you have traveled to the ends of the Earth to get information about the ozone layer?
A: Yes, I guess it is. My colleagues and I have studied the ozone hole in Antarctica, and we've measured and documented ozone chemistry above Greenland. But it's not all adventure. When I'm not visiting one of the poles, I run computer simulations of the atmosphere and study data at the National Oceanic and Atmospheric Administration (NOAA) in Boulder, Colorado.

Q: What is the significance of discoveries regarding the ozone hole?
A: Before British scientists discovered the ozone hole in Antarctica, no one was sure about ozone changes in the atmosphere. The popular belief was that in 100 years there might be 5 percent less ozone. So there were questions about whether it was a serious environmental problem. But when the British researchers released data that showed 50 percent less ozone over Antarctica in 1985 than was present 20 years earlier, the research raised our awareness that the problem was far more serious than previously thought.

Q: How have you contributed to the study of ozone?
A: Well, when the British data was first released, no one had much of an explanation about what was causing the destruction of the ozone layer. I thought about the problem a lot. Later that year, I sat in on a lecture about types of clouds called *polar stratospheric clouds*. These are beautifully colored clouds that are known for their iridescence. While I was looking at these clouds, which are common in the Antarctic but rare elsewhere, it occurred to me that they may have something to do with ozone depletion. Perhaps they provide a surface for chemical reactions that activate reactive chlorine from CFCs (human-made chlorofluorocarbons). If so, once activated, the chlorine could contribute to reactions that destroy ozone.

Q: Did you get the chance to test your hypothesis?
A: Yes, the next year the National Science Foundation chose me to lead a group of 16 scientists for a nine-week expedition in Antarctica. We were the first team of scientists from the United States sent to the Antarctic to study the ozone hole. Within one month we could see that unnaturally high levels of chlorine dioxide did occur in the stratosphere during ozone depletion. This discovery was very exciting because it seemed that we were on the right track. We kept collecting data that year and collected more data during a second trip the next year. Pretty soon, the evidence seemed to support my hypothesis that CFCs and ozone depletion are linked.

▶ The ozone hole can be seen in this satellite image. The hole is the pale blue and black region immediately above Solomon's shoulder.

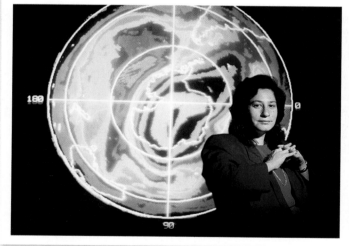

▶ Polar stratospheric clouds like these led Susan Solomon to make important discoveries about the cause of ozone depletion.

▶ Solomon has braved freezing polar temperatures to gather data about the ozone hole.

Q: How did this discovery make you feel?
A: On the one hand, it's very exciting scientifically to be involved in something like this. On the other hand, sometimes I think it's a little depressing. It would be nice to be involved in something more positive, to bring people good news. So far, we've brought nothing but bad news. We were hoping that we wouldn't find the same ozone chemistry in the Arctic that we found in the Antarctic. Unfortunately, we did. We hope for a positive result for the planet, but we don't always get it.

Q: How has your research helped to make a difference in our world?
A: Since our findings and others were announced, many of the world's countries have agreed to restrict or ban the use of CFCs. As a result, the ozone hole will eventually go away, but it will take a very long time. So although most countries have slowed their use of CFCs, CFCs from years past will still be hanging around in our atmosphere for the next 50 to 100 years. But I think our work has led in a small way to the realization that our actions do have consequences, and this realization should bring positive change.

Dr. Solomon has received international recognition for her work on the ozone hole over the Antarctic. She is a member of the U.S. National Academy of Sciences, the European Academy of Sciences, and the Académie des Sciences de France. In 2000, Dr. Solomon was awarded the National Medal of Science and the American Meteorological Society's Carl-Gustav Rossby Medal. In April 2002, she was nominated co-chair of the United Nations Climate Change Working Panel.

For More Information

If you would like free information about the ozone layer and what you can do to protect it, contact the Environmental Protection Agency, Public Outreach, 401 M St. SW, Washington, DC 20460.

What Do You Think?

If Susan Solomon had not sat in on the lecture about polar stratospheric clouds and did not realize the role that these clouds play in ozone destruction, where do you think our current understanding of the ozone hole would be? How does this reinforce the idea that a single person can make a tremendous contribution to humankind?

353

Answers to What Do You Think?

Without Susan Solomon's insight into the role of polar stratospheric clouds and CFCs in the destruction of ozone, the near-global push to restrict or ban the use of CFCs would not have occurred. As a result, the ozone hole would most likely have continued to grow and the detrimental effects to life from UV radiation would have become more pronounced.

Land
Chapter Planning Guide

PACING	CLASSROOM RESOURCES	LABS, DEMONSTRATIONS, AND ACTIVITIES
BLOCKS 1 & 2 · 90 min pp. 354–357 **Chapter Opener**		
Section 1 How We Use Land	**OSP** Lesson Plan * **CRF** Active Reading * **BASIC** **TT** Bellringer * **TT** Urban Vs. Rural Population by World Region *	**TE** Activity Where Do We Live? p. 355 **GENERAL**
BLOCKS 3, 4 & 5 · 135 min pp. 358–362 **Section 2** Urban Land Use	**OSP** Lesson Plan * **CRF** Active Reading * **BASIC** **TT** Bellringer * **TT** GIS Views of Seattle, Washington *	**TE** Activity Megalopolis, p. 358 **GENERAL** **SE** Field Activity Local Urban Sprawl, p. 359 **TE** Activity Identifying the Urban Crisis, p. 359 ◆ **BASIC** **TE** Group Activity Suburban Sprawl, p. 359 **ADVANCED** **SE** Inquiry Lab Creating a Land-Use Model, pp. 374–375 **CRF** Datasheets for In-Text Labs *
BLOCKS 6 & 7 · 90 min pp. 363–369 **Section 3** Land Management and Conservation	**OSP** Lesson Plan * **CRF** Active Reading * **BASIC** **TT** Bellringer * **TT** United States National Parks *	**TE** Internet Activity Putting Knowledge into Positive Action, p. 364 **GENERAL** **SE** QuickLab Measuring Soil Depth and Compaction, p. 365 **TE** Group Activity Balancing Recreation and Conservation, p. 368 **ADVANCED** **CRF** Consumer Lab * ◆ **BASIC** **CRF** Long-Term Project * ◆ **BASIC** **CRF** Design Your Own Lab * **GENERAL** **CRF** Modeling Lab * **ADVANCED** **CRF** Modeling Lab * **GENERAL**

BLOCKS 8 & 9 · 90 min

Chapter Review and Assessment Resources

- **SE** Chapter Review pp. 371–373
- **SE** Standardized Test Prep pp. 662–663
- **CRF** Study Guide * ■ **GENERAL**
- **CRF** Chapter Test * ■ **GENERAL**
- **CRF** Chapter Test * **ADVANCED**
- **CRF** Concept Review * ■ **GENERAL**
- **CRF** Critical Thinking * **ADVANCED**
- **OSP** Test Generator
- **CRF** Test Item Listing *

Online and Technology Resources

 Holt Online Learning

Visit **go.hrw.com** for access to Holt Online Learning or enter the keyword **HE6 Home** for a variety of free online resources.

 **One-Stop Planner® CD-ROM**

This CD-ROM package includes
- Lab Materials QuickList Software
- Holt Calendar Planner
- Customizable Lesson Plans
- Printable Worksheets
- ExamView® Test Generator
- Interactive Teacher Edition
- Holt PuzzlePro® Resources
- Holt PowerPoint® Resources

 Holt Environmental Science Interactive Tutor CD-ROM

This CD-ROM consists of interactive activities that give students a fun way to extend their knowledge of environmental science concepts.

KEY

TE	Teacher Edition	CRF	Chapter Resource File	*	Also on One-Stop Planner
SE	Student Edition	TT	Teaching Transparency	■	Also Available in Spanish
OSP	One-Stop Planner	CD	Interactive CD-ROM	◆	Requires Advance Prep

ENRICHMENT AND SKILLS PRACTICE	SECTION REVIEW AND ASSESSMENT	CORRELATIONS
SE **Pre-Reading Activity**, p. 354 TE **Using the Figure**, p. 354 `GENERAL`		**National Science Education Standards**
TE **Inclusion Strategies**, p. 355 TE **Skill Builder** Graphing, p. 356 `GENERAL` SE **Math Practice** Ecosystem Services, p. 357	SE **Section Review**, p. 357 TE **Reteaching**, p. 357 `BASIC` TE **Quiz**, p. 357 `GENERAL` TE **Alternative Assessment**, p. 357 `GENERAL` CRF **Quiz** * ■ `GENERAL`	SPSP 4a
TE **Skill Builder** Math, p. 360 `ADVANCED` TE **Using the Figure** Drawing Conclusions from GIS Maps, p. 361 `GENERAL` TE **Career** City Planner, p. 361	SE **Section Review**, p. 362 TE **Reteaching**, p. 362 `BASIC` TE **Quiz**, p. 362 `GENERAL` TE **Alternative Assessment**, p. 362 `GENERAL` CRF **Quiz** * ■ `GENERAL`	SPSP 3b SPSP 5b
SE **Graphic Organizer** Venn Diagram, p. 364 `GENERAL` SE **Case Study** California's Wilderness Corridors, pp. 366–367 TE **Inclusion Strategies**, p. 367 SE **Making a Difference** Restoring the Range, pp. 376–377 CRF **Map Skills** Land Use * `GENERAL`	SE **Section Review**, p. 369 TE **Reteaching**, p. 369 `BASIC` TE **Quiz**, p. 369 `GENERAL` TE **Alternative Assessment**, p. 369 `ADVANCED` CRF **Quiz** * ■ `GENERAL`	SPSP 3b SPSP 4a SPSP 5b SPSP 6e

Guided Reading Audio CDs

These CDs are designed to help auditory learners and reluctant readers. (Audio CDs are also available in Spanish.)

www.scilinks.org

Maintained by the **National Science Teachers Association**

TOPIC: Ecosystem Services
SciLinks code: HE4026

TOPIC: Land Use
SciLinks code: HE4059

TOPIC: Range Management
SciLinks code: HE4091

CNN Videos

Each video segment is accompanied by a Critical Thinking Worksheet.

Earth Science Connections

Segment 3 Satellite Farming

CHAPTER

14

Chapter Enrichment

This Chapter Enrichment provides relevant and interesting information to expand and enhance your classroom instruction of the chapter material.

How We Use Land

▶ An urban street scene in New York City

Differences in Land Use

More-industrialized countries, such as those of western Europe, tend to have more urban land than do less industrialized countries, such as the islands of the South Pacific. For example, the land area of the Netherlands breaks down as follows: 25 percent arable land, 3 percent permanent crops, 25 percent permanent pastures, 8 percent forests and woodland, and 39 percent other uses (primarily cities). In Papua New Guinea, on the other hand, arable crops make up 0.1 percent of the land; permanent crops, 1 percent; forests and woodlands, 92.9 percent; and other uses, 6 percent.

Changing Land Use

Scientists at the University of Wisconsin's Center for Sustainability and the Global Environment are working to understand how human land use has changed over long periods of time. The scientists combine information from satellite images with historical data from census records to construct digital maps of cropland and pasture land at different historical periods. They have created an animated movie that shows how the coverage of Earth by crops and pastures has changed over time.

Urban Land Use

The Development of Urban Life

Humans were initially nomadic hunters and gatherers, constantly on the move in search of food. With time, nomadic hunter-gatherers began to form semi-permanent camps, which allowed them to gather resources more efficiently. With the advent of farming, however, people adopted a more sedentary lifestyle. Agriculture also enabled people to produce food surpluses. The world's first cities were permanent agricultural villages that appeared in the Near East in the Mesopotamian region about 10,000 years ago.

Agricultural surplus contributed to the development of urban life. Excess food either coincided with or spurred the growth of a stratified social system. Food surpluses also allowed for an increase in the division of labor. People were able to specialize in a trade, craft, or service that was not directly involved with agricultural production.

Urban Crowding in Ur, Mesopotamia

Urban crowding has existed as long as there have been cities. Many early cities were also fortresses, so the population could not expand beyond the city walls. In the ancient Mesopotamian city of Ur, houses were cramped quarters located on narrow streets with no drainage. Instead of removing their trash, people threw it into the streets, where it accumulated. Archaeological excavations show that garbage levels rose so high in Ur that residents had to cut entrances into the second stories of their homes.

▶ Substandard housing in Hong Kong

BRAIN FOOD

The world's cities occupy only 2 percent of the Earth's surface yet account for roughly 78 percent of carbon emissions, 76 percent of industrial wood use, and 60 percent of water use.

Checkerboard Landscapes

Suburbs are often built in checkerboard patterns as development converts farmland adjacent to a city. This method of development increases expenses, because services become more inefficient the farther people live from service centers. For example, public transportation is scarce due to the high cost of operating trains or buses in areas with low population densities. Consequently, more residents use cars, expending more fossil-fuel resources. Checkerboard suburbs also contribute to the loss of family farms. Farmers on small tracts of land within this checkerboard pattern are often highly taxed, which encourages them to sell to developers, who convert this land into housing tracts.

SECTION 3 Land Management and Conservation

Resource Availability

Several factors influence the ability of an area to produce food. An area must have arable land (land on which crops can be grown) and a consistent supply of water. Energy and labor to till the soil and harvest crops must also be available. However, resources are distributed unevenly across the globe. Temperate regions such as Europe have more of the resources necessary to produce

BRAIN FOOD

Traffic studies in midtown New York City show the effects of modern traffic congestion. On average, cars move at only 6 miles an hour today. However, in the early 1900s, horse-drawn vehicles could reportedly pass through the same area at over 11 miles an hour!

food than arid regions such as the Middle East or North Africa. More than 40 of the world's nations do not have adequate water supplies to grow enough food for their populations. Furthermore, scientists estimate that human mismanagement has degraded roughly 17 percent (about 4.9 billion acres) of the Earth's vegetated surface. Irrigation, fertilization, overgrazing, and poor plowing practices have degraded 13.6 million acres in less than 65 years. This amounts to 38 percent of the world's total cultivated land.

Farming Losses

Since the mid-1900s, the number of farmers has declined worldwide. It has become more difficult for small farms to compete with large farms. Thus, many farmers are forced off their farms and migrate to cities to find work. Farms are also being developed for housing, industry, and roads at a rapid pace. The loss of small farms and family farmers means that less information about agricultural practices will be handed down through the generations.

Property Rights

In general, the United States grants unusually strong property rights to landowners. For example, European countries typically require landowners to observe strict land-use guidelines and to permit at least limited access to the public where appropriate. Some people would like to see similar guidelines established in the United States. Supporters of strong property rights argue that private ownership promotes sound land-management practices because people generally take better care of property that they own. Critics of strong property rights argue that the profit motive drives landowners to seek maximum possible short-term yield from the land at the expense of its long-term health.

▶ Rangeland in the western United States

Overview

Tell students that humans use land for many purposes, including farmland to grow crops, rangeland to feed livestock, forest land for wood, cities to live and conduct business, and parks for recreational enjoyment. Tell students that understanding these uses and their implications can make us better stewards of our environment.

Using the Figure — **GENERAL**

In the opening photograph in this chapter, suburban housing is encroaching on desert ecosystem in Southern California's Coachella Valley. During the 1990s, the population of the Coachella Valley grew 38 percent—almost three times the state average. This rapid population growth has put pressure on a number of resources, particularly water. The aquifer that supplies water to the valley fell as much as 23 feet in some places during the 1990s—bad news for a water supply that has been declining for more than 20 years. **LS** Visual

VIDEO SELECT

For information about videos related to this chapter, go to **go.hrw.com** and type in the keyword **HE4 LANV.**

Land

1 **How We Use Land**

2 **Urban Land Use**

3 **Land Management and Conservation**

PRE-READING ACTIVITY

FOLDNOTES

Layered Book
Before you read this chapter, create the **FoldNote** entitled "Layered Book" described in the Reading and Study Skills section of the Appendix. Label the tabs of the layered book with "Rural Land Use," "Urban Land Use," "Land Management," and "Conservation." As you read the chapter, write information you learn about each category under the appropriate flap.

On the edge of Palm Springs, California, suburban housing has been built on what was once a desert ecosystem.

354

Chapter Correlations | *National Science Education Standards*

SPSP 3b The earth does not have infinite resources; increasing human consumption places severe stress on the natural processes that renew some resources, and it depletes those resources that cannot be renewed. **(Section 2 and Section 3)**

SPSP 4a Natural ecosystems provide an array of basic processes that affect humans. Those processes include maintenance of the quality of the atmosphere, generation of soils, control of the hydrologic cycle, disposal of wastes, and recycling of nutrients. Humans are changing many of these basic processes, and the changes may be detrimental to humans. **(Section 1 and Section 3)**

SPSP 5b Human activities can enhance potential for hazards. Acquisition of resources, urban growth, and waste disposal can accelerate rates of natural change. **(Section 2 and Section 3)**

SPSP 6e Humans have a major effect on other species. For example, the

influence of humans on other organisms occurs through land use—which decreases space available to other species—and pollution—which changes the chemical composition of air, soil, and water. **(Section 3)**

SECTION 1
How We Use Land

Some years ago, officials in California decided to find out how land was being used in the state. Measurements were made using maps, aerial photographs, field surveys, and a computerized mapping system. The results were startling—in just eight years (between 1984 and 1992) nearly 84,000 hectares (about 210,000 acres) of farmland, rangeland, and woodland had been converted into suburbs and cities. This change is happening in many communities all over the world.

Land Use and Land Cover

We use land for many purposes, including farming, mining, building cities and highways, and recreation. Land cover is what you find on a patch of land, and it often depends on how the land is used. For example, land cover might be a forest, a field of grain, or a parking lot. There are different types of land cover and different human uses for each cover type, as shown in **Table 1.**

Land that is covered mainly with buildings and roads is called **urban** land. For the purposes of determining land use and residence trends, the U.S. Census Bureau defines an urban area as an area that contains 2,500 or more people and usually has a governing body, such as a city council. Any population not classified as urban is considered rural. Land that contains relatively few people and large areas of open space are **rural** areas. **Figure 1** shows the relative proportion of each of the types of land cover defined in **Table 1.** As the table shows, most land provides one or more resources that humans consume. These resources include wood in forests, crops in farmland, and mineral resources.

Table 1 ▼

Primary Land-Use Categories	
Land cover type	**Human use of land**
Rangeland	land used to graze livestock and wildlife
Forest land	land used for harvesting wood, wildlife, fish, nuts, and other resources
Cropland	land used to grow plants for food and fiber
Parks and preserves	land used for recreation and scenic enjoyment and for preserving native animal and plant communities and ecosystems
Wetlands, mountains, deserts, and other	land that is difficult to use for human purposes
Urban land	land used for houses, businesses, industry, and roads

Objectives

► Distinguish between urban and rural land.
► Describe three major ways in which humans use land.
► Explain the concept of ecosystem services.

Key Terms

urban
rural
ecosystem services

Land Use in the United States

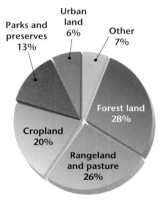

Source: Natural Resources Conservation Service.

Figure 1 ► The graph above shows the percentage of each land cover type in the United States.

355

Chapter Resource File

• **Lesson Plan**
• **Active Reading** BASIC
• **Section Quiz** GENERAL

Transparencies

TT Bellringer

INCLUSION Strategies

• *Attention Deficit Disorder*
• *Gifted and Talented*

Have students construct a chart similar to **Figure 1** for the state in which they live. Students may use the Natural Resources Conservation Service as a resource. After completing the chart, students should prepare a report comparing their state with another state. The report and chart should be presented to the class or to a small group of students.

Focus

Overview

Before beginning this section, review with your students the Objectives in the Student Edition. In this section, students will learn that land can be used for cities, agriculture, forestry, pasture, recreation, and preservation.

🔘 Bellringer

Ask students what percentage of people in the United States live in cities. Allow them time for discussion, and then tell them that more than three-quarters of the U.S. population lives in cities. Discuss why such a large percentage of the population lives in cities. **LS Interpersonal**

Motivate

Activity ——— GENERAL

Where Do We Live? Take the class outside and measure an area that is 100 m². Mark the area off with string and stakes. This area represents the surface area of Earth. Have the class help you measure a 70 m² area within the staked area. This area is the surface area of the Earth that is covered by oceans. Have the students move onto the land area. Next, mark an area that is 5 m² to represent the surface area of North America. To represent the surface area of the United States, mark an area that is 2 m². Tell students that about 77 percent of the U.S. population lives in urban areas, yet urban areas account for only about 2 percent of U.S. surface area. In the model, 400 cm² is equivalent to total urban land area in the U.S. **LS Kinesthetic/Visual**

Graphing Have students construct a pie graph similar to **Figure 1** that reflects the land distribution pattern in their state. Ask students, "How does state land use compare with national land use?" If there is a difference between state land use and national land use, ask students to discuss possible reasons for the difference. **LS** Visual

HISTORY —
CONNECTION   GENERAL

Militarism and Urbanization
The Roman word *castra*, meaning "military camp," is the basis for the names of several English urban areas, including Lancaster, Leicester, and Manchester. These places were originally military camps established by the Roman army as the army advanced across the land. As time passed, these military camps later developed into permanent settlements. Have students research the origin of other place-name suffixes such as *-burg*, *-ford*, and *-ville*. (Sample answer: *Ville* is a French word that means city or town.)

Figure 2 ▶ The photo on the left, of New York City, shows a typical urban scene, whereas the photo on the right, of the Connecticut River Valley, shows a typical rural scene.

Where We Live

Until about 1850, most people lived in rural areas. Many of them were farmers, who grew crops and raised livestock for food, clothing, and manufacturing. Other people managed the forests, worked in local mines or mills, or manufactured the necessities of life for a town.

The Industrial Revolution changed this pattern. Machinery was built that made it possible for fewer people to operate a farm or a grain mill. In addition, better transportation allowed manufacturers to be located far from their customers. So thousands of jobs in rural areas were eliminated. Many people had to move to cities to find jobs. As a result, urban areas grew rapidly during the 20th century and spread over more land. **Figure 3** shows that today, most people throughout the world live in urban areas. The movement of people from rural areas to urban areas happened rapidly in developed countries between about 1880 and 1950. Now, this movement is occurring rapidly in developing countries.

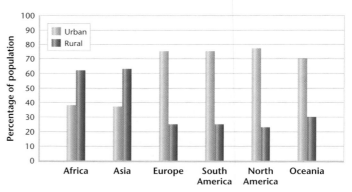

Figure 3 ▶ This graph shows the proportion of people living in urban areas and rural areas in different parts of the world.

Source: Population Reference Bureau.

BRAIN FOOD

Urban Explosion With 26.4 million inhabitants, ultra-dense Tokyo is the largest urban area in the world. The following are major world cities and their population counts in millions of people: Mexico City, Mexico (18.4); Bombay, India (18.0); São Paulo, Brazil (17.8); New York, New York (16.6); Lagos, Nigeria (13.4); Los Angeles, California (13.1); Calcutta, India (12.9); and Buenos Aires, Argentina (12.6). It is predicted that Bombay will become the second most populous city by 2015, with 26.1 million people. Lagos is predicted to be the third-largest city with 23.2 million people. According to the Worldwatch Institute, the cities of the developing world grew by 263 million people between 1990 and 1995. This number may actually be much higher, because millions of squatters living on city fringes are often not included in population counts. There are differences in published population counts because demographers delineate urban areas in different ways.

The Urban-Rural Connection

Whether people live in cities or in the countryside, people are dependent on resources produced in rural areas. These resources include clean drinking water, fertile soil and land for crops, trees for wood and paper, and much of the oxygen we breathe, which is produced by plants. The resources that are produced by natural and artificial ecosystems are called **ecosystem services.** Some examples of ecosystem services are shown in **Table 2.**

Supporting Urban Areas The area of rural land needed to support one person depends on many factors, such as the climate, the standard of living, and how efficiently resources are used. In a wet climate, for example, most agriculture depends on rain and does not depend on areas of lakes and rivers for irrigation. Each person in a developed country uses the ecosystem services provided by about 8 hectares of land and water. In the United States each person uses the ecosystem services from more than 12 hectares, whereas each person in Germany uses about 6 hectares' worth. Many people in developing nations do not have access to all the resources for a healthy life. They may use ecosystem services from less than a hectare of land per person.

Table 2 ▼

Examples of Ecosystem Services
purification of air and water
preservation of soil and renewal of soil fertility
prevention of flood and drought
regulation of climate
maintenance of biodiversity
movement and cycling of nutrients
detoxification and decomposition of wastes
aesthetic beauty

MATHPRACTICE

Ecosystem Services
Earth contains about 12.4 billion hectares of productive land—cropland, grazing land, forest, fresh water, and fisheries. In 1996, the world population was about 5.7 billion people, for a mean of 2.18 hectares of productive land per person. The world population in 2010 is projected to be 6.8 billion. On average, how much productive land per person will there be in 2010?

internet connect

www.scilinks.org
Topic: Ecosystem Services
SciLinks code: HE4026

SCiLINKS. Maintained by the National Science Teachers Association

SECTION 1 Review

1. **Explain** how ecosystem services link rural lands with urban lands.

2. **Describe** three main ways in which humans use land. Write a paragraph to explain your answer.
   WRITING SKILLS

3. **Distinguish** between rural lands and urban lands, and provide an example of each.

CRITICAL THINKING

4. **Making Decisions** What could individuals do to reduce the loss of ecosystem services per person as the human population grows?

5. **Making Inferences** How does the movement of people from rural lands to urban lands affect people's relationship with natural resources?

357

Answers to Section Review

1. Ecosystem services, such as clean drinking water and trees for paper and wood, are resources generated in rural lands that are used intensively to support urban lands.

2. Answers may vary. Humans create urban areas in which to build residences, businesses, and roads. We use pasture and rangeland to graze livestock. We use cropland to grow plants for food and fiber. We manage forests to produce wood and conserve wildlife.

3. Urban lands refer to areas that contain more than 2,500 people and usually have a governing body. Rural lands have limited population density and provide various ecosystem services. Sample answers: Chicago is urban; forested areas of the Pacific Northwest are rural.

4. Answers may vary. Humans could reduce their consumption by using fewer natural resources and fewer products that utilize such resources.

5. Answers may vary. The movement of people from rural lands to urban lands creates both physical and psychological distance between people and the natural resources on which they depend.

MATHPRACTICE

Answers
12.4 billion hectares of land divided by 6.8 billion people = 1.82 hectares productive land per person in 2010

Close

Reteaching ——— **BASIC**

Concept Mapping Write the following terms on the board: *urban, urban-rural connection, forests, farmland, ecosystem services, parks,* and *infrastructure.* Work as a class to make a concept map using these terms.
LS Logical   English Language Learners

Quiz ——— **GENERAL**

1. By United States Census standards, would a population of 1,000 constitute a city? (No, a city is defined as an area with more than 2,500 people.)

2. In North America, are there more people living in rural areas or urban areas? (urban areas) Is the world population mainly urban or rural? (Overall, the world's population is mainly urban.)

Alternative Assessment ——— **GENERAL**

Designing a Community Ask students to design a model community that is to be home for 1,000 people. Students may select a specific geographic area in which to locate the community, or they may invent one. Have students design the community to anticipate future population growth in the region. In addition, be sure students indicate where resources will come from and how community services such as electricity, water, trash collection, and police and fire protection will be delivered. Also have students consider the types of open space and alternative transportation routes (such as bike paths) they would include in their community.
LS Interpersonal   Co-op Learning

Overview

Before beginning this section, review with your students the Objectives in the Student Edition. This section discusses urban areas and the challenges of urbanization. Students will also learn how city planning is used to maximize the land resources of an area to accommodate its population.

Bellringer

Ask students to think of the features that can make a city a pleasant place to live and the features that can make a city an unpleasant place to live. Have them record their thoughts in their *EcoLog*.

LS Intrapersonal

Motivate

Activity ——————— GENERAL

Megalopolis, USA Tell students that while early cities were contained within the limits of their city walls, most modern cities sprawl in all directions. When one city runs into the next, it creates a large complex called a *megalopolis*. Have students examine a map of the megalopolis that extends from Washington, D.C., to Boston, Massachusetts. Have students look at a map of the United States and identify other areas of emerging megalopolises, such as the coast of southern California from Santa Barbara to San Diego. **LS** Visual

---

SECTION 2
Urban Land Use

Objectives

▶ Describe the urban crisis, and explain what people are doing to deal with it.
▶ Explain how urban sprawl affects the environment.
▶ Explain how open spaces provide urban areas with environmental benefits.
▶ Explain the heat-island effect.
▶ Describe how people use the geographic information system as a tool for land-use planning.

Key Terms

urbanization
infrastructure
urban sprawl
heat island
land-use planning
geographic information system
  (GIS)

Figure 5 ▶ The Washington, D.C.–Baltimore area has grown larger and more densely populated over the years. Red areas indicate urban development.

People live where they can find the things that they need and want, such as jobs, schools, and recreational areas. For most people today, this means living in an urban area.

Urbanization

The movement of people from rural areas to cities is known as **urbanization.** People usually leave rural areas for more plentiful and better paying jobs in towns and cities. In developed countries, urbanization slowed in the second half of the 20th century. In 1960, 70 percent of the U.S. population was classified as urban. By 1980, this percentage had increased only slightly to 75 percent. As urban populations have grown, many small towns have grown together and formed larger urban areas. The U.S. Census Bureau calls these complexes metropolitan areas. Some examples are Denver-Boulder in Colorado and Boston-Worcester-Lawrence in Massachusetts. **Figure 5** shows the expansion of the Washington, D.C.–Baltimore metropolitan area over the years. These maps were created using data from the U.S. Census Bureau.

Urban areas that have grown slowly are often relatively pleasant places to live. Roads and public transportation in these areas have been built to handle the growth, so that traffic flows freely. Buildings, roads, and parking lots are mixed in with green spaces and recreational areas. These green spaces may provide these urban areas with much needed ecosystem services such as moderation of temperature, infiltration of rainwater runoff, and aesthetic value.

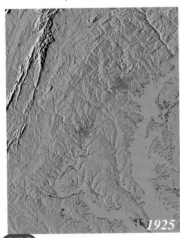
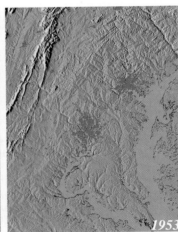
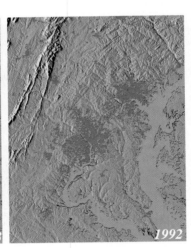

1925   *1953*   *1992*

MISCONCEPTION ALERT

Cultural Notions of Cities What makes a city? Different countries have unique ideas about what constitutes a city. The United States Census Bureau defines a city as an area with a population of 2,500 or more people. In India, a city is classified as an area that has greater than 5,000 people with a significant adult male population working in the professional or trade sector and not agriculture. In South Africa, a settlement of 500 people is a city. Have students research other official criteria that define cities around the world.

Chapter Resource File

• Lesson Plan
• Active Reading **BASIC**
• Section Quiz **GENERAL**

Transparencies

TT Bellringer

The Urban Crisis When urban areas grow rapidly, they often run into trouble. A rapidly growing population can overwhelm the infrastructure and lead to traffic jams, substandard housing, and polluted air and water. **Infrastructure** is all of the things that a society builds for public use. Infrastructure includes roads, sewers, railroads, bridges, canals, fire and police stations, schools, libraries, hospitals, water mains, and power lines. When more people live in a city than its infrastructure can support, the living conditions deteriorate. This growth problem has become so widespread throughout the world that the term *urban crisis* was coined to describe the problem. **Figure 6** shows an example of urban crisis in Hong Kong. The hillside is covered with substandard housing in an area that lacks the necessary infrastructure for people to live in healthy conditions.

Urban Sprawl Rapid expansion of a city into the countryside around the city is called **urban sprawl.** Much of this growth results in the building of suburbs, or housing and associated commercial buildings on the boundary of a larger town. People living in the suburbs generally commute to work in the city by car. Many of these suburbs are built on land that was previously used for food production, as shown in **Figure 7**. In 2000, more Americans lived in suburbs than in cities and the countryside combined. Each year suburbs spread over another 1 million hectares (2.5 million acres) of land in the United States.

Figure 6 ▶ Rapid urban growth has led to substandard housing on the hillsides above Hong Kong.

FIELD ACTIVITY

Local Urban Sprawl On your way home from school, observe your surroundings. In your *EcoLog*, write down any signs of urban sprawl that you observed. What criteria did you use for making this assessment?

Figure 7 ▶ This photograph shows suburban development spreading out around Maui, Hawaii.

Activity ——— BASIC

Identifying the Urban Crisis
Organize the class into small groups, and give each group a few current local newspapers or magazines. Have students search for articles that describe the urban crisis in some way. Typical articles that fall into this category include stories about increasing crime, homelessness, tax increases, and traffic congestion. If you do not live in an urban area, use newspapers from a large city. Have a class discussion about what the urban crisis is and what is being done to alleviate it. Then ask students to work together to make an informative bulletin-board display or a poster about the urban crisis in different areas.
LS Interpersonal

Co-op Learning | English Language Learners

Group Activity ——— ADVANCED

Suburban Sprawl To control the growth of cities, some states and counties have imposed urban limit lines around growing population centers. Urban limit lines are simply boundaries beyond which a municipality is not permitted to grow. These limit lines can effectively contain urban sprawl and force the full development and redevelopment of the urban areas they contain. Have students research urban limit lines and incorporate their findings into a written or videotaped news report. If you live in an urban or suburban area, suggest that students contact city planners to find out if urban limit lines exist for the area. LS Interpersonal
Co-op Learning

Notable Quotes ——— GENERAL

"A continent ages quickly once we come."

—Ernest Hemingway

Have students discuss what Hemingway might have meant, incorporating the concepts presented in this chapter. (Sample answer: When humans populate an area, they tend to use resources in an unsustainable way, which "ages" the land.)
LS Logical

BRAIN FOOD

20th-Century City Boom In 1910, there were 31 cities in the United States with populations between 100,000 and 250,000 people and 19 cities with populations in excess of 250,000 people. By 1990, there were 131 cities in the former category and 63 cities in the latter category.

Figure 8 ▶ The search for ocean views lead people to build these homes on the California coastline, which is giving way as a result of erosion.

The Challenges of Housing Development Have students read Chapter 3, "Los Angeles Against the Mountains," in *The Control of Nature* by Pulitzer Prize-winning author John McPhee. The book documents strategies people have used to try to control nature. Chapter 3 discusses the construction of basins built to catch debris flows that endanger Los Angeles area communities that lie at the foot of the San Gabriel Mountains. Instruct students to write an essay about the lessons learned in Los Angeles and about how these lessons might apply to development in their own area. **LS Intrapersonal**

SKILL BUILDER — ADVANCED

Math Have students do library, Internet, or phone research to determine the growth rate of their town or city. Ask them to find the current population and the population every 5 years before that, going back 25 years. Have them discuss how the growth or decline of the population is affecting land use. **LS Logical**

Connection to Geology

Flood Plains Rivers naturally flood their banks every so often. The potentially flooded area near a river is known as the flood plain. People increasingly build on flood plains, such as around the Mississippi River. Damage to buildings on flood plains often runs into the billions of dollars every year.

Figure 9 ▶ The urban heat island over Atlanta is shown in this computer-enhanced aerial view. Areas with higher temperatures appear red.

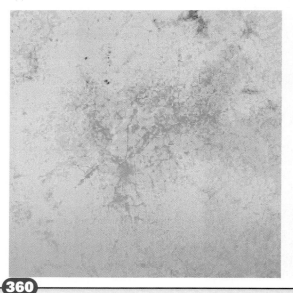

360

Development on Marginal Lands Many cities were first built where there was little room for expansion. As the cities grew, suburbs were often built on *marginal land*—land that is poorly suited for building. For example, Los Angeles and Mexico City are built in basins. These cities have expanded up into the surrounding mountains where the slopes are prone to landslides. The houses shown in **Figure 8** were built on land that is unsuited for development because of the natural processes of erosion along the coastline. Structures built on marginal land can become difficult or impossible to repair and can be expensive to insure.

Other Impacts of Urbanization Environmental conditions in the center of a city are different from those of the surrounding countryside. Cities both generate and trap more heat. The increased temperature in the city is called a **heat island.** Heat is generated by the infrastructure that makes a city run. Roads and buildings absorb more heat than vegetation does. They also retain heat longer. Atlanta, Georgia, is an example of a city that has a significant heat island, as shown in **Figure 9.**

Scientists are beginning to see that heat islands can affect local weather patterns. Hot air rises over a city, cooling as it rises, and eventually produces rain clouds. In Atlanta and many other cities, increased rainfall is a side effect of the heat island. The heat-island effect may be moderated by planting trees for shade and by installing rooftops that reflect rather than retain heat.

Hydrology and the Heat Island Effect
The majority of an urban center often includes impermeable surface cover. When rain falls, it is carried into storm drains rather than left as standing water in vegetation and soil. Evaporation from standing water cools the atmosphere, so when less moisture is available for evaporation, air temperatures remain higher. Revegetating cities helps increase evaporation and lowers temperatures in urban areas. Encourage students to do research to learn more about how cities are minimizing the heat-island effect. Reflective roofing, the use of light-colored construction materials, and planting shade trees are some methods that cities are using to mitigate the heat-island effect. **LS Intrapersonal**

Urban Planning

Land-use planning is determining in advance how land will be used—where houses, businesses, and factories will be built, where land will be protected for recreation, and so on. Land-use planners determine the best locations for shopping malls, sewers, electrical lines, and other infrastructure.

In practice, making land-use plans is complex and often controversial. The federal government requires developers to prepare detailed reports assessing the environmental impact of many projects. And the public has a right to comment on these reports. Developers, city governments, local businesses, and citizens often disagree about land-use plans. Projects that affect large or environmentally sensitive areas are often studied carefully and even bitterly debated.

Intelligent Design Land-use planners have sophisticated methods and tools available to them today. The most important technological tools for land-use planning involve using the geographic information system.

A **geographic information system (GIS)** is a computerized system for storing, manipulating, and viewing geographic data. GIS software allows a user to enter different types of data about an area, such as the location of sewer lines, roads, and parks, and then create maps with the data. **Figure 10** shows several GIS images of Seattle, Washington. Each image corresponds to a different combination of information. The power of GIS is that it allows a user to display layers of information about an area and to overlay these layers, like overhead transparencies, on top of one another.

Figure 10 ▶ The images below are of Seattle, Washington. Each image represents a different GIS layer, each with specific information.

361

Close

Reteaching ——— BASIC
Illustrating Land-Use Issues
Ask students to summarize the main points of this section in their own words. English language learners can draw pictures to illustrate each of the main points. Encourage students to include the drawings in their **Portfolio.** **English Language Learners**
LS Visual

Quiz ——— GENERAL

1. What infrastructure or services are most important for cities to provide? (Sample answers: housing, sanitation, transportation, and water)

2. Why shouldn't cities be allowed to simply spread out? (Urban and suburban sprawl create environmental problems by reducing ecosystem services and increasing the cost of city services.)

Alternative Assessment ——— GENERAL

Inner-City Renovation Tell students to imagine they have been assigned to a planning commission in charge of renovating an inner-city area. Explain that their planning options are unlimited but should include the use of incentives to encourage new investment and new residents. As a class, establish a budget for each project. Ask students to form groups of three or four to consider the problem and develop a plan that involves at least five renovation ideas. If appropriate, assign each group to a specific area of your town or city. **LS Interpersonal** **Co-op Learning**

Figure 11 ▶ This mass transit system in California's San Francisco Bay Area moves thousands of people a day with much less environmental impact than if the people took individual cars instead.

internet connect
www.scilinks.org
Topic: Land Use
SciLinks code: HE4059
SCiLINKS. Maintained by the National Science Teachers Association

Transportation Ask any urban dweller to name the main annoyance of big-city life, and the answer is likely to be "traffic." Most cities in the United States are difficult to travel in without a car. Many U.S. cities were constructed after the invention of the automobile. In addition, availability of land was not a limiting issue, so many American cities sprawl over large areas. By contrast, most cities in Europe were built before cars, have narrow roads, and are compact.

In many cities, *mass transit systems* were constructed in order to get people where they wanted to go. Mass transit systems, such as the one shown in **Figure 11,** use buses and trains to move many people at one time. Mass transit systems save energy, reduce highway congestion, reduce air pollution, and limit the loss of land to roadways and parking lots. Where the construction of mass transit systems is not reasonable, carpooling is an important alternative.

Open Space *Open space* is land within urban areas that is set aside for scenic and recreational enjoyment. Open spaces include parks, public gardens, and bicycle and hiking trails. Open spaces left in their natural condition are often called *greenbelts*. These greenbelts provide important ecological services.

Open spaces have numerous environmental benefits and provide valuable functions. The plants in open spaces absorb carbon dioxide, produce oxygen, and filter out pollutants from air and water. The plants even help keep a city cooler in the summer. Open spaces, especially those with vegetation, also reduce drainage problems by absorbing more of the rainwater runoff from building roofs, asphalt, and concrete. This ecological service results in less flooding after a heavy rain. These open spaces also provide urban dwellers with much-needed places for exercise and relaxation.

SECTION 2 Review

1. **Describe** the urban crisis, and explain how people are addressing it.

2. **Explain** how urban areas create heat islands.

3. **Explain** how open spaces provide environmental benefits to urban areas.

4. **Describe** how a GIS system can be used as a land-use tool.

CRITICAL THINKING

5. **Identifying Relationships** Write a short paragraph in which you describe the benefits of using a geographic information system for land-use planning.
WRITING SKILLS

6. **Making Decisions** Describe the environmental implications of urban sprawl.

362

Answers to Section Review

1. An urban crisis occurs when urban areas grow so rapidly that the needs of the population become greater than the resources available to meet these needs. People are addressing the crisis by improving infrastructure and services.

2. Large urban areas create heat islands because the infrastructure of the large areas generates heat and because the roads and buildings of these areas absorb more heat than the vegetation in the surrounding countryside does.

3. Open spaces support trees and other plants, which absorb carbon dioxide, produce oxygen, filter pollutants, and help cool the city in the

summer. Open spaces also reduce drainage problems by absorbing rainwater runoff.

4. Urban sprawl covers land once used for food production, destroys ecosystems, causes erosion through the development of marginal lands, and contributes to heat islands.

5. Answers may vary. These systems allow planners to visually compare land use, populations, and infrastructure to determine if land is being used most efficiently.

6. Answers may vary.

Land Management and Conservation

As the human population grows, the resources of more rural land are needed to support the population. The main categories of rural land are farmland, rangeland, forest land, national and state parks, and wilderness. Throughout our history, we have sometimes managed these lands sustainably so that they will provide resources indefinitely. We have also sometimes reduced their productivity by overusing or polluting them. The condition of rural land is important because of the ecological services that it provides. These services are especially important for the urban areas that rely on the productivity of rural land.

Farmlands

Farmland, such as that shown in **Figure 12**, is land that is used to grow crops and fruit. The United States contains more than 100 million hectares of prime farmland. However, in some places, urban development threatens some of the most productive farmland. Examples of places where farmland is threatened are parts of North Carolina's Piedmont region and the Twin Cities area of Minnesota. In 1996, the U.S. government established a national Farmland Protection Program to help state, county, and local governments protect farmland in danger of being paved over or otherwise developed.

Objectives

▶ Explain the benefits of preserving farmland.
▶ Describe two ways that rangeland can be managed sustainably.
▶ Describe the environmental effects of deforestation.
▶ Explain the function of parks and of wilderness areas.

Key Terms

overgrazing
deforestation
reforestation
wilderness

Ecofact

Hedgerows Farmland forms an important habitat for wildlife in Great Britain, which has relatively few remaining natural areas. Fields are separated by rows of bushes called hedgerows, which provide shelter for a variety of birds, mammals, reptiles, and insects.

Figure 12 ▶ This farmland next to the suburbs of Mililani, Hawaii, is used to grow a variety of crops.

363

Focus

Overview

Before beginning this section, review with your students the Objectives in the Student Edition. Rural land management issues are discussed in this section. The conservation and sustainability of farmland, rangeland, forest land, public land, and wilderness are addressed in this section.

Bellringer

Ask students, "Why does wilderness need to be preserved? Why shouldn't we consider the needs of humans first? Why don't we consider only our short-term needs and worry about the consequences later?"

Motivate

Discussion ——— GENERAL

Measuring Paper Use Tell your class that you would like to figure out how much paper was used in your school last year and how many trees were required to produce that much paper. Suggest a figure of 100 pages of typical letter-sized paper per student per month. Have students multiply this figure by the number of students in your school. A 5-foot section of a typical tree yields about 10,000 sheets, and the typical tree has about 30 feet of usable trunk. So each tree yields about 60,000 sheets. Ask students to figure out how many trees were required to produce the paper used in your school last year, assuming a nine-month school year. **LS Logical**

Chapter Resource File

• Lesson Plan
• Active Reading BASIC
• Section Quiz GENERAL

Transparencies

TT Bellringer

Discussion ——— ADVANCED

Energy Transfer Overgrazing has largely destroyed the native grass cover of much of the American West and has allowed nuisance plants such as mesquite, juniper, and cactus to invade. Ask students to discuss how this might change the flow of energy through the ecosystem. (There is less energy flow from plants to animals because fewer animals are able to feed on the nuisance species than on grass.)
LS Verbal

HISTORY ———
● CONNECTION — GENERAL

Sagebrush Rebellion In the late 1970s and the early 1980s, a movement that came to be known as the Sagebrush Rebellion sought to reduce federal land ownership in the West. Sagebrush "rebels" argued that public management by distant bureaucracies was invariably mishandled. They contended that only a private landowner with a vested interest could be expected to properly manage the land. The Sagebrush Rebellion ultimately failed, but the Wise-Use Movement, a private property-rights organization, has adopted many of its ideas.

Graphic
(**Organizer**) GENERAL

Venn Diagram
You may want to have students work in groups to create this Venn Diagram. Have one student draw the map and fill in information provided by other students from the group.

internet connect
www.scilinks.org
Topic: Range Management
SciLinks code: HE4091
SCi*LINKS*® Maintained by the National Science Teachers Association

Graphic
(**Organizer**) **Venn Diagram**
Create the **Graphic Organizer** entitled "Venn Diagram" described in the Appendix. Label the circles with "Rangelands Land Management" and "Forest Land Management." Then, fill in the diagram with characteristics that each type of land management shares with the other.

Figure 13 ▶ The photo below shows productive rangeland in the western United States.

364

Rangelands

Land that supports different vegetation types like grasslands, shrublands, and deserts and that is not used for farming or timber production is called *rangeland*. Rangelands can be arid, like rangelands in the desert Southwest, or relatively wet, like the rangelands of Florida. The most common human use of rangeland is for the grazing of livestock, as shown in **Figure 13**. The most common livestock are cattle, sheep, and goats, which are valued for their meat, milk, wool, and hides. Native wildlife species also graze these lands. Like farmland, rangeland is essential for maintaining the world's food supply. World population growth may require a 40 percent increase in the food production of rangeland from 1977 to 2030.

Problems on the Range Some rangelands in the United States have become degraded by poor land management strategies. Most damage to rangeland comes from overgrazing, or allowing more animals to graze in an area than the range can support. When animals overgraze, too many of the plants are eaten, and the land can become degraded. Overgrazing often results in changes in the plant community. Less desirable plant species may invade the area and replace more-desirable plant species. In cases of severe overgrazing, all the vegetation that covers the land is eaten. Once the plants are gone, there is nothing to keep the soil from eroding.

Maintaining the Range Much of the rangeland in the United States is public land managed by the federal government, which leases the rangeland to ranchers. Much of the rangeland in the United States is degraded. The Public Rangelands Improvement Act of 1978 was enacted to reverse this trend and improve land management practices.

Sustaining the productivity of rangeland generally means reducing overgrazing by limiting herds to sizes that do not degrade the land. Rangeland is also left unused for periods of time so that the vegetation can recover. Improving rangeland that has been degraded by overgrazing often includes methods such as killing invasive plants, planting native vegetation, and fencing areas to let them recover to the state they were in before they were overgrazed. Ranchers also control grazing by digging enough small water holes that livestock do not overgraze the vegetation around a single water hole.

⚡ Internet Activity ——— GENERAL

Putting Knowledge into Positive Action
Encourage students to identify from this section a conservation issue that interests them. Have them research this topic on the Internet with the purpose of finding a way to get involved. Many conservation organizations have volunteer programs and letter-writing campaigns centered around land-use issues. Have each student write a letter about his or her issue and read it to the class. **LS Intrapersonal**

Notable Quotes ——— GENERAL

"The horizons are still mine. The ragged peaks, the cactus, the brush, the hard brittle plants, these are mine and yours. We must be humble with them."

—Simon Ortiz (Acoma Pueblo poet)

Have students discuss the meaning of the quote. Ask them to think of ways citizens can ensure that rangelands, which are owned by American taxpayers, are respected and protected by all who use them. **LS Verbal**

Forest Lands

Trees are harvested to provide products we use everyday, such as paper, furniture, and lumber and plywood for our homes. In addition to wood and paper, we also value forest products such as maple syrup and turpentine. There are many ecosystem services provided by forests; however, one of the most important is the removal of CO_2 from the air.

Harvesting Trees People use enormous amounts of wood. The worldwide average is 1,800 cm³ of wood used per person each day. However, on average, each person in the United States uses about 3.5 times this amount. This is the equivalent of each person in the United States cutting down a tree that is 30 m tall every year. About 1.5 billion people in developing countries depend on firewood as their main source of fuel.

The timber industry classifies forest lands into three categories—virgin forest, which is forest that has never been cut; native forest, which is forest that is planted and managed; and tree farms, which are areas where trees are planted in rows and harvested like other crops. The two most widely used methods of harvesting trees are clear-cutting and selective cutting. These methods are shown in **Figure 14**. *Clear-cutting* is the process of removing all of the trees from an area of land. Clear-cutting large areas destroys wildlife habitat and causes soil erosion. The main alternative is selective cutting, which is usually practiced on smaller areas owned by individuals. *Selective cutting* is the process of cutting and removing only middle-aged or mature trees. Selective cutting is more expensive than clear-cutting, but selective cutting is usually much less destructive.

QuickLAB

Measuring Soil Depth and Compaction

Procedure

1. Find a plot of **undisturbed soil** in a forest, meadow, park, or other undisturbed area near your school.
2. Press a **meterstick** down into the undisturbed soil as far as it will go. Record how deep the meterstick went into the soil. Record how soft the soil was and how easy it was to press the meterstick into the soil. Repeat this five times in the same plot of undisturbed soil.
3. Pour 1 L of **water** onto the undisturbed soil. Use a **stopwatch** to record how long it takes for the soil to fully absorb the water.
4. Repeat this procedure at a plot of **disturbed soil** in a bike path, dirt road, or other area where the soil is bare and vegetation has been cleared or trampled.

Analysis

1. How did the soil depth and hardness in the plot of undisturbed soil differ from that in the plot of disturbed soil?
2. Which plot absorbed water faster?
3. How might grazing cattle affect the depth and compaction of an undisturbed plot of land?
4. How might clear-cutting affect an undisturbed plot of land?

QuickLAB

Skills Acquired:
• Measuring
• Collecting Data
• Inferring

Answers

1. Disturbed soils (such as trails) will be more compact, especially at the surface. Undisturbed soils will have a more even density.
2. The undisturbed soil should absorb water faster.
3. Grazing cattle will decrease the depth and increase the compaction of soil.
4. Clear-cutting an undisturbed area of land will remove the root systems that hold soil in place, compact soil with heavy machinery, and leave the surface topsoil bare and prone to erosion.

365

Issues in African Forestry The United Nations Food and Agricultural Organization estimates that 13,000 km² of African forest are cleared each year. Forest is cleared for agriculture, logging, firewood collection, charcoal production, and human settlement. In central Africa, this deforestation has reduced the habitats of great apes and chimpanzees, causing populations of the animals to decline. Scientists think that if the decline continues at the current rate, great apes will become extinct within 20 years. Deforestation has also directly affected African people. In remote, arid areas of northern and West Africa, there are often few energy sources besides firewood. In many villages, gathering firewood consumes more time and a greater share of a family's income than acquiring food does. Diminishing forest resources force locals to travel farther and farther to reach the retreating front line of shrinking forests.

Balancing Human Needs and the Environment In the Amazon rain forest, which is the world's largest rain forest, satellite data show that burning and deforestation are on the rise. Rain-forest residents start fires to clear the land of trees for planting food and grazing cattle. Other landowners harvest mahogany trees and other desirable hardwoods for profit. The effects of cattle grazing on the environment are particularly harmful. The amount of land required to graze cattle is 20 times greater than the amount of land needed to plant many crops. Deforestation caused by cattle grazing, crop planting, and tree harvesting leads to soil erosion, flooding, unstable agriculture, and the reduction of species habitat. Scientists think the forest area in the Amazon River basin may have already been reduced by 25 percent and may shrink another 20 percent by the year 2020. Although government regulations in countries such as Brazil require landowners to preserve 80 percent of their land, enforcing such laws is difficult. Ask students to discuss what should be done to strike a balance between the needs of the area's inhabitants and the need to sustain the environment of the Amazon rain forest. **LS Verbal**

Ecofact

Burning Trees and CO₂ When trees are cut and burned, they release carbon dioxide. From 1850 to 1990, deforestation released more than 100 billion metric tons of carbon dioxide into the atmosphere worldwide. Some scientists think this additional CO_2 is contributing to an increase in global temperatures.

Deforestation The clearing of trees from an area without replacing them is called **deforestation.** Most countries become severely deforested as populations expand and the demand for forest products increases. Forests are cleared to convert the land into farmland. People also clear forests to make space for roads, homes, factories, and office buildings.

Deforestation reduces wildlife habitat, but it has other impacts, too. For example, when forests are cleared from hillsides, soil erosion usually results if the area is not quickly planted with a cover crop. Without tree roots to hold the soil in place, soil is easily washed or blown into the valley below. In New York, forests on hillsides were cleared for farmland during the 19th century. Plowing also increased the rate of erosion, and as much as 90 percent of the soil eroded. Then, during the Great Depression, which was in the 1930s, hundreds of farmers in this area went bankrupt. The state bought many of these abandoned farms, and let the forests regenerate. Today, many of these hilltops are covered with state forest, which is used for recreation.

The rate of deforestation is especially high in tropical rain forests, where the soil is relatively thin. Farmers who clear forests in these areas must always move from one plot of land to another and clear more forest each time they move. Whether forests are cleared for farming or wood, if trees are not replanted, natural resources are steadily depleted.

CASE STUDY

California's Wilderness Corridors

California has an extraordinary range of habitats, from coastal islands, where elephant seals breed, to mountains where salmon, cougars, bobcats, and badgers are found. Many of these animals live on wildlife preserves and other public lands or on private land.

Animals do not know that they are safest if they stay on preserves. Many animals naturally migrate at different times of the year. Young animals are often evicted from their territories by their parents and must search for territories of their own to survive. Many animals also leave their territories in search of mates from other populations. This process is

important because if animals reproduce only with members of the species that live nearby, the population becomes inbred and genetic defects become more common.

As California becomes more urbanized, however, migration routes between one population of animals and another population are increasingly blocked by highways and housing developments.

A possible solution is the acquisition of conservation corridors—thin strips of protected land that connect one preserve with another preserve. Conservation biologists have argued for years over whether such corridors are

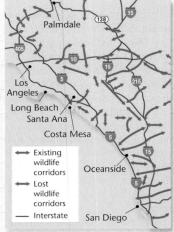

Source: Los Angeles Times

▶ This diagram shows suspected wildlife corridors around Los Angeles, California.

Cultural Awareness GENERAL

A Kenyan Conservationist Wangari Maathai is a Kenyan environmentalist who has won international recognition for her conservation and humanitarian efforts. As the founder of the Green Belt Movement, she not only has been instrumental in co-ordinating an enormous network of community reforestation projects in Kenya but also has been involved in starting similar movements in a dozen other countries, including the United States. Maathai's work has helped slow environmental degradation and has helped provide income and firewood for many poor rural people. At times a political dissident, Maathai is also a strong believer in the power of the individual: "When any of us feels she has an idea or an opportunity, she should go ahead and do it. . . . One person can make a difference." Encourage students to research and write an essay on the Green Belt Movement. Have them add this essay to their **Portfolio. LS Intrapersonal**

Reforestation Clear-cut forest can be replanted or allowed to regrow naturally, without human intervention. **Reforestation** is the process by which trees are planted to re-establish trees that have been cut down in a forest land. In some places, reforestation is happening faster than trees are being cut down. New England, for example, now contains more forest than it did a century ago. Much of the original forest had been cleared for farming, but the land was unable to sustain productive crops over a long period. As farming in this area became less economical, farms were abandoned and the forest regenerated. The same process has happened in places where steep hillsides were deforested for farming or development. The cost of deforestation, which caused soil erosion, landslides, and flooding, was too high. So forest has now been allowed to regenerate or has been replanted.

Some governments require reforestation after timber has been harvested from public land. A reforestation project is shown in **Figure 15**. Worldwide, more than 90 percent of all timber comes from forests that are not managed by an agency that monitors the health of forest ecosystems. Many governments are currently working to improve reforestation efforts and to promote less destructive logging methods. Private organizations have also established tree-planting programs on roadsides and in cities.

Figure 15 ▶ Tree seedlings have been planted to reforest this hillside as part of a reforestation project in the Fiji Islands.

▶ Cougars may need wildlife corridors in order to survive in parts of California.

effective in linking habitats and protecting animals.

One such corridor in California is the Tenaja corridor, which connects the Santa Rosa Plateau Ecological Preserve and the larger Cleveland National Forest in the Santa Ana Mountains southeast of

Los Angeles. Biologist Paul Beir studied the movements of a population of cougars in the Santa Ana Mountains by putting radio collars on more than 30 animals to track their movements. He found that the animals used the Tenaja corridor and avoided urban areas.

Now there is public pressure to preserve 232 of the corridors that link critical habitats. Voters have approved bond measures that will supply the money to buy some of the land, and the Nature Conservancy is also contributing land. The question of whether wildlife corridors preserve species may finally be answered by California's initiative.

CRITICAL THINKING

1. Applying Ideas California's state emblem is a grizzly bear, which is a species no longer found in the state. Why do you think the bears disappeared?

2. Expressing Viewpoints Should California spend state money to preserve habitats? Explain your answer.

CASE STUDY

California's Wilderness Corridors Use the information in the Case Study to locate the Tenaja Corridor. Have students conduct research on the area and create a map showing the corridor, the cougar's range, and nearby cities and highways. Have students propose ways to balance the needs of human communities with the need to preserve part or all of the corridor. **LS** Visual

Answers to Critical Thinking

1. California's grizzly bears disappeared largely because of human development of the state. Grizzly bears have wide ranges, which they exploit for food and use to find mates. As humans blocked the corridors used by the bears, the bears were no longer able to reach available habitats.

2. Answers may vary. Students could cite the importance of preserving ecosystems and species diversity, which necessitates spending state or federal money.

INCLUSION Strategies

• *Learning Disabled*
• *Attention Deficit Disorder*

Ask students to identify a national park close to where they live. They will create a brochure about the park to show information they leaned researching the park. The brochure should include the following information: the state in which the park is located, the date the park was established, the size of the park, the services that are available in the park, and the major sites to be seen in the park. The brochures may be displayed on a bulletin board in the classroom.

Group Activity — ADVANCED

Balancing Recreation and Conservation Tell students that a group of speleologists has discovered a huge cave with an ecosystem that contains a variety of organisms new to science. The landowner, Wilma Smith, has a dilemma. She can either sell the rights to the cave to Underground Resorts, Inc., which has made a generous offer, or sell the rights to the United States government, which has offered less money but will turn the site into a national park. Have the students hold a mock town hall meeting to discuss the decision. One student should play the mayor and serve as moderator of the meeting. Another student can play Wilma. Some students can represent the National Park Service, others can represent Underground Resorts, Inc., and the remainder can be concerned townspeople. At the end of the meeting, ask Wilma to announce her decision. **LS** Interpersonal Co-op Learning

LANGUAGE ARTS
● CONNECTION — GENERAL

Debating Conservation Policy In *Desert Solitaire*, Edward Abbey presents an interesting perspective on wilderness areas and the National Park Service. Have students read the book or select chapters to discuss. Set up a mock debate, and have students present their own views in support of or against Abbey's opinions of the park service and conservation of wilderness areas. **LS** Verbal

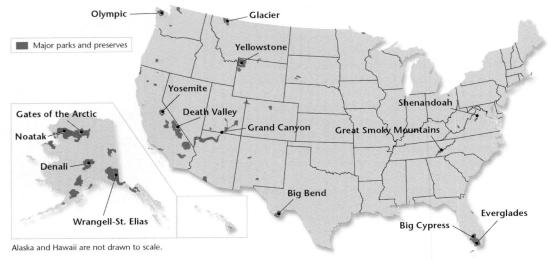

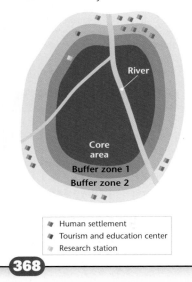

Figure 16 ▶ National parks in the United States are concentrated in the West.

Alaska and Hawaii are not drawn to scale.

Figure 17 ▶ Biosphere reserves are places where human populations and wildlife live side by side.

Human settlement
Tourism and education center
Research station

River

Core area

Buffer zone 1
Buffer zone 2

368

Parks and Preserves

In the 1870s, a group of explorers approached Congress with news of a magnificent expanse of land in Wyoming and Montana. These explorers expressed their concern that this land would be damaged by the development that had changed the northeastern United States. Congress agreed to protect this land by setting it aside for the public to use and enjoy, and the first national park—Yellowstone—was created. Today, the United States has about 50 national parks, as shown in **Figure 16.** If you are a U.S. citizen, you share ownership of these lands.

Public lands in the United States have many purposes. Most public lands are not as protected as the national parks are. Some public lands are leased to private companies for logging, mining, and ranching. Other public lands are maintained for hunting and fishing, as wildlife refuges, or to protect endangered species.

International efforts include the United Nations's Man and the Biosphere Program. This program has set up several hundred preserves throughout the world since 1976. These preserves are called *biosphere reserves* and are unusual in that they include people in the management plan of the reserves, as shown in **Figure 17.**

Wilderness The U.S. Wilderness Act, which was passed in 1964, designated certain lands as wilderness areas. **Wilderness** is an area in which the land and the ecosystems it supports are protected from all exploitation. Wilderness areas are found within several of the nation's public land systems. So far, 474 regions covering almost 13 million hectares (32 million acres) have been designated as wilderness in the United States. **Figure 18** shows an example of a wilderness area. These areas are open to hiking, fishing, boating (without motors), and camping. Building roads or structures and using motorized equipment is not allowed in wilderness areas.

HISTORY
● CONNECTION — BASIC

Birthing the Park System In 1872, Yellowstone National Park became the country's first national park, as decreed by President Ulysses S. Grant. In 1916, with over 60 designated sites in its possession, the National Park Service was signed into existence by an act created by President Woodrow Wilson. The National Park System comprises 384 areas that cover more than 83.3 million acres. National Parks are found in every state (except Delaware), the District of Columbia, American Samoa, Guam, Puerto Rico, and the Virgin Islands. The different types of parks include national parks, monuments, battlefields, military parks, historical parks, historic sites, lakeshores, seashores, recreation areas, scenic rivers and trails, and the White House. Have students use the Internet to find the national park that is located closest to your school. Ask them to find out when the site was designated as a national park and why the site is nationally important. If possible, take a class trip to the park. **LS** Intrapersonal

Benefits of Protected Areas Without national and private parks and preserves around the world, many more species would now be extinct. In a crowded world, these protected areas often provide the only place where unspoiled forests, deserts, or prairies remain. Without these areas, the plants and animals that can survive only in these ecosystems would disappear. These protected areas also provide recreation for people. People can camp, hike, fish, and watch birds and other wildlife in these areas. Wilderness areas also serve as outdoor classrooms and research laboratories where people can learn more about the natural world.

Threats to Protected Areas There is a constant battle in our world between our conservation efforts and the growing and increasingly mobile population. Around the world, more people visit national parks and wilderness areas each year and leave their mark on the land. Litter and traffic jams that have plagued our cities now plague many of our national parks. Rangelands, mining and logging sites, oil and gas drilling operations, factories, power plants, and urban areas are often close enough to the parks to affect the parks. In addition, preserved areas are affected by climate change and by air and water pollution, as are most other parts of the world.

In attempts to protect wilderness from damage, limits have been set in some areas on the number of people permitted in the area at any given time. Some areas are completely closed to visitors to allow wild animals to breed. In addition, volunteer programs are now active in many wilderness areas. Volunteers help pick up trash, build trails, control invading or exotic species, and help educate the visiting public.

Figure 18 ▶ In the United States, wilderness areas, such as the High Uintas Wilderness area shown here, are supposed to be preserved untouched for our own and future generations.

SECTION 3 Review

1. **Explain** what reforestation is and why it is important.

2. **List** and explain two methods of managing rangelands sustainably.

3. **Describe** the function of parks and of wilderness.

4. **Describe** the environmental effects of deforestation.

CRITICAL THINKING

5. **Recognizing Relationships** Read the first paragraph under the head "Threats to Protected Areas." Why do you suppose that some of our nation's national parks and wilderness areas are degraded?
READING SKILLS

6. **Recognizing Relationships** What are the benefits of preserving farmland?

369

Answers to Section Review

1. Reforestation involves replacing trees that have died or been cut down. It helps restore wildlife habitat and prevents soil erosion, because tree roots hold soil in place.

2. Rangeland can be managed by grazing fewer animals in a given area, which helps avoid overgrazing. Also, land can be left untouched for a period of time to give grasses and other plants time to recover.

3. Parks and wilderness protect certain species from extinction by preserving their habitat. Types of ecosystems that may not exist elsewhere in the country or world may be

preserved. Parks and wilderness also support recreation, research, and outdoor education.

4. Answers may vary. Deforestation leads to destruction of wildlife habitat and soil erosion.

5. Many parks are overcrowded, which leads to problems such as pollution and littering. Also, many parks are located close enough to urban areas, oil and gas drilling operations, logging sites, factories, or power plants to be affected by them.

6. Preserving farmland ensures that arable land that may be needed to produce crops in the future will be available.

Close

Reteaching ——————— BASIC

Section Review Work with individuals or small groups of English language learners to recap the main points of each subsection. Ask students to draw concept maps illustrating the main points and the ways the points relate to each other. Quiz students orally to ensure their understanding. | English Language Learners |

Quiz ——————— GENERAL

1. What type of land is usually used for ranching? (People generally use land that supports different vegetation types and that is not being used for farming or timber production for ranching.)

2. What are the consequences of deforestation? (habitat loss, soil erosion, and local climate change due to vegetation loss)

Alternative Assessment ——————— ADVANCED

Candidates for Conservation
Organize students into groups, and ask them to prepare a statement nominating an area, either a real or imaginary place, for national park designation. Have each group identify the qualities that would make the area a likely candidate for special protection. Also have students identify the type of park designation that would best fit the area. Tell them that the National Park Service Web site has a complete list of designation descriptions. Finally, have students determine recreational and commercial uses of their park. Ask the groups to present their statement to the rest of the class. Then have the whole class choose the area most deserving of the national park designation. **LS** Interpersonal | Co-op Learning |

Transparencies
TT United States National Parks

Alternative Assessment — GENERAL

Resource Distribution Divide the class into groups of 5 to 10 students. Each group will represent a country. Hand each group sets of cards or pieces of paper that represent different resources (blue cards for water, yellow cards for electricity, green cards for food, brown cards for housing availability, and so forth). Tell students that each card represents enough of that resource for one person. Distribute the cards unevenly so groups receive different amounts of resources. Then, have the groups discuss among themselves how they will structure their "country," how they will make use of their resources, and how many people they believe the resources of their country can support. Have each group discuss their results with the class. If there is additional time, have the countries interact with each other to discuss how resources might be shared in order to meet the needs of all groups. **LS** Interpersonal

Co-op Learning

Political Solutions Tell students to imagine that they are candidates running for city manager of their community or of a community of their choice. Have each student compose and deliver a campaign speech in which he or she cites at least three specific real-world land-use problems and provides potential solutions to each. **LS** Verbal

1 How We Use Land

Key Terms

urban, 355
rural, 355
ecosystem services, 357

Main Ideas

▶ Land is covered with forest, farm fields and pastures, roads, and towns.

▶ Urban areas are mostly covered with houses, roads, businesses, and industrial and municipal structures. Rural areas have less dense human populations and include forest land, cropland, rangeland, and other land cover types.

▶ Urban areas need very large areas of rural ecosystems to supply them with water, food, wood, and other ecosystem services.

2 Urban Land Use

Key Terms

urbanization, 358
infrastructure, 359
urban sprawl, 359
heat island, 360
land-use planning, 361
geographic information system (GIS), 361

Main Ideas

▶ Urbanization is the migration of people from rural to urban areas.

▶ When cities grow more rapidly than infrastructure can be built, they tend to suffer from substandard housing and traffic problems.

▶ Unplanned growth of a city results in urban sprawl, as low-density development spreads into the surrounding countryside.

▶ Land-use planning is essential if urban areas are to be pleasant places to live.

3 Land Management and Conservation

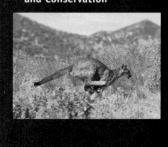

Key Terms

overgrazing, 364
deforestation, 366
reforestation, 367
wilderness, 368

Main Ideas

▶ Farmland is used to raise crops and livestock.

▶ Rangeland is land used primarily for grazing livestock. Rangeland is easily degraded by overgrazing.

▶ Trees are harvested for many purposes. Deforestation can cause soil erosion and may threaten forest plants and animals with extinction.

▶ National lands are used for many purposes, including lumber, mining, and recreation. Wilderness is national land that is protected from all exploitation for the benefit of future generations.

370

Chapter Resource File

- **Chapter Test** GENERAL
- **Chapter Test** ADVANCED
- **Concept Review** GENERAL
- **Critical Thinking** ADVANCED
- **Test Item Listing** GENERAL
- **Design Your Own Lab** GENERAL
- **Modeling Lab** ADVANCED
- **Modeling Lab** GENERAL
- **Consumer Lab** BASIC
- **Long-Term Project** BASIC

Using Key Terms

Use each of the following terms in a separate sentence.

1. *rangeland*
2. *infrastructure*
3. *urbanization*
4. *ecosystem services*
5. *geographic information system*

For each pair of terms, explain how the meanings of the terms differ.

6. *heat island* and *urban sprawl*
7. *overgrazing* and *deforestation*
8. *urban* and *rural*
9. *selective cutting* and *clear-cutting*

> ✔ **STUDY TIP**
>
> **Flash Cards** With a partner, make flash cards for the key words and most important ideas in the chapter. Take turns quizzing each other about the content of the course. Do another round, and this time the person being asked questions should try to use each key word and idea in a complete sentence.

Understanding Key Ideas

10. Building a mass transit system is likely to have which of the following effects?
    a. increasing air pollution
    b. traffic congestion
    c. increasing the temperature of the urban heat island
    d. none of the above

11. National parks and wilderness areas are designed to do which of the following?
    a. provide recreation
    b. protect wildlife
    c. preserve natural areas
    d. all of the above

12. Which of the following is *not* an example of urbanization?
    a. Immigrants settle in New York City.
    b. A farmer who can no longer afford to lease farmland moves to a city.
    c. A drop in timber prices in Oregon causes a lumberjack to lose his job and he moves to Portland.
    d. An Indian family moves to the city of Calcutta after a landslide destroys their village.

13. Which of the following is *not* an example of infrastructure?
    a. a railroad
    b. a school
    c. a telephone line
    d. a dairy farm

14. Which of the following is a likely result of deforestation?
    a. The amount of carbon dioxide removed from the atmosphere is reduced.
    b. Wind blows soil away because its plant cover has been removed.
    c. Water runs off the land more rapidly and causes floods.
    d. all of the above

15. Which of the following is *not* likely to cause the degradation of rangeland?
    a. adding more animals to a herd grazing on rangeland
    b. a drought in which rainfall is lower than usual for three years
    c. planting grass seed on the land
    d. driving a vehicle off-road

16. Which of the following is an example of reforestation?
    a. replanting forest land that has been clear-cut
    b. planting a cherry tree in your backyard
    c. planting oak trees in a city
    d. all of the above

17. Which of the following is *not* an ecosystem service provided by rural lands?
    a. oxygen in the air
    b. food
    c. aesthetic beauty
    d. wood for making paper

371

Assignment Guide

Section	Questions
1	4, 8, 17, 26
2	2, 3, 5, 6, 10, 12, 13, 22, 26, 31, 34
3	1, 7, 9, 11, 14–16, 18–21, 23–25, 27–30, 32, 33

ANSWERS

Using Key Terms

1. Sample answer: On rangeland, some of the animals you would expect to see are cows, sheep, or buffalo.

2. Sample answer: Roads, schools, and bridges are part of the infrastructure of a city.

3. Sample answer: Urbanization leads to problems when there aren't enough resources for all the people who live in a city.

4. Sample answer: Ecosystem services include water, oxygen, fertile soil, and other products generated by healthy ecosystems.

5. Sample answer: The urban planners used a geographic information system map to generate useful information about the city they were planning.

6. Heat islands are areas of increased temperature over cities. Urban sprawl is the expansion of the city into the surrounding countryside.

7. Overgrazing refers to the damage caused by more animals grazing an area than the land can support. Deforestation is cutting down trees without replacing them.

8. Urban areas contain more than 2,500 people and are characterized by a formal structure or government. Rural areas are sparsely populated and contain forest and farmland.

9. Selective cutting is the removal of only middle-aged and mature trees from a forest. Clear-cutting is the complete removal of trees in a given area of land.

Understanding Key Ideas

10. d
11. d
12. a
13. d
14. d
15. c
16. a
17. c

Short Answer

18. Rangeland may be degraded by overgrazing. Overgrazing can allow undesirable plant species to invade and may lead to erosion.

19. Answers may vary. National parks and forests in the United States do not necessarily protect ecosystems from human activities. They are often leased for mining, logging, and ranching. Even lands that are not leased may be located close enough to mining, oil and gas drilling, or industrial operations to be affected by them.

20. A national park supports more uses than a wilderness area does. Wilderness areas are protected from all exploitation but can be used for hiking, fishing, and nonmotorized boating. National parks can be leased for mining, logging, and ranching, and motorized activities such as snowmobiling are allowed.

21. No, national parks are located in many countries.

22. Mass transit systems decrease pollution and traffic by lessening the number of cars on the road.

Interpreting Graphics

23. Answers may vary. Sample answer: The reserve's research station is located in buffer zone 1. It is most likely located there to allow researchers easy access to the heart of the reserve.

24. The reserve is likely an ecotourism destination because there is a tourism and education center.

25. Buffer zone 2 forms a barrier between the human settlements and the rest of the reserve.

Short Answer

18. Explain one way rangeland can be degraded.

19. Do national parks and forests in the United States protect ecosystems from human activities? Explain your answer.

20. What is the difference between a U.S. wilderness area and a national park?

21. Are national parks located only in the United States?

22. How can building a mass transit system improve living standards in an urban area?

Interpreting Graphics

The map below shows a typical UN Biosphere Reserve. Use the map to answer questions 23–25.

23. Where is the reserve's research station located, and why has it been placed there rather than anywhere else in the reserve?

24. What indicators can you see that this reserve might be an ecotourism destination?

25. What does the map tell you about the function of buffer zone 2?

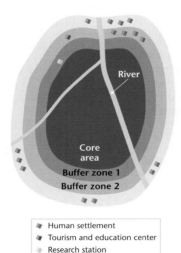

River

Core area

Buffer zone 1
Buffer zone 2

- ■ Human settlement
- ■ Tourism and education center
- ■ Research station

Concept Mapping

26. Answers to the concept mapping questions are on pp. 467–472.

Critical Thinking

27. Answers may vary. Sample answer: Clear-cutting a hillside may cause rapid soil erosion. Trees and other plants hold soil, so when they are removed, soil is more easily blown away by wind or washed away by heavy rains.

Concept Mapping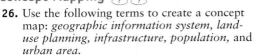

26. Use the following terms to create a concept map: *geographic information system, land-use planning, infrastructure, population,* and *urban area.*

Critical Thinking

27. **Recognizing Relationships** Read about clear-cutting under the head "Harvesting Trees." What effects does clear-cutting a hillside have on the environment? **READING SKILLS**

28. **Drawing Inferences** If we see many invasive plant species and large areas of bare soil on rangeland, what conclusions can we draw about the land management practices on this rangeland? Explain your answer.

29. **Evaluating Assumptions** We tend to think that the main use of livestock is for meat. However, the Masai herders of Africa do not slaughter their cattle. They use the milk. They also bleed the cattle and use to the blood to make a protein-rich sausage. What other uses for livestock can you think of that do not involve killing the animals?

Cross-Disciplinary Connection

30. **History** Find out how deforestation has affected a community. If you live in a forest biome, you can document the effects of deforestation on local rivers and farmland. If not, you will probably have to find an example on the Internet or in a magazine. Write a paragraph for your answer, using at least three key terms from this chapter. **WRITING SKILLS**

Portfolio Project

31. **Research** Diagram the growth of your community over the last 100 years. Express this as a graph that shows the growth of the population and a map that shows the area of ground the community covers. There are various possible sources for the data you will need. If there is a local historical society, this is probably the best source. Otherwise, city hall or the local newspaper will probably have the information.

28. We can conclude that the land has been over-grazed. This destroys native vegetation and allows invasive species to populate the area.

29. Answers may vary. The fur of sheep can be shorn for making clothing, the milk of cows and goats can be used for cheese, and oxen can be used to pull plows.

Cross-Disciplinary Connection

30. Answers may vary.

Portfolio Project

31. Answers may vary.

MATH SKILLS

The graph below shows land cover in the United States in 1997. Use the graph below to answer questions 32–33.

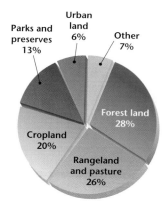

Parks and preserves 13%

Urban land 6%

Other 7%

Forest land 28%

Cropland 20%

Rangeland and pasture 26%

32. Analyzing Data If the percentage of cropland increased to 25 percent, and all other land cover categories except for rangeland and pasture remained the same, what percentage would rangeland and pasture be?

33. Making Calculations What percentage of the United States is planted in crops if 11 percent of cropland is idle (unplanted) at any one time?

WRITING SKILLS

34. Communicating Main Ideas In what ways does urban sprawl reduce the quality of life for people in the suburbs as well as in the town or city?

STANDARDIZED TEST PREP

For extra practice with questions formatted to represent the standardized test you may be asked to take at the end of your school year, turn to the sample test for this chapter in the Appendix.

READING SKILLS

Read the passage below, and then answer the questions that follow.

When more people live in a city than its infrastructure can support, living conditions deteriorate. For example, many people do not have access to clean water for drinking and washing. In addition, overcrowding causes the prices of existing houses and apartments to rise above the reach of many workers. Increasing numbers of people become homeless. These problems have become so widespread throughout the developed and developing world, that the term *urban crisis* was coined to describe the problem. According to the United Nations, the crisis is so bad that almost one-fourth of the world's city dwellers could be homeless by the year 2020.

1. According to the passage, which of the following statements is true?
  a. Cities have grown so rapidly that the rural areas that supply the cities with food and fuel cannot do so fast enough.
  b. Overcrowded cities lead to unaffordable housing and homelessness.
  c. The urban crisis does not usually involve water pollution.
  d. Living conditions usually improve when the population of a city increases.

2. Which of the following is an example of the urban crisis, according to the passage?
  a. To make space for more cars, roads into and through a city are widened.
  b. Deforestation in the countryside surrounding a city causes the water level in a river that flows through the city to increase when it rains.
  c. Homelessness is increasing rapidly in cities in both developed and developing countries.
  d. Inadequate planning for a growing number of commuters results in traffic jams on roads leading into the city in rush hour.

Math Skills

32. The percentage of pasture and rangeland would decrease by 5 percent to become 21 percent of land use.

33. Only 17.8 percent of the United States would be planted in crops. (If 11 percent of cropland is idle, then 89 percent of cropland is planted. Thus, 0.89 x 20% = 17.8% of the United States is planted.)

Writing Skills

34. Answers may vary. Urban sprawl reduces the quality of life because it eliminates ecosystem services used by people in the suburbs as well as in the city. Services are disrupted through deforestation, pollution, and soil degradation. Urban sprawl also increases the travel and commute times as well as the associated gasoline costs for suburban dwellers. Sprawl also separates suburban dwellers by long distances from goods and services they need.

Reading Skills

  1. b
  2. c

CREATING A LAND-USE MODEL

Teacher's Notes

Time Required
one 45-minute class period

Lab Ratings

EASY —————————→ HARD

TEACHER PREPARATION 🧪
STUDENT SETUP 🧪🧪🧪
CONCEPT LEVEL 🧪🧪
CLEANUP 🧪

Skills Acquired
- Constructing Models
- Identifying and Recognizing Patterns
- Interpreting
- Organizing and Analyzing Data

The Scientific Method
In this lab, students will:
- Ask Questions
- Analyze the Results
- Draw Conclusions

Materials
The materials on the student page are enough for a group of four students.

Objectives
▶ **Create** a simulated land-use model.
▶ **Recognize** conflicts of interest that arise during a negotiation.
▶ USING SCIENTIFIC METHODS **Analyze** and draw conclusions about the effect of compromise on the desired outcome for each interested party in a land-use plan.

Materials
colored pencils
graph paper
pens

Laws
- At least 10 percent of each type of habitat must be preserved.
- Landfills must be at least two acres away from all housing, wetlands and fresh water sites.
- Roads and bridges may cross rivers and wetlands but they must go around large natural areas.
- Roads must be connected to all developed areas of the city.
- There must be no building over wetlands, slopes and fresh water. Only parks may partially cover these habitats and roads/bridges may cross them.

Creating a Land-Use Model
Land-use plans are drawn up by planners, but they are created with the combined input of various members of a community. Along with three other people, you are meeting to plan the development of 400 acres of land for your growing city. Your team is composed of the following four members:

Team Members
• The **Planner** is concerned with creating a plan that encourages the sort of growth that will attract businesses and new citizens to the area.
• The **Developer** bought the land from the city and is interested in the right to build housing and a shopping center.
• The **Conservationist** is interested in preserving open space and natural areas from further development.
• The **Law Enforcer** ensures that all of the laws and regulations are met for any new development project.

Procedure
1. Have each team member select one of the four jobs above.
2. Use all or part of a large piece of graph paper as your map. Mark off an area that will represent 400 acres. Determine the approximate scale, and label the sides of your area accordingly.
3. The planner will color in the map as follows:
   a. 40 acres will be fresh water (rivers and/or lakes) and is colored light blue.
   b. 80 acres will be wetlands that are right next to some of the fresh water and are colored light purple or lavender.
   c. 40 acres will be land that is too sloped for building and will be colored tan.
   d. 240 acres is land that is good for development and will be colored light green.
4. Once the land is colored in, it cannot be altered. That will be the land you work with.
5. After the area is colored in, the group must discuss how and where to put the following items:
   a. 40 acres for a landfill
   b. 20 acres for utilities such as power plants, water treatment facilities, etc.
   c. 40 acres for parks and wildlife
   d. 40 acres for housing. Try to put the houses near a beautiful area.

374

Tips and Tricks
Make sure students understand and follow all of the guidelines as they go. The guidelines restrict various options and choices. If students forget to follow some of the guidelines, they will have to redo part or all of the exercise. This restriction mirrors the real-life situation people face when planning a community.

e. 40 acres for shopping

f. 20 acres for anything that the group agrees to add. For example, you could add a few acres for community gardens or for sports and playing fields. The law enforcer cannot suggest anything, but if the group can't agree on what to add, the law enforcer may cast the deciding vote.

g. 40 acres of roads and bridges (you can divide an acre up so that you can build long, thin roads rather than create short, fat roads that are an entire acre thick). Make sure at least one road goes into and out of town.

6. The law enforcer should make sure that the plans abide by the planning regulations by checking the map for violations.

7. Use the key under the map to mark which areas are which. For example, an R denotes a road or bridge. Use a pencil and write in the things softly at first in case changes are to be made. You may need a second copy of the map in case you make mistakes the first time.

Analysis

1. **Describing Events** Did everyone on your team agree on the plan, or were there conflicts of interest? Explain.

2. **Describing Events** Were you able to get everything your team wanted into the plan or did it face any problems? Describe what happened.

3. **Identifying Patterns** How did the features of the land constrain the plan that you made? Did you encounter any problems?

Conclusions

4. **Evaluating Results** Does the plan your group created meet the needs of all of the group members? Does it allow for development while preserving the environment?

5. **Evaluating Models** How do you think this land planning "simulation" compares to the real-life process of land-use planning?

Extension

1. **Research and Communications** Look in the newspaper or on the Internet for a story about a land-use debate in your area. Identify the different members involved. Role-play with your team to see what forces will bear on this controversy.

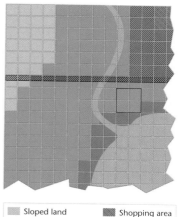

Sloped land	Shopping area
Wetlands	Housing
Fresh water	Utilities
For development	Road

▶ **Example Map** This is an example of what your land-use model might look like.

375

Chapter Resource File

- **Datasheets for In-Text Labs**
- **Lab Notes and Answers**

Sharon Harris
Mother of Mercy
High School
Cincinnati, Ohio

RESTORING THE RANGE

Background

The Bamberger ranch is just one example of a private effort to preserve species and habitat. Haliburton Forest and Wildlife Reserve Ltd. in Ontario, Canada, was purchased from a logging company in 1962 by German Baron von Fuerstenberg. The 60,000-acre depleted forest has since been restored and became Canada's first "certified sustainable forest." The privately owned Haliburton Forest is still used for logging, but now these practices are conducted in a manner that serves as a model for forest management. Streams and lakes are preserved, bird nests are avoided, and the rare red spruce is protected. Haliburton Forest serves to educate the public and provide recreational opportunities while still protecting the environment.

GEOGRAPHY
CONNECTION — **GENERAL**

A Costa Rican Preserve Punta Leona is home to a private preserve on Costa Rica's Pacific coast. The reserve is 27 years old and is one of the few remaining forests in the transitional life zone between the dry forest of the northwest and the rain forest of the southwest. The reserve has primarily served as a center for research. In 1996, biologist Christopher Vaughn began a program at Punta Leona to help protect the scarlet macaw. The effort received national attention. Punta Leona houses a hotel and attracts many tourists.

Making a difference

RESTORING THE RANGE

When Ohioan J. David Bamberger first moved to San Antonio, Texas as a vacuum cleaner sales representative, he was charmed by the dry, grass-covered rangeland of the Texas Hill Country. But much of the land was degraded. It had been overgrazed by cattle and was left with thin soil and dried up creeks.

Bamberger became intrigued by the idea of restoring some of the range to its original beauty. He was inspired by a book his mother gave him called *Pleasant Valley,* by Louis Bromfield. Long before it was popular, Bromfield had theories about how degraded habitats could be restored and how they could then be managed in a sustainable manner. Bamberger was intrigued by the idea of putting Bromfield's theories into action.

The Bamberger Ranch

In 1959, David Bamberger bought his first plot of land near Johnson City. Since then, David and Margaret Bamberger have expanded the ranch to nearly 2,300 hectares (5,500 acres). It is one of the largest habitat restoration projects in Texas, and shows the beauty of this area before it was damaged by human activities.

In its natural state, the ranch should have been grassland, with woody shrubs only near creeks. Instead, it had become overgrown with juniper shrubs and trees (often called cedar, *Juniperus ashei*), which can grow in poor soil and choke out other plants.

Bamberger read everything he could find on the degradation and restoration of rangeland. He found that two main things destroy the range: overgrazing and the suppression of wildfires.

Overgrazing causes soil erosion. The lack of fires permits the growth of shrubs that shade out grasses and wildflowers.

The Bambergers set to work to restore the property. They cleared most of the junipers, which left more water in the soil. They planted native trees, wildflowers, and grasses, and they controlled the grazing.

Grazing is necessary for healthy grassland. The American prairies were home to huge herds of bison (buffalo), which cropped the grass and fertilized the soil with their droppings. The Bambergers combined the grazing they needed with the preservation of an endangered species. San Antonio Zoo asked the Bambergers if they could help preserve the endangered scimitar-horned oryx, an antelope with thin, curved horns that is native to North Africa. Only a few small herds of this species remained, and the zoo feared that the oryx were becoming inbred, with too little genetic diversity. The Bambergers agreed, and the ranch is now home to a large herd of oryx.

The Effects of Restoration

The change in the ranch since Bamberger first bought it is most obvious at the fence line bordering the ranch. Beyond the fence there is a small forest of junipers and little other vegetation. On Bamberger's side, the main plants are grasses and wildflowers, with shrubs and trees in canyons and gullies beside the creeks. When the Bambergers first arrived, they counted only 48 species of birds on the ranch. Now, there are more than 150 species because there are many more different plants on the ranch. In the early

► David Bamberger, founder of the Bamberger Ranch Preserve.

days, deer on the ranch weighed only about 20 kg. Now they weigh about 40 kg, thanks to the improved grazing.

In addition to deer and oryx, cattle and goats live on the ranch. Some of these are used for experiments on the effects of domestic animals on rangeland. Students and faculty from nearby universities are studying this question by using exclosures. These are fences that keep large animals out of an area. The vegetation inside an exclosure is invariably taller than that outside because grazing animals are excluded. But in addition, the plant mix inside the exclosure is different from that outside. This is because grazing mammals eat only a few nutritious species and leave the others.

The Distribution of Water

One important change in the ranch under the Bambergers' management has been the change in water distribution. Water is very important in rangeland, which naturally gets little rainfall. Many of the creeks dry up between rainy periods, but water remains in the soil and underground. Grasses have spreading root systems that absorb water from a wide area. Poor management changes this balance by allowing junipers to take over the land. A juniper can take up 10 L of water a day from the soil, leaving too little for nearby grasses and wildflowers to survive. Then, when it rains heavily, the junipers cannot absorb all the water and it runs off the land. With no grass roots to hold the soil in place, the soil erodes into the creeks.

When the Bambergers arrived at the ranch, it was degraded rangeland. They drilled wells 150 m deep (500 ft) and did not reach the water table. Now, with the restoration of grassland, soil erosion has been reduced and much more water remains in the soil. Creeks and lakes contain water for most of the year, and a dry spell is not a disaster. The water in the creeks and lakes is clear and full of fish, instead of muddy because it is full of soil.

Sustainability

The Bamberger Ranch is a working ranch, raising and selling livestock, but it is also home to dozens of other projects. Bamberger consultants advise others who are interested in managing rangeland in a sustainable fashion. Volunteers help by building and repairing nature trails and performing all kinds of maintenance work. The ranch hosts research on grasslands and range management, conferences on habitat restoration, educational workshops, as well as vacations for those interested in all aspects of nature.

▶ At nearly 2,300 hectares, the Bamberger Ranch is one of the largest habitat restoration projects in Texas. This is a photo of a portion of the Bamberger Ranch used for sustainable ranching.

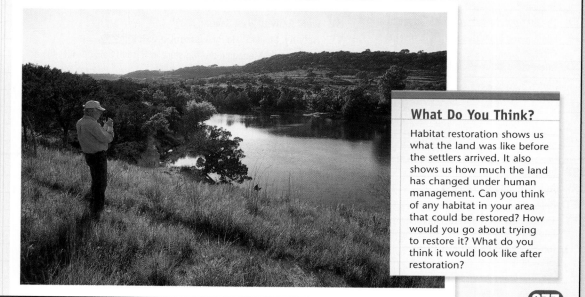

What Do You Think?

Habitat restoration shows us what the land was like before the settlers arrived. It also shows us how much the land has changed under human management. Can you think of any habitat in your area that could be restored? How would you go about trying to restore it? What do you think it would look like after restoration?

Answers to What Do You Think?

Answers may vary.

⏏ **Internet** Activity — GENERAL

Gorilla Haven In 1999, Steuart and Jane Dewar began construction of a project called Gorilla Haven in the mountains of northern Georgia. When completed, Gorilla Haven will provide a temporary facility for zoo gorillas in need. Gorilla Haven will accommodate gorillas when a zoo is unable to do so. Circumstances that will lead to the housing of gorillas at Gorilla Haven include a surplus in male gorilla zoo populations, genetic redundancy, personality disorders, health concerns, and zoo renovations. This facility will work with zoos to ensure the long-term future of zoo gorilla populations. Gorilla Haven is still under construction, but there are plans to begin accepting gorillas in the near future. Have students research Gorilla Haven and similar projects on the Internet. Encourage students to include their findings in their **Portfolio.** Ⓛ **Intrapersonal**

Food and Agriculture
Chapter Planning Guide

PACING	CLASSROOM RESOURCES	LABS, DEMONSTRATIONS, AND ACTIVITIES
BLOCKS 1 & 2 · 90 min pp. 378–383 **Chapter Opener**		
Section 1 Feeding the World	**OSP** Lesson Plan * **CRF** Active Reading * **BASIC** **TT** Bellringer * **TT** Total Calorie Supply, per Person, per Day * **TT** World Grain Production Vs. Grain Production per Person *	**TE** Demonstration Food Choices, p. 379 ◆ **BASIC** **TE** Demonstration Food Distribution, p. 380 ◆ **GENERAL** **CRF** Consumer Lab * ◆ **BASIC**
BLOCKS 3, 4 & 5 · 135 min pp. 384–394 **Section 2** Crops and Soil	**OSP** Lesson Plan * **CRF** Active Reading * **BASIC** **TT** Bellringer * **TT** The Structure and Composition of Soil * **TT** Worldwide Vulnerability of Soils to Water Erosion * **TT** The Steps of Integrated Pest Management * **TT** Engineering Plant Resistance to Insect Pests * **CD** Interactive Tutor Biosphere	**TE** Group Activity Soil Qualities, p. 385 ◆ **GENERAL** **SE** QuickLab Preventing Soil Erosion, p. 386 ◆ **TE** Activity World Farming Methods, p. 387 **BASIC** **TE** Internet Activity The Land Grant System, p. 387 **GENERAL** **TE** Demonstration Unexpected Pesticides, p. 389 **GENERAL** **TE** Group Activity Role-Playing Game, p. 390 **GENERAL** **SE** Field Activity Pest Search, p. 391 **TE** Actvitiy Small-Scale Pest Control, p. 391 **GENERAL** **SE** Inquiry Lab Managing the Moisture in Garden Soil, pp. 404–405 ◆ **CRF** Datasheets for In-Text Labs * **CRF** Modeling Lab * ◆ **GENERAL** **CRF** Observation Lab * ◆ **ADVANCED** **CRF** Research Lab * ◆ **GENERAL**
BLOCKS 6 & 7 · 90 min pp. 395–399 **Section 3** Animals and Agriculture	**OSP** Lesson Plan * **CRF** Active Reading * **BASIC** **TT** Bellringer * **TT** The Collapse of the North Atlantic Cod Fishery * **CD** Interactive Exploration 6 How's it Growing?	**TE** Internet Activity Fish Harvests and Aquaculture, p. 396 **ADVANCED** **TE** Group Activity Overlooked Food, p. 397 **GENERAL** **CRF** Long-Term Project * ◆ **GENERAL**

BLOCKS 8 & 9 · 90 min

Chapter Review and Assessment Resources

- **SE** Chapter Review pp. 401–403
- **SE** Standardized Test Prep pp. 664–665
- **CRF** Study Guide * ■ **GENERAL**
- **CRF** Chapter Test * ■ **GENERAL**
- **CRF** Chapter Test * **ADVANCED**
- **CRF** Concept Review * ■ **GENERAL**
- **CRF** Critical Thinking * **ADVANCED**
- **OSP** Test Generator
- **CRF** Test Item Listing *

Online and Technology Resources

Visit **go.hrw.com** for access to Holt Online Learning or enter the keyword **HE6 Home** for a variety of free online resources.

 **One-Stop Planner® CD-ROM**

This CD-ROM package includes
- Lab Materials QuickList Software
- Holt Calendar Planner
- Customizable Lesson Plans
- Printable Worksheets
- ExamView® Test Generator
- Interactive Teacher Edition
- Holt PuzzlePro® Resources
- Holt PowerPoint® Resources

 Holt Environmental Science Interactive Tutor CD-ROM

This CD-ROM consists of interactive activities that give students a fun way to extend their knowledge of environmental science concepts.

KEY

TE Teacher Edition	**CRF** Chapter Resource File	*****	Also on One-Stop Planner
SE Student Edition	**TT** Teaching Transparency	■	Also Available in Spanish
OSP One-Stop Planner	**CD** Interactive CD-ROM	◆	Requires Advance Prep

ENRICHMENT AND SKILLS PRACTICE	SECTION REVIEW AND ASSESSMENT	CORRELATIONS
SE Pre-Reading Activity, p. 378 **TE** Using the Figure, p. 378 GENERAL		National Science Education Standards
TE Using the Figure Meat Consumption, p. 380 GENERAL **SE** MathPractice Extra Calories, p. 381 **TE** Interpreting Statistics Production Vs. Population, p. 382 BASIC **TE** Skill Builder Writing, p. 382 BASIC	**SE** Section Review, p. 383 **TE** Reteaching, p. 383 BASIC **TE** Quiz, p. 383 GENERAL **TE** Alternative Assessment, p. 383 GENERAL **CRF** Quiz * ■ GENERAL	LS 4e SPSP 4a SPSP 5c
TE Student Opportunities Soil Surveys, p. 385 **TE** Using the Figure Vulnerable Soils, p. 386 GENERAL **TE** Career Agricultural Specialist, p. 389 **TE** Using the Figure Integrated Pest Management, p. 392 ADVANCED **TE** Student Opportunities Community Gardens and Farmer's Markets, p. 392 ADVANCED **TE** Using the Figure Engineering Plant Resistance, p. 393 ADVANCED **TE** Inclusion Strategies, p. 393	**SE** Mid-Section Review, p. 388 **TE** Homework, p. 390 GENERAL **SE** Section Review, p. 394 **TE** Reteaching, p. 394 BASIC **TE** Quiz, p. 394 GENERAL **TE** Alternative Assessment, p. 394 GENERAL **CRF** Quiz * ■ GENERAL	LS 4e SPSP 4a SPSP 5c
TE Inclusion Strategies, p. 395 **SE** Case Study Menhaden, pp. 396–397 **SE** Graphic Organizer Comparison Table, p. 397 GENERAL **TE** Student Opportunities Heifer International, p. 396 **TE** Using the Table Livestock Populations, p. 398 GENERAL **SE** Points of View Genetically Engineered Foods, pp. 406–407 **CRF** Map Skills U.S. Crops * GENERAL	**TE** Homework, p. 398 GENERAL **SE** Section Review, p. 399 **TE** Reteaching, p. 399 BASIC **TE** Quiz, p. 399 GENERAL **TE** Alternative Assessment, p. 399 GENERAL **CRF** Quiz * ■ GENERAL	SPSP 4a SPSP 5c

 Guided Reading Audio CDs

These CDs are designed to help auditory learners and reluctant readers. (Audio CDs are also available in Spanish.)

 www.scilinks.org

Maintained by the **National Science Teachers Association**

TOPIC: Food and Diet
SciLinks code: HE4042

TOPIC: Green Revolution
SciLinks code: HE4050

TOPIC: Soil Erosion
SciLinks code: HE4100

TOPIC: Genetic Engineering
SciLinks code: HE4047

TOPIC: Livestock Production
SciLinks code: HE4086

 CNN Videos

Each video segment is accompanied by a Critical Thinking Worksheet.

Science, Technology & Society
Segment 16 Homemade Dirt

CHAPTER 15

Chapter Enrichment

This Chapter Enrichment provides relevant and interesting information to expand and enhance your classroom instruction of the chapter material.

SECTION 1 Feeding the World

The Origins of Agriculture

The practice of farming began at different times in different places in the world. It is thought to have first occurred in the Near East about 12,000 years ago. Most of today's major crops were in cultivation by Roman times. All modern crops were derived from wild species. Often, the wild cousins of our modern foodstuffs are unpalatable or poisonous. Wild almonds, for example, contain enough cyanide to kill a human. The wild ancestors of potatoes, eggplants, and lima beans were also bitter or poisonous. Early humans influenced the evolution of more-palatable varieties by selectively eating and spreading the seeds of those fruits and nuts with more-desirable characteristics.

Some plants were easier to domesticate than others. Barley, wheat, and peas were domesticated 8,000–10,000 years ago in the Near East. These crops were good candidates for domestication because they were already edible and productive in the wild, produced within a year of planting, were less perishable in storage, and were self-pollinating. Fruit and nut crops, such as olives, figs, dates, and grapes, were the next crops to be domesticated in the Near East around 4,000 BCE. These woody plants do not grow annually like cereals do and only become productive about 10 years after planting. Later, the Chinese perfected the deliberate and difficult process of tree-grafting to cultivate fruits such as apple, pear, plum, and cherry. Strawberries were domesticated in the Middle Ages, and pecans were domesticated in the mid-1800s.

HIGH-YIELDING DEMONSTRATION
GUANG-YOU-QING

▶ An experimental rice farm in China

Scurvy

Probably the first disease to be definitely associated with a dietary deficiency, scurvy results from a lack of vitamin C (ascorbic acid). Scurvy causes weakness, pain in the limbs, gum disease, skin hemorrhages, and sometimes death. Scurvy was a serious problem in the past whenever fresh fruits and vegetables were not available during the winter. It was especially common among sailors in the days before perishable foods could be stocked on ships. In the 1700s, the British navy began to stock rations of lime juice on long sea voyages. This is the origin of the term *limeys* for British sailors. (Encourage students to provide reasons for their choices.) Inform students that each loaf of bread contains 1190 Calories and 32 g of protein and the hamburger patty contains 242 Calories and 20 g of protein. However, the resources and energy that went into producing the hamburger patty were probably about equal to those that went into producing the 3 loaves of bread.

BRAIN FOOD

With current food production methods, many food experts calculate that 10 billion people could be fed a nutritious—but largely vegetarian—diet. But current production could not provide an average American diet to everyone on Earth today. The human population is expected to increase to about 9 billion in the next 50 years.

SECTION 2 Crops and Soils

Amazonian Soil

The town of Yurimaguas in the Peruvian Amazon has had limited potential for farming because the local soils have a low nutrient content, high acidity, and aluminum toxicity. The local people were importing expensive, high-quality tomatoes from the Pacific coast of Peru. Recently, scientists found a way to use thousands of tons of sawdust waste from Amazonian timber industries. By combining this wasted resource with earthworms, the scientists were able to create soil beds that produce high-quality tomatoes for local consumption. New methods allowed a certain species of earthworm to multiply rapidly in a bed containing 25 percent composted sawdust and 75 percent soil. Earthworm populations increased 15-fold in just four months. Their activity increased the availability of calcium, magnesium, and potassium in the soil and reduced aluminum toxicity.

► Traditional farming, using livestock to pull a plow

Norman Borlaug

Norman Ernest Borlaug is an American agricultural scientist who received the Nobel Peace Prize in 1970 for developing high-yield strains of wheat that reduced the threat of famine in many developing countries. Borlaug was thus considered one of the pioneers of the green revolution. A Nobel Prize committee member said of Borlaug, "More than any other single person of this age, he has helped to provide bread for a hungry world." In his acceptance speech, Borlaug stated that the green revolution has given humans only temporary success in the war against hunger. He admitted that the revolution did not solve all of the problems of food production and distribution. But he said, "It is far better for mankind to be struggling with new problems caused by abundance rather than with the old problem of famine."

Farming for Wildlife

California's rice farms are taking on the role of the state's dwindling natural wetlands. After harvest, the rice fields provide birds migrating along the Pacific coast with wintertime food and shelter. Each year, waterfowl that visit the state's rice fields may feast on 200 million pounds of leftover rice as well as other seeds, plants, and invertebrates. About 60 percent of Pacific-migrating waterfowl spend the winter in the Sacramento Valley. The Valley is also home to the California Ricelands Habitat Partnership, a cooperative effort among farmers, conservation groups, and government agencies. The program promotes farming methods that benefit wetlands-dependent species. Under the partnership's plan, rice fields are flooded during the winter. Foraging birds benefit the farmers by trampling the leftover rice straw so that tilling is not necessary. Project organizers believe the partnership may serve as a national model for agriculture working with the environment.

BRAIN FOOD

Conventional farming, which relies on pesticides and synthetic chemicals, receives billions of dollars of support through the U.S. Department of Agriculture (USDA) and the land grant university system. Currently, only a small proportion of USDA research funds are directed toward organic farming practices. Organic farmers contend that organic farming has the potential to produce yields matching or surpassing those of conventional crops.

SECTION

3 Animals and Agriculture

Agriculture

The word *agriculture* comes from the Latin words *ager*, which means "field," and *cultura*, which means "cultivation." Agriculture originally meant the cultivation of crops in fields. However, by the mid-1800s, the concept of "agriculture" began to include the science of cultivating the soil, producing crops, and raising livestock. The United States Department of Agriculture (USDA) was established in 1862. The goal of the USDA was to collect and study seeds, plants, soils, and animals; to test new farm equipment; and to advise farmers on agricultural matters.

Early Animal Domestication

Throughout history, the domestication of animals has varied from region to region. In Eurasia, Neolithic people domesticated dogs, sheep, goats, cattle, pigs, chickens, ducks, and water buffalo. In the Americas, people domesticated dogs, turkeys, llamas, alpacas, and guinea pigs. In Africa, people raised cattle, sheep and goats, but these were probably domesticated in the Near East.

► The largest fish market in the world, in Tokyo, Japan

Overview

Tell students that this chapter will discuss food production, maintenance of soil productivity, and the challenges of feeding the world. Various alternative methods of planting and harvesting crops, controlling pests, and raising livestock are examined.

Using the Figure — GENERAL

Ask students to imagine being the owner of this farmland in rural Pennsylvania. What might their daily life involve? What concerns and needs would they have in order to maintain their income and way of life? Tell students that today less than 10 percent of agricultural land in the United States is owned and operated by families. Ask them to brainstorm possible reasons for this situation. (Answers may vary. Two reasons are that agriculture is becoming more industrialized and that the costs of farming are increasing.)

PRE-READING ACTIVITY

Have pairs of students use their FoldNotes to study key terms from the chapter. Instruct one student to use the FoldNote to provide the key term, and have the other student give the definition. Have the student who provides the key term correct the other student's definition

For information about videos related to this chapter, go to **go.hrw.com** and type in the keyword **HE4 FOOV**.

Food and Agriculture

1 Feeding the World

2 Crops and Soil

3 Animals and Agriculture

PRE-READING ACTIVITY

FOLDNOTES

Key-Term Fold
Before you read this chapter, create the **FoldNote** entitled "Key-Term Fold" described in the Reading and Study Skills section of the Appendix. Write a key term from the chapter on each tab of the key-term fold. Under each tab, write the definition of the key term.

This farmland in rural Pennsylvania is used to grow alfalfa, corn, and soybeans. Agriculture can be thought of as one of the most important relationships people have with the environment.

378

Chapter Correlations *National Science Education Standards*

LS 4e Human beings live within the world's ecosystems. Increasingly, humans modify ecosystems as a result of population growth, technology, and consumption. Human destruction of habitats through direct harvesting, pollution, atmospheric changes, and other factors is threatening current global stability, and if not addressed, ecosystems will be irreversibly affected. **(Section 1 and Section 2)**

SPSP 4a Natural ecosystems provide an array of basic processes that affect humans. Those processes include maintenance of the quality of the atmosphere, generation of soils, control of the hydrologic cycle, disposal of wastes, and recycling of nutrients. Humans are changing many of these basic processes, and the changes may be detrimental to humans. **(Section 1, Section 2, and Section 3)**

SPSP 5c Some hazards, such as earthquakes, volcanic eruptions, and severe weather, are rapid and spectacular. But there are slow and progressive changes that also result in problems for individuals and societies. For example,

change in stream channel position, erosion of bridge foundations, sedimentation in lakes and harbors, coastal erosions, and continuing erosion and wasting of soil and landscapes can all negatively affect society. **(Section 1, Section 2, and Section 3)**

Feeding the World

In 1985, lack of rain, loss of soil, and war had caused crops to fail in Ethiopia. This resulted in **famine,** widespread starvation caused by a shortage of food. Events like those in Ethiopia present a frightening picture of the difficulty of feeding the Earth's growing population. Modern agricultural practices provide most of the world's population with enough food to survive. However, some of these practices can cause environmental damage that eventually makes growing food crops more difficult. In this chapter, you will learn why feeding all of the world's people is so difficult and about efforts to increase food production.

Humans and Nutrition

The human body uses food both as a source of energy and as a source of materials for building and maintaining body tissues. The amount of energy that is available in food is expressed in *Calories*. One Calorie (Cal) is equal to 1,000 calories, or one kilocalorie. As shown in **Table 1,** the major nutrients we get from food are carbohydrates, proteins, and lipids. Our bodies need smaller amounts of vitamins and minerals to remain healthy.

Malnutrition is a condition that occurs when people do not consume enough Calories or do not eat a sufficient variety of foods to fulfill all of the body's needs. There are many forms of malnutrition. For example, humans need to get eight essential amino acids from proteins. This is easily done if a variety of foods are eaten. However, in some parts of the world, the only sources of food may be corn or rice. Each of these foods contains protein but lacks one of the essential amino acids. A type of malnutrition called *amino acid deficiency* can result from such a limited diet.

Objectives

▶ Identify the major causes of malnutrition.
▶ Compare the environmental costs of producing different types of food.
▶ Explain how food distribution problems and drought can lead to famine.
▶ Explain the importance of the green revolution.

Key Terms

famine
malnutrition
diet
yield

Table 1 ▼

Major Nutrients in Human Foods				
Nutrient	Composition	Sources	Energy yield	Function
Carbo-hydrates	sugars	wheat, corn, and rice	4 Cal/g	is the main source of the body's energy
Lipids (oils and fats)	fatty acids and fatty alcohols	olives, nuts, and animal fats	9 Cal/g	helps form membranes and hormones
Proteins	amino acids	animal food and smaller amounts from plants	about 4 Cal/g	helps build and maintain all body structures

internet connect

www.scilinks.org
Topic: Food and Diet
SciLinks code: HE4042
SCI*LINKS*® Maintained by the National Science Teachers Association

Connection to ▶ Biology

Essential Amino Acids Animals make their own proteins from amino acids. Essential amino acids are those that must be supplied in the diet because the body needs them but cannot make them from other amino acids. A lack of essential amino acids in the diet can lead to the human diseases kwashiorkor and marasmus, which can cause brain damage in children.

379

Chapter Resource File

• **Lesson Plan**
• **Active Reading** BASIC
• **Section Quiz** GENERAL

Transparencies

TT Bellringer

Focus

SECTION 1

Overview

Before beginning this section, review with your students the Objectives in the Student Edition. In this section, students learn about the nutritional needs of humans, methods used to feed the world's population, and reasons why providing food for all of the world's people is difficult.

Bellringer

Begin class by asking students if there is enough food available in the world to feed everyone. Ask them to guess how many people in the world do not get enough to eat each day. (There may be not enough food available to feed everyone fully nutritious diets, but there may be enough to prevent starvation. About 800 million people are undernourished each day out of a human population of about 6.1 billion.) Ask them to hypothesize why everyone does not get enough food. (Reasons include unequal distribution of resources and environmental factors, such as drought, that may or may not be related to human activities.)
LS Logical

Motivate

Demonstration —— BASIC

Food Choices Bring in or show pictures of 3 loaves of whole grain bread and 1 hamburger patty. Ask students, "If, for the same amount of money, you could buy these three loaves of bread or this one hamburger patty, which would you choose?" (Answers may vary).

Meat Consumption Have students look at **Figure 1** and **Figure 2** to compare the relative amounts of meat produced and consumed around the world. (Meat forms a small percentage of total world food production.) Ask students to discuss why they think proteins and fats form a larger percentage of the diets of western, developed countries, and less so elsewhere. (Meat requires more resources to produce and is thus more available to people in developed countries.) **LS** Logical

Demonstration — GENERAL

Food Distribution Organize students into two groups as follows. Group A will represent the United States, the world's third-largest country, with a population of 278 million. Group B will represent India, the world's second-largest country, with a population of 1,030 million. For each student in group A, put three students in group B. Then, give 5 1/3 cups of dried rice to each student in group A and give 3 1/2 cups to each student in group B. Explain that this arrangement is an approximation of the current distribution of food and people in these two countries. The rice represents the daily supply of Calories per person. Three cups of rice provide about 2,100 Calories, the average minimum daily intake recommended by the U.S. government. In the least developed countries, the food supply is usually less than 2,100 Cal/day/person. Have students discuss their ideas about international food distribution and abundance. **LS** Kinesthetic/Logical

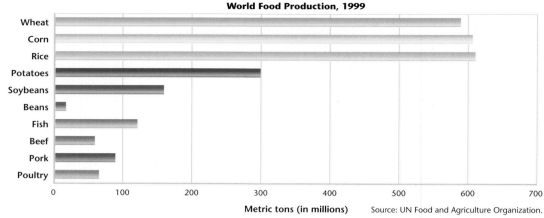

World Food Production, 1999

Metric tons (in millions)   Source: UN Food and Agriculture Organization.

Figure 1 ▶ This bar graph shows that in 1999, grains (wheat, corn, and rice) were produced in greater amounts than was any other food. Wheat and corn are eaten by humans and are fed to farm animals.

Sources of Nutrition A person's **diet** is the type and amount of food that he or she eats. A healthy diet is one that maintains a balance of the right amounts of nutrients, minerals, and vitamins. In most parts of the world, people eat large amounts of food that is high in carbohydrates, such as rice, potatoes, and bread. As shown in **Figure 1**, the foods produced in the greatest amounts worldwide are *grains*, plants of the grass family whose seeds are rich in carbohydrates. Besides eating grains, most people eat fruits, vegetables, and smaller amounts of meats, nuts, and other foods that are rich in fats and proteins.

Diets Around the World People worldwide generally consume the same major nutrients and eat the same basic kinds of food. But diets vary by region, as shown in **Figure 2**. People in more-developed countries tend to eat more food and a larger proportion of proteins and fats than people eat in less developed countries. For example, in the United States, almost half of all Calories people consume come from meat, fish, and oil. The Japanese, whose diet traditionally included a mix of rice, vegetables, and seafood, have started to consume more beef in recent decades.

Figure 2 ▶ People in developed countries generally eat more food and more proteins and fats than people in less developed countries eat.

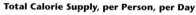

Total Calorie Supply, per Person, per Day

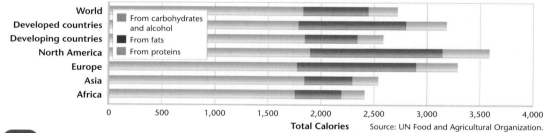

- From carbohydrates and alcohol
- From fats
- From proteins

Total Calories   Source: UN Food and Agricultural Organization.

Transparencies

TT Total Calorie Supply, per Person, per Day

Cultural Awareness — GENERAL

Spud Diversity Scientists and sociologists are making efforts to record and preserve the nearly 4,000 varieties of potato that have been grown in the Andes. These potatoes come in a variety of shapes and colors. There is also diversity in the conditions under which the plants can grow, nutritional values, and resistance to rotting during storage. Researchers speculate that for traditional Andean communities, such diversity served both as a means of food security and as a way to add variety to a potato-based diet.

The Ecology of Food

As the human population grows, farmland replaces forests and grasslands. Feeding everyone while maintaining natural ecosystems becomes more difficult. Different kinds of agriculture have different environmental impacts and different levels of efficiency.

Food Efficiency The *efficiency* of a given type of agriculture is a measure of the quantity of food produced on a given area of land with limited inputs of energy and resources. An ideal food crop is one that efficiently produces a large amount of food with little negative impact on the environment.

On average, more energy, water, and land are used to produce a Calorie of food from animals than to produce a Calorie of food from plants. Animals that are raised for human use are usually fed plant matter. Because less energy is available at each higher level on a food chain, only about 10 percent of the energy from the plants gets stored in the animals. Thus, a given area of land can usually produce more food for humans when it is used to grow plants than when it is used to raise animals. The efficiency of raising plants for food is one reason why diets around the world are largely based on plants. However, meat from animals generally provides more nutrients per gram than most food from plants.

Old and New Foods Researchers hope to improve the efficiency of food production by studying plants and other organisms that have high **yield**—the amount of food that can be produced in a given area. Researchers are interested in organisms that can thrive in various climates and that do not require large amounts of fertilizer, pesticides, or fresh water. As shown in **Figure 3**, some organisms have been a source of food for centuries, while other sources are just being discovered.

MATHPRACTICE

Extra Calories An active man who weighs 70 kg maintains his weight if he eats 2,700 Cal per day. Unused Calories are converted into stored fat at the rate of 1 kg of fat per 9,000 Cal that are unused. If this active man consumes 3,600 Cal per day, how much weight does he gain each year?

Figure 3 ▶ Glasswort (top) is a salad green that may become an important food source in the future because it can grow in salty soil. Seaweed (bottom) is a multicellular organism called a *protist* and has been harvested and eaten by humans for centuries.

381

Cultural Awareness **GENERAL**

Staple Foods Around the World In many countries around the world, meals consist of one staple food that is eaten with seasoned vegetables and, sometimes, small amounts of meat. The most important of these staples is rice. Rice is eaten by over half of the world's population, especially in Asia. In parts of China and India, however, wheat is the preferred staple, in the form of noodles, buns, and flat breads. In Central America, cornmeal is more important. In European countries, the staples are potatoes and bread grains such as wheat and rye. In Ethiopia, the staples include millet (a cereal grass) and buckwheat, made into a fermented bread called *injera*.

Interpreting Statistics — BASIC

Production Vs. Population Have students examine **Figure 5** and notice the two trends represented. Ask students to explain how grain production could increase while grain per person decreases. (Both grain production and the human population increase in size, but the human population grows faster.) Ask students to identify the time when the human population began increasing faster than grain production did. (around 1987) **LS** Visual/Logical

REAL-LIFE —
• CONNECTION — GENERAL

Fungus: It's What's for Dinner
In the ongoing search to develop alternative food sources, one British company has developed a new product from the fungus *Fusarium venenatum*, which it identified in a local meadow. The fungus is grown in fermentation vats, like yogurt. Then it is mixed with eggs and flavorings and turned into an imitation-beef or imitation-chicken product called *Quorn*™. Compared to beef, the product has a similar ratio of Calories from protein, a lower Calorie density, and less fat. Quorn has been fashioned into numerous products, including "chicken" nuggets, lasagna, and ground "beef." The manufacturers of Quorn claim that it is "mushroom in origin," but detractors claim that even though it is indeed a fungus, it is nonetheless not a true mushroom and should thus require further testing.

Figure 4 ▶ Malnourished citizens in Bangladesh (a country in Asia) wait for food assistance.

internet connect
www.scilinks.org
Topic: Green Revolutions
SciLinks code: HE4050
SCiLINKS® Maintained by the National Science Teachers Association

World Food Problems

As shown in **Figure 4**, some people become malnourished because they simply do not get enough food. More food is needed each year to feed the world's growing population. As shown in **Figure 5**, world food production has been increasing for decades, but now food production is not increasing as fast as the human population is increasing.

Unequal Distribution If all the food in the world today were divided equally among the human population, no one would have quite enough food for good health. But food is not divided equally. And malnutrition is largely the result of poverty. Even in the United States, many poor people suffer from malnutrition. Wars and political strife can also lead to malnutrition because they interrupt transportation systems. During wars, even if food is available, it often cannot be transported to the people who need it.

Droughts and Famines A *drought* is a prolonged period during which rainfall is below average. Crops grown without irrigation may produce low yields or fail entirely. A drought is more likely to cause famine in places where most food is grown locally than in places where most food is imported. If a drought occurs, there may be no seed to plant crops the following year. The effects of a drought can continue for years.

People in a given area can usually survive one crop failure. They may have saved enough food from previous seasons, or they may have systems for importing food from elsewhere. But several years of drought cause severe problems for any area of the world. After a long drought, the soil may be less able to support the production of food crops.

382

Notable Quotes — ADVANCED

"In simplest terms, agriculture is an effort by man to move beyond the limits set by nature."

—Lester R. Brown and Gail W. Finsterbusch, authors of *Man and His Environment: Food*

Have students discuss the limits they think the authors are referring to and ways that humans overcome these limits. (Answers may vary. Limits are the amounts of food that grow naturally in given areas.) **LS** Verbal

SKILL BUILDER — BASIC

Writing Ask students to imagine living on a farm during a period of severe drought. Have them write a short story, in the first person, describing their daily experiences and major events during the drought. Some students may have actually had this experience or know someone who has. **LS** Verbal/Intrapersonal

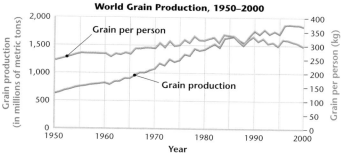

World Grain Production, 1950–2000

Grain production (in millions of metric tons) / Grain per person (kg)

Source: U.S. Department of Agriculture.

Figure 5 ▶ Worldwide grain production has increased steadily, but not as rapidly as the population has grown.

The Green Revolution

Between 1950 and 1970, Mexico increased its production of wheat eight-fold and India doubled its production of rice. Worldwide, increases in crop yields resulted from the use of new crop varieties and the application of modern agricultural techniques. These changes were called the *green revolution*. An example of one of the new varieties of grain is shown in **Figure 6**. Since the 1950s, the green revolution has changed the lives of millions of people.

However, the green revolution also had some negative effects. Most new varieties of grain produce large yields only if they receive large amounts of water, fertilizer, and pesticides. In addition, the machinery, irrigation, and chemicals required by new crop varieties can degrade the soil if they are not used properly. As a result of the overuse of fertilizers and pesticides, yields from green revolution crops are falling in many areas. Grain production in the United States has decreased since 1990, partly because the amount of water used for irrigation has decreased during the same period.

In addition, the green revolution had a negative impact on *subsistence farmers*—farmers who grow only enough food for local use. Before the green revolution, subsistence farmers worked most of the world's farmland. But they could not afford the equipment, water, and chemicals needed to grow the new crop varieties.

Figure 6 ▶ New rice varieties and farming methods developed during the green revolution are used to increase yield in this experimental farm in China.

SECTION 1 Review

1. **Identify** the major causes of malnutrition.

2. **Compare** the environmental costs of producing different types of foods.

3. **Explain** how drought and the problems of food distribution can lead to famines.

4. **Describe** the importance and effects of the green revolution.

CRITICAL THINKING

5. **Identifying Relationships** Study the graph in Figure 5. World grain production peaked in the mid-1990s. Why was the amount of grain per person declining?

6. **Inferring Relationships** Write a short paragraph that explains how a decrease in the production of grain worldwide could lead to a shortage of other food sources. **WRITING SKILLS**

383

Answers to Section Review

1. Malnutrition occurs when people do not get enough Calories or do not eat a sufficient variety of foods to fulfill all of the body's nutritional needs. All over the world, many people have poor diets or do not get enough food.

2. Compared to meat, plant foods generally require less land, energy, and water to produce the same number of food Calories.

3. A drought can decrease crop yields and leave no seeds for the next year, causing famine in places where most food is grown locally. When poverty, war, or political strife disrupt distribution systems, food may become unavailable in an area, which can lead to famine.

4. The green revolution caused worldwide crop yield increases, helping feed millions of people. However, the crop varieties and methods of the green revolution required the use of large amounts of fertilizers, pesticides, and irrigation, causing soil degradation. It also created hardship for farmers who could not afford the expensive inputs.

5. Less grain was available per person because population had increased at a greater rate than grain production had increased.

6. Answers may vary. Students should note that humans feed grains to the animals that produce meat, eggs, and milk.

Overview

Before beginning this section, review with your students the Objectives in the Student Edition. This section focuses on soil and its role in agriculture. Farming methods that degrade soil and those that preserve or restore soil will be contrasted. The section also explores methods of pest control.

 **Bellringer**

Ask students to imagine that they are farmers. Let them discuss what type of crop would be appropriate for their region. Then, have them discuss the things that would be needed for a healthy crop. (Examples include healthy soil, reliable water sources, and a pest-management system.) **LS Verbal**

Motivate

Identifying Preconceptions —— GENERAL

The Fertile Crescent Ask students, "What kind of landscape do you think exists in the area of Iraq, Syria, Jordan, and Lebanon?" Point out the area on a world map. Tell students that these countries were once part of an area called the *Fertile Crescent*. Explain that the area is still relatively fertile, but the countries mentioned above no longer have the same lush vegetation. Ask students, "What might have caused this dramatic change in landscape?" (This was one of the first areas of the world to discover irrigation. Over the years, the area was overirrigated, causing soil problems such as salinization.) **LS Visual/ Intrapersonal**

SECTION 2
Crops and Soil

Objectives

▶ Distinguish between traditional and modern agricultural techniques.
▶ Describe fertile soil.
▶ Describe the need for soil conservation.
▶ Explain the benefits and environmental impacts of pesticide use.
▶ Explain what is involved in integrated pest management.
▶ Explain how genetic engineering is used in agriculture.

Key Terms

arable land
topsoil
erosion
desertification
compost
salinization
pesticide
biological pest control
genetic engineering

Figure 7 ▶ In modern agriculture, machinery is used to do much of the work previously performed by humans and animals.

The Earth has only a limited area of **arable land,** land that can be used to grow crops. As our human population continues to grow, the amount of arable land per person decreases. In this section, you will learn how food is produced, how arable land can become degraded, and how we can ensure that we will continue to grow the crops we need in the future.

Agriculture: Traditional and Modern

The basic processes of farming include plowing, fertilization, irrigation, and pest control. Traditionally, plows are pushed by the farmer or pulled by livestock. Plowing helps crops grow by mixing soil nutrients, loosening soil particles, and uprooting weeds. Organic fertilizers, such as manure, are used to enrich the soil so that plants grow strong and healthy. Fields are irrigated by water flowing through ditches. Weeds are removed by hand or machine. These traditional techniques have been used since the earliest days of farming, centuries before tractors and pesticides were invented.

In most industrialized countries, the basic processes of farming are now carried out using modern agricultural methods. Machinery powered by fossil fuels is now used to plow the soil and harvest crops, as shown in **Figure 7.** Synthetic chemical fertilizers are now used instead of manure and plant wastes to fertilize soil. A variety of overhead sprinklers and drip systems may be used for irrigation. And synthetic chemicals are used to kill pests.

Chapter Resource File

• **Lesson Plan**
• **Active Reading** BASIC
• **Section Quiz** GENERAL

 **Transparencies**

TT Bellringer

 *Notable Quotes* —— GENERAL

"The soil is the source of life, creativity, culture, and real independence."

—David Ben-Gurion, former prime minister of Israel

Have students discuss the relationship between good soil and a thriving civilization. (Good soil allows countries to feed their citizens. When people are well nourished, they can spend more energy on work and creative activities.) **LS Verbal**

Fertile Soil: The Living Earth

Soil that can support the growth of healthy plants is called *fertile soil*. Plant roots grow in **topsoil**, the surface layer of soil, which is usually richer in organic matter than the subsoil is. Fertile topsoil is composed of living organisms, rock particles, water, air, and organic matter, such as dead or decomposing organisms.

Most soil forms when rock is broken down into smaller and smaller fragments by wind, water, and chemical weathering. *Chemical weathering* happens when the minerals in the rock react chemically with substances such as water to form new materials. Temperature changes and moisture cause rock to crack and break apart, which creates smaller particles on which the seeds of pioneer plants fall and take root. It can take hundreds or even thousands of years for these geological processes to form a few centimeters of soil.

Other processes also help to produce fertile topsoil. The rock particles supply mineral nutrients to the soil. Fungi and bacteria live in the soil, and they decompose dead plants as well as organic debris and add more nutrients to the soil. Earthworms, insects, and other small animals help plants grow by breaking up the soil and allowing air and water into it. One way to tell whether soil is fertile is to see if it contains earthworms. **Table 2** lists some of the organisms that live in fertile soil.

As you can see in **Figure 8**, several layers of soil lie under the topsoil. The bottom layer is bedrock, which is the solid rock from which most soil originally forms.

Table 2 ▼

Numbers of Organisms in Average Farm Soil	
Organisms	Quantity
Insects	23 million per hectare
All arthropods (including insects)	725 million per hectare
Bacteria	2.5 billion per gram
Algae	50,000 per gram
Earthworms	6 million per hectare

Note: One hectare equals about 2.47 acres.
Source: US Department of Agriculture.

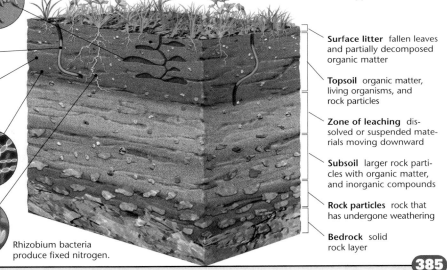

Ants and earthworms break up and aerate the soil.

Bacteria and fungi decompose organic matter.

Rhizobium bacteria produce fixed nitrogen.

Figure 8 ▶ Soil is made of rock particles, air, water, and dead and living organisms. The number and characteristics of the soil layers may be different in different types of soil.

Surface litter fallen leaves and partially decomposed organic matter

Topsoil organic matter, living organisms, and rock particles

Zone of leaching dissolved or suspended materials moving downward

Subsoil larger rock particles with organic matter, and inorganic compounds

Rock particles rock that has undergone weathering

Bedrock solid rock layer

385

Student Opportunities

Soil Surveys Local, state, or federal agencies frequently conduct surveys of agricultural or forest soils. Encourage students who may be interested in learning more about soils to inquire about summer job or internship opportunities with these agencies.

Teach

Group Activity — GENERAL

Soil Qualities Obtain some soil-testing kits, which measure pH and the levels of several key nutrients. Inexpensive kits are available at most nurseries or through a local agricultural extension service. Have students work in groups to collect one soil sample from an area that they think will be fertile and another from an area that they think will be infertile. Tell students to try to keep the layers of their soils intact. Also, have them collect subsoil samples. (Make sure students obtain permission before digging.) Back in the classroom, have groups compare their soil samples to the cross section in **Figure 8.** They should take note of similarities to the figure and then note each soil's color, texture, moisture content, organic content (a hand lens will be helpful), and consistency. Have them test their soils using the kits. Ask them to record in their *EcoLog* which soil sample was the most fertile, which was the least fertile, and which were in between. Have the groups share their findings with the rest of the class. **LS Kinesthetic** Co-op Learning

REAL-LIFE CONNECTION — ADVANCED

Learning About Local Soils The USDA Soil Conservation Service provides soil surveys for each state, county, and district. These surveys contain maps, tables of information on the properties and capabilities of each soil type, and aerial photographs of the land. Suggest that students find soil survey information for their local area (some of this information should be available on the Internet). Have them study the information and then report on the soil conditions in your area. Interested students can include their findings in their **Portfolio. LS Intrapersonal**

Using the Figure — GENERAL

Vulnerable Soils Have students use **Figure 9** to find the areas of the world where soils are most vulnerable to erosion by water. Ask students to discuss why they think these areas are vulnerable. (Areas may be vulnerable because of naturally poor soils, sloping landscapes, deforestation, agricultural mismanagement, and/or successive droughts.) **LS** Logical

QuickLAB

Skills Acquired:

• Constructing Models
• Experimenting
• Measuring
• Inferring

Answers to Analysis

1. The tray with topsoil probably had the most erosion and runoff because there were no roots or plant matter to absorb water or hold soil in place. The tray with sod or mulch probably had the least amount of runoff and erosion. Planting a cover crop can prevent erosion. Mulch can also prevent erosion but can be washed away on slopes.

Teacher's Note: Have students cover textbooks with plastic to avoid getting them wet or dirty.

Transparencies

TT Worldwide Vulnerability of Soils to Water Erosion

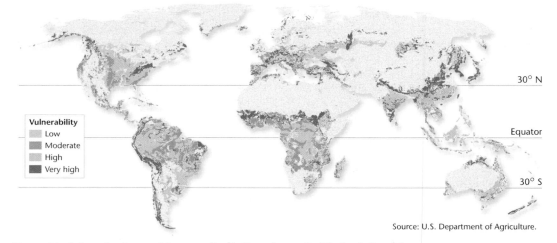

Vulnerability
- Low
- Moderate
- High
- Very high

30° N

Equator

30° S

Source: U.S. Department of Agriculture.

Figure 9 ► Soil erosion is one of the most serious environmental problems the world faces. This map shows the vulnerability of soils worldwide to erosion by water.

QuickLAB

Preventing Soil Erosion

Procedure

1. Obtain three **trays**, and fill one with **sod**, one with **topsoil**, and one with a type of mulch, such as **hay**.

2. Place each tray at an angle by creating a surface that resembles a hill by using **doorstoppers** and **textbooks**. Place a **large bowl** at the bottom of each tray to catch the runoff.

3. Sprinkle **2L of water** slowly on each tray to simulate heavy rainfall.

4. Use a **scale** to weigh the runoff of soil and water that collected in each bowl.

Analysis

1. Which tray had the most soil erosion and water runoff? Which tray had the least? Why? What does this lab demonstrate about soil erosion?

Soil Erosion: A Global Problem

Erosion is the wearing away of rock or soil by wind and water. In the United States, about half of the original topsoil has been lost to erosion in the past 200 years. **Figure 9** shows potential soil erosion worldwide. Without topsoil, crops cannot be grown.

Almost all farming methods increase the rate of soil erosion. For example, plowing loosens topsoil and removes plants that hold the soil in place. The topsoil is then more easily eroded by wind or rain.

Land Degradation

Land degradation happens when human activity or natural processes damage the land so that it can no longer support the local ecosystem. In areas with dry climates, desertification can result. **Desertification** is the process by which land in arid or semiarid areas becomes more desertlike because of human activity or climatic changes. This process is causing some of our arable land to disappear.

Desertification has happened in the Sahel region of northern Africa. In the past, people who lived in the drier part of the Sahel grazed animals, whereas people who lived in areas of the Sahel with more rainfall planted crops. The grazing animals were moved from place to place to find fresh grass and leaves. The cropland was planted for only a few years, and then the land was allowed to lie *fallow,* or to remain unplanted, for several years. These methods of farming and grazing allowed the land to adequately support the people in the Sahel. But the population in the region has grown, and the land is being farmed, grazed, and deforested faster than it can regenerate. Now, too many crops are planted too frequently, and fallow periods are being shortened or eliminated. As a result, the soil is losing its fertility and productivity. Because of overgrazing, the land has fewer plants to hold the topsoil in place. So, large areas have become desert and can no longer produce food.

LANGUAGE ARTS
CONNECTION — GENERAL

The Dust Bowl and *The Grapes of Wrath* John Steinbeck's *The Grapes of Wrath* describes some of the conditions of the Dust Bowl, a reference to the Great Plains region of the United States during the 1930s. *The Dust Bowl* describes the conditions that resulted when poor agricultural practices combined with years of severe drought to set the stage for massive soil erosion by wind. The book follows the fortunes of the Joads, one of the displaced Oklahoma farming families that moved west to California to seek a better life. Readings from the novel may help students understand the terrible impact of the Dust Bowl on the people who lived in the region. Tell students that today, as a result of the Soil Conservation Act passed after the Dust Bowl events, U.S. farmers are more likely to use planting and grazing techniques that protect the soil from erosion. **LS** Intrapersonal

Soil Conservation

There are many ways of protecting and managing topsoil and reducing erosion. Soil usually erodes downhill, and many soil conservation methods are designed to prevent downhill erosion, as shown in **Figure 10**. Building soil-retaining terraces across a hillside may be cost-effective for producers of valuable crops, such as wine grapes and coffee. On gentler slopes, *contour plowing* is used. This method includes plowing across the slope of a hill instead of up and down the slope. An even more effective method of plowing is leaving strips of vegetation across the hillside instead of plowing the entire slope. These strips catch soil and water that run down the hill. Still, many areas of land that have hills are not suited to farming, but may be better used as forest or grazing land.

In traditional farming, after a crop is harvested, the soil is plowed to turn it over and bury the remains of the harvested plants. In *no-till farming*, shown in **Figure 11**, a crop is harvested without turning over the soil. Later, the seeds of the next crop are planted among the remains of the previous crop. The remains of the first crop hold the soil in place while the new crop develops. No-till farming saves time compared with conventional methods. This method can also reduce soil erosion to one-tenth of the erosion caused by traditional methods. However, no-till farming may not be suitable for some crops. Other disadvantages to this method can include soil that is too densely packed and lower crop yields over time.

Figure 10 ▶ Terracing (left) keeps soil in multiple, small, level fields. Contour plowing (right) follows the natural contours of the land. Both methods prevent soil erosion by keeping water from running directly downhill.

internet connect

www.scilinks.org
Topic: Soil Erosion
SciLinks code: HE4100

SCiLINKS. Maintained by the National Science Teachers Association

Figure 11 ▶ This farmer is practicing no-till farming. The tractor plants a new crop by poking seeds into the soil through the remains of the old crop.

387

📶 **Internet Activity** ———— **GENERAL**

The Land Grant System Public universities did not exist in the United States before the mid-1800s. There were only private institutions with tuitions that were too high for the average family. In 1862, President Abraham Lincoln signed the first Morrill Act, which gave 10,000 acres of federal land to each state to sell and use the proceeds for the creation of a public university. Today, every state has a land grant university, which bears the major responsibility for agricultural research and teaching and has a mission of extension education and public outreach. Have students find out where their state's land grant university is and look up its agricultural programs on the Internet. Have some students find Internet sites for other land grant universities in other states. Have them research the curricula for students specializing in agriculture, learn about methods of agriculture being studied, and research how those methods differ regionally. Ask students to share their findings and include them in their **Portfolio.**
LS Intrapersonal

Activity ———— **BASIC**
World Farming Methods Ask each student to choose a foreign country and then to create a poster about its farming methods. If students in your class are from other countries, ask them to create a poster about the dominant type(s) of farming in their native land. Students should include information about crops planted, farming equipment used, and popular farming techniques used. Encourage students to use visuals (photographs or drawings) to supplement any text. Have students present their posters to the rest of the class.   **English Language Learners**
LS Visual

REAL-LIFE ————
● **CONNECTION** — **GENERAL**

The Monitoring Project Farming is a complex enterprise. Farmers must manage finances while considering ecosystem interactions and their own quality of life. To assess total farm quality and success, a group of farmers teamed up with the Minnesota Institute for Sustainable Agriculture to create The Monitoring Project. The project develops user-friendly monitoring techniques for farmers, drawing on the collaborative expertise of farmers, biologists, economists, sociologists, and government agency staff. Techniques for gauging indicators of farm quality include wildlife watching (birds, frogs, and worms), soil testing, water analysis (streams and groundwater), quality of life analysis, and finance tracking. *The Monitoring Tool Box* offers practical instructions, such as how to set up stream- or bird-monitoring stations.

HISTORY
CONNECTION · GENERAL

George Washington Carver

George Washington Carver (1856–1943) was an internationally known African-American scientist. Carver sought to teach soil conservation and techniques for improving crop production to Southern farmers—particularly African Americans. He did so by writing literature about applied agriculture, setting up exhibits at conferences, and giving public lectures. Carver first learned about agriculture as a slave in Diamond Grove, Missouri. He went on to obtain a master's degree in agricultural science, teach at two universities, head a university agriculture department, and direct a state agricultural station. As a member of the Commission on Inter-Racial Cooperation and the Young Men's Christian Association (YMCA), Carver also spent much time working to improve race relations. Carver received many awards in his lifetime, including a fellowship to London's Royal Society of Arts and the Theodore Roosevelt Medal for valuable contributions to science.

BRAIN FOOD

Salty Vengeance It is said that after the Romans defeated the Carthaginians, they salted the Carthaginians' soil to make the land unsuitable for growing crops.

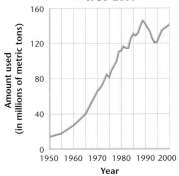

World Fertilizer Use, 1950–2000

Source: UN Food and Agriculture Organization.

Figure 12 ▶ The use of inorganic fertilizers has increased dramatically worldwide since 1950.

Connection to Geology

Soil Formation Over Time
Most rock breaks down into finer particles over time and changes from gravel to sand to clay. You can tell the age of soil by looking at its rock particles. Young soil is sandy or gravelly, and it falls apart when you squeeze it in your hand. Older soils contain clay, and damp clay stays together in lumps when you squeeze it in your hand.

Enriching the Soil

Soil was traditionally fertilized by adding organic matter, such as manure and leaves, to the soil. As organic matter decomposes, it adds nutrients to the soil and improves the texture of the soil. However, inorganic fertilizers that contain nitrogen, phosphorus, and potassium have changed farming methods. Without these fertilizers, world food production would be less than half of what it is today. Over the past 50 years, the use of such fertilizers has increased rapidly, as shown in **Figure 12.** If erosion occurs in areas where the soil has been fertilized with inorganic chemicals, fertilizers and pesticides may pollute waterways.

A modern method of enhancing the soil is to use both organic and inorganic fertilizers by adding compost and chemical fertilizers to the soil. **Compost** is partly decomposed organic material. Compost comes from many sources. For example, you can buy composted cow manure in a garden store. Also, many cities and industries now compost yard waste and crop wastes. This compost is sold to farmers and gardeners, and the process is saving costly landfill space.

Salinization

The accumulation of salts in the soil is known as **salinization** (SAL uh nie ZAY shuhn). Salinization is a major problem in places such as California and Arizona, which have low rainfall and naturally salty soil. In these areas, irrigation water comes from rivers or groundwater, which is saltier than rainwater. When water evaporates from irrigated land, salts are left behind. Eventually, the soil may become so salty that plants cannot grow.

Irrigation can also cause salinization by raising the groundwater level temporarily. Once groundwater comes near the surface, the groundwater is drawn up through the soil like water is drawn up through a sponge. When the water reaches the surface, the water evaporates and leaves salts in the soil. Salinization can be slowed if irrigation canals are lined to prevent water from seeping into the soil, or if the soil is watered heavily to wash out salts.

SECTION 2 Mid-Section Review

1. **Explain** the differences between traditional and modern farming methods.

2. **Describe** the structure and composition of fertile soil.

3. **Explain** why the presence of plants helps prevent soil erosion.

4. **Explain** why soil conservation is an important agricultural practice.

CRITICAL THINKING

5. **Inferring Relationships** Study the graph in Figure 12. What do you think might have happened to food production between 1990 and 1995?

6. **Applying Ideas** Erosion is a natural process. Why has it become such a serious environmental problem? Write a paragraph that explains your reasoning.
WRITING SKILLS

388

Mid-Section Review

1. Traditional farming is generally done by small groups of people on small tracts of land, using relatively low-technology tools. Modern industrial agriculture usually relies on heavy machinery and pesticides to farm large tracts of land for high yields.

2. Fertile soil usually exists as a series of layers. The rich topsoil contains living organisms, rock particles, water, air, and organic matter.

3. Plants prevent erosion by shielding the ground from wind and rain and by holding the soil in place with their roots.

4. Soil conservation protects topsoil and reduces erosion, which helps people farm sustainably.

5. Answers may vary. Food production probably declined as fertilizer use declined.

6. Answers may vary. Erosion has become a serious problem because human activities have dramatically increased erosion rates, which has polluted air and water and reduced the amount of land available to grow crops and trees.

Figure 13 ▶ **Examples of major crop pests** include ❶ weeds, ❷ plant-eating insects, and ❸ fungi.

Pest Control

In North America, insects eat about 13 percent of all crops. Crops in tropical climates suffer even greater insect damage because the insects grow and reproduce faster in these climates. In Kenya, for example, insects destroy more than 25 percent of the nation's crops. Worldwide, pests destroy about 33 percent of the world's potential food harvest.

As shown in **Figure 13**, insects are one of several types of organisms considered pests. A *pest* is any organism that occurs where it is not wanted or that occurs in large enough numbers to cause economic damage. Humans try to control populations of many types of pests, including many plants, fungi, and microorganisms.

Wild plants often have more protection from pests than crop plants do. Wild plants grow throughout a landscape, so pests have a harder time finding and feeding on a specific plant. Crop plants, however, are usually grown together in large fields, which provides pests with a one-stop source of food. Wild plants are also protected from pests by a variety of pest predators that live on or near the plants. Some wild plants have also evolved defenses to many pests, such as poisonous chemicals that repel pests.

Pesticides

Many farmers rely on pesticides to produce their crops. **Pesticides** are chemicals used to kill insects, weeds, and other crop pests. During the last 50 years, scientists invented many new pesticides. The pesticides were so effective that farmers began to rely on them almost completely to protect their crops from pests. However, pesticides can also harm beneficial plants and insects, wildlife, and even people.

Ecofact

Crop Rotation Farmers and gardeners have known for centuries that you get higher yields and less pest damage if you plant different crops each year on a piece of land. This method works because most pests are specialists and will only eat one or a few types of plants. The tomato hornworm is an example of one of these pests. If you plant tomatoes in one place every year, the hornworm population grows rapidly and will destroy the crop. If beans are planted in place of the tomatoes in alternate years, the hornworms cannot find food and die.

Demonstration ── GENERAL

Unexpected Pesticides Bring in a selection of some canned fruits and fruit products sold in grocery stores. Have students look at the labels, to determine where each of the products was grown. (Students may find that many fruit products are imported from other countries.) Tell students that pesticides that are not legal to use in the U.S. may still be produced and exported to other countries by pesticide companies. It is possible that crops treated with these pesticides are then being sold to consumers in this country. Ask students: What can consumers, such as yourselves, do about it? Encourage the class to discuss and identify ways to find out more about possible pesticide residues in food imports. **LS** Intrapersonal

Cultural Awareness GENERAL

Food or Disease? Corn crops often become infected by the fungus *Ustilago maydis*, a corn smut, depicted in **Figure 13**. In the United States, a crop infected with corn smut is usually destroyed, causing economic loss. However, this fungus has been a delicacy in Mexico since the days of the Aztecs. A crop of corn smut may actually earn the Mexican farmer more income than a crop of corn. Demand for corn smut, called *cuitlacoche* or *huitlacoche* in Mexico, is so high that it is canned and sold in supermarkets.

389

HISTORY ──
● CONNECTION ── BASIC

Irish Potato Famine In the early 1800s, much of Ireland's population was poor, and potatoes were the primary means of sustenance. When a fungal disease wiped out virtually all of Ireland's potato crops between 1845 and 1847, nearly 1 million people died of starvation or disease. Ask students, "What could Irish leaders have done to help prevent this disaster?" (Answers may vary. They could have encouraged a more diverse crop base, developed fungicides, or explored alternative methods of farming and food distribution.) **LS** Logical

Notable Quotes ── GENERAL

"One for the maggot, one for the crow, one for the cutworm, and one to grow."

—*Farmer's Almanac*

Discuss the quote with the class. Have them think about what the quote implies about the natural relationship between plants and other organisms in their environment. (In nature, plants and seeds experience heavy animal predation. In a natural system of agriculture, where no pesticides are used, a farmer might expect to lose many seeds to other organisms.) **LS** Logical

Career

Agricultural Specialist Invite a farmer, a county agricultural extension agent, a high school or college agriculture teacher, or another agricultural specialist to speak to the class about farming methods used in your area. Suggest that students prepare a list of questions in advance. If your speaker is willing, have him or her lead a field trip to a local farm to observe current farming practices. **LS** Intrapersonal

Group Activity ── GENERAL

Role-Playing Game Tell the class you are going to play a game called "farmers versus pests." Divide the class into two teams. Tell students that each team's turn will consist of deciding on a single, realistic action, which must be written on a note card and then read aloud to the class. Have each team elect a scientific "judge" to vote to settle any disputes. Allow the farmers to start by choosing which crop(s) to plant. Then, allow the pests a chance to eat or damage the crop. Allow several rounds of the game and then discuss the ideas generated. Students may find that compromise between the two sides is possible, that there are many creative ways to combat pests, or that the "arms race" escalates outrageously. The class may wish to attempt the game with different rules for "winning."

LS **Verbal/Logical** Co-op Learning

LAW ──
● CONNECTION ── GENERAL

Degree of Exposure Studies by the National Academy of Sciences showed that the Federal Insecticide, Fungicide, and Rodenticide Act (FIFRA) of 1972 was the most poorly enforced environmental law in our legal system. FIFRA was controversial because it banned pesticides that caused cancer in lab animals, regardless of the pesticides' concentration. The government amended FIFRA in 1996 to add the consideration of degree of exposure as a component of evaluating pesticide safety. Have students research current regulations on pesticide use. **LS** **Verbal/Logical**

Figure 14 ▶ A cropduster sprays pesticide on a field of pineapples in Hawaii. Cropdusting is an easy way to apply pesticide to a large area.

Pesticide Resistance You might think that the most effective way to get rid of pests is to spray often with large amounts of pesticide, as shown in **Figure 14.** However, over time, this approach usually makes the pest problem worse. Pest populations may evolve *resistance*, the ability to survive exposure to a particular pesticide. More than 500 species of insects have developed resistance to pesticides since the 1940s.

Human Health Concerns Pesticides are designed to kill organisms, so they may also be dangerous to humans. For example, in some areas fruit and vegetable farmers use large amounts of pesticides on their crops. Cancer rates among children in those areas are sometimes higher than the national average, and nervous system disorders may be common. Workers in pesticide factories may also become ill. And people who live near these factories may be endangered by accidental chemical leaks. People who apply pesticides need to follow safety guidelines to protect themselves from contact with these chemicals.

Pollution and Persistence The problem of pesticides harming people and other organisms is especially serious with pesticides that are persistent. A pesticide is *persistent* if it does not break down easily or quickly in the environment. Persistent pesticides do not break down rapidly into harmless chemicals when they enter the environment. As a result, they accumulate in the water and soil. Some persistent pesticides have been banned in the United States, but many of them remain in the environment for many years. DDT, a persistent pesticide banned in the United States in the 1970s, can still be detected in the environment and has even been found in women's breast milk.

Connection to **Law**

Pesticide Regulation The only pesticides that are fully regulated in the United States are newly introduced pesticides designed for use on some food crops. Many older pesticides in use have not been adequately tested for toxicity and are not effectively regulated. According to the National Academy of Sciences, much of the cancer risk from pesticides in our diet comes from older pesticides used on foods such as tomatoes, potatoes, and oranges.

390

Homework ──────── GENERAL

Pesticides and Produce Ask students whether they or their families rinse and scrub their produce before eating it. Then, share with students the following information: Most studies have indicated that pesticide residues on produce rarely exceed regulatory guidelines. However, these guidelines are made with the presumption that pesticide residues will be further reduced by "properly" rinsing and scrubbing the surfaces of produce before it is eaten. Have students use the library or the Internet to research the levels of pesticide residue that are allowed on the produce that they buy. Also have students research alternatives to simply rinsing their produce with water. **LS** **Logical/Intrapersonal**

Figure 15 ▶ A parasitic wasp injects its eggs into an aphid (left). A predatory mite is attacked by another species of mite (right).

Biological Pest Control

Most farmers practice some form of *pest management*. **Biological pest control** is the use of living organisms to control pests. Every pest has enemies in the wild, and these enemies can sometimes be used to control pest populations, as shown in **Figure 15**. One of the first recorded examples of biological control was in India in the mid-1800s. American prickly pear cactus had been introduced into India to feed insects that are used to make a valuable red dye. Because the cactus had no natural enemies in India, the cactus grew and spread. The plants were finally controlled by the introduction of an American beetle that eats the cactus.

Pathogens Organisms that cause disease, called *pathogens* (PATH uh juhnz), can also be used to control pests. One of the most common pathogens used to control pests is the bacterium *Bacillus thuringiensis* (buh SIL uhs THUHR in JIEN sis), often abbreviated *Bt*. This bacterium can kill the caterpillars of moths and butterflies that we consider to be pests.

Plant Defenses Scientists and farmers have bred plant varieties that have defenses against pests. For example, if you buy tomato plants or seeds, you may see that they are labeled "VNT." This label means they are resistant to certain fungi, worms, and viruses. Examples of plant defenses include chemical compounds that repel pests and physical barriers, such as tougher skin.

Chemicals From Plants Another type of biological pest control also makes use of plants' defensive chemicals. For example, chemicals found in chrysanthemum plants are now sold as pesticides. Most insect sprays that contain these chemicals are designed for use in the home because they are less harmful to humans and pets. These products are biodegradable, which means that they are broken down by bacteria and other decomposers.

FIELD ACTIVITY

Pest Search Make a list of the pests you can find in your area. Look for weeds and insects. What evidence can you find that these organisms are pests? You will not be able to see pests in the soil or microscopic bacteria, fungi, or viruses, but you may be able to see the damage the microscopic pests cause—black spots or dead patches on leaves. Can you think of a way to decrease the damage that is caused by these pests that involves the use of biological pest control? Record your observations in your *EcoLog*.

Activity —— GENERAL

Small-Scale Pest Control Ask students to suggest organisms that might help to control pests on crops. (Sample answers: some kinds of birds, lizards, amphibians, spiders, praying mantises, ladybugs, ants, wasps, snakes, hawks, owls, foxes, coyotes, bacteria, and viruses.) Encourage students to draw a picture of some of these beneficial predators in action. **LS** Intrapersonal

BIOLOGY
CONNECTION —— GENERAL

Natural Disease Control Interplanting, also called *companion planting*, is another example of a biological method of pest control that can also improve crop yields. For example, conventional tomato growers must rotate the location of their crop every year because disease pathogens accumulate in the soil and destroy the roots of the second year's crop. Farmers usually rotate less profitable crops for two to five years before replanting tomatoes in that same plot. Yet, organic growers have produced good yields planting consecutive tomato crops. The farmers plant an onion, leek, or garlic plant (all members of the *Allium* genus) alongside each tomato. The roots of these *Allium* plants support a type of bacteria that inhibits pathogens that might otherwise infect the tomato plants' roots.

391

MICROBIOLOGY
CONNECTION —— ADVANCED

Friendly Fungi VAM (vesicular-arbuscular mycorrhizae) are a beneficial type of fungi that attach to the roots of many crops. The filaments of these fungi spread out into the soil and reach moisture and nutrients beyond the reach of the crop roots. In this symbiotic relationship, VAM deliver water and nutrients to the crop in exchange for extracting small amounts of sugar photosynthesized by the leaves of the crop. The mass of VAM on a given plant can be up to 100 times greater than the mass of the plant's roots. Crop rotation and other organic practices encourage VAM growth, resulting in crops that can tolerate greater stress from heat and drought. This symbiosis partially explains how organic farms can have higher crop yields than otherwise predicted by conventional agronomists.

Using the Figure — ADVANCED

Integrated Pest Management
Review with students the steps involved in IPM as shown in **Figure 16**. Try to brainstorm examples of what farm managers might consider at each step. For example, "Economic considerations" might include the amount of crop loss that a farm can financially afford; "Crop and pest ecology" might involve the study of relationships between a particular crop, a pest of that crop, and the local environment; and "Set action thresholds" might mean deciding on the amount of crop loss above which the farmer will take action. **LS** Visual/Verbal

Opportunities

Community Gardens and Farmer's Markets Many communities have open markets where local small-scale farmers sell their produce. Some communities have garden clubs or programs which offer gardening advice, materials, and sometimes free garden plots available for local residents. Give students incentives to pursue such opportunities in your community.

Transparencies

TT The Steps of Integrated Pest Management

Disrupting Insect Breeding *Growth regulators* are chemicals that interfere with some stage of a pest's life cycle. If you have a dog, you may feed it a pill once a month to keep it free of fleas. The pill contains a growth regulator that prevents flea eggs from developing. When a flea sucks the dog's blood, the flea ingests the growth regulator. The regulator stops the flea's eggs from developing into adult fleas.

Pheromones (FER uh MOHNZ), chemicals produced by one organism that affect the behavior of another organism, can also be used in pest control. For example, female moths release pheromones that attract males from miles away. By treating crops with pheromones, farmers can confuse the male moths and interfere with the mating of the moths. Another way to prevent insects from reproducing is to make it physically impossible for the males to reproduce. For example, male insects are treated with X rays to make them sterile and then are released. When they mate with females, the females produce eggs that do not develop.

Integrated Pest Management

Integrated pest management is a modern method of controlling pests on crops. The steps involved in integrated pest management are shown in **Figure 16**. The goal of integrated pest management is not to eliminate pest populations but to reduce pest damage to a level that causes minimal economic damage. A different management program is developed for each crop. The program can include a mix of farming methods, biological pest control, and chemical pest control. Each of these methods is used at the appropriate time in the growing season. Fields are monitored from the time the crops are planted. When significant pest damage is found, the pest is identified. Then a program to control the pest is created.

Figure 16 ▶ This flow diagram shows the steps involved in integrated pest management.

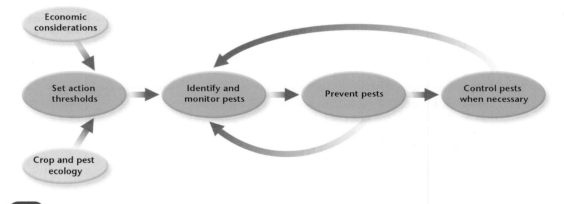

MISCONCEPTION ALERT

The Meaning of Organic Students may find the use of the term *organic* confusing. *Organic agriculture* generally means farming that depends on ecological interactions—instead of synthetic chemicals—to maintain yields, control pests, and manage soil fertility. In 2002, the U.S. Department of Agriculture (USDA) defined new nationwide standards by which foods could be labeled organic. Prior to this, various states and organizations had created more than 40 different standards. The new *national organic standards* govern the labeling of food products and the certification and monitoring of how food is grown, handled, and processed. The national organic standards prohibit genetically modified organisms, food irradiation, and the use of sewage sludge as fertilizer. For livestock production, organic standards prohibit the routine use of growth hormones and antibiotics and require that animals have access to the outdoors. Many consumer and health-food advocates praised the USDA for creating the national standards.

① Scientists isolate the gene from *Bt* that directs a cell to produce a toxin. The *Bt* gene is then joined to a "marker gene" that enables a cell to break down an antibiotic.

Toxin gene from *Bt*

Bacillus thuringiensis (Bt)

Marker gene

② The two genes are inserted into corn plant cells.

③ Scientists grow the corn cells and expose them to an antibiotic. Only those cells that have incorporated the inserted genes survive.

Antibiotic

④ The surviving cells grow into corn plants. These plants produce the *Bt* toxin, which kills caterpillars.

Figure 17 ▶ Genetic Engineering
This diagram shows the main steps used to produce a genetically modified plant—in this case, corn that produces its own insecticide.

Biological methods are the first methods used to control the pest. So, natural predators, pathogens, and parasites of the pest may be introduced. Cultivation controls, such as vacuuming insects off the plants, can also be used. As a last resort, small amounts of insecticides may be used. The insecticides are changed over time to reduce the ability of pests to evolve resistance.

Engineering a Better Crop

Plant breeding has been used since agriculture began. Farmers select the plants that have the tastiest tomatoes and the least pest damage. They save seeds from these plants to use in planting the next crop. The selected seeds are more likely to contain the genes for large, tasty fruits and for pest resistance than seeds from other plants are.

A faster way of creating the same result is to use **genetic engineering,** the technology in which genetic material in a living cell is modified for medical or industrial use. Genetic engineering involves isolating genes from one organism and implanting them into another. Scientists may use genetic engineering to transfer desirable traits, such as resistance to certain pests. The plants that result from genetic engineering are called *genetically modified* (GM) plants.

Figure 17 shows an example of the steps used to produce a GM plant. In this case, the gene introduced into the plant is not a plant gene. It is an insecticide gene from *Bt*, a bacterium that produces a chemical that kills plant-eating caterpillars but does not harm other insects. Plants that have the *Bt* gene make this insecticide within their leaves. Hundreds of gene transfers have now been performed to create many other GM crops.

internet connect

www.scilinks.org
Topic: Genetic Engineering
SciLinks code: HE4047

SCiLINKS® Maintained by the National Science Teachers Association

393

Reteaching ——— BASIC

Agricultural Concerns Tell students to imagine they are a farmer or gardener. Ask them to suggest the kinds of things that they should know about and pay attention to as farmers. Build a concept map with their responses. Then, have them review the headings in this section to see if they have left out any important issues. **LS** Visual/Verbal

Quiz ——— GENERAL

1. Why are earthworms, ants, and other burrowing animals important to fertile, productive soil? (They break up and aerate the soil.)

2. What sort of natural conditions make salinization a major problem? (low rainfall and naturally salty soil, such as occur in deserts)

3. What are the problems with persistent pesticides? (Persistent pesticides do not break down but accumulate in water, soil, and organisms.)

Alternative Assessment ——— GENERAL

Maintaining Soil Health Tell students that they have just been called in as consultants to a major agricultural company. The firm is concerned about maintaining soil quality. Have students create a presentation explaining some methods the company could use to help maintain soil fertility and prevent problems such as erosion. Encourage students to include their presentation in their **Portfolio. LS** Intrapersonal

Ecofact

Nitrogen Fixation One of the most valuable families of crop plants is the legumes (LEG YOOMZ), which include peas and beans. Legumes produce higher grade proteins than most plants produce, so legumes are part of diets in many parts of the world. Planting legumes also improves the soil. Their roots have nodules containing bacteria that take nitrogen gas from the air and that convert the nitrogen into a form other plants can use to build proteins.

Implications of Genetic Engineering In the United States, we now eat and use genetically engineered agricultural products every day. Many of these products have not been fully tested for their environmental impacts, and some scientists warn that these products will cause problems in the future. For example, genes are sometimes transferred from one species to another in the wild. Suppose a corn plant that was genetically engineered to be resistant to a pesticide were to pass the resistance genes to a wild plant. That wild plant might be a pest that could not be killed by that pesticide.

Sustainable Agriculture

Farming that conserves natural resources and helps keep the land productive indefinitely is called *sustainable agriculture*. Also called *low-input farming*, sustainable agriculture minimizes the use of energy, water, pesticides, and fertilizers. This method involves planting productive, pest-resistant crop varieties that require little energy, pesticides, fertilizer, and water. **Figure 18** shows an experimental farm where new sustainable agriculture techniques are being researched.

Figure 18 ▶ At the Land Institute in Salina, Kansas, sustainable agriculture techniques are being used to increase seed quantity in wheatgrass (background) and to increase yield in sunflowers (foreground).

SECTION 2 Review

1. **Define** the term *pest*.

2. **Compare** the benefits and environmental impact of pesticide use.

3. **Describe** how biological pest control is part of integrated pest management.

4. **Describe** how genetic engineering is used in agriculture.

CRITICAL THINKING

5. **Inferring Relationships** Write a paragraph to explain the similarities and differences between traditional plant breeding and genetic engineering. **WRITING SKILLS**

6. **Predicting Consequences** Read the description of integrated pest control in this section. Why do you think this pest control technique is not practiced everywhere? **READING SKILLS**

394

Answers to Section Review

1. A pest is any organism that occurs where it is not wanted or that causes economic damage.

2. Pesticides can kill insects, weeds, and pests that destroy crops. However, pesticides can threaten human health, pollute ecosystems, and prompt the evolution of pest resistance.

3. Biological pest control makes use of natural living enemies to control pests, and as such is part of one of the steps of IPM.

4. Scientists use genetic engineering to transfer genes for desirable traits, such as pest resistance, into crop plants from other organisms.

5. Answers may vary. Traditional plant breeding is similar to genetic engineering because both practices select for desired traits. In genetic engineering, genes from other species may be spliced into crop genes to produce a desired trait.

6. Answers may vary. Integrated pest control requires an understanding of all of the variables that affect pest populations and when they should be controlled. It requires education and a willingness to proactively manage crop pests.

Animals and Agriculture

We have seen that the total energy needed to grow plants for food is much less than the energy needed to raise animals as food. However, most animal proteins contain more essential amino acids than proteins found in plants do, and most humans include some animal products in their diet. Food from animals has been the basis of life for some human populations for centuries. For example, many human populations have traditionally obtained most of their protein from fish and seafood.

Our ancestors obtained animal protein by hunting and fishing, but today most people get animal protein from domesticated species. About 50 animal species have been **domesticated,** which means that they are bred and managed for human use. Domesticated animals include chicken, sheep, cattle, honey bees, silkworms, fish, and shellfish. In many parts of the world, goats, pigs, and water buffalo are also important domesticated animals.

Food from Water

Because fish are an important food source for humans, the harvesting of fish has become an important industry worldwide. However, as shown in **Figure 19,** when too many fish are harvested over a long period of time, ecological systems can be damaged.

Overharvesting Catching or removing from a population more organisms than the population can replace is called **overharvesting.** Many governments are now trying to stop overharvesting. They have created no-fishing zones, so that fish populations can recover. Research shows that fishing in areas surrounding no-fishing zones improves after no-fishing zones have existed for a few years. In some areas of the world, such restrictions are necessary if fish markets, such as the one shown in **Figure 20,** are to prosper.

Objectives

▶ Explain how overharvesting affects the supply of aquatic organisms used for food.

▶ Describe the current role of aquaculture in providing seafood.

▶ Describe the importance of livestock in providing food and other products.

Key Terms

domesticated
overharvesting
aquaculture
livestock
ruminant

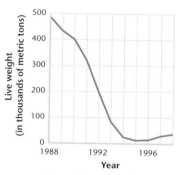

Figure 19 ▶ The North Atlantic cod fishery has collapsed because of overharvesting.

Figure 20 ▶ Whole, fresh tuna are one of the many types of seafood for sale at the Tokyo fish market, the largest fish market in the world.

395

INCLUSION Strategies

• *Learning Disabled*
• *Attention Deficit Disorder*

Have students list and define the Key Terms from the section or the chapter using a word processor or pencil and paper. Students should then mix-up the definitions and terms to create a matching activity. Students should exchange papers and try to complete the activity. Completed activities can be used as a study guide.

Chapter Resource File

• **Lesson Plan**
• **Active Reading** BASIC
• **Section Quiz** GENERAL

Transparencies

TT Bellringer
TT The Collapse of the North Atlantic Cod Fishery

SECTION 3

Focus

Overview

Before beginning this section, review with your students the Objectives in the Student Edition. This section introduces the ways in which we manage and use animals for food. Methods of raising fish, livestock, and poultry are discussed.

 **Bellringer**

Ask students to share what types of meat they consume. (Sample answer: beef, chicken, fish.) Ask them to consider how that meat was produced and to list ways that the conditions in which animals are raised could affect meat quality. Have them write their responses in their *EcoLog.* (Sample answers: Crowded conditions allow diseases to spread easily; lack of exercise increases the fat content of meat; animals may be given drugs to combat disease or to grow more quickly, and these drugs might remain in the meat.) **LS Logical/Intrapersonal**

Motivate

Discussion ——— BASIC

Gone Fishing Ask students if they go fishing or know someone who does. Have them share descriptions of the experience, focusing on what they have learned about fish biology and meat preparation. Ask if they have noticed or heard of any trends in the number, type, or size of fish that can be caught. (Students may have heard that some fish catches are getting smaller, that more anglers are practicing "catch-and-release," that certain species are becoming more or less common, that some fish and water sources are considered dangerously contaminated, or that regulations on fishing are becoming stricter.) **LS Verbal/Intrapersonal**

Fish Harvests and Aquaculture
Assign a coastal country to each pair of students. Have them search the Internet to find out how much fish and seafood is harvested, consumed, and exported by that country. Have them also try to find statistics on the amount and types of aquaculture practiced in that country. Create a large table on chart paper to collect this information, and display it in the classroom. Have students add the information for their country to the table. As an alternative or extension, have students create graphs of the statistics and post them next to a world map, using pinned strings to make a connection to each country.
LS Verbal/Visual/Logical

Student Opportunities

Heifer International This organization combats hunger and poverty by donating appropriate livestock, training, and related services to small-scale farmers worldwide. The organization began in the 1930s when a U.S. farmer ladling out milk for victims of the Spanish Civil War decided that they needed "not a cup, but a cow." He asked his American friends to donate young cows, and the charity began. Current projects are selected based on 12 *Cornerstones for Just and Sustainable Development*. Each recipient of an animal must give away some of the offspring and develop self-sufficiency. The organization offers hands-on learning and volunteer opportunities.

Figure 21 ▶ This oyster farm in Washington State shows how aquaculture concentrates seafood production.

Aquaculture

Fish and other aquatic organisms provide up to 20 percent of the animal protein consumed worldwide. But overharvesting is reducing the amount of fish and other organisms in the world's oceans. One solution to this problem may be a rapid increase in **aquaculture** (AK wuh KUHL chuhr), the raising of aquatic organisms for human use or consumption.

Aquaculture is not a new idea. This practice probably began in China about 4,000 years ago. Today, China leads the world in using aquaculture to produce freshwater fish.

There are a number of different methods of aquaculture. The oyster farm shown in **Figure 21** represents one such method. The most common method is known as a fish farm. Fish farms generally consist of many individual ponds that each contain fish at a specific stage of development. Clean water is circulated through the ponds and brings in oxygen while sweeping away carbon dioxide and fecal wastes. The fish grow to maturity in the ponds and then are harvested.

Another type of aquaculture operation is known as a ranch. In this method, fish such as salmon are raised until they reach a

Menhaden: The Fish Behind the Farm

One of the largest commercial catches in the United States each year is of a fish that most people have never heard of—the menhaden (men HAYD 'n). Menhaden are small, silver, oily fish in the herring family and are found in the Atlantic Ocean from Maine to Florida. More than one-third of the weight of commercial fish caught on the East Coast each year is menhaden. But menhaden are so full of bones that they are inedible. So why are they so important?

When the first colonists arrived in the area we now call New England, local Indians showed them how to fertilize their crops with menhaden. This is where the legend that the best corn is grown by planting a fish with each seed came from. Later, menhaden oil was used

in oil lamps, and ground menhaden were added to cattle feed.

The menhaden catch is processed to produce fishmeal and fish oil. The oil is used in cooking oils and margarine. The fishmeal has a high protein content, and it is added to the feed of pets, chickens, turkey, hogs, cattle, and farm fish. Menhaden is also used by recreational fishermen as bait for fish such as bluefin, striped bass, shark, and tuna.

Menhaden spawn in the ocean. The eggs hatch into larvae, which are carried into estuaries where they spend their first year. After the menhaden mature, they return to the ocean and usually live within 50 km of the coast. The Chesapeake Bay is one of the most important nurseries for menhaden.

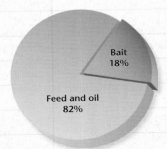

Source: Menhaden Resource Council.

▶ The enormous menhaden catch is used entirely to produce feed and oil and as bait for catching other fish.

Menhaden live in large schools near the surface, so they are easily caught with *purse seine* nets, nets that hang down from the surface. Boats towing the nets encircle the

396

BRAIN FOOD

The Benefits of Fish Although people typically think fish is beneficial as a source of protein, it is also an important source of the following nutrients: lysine (which helps prevent hardening of the arteries and hypertension); polyunsaturated fats; minerals such as calcium, phosphorus, and iron; vitamins A, B_1,

B_2, B_{12}, and D; and trace elements such as iodine and zinc. Fish is especially beneficial to underweight preschool children and those suffering from many vitamin and micronutrient deficiencies. Few other natural animal or vegetable protein sources can match the combined benefits of fish.

certain age and then are released. The salmon, for example, migrate downstream to the ocean, where they live until adulthood. When they are mature, the fish return to their birthplace to reproduce. When they return, they are captured and harvested.

Today, most of the catfish, oysters, salmon, crayfish, and rainbow trout eaten in the United States are the products of aquaculture. In the 1980s, domestic production of these species quadrupled, and imports of these species increased even faster. Worldwide, about 23 percent of seafood now comes from aquaculture.

However, as with other methods of food production, aquaculture can cause environmental damage if not managed properly. For example, the aquatic organisms can create a large amount of waste, which can be a source of pollution. Also, because aquaculture requires so much water, the process can deplete local water supplies. In a few cases, sensitive wetlands have been damaged when large aquaculture operations were located within the wetland. Despite these problems, aquaculture will continue to be an important source of protein for the human diet.

▶ A menhaden catch is unloaded from purse seine nets in Chesapeake Bay, Virginia.

fish, which are pumped out of the ocean into refrigerated containers.

An adult menhaden is an important member of the marine ecosystem. The fish are filter feeders that scoop up large mouthfuls of water and filter out the plankton for food. An adult menhaden can filter a million gallons of water in six months.

The Chesapeake Bay Ecological Foundation estimates that the menhaden population removes up to one-fourth of the nitrogen pollutants dumped into the Chesapeake Bay each year. Because nitrogen runoff from lawns and farms is a major pollutant of the Chesapeake Bay, this function of the fish is important. Sport fishermen also value menhaden as bait because they are the natural food of many sportfish.

Both environmentalists and the sport fishing industry were worried when the menhaden catch declined during the 1990s. The catch in 2000 was the second-lowest catch

on record. Both groups believe that overharvesting by commercial fishing boats was the reason for the reduced catch. As a result, the Atlantic Menhaden Management Board, which manages the menhaden fishery, has been restructured to have fewer members who represent the commercial fisheries.

CRITICAL THINKING

1. Applying Ideas Many different groups have potentially conflicting interests in the future of the menhaden fishery. Write a paragraph that explains the opposing points of view of two of these groups. **WRITING SKILLS**

2. Expressing Viewpoints If you were on the Atlantic Menhaden Management Board, what changes would you suggest to prevent the fishery from declining? Write a paragraph that explains these changes. **WRITING SKILLS**

Livestock Populations Ask students to examine **Table 3.** Ask, "Which types of livestock have increased the most in recent decades?" (chickens, pigs, and goats) If any students have direct experience with these type of animals, have them share their knowledge of how the animals might be raised. Ask students to speculate about why populations of these livestock have increased. (Answers may vary.) Suggest that interested students conduct further research and report back to the class.  Logical

Cultural Awareness GENERAL

Livestock Diversity Of 50,000 known mammal and bird species, only about 30 species have been domesticated and used extensively as livestock. Currently, most of the world's livestock consists of 15 species, although there are many breeds (or "races") of each. In this century, between 300 and 600 of over 4,000 unique livestock breeds have died out, and at least as many are critically endangered. Food development organizations see a danger in this trend and advocate using native and locally-bred animals as part of hunger prevention efforts. Indigenous species and locally-bred livestock are presumably adapted to the local environment, and thus are more sustainable to raise than imported breeds.

Figure 22 ▶ Modern livestock operations, such as this pig farm in North Carolina, are large and efficient.

Table 3 ▼

UN FAO Estimates of Animal Populations			
	Global Livestock Populations		
Species	1961	2001	Increase
Chickens	3.9 billion	14.8 billion	280%
Sheep	1 billion	1 billion	0%
Cattle	942 million	1.4 billion	53%
Pigs	406 million	928 million	129%
Goats	349 million	702 million	101%
Horses, donkeys, and mules	110 million	114 million	4%

Livestock

Domesticated animals that are raised to be used on a farm or ranch or to be sold for profit are called **livestock.** As shown in **Table 3,** populations of livestock have changed dramatically in the last 40 years. Large livestock operations, such as the pig farm shown in **Figure 22,** produce most of the meat that is consumed in developed countries. Meat production per person has increased worldwide since 1950, as shown in **Figure 23.** Livestock are also important in developing countries. In these countries, livestock not only provide leather, wool, eggs, and meat, but also serve many other functions. Some livestock are used as draft animals to pull carts and plows. Other livestock provide manure as the main source of plant fertilizer or as a fuel for cooking.

Ruminants Cattle, sheep, and goats are **ruminants** (ROO muh nuhnts), cud-chewing mammals that have three- or four-chambered stomachs. *Cud* is the food that these animals regurgitate from the first chamber of their stomachs and chew again to aid digestion. Ruminants also have microorganisms in their intestines that allow the animals to digest plant materials that humans cannot digest. When we eat the meat of ruminants, we are using them to convert plant material, such as grass stems and woody shrubs, into food that we can digest—such as beef.

Humans have created hundreds of breeds of cattle that are suited to life in different climates. Cattle are most common in North America, India, and Africa. But the cattle are not always slaughtered for meat. In Africa for example, traditional Masai herders drink milk and blood from their cattle, but the herders rarely kill them for meat. India has almost one-fifth of the world's cattle. However, many of these cattle are not killed or eaten because cows are sacred to Hindus, who make up a large part of India's population. These cattle instead produce milk and dung, and the cattle are used as draft animals.

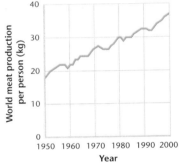

Figure 23 ▶ Worldwide meat production per person has increased significantly since 1950.

Source: Worldwatch Institute.

Homework —— GENERAL

Comparing Meat Alternatives Have students compare the nutritional value of some products made with meat to those made with a meat substitute. Meat substitutes are often marketed as "health foods." Have students read the ingredients and the nutritional label on each product and record the grams of protein, carbohydrates, and fat. Ask them to compare these to the U.S. recommended daily allowances for protein, carbohydrates, and fat. Have them create a summary of their findings and include their own conclusions about the products they studied.  Intrapersonal

BRAIN FOOD

Controlling Disease the Organic Way
In organic livestock farming, farmers avoid using synthetic medications and hormones. Diseases and parasites are avoided through preventative measures such as rotational grazing, carefully balanced diets, sanitary conditions, and gentle treatment of the animals.

Poultry Since 1961, the population of chickens worldwide has increased by a greater percentage than the population of any other livestock, as shown in **Table 3**. Chickens are a type of *poultry*, domesticated birds raised for meat and eggs, which are good sources of essential amino acids. In more-developed countries, chickens and turkeys are usually raised in factory farms, as shown in **Figure 24**. Some people have criticized this industry because the animals live in cramped, artificial environments.

Fewer ducks and geese are raised worldwide than chickens, but in some areas ducks and geese are economically important. For example, the Chinese use ducks not only for meat, but also as part of an integrated system that produces several types of food at one time. The ducks' droppings are used to fertilize fields of rice called *rice paddies*. The rice paddies are flooded several times per year with water from nearby ponds. Mulberry trees, which feed silkworms, are also irrigated by the ponds. Plant materials and filtered sewage are dumped in the ponds and serve as food for carp and other fish. The integrated system uses little fresh water, recycles waste, and produces ducks, silk, rice, and fish.

internet connect
www.scilinks.org
Topic: Livestock Production
SciLinks code: HE4086
SciLINKS. Maintained by the National Science Teachers Association

Figure 24 ▶ Modern chicken farms, such as this one, are often huge, industrial-scale operations.

SECTION 3 Review

1. **Explain** why the percentage of seafood produced by aquaculture is increasing so rapidly.

2. **Explain** how overharvesting affects the supply of fish such as salmon.

3. **Describe** the importance of livestock to cultures that consume no meat.

CRITICAL THINKING

4. **Inferring Relationships** Read the description of poultry above and explain why chickens are such an important source of food for humans. READING SKILLS

5. **Applying Ideas** Look at the graph in Figure 23. Write a short paragraph explaining why meat production has increased so rapidly. WRITING SKILLS

399

Answers to Section Review

1. Answers may vary. The percentage of seafood produced by aquaculture is increasing rapidly because the supply of fish and shellfish from the oceans is declining as a result of overfishing. Also, aquaculture provides controlled environments in which to raise seafood.

2. Overharvesting affects fish supplies by removing a large percentage of fish populations.

3. Even in cultures that eat no meat, livestock may serve as draft animals for crop production or as sources of milk, dung, and other materials.

4. Chickens are an important food source for humans because they produce meat and eggs, which are both good sources of essential amino acids.

5. Answers may vary. Human populations in countries whose citizens eat large amounts of meat have increased, leading to greater demand. Meat consumption has also increased in some countries because of changes in food preferences. Increases in meat production have been supported by new production methods.

Close

Reteaching ———— BASIC

Evaluating Food Choices Ask students to revisit the list they made during the *Bellringer* activity of what types of meat they consume. Have them create a table listing each type of meat, the way that meat was produced, and the benefits, risks, and environmental impacts associated with producing each type of meat. LS Verbal/Intrapersonal

Quiz ———— GENERAL

1. What environmental problems are associated with aquaculture? (Aquaculture operations generate large amounts of waste, which can pollute waterways. Aquaculture also uses large amounts of water and so can deplete local water supplies.)

2. Besides chickens, what types of poultry are important in other parts of the world? (ducks and geese)

3. What is the relationship between no-fishing zones and overharvesting? (No-fish zones allow fish populations to recover from the effects of overharvesting.)

Alternative Assessment ———— GENERAL

Applying Concepts Tell students they are the proud owners of a new chicken farm. They have the option of running their farm with any of the methods discussed in the chapter. Have each student write a description of her or his farm. Have students choose a name for their farm and explain the following: how big the farm is, how many animals they are raising, how they would care for the animals, and how they would manage the land. LS Intrapersonal

Chapter 15 · Food and Agriculture **399**

CHAPTER 15 Highlights

Alternative Assessment ——— GENERAL

Eco-Friendly Eating Divide the class into groups, and ask them to compile a list of their own diet tips that are both healthy and environmentally responsible. Ask groups to put the list in a poster, brochure, newspaper advice column, or other creative format for public display.

LS Interpersonal/Visual   Co-op Learning

Agriculture Case Study Have pairs or small groups of students research the history of a specific farm, ranch, or fishery operation. Suggest that they choose an operation within their state, or perhaps in another country. Have students create a poster or multimedia presentation depicting the history and including commentary on critical points where the managers of the operation made decisions that had environmental impacts. Have students present to their classmates and answer questions.

LS Intrapersonal/Verbal   Co-op Learning

Family Roots Assign this project to students who have any form of agriculture in their family history. Tell them to interview family members (and record the interview on audio or video tape if possible) and collect photographs and memorabilia of this history. Then, have them share these materials with the class.

LS Intrapersonal/ Verbal   English Language Learners

1 Feeding the World

Key Terms

famine, 379
malnutrition, 379
diet, 380
yield, 381

Main Ideas

▶ The foods produced in the greatest amounts worlwide are grains, the seeds of grass plants.

▶ Malnutrition is a condition that occurs when people do not consume enough Calories or do not eat a sufficient variety of foods to fulfill all of the body's needs.

▶ More food is needed each year to feed the world's growing population. Distribution problems and drought can lead to food shortages.

▶ The green revolution introduced new crop varieties with increased yields through the application of modern agricultural techniques.

2 Crops and Soils

Key Terms

arable land, 384
topsoil, 385
erosion, 386
desertification, 386
compost, 388
salinization, 388
pesticide, 389
biological pest control, 391
genetic engineering, 393

Main Ideas

▶ The basic processes of farming are plowing, fertilization, irrigation, and pest control. Modern agricultural methods have replaced traditional methods in much of the world.

▶ Fertile soil is soil that can support the growth of healthy plants. Soil conservation methods are important for protecting and managing topsoil and reducing erosion.

▶ Pests cause considerable crop damage. The use of pesticides has both positive and negative effects on the environment. Integrated pest management can minimize the use of chemical pesticides.

▶ Genetic engineering is the process of transferring genes from one organism to another. Plants that result from genetic engineering are called genetically modified plants.

3 Animals and Agriculture

Key Terms

domesticated, 395
overharvesting, 395
aquaculture, 396
livestock, 398
ruminant, 398

Main Ideas

▶ Overharvesting has reduced the populations of many aquatic organisms worldwide.

▶ Aquaculture, the raising of aquatic animals, may be a solution to the problem of overharvesting.

▶ Livestock are important for the production of food and other products. Worldwide meat production per person has increased greatly over the past several decades.

400

Chapter Resource File

- **Chapter Test** GENERAL
- **Chapter Test** ADVANCED
- **Concept Review** GENERAL
- **Critical Thinking** ADVANCED
- **Test Item Listing**
- **Modeling Lab** GENERAL
- **Observation Lab** ADVANCED
- **Research Lab** GENERAL
- **Consumer Lab** BASIC
- **Long-Term Project** GENERAL

Using Key Terms

Use each of the following terms in a separate sentence.

1. *overharvesting*
2. *erosion*
3. *livestock*
4. *yield*
5. *genetic engineering*

For each pair of terms, explain how the meanings of the terms differ.

6. *pesticide* and *biological pest control*
7. *arable land* and *topsoil*
8. *livestock* and *ruminant*
9. *malnutrition* and *famine*
10. *salinization* and *desertification*

✓ **STUDY TIP**

Making It a Habit Many people find that developing a routine helps them to study more effectively. Decide which time of day you feel most alert, and set it aside for studying. Make sure that any distractions around you will be minimal. When you regularly follow through with your study plan, you may find that you begin to learn more in less time.

Understanding Key Ideas

11. Malnutrition can be caused by
    a. a lack of enough Calories.
    b. a lack of carbohydrates.
    c. a lack of essential amino acids.
    d. All of the above

12. Humans need which of the following nutrients?
    a. carbohydrates and minerals
    b. lipids and vitamins
    c. proteins
    d. all of the above

13. Which of the following is *not* one the six most produced foods worldwide each year?
    a. potatoes
    b. beef
    c. rice
    d. wheat

14. Which of the following statements about human diets in all parts of the world is true?
    a. Most people eat pork.
    b. An adequate diet includes carbohydrates, proteins, and fats.
    c. Most people do not have protein in their diets.
    d. Most people are obese.

15. Malnutrition is largely a result of
    a. war.
    b. soil erosion.
    c. poverty.
    d. salinization.

16. Which of the following is *not* found in fertile soil?
    a. rock particles
    b. worms
    c. high concentrations of salts
    d. high concentrations of organic matter

17. Which of the following is *not* a soil conservation method?
    a. contour plowing
    b. salinization
    c. no-till farming
    d. terracing

18. Which of the following statements is a disadvantage of using chemical pesticides?
    a. Pesticides can pollute waterways.
    b. Pests evolve resistance to pesticides.
    c. Pesticides kill beneficial insects.
    d. all of the above

19. How do pesticides that are growth regulators work?
    a. They kill fleas.
    b. They disrupt the pest's life cycle.
    c. They attract predators of the pest.
    d. They prevent the pest from attacking the plant by poisoning its nervous system.

401

Assignment Guide	
Section	Questions
1	4, 9, 11–15, 20, 25–27, 29, 33, 35–38
2	2, 5–7, 10, 16–19, 21, 22, 24, 28, 30, 31, 34, 39, 40
3	1, 3, 8, 20, 23, 32

ANSWERS

Using Key Terms

1. Sample answer: Overharvesting has reduced the populations of many fish species.

2. Sample answer: Wind erosion and water erosion cause soil degradation.

3. Sample answer: Some types of livestock include cows, sheep, and goats.

4. Sample answer: The yield of a particular crop depends on factors such as soil fertility and water availability.

5. Sample answer: Genetic engineering allows scientists to incorporate the genes for desirable traits into crop plants.

6. Pesticides are inorganic chemicals used to control pests. Biological pest control involves using natural enemies to control pests.

7. Arable land refers to land that can be farmed. Topsoil is the uppermost layer of soil, in which crops grow.

8. Livestock are domestic animals raised for use on a farm or ranch or to be sold for profit. Ruminants are a type of livestock, such as cattle, sheep, and goats, that chew cud and have three- or four-chambered stomachs.

9. Malnutrition is a condition that results from the lack of proper nutrients necessary for human health. A famine is widespread starvation caused by a shortage of food.

10. Salinization and desertification are two ways in which fertile soil can be degraded. Salinization is an accumulation of salts in soil. Desertification is the process by which land becomes more desert-like because of human activities or climate change.

Understanding Key Ideas

11. d
12. d
13. b
14. b
15. c
16. c
17. b
18. d
19. b

CHAPTER 15 Review

Short Answer

20. More resources (energy, water, and land) are required to produce a given amount of meat than to produce the equivalent amount of plants.

21. Plowing soil digs up plants. Without roots holding the soil in place, the soil is more susceptible to erosion. Plowing soil also exposes a greater surface area of soil to wind and water.

22. Pests can quickly evolve resistance to chemical pesticides. Chemical pesticides generally kill many types of organisms, including beneficial ones.

23. Ruminants convert plant material that human cannot digest, such as grass stems, into food humans can digest.

24. Soil degradation is a loss of soil or soil fertility that renders land less suitable for growing crops. Often, this creates a worsening cycle of degradation that ends with desertification, salinization, or soil that is otherwise not arable.

Interpreting Graphics

25. 1981; 1983

26. About 76 million acres of corn were planted in 1991.

27. It is likely that some of the corn crop is lost each year.

Concept Mapping

28. Answers to the concept-mapping questions are on pages 667–672.

Short Answer

20. Why does it cost more to produce a kilogram of meat than to produce a kilogram of plants?

21. How does plowing soil increase soil erosion?

22. Why are biological controls for killing pests sometimes more effective than chemical pesticides are?

23. Why are ruminants valuable livestock?

24. Explain how soil degradation leads to loss of arable land.

Interpreting Graphics

Use the graph below to answer questions 25–27.

25. In which year was the most corn planted? In which year was the least corn harvested?

26. How many acres were planted with corn in 1991?

27. According to the graph, more acres of corn are planted than are harvested each year. Why?

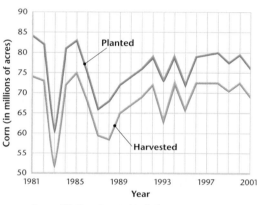

Source: U.S. Department of Agriculture.

Concept Mapping

28. Use the following terms to create a concept map: *contour plowing, no-till farming, organic farming, careful irrigation, soil erosion, nutrient depletion,* and *salinization.*

Critical Thinking

29. **Making Comparisons** Both wars and drought can lead to famine. In what ways do they have similar effects, and in what ways are their effects different? Write a short paragraph that explains your answer. **WRITING SKILLS**

30. **Analyzing Ideas** What incentives to conserve soil do farmers in developed nations have?

31. **Inferring Relationships** Read the text in this chapter under the heading, "Disrupting Insect Breeding." Are pheromones a type of pesticide? Explain your reasoning. **READING SKILLS**

Cross-Disciplinary Connection

32. **Social Studies** Thousands of tons of dead fish are shoveled back into the ocean each year from fishing vessels because the fish are species that consumers do not want to buy. Identify some ways that humans might be able to reuse this protein.

33. **Economics** Hundreds of thousands of people starve to death every year. How is this problem related to the problems of poverty? Explain your answer.

Portfolio Project

34. **Prepare a Report** Environmental degradation caused by farming is not a new problem. The Dust Bowl of the 1930s is an example of an environmental disaster caused by farming practices that we would now consider to be damaging. Investigate the Dust Bowl, and write a report about it. Include information about the farming practices, laws, and regulations that were introduced in the United States as a result of the lessons learned during the 1930s. **WRITING SKILLS**

Critical Thinking

29. Answers may vary. Wars and droughts can both take land out of production and disrupt food distribution networks. Droughts often affect large areas uniformly, whereas wars tend to affect some areas more intensely than others.

30. Answers may vary. One major incentive for conserving soil is to sustain annual crop yields.

31. Answers may vary. Pheromones could be considered a type of pesticide in that they are chemicals applied to crops to control insect pests. However, unlike synthetic pesticides, pheromones are naturally-occurring chemicals.

Cross-Disciplinary Connection

32. Answers may vary. The unwanted fish could be ground up and used to feed livestock or pets or to nutritionally supplement human food products. They could also be used as fertilizer or bait.

33. Answers may vary. People may simply lack money to buy enough food or lack the resources to grow food.

Portfolio Project

34. Answers may vary.

MATH SKILLS

Use the table below to answer questions 35–38.

World Food Production (in Millions of Tons)			
Food	1990	1995	1999
Total grains	1,700	1,800	1,900
Wheat	590	540	590
Rice	350	370	400
Legumes	58	55	59
Poultry	37	51	58
Milk	441	381	387

35. Analyzing Data Which foods had increased production in 1995 and 1999?

36. Analyzing Data Which foods had lower production in 1995 than in 1990?

37. Analyzing Data Taking into account the 1999 data, can you think of any possible reasons for the answer to question 36?

38. Analyzing Data The human population of the world grew by 15 percent between 1990 and 1999. By what percentage did total grain production increase during this time?

WRITING SKILLS

39. Communicating Ideas Explain how the way in which insects reproduce allows them to evolve pesticide resistance very rapidly.

40. Analyzing Ideas Explain why the pesticide DDT can still be detected in the environment even though its use was banned decades ago.

STANDARDIZED TEST PREP

For extra practice with questions formatted to represent the standardized test you may be asked to take at the end of your school year, turn to the sample test for this chapter in the Appendix.

READING SKILLS

Read the passage below, and then answer the questions that follow.

A large amount of energy is needed to produce food. In all parts of the world, the energy used to process, distribute, and cook food is greater than the energy used to grow it. In the United States, it is estimated that every Calorie of food on our dinner tables has required 9 Calories of energy to get there. Half a Calorie accounts for the energy used on the farm. The other 8.5 Calories account for energy for processing, packaging, distribution, and cooking. In rural India, twice as much energy goes into cooking a kilogram of rice as was invested in producing it. Energy shortages, such as a shortage of wood for cooking, have caused environmental problems such as deforestation. In poor countries, cooking may require more energy than is used by transportation, heating, and all other uses for energy combined.

1. According to the passage, which of the following statements about food is true?
   a. Most of the energy invested in food production goes into distributing the food.
   b. More energy is used to grow food on the farm than is used to cook the food.
   c. Most energy used to produce food goes into processing, distributing, and cooking the food.
   d. In developing countries only, cooking food requires more energy than growing food does.

2. Which of the following points is not discussed in this passage?
   a. Packaging is the least costly part of preparing food for sale to the customer.
   b. It takes more energy to cook a kilogram of rice than to grow a kilogram of rice.
   c. In some countries, cooking requires more energy than all other processes that use energy combined.
   d. Gathering sufficient wood to cook food has led to deforestation.

403

MANAGING THE MOISTURE IN GARDEN SOIL

Teacher's Notes

Time Required

one 45-minute class period

Lab Ratings

EASY ———————→ HARD

TEACHER PREPARATION 🧪🧪
STUDENT SETUP 🧪🧪🧪
CONCEPT LEVEL 🧪
CLEANUP 🧪🧪

Skills Acquired

• Forming hypotheses
• Interpreting data

The Scientific Method

In this lab, students will:
• Analyze Results
• Draw Conclusions

Materials

The materials listed are enough for a group of 3 to 4 students. Provide 200g of potting soil to each group. All other materials should be completely dry.

Safety Cautions

Supervise students closely while they are heating soil samples.

Objectives

▶ **USING SCIENTIFIC METHODS**
   Hypothesize how to reduce the amount of water a garden needs.
▶ **Compare** the amount of water different soil samples can hold.
▶ **Explain** how adding materials to a soil sample can help increase the sample's ability to hold water.

Materials

beaker, 250 mL
compost, 5 g
crucible (or other heat-safe container)
dry chopped grass clippings, 5 g
eyedropper
filter paper
funnel
heat source (hot plate or oven)
metric balance
sawdust, 5 g
soil sample, 50 g
stirring rod
tongs
watch (or clock)
water

▶ **Step 4** Fold the moist filter paper into quarters, and then open it to form a cup that fits in a funnel.

Managing the Moisture in Garden Soil

You work as a soil specialist with the Smith County Soil Conservation District. You are trying to help Mrs. Latisha Norton, a local resident, solve an agricultural problem. Mrs. Norton has found that she must water her vegetable garden very often to keep it healthy. As a result, her family's water bills have skyrocketed! Mrs. Norton and her family may have to give up their garden project because of the added expense.

You realize that the water is probably draining out of the garden soil too quickly. To solve this problem, you need to find out how much water the soil can hold. You visit her garden and collect several soil samples. (Your teacher will provide you with soil.)

Procedure

1. Dry your soil sample without burning any of the organic matter. To do this, place about 50 g of soil in a crucible or other heat-safe container. Using tongs, gently heat the sample over a hot plate or put the sample in an oven. Stir the sample occasionally with a stirring rod to ensure that the sample becomes completely dry.

2. After the sample is completely dry, weigh about 10 g of dry soil. Record the mass in a data table.

3. Dampen a circle of filter paper until it is thoroughly moist, but not dripping. Weigh the moist filter paper, and record its mass in a data table.

4. As shown below, fold the moist filter paper into quarters. Next, open the filter paper to form a cup that fits in a funnel. Place the cup-shaped filter paper in the funnel.

5. Place the dry soil sample on the filter paper in the funnel. Place the funnel in the beaker.

6. Add water to the soil sample one drop at a time until all of the soil is moist and water begins to drip out of the funnel. Stop adding water, and let the funnel sit for 5 min.

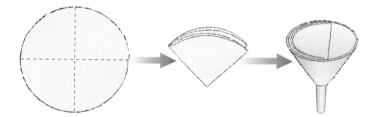

404

Tips and Tricks

Have students think about other ways that water can be retained in a garden. Have them research and then discuss how plants might be planted, soil might be arranged, or edging might be used to prevent water loss. Have students contact local farmers or gardeners to learn their techniques for managing moisture.

For the extension, students may try sand, clay, various organic materials, or soil additives that are available from commercial gardening

suppliers. Encourage creativity, and don't give too many hints, but be sure the students hypothesize why certain soils may hold water best and have reasons for their proposed hypotheses. Students should be able to test their hypotheses and provide appropriate data. Students may also want to experiment with the shape of their "garden" to try to reduce water runoff.

7. After 5 min, remove the filter paper and moist soil from the funnel, and weigh the paper and soil together. Record their mass in a data table.

8. Calculate the mass of the moistened soil sample by subtracting the mass of the damp filter paper from the mass of the completely moistened sample and the filter paper. Record the mass in a data table.

9. Calculate the amount of water that your soil sample can hold by subtracting the mass of the dry soil sample from the mass of the moistened soil sample. Record the result in a data table.

10. Calculate the percentage of water that your sample held. Divide the mass of water the soil held by the mass of the moistened soil sample, and multiply by 100. The higher the percentage is, the more water the soil can hold. Record the percentage in a data table.

11. Divide the remaining dry soil sample into three 5 g portions. To the first soil sample, add 5 g of dry compost. To the second soil sample, add 5 g of dry chopped grass clippings. To the third soil sample, add 5 g of dry sawdust. Weigh each mixed soil sample, and record the masses of the three samples in a data table.

12. Perform steps 3–10 for each of your mixed soil samples. Record your results in a data table.

Analysis

1. **Organizing Data** Compare your results with the results of your classmates. Which soil samples held water the best? Why?

2. **Analyzing Data** Which of the additional materials improved the soil's ability to hold water?

Conclusions

3. **Evaluating Methods** Based on your results as well as your research, what could you recommend to Mrs. Norton to reduce the amount of water her garden needs?

Extension

1. **Designing Experiments** With the help of your teacher, choose one more material in addition to the three materials you used in step 11. Combine two of these materials, and mix them with a soil sample. Combine the remaining two materials with another soil sample. Perform steps 3–10 for these two mixed soil samples. Compare your results with the results you gathered earlier in the lab. Which combination of materials in the soil samples held water the best?

▶ **Step 6** When adding water to the soil sample, add one drop at a time until all of the soil is moist and water begins to drip out of the funnel.

405

Sample Student Data Table:

Sample Contents	Mass of dry soil sample	Mass of damp filter paper	Mass of moistened soil sample on filter paper	Percent water-holding capacity
Plain soil	10g			
Soil + compost	10g			
Soil + grass clippings	10g			
Soil + sawdust	10g			

Jason Marsh
Moose Lake High School
Moose Lake, Minnesota

Answers to Analysis

1. The dry chopped grass clippings should improve the soil's water retention the most. This material can readily absorb water. It also separates soil particles, which allows more water to bond to the soil.

2. All of the materials probably improved the soil's ability to hold water.

Answers to Conclusions

3. Adding organic matter (such as dry chopped grass clippings or wood chips) should help Mrs. Norton's soil retain water and reduce the amount of water she needs to add to the garden.

Answers to Extension

1. Answers may vary. Students may find that a combination of materials may be most effective in helping soil retain moisture.

Student Opportunities

Soil and Water Conservation Official Invite the local soil and water conservation district manager into the classroom to help you with this exercise. Ask her or him to suggest better alternatives to the materials used in the experiment. Also, ask the guest to talk about soil management practices as they relate to water availability in your area.

Chapter Resource File

• Datasheets for In-Text Labs
• Lab Notes and Answers

POINTS of view

GENETICALLY ENGINEERED FOODS

Genetically engineered foods are now on sale in the world's supermarkets, and we do not recognize them because they are not labeled as such.

As the world's population continues to increase, food production must try to keep pace with the increase. Genetic engineering provides one way to develop new foods. Biotechnologists develop desirable characteristics in an organism by altering its genes or by inserting new genes into the organism's cells. For example, a gene that makes one plant species resistant to pests might be transferred to another plant species. The second plant species would then have the same resistance to pests.

In 1994, the first genetically modified food was offered for sale. It is a tomato called the Flavr Savr™ which softens slowly, so it can remain on grocery shelves longer before becoming soft and overripe. Biotechnologists developed the tomato by altering the gene that causes ripe tomatoes to soften. The Food and Drug Administration (FDA) said it was as safe as other tomatoes and cleared it for sale. Here are two points of view on genetically engineered foods.

The Benefits Outweigh the Risks

Scientists who support the development of genetically engineered foods view the process as simply an extension of previous plant-breeding techniques. Traditionally, farmers altered the genetic makeup of plants by crossbreeding different strains to combine the best traits of both plants. However, the direct manipulation of genes makes it possible to control genetic changes more precisely and efficiently.

▶ A scientist examines experimental samples of genetically modified fruit trees.

Biotechnologists say that their new products are as safe for consumers as plants developed through crossbreeding. Why shouldn't genetically engineered foods sit beside other foods on grocery store shelves?

The benefits of creating genetically engineered fruits and vegetables include keeping produce fresh longer, adding nutrients, and creating more-successful crops. For example, by inserting a gene that gives virus resistance to squash plants, scientists could boost the plants' resistance to viral infection. These resistant squash could produce five times the amount of squash per harvest as other squash does. Or scientists could increase the amino acids in a food product to give it more nutritional value.

Crops could be developed to grow faster and have higher yields. To combat world hunger, scientists may be able to develop seeds that can grow well in areas that have poor soil or poor water conditions. For more immediate relief, genetically engineered foods that would not spoil as quickly could be shipped to needy nations.

▶ The corn plants in this field have been genetically engineered to resist the effects of herbicides.

CROPLAN GENETICS

212

406

CHEMISTRY ———
• CONNECTION —— ADVANCED

Adding Nutritional Value? Genetic Modification (GM) techniques are also applied to alter the nutritional profiles of existing crops. Some of the techniques applied increase the content of vitamins, minerals and other micronutrients, modify fats and oils, alter starch and sugar content, alter the protein and amino acid profile, reduce levels of anti-nutritional or allergy factors, and enhance flavor. For example, Swiss researchers are working on "golden rice," which contains enhanced levels of beta-carotene and iron. This rice is intended to provide benefits to people in developing countries, where Vitamin A deficiency is a major cause of blindness.

The Risks Outweigh the Benefits

Critics of genetically engineered foods believe that these products are significantly different from foods developed through traditional methods. Genetic engineering allows genes from any living organism, including genes from animals or bacteria, to be placed into crops. Opponents are concerned about the safety of foods that contain these "foreign" genes.

Another safety concern is the possibility of allergic reactions. Some foods, such as peanuts and shellfish, cause allergic reactions in many people. If genes from these foods are placed in an entirely different product, people who eat these new products and do not know the products contain the foreign genes may have allergic reactions.

Other critics object because of religious or ethical reasons. Certain

▶ This farmer from Oaxaca, Mexico, holds up ears of traditional corn varieties. Some people fear that genes from genetically engineered varieties could accidentally be introduced to native varieties.

religions prohibit eating pork and other foods. People may object to the insertion of genes from pigs or other prohibited foods into foods they normally eat. Similarly, vegetarians might object to eating foods containing animal genes.

Some scientists are concerned that genetically engineered plant species may be accidentally introduced into the wild. Genetic engineering may give a new species an advantage over an existing wild species. If the new species thrives at the expense of the wild species, the wild species could become extinct.

▶ These people in Montreal, Quebec, are protesting the importation of genetically modified organisms (GMOs). Many countries have not accepted genetically engineered crops as much as the United States has.

What Do You Think?

Some people propose that genetically engineered foods should have labels that identify them as such. Could such a measure decrease criticism about the safety of genetically engineered foods? Based on what you have read, decide whether you would buy genetically engineered foods at the grocery store. Explain your reasoning.

This pit in Brazil is one of the world's largest iron ore mines. Mineral and energy resources are essential to human societies, but extracting and using these resources has environmental consequences.

Mining and Mineral Resources
Chapter Planning Guide

PACING	CLASSROOM RESOURCES	LABS, DEMONSTRATIONS, AND ACTIVITIES
BLOCKS 1 & 2 • 90 min pp. 410–414 **Chapter Opener**		
Section 1 Minerals and Mineral Resources	**OSP** Lesson Plan * **CRF** Active Reading * BASIC **TT** Bellringer * **TT** Minerals Used in the Lifetime of the Average U.S. Citizen * **TT** Common Elements and Their Ore Minerals * **TT** Environments of Mineral Formation * **CD** Interactive Tutor Mineral Identification **CD** Interactive Tutor Minerals in Our Daily Lives	**SE** Field Activity Identifying Objects Made of Minerals, p. 411 **TE** Demonstration Rocks and Minerals, p. 411 ◆ GENERAL **TE** Group Activity Unique Mineral Properties, p. 412 GENERAL **SE** Skills Practice Lab Extraction of Copper from Its Ore, pp. 430–431 ◆ **CRF** Datasheets for In-Text Labs * **CRF** Long-Term Project * ◆ ADVANCED **CRF** Design Your Own Lab * ◆ GENERAL **CRF** Consumer Lab * ◆ BASIC
BLOCKS 3 & 4 • 90 min pp. 415–420 **Section 2** Mineral Exploration and Mining	**OSP** Lesson Plan * **CRF** Active Reading * BASIC **TT** Bellringer *	**TE** Group Activity Mining Chocolate Minerals, p. 416 ◆ BASIC **TE** Activity Calculating the Percentage Composition of an Ore, p. 417 ◆ ADVANCED **SE** QuickLab Surface Coal Mining, p. 420 **CRF** Modeling Lab * ◆ GENERAL
BLOCKS 5, 6 & 7 • 135 min pp. 421–425 **Section 3** Mining Regulations and Mine Reclamation	**OSP** Lesson Plan * **CRF** Active Reading * BASIC **TT** Bellringer *	**TE** Group Activity Measuring the Impact of a Mine, p. 422 GENERAL **TE** Internet Activity Bioremediation, p. 422 GENERAL **TE** Activity Obtaining Permits for Mining, p. 424 ADVANCED **CRF** Observation Lab * ◆ GENERAL

BLOCKS 8 & 9 • 90 min

Chapter Review and Assessment Resources

- **SE** Chapter Review pp. 427–429
- **SE** Standardized Test Prep pp. 666–667
- **CRF** Study Guide * ■ GENERAL
- **CRF** Chapter Test * ■ GENERAL
- **CRF** Chapter Test * ADVANCED
- **CRF** Concept Review * ■ GENERAL
- **CRF** Critical Thinking * ADVANCED
- **OSP** Test Generator
- **CRF** Test Item Listing *

Online and Technology Resources

 Holt Online Learning

Visit **go.hrw.com** for access to Holt Online Learning or enter the keyword **HE6 Home** for a variety of free online resources.

 **One-Stop Planner® CD-ROM**

This CD-ROM package includes
- Lab Materials QuickList Software
- Holt Calendar Planner
- Customizable Lesson Plans
- Printable Worksheets
- ExamView® Test Generator
- Interactive Teacher Edition
- Holt PuzzlePro® Resources
- Holt PowerPoint® Resources

 Holt Environmental Science Interactive Tutor CD-ROM

This CD-ROM consists of interactive activities that give students a fun way to extend their knowledge of environmental science concepts.

KEY

TE	Teacher Edition	CRF	Chapter Resource File
SE	Student Edition	TT	Teaching Transparency
OSP	One-Stop Planner	CD	Interactive CD-ROM

* Also on One-Stop Planner
■ Also Available in Spanish
♦ Requires Advance Prep

ENRICHMENT AND SKILLS PRACTICE	SECTION REVIEW AND ASSESSMENT	CORRELATIONS
SE **Pre-Reading Activity,** p. 410 TE **Using the Figure,** p. 410 `GENERAL`		**National Science Education Standards**
TE **Skill Builder** Vocabulary, p. 411 `GENERAL` TE **Skill Builder** Vocabulary, p. 413 `GENERAL` TE **Student Opportunities** Rockhounds, p. 413 TE **Inclusion Strategies,** p. 413	SE **Section Review,** p. 414 TE **Reteaching,** p. 414 `BASIC` TE **Quiz,** p. 414 `GENERAL` TE **Alternative Assessment,** p. 414 `GENERAL` CRF **Quiz** * ■ `GENERAL`	SPSP 3a
SE **Case Study** Hydraulic Mining in the California Goldfields, pp. 418–419	SE **Section Review,** p. 420 TE **Reteaching,** p. 420 `BASIC` TE **Quiz,** p. 420 `GENERAL` TE **Alternative Assessment,** p. 420 `BASIC` CRF **Quiz** * ■ `GENERAL`	SPSP 3b
TE **Inclusion Strategies,** p. 421 SE **MathPractice** Volume, p. 423 SE **Graphic Organizer** Spider Map, p. 424 `GENERAL` SE **Maps in Action** Mineral Production in the United States, p. 432 SE **Society & the Environment** Coltan and the War in the Congo, p. 433 CRF **Map Skills** South Africa's Minerals * `GENERAL`	TE **Homework,** p. 424 `GENERAL` SE **Section Review,** p. 425 TE **Reteaching,** p. 425 `BASIC` TE **Quiz,** p. 425 `GENERAL` TE **Alternative Assessment,** p. 425 `GENERAL` CRF **Quiz** * ■ `GENERAL`	SPSP 3b SPSP 4b SPSP 4c SPSP 5b

Guided Reading Audio CDs

These CDs are designed to help auditory learners and reluctant readers. (Audio CDs are also available in Spanish.)

NSTA

www.scilinks.org

Maintained by the **National Science Teachers Association**

TOPIC: Minerals
SciLinks code: HE4067

TOPIC: Mining Minerals
SciLinks code: HE4069

TOPIC: Mining Reclamation
SciLinks code: HE4070

CNN Videos

Each video segment is accompanied by a Critical Thinking Worksheet.

Earth Science Connections
Segment 14 Yellowstone Hot Water

CHAPTER

16

Chapter Enrichment

This Chapter Enrichment provides relevant and interesting information to expand and enhance your classroom instruction of the chapter material.

SECTION 1 Minerals and Mineral Resources

► Gold in quartz matrix

Indian Turquoise Mines

Archeological evidence suggests that Native Americans were mining turquoise in the American Southwest by the 10th century. One ancient turquoise mine is in the Cerrillos Hills southwest of Santa Fe, New Mexico. Tools recovered from prehistoric turquoise mines include picks, hammers, grooved axes, and mauls. These tools were made from igneous rock, quartz, and quartzite. In order to shape the turquoise, Native Americans would heat the stone and then cool it quickly by quenching it in water. This would cause the stone to shatter, making it easier to work using lapidary stones and other tools. The Pueblo Indians mined turquoise in the Cerrillos Hills for hundreds of years. The turquoise was used for ornamentation and as an important trade item.

BRAIN FOOD

Zeolites are known for the ability to adhere thin layers of gas and water molecules to their surface and to release the water molecule coating. These characteristics make them useful for environmental pollution control. Zeolite minerals are particularly useful in the removal of metals from mine wastewater, radioactive waste treatment, air pollution control, aquaculture (fish and shrimp farming), gas purification and separation, and heat storage.

SECTION 2 Mineral Exploration and Mining

Mining Myths

Myths of lost mines and secret stashes of gold have become a part of American mining lore. One of the most famous of these myths is the tale of the Lost Dutchman Mine. Jacob Waltz, a Prussian immigrant, moved to California during the Gold Rush of 1849. He did not strike it rich, and during the 1870s, Waltz moved to Arizona to continue his search for gold. He and his partner, Jacob Weiser, allegedly found the lost Peralta mine in the Superstition Mountains with the help of a Peralta descendant. They purportedly protected their wealth by burying caches of gold in the vicinity of the rock formation known as Weaver's Needle. Waltz died in 1891 in apparent poverty, but his caretaker, Julia Thomas, reportedly quit her ice-cream business and went in search of the gold Waltz had buried years before. Thus, the legend of the Lost Dutchman Mine began, and the myth grew over the decades. The Lost Dutchman State Park was established in the Superstition Mountains to commemorate this myth. However, the park was closed in July 2002 as a result of budget cuts.

► A rotating shearer in use at a longwall coal mine

Mining Biotechnology

Bacteria are used in many mining applications. A common application is bioleaching, also known as biohydro-metallurgy (mining with living organisms and water). In the bioleaching process, ore is mounded in a safe pond area, and bacteria such as *Thiobacillus ferrooxidans* and *Leptospirillium ferrooxidans* are introduced to remove metal from metal sulfides. Although further studies are being performed to determine the environmental effects of bioleaching, the method currently appears to be a more environmentally safe and less expensive solution than cyanide heap leaching.

Robotics in Mining

Engineers are currently developing automated mining equipment that will be used in surface and subsurface mining environments. This new technology promises to improve productivity and at the same time increase mine safety. An example of this technology is a fully automated robotic excavator that will be used to load trucks with soil and other materials at surface mines and quarries. The excavator was developed at the Robotics Institute at Carnegie Mellon University in Pittsburgh. A pair of scanning laser range finders that work independently are mounted atop the excavator. One range finder is used to recognize, localize, and measure the dimensions of a truck. On-board software processes this information. The dump point in the truck is planned, and the bucket swings toward the truck. At the same time, the other range finder measures the soil on the dig face, and

the next location to dig is calculated. After the material is dumped into the truck, the first range finder scans the dig face for any obstacles—including humans—that may be in the way of the excavator. If an obstacle is recognized, the machine's motion is stopped. If not, the excavator continues to swing to the next dig site. It takes six 15- to 20-second passes to load one 22-ton truck.

SECTION 3 Mining Regulations and Mine Reclamation

Remedies for Acid Mine Drainage

Several methods for addressing and mitigating the problem of acid mine drainage (AMD) exist. Chemical treatments known as active systems are costly and present a long-term problem. Passive systems are being investigated to determine their ability to remove metals and reduce acidity from AMD. Some common features of passive systems are the circulation of AMD water through a semi-natural environment (artificial streams and wetlands), bacteria-rich areas, and the use of limestone to reduce acidity.

▶ Almost 3,000 kg of diamonds have been removed from this mine in South Africa

Overview

Tell students that this chapter discusses the economic importance of minerals, surface and subsurface mining techniques, the environmental ramifications of mining, and the laws regulating mining and mine reclamation. Point out that the information in this chapter directly affects students, because they use minerals in their everyday life.

Using the Figure — GENERAL

The photograph shows a large open-pit copper mine in Arizona. In open-pit mining, the ore is mined downward in layers. The stair-step excavation of the walls helps keep the sides of the mine from caving. The Bingham Canyon open-pit mine in Utah is one of only a few human-made objects that can be seen from outer space. Have students research the Bingham Canyon Mine on the Internet and report their findings to the class.
LS Visual/Verbal

PRE-READING ACTIVITY

FOLDNOTES
You may want to use this FoldNote in a classroom discussion to review material from the chapter. On the board, write each category from the Three-Panel Flip Chart. Then, ask students to provide information for each category. Under the appropriate category on the board, write the information that students provide.

VIDEO SELECT

For information about videos related to this chapter, go to **go.hrw.com** and type in the keyword **HE4 MMRV**.

Mining and Mineral Resources

CHAPTER 16

1 **Minerals and Mineral Resources**

2 **Mineral Exploration and Mining**

3 **Mining Regulations and Mine Reclamation**

PRE-READING ACTIVITY

FOLDNOTES
Three-Panel Flip Chart
Before you read this chapter, create the **FoldNote** entitled "Three-Panel Flip Chart" described in the Reading and Study Skills section of the Appendix. Label the flaps of the chart with "Mineral Resources," "Mining," and "Mining Regulations and Reclamation." As you read the chapter, write information you learn about each category under the appropriate panel.

Obtaining the minerals humans need requires removing large amounts of rock at mines like this open-pit copper mine in Arizona. The challenge is to satisfy the increasing demand for mineral resources while minimizing the cost to the environment.

410

Chapter Correlations | *National Science Education Standards*

SPSP 3a Human populations use resources in the environment in order to maintain and improve their existence. Natural resources have been and will continue to be used to maintain human populations. **(Section 1)**

SPSP 3b The earth does not have infinite resources; increasing human consumption places severe stress on the natural processes that renew some resources, and it depletes those resources that cannot be renewed. **(Section 2 and Section 3)**

SPSP 4b Materials from human societies affect both physical and chemical cycles of the earth. **(Section 3)**

SPSP 4c Many factors influence environmental quality. Factors that students might investigate include population growth, resource use, population distribution, overconsumption, the capacity of technology to solve problems, poverty, the role of economic, political, and religious views, and different ways humans view the earth. **(Section 3)**

SPSP 5b Human activities can enhance potential for hazards. Acquisition of resources, urban growth, and waste disposal can accelerate rates of natural change. **(Section 3)**

Minerals and Mineral Resources

Take a look at the human-made objects that surround you. As you may notice, almost every solid object you see is made of minerals. As shown in **Figure 1**, we depend on the use of mineral resources in almost every aspect of our daily lives. However, our dependence on minerals has not come without a price. The current challenge is to obtain the minerals that an ever-increasing world population demands at minimal cost to the environment. In this chapter you will learn about how we use our mineral resources and how we deal with the environmental effects of mining.

What Is a Mineral?

A **mineral** is a naturally occurring, usually inorganic solid that has a characteristic chemical composition, an orderly internal structure, and a characteristic set of physical properties. Minerals are made up of atoms of a single element, or of *compounds*—atoms of two or more elements chemically bonded together. The atoms that make up minerals are arranged in regular, repeating geometric patterns. The arrangement of the atoms, along with the strength of the chemical bonds between them, determine the physical properties of minerals.

The elements gold, silver, and copper are considered minerals. These types of minerals are called *native elements*. However, most minerals are compounds. For example, the mineral quartz is made up of silica, a compound consisting of one silicon atom and two oxygen atoms. When combined with other elements, silica forms most of the minerals that make up Earth's crust.

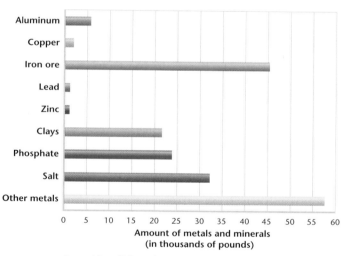

Amount of metals and minerals
(in thousands of pounds)

Source: Mineral Information Institute.

Objectives

▶ Define the term *mineral*.
▶ Explain the difference between a metal and a nonmetal, and give two examples of each.
▶ Describe three processes by which ore minerals form.

Key Terms

mineral
ore mineral

FIELD ACTIVITY

Identifying Objects Made of Minerals Take a walk around your neighborhood or through your home with a notebook and pencil. Pick an object such as a car, an appliance, or a computer. List as many materials that make up that object as you can. Be as specific as possible. Repeat the procedure for several other objects. Which of these objects do you think are made from minerals? Write to the company that made one of your objects, and ask what materials are used to make the object. Record your observations, along with the company's response, in your *EcoLog*.

Figure 1 ▶ Mineral consumption is greatest in developed countries, such as the United States. This graph shows the average amount of minerals a person in the U.S. will consume over his or her lifetime.

411

Teach

Group Activity — GENERAL

Unique Mineral Properties

Divide the class into four groups, and give each group a tray of minerals (for example, halite, galena, sulfur, optical grade calcite, crystalline quartz, hematite, gypsum, magnetite, and muscovite). Give each group tools to identify each mineral (a nail, an unglazed white tile, a magnet, and a hand lens) and a list of the minerals and their identifying characteristics. Ask students to look at the list and identify the minerals in the tray using the following characteristics: halite—cubic crystal structure, water soluble; galena—metallic luster with a high density; sulfur—smells like rotten eggs when scratched; calcite—creates a double image when placed over a dot on a piece of paper; quartz—six-sided prism not scratched by the nail; hematite—red streak when rubbed on the tile; gypsum—looks like calcite but doesn't have the optical property and can be scratched with a fingernail; magnetite—magnetic; and muscovite—peels apart in thin sheets. Ask students to list their findings, along with a description of what the mineral looks like, on a sheet of paper and include it in their *EcoLog*

LS Kinesthetic Co-op Learning

Transparencies

TT Common Elements and Their Ore Minerals

Figure 2 ▶ Certain minerals are mined because of the valuable metals they contain, as shown in the table. Wulfenite (above) is a minor ore of lead. Nice specimens of wulfenite are much sought after by mineral collectors.

internet connect

www.scilinks.org
Topic: Minerals
SciLinks code: HE4067

SCILINKS. Maintained by the National Science Teachers Association

Figure 3 ▶ Gold is one of the most economically important metallic minerals.

412

Table 1 ▼

Common Elements and Their Ore Minerals	
Element	**Important ore minerals**
Aluminum (Al)	gibbsite, boehmite, diaspore (bauxite)
Beryllium (Be)	beryl
Chromium (Cr)	chromite
Copper (Cu)	bornite, cuprite, chalcocite, chalcopyrite
Iron (Fe)	goethite, hematite, magnetite, siderite
Lead (Pb)	galena
Manganese (Mn)	psilomelane, pyrolusite
Mercury (Hg)	cinnabar
Molybdenum (Mo)	molybdenite
Nickel (Ni)	pentlandite
Silver (Ag)	acanthite
Tin (Sn)	cassiterite
Titanium (Ti)	ilmenite, rutile
Uranium (U)	carnotite, uraninite
Zinc (Zn)	sphalerite

Ore Minerals

Minerals that are valuable and economical to extract are known as **ore minerals.** As shown in **Table 1,** ore minerals contain elements, many of which are economically valuable. During the mining process, ore minerals, along with minerals that have no commercial value, or *gangue* (GANG) *minerals*, are extracted from the host rock. After extraction, mining companies use various methods to separate ore minerals from the gangue minerals. The ore minerals are then further refined to extract the valuable elements they contain. For mining to be profitable, the price of the final product must be greater than the costs of extraction and refining.

Metallic Minerals Ore minerals are either metallic or nonmetallic. Metals conduct electricity, have shiny surfaces, and are opaque. Many valuable metallic minerals are native elements such as gold, shown in **Figure 3.** Silver and copper are also important native elements. Other important ore minerals are compounds in which metallic elements combine with nonmetallic elements, such as sulfur or oxygen.

Nonmetallic Minerals Nonmetals tend to be good insulators, may have shiny or dull surfaces, and may allow light to pass through them. Nonmetallic minerals can also be native elements or compounds.

FINE ARTS
CONNECTION — GENERAL

Coloring the World Minerals are used as pigments in many of the fine arts, including pottery, photography, and glass blowing. Potters use glazes that, when fired, form a colorful glass-like surface to coat their wares. Glass blowers also use minerals to color glass. Dale Chihuly, a modern glass artist based in Seattle, Washington, has about 300 different colors of glass he works with to create pieces of art. Some minerals used as pigments include azurite and lapis—blue; malachite—green; hematite—red, orange, or black; and gold—cranberry. Have students research and prepare a report on an art technique that involves minerals. **LS** Logical

---

Placer Mining

When rock weathers and disintegrates, minerals within the rock are released. These minerals are concentrated by wind and water into surface deposits called **placer deposits.** The most important placers are stream placers. Streams transport mineral grains to a point where they fall to the streambed and are concentrated. Concentration occurs at places where currents are weak and the dense mineral grains can no longer be carried in the water. These stream placers often occur at bends in rivers, where the current slows.

Placer deposits may form along coastlines by heavy minerals that wash down to the ocean in streams. These heavy minerals are concentrated by wave action.

Placer gold, diamonds, and other heavy minerals are mined by dredging. As shown in **Figure 12**, a dredge consists of a floating barge on which buckets fixed on a conveyor are used to excavate sediments in front of the dredge. Gold, diamonds, or heavy minerals are separated from the sediments within the dredge housing. The processed sediments are discharged via a conveyor that is located behind the dredge.

Figure 12 ▶ This dredge is mining gold from placer deposits along a river on New Zealand's South Island.

▶ To this day, the mountainsides in the Sierra Nevada bear the scars of hydraulic mining.

troughs called *sluices*. The sluices were lined with a series of devices known as *riffles* to catch the gold. Mercury was also added to the riffles to help capture the gold. The muddy water and processed sediments were then discharged into adjacent rivers.

Hydraulic mining proved to be an environmental disaster. Muddy water and sediments polluted rivers and caused them to fill with silt. The silt from the hydraulic mines traveled as far downstream as San Francisco and into the Pacific Ocean. As much as 1.4 to 3.6 million kilograms of mercury may have been released downstream, which poisoned fish, amphibians, and invertebrates. Farmers in California's central valley sustained millions of dollars in damage as their fields were flooded when the sediment-choked Sacramento

River overflowed its banks. But the farmers fought back. In January 1884, Judge Lorenzo Sawyer ruled that mine tailings could no longer be discharged into the rivers. The Sawyer decision was the first environmental ruling to be handed down in the United States. This ruling closed the door on hydraulic mining in the Sierra Nevada goldfields, where 2 billion cubic meters of soil and rocks had been carved from the mountainsides in slightly over 30 years.

CRITICAL THINKING

1. Making Inferences What do you think were other environmental effects of hydraulic mining that were not mentioned in this article?

2. Analyzing Relationships Write a paragraph about how the mercury that was lost during hydraulic mining may still be affecting the environment today. **WRITING SKILLS**

419

QuickLAB

Surface Coal Mining

Procedure

1. Cut off the top part of a **2 L plastic soda bottle** to make a container that has an open end.
2. Spread a 5 cm layer of **soil** on the bottom of the bottle.
3. Spread a 0.75 cm layer of **rice** on top of the soil to represent a coal seam.
4. Spread a 12.5 cm layer of soil on top of the coal.
5. To excavate the coal, dig out the top layer of soil with a **spoon**, and place it in a **bowl**. Measure the volume of this soil by using a **graduated cylinder**. Record the volume.
6. Dig out the layer of coal, and place it in a second bowl. Measure and record the volume.

Analysis

1. What is the ratio of overburden to coal?
2. What are some factors that you would need to consider if you were going to surface-mine coal economically?

Figure 13 ▶ At a smelter, such as this aluminum smelter in Venezuela, ore is melted at high temperatures in a furnace to obtain a desired metal.

Smelting

In the process called **smelting**, crushed ore is melted at high temperatures in furnaces to separate impurities from molten metal. In the furnace, material called a *flux* bonds with impurities and separates them from the molten metal. The molten metal, which is desired, falls to the bottom of the furnace and is recovered. The flux and impurities, which are less dense, form a layer called *slag* on top of the molten metal. Gases such as sulfur dioxide form within the furnace and are captured, so they do not enter the environment. **Figure 13** shows a smelter in Venezuela.

Undersea Mining

The ocean floor contains significant mineral resources, which include diamonds, precious metals such as gold and silver, mineral ores, and sand and gravel. Since the late 1950s, several attempts have been made to mine the ocean. These attempts met with varying degrees of success. Competition with land-based companies that can mine minerals more cheaply and the great water depths at which some mineral deposits are found are two of the reasons undersea mining has been largely unsuccessful to date.

SECTION 2 Review

1. **List** the steps in mineral exploration.
2. **Describe** three methods of subsurface mining.
3. **Describe** two methods of surface mining.
4. **Describe** the steps involved in smelting ore.
5. **Define** the term placer deposit, and explain how placer deposits form.

CRITICAL THINKING

6. **Making Comparisons** Read about surface and subsurface mining techniques. What are some of the advantages and disadvantages of each technique?
   READING SKILLS
7. **Understanding Relationships** If a mining company were exploring a river for potential placer deposits, where are some likely places they would focus their exploration?

420

SECTION 3
Mining Regulations and Mine Reclamation

With an increase in U.S. energy requirements, particularly the demand for coal to fuel power plants that produce electricity, the scale of surface mining has grown. So, too, have the potential environmental effects of mining. For example, surface coal mining requires the removal of enormous amounts of soil and rock to reach near-surface coal seams.

Because of the potential environmental impacts of mining on such a large scale, mining has become one of the most heavily regulated industries in the United States. Mining companies now spend large amounts of money to preserve the environment. Reclaiming the land, or returning land to its original condition after mining is completed, is part of every surface coal mining operation. Before mining, companies develop a plan to reclaim the land. Even before mining is complete, this plan is put into action. With environmental preservation now a clear goal of mining companies, future generations of Americans will not have to view scars in the Earth, such as the one shown in **Figure 14.**

The Environmental Impacts of Mining

There are many potential environmental impacts of mining. In the United States, the federal and state governments and mining companies are spending billions of dollars to clean up abandoned mines. Much of this legacy has been left to U.S. citizens from a time when there was little regulation of mining and mineral processing.

Objectives

▶ Describe seven important potential environmental consequences of mining.

▶ Name four federal laws that relate to mining and reclaiming mined land.

▶ Define the term *reclamation*.

▶ Describe two ways in which state governments regulate mining.

Key Terms

subsidence
reclamation

Figure 14 ▶ At 215 m deep and 1.6 km in circumference, the "Big Hole" at the Kimberley Mine in South Africa is the largest hand-dug excavation in the world. By the time the mine closed in 1914, 22.5 million tons of rock had yielded almost 3,000 kg of diamonds.

421

Group Activity — GENERAL

Measuring the Impact of a Mine
Tell students that mining companies and environmental groups often work together to develop a plan that considers how a mine will affect the environment and how the area will be returned to its original condition. Have students work in groups to develop a list of issues that the two groups would need to discuss before a mining project could begin. (Sample answers: air and noise pollution, water protection, and minimizing the impact on wildlife) For each item on their list, have groups determine how the item would be measured. Have students consider the following questions: "What kinds of measurements would be taken to determine environmental impact? When would measurements need to be made?" (Students' responses should demonstrate an understanding that measurements will need to be made before, during, and after mining.)
LS Logical/Interpersonal

internet connect
www.scilinks.org
Topic: Mining Reclamation
SciLinks code: HE4070
SCiLINKS. Maintained by the National Science Teachers Association

Air and Noise Pollution Surface mining can cause both air pollution and noise pollution. At surface coal mines, removing, loading, hauling, and dumping soil and overburden produce dust. Wind that blows across unreclaimed soil and overburden storage areas also adds dust to the air. Loading, hauling, and unloading ore produce dust emissions at open-pit mines. Dust is also generated in open-pit mines when the ore is blasted apart with explosives.

Noise is created by the equipment that is used in a mine as well as by blasting. Whereas equipment noise may be a nuisance, blasting can cause physical damage to structures that are located near the mine.

Because of air and noise pollution, most surface mines are not located near urban populations. More important, regulations in the U.S. forbid mining operations to allow dust or noise to exit the area that is being mined.

Water Contamination Water resources can be negatively impacted by mining. Water that seeps into mines or through piles of excess rock can pick up or dissolve toxic substances like arsenic. These contaminants can wash into streams, where they can harm or kill aquatic life.

Coal or minerals that contain a lot of sulfur can cause a similar problem. When these materials react with oxygen and water, they form a dilute sulfuric acid. This acid can dissolve toxic minerals that remain in mines and excess rock. The contaminated water that results from this process is known as *acid mine drainage*, or AMD, which is shown in **Figure 15**. Mining regulation in the U.S. requires companies to dispose of acid-producing rock in such a way that water is not contaminated.

Figure 15 ▶ Copper mines have polluted the Queen River in Tasmania with acid mine drainage. This photo shows the river flowing past residential housing.

Displacement of Wildlife Removing soil from a surface mine site strips away all plant life. With their natural habitat removed, animals will leave the area. In addition, when mining is completed and the soil is returned to the mine site, different plants and animals may establish themselves, which creates an entirely new ecosystem. However, a good development plan to reclaim a mine site can ensure that the displacement of wildlife is merely temporary.

Dredging can negatively affect aquatic ecosystems and physically change the bottoms of rivers. Dredging disturbs river bottoms and destroys aquatic plant life in the dredged portion of the river. The disturbance of a riverbed can cause muddy sediments to contaminate a river for up to 10 km.

422

Internet Activity — GENERAL

Bioremediation Bioremediation uses living organisms to reduce or eliminate hazards from toxic chemicals or wastes. Have students search the Internet for articles that discuss possible ways in which bacteria may be used to remediate toxic mine waste. For example, researchers at the University of Missouri–Columbia have found a bacterium that can remove uranium from water. Have students report their findings to the class. **LS** Logical

GEOGRAPHY
CONNECTION — ADVANCED

GIS and Mining Geographic information systems (GIS) are an integral tool in the mining industry. Applications of GIS include identifying landownership and mineral claims, managing mine exploration and production, and determining evacuation routes and safety chamber sites in case of emergency. Ask students to research examples of GIS use in mining and write their findings in the form of a newspaper article. Have them include the article in their **Portfolio**. **LS** Verbal

Erosion and Sedimentation Excess rock from mines is sometimes dumped into large piles called *dumps*. Running water erodes unprotected dumps and transports sediments into nearby streams. These sediments may choke streams and damage water quality and aquatic life.

Soil Degradation Soil at a mine site is removed from the uppermost layer downward. When this soil is stored for later reuse, care must be taken to ensure that the upper soil layers are not buried beneath soil layers that were originally below them. In this way, the soil layers that are richest in important nutrients are not covered. If soil is not removed and stored in separate layers, the soil may be nutrient poor when it is reclaimed.

Minerals that contain sulfur may be found in deeper soil layers. If these minerals are exposed to water and oxygen in the atmosphere, chemical reactions result in the release of acid, which then acidifies the soil. When mining is completed and the soil is returned to the mine site, it may be difficult for plants to establish themselves if soil is acidified.

Subsidence The sinking of regions of the ground with little or no horizontal movement is called **subsidence** (suhb SIED'ns). Subsidence occurs when pillars that have been left standing in mines collapse or the mine roof or floor fails.

The locations of many abandoned mines are unknown. Buildings, houses, roads, bridges, underground pipelines, and utilities that are built over these mines could be damaged if the ground below them subsides. In November and December 2000, underground limestone mines that were several hundred years old collapsed in Edinburgh, Scotland. The collapse caused property damage and forced people to evacuate their homes. **Figure 16** shows the potential effects of mine subsidence.

Figure 16 ▶ A hole created by the subsidence of a gold mine has swallowed this house in New Zealand.

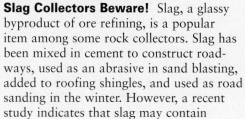

BRAIN FOOD

Slag Collectors Beware! Slag, a glassy byproduct of ore refining, is a popular item among some rock collectors. Slag has been mixed in cement to construct roadways, used as an abrasive in sand blasting, added to roofing shingles, and used as road sanding in the winter. However, a recent study indicates that slag may contain significant amounts of toxic materials such as lead, cadmium, barium, and arsenic. Research indicates these toxic substances may be released into the environment through weathering processes and could contaminate soils, surface water, and groundwater.

Teach, continued

Activity — ADVANCED
Obtaining Permits for Mining
Organize the students into groups, and provide them with copies of a guide to the Pennsylvania permitting process (this can be found on the Web site for the Pennsylvania Department of Environmental Protection). Each student group should represent a mining company that wants to excavate a subsurface coal mine in Pennsylvania. Have each team write up a brief mining proposal. (Sample answer: The mine will be dug in southeastern Bucks County, the coal refuse will be placed in PVC-lined shallow pits, the land will be reclaimed by filling the mine with tailings, and the area will be covered with soil.) Have students list the activities that require permits. (Sample answers: water discharge, underground mining of coal, and disposing of coal refuse) Assume the application is complete and the public has been notified. Then, tell groups that the public has a concern about the mine project (such as AMD, endangered species, or highway encroachments) and they need to write a brief response before permitting can continue. Once this is done, assume the process is complete and the permits are ready to be reviewed with a request to post bond. Have each group review and decide on the other group's proposal. **LS** Logical Co-op Learning

Ecofact
Bats and Mines Over the past century, human disturbance of traditional bat roosting sites, such as caves and trees, has caused bats to move into abandoned mines. At present, 30 of the 45 species of bats in the United States live in mines. Some of the largest populations of endangered bat species now live in abandoned mines.

Graphic Organizer
Spider Map
Create the **Graphic Organizer** entitled "Spider Map" described in the Appendix. Label the circle "Mining Regulations." Create a leg for each type of mining regulation. Then, fill in the map with details about each type of mining regulation.

Underground Mine Fires Fires that start in underground coal seams are one of the most serious environmental consequences of coal mining. Lightning, forest fires, and burning trash can all cause coal-seam fires. In addition, fires can start by themselves when minerals in the coal that contain sulfur are exposed to oxygen. These fires are hard to put out and are often left to burn themselves out, which may take decades or even centuries. For example, a fire that has been burning through an underground coal seam in an Australian mountain is estimated to be 2,000 years old! Underground fires that burn their way to the surface release smoke and gases that can cause respiratory problems. A fire in a coal seam is shown in **Figure 17**.

Mining Regulation and Reclamation
Mines on land in the United States are regulated by federal and state laws. To ensure that contaminants from mines do not threaten water quality, mining companies must comply with regulations of the Clean Water Act and the Safe Drinking Water Act. The release of hazardous substances into the air, soil, and water by mining is regulated by the Comprehensive Response Compensation and Liability Act. In addition, all mining operations must comply with the Endangered Species Act. This act ensures that mining activities will not affect threatened or endangered species and their habitats.

Reclamation The process of returning land to its original or better condition after mining is completed is called **reclamation.** The Surface Mining Control and Reclamation Act of 1977 (SMCRA) created a program for the regulation of surface coal mining on public and private land. The act set standards that would minimize the surface effects of coal mining on the environment. SMCRA also established a fund that is administered by the federal government and is used to reclaim land and water resources that have been adversely affected by past coal-mining activities.

424

Graphic Organizer — GENERAL

Spider Map
You may want to have students work in groups to create this Spider Map. Have one student draw the map and fill in the information provided by other students from the group.

Homework — GENERAL

Underground Mine Fire The town of Centralia, Pennsylvania, lies over an abandoned coal mine. In 1962, the state decided to use the mine as a landfill. That summer the garbage caught fire and ignited the surrounding coal. Between 1962 and 1978, state and federal attempts to contain the fire cost an estimated $14.78 million (adjusted for inflation). By 1985, a grant was ceded to the state to purchase homes and businesses in Centralia. Today, only 20 residents remain in the once thriving town, and the fire continues to burn. For information on the Centralia fire, direct students to the Pennsylvania Department of Environmental Protection, Bureau of Deep Mine Safety Web site. Have students report on the fire in a news narrative format and incorporate images of the fire in their report. **LS** Verbal/Visual

State Regulation of Mining States have created programs to regulate mining on state and private lands. Mining companies must obtain permits from state environmental agencies before mining a site. These permits specify certain standards for mine design and reclamation. In addition, some states have bond forfeiture programs. In a bond forfeiture program, a mining company must post funds, called a *bond*, before a mining project begins. If the company does not mine and reclaim a site according to the standards required by its permits, the company must give these funds to the state. The state then uses the funds to reclaim the site. A reclaimed surface coal mine is shown in **Figure 18**.

State agencies are also responsible for inspecting mines to ensure compliance with environmental regulations. Agencies issue violations to companies that do not comply with environmental regulations and assess fines for noncompliance. In addition, states such as Pennsylvania have begun large projects to reclaim abandoned mine lands. Acid mine drainage, mine fires, mine subsidence, and hazards related to open shafts and abandoned mining structures are all problems that these projects will attempt to correct.

Connection to History

Jihlava Jihlava is an ancient town in the Czech Republic. In the 1200s, silver was discovered in Jihlava. The rush that followed brought miners, merchants, and traders from all over Europe. As a result, Jihlava became very prosperous. In addition to creating municipal laws, the town passed its own mining laws. Jihlava's mining laws served as an example for other mining towns in central Europe.

Figure 18 ▶ Reclamation often includes seeding, planting, and irrigating to return the land to its original state.

SECTION 3 Review

1. **List** seven potential environmental impacts of mining.

2. **Name** four federal laws that regulate mining activities in the United States.

3. **Define** the term *reclamation.*

4. **Describe** two ways in which state governments regulate mining.

CRITICAL THINKING

5. **Making Decisions** Give examples of environmental concerns that would be taken into account by a mining company when it created a reclamation plan for a mine site.

6. **Making Decisions** Read about how topsoil is removed and stored for later reclamation under the heading "Soil Degradation." How can this process be implemented to keep soils from degrading?
   READING SKILLS

425

Answers to Section Review

1. air and noise pollution, water contamination, wildlife displacement, erosion and sedimentation, soil degradation, subsidence, and underground mine fires

2. Sample answer: Four federal laws that regulate mining practices are the Clean Water Act, the Safe Drinking Water Act, the Comprehensive Response Compensation and Liability Act, and the Surface Mining Control and Reclamation Act of 1977.

3. Reclamation is the process of returning land to its original condition or better after mining operations are completed.

4. Sample answer: State governments regulate mining with permitting processes and by assessing fines for noncompliance.

5. Answers may vary.

6. Answers may vary.

Alternative Assessment —— GENERAL

Mining Information Brochure

Have students create a brochure that describes the history, environmental impact, and economic influence of mining in your state or region. Students should use the information in each section to describe the minerals that are mined in your region, what type of mining is used, and how this activity has affected the natural environment and the economic status of your region. Encourage students to use text and images in their brochures. For example, students could include images of the minerals that are mined, as well as pictures of the mines and of any areas that have been reclaimed. Students can describe the history of mining in the region and speculate about what the future may hold for your area. **LS** Verbal/Visual

Chapter Resource File

- **Chapter Test** GENERAL
- **Chapter Test** ADVANCED
- **Concept Review** GENERAL
- **Critical Thinking** ADVANCED
- **Test Item Listing**
- **Design Your Own Lab** GENERAL
- **Modeling Lab** GENERAL
- **Observation Lab** GENERAL
- **Consumer Lab** BASIC
- **Long-Term Project** ADVANCED

CHAPTER 16 Highlights

1 Mineral and Mineral Resources

Key Terms

mineral, 411
ore mineral, 412

Main Ideas

▶ A mineral is a naturally occurring, usually inorganic solid that has a characteristic chemical composition, an orderly physical structure, and a characteristic set of physical properties.

▶ Minerals that are valuable and economical to extract are known as *ore minerals*.

▶ Ore minerals may form from the cooling of magma, the circulation of hot-water solutions through rocks, and the evaporation of water that contains salts.

▶ Metals are important economically because of their electrical and thermal conductivity, durability, and heat and corrosion resistance.

2 Mineral Exploration and Mining

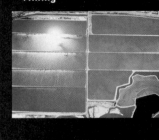

Key Terms

subsurface mining, 416
surface mining, 417
placer deposit, 419
smelting, 420

Main Ideas

▶ Mining companies conduct mineral exploration to identify areas where there is a high likelihood of finding valuable mineral resources in quantities worth mining.

▶ Room-and-pillar mining, longwall mining, and solution mining are subsurface mining methods.

▶ Open-pit mining, surface coal mining, quarrying, and solar evaporation are surface-mining methods.

▶ Minerals are concentrated by wind and water into surface deposits called *placer deposits*.

▶ Smelting is the process in which ore is melted at high temperatures to separate impurities from the molten metal.

3 Mining Regulations and Mine Reclamation

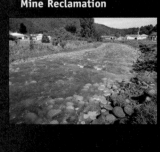

Key Terms

subsidence, 423
reclamation, 424

Main Ideas

▶ Some of the environmental consequences of mining may include air and noise pollution, water contamination, displacement of wildlife, erosion and sedimentation, soil degradation, subsidence, and underground mine fires.

▶ The U.S. government has enacted legislation that regulates mining and attempts to minimize the impact of mining on the environment.

▶ Federal and state agencies issue permits to mining companies, issue violations and assess penalties when mining companies do not comply with standards set by their permits, and ensure that abandoned mine lands are reclaimed.

Using Key Terms

Use each of the following terms in a separate sentence.

1. *mineral*
2. *placer deposit*
3. *smelting*
4. *subsidence*
5. *reclamation*

For each pair of terms, explain how the meanings of the terms differ.

6. *element* and *mineral*
7. *ore mineral* and *gangue mineral*
8. *placer deposit* and *dredging*
9. *subsurface mining* and *surface mining*

✓ STUDY TIP

Using Terms Work together with a study partner. Learn the definitions of both the boldfaced and italicized words that appear in this chapter. When both you and your partner feel confident in having learned the meanings of these terms, take out a piece of paper. On this paper, you and your partner will each write a one-page essay in which you use as many of these terms as possible. When you both are finished, exchange essays and review them for accuracy.

Understanding Key Ideas

10. Which of the following statements does *not* correctly describe a mineral?
    a. A mineral is a naturally occurring substance.
    b. A mineral is an organic substance.
    c. A mineral is a solid substance.
    d. A mineral has a characteristic chemical composition.

11. Gold, silver, and copper are
    a. nonmetallic minerals.
    b. native elements.
    c. compounds.
    d. gangue minerals.

12. Ore deposits form from
    a. the cooling of magma.
    b. the evaporation of water that contains salts.
    c. the circulation of hot-water solutions in rocks.
    d. All of the above

13. Which of the following economically important elements is *not* a metal?
    a. zinc
    b. titanium
    c. copper
    d. sulfur

14. Which of the following methods is *not* a subsurface mining method?
    a. quarrying
    b. solution mining
    c. longwall mining
    d. room-and-pillar mining

15. Which of the following mining methods would most likely be used to mine salt?
    a. solution mining
    b. open-pit mining
    c. solar evaporation
    d. both (a) and (c)

16. Dredging would *not* be used to mine
    a. diamonds.
    b. coal.
    c. heavy minerals.
    d. gold.

17. Which of the following elements in minerals causes soil to become acidified?
    a. potassium
    b. calcium
    c. sulfur
    d. barium

18. Which of the following pieces of federal legislation established a program for regulating coal mining on public and private lands?
    a. the Comprehensive Response and Liability Act
    b. the Clean Air Act
    c. the Clean Water Act
    d. the Surface Mining Control and Reclamation Act of 1977

427

Assignment Guide

Section	Questions
1	1, 6, 7, 10–13, 19, 25–26
2	2, 3, 8, 9, 14–16, 20–22, 27–30, 32, 33, 35
3	4, 5, 17, 18, 23, 24, 31, 34

ANSWERS

Using Key Terms

1. Sample answer: A mineral is a naturally occurring substance that comprises rocks and may have economic value.

2. Sample answer: Placer deposits are minerals that have been concentrated by wind or water.

3. Sample answer: A metal is removed from an ore by smelting.

4. Sample answer: Subsidence happens when mines collapse, causing the ground above to sink.

5. Sample answer: Mining companies must successfully use reclamation techniques during mining and after they close a mine in order to get their bond back from the state.

6. An element may be any of the substances listed in the periodic table; elements are composed of only a single type of atom and may be solid, liquid, or gas. A mineral is a solid that may be composed of atoms of a single element or of compounds.

7. An ore mineral is a mineral that contains an element or elements of economic value, and a gangue mineral is a mineral that has no economic importance.

8. A placer deposit is a deposit of wind- or water-concentrated minerals that is mined by a technique called *dredging*.

9. Subsurface mining occurs underground, and surface mining occurs on Earth's surface.

Understanding Key Ideas

10. b
11. b
12. d
13. d
14. a
15. d
16. b
17. c
18. d

Short Answer

19. Native elements are composed of one element, whereas compounds are made up of more than one element.

20. During solar evaporation, the sun's energy evaporates the water and leaves behind the concentrated solids.

21. Methods of subsurface mining are room-and-pillar mining and longwall mining. Surface mining methods are open-pit mining and surface coal mining.

22. Undersea mining has not been very profitable because water depths can be great and the cost is prohibitive.

23. Reclaimed soil can degrade if deeper soil layers are placed over shallower layers and if the deeper layers contain sulfur minerals. In sulfur-rich soil, exposure to water and oxygen can acidify the soil.

24. A state bond forfeiture program is meant to encourage proper reclamation techniques and to provide funds for pollution cleanup if mine reclamation fails.

Interpreting Graphics

25. ($14 billion in metals/$39 billion total mineral production) × 100 = 35.9% of the total U.S. production

26. ($10 billion in metals/$39 billion total mineral production) × 100 = 25.6% of the total U.S. production

Concept Mapping

27. Answers to the concept mapping questions are on pp. 667–672.

Short Answer

19. What is the difference between native elements and compounds?

20. Describe the solar evaporation process.

21. What are the surface and subsurface methods by which coal is commonly mined?

22. Explain why undersea mining has been largely unsuccessful to date.

23. Describe how reclaimed soil may become degraded.

24. Explain the purpose of a state bond forfeiture program.

Interpreting Graphics

The graph below shows total U.S. mineral production from 1995 to 1999. Use the graph to answer questions 25–26.

25. In 1995, metals accounted for $14 billion of the $39 billion total U.S. production of minerals. Metals accounted for what percentage of total U.S. production of minerals?

26. In 1999, metals accounted for only $10 billion of the $39 billion total U.S. production of minerals. Metals accounted for what percentage of total U.S. production of minerals?

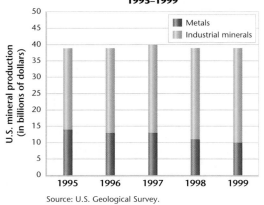

Changes in U.S. Mineral Production 1995–1999

Source: U.S. Geological Survey.

Concept Mapping

27. Use the following terms to create a concept map: *subsurface mining, surface mining, room-and-pillar mining, longwall mining, solution mining, open-pit mining, surface coal mining,* and *quarrying.*

Critical Thinking

28. **Analyzing Relationships** Read about the technological changes in the mining industry that are discussed in the introduction to Section 2. What method or methods of mining seem well suited for automation, particularly robotics? READING SKILLS

29. **Making Decisions** Mining companies use computer models to show them where high- and low-grade ores are located in the deposit that they are mining. If the price of the ore mineral that a company is mining suddenly increases, how would computer modeling help the company economically exploit the mineral deposit to take advantage of the increase in price?

Cross-Disciplinary Connection

30. **Social Studies** Fifteen to 20 miles southwest of Santa Fe, New Mexico, are a series of low hills known as Los Cerrillos. Native Americans mined the blue-green gemstone turquoise from narrow veins in rock from these hills for almost 1,000 years, beginning in about the year 875. Research Native American mining at Los Cerrillos, New Mexico. Write a short report about your findings. WRITING SKILLS

Portfolio Project

31. **Debate** A mining company has applied for permits to establish a surface mine on land that is located near a stretch of river in which an endangered species of fish lives. Assume that the ore to be mined is rare and has important new applications in cancer treatment. Weighing both sides of the argument, would you issue the permits? Make your case for or against issuing the mining permits in a debate with your classmates.

Critical Thinking

28. Answers may vary. Students should demonstrate comprehension of mining techniques that are hazardous to humans and that would benefit from automation.

29. Answers may vary. Students should show an understanding that when the price of an ore mineral, such as gold, is high, a mining company uses computer modeling to determine where the highest-grade ore is located in a deposit.

Cross-Disciplinary Connection

30. Answers may vary. Information about Native American mining techniques can be found on the Internet.

Portfolio Project

31. Answers may vary.

MATH SKILLS

32. Making Calculations Some low-grade gold ores that have been mined economically average about 0.1 oz of gold per ton of ore. Five tons of rock must be removed to obtain one ton of ore. How many tons of rock must be mined to obtain 1 oz of gold? How many pounds of ore must be processed to obtain 1 oz of gold?

WRITING SKILLS

33. Communicating Main Ideas One of the main ideas of this chapter is that the human need for minerals requires mining companies to continually find new deposits of minerals that can be extracted inexpensively. Extraction must be done in such a way that the environment is not severely affected. Using surface coal mining or open-pit mining as an example, explain why it is difficult to mine large ore deposits without affecting the environment.

34. Writing Persuasively A mining company is applying for permits to establish an open-pit mine near your home. Do research to determine what impact, if any, the operation will have on your quality of life, the environment, and the economics of your community. Summarize your findings in a concise one-page paper.

35. Outlining Topics You are an exploration geologist who works for a mining company. You are searching for a new deposit of an ore mineral. Outline the steps you would take to find a deposit and to determine whether that deposit would be economical to mine.

STANDARDIZED TEST PREP

For extra practice with questions formatted to represent the standardized test you may be asked to take at the end of your school year, turn to the sample test for this chapter in the Appendix.

READING SKILLS

Read the passage below, and then answer the questions that follow.

One of the only two rocks in which diamonds have been found is called *kimberlite*. Kimberlite is an uncommon kind of rock that forms cylindrical subsurface bodies called *kimberlite pipes*. Kimberlite pipes look very much like the vents that bring lava to the surface in volcanoes. Diamonds form deep in the Earth's mantle under enormous temperatures and pressures. Diamonds are believed to be carried to the surface in kimberlite pipes in very rapid, explosive events. However, not all kimberlite pipes contain diamonds.

Kimberlite is a soft, black, blue, or green rock that weathers rapidly when it reaches Earth's surface. Because kimberlite decomposes rapidly, it does not form rock outcrops. Instead, it forms circular depressions several feet below the surface of the ground. These depressions may be covered with a bluish kimberlite soil called *blue ground*. Iron-stained soils may also cover depressions. These soils are referred to as *yellow ground*.

1. Which of the following statements about kimberlite is *not* true?
   a. Diamonds are found in kimberlite.
   b. All kimberlite contains diamonds.
   c. Kimberlite weathers rapidly at Earth's surface.
   d. Kimberlite is an uncommon kind of rock.

2. If you were an exploration geologist searching for a deposit of diamonds, which of the following would *not* be a good surface indicator of the existence of a kimberlite pipe?
   a. a circular depression
   b. a bluish soil that fills a depression
   c. an iron-stained soil that fills a depression
   d. a large rock outcrop

Math Skills

32. 50 tons of rock are mined to obtain 1 oz of gold; 20,000 lb of ore are processed for 1 oz of gold

Writing Skills

33. Answers may vary.

34. Answers may vary.

35. Answers may vary.

Reading Skills

   1. b
   2. d

EXTRACTION OF COPPER FROM ITS ORE

Teacher's Notes

Time Required

one 45-minute class period

Lab Ratings

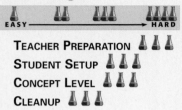

TEACHER PREPARATION ⚗⚗⚗
STUDENT SETUP ⚗⚗⚗
CONCEPT LEVEL ⚗⚗⚗
CLEANUP ⚗⚗⚗

Skills Acquired

• Constructing Models
• Experimenting
• Inferring

The Scientific Method

In this lab, students will:
• Make Observations
• Analyze the Results
• Draw Conclusions

Materials

The materials listed are enough for groups of 3 to 4 students. Additional materials, such as a spatula and a stop-watch, may be helful for students to use during the lab.

Copper carbonate, $CuCO_3$, is more readily available and purer than malachite. Excellent results are obtained using $CuCO_3$.

Objectives

▶ **Extract** copper from copper carbonate in much the same way that copper is extracted from malachite ore.

▶ **USING SCIENTIFIC METHODS**
Hypothesize how this process can be applied to extract other metallic elements from ores.

Materials

Bunsen burner
copper (cupric) carbonate
funnel
iron filings
sulfuric acid, dilute
test-tube holder
test-tube rack
test tubes, 13 mm x 100 mm (2)
water

▶ **Copper Ore** Malachite is a carbonate of copper that commonly forms in copper deposits. It is sometimes used as an ore of copper.

Extraction of Copper from Its Ore

Most metals are combined with other elements in the Earth's crust. A material in the crust that is a profitable source of an element is called an *ore*. Malachite (MAL uh KIET) is the basic carbonate of copper. The green corrosion that forms on copper because of weathering has the same composition that malachite does. The reactions of malachite are similar to those of copper carbonate.

In this investigation, you will extract copper from copper carbonate using heat and dilute sulfuric acid. The process you will be using will be similar to the process in which copper is extracted from malachite ore.

Procedure

1. CAUTION: Wear your laboratory apron, gloves, and safety goggles throughout the investigation. Fill one of the test tubes about one-fourth full of copper carbonate. Record the color of the copper carbonate.

2. Light the Bunsen burner, and adjust the flame.

3. Heat the copper carbonate by holding the tube over the flame with a test-tube holder, as shown in the figure on the next page. CAUTION: When heating a test tube, point it away from yourself and other students. To prevent the test tube from breaking, heat it slowly by gently moving the test tube over the flame. As you heat the copper carbonate, observe any changes in color.

4. Continue heating the tube over the flame for 5 min.

5. Allow the test tube to cool. Observe any change in the volume of the material in the test tube. Then, place the test tube in the test-tube rack. Insert a funnel in the test tube, and add

Tips and Tricks

Have students add 100–150 mL of sulfuric acid (1 M) to the heated copper carbonate. When copper oxide is redissolved in sulfuric acid (Step 6), the blue color indicates the presence of the hydrated copper (II) ion. There will probably be some CuO left in the test tube. If you wish to save it, you may direct students to wash it into a safe container.

Make sure students understand what they are observing in Step 8. When the iron filings are added to the test tube containing copper sulfate, the copper ions are reduced to solid copper, which forms around the iron filings.

Safety Cautions

Before attempting this activity, become familiar with the material safety data sheet for sulfuric acid. When working with caustic or poisonous chemicals, use extreme caution and allow only your most mature students to handle these materials. A functioning eyewash station should also be immediately accessible. Because an open flame is being used in part of the lab, please address fire hazards, and review how to use fire extinguishers and fire blankets. Make sure that students wear goggles, gloves, and lab aprons at all times.

dilute sulfuric acid until the test tube is three-fourths full. CAUTION: Avoid touching the sides of the test tube, which may be hot. If any of the acid gets on your skin or clothing, rinse immediately with cool water and alert your teacher.

6. Allow the test tube to stand until some of the substance at the bottom of the test tube dissolves. After the sulfuric acid has dissolved some of the solid substance, note the color of the solution.

7. Use a second test tube to add more sulfuric acid to the first test tube until the first test tube is nearly full. Allow the first test tube to stand until more of the substance at the bottom of the test tube dissolves. Pour this solution (copper sulfate) into the second test tube.

8. Add a small number of iron filings to the second test tube. Observe what happens.

9. Clean all of the laboratory equipment, and dispose of the sulfuric acid as directed by your teacher.

Analysis

1. **Explaining Events** Disregarding any condensed water on the test-tube walls, what do you call the substance formed in the first test tube? Explain any change in the volume of the new substance relative to the volume of the copper carbonate.

2. **Explaining Events** When the iron filings were added to the second test tube, what indicated that a chemical reaction was taking place? Explain any change to the iron filings. Explain any change in the solution.

Conclusions

3. **Drawing Conclusions** Why was sulfuric acid used to extract copper from copper carbonate?

▶ **Step 3** To heat the copper carbonate, hold the tube over the flame with a test-tube holder. Point the test tube away from yourself and other students.

Extension

1. **Analyzing Data** Suppose that a certain deposit of copper ore contains a minimum of 1 percent copper by mass and that copper sells for $0.30 per kilogram. Approximately how much could you spend to mine and process the copper from 100 kg of copper ore and remain profitable?

2. **Making Comparisons** How is the process used in this experiment similar to the cyanide heap-leaching process used to extract gold from low-grade ore?

431

Discussion

Before students begin this investigation, you may wish to discuss metallurgical processes. Metallurgy is the process whereby a metal is extracted from its ore and prepared for practical use. In their compounds, metals almost always exist in positive oxidation states. Therefore, the metal must be reduced (gain electrons) to obtain a metal from its ore. Ores that contain impurities are treated to concentrate the metal and to convert some metal compounds into substances that can be more easily reduced.

Answers to Analysis

1. The new substance is copper oxide. Because CO_2 gas is released during the reaction, there is less solid material left in the test tube.

2. Bubbles rose from the solution. The iron filings became red in color because the copper sulfate was reduced to solid copper, which formed around the iron filings. The solution became less blue because the copper sulfate that caused the blue color was being reduced and it precipitated out of the solution.

Answers to Conclusions

3. Sulfuric acid was used to extract the copper from copper carbonate because the chemical reaction that takes place forms water (H_2O) and carbon dioxide (CO_2), releasing the copper from the copper carbonate.

Answers to Extension

1. Less than $0.30 can be spent on mining and processing because only 1 kg of copper can be extracted from 100 kg of ore that is 1 percent copper and the selling price for 1 kg of copper is $0.30.

2. Cyanide heap leaching and this process both rely on an acid to chemically react with the desired element to extract that element from low-grade ore.

MINERAL PRODUCTION IN THE UNITED STATES

Activity ——————— GENERAL

Maps from Satellite Images
Organize students into groups of three or four, and give each group the following: tracing paper, some tape, colored pencils, and a satellite image (found on the Internet). Satellite images of large geologic features such as the Ouachita Mountains work well for this activity. Have each group tape a sheet of tracing paper over an interesting, relatively unpopulated area of the satellite image. Ask them to trace and color in areas where they see differences in color on the image. Have them color-code the areas with the colored pencils and make a map legend that shows which map color corresponds to a color on the satellite image. When finished, students should compare the results with a geologic map of the area (available online or from the USGS). The maps should match up relatively well because rock formations have distinctive colors and vegetation, which show up on satellite images. Tell students that this is an example of using remote sensing to map an area, similar to the way in which mining companies look for mineral-rich areas.

LS Visual/Kinesthetic   Co-op Learning

Transparencies
TT Mineral Production in the United States

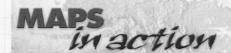

MINERAL PRODUCTION IN THE UNITED STATES

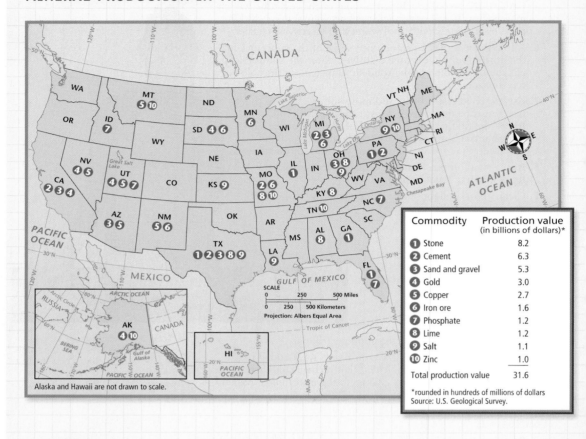

Commodity	Production value (in billions of dollars)*
① Stone	8.2
② Cement	6.3
③ Sand and gravel	5.3
④ Gold	3.0
⑤ Copper	2.7
⑥ Iron ore	1.6
⑦ Phosphate	1.2
⑧ Lime	1.2
⑨ Salt	1.1
⑩ Zinc	1.0
Total production value	31.6

*rounded in hundreds of millions of dollars
Source: U.S. Geological Survey.

MAP SKILLS

In 1999, the top 10 mineral commodities produced in the United States had a total value of about $31.6 billion. Over half of this production value came from the top three commodities: stone, cement, and sand and gravel. The map above shows the distribution of the production of these commodities by state. Use the map above to answer the following questions.

1. **Using a Key** Find your state on the map of mineral production. Which of the top 10 mineral commodities, if any, were produced in your state in 1999?

2. **Evaluating Data** Gold, copper, iron ore, and zinc are metals in the top 10 mineral commodities produced in 1999. What percentage of total 1999 production value do these metals represent? Which states were the producers of these metals in 1999?

3. **Evaluating Data** Stone, sand, and gravel are collectively known as *aggregates*. What percentage of total 1999 production value do aggregates represent? Which states were the major producers of aggregates in 1999?

4. **Using a Key** Which states produced salt in 1999?

432

Answers to Map Skills

1. Students should identify any mineral commodities produced in their state in 1999.

2. [(gold + copper + iron ore + zinc)/31.6] × 100 = 26.3% of total 1999 production; gold: Alaska, California, Nevada, South Dakota, and Utah; copper: Arizona, Montana, Nevada, New Mexico, and Utah; iron ore: Michigan, Minnesota, Missouri, New Mexico, and South Dakota; and zinc: Alaska, Missouri, Montana, New York, and Tennessee.

3. [(stone + sand and gravel)/31.6] × 100 = 42.7% of total 1999 production; Arizona, California, Florida, Georgia, Illinois, Michigan, Ohio, Pennsylvania, and Texas were main producers of aggregates.

4. New York, Ohio, Kansas, Louisiana, and Texas were major producers of salt in 1999.

COLTAN AND THE WAR IN THE CONGO

If you purchase a mobile phone, pager, or laptop computer, you may not be aware of the connection between these devices and politics in central Africa. Each of these products requires tantalum in its manufacture. Tantalum is a heat-resistant metal that can hold a high electric charge. Tantalum is ideal for the production of capacitors, which are used to regulate voltage in many of the electronics products in use today.

The main ore of tantalum is columbite-tantalite, which is often shortened to coltan. Eighty percent of the world's coltan reserves are found in the mountains of the eastern part of the Democratic Republic of the Congo (DRC). From the DRC, coltan makes its way into the world market, much of it illegally.

The 1996 Civil War

In 1996, hostilities between ethnic peoples caused civil war to break out in the DRC. Two years later, neighbors Rwanda and Uganda entered the conflict and backed two Congolese rebel movements. Shortly thereafter, Angola, Namibia, and Zimbabwe lent their support to the government of the DRC. Today, the Rwandan- and Ugandan-backed rebels have primary control of the coltan ore in the eastern DRC.

The war in the DRC is as much an economic war as an ethnic war. The price of coltan has reached prices as high as $400 per kilogram. Forces from neighboring Rwanda, Uganda, and Burundi have been accused of smuggling coltan out of the DRC and making enormous profits. This money is being used to help finance the continuing war efforts of these countries.

The Consequences of Civil War

Since 1998, almost 2.5 million people have died in the fighting in the

▶ Cell phones are just one of the electronic products in common use today that require tantalum in their production.

DRC. Government and rebel forces have attacked, killed, and tortured innocent civilians to maintain their rule. Almost half of the population of the DRC lacks safe drinking water. Access to health care is limited, and an estimated 2 million people suffer from HIV/AIDS.

Meanwhile, coltan mining has moved into coltan-rich national parks and reserves. Wildlife is being lost at an alarming rate as miners kill animals for food and elephants for ivory.

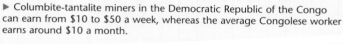

▶ Columbite-tantalite miners in the Democratic Republic of the Congo can earn from $10 to $50 a week, whereas the average Congolese worker earns around $10 a month.

What Do You Think?

Many electronics companies have stopped buying columbite-tantalite ore from central African countries. What effect, if any, do you think this action will have on present conditions in the DRC?

433

Society & the Environment

COLTAN AND THE WAR IN THE CONGO

Discussion ——— GENERAL

Diamond Conflicts Several civil wars and conflicts in Africa are funded with illegal diamonds. Diamonds are sold on the market, and the proceeds are used to purchase military supplies for the armies. In hopes of stopping the purchase of so-called conflict diamonds, the two major diamond-trading associations agreed to establish a system of certificates establishing a diamond's origin. Anyone trading in conflict diamonds will be expelled from the diamond trade, and any nation knowingly involved in such activities would lose its export accreditation. Ask students why diamonds would be a revenue source for civil wars. (Diamonds have high value, they are easily smuggled, and they are difficult to trace.) LS **Verbal/Logical**

Answers to What Do You Think?

Answers may vary. Sample answer: When companies stopped buying columbite-tantalite ore from central African countries, the money used to fund the civil wars probably decreased significantly. Additionally, the pressure to smuggle coltan probably also decreased because the smugglers were not as able to sell the illegal ore. Thus, a potential decrease in hostilities might be expected as a result of the removal of financial support and the lessening of incursions to steal coltan ore.

PACING	CLASSROOM RESOURCES	LABS, DEMONSTRATIONS, AND ACTIVITIES
BLOCKS 1, 2 & 3 • 135 min pp. 434–443		
Chapter Opener		
Section 1 Energy Resources and Fossil Fuels	**OSP** Lesson Plan * **CRF** Active Reading * BASIC **TT** Bellringer * **TT** How a Coal-Fired Power Plant Works * **TT** Fossil Fuels in the United States *	**SE** QuickLab Generating Electricity, p. 437 **TE** Activity Global Fossil Fuel Distribution, p. 438 ADVANCED **TE** Group Activity Simulating Coal Formation, p. 438 GENERAL **TE** Group Activity Making Plastic, p. 440 GENERAL **TE** Demonstration Peanut Power, p. 441 GENERAL **SE** Skills Practice Lab Your Household Energy Consumption, pp. 452–453 **CRF** Datasheets for In-Text Labs * **CRF** Modeling Lab * ◆ GENERAL **CRF** Observation Lab * ◆ BASIC **CRF** Consumer Lab * ◆ GENERAL **CRF** CBL™ Probeware Lab * ◆ GENERAL
BLOCKS 4 & 5 • 90 min pp. 444–447		
Section 2 Nuclear Energy	**OSP** Lesson Plan * **CRF** Active Reading * BASIC **TT** Bellringer * **TT** Fission & Fusion * **TT** How a Nuclear Power Plant Works *	**CRF** Long-Term Project * ADVANCED

BLOCKS 6 & 7 • 90 min

Chapter Review and Assessment Resources

SE Chapter Review pp. 449–451
SE Standardized Test Prep pp. 668–669
CRF Study Guide * ■ GENERAL
CRF Chapter Test * ■ GENERAL
CRF Chapter Test * ADVANCED
CRF Concept Review * ■ GENERAL
CRF Critical Thinking * ADVANCED
OSP Test Generator
CRF Test Item Listing *

Online and Technology Resources

 Holt Online Learning

Visit go.hrw.com for access to Holt Online Learning or enter the keyword **HE6 Home** for a variety of free online resources.

 **One-Stop Planner® CD-ROM**

This CD-ROM package includes
• Lab Materials QuickList Software
• Holt Calendar Planner
• Customizable Lesson Plans
• Printable Worksheets
• ExamView® Test Generator
• Interactive Teacher Edition
• Holt PuzzlePro® Resources
• Holt PowerPoint® Resources

 Holt Environmental Science Interactive Tutor CD-ROM

This CD-ROM consists of interactive activities that give students a fun way to extend their knowledge of environmental science concepts.

KEY

TE	Teacher Edition	CRF	Chapter Resource File	*	Also on One-Stop Planner
SE	Student Edition	TT	Teaching Transparency	▨	Also Available in Spanish
OSP	One-Stop Planner	CD	Interactive CD-ROM	◆	Requires Advance Prep

ENRICHMENT AND SKILLS PRACTICE	SECTION REVIEW AND ASSESSMENT	CORRELATIONS
SE **Pre-Reading Activity,** p. 434 TE **Using the Figure,** p. 434 `GENERAL`		**National Science Education Standards**
TE **Using the Figure** Power Plant Efficiency, p. 436 `GENERAL` TE **Reading Skill Builder** Prediction Guide, p. 436 `BASIC` TE **Interpreting Statistics** Energy Consumption, p. 437 `GENERAL` TE **Career** Working with Electricity, p. 437 TE **Reading Skill Builder** Reading Organizer, p. 437 `GENERAL` SE **Graphic Organizer** Chain-of-Events Chart, p. 438 `GENERAL` TE **Interpreting Statistics** Energy Production, p. 439 `GENERAL` TE **Skill Builder** Vocabulary, p. 439 `GENERAL` TE **Inclusion Strategies,** p. 439 SE **Case Study** Methane Hydrates, pp. 440–441 SE **MathPractice** World Energy Use, p. 442 TE **Using the Figure** Fossil Fuel Predictions, p. 442 `GENERAL`	TE **Reteaching,** p. 439 `BASIC` TE **Homework,** p. 441 `ADVANCED` SE **Section Review,** p. 443 TE **Reteaching,** p. 443 `BASIC` TE **Quiz,** p. 443 `GENERAL` TE **Alternative Assessment,** p. 443 `GENERAL` CRF **Quiz** * ▨ `GENERAL`	PS 3b SPSP 3a SPSP 3b
TE **Using the Figure** Comparing Nuclear and Fossil-Fuel Power Plants, p. 445 `GENERAL` TE **Skill Builder** Math, p. 445 `GENERAL` SE **Maps in Action** Nuclear Power in the United States, 1950–2002, p. 454 SE **Science & Technology** From Crude Oil to Plastics, p. 455 CRF **Map Skills** Coal Deposits * `GENERAL`	TE **Reteaching,** p. 445 `BASIC` TE **Homework,** p. 446 `ADVANCED` SE **Section Review,** p. 447 TE **Reteaching,** p. 447 `BASIC` TE **Quiz,** p. 447 `GENERAL` TE **Alternative Assessment,** p. 447 `GENERAL` CRF **Quiz** * ▨ `GENERAL`	PS 1c SPSP 3a SPSP 3b

Guided Reading Audio CDs

These CDs are designed to help auditory learners and reluctant readers. (Audio CDs are also available in Spanish.)

www.scilinks.org

Maintained by the **National Science Teachers Association**

TOPIC: Fossil Fuels
SciLinks code: HE4044

TOPIC: Nuclear Energy
SciLinks code: HE4075

CNN Videos

Each video segment is accompanied by a Critical Thinking Worksheet.

Science, Technology & Society
Segment 6 BioDiesel

CHAPTER

17

Chapter Enrichment

This Chapter Enrichment provides relevant and interesting information to expand and enhance your classroom instruction of the chapter material.

SECTION 1 Energy Resources and Fossil Fuels

Coal and China

The people of Guizhou Province in southwestern China have been burning coal for household use since the early 1900s. They use coal-burning stoves to dry corn, chili peppers, and other crops in their homes because it is too cold and damp to dry them outside. The problem is that the coal in this region contains unusually high concentrations of arsenic and fluorine, two very toxic chemicals. The concentration of arsenic in this coal can be over 35,000 parts per million, compared to the U.S. average of 22 parts per million (the U.S. concentration is not considered hazardous). When this coal is burned, arsenic and fluorine are released into the air, where they can be inhaled by people or absorbed by the food set out to dry. Many homes are poorly ventilated, so the toxic fumes have nowhere to go. Scientists in the United States and China are working together to map the most dangerous coal deposits and to develop test kits that the people can use to measure the concentrations of arsenic and fluorine in the coal they use.

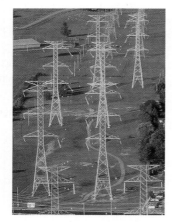
▶ **Pylons and transmission lines distribute electrical energy**

Clean Coal Technologies

Several methods are used to reduce the pollution caused by coal-burning power plants. Sulfur can be removed in two ways. One way to remove small particles of sulfur is to crush and wash the coal. However, some sulfur is chemically bonded to the carbon in the coal and cannot be washed out. In this case, a second method of removal is used. The sulfur is removed from the combustion gases before the gases leave the smokestack. This can be done using "scrubbers" that spray a solution of water and limestone powder into the combustion gases. The limestone reacts with the sulfur, which is collected in the form of a paste.

Nitrogen oxides, or NOx, are another air pollutant formed when coal is burned. Nitrogen makes up about 80 percent of the atmosphere, and at high temperatures it combines with oxygen to form nitrogen oxides. One way to reduce the formation of NOx is by using low-NOx burners. Coal is burned in chambers that contain more fuel than air so that the oxygen tends to combine with the fuel rather than with the nitrogen in the air. This method removes more than one-half the NOx from the combustion gases. Another way to remove NOx involves scrubbers. NOx scrubbers use a catalyst to break down the nitrogen oxides to nonpolluting gases before they leave the smokestack. This method can remove up to 90 percent of NOx but is more expensive than low-NOx burners.

BRAIN FOOD

One of the cleanest ways to burn coal is called *coal gasification*. When coal is heated and sprayed with hot water vapor, it turns into a gas that is a mixture of carbon monoxide and hydrogen. This gas can be burned to generate electricity. Coal gasification can remove 99.9 percent of the sulfur and particulates from power-plant emissions.

Oil Production

Crude oil is under tremendous pressure from the rock above it and from heat within the Earth, which expands any gases trapped within the oil. When a well is drilled into rock that contains an oil reservoir, the pressure is released and the oil is forced out of the rock. After this oil has been extracted from the reservoir, the remaining oil must be pumped out. If natural gas is found with the oil, one way to pump the oil out is to separate the natural gas, inject the gas back into the ground in order to build up the pressure again, and force more oil out. Oil companies can also flood the reservoir with water pumped through a separate injection well. The water washes the oil out of the rock, and the water-oil mixture is pumped out through the production well. However, this process still only extracts 5 to 10 percent of the original oil volume. The oil production process can leave about two barrels of oil in the ground

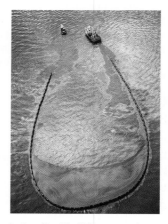
▶ **An oil spill off the coast of Southwest Wales**

for every one barrel that is pumped out. Scientists are investigating various methods to use microbes or chemicals called *surfactants* to help release the remaining oil.

BRAIN FOOD

There are few high-pressure natural gas reservoirs remaining, so other techniques are used to remove the gas from reservoirs. In one technique, fluids, such as water, are pumped into the rock under high pressure. The high-pressure fluids crack the rock and create fractures where the gas can flow. Materials such as sand or small glass beads are pumped in with the water to act as "propping agents," which hold the fractures open so that the gas can be released.

The Arab Oil Embargo

The Arab oil embargo of 1973 showed how vulnerable the United States is to disruptions of its oil supply. In retaliation for American support of Israel in the 1973 Yom Kippur War, several Arab nations suspended shipments of crude oil to the United States. This action, which reduced the oil supply in the United States by less than 10 percent, was extremely disruptive. Gasoline shortages were common. Many service stations closed, and long lines formed at the stations that remained open. Gasoline prices skyrocketed, and the federal government prepared to ration gasoline. Military action to restore the flow of oil was discussed. The embargo also strengthened a struggling conservation movement in America and led to remarkable improvements in American energy efficiency. More recently, the Persian Gulf War slightly disrupted the supply of Middle Eastern oil and sent oil prices sharply upward.

SECTION

2 Nuclear Energy

Breeder Reactors

Breeder reactors convert uranium-238, which is relatively common but not fissionable (usable as nuclear fuel), into plutonium-239, which is fissionable. Usable power is generated as the conversion takes place. Well-designed breeder reactors can actually produce more usable nuclear fuel than they consume. One major concern about breeder reactors is that the plutonium they produce is a key ingredient of atomic bombs, and some

people are concerned that terrorists or unstable governments could obtain plutonium from breeder reactors.

▶ A nuclear power plant in California

Fusion Technologies

After 40 years of experimentation, scientists have succeeded in carrying out fusion on a very limited scale. So far, however, usable amounts of energy have not been produced. There are two basic fusion-reactor designs: the magnetic-confinement method and the pellet-fusion method. The most successful magnetic-confinement system, called a *tokamak* (a Russian acronym for "toroidal magnetic chamber"), uses intense magnetic fields and a strong electric current to confine deuterium plasma in a toroidal (doughnut-shaped) field while heating it to ignition temperatures of 100 million degrees Celsius. The pellet-fusion method uses a spherical array of powerful, simultaneously firing laser beams to implode a small deuterium-enriched pellet with such force that its atoms fuse. Enormous amounts of energy are needed to achieve pellet fusion. The combined power of the lasers exceeds, for an instant, the power output of the entire United States. Both fusion methods are believed to be ultimately workable, but the technological challenges are great.

BRAIN FOOD

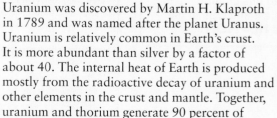

Uranium was discovered by Martin H. Klaproth in 1789 and was named after the planet Uranus. Uranium is relatively common in Earth's crust. It is more abundant than silver by a factor of about 40. The internal heat of Earth is produced mostly from the radioactive decay of uranium and other elements in the crust and mantle. Together, uranium and thorium generate 90 percent of Earth's heat.

Overview

Tell students that the purpose of this chapter is to introduce them to the fundamentals of energy use in our society. In the United States, energy consumption patterns produce a great demand for fuels in the transportation, industrial, residential, and commercial sectors. Our choice of fuels and our dependence on them has economic, environmental, and political consequences.

Using the Figure — GENERAL

The tall, thin towers in the photograph are fractionation towers. The towers separate petroleum into its various components based on their boiling points. Gasoline has a relatively low boiling point and is one of the first components to be separated. The flames at the top of the columns are the result of burning excess gases. Students can find out more about fractionation in the feature at the end of the chapter. Ask students to estimate the percentage of petroleum in the U.S. that is used for gasoline. (about 40 percent)
LS Visual

PRE-READING ACTIVITY

FOLDNOTES

You may want to assign this FoldNote activity as homework.
Collect the FoldNotes to check students' understanding of the material.

For information about videos related to this chapter, go to **go.hrw.com** and type in the keyword **HE4 NONV.**

Nonrenewable Energy

1 **Energy Resources and Fossil Fuels**

2 **Nuclear Energy**

PRE-READING ACTIVITY

FOLDNOTES

Booklet
Before you read this chapter, create the **FoldNote** entitled "Booklet" described in the Reading and Study Skills section of the Appendix. Label each page of the booklet with a main idea from the chapter. As you read the chapter, write what you learn about each main idea on the appropriate page of the booklet.

At this petroleum refinery in Germany, crude oil pumped from the Earth is changed into fuels, such as gasoline, and other products.

434

Chapter Correlations — *National Science Education Standards*

PS 1c The nuclear forces that hold the nucleus of an atom together, at nuclear distances, are usually stronger than the electric forces that would make it fly apart. Nuclear reactions convert a fraction of the mass of interacting particles into energy, and they can release much greater amounts of energy than atomic interactions. Fission is the splitting of a large nucleus into smaller pieces. Fusion is the joining of two nuclei at extremely high temperature and pressure, and is the process responsible for the energy of the sun and other stars. **(Section 2)**

PS 3b Chemical reactions may release or consume energy. Some reactions such as the burning of fossil fuels release large amounts of energy by losing heat and by emitting light. Light can initiate many chemical reactions such as photosynthesis and the evolution of urban smog. **(Section 1)**

SPSP 3a Human populations use resources in the environment in order to maintain and improve their existence. Natural resources have been and will continue to be used to maintain human populations. **(Section 1 and Section 2)**

SPSP 3b The earth does not have infinite resources; increasing human consumption places severe stress on the natural processes that renew some resources, and it depletes those resources that cannot be renewed. **(Section 1 and Section 2)**

Energy Resources and Fossil Fuels

How does a sunny day 200 million years ago relate to your life today? Chances are that if you traveled to school today or used a product made of plastic, you used some of the energy from sunlight that fell on Earth several hundred million years ago. Life as we know it would be very different without the fuels or products formed from plants and animals that lived alongside the dinosaurs.

The fuels we use to run cars, ships, planes, and factories and to produce electricity are natural resources. Most of the energy we use comes from a group of natural resources called *fossil fuels*. **Fossil fuels** are the remains of ancient organisms that changed into coal, oil, or natural gas. Fossil fuels are central to life in modern societies, yet there are two main problems with fossil fuels. First, the supply of fossil fuels is limited. Second, obtaining and using them has environmental consequences. In the 21st century, societies will continue to explore alternatives to fossil fuels but will also focus on developing more-efficient ways to use these fuels.

Fuels for Different Uses

Fuels are used for four main purposes: for transportation, for manufacturing, for heating and cooling buildings, and for generating electricity to run machines and appliances. The suitability of a fuel for each application depends on the fuel's energy content, cost, availability, safety, and byproducts of the fuel's use. For example, it's hard to imagine an airplane, such as the one shown in **Figure 1**, running on piles of coal. Although coal is readily available and inexpensive, to power an airplane using coal would require hundreds of tons of coal. Likewise, the people shown around the campfire are not warming themselves by burning airplane fuel, they are burning wood, which is a perfect fuel for their needs.

Objectives

▶ List five factors that influence the value of a fuel.

▶ Explain how fuels are used to generate electricity in an electric power plant.

▶ Identify patterns of energy consumption and production in the world and in the United States.

▶ Explain how fossil fuels form and how they are used.

▶ Compare the advantages and disadvantages of fossil-fuel use.

▶ List three factors that influence predictions of fossil-fuel production.

Key Terms

fossil fuels
electric generator
petroleum
oil reserves

Figure 1 ▶ Different Fuels, Different Purposes The airplane (left) is being refueled with a highly refined liquid fuel. Airplane fuel must have a high ratio of energy to weight. The campers (below) are keeping warm by burning wood, which is much safer than burning airplane fuel!

435

SECTION 1

Focus

Overview

Before beginning this section, review with your students the Objectives in the Student Edition. This section discusses energy use in our society and the important role of fossil fuels in providing energy. Students will learn how electricity is produced from fossil fuels as well as how fossil fuels are formed, their effects on the environment, and the potential for their continued use in the future.

Bellringer

Ask students to describe the photos in **Figure 1**. Have them list the characteristics of each fuel shown.

Motivate

Discussion ———— **GENERAL**

Life Without Fossil Fuels Explain that fossil fuels are materials such as coal, oil, and natural gas that are burned to supply heat or generate electricity. Suggest that students run through their daily routine from the moment they wake up in the morning to the time they go to sleep and think about how fossil fuels affect that routine. Ask students, "How many of the things that you do each day depend on fossil fuels in some way?" Then, encourage discussion about how life would be different if fossil fuel resources were more limited. **LS** Verbal

Chapter Resource File

• Lesson Plan
• Active Reading **BASIC**
• Section Quiz **GENERAL**

Transparencies

TT Bellringer

Using the Figure — GENERAL

Power Plant Efficiency Most of the electric power used in this country is generated at plants such as the one shown in **Figure 3**. Tell students that the typical power plant only converts about 33 to 40 percent of the energy contained in fossil fuels into electric energy. Ask students to examine the figure to see if it contains any clues as to where the "lost" energy goes. (The energy is dissipated as waste heat in the following ways. Heat from the burning fuel escapes through combustion chamber walls, smoke stacks, boiler walls, and through pipes carrying steam to the turbine. Some of the steam is released after it passes through the turbine. The turbine does not convert all of the kinetic energy of the steam into electrical energy. Friction in the turbine and generator causes the loss of some energy. Electrical resistance in the generator and associated circuitry dissipates energy.) **LS** Visual

READING SKILL BUILDER — BASIC

Prediction Guide Have students write down five questions about fossil fuels. Then have them read the section and write down the answers to their questions. Encourage them to research the answers that they did not find in this chapter. **LS** Intrapersonal

Transparencies

TT How a Coal-Fired Power Plant Works

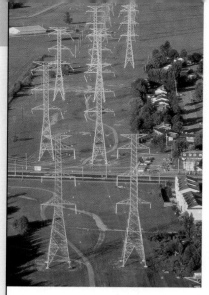

Figure 2 ▶ These pylons and wires are part of an electricity distribution grid in upstate New York.

Electricity—Power on Demand

The energy in fuels is often converted into electrical energy in order to power machines, because electricity is more convenient to use. Computers, for example, run on electricity rather than oil. Electricity can be transported quickly across great distances, such as an entire state, or across tiny distances, such as inside a computer chip. The electricity that powers the lights in your school was generated in a power plant and then carried to users through a distribution grid like the one shown in **Figure 2**. Two disadvantages of electricity are that it is difficult to store and other energy sources have to be used to generate it.

How Is Electricity Generated? An **electric generator** is a machine that converts mechanical energy, or motion, into electrical energy. Generators produce electrical energy by moving an electrically conductive material within a magnetic field. Most commercial electric generators convert the movement of a turbine into electrical energy, as shown in **Figure 3**. A *turbine* is a wheel that changes the force of a moving gas or a liquid into energy that can do work. In most power plants, water is boiled to produce the steam that turns the turbine. Water is heated by burning a fuel in coal-fired and gas-fired plants or is heated from the fission of uranium in nuclear plants. The turbine spins a generator to produce electricity.

Figure 3 ▶ How a Coal-Fired Power Plant Works

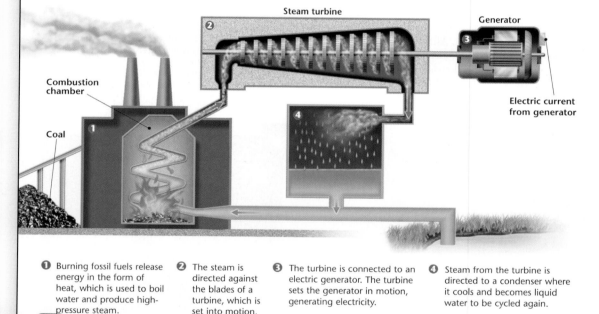

1 Burning fossil fuels release energy in the form of heat, which is used to boil water and produce high-pressure steam.

2 The steam is directed against the blades of a turbine, which is set into motion.

3 The turbine is connected to an electric generator. The turbine sets the generator in motion, generating electricity.

4 Steam from the turbine is directed to a condenser where it cools and becomes liquid water to be cycled again.

436

MISCONCEPTION ALERT

United States Petroleum Imports The United States relies on imported oil for the majority of its oil consumption, but this has not always been the case. In 1973, the United States imported only 28 percent of its oil, but today it imports about 60 percent. Students may think that most imported oil comes from the Middle East or the Persian Gulf. Saudi Arabia is the largest single exporter of oil to the United States, but they provide only 17 percent of total U.S. oil imports. In fact, the Persian Gulf countries account for only 29 percent of U.S. imports. After Saudi Arabia, the largest oil exporters to the United States are Mexico, Canada, and Venezuela. The oil from these three countries accounts for 45 percent of U.S. oil imports.

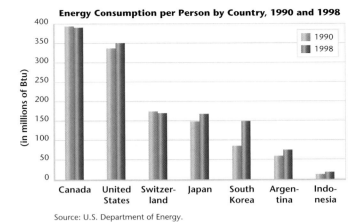

Energy Consumption per Person by Country, 1990 and 1998

(in millions of Btu)

Legend: 1990, 1998

Countries: Canada, United States, Switzerland, Japan, South Korea, Argentina, Indonesia

Source: U.S. Department of Energy.

Energy Use

Everything from the food you eat to the clothes you wear requires energy to produce. Furthermore, the price of nearly every product or service that you use reflects the cost of energy. When you purchase a plane ticket, for example, you purchase part of the fuel that will help you reach your destination. In 2000, airlines spent $5.4 billion on fuel—their second-highest expense after labor costs.

World Patterns There are dramatic differences in fuel use and efficiency throughout the world. People in developed societies use much more energy than people in developing countries do. However, energy use in some developing countries is growing rapidly. Even within the developed world there are striking differences in energy use. For example, **Figure 4** shows that a person in Canada or the United States uses more than twice as much energy as an individual in Japan or Switzerland does. Yet personal income in Japan and Switzerland is higher than personal income in Canada and the United States. One reason for this pattern lies in how energy is generated and used in those countries.

Energy Use in the United States The United States uses more energy per person than any other country in the world except Canada and the United Arab Emirates. Part of the reason that the United States uses so much energy is that, as **Figure 5** shows, the United States uses more than 25 percent of its energy resources to transport goods and people, mainly by trucks and personal vehicles. In contrast, Japan and Switzerland have extensive rail systems and they are relatively small, compact countries. The availability and cost of fuels also influence fuel use. Residents of the United States and Canada enjoy some of the lowest gasoline taxes in the world. There is little incentive to conserve gasoline when its cost is so low. Japan and Switzerland, which have minimal fossil-fuel resources, supplement a greater percentage of their energy needs with other energy sources, such as hydroelectric or nuclear power.

Figure 4 ▶ Canada and Switzerland slightly reduced their energy use per person during this nine-year period, while energy use per person in South Korea increased by almost 50 percent.

QuickLAB

Generating Electricity

Procedure

1. Tightly wrap 100 cm of **fine-gauge copper wire** around a small cardboard tube.
2. Attach a **galvanometer** or a **battery tester** to the ends of the wire.
3. Pass a **bar magnet** through the cardboard tube, and observe the galvanometer.

Analysis

1. What did you observe?
2. How could you increase the current you detected?
3. How does this lab model an electric power plant?

How Energy Is Used in the United States

Residential 19%
Industrial 38%
Commercial 16%
Transportation 27%

Source: International Energy Agency.

Figure 5 ▶ This graph shows the percentages of total energy use in the United States for different purposes.

437

Interpreting Statistics — GENERAL

Energy Consumption The energy consumption of several countries is shown in **Figure 4.** The vertical axis uses Btu, or British Thermal Units. One Btu is the amount of heat required to raise the temperature of one pound of water by one degree Fahrenheit. One wooden kitchen match can produce about 1 Btu. The average candy bar produces about 1,000 Btu, and one pound of coal produces approximately 15,000 Btu. Ask students to hypothesize why Canada and Switzerland decreased their energy usage while South Korea almost doubled their energy use during the same time period. (One possible reason is that Canada and Switzerland had very little population or industrial growth during this period, while South Korea had significant growth in commercial and industrial sectors. In addition, advances in efficiency and conservation may have reduced energy use in Switzerland and Canada.)

LS Visual/Logical

QuickLAB

Skills Acquired:

• Experimenting
• Measuring
• Inferring

Answers

1. Students should detect a current. If they do not, they should wrap a longer length of wire around the tube.
2. Wrapping more wire around the tube or using a stronger bar magnet would increase the current.
3. This lab is a simple model of a generator in an electric power plant.

READING SKILL BUILDER ——— GENERAL

Reading Organizer Before students read the rest of this section, have them create a table with the column headings "Coal," "Petroleum," and "Natural gas" and the row headings "Formation," "Uses," and "Environmental effects." Tell students to fill in the table with the information they learn as they read the remainder of this section. **LS** Logical

English Language Learners

Career

Working with Electricity Electric power plants require skilled personnel in order to operate efficiently and safely. Electrical and mechanical engineers design the power plants and the distribution grid to transmit the electricity. Power-plant operators oversee the daily operation of the power plant from the control room. Line workers maintain and restore power in the distribution lines. Have interested students research one of these careers. Invite someone in one of these fields to your class to discuss their work.

Activity — ADVANCED

Global Fossil Fuel Distribution
Have students produce a map similar to **Figure 6** for another region of the world, such as the Middle East, the Caspian Sea, or China. The map should show the distribution of coal, oil, and natural gas in that region. Ask students, "Based on your map, how would this region's natural resources affect its role in the global economy?" As an extension, suggest that they research the role of fossil fuels in the history of the region and make a timeline showing the history. Encourage them to present their findings to the class.
LS Visual/Interpersonal

Group Activity — GENERAL

Simulating Coal Formation Have small groups of students perform the following experiment: Place 4–6 inches of water in an aquarium. Add 2 inches of fine- to medium-grained sand. On top of the sand add fern fronds, leaves, and twigs. Let the aquarium stand for two weeks, and observe any changes that take place. Gently pour 2 inches of fine silt or mud on top. Let the experiment stand for another two weeks. After the two weeks of observation, remove the surface water and let the remaining ingredients dry for a week or two. Students can take a core sample by pushing a straw into the dry sediment and then cutting the straw open to remove the plug. They may also be able to see "fossils" of the original plant matter by gently breaking up the dry sediment in the aquarium.
LS Interpersonal   Co-op Learning

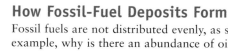

Graphic
Organizer **Chain-of-Events Chart**
Create the **Graphic Organizer** entitled "Chain-of-Events Chart" described in the Appendix. Then, fill in the chart with details about each step of the formation of fossil fuels.

How Fossil-Fuel Deposits Form

Fossil fuels are not distributed evenly, as shown in **Figure 6.** For example, why is there an abundance of oil in Texas and Alaska but very little in Maine? Why does the eastern United States produce so much coal? The answers to these questions lie in the geologic history of the areas.

Coal Formation   Coal forms from the remains of plants that lived in swamps hundreds of millions of years ago. Much of the coal in the United States formed about 300 million to 250 million years ago, when vast areas of swampland covered the eastern United States. As ocean levels rose and fell, these swamps were repeatedly covered with sediment. Layers of sediment compressed the plant remains, and heat and pressure within the Earth's crust caused coal to form. Coal deposits in the western United States also formed from ancient swamps, but those deposits are much younger. The abundant coal deposits in states such as Wyoming formed between 100 million and 40 million years ago.

Oil and Natural Gas Formation   Oil and natural gas result from the decay of tiny marine organisms that accumulated on the bottom of the ocean millions of years ago. After these remains were buried by sediments, they were heated until they became complex energy-rich carbon molecules. Over time, these molecules migrated into the porous rock formations that now contain them. Much of the oil and natural gas in the United States is located in Alaska, Texas, California, and the Gulf of Mexico.

Figure 6 ▶ This map shows coal, oil, and natural gas deposits in the United States.

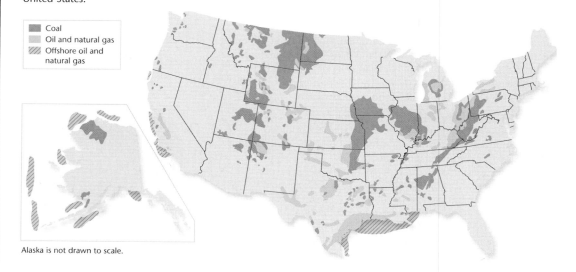

- ■ Coal
- ■ Oil and natural gas
- ▨ Offshore oil and natural gas

Alaska is not drawn to scale.

Graphic
Organizer GENERAL

Chain of Events Chart
You may want to use this Graphic Organizer to assess students' prior knowledge before beginning a discussion of each step of the formation of fossil fuels. You may also choose to use a similar activity as a quiz to assess students' knowledge of formation of fossil fuels or to assess students' knowledge after students have read the description of formation of fossil fuels.

MISCONCEPTION ///ALERT\\\

Oil Reservoirs Students may think that oil exists underground in large lakes or pools. However, most oil is trapped inside the rock in tiny spaces called *pores.* The word *petroleum* literally means "rock oil" in Latin: *petra* means "rock," and *–oleum* means "oil."

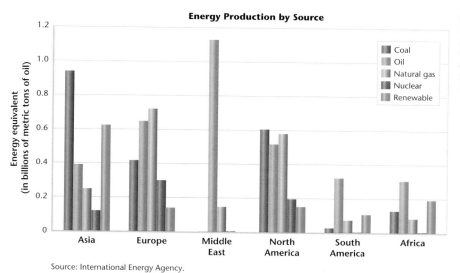

Energy Production by Source

Energy equivalent (in billions of metric tons of oil)

Legend:
- Coal
- Oil
- Natural gas
- Nuclear
- Renewable

Asia · Europe · Middle East · North America · South America · Africa

Source: International Energy Agency.

Figure 7 ▶ The Middle East produces the majority of the world's oil. Asia, however, produces the most coal.

Coal

Most of the world's fossil-fuel reserves are made up of coal. Asia and North America are particularly rich in coal deposits, as shown in **Figure 7.** Two major advantages of coal are that it is relatively inexpensive and that it needs little refining after it has been mined. More than half of the electricity generated in the United States comes from coal-fired power plants, as shown in **Figure 8.**

Coal Mining and the Environment The environmental effects of coal mining vary. Underground mines can have a minimal effect on the environment at the surface. However, surface coal-mining operations sometimes remove the top of an entire mountain to reach the coal deposit. In addition, if waste rock from coal mines is not properly contained, toxic chemicals can leach into nearby streams. A lot of research focuses on developing better methods of locating the most productive, clean-burning coal deposits and developing less damaging methods of mining coal.

Air Pollution The quality of coal varies. Higher-grade coals, such as bituminous coal, produce more heat and less pollution than a lower-grade coal, such as lignite. Sulfur, which is found in all grades of coal, can be a major source of pollution when coal is burned. When high-sulfur, low-grade coal is burned, it releases much more pollution than a low-sulfur bituminous coal does. The air pollution and acid precipitation that result from burning high-sulfur coal without adequate pollution controls are serious problems in countries such as China. However, clean-burning coal technology has dramatically reduced air pollution in countries such as the United States.

Figure 8 ▶ More than half of the electricity generated in the United States comes from burning coal. This power plant (bottom) in West Virginia is located close to the abundant coal deposits in that state.

How Electricity Is Generated in the United States

- Hydroelectricity 8%
- Natural gas 10%
- Oil 2%
- Nuclear 23%
- Coal 57%

439

Group Activity ——— GENERAL

Making Plastic Before petroleum was widely used, natural substances were used to make plastics. In 1897 Adolph Spitteler and W. Kirsche created a plastic called *casein-formaldehyde* from casein, a protein in milk, and formaldehyde. This plastic was used to make buttons, fountain pens, jewelry, and other objects. Students can make a plastic-like substance by separating the casein out of milk. Give the students the following instructions: Gently heat 1 cup of cream in a pan. When it is simmering but not boiling, add 1 tablespoon of vinegar or lemon juice. Stir until it begins to clump. Add more vinegar or lemon juice if necessary. Let it cool and then wash off the "curds." Mold the rubbery substance into different shapes and set it aside to dry for a few days. Tell students that the acid in the vinegar or lemon juice reacted with the casein to separate it from the rest of the cream.
LS Kinesthetic Co-op Learning

Discussion ——— GENERAL

Oil Wells and Rock Formations
Ask students to identify areas where many oil wells exist as well as the name and rock type of the main oil reservoir for that area. (Sample answer: Sandstone of the Frio Formation is the oil reservoir for the Corpus Christi, Texas, area.) Ask students why they think that formation makes a good reservoir. (Sample answer: the porosity and permeability of the rock and the presence of oil)

Connection to Chemistry

Catalytic Converters Catalytic converters are one of the most important emission-control features on cars. These devices use two separate catalysts—a *reduction catalyst* and an *oxidation catalyst*. The reduction catalyst uses platinum and rhodium to separate nitrous oxides, forming nitrogen and oxygen molecules. The oxidation catalyst uses platinum and palladium to burn—or oxidize—hydrocarbons and carbon monoxide, forming carbon dioxide, which is less harmful.

Petroleum

Oil that is pumped from the ground is also known as *crude oil*, or petroleum. Anything that is made from crude oil, such as fuels, chemicals, and plastics, is called a *petroleum product*. Much of the world's energy needs are met by petroleum products. In fact, petroleum accounts for 45 percent of the world's commercial energy use.

Locating Oil Deposits Oil is found in and around major geologic features that tend to trap oil as it moves in the Earth's crust. These features, which include folds, faults, and salt domes, are bound by impermeable layers of rock, which prevent the oil from escaping. Most of the world's oil reserves are in the Middle East. Large oil deposits also exist in the United States, Venezuela, the North Sea, Siberia, and Nigeria. Geologists use many different methods to locate the rock formations that could contain oil. When geologists have gathered all of the data that they can from the Earth's surface, exploration wells are drilled to determine the volume and availability of the oil deposit. If oil can be extracted at a profitable rate, wells are drilled and the oil is pumped or flows to the surface. After petroleum is removed from a well, it is transported to a refinery to be converted into fuels and other petroleum products.

Methane Hydrates— Fossil Fuel of the Future?

Deep under the waves in the Gulf of Mexico lies an untapped resource that could be the fuel of the future. It looks like ice, but it burns with a bright fire. This strange compound is called methane hydrate.

A methane hydrate is a cagelike lattice of ice that contains trapped molecules of methane. Methane is a natural gas made up of carbon and hydrogen. The methane in methane hydrates results from the bacterial decomposition of organic matter.

Methane hydrates have been known to exist since 1890. However, nobody knew that the hydrates formed in nature until 1964, when a large deposit was discovered by a Soviet crew that was

drilling for oil in Siberia. Today, huge deposits of this "solid" natural gas have been discovered around the edges of most continents. The deposits are often several hundred meters thick.

Methane hydrates form in geologic situations in which temperatures are stable and low and pressure is high. In places such as Siberia and Alaska, methane hydrates form below the tundra where permafrost extends down into shallow sediments. They also form under the ocean in water that is deeper than 500 m. In the United States, there are deposits of methane hydrates off the shores of Alaska, Washington, California, and

the Carolinas and in the Gulf of Mexico. If we could recover just one percent of the methane hydrate around the United States we could more than double our supply of natural gas, a clean-burning fuel that produces little pollution except for carbon dioxide.

Natural gas is in increasing demand for use in new electric power stations. These power stations are cheaper to build than other power stations and produce little air pollution. Natural gas is also used to fuel low-pollution vehicles.

Natural gas will become an increasingly important substitute for coal and petroleum as countries limit their emissions of the greenhouse

---

 **Cultural Awareness** GENERAL

Native Alaskans and Oil The Arctic National Wildlife Refuge (ANWR) is a region of wilderness along the northern coast of Alaska that oil companies are trying to open for oil and gas exploration. Two groups of native people in and around the ANWR, the Inupiat and the Gwich'in, have taken opposite positions regarding the future of the refuge. The Inupiat Eskimos live in Kaktovik, the only town inside the ANWR. The Inupiat traditionally have a subsistence culture but benefit from various economic developments, such as whaling and trapping. They originally opposed the drilling on environmental grounds but now think that oil drilling is needed for the money and the jobs it will generate. The Gwich'in live in 15 small towns across northeast Alaska and northwest Canada. Caribou is their main source of livelihood, and they rely on the Porcupine Caribou Herd, which uses the ANWR as a place to bear their young and feed. The Gwich'in want permanent protection of the ANWR in order to protect the caribou.

The Environmental Effects of Using Oil When petroleum fuels are burned, they release pollutants. Internal combustion engines in vehicles that burn gasoline pollute the air in many cities. These pollutants contribute to the formation of smog and cause health problems. Emissions regulations and technology such as catalytic converters have reduced air pollution in many areas. However, in developing countries, cars are generally older, and the gasoline that they burn contains significantly more sulfur, a pollutant that contributes to acid precipitation. Many scientists also think that the carbon dioxide released from burning petroleum fuels contributes to global warming.

Oil spills, such as the one shown in **Figure 9**, are another potential environmental problem of oil use. In recent years, new measures have been taken to prevent oil spills from tankers. These measures include requiring that new tankers be double-hulled so that puncturing the outer hull does not allow the oil to leak out. Also, response times to clean up oil spills have improved. While oil spills are dramatic, much more oil pollution comes from everyday sources, such as leaking cars. However, measures to reduce everyday contamination of our waterways from oil lag far behind the efforts made to prevent large spills.

Figure 9 ▶ This ship is attempting to contain the oil spilled from the *Sea Empress* in 1996.

▶ A researcher holds a chunk of burning methane hydrate (far left). A methane hydrate mound in the Gulf of Mexico (left).

gas carbon dioxide to reduce global warming. As a result, the use of fuel-cell technology in vehicles will also increase. Fuel cells need hydrogen as a power source, and natural gas is a convenient source of hydrogen. The first fuel-cell vehicles will probably be fueled at a gas station that also has a natural gas pump.

So far, we have no technology to recover or use this strange mixture of ice and methane. One potential idea is to pump heated water into a methane hydrate deposit to melt the ice and release the methane gas. The gas would then have to be pumped to a processing plant.

CRITICAL THINKING

1. Applying Processes What are the differences between the geologic processes by which petroleum and natural gas form and the way methane hydrates form?

2. Analyzing Relationships Methane is a very effective greenhouse gas. How might this factor into the extraction or use of methane hydrates in the future?

441

Fossil Fuel Predictions Make sure students realize that projections of fossil-fuel reserves, shown in **Figure 11,** are based largely on deposits not yet discovered. It is theoretically possible that little or no additional reserves will be found, which would greatly shorten the life span of the resource. On the other hand, it is possible that recoverable fossil-fuel deposits may exceed estimated amounts by as much as a factor of two or three. **LS** Visual

The Future of Fossil Fuels
Organize the class into two groups to debate the energy future of our country. One group will take the position of continued fossil fuel usage and further exploration for more reserves. Tell the second group to argue for reducing fossil fuel consumption and to advocate alternative energy sources. Have students base their arguments on what they learned in this chapter and on any research done before the debate takes place.
LS Verbal/Interpersonal

MATHPRACTICE

Answers

$100 \times (66.9$ million barrels/day $- 59.6$ million barrels/day$) \div$ 59.6 million barrels/day $=$ 12.2% increase in petroleum production

Figure 10 ▶ Except when it is refueling, a vehicle that runs on natural gas looks like one that runs on gasoline.

MATHPRACTICE

World Energy Use
In 1980, worldwide production of petroleum was 59.6 million barrels per day. In 1998, petroleum production was 66.9 million barrels a day. Calculate the percent increase in oil production during this period.

Figure 11 ▶ This graph shows past oil production and one prediction for the future.

Natural Gas About 20 percent of the world's nonrenewable energy comes from natural gas. Natural gas, or methane (CH_4), is a good example of how advances in technology can make a fuel more common. In the past, when natural gas was encountered in an oil well, it was burned off because it was considered a nuisance. As technology improved, transporting natural gas in pipelines and storing it in compressed tanks became more practical. Now, many more oil wells recover natural gas. Because burning natural gas produces fewer pollutants than other fossil fuels, vehicles that run on natural gas, such as the one in **Figure 10,** require fewer pollution controls. Electric power plants can also use this clean-burning fuel.

Fossil Fuels and the Future

Today, fossil fuels supply about 90 percent of the energy used in developed countries. Some projections suggest that by 2050 world energy demand will have doubled, mainly as a result of increased population and industry in developing countries. As the demand for energy resources increases, the cost of fossil fuels will likely increase enough to make other energy sources more attractive. Planning now for the energy we will use in the future is important because it takes many years for a new source of energy to make a significant contribution to our energy supply.

Predicting Oil Production Oil production is still increasing, but it is increasing much more slowly than it has in the past, as shown in **Figure 11.** Many different factors must be considered when predicting oil production. **Oil reserves** are oil deposits that can be extracted profitably at current prices using current technology. In contrast, some oil deposits are yet to be discovered or to become commercial. Predictions must also take into account changes in technology that would allow more oil to be extracted in the future.

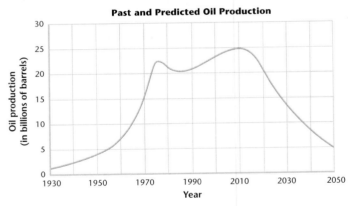

Past and Predicted Oil Production

Source: Petroconsultants S.A.

442

REAL-LIFE

CONNECTION — BASIC

Natural Gas Natural gas is an odorless gas that is composed largely of methane. Because natural gas leaks can be deadly, gas companies add *mercaptan,* a relatively harmless substance that smells like rotten cabbage, so that if there is a leak, it can be detected. Make sure that students can identify this smell and know what to do if they detect it.

Finally, all predictions of future oil production are guided by an important principle: the relative cost of obtaining fuels influences the amount of fossil fuels that we extract from the Earth. For example, as the supply of readily available oil decreases, we may begin to rely less on oil reserves and focus on using oil more selectively. At that time, oil will begin to be used more for applications in which it is essential. Cars and power plants, which can be powered in many ways, will begin to rely on other energy sources.

Future Oil Reserves No large oil reserves have been discovered in the past decade, and geologists predict that oil production from fields accessible from land will peak in about 2010. Additional oil reserves are under the ocean, but extracting oil from beneath the ocean floor is much more expensive. Currently, oil platforms can be built to drill for oil at depths greater than 1,800 m, but much of the oil in the deep ocean is currently inaccessible. Deep-ocean reserves may be tapped in the future, but unless oil-drilling technology improves, oil from the deep ocean will be much more expensive than oil produced on land.

Connection to Astronomy

Cosmic Oil Some scientists support a controversial hypothesis that some oil and natural gas deposits did not come from the remains of ancient life but were incorporated into the Earth during its formation from the solar nebula 4.5 billion years ago. Hydrocarbons are some of the most common compounds in the universe and large amounts of them could have been incorporated in the Earth as it formed. If this theory is correct, the Earth could contain vast reserves of "cosmic fossil fuels" that are not yet discovered.

Figure 12 ▶ This offshore oil rig is extracting petroleum from beneath the ocean floor.

SECTION 1 Review

1. **Describe** five factors that influence the value of a fuel.

2. **Describe** how fossil fuels are used to produce electricity, and explain how an electric generator works.

3. **Describe** how coal, oil, and natural gas form, how these fuels are used, and how using each fuel affects the environment.

CRITICAL THINKING

4. **Analyzing Relationships** What is the relationship between natural gas and petroleum?

5. **Making Comparisons** Read the description of how fossil-fuel deposits form. Are fossil fuels produced today by the same geologic processes as in the past?
   READING SKILLS

6. **Making Inferences** Examine Figure 11. What do you think accounts for the dramatic increase in the worldwide production of oil after 1950?

443

Overview

Before beginning this section, review with your students the Objectives in the Student Edition. The advantages and disadvantages of nuclear energy are examined in this section. Nuclear fusion is also discussed as a future energy source.

🔊 Bellringer

Ask students to think about how nuclear energy and nuclear power plants are portrayed in the media. Then, have them write about what they have seen on TV or in the movies and list any questions about nuclear power that they would like to answer. **LS** Visual

Motivate

Discussion ──────── GENERAL

A New Nuclear Power Plant in Your Community Tell the class to imagine that a nuclear power plant has been proposed to replace an aging coal-fired power plant in their community. Tell half the class that they are to defend the proposal while the other half will oppose it. Allow time for students to build a case for their position. Suggest that groups have each member assume the role of a person such as a utility company representative, a design engineer, an unemployed worker, a concerned citizen and parent, a property owner, or a business owner. **LS** Verbal Co-op Learning

---

SECTION 2
Nuclear Energy

Objectives

▶ Describe nuclear fission.
▶ Describe how a nuclear power plant works.
▶ List three advantages and three disadvantages of nuclear energy.

Key Terms

nuclear energy
nuclear fission
nuclear fusion

📶 **internet** connect

www.scilinks.org
Topic: Nuclear Energy
SciLinks code: HE4075

SCI *LINKS* Maintained by the National Science Teachers Association

Figure 13 ▶ Neutrons are released from the fission, or the splitting, of a uranium atom's nucleus. Some of these neutrons then cause other atoms to undergo nuclear fission in a process called a *chain reaction*.

In the 1950s and 1960s, nuclear power plants were seen as the power source of the future because the fuel they use is clean and plentiful. It was predicted that a nationwide network of nuclear power plants would provide electricity that was "too cheap to meter." But in the 1970s and 1980s, almost 120 planned nuclear power plants were canceled, and about 40 partially constructed nuclear plants were abandoned. What happened? In this section, you'll learn how nuclear power works and why about 17 percent of the world's electricity comes from nuclear power today.

Fission: Splitting Atoms

Nuclear power plants get their power from **nuclear energy**, the energy within the nucleus of an atom. The forces that hold together the nucleus of an atom are more than 1 million times stronger than the chemical bonds between atoms. In nuclear power plants, atoms of the element uranium are used as the fuel.

The nuclei of uranium atoms are bombarded with atomic particles called *neutrons*. These collisions cause the nuclei to split in a process called **nuclear fission**. A fission reaction is shown in **Figure 13**. Nuclear fission releases a tremendous amount of energy and more neutrons, which in turn collide with more uranium nuclei. If a fission reaction is allowed to continue, this chain reaction will escalate quickly. One example of an uncontrolled fission reaction is the explosion of an atomic bomb. In contrast, nuclear power stations are designed so that the chain reaction produces a controllable level of energy.

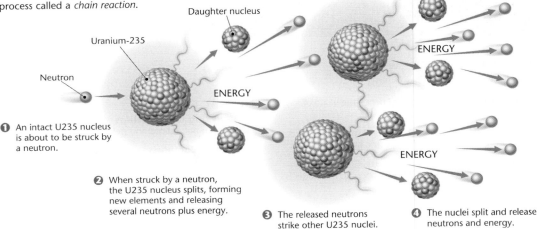

Daughter nucleus

Uranium-235

Neutron

ENERGY

ENERGY

ENERGY

ENERGY

❶ An intact U235 nucleus is about to be struck by a neutron.

❷ When struck by a neutron, the U235 nucleus splits, forming new elements and releasing several neutrons plus energy.

❸ The released neutrons strike other U235 nuclei.

❹ The nuclei split and release neutrons and energy.

444

Chapter Resource File

• **Lesson Plan**
• **Active Reading** BASIC
• **Section Quiz** GENERAL

Transparencies

TT Bellringer
TT Fission & Fusion

MISCONCEPTION
/// ALERT \\\

Uncontrolled Fission Reactions
Students may think that it is possible for a nuclear power plant to explode like an atomic bomb if the reaction is uncontrolled. Point out that atomic bombs use different fuel from that used in a power plant. Weapons-grade uranium is over 90 percent pure uranium (U-235), compared to 3.5 percent uranium in commercial fuel rods. The uranium fuel rods may melt, but they cannot explode.

How Nuclear Energy Works

A nuclear reactor is surrounded by a thick pressure vessel that is filled with a cooling fluid. The pressure vessel is designed to contain the fission products in case of an accident. Thick concrete walls also surround reactors as shown in **Figure 14.**

Inside a reactor, shown in **Figure 15,** metal fuel rods that contain solid uranium pellets are bombarded with neutrons. The chain reaction that results releases energy and produces more neutrons. The reactor core contains control rods, which are made of a material such as boron or cadmium that absorbs the neutrons to prevent an uncontrolled chain reaction. When the control rods are lowered between the fuel rods, they slow the fission reactions. If the control rods are lowered completely, they prevent fission and shut down the reactor.

The heat released during nuclear reactions is used to generate electricity in the same way that power plants burn fossil fuels to generate electricity. In a nuclear power plant, energy released from the fission reactions heats a closed loop of water that heats another body of water. As the water boils, it produces steam that drives a steam turbine, which is used to generate electricity.

Figure 14 ▶ Every year, the Diablo Canyon nuclear plant generates enough energy for 2 million Californian households—the energy equivalent of burning 20 million barrels of oil.

Figure 15 ▶ How a Typical Nuclear Power Plant Works

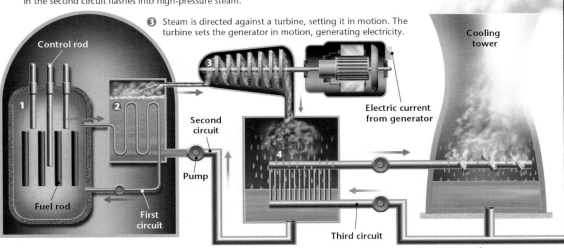

❷ The superheated water is pumped to a heat exchanger, which transfers the heat of the first circuit to the second circuit. Water in the second circuit flashes into high-pressure steam.

❸ Steam is directed against a turbine, setting it in motion. The turbine sets the generator in motion, generating electricity.

Control rod

Cooling tower

Electric current from generator

Second circuit

Pump

Fuel rod

First circuit

Third circuit

❶ Energy released by the nuclear reaction heats water in the pressurized first circuit to a very high temperature.

❹ A third circuit cools the steam from the turbine and the waste heat is released from the cooling tower in the form of steam.

445

BRAIN FOOD

Nuclear Power Production and Nuclear Weapons Some commercial nuclear reactors produce byproducts (plutonium and an isotope of uranium) that can be reprocessed into ingredients necessary for making nuclear weapons. This has led to a fear of nuclear weapons proliferation. Have students discuss the advantages and disadvantages of nuclear power in this light. Ask students to propose solutions to ensure that nuclear reactor byproducts are not misused. (Answers may vary. Students may suggest independent monitoring of the transfer of plutonium and uranium.) You may also want to compare this issue with the effects of dependence on imported oil. Point out that the use of most energy resources has ramifications in international politics. **LS Interpersonal**

Discussion ─────────── GENERAL

Nuclear Waste Disposal There are several options for nuclear waste disposal. These options include sending the waste into space; storing the waste in deep, isolated parts of the ocean; storing it in Antarctica; paying a developing country to dispose of it; storing the waste in a stable geologic formation in an isolated region; studying the issue further; and doing nothing. Have each student choose an option and explain their reasoning. **LS** **Verbal** Co-op Learning

---

MISCONCEPTION ///ALERT\\\

Nuclear Power and Radioactivity Modern, properly operated nuclear power plants release almost no radioactivity into the environment. Instead, the radioactive substances are contained in the plant's wastes. Coal naturally contains very small amounts of radioactive substances, such as uranium and thorium. When the coal is burned in coal-fired power plants, small quantities of these substances are released into the environment in the ash and exhaust. As a result, people living near coal-fired power plants are exposed to more nuclear radiation than people living near properly maintained nuclear plants.

Figure 16 ▶ Uranium is a very compact fuel. The uranium pellets (above) can generate as much electricity as the trainload full of coal does.

Geofact

Radon Uranium occurs naturally in rock and soil. When uranium undergoes radioactive decay, it gives off a number of products, including an invisible and odorless radioactive gas called radon. Radon can seep into buildings from the surrounding rock and soil, and if buildings are not ventilated properly, dangerous levels of radon can build up. It is estimated that radon causes 5,000 to 20,000 people in the United States to die from lung cancer each year.

446

The Advantages of Nuclear Energy

Nuclear energy has many advantages. Nuclear fuel is a very concentrated energy source, as shown in **Figure 16.** Furthermore, nuclear power plants do not produce air-polluting gases. When operated properly, nuclear plants release less radioactivity than coal-fired power plants do. Many countries with limited fossil-fuel reserves rely heavily on nuclear plants to supply electricity. France, for example, generates about three-fourths of its electricity from nuclear power. France produces less than one-fifth of the air pollutants per person than does the United States, which relies on fossil fuels for almost 70 percent of its electricity needs.

Why Aren't We Using More Nuclear Energy?

Building and maintaining a safe reactor is very expensive. As a result, nuclear power is no longer competitive with other energy sources in many countries. The last 20 nuclear reactors built in the United States cost more than $3,000 per kilowatt of electrical capacity. In contrast, wind power is being installed at less than $1,000 per kilowatt, and newer natural gas power plants can cost less than $600 per kilowatt. However, the actual cost of new nuclear power plants is uncertain, so it is difficult to predict whether investors will build new plants in the United States.

Storing Waste The difficulty of finding a safe place to store nuclear wastes is one of the greatest disadvantages of nuclear power. The fuel cycle of uranium produces fission products that can remain dangerously radioactive for thousands of years. Uranium mining and fuel development produce radioactive waste. In addition, the used fuel, liquids, and equipment from a reactor core are also considered hazardous wastes. Storage sites for nuclear wastes must be located in areas that are geologically stable for tens of thousands of years. The United States has spent over two decades studying a site called Yucca Mountain in southern Nevada as a place to store nuclear waste. Scientists are also researching a process called transmutation, that would recycle the radioactive elements in nuclear fuel.

Safety Concerns In a poorly designed nuclear plant, the fission process can potentially get out of control. This is what happened during the world's worst nuclear reactor accident, which occurred at Chernobyl in the Ukraine in 1986. Engineers turned off most of the reactor's safety devices to conduct an unauthorized test. This test caused explosions that destroyed the reactor and blasted tons of radioactive materials into the air. Hundreds of firefighters, residents, and workers died from radiation exposure. Areas of northern Europe and the Ukraine are still contaminated from the disaster. The Chernobyl reactor was an old design that, for safety reasons, is not used in the United States. The nuclear reactor had no containment building. In addition, the engineers at Chernobyl violated basic safety guidelines.

Homework ───────── ADVANCED

Nuclear Fusion Suggest that students research nuclear fusion and produce a poster, report, or short presentation describing the main methods of nuclear fusion and what a fusion power plant might look like. Students may want to focus on the incredibly complicated technical problems that are posed by fusion and to discuss how those problems are being solved. Suggest that students include their findings in their **Portfolio.** **LS** **Interpersonal**

MISCONCEPTION ///ALERT\\\

Radioactive Waste Students may not understand that the byproducts of nuclear fission do not remain radioactive forever. After the fission reaction, the waste products give off heat and radioactivity, cooling down and gradually becoming less hazardous. The amount of time this takes varies from minutes to thousands of years, depending on the element. Plutonium, one of the byproducts of breeder reactors, remains hazardous to humans for about 240,000 years.

In the United States, the most serious nuclear accident occurred in 1979 at the Three Mile Island nuclear power plant in Pennsylvania. Human error, along with blocked valves and broken pumps, was responsible for the accident at Three Mile Island. Fortunately, only a small amount of radioactive gas escaped. Since this accident, the U.S. Nuclear Regulatory Commission has required more than 300 safety improvements to nuclear power plants.

The Future of Nuclear Power

One possible future energy source is nuclear fusion. **Nuclear fusion** occurs when lightweight atomic nuclei combine to form a heavier nucleus and release tremendous amounts of energy. **Figure 17** illustrates the process of nuclear fusion. Nuclear fusion is the process that powers the stars, including our sun. It is potentially a safer energy source than nuclear fission is because it creates less dangerous radioactive byproducts.

Unfortunately, although the potential of fusion is great, so is the technical difficulty of achieving that potential. For fusion to occur, atomic nuclei must be heated to extremely high temperatures (about 100,000,000°C, or 180,000,000°F). The nuclei also must be maintained at very high concentrations and properly confined. Achieving all three of these conditions simultaneously is extremely difficult. The technical problems are so complex that building a nuclear fusion plant may take decades or may never happen.

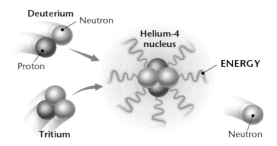

Connection to **History**

Three Mile Island The Three Mile Island accident was a wake-up call for the nuclear industry. Many reforms and safety measures were instituted throughout the industry after the accident occurred. In 1989, 10 years after the accident, the nuclear plant received the best INPO rating in the world. The rating was based on a measure of reliability, efficiency, and safety. In 1999, the plant set a world record after running continuously for 688 days.

Figure 17 ▶ During nuclear fusion, the nuclei of two forms of hydrogen (deuterium and tritium in this case) join to form helium, which releases large amounts of energy.

SECTION 2 Review

1. **Compare** a power plant that burns fossil fuels with a nuclear power plant.

2. **Describe** two advantages and two disadvantages of nuclear power plants.

3. **Explain** the difference between nuclear fission and nuclear fusion.

CRITICAL THINKING

4. **Applying Ideas** Read about the advantages of nuclear energy. Explain why countries such as France and Japan rely heavily on nuclear power. `READING SKILLS`

5. **Making Decisions** Which poses more of an environmental threat: transporting spent nuclear fuel or transporting toxic chemicals? Write your opinion in the form of a short essay. `WRITING SKILLS`

447

Close

Reteaching —————— `BASIC`

Section Review Have students work in pairs to quiz each other about the material in this section. Students should take turns asking and answering questions that they develop based on the text. `LS` **Verbal** | English Language Learners |

Quiz —————— `GENERAL`

1. What percentage of the world's electricity comes from nuclear power? (about 17 percent)

2. What is the fuel used in most nuclear power plants? (uranium)

3. What safety feature of nuclear power plants can slow down and stop the nuclear reaction? (control rods made of graphite or cadmium that absorb the extra neutrons)

Alternative Assessment —————— `GENERAL`

Nuclear Energy: Take a Side
Have students prepare a brochure or advertisement for either a pro-nuclear energy or an anti-nuclear energy organization. The document should be persuasive, but tell students to support their document with facts and sound reasoning, not with emotional appeals. Encourage students to include their work in their **Portfolio.** `LS` **Verbal**

Answers to Section Review

1. Both plants heat water to produce steam that drives an electric generator. A nuclear power plant generates heat by means of a nuclear reaction, while a fossil-fuel-burning power plant generates heat by burning fossil fuels.

2. Answers may vary. Some advantages are that nuclear fuel is a very concentrated energy source, it produces no air-polluting gases, and it provides an alternative energy source for countries with limited fossil-fuel supplies. Disadvantages include the high costs of construction and maintenance, the difficulty of storing the waste, and safety concerns.

3. In nuclear fission, atomic nuclei are split into neutrons and energy and many of the byproducts are highly radioactive. In nuclear fusion, lightweight atomic nuclei combine to form heavier nuclei, and radioactive byproducts are fewer and short-lived. The conditions for nuclear fusion are much more difficult to reach than the conditions for nuclear fission.

4. Answers may vary. Sample answers: There are few fossil-fuel resources in Japan and France. The two countries want an independent source of energy so that they don't have to rely on sources from politically unstable countries. They want to reduce air pollution.

5. Answers may vary. Students should consider the amount of fuel that is transported, the relative dangers of each material, and the safety of the transportation method.

CHAPTER 17 Highlights

Alternative Assessment — GENERAL

Nonrenewable Energy Current Events Have student groups create a newspaper or magazine focused on current news in the nonrenewable energy sector by researching related articles and pictures from periodicals. Suggest that they create different sections, such as business, politics, and lifestyle, and include appropriate material in each section. **LS** Visual/Interpersonal Co-op Learning

🔹 Internet Activity — GENERAL

The Cost of Gasoline Have students use the Internet to research the price per liter of gasoline in various countries. Ask students if these prices represent the true cost of gasoline. (no, because governments provide large subsidies to fossil-fuel companies, gasoline is taxed at different rates, and the environmental costs, such as air pollution problems, are not included in the prices.) If possible, have students find out what percentage of fuel prices is due to taxation and what the tax money is used for. Ask them, "What are the advantages and disadvantages of high gasoline taxes?" (Sample answers: Some advantages are that money can be raised for other government expenses and that people will be encouraged to conserve fuel. A disadvantage is that driving a gas-powered vehicle is more expensive.) **LS** Intrapersonal

1 Energy Resources and Fossil Fuels

Key Terms

fossil fuels, 435
electric generator, 436
petroleum, 440
oil reserves, 442

Main Ideas

▶ Most of the world's energy needs are met by fossil fuels, which are nonrenewable resources.

▶ Coal is abundant in North America and Asia. In the United States, coal is used primarily to produce electricity.

▶ Petroleum can be refined into fuels to power vehicles and machines. Petroleum can also be used to manufacture many other products.

▶ Natural gas is often found above oil deposits. In general, burning natural gas releases fewer pollutants than burning coal or oil.

▶ The extraction, transportation, and use of fossil fuels cause many environmental problems, including air and water pollution and habitat destruction.

▶ Calculations of fossil-fuel reserves predict that oil production will peak and then decline in the early 21st century.

2 Nuclear Energy

nuclear energy, 444
nuclear fission, 444
nuclear fusion, 447

▶ Nuclear energy is energy that exists within the nucleus of an atom. When uranium nuclei are bombarded with neutrons, they undergo fission and release large amounts of energy.

▶ In a nuclear power station, the heat generated by fission is used to heat water to form steam. The steam drives turbines that generate electricity.

▶ The main advantages of nuclear power are that the fuel is compact and the power stations generally do not pollute. The main disadvantage is that nuclear power produces radioactive waste, which will be dangerous for centuries.

448

Chapter Resource File

- Chapter Test GENERAL
- Chapter Test ADVANCED
- Concept Review GENERAL
- Critical Thinking ADVANCED
- Test Item Listing
- Modeling Lab GENERAL
- Observation Lab BASIC
- CBL™ Probeware Lab GENERAL
- Consumer Lab GENERAL
- Long-Term Project ADVANCED

Using Key Terms

Use each of the following terms in a separate sentence.

1. *fossil fuels*
2. *petroleum*
3. *oil reserves*
4. *nuclear fission*
5. *nuclear fusion*

For each pair of terms, explain how the meanings of the terms differ.

6. *petroleum* and *oil reserve*
7. *turbine* and *electric generator*
8. *nuclear fission* and *nuclear fusion*

✔ **STUDY TIP**

Get Organized Being organized can help make studying more efficient and less confusing. Start by reducing clutter and consolidating loose papers. Arrange your items by subject, and be sure to label your books, notebooks, and dividers. A planner, or agenda book, can help you balance schoolwork with other activities. It also can serve as reminder of upcoming deadlines and help you to prioritize multiple tasks.

Understanding Key Ideas

9. Which of the following statements provides a reason for the widespread use of fossil fuels?
   a. Fossil fuels are a renewable source of energy.
   b. Fossil fuels are readily available and inexpensive.
   c. Fossil fuels are not harmful to the environment.
   d. all of the above

10. Which of the following pairs are design features that nuclear power plants and coal-fired power plants share?
    a. fuel rods and containment buildings
    b. turbines and generators
    c. combustion chamber and reactor cores
    d. none of the above

11. The main reason for the worldwide slowdown in the construction of nuclear power plants is that
    a. we have run out of uranium fuel.
    b. the electricity from nuclear power is generally more expensive to produce than electricity from other sources.
    c. nuclear reactors are inherently unsafe.
    d. nuclear reactors release large quantities of greenhouse gases.

12. Which is an example of the direct use of fossil fuels?
    a. a nuclear reactor
    b. an oil-fired furnace
    c. a wood-burning stove
    d. an electric generator

13. Which of the following statements describes the process by which modern nuclear power plants use nuclear energy?
    a. Power plants use nuclear fusion to split uranium atoms and release nuclear energy.
    b. Power plants use nuclear fusion to combine atomic nuclei and release nuclear energy.
    c. Power plants use nuclear fission to split uranium atoms and release nuclear energy.
    d. Power plants use nuclear fission to combine atomic nuclei and release nuclear energy.

14. If fossil fuels are still forming today, why are they considered nonrenewable resources?
    a. Fossil fuels are broken down by natural processes faster than they form.
    b. We are depleting fossil fuels much faster than they form.
    c. The fossil fuels being formed today are deep under the ocean, where they cannot be reached.
    d. The only fossil fuels being produced are methane hydrates, which we cannot use yet.

15. Which of the following is *not* a concern about nuclear energy?
    a. the difficulty of safe storage of nuclear waste
    b. the high levels of air pollution produced
    c. the high costs of nuclear energy
    d. the possibility that a nuclear chain reaction can get out of control

449

ANSWERS

Using Key Terms

1. Sample answer: Fossil fuels are a nonrenewable resource.

2. Sample answer: Petroleum provides a large percentage of our fuel needs.

3. Sample answer: Knowledge of current oil reserves is important for predicting future oil production.

4. Sample answer: Nuclear fission involves splitting uranium atoms.

5. Sample answer: Nuclear fusion could be an energy source of the future.

6. Petroleum is oil that has been pumped from the ground. Oil reserves are oil deposits that can be extracted profitably at current prices using current technology.

7. A turbine is a wheel that changes the force of a moving gas or liquid into energy and sets a generator in motion. The generator converts this motion into electrical energy.

8. Nuclear fission is the splitting of atoms. Nuclear fusion is the combining of atoms.

Understanding Key Ideas

9. b
10. b
11. b
12. b
13. c
14. b
15. b

Assignment Guide	
Section	Questions
1	1–3, 6, 7, 9, 12, 14, 16, 19–29, 32, 33
2	4, 5, 8, 10, 11, 13, 15, 17, 18

Short Answer

16. Fossil fuels are readily available and are a more concentrated source of energy than, for example, fuelwood.

17. The Three Mile Island accident led to over 300 safety improvements in nuclear power plants in the United States.

18. Answers may vary. Safety controls, waste storage, and plant maintenance costs make nuclear power expensive.

19. An oil deposit is any known petroleum deposit on Earth. Oil reserves are oil deposits that can be extracted profitably at current prices using current technology.

Interpreting Graphics

20. approximately 1950

21. Answers may vary. The Industrial Revolution created more uses for coal, including coal-powered steam engines.

22. Answers may vary. Oil and natural gas production are driven by demand and price. Because oil and natural gas are commonly found in the same deposits, the demand for oil influences the production of natural gas.

23. Answers may vary. Coal is being used increasingly to power many of the power plants built to satisfy the nation's growing electricity demand.

24. Answers may vary. Although fossil fuels are used in many applications where wood was once used, people still use wood for heating. The increase in population may compensate for the relative decrease in the percentage of people using wood as fuel.

Short Answer

16. Why have fossil fuels become our primary energy resource?

17. How did the Three Mile Island accident affect nuclear safety in the United States?

18. What factors make nuclear power expensive?

19. What is the difference between oil reserves and oil deposits?

Interpreting Graphics

The graph below shows the different contributions of various fuels to the U.S. energy supply since 1850. Use the graph to answer questions 20–24.

20. When did oil first become a more important energy source than coal?

21. Why do you think the use of coal increased so rapidly between 1850 and 1920?

22. The data for oil and natural gas are nearly parallel—they rise and fall together. Why do you think this pattern exists?

23. Why do you think the use of coal is on the rise after having fallen in the 1950s?

24. Why do you think that the use of wood as a fuel has not significantly increased or decreased since about 1850?

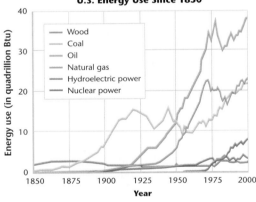

U.S. Energy Use Since 1850

Legend:
— Wood
— Coal
— Oil
— Natural gas
— Hydroelectric power
— Nuclear power

Y-axis: Energy use (in quadrillion Btu), 0 to 40
X-axis: Year, 1850 to 2000

Concept Mapping

25. Use the following terms to create a concept map: *oil well, petroleum, refinery, gasoline, natural gas, plastics,* and *oil reserve.*

Critical Thinking

26. **Demonstrating Reasoned Judgment** The invention of artificial plastics had a damaging effect on the environment because most plastics break down very slowly, so they remain in landfills and are dangerous to wildlife. However, the invention of plastics also affected the environment in many positive ways. List as many positive effects as you can.

27. **Analyzing Relationships** Read the description of how fossil-fuel deposits form. Explain why fossil fuels are a form of stored solar energy. **READING SKILLS**

28. **Analyzing Relationships** The United States currently imports about half of all the crude oil it uses. Why might this be a problem? Write a paragraph that describes the recommendations that you would make to U.S. lawmakers, manufacturers, and consumers to reduce the country's dependence on foreign oil. **WRITING SKILLS**

Cross-Disciplinary Connection

29. **Economics** What incentives could encourage automobile manufacturers in the United States to produce more fuel-efficient cars? The U.S. government could increase the requirements for fuel efficiency. However, at least two other strong forces are likely to change the types of vehicles that manufacturers produce. What do you think these forces are?

Portfolio Project

30. **Prepare a Display** Find out how petroleum, natural gas, coal, or uranium are extracted. For example, engineers have developed methods to drill sideways to reach oil deposits thousands of feet underground. Research one method and prepare a model or a posterboard display that communicates your findings. Be sure to include information about the environmental effects of the method you studied.

Concept Mapping

25. Answers to the concept mapping questions are on pp. 667–672.

Critical Thinking

26. Answers may vary. Sample answer: Plastics can replace products formerly obtained from endangered species, such as ivory.

27. Answers may vary. Fossil fuels were formed from the remains of plants and tiny marine organisms. Plants store energy that is absorbed during the process of photosynthesis. The marine organisms either use sunlight directly or consume organisms that do. The bodies of all of these organisms were ultimately formed using solar energy.

28. Answers may vary. Students may recommend exploring for new fossil-fuel deposits, developing technology to more efficiently extract fuel, developing new sources of energy, or conserving energy.

 **MATH SKILLS**

The graph below compares the contribution of each world region to world oil production. Use the graph to answer question 31.

31. Analyzing Data If the total sales of oil in 2002 were $500 billion, what is the value of the oil produced by each region?

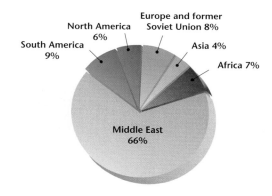

North America 6%

South America 9%

Europe and former Soviet Union 8%

Asia 4%

Africa 7%

Middle East 66%

WRITING SKILLS

32. Communicating Main Ideas How would our lives change if oil reserves became so depleted that gasoline is very expensive?

33. Recognizing Relationships Outline the major forms of environmental change that have resulted from fossil-fuel use. Include your thoughts on subjects such as habitat loss, pollution, and our use of land. Remember to include positive environmental changes.

STANDARDIZED TEST PREP

For extra practice with questions formatted to represent the standardized test you may be asked to take at the end of your school year, turn to the sample test for this chapter in the Appendix.

 READING SKILLS

Read the passage below, and then answer the questions that follow.

Paula Curtis became chief executive officer of Zaft Motors in 2002. She has a strong interest in the environment. Because she is chief executive of the country's second-largest auto manufacturer, she has an influence on the automobile industry. For instance, Zaft left the Global Climate Coalition, a group of companies that denied the scientific research proving global warming. Within four months, two other auto manufacturers also left. Zaft publishes a "corporate citizenship" report each year. In 2001, the report stated that Zaft's vehicles and factories emit 350 million metric tons of carbon dioxide annually and contribute to global warming. The report also stated that Zaft was committed to reducing this number.

However, Zaft has a long way to go to fulfill this goal. Zaft has failed to improve the fuel economy of its cars and trucks, so its new vehicles get fewer miles per gallon, on average, than the vehicles built in 1982. Modern technology for engines, transmissions, and aerodynamics could help Zaft achieve an average fuel economy of 40 mi/gal for its cars, pickups, and sport utility vehicles. As a result, the United States would save almost 1 million barrels of oil per day—over half as much as the country imports from Saudi Arabia.

1. Which of the following statements best describes the thesis of the article?
   a. Zaft Motors is jeopardizing its position as the country's second-largest automaker by enacting environmental controls.
   b. Zaft left the Global Climate Coalition because it acknowledged the scientific evidence for global warming.
   c. Although Zaft has taken some actions to be an environmentally responsible corporate citizen, the company still needs to improve the fuel efficiency of its vehicles.
   d. none of the above

451

Cross-Disciplinary Connection
29. Sample answers: consumer demand and the price of fuel.

Portfolio Project
30. Answers may vary.

Math Skills
31. Middle East: $330 billion; South America: $45 billion; Europe and former Soviet Union: $40 billion; Africa: $35 billion; North America: $30 billion; Asia: $20 billion

Writing Skills
32. Answers may vary.
33. Answers may vary.

Reading Skills
   1. c

YOUR HOUSEHOLD ENERGY CONSUMPTION

Teacher's Notes

Time Required
one-hour home activity and one 45-minute class period

Lab Ratings

EASY ———————→ HARD

TEACHER PREPARATION
STUDENT SETUP
CONCEPT LEVEL
CLEANUP

Skills Acquired
- Collecting Data
- Interpreting
- Organizing and Analyzing Data

The Scientific Method
In this lab, students will:
- Analyze Results
- Draw Conclusions

Materials and Equipment
The materials listed are enough for each student.

Safety Cautions
Students should be reminded of the danger of electric shock. When performing the home energy audit and reading the electric meter, students should not attempt to unplug or in any way alter the appliances or electric meter. Students should not touch the electric meter, circuit breaker boxes, or any of the wiring connected to these devices.

CHAPTER **17** Skills Practice Lab: CONSUMER

Objectives

▶ **USING SCIENTIFIC METHODS** **Identify** the ways in which electricity is consumed in your household.

▶ **Compute** the energy consumption of your household.

▶ **Interpret** an electric utility bill and an electric meter.

Materials
calculator
electric bill
notebook
pen or pencil

▶ **Keeping Track of Energy Use**
An electric meter (below) records the amount of electricity that a household uses. A utility bill (below) calculates the cost of the electricity used.

Your Household Energy Consumption

We use electricity for many activities at home, such as drying clothes, cooking food, and heating and cooling. The total amount of energy that we use depends both on how much energy each individual appliance consumes and on how long we use the appliance each day. In this lab, you will survey your household to determine how much electricity you consume and you will analyze an electric bill to calculate how much you pay for your electricity.

Procedure

1. Create a table similar to the one shown below. To determine daily energy consumption in kilowatt-hours, divide the wattage of an appliance by 1,000 and then multiply by the number of hours the item is used per day.

Appliance	Energy consumed in 1 hour (watts)	Hours used (per day)	Daily energy consumption (Kwh)

DO NOT WRITE IN THIS BOOK

2. Walk through your home, and identify all appliances and devices that use electricity. List each item in your table.

3. Fill in each column in your table. Determine the wattage of each item by referring to the table on the next page.

4. Find the electric meter. It may be on an outside wall of your house or apartment building. Record the current reading on the meter. The reading may change as you watch it. If so, electricity is currently being consumed in your household. If the reading is changing, write down an estimate of the current reading.

Electric Service	Meter#	Read Date		Reading
	141707	04/05/2002		87671.00
		03/07/2002		87503.00
		Read Difference		168.00
		Total Consumption in KWH		168

Billing Rate: Residential Service Winter
Customer Charge...$6.00
Energy Charge   168.00 @ $ .0355000 per KWH.................$5.96
Fuel Charge   168.00 @ $ .0177400 per KWH.................$2.98
Sales Tax...$0.15
TOTAL CURRENT CHARGES - Electric.................................$15.09

Tips and Tricks
Students may have appliances that are not listed in the table. If students include these in their survey, they could take the energy consumption value directly from the appliance (if it is listed there), research the appliance's energy consumption on the Internet, or choose a similar appliance from the table.

INCLUSION Strategies

- *Developmentally Delayed*
- *Learning Disabled*

Have students create a table that shows how they use electricity in their home. In the first column, students should list all of the rooms in their home or apartment. In the second column, students should indicate how electricity is used in each room. In the third column, students should indicate ways to reduce the amount of electricity used in each room.

Analysis

1. **Organizing Data** Add up the energy consumption per day for all items. This number is the total energy consumed by your household in one day.

2. **Organizing Data** On your electric bill, find the total number of kilowatt-hours consumed during this time period. An electric bill usually lists a meter reading for the beginning of the time period and for the end of the time period. The difference is the energy consumption in kilowatt-hours.

3. **Analyzing Data** Divide the number of kilowatt-hours from your electric bill by the number of days in the time period. This number reflects the average daily energy consumption for this time period.

4. **Analyzing Results** Compare the daily energy consumption that you calculated from your home survey with the average calculated from your electric bill. Is there a difference? If so, what could explain the difference?

5. **Analyzing Data** Find the cost of electricity per kilowatt-hour on your electric bill. How much does washing your clothes in a washing machine cost?

Conclusions

6. **Drawing Conclusions** What can you conclude about energy consumption in your home? What activities consume the most energy? How could you reduce the energy consumption in your home?

7. **Evaluating Methods** How could the energy survey be refined to estimate more accurately your daily energy consumption?

Extension

1. **Communicating Ideas** Even when an appliance is turned off, it can still consume electricity. This type of electricity consumption is called a *phantom load*. Find out about phantom loads and prepare a booklet that shows how people can reduce this type of energy use.

Energy Consumption for Common Household Appliances	
Appliance	Energy consumed in 1 hour (watts)
Ceiling fan	120
Clock radio	10
Clothes washer	425
Clothes dryer (electric)	3,400
Coffee maker	1,050
Dishwasher	1,800
Window fan	150
Hair dryer	1,500
Heater (portable)	1,100
Iron	1,400
Light bulbs	60, 75, 100
Microwave oven	900
Personal computer	270
Refrigerator (frost free, 16 ft³)	725
Stereo	400
Television (color)	130
Toaster	1,100
Toaster oven	1,225
Vacuum cleaner	1,200
VCR/DVD	19/22
Water heater (40 gal)	5,000
Water pump (deep well)	650
Window fan	150

Answers to Analysis

1. Answers may vary. The total energy consumption should be in kilowatt-hours.

2. Answers may vary.

3. Answers may vary.

4. Answers may vary. Students could have wrongly estimated the daily usage of appliances, or the values for energy consumption in the table may not match the actual appliances in the home. There may also be other appliances or devices that students did not consider.

5. Answers may vary. Students should use the value for the clothes washer's energy consumption from the table.

Answers to Conclusions

6. Answers may vary. Answers will depend both on the hours per day an appliance is used and on the hourly energy consumption of the appliance. Energy-intensive appliances include refrigerators, heaters, clothes dryers, and air conditioners.

7. Answers may vary. The energy consumption could be recorded directly from the appliance, usually from a sticker somewhere on the back. The students could also do a more detailed audit in which they monitor energy usage for a day or a week to get a more accurate estimate of appliance usage.

Answers to Extension

1. Answers will vary.

453

Chapter Resource File

• Datasheets for In-Text Labs
• Lab Notes and Answers

CLASSROOM TESTED & APPROVED

Dan Trockman
Hopkins High School
Minnetonka, Minnesota

NUCLEAR POWER IN THE UNITED STATES, 1950–2002

Background

The first controlled nuclear fission chain reaction was achieved at the University of Chicago in 1942 by Enrico Fermi's group. Electricity was first generated from an experimental reactor in the United States in 1951. Construction planning for nuclear reactors began on a large scale from 1966 to 1974. Although a total of 259 reactors were planned, 48 percent were never built. Only about 104 reactors are in operation today. The numbers and locations of nuclear power plants are based on the best available information at the time of publication.

Homework — GENERAL

Nuclear Waste, Fast Forward
Suggest that students write a fictional story set 10,000 years in the future. Tell students that archeologists have just discovered a strange stash of metals stored in canisters. Not realizing that their ancestors used nuclear energy as a source of power and then stored the wastes here, the society must deal with the consequences of their discovery. Have students describe the stories that the people invent to explain the discovery and what they think about the ancestors who built this structure. **LS** Intrapersonal

Transparencies

TT Nuclear Power in the United States, 1950–2002

MAPS *in action*

NUCLEAR POWER IN THE UNITED STATES, 1950–2002

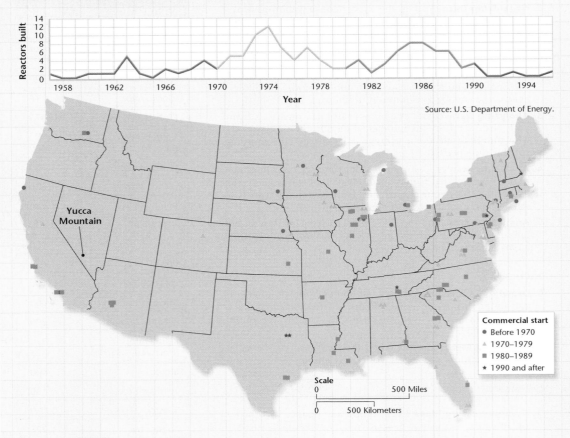

Source: U.S. Department of Energy.

Yucca Mountain

Commercial start
- ● Before 1970
- ▲ 1970–1979
- ■ 1980–1989
- ★ 1990 and after

Scale
0 ————— 500 Miles
0 ————— 500 Kilometers

MAP SKILLS

1. **Identifying Trends** During what time period did most of the nuclear plants in the United States begin operation?

2. **Using the Key** Use the symbols on the map to describe the history of nuclear power in the Northeast, the Southeast, the Great Plains states, and the western states.

3. **Comparing Areas** What region of the United States has the most nuclear power plants? What region has the fewest? Why?

4. **Identifying Relationships** Why do you think that many nuclear reactors are built close together?

5. **Calculating Problems** The U.S. government is considering storing most of its nuclear waste produced in the United States in a facility in Yucca Mountain in Nevada. This location is marked on the map. What are the average distances from the nuclear plants in California, Texas, Florida, and New York to this site?

454

Answers to Map Skills

1. 1970–1979

2. Sample answer: The large concentrations of people and of industries in the northeastern U.S. encouraged the early and abundant use of nuclear energy. The southeastern U.S. followed their example. In the western U.S., the population density was lower, so fewer nuclear power plants were built there. The prevalence of agriculture and the lower population in the Great Plains states may have affected the number of nuclear plants built. Nuclear power plants are usually built near large, reliable sources of water.

3. The eastern U.S. has the most nuclear power plants, and the west-central U.S. has the fewest.

This is probably due to the large concentration of people in the East.

4. Reactors are built close together to take advantage of the safest and most economically feasible areas available. In addition, once a nuclear power plant is built, adding additional reactors is more convenient and more cost-effective than building a new power plant in another location.

5. Yucca Mountain is approximately 375 km (250 mi) from California reactors, 1,950 km (1,200 mi) from Texas reactors, 3,200 km (2,000 mi) from Florida reactors, and 3,600 km (2,200 mi) from New York reactors.

FROM CRUDE OIL TO PLASTICS

Can you imagine using shampoo from a glass bottle? Your parents or grandparents might remember a time when a dropped shampoo bottle meant shards of glass on the shower floor. Today, plastic is in everything from shampoo bottles, car fenders, artificial limbs, refrigerators, and cameras, to snowboards. While these products are all vastly different, they started out in the same way, as petroleum, or crude oil.

Petroleum is the most common fuel used to power vehicles and heat homes. But you may not realize that this versatile fuel contains many different organic compounds, which are used to produce many other products and chemicals of modern society. However, before these compounds can be used, they must be separated from petroleum. At a refinery, petroleum is heated so that it separates into petroleum distillates during a process called fractional distillation. When petroleum is heated, petroleum distillates evaporate and condense at different temperatures. These different compounds are then used to make many products we use every day.

A Refined Resource

When the crude oil is heated, volatile compounds that have very low boiling points, such as those that make up gasoline and airplane fuel, evaporate first and are collected. Compounds that have very high boiling points are left behind. These compounds are used to make products such as diesel oil, heating oil, lubricants, asphalt, paraffin wax, and tar. Petroleum distillates are also used in pesticides, cleaning fluids, metal polishes, spot removers, lubricants, and many other products.

▶ A fractionation tower separates petroleum into its component compounds.

All petroleum distillates are toxic. When their vapors are breathed, they can cause chemical pneumonia, lung damage, or death. Therefore, all household products that contain 10 percent or more petroleum distillates are required to have hazard warnings. Products containing petroleum distillates should be used carefully.

Plastics

Plastic is one type of material that is manufactured from petroleum distillates. Before plastics can be made, however, the petroleum distillates must be processed. Ethane and propane, two petroleum distillates, are "cracked," or broken into smaller compounds called ethylene and propylene. These compounds are combined with a catalyst and other additives to create a powdered polymer. This polymer is melted and formed into small pellets. Manufacturers purchase the pellets, melt them, add color, and create many products with which you are familiar.

Plastics are also being used in ways you may not have thought of before. Scientists at Ohio State University recently developed a plastic that can withstand temperatures up to 800°F. Military aircraft may soon take advantage of this technology, and it may not be long before you find yourself in a car that has an engine made of plastic.

Lowest Boiling Point

Condensation cap

Gases

Gasoline

Airplane fuel

Kerosene

Heating oil

Diesel oil

Grease, lubricants, and wax

Asphalt

Heated Crude Oil

Highest Boiling Point

What Do You Think?

There are many possible substitutes for petroleum fuels to power vehicles and electric generators. However, finding substitutes for petroleum used as solvents and in other products is not easy. If petroleum supplies are limited, can you think of any substitutes for petroleum?

455

Answers to What Do You Think?

Answers may vary. There are many substitutes for petroleum products, including cleaners made from baking soda and vinegar, soap-based insecticides, water-based paints, plastics made from starch, lactic acid or bacteria, inks based on water or vegetable oil, and candles and crayons from soybean oil.

Science & Technology

FROM CRUDE OIL TO PLASTICS

Background

Crude oil is approximately 84 percent carbon and 14 percent hydrogen. The carbon and hydrogen atoms are bonded together in chains and other structures of various lengths called *hydrocarbons*. Hydrocarbons are useful because they store a great deal of energy, making them useful as fuels, and because they can be arranged in many different structures, producing everything from petroleum jelly to nylon. The refining process separates the different hydrocarbons based on their boiling point: the longer the carbon chain, the higher the boiling point. The lightest, gaseous component can have just four carbon atoms, whereas the thickest residue, and the last to be boiled off, may have 80 carbon atoms.

The various distillates are treated to remove impurities and then processed and blended to produce the final products, such as gasoline or lubricants. These products are kept in storage on-site until they can be transported to their final destination, where they may be used directly, as in the case of gasoline, or further processed to make other products, such as wax or plastics.

ECONOMICS
▶ CONNECTION   GENERAL

Petroleum and Transportation

The majority of petroleum is refined for use in the transportation sector. About 40 percent goes to produce gasoline alone. All fuels, including gasoline, jet fuel, and heating oil, comprise 86 percent of petroleum usage. Products such as lubricants, asphalt, and plastics account for the remainder. Point out to students the strong dependence of the U.S. transportation system on petroleum, not just for fuels but also for motor lubricants and asphalt to pave the roads.

LS Verbal

Renewable Energy
Chapter Planning Guide

PACING	CLASSROOM RESOURCES	LABS, DEMONSTRATIONS, AND ACTIVITIES
BLOCKS 1, 2, 3 & 4 • 180 min pp. 456–465 **Chapter Opener**		
Section 1 Renewable Energy Today	**OSP** Lesson Plan * **CRF** Active Reading * **BASIC** **TT** Bellringer * **TT** A Passive-Solar Home * **TT** Active Solar Energy & PV Cells * **TT** World Use of Woodfuels * **TT** How Hydropower Works * **TT** Geothermal Energy & Geothermal Heat Pumps * **CD** Interactive Exploration 8 The Generation Gap	**TE** Demonstration Solar Heater, p. 457 ◆ **GENERAL** **TE** Group Activity Cooking with the Sun, p. 458 **ADVANCED** **TE** Group Activity Measuring Wind Power Potential, p. 461 **GENERAL** **SE** Field Activity Biomass Survey, p. 462 **TE** Demonstration Hydropower, p. 463 ◆ **BASIC** **TE** Activity Micro-hydropower Systems, p. 464 **GENERAL** **TE** Activity Alternative Energy Possibilities Near You, p. 464 **GENERAL** **SE** Inquiry Lab Blowing in the Wind, pp. 476–477 ◆ **CRF** Datasheets for In-Text Labs * **CRF** Modeling Lab * ◆ **GENERAL** **CRF** Observation Lab * ◆ **GENERAL**
BLOCKS 5 & 6 • 90 min pp. 466–471 **Section 2** Alternative Energy and Conservation	**OSP** Lesson Plan * **CRF** Active Reading * **BASIC** **TT** Bellringer * **TT** Tidal Power & OTEC * **TT** Fuel Cells & Hybrid Cars *	**SE** QuickLab Hydrolysis, p. 468 **TE** Demonstration Inefficient Light Bulbs, p. 469 **GENERAL** **TE** Internet Activity Alternative Fuel Vehicles, p. 469 **GENERAL** **CRF** Observation Lab * ◆ **BASIC** **CRF** Consumer Lab * ◆ **GENERAL** **CRF** Long-Term Project * **ADVANCED**

BLOCKS 7 & 8 • 90 min

Chapter Review and Assessment Resources

SE Chapter Review pp. 473–475
SE Standardized Test Prep pp. 670–671
CRF Study Guide * ■ **GENERAL**
CRF Chapter Test * ■ **GENERAL**
CRF Chapter Test * **ADVANCED**
CRF Concept Review * ■ **GENERAL**
CRF Critical Thinking * **ADVANCED**
OSP Test Generator
CRF Test Item Listing *

Online and Technology Resources

Visit **go.hrw.com** for access to Holt Online Learning or enter the keyword **HE6 Home** for a variety of free online resources.

 One-Stop Planner® CD-ROM

This CD-ROM package includes
• Lab Materials QuickList Software
• Holt Calendar Planner
• Customizable Lesson Plans
• Printable Worksheets
• ExamView® Test Generator
• Interactive Teacher Edition
• Holt PuzzlePro® Resources
• Holt PowerPoint® Resources

 Holt Environmental Science Interactive Tutor CD-ROM

This CD-ROM consists of interactive activities that give students a fun way to extend their knowledge of environmental science concepts.

KEY

TE	Teacher Edition	**CRF**	Chapter Resource File	*	Also on One-Stop Planner
SE	Student Edition	**TT**	Teaching Transparency	■	Also Available in Spanish
OSP	One-Stop Planner	**CD**	Interactive CD-ROM	◆	Requires Advance Prep

ENRICHMENT AND SKILLS PRACTICE	SECTION REVIEW AND ASSESSMENT	CORRELATIONS
SE Pre-Reading Activity, p. 456 **TE** Using the Figure, p. 456 [GENERAL]		**National Science Education Standards**
TE Student Opportunities Design Contests, p. 457 **TE** Using the Figure Mesa Verde, p. 458 [GENERAL] **TE** Inclusion Strategies, p. 458 **SE** Case Study A Super-Efficient Home, pp. 458–459 **TE** Student Opportunities Renewable Energy Opportunities, p. 459 **TE** Using the Figure Solar Cells and Light, p. 460 [GENERAL] **TE** Interpreting Statistics The Cost of Wind Power, p. 461 [GENERAL] **TE** Inclusion Strategies, p. 461	**TE** Homework, p. 458 [ADVANCED] **TE** Homework, p. 463 [BASIC] **SE** Section Review, p. 465 **TE** Reteaching, p. 465 [BASIC] **TE** Quiz, p. 465 [GENERAL] **TE** Alternative Assessment, p. 465 [ADVANCED] **CRF** Quiz * ■ [GENERAL]	ES 1a SPSP 3a SPSP 3b
TE Career Renewable Energy, p. 467 **TE** Using the Figure Fuel Cell Usage, p. 468 [GENERAL] **TE** Skill Builder Writing, p. 469 [GENERAL] **SE** MathPractice Energy Efficiency, p. 470 **SE** Graphic Organizer Spider Map, p. 470 [GENERAL] **SE** Maps in Action Wind Power in the United States, p. 478 **SE** Science & Technology Back to Muscle Power and Springs, p. 479 **CRF** Map Skills Hydroelectric Power * [GENERAL]	**TE** Homework, p. 467 [GENERAL] **SE** Section Review, p. 471 **TE** Reteaching, p. 471 [BASIC] **TE** Quiz, p. 471 [GENERAL] **TE** Alternative Assessment, p. 471 [GENERAL] **CRF** Quiz * ■ [GENERAL]	SPSP 3a SPSP 3b

 Guided Reading Audio CDs

These CDs are designed to help auditory learners and reluctant readers. (Audio CDs are also available in Spanish.)

 www.scilinks.org

Maintained by the **National Science Teachers Association**

TOPIC: Renewable Sources of Energy
SciLinks code: HE4093

TOPIC: Fuel Cells
SciLinks code: HE4046

TOPIC: Mass Transit
SciLinks code: HE4063

TOPIC: Energy Conservation
SciLinks code: HE4033

 CNN Videos

Each video segment is accompanied by a Critical Thinking Worksheet.

Science, Technology & Society

Segment 14 Wind Power

Segment 24 Harnessing Sound Energy

Chapter Enrichment

This Chapter Enrichment provides relevant and interesting information to expand and enhance your classroom instruction of the chapter material.

1 Renewable Energy Today

Solar Cookers

Solar cookers come in many different types—box, panel, parabolic, and concentrator cookers are the most common. They can be made out of various materials and objects, from cardboard boxes and pizza containers to more durable substances such as wood. Some function like an oven, while others work like a stovetop. Simple box cookers can reach temperatures of 150°C (302°F), whereas commercial solar cookers can reach temperatures of about 232°C (450°F), comparable to conventional ovens. The simplest cookers take about twice as long to cook something as a conventional oven, but parabolic cookers cook foodstuffs in about the same amount of time as a stovetop burner.

BRAIN FOOD

European scientists have developed a prototype office building that generates electricity from turbines located between elevated walkways. The commercial version of the building will be 48 stories tall and shaped like a boomerang to direct wind toward the turbines. Three turbines will produce about 20 percent of the building's electricity.

▶ **Wind farm at Altamont Pass, California**

Concentrating Solar Power Systems

A solar technology called *concentrating solar power systems* can be used to generate large amounts of electricity. In these systems, solar energy is used to generate heat, which is then converted to electricity by a steam turbine. Several different mechanisms are used to generate the heat. The trough system utilizes parabolic mirrors to

▶ **Cliff dwellings in Mesa Verde, New Mexico**

focus sunlight on a tube containing liquid. The liquid is converted to steam which generates electricity. The central receiver system uses thousands of mirrors that track the sun and reflect sunlight onto a receiver on top of a tall tower. The solar energy heats molten salt inside the receiver. Molten salt, which can store energy for days, is used to generate steam to power a turbine. The dish system employs a parabolic mirror to focus sunlight on a receiver above the dish. The heat is used to generate electricity. Some solar power plants are hybrids, which combine solar power with fossil fuels so they can produce electricity 24 hours a day. Electricity from solar power plants can cost 9 to 12 cents per kilowatt-hour and less than 8 cents per kilowatt-hour if the plant is a fossil-fuel hybrid.

Biodiesel Fuel

Biodiesel is an alternative to petroleum diesel fuel. It is a non-toxic, biodegradable fuel that is not as damaging to waterways as petroleum diesel. Most biodiesel fuel is made from soybean oil, but fryer oil discarded from restaurants or any vegetable oil found in a grocery store can also be used. The oil must be treated in a two-step process with wood or grain alcohol and sodium hydroxide (lye) in order to remove the glycerin from the oil, but this is a simple process that can even be done at home. Vegetable oil works as a substitute for diesel because both compounds have similar chemical compositions.

Compared to petroleum diesel fuel, pure biodiesel emits nearly 50 percent less carbon monoxide, 68 percent fewer hydrocarbons, 40 percent less particulates and no sulfates. The increased levels of nitogen oxides emitted by biodiesel can be mitigated with emissions-control technology.

SECTION 2 Alternative Energy and Conservation

Hydrogen Energy

The first large-scale use of hydrogen fuel cells will most likely be to power vehicles. DaimlerChrysler and Ford are working on fuel cells for public transportation buses in Europe, and General Motors hopes to start mass-producing fuel cell automobiles by 2010. Honda has built a hydrogen fueling station in Torrance, California, that produces hydrogen from water by use of solar power. Germany, Japan, and Iceland are leading the transition to hydrogen energy with their strong support of hydrogen research. Iceland plans to be the first nation with a hydrogen fuel-based economy by replacing fossil fuels with hydrogen in buses, automobiles, and fishing boats within 30 to 40 years.

Greening Architecture

Glenn Murcott, an architect in Australia, designs site-specific houses that do not require electricity to maintain a comfortable atmosphere. Steadfastly refusing to use air conditioning in a region where temperatures can reach 105°F, he instead incorporates solar energy and energy conservation methods into his designs. First, he studies the local area, including prevailing breezes, water drainage patterns, and flora and fauna, and then creates a home integrating these observations into an environmentally and human-friendly place. His designs typically include a multi-layered north wall consisting of insect screens, moving glass panels, adjustable louvers, and thermal blinds that the owner adjusts periodically to provide maximum efficiency in heating and cooling the structure. His designs are so efficient that temperatures inside can be a chill 75°F in summer and a warm 62°F in winter when outside temperatures are 105°F in summer and 32°F in winter.

BRAIN FOOD

It's amazing what could be accomplished if energy efficiency were implemented on a large scale. For example, if all windows in the frost-belt (the New England, Mideast, Great Lakes, and Plains regions) were replaced with the most energy-efficient windows available today, the amount of oil saved would be more than the total annual oil production of Alaska.

Incandescent or Fluorescent?

Most households use incandescent light bulbs for lighting. These light bulbs produce light by passing electrical energy through a filament, which glows and produces light. Unfortunately, 90 percent of the electricity used to power incandescent bulbs is converted to heat and only 10 percent is converted to light. Therefore, the typical light bulb in a home is actually a better heat source than a light source. An alternative is the compact fluorescent light bulb (CFL), which is more energy-efficient. CFLs are gas-filled bulbs with a white phosphor coating on the inside. Electricity causes the gas in the bulb to emit ultraviolet light, which excites the phosphor coating inside the bulb. The phosphor coating then emits visible light into the room. Although they are more expensive initially than incandescent bulbs, CFLs use only one-quarter the electricity and last much longer.

▶ Bicycles in Denmark, free for public use

Overview

Tell students that renewable and alternative energy resources play an increasingly important role in reducing our dependence on non-renewable energy sources. However, energy conservation and efficiency may have an even greater effect on the reduction of fossil-fuel use.

Using the Figure — GENERAL

The windmills shown here are on the plains of La Mancha, Spain. The first windmills were probably built during the 7th century in the Sassanid Empire of Persia, located near the present-day border of Afghanistan and Iran. These windmills were built to grind grain and pump water for irrigation. The Crusades brought this technology to Europe, and windmills became a common sight by the 1100s. Seven centuries later, windmills were a vital source of renewable energy in America. At one time, more than 1 million windmills pumped water on American farms. Windmill use declined in the 1930s, but wind-power is making a comeback. Ask students if they can think of any uses for simple windmills today.
LS Visual/Logical

PRE-READING ACTIVITY

FOLDNOTES

Encourage students to use their FoldNote as a study guide to quiz themselves for a test on the chapter material. Students may want to create Double Door FoldNotes for different topics within the chapter.

VIDEO SELECT

For information about videos related to this chapter, go to **go.hrw.com** and type in the keyword **HE4 RENV**.

Renewable Energy

1 **Renewable Energy Today**
2 **Alternative Energy and Conservation**

PRE-READING ACTIVITY

FOLDNOTES

Double-Door Fold
Before you read this chapter, create the **FoldNote** entitled "Double-Door Fold" described in the Reading and Study Skills section of the Appendix. Write "Advantages of renewable energies" on one flap of the double door and "Disadvantages of renewable energies" on the other flap. As you read the chapter, compare the two topics, and write characteristics of each type of renewable energy on the inside of the appropriate flap.

The power of the wind is one of the oldest energy resources used by humans. These Spanish windmills were built to grind grain hundreds of years ago. Today, wind energy is a rapidly growing industry.

456

Chapter Correlations   *National Science Education Standards*

ES 1a Earth systems have internal and external sources of energy, both of which create heat. The sun is the major external source of energy. Two primary sources of internal energy are the decay of radioactive isotopes and the gravitational energy from the earth's original formation. **(Section 1)**

SPSP 3a Human populations use resources in the environment in order to maintain and improve their existence. Natural resources have been and will continue to be used to maintain human populations. **(Section 1 and Section 2)**

SPSP 3b The earth does not have infinite resources; increasing human consumption places severe stress on the natural processes that renew some resources, and it depletes those resources that cannot be renewed. **(Section 1 and Section 2)**

Renewable Energy Today

When someone mentions renewable energy, you may think of high-tech solar-powered cars, but life on Earth has always been powered by energy from the sun. **Renewable energy** is energy from sources that are constantly being formed. In addition to solar energy, renewable energy sources include wind energy, the power of moving water, and the Earth's heat.

Many governments plan to increase their use of renewable sources. For example, the European Union plans to produce 12 percent of their energy from renewable sources by 2010. Such a change will reduce the environmental problems caused by the use of nonrenewable energy. However, all sources of energy, including renewable sources, affect the environment.

Solar Energy—Power from the Sun

What does the space station shown in **Figure 1** have in common with a plant? Both are powered by energy from the sun. The sun is a medium-sized star that radiates energy from nuclear fusion reactions in its core. Only a small fraction of the sun's energy reaches the Earth. However, this energy is enough to power the wind, plant growth, and the water cycle. So nearly all renewable energy comes directly or indirectly from the sun. You use direct solar energy every day. When the sun shines on a window and heats a room, the room is being heated by solar power. Solar energy can also be used indirectly to generate electricity in solar cells.

Objectives

▶ **List** six forms of renewable energy, and compare their advantages and disadvantages.

▶ **Describe** the differences between passive solar heating, active solar heating, and photovoltaic energy.

▶ **Describe** the current state of wind energy technology.

▶ **Explain** the differences in biomass fuel use between developed and developing nations.

▶ **Describe** how hydroelectric energy, geothermal energy, and geothermal heat pumps work.

Key Terms

renewable energy
passive solar heating
active solar heating
biomass fuel
hydroelectric energy
geothermal energy

Figure 1 ▶ What does this plant have in common with a space station's solar panels? Both use energy from the sun.

457

Student Opportunities

Design Contests There are several renewable energy design contests for high school students. The Solar Decathlon is sponsored by the U.S. Department of Energy; students compete to design and create the most efficient and livable solar-powered home. The Carolina EV Challenge is a year-long event that includes converting a gasoline-powered car to an electric-powered car. The National Solar Design Contest gives students the chance to design solar-powered devices. Help students research these and other competitions on the Internet.

Focus

Overview

Before beginning this section, review with your students the Objectives in the Student Edition. This section describes renewable energy sources including solar, wind power, biomass fuels, hydroelectricity, and geothermal energy.

🔔 Bellringer

Ask students to think about how their great-grandparents met their energy needs. (Sample answers: public transportation or the use of farm animals for transportation, windmills on farms) Ask students which of these sources were renewable and which were non-renewable. (renewable: farm animals, windmills, muscle power; non-renewable: fossil fuels) **LS Logical**

Motivate

Demonstration —— GENERAL

Solar Heater Tell students that solar water heating systems are used in more than 1 million U.S. homes as an alternative to using electricity or burning natural gas. Make a simple solar water heater by painting a garden hose black and then place it in a sunny spot. Fill a container with water and measure the temperature of the water. Then pour the water into one end of the hose, allowing the water to run slowly through. Collect the water at the other end. Measure the temperature of the water as it comes out of the hose. The temperature at the end should be higher. Have students brainstorm ways to improve your simple water heater design. **LS Visual/Kinesthetic**

Group Activity — ADVANCED

Cooking with the Sun Organize students into small groups. Have each group research a solar cooker design on the Internet and then build a solar cooker based on that design. Have students use their solar cooker to cook foods for a class picnic. Ask students, "How might your geographic location affect the efficiency of your solar cooker?" (Lower latitudes, high altitudes, and low amounts of atmospheric moisture produce favorable conditions for solar cooking.)
LS Kinesthetic Co-op Learning

Using the Figure — GENERAL

Mesa Verde The southern orientation of the cliff dwellings in **Figure 2** is a passive solar design principle. In summer, the sun's path is higher in the sky so the cliffs shade the dwellings from the sun and the rooms remain cool. In winter, the sun's path is lower in the sky so the sun heats the rooms. Ask students, "How could these principles apply to any house?" (Sample answers: build a roof overhang on the south side, shade the south side of the building with trees, use building materials that provide efficient thermal storage) As an extension, initiate a passive solar design contest in which student groups build scale model houses out of common materials. Each design can be tested using a heat lamp and thermometer. Assign grades based on the temperature inside each house (in degrees Fahrenheit).

Figure 2 ▶ Seven hundred years ago, the Ancestral Puebloans, also called the Anasazi, lived in passive solar cliff dwellings in Mesa Verde, New Mexico.

Passive Solar Heating The cliff dwellings shown in **Figure 2** used passive solar heating, the simplest form of solar energy. **Passive solar heating** uses the sun's energy to heat something directly. In the Northern Hemisphere, south facing windows receive the most solar energy, so passive solar buildings have large windows that face south. Solar energy enters the windows and warms the house. At night, the heat is released slowly to help keep the house warm. Passive solar buildings must be well insulated with thick walls and floors in order to prevent heat loss.

Passive solar buildings are oriented according to the yearly movement of the sun. In summer, the sun's path is high in the sky and the overhang of the roof shades the building and keeps it cool. In winter, the sun's path is lower in the sky, so sunlight shines into the home and warms it. If there is reliable winter sunlight, an extremely efficient passive solar heating system can heat a house even in very cold weather without using any other source of energy. However, an average household could reduce its energy bills by using any of the passive solar features shown in **Figure 3**.

CASE STUDY — A Super-Efficient Home

Imagine a home located deep in the Rocky Mountains, where winter temperatures can plunge to −40°C (−40°F). The home has no furnace, yet it manages to stay comfortably warm even in the coldest weather. This home, built by energy experts Hunter and Amory Lovins in Snowmass, Colorado, is a prime example of a new generation of super-efficient structures.

Efficiency without sacrifice was the goal in designing the Lovins's home, which also houses the Rocky Mountain Institute (RMI), an energy-research organization. The structure uses one-tenth the electricity and one-half the water of a similar-sized conventional building. The building

cost more to build than a conventional structure, but that extra cost was recovered through energy savings in only three years.

Solar energy is the most important energy source for RMI. An abundance of south-facing windows lets in plenty of sunshine. As a result, little daytime lighting is required. Artificial lighting is provided by compact fluorescent lamps that draw only 18 W but provide as much light as standard 75 W incandescent bulbs. These lamps also last 10 to 13 times longer than ordinary bulbs. Motion sensors turn the lights off when a room is empty and turn them back on when someone enters the room.

Much of the building's electricity is provided by solar cells. If the building did not have equipment such as copiers and computers, it might not require any outside electricity at all. RMI staffer Owen Bailey said, "When the copier is not running, we actually send power back to the utility company."

Solar energy, plus the heat from appliances and human bodies, meets 90 percent of the heating needs. The other 10 percent is provided by two wood-burning stoves. The walls and roof of RMI are heavily insulated, greatly reducing heat loss. Also, the walls and windows are airtight, eliminating another common source of heat loss.

INCLUSION Strategies

• *Attention Deficit Disorder*

Have students research passive solar heating and active solar heating using reference books or the Internet. Students should find definitions of each type of heating, examples or models of buildings using the two types of heating, and estimates of how much money each method could save a homeowner in your area in one year. Students can give an oral or written report on their findings.

Homework — ADVANCED

Suncharts A sunchart is a map of the sun's annual path across the sky in a given location. These charts allow an architect or engineer to design buildings that best utilize the annual movement of the sun. Have students do research to create a sunchart for their region. Then, have them draw a profile of the southern horizon from the school or from their home on the sunchart. Have students indicate the best locations to place or remove trees, windows, shades, photovoltaic panels, and other passive- or active-solar elements.
LS Visual/Logical

Vent allows hot air to escape in summer.

Insulated drapes or window shades reduce nighttime heat loss in winter.

Ceilings are heavily insulated.

Summer sun

Shade trees help keep a home cool in the summer.

Winter sun

Thick walls and floors store heat in winter.

South-facing, double-paned windows let sunlight in but reduce heat loss on cold nights.

▶ **The Rocky Mountain Institute** uses the energy of the sun so efficiently that it can stay warm in the coldest Colorado winters.

During extended cloudy winter weather (with no solar heat input) the building loses only about 1°F per day. Nevertheless, the structure is well ventilated. It has specially designed air exchangers that vent stale air and warm the incoming fresh air.

The RMI structure shows that conservation does not require discomfort. The building is comfortable and spacious. As Amory Lovins said, "The main thing that the Institute demonstrates is that conservation . . . doesn't mean freezing in the dark."

CRITICAL THINKING

1. Inferring Relationships Specially designed homes in Colorado are able to meet most of their heating needs using passive solar heating. But in parts of Canada and Alaska where winter weather can be similar to the weather in Colorado, solar-heating systems are often inadequate. Use what you know about latitude and solar radiation and write an explanation for this. **WRITING SKILLS**

2. Applying Ideas Currently, only about 1 percent of the homes built in this country have energy-efficient designs. What could be done to increase this percentage?

459

Solar Cells and Light Explain to students that the power output of a photovoltaic module depends on the system voltage and the current produced. The voltage output of the cell shown in **Figure 5** is approximately 0.5 volts. When several solar cells are connected in series (the positive electrode of one cell is connected to the negative electrode of the next cell), the voltage is increased. Thus, a 36-cell module that uses 0.5 volt cells connected in series produces 18 volts. Ask students to list the factors that might affect the current produced by a solar cell. (surface area of the cell, sunlight intensity, and the efficiency of the cell) **LS Visual**

Discussion — GENERAL

Mandating Solar Water Heating If a government supports or requires solar water heaters, as is done in Japan and Israel, then they are more widely used. In Israel, building codes require all new buildings to have solar water heaters and, as a result, they are now found in seventy percent of all buildings. Ask students to debate the following question: "Should governments require measures, such as solar water heating in new buildings, as part of an energy efficiency plan?" **LS Verbal**

Transparencies

TT Active Solar Energy & PV Cells

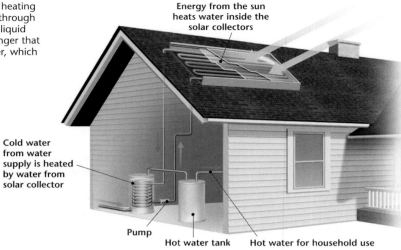

Figure 4 ▶ In a solar water heating system, a liquid is pumped through solar collectors. The heated liquid flows through a heat exchanger that transfers the energy to water, which is used in a household.

Energy from the sun heats water inside the solar collectors

Cold water from water supply is heated by water from solar collector

Pump

Hot water tank

Hot water for household use

Active Solar Heating Energy from the sun can be gathered by collectors and used to heat water or to heat a building. This technology is known as **active solar heating.** More than 1 million homes in the United States use active solar energy to heat water. Solar collectors, usually mounted on a roof, capture the sun's energy, as shown in **Figure 4.** A liquid is heated by the sun as it flows through the solar collectors. The hot liquid is then pumped through a heat exchanger, which heats water for the building. About 8 percent of the energy used in the United States is used to heat water; therefore, active solar technology could save a lot of energy.

Photovoltaic Cells Solar cells, also called *photovoltaic* (FOHT oh vahl TAY ik) *cells*, convert the sun's energy into electricity, as shown in **Figure 5.** Solar cells were invented more than 120 years ago, and now they are used to power everything from calculators to space stations. Solar cells have no moving parts, and they run on nonpolluting power from the sun. So why don't solar cells meet all of our energy needs? A solar cell produces a very small electrical current. So meeting the electricity needs of a small city would require covering hundreds of acres with solar panels. Solar cells also require extended periods of sunshine to produce electricity. This energy is stored in batteries, which supply electricity when the sun is not shining.

Despite these limitations, energy production from solar cells has doubled every four years since 1985. Solar cells have become increasingly efficient and less expensive. Solar cells have great potential for use in developing countries, where energy consumption is minimal and electricity distribution networks are limited. Currently, solar cells provide energy for more than 1 million households in the developing world.

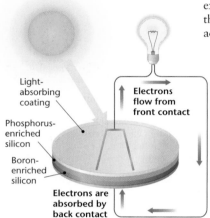

Light-absorbing coating

Phosphorus-enriched silicon

Boron-enriched silicon

Electrons flow from front contact

Electrons are absorbed by back contact

Figure 5 ▶ Sunlight falls on a semiconductor, causing it to release electrons. The electrons flow through a circuit that is completed when another semiconductor in the solar cell absorbs electrons and passes them on to the first semiconductor.

460

REAL-LIFE CONNECTION — GENERAL

A Million Solar Roofs Announced in June 1997, Million Solar Roofs Initiative (MSRI) is an initiative to install solar energy systems on one million U.S. buildings by 2010. The initiative includes solar-electric systems (photovoltaics) and solar-thermal systems. The main goal is to build a strong market for solar energy applications for buildings. The U.S. Department of Energy is leading the initiative and partnering with other federal agencies, local and state governments, utilities, energy service providers, the solar energy industry, the building industry, financial institutions, schools, and nongovernmental organizations. Have students find out if their school can participate in the MSRI through the Interstate Renewable Energy Council's "Schools Going Solar" program or through the "Solar Schools" program. Students may also find out about the Department of Energy's EnergySmart schools program.

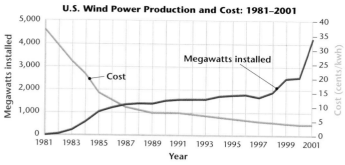

U.S. Wind Power Production and Cost: 1981–2001

Source: American Wind Energy Association.

Figure 6 ▶ The cost of wind power has been steadily falling as wind turbines have become more efficient.

Wind Power—Cheap and Abundant

Energy from the sun warms the Earth's surface unevenly, which causes air masses to flow in the atmosphere. We experience the movement of these air masses as wind. Wind power, which converts the movement of wind into electric energy, is the fastest-growing energy source in the world. New wind turbines are cost effective and can be erected in three months. As a result, the cost of wind power has declined dramatically, as shown in **Figure 6.** The world production of electricity from wind power quadrupled between 1995 and 2000.

Wind Farms Large arrays of wind turbines, like the one shown in **Figure 7,** are called *wind farms.* In California, large wind farms supply electricity to 280,000 homes. In windy rural areas, small wind farms with 20 or fewer turbines are also becoming common. Because wind turbines take up little space, some farmers can add wind turbines to their land and still use the land for other purposes. Farmers can then sell the electricity they generate to the local utility.

An Underdeveloped Resource Scientists estimate that the windiest spots on Earth could generate more than ten times the energy used worldwide. Today, all of the large energy companies are developing plans to use more wind power. Wind experts foresee a time when prospectors will travel the world looking for potential wind-farm sites, just as geologists prospect for oil reserves today. However, one of the problems of wind energy is transporting electricity from rural areas where it is generated to urban centers where it is needed. In the future, the electricity may be used on the wind farm to produce hydrogen from water. The hydrogen could then be trucked or piped to cities for use as a fuel.

Figure 7 ▶ California wind farms, such as this one in Altamont Pass, generate more than enough electricity to light a city the size of San Francisco.

461

Interpreting Statistics ──── GENERAL

The Cost of Wind Power Point out to students that the graph in **Figure 6** has two vertical axes: the axis on the left is used to read the line marked "Megawatts installed," and the axis on the right is used to read the line marked "Cost." The axes have different values and units, so it is important to match the data with the correct axis.

Group Activity ──── GENERAL

Measuring Wind Power Potential Have students work in small groups to construct simple wind speed indicators (anemometers). Supply each group with four small paper cups and have them color one cup with a marker. Then give students the following instructions: Cut out two stiff pieces of wire (approximately 25 cm) and join them together to form an "X." Then attach one cup to each end of the wires so that all cups face the same direction. Attach the center of the X to the top of a dowel with a push pin so that it spins freely. Students can measure the relative wind speed by counting the number of times the colored cup goes around in one minute. Have students measure the wind speed at various locations on the school grounds. Ask students to locate the best windpower sites at your school. **LS** Interpersonal/ Kinesthetic

Co-op Learning

 MISCONCEPTION //// **ALERT** \\\\

Wind Farms and Birds Students may think that many birds will be killed by colliding with wind turbines. The highly publicized deaths of golden eagles and other raptors at Altamont Pass in **Figure 7** prompted the windpower industry to reevaluate wind turbine designs and wind farm locations. Some studies show that birds become accustomed to the turbines and learn to avoid them. In fact, more birds are killed by high-voltage power lines than are killed by wind turbines.

INCLUSION Strategies

• *Learning Disabled*
• *Attention Deficit Disorder*

Have students draw or make a model of a wind farm. They may use **Figure 7** for reference or find images of wind farms on the Internet. The display should include a description of how many turbines are used in the wind farm, how the turbines generate electricity, and how many homes will be supplied with electricity. An oral presentation to a small group, teacher, or the class can be used to show understanding of the concept.

Renewable Energy in Nicaragua People who live in rural villages of Nicaragua often have no electricity, and it is expensive to expand the electric grid to reach them. As a result, they rely heavily on natural gas, firewood, kerosene, or diesel fuel for their electricity and fuel needs. The use of these fuels can lead to problems such as respiratory disease, deforestation, and air pollution. They are also expensive—a family may spend one-third of its income on fuel. Grupo Fenix, based out of the National Engineering University in Managua, was formed in 1996 to provide people with energy alternatives. Grupo Fenix builds solar ovens and teaches people how to use them. They also install solar panels on homes to power fluorescent light bulbs for a few hours in the evening. The solar panels and other necessary electronics are manufactured by victims of land mines remaining from the Sandinista-Contra war. Grupo Fenix also utilizes micro-hydropower to provide people with electricity from waterfalls.

Transparencies

TT World Use of Woodfuels

Share of Woodfuels in Energy Consumption

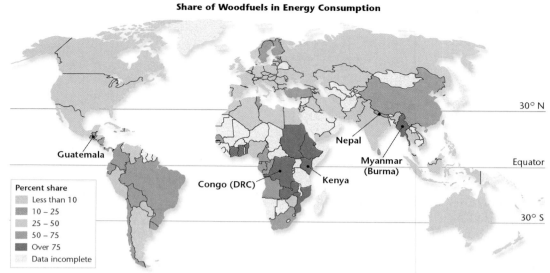

Percent share
- Less than 10
- 10 – 25
- 25 – 50
- 50 – 75
- Over 75
- Data incomplete

30° N

Nepal

Guatemala

Myanmar (Burma)

Equator

Congo (DRC)   Kenya

30° S

Figure 8 ▶ The consumption of wood as an energy source has increased by nearly 80 percent since 1960. In developing countries such as Nepal, Burma, Guatemala, Congo (DRC), and Kenya, the use of fuelwood places an enormous burden on local environments.

FIELD ACTIVITY

Biomass Survey Walk around your neighborhood, and list as many sources of biomass fuel as you can find. Are any of these (such as a pile of firewood) large enough to be used as fuel sources? What do you think the advantages and disadvantages of using biomass as a fuel in your area would be? Record your observations in your *EcoLog*.

Biomass—Power from Living Things

Plant material, manure, and any other organic matter that is used as an energy source is called a **biomass fuel.** While fossil fuels are organic and can be thought of as biomass energy sources, fossil fuels are nonrenewable. Renewable biomass fuels, such as wood and dung, are major sources of energy in developing countries, as shown in **Figure 8.** More than half of all wood cut in the world is used as fuel for heating and cooking. Although wood is a renewable resource, if trees are cut down faster than they grow, the resulting habitat loss, deforestation, and soil erosion can be severe. In addition, harmful air pollution may result from burning wood and dung.

Methane When bacteria decompose organic wastes, one by-product is methane gas. Methane can be burned to generate heat or electricity. In China, more than 6 million households use bio-gas digesters to ferment manure and produce gas used for heating and cooking. In the developed world, biomass that was once thought of as waste is being used for energy. In 2002, Britain's first dung-fired power station started to produce electricity. This power station uses the methane given off by cow manure as fuel. Similarly, some landfills in the United States generate electricity by using the methane from the decomposition of trash.

Alcohol Liquid fuels can also be derived from biomass. For example, ethanol, an alcohol, can be made by fermenting fruit or agricultural waste. In the United States, corn is a major source of ethanol. Cars and trucks can run on ethanol or *gasohol*, a blend of gasoline and ethanol. Gasohol produces less air pollution than fossil fuels do. For this reason, some U.S. states require the use of gasohol in vehicles as a way to reduce air pollution.

BRAIN FOOD

Generating Electricity from Airplane Food Los Angeles International Airport has implemented a pilot project to use bacteria to break down food waste from airline meals to generate electricity. The methane produced by the bacteria is transferred to a power plant, where it is burned to generate electricity.

Notable Quotes ——— GENERAL

"People and nations behave wisely— once they have exhausted all other alternatives."

—Abba Eban, Israeli diplomat and writer.

Ask students how they think this quote applies to energy sources. **LS** Logical

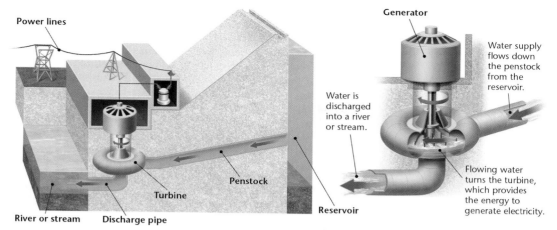

Power lines

Generator

Water supply flows down the penstock from the reservoir.

Water is discharged into a river or stream.

Penstock

Turbine

River or stream

Discharge pipe

Reservoir

Flowing water turns the turbine, which provides the energy to generate electricity.

Hydroelectricity—Power from Moving Water

Energy from the sun causes water to evaporate, condense in the atmosphere, and fall back to the Earth's surface as rain. As rainwater flows across the land, the energy in its movement can be used to generate electricity. **Hydroelectric energy,** which is energy produced from moving water, is a renewable resource that accounts for about 20 percent of the world's electricity. The countries that lead the world in hydroelectric energy are, in decreasing order, Canada, the United States, Brazil, China, Russia, and Norway.

Figure 9 shows how a hydroelectric power plant works. Large hydroelectric power plants have a dam that is built across a river to hold back a reservoir of water. The water in the reservoir is released to turn a turbine, which generates electricity. The energy of this water is evident in **Figure 10,** which shows the spillway of the world's largest hydroelectric dam.

The Benefits of Hydroelectric Energy Although hydroelectric dams are expensive to build, they are relatively inexpensive to operate. Unlike fossil fuel plants, hydroelectric dams do not release air pollutants that cause acid precipitation. Hydroelectric dams also tend to last much longer than fossil fuel-powered plants. So the importance of hydroelectric energy is clear when you consider that nearly a quarter of the world's electricity is generated from this nonpolluting, renewable energy source. Dams also provide other benefits such as flood control and water for drinking, agriculture, industry, and recreation.

Figure 9 ▶ Hydroelectric dams convert the *potential energy,* or stored energy, of a reservoir into the *kinetic energy,* or moving energy, of a spinning turbine. The movement of the turbine is then used to generate electricity.

Figure 10 ▶ The Itaipu Dam in Paraguay supplies about 75 percent of the electricity used by Paraguay and 25 percent of the electricity used by Brazil.

463

Homework ──────── **BASIC**

Crossword Puzzles Suggest that students create a crossword puzzle using terms discussed in this section. Provide them with a puzzle from a newspaper to remind them of its format. Each term in the puzzle should have the appropriate clue listed in the "Down" or "Across" list. After they are done, have students exchange crossword puzzles to see how well they do in solving their classmates' puzzles. **LS Verbal**

English Language Learners

GEOLOGY
▶ **CONNECTION** ── **GENERAL**

Silt and Hydroelectric Dams Silt accumulates behind a dam as particles settle out of the water. This silt reduces the storage capacity of the reservoir. It can take a reservoir between 50 and several hundred years to silt up, depending on how much silt is in the water. Silt must be removed from the reservoir, usually by dredging, because once the reservoir fills up it can no longer store as much water and the generating capacity of the hydroelectric plant is reduced.

Demonstration ──── **BASIC**

Hydropower Cut the top off a half gallon milk carton, then punch holes of equal size in one side at 1.5 cm, 3 cm, 5 cm, and 10 cm intervals from the carton bottom. Seal all the holes with a single piece of tape. Fill the carton with water. While holding the carton over a sink, quickly remove the piece of tape from all the holes. Ask students to explain what they observed. (The water coming out of the bottom holes travels farther than the water from the top holes. The water at the bottom is under higher pressure, and so it shoots out at a greater force.) Ask students where they think the turbines should be placed in a hydroelectric dam. (Hydroelectric turbines are located at the bottom of a dam rather than at the top because the water flowing from the bottom of a dam is under higher pressure.) **LS Visual**

English Language Learners

Teaching Tip ──── **GENERAL**

Turbines and Electricity Generation Remind students that most power plants use turbines to produce electricity. List types of power plants, and ask students how the turbine is turned in each type of plant. (Sample answer: A coal power plant may use steam from heated water.) Have students brainstorm some of the causes of inefficiency in power plants. (Sample answer: heat that escapes from the boilers and turbine, friction in the shaft that turns the turbine) Ask students, "What kind of power plant can produce electricity without a turbine?" (Sample answers: a plant with photovoltaic cells or hydrogen cells) **LS Logical/Verbal**

Transparencies

TT How Hydropower Works

Micro-hydropower Systems
Have students research how micro-hydropower works and where it is used in the world, then have them design their own micro-hydropower generator. The Internet is a good resource for this activity. Suggest that they include the design in their **Portfolio.** **Intrapersonal**

Alternative Energy Possibilities Near You Ask students to determine which alternative energy sources are suitable for your area and have them explain their reasoning. Tell students to assume the following information and incorporate it into their reasoning: 200 days of sunshine per year are required for solar energy systems, wind energy systems require an average wind speed greater than 13 mph, geothermal energy is generally an option only in areas with geothermal deposits, and hydropower is an option only in regions with abundant flowing water and significant changes in elevation. Students can use an atlas or information from the weather service to determine the meteorological data for your area. Maps of geothermal potential can be obtained from encyclopedias or from the Internet. **LS Logical**

Transparencies

TT Geothermal Energy & Geothermal Heat Pumps

Figure 11 ▶ Geothermal power plants generate electricity using the following steps: ❶ steam rises through a well; ❷ steam drives turbines, which generate electricity; ❸ leftover liquid water is pumped back into the hot rock.

Heated water

Hot rock

Disadvantages of Hydroelectric Energy A dam changes a river's flow, which can have far-reaching consequences. A reservoir floods large areas of habitat above the dam. The water flow below the dam is reduced, which disrupts ecosystems downstream. For example, many of the salmon fisheries of the northwestern United States have been destroyed by dams that prevent the salmon from swimming upriver to spawn. When the land behind a dam is flooded, people are often displaced. An estimated 50 million people around the world have been displaced by dam projects. Dam failure can be another problem—if a dam bursts, people living in areas below the dam can be killed.

Dams can also affect the land below them. As a river slows down, the river deposits some of the sediment it carries. This fertile sediment builds up behind a dam instead of enriching the land farther down the river. As a result, farmland below a dam can become less productive. Recent research has also shown that the decay of plant matter trapped in reservoirs can release large amounts of greenhouse gases—sometimes more than a fossil-fuel powered plant.

Modern Trends In the United States, the era of large dam construction is probably over. But in developing countries, such as Brazil, India, and China, the construction of large dams continues. A modern trend is *micro-hydropower,* which is electricity produced in a small stream without having to build a big dam. The turbine may even float in the water, not blocking the river at all. Micro-hydropower is much cheaper than large hydroelectric dam projects, and it permits energy to be generated from small streams in remote areas.

Geothermal Energy—Power from the Earth

In some areas, deposits of water in the Earth's crust are heated by energy within the Earth. Such places are sources of **geothermal energy**—the energy from heat in the Earth's crust. As **Figure 11** shows, this heat can be used to generate electricity. Geothermal power plants pump heated water or steam from rock formations and use the water or steam to power a turbine that generates electricity. Usually the water is returned to the Earth's crust where it can be heated and used again.

The United States is the world's largest producer of geothermal energy. The world's largest geothermal power plant is The Geysers, in California, which produces electricity for about

---

GEOLOGY
●—[CONNECTION]—(GENERAL)

Earth's Heat Two sources of geothermal energy are the decay of radioactive elements in Earth's interior and gravitational energy from Earth's formation. Over billions of years this thermal energy will dissipate and Earth will cool off. So, on a human time scale, geothermal energy may be considered renewable. However, the underground water deposits that are used by most geothermal power plants can be exhausted.

MISCONCEPTION
////ALERT\\\\

Environmental Effects of Renewable Energy Students may think that using renewable energy does not harm the environment. However, there are environmental consequences associated with any form of energy. For example, toxic substances are used in the manufacture of photovoltaic cells, and the battery systems that store the energy use toxic heavy metals such as lead. Ask students to choose a form of renewable energy and list ways to minimize the harmful effects of that technology.

The ground is warmer than the air in winter.

Heat is transferred from the ground to warm the house.

The ground is cooler than the air in summer.

Heat is transferred from the house to the ground to cool the house.

Figure 12 ▶ In the winter (left), the ground is warmer than the air is. A fluid is circulated underground to warm a house. In the summer (right), the ground is cooler than the air is, and the fluid is used to cool a house.

1.7 million households. Other countries that produce geothermal energy include the Philippines, Iceland, Japan, Mexico, Italy, and New Zealand. Although geothermal energy is considered a renewable resource, the water in geothermal formations must be managed carefully so that it is not depleted.

Geothermal Heat Pumps: Energy for Homes More than 600,000 homes in the United States are heated and cooled using geothermal heat pumps such as the one shown in **Figure 12**. Because the temperature of the ground is nearly constant year-round, a *geothermal heat pump* uses stable underground temperatures to warm and cool homes. A heat pump is simply a loop of piping that circulates a fluid underground. In warm summer months, the ground is cooler than the air, and the fluid is used to cool a home. In the winter, the ground is warmer than the air, and the fluid is used to warm the home.

🖳 internet connect

www.scilinks.org
Topic: Renewable Sources of Energy
SciLinks code: HE4093

SCiLINKS Maintained by the National Science Teachers Association

SECTION 1 Review

1. **List** six forms of renewable energy, and compare the advantages and disadvantages of each.

2. **Describe** the differences between passive solar heating, active solar heating, and photovoltaic energy.

3. **Describe** how hydroelectric energy, geothermal energy, and geothermal heat pumps work.

4. **Explain** whether all renewable energy sources have their origin in energy from the sun.

CRITICAL THINKING

5. **Making Decisions** Which renewable energy source would be best suited to your region? Write a paragraph that explains your reasoning. **WRITING SKILLS**

6. **Identifying Trends** Identify a modern trend in hydroelectric power and in wind energy.

7. **Analyzing Relationships** Write an explanation of the differences in biomass fuel use between developed and developing countries. **WRITING SKILLS**

465

Close

Reteaching ——— BASIC

A Sustainable Energy Future As a class, construct a concept map that identifies the differences between renewable and nonrenewable energy sources. Then lead a discussion about a sustainable energy future. Explain that a sustainable energy future is one in which energy resources can be maintained indefinitely. **LS** Verbal

English Language Learners

Quiz ——— GENERAL

1. List two forms of renewable energy that have promising applications in developing countries. (Sample answers: solar cells and micro-hydropower)

2. What percentage of the world's electricity is produced from hydroelectric energy? (about 20 percent)

3. What is the fastest growing energy source in the world? (wind energy)

Alternative Assessment ——— ADVANCED

Writing a Proposal Ask your students to write a hypothetical proposal for an alternative energy system to serve their community. Students may choose from any of the systems described in this section. The energy source should be appropriate for the area and economically feasible. Encourage students to use visual aids to enhance their proposal. **LS** Verbal/Visual

Answers to Section Review

1. Solar cells are non-polluting and have no moving parts, but they are inefficient and expensive. Passive solar energy requires no fuels, but efficiency depends on location. Wind energy is inexpensive, but transmitting electricity from remote sites is difficult. Biomass produces less pollution than fossil fuels, but it can produce environmental problems if used in a nonrenewable manner. Hydroelectricity facilities are inexpensive to operate and produce no air pollution, but they can have adverse effects on the environment and the surrounding population. Geothermal plants are less polluting than fossil-fuel plants and require no fuels, but hydrothermal water deposits can be exhausted and facilities can be expensive to build.

2. Passive solar heating uses the sun's energy directly by incorporating features into a building's construction. Active solar heating stores the sun's energy and uses it indirectly to heat water or a building. Photovoltaic energy uses the sun's energy indirectly to generate electricity.

3. To produce hydroelectric energy, flowing water turns a turbine, which generates electricity. Geothermal energy is generated when heated water or steam is pumped up from inside Earth to power a turbine that generates electricity. Geothermal heat pumps circulate fluid in a pipe underground and back to the surface to heat or cool a house.

4. All of the sources have their origin in solar energy except geothermal power.

5. Answers may vary.

6. Answers may vary. Two trends are micro-hydropower and increased use of windpower.

7. Developing countries use woodfuels and dung as energy sources. Developed countries use methane or alcohol from wastes.

Focus

Overview

Before beginning this section, review with your students the Objectives in the Student Edition. This section describes three forms of alternative energy and discusses energy efficiency and energy conservation.

Bellringer

Have students examine the classroom to see if they can find any instances where energy is being wasted. Have them write down what they observe and list ways that energy could be saved. As an extension, have students map the room and then use a thermometer or temperature probe to detect and map temperature zones in the room. **LS Logical/Kinesthetic**

Motivate

Identifying Preconceptions — GENERAL

Energy Conservation Ask students what the phrase "energy conservation" means to them. Students may mention ideas such as turning off lights or saving energy. Ask them what the benefits of conservation would be. (saving energy resources and money, and reducing pollution) Ask them if they see any disadvantages to energy conservation. Explain that energy conservation often means changing old habits, which can be difficult or uncomfortable for some people. It may require self-discipline, lifestyle changes, or changes in buying habits, but the benefits can be significant. **LS Intrapersonal**

Alternative Energy and Conservation

Objectives

▶ Describe three alternative energy technologies.
▶ Identify two ways that hydrogen could be used as a fuel source in the future.
▶ Explain the difference between energy efficiency and energy conservation.
▶ Describe two forms of energy-efficient transportation.
▶ Identify three ways that you can conserve energy in your daily life.

Key Terms

alternative energy
ocean thermal energy
  conversion (OTEC)
fuel cell
energy efficiency
energy conservation

To achieve a future where energy use is sustainable, we must make the most of the energy sources we already have and develop new sources of energy. **Alternative energy** describes energy sources that are still in development. Some renewable energy sources that we use now, such as geothermal power, were once considered alternative energy. For an alternative energy source to become a viable option for the future, the source must be proven to be cost effective. Also, the environmental effects of using the energy source must be acceptable. Government investment is often the only way to research some of these future energy possibilities.

Tidal Power

Tides are the movement of water in the oceans and seas caused by gravitational attraction between the sun, Earth, and moon. The tides, which happen twice each day, are marked by the rising and falling of the sea level. The energy of the tides was used nearly a thousand years ago to power mills in France and Britain. Today, tidal power is used to generate electricity in countries such as France, Russia, and Canada.

As **Figure 13** shows, a tidal power plant works much like a hydroelectric dam. As the tide rises, water flows behind a dam; when the sea level falls, the water is trapped behind the dam. When the water in the reservoir is released, it turns a turbine that generates electricity. Although tidal energy is renewable and non-polluting, it will not become a major energy source in the future. The cost of building and maintaining a tidal power plant is high, and there are few locations that are suitable.

Figure 13 ▶ As the tide rises, water enters a bay behind a dam. The gate then closes at high tide. At low tide, the gate opens and the water in the bay rushes through, spinning a turbine that generates electricity.

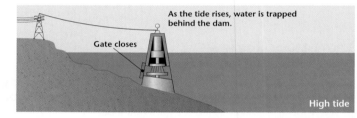
Gate closes
As the tide rises, water is trapped behind the dam.
High tide

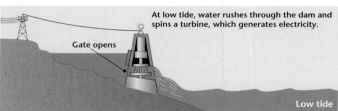
Gate opens
At low tide, water rushes through the dam and spins a turbine, which generates electricity.
Low tide

Chapter Resource File

• **Lesson Plan**
• **Active Reading** BASIC
• **Section Quiz** GENERAL

Transparencies

TT Bellringer
TT Tidal Power & OTEC

Ocean Thermal Energy Conversion

In the tropics, the temperature difference between the surface of the ocean, which is warmed by solar energy, and deep ocean waters can be as much as 24°C (43°F). An experimental power station off the shores of Hawaii uses this temperature difference to generate electricity. This technology, which is shown in **Figure 14**, is called **ocean thermal energy conversion (OTEC)**. In this system, warm surface water is used to boil sea water. This is possible because water boils at low temperatures when it is at low pressure in a vacuum chamber. The boiling water turns into steam, which spins a turbine. The turbine runs an electric generator. Cold water from the deep ocean cools the steam, turning the steam into water that can be used again.

Japan has also experimented with OTEC power, but so far, no project has been able to generate electricity cost effectively. One problem with OTEC is that the power needed to pump cold water up from the deep ocean uses about one-third of the electricity the plant produces. Therefore, the OTEC plants are inefficient. The environmental effects of pumping large amounts of cold water to the surface are also unknown.

Hydrogen—A Future Fuel Source?

The fuel of the future might be right under our noses. Hydrogen, the most abundant element in the universe, can be burned as a fuel. Hydrogen is found in every molecule of living things, and it is also found in water. Hydrogen does not contain carbon, so it does not release pollutants associated with burning fossil fuels and biomass. When hydrogen is burned in the atmosphere, it combines with oxygen to produce water vapor, a harmless byproduct, and small amounts of nitrogen oxides. Hydrogen gas (H_2) can be produced by using electricity to split molecules of water (H_2O). Or, in the future, we may also be able to grow plants to produce hydrogen cost effectively, as shown in **Figure 15**.

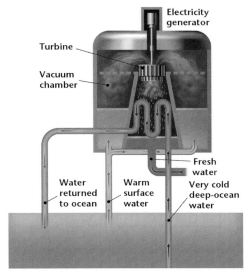

Figure 14 ▶ In an open cycle OTEC plant, warm surface water is brought to boil in a vacuum chamber. The boiling water produces steam to drive a turbine that generates electricity. Cold deep-ocean water is pumped in to condense the steam. Fresh water is a byproduct of this type of OTEC plant.

🔲 **internet** connect

www.scilinks.org
Topic: Fuel Cells
SciLinks code: HE4046

SCiLINKS. Maintained by the National Science Teachers Association

Figure 15 ▶ Hydrogen fuel can be made from any material that contains a lot of hydrogen, including the experimental plot of switchgrass shown here.

467

Career

Renewable Energy Careers in renewable energy cover a wide range of disciplines. Examples include "windsmiths" to operate wind turbines and people that install photovoltaic systems. Geologists, physicists, and material scientists research ways to utilize renewable energy resources more efficiently. The space program has also made great advances in renewable energy and conservation.

As a class, interview someone who works in the field of renewable energy. Prepare a list of questions beforehand. Suggest that students ask the interviewee about the education required, his or her daily duties, what he or she enjoys about the job, and the challenges of working in renewable energy. Then have students transcribe the interview in their **Portfolio.** 🅛 **Verbal/Interpersonal**

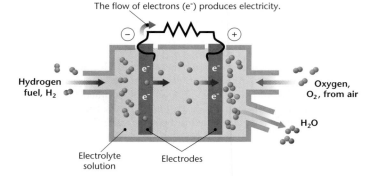

The flow of electrons (e⁻) produces electricity.

Hydrogen fuel, H₂

Oxygen, O₂, from air

H₂O

Electrolyte solution

Electrodes

Teach, *continued*

Using the Figure — GENERAL

Fuel Cell Usage One promising application of fuel cells, such as the one in **Figure 16,** is as a fuel source for automobiles. Have students research how fuel cells could be used in automobiles. Suggest that they create a poster or three-dimensional model of a fuel-cell engine to show how it operates. Have them include this work in their **Portfolio.** LS Verbal

QuickLAB

Skills Acquired:

• Experimenting
• Organizing and Analyzing

Safety Caution: Both hydrogen and oxygen are flammable. Make sure students exercise caution when removing the test tubes. No open flames should be in the room when performing this activity. Perform this activity for 15 minutes only. Examine batteries carefully for leaks.

Answers

1. There are two atoms of hydrogen for every atom of oxygen in a molecule of water, so the volumes of gas collected will not be equal.

Transparencies

TT Fuel Cells & Hybrid Cars

QuickLAB

Hydrolysis

Procedure

1. Coat a **9 V cell** with **petroleum jelly.** Be careful not to get any on the terminals.
2. Mix **1 Tbsp** of **salt** in a **600 mL beaker** of **water.**
3. Fill **two test tubes** with the saltwater solution, and invert them in the beaker, making sure to cover the ends of the test tubes. No air should be trapped in the test tubes.
4. Place the 9 V cell upright in the beaker. Position a battery terminal under the open mouth of each test tube. You will observe hydrogen gas collecting in the test tube located over the negative terminal and oxygen gas collecting over the positive terminal.

Analysis

1. Did you collect the same volume of hydrogen as oxygen? Explain why or why not.

The Challenge of Hydrogen Fuel So why is hydrogen the fuel of the future and not the fuel of today? One difficulty is that hydrogen takes a lot of energy to produce. If this energy comes from burning fossil fuels, generating hydrogen would be expensive and polluting. One alternative is to use electricity from solar cells or wind power to split water molecules to produce hydrogen. Hydrogen could then be stored in pressurized tanks and transported in gas pipelines. Or hydrogen might not be stored at all—it might be used as it is produced, in fuel cells.

Fuel Cells Fuel cells, like the one in **Figure 16,** may be the engines of the future. Like a battery, a **fuel cell** produces electricity chemically, by combining hydrogen fuel with oxygen from the air. When hydrogen and oxygen are combined, electrical energy is produced and water is the only byproduct. Fuel cells can be fueled by anything that contains plenty of hydrogen, including natural gas, alcohol, or even gasoline. The space shuttles have used fuel cells for years. In the change from cars powered by internal combustion engines to those powered by fuel cells, vehicles may get hydrogen from gasoline so that they can be refueled at existing gas stations. By 2010, portable devices such as phones and video-game players may be powered by micro fuel cells. These fuel cells would be fueled with alcohol and may end the problem of charging or changing batteries.

Energy Efficiency

There are two main ways to reduce energy use—lifestyle changes and increases in energy efficiency. Lifestyle changes might include walking or biking for short trips and using mass transit. **Energy efficiency** is the percentage of energy put into a system that does useful work. Energy efficiency can be determined using this simple equation: energy efficiency (in %) = energy out/energy in × 100. Thus, the efficiency of a light bulb is the proportion of electrical energy that reaches the bulb and is converted into light energy rather than into heat. Most of our devices are fairly inefficient. More than 40 percent of all commercial energy used in the

BRAIN FOOD

Electricity from Bacteria Scientists have discovered bacteria on the ocean floor that remove electrons from carbon atoms to produce carbon dioxide. When an electrode is planted in the ocean floor, the electrons can be gathered to produce an electric current. Scientists are investigating whether these bacteria could be used to generate electricity on a larger scale.

CHEMISTRY — CONNECTION — GENERAL

Hydrogen Dangers Hydrogen can ignite at very low concentrations from a small electric spark. Care must be taken during transportation and storage to ensure its containment. Hydrogen is normally transported as a super-cooled liquid. Hydrogen could also be transported in pipelines, as this has proven to be a safe method of transporting natural gas. Have students research the properties of hydrogen and the history of its use as a fuel.

United States is wasted. Most of it is lost from inefficient fuel-wasting vehicles, furnaces, and appliances and from leaky, poorly insulated buildings. We could save enormous amounts of energy by using fuel cells instead of internal combustion engines in cars, and by changing from incandescent to fluorescent light bulbs, as shown in **Table 1**. However, many increases in efficiency involve sacrifices or investments in new technology.

Efficient Transportation Nothing would increase the energy efficiency of American life more than developing efficient engines to power vehicles and increasing the use of public transportation systems. The internal combustion engines that power most vehicles use fuel inefficiently and produce air pollution. The design of these engines has hardly changed since 1900, but they may change radically in the next 50 years. However, in the United States, gasoline prices are currently so low that there is little demand for fuel-efficient vehicles, which are more common in other countries.

Hybrid Cars Hybrid cars, such as the one shown in **Figure 17**, are examples of energy-efficient vehicles currently in use. You have probably seen hybrid cars on the road. Hybrid cars use a small, efficient gasoline engine most of the time, but they also use an electric motor when extra power is needed, such as while accelerating. Hybrid cars feature many other efficient technologies. They convert some of the energy of braking into electricity and they store this energy in the battery. To save fuel, hybrid cars sometimes shut off the gasoline engine, such as when the car is stopped at a red light. Hybrid cars are also designed to be aerodynamic, and they are made of lightweight materials so that they need less energy to accelerate. Hybrid cars do not cost much more than conventional vehicles, they cost less to refuel, and they produce less harmful emissions. These benefits are leading several top auto makers to design many hybrid car models, including hybrid trucks and SUVs.

Table 1 ▼

Energy Efficiency of Common Conversion Devices	
Device	Efficiency
Incandescent light bulb	5%
Fluorescent light bulb	22%
Internal combustion engine (gasoline)	10%
Human body	20%–25%
Steam turbine	45%
Fuel cell	60%

internet connect

www.scilinks.org
Topic: Mass Transit
SciLinks code: HE4063
Topic: Energy Conservation
SciLinks code: HE4033

SC*LINKS* Maintained by the National Science Teachers Association

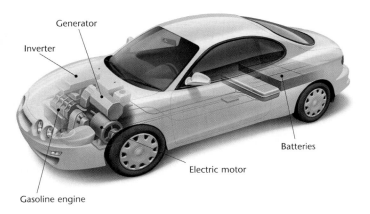

Figure 17 ▶ A hybrid car has a gasoline engine and an electric motor.

Generator
Inverter
Batteries
Electric motor
Gasoline engine

469

MATH**PRACTICE**

Answers

A person in the United States would save 0.02 × 459 gallons = 9.18 gallons. A person in Germany would save 0.02 × 140 gallons = 2.8 gallons.

ARCHITECTURE
CONNECTION **GENERAL**

Green Building Building construction that incorporates energy-efficient and environmentally friendly design features, such as passive solar heating or local building materials, is known as *green building*. Encourage students to investigate green building techniques in their region and have them write a report on their findings. You could also invite a professional, such as an architect specializing in green design (ask the local chapter of the American Institute of Architects for a list) or a representative from a city or state green building program, to come and discuss their work with the class. **LS Interpersonal**

Graphic
Organizer **GENERAL**

Spider Map
You may want to have students work in groups to create this Spider Map. Have one student draw the map and fill in the information provided by other students from the group.

Graphic
Organizer **Spider Map**
Create the Graphic Organizer entitled "Spider Map" described in the Appendix. Label the circle "Ways to Conserve Energy At Home." Create a leg for each way to conserve energy at home. Then, fill in the map with details about each way to conserve energy at home.

MATH**PRACTICE**

Energy Efficiency In the United States, each person uses an average of 459 gallons of gasoline per year. In Germany, each person uses an average of 140 gallons a year. Auto manufacturers estimate that vehicles would use 2 percent less gasoline if everyone kept their tires inflated to the correct pressure. How much gasoline would a person in the United States save and a person in Germany save each year if their tires were kept inflated to the correct pressure?

Figure 18 ▶ In Copenhagen, Denmark, companies provide free bicycles in exchange for publicity. Anyone wishing to use a bike is free to borrow one after paying a refundable deposit. The program helps cut down on pollution and auto traffic.

470

REAL-LIFE
CONNECTION **GENERAL**

Energy Audits Many utility companies offer energy audits, in which trained technicians survey a home or business to determine its energy efficiency and suggest improvements. If such a service is available in your community, invite an energy auditor to come to your class and describe an energy audit. Invite him or her to perform an energy audit of your classroom.

Cogeneration

One way to use fuel more efficiently is *cogeneration*, the production of two useful forms of energy from the same fuel source. For example, the waste heat from an industrial furnace can power a steam turbine that produces electricity. The industry may use the electricity or sell it to a utility company. Small cogeneration systems have been used for years to supply heat and electricity to multiple buildings at specific sites. Small units suitable for single buildings are now available in the United States.

Energy Conservation

Energy conservation means saving energy. It can occur in many ways, including using energy-efficient devices and wasting less energy. The people in **Figure 18** are conserving energy by bicycling instead of driving. Between 1975 and 1985, conservation made more energy available in the United States than all alternative energy sources combined did.

Cities and Towns Saving Energy The town of Osage, Iowa, numbers 3,600 people. You might not think that a town this small could make much of a difference in energy conservation. Yet the town adopted an energy conservation plan that saves more than $1 million each year. The residents plugged the leaks around windows and doors where much of the heat escapes from a house. In addition, they replaced inefficient furnaces and insulated their hot water heaters. Businesses in Osage also found ways to conserve energy. In addition to saving energy, the town has greatly improved its economy through energy conservation. Businesses have relocated to the area in order to take advantage of low energy costs. Unemployment rates have also declined. This small town in Iowa is just one example of the dramatic benefits of energy conservation.

Conservation Around the Home The average household in the United States spends more than $1,200 on energy bills each year. Unfortunately, much of that energy is wasted. Most of the energy lost from homes is lost through poorly insulated windows, doors,

MISCONCEPTION
///**ALERT**\\\

The Cost of Energy Conservation
Many people think that it is expensive to make a home energy efficient. Although some measures, such as new appliances, are expensive, many are free or inexpensive. It costs nothing to turn off unused lights, clean refrigerator coils, or air-dry clothes. These methods are inexpensive and will pay for themselves in less than one year: wrap your hot water heater ($12), install a programmable thermostat ($25), change air filters on heating and cooling systems ($12).

walls, and the roof. So a good way to increase energy efficiency is to add to the insulation of a home. Replacing old windows with new high efficiency windows can reduce your energy bill by 15 percent. Two of the best places to look for ways to conserve energy are doors and windows. Much of the energy lost from a house escapes as hot air in winter or cold air in summer passes through gaps around doors and windows. Hold a ribbon up to the edges of doors and windows. If it flutters, you've found a leak. Sealing these leaks with caulk or weather stripping will help conserve energy. There are dozens of other ways to reduce energy use around the home. Some of these are shown in **Figure 19.**

Conservation in Daily Life There are many simple lifestyle changes that can help save energy. First, remember that using less of any resource usually translates into saving energy. For example, washing your clothes in cold water uses only 25 percent of the energy needed to wash your clothes in warm water. **Table 2** lists a few ways that you can conserve energy every day. Can you think of other ways?

Table 2 ▼

Energy Conservation Tips
• Walk or ride a bicycle for short trips.
• Carpool or use public transportation whenever possible.
• Drive a fuel-efficient automobile.
• Choose ENERGY STAR® products.
• Recycle and choose recycled products whenever possible.
• Set computers to "sleep" mode when they are not in use.

Figure 19 ▶ Ways to Save Energy Around the House

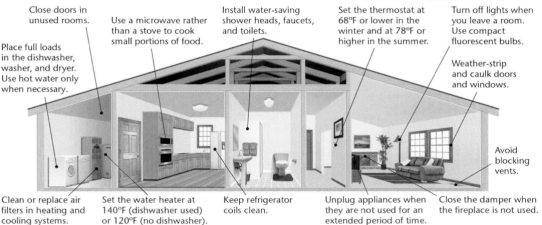

Close doors in unused rooms.

Use a microwave rather than a stove to cook small portions of food.

Install water-saving shower heads, faucets, and toilets.

Set the thermostat at 68°F or lower in the winter and at 78°F or higher in the summer.

Turn off lights when you leave a room. Use compact fluorescent bulbs.

Place full loads in the dishwasher, washer, and dryer. Use hot water only when necessary.

Weather-strip and caulk doors and windows.

Avoid blocking vents.

Clean or replace air filters in heating and cooling systems.

Set the water heater at 140°F (dishwasher used) or 120°F (no dishwasher).

Keep refrigerator coils clean.

Unplug appliances when they are not used for an extended period of time.

Close the damper when the fireplace is not used.

SECTION 2 Review

1. **Describe** three alternative energy technologies, and identify two ways that hydrogen could be used as a fuel source in the future.

2. **List** as many ways as you can for individuals and communities to conserve energy.

3. **Describe** the difference between energy conservation and energy efficiency.

CRITICAL THINKING

4. **Making Inferences** What factors influence a person's choice to conserve energy?

5. **Making Comparisons** Read the description of hydrogen fuel cells and explain why hydrolysis (splitting water molecules with electricity to produce hydrogen and oxygen) is the opposite of the reaction that occurs in a hydrogen fuel cell. **READING SKILLS**

471

Answers to Section Review

1. Tidal power uses the energy of the tides to power a turbine that generates electricity. OTEC produces electricity by driving a turbine with steam that has been heated by warm surface water. The steam is turned back to liquid again by cooling with cold water from the deep ocean. Hydrogen can be used directly by burning it as fuel, or it can be stored in a fuel cell and then combined with oxygen to produce electricity.

2. Answers will vary.

3. Energy conservation involves reducing the amount of energy we use. Energy efficiency involves getting more useful work out of the energy that we use.

4. Answers may vary.

5. During hydrolysis water molecules are split into hydrogen and oxygen using electricity. In a fuel cell, hydrogen and oxygen are combined to produce electricity and water is produced as a byproduct.

Alternative Assessment — GENERAL

Design an Energy Policy Tell groups of students that a 51st state has been created from existing U.S. territory. Their responsibility is to create the energy policy for the state, and they must decide how the state will generate its energy (from fossil fuels, nuclear energy, and/or renewables) and then how it will consume that energy (for uses such as heating, cooking, building construction, and transportation). Students should consider all sources of energy and base their decisions on their advantages and disadvantages. Ask them to write up their energy policy and include diagrams or pictures to illustrate their policy. Have them include their work in their **Portfolio.** LS Logical

Debate: Energy Research Funding Organize the class into small groups. Assign each group either a renewable or nonrenewable energy source discussed in this chapter. Tell them that they are competing for $1 billion in research grants that will be awarded for energy research and development. Have them debate how the money should be awarded. Ask, "Does it make more sense to focus on one research area, or divide the money between several different areas? Would it be better to spend a large amount of money investigating new, unproven techniques, or spend the money to improve the efficiency and environmental effects of the energy sources we currently use?" LS Verbal

CHAPTER 18 Highlights

1 Renewable Energy Today

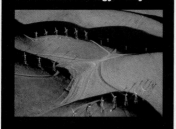

Key Terms

renewable energy, 457

passive solar heating, 458

active solar heating, 460

biomass fuel, 462

hydroelectric energy, 463

geothermal energy, 464

Main Ideas

▸ Renewable energy sources are forms of energy that are constantly being formed from the sun's energy.

▸ Solar energy can be used to heat a house directly or to heat another material, such as water, which can then be used to heat a house. Solar cells can also be used to generate electricity.

▸ Wind power is the fastest growing source of energy in the world.

▸ Many people in developing countries get most of their energy from biomass such as fuelwood. Biomass is increasingly used in developed countries to generate electricity.

▸ Hydroelectric energy is electricity generated by the energy of moving water.

▸ Geothermal energy, the heat within the Earth, can be used to generate electricity.

2 Alternative Energy and Conservation

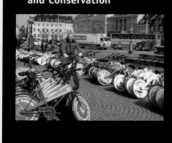

Key Terms

alternative energy, 466

ocean thermal energy conversion (OTEC), 467

fuel cell, 468

energy efficiency, 468

energy conservation, 470

Main Ideas

▸ Alternative energy sources are energy sources that are still in development.

▸ Ocean thermal energy conversion (OTEC) uses the temperature difference between layers of ocean water to generate electricity.

▸ Hydrogen may be one of the fuels of the future. It can be made from any organic material and produces only water as a waste product when burned.

▸ Hydrogen fuel cells may be the engines of the future. Many experiments with them are now underway.

▸ Energy efficiency is the percentage of energy put into a system that does useful work. Energy conservation means saving energy.

472

Chapter Resource File

- Chapter Test GENERAL
- Chapter Test ADVANCED
- Concept Review GENERAL
- Critical Thinking ADVANCED
- Test Item Listing
- Modeling Lab GENERAL
- Observation Lab BASIC
- Observation Lab GENERAL
- Consumer Lab GENERAL
- Long-Term Project ADVANCED

Using Key Terms

Use the correct key term to complete each of the following sentences.

1. Much of the energy needs of the developing world are met by _____, such as fuel-wood.

2. A _____ converts the potential energy of a reservoir into the kinetic energy of a spinning turbine.

3. Turning off the lights when you leave a room is an example of _____.

Use each of the following terms in a separate sentence.

4. *renewable energy*

5. *geothermal energy*

6. *alternative energy*

7. *energy conservation*

> ✔ **STUDY TIP**
>
> **Get Some Exercise** Ride a bike, go for a walk, or play Frisbee or basketball. Try to get at least a half hour of exercise before you begin studying. Then when you study you will be more relaxed and you will be able to focus on the subject you want to learn. As you study, take a moment to notice if the exercise helped. Scientists have proven that regular physical exercise helps fight memory loss.

Understanding Key Ideas

8. Which of the following forms of renewable energy uses the sun's energy most directly?
   a. biomass fuel
   b. passive solar heating
   c. geothermal energy
   d. a hydrogen fuel cell

9. Which of the following energy sources is useful in most parts of the world?
   a. tidal power
   b. OTEC
   c. geothermal energy
   d. active solar energy

10. A house that uses passive solar heating in the Northern Hemisphere will
   a. be built of a material such as concrete or adobe that stores heat well.
   b. have little insulation.
   c. have large north-facing windows.
   d. have an overhang to shade the house from direct winter sun.

11. A passive solar house in the Southern Hemisphere will face
   a. north.
   b. south.
   c. east.
   d. west.

12. Photovoltaic cells convert the sun's energy into
   a. heat.
   b. fuel.
   c. electricity.
   d. light.

13. In a developing country, you are most likely to find biomass used
   a. to generate electricity.
   b. for manufacturing.
   c. for heating and cooking.
   d. as a source of hydropower.

14. Which of the following is *not* true of fuel cells?
   a. They produce electricity.
   b. They will work with many different fuels.
   c. They are more energy efficient than most engines used today.
   d. They cannot be fueled by hydrogen.

15. Which renewable energy source is the fastest growing energy source in the world?
   a. oil
   b. wind
   c. biomass
   d. photovoltaic cells

16. Which statement describes why geothermal heat pumps work?
   a. They are located in areas with abundant geothermal energy.
   b. The ground is warmer than the air in summer and colder than the air in winter.
   c. The ground is colder than the air in summer and warmer than the air in winter.
   d. They run on hydrogen fuel cells.

473

ANSWERS

Using Key Terms

1. biomass fuel

2. hydroelectric power plant

3. energy conservation

4. Sample answer: Renewable energy is produced from sources that are constantly being renewed.

5. Sample answer: Geothermal energy is only useful in areas where geothermal heat sources are close to Earth's surface.

6. Sample answer: Alternative energy sources need development and research before they can be implemented in a large scale.

7. Sample answer: Energy conservation may mean changing lifestyle habits in order to use less energy.

Understanding Key Ideas

8. b
9. d
10. a
11. a
12. c
13. c
14. d
15. b
16. c

Assignment Guide

Section	Questions
1	1, 2, 4, 5, 8–13, 15–17, 19–23, 28–31
2	3, 6, 7, 14, 18, 24–29, 32, 33

Short Answer

17. The sun is the original source of hydroelectric energy.

18. Corrosion will increase the cost of tidal power.

19. Micro-hydropower plants are less expensive to build and they can be built in remote areas on small streams. The environmental effects of micro-hydropower are less than the effects of large-scale hydropower projects.

Interpreting Graphics

20. In the winter, the water is warmed underground and then pumped up to the house, where the water heats the house. The water is cooled in the house.

21. In the summer, the water is cooled underground and then pumped up to the house, where the water cools the house. The water is warmed in the house.

22. In the summer, the closed loop is the coolest and the air is the warmest. In the winter, the closed loop is the warmest and the air is the coolest.

Concept Mapping

23. Answers to the concept mapping questions are on pp. 667–672.

Short Answer

17. Rivers are recharged by the water cycle, so what is the original source of hydroelectric energy?

18. Salt water corrodes metals rapidly. What effect is this likely to have on the cost of electricity produced from tidal power?

19. Why is it likely that hydroelectric energy will be generated increasingly by micro-hydropower plants rather than by large hydroelectric dams?

Interpreting Graphics

Use the information in the figure below to answer questions 20–22.

20. Describe the path of the water in the loop during winter. Where is the water warmed? Where is the water cooled?

21. Describe the path of the water in the loop during summer. Where is the water warmed? Where is the water cooled?

22. What is the difference in the temperature between the house, the closed loop, and the air in the summer? What is the temperature difference in the winter?

Concept Mapping

23. Use the following terms to create a concept map: *sun, hydroelectric energy, solar energy, passive solar heating, active solar heating, water cycle, biomass fuel, wind energy, photovoltaic cell,* and *electric current.*

Critical Thinking

24. **Making Comparisons** Read the description of energy efficiency and energy conservation in this chapter. How are the two concepts related? Give several examples. **READING SKILLS**

25. **Analyzing Ideas** Does the energy used by fuel cells come from the sun? Explain your answer.

26. **Analyzing Ideas** Explain whether you think the most important advances of the 21st century will be new sources of energy or more efficient use of sources that already exist.

27. **Drawing Inferences** Don Huberts of Shell Hydrogen said, "The Stone Age didn't end because the world ran out of stones." He was talking about the future of fossil fuels. Write a short essay that explains what he meant. **WRITING SKILLS**

Cross-Disciplinary Connection

28. **Geography** Create a world map that shows at least 10 renewable energy or alternative energy projects currently in operation. Annotate your map with details and photographs of each project.

Portfolio Project

29. **Energy Timeline** The first energy source used by human societies was human muscle. It was used to build houses, make clothing, and shape tools that could be used to dig up plants and kill animals for food. What was the next source of energy? Make a timeline of the energy sources that humans began to use at various times in history. Add interesting facts and images that relate to each energy source on your timeline. Continue your timeline into the future. What energy sources do you think we will use in the future?

Critical Thinking

24. Energy conservation involves reducing energy use, whereas energy efficiency involves getting more work from the energy used. Both concepts are related to saving energy resources. Turning off lights that are not in use is an example of energy conservation, and a refrigerator that uses less energy to achieve the same amount of cooling is an example of energy efficiency.

25. If the source of hydrogen for fuel cells is organic, then the energy originates from the sun. If the source of hydrogen is inorganic, then the energy might not have come from the sun.

26. Answers may vary.

27. Answers may vary. The quote implies that we will stop using fossil fuels because we will transition to other sources of energy that are more useful or practical, not because we will run out of fossil fuels.

Cross-Disciplinary Connection

28. Answers may vary.

Portfolio Project

29. Answers may vary.

MATH SKILLS

The pie graph below shows electric generating capacity from renewable sources in the United States in 1998. Use the data to answer questions 30–31.

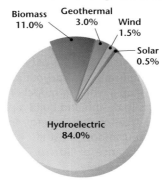

Biomass
11.0%

Geothermal
3.0%

Wind
1.5%

Solar
0.5%

Hydroelectric
84.0%

30. **Making Calculations** How much generating capacity came from biomass, geothermal, wind, and solar combined?

31. **Making Calculations** In 1998, the United States had a total of 94,822 MW of electric generating capacity from renewable energy. How much of that capacity came from biomass? How much came from wind power?

WRITING SKILLS

32. **Communicating Main Ideas** Explain why scientists are working to reduce the use of the two main sources of energy people use today—fossil fuels and biomass.

33. **Writing Persuasively** Write a guide that encourages people to conserve energy and offers practical tips to show them how.

STANDARDIZED TEST PREP

For extra practice with questions formatted to represent the standardized test you may be asked to take at the end of your school year, turn to the sample test for this chapter in the Appendix.

READING SKILLS

Read the passage below, and then answer the questions that follow.

Aluminum is refined from the ore *bauxite*, which is deposited in a thin layer at the Earth's surface. Worldwide, bauxite strip mines cover more of the Earth's surface than any other type of metal ore mine. Aluminum production uses so much electrical energy that the metal has been referred to as "congealed electricity." Producing six aluminum cans takes the energy equivalent of 1 L of gasoline. For this reason, aluminum smelters are located close to cheap and reliable energy sources, such as hydroelectric dams in the Pacific Northwest, Quebec, and the Amazon. When the environmental effects of producing new aluminum are considered, the importance of recycling becomes clear. Recycling one aluminum can saves enough energy to run a television set for 4 hours! Currently, the United States obtains about 20 percent of its aluminum from recycling.

1. Why is aluminum referred to as "congealed electricity"?
   a. Smelting aluminum requires a different form of electrical energy.
   b. Aluminum has an electric charge.
   c. Like electrical energy, aluminum can also be recycled.
   d. So much electrical energy is required to produce aluminum that it is almost as if aluminum were solidified electricity.

2. Which of the following statements describes the author's main point?
   a. Hydroelectricity is a cheap, reliable source of energy.
   b. Recycling aluminum can make a significant contribution to energy conservation.
   c. Aluminum is available in many places, so there is no need to conserve it.
   d. The environmental effects of hydroelectric dams are not related to the consumption of aluminum.

Math Skills

30. 11.0 percent + 3.0 percent + 1.5 percent + 0.5 percent = 16.0 percent

31. 0.11 × 94,822 MW = 10,430 MW from biomass; 0.015 × 94,822 MW = 1,422 MW from wind power

Writing Skills

32. Answers may vary. Both fossil fuels and biomass (in the form of wood) produce greenhouse gases and contribute to global warming.

33. Answers may vary.

Reading Skills

1. d
2. b

475

BLOWING IN THE WIND

Teacher's Notes

Time Required

two 45-minute class periods

Lab Ratings

EASY ———————————→ HARD

TEACHER PREPARATION ▲▲
STUDENT SETUP ▲▲▲
CONCEPT LEVEL ▲▲
CLEANUP ▲▲

Skills Acquired

- Designing Experiments
- Constructing Models
- Experimenting
- Collecting Data
- Organizing and Analyzing Data
- Communicating

The Scientific Method

In this lab, students will:
- Make Observations
- Ask Questions
- Test the Hypothesis
- Analyze the Results
- Draw Conclusions
- Communicate the Results

Materials

The materials listed are enough for a group of 2 to 4 students. A high-speed fan can be used in place of the hair dryer. Provide a wide variety of scrap materials. Assorted sizes of dowels, pulleys, string, and cardboard are all useful items. No commercially-available wind-catching devices, such as pinwheels, should be used.

Objectives

▶ **USING SCIENTIFIC METHODS** **Prepare** a detailed sketch of your solution to the design problem.

▶ **Design and build** a functional windmill that lifts a specific weight as quickly as possible.

Materials

blow-dryer, 1,500 W
dowel or smooth rod
foam board
glue, white
paper clips, large (30)
paper cup, small (1)
spools of thread, empty (2)
string, 50 cm

optional materials for windmill blades: foam board, paper plates, paper cups, or any other lightweight materials

▶ **Windmill Base** Your windmill base should allow the dowel to spin as freely as possible. The pinwheel shown at the end of the dowel is a suggested design for your windmill blades.

Tips and Tricks

By the end of the first class period, students should have a detailed sketch available for approval. The sketch should be evaluated in terms of how well the design problem is addressed. Students can then start the initial construction. During the second class period, students should have a final working model of their windmill. It should have been tested several times in order for an average speed to be obtained.

Blowing in the Wind

MEMO

To: Division of Research and Developers

Quixote Alternative Energy Systems is accepting design proposals to develop a windmill that can be used to lift window washers to the tops of buildings. As part of the design engineering team, your division has been asked to develop a working model of such a windmill. Your task is to design and build a prototype of a windmill that can capture energy from a 1,500 W blow-dryer. Your model must lift 30 large paper clips a vertical distance of 50 cm (approximately 2 ft) as quickly as possible.

Procedure

1. Build the base for your windmill (shown below). Begin by attaching the two spools to the foam board using the glue. Make sure the spools are parallel before you glue them.

2. Pass a dowel or a smooth rod through the center of the spools. The dowel should rotate freely. Attach one end of the string securely to the dowel between the two spools.

3. Poke a hole through the middle of the foam board to allow the string to pass through.

4. Attach the cup to the end of the string. You will use the cup to lift the paper clips.

5. Place your windmill base between two lab tables or in any other area that will allow the string to hang freely.

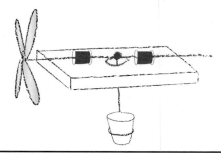

Safety Cautions

Have students use the blowdryer on a cool setting, not hot. To maintain a safe distance between the blowdryer and the wind turbines, a 30 cm dowel can be attached to the blow dryer. Remind students to use care and eye protection if they use a drill.

6. Prepare a sketch of your prototype windmill blades based on the objectives for this lab. Include a list of the materials that you will use and safety precautions (if necessary).

7. Have your teacher approve your design before you begin construction.

8. Construct a working prototype of your windmill blades. Test your model several times to collect data on the speed at which it lifts the paper clips. Record your data for each trial.

9. Vary the type of material used for construction of your windmill blades. Test the various blades to determine whether they improve the original plan.

10. Vary the number and size of the blades on your windmill. Test each design to determine whether the change improves the original plan.

Analysis

1. **Summarize Results** Create a data table that lists the speed for each lift for several trials. Include an average speed.

2. **Graphing Data** Prepare a bar graph that shows your results for each blade design.

Conclusions

3. **Evaluating Methods** After you observe all of the designs, decide which ones you think best solve the problem and explain why.

4. **Evaluating Models** Which change improved your windmill the most—varying the materials for the blades, varying the number of blades, or varying the size of the blades? Would you change your design further? If so, how?

Extension

1. **Research** Windmills have been used for more than 2,000 years. Research the three basic types of vertical-axis machines and the applications in which they are used. Prepare a report of your findings.

2. **Making Models** Adapt your design to make a water wheel. You'll find that water wheels can pull much more weight than a windmill can. Find designs on the Internet for micro-hydropower water wheels such as the Pelton wheel, and use the designs as inspiration for your models. You can even design your own dam and reservoir.

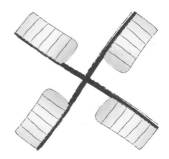
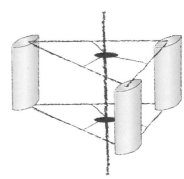
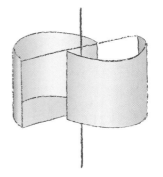

▶ **Sample Windmill Blade Designs**

477

Chapter Resource File
- Datasheets for In-Text Labs
- Lab Notes and Answers

CLASSROOM TESTED & APPROVED

Stu Lipsky
Seward Park High School
New York, New York

WIND POWER IN THE UNITED STATES

Background

Along with the colt revolver and barbed wire, windmills played an important role in the settlement of the American West. Over 6 million windmills were installed on U.S. farms between 1850 and 1970, mostly to pump water. Sears Roebuck and Company sold a wide variety of mass-produced windmills and at one time included a windmills section in its catalogs that was over 100 pages long. By the mid-1920s windmills were also used to generate electricity in rural areas, but in the 1930s and 1940s the Rural Electrification Program extended the nation's electric distribution grid. This resulted in the rapid decline of windmills.

SKILL BUILDER — BASIC

Graphing Have students count all the wind power projects in each category shown in the key and create a bar graph to display this data.
 Visual | English Language Learners |

Homework — GENERAL

Wind Power Around the World
Ask students to research and report on wind energy in Denmark, the Netherlands, or Germany.
Intrapersonal | English Language Learners |

Transparencies

TT Wind Power in the United States

MAPS in action

WIND POWER IN THE UNITED STATES

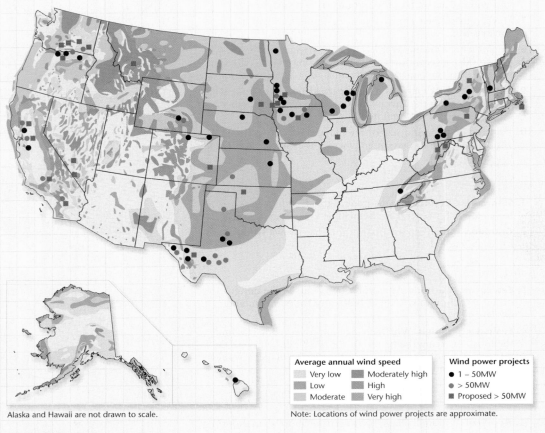

Alaska and Hawaii are not drawn to scale.

Average annual wind speed
- Very low
- Low
- Moderate
- Moderately high
- High
- Very high

Wind power projects
- ● 1 – 50MW
- ● > 50MW
- ■ Proposed > 50MW

Note: Locations of wind power projects are approximate.

MAP SKILLS

1. **Analyzing Data** Why are most of the wind farms located in the western and central United States and not in the eastern United States?

2. **Understanding Topography** Examine Idaho, Wyoming, Montana, and Colorado. What landscape feature might account for the strong winds in those states?

3. **Using the Key** Use the wind power key to locate where you would plan five wind power projects that are larger than 50 MW.

4. **Using the Key** The Great Plains states have been called the "Saudi Arabia of wind energy." Use the key to explain what this statement means.

5. **Finding Locations** The first offshore wind farm in the United States is proposed off the East Coast. Find where the proposed wind farm will be located, and describe the wind conditions in that area.

6. **Using the Key** Use the map to determine which state has the greatest unused potential for wind energy. Explain your reasoning.

478

Answers to Map Skills

1. Answers will vary. The western and central United States have a greater number of areas with high wind power, as well as more open land on which to build wind farms. Wind speeds are generally higher in the central and western United States.

2. the Rocky Mountains

3. Suitable areas are those with the highest wind speed; for example, Montana, Wyoming, and Nebraska.

4. Large areas of the Great Plains have high wind speed and therefore a large potential for energy production.

5. The wind farm will be located off the coast of Massachusetts, where the wind power capacity is among the highest in the country.

6. Answers will vary. Sample answer: Wyoming has the largest unused wind power potential because wind power capacity is high but there are few wind power projects.

BACK TO MUSCLE POWER AND SPRINGS

In 2001, U.S. forces dropped radios into remote parts of Afghanistan so that people could hear news broadcasts. The radios did not contain batteries, and there are few electrical outlets in rural Afghanistan. So how were people supposed to power the radios? The answer is surprisingly simple. The radios use our oldest form of energy—human muscle. Thirty seconds of cranking a handle on one of these radios stores enough energy for an hour of listening.

All Wound Up
The windup radios were invented by Trevor Bayliss in a London garden shed. They were first marketed in 1995 by Freeplay Energy. Now there are also flashlights and electric generators powered by cranks. Some people use them for boating and camping trips and for times when the power goes out. In developing countries, these devices are used in areas where there is no power supply.

These windup devices have several advantages over battery-powered devices. One main advantage is that there is no hazardous waste to dispose of in the form of used batteries. Also, batteries are heavy. Replacing batteries with longer-lasting, lighter sources of electricity has long been a goal of inventors.

The Secret Is in the Design
The first Freeplay radios worked like clockwork toys. The user would turn the handle, winding up a long spring, which then slowly unwound, releasing energy. In a newer model, cranking drives an alternator that charges a tiny battery.

How are these devices different from older clockwork devices, such as the first record players? They are much lighter and much easier to crank. The secret, says Freeplay, is enormously strong, lightweight components that survive for a long time.

An Inspiring Invention
This new technology has captured the imagination of many groups with different goals. There are dozens of experiments going on all over the world. A new line of windup generators will power everything from computers and mobile phones to land-mine detectors and water purification systems.

Watchmakers have developed watches whose batteries never need to be changed. These watches are powered by movement of the wrist. Typing your term paper or playing volleyball powers your watch. Another watch is powered by the heat of your body.

A shoe company is investigating ways to charge small batteries by walking. One design contains a material that generates electricity every time your heel hits the ground.

Defense agencies are studying ways to convert mechanical energy and heat energy into electricity. The goal is to make lighter versions of equipment that has to be carried. Soldiers would no longer have to carry heavy batteries in their packs. The possible applications for these new energy technologies are almost endless.

▶ The Freeplay Radio (below) was invented in a London garden shed, and now the idea is being used all over the world. The flashlight (right) is powered by shaking.

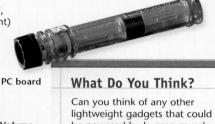

Antenna

Solar panel

Alternator

Rechargeable battery

PC board

Volume knob

DC input jack

Earphone socket

Tuning knob

Winder handle   Input gear   Secondary gear

What Do You Think?
Can you think of any other lightweight gadgets that could be powered by human muscle or body heat instead of by batteries? In what ways are the devices described here more environmentally friendly than the devices they replace?

479

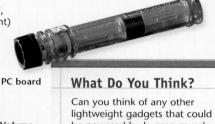

BRAIN FOOD
Human-Powered Washing Machines
It is possible to wash your clothes and get exercise at the same time. The Campus Center for Appropriate Technology at Humboldt State University in California developed a way to hook up an exercise bicycle to a washing machine using a pulley system. As you pedal the bicycle, a fan belt connected to the bicycle wheel turns a gear in the washing machine.

Answers to What Do You Think?
Answers may vary. Examples of lightweight devices include clocks and telephones. The devices described are more environmentally friendly than the products they replace because batteries require toxic materials for their construction and they present a disposal problem.

Cultural Awareness GENERAL
Water Pumps and Seesaws
In the arid plains of eastern Columbia, a village called Gaviotas is an international experiment in sustainable development. Many innovative ideas and technologies have been developed in the past 30 years in Gaviotas. For example, the schoolyards have seesaws that are water pumps. As the children ride the seesaw up and down, water is pumped from the ground into a tank. Instead of using fuel for a basic necessity, the townspeople have combined water pumping with an activity that children do for fun. Students can find out more about Gaviotas on the Internet or in the book, *Gaviotas: A Village to Reinvent the World*. Have students find out more about the sustainable technologies used in Gaviotas and build models of some of the simple devices used in the village.
LS Kinesthetic   English Language Learners

Waste
Chapter Planning Guide

PACING	CLASSROOM RESOURCES	LABS, DEMONSTRATIONS, AND ACTIVITIES

BLOCKS 1, 2 & 3 · 135 min pp. 480–487

Chapter Opener

Section 1
Solid Waste

- **OSP** Lesson Plan *
- **CRF** Active Reading * BASIC
- **TT** Bellringer *
- **TT** Sanitary Landfills *
- **TT** How a Solid-Waste Incinerator Works *

- **TE** Activity Packaging Analysis, p. 481 ◆ GENERAL
- **TE** Activity Landfills, p. 485 ◆ BASIC
- **SE** Skill Practice Lab Solid Waste in Your Lunch, pp. 504–505 ◆
- **CRF** Datasheets for In-Text Labs *
- **CRF** Modeling Lab * ◆ ADVANCED

BLOCKS 4 & 5 · 90 min pp. 488–492

Section 2
Reducing Solid Waste

- **OSP** Lesson Plan *
- **CRF** Active Reading * BASIC
- **TT** Bellringer *

- **TE** Demonstration A New Beginning, p. 488 ◆ GENERAL
- **SE** Field Activity Is It Really Recyclable? p. 489
- **TE** Activity Creating a School Recycling Program, p. 490 ADVANCED
- **TE** Group Activity Recycling Education Campaign, p. 491 GENERAL
- **CRF** Design Your Own Lab * ◆ GENERAL
- **CRF** Modeling Lab * GENERAL

BLOCKS 6 & 7 · 90 min pp. 493–499

Section 3
Hazardous Waste

- **OSP** Lesson Plan *
- **CRF** Active Reading * BASIC
- **TT** Bellringer *
- **TT** Superfund Sites *
- **TT** Hazardous-Waste Deep-Well Injection *

- **SE** QuickLab Neutralizing Hazardous Waste, p. 495
- **TE** Activity Energy in Hazardous Waste, p. 495 ADVANCED
- **TE** Demonstration Hazardous Waste Spills, p. 496 GENERAL
- **TE** Internet Activity Hazardous Substances, p. 497 GENERAL
- **TE** Group Activity Household Hazardous Waste Disposal, p. 498 GENERAL
- **CRF** Consumer Lab * ◆ GENERAL
- **CRF** Long-Term Project * ◆ BASIC

BLOCKS 8 & 9 · 90 min

Chapter Review and Assessment Resources
- **SE** Chapter Review pp. 501–503
- **SE** Standardized Test Prep pp. 672–673
- **CRF** Study Guide * ■ GENERAL
- **CRF** Chapter Test * ■ GENERAL
- **CRF** Chapter Test * ADVANCED
- **CRF** Concept Review * ■ GENERAL
- **CRF** Critical Thinking * ADVANCED
- **OSP** Test Generator
- **CRF** Test Item Listing *

Online and Technology Resources

 Holt Online Learning

Visit **go.hrw.com** for access to Holt Online Learning or enter the keyword **HE6 Home** for a variety of free online resources.

 **One-Stop Planner® CD-ROM**

This CD-ROM package includes
- Lab Materials QuickList Software
- Holt Calendar Planner
- Customizable Lesson Plans
- Printable Worksheets
- ExamView® Test Generator
- Interactive Teacher Edition
- Holt PuzzlePro® Resources
- Holt PowerPoint® Resources

 Holt Environmental Science Interactive Tutor CD-ROM

This CD-ROM consists of interactive activities that give students a fun way to extend their knowledge of environmental science concepts.

Compression guide:
To shorten from 9 to 7 45 min
blocks, eliminate blocks 2 and 3.

KEY

TE	Teacher Edition	CRF	Chapter Resource File		*	Also on One-Stop Planner
SE	Student Edition	TT	Teaching Transparency		■	Also Available in Spanish
OSP	One-Stop Planner	CD	Interactive CD-ROM		◆	Requires Advance Prep

ENRICHMENT AND SKILLS PRACTICE	SECTION REVIEW AND ASSESSMENT	CORRELATIONS
SE Pre-Reading Activity, p. 480 TE Using the Figure, p. 480 `GENERAL`		National Science Education Standards
TE Inclusion Strategies, p. 481 TE Using the Figure NIMBY, p. 482 `GENERAL` TE Interpreting Statistics Municipal Solid Waste, p. 482 `GENERAL` TE Skill Builder Math, p. 482 `GENERAL` TE Career The Wide World of Waste, p. 483 SE MathPractice Municipal Solid Waste, p. 484 TE Interpreting Statistics Municipal Solid Waste, p. 484 `GENERAL` TE Skill Builder Graphing, p. 484 `GENERAL` TE Skill Builder Writing, p. 486 `GENERAL` TE Reading Skill Builder Paired Summarizing, p. 486 `GENERAL`	TE Homework, p. 483 `GENERAL` TE Homework, p. 483 `ADVANCED` TE Homework, p. 485 `GENERAL` SE Section Review, p. 487 TE Reteaching, p. 487 `BASIC` TE Quiz, p. 487 `GENERAL` TE Alternative Assessment, p. 487 `GENERAL` CRF Quiz * ■ `GENERAL`	LS 4e SPSP 2a SPSP 3b SPSP 3c SPSP 6d
TE Using the Figure The Recycling Cycle, p. 489 `GENERAL` SE Case Study Paper or Plastic? pp. 490–491 SE Graphic Organizer Chain-of-Events Chart, p. 492 `GENERAL`	TE Homework, p. 491 `GENERAL` SE Section Review, p. 492 TE Reteaching, p. 492 `BASIC` TE Quiz, p. 492 `GENERAL` TE Alternative Assessment, p. 492 `GENERAL` CRF Quiz * ■ `GENERAL`	LS 4e SPSP 3b SPSP 3c SPSP 6d
TE Inclusion Strategies, p. 493 SE Case Study Love Canal, pp. 496–497 TE Student Opportunities Community Involvement, p. 497 SE Points of View Should Nuclear Waste Be Stored at Yucca Mountain? pp. 506–507 CRF Map Skills Recycling Centers * `GENERAL`	TE Homework, p. 498 `GENERAL` SE Section Review, p. 499 TE Reteaching, p. 499 `BASIC` TE Quiz, p. 499 `GENERAL` TE Alternative Assessment, p. 499 `GENERAL` CRF Quiz * ■ `GENERAL`	LS 4e SPSP 3b SPSP 6d

Guided Reading Audio CDs

These CDs are designed to help auditory learners and reluctant readers. (Audio CDs are also available in Spanish.)

www.scilinks.org

Maintained by the **National Science Teachers Association**

TOPIC: Solid Waste
SciLinks code: HE4102

TOPIC: Biodegradable and Nonbiodegradable Materials
SciLinks code: HE4128

TOPIC: Waste Prevention
SciLinks code: HE4120

TOPIC: Hazardous Waste
SciLinks code: HE4054

CNN Videos

Each video segment is accompanied by a Critical Thinking Worksheet.

Earth Science Connections
Segment 10 Mercury Disposal

CHAPTER

19

Chapter Enrichment

This Chapter Enrichment provides relevant and interesting information to expand and enhance your classroom instruction of the chapter material.

1 Solid Waste

Solid Waste Milestones

The first recorded municipal dump in the Western world is in Athens, Greece, and dates to 500 BCE. The dump is one mile outside of what was then the city. England built the first systematic incinerator in 1874. Until this time, residents often burned garbage in their backyards. The first incinerator in the United States was built in New York in 1885. Four years later, the U.S. government declared that we were running out of room for waste disposal. In the 1920s, the United States began using filled wetlands to dispose of waste. The first federal solid waste management laws were passed in 1965.

BRAIN FOOD

There was no garbage-collection system in 18th-century New York City. People threw their garbage directly into the streets. Free-running hogs then came along and ate the garbage.

Piggeries

In 1900, "piggeries" were established in the United States for waste management. Fresh or cooked garbage was fed to pigs for disposal. In the mid-1950s, a group of pigs got sick from eating raw garbage, so a law was passed stating that the pigs could only eat cooked garbage. The method of cooking was expensive and ineffective, and piggeries were phased out in the late 1960s. In some communities, food waste from institutional or commercial kitchens is collected and fed to pigs on local farms as a method of reducing food waste.

Waste and Disease

Throughout human history, trash has provided breeding grounds for vectors of diseases such as bubonic plague, cholera, and typhoid fever. In towns, garbage was often thrown out of windows into the streets, where it supported rats and contaminated the water supply. In 1842, the "age of sanitation" finally began when Edwin Chadwick linked higher mortality rates to unsanitary conditions in England.

Plastics in the Environment

Plastics are petroleum-based polymers that consist of long chains of molecules. Most of these molecular chains are very difficult to break. In fact, most plastics used today are not readily biodegradable; they take 200 to 400 years to degrade. In 1997, the United States used 47 million barrels of oil to produce plastic packaging alone. In a landfill, toxic compounds that are sometimes added to plastics can leach out. If the landfill leaks, these compounds can contaminate groundwater and surface water. The petrochemicals from which plastics are made release toxins into the air when incinerated.

▶ **Some fast-food cartons are made out of plastic**

2 Reducing Solid Waste

Appliances and Landfills

More than 8 million appliances, such as refrigerators and clothes washers and dryers, are thrown away in the United States each year. Eighty-one percent of these appliances are recycled, mostly for their steel content. Eighteen states ban appliances from landfills, forcing consumers and manufacturers to recycle them.

BRAIN FOOD

Today, the most efficiently recycled material is aluminum. Recycling aluminum saves 95 percent of the energy needed to process virgin aluminum. About 2,200 aluminum cans are recycled every second in the United States. However, enough aluminum is thrown out to rebuild the entire United States commercial air fleet every three months.

The Myths of Plastics Recycling

When plastics are recycled, the quality of the plastic is usually reduced. Plastic is usually recycled to make secondary products that are not recyclable, such as certain textiles. Thus, the plastic's life cycle is extended, but the materials eventually end up in landfills, which is not the desired outcome of recycling. Plastics recycling has not significantly reduced the consumption of natural resources because a "closed loop" system has not been established.

In most areas, only air-blown plastics, such as narrow-necked plastic bottles, are accepted for recycling. Injection-molded plastics, such as caps or margarine containers, are rarely accepted. For example, yogurt cups with number 2 on them are not recyclable even though they are made from one of the most commonly recycled plastics. The chemicals that are added to the plastic to form its shape substantially change its composition so that it is no longer similar to the end products of other number 2 plastics.

Curbside Recycling

Curbside recycling began in the city of Baltimore, Maryland, in 1974. By 2000, there were more than 9,200 curbside recycling programs in the United States. Many programs arose because municipalities began to run out of landfill space. One of the greatest problems faced by recycling programs today is finding buyers for the products made from the recycled materials, because these products are sometimes more expensive.

▶ Recycling begins with collecting and sorting materials

Composting

Each year in the United States, people bag 18 billion kg (20 million tons) of leaves and grass clippings and send them to landfills. At the same time, farmers and home-owners buy approximately 20 billion kg (22 million tons) of commercial synthetic fertilizers to add to their crops and lawns. Many of these fertilizers are made from petrochemicals and can be toxic to the local environment. If yard wastes were composted, the byproducts could add valuable nutrients to topsoil and serve as a soil enricher, thereby reducing the amount of synthetic fertilizer necessary. Also, the yard clippings would no longer take up dwindling landfill space.

3 Hazardous Waste

BRAIN FOOD

Bioremediation relies on microorganisms to break down wastes. Some microbes can metabolize organic pollutants, such as petroleum compounds. Bioremediation techniques are now being used to clean up environmental pollutants such as crude oil, sewage effluent, chlorinated solvents, pesticides, and gasoline.

Cell Phones and Hazardous Waste

By 2005, about 130 million cell phones will be thrown away each year in the United States. This represents 59,000 metric tons (65,000 tons) of plastic and metal laced with toxic substances that could contaminate the environment. Cell phones contain antimony, arsenic, beryllium, cadmium, copper, lead, nickel, and zinc, which persist in the environment, accumulate in fatty tissues, and can potentially cause severe health problems. The European Union plans to implement take-back programs for cell phones. At this time, no such program exists in the United States.

Overview

Tell students that the purpose of this chapter is to introduce them to our society's waste problems. Society produces many different kinds of solid wastes, and these wastes must be disposed of properly. Producing less waste, recycling, buying recycled products, composting, and changing the types of materials we use can help alleviate the waste problem.

Using the Figure — GENERAL

A modern landfill is different from what students may think of as a dump, which is just an open pile of trash with no sanitary management. A landfill is designed to separate the trash from the surrounding environment using a bottom liner of plastic or clay and a layer of soil on top. The storage area for the waste may take up only one-third of the area set aside for the landfill. The remaining area is used for leachate and runoff-collection ponds, as a buffer zone, and as a soil source. Ask students if we could use landfills to store all our waste. (No; We could eventually run out of space, and some items, such as motor oil, tires, and lead-acid batteries, are toxic and require special disposal.) **LS** Visual

PRE-READING ACTIVITY

FOLDNOTES

You may want to use this FoldNote in a classroom discussion to review material from the chapter. Ask students to provide information for each category from the FoldNote. Write the information that students provide on the board.

VIDEO SELECT

For information about videos related to this chapter, go to **go.hrw.com** and type in the keyword **HE4 WASV**.

Waste

1 Solid Waste
2 Reducing Solid Waste
3 Hazardous Waste

PRE-READING ACTIVITY

FOLDNOTES

Three-Panel Flip Chart
Before you read this chapter, create the **FoldNote** entitled "Three-Panel Flip Chart" described in the Reading and Study Skills section of the Appendix. Label the flaps of the chart with "Solid Waste," "Reducing Solid Waste," and "Hazardous Waste." As you read the chapter, write information you learn about each category under the appropriate panel.

This landfill in New Jersey stores municipal solid waste that people throw away on a day-to-day basis. Every year, the United States generates more than 210 million metric tons of municipal solid waste.

480

Chapter Correlations   *National Science Education Standards*

LS 4e Human beings live within the world's ecosystems. Increasingly, humans modify ecosystems as a result of population growth, technology, and consumption. Human destruction of habitats through direct harvesting, pollution, atmospheric changes, and other factors is threatening current global stability, and if not addressed, ecosystems will be irreversibly affected. **(Section 1, Section 2, and Section 3)**

SPSP 2a Populations grow or decline through the combined effects of births and deaths, and through emigration and immigration. Populations can increase through linear or exponential growth, with effects on resource use and environmental pollution. **(Section 1)**

SPSP 3b The earth does not have infinite resources; increasing human consumption places severe stress on the natural processes that renew some resources, and it depletes those resources that cannot be renewed. **(Section 1, Section 2, and Section 3)**

SPSP 3c Humans use many natural systems as resources. Natural systems have the capacity to reuse waste, but that capacity is limited. Natural systems can change to an extent that exceeds the limits of organisms to adapt naturally or humans to adapt technologically. **(Section 1 and Section 2)**

SPSP 6d Individuals and society must decide on proposals involving new research and the introduction of new technologies into society. Decisions involve assessment of alternatives, risks, costs, and benefits and consideration of who benefits and who suffers, who pays and gains, and what the risks are and who bears them. Students should understand the appropriateness and value of basic questions—"What can happen?"—"What are the odds?"—and "How do scientists and engineers know what will happen?" **(Section 1, Section 2, and Section 3)**

Solid Waste

It is lunchtime. You stop at a fast-food restaurant and buy a burger, fries, and a soda. Within minutes, the food is gone, and you toss your trash into the nearest wastebasket. **Figure 1** shows what might be in your trash: a paper bag, a polystyrene burger container, the cardboard carton that held the fries, a paper cup with a plastic lid, a plastic straw, a handful of paper napkins, and several ketchup and mustard packets. Once you throw away your trash, you probably do not give it a second thought. But where does the trash go?

The trash from the wastebasket probably will be picked up by a collection service and taken to a landfill, where the trash will be dumped with thousands of tons of other trash and covered with a layer of soil. That trash will not bother anyone anymore, will it? Maybe not, unless the landfill fills up next year and the city has no place to put the garbage. What would happen if rainwater ran down into the landfill, and leached a harmful chemical, such as paint thinner, and it seeped into the groundwater? Suddenly, the trash that was not bothering anyone is causing an environmental problem.

The Generation of Waste

Imagine multiplying the waste disposal problems that come with your lunch by the number of things that you and everyone else throw away each day. Every year, the United States generates more than 10 billion metric tons of solid waste. **Solid waste** is any discarded solid material. Solid waste includes everything from junk mail to coffee grounds to cars. Many products that we buy today are used once and then thrown away. As a result, the amount of solid waste each American produces each year has more than doubled since the 1960s.

Objectives

▶ **Name** one characteristic that makes a material biodegradable.

▶ **Identify** two types of solid waste.

▶ **Describe** how a modern landfill works.

▶ **Name** two environmental problems caused by landfills.

Key Terms

solid waste
biodegradable
municipal solid waste
landfill
leachate

internet connect

www.scilinks.org
Topic: Solid Waste
SciLinks code: HE4102

SCiLINKS Maintained by the National Science Teachers Association

Figure 1 ▶ Where does your trash go when you throw it away?

INCLUSION Strategies

• *Learning Disabilities*
• *Attention Deficit Disorder*

Have students give a report on the total amount of waste produced by the class in one day. Students should identify waste that is biodegradable, nonbiodegradable, and waste that can be recycled. Students can include a plan for the class to reduce the amount of waste that would go to a land-fill. Students could report findings and recommendations as though they were reporting for a television news show.

Focus

Overview

Before beginning this section, review with your students the Objectives in the Student Edition. This section encourages students to think about where trash comes from and what happens to trash after people dispose of it. Students also learn how the disposal of certain kinds of human-produced materials, such as plastics, affects the environment.

📀 Bellringer

Have students imagine that they are on an archeological expedition 500 years from now and have just come across a garbage site from the 21st century. Ask them to list the types of objects they think they would find. (Answers may vary. Students should list objects that have not decomposed, such as those made from plastic, metal, or glass.)
LS Logical

Motivate

Activity ——————— GENERAL

Packaging Analysis Organize the class into several groups. Give each group a variety of packaged products. Have students calculate, for each product, the percentage of the total mass of the consumable product and its packaging. Then, ask them to compare the percentages. Finally, encourage students to discuss the benefits and drawbacks of each type of packaging.

Chapter Resource File

• **Lesson Plan**
• **Active Reading** BASIC
• **Section Quiz** GENERAL

Transparencies

TT Bellringer

Teach

Using the Figure — GENERAL

NIMBY Point out to students that the situation described in **Figure 2** is an example of the NIMBY phenomenon. Explain that NIMBY stands for "Not in My Backyard" and that it represents the mind-set of not wanting a landfill near one's home. Ask students why a landfill is considered undesirable by some. (Sample reasons are the possible sights and smells, declining property values, and possible groundwater contamination.) Then ask, "What would happen if everyone took the NIMBY point of view?" (Landfills would have to be located far from people's homes and businesses. Garbage trucks would have to transport the waste farther from its source, causing additional environmental problems.) **LS** Visual

Interpreting Statistics — GENERAL

Municipal Solid Waste Use **Figure 3** to help students visualize the tremendous increase in municipal solid waste between 1960 and 1999. Have students copy the graph onto a sheet of paper. Tell them to assume a similar rate of increase until the year 2010. Then ask, "How much total waste did Americans generate in 1980?" (about 150 million tons) "How much will we generate in the year 2010?" (about 250 million tons) Ask students, "Where will we put all this trash?" (Answers may vary. Most waste will go in a growing number of landfills or be incinerated.) Encourage students to brainstorm ways to reduce the volume of trash we produce. (reduce, reuse, and recycle) **LS** Logical

Figure 2 ▶ The barge *Mobro* (right) from Islip, New York, sailed up and down the East Coast and to the Gulf of Mexico for five months looking for a place to dump its load of garbage. The map below shows its route.

Space and Waste Today, many towns are running out of space to dispose of the amounts of waste that people create. For example, in 1987, the barge shown in **Figure 2** was loaded with 3,200 tons of garbage and left the town of Islip, New York, in search of a place to unload its waste. The barge sailed along the Atlantic coast to the Gulf of Mexico for more than five months in search of a state that would be willing to dispose of the waste. When no one would accept the garbage, it was finally burned in New York, and the 430 tons of ash was sent to Islip to be buried.

Population and Waste While the Earth's human population and the amount of waste we produce grows larger, the amount of land available per person becomes smaller. Thousands of years ago, in the time of hunter-gatherer societies, the human population was smaller and the waste created consisted mostly of animal and vegetable matter. This type of waste combined with a large amount of land made disposing the waste much easier. However, today, the average person living in the United States produces 4.4 pounds of solid waste per day, as shown in **Figure 3**. Because the human population and the amount of waste we create is increasing and the amount of land available is decreasing, it is getting harder to dispose of the waste we create.

Figure 3 ▶ The total amount of municipal solid waste generated in the United States has more than doubled in the past 40 years.

Municipal Solid Waste Generation (U.S.)

Source: U.S. Environmental Protection Agency.

SKILL BUILDER — GENERAL

Math Have students calculate the number of days it would take them to generate the amount of solid waste equivalent to their body weight. (Answers may vary. Sample answer: 140 lb/(4.4 lb/day) = 32 days for a 140 lb person) Then, have them calculate, in units of their body weight, the amount of solid waste they produce in one year. (Sample answer: 365 days/32 days = 11.4, so a 140 lb person would produce 11.4 times his or her body weight in solid waste in one year.) **LS** Logical

Notable Quotes — GENERAL

"But our waste problem is not the fault only of producers. It is the fault of an economy that is wasteful from top to bottom . . . and all of us are involved in it."

—Wendell Berry

Ask students to explain this quote. Ask where they think the responsibility for our waste problems lies—with manufacturers or with the consumer. (Answers may vary.) **LS** Verbal

Not All Wastes Are Equal

Problems are caused not only by the amount of solid waste but also by the type of solid waste. There are two basic materials that wastes are made of: wastes made of biodegradable materials and wastes made of nonbiodegradable materials. A material is **biodegradable** if it can be broken down by biological processes. Plant and animal matter are examples of biodegradable materials that can be broken down and absorbed by the environment. Products made from natural materials are usually biodegradable. Examples of biodegradable products include newspapers, paper bags, cotton fibers, and leather.

Many products made from synthetic materials are not biodegradable. A *nonbiodegradable material* cannot be broken down by biological processes. Synthetic materials are made by combining chemicals to form compounds that do not form naturally. Some examples of synthetic materials are polyester, nylon, and plastic.

Plastic Problems Plastics illustrate how nonbiodegradable materials can cause problems. Plastics are made from petroleum or natural gas. Petroleum and natural gas consist mostly of carbon and hydrogen, which are the same elements that make up most molecules found in living things. But in plastics, these elements are put together in molecular chains that are not found in nature. Over millions of years, microorganisms have evolved the ability to break down nearly all biological molecules. However, microorganisms have not developed ways to break down the molecular structures of most plastics. Therefore, some plastics that we throw away may accumulate and last for hundreds of years.

Types of Solid Waste

Most of what we throw out on a day-to-day basis is called municipal solid waste. Manufacturing waste, such as the computers shown in **Figure 4**, and mining waste make up about 70 percent of the other types of solid waste produced in the United States.

internet connect

www.scilinks.org
Topic: Biodegradable and Nonbiodegradable Materials
SciLinks code: HE4128

SCI LINKS. Maintained by the National Science Teachers Association

Ecofact

Breaking Down Biodegradable Material Decomposers, such as fungi and bacteria, are examples of organisms that break down biodegradable material. Once these materials are broken down, they can be reused by other organisms. Scavengers, such as vultures, and insects, such as dung beetles, also help recycle organic waste.

Figure 4 ▶ These discarded computers have been exported from the United States and disposed of in China. Unwanted computers, televisions, audio equipment, and printers, are types of electronic waste.

483

Discussion ——— GENERAL

Biodegradable Packaging Some packaging manufacturers argue that it does not make sense to create biodegradable packaging given our current municipal waste management system. The majority of municipal solid waste is either sent to landfills, where it usually does not decompose very quickly, because of lack of water and air, or the waste is burned in incinerators. Until municipal composting makes up a larger part of waste management, these packaging manufacturers argue, creating packaging that biodegrades does not make sense. Some manufacturers also argue that biodegradable packaging could increase litter, because consumers will be less motivated to properly dispose of waste if they think it will decompose. Discuss this position with students. Ask them if they think packaging should be biodegradable and why or why not. **LS Verbal**

Homework ——— GENERAL

Recycling Computers Have students find out if there is a business or organization in their community that recycles computers and, if so, what products it makes from the computers. If there is no computer recycling facility in your community, have them write a computer manufacturer to ask if it recycles its own computers. **LS Interpersonal**

Career

The Wide World of Waste Waste management offers some interesting possibilities for careers that students may want to explore. City recycling coordinators are needed to design and oversee municipal recycling programs. Attorneys who specialize in environmental law provide advice on landfill siting and compliance with environmental laws. There are also entrepreneurs who recycle waste and produce new, often imaginative products. If a student expresses an interest in waste management, suggest that he or she interview one of these professionals to get more information.
LS Interpersonal/Intrapersonal

Homework ——— ADVANCED

Plastics Plastics production has increased steadily from less than 1 percent of solid waste in 1960 to 10.7 percent of solid waste in 2000. Have students research the seven types of plastics, which correspond to the numbers on the bottoms of plastic containers. These numbers represent the type of resin used to produce the plastic. Have students investigate the manufacturing method, ingredients, uses, the potential for recycling, biodegradability, and other characteristics of each plastic. Suggest that students present their findings to the class. **LS Verbal**

Answers

(229.9 million tons − 223 million tons)/223 million tons × 100 = about 3 percent

Interpreting Statistics — GENERAL

Municipal Solid Waste The pie graph in **Figure 5** shows the percentages of municipal solid waste by weight generated in the United States. Ask students the following questions: "What percentage of municipal solid waste can be composted?" (12.1% + 10.9% = 23%) Tell students, "The materials that are the best candidates for recycling are paper, metal, and glass. If all of these three wastes were recycled, what percentage of waste would be prevented?" (51.4 percent) "What percentage of waste would be left over after composting and recycling these materials?" (25.6 percent) **LS** Logical

SKILL BUILDER — GENERAL

Graphing Based on the information on this page, have students create a pie graph showing the percentages of all solid wastes produced in the United States. Students will have to calculate the percentage of mining waste. Have them put waste that is unaccounted for in the category "other." (manufacturing: 56 percent, mining: 14 percent, agricultural: 9 percent, municipal: 2 percent, other: 14 percent) **LS** Visual

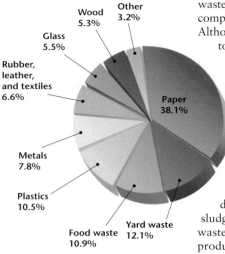

United States Municipal Solid Waste (Percentage by Weight)

Other 3.2%
Wood 5.3%
Glass 5.5%
Rubber, leather, and textiles 6.6%
Paper 38.1%
Metals 7.8%
Plastics 10.5%
Food waste 10.9%
Yard waste 12.1%

Source: U.S. Environmental Protection Agency.

Figure 5 ▶ Paper makes up most of the municipal solid waste in the United States. How much of the waste shown in this graph could be recycled?

Municipal Solid Waste
The United States generated approximately 229.9 million tons of municipal solid waste in 1999. In 1998, the United States generated approximately 223 million tons of municipal solid waste. What was the percent increase in municipal solid waste generation from 1998 to 1999?

Figure 6 ▶ Modern landfills are lined with clay and plastic and have a system for collecting and treating liquid that passes through the compacted solid waste.

Municipal Solid Waste About 2 percent of the total solid waste in the United States is made up of **municipal solid waste,** which is the waste produced by households and businesses. **Figure 5** shows the composition of municipal solid waste in the United States. Although municipal solid waste makes up only 2 percent of the total solid waste in the United States, this amounts to more than 210 million metric tons each year. That is enough waste to fill a convoy of garbage trucks that would stretch around the Earth about six times. Furthermore, the amount of municipal solid waste is growing much faster than the amount of mining or agricultural waste.

Solid Waste from Manufacturing, Mining, and Agriculture Solid waste from manufacturing, mining, and agriculture make up the rest of the total solid waste produced in the United States. Solid waste from manufacturing makes up 56 percent of the total solid waste produced and includes items such as scrap metal, plastics, paper, sludge, and ash. Although consumers do not directly produce waste from manufacturing, they indirectly create it by purchasing products that have been manufactured.

Waste from mining consists of the rock and minerals that are left over from excavation and processing. This waste is left exposed in large heaps, is dumped in oceans or rivers, or is disposed of by refilling and landscaping abandoned mines. Agricultural waste makes up 9 percent of the total solid waste produced and includes crop wastes and manure. Because agricultural waste is biodegradable, it can be broken down and returned to the soil. However, the increasing use of fertilizers and pesticides may cause agricultural waste to become more difficult to dispose of because the waste may be harmful if returned to the soil.

Solid Waste Management

Most of our municipal waste in the United States is sent to landfills such as the one shown in **Figure 6**. However, some of our waste is incinerated, and more than 28 percent of our waste is recycled. By comparison, in 1970, we recycled only 6.6 percent of our waste.

REAL-LIFE CONNECTION — GENERAL

What Do You Throw Away? Give each student a large garbage bag. Tell them that they are to carry it around with them for the next 24 hours and put in the bag all the waste that they would normally throw away. They should also write down on paper each object they put in the bag. At the end of the 24 hours, have them weigh the bag. Have them compare the weight of the garbage with the 2 kg (4.4 lb) generated by the average U.S. consumer in one day. Ask them how they could reduce the waste they generate. (Answers may vary.) **LS** Interpersonal

Cultural Awareness — GENERAL

Construction Waste in Japan Construction materials often take up an enormous amount of landfill space. Often, these materials could be used again. For example, the construction of a tunnel in an urban area of Japan removed massive quantities of earth and rocks. Ordinarily, this material would be transported a long distance to a rapidly filling landfill. Instead, it was used to reclaim land that was previously unusable by the city. This provided two valuable services: creating urban land space that could be used as parkland or for development, and saving precious landfill space.

Landfills More than 50 percent of the municipal and manufacturing solid waste created in the United States ends up in landfills as shown in **Table 1.** A **landfill** is a permanent waste-disposal facility where wastes are put in the ground and covered each day with a layer of soil, plastic, or both. The parts of a modern landfill are shown in **Figure 7.** The most important function of a landfill is to contain the waste that is buried inside and to keep the waste from causing problems with the environment. Most importantly the waste inside a landfill must not come into contact with the soil and groundwater that surrounds the landfill.

Problems with Landfills One problem with landfills is leachate. **Leachate** is a liquid that has passed through compacted solid waste in a landfill. Leachate forms when water seeps down through a landfill and contains dissolved chemicals from decomposing garbage. Leachate may contain chemicals from paints, pesticides, cleansers, cans, batteries, and appliances. Landfills typically have monitoring wells and storage tanks to measure and store leachate. Stored leachate can then be treated as waste water. However, if landfills are not monitored properly, leachate can flow into groundwater supplies and make water from nearby wells unsafe to drink.

Another problem with landfills is methane. As organic waste decomposes deep in the landfill where there is no oxygen, it produces methane, a highly flammable gas. Methane gas is usually pumped out of landfills and used as fuel. However, if methane gas production is not monitored safely, it may seep through the ground and into basements of homes up to 300 m from a landfill. If the methane is ignited by a spark, it can cause dangerous explosions.

Table 1 ▼

Where Waste in the United States Goes	
Waste-disposal method	Percentage of waste by weight
Stored in landfills	57
Recycled	28
Incinerated	15

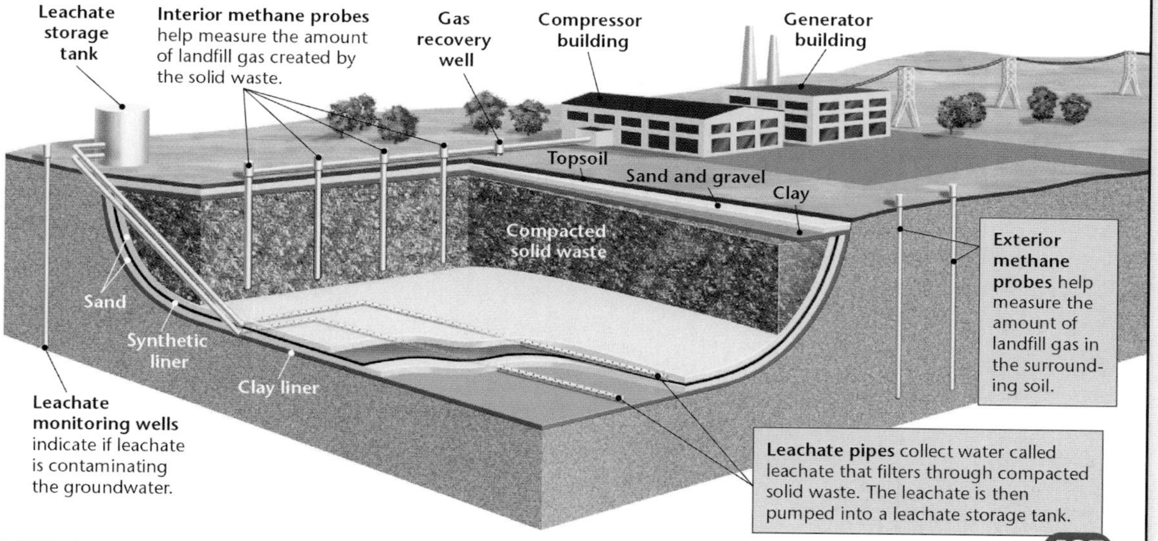

Figure 7 ► This landfill generates electricity by burning methane gas that is produced by decomposing garbage.

Leachate storage tank

Interior methane probes help measure the amount of landfill gas created by the solid waste.

Gas recovery well

Compressor building

Generator building

Topsoil

Sand and gravel

Clay

Compacted solid waste

Sand

Synthetic liner

Clay liner

Leachate monitoring wells indicate if leachate is contaminating the groundwater.

Exterior methane probes help measure the amount of landfill gas in the surrounding soil.

Leachate pipes collect water called leachate that filters through compacted solid waste. The leachate is then pumped into a leachate storage tank.

485

Mail-Order Catalogs Every year 19.5 billion mail-order catalogs are produced in the United States, which amounts to 71 catalogs per person. This represents 3.3 million metric tons (3.6 million tons) of waste that ends up in landfills. Ask students how they can help reduce this waste. (Sample answer: Have my address removed from mailing lists and urge catalog companies to use recycled paper.) **LS Verbal**

Debate — GENERAL

New Incinerator in Town Have one student portray a radio talk-show host doing a program on a new solid-waste incinerator to be built near a residential part of town. The other students should prepare arguments for or against the incinerator and then "call in" their opinions to the radio station. Encourage students to take a variety of viewpoints. **LS Verbal/Interpersonal**

Co-op Learning

SKILL BUILDER — GENERAL

Writing Tell students to imagine that all the landfills are full and that no more land is available to build new landfills. Have them write a short story describing a scenario that they think would be likely to occur. Stories should describe what people would do with their waste and should consider economic, political, and environmental consequences. **LS Interpersonal**

Figure 8 ▶ Biodegradable materials do not degrade quickly in modern landfills. This newspaper was put in a Tempe, Arizona landfill in 1971 and was removed in 1989.

Figure 9 ▶ The map below shows the number of years until landfill capacity is reached in each state.

Safeguarding Landfills The Resource Conservation and Recovery Act, passed in 1976 and updated in 1984, requires that new landfills be built with safeguards to reduce pollution problems. New landfills must be lined with clay and a plastic liner and must have systems for collecting and treating leachate. Vent pipes must be installed to carry methane out of the landfill, where the methane can be released into the air or burned to produce energy.

Adding these safeguards to landfills increases the cost of building them. Also, finding acceptable places to build landfills is difficult. The landfills must be close to the city producing the waste but must be far enough from residents who object to having a landfill near their homes. Any solution is likely to be expensive, either because of the legal fees a city must pay to fight residents' objections or because of the cost of transporting garbage to a distant site.

Building More Landfills Although we can build safer landfills, we are currently running out of space that we are willing to develop for new landfills. The materials we bury in landfills are not decomposing as fast as we can fill landfills. Even biodegradable materials, such as the newspaper in **Figure 8**, take several years to decompose. The total number of active landfills in the United States in 1988 was 8,000. By 1999, the total number of active landfills decreased to 2,300 because many landfills had been filled to capacity. The U.S. Environmental Protection Agency (EPA) estimates that the active landfills in 20 states will be filled to capacity within 10 years as shown in **Figure 9**.

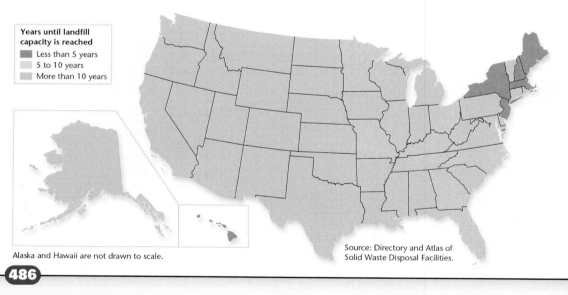

Years until landfill capacity is reached
- Less than 5 years
- 5 to 10 years
- More than 10 years

Alaska and Hawaii are not drawn to scale.

Source: Directory and Atlas of Solid Waste Disposal Facilities.

READING SKILL BUILDER — GENERAL

Paired Summarizing After students have read this section, have them summarize the main points with a partner. Ask them to list the different types of solid waste and the different options for waste disposal along with the advantages and disadvantages of each. **LS Verbal**

Cultural Awareness — GENERAL

Manufacturer Responsibility Extended producer responsibility (EPR), also known as product stewardship or take-back, began in Germany in 1991 when the country began running out of room for waste disposal. German manufacturers are required to take back their products' packaging and reuse or recycle it. This gives manufacturers an incentive to design packaging that is easily recyclable or reusable, and it puts the financial burden on the manufacturer instead of on the consumer or the government. Many other countries have similar programs.

Incinerators One option for reducing the amount of solid waste sent to landfills is to burn it in incinerators, as shown in **Figure 10.** In 1999, the United States had 102 operational incinerators that were capable of burning up to 94,000 metric tons of municipal solid waste per day. However, the waste that is burned does not disappear. Although incinerators can reduce the weight of solid waste by 75 percent, incinerators do not separate materials that should not be incinerated before burning the waste. So, some materials such as cleansers, batteries, and paints that should not be burned, end up in the air as polluting gases. The rest of the solid waste is converted into ash that must be disposed of in a landfill.

Incinerated materials take up less space in landfills, but the incinerated material can be more toxic than it was before it was incinerated. Even incinerators that have special air pollution control devices release small amounts of poisonous gases and particles of toxic heavy metals into the air.

Figure 10 ▶ A solid-waste incinerator reduces the amount of trash that goes to landfills and can be used to generate electricity. However, the material that is created by the incinerator can be toxic.

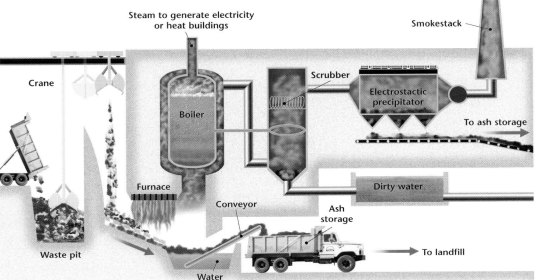

SECTION 1 Review

1. **Explain** what makes a material biodegradable.

2. **Compare** municipal solid waste and manufacturing solid waste.

3. **Describe** how a modern landfill works. Write a short paragraph to explain your answer. List two environmental problems that can be caused by landfills.
   WRITING SKILLS

4. **Describe** one advantage and one disadvantage of incinerating solid waste.

CRITICAL THINKING

5. **Identifying Relationships** Name two non-biodegradable products that you use. What makes these products nonbiodegradable? Name two biodegradable products that you can use instead.

6. **Identifying Alternatives** What can you do to help reduce the amount of solid waste that you throw away? What can you do to help people in your neighborhood reduce the amount of solid waste that is thrown away?

487

Answers to Section Review

1. A biodegradable material can be broken down by biological processes.

2. Municipal solid waste is produced by households and businesses and is mostly paper, yard waste, food waste, and plastics. Manufacturing solid waste, which includes scrap metal, paper, sludge, and ash, makes up a much larger percentage of the total solid waste produced.

3. Answers may vary but should describe how landfills work. Two potential environmental problems are groundwater contamination from leachate and explosions from methane leaks.

4. Incinerators reduce the volume and weight of solid waste, but the waste is more toxic than it was before being incinerated.

5. Answers may vary. Sample answer: Two non-biodegradable products are plastic bottles and polyester clothing. Instead of these products, you could use paper cartons and cotton clothing.

6. Answers may vary. To reduce waste, you could use products that can be used multiple times or you could start a neighborhood recycling program.

Reteaching — BASIC

Categorizing Trash Have students help you collect trash discarded in the classroom and set up a display. Label the items as natural or synthetic, biodegradable or nonbiodegradable, and renewable or nonrenewable. **LS Interpersonal**

Quiz — GENERAL

1. Define *solid waste*. (any discarded solid material)

2. How much solid waste does the average person in the U.S. produce per day? (4.4 lb)

3. Which sector of society produces the largest percentage of solid waste? (manufacturing)

4. What are three ways to dispose of solid waste? (recycle, put in landfills, burn in incinerators)

5. Why is incinerated waste often more toxic than the original waste? (Toxic products, such as batteries, are often not separated out before the waste is incinerated. These products are incinerated and end up polluting the area with toxic gases and heavy metals. Remaining toxic substances are often concentrated in the ash.)

Alternative Assessment — GENERAL

Natural and Synthetic Materials Have students create an illustration, a poem, or a short story to contrast natural and synthetic materials. Suggest that students include their work in their **Portfolio. LS Intrapersonal**

Transparencies

TT How a Solid-Waste Incinerator Works

Focus

Overview

Before beginning this section, review with your students the Objectives in the Student Edition.
This section discusses ways to reduce the amount of waste sent to landfills and incinerators. These options include producing less waste, recycling, buying recycled products, composting, and changing the types of materials we use.

🔊 Bellringer

Ask students to list all the disposable products (such as razors or soap bottles) that they bought in the last month. Ask students to write down any non-disposable or durable alternatives to these products. Then, ask them why they buy disposable products instead of those that are more durable or have less packaging.
LS Logical/Intrapersonal

Motivate

Demonstration —— GENERAL

A New Beginning Bring the following items to class: a sheet of white office paper, a plastic soft-drink bottle, a glass container, and a bicycle tire. Hold up the sheet of paper, and tell students, "This could become part of a home's insulation, bedding for farm animals, or toilet paper." Hold up the other items and say, "This bottle could become carpet, insulation in a jacket, stuffing in a pillow, or part of a car's interior. This glass could be added to asphalt to make new streets, and this tire could become part of the soles on your shoes."

Reducing Solid Waste

Objectives

▶ Identify three ways you can produce less waste.
▶ Describe how you can use your consumer buying power to reduce solid waste.
▶ List the steps that an item must go through to be recycled.
▶ List two benefits of composting.
▶ Name one advantage and one disadvantage to producing degradable plastic.

Key Terms

source reduction
recycling
compost

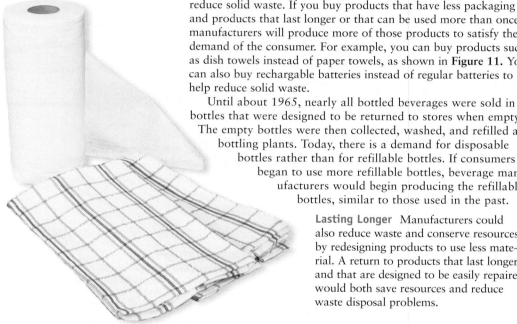

Figure 11 ▶ You can help reduce solid waste by purchasing items that have less packaging. Purchasing items that last longer, such as dish towels, can also reduce solid waste.

488

If landfills and incinerators can pollute the environment and are expensive to operate, what else can we do to safely reduce solid waste? This section examines ways to reduce solid waste through producing less waste, recycling, and changing the materials and products we use. All of these techniques help reduce waste before it is delivered to landfills or incinerators. This method of reducing solid waste is known as source reduction. **Source reduction** is any change in design, manufacture, purchase, or use of materials or products to reduce their amount or toxicity before they become municipal solid waste.

Reducing Solid Waste

If we produce less waste, we will reduce the expense and difficulty of collecting and disposing of it. Many ideas for reducing waste are common sense, such as using both sides of a sheet of paper and not using unneeded bags, napkins, or utensils at stores and restaurants.

Buying Less As a consumer, you can influence manufacturers to reduce solid waste. If you buy products that have less packaging and products that last longer or that can be used more than once, manufacturers will produce more of those products to satisfy the demand of the consumer. For example, you can buy products such as dish towels instead of paper towels, as shown in **Figure 11.** You can also buy rechargable batteries instead of regular batteries to help reduce solid waste.

Until about 1965, nearly all bottled beverages were sold in bottles that were designed to be returned to stores when empty. The empty bottles were then collected, washed, and refilled at bottling plants. Today, there is a demand for disposable bottles rather than for refillable bottles. If consumers began to use more refillable bottles, beverage manufacturers would begin producing the refillable bottles, similar to those used in the past.

Lasting Longer Manufacturers could also reduce waste and conserve resources by redesigning products to use less material. A return to products that last longer and that are designed to be easily repaired would both save resources and reduce waste disposal problems.

Chapter Resource File

• **Lesson Plan**
• **Active Reading** BASIC
• **Section Quiz** GENERAL

🖥 Transparencies

TT Bellringer

REAL-LIFE ——
●━ CONNECTION ━ GENERAL

Closing the Loop If recycling is to be effective, products made from recycled materials must be bought and used by consumers. This is known as "closing the loop." In a sustainable loop, materials are recycled to produce the original product. For example, glass bottles are melted down and made into new bottles. Have students identify the "loops" discussed as part of the Demonstration on this page. Ask them which of the loops are sustainable and which are not. (Paper recycled back to toilet paper is sustainable. Plastic is an example of a nonsustainable loop.)

Recycling

In addition to reducing waste, we need to find ways to make the best use of all the materials we throw away. **Recycling** is the process of reusing materials or recovering valuable materials from waste or scrap. Making products from recycled materials usually saves energy, water, and other resources. For example, 95 percent less energy is needed to produce aluminum from recycled aluminum than from ore. About 75 percent less energy is needed to make steel from scrap than from ore. And about 70 percent less energy is needed to make paper from recycled paper than from trees.

Recycling: A Series of Steps When most people think about recycling, they probably think of only the first step of bringing their bottles, cans, and newspapers to a recycling center or putting these things at the curb in specially marked containers. However, as shown in **Figure 12**, recycling actually involves a series of steps that must happen for recycling to work.

First, the discarded materials must be collected and sorted by type. Next, each type of material must be taken to a facility where it can be cleaned and made ready to be used again. For example, glass is sorted by color and is crushed, and paper is sorted by type and made into a pulp with water. Then the materials are used to manufacture new products. Finally, the new products are sold to consumers. If more people buy products made from recycled materials, there will be an increase in the demand for these products. This demand encourages manufacturers to build facilities to make recycled products. When such facilities are built, it becomes easier for communities to sell the materials they collect from residents for recycling.

FIELD ACTIVITY

Is It Really Recyclable?
Conduct a survey of the plastic containers in your household that are recyclable. Note the number of plastic containers found in your household. Now look at the number printed on the bottom of each container. The plastics industry has established a system of designating which plastics are recyclable. Types 1 and 2 are most commonly recycled by most communities. Type 4 is less commonly recycled, and types 3, 5, 6, and 7 are most likely not to be recycled. In your *EcoLog*, record the total number of plastic containers for each type of plastic that you find in your household. How many Type 1 and Type 2 plastic containers did you find in your household?

Figure 12 ▶ The steps of recycling include ❶ collecting and sorting discarded materials by type, ❷ taking the materials to a recycling facility, ❸ cleaning the discarded materials so that they can be shredded or crushed, and ❹ reusing the shredded or crushed material to manufacture new products.

489

MISCONCEPTION ALERT

Recycling Students may think that the best thing they can do to reduce waste is to recycle, but in fact, recycling should be a last resort. Students may have heard of the three Rs: reduce, reuse, and recycle. Explain to students that reducing your waste production has the greatest impact. The next best thing is to reuse items, which decreases the rate at which natural resources are consumed. Finally, items should be recycled if possible.

Teach

Discussion ── GENERAL

Recycling Newspapers Ask students, "Does your family have a daily newspaper delivered to your home? If so, how long does it take for those papers to pile up in a wastebasket or recycling stack?" (Most students will have noticed that the newspapers pile up quickly.) Tell students to imagine newspapers piling up in the homes of the millions of other people across the country who receive a daily newspaper. Ask students, "How might newspapers contribute to the nation's waste-disposal problems?" (Students should recognize that a newspaper may seem to be an insignificant item, but when the entire population of newspaper readers is considered, the environmental impact of newspapers becomes significant.) Give students the following example: Newspapers take up about 20 percent of landfill space. On the other hand, less than 1 metric ton of recycled newspaper saves 17 trees and 2.4 m³ (3 cubic yards) of landfill space. And if each person recycled his or her daily paper, four trees would be saved per person each year. **LS Verbal**

Using the Figure ── GENERAL

The Recycling Cycle Call on a student to describe the "life cycle" of a material as the material is recycled by following the steps shown in **Figure 12.** Ask students to discuss what effect they think an increase in recycling will have on jobs. (Answers may vary. Many jobs will be created in the recycling industry. However, some jobs, including those involved in managing landfills or in obtaining raw materials, such as wood and mined metals, will also be lost.) **LS Visual**

Activity — ADVANCED

Creating a School Recycling Program Have students interview your school's custodian or administrative staff to find out if the school has a recycling program. If there is no program, suggest that students write a proposal to start one. The proposal should include the types of materials that could be recycled from the school's waste, a list of the companies available within the area that have pick-up or drop-off recycling programs, and a description of how the program could be initiated and maintained. If a recycling system is already in place, students can verify the efficiency of the system and make recommendations for its improvement, if necessary. **LS** Interpersonal

CHEMISTRY — CONNECTION — GENERAL

Composting Processes Explain to students that there are two types of decomposition: aerobic and anaerobic. In aerobic decomposition, organic matter combines with oxygen and water to produce carbon dioxide, water, heat, and energy. In the anaerobic process, organic matter combines only with water to produce carbon dioxide, methane, hydrogen sulfide, and energy. Suggest that students build a compost pile at school or at home to study the process. If the pile starts to smell rotten, ask students why. (Anaerobic decomposition is taking place. The pile needs more oxygen, provided by turning the pile, or less water.)
LS Kinesthetic

Table 2 ▼

Benefits of Composting

- keeps organic wastes out of landfills
- provides nutrients to the soil
- increases beneficial soil organisms, such as worms and centipedes
- suppresses some plant diseases
- reduces the need for fertilizers and pesticides
- protects soil from erosion

Composting Yard waste often makes up more than 15 percent of a community's solid waste. None of this waste has to go to a landfill. Because yard waste is biodegradable, it can decompose in a compost pile. Many people also put fruit and vegetable trimmings and table scraps in their compost piles. The warm, moist, dark conditions inside a large pile of biodegradable material allow bacteria to grow and break down the waste rapidly. Eventually the material becomes **compost,** a dark brown, crumbly material made from decomposed plant and animal matter that is spread on gardens and fields to enrich the soil. Compost is rich in the nutrients that help plants grow. More benefits of composting are listed in **Table 2.**

Some cities collect yard waste from homes and compost it at a large, central facility. Although most city composting facilities in the United States collect only yard wastes, several European cities also collect and compost food wastes in municipal facilities. Composting can also be an effective way of handling waste from food-processing plants and restaurants, manure from animal feedlots, and municipal sewage sludge. If all biodegradable wastes were composted, the amount of solid waste going to landfills could be reduced.

CASE STUDY

Paper or Plastic?

The following question may sound familiar: Do you want paper or plastic? If you have ever stood in the checkout line of a grocery store, it probably is. Almost every grocery store today offers a choice between either paper or plastic bags for sacking grocery items. Many people make their choice based on convenience. But what is the best choice for someone who is concerned about the environment?

On the surface, it may seem that paper is the better choice. Paper comes from a renewable resource—trees—and is biodegradable. Plastic, on the other hand, comes from petroleum or natural gas, which are usually considered nonrenewable resources. In addition, plastic bags are not biodegradable.

Upon closer examination, however, the decision may not be as simple as it seems. Removing large numbers of trees from forests to manufacture paper can disrupt woodland ecosystems. Plus, a tremendous amount of energy is required to convert trees into pulp and then manufacture paper from the pulp.

To make the best decision about which product is better for the environment, the following questions should be considered.

- How much raw material, energy, and water is needed to manufacture each bag?
- What waste products will result from the manufacture of each bag, and what effect will those wastes have on water, the atmosphere, and land?

▶ Making an educated decision at the grocery store will help reduce solid waste.

- Can recycled materials be used in the manufacture of the bag? If so, to what degree will the use of recycled materials reduce the amount of raw materials, energy,

HISTORY — CONNECTION — GENERAL

Recycling During World War II Tell students that the last time in history there was as much recycling done in the United States was during World War II, when the population was about half of what it is today. Explain that cooking fat was saved and used for making explosives and that aluminum chewing-gum wrappers were saved for airplane fuselages. Ask, "Why was recycling so important during World War II, and why are people once again beginning to recycle?" (During WWII, people were faced with shortages of certain products. They reacted by reusing the useful materials in discarded products. Today, largely because of increasing human populations, many countries are worrying about national and worldwide shortages of raw materials and thus are establishing recycling programs.) Suggest that interested students research WWII salvage operations and prepare a written, visual, or audiovisual report to reflect their findings. Have students include their report in their **Portfolio. LS** Visual

Changing the Materials We Use

Simply changing the materials we use could eliminate much of the solid waste we produce. For example, single-serving drink boxes are made of a combination of foil, cardboard, and plastic. The drink boxes are hard to recycle because there is no easy way to separate the three components. More of our waste could be recycled if such products were no longer made and if all drinks came in recyclable glass, cardboard, or aluminum containers.

Recycling other common household products into new, useable products could also help eliminate solid waste. For example, newspapers can be recycled to make cardboard, egg cartons, and building materials. Telephone books, magazines, and catalogs can also be recycled to make building materials. Used aluminum beverage cans can be recycled to make new beverage cans, lawn chairs, aluminum siding for houses, and cookware. Used glass jars and bottles can be recycled to make new glass jars and bottles. Finally, plastic beverage containers can be recycled to make nonfood containers, insulation, carpet yarn, textiles, fiberfill, scouring pads, toys, plastic lumber, and crates.

▣ internet connect ▤

www.scilinks.org
Topic: Waste Prevention
SciLinks code: HE4120

SCi
LINKS   Maintained by the
         National Science
         Teachers Association

and water used and wastes produced in making the bag?
- How will the bag decompose, and what will the environmental impact be if it is incorrectly disposed of?

Although several studies have analyzed these questions, most have been conducted by parties with a vested interest, such as plastic or paper manufacturing companies. As you might expect, the studies done by plastic manufacturers conclude that plastic bags have the least environmental impact, while studies done by paper producers conclude that paper bags have the least environmental impact. Often, these researchers fail to study all of the important factors listed above.

But the plastic versus paper debate has caused both industries to improve the way their products affect the environment. For example, paper bags recently outsold plastic bags because they were considered

▶ A reusable canvas shopping bag may be the best response to the paper-or-plastic question.

stronger, better for reusing or recycling, and less harmful in a landfill.

Then, new technology allowed the plastics industry to gain a larger market share. By incorporating recycled plastic into the bags, manufacturers improved the image of plastic bags.

Therefore, the debate continues and environmentally conscious people are still wondering which is better. Right now there seems to be no right answer. However, the following are environmentally sound options.

- Carry your groceries in bags brought from home (paper, plastic, or canvas bags).
- Choose the bag you are most likely to reuse in the future.
- If you have only one or two small items do not use a bag.

CRITICAL THINKING

1. Identifying Relationships
Explain how environmentally conscious shoppers have helped improve paper and plastic bag manufacturing in this country.

2. Understanding Concepts
Why should a person care which bag he or she is given at the grocery store?

491

Group Activity ───── GENERAL
Recycling Education Campaign
Working together in small groups, students could develop a campaign to educate the rest of the school about the importance of recycling. Encourage students to think of innovative ways to reduce the number of resources they need to complete their project. For example, a pamphlet could be copied onto recycled paper. Then, rather than distributing one pamphlet to each student, students could install a "take-one" display that includes limited numbers of the pamphlet at various locations in the school. Suggest that students include something from their project in their **Portfolio.** LS Interpersonal

Co-op Learning

CASE STUDY

Paper or Plastic? After students have read the Case Study, encourage the class to discuss it. Then, lead students in a discussion of why "scientific" research done by advocacy groups often cannot be trusted. Encourage students examining research data to find out if the researcher had any special interest in the outcome. Students should also find out as much as possible about the researcher's methods and conclusions.

Answers to Critical Thinking

1. Answers may vary. For example, consumer pressure on plastic-bag manufacturers led them to include recycled plastic in their bags.
2. Answers may vary. The choice of paper or plastic has environmental and economic consequences.

Homework ───────── GENERAL

Making Paper Suggest that students find out how to make recycled paper and then make some. Detailed instructions can be found in many activity books on recycling. In general, the process involves shredding used paper, mixing it with water in a blender, draining the mixture, and spreading it over a mesh screen that is placed on a cloth towel. When dry, the mixture can be lifted from the screen in a sheet, and this sheet is the recycled paper. Encourage students to include their recycled paper in their **Portfolio.** LS Kinesthetic

MISCONCEPTION
///**ALERT**\\\

Recycling Symbols The "chasing arrow" symbol on the bottom of plastic containers does not mean that the plastic is recyclable. It is merely intended to highlight the number inside, which indicates the general type of resin that the plastic was made from. In general, plastics with the same number can be recycled together (if they are recyclable at all), although the dyes, softeners, and other chemicals added to produce the final product can lead to incompatibility.

Reteaching ─────── BASIC
Visit a Recycling Center
Arrange a class visit and tour of a local recycling center or an establishment that makes new products out of recycled materials. Ask the tour guide to discuss the cycle of a product from its original state to a recycled product and the importance of consumer demand for recycled products. Encourage students to prepare questions in advance. **LS** Verbal

Quiz ─────── GENERAL
1. What are three ways to reduce solid waste? (produce less waste, recycle, and change the materials and products we use)
2. What are two types of degradable plastics? (photodegradable plastic and green plastic)

Alternative Assessment ─── GENERAL
Waste Disposal Concept Map
Ask students to draw a concept map showing some solid-waste disposal concerns and options for addressing those issues in the future. (Answers will vary. Some concerns include running out of landfill space and the contamination of water supplies by leachate. Some options for addressing those concerns include producing less waste, recycling, composting, and changing the materials we use.) **LS** Visual

Graphic
Organizer ─ GENERAL
Chain of Events Chart
You may want to use this Graphic Organizer to assess students' prior knowledge before beginning a discussion of each step of the degradation of degradable plastics. You may also choose to use a similar activity as a quiz to assess students' knowledge of the degradation of degradable plastics or to assess students' knowledge after students have read the description of the degradation of degradable plastics.

Graphic
Organizer — Chain-of-Events Chart
Create the **Graphic Organizer** entitled "Chain-of-Events Chart" described in the Appendix. Then, fill in the chart with details about each step of the degradation of degradable plastics.

Figure 13 ▶ Green plastics made from living things are biodegradable. The plastic fork below has been engineered to degrade within 45 days of disposal.

| DAY 0 | DAY 12 | DAY 33 | DAY 45 |

Degradable Plastics As you read earlier, most plastics are not biodegradable. To make plastic products more appealing to people who are concerned about the environment, several companies have developed new kinds of plastics that they say are degradable. One type, called *photodegradable plastic,* is made so that when it is left in the sun for many weeks, it becomes weak and brittle and eventually breaks into pieces.

Another type of degradable plastic, called *green plastic,* is made by blending the sugars in plants with a special chemical agent to make plastic. Green plastics are labeled as green because they are made from living things and are considered to be more environmentally friendly than other plastics. The production of green plastics requires 20 to 50 percent less fossil fuel than the production of regular plastics does. The fork in **Figure 13,** is made of green plastic. This plastic has been engineered to degrade within 45 days of being thrown away. When this plastic is buried, the bacteria in the soil eat the sugars and leave the plastic weakened and full of microscopic holes. The chemical agent then gradually causes the long plastic molecules to break into shorter molecules. These two effects combine to cause the plastic to eventually fall apart into small pieces.

Problems with Degradable Plastics The main problem with these so-called degradable plastics is that although they do break apart and the organic parts can degrade, the plastic parts are only reduced to smaller pieces. This type of plastic can help reduce the harmful effects that plastic litter has on animals in the environment, because the plastic pieces will be too small to get caught in their throats or around their necks. However, the small pieces of plastic will not disappear completely. Instead, the pieces of plastic will be spread around. So, these biodegradable plastics can remain in landfills for many years, just as regular plastics can.

SECTION 2 Review

1. **Name** three things you could do each day to produce less waste.
2. **Explain** how buying certain products can help reduce solid waste.
3. **Describe** the steps it takes to recycle a piece of plastic.
4. **List** two benefits of composting.

CRITICAL THINKING
5. **Analyzing Methods** What are the advantages and disadvantages to producing degradable plastics?
6. **Demonstrating Reasoned Judgement** Read the Case Study in this section and decide which type of bag you would choose the next time you go shopping. Explain why you made this choice. What are other uses of the bag you chose? **READING SKILLS**

492

Answers to Section Review
1. Answers may vary. For example, one could purchase items with less packaging, use both sides of a sheet of paper, or buy in bulk.
2. Answers may vary. For example, rechargeable batteries last longer, so a consumer does not need to buy them as often.
3. The discarded plastic must be collected, sorted, cleaned, and processed. Then, the materials are used to manufacture new products.
4. Answers may vary. Sample answer: Composting keeps organic wastes out of landfills and suppresses some plant diseases.
5. The advantages are that they reduce harmful effects on wildlife and that they may require fewer fossil fuels for manufacturing. The disadvantages are that they merely degrade into smaller pieces that can be spread around, and the plastic itself does not decompose.
6. Answers may vary. Students should base their logic on the arguments presented in the Case Study.

Hazardous Waste

Many of the products we use today, from laundry soap to computers, are produced in modern factories that use thousands of chemicals. Some of these chemicals make up parts of the products, while other chemicals are used as cleansers or are used to generate electricity for the factories. Large quantities of the chemicals used are often leftover as waste. Many of these chemicals are classified as **hazardous waste,** which is any waste that is a risk to the health of humans or other living things.

Types of Hazardous Waste

Hazardous wastes may be solids, liquids, or gases. Hazardous wastes often contain toxic, corrosive, or explosive materials. Some examples of hazardous wastes include substances such as cleansers used to disinfect surfaces or lubricants used to help machines run smoothly. More examples of hazardous wastes are listed in **Table 3.**

The methods used to dispose of hazardous wastes often are not as carefully planned as the manufacturing processes that produced them. One case of careless hazardous waste disposal that had horrifying results occurred at Love Canal, in Niagara Falls, New York. At Love Canal, homes and a school were built on land that a chemical company had used as a site to dump toxic waste. Problems started when the toxic waste began to leak from the site.

The events at Love Canal shocked people into paying more attention to how hazardous wastes were being disposed of and stored throughout the United States. In other places throughout the country, improperly stored or discarded wastes—such as those shown in **Figure 14**—were leaking into the air, soil, and groundwater. Federal laws were passed to clean up old waste sites and regulate future waste disposal.

Objectives

▶ Name two characteristics of hazardous waste.
▶ Describe one law that governs hazardous waste.
▶ Describe two ways in which hazardous waste is disposed.

Key Terms

hazardous waste
deep-well injection
surface impoundment

Table 3 ▼

Types of Hazardous Waste

- dyes, cleansers, and solvents
- PCBs (polychlorinated biphenyls) from older electrical equipment, such as heating systems and television sets
- plastics, solvents, lubricants, and sealants
- toxic heavy metals, such as lead, mercury, cadmium, and zinc
- pesticides
- radioactive wastes from spent fuel that was used to generate electricity

Figure 14 ▶ An improperly maintained hazardous waste site can leak toxic wastes into the air, soil, and groundwater.

493

Chapter Resource File

- Lesson Plan
- Active Reading BASIC
- Section Quiz GENERAL

Transparencies

 TT Bellringer

 INCLUSION Strategies GENERAL

- *Attention Deficit Disorder*
- *Learning Disabled*

Help students call a local automotive repair shop, a paint store, a fertilizer dealer, a computer store, and a cellular telephone dealer to ask questions about the proper disposal of hazardous products associated with their businesses. The students can list the products and the answers they were given in their telephone interviews. Responses can be reported to the class or printed as a flyer for the class.

SECTION 3

Focus

Overview

Before beginning this section, review with your students the Objectives in the Student Edition. In this section, students learn what constitutes hazardous waste. Through a discussion of the Superfund Act, they discover how difficult it is to dispose of hazardous waste. Finally, students learn how hazardous wastes are best managed.

Bellringer

Have students write down their definition of hazardous waste, and have them list substances that they consider hazardous. (Answers may vary.) LS Logical

Motivate

Discussion ———— GENERAL

Proper Disposal Bring in the following common household products: a can of paint, furniture oil, nail-polish remover, motor oil, and weedkiller. Ask students if they have ever used any of these products. If so, ask, "How did you dispose of the product when you were finished with it?" Tell students that most people simply toss the empty or partially empty container into the trash. Ask students, "Why would anyone take these common materials to a hazardous waste collection facility?" Encourage students to look at the labels of the products, most of which probably have warning labels and hazardous-product symbols. Ask, "What possible effects could those chemicals, when discarded, have on the environment?" (Sample answer: They could contaminate local water sources, possibly harming aquatic life, birds, pets, and other organisms.) LS Verbal

Teach

Debate — GENERAL

Superfund in Trouble The basis of the Superfund program was the idea that "the polluter pays." Polluting industries, mostly the chemical and oil industries, paid special taxes that went directly to the fund, amounting to a total of about $1 billion per year. But industry complained that the system was poorly managed. In 1995, Congress failed to renew the corporate taxes funding the Superfund program, and the amount of money available in the fund for cleanup has been decreasing ever since. The fund is expected to have only $28 million in 2003 compared to a high of $3.8 billion in 1996. If the corporate taxes are not reinstated, taxpayers will have to pay to clean up abandoned sites. Have students stage a debate about who should pay for the clean-up of severely contaminated hazardous waste sites. Make sure students consider the following three scenarios: the company has abandoned the site, the company refuses or is unable to pay for clean-up, and the company has gone out of business. **LS Verbal**

Transparencies

TT Superfund Sites

Connection to Law

Resource Conservation and Recovery Act The Resource Conservation and Recovery Act (RCRA) was passed by Congress in 1976 and amended in 1984. The RCRA created the first significant role for federal government in waste management. The act was established to regulate solid and hazardous waste disposal and to protect humans and the environment from waste contamination.

The primary goals of the RCRA include protecting human health from the hazards of waste disposal, conserving energy and natural resources by recycling and recovering, reducing or eliminating waste, and cleaning up waste, which may have spilled, leaked, or been improperly disposed of.

Figure 15 ▶ This map shows the number of approved and proposed Superfund sites as of 2001. These sites are some of the most hazardous areas in the United States.

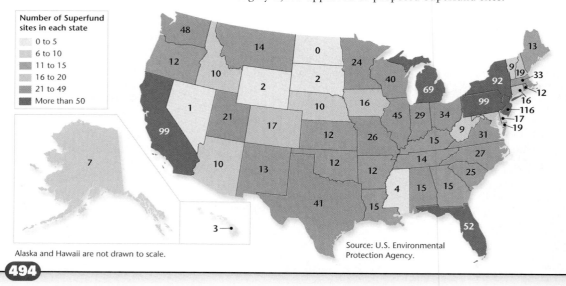

Number of Superfund sites in each state

- 0 to 5
- 6 to 10
- 11 to 15
- 16 to 20
- 21 to 49
- More than 50

Alaska and Hawaii are not drawn to scale.

Source: U.S. Environmental Protection Agency.

494

Resource Conservation and Recovery Act

The Resource Conservation and Recovery Act (RCRA) requires producers of hazardous waste to keep records of how their wastes are handled from the time the wastes are made to the time the wastes are placed in an approved disposal facility. If the wastes cause a problem in the future, the producer is legally responsible for the problem. RCRA also requires all hazardous waste treatment and disposal facilities to be built and operated according to standards that are designed to prevent the facilities from polluting the environment.

The Superfund Act

Because the safe disposal of hazardous wastes is expensive, companies that produce hazardous wastes may be tempted to illegally dump them to save money. In 1980, the U.S. Congress passed the Comprehensive Environmental Response, Compensation, and Liability Act, more commonly known as the Superfund Act. The Superfund Act gave the U.S. Environmental Protection Agency (EPA) the right to sue the owners of hazardous waste sites who had illegally dumped waste. Also, the EPA gained the right to force the owners to pay for the cleanup. The Superfund Act also created a fund of money to pay for cleaning up abandoned hazardous waste sites.

Cleaning up improperly discarded waste is difficult and extremely expensive. At Love Canal alone, $275 million was spent to put a clay cap on the site, to install a drainage system and treatment plant to handle the leaking wastes, and to relocate the residents. Now, more than 20 years after Love Canal was evacuated, many Superfund sites still need to be cleaned up, as shown in **Figure 15.** Cleanup has been completed at only 75 of the roughly 1,200 approved or proposed Superfund sites.

REAL-LIFE CONNECTION — GENERAL

Superfund Sites in Your Area Direct students' attention to **Figure 15.** Suggest that students form teams to investigate whether there are any Superfund sites near the school. They could read regional newspapers and other periodicals and contact the nearest office of the Environmental Protection Agency for information. Students could present their findings in the form of a news report, video, or poster. Suggest that students include something from the project in their **Portfolio.** Note: If there is not a Superfund site nearby, have students choose the closest site to research. **LS Interpersonal**

BRAIN FOOD

Reporting Hazardous Waste Sites

Under the Superfund Act, anyone can report a hazardous waste site they think needs to be cleaned up. The EPA then investigates the site and decides if it needs to be cleaned up and, if so, who takes charge: the federal government (Superfund), state agencies, or others. Although there are only about 1,200 proposed or approved Superfund sites, the EPA has assessed over 41,400 reported hazardous waste sites in the United States.

Figure 16 ▶ Safely transporting hazardous waste is an important part of hazardous waste management.

Hazardous Waste Management

Each year, the United States produces about 252 million metric tons of hazardous waste, and this amount is growing each year. It is difficult to guarantee that the disposal techniques used today will not eventually pollute our air, food, or water.

Preventing Hazardous Waste One way to prevent hazardous waste is to produce less of it. In recent years, many manufacturers have discovered that they can redesign manufacturing methods to produce less or no hazardous waste. For example, some manufacturers who used chemicals to clean metal parts of machines have discovered that they can use tiny plastic beads instead. The beads act like a sandblaster to clean the parts, can be reused several times, and are not hazardous when disposed. Often, such techniques save the manufacturers money by cutting the cost of materials as well as in cutting the cost of waste disposal.

Another way to deal with hazardous waste is to find a way to reuse it. In the United States, more than 50 programs have been set up to help companies work with other companies that can use the materials that they normally throw away. For example, a company that would usually throw away a cleaning solvent after one use can instead sell it to another company. The company that buys the used solvent may produce a product that is not harmed by small amounts of contamination in the solvent.

Conversion into Nonhazardous Substances Some types of wastes can be treated with chemicals to make the wastes less hazardous. For example, lime, which is a base, can be added to acids to neutralize them. A base is a compound that can also react with acids to convert acids into salts, which are less harmful to the environment. Also, cyanides, which are extremely poisonous compounds, can be combined with oxygen to form carbon dioxide and nitrogen. In other cases, wastes can be treated biologically. Sludge from petroleum refineries, for example, may be converted by soil bacteria into less harmful substances.

internet connect

www.scilinks.org
Topic: Hazardous Waste
SciLinks code: HE4054

SCi LINKS. Maintained by the National Science Teachers Association

QuickLAB

Neutralizing Hazardous Waste

Procedure

1. Using a **measuring spoon,** obtain about a teaspoon of **baking soda,** and place it in a **500 mL beaker.** The baking soda will act as the base which will neutralize the acid.

2. In a separate **500 mL beaker,** pour approximately **200 mL of vinegar.** The vinegar is a weak acid.

3. Add the vinegar (acid) to the baking soda (base).

Analysis

1. What happened when you added the vinegar to the baking soda?

2. How is this lab similar to the technique used to convert some hazardous wastes into nonhazardous substances?

495

QuickLAB

Skills Acquired:

• Experimenting
• Identifying and Recognizing Patterns
• Interpreting

Answers

1. The vinegar reacted with and dissolved the baking soda.

2. In both situations, one chemical is added to another chemical to neutralize it.

Activity ——— **ADVANCED**

Energy in Hazardous Wastes

Ask students, "Can the potential energy in discarded hazardous wastes be harnessed?" (Toxic chemicals are capable of reacting with other chemicals to produce energy. In addition, when a toxic waste is burned, or incinerated, the burning process produces energy. Both kinds of energy can potentially be harnessed to create new chemical compounds or to generate electricity.) Ask students to find out what new technologies are needed to enable us to safely use the potential energy of hazardous wastes. Suggest that they present their findings in a creative way in their **Portfolio.** 🄻 **Verbal/Interpersonal**

BIOLOGY ———
● **CONNECTION** ▶ **GENERAL**

Phytoremediation One method of biologically treating hazardous waste is called *phytoremediation.* Certain plants are grown on soil that contains hazardous waste, such as an area with a chemical spill. The plants absorb one or more hazardous substances and are then harvested and disposed of as hazardous waste. Some plants convert the contaminant into a nonhazardous substance; in this case, the plants can be used for animal feed or paper. Other plants absorb and concentrate the contaminant, as is the case with heavy metals such as mercury and lead. Sunflowers have been planted at the Chernobyl site to absorb radioactive cesium-137 and strontium-90.

Demonstration —— GENERAL

Hazardous Waste Spills Fill three clear glasses or beakers with water. Pour a small amount of vegetable oil in one of the beakers. Ask students to describe what they see. Explain to them that the oil behaves like jet fuel: it floats on top of the water and does not dissolve. Then, pour a small amount of food coloring into another beaker. Tell them this is how gasoline behaves: it is water soluble and mixes with water. Finally, pour a small amount of maple syrup or molasses into the third beaker. Ask students what they observe. The syrup behaves like the hazardous solvent trichloroethylene (TCE) or like polychlorinated biphenyls (PCBs). The syrup sinks below water but does not mix with the water. **LS Kinesthetic/Visual**

Connection to Chemistry

Hazardous Chemical Reactions After a material is thrown away, it may become more hazardous as a result of a chemical reaction with other discarded wastes. For example, metallic mercury is considered to be toxic. Metallic mercury is often used in thermometers and computers. If it is buried in a landfill, the bacteria in a landfill can cause it to react with methane to form methyl mercury. Methyl mercury, which is more toxic than metallic mercury, can cause severe nerve damage.

Land Disposal Most of the hazardous waste produced in the United States is disposed of on land. One land disposal facility, illustrated in **Figure 17**, is called deep-well injection. During **deep-well injection,** wastes are pumped deep into the ground, where they are absorbed into a dry layer of rock below the level of groundwater. After the wastes are buried below the level of groundwater, the wastes are covered with cement to prevent contamination of the groundwater. Another common land disposal facility is a **surface impoundment,** which is basically a pond that has a sealed bottom. The wastes accumulate and settle to the bottom of the pond, while water evaporates from the pond and leaves room to add more wastes.

Hazardous wastes in concentrated or solid form are often put in barrels and buried in landfills. Hazardous waste landfills are similar to those used for ordinary solid waste, but these landfills have extra safety precautions to prevent leakage.

In theory, if all of these facilities are properly designed and built, they should provide safe ways to dispose of hazardous wastes. However, if they are not properly maintained, they can develop leaks that may result in contamination of the air, soil, or groundwater.

Love Canal: A Toxic Nightmare

To someone who has never heard of it, Love Canal may sound like a pleasant place for a picnic. But in the minds of people familiar with the abandoned canal site in Niagara Falls, New York, the area is synonymous with chemicals, disease, and financial loss.

The problems began in 1942 when a chemical company bought the area and used it as a dump for toxic wastes. Over the next 11 years, the company buried almost 20,000 metric tons of hazardous chemicals in the canal. At the time, disposing of chemical wastes in this way was legal. It was thought that the thick clay that lined the canal would prevent the wastes from escaping into the surrounding soil.

By 1953, the site was full. It was covered with a cap of clay and soil and was sold to the school board of Niagara Falls. The school board ignored warnings from the chemical company and built an elementary school and playgrounds on top of the canal. In addition, hundreds of homes were built on the canal. Roads and sewer lines were also constructed across the site, which disturbed its clay cap and occasionally exposed barrels of waste. The new homes attracted many new residents, who were not warned about the hazardous waste dump nearby.

By the late 1950s, problems started occurring. Children playing near the school were burned by chemicals that leaked from the

▶ Toxic waste that leaked from the barrels buried at Love Canal leaked into this man's basement.

BRAIN FOOD

Leaking Waste Sites Both municipal landfills and hazardous waste landfills can leak toxic substances that can produce adverse health effects. The most common health effects are lower birth weight in babies and birth defects. Scientists do not know whether air pollution or water pollution from landfills poses the greater risk.

LANGUAGE ARTS
CONNECTION —— GENERAL

Analyzing Environmental Reporting Suggest that students read "Child's Plague," found in *Sierra*, Nov.–Dec. 1997. This article discusses the possible effects of environmental toxins on children. Have interested students assume the role of investigative reporters to determine whether the article is written fairly and accurately. Have them write an assessment of their findings. Students' research should include other articles and books. Students may also wish to contact the EPA, epidemiologists, pediatricians, and manufacturers of hazardous wastes.

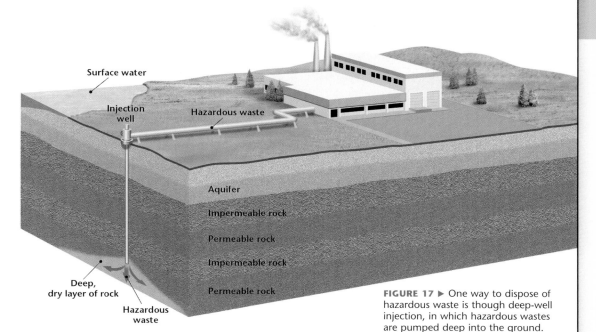

Surface water

Injection well

Hazardous waste

Aquifer

Impermeable rock

Permeable rock

Impermeable rock

Permeable rock

Deep, dry layer of rock

Hazardous waste

FIGURE 17 ▶ One way to dispose of hazardous waste is though deep-well injection, in which hazardous wastes are pumped deep into the ground.

Discussion ———— GENERAL

Risk Assessment The risk that a hazardous waste storage facility poses to a community depends on factors such as the geologic characteristics of the site, the types of wastes stored there, and the probability that contaminants will be exposed to the environment. Ask students how contaminants could be spread from a site. (Contaminants could be spread through runoff from rainfall or by leaching into groundwater.) Ask them how humans, other animals, and plants could be exposed to contaminants. (Answers will vary.) Ask students if they can think of any other factors that would affect the risk a site poses. (Concentration of contaminants and extent of degradation are examples.) **LS** Logical/Verbal

🖳 Internet Activity — GENERAL

Hazardous Substances Have each student use the Internet to research one hazardous substance. Suggest that they begin their search with the EPA Web site, which may have information on safe exposure levels for the substance. Tell students to find out the following information: products that contain the substance, uses, hazards, safe storage methods, and any environmental effects. **LS** Interpersonal

Student Opportunities

Community Involvement The solution to waste management problems often begins with citizen participation. Suggest that students research waste management problems or hazardous waste problems in their community. Have them contact local community organizations to see what, if anything, is being done about these problems. Then, have students design a plan to get involved in addressing one of the problems.

▶ This chemical plant buried almost 20,000 metric tons of hazardous chemicals in Love Canal.

After years of protests and court cases, a federal judge ruled that the chemical company was responsible for the Love Canal disaster. The company agreed to pay $98 million to the state of New York and agreed to reimburse the state and federal governments for the cleanup.

ground. In the 1960s and 1970s the chemical leaks became more obvious. Puddles of chemicals appeared in backyards. Thick, black sludge oozed into basements. Health problems, such as asthma, dizziness, blurred vision, seizures, miscarriages, stillbirths, and birth defects, became more common among the residents.

Local, state, and federal officials began to take notice of the prob-lems at Love Canal in the mid-1970s. Water-, soil-, and air-quality tests showed chemical contamina-tion. In 1978, the governor of New York ordered the 239 families living closest to the chemical dump to evacuate. The state purchased their homes and paid for their relocation. In 1980, Love Canal was declared a federal disaster area, and another 710 families were relocated.

CRITICAL THINKING

1. Analyzing Ideas Use the Love Canal situation to explain why when we throw something away, it is never really gone.

2. Evaluating Conclusions Now that you have read about Love Canal, how might you change the ways in which we dispose of hazardous waste?

497

CASE STUDY

Love Canal Toxic chemicals began to resurface in Love Canal in 1958. However, officials did not take notice until 1978, after strong public pressure from residents. Ask students, "What is the ethical duty of a public official in a case like this? Is it possible to take action based only on the scientific evidence of the threat to public health, regardless of the economic and political consequences?" As an extension, have students research this case in more detail, paying close attention to the conflicts that developed between the residents, public health officials, politicians, and scientists involved.

Answers to Critical Thinking

1. When we throw something away, it only leaves our possession. Even if it is chemically transformed, the resulting material ends up somewhere else.

2. Answers may vary. Sample answer: Designate hazardous waste sites as permanent storage sites to prevent others from building on them.

Transparencies

TT Hazardous-Waste Deep-Well Injection

Teaching Tip ———— BASIC

Videos on Waste Contact your city or state environmental agency to find out if it has any films or videos about hazardous waste in your area. If not, obtain a video from a local library or university.
LS Visual

Group Activity ———— GENERAL

Household Hazardous Waste Disposal Have students work in small groups to create a brochure or poster showing how to dispose of various hazardous household wastes in your area. For example, students could create a table listing the name of the material, its uses, hazards, first-aid treatment, and proper means of disposal. They might also include a column listing nonhazardous alternatives.
LS Interpersonal Co-op Learning

Figure 18 ▶ Chemicals can be used to clean up hazardous wastes. This tractor is spreading chemicals to help break down the oil from an oil spill on a beach in Wales, United Kingdom.

Geofact

Biomining Bacteria are not only used to break down hazardous wastes, but they are also used to extract copper and gold from ore. This technique is called *biomining*. Currently, 25 percent of the world's copper is produced through bio-mining. Today, scientists are attempting to bioengineer bacterial strains that can mine poisonous heavy metals such as arsenic, cad-mium, and mercury from ore.

Biologically Treating Hazardous Waste Some hazardous wastes can be absorbed, broken down, or their toxic-ity can be reduced when they are treated with biological and chemical agents. Certain bacteria can be used to clean up an area in the environment that has been contaminated with haz-ardous substances, such as crude oil, PCBs, and cyanide. Scientists can grow bacteria in a lab and apply the bacteria to a contaminated area in the environ-ment to break down the hazardous sub-stances. Flowering plants and trees that absorb heavy metals can also be planted in contaminated areas. Chemicals can also be used to neutralize and absorb hazardous wastes. In **Figure 18**, chemicals were applied to an oil spill to help absorb the oil and help pre-vent harm to the plants and animals that live in and around the beach by hazardous waste.

Incinerating Hazardous Waste Some hazardous wastes are dis-posed of by burning, often in specially designed incinerators. Incinerators can be a safe way to dispose of waste, but they have several problems. Incineration is generally the most expensive form of waste disposal because they require a lot of energy to operate. Incinerators also need pollution-control devices and need to be carefully monitored so that hazardous gases and particles are not released into the air. Also, after hazardous waste is incinerated, the leftover ash needs to be buried. This ash is usually buried in a haz-ardous waste landfill.

When we put hazardous waste into disposal facilities for long-term storage the wastes do not disappear. Instead, they must be closely monitored. For example, disposal of radioactive wastes from nuclear reactors is an especially difficult storage problem. The only way to make the radioactive wastes nonhazardous is to let them sit for thousands of years until the radioactivity decreases to safe levels. Therefore, engineers and geologists search for disposal sites that probably will not be damaged by move-ments of the Earth for thousands of years.

Exporting Hazardous Waste Until recently, only local laws regu-lated waste disposal in the United States. Companies would often get rid of hazardous wastes by sending them to landfills in other states, especially the less populated southern states. In the 1980s, as southern populations grew, these southern states began to refuse hazardous wastes from other states.

Hazardous wastes are also exported through international trade agreements. Some hazardous wastes are exported to other countries because there may be a facility in another country that specializes in treating, disposing of, or recycling a particular hazardous waste.

498

MISCONCEPTION
///ALERT\\\

Storm Drains Students may think that there is nothing wrong with disposing of hazardous materials, such as paint or motor oil, by pouring them down storm drains along the street. However, storm-drain water is not treated at a wastewater plant; it goes directly to local streams, rivers, and lakes. Emphasize to students that hazardous substances should never be poured into gutters. Tell them to take hazardous materials to the local hazardous waste collection facility.

Homework ———— GENERAL

Mapping Your Community Have students create a map of solid-waste activities in their community. Suggest that they include the following facilities: landfills, incinerators, composting facilities, hazardous waste sites, recycled waste processing and collection centers, businesses that create products from recycled material, businesses that sell recycled products, and the municipal solid-waste department. Ask them if any of these activities are clustered in particular areas of the commu-nity. If so, ask them to brainstorm possible reasons for the concentration. **LS** Visual

Hazardous Wastes at Home

You may think of hazardous waste management as a problem that only big industries face. However, everyday household products can also create hazardous waste. Chemicals, including house paint, pesticides, and batteries all create hazardous waste and are used in homes, schools, and businesses. Additional hazardous household products are listed in **Table 4.** Hazardous materials poured down the drain or put in the trash end up in solid-waste landfills. These hazardous wastes should instead be disposed of in a specially designed hazardous waste landfill.

Disposing of Household Hazardous Waste To make sure that household hazardous waste is disposed of properly, more and more cities around the country have begun to provide collection for household hazardous waste. Some cities collect materials only once or twice a year, while other cities have permanent facilities where residents can drop off hazardous waste. Trained workers sort the hazardous materials and send some materials for recycling and pack other materials into barrels for disposal. Used batteries and motor oil are recycled. Paint may be blended and used for city park maintenance or to clean up graffiti.

Motor Oil If you have ever changed the oil in your car yourself, you have probably wondered what to do with the old, dirty oil. It is illegal to pour it on the ground or throw it in the trash. But, you may be surprised to find out that people in the United States throw away about 700 million liters (185 million gallons) of used motor oil every year. This amount does not include the oil disposed of by service stations and automobile repair shops.

So what can people do with the oil? One option is to take it to an automobile service station, where it will be turned in for recycling. Some cities have designated oil-collection receptacles as shown in **Figure 19.** These cities recycle the used oil turned in by citizens. If you do not know what services your community provides, you can call your local city government and find out.

Table 4 ▼

Common Hazardous Household Products

- motor oil
- paints
- batteries
- computers
- mobile phones
- pesticides
- fertilizers
- cleaners
- antifreeze

Figure 19 ▶ Used motor oil should be disposed of at an automobile service station or in an oil-collection receptacle.

SECTION 3 Review

1. **Name** two characteristics of hazardous waste.

2. **Identify** one law that governs hazardous waste.

3. **Describe** two common ways to dispose of hazardous waste in the United States. What is one advantage and one disadvantage of one of these methods?

4. **Describe** how bacteria could be used to degrade hazardous wastes. Write a short paragraph to explain your answer. **WRITING SKILLS**

CRITICAL THINKING

5. **Evaluating Ideas** Suppose that a surface impoundment site for hazardous waste is planned for your community. Would you oppose locating the site in your community? Explain your answer.

6. **Applying Ideas** Suppose someone dumped left-over motor oil on a driveway. Could this disposal method contaminate the air, water, or soil? Explain your answer.

499

Answers to Section Review

1. Hazardous waste is toxic, corrosive, or explosive.
2. Answers may vary. The Resource Conservation and Recovery Act regulates solid and hazardous waste disposal, and the Superfund Act established the means to clean up highly contaminated hazardous waste sites.
3. Land disposal is a common way of disposing of hazardous waste. One advantage is relatively cheap operating costs. A disadvantage is that toxic substances can leach into soil or groundwater. Another common way is incin-

eration. Incineration reduces the volume of waste but releases hazardous gases and particles.

4. Answers may vary. Bacteria are grown in a lab and then applied to a contaminated area, where they break down the hazardous substances.
5. Answers may vary.
6. It could contaminate the soil by running off the driveway into the yard, and it could contaminate the water if it runs off into a nearby gutter.

Close

Reteaching ———— BASIC

Reviewing Hazardous Wastes
Ask students to define hazardous waste. (waste that is a risk to the health of humans or other organisms) Have them give examples of hazardous wastes. (Sample answers: pesticides, acids, and batteries) Then, ask them how each of these should be safely disposed of. (They should be taken to a hazardous waste collection center.) **LS Verbal**

Quiz ———— GENERAL

1. What percentage of Superfund sites have been cleaned up? ($75 \div 1,200 \times 100 = 6.25$ percent)

2. What are three ways of dealing with hazardous waste that do not involve storage? (redesigning manufacturing methods to reduce waste, recycling the waste by using it for another purpose, and chemically treating the waste to convert it to a nonhazardous substance)

3. What are two types of land disposal techniques for hazardous waste? (deep-well injection and surface impoundment)

4. Name three types of organisms used to treat hazardous waste. (bacteria, certain flowering plants, and certain trees)

5. Who can you call to find out how to dispose of household hazardous waste? (your local city government)

Alternative Assessment ———— GENERAL

Alternatives to Hazardous Cleaning Products Have students compare a hazardous household cleaning product with one of its nonhazardous alternatives. Suggest that groups of students test the effectiveness of each product and write a report summarizing their findings. Make sure students follow all safety cautions on the products. **LS Interpersonal**

Alternative Assessment —— GENERAL

Alternatives to Hazardous Household Products Give students the following list of hazardous household products, and have them research alternatives to each one. Suggest that they make a poster listing each product and its alternatives.

Top Hazardous Household Products

- Air fresheners (contain formaldehyde and phenol)
- Ammonia
- Bleach
- Carpet and upholstery shampoo (contain perchloroethylene and ammonium hydroxide)
- Dishwasher detergents (usually contain chlorine)
- Drain cleaner (contains lye, hydrochloric acid, or trichloroethane)
- Furniture polish (contains petroleum distillates, phenol, and nitrobenzene)
- Mold and mildew cleaners (contain sodium hypochlorite and formaldehyde)
- Oven cleaner (contains sodium hydroxide)
- Antibacterial cleaners (may contain triclosan)
- Laundry room products (contain sodium or calcium hypochlorite, linear alkylate sulfonate, and sodium tripolyphosphate)
- Toilet bowl cleaners (contain hydrochloric acid and hypochlorite bleach)

1 Solid Waste

Key Terms

solid waste, 481
biodegradable, 483
municipal solid waste, 484
landfill, 485
leachate, 485

Main Ideas

▶ Every year, people in the United States generate more than 10 billion metric tons of soild waste.

▶ Materials that are biodegradable, such as newspapers and cotton fibers, can be broken down by biological processes. Materials that are not biodegradable such as plastics, are a major cause of disposal problems.

▶ Municipal solid waste makes up only a small fraction of the total solid waste generated, but it still amounts to over 210 million metric tons per year.

▶ Landfills and incinerators are two facilities used for disposing solid waste.

2 Reducing Solid Waste

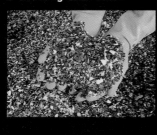

Key Terms

source reduction, 488
recycling, 489
compost, 490

Main Ideas

▶ Source reduction is a method by which we can produce less waste, recycle, and reuse materials.

▶ Recycling is the process of reusing materials or recovering valuable materials from waste or scrap.

▶ A compost pile made from plant and animal matter can be spread on gardens and fields to enrich the soil.

▶ Degradable plastic is a type of plastic that is partially made from living things.

3 Hazardous Waste

Key Terms

hazardous waste, 493
deep-well injection, 496
surface impoundment, 496

Main Ideas

▶ Hazardous waste is any waste that is a risk to the health of humans or other living things.

▶ The Resource Conservation and Recovery Act (RCRA) and the Superfund Act were established to regulate solid and hazardous waste disposal and to protect humans and the environment from waste contamination.

▶ Activities at home can create hazardous waste. Household hazardous wastes should be properly disposed of at designated collection sites.

Chapter Resource File

- **Chapter Test** GENERAL
- **Chapter Test** ADVANCED
- **Concept Review** GENERAL
- **Critical Thinking** ADVANCED
- **Test Item Listing**
- **Design Your Own Lab** GENERAL
- **Modeling Lab** ADVANCED
- **Modeling Lab** GENERAL
- **Consumer Lab** GENERAL
- **Long-Term Project** BASIC

Using Key Terms

Use each of the following terms in a separate sentence.

1. *source reduction*
2. *leachate*
3. *municipal solid waste*
4. *biodegradable*
5. *recycling*

Use the correct key term to complete each of the following sentences.

6. _____ is any waste that is a risk to the health of humans or other living things.

7. A dark brown, crumbly material made from decomposed vegetable and animal matter is called _____.

8. A _____ is a waste disposal facility where wastes are put in the ground and covered each day with a layer of dirt, plastic, or both.

> ✔ **STUDY TIP**
>
> **Increase Your Vocabulary** To learn and remember vocabulary words, use a dictionary for words you do not understand and become familiar with the glossaries of your textbooks.

Understanding Key Ideas

9. Solid waste includes all of the following *except*
   a. newspaper and soda bottles.
   b. food scraps and yard clippings.
   c. ozone and carbon dioxide.
   d. junk mail and milk cartons.

10. If your shirt is made of 50 percent cotton and 50 percent polyester, what part is biodegradable?
   a. cotton
   b. polyester
   c. both (a) and (b)
   d. none of the above

11. Microorganisms are unable to break down plastics because plastics
   a. are made from oil.
   b. are too abundant.
   c. are made of elements not found in any other substance.
   d. do not occur in nature.

12. Municipal solid waste is approximately what percentage of all solid waste?
   a. 2 percent
   b. 20 percent
   c. 60 percent
   d. 90 percent

13. Leachate is a substance that
   a. is produced in a compost pile.
   b. is a byproduct of bacterial digestion.
   c. is produced by incinerators.
   d. contains dissolved toxic chemicals.

14. Which of the following is not a benefit of incinerating waste?
   a. It reduces the amount of material sent to the landfill.
   b. It produces energy in the form of heat.
   c. It can be used to produce electricity.
   d. It neutralizes all of the toxic materials in the waste.

15. Manufacturers could reduce waste and conserve resources by making products that
   a. use more materials.
   b. are more durable.
   c. are difficult to repair.
   d. are disposable.

16. Which of the following is one way to reduce an over-supply of recyclable materials?
   a. build more recycling plants
   b. limit the amount of recyclable materials that can be collected
   c. increase the demand for products made from recycled materials
   d. put the excess materials in landfills

17. Most of the municipal solid waste in the United States is
   a. stored in landfills.
   b. recycled.
   c. incinerated.
   d. None of the above

501

ANSWERS

Using Key Terms

1. Sample answer: Changing the materials and products we use in order to reduce waste is an example of source reduction.

2. Sample answer: Groundwater contamination from leachate is one problem with some landfills.

3. Sample answer: Municipal solid waste makes up 2 percent of all solid waste.

4. Sample answer: Biodegradable products are environmentally friendly because they can be absorbed back into the environment.

5. Sample answer: Recycling is effective only if people buy products made from recycled materials.

6. Hazardous waste

7. compost

8. landfill

Understanding Key Ideas

9. c
10. a
11. d
12. a
13. d
14. d
15. b
16. c
17. a

Assignment Guide	
Section	**Questions**
1	2–4, 8–14, 17–19, 22–26
2	1, 5, 7, 15, 16, 20, 27, 28, 30, 32–36
3	6, 21, 25, 29, 31, 37

Short Answer

18. Answers may vary, but students should recognize that glass and metal wastes are easy to recycle.

19. They can help prevent toxins from leaking into the surrounding soil and water.

20. Compost is made of yard and food waste. Some benefits of composting are that it keeps organic wastes out of landfills, it provides nutrients to soil, and it protects soil from erosion.

21. The EPA can sue companies who illegally dispose of hazardous wastes.

Interpreting Graphics

22. Approximately 7,900 landfills existed in 1988, and about 2,300 landfills existed in 1998.

23. The overall number of landfills decreased. This decrease probably occurred because landfills filled up and few new ones were built.

24. The graph for 2028 would show few, if any, active landfills.

Concept Mapping

25. Answers to the concept mapping questions are on pp. 667–672.

Critical Thinking

26. Answers may vary. Because of the recession, households and businesses probably decreased their consumption, which also decreased the amount of municipal waste produced.

27. Answers may vary. A recycled product reduces the amount of waste produced. A brand-new product may be cheaper.

Short Answer

18. Do you think incineration is an efficient disposal method for glass and metal wastes? Write a short paragraph that explains why or why not. **WRITING SKILLS**

19. How do plastic liners and layers of clay help protect the environment around a landfill?

20. What are the materials that make up compost? List at least three benefits of composting.

21. How does the Superfund Act allow the federal government to ensure proper disposal of hazardous waste?

Interpreting Graphics

The graph below shows the number of landfills in the United States from the year 1988 to the year 1998. Use the graph to answer questions 22–24.

22. Approximately how many landfills existed in 1988? in 1998?

23. During the span of 10 years, did the overall number of landfills increase or did the number decrease? What may have caused this change? Explain your answer.

24. If this trend continues, what might the graph look like for the year 2028?

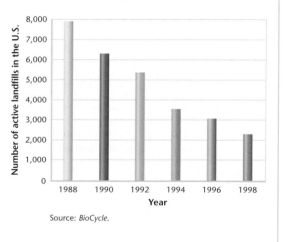

Source: *BioCycle.*

Concept Mapping

25. Use the following terms to create a concept map: *solid waste, hazardous waste, landfills, types of waste, surface impoundment, methods of waste disposal, incineration,* and *deep-well injection.*

Critical Thinking

26. **Understanding Concepts** During the 1970s, the production of municipal solid waste decreased. An economic recession was also occurring. How might the reduction in waste have been related to the recession?

27. **Making Comparisons** Read the description of recycling in this chapter and compare the benefits of buying a product that has been recycled to the benefits of buying a brand new product. Which product would you prefer to buy? Explain your answer. **READING SKILLS**

28. **Evaluating Information** How would a ban on the production of plastics affect both the environment and society?

29. **Identifying Relationships** When we purchase hazardous household products, such as motor oil, bleach, and pesticides, what happens to the containers when they are empty? What happens to the hazardous waste that these products create?

30. **Predicting Consequences** How might a person's current shopping habits affect the quality of the environment 100 years in the future?

Cross-Disciplinary Connection

31. **Social Studies** Use an almanac to determine which five states have the greatest number of hazardous waste sites. What factors do you think might account for the number of hazardous waste sites located in a state?

Portfolio Project

32. **Make a Display** Do a special project about recycling in your community. Determine what types of materials are collected, where they are taken for processing, how they are recycled, and what products are made from them. Display your findings on a poster.

28. Answers may vary. The ban would be beneficial to the environment because materials that persist for hundreds of years would no longer be produced. The ban would most likely cause loss of jobs in the plastics manufacturing industry, but this job loss might be compensated for by job creation in industries that replace plastics manufacturing.

29. Answers may vary. If the containers are thrown in the municipal trash, the hazardous waste can end up in a landfill and possibly leach out into waterways and soil.

30. Answers may vary. If many people buy plastic products, these products will still be around 100 years from now, piled in landfills or other storage sites.

Cross-Disciplinary Connection

31. Answers may vary. Available land and political influence might account for the number of hazardous waste sites.

Portfolio Project

32. Answers may vary.

MATH SKILLS

Use the table below to answer questions 33–35.

Paper Products in Municipal Solid Waste		
Product	Generation (tons)	Percentage recycled
Newspapers	13,620	56.4
Books	1,140	14.0
Magazines	2,260	20.8
Office papers	7,040	50.4

33. Evaluating Data How many tons of paper products were generated according to the table?

34. Making Calculations How many tons of newspapers were recycled? How many tons of newspapers were not recycled?

35. Making Calculations How many tons of office papers were recycled? How many tons of office papers were not recycled?

WRITING SKILLS

36. Writing Persuasively Pretend that you work for a company that sells degradable plastics. Write an advertising campaign that would persuade consumers to buy materials made from your company's brand of degradable plastic.

37. Outlining Topics Describe the various ways in which hazardous waste can be disposed. List the advantages and disadvantages of each way.

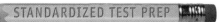

STANDARDIZED TEST PREP

For extra practice with questions formatted to represent the standardized test you may be asked to take at the end of your school year, turn to the sample test for this chapter in the Appendix.

READING SKILLS

Read the passage below, and then answer the questions that follow.

All organisms need nitrogen to make proteins and nucleic acids. The complex pathway that nitrogen follows within an ecosystem is called the nitrogen cycle. Most living things cannot use nitrogen gas directly from the atmosphere. The process of converting nitrogen gas to compounds that organisms can use is called nitrogen fixation. Organisms rely on the actions of bacteria that are able to transform and "fix" nitrogen gas into these compounds. Nitrogen-fixing bacteria convert nitrogen gas into ammonia, which plants can absorb and use to make proteins. Nitrogen-fixing bacteria live in the soil and in the roots of some kinds of plants, such as beans, peas, clover, and alfalfa.

Decomposers break down the wastes of organisms and release the nitrogen they contain as ammonia. This process is known as ammonification. Through ammonification, nitrogen that would otherwise be buried is reintroduced into the ecosystem.

1. After nitrogen-fixing bacteria convert nitrogen gas into ammonia,
   a. nitrogen fixation occurs.
   b. plants can absorb the ammonia to make proteins.
   c. nitrogen-fixing bacteria absorb the ammonia.
   d. decomposers absorb the ammonia.

2. If decomposers did not break down the waste that organisms create,
   a. nitrogen would be released into the atmosphere as ammonia.
   b. ammonification would occur.
   c. nitrogen-fixing bacteria would not convert nitrogen gas into ammonia.
   d. nitrogen would not be released into the soil as ammonia.

Math Skills

33. 24,060 tons

34. 7,682 tons were recycled; 5,938 tons were not recycled.

35. 3,548 tons were recycled; 3,492 tons were not recycled.

Writing Skills

36. Answers may vary.

37. Answers may vary.

Reading Skills

1. b

2. d

SOLID WASTE IN YOUR LUNCH

Teacher's Notes

Time Required

one 45-minute class period

Lab Ratings

EASY ——————→ HARD

TEACHER PREPARATION
STUDENT SETUP
CONCEPT LEVEL
CLEANUP

Skills Acquired

• Predicting
• Measuring
• Collecting Data
• Interpreting
• Inferring
• Organizing and Analyzing Data
• Communicating

The Scientific Method

In this lab, students will:
• Make Observations
• Analyze the Results
• Draw Conclusions
• Communicate the Results

Objectives

▶ **Recognize** various categories and amounts of solid waste produced.

▶ **Compute** percentages of waste, by category, produced per person in a single meal.

▶ **Generalize** data from a small sample for a large population using calculations.

▶ USING SCIENTIFIC METHODS **Infer** from small data samples the impact that waste production has on a large population.

▶ USING SCIENTIFIC METHODS **Evaluate** how waste data can be used to communicate results and offer solutions.

Materials

balance, triple beam or
  electronic calculator
paper towels
plastic bags
ruler

Solid Waste in Your Lunch

Are you aware of how much waste you produce during one meal? Various government and private agencies study the amount and types of food waste we produce and are continuously working to solve the problems of waste disposal. In this lab activity, you will determine how much solid waste you produce during a typical lunch. You will also predict through calculations how much solid waste your school population produces during lunch.

Procedure

1. Collect all your lunch waste on the day of the lab activity or the day before the lab activity depending on whether your class meets before or after lunch. Put all of your lunch waste in a plastic bag, including leftover food items, wrappers, napkins, straws, unopened containers of condiments, and disposable trays.

2. Each lab group member should place his or her plastic bag of waste on the worktable. Each member should separate his or her waste on a paper towel into the following categories: paper and cardboard, plastic, metal, glass, wood, and food.

3. Determine the mass of the waste in grams produced for each category for each person in the group. Create a data table similar to the one shown below and record the masses.

4. Determine the total mass for each category for the lab group. Then, determine the average mass of solid waste per student for each category. Finally, determine the overall total amount of solid waste produced for each student.

Waste category	Student 1	Student 2	Student 3	Total mass of lab group	Average mass/student
Paper and cardboard					
Plastic					
Metal					
Glass					
Wood					
Food					
Total					

DO NOT WRITE IN THIS BOOK

Safety Cautions

You may want to provide students with surgical masks when sifting through the collected waste. Be sure to caution students to be careful handling open soda cans or other open metal cans with sharp edges. Remind them of cleanup procedures concerning broken glass. Remind students not to eat or drink in the lab. Also, be sure to remind them to clean their work areas and wash their hands after completing the activity.

Tips and Tricks

If your class period is before lunch, have students collect lunch waste the day before performing this activity. So that students do not have to carry their lunch waste with them during the day, provide in the school cafeteria a box or container in which students can drop off their bag of waste. Their names should be on their bags.

Demonstrate to students how to use a balance. Also, you may have to review the steps in determining percentages.

Analysis

1. **Organizing Data** Use the equation below to determine the percentage for each waste category that makes up your total waste as an individual. Add another column to your data table to record this value.

$$\frac{\text{Mass (in grams) of waste category}}{\text{Mass (in grams) of total waste}} \times 100 = \begin{array}{l}\text{waste category's}\\ \text{percentage of}\\ \text{total waste}\end{array}$$

2. **Organizing Data** Use the equation above to determine the percentage for each waste category that makes up the total waste for your lab group. Divide the total waste for each category from the table on the previous page by the grand total and multiply by 100. Add another column to your data table to record these values.

3. **Examining Data** Compare your averages for each category and the total with other groups in the class. How and why are the data different or similar?

4. **Examining Data** Which category of waste makes up the greatest percentage of the total waste? Explain your answer.

Conclusions

5. **Making Predictions** How can you calculate the lunch waste produced in each category and overall by your entire school's student body in a day? Use your equation to make this calculation.

6. **Applying Conclusions** How can you use the knowledge you have acquired by doing this calculation exercise to reduce the amount of waste you produce?

Extension

1. **Research and Communications** Write a letter to the editor of your school's newspaper, the editor of the local newspaper, or your school's principal or cafeteria manager sharing the data your class has gathered and calculated. Offer creative solutions to eliminate and reduce some of the waste.

▶ **Step 4** Determine the mass of the waste produced in grams for each category of waste.

Answers to Analysis

1. Answers may vary. Check to be sure that the correct equation was used and that the results make sense.

2. Answers may vary. Check to be sure that the correct equation was used and that the results make sense.

3. Similarities are due to similar lunches (bought in cafeteria), and differences are due to different items purchased or brought from home.

4. Answers may vary.

Answers to Conclusions

5. Answers may vary. Students should explain that the average waste per person in each category may be multiplied by the total school enrollment.

6. Answers may vary.

Answers to Extension

1. Answers may vary.

Chapter Resource File

- Datasheets for In-Text Labs
- Lab Notes and Answers

CLASSROOM TESTED & APPROVED

Jane Frailey
Hononegah High School
Hononegah, Illinois

SHOULD NUCLEAR WASTE BE STORED AT YUCCA MOUNTAIN?

Background

Yucca Mountain, Nevada, is located 145 km (90 mi) northwest of Las Vegas, a city that has 1.4 million inhabitants. Local residents oppose the planned nuclear waste repository at Yucca Mountain, and the state of Nevada has filed five lawsuits to stop the project.

The arguments over the environmental suitability of the site still have not been settled. One report by independent scientists stated that the flow of water at the site is not sufficiently understood and that it is not possible to rule out groundwater contamination if the metal containers containing the nuclear waste corrode. As a precaution, the metal containers are being designed with umbrella-like features to keep out water. The geologic stability of the site is also questionable. Since 1976, there have been 621 earthquakes of magnitude 2.5 or greater within 80 km (50 mi) of Yucca Mountain. The largest earthquake had a magnitude of 5.6 and occurred in 1992. There are other unknowns, such as the thermal effects of radiation on the surrounding environment, the potential effects of global warming on the level of the water table, and the possibility of volcanic activity. These uncertainties are some of the problems in planning a structure that must safely store nuclear waste for at least 10,000 years.

POINTS of view

SHOULD NUCLEAR WASTE BE STORED AT YUCCA MOUNTAIN?

Yucca Mountain, in Nevada, has been chosen as the location for the nation's first permanent storage site for nuclear waste. Nuclear fuel is used to generate electricity. Nuclear waste is created after nuclear fuel can no longer be used to generate electricity. This waste is called high-level radioactive waste. High-level radioactive waste includes solids, liquids, and gases that contain a high concentration of radioactive isotopes that take thousands of years to decay. The idea is to seal 77,000 tons of radioactive waste in steel canisters and store the canisters in underground tunnels designed to last 10,000 years. Yucca Mountain is scheduled to receive its first shipment of nuclear waste by 2010.

Construction of the facility has already begun. But the debate continues about whether it would be safer to store radioactive wastes at Yucca Mountain or to keep them where they are now—in temporary storage facilities at each nuclear power plant.

For the Yucca Mountain Site

Those who support construction of the facility point out that there are two major advantages to the plan. First, Yucca Mountain is located in a remote region that is far from large populations of people. Second, the climate is extremely dry. Yucca Mountain usually receives less than 20 cm of precipitation a year, most of which evaporates before it can soak into the ground. Therefore, this dry climate means that precipitation is unlikely to cause the water table to rise and come in contact

▶ Supporters of the Yucca Mountain storage facility think that this isolated spot in Nevada is a suitable place for permanent nuclear-waste disposal.

with the stored nuclear waste. Water is the primary way by which radioactive material could move from the storage facility.

Many opponents of the site worry that changes in the climate might cause the water table to rise. They say groundwater could then reach the stored nuclear waste and become contaminated. However, supporters of the site point to several scientific studies, which determined that no significant rise or fall of the water table has occurred in the past 100,000 years.

Operators of nuclear power plants are anxious for the Yucca Mountain facility to be completed. Currently, each power plant stores its nuclear waste near the plant. Many of these storage sites have been in use for decades and are approaching their maximum capacity.

Some people think that storing wastes in one location will be safer than storing them at the individual power plants. In addition, some of the nuclear waste containers are stored in pools of water rather than in surface or underground storage

areas. Some people fear that the hazardous wastes could leak into neighborhoods around the country.

Supporters of the Yucca Mountain storage facility think that this isolated location in Nevada is a suitable place for permanent nuclear-waste disposal.

Against the Yucca Mountain Site

Perhaps the fiercest outcry against the Yucca Mountain site comes from Nevada residents. They fear that if tons of highly toxic waste are stored in one place, some of it might eventually leak. Because some of this waste is so toxic that a tiny amount could be lethal, some people think a major environmental disaster could result if small quantities of waste reach the environment.

Some people are concerned that the radioactive waste might leak into the groundwater. The waste containers are expected to last 500 to 1,000 years, but they will have to remain isolated and not come into contact with water for 10,000 years. Opponents of the plan say

CHEMISTRY CONNECTION — GENERAL

Container Materials One possibility for nuclear waste storage containers is a new group of ceramic materials called *complex oxides*. They have a structure similar to the mineral fluorite, and their advantage is that the atoms can move around relatively easily because the structures are not highly ordered. When defects caused by radiation occur, the material can adjust and is not weakened. Containers made from these ceramics could last thousands of years.

that nobody can guarantee that the containers will remain isolated for that long.

If radioactive waste leaked out of the facility, the waste could contaminate the water in wells, springs, and streams. In time, the contamination could spread from the site and into the environment.

Another worry is that transporting nuclear waste across vast distances to Yucca Mountain is riskier than leaving the material near the facilities where it is produced. Any accident along the way could release radioactivity into the environment.

Most opponents of the Yucca Mountain site agree that current methods of storing nuclear waste are dangerous and should be improved. They suggest that by transferring the waste to steel and concrete containers, the waste could be safely stored at each nuclear power facility for 75 to 100 years. By that time, they suggest, more will be known about how to store the wastes safely for thousands of years.

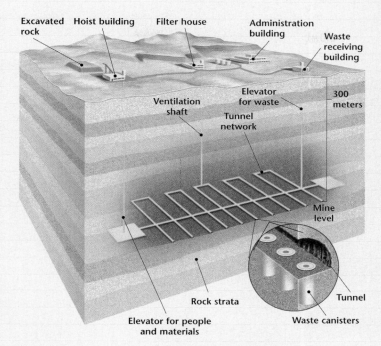

▶ The preliminary plan for the Yucca Mountain nuclear-waste storage facility shows radioactive materials carefully packaged and buried in tunnels deep underground.

▶ This map shows the nuclear power plants around the country that are possible sources of nuclear waste for the Yucca Mountain facility.

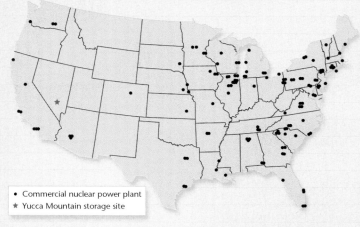

- Commercial nuclear power plant
- ★ Yucca Mountain storage site

What Do You Think?

There are over 100 nuclear power facilities in the United States. Using the Internet, research to find a nuclear power facility near your community. If there is not one near your community, how close is the nearest nuclear power facility? Is this nuclear power plant still in operation? After researching, would you be for or against the Yucca Mountain site?

507

CHAPTER 20
The Environment and Human Health

CHAPTER 21
Economics, Policy, and the Future

The casuarina tree is native to Australia and is one of the few pine trees that grow in nutrient-poor, sandy areas. This casuarina plantation on the coast of South Africa was established to hold sand dunes in place and to serve as a local source for wood fuel.

The Environment and Human Health
Chapter Planning Guide

PACING	CLASSROOM RESOURCES	LABS, DEMONSTRATIONS, AND ACTIVITIES
BLOCKS 1, 2 & 3 · 135 min pp. 510–518 **Chapter Opener**		
Section 1 Pollution and Human Health	**OSP** Lesson Plan * **CRF** Active Reading * `BASIC` **TT** Bellringer * **TT** A Typical Dose-Response Curve * **TT** A Model of Air-Pollution Movement Through a City *	**TE** Internet Activity ECOTOX and IRIS, p. 512 `ADVANCED` **TE** Group Activity Eliminating Everyday Mercury, p. 513 `GENERAL` **TE** Group Activity Testing the Air, p. 514 `GENERAL` **SE** Field Activity Sources of Pollution, p. 515 **TE** Activity Pollution Diary, p. 515 `GENERAL` **SE** Skills Practice Lab Lead Poisoning and Mental Ability, pp. 528–529 ◆ **CRF** Datasheets for In-Text Labs * **CRF** Research Lab * ◆ `BASIC` **CRF** CBL™ Probeware Lab * ◆ `ADVANCED` **CRF** Long-Term Project * ◆ `GENERAL`
BLOCKS 4 & 5 · 90 min pp. 519–523 **Section 2** Biological Hazards	**OSP** Lesson Plan * **CRF** Active Reading * `BASIC` **TT** Bellringer * **TT** A Model of Increase in Malaria Risk Due to Climate Change * **CD** Interactive Exploration 2 What's Bugging You?	**TE** Demonstration Tiny but Deadly, p. 519 `BASIC` **SE** QuickLab Simulating an Epidemic, p. 520 ◆ **TE** Activity Forest Clearing and Disease, p. 521 `GENERAL` **TE** Group Activity Asking About Antibiotic Use, p. 521 `GENERAL` **CRF** Consumer Lab * ◆ `GENERAL`

BLOCKS 6 & 7 · 90 min

Chapter Review and Assessment Resources

- **SE** Chapter Review pp. 525–527
- **SE** Standardized Test Prep pp. 674–675
- **CRF** Study Guide * ■ `GENERAL`
- **CRF** Chapter Test * ■ `GENERAL`
- **CRF** Chapter Test * `ADVANCED`
- **CRF** Concept Review * ■ `GENERAL`
- **CRF** Critical Thinking * `ADVANCED`
- **OSP** Test Generator
- **CRF** Test Item Listing *

Online and Technology Resources

 Holt Online Learning

Visit **go.hrw.com** for access to Holt Online Learning or enter the keyword **HE6 Home** for a variety of free online resources.

 **One-Stop Planner® CD-ROM**

This CD-ROM package includes
- Lab Materials QuickList Software
- Holt Calendar Planner
- Customizable Lesson Plans
- Printable Worksheets
- ExamView® Test Generator
- Interactive Teacher Edition
- Holt PuzzlePro® Resources
- Holt PowerPoint® Resources

 Holt Environmental Science Interactive Tutor CD-ROM

This CD-ROM consists of interactive activities that give students a fun way to extend their knowledge of environmental science concepts.

Compression guide:
To shorten from 7 to 5 45 min
blocks, eliminate blocks 2 and 3.

KEY

TE	Teacher Edition	**CRF**	Chapter Resource File	*	Also on One-Stop Planner
SE	Student Edition	**TT**	Teaching Transparency	■	Also Available in Spanish
OSP	One-Stop Planner	**CD**	Interactive CD-ROM	◆	Requires Advance Prep

ENRICHMENT AND SKILLS PRACTICE	SECTION REVIEW AND ASSESSMENT	CORRELATIONS
SE Pre-Reading Activity, p. 510 **SE** Reading Warm-Up, p. 510 **TE** Using the Figure, p. 510 GENERAL		**National Science Education Standards**
TE Skill Builder Graphing, p. 512 GENERAL **TE** Inclusion Strategies, p. 512 **SE** Graphic Organizer Spider Map, p. 513 GENERAL **TE** Inclusion Strategies, p. 514 **SE** MathPractice Concentration, p. 516 **SE** Case Study Chemicals That Disrupt Hormones, pp. 516–517 **TE** Skill Builder Vocabulary, p. 516 GENERAL	**SE** Section Review, p. 518 **TE** Reteaching, p. 518 BASIC **TE** Quiz, p. 518 GENERAL **TE** Alternative Assessment, p. 518 GENERAL **CRF** Quiz * ■ GENERAL	SPSP 3c SPSP 5d SPSP 6c
TE Interpreting Statistics Risks of Transmission, p. 522 BASIC **SE** Maps in Action Lyme Disease Risk, p. 530 **SE** Society & the Environment Toxic Mold, p. 531 **CRF** Maps Skills Disease Distribution * GENERAL	**SE** Section Review, p. 523 **TE** Reteaching, p. 523 BASIC **TE** Quiz, p. 523 GENERAL **TE** Alternative Assessment, p. 523 GENERAL **CRF** Quiz * ■ GENERAL	SPSP 3c SPSP 5d SPSP 6c

Guided Reading Audio CDs

These CDs are designed to help auditory learners and reluctant readers. (Audio CDs are also available in Spanish.)

www.scilinks.org

Maintained by the **National Science Teachers Association**

TOPIC: Toxicology
SciLinks code: HE4082

TOPIC: Toxic Waste
SciLinks code: HE4114

TOPIC: Emerging Viruses
SciLinks code: HE4030

 CNN Videos

Each video segment is accompanied by a Critical Thinking Worksheet.

Science, Technology & Society
Segment 5 Radioactive Medicine

CHAPTER

20

Chapter Enrichment

This Chapter Enrichment provides relevant and interesting information to expand and enhance your classroom instruction of the chapter material.

Pollution and Human Health

▶ Motor-vehicle emissions in an urban area

Paracelsus and the Dose-Response Relationship

In 1567, Swiss doctor Philippus von Hohenheim-Paracelsus stated the following: "All substances are poisons; there is none which is not a poison. The right dose differentiates a poison and a remedy." Thus, Paracelsus identified that there is a dose-response relationship between chemicals and organisms. Paracelsus's observation is the basis for the toxicologists' maxim, "The dose makes the difference." Paracelsus went on to state that therapeutic and toxic properties of a chemical could be determined by careful experimentation.

BRAIN FOOD

Asthma is a disease of the respiratory system that a person has for life. Asthma attacks are periods during which an affected person has difficulty breathing because the small airways in the lungs constrict, swell, and secrete mucus. Most of the common triggers of asthma attacks are environmental in origin and include smoke, allergens (such as pollen), and infectious diseases, such as colds.

Allergies and Animals

Many people have respiratory ailments that are triggered by animal dander. Scientists used to think that pets in the home increased a child's risk for developing allergies, but recent studies have found that the opposite is true. New research indicates that children who grow up around dogs and cats are less likely to develop many types of allergies. In fact, children exposed to two or more pets from birth are 70 percent less likely to develop a pet allergy. Scientists believe that early exposure to dander actually immunizes children against future allergies. Unfortunately for those already allergic to dander, once an allergy develops, the presence of a pet usually increases allergy symptoms.

Clancy, the Mercury Detecting Dog

As a part of the Minnesota Pollution Control Agency's Mercury-Free Zone program, a chocolate labrador retriever has been trained to detect mercury. Clancy, the mercury-detecting dog, visits schools to help sniff out mercury spills or products in the classroom that contain mercury. With Clancy's help, MPCA officials have been able to remove over 210 pounds of mercury or mercury-bearing material from Minnesota classrooms. Clancy also helps officials educate students and teachers about the dangers of mercury and the proper way to dispose of mercury-containing products. Mercury-detecting dogs were first used in Sweden, where they were able to help officials remove an estimated 1.4 tons of mercury from schools. Health risks to the dogs are minimal, and the dogs' blood is checked frequently for mercury levels. Clancy and his handlers will soon be expanding their duties to include classrooms in Wisconsin, Illinois, Michigan, and Ohio. With Clancy's help, classrooms in those states will soon be safe from the dangers of mercury.

Mad as a Hatter

Mercury intoxication, which leads to psychosis and death, used to be common among people who made hats, called *hatters*. To manufacture a hat made of rabbit fur (a cheap alternative to beaver), a hatter applied mercurous nitrate to the fur so that the fibers would mat more easily. These fibers were then shaved off, turned into felt, and hardened by boiling. The hatter then steamed and ironed the hat to shape it. In a poorly ventilated shop, hatters would breathe mercury vapors during each step of the hat-making process. Mercury accumulated in their tissues, causing kidney and brain damage. Some symptoms of mercury poisoning include palsy, memory loss, personality changes, and slurred speech.

SECTION 2 Biological Hazards

▶ A poultry market in Hong Kong

Bacteria and Children

Bacterial pathogens, such as *Salmonella* and *Listeria* (a bacterium that grows in moist areas), can cause acute intestinal discomfort in healthy adults, but most adults recover quickly. However, these microbes can be deadly to small children, pregnant women, and the elderly. Diarrhea associated with these food-borne pathogens can cause dehydration in young children and senior citizens, which can lead to death. *Listeria* can cause the spontaneous abortion of an otherwise healthy fetus. Because of these significant health risks, the United States Department of Agriculture has programs to help keep pathogens out of the food supply.

BRAIN FOOD

Cholera is a disease caused when the bacterium *Vibrio cholerae* infects the human intestine. Cholera infections are often asymptomatic. However, about one in 20 infected people develops severe symptoms, which can be life-threatening. These symptoms include watery diarrhea, vomiting, and leg cramps. The greatest health threat is from rapid dehydration and associated shock. The most effective treatment is rapidly replacing fluids and salts.

Super Germs

In 1999, one in 10 Russian prisoners was infected with tuberculosis (TB). Incomplete treatment of those affected has led to the evolution of a new superstrain of TB that is resistant to most antibiotics. Russian officials are worried that as prisoners spread this strain of tuberculosis to guards and prison medical personnel, those people will infect the public.

BRAIN FOOD

Prions are tiny infectious proteins that cause neurodegenerative wasting diseases, such as mad cow disease, scrapie (a disease of sheep), and Creutzfelt-Jacob disease. Prions cause illness by entering brain cells and converting normal proteins into other prions. Brains of infected people become spongelike, causing mental illness. Scientists think that prions can cross between species. For example, mad cow disease is transmitted to humans through infected meat.

Transgenic Mosquitoes

Molecular biologists are developing "transgenic mosquitoes" that have been genetically modified so that they are unable to transmit malaria and other tropical diseases, such as dengue fever. Biologists hope to introduce these mosquitoes to the wild, where their genes for resistance would spread throughout mosquito populations, potentially eradicating malaria and saving millions of lives. But ecologists question whether the proposal would work. For example, they have asked, "Would the resistance genes really spread throughout mosquito populations? Would the malaria parasite develop resistance to the genes? Would all of the malaria species that now spread these diseases need to be treated?" Ecologists are now designing studies to answer these questions.

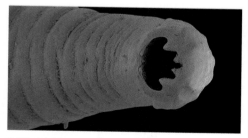

▶ Worldwide, an estimated 500 million people are hosts for hookworms

Overview

Tell students that this chapter will discuss how human health is affected by environmental conditions. This chapter examines human health problems that are related to natural and human-produced pollution, as well as the connection between human diseases and organisms that serve as pathogens and vectors.

Using the Figure — GENERAL

In the photo, a Nepali woman uses contaminated water to clean dishes. According to the United Nations, a lack of clean drinking water is the number-one health issue facing developing countries. Ask students to think of diseases spread by contaminated water and discuss how water could be made safer for human use. (Simple filters could be used, water could be boiled, and villages could set up a small municipal water supply that is separate from areas containing latrines.) **LS** Logical

PRE-READING ACTIVITY

You may want to collect students' FoldNotes to determine their prior knowledge. You may also want to modify your lesson plan to answer questions that students list in the "Want" column of their table.

VIDEO SELECT

For information about videos related to this chapter, go to **go.hrw.com** and type in the keyword **HE4 EHHV**.

The Environment and Human Health

CHAPTER 20

1 Pollution and Human Health
2 Biological Hazards

PRE-READING ACTIVITY

FOLDNOTES

Tri-Fold
Before you read this chapter, create the **FoldNote** entitled "Tri-Fold" described in the Reading and Study Skills section of the Appendix. Write what you know about environmental effects on human health in the column labeled "Know." Then, write what you want to know in the column labeled "Want." As you read the chapter, write what you learn about biodiversity in the column labeled "Learn."

This woman is washing a pot in a polluted river in Kathmandu, Nepal. Pollution is only one way the environment affects our health.

510

Chapter Correlations | *National Science Education Standards*

SPSP 3c Humans use many natural systems as resources. Natural systems have the capacity to reuse waste, but that capacity is limited. Natural systems can change to an extent that exceeds the limits of organisms to adapt naturally or humans to adapt technologically. **(Section 1 and Section 2)**

SPSP 5d Natural and human-induced hazards present the need for humans to assess potential danger and risk. Many changes in the environment designed by humans bring benefits to society, as well as cause risks. Students should understand the costs and trade-offs of various hazards—ranging from those with minor risk to a few people to major catastrophes with major risk to many people. The scale of events and the accuracy with which scientist and engineers can (and cannot) predict events are important considerations. **(Section 1 and Section 2)**

SPSP 6c Progress in science and technology can be affected by social issues and challenges. Funding priorities for specific health problems serve as examples of ways that social issues influence science and technology. **(Section 1 and Section 2)**

SECTION 1
Pollution and Human Health

If you have ever coughed from breathing car exhaust, you have experienced a mild health effect of air pollution. Pollution of air, water, and soil is frequently in the news. Because people in the United States are so concerned about pollution, our country enjoys a relatively clean environment. But this situation is also due to the efforts of scientists who have studied the relationship between pollution and human health. Scientists are also beginning to understand the broader relationships between health and the environment.

Environmental Effects on Health

Pollution causes illness in two main ways. First, it may cause illness directly by poisoning us, as in the cases of lead poisoning and lung cancer. Second, pollution may cause illness indirectly because many infectious diseases spread in polluted environments. Examples of these diseases include cholera and river blindness, diseases caused by organisms found in polluted water.

The World Health Organization (WHO) has begun to collect data on how the environment affects human health. **Figure 1** shows the WHO's estimate of poor health by world region. Poor health is measured by the estimated number of days of healthy life lost to death and disease. The WHO graph shows that, in general, people in developed countries suffer fewer health impacts due to environmental causes than people in developing countries do. The main factor behind this situation is the enormous role of infectious diseases, such as tuberculosis and cholera, which are more common in crowded areas with poor sanitation.

Objectives

▶ List five pollutants, their sources, and their possible effects on human health.

▶ Explain how scientists use toxicology and epidemiology.

▶ Explain how pollution can come from both natural sources and human activities.

▶ Describe the relationship between waste, pollution, and human health.

Key Terms

toxicology
dose
dose-response curve
epidemiology
risk assessment
particulates

Figure 1 ▶ This bar graph shows the environment's contribution to disease in different parts of the world. Regions that suffer from poor health generally also suffer more from environmental causes of poor health, such as infectious diseases.

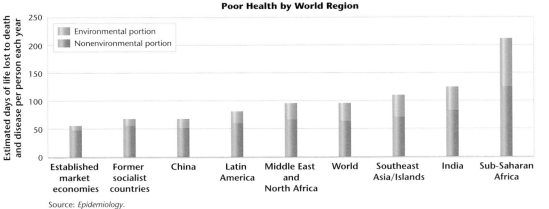

Poor Health by World Region

Source: *Epidemiology*.

511

Chapter Resource File

• Lesson Plan
• Active Reading (BASIC)
• Section Quiz (GENERAL)

Transparencies

TT Bellringer

REAL-LIFE
CONNECTION — GENERAL

Lost Days of Healthy Life In **Figure 1**, each country or region is compared using the number of annual days of healthy life lost per person. Potential days of healthy life lost is a common unit of measurement in the field of public health. It is an estimate of the days of healthy life lost prior to age 70 from causes that are considered preventable given appropriate healthcare interventions or preventive measures. Therefore, people in a country with less pollution but worse healthcare might have more days lost to illness than people in a country with more pollution but excellent healthcare.

SECTION 1

Focus

Overview

Before beginning this section, review with your students the Objectives in the Student Edition. The first section focuses on human health problems related to natural and human-produced pollution. Methods of determining risk from pollutants are discussed.

📻 Bellringer

Ask students to brainstorm all the possible pollutants that they think might affect human health. Tell them not to forget natural pollutants. Also ask them to list the sources of these pollutants. (Pollutants may include pesticides from farming or landscaping; heavy metals in soils or from industrial pollution; radioactive materials [radon gas and nuclear waste]; particulate matter from dust storms, volcanoes, industry, and cars; and bacteria from animal waste products.) Have them record their list in their *EcoLog*.
LS Logical

Motivate

Identifying
Preconceptions — GENERAL

Lost Days Before students read this section, ask them to guess how many days of poor health, on average, people in the United States suffer in a year. Then, ask them to estimate the average days of poor health per person per year people experience in China, Russia, sub-Saharan Africa, and Europe. When students finish guessing, have them look at **Figure 1.** Students may be surprised to learn that people in developing countries lose many days to poor health. Have students discuss some of the environmental factors that could contribute to poor health. (Factors include poor sanitation, discharge of chemicals into the environment, and natural pollutants, such as dust.) **LS** Visual

Table 1 ▼

Pollutant Types and Effects		
Pollutant	**Source**	**Possible effects**
Pesticides	use in agriculture and landscaping	nerve damage, birth defects, and cancer
Lead	lead paint and gasoline	brain damage and learning problems
Particulate matter	vehicle exhaust, burning waste, fires, and tobacco smoke	respiratory damage (asthma, bronchitis, cancer)
Coal dust	coal mining	black lung disease
Bacteria in food	poor sanitation and poor food handling	gastrointestinal infections

Teach

SKILL BUILDER — GENERAL

Graphing Ask each student to graph the following data:

Dose (ppm)	Mortality (%)
0	3
1	5
10	15
20	30
40	65

Have students determine the dose that would most likely cause 50 percent mortality. (around 32 ppm)
LS Visual

Internet Activity — ADVANCED

ECOTOX and IRIS Toxicology professionals need to gather information on environmental toxins to determine whether a certain concentration of a pollutant is dangerous to humans. Experiments are expensive and time-consuming, so scientists share data. Some Web sites coordinate data from many experiments so that scientists can compare results to determine how harmful a chemical is. ECOTOX is the EPA's ecotoxicology Web site. Its focus is on the effects of toxins on plants and wildlife. IRIS is the EPA's Integrated Risk Information System, which focuses on human health risks related to pollutants found in the environment. Have students use both databases to get a feeling for how toxicologists gather and process information.

🔲 internet connect
www.scilinks.org
Topic: Toxicology
SciLinks code: HE4082
SCiLINKS. Maintained by the National Science Teachers Association

Figure 2 ▶ A dose-response curve shows the response of an organism to different concentrations of a substance.

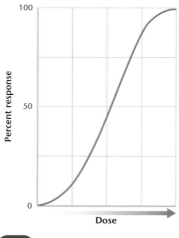

Toxicology

The word *toxic* means poisonous. **Toxicology** is the study of the harmful effects of substances on organisms. **Table 1** lists some important pollutants and their toxic effects.

Toxicity: How Dangerous Is It? We are exposed to small amounts of chemicals every day, in food, in the air we breathe, and sometimes in the water we drink. Almost any chemical can be harmful if taken in, or *ingested*, in large enough amounts. The question is whether the concentration of any particular chemical in the environment is high enough to be harmful.

To determine the effect of a pollutant on health, we need to know several things. We need to know how much of the pollutant is in the environment and how much gets into the body. Then we need to determine what concentration of the toxin damages the body. The amount of a harmful chemical to which a person is exposed is called the **dose** of that chemical. The damage to health that results from exposure to a given dose is called the *response*.

Whether a chemical has a toxic effect depends in part on the dose. The response also depends on the number of times a person is exposed, the person's size, and how well the person's body breaks down the chemical.

A persistent chemical is a chemical that breaks down slowly in the environment. The pesticide DDT is an example of a persistent chemical. Persistent chemicals are dangerous because more people are likely to come into contact with them, and these chemicals are more likely to remain in the body.

Dose-Response Curves The toxicity of a chemical can be expressed by a dose-response curve, as shown in **Figure 2**. A **dose-response curve** shows the relative effect of various doses of a drug or chemical on an organism or organisms as determined by experiments. Sometimes, we find that there is a *threshold dose*. Exposure to any amount of chemicals less than the threshold dose has no adverse effect on health. Exposure to levels above the threshold dose usually leads to worse effects.

512

INCLUSION Strategies

• *English Language Learner*
• *Learning Disabled*

Ask students to use **Table 1** to write warning statements for each of the pollutants listed in the table. Each warning statement should include the type of pollutant and the possible effects of exposure. Students can invent names of the products and design the container label with the warning included. Examples of actual products with warning labels could be displayed.

HISTORY
● CONNECTION — GENERAL

Plumbing and the Roman Empire Ask students if they know the chemical symbol for lead. Many may know it is Pb. Ask them why it is not a symbol such as Ld. Tell them that lead has been used in plumbing for centuries, so the element lead is represented by Pb, which is short for *plumbum*, the Latin word for "plumbing." The Romans used lead pipes to distribute water within their cities. As poisonous lead leached into the water, Romans experienced an increase in dementia. The use of lead in wine making also contributed to lead poisoning.

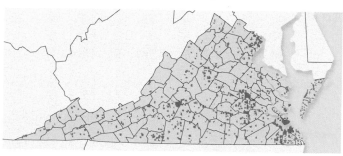

Source: Virginia Department of Health.

Figure 3 ▶ This map shows the location of cases of mercury poisoning in Virginia. Patterns point scientists toward areas of mercury pollution.

Epidemiology

When an epidemic occurs, such as a widespread flu infection, health officials use their knowledge of epidemiology to take action. **Epidemiology** (EP uh DEE mee AHL uh jee) is the study of the spread of diseases. Epidemiologists collect data from health workers on when and where cases of the disease have occurred. This information can be used to produce a map like the one in **Figure 3**.

Then scientists trace the disease to try to find its origin and how to prevent it from spreading. For example, in a case of mercury poisoning, health officials may ask questions such as: What did the people with mercury poisoning have in common? Were they all exposed to the same chemicals?

Risk Assessment In order to safeguard the public, health officials determine the risk posed by particular pollutants. Recall that risk is the probability of a negative outcome. In the case of human health, risk is the probability of suffering a disease, injury, or death.

Scientists and health officials work together on risk assessments for pollutants. A **risk assessment** is an estimate of the risk posed by an action or substance. During the process of risk assessment, scientists first compile and evaluate existing information on the substance. Then they determine how people might be exposed to it. **Figure 4** shows a diagram, created by a computer model, of how air pollutants might travel through a city area. The third step is determining the toxicity of the substance. Finally, scientists characterize the risk that the substance poses to the public. Risk assessments may lead to government regulations on how and where the substance can be used. In the United States, the Environmental Protection Agency (EPA) formulates these regulations.

Figure 4 ▶ Air flow models like this one help scientists predict the path that air pollutants may follow through a city. The bright orange areas are receiving the most pollutants.

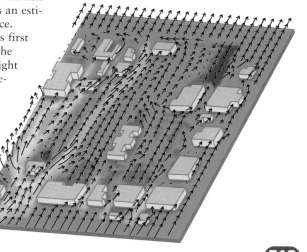

513

Graphic Organizer **Spider Map**
Create the **Graphic Organizer** entitled "Spider Map" described in the Appendix. Label the circle "Environmental Factors That Cause Disease." Create a leg for each type of environmental factor that causes disease. Then, fill in the map with details about each type of environmental factor that causes disease.

Group Activity —— GENERAL
Eliminating Everyday Mercury
On the board, list products that typically contain mercury. These products include thermostats, thermometers, fluorescent lights, light switches in older car trunks, some batteries, blood pressure gauges, and dental amalgam (the compound used in dental fillings). Then, have students create a poster that lists each item that contains mercury and suggests alternative materials. Also have them include tips for proper disposal of the items that contain mercury on the poster. Hang the poster in the hallway, or have students visit other classrooms to explain the hazards of mercury and the proper way to dispose of mercury-containing products. Also ask each student to check his or her house for these items, and encourage students to talk to their parents about proper mercury disposal.
LS Visual

Co-op Learning   English Language Learners

Graphic Organizer **GENERAL**

Spider Map
You may want to have students work in groups to create this Spider Map. Have one student draw the map and fill in the information provided by other students from the group.

REAL-LIFE CONNECTION —— GENERAL

Identifying Outbreaks Sometimes, outbreaks of disease are difficult to identify, so many people become ill before public health officials can take measures to prevent further spread of a disease. To identify outbreaks within days instead of weeks, a National Electronic Disease Surveillance System (NEDSS) is being created. This system will allow clinics to enter and share critical information about diseases on the Web. Emergency room visits and positive lab tests for certain diseases can be tracked, which will allow epidemiologists to identify patterns of infection. This type of system has helped identify outbreaks on a smaller scale. In New York City, cases of giardia and cryptosporidiosis (two diseases associated with polluted water) have been monitored successfully using the following methods: pharmacies report weekly sales information from over-the-counter antidiarrheal medications; clinical laboratories perform and report findings from stool samples; and nursing homes report new cases of gastrointestinal disease daily.

Transparencies

TT A Typical Dose-Response Curve
TT A Model of Air-Pollution Movement Through a City

Testing the Air Have your class monitor the air for particulate matter. Bring in 4 × 6-inch unlined index cards, petroleum jelly, rulers, toothpicks, string, and magnifying glasses. Have the class brainstorm places to put particulate collectors. You may want to place collectors indoors and outdoors. Have students draw four 2 × 3-inch squares on each card. On the back of the card, students should write the following: site number, description of the site, wind direction and the approximate wind speed, height above ground, and possible pollution sources. They should hang the card with string. Then, have them apply a thin layer of petroleum jelly with a toothpick to the squares on the card. Leave the collectors out for 24 hours. When the experiment is finished, have students examine each collector with a magnifying glass. Ask students to count the particles within each box, and to record those numbers on the card. Back in the classroom, discuss which areas had more pollution than others and why.

LS Interpersonal

| Co-op Learning | English Language Learners |

Figure 5 ▶ A dust storm descends upon Marrakesh, Morocco, in the photo above. Dust is perhaps the most common natural pollutant.

Figure 6 ▶ A town is coated with ash after the 1991 eruption of Mount Pinatubo, in the Philippines.

514

Pollution from Natural Sources

You may think of pollution as being entirely caused by people, but some pollutants occur naturally in the environment. Naturally occurring pollutants usually become hazardous to health when they are concentrated above their normal levels in the environment. One example is the radioactive gas radon. In some areas, radon from granite bedrock may seep into buildings, where it becomes concentrated. Because it is an odorless gas, people may unknowingly breathe it in. Radon causes an estimated 5,000 to 20,000 cancer deaths every year in the United States.

Particulates The most common pollutants from natural sources are dust, soot, and other particulates. **Particulates** (pahr TIK yoo lits) are particles in the air that are small enough to breathe into the lungs. These particles become trapped in the tiny air sacs in our lungs and cause irritation. This irritation can make lung conditions, such as chronic bronchitis and emphysema, worse. **Figure 5** shows particulate pollution from a dust storm, whereas **Figure 6** shows pollution from a volcanic eruption. Wildfires also produce large amounts of particulates.

Heavy Metals Another important type of pollution from natural sources are the so-called *heavy metals*. Dangerous heavy metals include the elements arsenic, cadmium, lead, and mercury. These metals occur naturally in rocks and soil. Most of these elements cause nerve damage when they are ingested beyond their threshold dose. Selenium, also found naturally in many soils, is actually a beneficial element when taken in very small quantities. But larger doses pose health risks to humans.

Pollution from Human Activities

Human activities release thousands of types of chemicals into the environment, but we know surprisingly little about the health effects of most of them. Only about 10 percent of commercial chemicals have been tested for their toxicity, and about 1,000 new chemicals are introduced every year. **Figure 7** shows the introduction of pollutants into the environment by human activities.

Recent Improvements In the United States, regulations have helped reduce our exposure to pollutants. Most vehicles and factories now have pollution-control devices. As a result, people living in the United States contain lower levels of some toxic chemicals in their bodies, on average, than they did in the recent past. In 2001, the U.S. Centers for Disease Control and Prevention (CDC) released a study on chemical residues in 3,800 people. Levels of nicotine (from smoking), lead, and several other toxic chemicals were considerably lower in these peoples' tissues than they had been 10 years earlier.

Because we know so little about the effects of chemicals on our health, new health risks are discovered frequently. For example, scientists now think that chemical pollution may be at least part of the cause of Parkinson's disease and Alzheimer's disease.

Burning Fuels Despite the very real advances in public health resulting from pollution control, air pollution is still a major health problem. Burning fuels in vehicles, home furnaces, power plants, and factories introduces enormous amounts of pollutants into the air. These pollutants include the gas carbon monoxide and many kinds of particulates. Gasoline and coal burning contribute to the many premature deaths each year from asthma, heart disease, and lung disorders. In fact, it may be possible to predict an area's death rate based on the amount of pollution. A recent study found that long-term exposure to air contaminated with soot particles raises a person's risk of dying from lung cancer or other lung and heart diseases.

FIELD ACTIVITY

Sources of Pollution Walk around your neighborhood, and record potential sources of pollution. Suggest ways in which the amount of pollution from each source might be reduced. Write your observations, suggestions, and any evidence that supports your analysis in your *EcoLog*.

Figure 7 ▶ Human activities can pollute air and water. Paper mills contribute pollutants to rivers (above). Vehicle emissions cloud the air in urban areas worldwide (left).

515

Activity — **GENERAL**

Pollution Diary After students learn about human-caused pollutants, have them keep a diary of the actions they take to reduce pollution. Encourage students to involve their families in these activities. Also encourage students to think of fun, creative ways to reduce the pollution they generate.
LS Intrapersonal

Cultural Awareness **GENERAL**

Wells in Bangladesh Surface water in Bangladesh is contaminated with many pathogens. Programs established to provide safe drinking water promoted the drilling of five million tube wells, which helped decrease the incidence of waterborne diseases. However, arsenic has been discovered in groundwater in almost 30 percent the wells tested so far, and almost 8,000 cases of arsenic poisoning have been diagnosed up to the present. Healthcare professionals are now looking for alternative methods of supplying clean drinking water to the Bangladeshi people.

REAL-LIFE CONNECTION **ADVANCED**

Household Pollutants Levels of industrial pollutants in waterways have been tracked for decades, but until recently, few studies have monitored the levels of common household chemicals, hormones, and antibiotics in water supplies. In 2002, the USGS released "Pharmaceuticals, Hormones, and Other Wastewater Contaminants in U.S. Streams, 1999–2000." The study found pharmaceuticals, hormones, and other organic wastewater pollutants at very low concentrations in streams across the United States. Standards for most of the 95 chemicals studied do not exist for drinking water. Common pollutants included steroids, detergent metabolites, nonprescription drugs, insect repellents, caffeine, disinfectants, antibiotics, and fire retardants. Have students read the report online to find data about streams in their area.

MATH**PRACTICE**

Answers

If 1 tsp in 2 gal of water produces a concentration of 1,000 ppm, then 1 tsp in 1 gal of water is double that concentration (2,000 ppm). If you divide that concentration by five (by adding 1 tsp to five gal), the concentration is 400 ppm.

SKILL BUILDER — GENERAL

Vocabulary Hormone disruptors that mimic estrogen can be called by a number of names: *environmental estrogen*; *ecoestrogen*; *xenoestrogen*, which means "foreign estrogen"; and *phytoestrogen*, which means "plant estrogen."

LS Verbal | English Language Learners

MATH**PRACTICE**

Concentration Concentrations of chemicals in the environment are often expressed in parts per million (ppm) or parts per billion (ppb). One teaspoon of salt in two gallons of water produces a salt concentration of 1,000 ppm. What salt concentration, in ppm, would result from dissolving one teaspoon of salt in five gallons of water?

Pesticides *Pesticides* are chemicals designed to kill unwanted organisms such as insects, fungi, or weeds. Pesticides are beneficial in that they allow us to grow more food by reducing pest damage. Many of the increases in food production in the past 60 years are partly due to the development and use of more effective pesticides.

But because pesticides are designed to kill organisms, they are often dangerous to humans in large enough doses. Although we are exposed to pesticide residues on fruits and vegetables, the amounts consumed by most people pose little danger.

Most modern pesticides, such as most of those used in the United States, break down quickly in the environment into harmless substances. Widely used *organophosphate* pesticides have replaced more persistent pesticides, such as DDT. But organophosphates are very toxic, causing nerve damage and perhaps cancer. In 1999, U.S. poison centers reported more than 13,000 cases of organophosphate poisoning. Most cases of pesticide poisoning affect the people applying the chemicals.

Persistent chemicals are still used in many developing countries. Such pesticides pose the greatest risk to children, whose internal organs are still developing and who eat and drink more in relation to their body weight than adults do.

CASE STUDY

Chemicals That Disrupt Hormones

In recent years, scientists have collected evidence that many pollutants disrupt the endocrine system. The glands that make up the human endocrine system produce hormones. *Hormones* are chemicals that circulate in the bloodstream and control many life processes, such as the development of muscles and sex organs.

Some pollutants, called *hormone mimics,* behave like natural hormones. Other pollutants are *hormone disrupters,* which prevent natural hormones from functioning normally. Even low levels of these kinds of pollutants can affect developing embryos and infants.

Hormone mimics were first discovered in fish in Europe. Researchers in England and France found that male trout and eels downstream

from sewage treatment plants contained egg-yolk proteins usually produced only by females. Lab experiments showed that the water contained estrogen-like chemicals and that these chemicals induced the male fish to make proteins usually produced only by females. The chemicals are believed to have come from detergents and from the urine of women taking contraceptive pills.

Most hormone disrupters interfere with the sex hormones. They prevent normal production of testosterone in males or increase the chances of sexual abnormality in females. Examples of hormone disrupters include phthalates, which are widely used in cosmetics, such as hair dyes and fingernail polish. Polychlorinated biphenyls (PCBs),

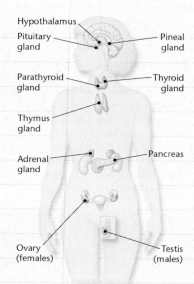

► The diagram above shows the major organs and glands of the human endocrine system.

516

BRAIN FOOD

Imposex in Snails In marine snails, imposex is a condition in which females develop male sexual organs, including a penis. This condition can cause sterility and steep declines in marine snail populations. Researchers at the Plymouth Marine Laboratory in England were able to link the chemical tributyltin (TBT) to the development of imposex in dogwhelks, a type of marine snail. TBT had been a common

additive in antifouling paints. As TBT leached into the water, it dramatically affected the snails. When TBT was banned for yachts, small boats, and fish cages, dogwhelk populations bounced back. Imposex in common whelk populations, however, could not be linked to TBT. Researchers are searching for reasons why these snails still mysteriously "change sex."

Industrial Chemicals Railroad tankers carrying industrial solvents overturned near Rochester, New York, in 2002. Two solvents reacted to cause a fire that destroyed several houses. Several people were treated for breathing the fumes.

We are exposed to low levels of industrial chemicals every day, particularly inside new buildings that have new furnishings. Toxic chemicals are used to make building materials, carpets, cleaning fluids, and furniture. Older buildings, like the one shown in **Figure 8**, were often painted using lead-based paint. Lead is directly linked to brain damage and learning disabilities.

Often, industrial chemicals are not known to be toxic until they have been used for many years. For example, polychlorinated biphenyls (PCBs) are oily fluids that have been used for years as insulation in electrical transformers. PCBs do not break down in the environment. In 1996, studies showed that children exposed to PCBs in the womb can develop learning problems and IQ deficits. The waters of the Great Lakes are polluted by PCBs, and doctors warn pregnant women not to eat fish from these lakes. Studies have shown that adults with high concentrations of PCBs in their tissues have more memory problems than adults who do not.

Figure 8 ▶ Lead poisoning in children is most often due to direct exposure to lead-based paint.

▶ The fertility of American alligators, such as this one, has been reduced by their exposure to hormone disrupting pollutants.

some pesticides, lead, and mercury may also act as hormone disrupters.

Many cases of pollution by hormone disrupters have now been found in the United States. For example, alligators in a Florida lake that was polluted with local hazardous waste had such abnormally small penises and low testosterone levels that they could not reproduce. In 2002, scientists reported that even small amounts of the widely used herbicide atrazine disrupt the sexual development of frogs.

During the past 50 years, there has been a large increase in cancers of the prostate, testicles, ovaries, and breasts in most industrialized countries. All of these forms of cancer can be accelerated by abnormal levels of sex hormones. A recent analysis of sperm counts among men in industrialized countries shows that sperm counts have fallen by 50 percent in the last 50 years.

Scientists do not yet have concrete evidence that hormone disrupters in the environment are actually causing these human health problems. Research into these questions has accelerated since an international conference on hormone disrupters was held in 1996.

CRITICAL THINKING

1. Reading Comprehension Explain the difference between hormone mimics and hormone disrupters.

2. Analyzing Relationships If humans are increasingly exposed to these pollutants, what are some possible results?

517

Close

Reteaching ——— BASIC

Crossword Puzzles Have each student use the terms in this section to create a crossword puzzle. Encourage students to include as many terms as possible in their puzzle. Have students exchange puzzles and try to complete them.
LS Visual/Interpersonal

Quiz ——— GENERAL

1. Give two examples of how hormone disruptors have caused physical changes in some animals. (Accept two of the following answers: male trout and eels downstream from a sewage treatment plant contained egg-yolk proteins normally found only in females; alligators exposed to hazardous waste had abnormally small penises and low testosterone levels; and atrazine exposure disrupted sexual development in frogs.)

2. Where do heavy metals occur naturally? (Heavy metals occur naturally in some rocks and soil.)

Alternative Assessment ——— GENERAL

The Case of the Mysterious Outbreak Have students write a detective story about a disease that was caused by environmental factors. Have students make sure that their detective uses epidemiological techniques to solve the case. Encourage students to add this story to their **Portfolio.**
LS Intrapersonal

Figure 9 ▶ Waste that is not disposed of properly can pollute beaches, where it can pose a threat to swimmers and sunbathers.

internet connect

www.scilinks.org
Topic: Toxic Waste
SciLinks code: HE4114

SCI LINKS. Maintained by the National Science Teachers Association

Waste Disposal Much of the pollution in our environment is a byproduct of inadequate waste disposal. **Figure 9** shows the pollution of a beach with solid waste. Wastewater from cities can carry oil and dozens of toxic chemicals into our waterways. Waste incineration plants can emit toxic products into the air, and mining can release toxic contaminants into streams and rivers.

One of the reasons that our air and water is less polluted in many areas than it was 50 years ago is that methods of disposing of waste have improved. However, problems remain. Many old landfills are leaking. And many communities still have sewage treatment plants that release raw sewage into a river or the ocean after heavy rains. In addition, laws regulating waste disposal are not always enforced.

The United States government has not decided how it will dispose of radioactive waste from nuclear power plants. Meanwhile, the waste remains in barrels at or near the plants, and small quantities of radioactive iodine, cesium, and other elements leak into nearby waterways.

SECTION 1 Review

1. **List** five pollutants, their sources, and their possible effects on human health.

2. **Explain** how pollution can arise from both natural sources and from human activities.

3. **Describe** the relationship between waste, pollution, and human health.

CRITICAL THINKING

4. **Making Comparisons** Write a short paragraph that explains the relationship between toxicology and epidemiology. WRITING SKILLS

5. **Analyzing Relationships** In what ways do human activities increase the health risks from natural pollutants?

518

Answers to Section Review

1. PCBs are found in electrical transformers and cause learning problems. Organophosphate pesticides are found on and near crop fields and cause nerve damage. Particulates come from various sources (dust, pollen, and combustion) and can affect lung functions. Heavy metals occur in some rocks and soil and can cause nerve damage. Radon comes from granite bedrock and can cause cancer.

2. Pollution can come from natural sources such as volcanoes and dust storms. Heavy metals and radon also occur naturally in some rocks. Human sources of pollution include human and domestic animal wastes in drinking-water

supplies and chemical pollution from industrial activities.

3. Answers may vary. Students should stress that improper reuse or disposal of waste creates pollution that can then affect human health.

4. Answers may vary but should include the idea that epidemiologists use information from toxicological studies to determine the types of pollutants that may be causing the rapid spread of a disease.

5. Answers may vary. Industrial activities sometimes concentrate natural heavy metals. Human carelessness can also start wildfires, which release particulates.

Biological Hazards

Some of the damage to human health in which the environment plays a role is not caused by toxic chemicals but by organisms that carry disease. Today, we have outbreaks of diseases that did not exist or that few people had heard of 100 years ago, such as AIDS, ebola, West Nile virus, hanta virus, and mad cow disease. In addition, diseases that have killed people for centuries, such as malaria, tuberculosis, yellow fever, and hookworm, kill many more people today than they did 50 years ago. All these diseases are caused by organisms. One of the reasons these diseases are now widespread is that we have altered our environment in ways that encourage them to spread.

The Environment's Role in Disease

Infectious diseases are caused by **pathogens,** organisms that cause disease. Some of these diseases, such as tuberculosis and whooping cough, are spread from person to person through the air. Other diseases are spread by drinking water that contains the pathogen. Still other diseases are transmitted by a secondary host, such as a mosquito. A **host** is an organism in which a pathogen lives all or part of its life. **Table 2** lists the most deadly infectious diseases worldwide.

Objectives

▶ Explain why the environment is an important factor in the spread of cholera.

▶ List two changes to the environment that can lead to the spread of infectious diseases.

▶ Explain what scientists mean when they say that certain viruses are emerging.

Key Terms

pathogen
host
vector

Table 2 ▼

Deaths from Infectious Diseases in 2000, Estimated by the World Health Organization		
Disease and examples	Cause	Estimated deaths per year (in millions)
Total infectious and parasitic diseases	bacteria, viruses, and parasites	10.5
Respiratory infections (pneumonia, influenza, and whooping cough)	bacteria, viruses	4.0
AIDS	virus	2.9
Diarrheal diseases (cholera, typhus, typhoid, and dysentery)	bacteria, viruses, parasites	2.1
Tuberculosis	bacteria	1.7
Childhood diseases (measles and diphtheria)	virus	1.5
Malaria	parasitic protist	1.1
Tetanus	bacteria	0.3
Tropical diseases (trypanosomiasis, Chagas' disease, schistosomiasis, and leishmaniasis)	bacteria, viruses, and parasites	0.1

519

Chapter Resource File

- **Lesson Plans**
- **Active Reading** BASIC
- **Section Quiz** GENERAL

Transparencies

TT Bellringer

BIOLOGY

CONNECTION ADVANCED

What Is a Retrovirus? The human immunodeficiency virus (HIV) is a retrovirus that causes AIDS (acquired immune deficiency syndrome). Retroviruses are infectious, nonliving particles filled with RNA instead of DNA. A retrovirus is able to inject RNA into a cell along with an enzyme that allows the virus to make copies of itself. In this way, a retrovirus uses a human cell's machinery to multiply. Have interested students research the way that retroviruses multiply and spread. Ask students to put together a diagram of the retrovirus infection cycle and to present their diagram to the class.

Focus

Overview

Before beginning this section, review with your students the Objectives in the Student Edition. This section explores the connection between biological pathogens, such as bacteria and viruses, and human diseases. The section also examines how environmental changes have caused certain diseases to spread.

Bellringer

Have students review the "Estimated deaths . . ." column in **Table 2.** Ask students to discuss why the number of deaths from respiratory infections is greater than deaths from other diseases. (Respiratory infections are passed through the air or through casual contact, so they are easier to spread than blood-borne or waterborne diseases are. Because they are spread so easily, respiratory infections affect and kill more people than other infections do.) **LS Logical**

Motivate

Demonstration —— BASIC

Tiny but Deadly Show students the head of a pin, and tell them that it is about 2 mm in diameter. Draw a circle with a diameter of 2 m to represent the head of a pin enlarged 1,000 times. Tell students that if the head of the pin were this size, you would begin to be able to see bacteria on it. Then, tell students that you would be able to see viruses on the pin if its head were 20 m in diameter, or 10,000 times larger. It might help students to visualize this ratio if they measure 20 m of string and then stretch it out in an open area. Explain that seeing viruses in detail would require the head of the pin to be between 200 and 2,000 m wide (100,000 to 1,000,000 times larger). Viruses are so small that we must study them using powerful electron microscopes. **LS Visual/ Kinesthetic** | English Language Learners

Teach

QuickLAB

Simulating an Epidemic

Procedure

1. Obtain one **test tube** of **water** from your teacher. Your teacher has "contaminated" one of the test tubes with an invisible substance.

2. Pour half your water into the test tube of a classmate. Your classmate will then pour an equal amount back into your test tube. Exchange water with three classmates in this way.

3. Your teacher will now put a small amount of a **test chemical** into your test tube. If your water turns cloudy, you have been "contaminated."

Analysis

1. Who had the test tube that started the "infection"?

2. Name a disease that could be spread in this way. Explain your answer.

Figure 10 ▶ This child is undergoing rehydration therapy during a cholera epidemic in South Africa.

Waterborne Disease

Nearly three-fourths of infectious diseases are transmitted through water. In developing countries, where there is not enough water for basic needs, the local water supply is often used for drinking, washing, and sewage disposal. So, the water is usually very polluted and is a good breeding ground for pathogens. The pathogens breed in water and transfer diseases directly to humans through water, or organisms that carry the pathogens transfer them to humans through the water. Organisms, such as mosquitoes, that transmit diseases to people are called **vectors** of the disease. The widespread construction of irrigation canals and dams, particularly in the tropics, has increased the habitat for vectors. For example, it is dangerous to bathe in tropical ponds. The water might contain the snail vector for schistosomiasis, an incurable disease that kills thousands of people each year.

Cholera The deadliest waterborne diseases come from drinking water polluted by human feces. Pathogens, such as those that cause *cholera* and *dysentery,* enter the water in human feces. These diseases cause the body to lose water by diarrhea and vomiting. These diseases cause most of the infant mortality around the world. A baby's body has less water than an adult's body and therefore suffers more from dehydration. **Figure 10** shows a child being treated for dehydration.

Malaria Another waterborne disease called *malaria* was once the world's leading cause of death. The disease is caused by parasitic protists and is transmitted by a bite from females of many species of mosquitoes. The mosquito vector lays her eggs in stagnant fresh water, which is where the mosquito larva develops. No effective vaccine for malaria exists, but preventative measures are used to control mosquitoes.

520

BIOLOGY
CONNECTION — GENERAL

Cholera and Copepods The bacterium that causes cholera, *Vibrio cholerae,* thrives in water polluted by sewage. Until recently, however, it was a mystery how these bacteria arrived in rivers and ponds. Research in the 1990s revealed a link between cholera and copepods, small crustaceans found in both marine and freshwater environments. Studies in Bangladesh showed that cholera bacteria hitch a ride on traveling marine copepods.

These copepods follow phytoplankton as it moves with tides and ocean currents. In areas polluted by sewage effluent, such as the mouths of rivers, phytoplankton blooms provide a feast for copepods. As copepods populate the polluted area, so do *Vibrio cholerae,* which can lead to an outbreak of cholera. Have students create a visual display that shows the life cycle of a common pathogen.
LS Visual

Environmental Change and Disease

Many ways in which we alter the environment make the environment more suitable for pathogens to live and reproduce. For example, soil is often polluted with chemicals and pathogens. When soil erodes, these pollutants blow away and wash away with the soil and may contaminate areas thousands of miles away. Many parasites are spread through soil that is contaminated with feces. Hookworm, which causes acute exhaustion, was once common in the United States. People are infected by walking barefoot on soil that contains human and animal feces or by consuming contaminated food or water. **Figure 11** shows soil erosion in Nepal. In 1984, 87 percent of the population was found to be infected by parasitic worms, which people were exposed to due to widespread soil erosion.

Antibiotic Resistance Our actions cause pathogens to evolve resistance to antibiotics that are used to kill them. For example, in the United States, large quantities of antibiotics are fed to livestock each year to speed their growth. As a result, *Salmonella, Escherichia coli (E. coli)*, and other bacteria that live in livestock evolve resistance to antibiotics. These bacteria now make thousands of U.S. citizens sick each year when they eat contaminated meat that has been improperly refrigerated or undercooked.

We also use enormous amounts of antibiotics to treat human illnesses. In 1979, 6 percent of European strains of pneumonia bacteria were resistant to antibiotics. Ten years later, 44 percent of the strains were resistant. Tuberculosis (TB) is another illness treated with antibiotics. The spread of TB in recent years is mostly due to the evolution of antibiotic resistance in the bacterium that causes TB.

Figure 11 ▶ Soil erosion in Nepal (top) leads to the spread of parasites such as the hookworm (bottom).

Ecofact

Suburbs Spread Lyme Disease
Lyme disease is the most widespread vector-borne disease in the United States. It is caused by a bacterium similar to the one that causes the sexually transmitted disease syphilis. The vector is a tick found on white-tailed deer. The suburbs are a suitable place for deer to grow and reproduce, and their populations have exploded as suburbs have expanded. Lyme disease was first described in 1976. It now infects more than 13,000 people a year in the United States.

521

Interpreting Statistics ——— BASIC

Risks of Transmission Have students examine **Figure 12,** which shows the predicted change in the risk of malaria transmission as global warming increases. Areas in Europe and Russia show the greatest probability for increased risk. Ask students to discuss whether Russia and Europe would soon have the most cases of malaria under this scenario. (They would not. Because there are now few cases of malaria in these countries, a doubling of risk translates into a relatively small increase in the number of cases.) **LS** Logical

ECONOMICS ———
● **CONNECTION** ——— GENERAL

HIV in Africa The spread of HIV and AIDS in sub-Saharan Africa is a human health tragedy that has resulted in the infection of about 15 percent of the adult population in eight African nations. This infection rate has had a significant effect on the African economy. The United Nations Development Program estimated that as of 1993, the HIV epidemic had cost the nations of Africa about $30 billion. Estimates by the United Nations indicate that Africa will lose 25 percent of its labor force to AIDS by 2010. Have students research the potential economic effects of HIV/AIDS in Africa or other parts of the world. **LS** Intrapersonal

Transparencies

TT A Model of Increase in Malaria Risk Due to Climate Change

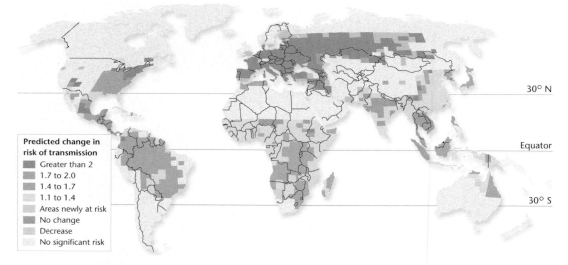

Predicted change in risk of transmission

- Greater than 2
- 1.7 to 2.0
- 1.4 to 1.7
- 1.1 to 1.4
- Areas newly at risk
- No change
- Decrease
- No significant risk

30° N

Equator

30° S

Figure 12 ▶ This computer model shows how malaria might spread under specific global warming conditions.

internet connect

www.scilinks.org
Topic: Emerging Viruses
SciLinks code: HE4030

SCI LINKS Maintained by the National Science Teachers Association

Connection to ▶ **Biology**

The Viral Advantage Antibiotics kill bacteria but not viruses, such as those that cause colds and flu. Antibiotics kill bacteria by interfering with their cellular mechanisms. Viruses do not have cellular mechanisms. Many antibiotics destroy the system a bacterium uses to make proteins. Viruses do not make their own proteins. Instead, they take over the cellular machinery of the cells they invade and use the cell to make proteins.

522

Malaria on the March Insects that breed in water are the secondary hosts that transmit malaria. The mosquitoes that transmit malaria are found in the warmer parts of the world. Epidemiologists believe that global warming may increase the area where malaria occurs. **Figure 12** shows that malaria might spread across large areas of Central America, South America, Africa, and Asia.

Malaria was common in much of the United States and Europe before the days of mosquito control. Now, it is most common in tropical countries. Historically, malaria was controlled by draining marshes and rice paddies where the mosquitoes breed and by spraying with pesticides. Since the 1970s, however, mosquitoes have evolved resistance to most of the pesticides. Newer methods for controlling mosquitoes involve spreading growth regulators that prevent mosquito larvae from maturing into adults or that sterilize the female mosquitoes.

Emerging Viruses In recent years, medical scientists have been focusing on previously unknown viruses, the so-called emerging viruses that were unknown 100 years ago. One example is AIDS (acquired immune deficiency syndrome), which is caused by HIV (human immune deficiency virus). Other examples of emerging viruses include the hanta virus, the ebola virus, and the West Nile virus. Most viral diseases spread directly from one person to another. Often, the virus invades the body through a cut or through mucus membranes. We do not have many effective drugs to treat viral diseases, and the drugs that we have are only effective against a certain virus. Our main defense against viral diseases is vaccination. The problem with vaccines is that they are very specific, and viruses evolve rapidly, so when a new strain of a viral pathogen evolves, a new vaccine must be developed.

GENETICS ———
● **CONNECTION** ——— GENERAL

Sickle Cell Anemia and Malaria Sickle cell anemia is an inherited disease that affects Africans and African Americans. A person who inherits the sickle cell gene from both parents produces abnormal hemoglobin, which deforms red blood cells. The deformed cells become trapped in small blood vessels, causing pain, organ damage, and even premature death. Because the disease used to kill people before they were able to reproduce, one would think that the gene would have been quickly lost when those carrying the disease died. But the gene has persisted, primarily in areas where malaria is widespread. In these areas, a carrier with one sickle cell gene and one normal gene is somewhat protected from malaria infection. That person may have a better chance of surviving to pass on their genes than does a person who does not have the sickle cell gene.

Cross-Species Transfers In recent years, scientists have discovered an increasing number of pathogens that have made a *cross-species transfer,* or have moved from one species to another. For example, HIV and West Nile virus fall into this category. The pathogens that cause these diseases have lived for centuries in some species of wild animals and have done little damage. When the pathogens invade humans, the pathogens cause serious diseases. Some ecologists think that the ways in which we are altering the environment and destroying habitats ensure that diseases like these will become more common in the future.

Examples of Cross-Species Transfers One example of pathogens that made a cross-species transfer occurred in Argentina. Herbicides were sprayed on crops in Argentina. The herbicide killed the native grasses and allowed other plants to invade the farmland. These new plants attracted a species of rodent that feeds on them. The rodents were carrying viruses for a hemorrhagic fever, which infected many of the agricultural workers. Hemorrhagic fevers cause hemorrhages, or internal bleeding, by breaking blood vessels. Hanta virus is an example of a virus that causes hemorrhagic fever.

Influenza, or flu, is highly contagious. The flu virus passes from humans to animals (particularly birds) and back to humans again. Hong Kong flu gets its name from the fact that the virus was transmitted to humans from ducks bred in Hong Kong for food. **Figure 13** shows a poultry market, where the Hong Kong flu virus probably transferred from birds to people. Because flu is so easily spread from one person to another, epidemiologists predict that the greatest threat to human health may be the outbreak of a new, very virulent strain of influenza virus, which would spread rapidly through crowded urban populations.

Figure 13 ▶ Poultry markets, such as this one in Hong Kong, can contribute to the cross-species transfer of viruses from birds to humans.

SECTION 2 Review

1. **List** two changes to the environment that can lead to the spread of infectious diseases.
2. **Explain** why some diseases are likely to spread as a result of global warming.
3. **Explain** why the environment is an important factor in the spread of cholera.
4. **Explain** the term *emerging virus.*

CRITICAL THINKING

5. **Understanding Concepts** Read the information under the heading "Antibiotic Resistance." How is the use of antibiotics by humans increasing antibiotic resistance in pathogens? **READING SKILLS**

6. **Analyzing Relationships** How do human activities cause pathogens to move from one species to another? Give examples of cross-species transfer to help explain your answer.

523

Answers to Section Review

1. Sample answers: soil erosion, the creation of irrigation canals and dams, the use of antibiotics, and habitat destruction which puts humans in greater contact with wild animals

2. Widespread warming may allow mosquitoes to flourish in areas that they do not currently occupy. The mosquitoes would likely bring with them a variety of disease organisms.

3. Cholera is transmitted to humans through polluted water sources in the environment.

4. The term *emerging virus* refers to a virus that was unknown in the recent past but that now threatens humans.

5. Bacteria in livestock have developed a resistance to antibiotics due to the large quantities of antibiotics that are fed to livestock to speed their growth. The evolution of antibiotic resistance in bacteria that cause human disease can be traced to the enormous numbers of antibiotics that are used to treat human illnesses.

6. The way in which humans are altering the environment is one of the main reasons why pathogens move from one species to another. The emergence of hemorrhagic fever in Argentina is an example of how human change to the environment has caused a cross-species transfer.

Alternative Assessment GENERAL

Water and Human Health Have each student write an essay that evaluates whether the lack of access to uncontaminated water is the most important human health problem in the world today. Have students include information about the effects of waterborne microbes, pesticides, hormone disrupters, and heavy metals. Have students include data, graphs, and tables (with sources) to defend their thesis. **LS Intrapersonal**

Microbe Models Have students look for pictures of different types of pathogens on the Internet. Have students work together to make clay models of each organism. If possible, have students size the models proportionately. Students should label the features of each pathogen and include a short description of its effects on human health. As an extension, students can make comic book-style pamphlets that show the life cycle of each pathogen. **LS Kinesthetic/Visual**

Chapter Resource File

- Chapter Test GENERAL
- Chapter Test ADVANCED
- Concept Review GENERAL
- Critical Thinking ADVANCED
- Test Item Listing
- Research Lab BASIC
- CBL™ Probeware Lab ADVANCED
- Consumer Lab GENERAL
- Long-Term Project GENERAL

CHAPTER 20 Highlights

1 Pollution and Human Health

Key Terms

toxicology, 512
dose, 512
dose-response curve, 512
epidemiology, 513
risk assessment, 513
particulates, 514

Main Ideas

▶ Toxic chemicals from both natural sources and human activities that pollute air, soil, water, and food may damage human health.

▶ Toxicology is used to determine how poisonous a substance is.

▶ After an outbreak of illness occurs, epidemiologists attempt to find its origin and try to find ways to prevent future epidemics.

▶ Most pollutants come from human activities, but some pollutants occur naturally.

▶ Improperly disposed of wastes may leak hazardous pollutants into the environment.

2 Biological Hazards

Key Terms

pathogen, 519
host, 519
vector, 520

Main Ideas

▶ Most human diseases that have an environmental component are caused by pathogens.

▶ The environment provides breeding grounds for pathogens and for their secondary hosts and vectors.

▶ The transmission of many infectious diseases includes water. We increase the areas where organisms that carry these diseases can reproduce when we create irrigation canals and inadequate sewage systems.

▶ Environmental changes that help spread infectious diseases include global warming and the spread of suburbs and farmland.

▶ Many emerging diseases are caused by pathogens that have made cross-species transfers from animals to humans.

524

Student Opportunities

Toxicology Internships Government agencies such as the Environmental Protection Agency or your state's pollution control agency generally have summer internships available to students. Internships may involve performing dose-response experiments or monitoring pollutants in the environment. Encourage students who are interested in toxicology to contact one of these agencies about summer internships via phone or the Internet.

Using Key Terms

Use each of the following terms in a separate sentence.

1. *dose*
2. *vector*
3. *risk assessment*
4. *particulates*
5. *epidemiology*

For each pair of terms, explain how the meanings of the terms differ.

6. *pathogen* and *host*
7. *response* and *dose*
8. *toxicology* and *epidemiology*

> ✔ **STUDY TIP**
>
> **Vocabulary Practice** To practice vocabulary, write the terms and definitions on a piece of paper and fold the paper lengthwise so that the definitions are covered. First, see how many definitions you already know. Then, write the definitions you don't know on another piece of paper, and practice again until you know all of them.

Understanding Key Ideas

9. Which of the following is *not* a true statement about the effects of pollution on health?
   a. It is difficult to determine how pollution affects health because many factors often contribute to a disease.
   b. The toxic effects of a pollutant depend upon the dose to which you are exposed.
   c. Many pollutants cause chronic diseases that result from exposure to the pollutant over the course of many years.
   d. Persistent chemicals are less toxic than chemicals that break down rapidly.

10. Which of the following is an emerging disease that was unknown 50 years ago?
    a. malaria
    b. dengue fever
    c. Lyme disease
    d. schistosomiasis

11. Cholera is usually transmitted from person to person by water because
    a. it is caused by a snail that breeds in water.
    b. it is usually contracted by someone drinking water polluted with human feces that contain the cholera pathogen.
    c. it is transmitted by mosquitoes.
    d. it is caused by a virus.

12. Tuberculosis (TB), which was once almost eradicated, is becoming more common, even in developed countries, because
    a. new varieties of the tuberculosis pathogen have evolved in rodents.
    b. livestock are given antibiotics.
    c. the pathogen that causes TB breeds in polluted water.
    d. some populations of the pathogen that causes TB are resistant to the antibiotics.

13. Which of the following statements about environmental pollutants is true?
    a. Our environment contains fewer toxic chemicals than it did 50 years ago.
    b. Hormone mimics in our water supply pose no danger to humans
    c. There is no health risk from pollutants in indoor air.
    d. The bodies of people who live in the United States contain lower levels of some toxic chemicals than they did 20 years ago.

14. Which of the following actions is most likely to prevent yellow fever, which is transmitted by mosquitoes, from becoming epidemic?
    a. preventing dehydration in patients by treating them with oral rehydration therapy
    b. taking antibiotics
    c. encouraging people to empty water out of old cans, tires, plant saucers, and other areas that contain standing water
    d. spraying the area repeatedly with pesticides

525

ANSWERS

Using Key Terms

1. Sample answer: The amount of a harmful substance to which a person is exposed is called a *dose*.

2. Sample answer: Certain mosquitoes are vectors that transmit to humans the organism that causes malaria.

3. Sample answer: Risk assessment is done to try to determine how potentially dangerous a substance is to people.

4. Sample answer: The forest fire produced huge amounts of particulates, which could be seen as smoke from miles away.

5. Sample answer: Healthcare workers use their knowledge of epidemiology to stop the spread of disease.

6. A pathogen is a disease-causing organism, whereas a host is an organism in which a pathogen lives all or part of its life.

7. The response is the damage to health that results from exposure to a given dose, or amount, of a substance.

8. Toxicology is the study of the harmful effects of substances on organisms, whereas epidemiology is the study of how a disease spreads.

Understanding Key Ideas

9. d
10. c
11. b
12. d
13. d
14. c

Assignment Guide	
Section	**Questions**
1	1, 3–5, 7–9, 13, 15, 18, 19–21, 23, 24, 27, 31
2	2, 6, 10–12, 14, 16, 17, 22, 23, 25, 26, 28–30

CHAPTER 20 **Review**

Short Answer

15. Scientists expose organisms to different doses of a chemical to create dose-response curves and to determine which concentrations of the chemical may produce a harmful effect.

16. Answers may vary. Deforestation or desertification may lead to soil erosion, which can expose people to disease-causing organisms in the soil. The construction of irrigation canals can increase the spread of malaria.

17. Answers may vary. Many disease vectors increase in areas where there is standing water. Also, biotic interactions in the environment can allow pathogens to be passed from primary to secondary hosts.

18. There may be more particulate matter, because of industrial activity and heavy traffic, in a large urban area.

Interpreting Graphics

19. The chemical represented by the top line is more toxic at a lower dose.

20. Both chemicals appear equally toxic at a very large dose.

21. Assuming that both chemicals are found in the environment in equal concentrations and accumulate in the food chain in similar ways, the first chemical would probably be more of a problem if it persisted in the environment.

Concept Mapping

22. Answers to the concept mapping questions are on pp. 667–672.

Short Answer

15. How do scientists determine the toxicity of a chemical?

16. How can land use change contribute to the spread of infectious disease?

17. What role does the environment play in the transmission of infectious diseases?

18. Why would lung disease be more common in a large urban area than in a remote rural area?

Interpreting Graphics

The graph below shows the dose-response curves for two chemicals. Use the graph to answer questions 19–21.

19. Which chemical is more toxic at a lower dose?

20. Which chemical is more toxic at a very large dose?

21. Can you tell from the graph which chemical is more likely to be a problem if it persists in the environment?

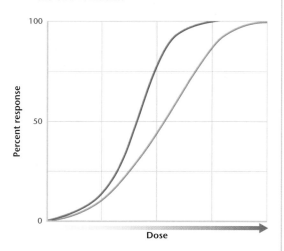

Concept Mapping

22. Use the following terms to create a concept map: *habitat destruction, pathogen, animal, vector,* and *human disease.*

Critical Thinking

23. **Making Comparisons** In what ways does a disease such as lung cancer, which is caused by breathing pollutants over a long period of time, differ from a disease such as malaria, which is caused by a pathogen?

Cross-Disciplinary Connection

24. **History** In 1775, Percival Pott noted that chimney sweeps had a high rate of cancer of the scrotum. What further investigations might be performed to find out what occupational hazard might be causing the cancer? How many of these would have been possible at the time, and how many require modern technology?

25. **Economics** Write a proposal to reduce the mosquito population of an area. How might you encourage the public to assist in this effort? **WRITING SKILLS**

26. **Biology** Read about mosquitoes under the heading "Malaria." How would you design an irrigation system to minimize the chances that mosquitoes would breed in it? **READING SKILLS**

Portfolio Project

27. Collect half a dozen pesticide containers that still have their labels. Make a table that has three columns. List the names of the pesticides in one column. Then read the label on each container. Use this information to decide which pesticide is the most dangerous and which pesticide is the least dangerous. In the second column, label the pesticides as most to least dangerous. In the third column, list the most important safety precautions required of anyone who uses the pesticide. **CAUTION:** Do not get pesticide on your face, and wash your hands thoroughly after handling the pesticide cans.

Critical Thinking

23. Answers may vary. A disease such as lung cancer takes many years to develop, is not infectious, and is usually caused by long-term exposure to inorganic substances. A disease such as malaria develops relatively soon after exposure and is infectious.

Cross-Disciplinary Connection

24. Answers may vary. The components of creosote (the soot in chimneys) could be analyzed and dose-response data for each component could be gathered by research or by experiments. The components of creosote

and some of their effects might have been known in 1775, but dose-response experimentation was not used at that time. Determining specific doses would require more modern technology.

25. Answers may vary. Students could encourage the public to empty standing water out of old tires or to frequently change the water in birdbaths.

26. Answers may vary. The system would need to minimize the amount of standing water. Water could be delivered slowly, so that it is absorbed by plant roots at the time it is delivered.

MATH SKILLS

The table below shows four diseases and the number of cases of each disease that were reported to the United States Centers for Disease Control in 1990 and 1998. Use the table below to answer questions 28–29.

Disease	1990	1998
Cryptosporidiosis	2	3,793
Lyme disease	2	16,801
Malaria	1,292	1,611
Typhoid fever	552	375

28. **Analyzing Data** Malaria cases increased between 1990 and 1998. What other facts would you want to know before deciding that the United States has a growing malaria problem?

29. **Making Calculations** By what percentage did the number of typhoid fever cases decline between 1990 and 1998?

WRITING SKILLS

30. **Communicating Main Ideas** Why do sewage systems that overflow when it rains need to be replaced with modern systems that do not overflow?

31. **Writing Persuasively** Write a letter to a newspaper. In the letter, argue either for or against homeowners' use of pesticides on their lawns and gardens.

STANDARDIZED TEST PREP

For extra practice with questions formatted to represent the standardized test you may be asked to take at the end of your school year, turn to the sample test for this chapter in the Appendix.

READING SKILLS

Read the passage below, and then answer the questions that follow.

Dehydration is a serious threat to human survival—as dangerous as a high fever. However, as any athlete knows, drinking water alone is often not an adequate cure for dehydration. Sports drinks contain sugar and electrolytes (minerals) as well as water. This principle also underlies oral rehydration therapy, which is used to treat people suffering from diseases such as cholera and dysentery. These diseases cause water loss from diarrhea and vomiting. Severe dehydration often causes death, particularly in small children. Patients being treated for dehydration are fed a solution of salt, sugar, and water. The sugar and salt help the body absorb the water from the stomach. Sugar and salt also add electrolytes to the body fluids so that these are not diluted. Millions of lives have been saved by rehydration therapy.

1. According to the passage, which of the following statements about oral rehydration therapy is *not* true?
   a. A solution containing sugars and salts is absorbed by the stomach more rapidly than water alone.
   b. The salts replace electrolytes in the bloodstream so that these are not diluted by the water.
   c. Any source of water is adequate to make up the solution of salts and sugar.
   d. Millions of lives have been saved by oral rehydration therapy.

2. According to the passage, which of the following statements about dehydration is *not* true?
   a. It may be fatal.
   b. It is especially dangerous to small children.
   c. It may be caused by diarrhea and vomiting from diseases such as cholera.
   d. It is not often caused by exercising on a hot day without drinking.

527

Portfolio Project
27. Answers may vary.

Math Skills
28. You would want to know how much the United States population increased during this time to see if the rate of infection actually changed. You would want to see the number of cases each year between 1990 and 1998 to see if the number of cases fluctuated or increased steadily. You would also want to see if rainfall totals were higher during these years to determine if there is a correlation between higher average rainfall and the incidence of malaria.

29. $(552 - 375)/552 = 0.32 = 32$ percent

Writing Skills
30. Answers may vary. Sample answer: When sewage overflows, it may be deposited in the soil. Pathogens can then be spread by wind or direct human contact. Overflowing sewage systems can deposit raw sewage into the drinking water supply, which can also cause disease. Replacing old sewer systems with modern systems that do not overflow will therefore reduce the risk of spreading waterborne disease.

31. Answers may vary.

Reading Skills
1. c
2. d

CHAPTER **20** Skills Practice Lab: MATH/GRAPHING

LEAD POISONING AND MENTAL ABILITY

Teacher's Notes

Time Required

one 45-minute class period

Lab Ratings

EASY ————————————→ HARD

TEACHER PREPARATION ▲
STUDENT SETUP ▲▲
CONCEPT LEVEL ▲▲
CLEANUP ▲

Skills Acquired

- Designing Experiments
- Interpreting
- Organizing and Analyzing Data

The Scientific Method

In this lab, students will:
- Test the Hypothesis
- Analyze the Results
- Draw Conclusions

Materials

The materials listed are enough for groups of 3 to 4 students. Students may want to use colored pencils to make the graph easier to read.

Objectives

▶ USING SCIENTIFIC METHODS **Analyze** the relationship between lead poisoning and children's IQ.
▶ **Graph** experimental data.
▶ **Interpret** graphical data.

Materials

notebook
pen or pencil

▶ **Effects of Lead** Lead smelters, such as the one below in Yugoslavia, can cause air pollution and lead poisoning.

Lead Poisoning and Mental Ability

People are usually exposed to lead in old buildings that were painted with lead paint. The lead can enter your body as dust when you breathe and can permanently damage the brain and nervous system. Lead poisoning can cause aggressive behavior, hyperactivity, headaches, and hearing loss. At high levels, it can cause seizures, coma, and even death. The Centers for Disease Control and Prevention (CDC) state that a lead level of only 10 micrograms per deciliter in the blood can be harmful. (A microgram is one-millionth of a gram, and a deciliter is one-tenth of a liter.) In this lab, you will explore the effect of lead poisoning on the mental ability of children. The children all grew up near a lead smelter, a factory where raw lead ore is processed. Scientists measured the concentration of lead in the children's blood over time. Psychologists also performed tests on the children to determine their IQ. You will analyze the data to see if you can find a pattern.

Procedure

1. Design a hypothesis for the relationship between the lead concentration in the blood, the IQ, and the age of the children. As the blood-lead concentration increases, how would you expect the person's IQ to change? How do you think this relationship would change as the children grow older?

2. The table on the next page lists the blood-lead concentration and IQ data for a group of 494 children. The children were measured five times as they grew up. The first measurement was made when they were six months old, and the last measurement was made when they were seven years old. The children were divided into four groups according to the amount of lead in their blood. Group 1 had the lowest concentration of lead, and group 4 had the highest concentration of lead. Prepare a graph for the data in the table. Plot lead concentration on the *x*-axis and IQ on the *y*-axis. Label each axis with the correct units. Choose an appropriate scale for each axis so that the entire range of data in the table will fit on the graph.

3. Plot the data from the table on your graph. Connect all data points for a single age group with a single line. You should have five lines of data on your graph and have one line for each age group.

528

Tips and Tricks

Have students choose null and alternative hypotheses before they begin graphing data. The alternative hypothesis would be what they expect to find. In this case, students might propose the following alternative hypothesis in the following format: H_1— As lead concentration in a person's blood increases, his or her IQ decreases. The appropriate null hypothesis, or hypothesis of "no difference," would be as follows: H_0— Lead concentration will have no effect on IQ. Null

hypotheses are used because it is easier to disprove a hypothesis than it is to prove it. The second pair of hypotheses could be the following: H_1—Exposure to lead over time will lower IQ; H_0—Exposure to lead over time will not affect IQ.

Students could graph the data points at home by using a computer. This lab would give them important practice in creating computer graphs.

Group of children		Average blood-lead concentration (micrograms per deciliter)	Average IQ score
6 mo	1	8.3	109.4
	2	12.6	104.7
	3	16.8	102.9
	4	24.2	100.0
15 mo	1	11.8	109.3
	2	18.6	106.5
	3	24.4	102.9
	4	34.4	101.3
3 yr	1	11.6	110.2
	2	17.4	106.5
	3	22.4	102.2
	4	30.2	100.0
5 yr	1	8.3	109.3
	2	12.6	106.1
	3	17.2	104.1
	4	23.6	98.8
7 yr	1	6.6	109.6
	2	10.1	107.7
	3	13.7	102.7
	4	20.0	98.7

▶ **Lead Paint** Dust from lead paint peelings can cause lead poisoning.

Analysis

1. **Analyzing Data** For a single age group, how does IQ vary with lead concentration? Is this true for all age groups?

2. **Analyzing Data** How does the relationship between lead concentration and IQ change as a child grows older?

Conclusions

3. **Drawing Conclusions** What conclusions can you draw from your analysis about the effect of lead on IQ?

4. **Applying Conclusions** Based on your conclusions, what long-term effects might lead poisoning have on a community?

Extension

1. **Analyzing a Viewpoint** Based on the data presented in this lab, do you think the CDC's limit of 10 micrograms per deciliter is reasonable? Explain your answer.

Answers to Analysis

1. For each age group, IQ varies negatively with blood lead concentration. This means that IQ decreases as blood lead concentration increases.

2. The negative relationship between IQ and lead concentration becomes stronger as a child grows older.

Answers to Conclusions

3. Lead concentrations and IQ are negatively correlated. Higher concentrations of lead in human blood cause a consistent decrease in IQ.

4. Lead poisoning could cause brain and nervous system damage in community residents over time, which would decrease residents' quality of life.

Answers to Extension

1. In children that had approximately 10 micrograms per deciliter in their blood for the entire study (group 1), IQ scores remained fairly steady. It appears from these data that the CDC's limit is reasonable. However, we do not have a control group of children with extremely low lead levels. Comparing the IQ scores of a control group and group 1 would be useful.

Chapter Resource File

- Datasheets for In-Text Labs
- Lab Notes and Answers

REAL-LIFE
● **CONNECTION** — GENERAL

Sources of Lead Poisoning Even people who do not live near a smelter or manufacturing plant that uses lead can be exposed to lead. Lead crystal glasses and some ceramics also contain lead that may leach into food. Areas along roadsides may contain residues from leaded gasoline, which was largely discontinued in the 1980s and 1990s. And lead-based paints may still be found in older houses. The EPA Web site is a good source for more information about lead.

TEACHER TESTED & APPROVED

Linda Culp
Thorndale High School
Thorndale, Texas

Demonstration — GENERAL

Forests and Lyme Disease
Obtain a map that shows the biomes of the United States. Have students compare areas of predicted Lyme disease risk to different biomes. Because ticks obtain the disease from deer and deer are abundant in deciduous forests, areas at high risk for disease transmission will probably correspond to the deciduous forest biome.
LS Visual

BRAIN FOOD

Rocky Mountain Spotted Fever Rocky Mountain spotted fever is another illness caused by a tick-borne pathogen. This disease causes a rash of black spots and was often fatal before the discovery of antibiotics. Rocky Mountain spotted fever is actually not common in the Rocky Mountain area of the United States, but is mainly a disease of the southeastern United States. Interestingly, Rocky Mountain spotted fever is rare in states with high incidences of Lyme disease.

Transparencies

TT Lyme Disease Risk

MAPS in action

LYME DISEASE RISK

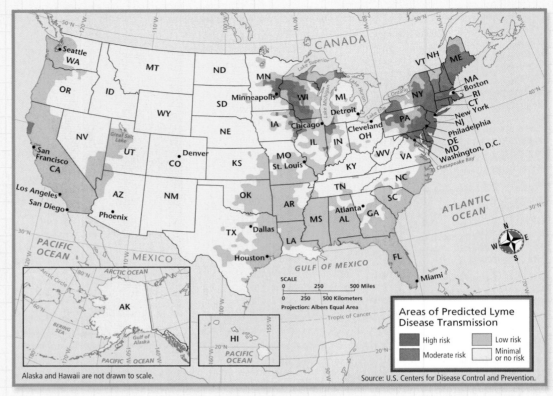

▶ The map above shows the risk of contracting Lyme disease by geographic location inside the United States.

MAP SKILLS

Use the Lyme disease risk map for the United States to answer the questions below.

1. **Using a Key** Using the map above, determine the risk of contracting Lyme disease in your city or town.

2. **Using a Key** In what general region of the United States is the risk of contracting Lyme disease greatest?

3. **Analyzing Relationships** Can you determine the relationship between the risk of contracting Lyme

disease and the concentration of ticks that act as vectors for the disease? Explain your answer.

4. **Analyzing Data** What is the difference between the risk of contracting Lyme disease in rural Massachussetts and the risk of contracting Lyme disease in rural Nevada?

5. **Forming a Hypothesis** What factors might account for the relatively high risk of contracting Lyme disease in the Northeast?

530

Answers to Map Skills

1. Students should accurately read the risk for their city or town.

2. The risk of contracting Lyme disease is greatest in the northeastern United States.

3. Answers may vary. You cannot determine this relationship from the map. But you can assume that because ticks are essential Lyme disease vectors, the ticks will occur in the same relative concentrations as they do in the predicted areas of disease transmission.

4. The difference is large—the difference between "high risk" and "minimal or no risk." But

percentages of infection are not given in the map key, so an exact percentage difference cannot be determined.

5. Ticks spread Lyme disease, but they acquire the disease pathogen by biting deer. The Northeast has large numbers of deer, which could support and infect large numbers of ticks. The Northeast also has large concentrations of people living in a relatively small, wooded region, which increases the probability that a person will be bitten by an infected tick.

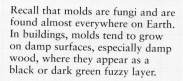
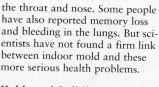
TOXIC MOLD

You may have seen stories in the news with titles such as "Mold Closes Schools" or "Homes Infested with Toxic Mold." In the past 10 years, news stories have reported on school evacuations, strange illnesses, and multimillion dollar lawsuits, all due to mold. What is toxic mold, and why is it a problem?

"Toxic mold" is a popular term for molds that grow indoors and that are suspected of making people sick who are exposed to them.

▶ The mold pictured below, *Stachybotrys,* is commonly implicated in toxic building episodes.

▶ This health worker is spraying portions of a house infested with mold that may be toxic.

Recall that molds are fungi and are found almost everywhere on Earth. In buildings, molds tend to grow on damp surfaces, especially damp wood, where they appear as a black or dark green fuzzy layer.

Are Molds Toxic?

Some species of mold produce toxins that they use mainly to compete with other molds. These toxins can also harm people. The toxins become airborne attached to spores, the mold's reproductive particles, or on tiny mold fragments. Once the toxins are airborne, people can breathe them in.

The most commonly mentioned toxic mold is *Stachybotrys chartarum.* This mold produces several potent toxins that affect the immune system and cause hemorrhaging in toxicology tests on mice.

People handling material contaminated with *Stachybotrys* report coughing and burning sensations of the throat and nose. Some people have also reported memory loss and bleeding in the lungs. But scientists have not found a firm link between indoor mold and these more serious health problems.

Molds and Buildings

Molds are especially common in areas with high rainfall. If wood or paper stays damp for any length of time, odds are that mold spores will land on it and begin to grow.

Most of the problems with indoor mold occur in areas between walls or in other rarely seen places. The solution is to fix leaks as soon as possible and improve air circulation so that damp areas dry out.

Scientific Uncertainty

There are only a small number of well-studied cases of toxic mold poisoning, and even in these there exists the possibility of other causes for the illnesses. Scientists see toxic mold as an example of the difficulty of linking environmental exposure with human health. "You can't prove causation from epidemiological studies," notes one doctor. "All you can do is show that there is a correlation."

What Do You Think?

School districts and homeowners have spent millions of dollars replacing moldy parts of buildings. But in many cases the mold had not caused health problems. Should the government require schools and homeowners to repair moldy buildings? Who should pay for it?

531

TOXIC MOLD

Background

In houses, the interior walls of corner rooms are the most likely places for mold to grow. Because these areas are generally colder than other locations inside a house, they tend to have a higher relative humidity. Warmer rooms in the same house will have lower relative humidity and be less susceptible to mold formation.

REAL-LIFE CONNECTION   GENERAL

Asthma and Mold Whether mold is toxic or not, it is still a common allergen. Exposure to mold can lead to asthma attacks and respiratory distress in some people. The Environmental Protection Agency recommends that asthma sufferers, who are often sensitive to mold, maintain low humidity indoors by fixing leaky pipes, emptying refrigerator and dehumidifier pans, venting clothes dryers, installing fans to disperse moisture, and keeping surfaces dry.

Answers to What Do You Think?

Answers may vary. Students may suggest that the government require moldy areas to be treated only if spore counts become high and suspected toxic molds are found.

Economics, Policy, and the Future

CHAPTER 21

Chapter Planning Guide

PACING	CLASSROOM RESOURCES	LABS, DEMONSTRATIONS, AND ACTIVITIES
BLOCKS 1 & 2 • 90 min pp. 532–538 **Chapter Opener**		
Section 1 Economics and International Cooperation	**OSP** Lesson Plan * **CRF** Active Reading * `BASIC` **TT** Bellringer * **TT** Major International Environmental Agreements * **TT** Economic Systems Within the Biosphere *	**TE** Internet Activity UN Documentation, p. 535 `ADVANCED` **TE** Internet Activity Making Conservation Profitable, p. 537 `GENERAL` **CRF** Consumer Lab * `BASIC`
BLOCKS 3 & 4 • 90 min pp. 539–543 **Section 2** Environmental Policies in the United States	**OSP** Lesson Plan * **CRF** Active Reading * `BASIC` **TT** Bellringer * **TT** U.S. Agencies and Environmental Laws *	**TE** Activity Whose Issue? p. 539 `GENERAL` **TE** Internet Activity U.S. Agencies, p. 540 `GENERAL` **TE** Activity Advertising Campaign, p. 541 `BASIC` **TE** Group Activity The Impact of an EIS, p. 541 `ADVANCED` **SE** Field Activity Local Policies, p. 542 **TE** Activity Local Politics, p. 542 `GENERAL` **TE** Group Activity How Does a City Council Work? p. 542 `ADVANCED` **CRF** Modeling Lab * ◆ `GENERAL` **CRF** Research Lab * `GENERAL`
BLOCKS 5, 6, 7 & 8 • 180 min pp. 544–547 **Section 3** The Importance the Individual	**OSP** Lesson Plan * **CRF** Active Reading * `BASIC` **TT** Bellringer *	**TE** Demonstration Product Packaging, p. 545 ◆ `GENERAL` **TE** Internet Activity Researching Influential Individuals, p. 545 `GENERAL` **SE** QuickLab Making a Decision, p. 546 **SE** Inquiry Lab Be an Environmental Scientist, pp. 552–553 ◆ **CRF** Datasheets for In-Text Labs * **CRF** Research Lab * ◆ `ADVANCED` **CRF** Long-Term Project * ◆ `GENERAL`

BLOCKS 9 & 10 • 90 min

Chapter Review and Assessment Resources

- **SE** Chapter Review pp. 549–551
- **SE** Standardized Test Prep pp. 676–677
- **CRF** Study Guide * ▪ `GENERAL`
- **CRF** Chapter Test * ▪ `GENERAL`
- **CRF** Chapter Test * `ADVANCED`
- **CRF** Concept Review * ▪ `GENERAL`
- **CRF** Critical Thinking * `ADVANCED`
- **OSP** Test Generator
- **CRF** Test Item Listing *

Online and Technology Resources

Holt Online Learning
Visit **go.hrw.com** for access to Holt Online Learning or enter the keyword **HE6 Home** for a variety of free online resources.

One-Stop Planner® CD-ROM
This CD-ROM package includes
- Lab Materials QuickList Software
- Holt Calendar Planner
- Customizable Lesson Plans
- Printable Worksheets
- ExamView® Test Generator
- Interactive Teacher Edition
- Holt PuzzlePro® Resources
- Holt PowerPoint® Resources

Holt Environmental Science Interactive Tutor CD-ROM
This CD-ROM consists of interactive activities that give students a fun way to extend their knowledge of environmental science concepts.

531A   **Chapter 21 • Chapter Planning Guide**

KEY

TE	Teacher Edition	CRF	Chapter Resource File	*	Also on One-Stop Planner
SE	Student Edition	TT	Teaching Transparency	■	Also Available in Spanish
OSP	One-Stop Planner	CD	Interactive CD-ROM	◆	Requires Advance Prep

ENRICHMENT AND SKILLS PRACTICE	SECTION REVIEW AND ASSESSMENT	CORRELATIONS
SE Pre-Reading Activity, p. 532 **TE** Using the Figure, p. 532 **GENERAL**		**National Science Education Standards**
TE Skill Builder Writing, p. 533 **ADVANCED** **TE** Student Opportunities International Exchange, p. 534 **TE** Reading Skill Builder Concept Mapping, p. 534 **BASIC** **SE** Case Study International Whaling, pp. 536–537 **TE** Using the Figure Economic Model, p. 536 **GENERAL** **TE** Inclusion Strategies, p. 537 **SE** MathPractice Nature Conservancy Assets, p. 538	**SE** Section Review, p. 538 **TE** Reteaching, p. 538 **BASIC** **TE** Quiz, p. 538 **GENERAL** **TE** Alternative Assessment, p. 538 **GENERAL** **CRF** Quiz * ■ **GENERAL**	SPSP 3a SPSP 4c SPSP 6b
TE Using the Table U.S. Agencies, p. 540 **BASIC** **TE** Inclusion Strategies, p. 541 **TE** Career Environmental Management and Policy, p. 542	**TE** Homework, p. 541 **GENERAL** **TE** Homework, p. 542 **ADVANCED** **SE** Section Review, p. 543 **TE** Reteaching, p. 543 **BASIC** **TE** Quiz, p. 543 **GENERAL** **TE** Alternative Assessment, p. 543 **GENERAL** **CRF** Quiz * ■ **GENERAL**	SPSP 3a SPSP 4c SPSP 6b SPSP 6c
TE Skill Builder Writing, p. 545 **BASIC** **SE** Graphic Organizer Chain-of-Events Chart, p. 547 **GENERAL** **SE** Making a Difference Student Club Saves Eagles and More, pp. 554–555 **CRF** Map Skills National Parks * **GENERAL**	**SE** Section Review, p. 547 **TE** Reteaching, p. 547 **BASIC** **TE** Quiz, p. 547 **GENERAL** **TE** Alternative Assessment, p. 547 **ADVANCED** **CRF** Quiz * ■ **GENERAL**	SPSP 4c SPSP 6c

 Guided Reading Audio CDs

These CDs are designed to help auditory learners and reluctant readers. (Audio CDs are also available in Spanish.)

 www.scilinks.org

Maintained by the **National Science Teachers Association**

TOPIC: Sustainable Development
SciLinks code: HE4108

TOPIC: United Nations
SciLinks code: HE4117

TOPIC: U.S. National Parks
SciLinks code: HE4116

TOPIC: Earth Day
SciLinks code: HE4022

 CNN Videos

Each video segment is accompanied by a Critical Thinking Worksheet.

Earth Science Connections

Segment 11 Solar Nomads

CHAPTER

21

Chapter Enrichment

This Chapter Enrichment provides relevant and interesting information to expand and enhance your classroom instruction of the chapter material.

Economics and International Cooperation

What Makes a Sustainable Community?

Former U.S. Senator and Earth Day founder Gaylord Nelson defines a sustainable community based on the following five characteristics:

- **Renewability:** A sustainable community does not use renewable resources faster than the resources can be replaced.

- **Substitution:** A sustainable community tries to use renewable resources instead of nonrenewable resources.

- **Interdependence:** A sustainable community does not import other resources in a way that will tax another community, nor does it export wastes to other communities.

- **Adaptability:** A sustainable community is flexible enough to adapt to changes and new opportunities.

- **Institutional Commitment:** A sustainable community supports sustainability through laws, systems, processes, and education.

Economics and the Environment

Environmental concerns have an increasing influence on economics. We can see this influence in changing ways of doing business, new public policies, and evolving consumer behavior. Slowly, product packaging and design, manufacturing, investment methods, tax policies, and many forms of technology have begun to reflect environmental awareness. New industries have developed to provide "eco-friendly" products and services. Companies have changed their ways of doing business to meet new regulations and the demands of environmentally conscious buyers. Many social planners and leaders now recognize that economic actions and the environment are deeply linked.

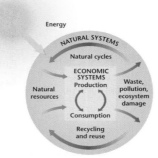

▶ **A model of economic systems operating within the environment**

BRAIN FOOD

Psychologist Abraham Maslow's *Hierarchy of Needs* provides a way of conceptualizing how humans prioritize needs. The hierarchy is usually illustrated as a pyramid. Physiological needs are at the base of the pyramid. The need for safety is next in line, love is third, and esteem and self-actualization are at the top. The pyramidal hierarchy illustrates the notion that the fulfillment of certain needs, such as having adequate food and shelter, take higher priority than the fulfillment of less critical needs or altruistic interests. In the past, concern for the environment may have been viewed as a low-priority need. However, when drinking water, farmlands, and air become degraded, environmental concerns register closer to the base of the pyramid.

A recent study on happiness by Oxford psychologist Michael Argyle led to this conclusion: "Above the poverty level, the relationship between income and happiness is remarkably small." Argyle found that the main factors that influence happiness are familial relationships, work, leisure time, and friendships.

Environmental Policies in the United States

Colorado River and Glen Canyon Dam

The Glen Canyon Dam has been a topic of controversy for many years due to its impact on the entire Colorado River. The Colorado is a major source of water for seven states—Wyoming, Utah, Colorado, New Mexico, California, Arizona, and Nevada—as well as for Mexico. In 1922, these seven states created the Colorado River Compact, an agreement to divide up and regulate the uses of the river's water. Glen Canyon Dam was built in the 1960s to store water by creating Lake Powell, the second-largest human-made reservoir in the United States. Water releases into the lower canyons are now controlled by the dam and are planned on a monthly basis to manage the sometimes-conflicting interests of recreation, power generation, flood control, and fish and wildlife habitat. Eighty-five percent of the water goes to agricultural production, and some is used in urban areas.

Since Glen Canyon Dam was built, the public and federal and state agencies have expressed increasing concern about how dam operations may be adversely affecting the downstream environment. The dam flooded a unique canyon and altered the entire Colorado River ecosystem. Before the dam was built, water flows and temperatures in the river fluctuated seasonally (mostly because of melting mountain snow). Also, the river was frequently filled with silt and sediment, and spring floods shaped the beaches, sandbars, and vegetation along the canyons. Now, sediment is trapped in Lake Powell, the dam prevents high river flows, many native fish have disappeared from the river, and five fish species are endangered. From 1989 to 1995, an Environmental Impact Statement (EIS) was prepared on the operation of Glen Canyon Dam. The final EIS proposed a process of "adaptive management" whereby the effects of dam operations on downstream resources would be monitored and assessed to guide future operations. As part of this process, a water flow of flood proportions was experimentally released from the dam in 1996 so that the effects could be studied.

▶ The Grand Canyon, downstream from Lake Powell and the Glen Canyon Dam

The United States Environmental Protection Agency

The role of the United States Environmental Protection Agency (EPA) is to coordinate efforts and enforce environmental laws. In 1970, before the EPA was established, the national approach to pollution control was unfocused. By creating the EPA, President Nixon enabled pollution issues across the country to be addressed as "a single, interrelated system." In a memo to Congress, Nixon stated that:

"A far more effective approach to pollution control would: identify pollutants; trace them through the entire ecological chain, observing and recording changes in form as they occur; determine the total exposure of man and his environment; examine interactions among forms of pollution; and identify where in the ecological chain interdiction would be most appropriate."

SECTION 3 The Importance of the Individual

Youth Voting

Voter turnout in American elections has plunged in recent years, especially among young people. Less than 50 percent of American adults voted in the 1996 presidential election, and only 36 percent voted in the November 1998 congressional elections. Only among senior citizens has voter turnout increased since 1972. Turnout dropped from 50 to 32 percent among citizens aged 18 to 24 and from 71 to 49 percent among citizens aged 25 to 44. Turnout by young African Americans increased until the late 1970s but has decreased since. Turnout by young Hispanics has declined slightly. However, young women have become more likely to vote than young men.

One survey indicated that "efficacy," the sense that one can make a difference, predicts young people's willingness to vote and to volunteer for civic causes. Other important factors are parental influence and the feeling among the youth that candidates are trying to reach them. Another survey showed an increase in young Americans' interest in public service careers since 1997.

▶ Youth voters, increasingly rare at polling places

Overview

Tell students that this chapter will focus on the many ways in which governments and people can affect environmental issues. Environmental decision making occurs at the level of the individual, the community, state or national goverment, or internationally.

Using the Figure — GENERAL

The Chimanimani Mountains lie along the border between Zimbabwe and Mozambique. Mozambique is among the poorest countries in the world because of decades of war and economic problems. Recently, organizations such as the Ford Foundation and the World Bank have begun working to support sustainable and equitable resource management in the region. The objectives are for local communities to establish democratic councils and to make decisions over the development and management of their local natural resources. In the Chimanimani area, there is an initiative to develop a community-managed Biosphere Reserve. Ask students to imagine that they have been assigned to assist with the Chimanimani Biosphere project, and to discuss how they might approach this assignment. **LS** Verbal/Visual

PRE-READING ACTIVITY

Encourage students to use the FoldNotes as a study guide to quiz themselves for a test on the chapter material. Students may want to create Four-Corner FoldNotes for differenet topics within the chapter.

VIDEO SELECT

For information about videos related to this chapter, go to **go.hrw.com** and type in the keyword **HE4 EPFV**.

Economics, Policy, and the Future
CHAPTER **21**

1 **Economics and International Cooperation**

2 **Environmental Policies in the United States**

3 **The Importance of the Individual**

PRE-READING ACTIVITY

FOLDNOTES **Four-Corner Fold**
Before you read this chapter, create the **FoldNote** entitled "Four-Corner Fold" described in the Reading and Study Skills section of the Appendix. Label each flap of the four-corner fold with a topic. Write what you know about each topic under the appropriate flap. As you read the chapter, add other information that you learn.

In the Chimanimani mountains of Zimbabwe, international groups are working with local residents to manage the natural resources of the area while improving the economic and political status of the residents.

532

Chapter Correlations *National Science Education Standards*

SPSP 3a Human populations use resources in the environment in order to maintain and improve their existence. Natural resources have been and will continue to be used to maintain human populations. **(Section 1 and Section 2)**

SPSP 4c Many factors influence environmental quality. Factors that students might investigate include population growth, resource use, population distribution, overconsumption, the capacity of technology to solve problems, poverty, the role of economic, political, and religious views, and different ways humans view the earth. **(Section 1, Section 2, and Section 3)**

SPSP 6b Understanding basic concepts and principles of science and technology should precede active debate about the economics, policies, politics, and ethics of various science and technology related challenges. However, understanding science alone will not resolve local, national, or global challenges. **(Section 1 and Section 2)**

SPSP 6c Progress in science and technology can be affected by social

issues and challenges. Funding priorities for specific health problems serve as examples of ways that social issues influence science and technology. **(Section 2 and Section 3)**

Economics and International Cooperation

More than six billion people are living on Earth, supported by unprecedented levels of human resource use, productivity, and scientific knowledge. On average, people live longer and have more education than they did 100 years ago. We continue to advance our understanding of human biology, social behavior, and our environment. But we still face many unknowns. Scientists do not agree on how humans are affecting the planet's ecosystems. People worldwide are worried about running short of resources such as fertile soil and fresh water. And many people disagree about how environmental problems should be addressed.

An important question is whether the present human condition is sustainable. **Sustainability** is the condition in which human needs are met in such a way that a human population can survive indefinitely. To plan for a sustainable society, one must understand economics and politics as well as environmental science.

International Development and Cooperation

We live in a time of *globalization*, when environmental and social conditions are linked across political borders worldwide. People cross borders in search of economic opportunities and a better quality of life. Increasingly, governments, organizations, and businesses around the world have a need to work together.

However, governments do not always agree on how to solve environmental problems. Within and between governments, people debate about who is responsible for environmental problems. People also debate about whether current levels of population growth and resource use are sustainable. Despite these different views, international leaders often meet to identify common goals and address problems, as shown in **Figure 1**.

Objectives

▶ **Describe** some of the challenges to achieving sustainability.

▶ **Describe** several major international meetings and agreements relating to the environment.

▶ **Explain** how economics and environmental science are related.

▶ **Compare** two ways that governments influence economics.

▶ **Give** an example of a private effort to address environmental problems.

Key Terms

sustainability
economics

Figure 1 ▶ At the 2000 Millennium Summit in New York, world leaders agreed on principles to guide the United Nations in the 21st century. Sustainable development is a shared goal among most nations.

533

Chapter Resource File

• **Lesson Plans**
• **Active Reading** BASIC
• **Section Quiz** GENERAL

Transparencies

TT Bellringer
TT Major International Environmental Agreements

SKILL BUILDER ——————— ADVANCED

Writing Ask students to write a proposal for a new scientific body that would conduct research on and suggest environmental policy options for all the nations of the world. Students should describe the organization's responsibilities, the way its members would be selected, and the skills and information needed by members of the organization to do the job. Ask students to detail the procedures the organization would use to make decisions. Have students put the proposal in their **Portfolio.** LS Intrapersonal

---

SECTION 1

Focus

Overview

Before beginning this section, review with your students the Objectives in the Student Edition. This section describes how nations can work together to address global environmental issues, and explains the relationship between economics and the environment.

Bellringer

On the board, write a list of environmental issues that cross international boundries. You might use chapter or section titles from this book to help form the list. Ask students to consider which are the most important issues to them and to choose two top priorities. Tell students to write in their *EcoLog* a brief description of why enacting international agreements on these issues might be difficult. LS Logical

Motivate

Discussion ——————— GENERAL

Crossing Borders Many environmental problems are international in scope and require international cooperation to be addressed. Air pollution, emission of greenhouse gases, loss of topsoil, disposal of nuclear waste, ocean pollution, and depletion of forests are just a few problems that cross international boundaries. A dramatic example is the Chernobyl disaster, which threw radioactive particles into the air and contaminated food crops and milk supplies in several northern European nations. Have students call out other examples to help you create a list on the board. Ask students to consider how the actions of one nation can affect the environmental health of many others. LS Visual

Table 1 ▼

Teach

Table 1 ▼

International Organizations, Meetings, and Agreements	
Related to Sustainable Development	
The World Conservation Union (IUCN), established 1948	a worldwide partnership of States, government agencies, private and nonprofit organizations, and scientists and experts from 140 countries; encourages and assists in conservation as well as equitable and sustainable use of natural resources
UN Conference on Human Environment, Stockholm, 1972	first international meeting to consider global environment and development needs; led to the formation of the UN Environment Programme (UNEP)
UN Conference on Environment and Development (UNCED or Earth Summit), Rio de Janeiro, 1992	meeting that produced Agenda 21 and the Rio Declaration (Earth Charter), which outlined key policies for sustainable development; established the UN Commission on Sustainable Development (UNCSD)
World Summit on Sustainable Development, Johannesburg, 2002	meeting to review 10-year progress of Agenda 21 and to consider several major treaties
Related to Climate and Atmosphere	
Intergovernmental Panel on Climate Change (IPCC), established 1988	group of scientists from around the world that studies the scientific, social, and economic aspects of human-induced climate change
Framework Convention on Climate Change, Rio de Janeiro, 1992	agreement that established international recognition of the problems of climate change; proposed strategies to limit greenhouse gases
Montreal Protocol on Substances That Deplete the Ozone Layer, 1987	agreement by many countries to eliminate substances, such as CFCs, that damage the atmosphere's protective ozone layer
Kyoto Protocol on Climate Change, 1997	agreement to reduce worldwide emissions of greenhouse gases; requires larger reductions by developed countries; allows trading of permitted levels of emissions; promotes pollution-free development

Debate — GENERAL

What Is Sustainable? For the sake of class debate, ask students to suggest working definitions of *sustainable* and *development*. After the class has established definitions, ask students to consider whether sustainable development is a possibility and whether it is worth trying to achieve. Also, ask them to consider how countries might aspire toward a path of sustainable development. Then, randomly assign teams to debate if and how sustainable development might be achieved. **LS Verbal/Interpersonal**

HISTORY CONNECTION — ADVANCED

Sustainable Development
The landmark 1987 UN World Commission on Environment and Development (WCED) defined sustainable development as "development which meets the needs of the present without compromising the ability of future generations to meet their own needs." The conference brought the concept of sustainable development from an "antigrowth" stance to a key principle of international political, economic, and social reform and cooperation. The WCED showed that the ability to sustain economic activity into the future is tied to wider social and environmental concerns.

Connection to Law

Small Islands, Global Issues
The United Nations Global Conference on the Sustainable Development of Small Island Developing States met in Barbados in 1994. The conference produced a declaration that included the following statement: "While small island developing States are among those that contribute least to global climate change and sealevel rise, they are among those that would suffer most from the adverse effects of such phenomena and could in some cases become uninhabitable."

534

Sustainable Development Many meetings and agreements among international governments have dealt with environmental concerns along with economic and political concerns. Some important examples are listed in **Table 1.** The Earth Summit of 1992 in Rio de Janeiro, Brazil, was a sign of new levels of international environmental awareness and cooperation. Representatives from around the world drew up several agreements. One of these was Agenda 21, a general plan to address a range of environmental problems while allowing continued economic development.

Climate and Atmosphere International organizations and agreements related to climate and the atmosphere are also listed in **Table 1.** One treaty, the Montreal Protocol, successfully reduced the amount of ozone-destroying chemicals in the atmosphere. However, not all agreements are successful. Any country may choose not to sign, enforce, or provide funding to implement an agreement.

For example, the Kyoto Protocol attempts to avoid or slow down global warming by reducing greenhouse-gas emissions around the world. Most of the developed countries have promised to reduce their emissions by about 5 percent by 2012. However,

Student Opportunities

International Exchange Organize a research project or a means of exchanging ideas and information with students from another part of the country or from another nation. The Global Rivers Environmental Education Network (GREEN) and the Center For Improved Engineering and Science Education (CIESE) are two organizations that connect classrooms around the world to cooperatively analyze water quality. The EarthWatch Institute, KidLink, Global Nomads Group, ePALS, and ThinkQuest also organize international youth collaborations.

READING SKILL BUILDER — BASIC

Concept Mapping Have groups of students identify broad environmental issues addressed by the agreements, meetings, and organizations in **Tables 1** and **2.** For example, students might list ocean pollution, endangered species, and forest conservation. Then, ask each group to create a concept map that links the various treaties and issues. Display the maps in the room, and have students add details as they learn more about international agreements related to these issues.
LS Visual/Verbal

Co-op Learning   English Language Learners

an argument against the Kyoto Protocol is that it would be costly to implement, even though it does not guarantee a stable climate. Another argument is that the treaty allows developing countries to continue to increase their use of fossil fuels, while it requires reductions in use of fossil fuels by the developed countries. Mainly for these reasons, the United States did not sign the treaty. However, U.S. corporations doing business in other parts of the world may still be subject to the treaty's requirements.

Other Agreements Hundreds of other international agreements have been made as new environmental issues have emerged. Sometimes, the results make news. For example, you may hear on the news that a cruise line was barred from a port or fined millions of dollars for dumping garbage at sea. The ship would be fined because its actions violate an agreement commonly known as MARPOL. (MARPOL refers to *marine pollution*.) Under MARPOL, large ships cannot dump garbage close to shore. MARPOL also regulates the practice of oil tankers washing out their tanks. As a result, beaches around the world are less polluted with tar despite the increasing volume of oil carried by tankers.

Table 2 ▼

Other International Organizations, Meetings, and Agreements Related to the Environment	
Antarctic Treaty and Convention, 1959	agreement to use Antarctica solely for peaceful purposes "in the interest of all mankind" and to cooperate in scientific research there
International Convention for the Prevention of Pollution from Ships (MARPOL), 1973; modified 1978	agreement that regulates disposal of wastes by ships on the ocean: specifies where and how different types of garbage, oil, sewage, and toxic wastes may be dumped
Convention on International Trade in Endangered Species (CITES), 1973	agreement that classifies endangered and threatened species worldwide and monitors international trade of these species; widely adopted and successful for many listed species
Convention on Migratory Species (CMS), 1979	agreement that protects wild animal species that migrate across international borders
Law of the Sea, 1982	agreement that addresses ocean pollution from land runoff, ocean dumping, hazardous materials, oil exploration, mining, and air pollution; designates deep-sea resources as "the common heritage of mankind"
Basel Convention, 1989	agreement that regulates transportation and disposal of hazardous wastes
Convention on Biological Diversity (CBD), 1992	agreement to inventory and protect endangered and threatened species; nations compensate each other for use of organisms in products
Intergovernmental Forum on Chemical Safety (IFCS), 1994	panel that facilitates cooperation among governments for environmentally sound management of chemicals
Cartagena Protocol on Biosafety, 2000	agreement that addresses transportation and use of genetically modified organisms
UN Forum on Forests (UNFF), 2001	panel that promotes the management, conservation, and sustainable development of all types of forests

535

"Act, act, act. You can't just watch."

—Angeles Serrano (grandmother and activist in Manila, Philippines)

Ask students: "How would you describe the tone of the statement?" (Urgent) "How does Serrano's statement apply to the environment?" (Sample answer: It conveys the message that we must solve environmental problems proactively.) Ask, "In the context of the quote, what does the word *act* mean to you?" (Answers may vary.) **LS** Verbal

---

HISTORY ──
● CONNECTION ── ADVANCED

Extreme Activism Have students research the international activities of a "direct-action" environmental group such as Greenpeace or the Sea Shepherd society. Have them create a "profile" or historical report of the group. Tell them to include the group's origins, the kinds of issues it has been involved with, the types of activism it has engaged in, and the reactions of the media and public to these actions. Ask students to conclude with a statement of their own opinion of the group's effectiveness.
LS Verbal/Intrapersonal

▲ **Internet Activity** ── ADVANCED

UN Documentation Most of the agreements, meetings, and organizations in **Tables 1** and **2** are well documented on the Web sites of the United Nations and its affiliate organizations. Have pairs of students research an item from the tables or a similar item of interest. Ask students to look for information about the history and progress of the item, which countries (often called "parties") have or have not cooperated, and how decisions have been made. Students may be somewhat overwhelmed by the complexity of the processes and by the volume of documentation produced in these international forums. Point out that this is indicative of the complexity of international cooperation. However, direct students to look for the summary and overview reports that are usually created. **LS** Verbal/Logical

GEOGRAPHY ──
● CONNECTION ── GENERAL

Climate Change and Himalayan Glaciers Approximately 2,000 glaciers have disappeared from the eastern Himalayas in the last 100 years. As these glaciers melt, the water that formed them runs into river valleys, like the Pearl River valley in China or the Ganges River valley in India. Over the past three decades, the frequency of floods from melting glaciers has risen, and tens of thousands of residents are at risk from the floods. According to a 2002 UNEP report, at least 44 glacial lakes in Bhutan and Nepal could burst their banks in five years. About 12 glacier incidents have been recorded in Tibet since 1935. The latest one took place in 1981 and destroyed three concrete bridges and crippled a long section of the Nepal-China Highway. Other glacial areas in the Himalayas and across the world are in a similar critical state. **LS** Logical

Economic Model Ask students to copy **Figure 2** into their *EcoLog* and to add notes during the following discussion. Ask students to explain, what is happening in each part of the diagram. Have them indicate where "common" resources and "shared" costs occur and where governments may try to influence the costs and benefits or inputs and outputs of an economic system. Point out that the resources drawn in, and the problems and wastes generated would still be part of the "shared" natural systems. Tie the diagram in to the topic of whaling discussed in the Case Study.
LS Visual

ECONOMICS

● CONNECTION — GENERAL

Free-Market Pollution Control
When governments are trying to reduce pollution, one alternative is a free-market approach. For example, a government could give each company a license to release a certain amount of pollutants. Companies could then buy, sell, and trade these permits on the open market. Thus, it might be more profitable for some companies to pollute as little as possible. Negotiations related to the Kyoto Protocol have proposed such a system, in which certain levels of greenhouse gas emissions could be traded between companies and between countries. A system like this is used by the U.S. Acid Rain Program to reduce SO_2 emissions.

Transparencies

TT Economic Systems Within the Biosphere

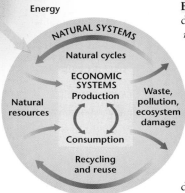

Energy

NATURAL SYSTEMS

Natural cycles

ECONOMIC SYSTEMS
Production

Natural resources

Waste, pollution, ecosystem damage

Consumption

Recycling and reuse

Figure 2 ▶ A complete economic model shows that economic systems operate within natural systems.

Economics and the Environment

Economics is the study of the choices people make as they use and distribute limited resources. In the traditional model of economics, *markets* are seen as self-contained economic systems, in which money and products flow in cycles. People within a market will decide the *value* of something by comparing the costs and benefits from their own perspective. For example, people decide how much they will pay for a product or how much they must be paid to do a certain job. These values change over time as people see changes in the costs or benefits of their actions.

Economists say that an economic system is successful when there is *economic growth*, an increase in the flow of money and products within a market. However, economic systems may not account for external factors that do not have a direct economic value, such as air or wildlife. As the fields of economics and various sciences share knowledge, economists develop more complex and realistic models of resource use. The example in **Figure 2** shows that economic systems are contained within and dependent upon the environment. Economies draw resources from the environment and may return waste or cause damage.

International Whaling: Conflict and Cooperation

Because no country controls the open ocean, the ocean has been treated as if the resources it contains are free for anyone to take. However, people around the world have noticed the disappearance of species and the pollution of their shores. The history of agreements between countries to regulate whaling illustrates both the problems and successes of international cooperation.

Whales were once hunted for their fat, which was used for lamp oil, and they are still hunted for meat. By the 20th century, most large whale species were endangered. So, countries have had to negotiate with each other to hunt for whales and to save whales from extinction.

In 1949, the International Whaling Commission (IWC) voted to limit commercial whaling to a nation's territorial waters. France objected and used a special provision to opt out of IWC rules. France was the first of many countries to use this loophole, which weakens the IWC's ability to create regulations.

The 1949 agreement also established quotas to limit the number of whales a nation could harvest. However, because the quotas specified the number of whales but not the type of whales, the quotas did not prevent the killing of endangered whale species. As a result, blue whales, fin whales, humpback whales, and sei whales were hunted nearly to extinction. Then in 1960,

the IWC suspended quotas entirely. What followed was the largest whale catch in the history of the IWC.

Whales are intelligent mammals, and many people have an emotional desire to save whales from extinction. Because of public pressure to save the whales, the IWC reestablished a quota in 1967. And in 1972, the IWC allowed observers from member nations to monitor the whale harvest of other member nations. In 1977, the IWC created more restrictions on whaling, and passed a resolution urging nations to stop importing whale products. Finally, the IWC called for a total ban on whaling that was to begin in 1984. However, three countries with large whaling industries opted out

536

Notable Quotes — ADVANCED

"The environmental crisis is an outward manifestation of a crisis of mind and spirit. There could be no greater misconception of its meaning than to believe it is concerned only with endangered wildlife, human-made ugliness, and pollution. These are part of it, but more importantly, the crisis is concerned with

the kind of creatures we are and what we must become in order to survive."

—Lynton K. Caldwell

Caldwell is Professor Emeritus of Political Science at the Indiana University School of Public and Environmental Affairs. Have interested students read excerpts from one of Caldwell's many books, such as *Environment as a Focus for Public Policy*. **LS** Verbal

Economists see environmental problems as *market failures.* The market has failed if the price of something does not reflect its true cost. For example, the price of gasoline does not reflect the other expenses caused by auto emissions. Illnesses caused by air pollution cost society billions of dollars a year. In a balanced economic system, the price of gasoline should reflect these costs. One difficulty in pricing is that sometimes we do not know environmental costs. An economic system can include only those costs that are understood at the time people make decisions.

Regulation and Economic Incentives Governments often try to influence economic systems. Governments may do this by creating regulations or punishing people with fines and jail sentences. Governments may also create *economic incentives* by paying out money for actions that benefit society or charging taxes on actions that have a social cost. For example, some governments offer rebates to people who purchase energy-saving appliances.

Governments have tried many ways to regulate environmental damage such as pollution. However, regulations are criticized when they are difficult to enforce, do not distribute costs evenly, or do not control environmental damage. Governments and economists continue to work on ways to link economic decisions with environmental effects.

Eco fact

Environmental Ratings Each year, the World Economic Forum ranks countries on an Environmental Sustainability Index. In 2001, the top five countries were Finland, Norway, Sweden, Canada, and Switzerland. The United States ranked 51st of 142 countries studied. The study also concluded that no country is on a truly sustainable path.

► This Icelandic whaling ship (left) is harvesting fin whales. Blue whales (right) are among the many endangered whale species.

CRITICAL THINKING

1. Expressing Opinions Write a paragraph describing your views about the issue of whaling.
WRITING SKILLS

2. Predicting Outcomes Demand for whale meat in Japan has been decreasing in recent years. Why might this change be happening, and what might be the results of this change?

of this agreement—Norway, Japan, and the former Soviet Union.

International debate over these issues has continued. Populations of a few whale species have recovered since whaling was restricted. Other species, such as the right whale, breed so slowly and have such small populations that extinction is likely.

In the 1990s, Norway and Japan continued to claim exceptions to the IWC's rules. Both countries harvested hundreds of minke whales each year, claiming that the minke population was large enough to survive limited hunting. Norway and Japan have also hunted in the IWC's designated whale sanctuary in the Antarctic Ocean. Japan has claimed that the IWC rules allow the country to harvest whales for research, although the whale meat is then sold as food in Japan.

537

Close

Reteaching ——— BASIC

Timeline Have students work together to create a large timeline of international meetings and agreements. Hang butcher paper on the wall, and supply markers. Have students consult **Tables 1** and **2** or other reference materials and illustrate the timeline.

Co-op Learning   **English Language Learners**

Quiz ——— GENERAL

1. How does the Nature Conservancy preserve ecosystems? (It creates large preserves using donated land and money, and allows local organizations to manage the preserves.)

2. Name two countries that have gotten around or opted out of whaling regulations. (Answers may include France, the former Soviet Union, Norway, and Japan.)

Alternative Assessment ——— GENERAL

Summit Report Have students create a news-style report or commentary on a major international summit or conference. Students may wish to transcribe or videotape their presentation for inclusion in their **Portfolio.**

LS Intrapersonal

MATH PRACTICE

Nature Conservancy Assets In 2000, the Nature Conservancy owned land worth a total of $1.3 billion. In 2001, it gained ownership of additional land worth $322 million. In the same year, it also sold land worth $88 million and gave away land worth $12 million to governments and other groups. What was the value of land held by the Nature Conservancy at the beginning of 2002?

Figure 3 ▶ The area around Mount Kilimanjaro in Kenya is an important home to wildlife such as elephants and giraffes. Several governments and organizations are working with local residents to manage the area for both wildlife preservation and sustainable economic development.

Private Efforts Businesses and private organizations also play a role in addressing environmental problems. Businesses may donate land for parks or preserves or donate money to environmental causes. Many businesses have found that recycling their wastes can save costs and improve their public image.

Private organizations often cooperate with each other and with governments. Such cooperation may include conducting research or creating plans for environmental management. **Figure 3** shows an area of Africa that several governments and private organizations are working together to manage. Local residents are also included in the process of planning for the area.

The Nature Conservancy is a nonprofit organization that uses a simple economic strategy to preserve ecosystems. This organization collects donations of money and land. If the donated land is not targeted for preservation, the organization trades or sells the land. Large preserves are put together by a combination of donations, exchanges, and purchases of land. The organization has created preserves in all 50 states and in 28 other countries.

SECTION 1 Review

1. **Describe** some of the challenges to achieving sustainability.

2. **Describe** three major international meetings or agreements relating to the environment.

3. **Compare** two ways that governments influence economics.

4. **Give an example** of a private effort to address environmental problems.

CRITICAL THINKING

5. **Analyzing Processes** Write a paragraph that explains why a local government might use tax money to purchase park lands. **WRITING SKILLS**

6. **Applying Ideas** Read about interactions of economics and the environment. List some ways that both governments and organizations could encourage people to conserve resources. **READING SKILLS**

538

Answers to Section Review

1. Answers may vary. Challenges include understanding environmental science, economics, and politics and balancing these things in decision making.

2. Answers may vary. Students may include descriptions of agreements discussed in the section.

3. Answers may vary. Two different approaches to regulation include motivating people with economic incentives or fining people who break rules.

4. Answers may vary. Examples include the strategies of the Nature Conservancy's and the cooperative efforts near Mount Kilimanjaro or in the Chimanimani mountains.

5. Answers may vary. Possible economic reasons include providing opportunities for recreation and protecting public water supplies from contamination. A public vote may have supported this decision, thus showing that a majority of citizens value public parklands.

6. Answers may vary. Governments can encourage conservation by providing economic incentives or by enforcing regulations, such as by charging fines. Organizations can cooperate with each other and with governments to educate citizens and to establish conservation programs.

Environmental Policies in the United States

Many people in the United States have demonstrated a concern about environmental problems. In both local and national elections in the United States, candidates often talk about environmental issues in their campaigns. Each year, millions of dollars are donated to environmental causes by U.S. citizens and businesses, and billions of federal tax dollars are spent to uphold environmental policies and to manage resources. In recent decades, the United States has reduced many types of pollution and improved water quality in many places. But the United States is still struggling to use its resources in a sustainable way and to preserve its unique ecosystems.

History of U.S. Environmental Policy

During the 1800s, people in the United States made use of the country's vast resources. Prairies were turned into cropland, ancient forests were cut down, and several species of animals were hunted to extinction. By the 1900s, citizens began to realize the consequences of these actions, and the citizens' attitudes started to change. Leaders such as President Theodore Roosevelt and conservationist John Muir, shown in **Figure 4,** called for increased protection and management of the nation's resources. Many national forests and parks, and agencies to manage them, were established around the early 1900s.

Objectives

▶ **Describe** two major developments in U.S. environmental history.
▶ **Give** examples of three federal agencies that have environmental responsibilities.
▶ **Explain** the purpose of Environmental Impact Statements.
▶ **Give** an example of how citizens can affect environmental policy at each level of government—local, state, and national.
▶ **Evaluate** the media as a source of information about the environment.

Key Terms

Environmental Impact Statement
lobbying

internet connect

www.scilinks.org
Topic: U.S. National Parks
SciLinks code: HE4116

SCiLINKS® Maintained by the National Science Teachers Association

Figure 4 ▶ In the late 1800s and early 1900s, President Theodore Roosevelt (on left) and naturalist John Muir (on right) were leaders in the conservation of natural areas. They are shown here at Yosemite National Park, one of the first national parks.

539

Chapter Resource File

• Lesson Plans
• Active Reading **BASIC**
• Section Quiz **GENERAL**

Transparencies

TT Bellringer

MISCONCEPTION
**///ALERT **

Legislation Is Not Permanent Public opinions about environmental issues change as new scientific information becomes available or as people perceive new threats. Sometimes, weaknesses and loopholes in a law become apparent and the laws are made stricter; at other times, arguments are made against laws that are very strict or costly. Laws may also be challenged and overturned in local, state, or national courts. Some laws are enacted with a requirement for periodic review and revision before they can be renewed.

SECTION 2

Focus

Overview

Before beginning this section, review with your students the Objectives in the Student Edition. In this section, students are introduced to environmental laws and policies of the United States. They also learn how people can influence policy at the national, state, and local levels.

Bellringer

Ask students to imagine that they live within 25 miles of a national park. Biologists have recommended expanding the boundaries of the park to preserve certain wildlife species. However, local industries that currently use these lands would have to move or stop operations if the boundaries were expanded. Ask students, "Who should decide the fate of the land: federal, state, or local government? Why?" (Answers may vary. Ideally, all levels should try to work together.) Ask them to write their answers in their *EcoLog.*
LS Logical

Motivate

Activity ———— **GENERAL**

Whose Issue? Ask students to think about the environmental issues discussed in this course. Ask students, "What is an environmental issue that is affecting your everyday life right now? Which level of government—local, state, or federal—is likely to be the most involved in making decisions about that issue?" Then, have students write a simple, short letter addressed to a government official at that level. Have students include a brief summary of the issue, a statement explaining their position on the issue, and a request for action from the official. Encourage students to include examples of ways the issue currently affects their lives. Suggest that students include their letter in their **Portfolio. LS Intrapersonal**

Using the Table — BASIC

U.S. Agencies Ask students to bring in old issues of periodicals and newspapers from home. Have student groups look through the reading materials, noting articles that relate to an environmental problem. Then, have students choose two articles and use **Table 3** to determine the federal agency or law that might address the reported issue or problem. Encourage group discussion by asking questions such as the following: "Is the issue or problem due to someone breaking a law? Is the law effective? Are people supporting or criticizing the law?" (Answers may vary.) **LS** Verbal

HISTORY —
● CONNECTION — GENERAL

Policy Changes Point out to students that environmental policies in the United States have not always grown stronger and stricter. Economic and political shifts sometimes result in the undoing of a previous protection or effort. For example, during the early part of the 2000s, a landmark decision was made to halt the recycling of glass and plastic in New York City. During the late 1990s, several environmental protection laws, such as the Endangered Species Act, were substantially altered. Have students research such a decision and then discuss the arguments for and against the decision. **LS** Logical

Transparencies

TT U.S. Agencies and Environmental Laws

Connection to ▶ Law

Inherited Laws In parts of the United States that were previously under the control of European countries, some of the old laws regarding property and land use are still in effect. In Texas and California, many provisions of Spanish land law still apply to the states' water sources. Most rivers and creeks in these states are public property. Also, Texas has ownership of coastal areas stretching 10.4 mi from its shores. This gives Texas the ownership of many offshore oil deposits. Other coastal states own only 3 mi, as established by English common law in those states.

Environmental Agencies and Laws Throughout the 1900s, U.S. citizens became more aware of environmental problems. Widespread crop disasters in the 1930s showed the country that poor farming practices were causing soil erosion and poverty. Policies to encourage soil conservation were adopted. People objected to living near smelly garbage dumps, so research on better methods of waste disposal began. The public began to complain about pollution. The first Earth Day, celebrated around the world in 1970, was a sign of widespread environmental awareness. In the same year, the U.S. Environmental Protection Agency (EPA) was created.

U.S. lawmakers have created many policies and federal agencies to manage environmental affairs, as shown in **Table 3**. For example, the EPA enforces the Clean Air Act and the Clean Water Act. These acts set standards for acceptable levels of pollutants in air and water. The EPA uses regulations and economic incentives to encourage individuals and businesses to meet these standards. Many of these laws continue to cause debate among citizens. Some citizens debate whether economic and personal freedoms are equal to long-term social interests. Some debate whether policies are based on science or on political trends. Meanwhile, many citizens are equally concerned about international politics, migration, and trade.

Table 3 ▼

U.S. Federal Agencies and Their Environmental Responsibilities	
Department or Agency	**Responsibilities**
Environmental Protection Agency	enforces National Environmental Policy Act; Clean Water Act; Clean Air Act; Solid Waste Disposal Act; Superfund; Federal Insecticide, Fungicide, and Rodenticide Control Act; Waste Reduction Act; Toxic Substances Control Act
Department of the Interior	enforces Wild and Scenic Rivers Act (managed across several agencies)
U.S. Fish and Wildlife Service	enforces Endangered Species Act, National Wildlife Refuge System Act, Alaska National Interest Lands Conservation Act, Species Conservation Act, Fish and Wildlife Improvement Act, Fish and Wildlife Conservation Act
Bureau of Land Management	enforces Federal Land Policy and Management Act, Taylor Grazing Act
National Parks Service	manages national parks
Office of Surface Mining Reclamation and Enforcement	enforces Surface Mining Control and Reclamation Act
Department of Agriculture	enforces Soil and Water Conservation Act, National Forests Management Act
Department of Commerce	
National Oceanic and Atmospheric Administration	monitors international atmosphere, climate, and oceans
National Marine Fisheries Service	enforces Marine Mammal Protection Act
Nuclear Regulatory Commission	regulates nuclear power stations and nuclear waste
Department of Energy	enforces National Energy Act, Public Utility Regulatory Policies Act

540

⚡ Internet Activity — GENERAL

U.S. Agencies Have students find the Web site for one of the agencies listed in **Table 3.** You may also suggest other agencies that are not listed but that have some role in U.S. environmental affairs. Examples include the USDA's Animal, Plant, and Health Inspection Service (APHIS) and the Department of Justice (which makes rulings on lawsuits involving environmental law). Have students find out when the agency was established; what the agency's purpose and primary responsibilities are; who the current director is, and what the most important current activities or issues involving this agency are. **LS** Verbal

Environmental Impact Statements Most government agencies are required to file an **Environmental Impact Statement** (EIS) for any proposed project or policy that would have a significant effect on the environment. Proposals for the construction of dams, highways, airports, and other projects that the federal government controls or funds must be evaluated with an EIS.

An EIS states the need for a project, the project's impact on the environment, and how any negative impact can be minimized. The public can comment on an EIS. For example, if a new dam is proposed, scientists and citizens may comment on any problems they foresee. Although public comment on an EIS rarely stops a project, the feedback may cause changes in the project's plans.

Federal agencies may also conduct an EIS when they plan changes in the regulation of public resources. Usually, several alternative actions are evaluated. For example, an EIS was conducted in the 1980s to evaluate alternative ways to release water from Glen Canyon Dam. Federal agencies were looking for ways to restore natural conditions downstream in the Grand Canyon, shown in **Figure 5**.

Unfunded Mandates and Economic Impacts Some limits have been placed on federal government's power to pass environmental laws. In 1995, Congress passed a law to prevent *unfunded mandates*, which are federal regulations that do not provide funds for state or local governments to implement the regulations. The federal government must now provide funding for any new laws that will cost more than 50 million dollars to implement. Congress can no longer pass laws such as the Clean Water Act, which requires local communities to conduct their own tests of public water supplies. Another limit being placed on many federal agencies requires the agencies to evaluate both the economic and environmental impacts of their policies.

Figure 5 ▶ The Grand Canyon ecosystem was changed when the Glen Canyon Dam was built upstream in 1962. An Environmental Impact Statement in the 1980s evaluated alternative ways to operate the dam.

U.S. Public Lands Twenty-eight percent of the area of the United States is publicly owned. This means that local, state, or federal governments hold the land in the public interest. Most of this public land is federally controlled and is in the western states. 80 percent of Nevada is publicly owned, and more than 60 percent of Alaska, Utah, and Idaho are publicly owned.

541

Activity ——— BASIC
Advertising Campaign Have students create a bumper sticker, an advertising slogan, or a brochure that relates to an environmental problem of interest to them. Their projects can be in English, their native language, or both. Suggest that students include their project in their **Portfolio.** English Language Learners
LS Visual

Group Activity ——— ADVANCED
The Impact of an EIS Have teams of students research an EIS that was conducted in their state or local community. They should be able to obtain a copy of the report from the agencies involved. Have students create a poster, video, or news article reporting on the process and outcome of the EIS. Suggest that students include their report in their **Portfolio.**
LS Interpersonal Co-op Learning

INCLUSION Strategies

• *Learning Disabled*
• *Developmentally Delayed*
• *Attention Deficit Disorder*

Using a word processor, pencil and paper, or a poster board, have students make a list of the National Parks of the United States. The list should include the name of each park, when it was established, and the state where the park is found. Students can then interview other students and teachers to find those who have visited a National Park on the list. Additionally, students may write to the National Park Service and ask for information about the parks or visit the web sites to get more information.

Homework ——— GENERAL
Publicly-Owned Land Have each student choose a state and determine the amount of land that is publicly owned in that state by contacting government agencies or using the library or Internet. Ask students to determine which level of government owns the land, how the land is regulated or managed, what private and public uses are allowed, and what role the land plays in local economies. In class, have students share their information.
LS Intrapersonal

REAL-LIFE ——
● CONNECTION — GENERAL

Citizen Participation Invite a city council representative or community member to speak to your class about citizen participation in local government. Have the speaker explain to students what they can do as individuals and as a group to affect environmental policies in their community. Encourage students to prepare a list of questions in advance.

Activity ——————— GENERAL

Local Politics Study the agenda for local town or city council meetings, and ask students to attend a meeting or watch or listen to a broadcast, if available. Often, decisions that involve environmental issues will be discussed, though perhaps not explicitly. Have students write a short report on the meeting, including what was discussed, what was decided, and what role environmental concerns played in the decision-making process, if any. They should also identify issues that perhaps ought to have considered environmental impacts but did not. Students might want to exercise their right to make public comments in these meetings. Find out if your whole class may be able to attend a council meeting as a field trip. **LS** Interpersonal/Verbal

Group Activity ——— ADVANCED

How Does a City Council Work?
Ask students who have experience in debate, Honor Society, or Model UN to role-play a discussion about a local environmental issue at a city council meeting. Encourage the "councilors" to discuss the issue, and also encourage the other students to formulate and ask questions as "citizens." **LS** Verbal/Interpersonal

FIELD ACTIVITY

Local Policies Use newspapers, TV, or the Internet to find out about a local environmental issue or policy that is currently being debated. Write a short newspaper article about this issue. Make sure to describe the problem, give some factual information, and state both sides of the debate or argument. Complete the article with your own opinion about what should be done. Write the article in your *EcoLog*. Make sure to consider what individuals should do as well as what the government, businesses, and the legal system should do. You might want to send the article to your school newspaper or to a local newspaper for publication.

Figure 6 ▶ Many environmental decisions are made at the local level. Citizens can participate in local government at public meetings (below). Some communities set aside local wildlife habitat and green spaces, such as the Barton Creek Greenbelt in Austin, Texas (right).

Influencing Environmental Policy

You can have more influence on environmental policy than you may realize. For example, as a citizen, you can contact your elected representatives to tell them your opinion on issues. There are also many other ways that consumers, businesses, the media, and organizations can influence policy at all levels of government.

Many laws related to the environment are created at the national level. However, there are also many state and local laws that affect the environment. It is easier for an individual to influence policy at the local level than at the national level. Local government is more responsive to citizen input than state or federal government is. It is also usually easier for citizens to organize and contact their representatives at the local level.

Local Governments Local governments and planning boards make many decisions for their communities. City councils and governmental agencies hold public meetings in which citizens may participate, such as the meeting shown in **Figure 6.** Local governments can decide how land may be used and developed, and where businesses and housing may be located. Local governments and agencies also create plans for public facilities, for waste disposal and recycling, and for many other facets of local life.

One common problem with local environmental planning is that communities often do not coordinate planning with each other. For example, your community may try to plan for clean air or water, but a neighboring community may allow development that creates pollution in your area. On the other hand, sometimes several local communities work together. For example, towns along the Hudson River in New York are cooperating to provide a "greenway" of natural areas for public use that stretch hundreds of miles along the river.

Homework ——————— ADVANCED

Green-Mapping Ask students to research a community project such as the Hudson River Valley Greenway or Barton Creek Greenbelt mentioned in this section. Provide maps to help them get started. Ask students to create a poster or presentation that includes the following: how these projects were started; which parties cooperated; how the local economy and local businesses are affected; which areas have been turned into public lands; how the areas are managed; and what activities are allowed. **LS** Visual

Career

Environmental Management and Policy
Career opportunities continue to grow in environmental management areas such as pollution control, drinking-water treatment, waste disposal, remediation, and risk assessment. These types of industry generate billions of dollars in revenue and employ millions of people worldwide. Many nonprofit environmental organizations have also experienced growth. Have students research job opportunities at an environmental management-related company, agency, or nonprofit organization within their county or state.

State Governments Environmental policy is also strongly influenced at the state level. The federal government passes laws that set environmental standards, but often these laws are minimum standards. Individual states are usually free to create laws that set higher standards. California's vehicle emission standards are higher than the federal standards because that state wants to control its problems with air pollution caused by traffic. States also have a lot of independent control over how to implement laws and manage public resources. For example, Ohio's Department of Natural Resources has used the state's endangered plant law to acquire habitats and to educate the public about the state's 350 endangered plant species. Many Ohio citizens voluntarily contribute to these conservation efforts as part of their income tax payments.

Lobbying Lawmakers are heavily influenced by lobbying on many sides of issues. **Lobbying** is an organized attempt to influence the decisions of lawmakers. Both environmental and industry groups hire lobbyists to provide information to lawmakers and urge them to vote a certain way. One way to influence policy is to support an organization that lobbies for the policies that you agree with.

The Media and Sources of Information The media, especially television news, as shown in **Figure 7**, is the main source of information about environmental topics for most of us. Popular TV, radio, and newspapers tell us, for example, when Congress is debating about oil drilling in the Arctic National Wildlife Refuge or when our local government is planning to build a new sewage plant. However, media reports are usually brief and leave out information.

If you want to understand environmental problems, you will want to find information from sources other than popular media. Many other sources are available, and you should evaluate all sources for bias and accuracy. Scientists and others who are familiar with environmental issues produce reports, magazines, and Web sites that contain in-depth information. Local organizations hold public meetings and produce newsletters. And through the Internet, you can get first-hand information from people all over the world.

Figure 7 ▶ A news broadcast may be the only way that many people learn about an environmental problem. From what other sources can people get information?

SECTION 2 Review

1. **Describe** two major developments in U.S. environmental history from each of the past two centuries.

2. **Give examples** of at least three federal agencies with environmental responsibilities.

3. **Explain** the purpose of Environmental Impact Statements. In what ways are citizens allowed to respond to an Environmental Impact Statement?

CRITICAL THINKING

4. **Relating Concepts** Describe three environmental issues that are important to people in your community.

5. **Expressing Viewpoints** Read about the ways of influencing environmental policy. Explain which of these ways you think is most effective. `READING SKILLS`

6. **Evaluating Information** Write a paragraph that evaluates an environmental news story from a newspaper, the radio, or TV. `WRITING SKILLS`

543

Close

Reteaching ──────── `BASIC`

Levels of Government Invite a social studies or civics teacher to your classroom to discuss the roles of each level of government (local, state, and national). When the guest has finished his or her discussion, work with the class to create a table with examples of environmental laws and regulations at each level. **LS** Verbal/Visual

Quiz ──────────── `GENERAL`

1. What is lobbying, and who does it? (Answers may vary. Lobbying is an attempt to influence lawmakers. Organized groups often hire lobbyists to represent the interests of their members.)

2. In what way is Congress limited in passing new laws such as the Clean Water Act? (The federal government must provide funding for any laws that cost more than $50 million to implement.)

Alternative Assessment ──── `GENERAL`

You Can Make a Difference Have students create a brochure or advertisement that explains how individuals their age can influence policy. **LS** Visual/Interpersonal

Answers to Section Review

1. Answers may vary. Sample answer: In the 1800s, prairies were turned into croplands, ancient forests were cut, and many animals were hunted to extinction; in the early 1900s, the national park system was created and people learned the value of soil conservation; in the late 1900s, the Clean Air Act was passed, and the Environmental Protection Agency was established.

2. Answers may vary but should reflect information from Table 3.

3. An EIS states the need for a project, the project's impact on the environment, and ways that any negative impact can be minimized. The public can comment on an EIS.

4. Answers may vary but may include land preservation, air quality, water quality, soil erosion, and endangered species.

5. Answers may vary.

6. Answers may vary.

Focus

Overview

Before beginning this section, review with your students the Objectives in the Student Edition. In this section, students are asked to consider the power of the individual. Students learn how individuals can influence environmental affairs through public and private actions.

Bellringer

Ask students to consider three of the most influential individuals they know of. Tell them to write down their thoughts in their *EcoLog.* Assure them that they won't have to share this information with the class. Students might choose people from the community, historical figures, or people who are discussed in the media. Ask what these people have in common in terms of their character. Have students imagine and then write down how their lives might be different without these individuals. **LS Intrapersonal**

Motivate

Discussion ——— GENERAL

What Difference Can I Make?
Ask for volunteers to begin a brainstorming session about what makes an individual stand out as a leader, hero, or someone who changes the world in some way. Lead students into a discussion of what kind of a difference a high school student, a teacher, or any other person can make. **LS Verbal**

Objectives

▶ Give examples of individuals who have influenced environmental history.
▶ Identify ways in which the choices that you make as an individual may affect the environment.

It is easy to feel that one person does not make much difference to the environment, but we all affect the environment with our daily actions. By learning about environmental problems and solutions, we are able to make responsible decisions. In addition, history has shown that one individual can have an influence on many others.

Influential Individuals

Some individuals who have influenced thinking about the environment in the United States are listed in **Table 4**. These people are famous because they brought attention to problems or convinced many people to think about new ideas. Many of these individuals wrote best-selling books about the subjects they knew well. These books were easy to understand and inspired people to think about environmental problems in a new way.

The 1960s Decade During the 1960s, environmental issues became widely known. It was then that biologists such as Paul Ehrlich, Barry Commoner, Rachel Carson, and Garrett Hardin drew public attention to environmental problems such as pollution, rapid population growth, and resource depletion.

Table 4 ▼

Some People Who Influenced Environmental Thinking in the United States	
Henry D. Thoreau (1817–1862) was a conservationist and writer who is best known for his essays about his stay in a cabin at Walden Pond in Massachusetts.	**Paul Ehrlich (1932–)** is a Stanford ecologist who warned of the dangers of rapid population growth with his 1968 book, *The Population Bomb.*
John Muir (1838–1914) was a Scottish-born naturalist and writer who founded the Sierra Club, explored the American West, and was a famous advocate for preserving western lands as wilderness.	**Jane Goodall (1934–)** studied chimpanzees in Tanzania's Gombe Stream National Park. Her books raised awareness of the plight of several endangered species and prompted new thinking about primate behavior.
Theodore "Teddy" Roosevelt (1858–1919) was the first American president to strongly support conservation. He founded the Forest Service, expanded national forests by 400 percent, and created the first National Monuments.	**Marion Stoddart (1928–)** led efforts to save the Nashua River in Massachusetts from pollution and development. *A River Ran Wild* is a book about her efforts. She is still active in protecting the Nashua River.
Alice Hamilton (1869–1970) was the first American expert on diseases caused by working with chemicals. In the early 1900s, she warned workers about exposure hazards and opposed the addition of lead to gasoline.	**Jacques Cousteau (1910–1997)** was a world-famous French oceanographer who produced many popular books, films, and TV programs that documented over four decades of his undersea explorations.
Rachel Carson (1907–1964) was a biologist with the U.S. Forest Service who raised awareness of toxic pesticides with her 1962 book, *Silent Spring.*	**Garrett Hardin (1915–)** is a distinguished professor of human ecology who is best known for his 1968 essay "The Tragedy of the Commons."

544

Chapter Resource File

- **Lesson Plans**
- **Active Reading** BASIC
- **Section Quiz** GENERAL

Transparencies

TT Bellringer

Notable *Quotes* ——— ADVANCED

"I know no safe depositary of the ultimate powers of the society but the people themselves; and if we think them not enlightened enough to exercise their control with a wholesome discretion, the remedy is not to take it from them, but to inform their discretion by education."

—Thomas Jefferson

Have students discuss how this quotation applies to environmental responsibility.

In *Silent Spring*, Rachel Carson argued that many public lands and resources were not adequately protected. She argued that resources such as water had to be protected and kept in natural, unpolluted conditions. Partly as a result of Carson's book, in 1964 Congress passed the Wilderness Act. This allowed the government to designate some federal lands as wilderness areas. These areas may only be used for low-impact recreation such as hiking and camping, and the number of visitors is limited.

Rising Awareness Also in the 1960s, several environmental disasters made headlines. Air pollution in New York City was blamed for 300 deaths. The bald eagle became endangered as a result of the widespread use of DDT. There was a massive oil spill near Santa Barbara. Lake Erie became so polluted that many of its beaches had to be closed. Eventually, pressure from the public led to new laws and efforts to reduce environmental damage. The first Earth Day, in 1970, was a historic demonstration of public concern for environmental issues.

Connection to **History**

Historical Writers Americans have been influenced by descriptions of America written by early explorers. An example is this passage written in 1805 by Meriwether Lewis, from his journal of the famous Lewis and Clark expedition:

"I beheld the Rocky Mountains for the first time . . . these points of the Rocky Mountains were covered with snow and the sun shone on it in such manner as to give me the most plain and satisfactory view. While I viewed these mountains I felt a secret pleasure in finding myself so near the head of the heretofore conceived boundless Missouri."

Figure 8 ▶ Examples of individuals who have brought attention to environmental issues: ❶ Rachel Carson, ❷ Marion Stoddart, and ❸ Jacques Cousteau.

545

Teach

Demonstration —— GENERAL
Product Packaging Have students bring in examples of products (or simply the packages) that have been bought and used by their families. Hold up each product and look on the packaging for claims or indicators of the relative environmental impact of each product. Note ingredients, whether the package contains recycled content or is recyclable, and claims about how the product was produced. Ask students to evaluate and discuss which products they think would be better to buy and why. **LS Visual**

⚡ Internet Activity — GENERAL

Researching Influential Individuals Have students research the lives of individuals such as those in **Table 4.** Lists of such individuals can be found in the *New York Public Library Science Desk Reference*, or students may research influential people in their community. In their reports, students should describe how the individuals became involved in environmental issues and what steps they took to reach their goals. Suggest that students include their report in their **Portfolio. LS Intrapersonal**

SKILL BUILDER — BASIC

Writing Ask students to list in their *EcoLog* individuals who could be added to **Table 4.** Perhaps they will choose someone they have learned about in the course or have met in real life. Ask students to write reasons why these individuals should be included in the table. **LS Intrapersonal**

Notable *Quotes* —— GENERAL

"Never doubt that a small group of thoughtful, committed citizens can change the world; indeed it is the only thing that ever has."

—Margaret Mead

Call on a student to read the quotation aloud. Have a class discussion about what the quote means to each student. Also, ask students to think of examples in which a small group has made a difference in the local community. **LS Verbal/Intrapersonal**

QuickLAB

Making a Decision

Procedure

1. Apply the following decision-making model to an environmental issue that interests you. After choosing an issue, find sources of information and opinions on different sides of the issue. Make notes about the ideas that you consider.
2. Consider which values apply to the issue. Consider scientific, economic, health, ethical, and cultural values. Which value is most important to you? to your community? Who else is involved, and how might they feel?
3. Explore the consequences of different actions. What are some possible outcomes? What are the pros and cons of each? How reliably can the outcomes be predicted?
4. Make your decision. Explain your reasoning in terms of the above considerations.

Analysis

1. Share your decision with a partner or group. Do not debate the issue; try to understand each others' reasoning, and give each other feedback about how carefully you applied the decision-making process.

Figure 9 ▶ Voting is an opportunity to make a decision that affects the environment.

Applying Your Knowledge

What will you be in the future? At the very least, you can expect to be a citizen who has the right to vote, a consumer who has choices of how to spend your money, and a member of the human race who has a role in the global environment. To make the decisions you will face, you can draw on your knowledge of environmental science.

Voting One of the most important decisions you may make is in the act of voting, as shown in **Figure 9**. The people we elect will make decisions that affect our environmental future. You have the right to support the candidates and laws that you think are best in both local and national elections. You can easily find out what a candidate thinks about environmental issues before an election. You can find information about candidates through the media, voter organizations, and Web sites.

One way to take action on environmental problems is as part of a group of people who share your concerns and interests. You can find many groups in your community asking for volunteers for activities such as planting trees, picking up trash, or maintaining trails. Many large nonprofit organizations hold meetings, educational activities, and trips to natural areas all over the country.

Weighing the Evidence A popular environmental slogan is to "think globally, act locally." This slogan reminds us that our everyday actions have broader effects. For example, every time we walk the dog, change the oil in a car, or toss aside a food wrapper, we may produce pollutants that will be washed into our drinking water supply by the next rain. Being aware of the effects of our actions is an important step in making decisions that affect the environment. What choices of action could you make today that will affect your environment?

Each of us has the responsibility to educate ourselves as we make the decisions that affect the world around us. There is a wealth of information about environmental issues on the Internet, in libraries, and in the media. When you research a topic, use reliable sources for statistics and information. Do not be misled by information that may look convincing but that has no supporting evidence.

Consumer Choices Another environmental slogan you may have heard is "reduce, reuse, recycle." As consumers, we can reduce the amount of things we buy and use, we can reuse things that are often used only once, and we can recycle many materials. How many examples can you think of to apply these ideas in your everyday life?

As a consumer, you may choose to buy products that are produced sustainably or that do less damage to the environment. It is not always easy to tell which products meet this standard. But as you learn more about environmental science, you'll be prepared to make decisions that guarantee that your impact on the environment will be a positive one.

internet connect

www.scilinks.org
Topic: **Earth Day**
SciLinks code: **HE4022**

SCI*LINKS*₀ Maintained by the National Science Teachers Association

Graphic

Organizer **Chain-of-Events Chart**

Create the **Graphic Organizer** entitled "Chain-of-Events Chart" described in the Appendix. Then, fill in the chart with details about how an individual choice at the grocery store can affect the environment.

Figure 10 ▶ As consumers, we make many choices that affect the environment. What choices could you make today that will affect your environment?

SECTION 3 Review

1. **Give examples** of at least three individuals in history who have had an impact on environmental thinking. What do they have in common?

2. **Identify** at least three ways individual citizens can influence their environment.

3. **List** five choices that you could make today that would have some kind of effect on the environment.

CRITICAL THINKING

4. **Identifying Relationships** Think of one activity that you do often. Write a paragraph explaining all the environmental effects, positive or negative, that this activity might have over time. **WRITING SKILLS**

5. **Predicting Consequences** Choose one environmental issue that you have learned about in this book and describe all the ways that you could make a difference on this issue.

547

Answers to Section Review

1. Answers may vary. Traits may include intelligence, foresight, hindsight, tenacity, confidence, and appreciation of nature.

2. Answers may vary. Individuals have an influence on the environment through voting, buying and using resources, or influencing others.

3. Answers may vary. Possible choices include what to buy, what to eat, whether to recycle or reuse a container, and whether to walk or drive somewhere.

4. Answers may vary.

5. Answers may vary.

Close

Reteaching — **BASIC**

Section Review Have students work in pairs to express the main ideas in this section as a series of sentences. Tell them to write the sentences down, leaving several blank lines between each sentence. Then have the pairs link these ideas together with sentences in each blank space. Have them repeat this exercise until they have crafted a narrative of the section. **LS** Verbal

English Language Learners

Quiz — **GENERAL**

1. Name one event that is at least partially attributed to the publication of Rachel Carson's *Silent Spring*. (Congressional passage of the Wilderness Act in 1964)

2. When was the first Earth Day? (The first Earth Day was in 1970.)

3. What does "think globally, act locally" mean? (Everyday actions have broad effects.)

Alternative Assessment — **ADVANCED**

Biography Have students do research and write biographies of someone listed in **Table 4** or of a similar person. Ask them to focus on the accomplishments of the person and to speculate about what the person must have been like as a friend, family member, co-worker, or acquaintance. The biography should include characteristics and actions that enabled the person to be influential. **LS** Intrapersonal

Graphic

Organizer **GENERAL**

Chain of Events Chart
You may want to use this Graphic Organizer, or a quiz, to assess students' prior knowledge before beginning a discussion, to assess students' knowledge after students have read how an individual choice at the grocery store can affect the environment.

CHAPTER 21 Highlights

Alternative Assessment — GENERAL

Environmental Milestones Have students make a timeline of major environmental events they have learned about in this chapter or throughout this book. Additional milestones in the history of environmental science are listed in the *New York Public Library Science Desk Reference*. Have students illustrate the timeline with art or photos and include it in their **Portfolio.** LS Logical/Visual

Environmental Stories Have teams of students create a children's book that deals with a theme from this chapter. Tell students to fully illustrate their book. Encourage students to share their books with younger siblings. Or take a class trip to a nearby elementary school and have students read their books to the children. Suggest that students include their children's book in their **Portfolio.** LS Interpersonal/Visual | Co-op Learning

Environmental Success Case Study Have teams of students research a successful effort to conserve resources or resolve an environmental issue at a local, national, or international level. This chapter has provided several examples. Ask students to create a "case study" about this effort and its "secret to success." Ask them to highlight the multiple parties involved, the processes used, any differences of opinion, and ways in which differences were addressed. Have students create a report or presentation that they can share with the class and include in their **Portfolio.** LS Interpersonal/Verbal | Co-op Learning

1 Economics and International Cooperation

Key Terms

sustainability, 533
economics, 536

Main Ideas

▶ To achieve sustainability will require cooperation and communication at many levels of society.

▶ Some international agreements on the environment have been achieved and successfully implemented. In some cases, goals have been set but not yet achieved.

▶ Economic systems operate within the environment by using resources and by returning both desired and undesired results. Economic systems sometimes fail to balance all the costs and benefits of people's actions.

2 Environmental Policies in the United States

Key Terms

Environmental Impact Statement, 541
lobbying, 543

Main Ideas

▶ In the last century, the U.S. government has developed policies to address environmental problems and has established agencies to implement those policies.

▶ Citizens can influence policy at all levels of government but especially at the local level.

▶ Lobbying and the media also influence policy and public opinion.

3 The Importance of the Individual

▶ Individuals can have an effect on environmental interactions through leadership and education. Many environmental problems were brought to the public's attention by a few individuals.

▶ You make important decisions about the environment every day. How you choose to spend money, vote, and use resources will have an impact on the environment.

▶ You can apply scientific thinking and knowledge to any decisions that you may face.

548

Chapter Resource File

- Chapter Test GENERAL
- Chapter Test ADVANCED
- Concept Review GENERAL
- Critical Thinking ADVANCED
- Test Item Listing
- Modeling Lab BASIC
- Research Lab GENERAL
- Research Lab ADVANCED
- Consumer Lab BASIC
- Long-Term Project GENERAL

Using Key Terms

Use each of the following terms in a separate sentence.

1. *sustainability*
2. *economics*

Use the correct key term to complete each of the following sentences.

3. Every federal project must complete a(n) _____.

4. Many groups try to influence government policies through _____.

STUDY TIP

Preparing for a Debate Participating in a debate can help you analyze an issue. To support a point of view, you must also understand opposing views. For practice, choose an issue discussed in this chapter or elsewhere in this book. At the top of a sheet of paper, state the basic problem. Draw two or more columns, and summarize different points of view at the top of each column. Then list the arguments in favor of each view. Try to find arguments that can be made against each other on similar points.

Understanding Key Ideas

5. Which of the following trends is *not* a challenge to achieving sustainability?
   a. the increasing human population
   b. the decreasing supply of fresh water in the world
   c. disagreement among governments
   d. advancement of scientific understanding

6. At the 1992 Earth Summit, representatives from around the world
   a. created the Kyoto Protocol.
   b. tried to balance economic development with environmental sustainability.
   c. could not reach agreement on anything important.
   d. talked about environmental problems for the first time ever.

7. International environmental agreements include
   a. the Montreal Protocol on Ozone.
   b. Earth Day.
   c. the World Trade Organization.
   d. the Wilderness Act of 1964.

8. Economic systems
   a. do not depend on limited natural resources.
   b. always balance the costs and benefits of every action.
   c. should not include the costs of pollution with the costs of an action.
   d. must operate within the environment.

9. Which of the following statements about U.S. environmental policy is *not* true?
   a. During most of the 19th century, most Americans were not concerned about environmental consequences.
   b. During the 1960s, several individuals had strong effects on public thinking about environmental issues.
   c. Before Earth Day 1970, no one in the United States cared about the environment.
   d. The Environmental Protection Agency was established at a time of increasing public awareness of environmental problems.

10. State and local environmental regulations
    a. cannot be influenced by individuals.
    b. simply enforce federal standards.
    c. do not have to follow federal standards.
    d. are often more strict than federal standards.

11. The main function of an Environmental Impact Statement is
    a. to predict the effect a federal project might have on the environment.
    b. to produce a record of environmental change throughout history.
    c. to satisfy the requirements of international agreements.
    d. to limit real estate development and the activities of businesses.

12. Local governments do not regulate
    a. recycling.
    b. sewage treatment.
    c. garbage disposal.
    d. Environmental Impact Statements.

549

ANSWERS

Using Key Terms

1. Sample answer: Sustainability is a way of living that can continue indefinitely.

2. Sample answer: Economics is the study of how people make decisions about resources by weighing costs and benefits.

3. Environmental Impact Statement

4. lobbying

Understanding Key Ideas

5. d
6. b
7. a
8. d
9. c
10. d
11. a
12. d

Assignment Guide

Section	Questions
1	1, 2, 5–8, 13, 14, 25, 26, 28
2	3, 4, 9–12, 15, 18–21, 24
3	16, 17, 22, 23, 27, 29

CHAPTER **21** Review

Short Answer

13. Leaders meet to create treaties that address environmental issues. Often leaders strive to allow for continued economic development.

14. Answers may vary. Any country may choose not to sign, enforce, or provide funding to implement a treaty.

15. State and local regulations are on a smaller scale than federal regulations and can therefore be more strict than federal regulations. Community decisions that pertain to development, waste disposal, and recycling are often made at a state or local level.

16. Citizens can comment on an EIS, write to representatives, and participate in decision-making meetings.

17. A consumer can affect the environment by carefully choosing what to buy or use. Reducing, reusing, and recycling materials makes a big difference.

Interpreting Graphics

18. The appliance on the right likely has an "energy star" on it because it is more energy-efficient than most similar models.

19. The most important quantity is how many kilowatt-hours per year the unit will consume.

20. Sample answer: The government has required these labels so that consumers can make economic and environmental choices based upon the energy efficiency of the products they may buy.

Short Answer

13. What do world leaders do at gatherings such as the Earth Summit?

14. Why are some treaties not successful?

15. In what ways do state or local regulations differ from federal regulations?

16. Describe several ways that citizens can influence environmental policy.

17. How can a consumer affect the environment?

Interpreting Graphics

The figures below show a type of label that is required by law to be placed on all new appliances. Use the figures to answer questions 18–21.

18. What is the most likely reason that the tag on the right has an "energy star" symbol?

19. Which quantity on the tags is the most important piece of information about these appliances?

20. Why do you think the government has required such labels to be placed on all new appliances?

21. There are two types of refrigerators represented on these labels: top-freezer and side-by-side. Which type is generally more efficient? How can you tell?

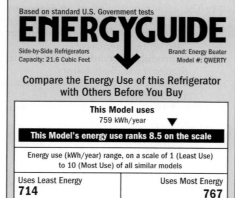

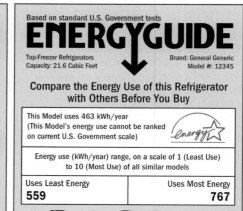

Concept Mapping

22. Use the following terms to create a concept map: *groups*, *individuals*, *lobbying*, *state laws*, *federal laws*, and *voting*.

Critical Thinking

23. **Expressing a Viewpoint** Read the section about influential individuals in this chapter. Describe at least one effect that one of these individuals may have had on your life. **READING SKILLS**

24. **Making Predictions** What might the effects be if the United States doubled the tax on gasoline over the next 10 years?

Cross-Disciplinary Connection

25. **History** Some people argue that developing nations should be allowed to create polluting industries in order to develop economically, just as the developed nations did in the past. Explain your opinion of this argument.

Portfolio Project

26. **An International Treaty** Write a proposal for a new international treaty that would address a pressing environmental problem and that you think could be agreed upon by many nations. **WRITING SKILLS**

21. Top-freezer refrigerators are generally more efficient. The refrigerator type is listed in the upper left of the tag and the scales on the tags indicate that some models of top-freezer refrigerators use less energy than any models of side-by-side refrigerators do.

Concept Mapping

22. Answers to the concept-mapping questions are on pp. 667–672.

Critical Thinking

23. Answers may vary.

24. Answers may vary. People would probably use less gas, and the cost of many goods would rise.

Cross-Disciplinary Connection

25. Answers may vary.

Portfolio Project

26. Answers may vary

MATH SKILLS

Use the graphs below to answer question 27.

27. **Analyzing Data** The graph on the left shows the proportions of federal money that the public thinks should be spent on different types of research and development, based on a 1993 poll. The graph on the right shows how the money was actually spent. Which types of spending show close agreement between government spending and public opinion? Which types show the greatest difference?

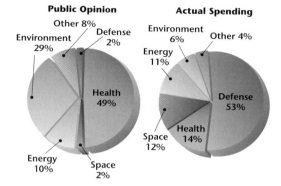

Public Opinion

Other 8%
Defense 2%
Environment 29%
Health 49%
Energy 10%
Space 2%

Actual Spending

Environment 6%
Other 4%
Energy 11%
Defense 53%
Health 14%
Space 12%

WRITING SKILLS

28. **Communicating Main Ideas** Describe some signs that the world may be progressing toward a sustainable future. What are some likely challenges ahead?

29. **Expressing Original Ideas** Describe your vision of a sustainable future. Consider lifestyles, technology, forms of government, economic systems, and social organizations.

STANDARDIZED TEST PREP

For extra practice with questions formatted to represent the standardized test you may be asked to take at the end of your school year, turn to the sample test for this chapter in the Appendix.

READING SKILLS

Read the passage below, and then answer the questions that follow.

Theodore Roosevelt was a unique and memorable president. A 1902 photograph shows him riding a moose across a river. Roosevelt thought of public lands as economic assets, to be used for timber harvesting, mining, and recreation. He felt that natural resources should be regulated and managed for the public benefit. He did not see these lands as refuges for threatened plants and animals. Roosevelt was considered to be the first conservationist president.

Roosevelt had some conflicts with preservationists such as John Muir. Preservationists believe in preserving public lands as untouched wilderness for future generations to study and enjoy. These conflicting views over how to use U.S. public lands continue today.

1. According to the passage, what is meant by the term *conservationist*?
   a. the same thing as *preservationist*
   b. one who believes in managing natural resources for the public benefit
   c. one who believes nature should be preserved untouched
   d. the same thing as *environmentalist*

2. According to the passage, which of these ideas about the uses of U.S. public lands was debated about 100 years ago?
   a. Mining should be legal.
   b. Wolves should be reintroduced into Yellowstone.
   c. Motors should be allowed in parks.
   d. Public lands should be preserved.

3. With which of the following opinions did both Roosevelt and Muir most likely agree?
   a. Public lands should not be used for mining or timber harvesting.
   b. The United States should own and regulate public lands.
   c. Wolves should be exterminated.
   d. Roosevelt was a preservationist.

551

Math Skills

27. The agreement between government spending and what the public thinks is most similar in the area of energy. Public opinion and actual spending differ most in the area of defense.

Writing Skills

28. Answers may vary. Many world leaders seem to be interested in sustainability and willing to meet and discuss policies that would lead to sustainability. Likely future challenges include agreeing upon a definition of sustainability and getting all levels of all societies to cooperate.

29. Answers may vary.

Reading Skills

1. b
2. d
3. b

CHAPTER 21 Inquiry Lab: RESEARCH

BE AN ENVIRONMENTAL SCIENTIST

Teacher's Notes

Time Required

two 45-minute class periods

Lab Ratings

EASY ———————————→ HARD

TEACHER PREPARATION ▲ ▲
STUDENT SETUP ▲ ▲ ▲
CONCEPT LEVEL ▲ ▲ ▲
CLEANUP ▲

Skills Acquired

• Collecting Data
• Classifying
• Interpreting
• Organizing and Analyzing Data
• Communicating

The Scientific Method

In this lab, students will:
• Ask Questions
• Analyze the Results
• Draw Conclusions
• Communicate the Results

Materials

Give students access to the library, Internet, note cards, file folders, and other research aids. Provide sample reports, such as an EIS, made by scientists for government and community agencies. You may also provide presentation materials such as poster board, overhead transparencies, markers, or software for word-processing, graphics, and "slideshow" presentations.

Objectives

▶ **USING SCIENTIFIC METHODS** **Research** a current environmental issue that requires an informed policy decision.

▶ **USING SCIENTIFIC METHODS** **Prepare and present** a report that is intended to inform the appropriate policy decision makers.

Materials

file folders
markers or colored pencils
note cards
posterboard
paper

optional materials: **computer for word processing, graphing, or making a presentation**

▶ **Scientific Reports** Scientists often present their findings in reports that can be reviewed by other scientists and by the public. Try to give your report a style and organization like that used by professional scientists.

Be an Environmental Scientist

Are you ready to put your knowledge of environmental science to work? Environmental scientists are often asked to help decision makers in government when there is a policy decision to be made that may affect the environment. Decision makers often want to make an informed decision based on a scientific analysis of a situation. Environmental scientists may be asked to study a situation or predict the results of an action and present their findings to the interested decision makers.

In this lab, you will be an environmental scientist who has been asked to prepare a report. The purpose of the report is to inform a group of decision makers of the possible results of their choices. You are expected to prepare an unbiased, thorough, and accurate report. And like a professional scientist's report, your report will be reviewed by your peers.

Procedure

1. Choose a current environmental issue that requires an informed policy decision. You might research legislation that is being considered in national or state governments. Or find out if any local projects or laws that have environmental effects are being debated in your community.

2. Do some simple beginning research to become familiar with the issue. Start in the library and then also try to find information from government agencies, scientific publications, and any private groups that are involved with the issue.

3. Write a brief description of the issue or proposal and a plan of how you will research and present your findings. Get your teacher's approval before proceeding.

4. Carry out your research. Don't forget to get help from librarians. Be sure to keep track of your sources of information, and check that they are reliable sources. Keep your teacher informed of your progress, and ask for help if you need it.

5. Create an outline of your report. Get your teacher's approval before proceeding.

Tips and Tricks

Help students visualize the result of this project by choosing sample reports to demonstrate to the class. Highlight for students the ways in which issues are presented and options are examined. Note any biases you observe, and emphasize the roles of tone and style in communicating ideas to the reader. You may prepare a list of issues that students can choose from or have a brainstorming session at the beginning of class to direct students toward suitable topics. You may also prepare and discuss with students a rubric for evaluating each other's reports by using the guidelines on the Student Edition page. If students are having trouble selecting an issue to focus on, tell them that many nonprofit groups provide information on recent and upcoming environmental legislation and policy decisions. Alternatively, you could allow students to choose one of the pieces of environmental legislation that is mentioned in this chapter.

6. Create the report. Be sure to do the following:

  a. Present the major options or different opinions being considered and the main arguments or reasons for each.

  b. For each option or potential action, explain the effects or consequences that might result.

  c. Create diagrams, tables, graphs, or other representations of the science involved.

  d. Provide citations of sources in a bibliography or other format, as approved by your teacher.

  e. Give a citation for each fact you present. Think critically about all sources of information you use. Try to find more than one source for information that seems doubtful.

  f. Be clear about how much data is available. Explain when there does not seem to be enough data to make a conclusion or establish a fact.

7. Present your report to your classmates and teacher.

▶ **Further Research** Like a real scientist, your research may lead you to new questions. You may wish to propose or conduct further research into the issue you have studied.

Analysis

1. **Analyzing Results** Read and listen to your classmates' reports, and evaluate them as described below.

Conclusions

2. **Evaluating Results** Evaluate your classmates' reports or presentations. Use the following criteria:

  a. What evidence or research did the scientist present to support his or her facts and conclusions?

  b. Was every conclusion supported by data or by scientific opinion?

  c. Was every fact or piece of data documented and supported by other sources?

  d. Were the concepts presented clearly? Did the report/presentation flow logically?

  e. Did the diagrams help you understand ideas?

  f. Was the report unbiased, or did the presenter show his or her opinion on the subject?

Extension

1. **Communications** Present or submit your report to a group that is making decisions about the issue you studied.

553

Chapter Resource File

• Datasheets for In-Text Labs
• Lab Notes and Answers

Alonda Droege
Evergreen High School
Seattle, Washington

Answers to Analysis

1. Students should present reports to or share reports with classmates.

Answers to Conclusions

2. Students should analyze each other's presentations using the criteria on the Student Edition page.

Answers to Extension

1. Answers may vary.

STUDENT CLUB SAVES EAGLES AND MORE

Background

Ask students, "What do you think these students learn other than information about the particular owl or raptor that they are working with?" (Sample answers: They would learn about animal behavior and biology, about working with other students toward a common goal, about being organized, and about taking responsibility.) Ask other students to share stories of socially-responsible activities they—or others they know—are involved in.

Emphasize that the SWCC students receive special training and permits for their dangerous work with rare and legally protected animals. Students in the club report that they must manage a delicate relationship with the birds. Each student "adopts" a bird so that the students and the bird can become familiar enough with each other to make safe physical contact. The students say that "the birds choose you" because each bird seems to be receptive to handling only from certain people. Students often feel they form a bond with the bird but must be careful to preserve the bird's wild instincts and also to stay emotionally detached, because the bird must later be released.

.

Making a difference

STUDENT CLUB SAVES EAGLES AND MORE

Not many people get to see a bald eagle up close, and most people never get to hold one. When Jeremy had the chance to care for an injured bald eagle and then return it to the wild, he felt "it was a life experience." Jeremy is just one of the hundreds of current and former members of the Southwestern High School Conservation Club (SWCC) in Somerset, Kentucky.

This club is unique in many ways. The SWCC's core mission is to help students understand the natural world through hands-on activities. In many ways, the club is an extension of a variety of environmental science classes offered at the school. However, the club is very busy and is involved in a wide variety of activities. Most important, the club members all feel that they are learning responsibility and important skills while making a difference in their environment.

Hands-On Science

Students at Southwestern High School (SWHS) can choose from six different environmental science courses, such as raptor biology or greenhouse management. In all of these classes, students spend more time getting their hands dirty than they spend using pencils and paper.

One SWHS science teacher said, "Biology is out there, beyond the classroom; you have to get outside to fully study it." Thanks to the hard work and leadership of students and teachers, the science facilities at SWHS now include a working greenhouse, native plant landscapes, a nature trail, an outdoor amphitheater, computer labs, and a weather station. The most exciting and unique facility at the school is the Raptor Rehabilitation Center, where SWCC members work every afternoon.

554

A Second Chance for Raptors

Raptors are birds of prey, such as owls, hawks, vultures, and eagles. Raptors in the United States have suffered many threats to their existence, from pollution to injury by cars or gunshots. Several federal and state laws are intended to protect raptors from such threats, but at least 300 raptors have been rescued from these threats by the students of SWHS.

Injured birds and orphaned fledglings are brought to the Raptor Rehabilitation Center from across the United States. The school has a special license to keep and care for raptors. The rehabilitation program requires veterinary equipment and supplies, specially designed cages, and a professional level of knowledge and training.

SWHS teacher and club sponsor Frances Carter started the rehabilitation program when the high school opened in 1993. She had previous experience with raptors and knew professionals in the field of wildlife management. These professionals asked her to help care for some of the birds

that were being found because there were no other raptor facilities in the area.

A typical day in a raptor biology class involves about 30 min of instruction and an hour of bird maintenance. Additional bird care is done by SWCC members during nonschool hours. For example, club members Jeremy, Ben, and Grant spent hours each day for months—including holidays—working with two bald eagles.

Maintenance on a live bird usually involves grinding down excess growth on the beak, trimming the talons, exercising the bird, conducting a physical exam, and giving medications. Cages have to be cleaned weekly. Special diets have to be prepared. All of these tasks can be dangerous and messy. Handling the large birds requires training and skill.

They Say the Birds Choose You

The first goal of raptor rehabilitation is to be able to return the birds to the wild. This goal involves a tricky balance between building trust with each bird and

▶ **An eagle named Justice** was the first bald eagle to be rehabilitated and released in Kentucky, thanks to the SWCC. "I got chills," said one of the club members who was present, "It was gorgeous. That's a feeling you only get a few times in your life."

🔲 Internet Activity ──────── GENERAL

Student Environmental Networks Have students search the Internet for the following organizations or networks of students interested in environmental issues:

- Student Environmental Action Coalition (SEAC)
- EarthTeam
- EarthForce
- UN's CyberSchoolBus "Atlas of Student Action for the Planet"
- The UN Environment Programme's "Children and Youth Programme"

preserving its wild instincts. The raptor program has succeeded in releasing more than 30 percent of its birds. Yet for many students, letting the birds go is the hardest part of the work.

Imagine being handed three fuzzy, softball-sized baby owls, squawking for their lost mother. You might guess that these young great horned owls stole the hearts of the students at Southwestern. The students named them Bert, Ernie, and Elmo. Club members Amy and Valerie virtually adopted these baby owls and even took the birds home over vacations and gave them round-the-clock care. Amy recalls 3 A.M. feedings of dead mice or chicken livers rolled in calcium. "Killing mice by myself—nasty stuff. I thought it was the grossest thing I ever had to do in my life," she said.

However, Amy learned that the owls depended on her, "and that was the only thing that mattered." Eventually, the three owls grew to be healthy adults and were released. Amy reflected, "It was hard to watch the 'babies' fly off, never to be seen again."

Responsibility and Reward
In addition to the news-making events when the club releases another raptor into the wild, the club makes many efforts to educate their community. Presentations in which live raptors are perched on a student's hand are popular events at other schools, community fairs, and teacher training workshops. These presentations are also opportunities to educate people about wildlife and environmental issues. The school provides leadership, resources, and workshops for over 40 other schools in Kentucky. Also, club

▶ **Students in SWCC** are multitalented. They do everything from cleaning birdcages to making presentations around the nation and from picking up litter to competing in the national Envirothon competition. The Club's latest plan is to build a nature center.

members participate in a yearly environmental science competition called the *Envirothon*. The club advanced to national level in 2001.

The raptor rehabilitation program, greenhouse, nature trail, weather stations, and computer lab sound like a lot of fun, but most club members say that responsibility and making a difference are the important reasons to be in the club. Everyone in the club has a job title, from Club Reporter to Greenhouse Manager to Webmaster.

What Will They Do Next?
For the club and the teachers at SWHS, there is always more to be done. Club members do most of the planning, fundraising, and manual labor in a variety of projects. Recent projects include landscaping the school grounds with native plants, creating composting and recycling centers, and expanding the school's facilities. The school hopes to unite these facilities into a complete nature center that would educate tourists,

students, and scientists from around the nation.

Club members have developed career interests in veterinary medicine, wildlife biology, conservation, or environmental science fields. Jeremy may be a firefighter or a civil engineer. Amy and Cara may go into journalism. They all say that the club has given them unique opportunities and that they will never forget the experience.

What Do You Think?

Are there any groups like the SWCC in your area? Would you like to be like these students? What other ways can students make a difference for the environment? Explain your answers.

Student Opportunities

HawkWatch Raptors are an essential part of healthy, functioning ecosystems. HawkWatch International (HWI) is a nonprofit organization whose mission is to monitor and protect raptors and their environments through research, education, and conservation. The core of HWI's research program focuses on the maintenance of a large database of raptor population numbers and trends. HWI collects data primarily by leading a network of standardized migration count projects at points along migration routes in North America and also internationally. HWI also uses trapping and banding, satellite telemetry, stable isotope analysis, and other methods to obtain raptor movement data. HWI's projects contribute to scientific knowledge of raptor populations in the U.S., and HWI's data is used by state and federal agencies, academia, and non-profit conservation organizations for setting raptor conservation and management priorities. HWI conducts education programs using live raptors and educates over 10,000 visitors each year at their migration study sites on topics such as raptor migration dynamics, the importance of raptor conservation, and raptor identification skills. Have students look into volunteering for HawkWatch's migration count projects or education programs.

Answers to What Do You Think?
Answers may vary.

Career

Biology and Beyond A variety of unique career opportunities await the SWCC members, who will graduate from SWHS with many proven skills and achievements. Amy Merrick started high school wanting to be "the next Barbara Walters," and ended up being known as "the bird lady." As Club president, Amy found new passions and horizons for her talents. For example, she wrote news articles and funding proposals for the Club. In her last year of high school, Amy had plans to be a "crusader" in college, majoring in public relations with a minor in Environmental Science.

Point out to students that participating in clubs and community service activities often helps young people find out what kind of work they might—or might not—enjoy and excel at in the future. Have students list their current career interests, and brainstorm examples of classes and activities that might give them more experience in their areas of interest. Encourage each student to find and try out at least one new club or community service opportunity.

Answers to Appendix Practice Questions

Reading and Study Skills

How to Make Power Notes

1. Sample answer:

> The Experimental Method
>
> <u>Power 1</u>: observing
>> <u>Power 2</u>: observation
>
> <u>Power 1</u>: hypothesizing and predicting
>> <u>Power 2</u>: hypothesis
>> <u>Power 2</u>: prediction
>
> <u>Power 1</u>: experimenting
>> <u>Power 2</u>: experiment
>>> <u>Power 3</u>: variable
>>> <u>Power 3</u>: experimental group
>>> <u>Power 3</u>: control group
>
> <u>Power 1</u>: organizing and analyzing data
>> <u>Power 2</u>: data
>
> <u>Power 1</u>: drawing conclusions
>
> <u>Power 1</u>: repeating experiments
>
> <u>Power 1</u>: communicating results

How to Make KWL Notes

1. **a.** The first step is observing.
   b. A hypothesis is more than a guess. It must be based on observations and testable by experiment.
   c. A good experiment has a single variable and a control group.

How to Make a Concept Map

1. **a.** core
   b. mesosphere
   c. inner core
2. Sample answers:
   - Earth is divided into layers.
   - The layers are categorized based on composition or structure.
   - The composition layers are the crust, the mantle, and the core.
3. **a.** concept
   b. linking words
   c. linking words
   d. concept

Math Skills Refresher

Order of Operations

1. 24
2. 7

Geometry

1. 225,000 m^2
2. 1,230 cm^3 (rounded to three significant figures)
3. 96 cm^2

Exponents

1. **a.** 9
   b. 14,348,907
   c. 537,824
   d. 1

Algebraic Rearrangements

1. **a.** $x = 20$
   b. $a = -1.75$
   c. $y = -6.3$
   d. $m = -4$
   e. $z = 2$
   f. $b = 5$

Scientific Notation

1. **a.** 1.23×10^7 m/s
   b. 4.5×10^{-12} kg
   c. 6.53×10^{-5} m
   d. 5.5432×10^{13} s
   e. 2.7315×10^2 K
   f. 6.2714×10^{-4} kg

Significant Digits

1. **a.** 4
   b. 5
   c. 4
   d. 3
2. **a.** 0.129 dm
   b. 2700 m/s
   c. 9.84 m^2
   d. 0.98 g

Appendix

CONTENTS

Lab Safety

General Guidelines For Laboratory Safety

In the laboratory, you can engage in hands-on explorations, test your scientific hypotheses, and build practical laboratory skills. However, while working in the laboratory or in the field, it is your responsibility to protect yourself and other students by conducting yourself in a safe manner. You will avoid accidents in the labroratory by following directions, handling materials carefully, and taking your work seriously. Read the following general safety guidelines and review the descriptions of the safety symbols on pp. xvii–xix before working in the laboratory.

Eye Protection   Clothing Protection   Caustic Substances   Chemical Safety

Animal Safety   Plant Safety   Electrical Safety   Heating Safety

Sharp Object Safety   Hand Safety   Fire Safety   Gas Safety

Glassware Safety   Waste Disposal   Hygienic Care/ Clean Hands

Before You Begin...

◆ Be prepared. Study assigned experiments before class. Resolve any questions about procedures before starting work.

◆ Keep your work area uncluttered. Store books, backpacks, jackets, or other items you do not need out of the way.

◆ Arrange the materials you are using for an experiment in an orderly fashion on your work surface. Keep laboratory materials away from the edge of the work surface.

◆ Tie back long hair and remove dangling jewelry. Roll up sleeves and secure loose clothing.

◆ Do not wear contact lenses in the laboratory. Chemicals could get between the contact lenses and your eyes and cause irreparable eye damage. If your doctor requires that you wear contact lenses instead of glasses, then you should wear eyecup safety goggles—similar to goggles that are worn for underwater swimming—in the laboratory.

◆ Do not wear open-toed shoes, sandals, or canvas shoes in the laboratory because they will not protect your feet if any chemical, glassware, or other object is dropped on them.

◆ Know the location of the nearest phone. Find out where emergency telephone numbers, such as the number for the nearest poison control center, can be found.

◆ Know the location of safety equipment such as eyewash stations and fire extinguishers. Know how to operate this equipment.

◆ Know the fire evacuation routes established by your school.

◆ Before you begin the experiment, review the supplies you will be using and the safety issues you should be concerned about. Be on the alert for the safety symbols shown on this page and those that appear in your experiment.

While You Are Working...

◆ Do not play in the lab. Take your lab work seriously, and behave appropriately in the laboratory. Be aware of your classmates' safety as well as your own at all times.

◆ Never perform an experiment not authorized by your teacher.

◆ Never work alone in the laboratory.

◆ Always wear safety goggles and a lab apron when you are working in the lab.

Laboratories contain chemicals that can damage your clothing, skin, and eyes.

◆ Wear protective gloves when working with an open flame, chemicals, solutions, wild or unknown plants, or other items as directed by your teacher.

◆ Never look directly at the sun through any optical device or use direct sunlight to illuminate a microscope. The focused light can seriously damage your eyes.

◆ When heating substances in a test tube, always point the test tube away from yourself and others.

◆ Keep your hands away from the sharp or pointed ends of scalpels, scissors, and other sharp instruments.

◆ Observe all of the safety symbols that accompany the procedural steps of the experiment. Be sure to follow the safety practices that are called for in the experiment.

◆ Never put anything in your mouth, and never touch or taste substances in the laboratory unless your teacher instructs you to do so.

◆ If your teacher instructs you to smell a chemical in the laboratory, follow the correct procedure. The correct method is to gently fan your hand over the substance, waving its

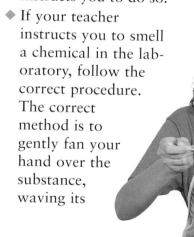

vapors toward your nose. Do not put your nose directly over the substance.

◆ Never eat, drink, chew gum, or apply cosmetics in the laboratory. Do not store food or beverages in the lab area.

◆ Report any accident, chemical spill, or unsafe incident to your teacher immediately.

◆ Check labels on containers of chemicals to be certain you are using the right material.

◆ When diluting an acid or base with water, always add the acid or base to water. Do NOT add water to the acid or base.

◆ Dispose of chemicals according to your teacher's instructions.

◆ Never return unused chemicals to the containers you obtained them from. Do not put any object into a bottle containing a laboratory chemical.

Emergency Procedures

Don't panic. In the event of a laboratory emergency follow these instructions.

◆ In the event of a fire, alert the teacher and leave the laboratory immediately.

◆ If your clothes catch fire, STOP, DROP, and ROLL! The quickest way to smother a fire is to stop immediately, drop to the floor, and roll.

◆ If your lab partner's clothes or hair catches fire, grab the nearest fire blanket and use it to extinguish the flames. Inform your teacher immediately.

◆ If a chemical spills on your skin or clothing, wash it off immediately with plenty of water, and notify your teacher.

Lab Safety

- If a chemical gets into your eyes or on your face, go to an eyewash station immediately, and flush your eyes (including under the eyelids) with running water for at least 15 minutes. Hold your eyelids open with your thumb and fingers, and roll your eyeball around. While doing this, have another student notify your teacher.
- If a chemical spills on the floor, do not clean it up yourself. Keep your classmates away from the area, and alert your teacher immediately.
- If you receive a cut, even if it is just a small one, notify your teacher.

Safety With Animals in the Laboratory

Observing and experimenting with animals can enrich your understanding of environmental science. However, you must use extreme caution to assure your own safety as well as the safety and comfort of animals you work with. When working with animals in the laboratory be sure to follow these guidelines.

- Do not touch or approach any animal unless your teacher specifically gives you permission.
- Handle animals only as your teacher directs. Mishandling or abusing any animal will not be tolerated.
- Do not bring any animal into the laboratory without your teacher's permission.
- Wear gloves or other appropriate protective gear when working with animals.
- Wash your hands after touching any animal.
- Inform your teacher immediately if you are scratched, bitten, stung, or otherwise harmed by an animal.
- Always follow your teacher's instructions regarding the care of laboratory animals. Ask questions if you do not clearly understand what you are supposed to do.
- Keep each laboratory animal in a suitable, escape-proof container in a location where the animal will not be frequently disturbed.

Animal containers should provide adequate ventilation, warmth, and light.

- Keep each laboratory animal's container clean. Clean cages of small birds and mammals daily.
- Provide each laboratory animal with water at all times.
- Feed animals regularly, according to their individual needs.
- If you are responsible for the care or feeding of animals, arrange for necessary care on weekends, holidays, and during vacations.
- No study that involves inflicting pain on a vertebrate animal should ever be conducted.
- Vertebrate animals must not be exposed to excessive noise, exhausting exercise, over-crowding, or other distressing stimuli.
- When an animal must be removed from the laboratory, your teacher will provide a suitable method.

Safety With Plants in the Laboratory

Some plants or plant parts are poisonous to the point of fatality, depending on the weight of the person and the amount of plant material ingested. Therefore, many plants or plant parts can present a safety hazard to you. When working with plants in the laboratory be sure to follow these guidelines.

- Never place any part of any plant in your mouth unless instructed to do so by your teacher. Seeds obtained from commercial growers can be particularly dangerous because such seeds may be coated with hormones, fungicides, or insecticides.
- Do not rub sap or juice of fruits on your skin or into an open wound.
- Never inhale or expose your skin or eyes to the smoke of any burning plant or plant parts.
- Do not bring unknown wild or cultivated plants into the laboratory.

- Do not eat, drink, or apply cosmetics after handling plants without first washing your hands.
- Provide adequate light and water and appropriate soil and temperature for plants growing in the laboratory.
- If you are responsible for plants, make necessary arrangements for their care on weekends, holidays, and during vacations.

Finishing up in the Laboratory...

- Broken glass, chemicals, and other laboratory waste products should be disposed of in separate special containers. Dispose of waste materials as directed by your teacher.
- Clean tables and sinks as directed by your teacher.
- Make sure all water faucets, gas jets, burners, and electrical appliances are turned off.
- Return all laboratory materials and equipment to their proper places.
- Wash your hands thoroughly with soap and water after completing each experiment.

Safe and Successful Fieldwork

Environmental scientists conduct much of their research in the field. For environmental scientists—and environmental science students such as you—there are three important issues to consider when working in the field. One issue is your personal safety. Another issue is the successful completion of the scientific work you set out to do. The third consideration is protection of the environment you have come to study. The following guidelines will help you address these three issues.

- Dress in a manner that will keep you comfortable, warm, and dry. Wear long pants rather than shorts or a skirt. Wear sturdy shoes that have closed toes. Do not wear sandals or heels. Wear waterproof shoes if you will be working in wetlands.

- Bring rain gear if there is any possibility of rain.
- Bring sunglasses, sunscreen, and insect repellent as needed.
- Do not go alone beyond where you can be seen or heard; travel with a partner at all times.
- Do not approach wild mammals, snakes, snapping turtles, or other animals that may sting, bite, scratch, or otherwise cause injury.
- Do not touch any animal in the wild without specific permission from your teacher.
- Find out whether poisonous plants or dangerous animals are likely to be where you will be going. Learn how to identify any hazardous species.
- Do not pick wildflowers or touch plants or plant parts unless your teacher gives you permission. Do not eat wild plants.
- Immediately report any hazard or injury to your teacher.
- Be sure you understand the purpose of your field trip and any assignments you have been given. Bring all needed school supplies, and keep them organized in a binder, backpack, or other container.
- Be aware of the impact you are having on the environments you visit. Just walking over fragile areas can harm them, so stay on trails unless your teacher gives you permission to do otherwise.
- Sketching, photographing, and writing field notes are generally more appropriate than collecting specimens for observation. Collecting from a field site may be permitted in certain cases, but always obtain your teacher's permission first.
- Do not leave garbage at the field site. Strive to leave natural areas just as you found them.

Reading and Study Skills

How to Make Power Notes

Power notes help you organize the environmental science concepts you are studying by distinguishing main ideas from details. Similar to outlines, power notes are linear in form and provide you with a framework of important concepts. To make power notes, you assign a *power* of 1 to each main idea and a 2, 3, or 4 to each detail. You can use power notes to organize ideas while reading your text or to restructure your class notes for studying purposes. Practice first by using simple concepts. For example, start with a few headers or boldfaced vocabulary terms from this book. Later you can strengthen your notes by expanding these simple words into more-detailed phrases and sentences. Use the following general format.

> <u>Power 1</u>: Main idea
>> <u>Power 2</u>: Detail or support for Power 1 idea
>>> <u>Power 3</u>: Detail or support for Power 2 concept
>>>> <u>Power 4</u>: Detail or support for Power 3 concept

1 Pick a Power 1 word or phrase from the text

The text you choose does not have to come from your environmental science textbook. You may make power notes from your lecture notes or from another source. We'll use the term *environmental problems* as an example of a main idea.

> <u>Power 1</u>: environmental problems

2 Using the text, select some Power 2 words to support your Power 1 word

We'll use the terms *resource depletion*, *pollution*, and *extinction*, which are the three main types of environmental problems.

> <u>Power 1</u>: environmental problems
>> <u>Power 2</u>: resource depletion
>> <u>Power 2</u>: pollution
>> <u>Power 2</u>: extinction

3 Select some Power 3 words to support your Power 2 words

We'll use the terms *renewable resources* and *nonrenewable resources*. These two terms are types of *resource depletion*, which is one of the Power 2 concepts.

Power 1: environmental problems
    Power 2: resource depletion
        Power 3: renewable resources
        Power 3: nonrenewable resources
    Power 2: pollution
    Power 2: extinction

4 Continue to add powers to support and detail the main idea as necessary

There are no restrictions on how many power numbers you can add to help you extend and organize your ideas. Words having the same power number should have a similar relationship to the power above, but do not have to be related to each other.

Power 1: environmental problems
    Power 2: resource depletion
        Power 3: renewable resources
        Power 3: nonrenewable resources
    Power 2: pollution
        Power 3: degradable pollutants
        Power 3: nondegradable pollutants
    Power 2: extinction
        Power 3: pollution
        Power 3: habitat loss

Practice

1. Use this book's lesson on the experimental method and power notes structure to organize the following terms: *observing, hypothesizing and predicting, experimenting, organizing and analyzing data, drawing conclusions, repeating experiments, communicating results, observation, hypothesis, prediction, experiment, variable, experimental group, control group,* and *data.*

How to Make KWL Notes

KWL stands for "what I Know—what I Want to know—what I Learned," The KWL strategy is somewhat different from other learning strategies because it prompts you to brainstorm about the subject matter before reading the assigned material. Relating new ideas and concepts with those that you have learned will help you to understand and apply the knowledge you obtain in this course. The section objectives throughout your text are ideal for using the KWL strategy. Read the objectives before reading each section, and follow the instructions in the example below.

1 Read the section objectives.

You may also want to scan headings, boldfaced terms, and illustrations in the section. We'll use a few of the objectives from a section of this book as examples.
- List and describe the steps of the scientific method.
- Describe why a good hypothesis is not simply a guess.
- Describe the two essential parts of a good experiment.

2 Divide a sheet of paper into three columns. Label the columns "What I know," "What I want to know," and "What I learned".

Here is an example table:

What I know	What I want to know	What I learned

3 Brainstorm about what you know about the information in the objectives, and write these ideas in the first column.

Because this table is designed to help you blend your own knowledge with new information, it is not necessary to write complete sentences.

4 Think about what you want to know about the information in the objectives. Write these ideas in the second column.

You'll want to know the information you will be tested on, so include information from both the section objectives and any other objectives your teacher has given you.

5 Use the third column to write down the information you learned. Do this while you read the section, or just after reading.

While you read, pay close attention to any information about the topics you wrote in the "What I want to know" column. If you do not find all of the answers you are looking for, you may need to reread the section or reference a second source. Be sure to ask your teacher if you still cannot find the information after reading the section a second time.

When you have completed reading the section, review the ideas you brainstormed. Compare your ideas in the first column with the information you wrote down in the third column. If you find that some of the ideas are incorrect, cross them out. Before you begin studying for your test, identify and correct any misconceptions you had prior to reading.

Here is an example of what your notes might look like after using the KWL strategy:

What I know	What I want to know	What I learned
• The experimental method is: predict, test, and conclude.	• What are the steps of the experimental method?	• The steps of the experimental method are: observing, hypothesizing, experimenting, organizing and analyzing data, drawing conclusions, communicating results, and repeating experiments.
• A hypothesis is like a guess, but you have an idea of what might happen.	• Why is a hypothesis not a guess?	• A hypothesis is more than a guess. You have to base it on observations and really think about what you are trying to prove. You should also design an experiment that can test if your hypothesis is wrong, but you cannot prove that it is correct.
• A good experiment includes a hypothesis and a lot of equipment.	• What are the two important parts of a good experiment?	• The two important parts of a good experiment are a single variable and a control group.

Practice

1. Use the third column from the table above to identify and correct any misconceptions in the following list of ideas.
   a. The first step of the experimental method is to predict.
   b. A hypothesis is like a guess.
   c. A good experiment includes a lot of equipment.

How to Make Two-Column Notes

Two-column notes can be used to learn and review definitions of vocabulary terms, examples of multiple-step processes, or details of specific concepts. The two-column-note strategy is simple: write the term, main idea, step-by-step process, or concept in the left-hand column, and the definition, example, or detail on the right.

One strategy for using two-column notes is to organize main ideas and their details. The main ideas from your reading are written in the left-hand column of your paper and can be written as questions, key words, or a combination of both. Details describing these main ideas are then written in the right-hand column of your paper.

1 | **Identify the main ideas.**

The main ideas for a chapter are listed in the section objectives. However, you decide which ideas to include in your notes. The example below shows some main ideas from the objectives in a section of this book.

- Define environmental science and compare environmental science with ecology.
- List the six major fields of study that contribute to environmental science.

2 | **Divide a blank sheet of paper into two columns, and write the main ideas in the left-hand column.**

Remind yourself that your two-column notes are precisely that—notes. Do not copy whole phrases out of the book or waste your time writing ideas in complete sentences. Summarize your ideas using quick phrases that are easy for you to understand and remember. Decide how many details you need for each main idea, and write that number in parentheses under the main idea.

<u>Main idea</u>	<u>Detail notes</u>
Environmental science (two definitions)	
Goals of environmental science (one main goal)	
What is studied (two main areas)	
Related fields of study (four major fields)	

3 Write the detail notes in the right-hand column

List as many details as you designated in the main-idea column.

Main idea	Detail notes
Environmental science (two definitions)	the environment is everything around us • includes the natural world and things produced by humans • is a complex web of connections environmental science is the study of how humans interact with the environment
Goals of environmental science (one main goal)	to understand and solve environmental problems
What is studied (two main areas)	interactions between humans and their environment • the ways we use natural resources, such as water and plants • how our actions alter our environment
Related fields of study (four major fields)	biology is the study of living organisms • includes zoology, botany, microbiology, ecology earth science is the study of the Earth's nonliving systems and the planet as a whole • includes geology, climatology, paleontology, hydrology chemistry is the study of chemicals and their interactions • includes biochemistry social sciences are the study of human populations • include geography, anthropology, sociology and demographics

You can use two-column notes to study for a short quiz or for a test on the material in an entire chapter. Cover the information in the right-hand column with a sheet of paper. Recite what you know and then uncover the notes to check your answers. Then, ask yourself what else you know about that topic. Linking ideas in this way will help you gain a more complete picture of environmental science.

Reading and Study Skills

How to Make a Concept Map

Making concept maps can help you understand new ideas and decide what is important as you read. A concept map is a simple drawing that shows how concepts are connected to each other. Making your own concept maps is a good way to study and to test your understanding of what you have learned. Concept maps are especially helpful if you have a visual learning style, because you can literally draw connections between words and ideas.

To make a concept map, you write main ideas in a few words, and draw lines to show relationships between them. Concept maps can be based on key vocabulary terms from a text. These terms are usually nouns, which make good labels for major concepts. You may add linking words to explain relationships. A group of connected words and lines together show a proposition. A proposition is another way of stating a main idea or explaining a concept. For example, "matter is changed by energy" is a proposition.

1 List all the important concepts

We will use some of the terms from a chapter in this book as an example.

Earth	layers
composition	structure
crust	mantle
core	lithosphere
aesthenosphere	mesosphere
outer core	inner core

2 Select a main concept for the map

We will use Earth as the main concept for this map.

3 Build the map by placing the concepts under the main concept, according to their importance

One way of arranging the concepts is shown in the following map.

4 Add linking words to give meaning to the arrangement of the concepts

When adding the links, be sure that each proposition makes sense. To distinguish concepts from links, place your concepts in circles, ovals, or rectangles. Then add cross-links with lines connecting concepts across the map. Note how the following map has represented the ideas found in step 1.

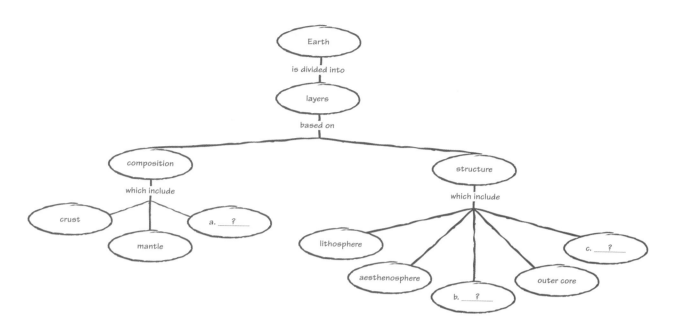

Practice mapping by making concept maps about topics you know, such as sports or hobbies. By perfecting your skills with information that you know very well, you will begin to feel more confident about making maps from the information in a chapter. Making maps might seem difficult at first, but the process gets you to think about the meanings and relationships among concepts. If you do not understand those relationships, you can get help early on.

Many people find it easier to study by looking at a concept map, rather than flipping through a book. Concept maps can be used to organize the information in a chapter by isolating the key concepts and making the relationships among ideas easy to see and understand. Another useful strategy is to trade concept maps with a classmate. Everybody organizes information in a slightly different way, and something your classmate may have done may help you understand the content better. Although concept mapping may take a little extra time, the time will pay off when you are reviewing for tests.

Practice

1. Use concepts from this book to fill in the rest of the blanks on the concept map above.
2. Write three propositions from the completed map.
3. Classify each of the following as either a concept or linking word(s).
   a. Earth
   b. is divided into
   c. which include
   d. crust

Reading and Study Skills

How to Make FoldNotes

Have you ever tried to study for a test or quiz but didn't know where to start? Or have you read a chapter and found that you can remember only a few ideas? Well, FoldNotes are a fun and exciting way to help you learn and remember the ideas you encounter as you learn science!

FoldNotes are tools that you can use to organize concepts. One Fold-Note focuses on a few main concepts. FoldNotes help you learn and remember how the concepts fit together. FoldNotes can help you see the "big picture." Below you will find instructions for building 10 different FoldNotes.

Pyramid

1. Place a **sheet of paper** in front of you. Fold the lower left-hand corner of the paper diagonally to the opposite edge of the paper.

2. Cut off the tab of paper created by the fold (at the top).

3. Open the paper so that it is a square. Fold the lower right-hand corner of the paper diagonally to the opposite corner to form a triangle.

4. Open the paper. The creases of the two folds will have created an **X.**

5. Using **scissors,** cut along one of the creases. Start from any corner, and stop at the center point to create two flaps. Use **tape** or **glue** to attach one of the flaps on top of the other flap.

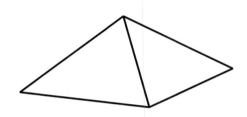

Double-Door Fold

1. Fold a **sheet of paper** in half from the top to the bottom. Then, unfold the paper.

2. Fold the top and bottom edges of the paper to the center crease.

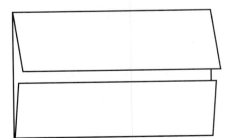

Table Fold

1. Fold a **piece of paper** in half from the top to the bottom. Then, fold the paper in half again.

2. Fold the paper in thirds from side to side.

3. Unfold the paper completely. Carefully trace the fold lines by using a **pen** or **pencil**.

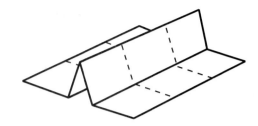

Booklet

1. Fold a **sheet of paper** in half from left to right. Then, unfold the paper.

2. Fold the sheet of paper in half again from the top to the bottom. Then, unfold the paper.

3. Refold the sheet of paper in half from left to right.

4. Fold the top and bottom edges to the center crease.

5. Completely unfold the paper.

6. Refold the paper from top to bottom.

7. Using **scissors,** cut a slit along the center crease of the sheet from the folded edge to the creases made in step 4. Do not cut the entire sheet in half.

8. Fold the sheet of paper in half from left to right. While holding the bottom and top edges of the paper, push the bottom and top edges together so that the center collapses at the center slit. Fold the four flaps to form a four-page book.

Layered Book

1. Lay **one sheet of paper** on top of **another sheet.** Slide the top sheet up so that 2 cm of the bottom sheet is showing.

2. Holding the two sheets together, fold down the top of the two sheets so that you see four 2 cm tabs along the bottom.

3. Using a **stapler,** staple the top of the Fold-Note.

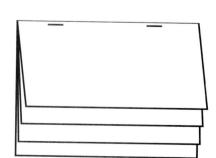

Two-Panel Flip Chart

1. Fold a **piece of paper** in half from the top to the bottom.

2. Fold the paper in half from side to side. Then, unfold the paper so that you can see the two sections.

3. From the top of the paper, cut along the vertical fold line to the fold in the middle of the paper. You will now have two flaps.

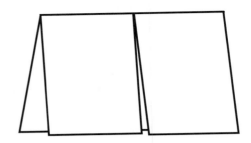

Key-Term Fold

1. Fold a **sheet of lined notebook paper** in half from left to right.

2. Using **scissors,** cut along every third line from the right edge of the paper to the center fold to make tabs.

Four-Corner Fold

1. Fold a **sheet of paper** in half from left to right. Then, unfold the paper.

2. Fold each side of the paper to the crease in the center of the paper.

3. Fold the paper in half from the top to the bottom. Then, unfold the paper.

4. Using **scissors,** cut the top flap creases made in step 3 to form four flaps.

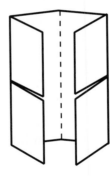

Three-Panel Flip Chart

1. Fold a **piece of paper** in half from the top to the bottom.
2. Fold the paper in thirds from side to side. Then, unfold the paper so that you can see the three sections.
3. From the top of the paper, cut along each of the vertical fold lines to the fold in the middle of the paper. You will now have three flaps.

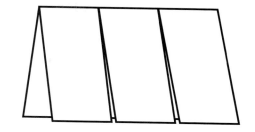

Tri-Fold

1. Fold a **piece a paper** in thirds from the top to the bottom.
2. Unfold the paper so that you can see the three sections. Then, turn the paper sideways so that the three sections form vertical columns.
3. Trace the fold lines by using a **pen** or **pencil.** Label the columns "Know," "Want," and "Learn."

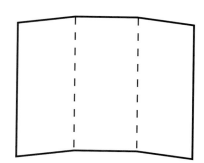

How to Make Graphic Organizers

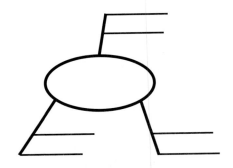

Have you ever wished that you could draw the many concepts you learn in your science class? Sometimes, being able to see how concepts are related helps you remember what you've learned. Graphic Organizers help you see the concepts! They are a way to draw or map out concepts.

You need only a piece of paper and a pencil to make a Graphic Organizer. Below, you will find instructions for five different Graphic Organizers that are designed to help you organize the concepts you'll learn in this book.

Spider Map

1. Draw a diagram like the one shown. In the circle, write the main topic.

2. From the circle, draw legs to represent different categories of the main topic. You can have as many categories as you want.

3. From the category legs, draw horizontal lines. As you read the chapter, write details about each category on the horizontal lines.

Comparison Table

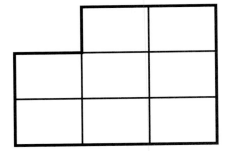

1. Draw a chart like the one shown. Your chart can have as many columns and rows as you want.

2. In the top row, write the topics that you want to compare.

3. In the left column, write characteristics of the topics that you want to compare. As you read the chapter, fill in the characteristics for each topic in the appropriate boxes.

Chain-of-Events-Chart

1. Draw a box. In the box, write the first step of a process or the first event of a timeline.

2. Under the box, draw another box, and use an arrow to connect the two boxes. In the second box, write the next step of the process or the next event in the timeline.

3. Continue adding boxes until the process or timeline is finished.

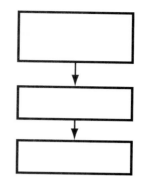

Venn Diagram

1. Draw a diagram like the one shown. You may have two or three circles depending on the number of topics. Make sure the circles overlap with each other.

2. In each circle, write a topic that you want to compare with a topic in another circle.

3. In the areas of the diagram where circles overlap, fill in characteristics that the topics in the overlapping circles share.

4. In the areas of the diagram where circles do not overlap, fill in characteristics that are unique to the topic of the particular circle.

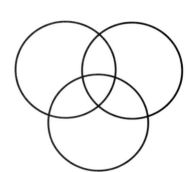

Cause-and-Effect Map

1. Draw a box and write a cause in the box. You can have as many cause boxes as you want. The diagram shown here is one example of a cause-and-effect map.

2. Draw another box to represent an effect of the cause. You can have as many effect boxes as you want. Draw a line from each cause to the effect(s).

3. In the cause boxes, write a description, explanation, or details about the cause. In the effect boxes, explain the effects that result from the process or factor identified in the cause box.

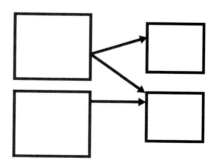

Reading and Study Skills

Analyzing Science Terms

You can often unlock the meaning of an unfamiliar science term by analyzing its word parts. Many parts of scientific word carry a meaning that derives from Latin or Greek. The parts of words listed below provide clues to the meanings of many science terms.

Word part	Meaning	Example
a-	not, without	abiotic
acr-, agr-	field	agriculture, acre
amphi-	both	amphibian
anti-	against	antibiotic
atmos-	vapor	atmosphere
auto-	self, same	autotrophic
benth-	depth	benthic, benthos
bio-	life	biology, biosphere, biotic
chloro-	green	chlorophyll
-cide	kill	insecticide, fungicide
co-, con-	with, together	coevolution, cooperation, commensalism
dem-, demo-	people	demography, epidemic
-duct-	to lead, draw	reproduction
e-, ec-, ex-	out, away from, outside	extinction, experiment
eco-	home, environment	ecology, ecosystem, economics
eu-	good, well	eutrophic
evolu-	to unroll	evolution
gen-	to give birth, produce	genetic, generation, genus
geo-	earth	geology, geosphere
hetero-	different	heterotroph
hydro-	water	hydrosphere, hydroelectric
im-, in-, ir-	not, without or in, into	invertebrate, immigration, irrigation
-ion	the act of	pollution, destruction
lith-	stone	lithosphere
-log-	to study	ecology, geology
-lu-, -lue-	dirt, impurity	pollution
mar-	sea	marine
micro-	small	microscopic, microorganisms
nutri-	food, nourishment	nutrient
organ-	tool, instrument	organic, organism
per-	through	permeable
photo-	light	photosynthesis
phyto-, -phyte	plant	phytoplankton, epiphyte
pre-	before, in front of	predator, prey, precipitation
pro-	forward	reproduction
re-	back, again	recycle, reproduce
spec-	appearance or shape	species, spectrum
-sphere	ball, globe	geosphere, ecosphere
stat-	position, standing	statistics, status
strati-, strato-	spread, layer	stratosphere
temper-	to measure or regulate	temperate, temperature
terra-, terre-	earth, land	terrain, terrestrial
thermo-	heat	thermosphere, thermal
-troph-	to feed, gather	eutrophic, autotroph

Math Skills Refresher

Geometry

A useful way to model the objects and substances studied in science is to consider them in terms of their shapes. For example, many of the properties of a wheel can be understood by pretending that the wheel is a perfect circle.

When using shapes as models, your ability to calculate the area or volume of shapes is a useful skill. The table below provides equations for the area and volume of several geometric shapes.

Geometric Areas and Volumes	
Geometric shape	**Useful equations**
Rectangle	Area $= lw$
Circle	Area $= \pi r^2$ Circumference $= 2\pi r$
Triangle	Area $= \frac{1}{2}bh$
Sphere	Surface area $= 4\pi r^2$ Volume $= \frac{4}{3}\pi r^3$
Cylinder	Volume $= \pi r^2 h$
Rectangular box	Surface area $= 2(lh + lw + hw)$ Volume $= lwh$

Practice

1. Calculate the area of a triangle that has a base of 900.0 m and a height of 500.0 m.
2. What is the volume of a cylinder that has a diameter of 14 cm and a height of 8 cm?
3. Calculate the surface area of a 4 cm cube.

Exponents

An exponent is a number that is superscripted to the right of another number. The best way to explain how an exponent works is with an example. In the value 5^4, the 4 is the exponent of the 5. The number with its exponent means that 5 is multiplied by itself 4 times:

$$5^4 = 5 \times 5 \times 5 \times 5 = 625$$

You will frequently hear exponents referred to as *powers*. Using this terminology, the above equation could be read as "five to the fourth power equals 625," or "five to the power of four equals 625." Keep in mind that any number raised to the power of 0 is equal to 1: $5^0 = 1$. Also, any number raised to the power of 1 is equal to itself: $5^1 = 5$.

A scientific calculator is a must for solving most problems involving exponents. Many calculators have dedicated keys for squares and square roots, but scientific calculators usually have a special key shaped like a caret, ^, for entering exponents. If you type in "5^4" and then hit the "=" key or the "Enter" key, the calculator will determine that 54 = 625 and display that answer.

Exponents		
	Rule	**Example**
Zero power	$x^0 = 1$	$7^0 = 1$
First power	$x^1 = x$	$6^1 = 6$
Multiplication	$(x^n)(x^m) = (x^{n+m})$	$(x^2)(x^4) =$ $x^{(2+4)} = x^6$
Division	$\frac{x^n}{x^m} = x^{(n-m)}$	$\frac{x^8}{x^2} = x^{(8-2)} = x^6$
Exponents raised to a power	$(x^n)^m = x^{nm}$	$(5^2)^3 =$ $5^6 = 15,625$

Practice

1. Perform the following calculations:
   - **a.** $9^1 =$
   - **b.** $(3^3)^5 =$
   - **c.** $(14^2)(14^3) =$
   - **d.** $11^0 =$

Math Skills Refresher

Order of Operations

Use this phrase to remember the correct order for long mathematical problems: "Please Excuse My Dear Aunt Sally" (some people just remember the acronym "PEMDAS"). This acronym stands for **p**arentheses, **e**xponents, **m**ultiplication, **di**vision, **a**ddition, and **s**ubtraction. This is the correct order in which to complete operations. These rules are summarized in the table below.

Order of Operations

1 Simplify groups inside parentheses. Start with the innermost group and work out.

2 Simplify all exponents.

3 Perform multiplication and division in order from left to right.

4 Perform addition and subtraction in order from left to right.

Look at the following example.

$$4^3 + 2 \times [8 - (3 - 1)] = ?$$

First, simplify the operations inside parentheses. Begin with the innermost parentheses:

$$(3 - 1) = 2$$
$$4^3 + 2 \times [8 - 2] = ?$$

Then, move on to the next-outer parentheses:

$$[8 - 2] = 6$$
$$4^3 + 2 \times 6 = ?$$

Now, simplify all exponents:

$$4^3 = 64$$
$$64 + 2 \times 6 = ?$$

Next, perform the remaining multiplication:

$$2 \times 6 = 12$$
$$64 + 12 = ?$$

Finally, perform the addition:

$$64 + 12 = 76$$

Practice

1. $2^3 \div 2 + 4 \times (9 - 2^2) =$
2. $\dfrac{2 \times (6 - 3) + 8}{4 \times 2 - 6}$

Algebraic Rearrangements

Algebraic equations contain *constants* and *variables*. Constants are simply numbers, such as 2, 5, and 7. Variables are represented by letters such as x, y, z, a, b, and c. Variables are unspecified quantities and are also called the *unknowns*. Often, you will need to determine the value of a variable in an equation that contains algebraic expressions.

An algebraic expression contains one or more of the four basic mathematical operations: addition, subtraction, multiplication, and division. Constants, variables, or terms made up of both constants and variables can be involved in the basic operations.

The key to finding the value of a variable in an algebraic equation is that the total quantity on one side of the equals sign is equal to the quantity on the other side. If you do the same operation on either side of the equation, the results will still be equal. To determine the value of a variable in an algebraic expression, you try to reduce the equation into a simple one that tells you exactly what x (or some other variable) equals.

Look at the simple problem below.

$$8x = 32$$

If we wish to solve for x, we can multiply or divide each side of the equation by the same factor. You can perform any operation on one side of an equation as long as you do the same thing to the other side of the equation. In this example, if we divide both sides of the equation by 8, we have:

$$\frac{8x}{8} = \frac{32}{8}$$

The 8s on the left side of the equation cancel each other out, and the fraction $\frac{32}{8}$ can be reduced to give the whole number, 4. Therefore, $x = 4$.

Next, consider the following equation.

$$2x + 4 = 16$$

If we divide each side by 2, we are left with $x + 2$ on the left and 8 on the right:

$$x + 2 = 8$$

Now, we can subtract 2 from each side of the equation to find that $x = 6$. In all cases, whatever operation is performed on the left side of the equals sign must also be performed on the right side.

Practice

1. Rearrange each of the following equations to give the value of the variable indicated with a letter.
   a. $8x - 32 = 128$
   d. $-2(3m + 5) = 14$
   b. $6 - 5(4a + 3) = 26$
   e. $\left[8 \frac{(8 + 2z)}{32}\right] + 2 = 5$
   c. $-3(y - 2) + 4 = 29$
   f. $\frac{(6b + 3)}{3} - 9 = 2$

Powers of 10	
Power of 10	Decimal Equivalent
10^4	10,000
10^3	1,000
10^2	100
10^1	10
10^0	1
10^{-1}	0.1
10^{-2}	0.01
10^{-3}	0.001

Scientific Notation

Many quantities that scientists deal with are very large or very small values. For example, light travels at about 300,000,000 meters per second, and an electron has a mass of about 0.000 000 000 000 000 000 000 000 000 9 g. Obviously, it is difficult to read, write, and keep track of numbers like these. We avoid this problem by using a method dealing with powers of the number 10.

Study the positive powers of 10 shown in the following table. You should be able to check these numbers using what you know about exponents. The number of zeros in the equivalent number corresponds to the exponent of the 10, or the power to which the 10 is raised. The equivalent of 10^4 is 10,000, so the number has four zeros.

But how can we use the powers of 10 to simplify large numbers such as the speed of light? The speed of light is equal to $3 \times 100,000,000$ m/s. The factor of 10 in this number has 8 zeros, so it can be rewritten as 10^8. So, 300,000,000 can be expressed as 3×10^8.

Negative exponents can be used to simplify numbers that are less than 1. Study the negative powers of 10 in the table. In these cases,

the exponent of 10 equals the number of decimal places you must move the decimal point to the right so that there is one digit just to the left of the decimal point. In the case of the mass of an electron, the decimal point has to be moved 28 decimal places to the right for the numeral 9 to be just to the left of the decimal point. The mass of the electron, about 0.000 000 000 000 000 000 000 000 000 9 g, can be rewritten as about 9×10^{-28} g.

Scientific notation is a way to express numbers as a power of 10 multiplied by another number that has only one digit to the left of the decimal point. For example, 5,943,000,000 is 5.943×10^9 when expressed in scientific notation. The number 0.000 083 2 is 8.32×10^{-5} when expressed in scientific notation.

Practice

1. Rewrite the following values using scientific notation.
   a. 12,300,000 m/s
   b. 0.000 000 000 004 5 kg
   c. 0.000 065 3 m
   d. 55,432,000,000,000 s
   e. 273.15 K
   f. 0.000 627 14 kg

Math Skills Refresher

Significant Digits

The following list can be used to review how to determine the number of *significant digits* (also called *significant figures*) in a given value or measurement.

Rules for Significant Digits:

1. All nonzero digits are significant. For example, 1,246 has four significant digits (shown in red).

2. Any zeros between significant digits are also significant. For example, 1,206 has four significant figures.

3. If the value does not contain a decimal point, any zeros to the right of a nonzero digit are not significant. For example, 1,200 has only two significant digits.

4. Any zeros to the right of a significant digit and to the left of a decimal point are significant. For example, 1,200.0 has four significant digits.

5. If a value has no significant digits to the left of a decimal point, any zeros to the right of the decimal point and also to the left of a significant digit are not significant. For example, 0.0012 has only two significant digits.

6. If a value ends with zeros to the right of a decimal point, those zeros are significant. For example, 0.1200 has four significant digits.

After you have reviewed the rules, use the following table to check your understanding of the rules. Cover up the second column of the table, and try to determine how many significant digits each number in the first column has. If you get confused, refer to the rule given.

When performing mathematical operations with measurements, you must remember to keep track of significant digits. If you are adding or

Significant Digits		
Measurement	Number of significant digits	Rule
12,345	5	1
2,400 cm	2	3
305 kg	3	2
2,350. cm	4	4
234.005 K	6	2
12.340	5	6
0.001	1	5
0.002 450	4	5 and 6

subtracting two measurements, your answer can only have as many decimal positions as the value that has the fewest number of decimal places. When multiplying or dividing measurements, your answer can only have as many significant digits as the value with the fewest number of significant digits.

Practice

1. Determine the number of significant digits in each of the following measurements:
   a. 65.04 mL   c. 0.007 504 kg
   b. 564.00 m   d. 1,210 K

2. Perform each of the following calculations, and report your answer with the correct number of significant digits and units:
   a. 0.004 dm + 0.12508 dm
   b. 340 m ÷ 0.1257 s
   c. 40.1 m × 0.2453 m
   d. 1.03 g − 0.0456 g

Graphing Skills

Line Graphs

In laboratory experiments, you will usually be controlling one variable and seeing how it affects another variable. Line graphs can show these relations clearly. For example, you might perform an experiment in which you measure the growth of a plant over time to determine the rate of the plant's growth. In this experiment, you are controlling the time intervals at which the plant height is measured. Therefore, time is the independent variable. The height of the plant is the dependent variable. The table below gives some sample data for an experiment that measures the rate of plant growth.

The independent variable is plotted on the x-axis. This axis will be labeled "Time (days)" and will have a range from 0 to 35 days. Be sure to properly label each axis, including the units.

The dependent variable is plotted on the y-axis. This axis will be labeled "Plant Height (cm)" and will have a range from 0 to 5 cm.

| Experimental Data for Plant Growth Vs. Time ||
Time (days)	Plant height (cm)
0	1.43
7	2.16
14	2.67
21	3.25
28	4.04
35	4.67

Think of your graph as a grid with lines running horizontally from the y-axis and vertically from the x-axis. To plot a point, find the x value on the x axis. For the example above, plot each value for time on the x axis. Follow the vertical line from the x axis until it intersects the horizontal line from the y-axis at the corresponding y value. For the example, each time value has a corresponding height value. Place your point at the intersection of these two lines.

The line graph below shows how the data in the table might be graphed.

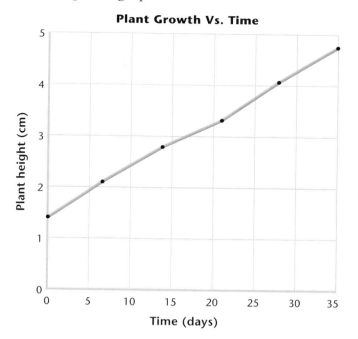

Bar Graphs

Bar graphs are useful for comparing data values. If you wanted to compare the area or depth of the major oceans, you might use a bar graph. The table below gives the data for each of these quantities.

| Depth of the Major Oceans ||
Ocean	Depth (m)
Pacific Ocean	4,028
Atlantic Ocean	3,926
Indian Ocean	3,963
Arctic Ocean	1,205

To create a bar graph from the data in the table, begin on the x-axis by labeling four bar positions with the names of the four oceans. Label the y-axis "Depth (m)." Be sure the range on your y-axis encompasses 1,205 m and 4,028 m. Then, draw the bars to represent the area of each ocean, with a bar height on the *y*-axis that matches each ocean's area value, as shown in the bar graph below.

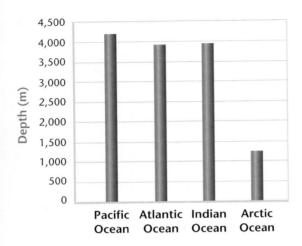

Pie Graphs

Pie graphs are an easy way to visualize how many parts make up a whole. Frequently, pie graphs are made from percentage data. For example, you could create a pie graph showing percentage of different materials that make up the waste generated in cities of the United States. Study the example data in the table below.

United States Municipal Solid Waste	
Material	**Percentage of total waste**
Paper	38%
Yard waste	12%
Food waste	11%
Plastics	11%
Metals	8%
Rubber, leather, and textiles	7%
Glass	6%
Wood	5%
Other	3%

To create a pie graph from the data in the table, begin by drawing a circle to represent the whole, or total. Then, imagine dividing the circle into 100 equal sections, to represent 100 percent. Shade in 38 consecutive sections, and label that area "Paper." Continue to shade sections with other colors until the entire pie graph has been filled in and until each type of waste has a corresponding area in the circle, as shown in the pie graph below.

United States Municipal Solid Waste (Percentage by Weight)

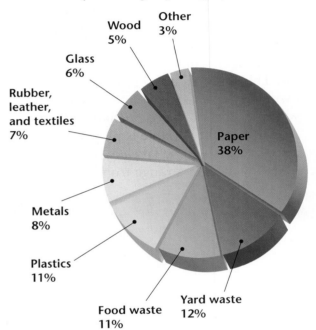

Chemistry Refresher

Atoms and Elements

Every object in the universe is made up of particles of matter. Matter is anything that has mass and takes up space. An element is a substance that cannot be separated into simpler substances by chemical means. Elements cannot be separated in this way because each element consists of only one kind of atom. An atom is the smallest unit of an element that maintains the properties of that element.

Atomic Structure Atoms are made up of small particles called *subatomic particles*. The three major types of subatomic particles are **electrons, protons,** and **neutrons.** Electrons have a negative electrical charge, protons have a positive charge, and neutrons have no electrical charge. The protons and neutrons are packed close to one another and form the **nucleus.** The protons give the nucleus a positive charge. The electrons of an atom are located in a region around the nucleus known as an **electron cloud.** The negatively charged electrons are attracted to the positively charged nucleus. An atom may have several energy levels in which electrons are located.

Atomic Number To help in the identification of elements, scientists have assigned an **atomic number** to each kind of atom. The atomic number is equal to the number of protons in the atom. Atoms with the same number of protons are all of the same element. In an uncharged, or electrically neutral, atom there are an equal number of protons and electrons. Therefore, the atomic number also equals the number of electrons in an uncharged atom. The number of neutrons, however, can vary for a given element. Atoms that have different numbers of neutrons but are of the same element are called **isotopes.**

Periodic Table of the Elements A periodic table of the elements is shown on the next page. In a periodic table, the elements are arranged in order of increasing atomic number. Each element in the table is found in a separate box. In each horizontal row of the table, each element has one more

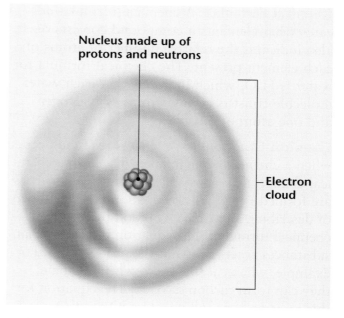

Nucleus made up of protons and neutrons

Electron cloud

▶ The nucleus of the atom contains the protons and neutrons. The protons give the nucleus a positive charge. The negatively charged electrons are in the electron cloud surrounding the nucleus.

electron and one more proton than the element to its left. Each row of the table is called a **period.** Changes in chemical properties across a period correspond to changes in the elements' electron arrangements. Each vertical column of the table, known as a **group,** contains elements that have similar properties. The elements in a group have similar chemical properties because they have the same number of electrons in their outer energy level. For example, the elements helium, neon, argon, krypton, xenon, and radon all have similar properties and are known as the noble gases.

Molecules and Compounds When the atoms of two or more elements are joined chemically, the resulting substance is called a **compound.** A compound is a new substance with properties different from those of the elements that compose it. For example, water (H_2O) is a compound formed when atoms of hydrogen (H) and oxygen (O) combine. The smallest complete unit of a compound that has all of the properties of that compound is called a **molecule.**

Chemical Formulas A chemical formula indicates what elements a compound consists of. It also indicates the relative number of atoms of each element present. The chemical formula for water is H_2O, which indicates that each water molecule consists of two atoms of hydrogen and one atom of oxygen.

Chemical Equations A chemical reaction occurs when a chemical change takes place. (In a chemical change, new substances with new properties are formed.) A chemical equation is a useful way of describing a chemical reaction by means of chemical formulas. The equation indicates what substances react and what the products are. For example, when carbon and oxygen combine, they can form carbon dioxide. The equation for this reaction is as follows: $C + O_2 \rightarrow CO_2$.

Acids, Bases, and pH An ion is an atom or group of atoms that has an electrical charge because it has lost or gained one or more electrons. When an acid, such as hydrochloric acid (HCl), is mixed with water, it separates into ions. An **acid** is a compound that produces hydrogen ions (H^+) in water. The hydrogen ions then combine with a water molecule to form a hydronium ion (H_3O^+). A solution that contains hydronium ions is an acidic solution. A **base**, on the other hand, is a substance that produces hydroxide ions (OH^-) in water.

To determine whether a solution is acidic or basic, scientists measure pH. **pH** is a measure of how many hydronium ions are in solution. The pH scale ranges from 0 to 14. The middle point, pH = 7, is neutral, neither acidic nor basic. Acids have a pH of less than 7; bases have a pH of more than 7. The lower the number, the stronger the acid. The higher the number, the stronger the base. A pH scale is shown in Figure 12, on page 314.

internet connect

Topic: Periodic Table
Go To: go.hrw.com
Keyword: Holt Periodic

Visit the HRW Web site for updates on the periodic table

THE PERIODIC TABLE OF THE ELEMENTS

Metals	**Nonmetals**
Alkali metals	Hydrogen
Alkaline-earth metals	Semiconductors
Transition metals	Halogens
Other metals	Noble Gases
	Other nonmetals

6 — Atomic number
C — Symbol
Carbon — Name

Group 18

1 **H** Hydrogen																	2 **He** Helium
Group 1	Group 2											Group 13	Group 14	Group 15	Group 16	Group 17	
3 **Li** Lithium	4 **Be** Beryllium											5 **B** Boron	6 **C** Carbon	7 **N** Nitrogen	8 **O** Oxygen	9 **F** Fluorine	10 **Ne** Neon
11 **Na** Sodium	12 **Mg** Magnesium	Group 3	Group 4	Group 5	Group 6	Group 7	Group 8	Group 9	Group 10	Group 11	Group 12	13 **Al** Aluminum	14 **Si** Silicon	15 **P** Phosphorus	16 **S** Sulfur	17 **Cl** Chlorine	18 **Ar** Argon
19 **K** Potassium	20 **Ca** Calcium	21 **Sc** Scandium	22 **Ti** Titanium	23 **V** Vanadium	24 **Cr** Chromium	25 **Mn** Manganese	26 **Fe** Iron	27 **Co** Cobalt	28 **Ni** Nickel	29 **Cu** Copper	30 **Zn** Zinc	31 **Ga** Gallium	32 **Ge** Germanium	33 **As** Arsenic	34 **Se** Selenium	35 **Br** Bromine	36 **Kr** Krypton
37 **Rb** Rubidium	38 **Sr** Strontium	39 **Y** Yttrium	40 **Zr** Zirconium	41 **Nb** Niobium	42 **Mo** Molybdenum	43 **Tc** Technetium	44 **Ru** Ruthenium	45 **Rh** Rhodium	46 **Pd** Palladium	47 **Ag** Silver	48 **Cd** Cadmium	49 **In** Indium	50 **Sn** Tin	51 **Sb** Antimony	52 **Te** Tellurium	53 **I** Iodine	54 **Xe** Xenon
55 **Cs** Cesium	56 **Ba** Barium	57 **La** Lanthanum	72 **Hf** Hafnium	73 **Ta** Tantalum	74 **W** Tungsten	75 **Re** Rhenium	76 **Os** Osmium	77 **Ir** Iridium	78 **Pt** Platinum	79 **Au** Gold	80 **Hg** Mercury	81 **Tl** Thallium	82 **Pb** Lead	83 **Bi** Bismuth	84 **Po** Polonium	85 **At** Astatine	86 **Rn** Radon
87 **Fr** Francium	88 **Ra** Radium	89 **Ac** Actinium	104 **Rf** Rutherfordium	105 **Db** Dubnium	106 **Sg** Seaborgium	107 **Bh** Bohrium	108 **Hs** Hassium	109 **Mt** Meitnerium	110 **Ds** Darmstadtium	111 **Uuu*** Unununium	112 **Uub*** Ununbium	113 **Uut*** Ununtrium	114 **Uuq*** Ununquadium	115 **Uup*** Ununpentium			

A team at Lawrence Berkeley National Laboratories reported the discovery of elements 116 and 118 in June 1999. The same team retracted the discovery in July 2001. The discovery of elements 113, 114, and 115 has been reported but not confirmed.

* The systematic names and symbols for elements greater than 110 will be used until the approval of trivial names by IUPAC.

58 **Ce** Cerium	59 **Pr** Praseodymium	60 **Nd** Neodymium	61 **Pm** Promethium	62 **Sm** Samarium	63 **Eu** Europium	64 **Gd** Gadolinium	65 **Tb** Terbium	66 **Dy** Dysprosium	67 **Ho** Holmium	68 **Er** Erbium	69 **Tm** Thulium	70 **Yb** Ytterbium	71 **Lu** Lutetium
90 **Th** Thorium	91 **Pa** Protactinium	92 **U** Uranium	93 **Np** Neptunium	94 **Pu** Plutonium	95 **Am** Americium	96 **Cm** Curium	97 **Bk** Berkelium	98 **Cf** Californium	99 **Es** Einsteinium	100 **Fm** Fermium	101 **Md** Mendelevium	102 **No** Nobelium	103 **Lr** Lawrencium

Economics Concepts

You may think that economics is about the complicated numbers of stock markets and interest rates, but the field of economics is based on simple concepts. *Economics* is the study of how people make decisions about the production, distribution, and consumption of limited resources as they attempt to fulfill their needs and wants. While economics can be a complex subject to study, it is a key part of understanding the relationship of humans and their environment. Here we will present some of the most basic concepts of economics.

Resources and Value

Resources that people use to create useful and desirable products are called *economic resources* or *capital*. Products and capital may exist in the form of *goods* or *services*. There are three general types of capital: natural, manufactured, and human. *Natural resources,* sometimes called *earth capital,* are resources such as land, fertile soil, air and water, oceans, wildlife, and minerals. *Manufactured capital* includes tools, machines, buildings, and other things that are made from natural resources and that are used to produce goods and services. *Human capital* includes the mental and physical abilities for which people may be paid wages or salaries.

A given resource or product has a specific *value* to a given group of people at a particular time. Generally, the *value* of something is the amount of money most people are willing to pay for it. For example, if many people are willing to pay $15 for a music CD, the value of that CD is $15. But value is not always the same as price. Some stores might sell the CD for $5. If the CD's price were lower than its value, an economist would say the CD is *undervalued*.

Economic activities that produce goods and services are called *industries*. A basic industry involves people using natural resources directly. The highest level of industry involves people working mostly with information instead of goods. The degree to which a single business or economic system has activities at multiple levels of industry is called *economic diversity*. For example, the United States has a high degree of economic diversity, but a small island that subsists on fishing and tourism has low economic diversity.

Economic Systems and Governments

Most societies use some type of *economic system* to decide what to produce, how to produce, and for whom to produce. The main difference between economic systems is in how much the government regulates the activities of businesses or controls access to resources.

In a *market economy* (also called a *capitalist, free-market,* or *free-enterprise system*), people own businesses and make their own decisions about what to make, sell, or buy. The theory of a market economy is that competition in open markets will result in the highest-quality goods being produced in the most efficient way and for the lowest price. In market theory, individuals acting in self-interest will efficiently decide what goods and services to buy or sell, so supply will be balanced with demand.

In a *command economy,* or a *centrally planned economy,* the government controls production and determines the amount and price of goods and services produced. Command economies are typically practiced by communist governments. A few countries and cultural groups still practice a *traditional economy,* which means they make economic decisions based on local customs or traditions.

However, economic systems are rarely practiced exactly according to theory. Most countries have a *mixed economy* in which a combination of government control and free markets exist. Governments may produce goods and services or may try to influence the flow of goods and services by charging taxes, paying out subsidies, or making regulations.

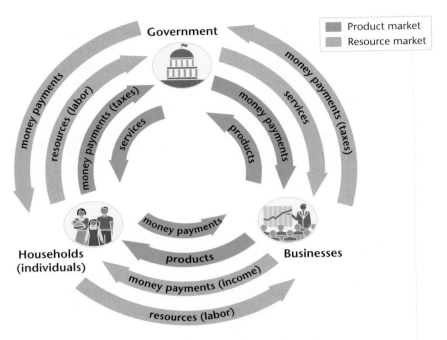

Product market
Resource market

Government

Households (individuals)

Businesses

money payments

resources (labor)

money payments (taxes)

services

services

money payments

products

money payments (taxes)

money payments

products

money payments (income)

resources (labor)

▶ **An Economic System Model** This *circular-flow model* illustrates the exchange of resources, products, and money payments in the economic system of nations such as the United States.

Economic Growth and Development

The economic growth within a country is usually measured by looking at the country's *gross domestic product* (GDP), or by looking at the *gross national product* (GNP). A country's GDP is the total value of all goods and services produced within the country in a year. The GNP is like the GDP, but the GNP includes income from outside of the country generated by individuals or companies based within the country. To represent each person's part in the economy, economists calculate the GNP or GDP *per capita,* which means the average GNP or GDP per person in the country.

Economists and social scientists often categorize countries based on indicators of their economic and social development. Countries that have high average incomes, slow population growth, diverse economies, and strong social support systems are considered to be *more devel-*

oped. Countries that have low average incomes, simple economies, and rapid population growth are considered to be *less developed.* However, these categories are difficult to apply because countries may develop in different ways and because the economies of different countries are interconnected as people and goods move between countries.

The economies of the world are now so interconnected that economists often refer to the *global economy.* In the 20th century, most countries became more developed and tended toward market system economies. Also, international trade continued to increase. Many countries now work together to help manage the global economy. International organizations such as the World Bank, the World Trade Organization, or the European Union have become as influential as national governments.

ENVIRONMENTAL EDUCATOR

As a child who watched *The Underwater World of Jacques Cousteau* on public television every chance she had, **Niki Espy** dreamed of one day studying aquatic mammals for a living. She went to college with the intent of continuing on to graduate school to focus on behavioral studies in marine biology. But while pursuing a bachelor's degree in biology, she interned as a naturalist. Today, Niki works for the Milwaukee Public Museum. She provides educational programs for children, adults, and families and is responsible for developing and implementing school programs that focus on cultural and natural history. Niki also facilitates training for educators, including student teachers, active teachers, museum volunteers, and museum docents.

> **If we look at humans as a separate component of the world, we will not be able to truly reach sustainability.**

Q: How does your current job relate to environmental education?

Niki Espy: I use the principles of environmental education to teach about natural and cultural history. The basics of awareness, appreciation, knowledge, and action assist me daily in my educational endeavors. I believe that if we don't have an understanding of the world, we can't begin to value or protect our resources. The museum's educational programs lead students to question, explore, analyze, evaluate, and discuss how the introductions of exotic plants and animals and the urbanization of the Milwaukee area have affected biodiversity. While interpreting the plant and animal changes, we don't forget the people and how indigenous groups used the land.

Q: What is the importance of including people in a discussion on biodiversity and environmental impact?

Niki Espy: If we look at humans as a separate component of the world, we will not be able to truly reach sustainability. By placing people in the equation, we can look at our behaviors and our impact on local and global ecoregions, economies, and social systems and can obtain the answers we need to create a sustainable future.

MORE ON THIS CAREER

Many museums have volunteer programs in which volunteers work directly with the public or in different administrative or scientific departments. For example, volunteers at the Milwaukee Public Museum may provide assistance at the information desk, give tours to the public through the exhibit galleries, demonstrate objects visitors can touch, help educate visitors about special exhibits, and work at special events. In addition, volunteers may work behind the scene in research areas such as anthropology, archeology, botany, geology, paleontology, and zoology. For more information on volunteer programs, contact a museum located near you.

▶ One of Niki Espy's goals as an environmental educator is to increase awareness and knowledge of the natural world.

Environmental Careers

ENVIRONMENTAL ENGINEER

John Roll began college studying chemistry and biology, and then halfway through his undergraduate degree he switched to agricultural engineering with a focus on water-quality issues. Roll completed his master's degree in agricultural engineering and expanded his environmental background to include livestock waste management. He spent the first three years after graduation on a project involving treated solids from a municipal wastewater treatment facility. For the next 14 years, John was manager of land reclamation and environmental permits for a surface coal mining company in Illinois. In 1990, Roll entered Oklahoma State University, where he studied groundwater transport of contaminants and received a Doctorate in Biosystems Engineering.

Q: What does an environmental engineer study in college?

John: First and foremost, a student must obtain an engineering degree. The individual must have a desire to study hard, and it helps to possess an aptitude for the math and hard-science courses (physics, chemistry, and mechanics) required by engineering programs. An environmental engineer can come from different engineering study areas, but all individuals should share a common desire to apply engineering principles to an aspect of the environment that is interesting to them. Agricultural engineering, chemical engineering, civil engineering, general engineering, geological engineering, and mechanical engineering programs routinely graduate individuals who work on environmental issues specific to their discipline.

Q: What kind of jobs does an environmental engineer do?

John: The range of jobs performed by an environmental engineer is extremely varied. A chemical engineer may develop new manufacturing methods that remove toxic contaminants from a product. A civil engineer may be involved in the design of water and wastewater treatment plants, the development of better methods to treat wastes, the development of road-building processes that

I have found out through firsthand experience that environmental issues require very careful communications skills.

▶ John Roll is shown conducting a survey of plant cover on reclaimed mined land.

are more environmentally friendly, and the design of groundwater treatment schemes. Mechanical and general engineering graduates may work on controlling air pollution from factories and producing changes in manufacturing methods to create less waste. Agricultural engineers often work on environmental issues involving livestock waste, runoff-water quality, erosion control, and application methods to lower the quantity of fertilizer, herbicides, and insecticides used to grow crops.

Q: What is the most important skill an environmental engineer should possess?

John: An environmental engineer must be skilled in the application of science and engineering principles to help solve a problem. Using a team approach to an environmental problem will yield a broader view on the issue. Team members usually have expertise in different environmental disciplines, and this results in multiple views on how to solve the problem at hand. Therefore, probably the least expected but the most important skill for an environmental engineer is the ability to communicate clearly through written and spoken words.

Q: Do you feel that environmental issues are often misunderstood?

John: I found out through first-hand experience that environmental issues require very careful communication skills. Environmental issues are often controversial. However, open communication between all interested parties, including those individuals who are against a project, can prevent misunderstanding. For example, the plans for the Industry Coal Mine were finalized after discussions with governmental agencies, local citizens, and authorities. The planning and public meetings lasted almost three years, and during this time everyone had a chance to question the coal company about its plans and to express their views. The public opinion ranged from very favorable to a few individuals who were totally against the project. By addressing the issues with good faith, a reclamation plan was developed that was ultimately approved by all state and federal agencies, local county officials, and zoning boards.

Q: What is the future need for environmental engineers?

John: My feeling is that the future will be a good one for environmental engineers. Since 1970, the environment has been an important focus for many people. Congress passed new laws and created new agencies such as the Environmental Protection Agency (EPA) to specifically address environmental problems. The agencies wrote regulations based on laws passed by Congress and approved by the President, and this resulted in new or additional permits, approvals, and public comment requirements for activities that might harm the environment. In order to enforce the regulations, new agencies were created in the states as well as the federal government. Industry and government currently hire many environmental engineers to meet regulatory requirements.

MORE ON THIS CAREER

For more information on environmental engineering as a career, contact the American Academy of Environmental Engineers, 130 Holiday Court, Suite 100, Annapolis, Maryland 21401, or call (410) 266-3311.

▶ John Roll managed the reclamation of this surface coal mine in Illinois. Land that has been reclaimed is seen to the right of the cut that is being mined.

ENVIRONMENTAL ARCHITECT

To **Michael Reynolds,** a house is not just a home, and old tires and empty soda cans are not just trash. For almost 30 years, this Taos, New Mexico, architect has been designing and building energy-efficient houses out of automobile tires, cans, and other discarded items. These houses, which Michael now calls "Earthships," not only provide a comfortable, affordable place for people to live but also contribute to a sustainable future for our planet.

The Origin of the Earthship Design

In 1970, a TV report about the growing number of beverage cans littering the streets and fields of America started Michael thinking about ways that trash could be used to build houses. Through many years of experimentation, he found that sturdy walls could be built by packing soil into old tires, stacking the tires like bricks, and covering them with cement or adobe, a heavy clay often used in buildings in the Southwest. Michael had this design tested by structural engineers to ensure that the walls would meet or surpass any existing building code requirements. One engineer even commented that the design could be used to construct dams!

Building an Earthship

The tire-stack design is used for three of the outside walls of an Earthship. These walls are approximately 1 m (3 ft) thick, and this large mass causes the walls to act like a battery, storing energy from the sun and releasing the energy when needed. Also, the base of the Earthship is built below the frost line (the deepest level to which the ground freezes). Below this line, the ground maintains a constant temperature— around 15°C (59°F)—and walls anchored below the frost line usually stay at that temperature, too. The fourth wall, which faces south, is constructed completely of glass to capture as much sunlight as possible. In the winter, the tire-stack walls hold in the sun's warmth. In the summer, cool air enters through windows in the front while warm air escapes through a skylight in the back.

To Michael, an Earthship is not just a home—it's a lifestyle.

▶ Earthships, above, often look more like natural land formations than like houses. Michael Reynolds, right, uses discarded materials, such as used soda cans, to construct environmentally friendly houses.

▶ A greenhouse, built along the Earthship's southern glass wall, can provide residents with a sustainable food source.

Even the soil that is excavated for the site of the house is used to build the house. Some of the soil is pounded into the tires to construct the walls, and the remaining soil is piled against the outside of these walls and on top of the roof (constructed of beams) for further insulation. The most suitable location for this design is a south-facing slope of a hill, where the Earthship can simply be built into the hill. Often, Earthships look more like natural formations of land than houses.

Inside the house, walls between rooms are constructed by embedding empty beverage cans into mortar or mud. When these walls are covered with cement and then painted with latex paint or some other durable finish, they look just like walls constructed with conventional materials. Other inside surfaces, including stairs and even bathtubs, can be built using the beverage-can technique. Because the cans are so lightweight, this method can even be used to create dynamic interior structures such as arches and domes.

The Environmental Impact of the Earthship Design

Earthships are typically built to obtain electricity from photovoltaic cells that convert sunlight to electricity. All household water is supplied by rainwater that is collected on the roof. Wastewater from sinks, tubs, and the laundry room is recycled to nourish plants in the greenhouse, which can provide a sustainable source of food. With these features, people who live in Earthships use fewer of the Earth's resources and often have no utility bills.

Because Earthships are environmentally friendly and inexpensive to buy and maintain, more and more people are choosing them instead of conventional homes. More than 1,000 Earthships have already been built, mostly in the Southwest. However, the design can be used anywhere. In fact, Earthships have been built in Florida, Vermont, Canada, Mexico, Bolivia, and Japan. Wetter environments simply require that the house is built entirely above the ground and uses more cans and tires.

More tires in the design certainly wouldn't be a problem. According to the Environmental Protection Agency, more than 250 million tires are discarded in the United States every year. But most landfills do not accept tires because of their tendency to rise to the surface even when the landfill is covered over. Tire dealers usually pay to have used tires hauled away to stockpile areas, where they sit indefinitely. Earthships provide one way to diminish the stockpiles.

Michael enthusiastically shares his Earthship concept with others. To Michael, an Earthship is not just a home—it's a lifestyle. His dedication to designing Earth-friendly homes is a result of his commitment to "reducing the stress involved in living on the Earth, for both humans and the planet."

▶ The tire-stack design of the outer walls accounts for much of the Earthship's energy efficiency. These tire stacks will be covered with cement or adobe for a finished exterior.

MORE ON THIS CAREER

For more information on environmentally friendly, energy-efficient housing, use the Internet to locate government and nonprofit organizations that are involved in "green" building projects.

ENVIRONMENTAL FILMMAKER

Haroldo Castro considers himself a "citizen of the planet." It's easy to see why: he was born in Italy to a Brazilian father and a French mother, he was educated in France, he speaks five languages, and he has visited more than 80 countries. Furthermore, Haroldo has devoted his life to improving the planet's well-being. He has accomplished this by taking photographs, writing books and articles, and producing award-winning video documentaries. Haroldo works for Conservation International (CI), an environmental organization that establishes partnerships with countries all over the world to develop and implement ecosystem conservation projects.

Q: What do you do at CI?

Haroldo: I am the International Communications Project Director. What I do is make documentaries and take photographs of CI's conservation projects. These videos and photos are designed

▶ Haroldo Castro, below left, is shown directing a video crew in Rio de Janeiro, Brazil.

If you are trying to deliver an important message to people of a different culture, it's important to step into their shoes and deliver it from their point of view.

to teach people how to better interact with their local environment. Most of our work is done in countries that have tropical rain forests, such as those in Latin America, Asia, and Africa.

Q: Can you describe one of your documentaries?

Haroldo: Sure. We made a documentary in Guatemala about products that local people can sustainably harvest from the northern tropical forests.

After one year of production, we completed a half-hour documentary called *Between Two Futures*. CI then distributed the video to government officials, environmental organizations, university professors, and teachers. We also encouraged its broadcast on TV channels in Guatemala and other Latin American countries.

The film has been a real success story. I think our ability to be culturally sensitive to the Guatemalan people contributed in large part to the film's success. Each of us who worked on the project had a Latin American background. We worked closely with the Guatemalan people; we had a Guatemalan narrator, and we used only Guatemalan music. If you are trying to deliver an important message to people of a different culture, it's important to step into their shoes and deliver it from their point of view.

Q: What is your educational background and experience?

Haroldo: Although I do have a degree in economics, my best education and training has definitely come from traveling and other real-life experiences. I learn by studying the diverse cultures around the world.

Once I spent two years traveling around Latin America by van; another time I drove from Europe to India in six months. These experiences are my education. When my friends say that it is necessary to have a master's degree or doctorate to gain respect, I respond by saying that I have a Ph.D. in "Travelology." That's a degree I think my real-world experience on the road has earned me.

Q: Do you ever have to deal with crisis situations?

Haroldo: [laughter] If there is not a crisis when I'm traveling, I'm worried—it usually means there will be a disaster later! Anyone who travels a lot has to deal with crises, such as getting sick on local food or getting robbed. I've had equipment stolen from Lebanon to Peru!

I would like to tell you a story. Several years ago we were working in a remote rain-forest region of Mexico for 10 days. When we were ready to leave, we boarded a small plane and set out for the nearest commercial airport, only to learn that it had been closed. We were forced to go to a nearby military airport instead.

When we landed and began to unload our large boxes of equipment, the military personnel got very nervous. We looked pretty grungy and unshaven and covered with mud. It was obvious that we'd been in the rain forest awhile. They thought we were terrorist guerrillas and surrounded us with machine guns. For three hours we pleaded our case, and finally they let us go. I think you might call that a crisis situation!

Q: If a high school student were to ask you what he or she could do to help the environment, what would your answer be?

Haroldo: I would say . . . Learn all you can, appreciate the world around you, and follow

▶ Haroldo Castro is filming slash-and-burn agriculture.

your passion. If you like photography, go out and take pictures of things that leave you with good and bad impressions. If you like gardening, start experimenting with seedlings. Whatever your interest, my advice is, just go for it!

MORE ON THIS CAREER

Many government offices, publishers, and environmental organizations have in-house communications departments for producing films or photographs. Have a librarian help you make a list of such places, and then call these places for more information and for possible volunteer or internship ideas.

While you're at the library, look through *The Guide to International Film and Video Festivals* for any mention of environmental film festivals in your area. Haroldo recommends attending a film festival if at all possible. "Doing so," he said, "would give you the invaluable opportunity to see some of the best films produced and to talk to the people who made them." If you can't find the guide or would like further information, contact the **Association of Independent Video & Filmmakers** at 304 Hudson Street, 6th Floor, New York, NY, 10013.

▶ For Haroldo, capturing images such as this Guatemalan girl holding a hummingbird on film allows him to recall rich travel experiences.

ENVIRONMENTAL LAWYER

Jana L. Walker used to be a nurse, but now she's showing her concern for individuals and their safety in a different way. She owns her own law practice that focuses on environmental protection and Native American issues.

Jana is a member of the Cherokee nation. She supports the "Great Law" of the Six Nations Iroquois Confederacy: "In our every deliberation, we must consider the impact of our decisions on the next seven generations." According to Jana, the Great Law is particularly relevant to environmental issues. She says the Great Law is relevant because it reflects the need to establish laws to protect our natural resources for future generations now, before lands and waters are permanently damaged and species are driven to extinction.

The environmental movement that took off in other parts of the country during the 1970s is only now reaching many Indian lands.

▶ The work Jana does in her New Mexico office is improving the quality of life for many tribal people.

Q: **What inspired you to change from the nursing field to law?**

Jana: Well, I'd always been interested in law, but I guess I never thought I'd be able to do it. But after seven years of nursing, I was really ready for a more independent career. So I made getting through law school my goal. Now I know that it's never too late to get additional education or to fulfill a personal goal.

After law school, I worked at a couple of different law firms—one large and one small. Then I decided that what I really wanted to do was practice law on my own. So I started a solo law practice to focus on Indian and environmental law issues. These issues are very important to me as an inhabitant of the planet, as an attorney, and as an Indian.

Q: **What is the relationship between environmental law and the Indian nations?**

Jana: Well, tribal lands have suffered from many environmental problems. You see, although the first federal environmental laws were enacted several decades ago, those laws did not address Indian tribes and reservations. And the tribes lacked the money to start these programs on their own. As a result, there are now over 53 million acres of tribal lands that have had little or no environmental protection for many years. So the environmental movement that took off in other parts of the country during the 1970s is only now reaching many Indian lands.

Q: **What kinds of environmental problems do you encounter?**

Jana: The problems range from leaking underground storage tanks to acid rain to radioactive contamination to water pollution to illegal, or "wildcat," dumping of trash.

Q: **Those are tough problems. What can you do about them?**

Jana: I help tribes set up regulatory programs to protect the wildlife, land, air, and water resources of the reservations. I also review environmental bills that could affect Indian lands to determine whether they would have a positive or negative impact. Then I lobby for those bills that would help tribal programs. It's an awful lot of reading and writing—definitely not what a television lawyer does!

Q: **What are the most frustrating aspects of your work?**

Jana: It's frustrating to see a tribe begin to move forward with environmental regulation and then have its efforts challenged by a neighboring community. For example, I know of a case in which the tribe wanted to establish water-quality standards for a large river that ran through its reservation. The river is listed as one of the 10 most endangered rivers in America because of severe pollution. The Environmental Protection Agency (EPA) approved of the tribal standards. Then officials for a large city upriver learned that the new standards would limit their use of the river for municipal waste discharge. So the city planners disputed the EPA's approval of the new standards. As a result, the improvements in water quality are again delayed for the tribal people as well as for other communities downstream from this city. This is frustrating! Persistence and the ability to cooperate with government authorities are necessary tools in such a situation.

Q: **What personal qualities do you think are most important in your field?**

Jana: Determination and self-motivation are musts. It's a long haul getting a law degree. Then, once you're a lawyer, the law is constantly changing. That means you must be willing to continue to learn and to study these changes. Creativity is also essential, because many times a law may not directly address your client's problem or need. As a result, you often have to weave together several legal theories to address a particular situation.

▶ Jana L. Walker, shown here next to the Río Grande in New Mexico, is working to improve the quality of river water that runs through tribal lands.

Q: **What message would you like to send to high school students today?**

Jana: I'd like to emphasize that protecting the Earth is everybody's job. But before we can tackle the work, we must become aware of the environment and how we fit into this world. And often that's not something you can learn from a book. It's only after we become truly conscious of nature and the environment that we can begin to see how our actions affect it and what steps must be taken to protect the Earth. So my advice is, go out and enjoy the natural world, and develop a real appreciation for it!

MORE ON THIS CAREER

If you're interested in learning more about a lawyer's work, check with your high school guidance counselor. You may be able to get a part-time or summer job in a law office. Or for more information, contact the American Bar Association, Law Student Division, 750 North Lake Shore Drive, Chicago, IL, 60611, or call (312) 988-5000.

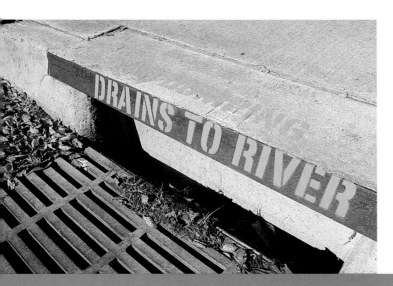

▶ Jana hopes that actions taken because of her work and the work of others will increase awareness about and help solve serious environmental problems, such as water pollution.

Environmental Careers

CLIMATE RESEARCHER

One summer when **Dr. Richard Somerville** was just a child, he built a weather station in his backyard. His creation grew out of a fascination for the great power of weather—a phenomenon that affects everyone every day. So using instruments made out of coffee cans, balloons, and rubber bands, Richard began keeping track of daily weather conditions and questioning how the world's weather systems worked. As time went on, he began to question more than just the weather—he looked at clouds, oceans, and the world of living things as well. These pursuits led Richard to the prestigious Scripps Institution of Oceanography. Today he is a professor of meteorology at the Scripps Institution, which is part of the University of California at San Diego in La Jolla, California.

There are two general classes of technology that are most important to my work: satellites and computers.

Q: **What exactly is meteorology?**

Richard: Simply put, it is the science of the atmosphere—especially the study of weather and weather forecasting.

Q: **What most appeals to you about your job?**

Richard: Probably the most exciting aspect of any scientist's work involves those few, rare "Eureka!" moments when you realize that you've discovered something that no one else on Earth knows about. That's quite a feeling. It's also rewarding to know that you're adding to the knowledge of others, transferring important pieces of information to important people who can use that information to improve this world.

Q: **What does your research involve?**

Richard: Well, I do research on the greenhouse effect, on climate changes in general, and on the effects of long-range climate changes. I also study El Niño events and Indian monsoons. I see how these events and phenomena affect people—such as people involved in agriculture. The climate really affects the way people live!

I'm also researching whether the activities of humans are affecting the atmosphere. For example, each year, the world's growing population uses more and more energy by burning coal, oil, natural gas, and wood. When all of these substances are burned, they add carbon dioxide to the atmosphere. So I study the atmosphere to see how much the added carbon dioxide is intensifying the greenhouse effect. Then I try to determine how those changes will affect humans. You see, the more we know about the atmosphere, the better we can predict what will happen next.

Q: **How is your research data used?**

Richard: Many of my findings can affect public policy. For instance, How should the energy of the world be generated? I can help policymakers

▶ This computer simulation of increasing global temperatures allows Richard to study the possible effects of a changing climate on our planet.

▶ This scientist uses state-of-the-art equipment to gather information about changes in ocean temperatures over time.

explore this question by providing them with data about the effects of fuels such as coal, oil, and gas on the atmosphere. Then I can recommend that they establish policies to reduce human reliance on those fuel sources. I can also encourage the use of resources such as solar, wind, and hydroelectric power.

Q: What tools do you use to obtain your data?

Richard: There are two general classes of technology that are most important to my work: satellites and computers. Together these two items have virtually revolutionized this field by hugely expanding what we've been able to observe and understand. Satellites, for example, can provide us with a whole different perspective of our world. The photographs generated by a satellite allow us to look at global temperatures as well as specific weather and sea conditions. Data are also collected on clouds, soil, and vegetation. By analyzing these observations, we can monitor changing conditions and identify possible problem areas.

Computers help us make sense of the data. Computer equipment in the satellites helps to answer our questions and helps us to better visualize the data. Personal computers help us record and summarize our findings. Then we have "super computers," which can simulate the motions of the atmosphere and the ocean, and thereby help us to answer questions and make predictions.

We also have access to ships and airplanes that are loaded with highly specialized equipment. These research platforms can be sent to specific areas of the world to gather more information about a situation or condition.

Q: What are the most frustrating aspects of your job?

Richard: Other demands that limit the time I spend doing research. There's a large fraction of time and energy that must be spent making research possible—you have to find money, so you spend lots of time writing proposals and doing other administrative work.

Q: What school subjects turned out to be the most important for your career?

Richard: You might be surprised. Math and science classes are essential, but in retrospect I value my English courses the most. Scientists are writers—the final products of their research are shown in published papers.

Q: What personal qualities do you think are most essential for a successful person in your field?

Richard: There are an enormous variety of scientists—some are sloppy, some are organized, some like to work alone, and some in teams. One thing all good scientists have in common, though, is dedication—they all want to do science above anything else. I think Thomas Edison's famous quotation, "Genius is 1 percent inspiration and 99 percent perspiration," is really on the mark. Not everyone can be born a genius, but anyone who is really dedicated can have a good career in science.

MORE ON THIS CAREER

If you are interested in learning more about a career in meteorology, contact the **American Meteorological Society,** 45 Beacon Street, Boston, MA, 02108.

RESEARCH WILDLIFE BIOLOGIST

Many people imagine wildlife biologists wrestling large game animals to the ground, slapping radio collars around their necks, and then creeping through the forest for weeks on end to study the creatures. According to **Mariko Yamasaki,** research wildlife biologist for the U.S. Department of Agriculture, Forest Service, there is a lot more to wildlife biology than that. To Mariko, "Nature is fascinating on many, many levels, from the tiniest ant all the way up to charismatic animals such as bears and wolves. We have to get away from the notion that animals with feathers or fur and big brown eyes are more important than slimy, scaly creatures with beady eyes. All organisms have a role—we must be sure that their contribution to the big picture is recognized."

Q: What is your educational background and experience?

Mariko: My background is basically a long and colorful stringing together of different experiences. I have bachelor's degrees in anthropology and zoology and a master's degree in natural resources (specific to wildlife). By the time I got out of school in the late 1970s, I came up against a surprising attitude—people in my home state really couldn't conceive of having female biologists supervising in the field. So I looked outside my home state. I ended up studying bald eagles for the Bureau of Land Management out West. This sort of snowballed into a permanent appointment in Washington as a wildlife biologist for the Bureau of Land Management. Today I work at the Northeastern Forest Experiment Station, where I do research in forested lands that cover a 200 mi radius, including parts of Maine and New Hampshire.

▶ To Mariko Yamasaki, every creature, no matter how small or seemingly insignificant, has an important role in this biosphere. Below, she is searching for salamanders in the wild.

Q: What organisms are you studying in the field right now?

Mariko: I'm studying small mammals, such as mice, shrews, voles, and squirrels. My colleagues and I also study insectivorous bats, migratory birds, and terrestrial salamanders. These are animals that we know something about, such as their basic biology, but we don't know how they respond to forest management. We're looking at these critters to get a sense of how they fit into the bigger picture.

Q: What types of questions are you trying to answer about these animals?

Mariko: One question my colleagues and I are trying to answer right now is how terrestrial salamanders respond to "even-aged management" of northern hardwoods. Even-aged management involves harvesting a large area of trees whose ages are within 20 years of each other.

We have to get away from the notion that animals with feathers or fur and big brown eyes are more important than slimy, scaly creatures with beady eyes.

► Mariko wants to know if the way in which trees are harvested from forested areas like this one affects the survival of terrestrial salamanders.

Foresters often use even-aged management because it is an efficient means of harvesting large amounts of timber at one time. My hypothesis is that when a large area of trees has been harvested, the ground temperature might change because the area is suddenly exposed to direct sunlight. This might affect the population and distribution of terrestrial salamanders in a negative way. I use the data I gather to make recommendations to forest managers about how they can manage tracts of forest to best support the needs of salamanders and other wildlife.

Q: **Do you work with other people a lot?**

Mariko: There's an old stereotype that a wildlife biologist leads a solitary life studying nature. This simply isn't true—it's important to know how to work with people and how to understand and deal with a variety of viewpoints. There is rarely a day that I sit alone in my office. But I will say that a wildlife biologist does have some control over the matter—generally, you can work with people as much or as little as you want.

Q: **Do you ever have to deal with crisis situations?**

Mariko: Not really, but I do see a lot of controversy, particularly related to wildlife and the use of natural resources. My work has often become the object of heated debate. Some people will support my findings wholeheartedly, while others call them worthless. There are any number of ways of dealing with this kind of pressure. I've found that it's real important to get my information together and analyze it as thoroughly as possible so that I can really stand behind what I'm saying. It's also important to realize that everyone is entitled to an opinion.

Q: **What are the most interesting or exciting aspects of your work?**

Mariko: Oh heavens! Being out and observing the natural world. Being able to test hypotheses. Being up real early on a bird survey. It's never the same twice. I also enjoy discovering something new—there's nothing any more special than that. There's a lot out there! The scale of things to observe and study is mind-boggling.

Q: **What advice might you give to someone who is searching for a career?**

Mariko: I think it's important to do something you are really interested in. My career, just like anybody else's, is not always a bed of roses. But if you really care about what you do, you can get beyond the problems and complications inherent to any job. It's also important to think that you've got something to contribute. I think that I can help contribute to the way people view wildlife, and that's important to me.

► This group of community leaders, politicians, and scientists is discussing how best to use the natural resources of a forested region in Maine.

MORE ON THIS CAREER

If you are interested in learning more about a career in wildlife biology, contact **The Wildlife Society,** 5410 Grosvenor Lane, Bethesda, MD 20814, or the **American Institute of Biological Sciences,** 1444 Eye Street NW, Suite 200, Washington, D.C., 20005.

ANY JOB CAN BE ENVIRONMENTAL

You don't have to be in an environmental career to make a positive impact on the environment. Gun Denhart is an excellent example of how you can make a difference through your career, even if your career doesn't directly involve the environment.

Hanna Andersson

Gun Denhart is the cofounder of Hanna Andersson, in Portland, Oregon. Hanna Andersson is a company that specializes in selling baby clothing and children's clothing through a mail-order catalog service. The company began in 1983 as an in-home enterprise, in which a spare room was used as the company office and the garage served as the warehouse. One-inch-square fabric samples were cut and pasted into each of the 75,000 catalogs that were mailed that first year. Hanna Andersson has grown enormously since 1983. An adult line of clothes has been added, several retail stores were opened, and a Web site was established.

The Used Clothing Recycling Program

As a parent, Gun Denhart realized that children outgrow their clothing very quickly and that clothing purchased at Hanna Andersson will last for more than a single child. Rather than waste clothing, she reasoned, why not pass these clothes on to children in need? So Gun instituted a program called Hannadowns®. The purpose of the program is to encourage the purchasers of Hanna Andersson clothing to recycle their used clothes. Gun says, *"You can make a critical difference. Most often, my clothes last for more than one child's use, and it's a great feeling to pass them on to younger children in your family, to your friends, or to charitable organizations. It is heartbreaking to realize how many children live at risk—an unbelievable 22 percent of children live below the poverty level in America. Providing them with nourishing food and warm clothing is a never-ending job. Fortunately, there are organizations that offer clothing and supportive services. To help them make a critical difference in children's lives, please send your outgrown children's clothes in good condition to these organizations."*

In the first 16 years of the company's existence, Hanna Andersson customers have donated over one million pieces of recycled clothes to children in need. These clothes could have ended up in landfills but have instead clothed kids all over the world.

In the first 16 years of the company's existence, Hanna Andersson customers have donated over one million pieces of recycled clothes to children in need.

▶ This clothing will be recycled because of an innovative program designed by Gun Denhart.

BOOSTING YOUR HOME'S
ENERGY EFFICIENCY

Many people don't realize the impact that energy production has on the environment. No matter what kind of energy plant serves your area, the production of that energy carries with it certain environmental risks. For example, when we burn coal to create electricity, many pollutants are released into the air. These pollutants may cause environmental problems such as global warming and acid rain. The more energy each of us uses, the more we contribute to these problems. So it makes environmental sense to conserve energy. Conservation is also a good way to save money—just a few energy-saving measures can substantially lower an energy bill.

Could the energy efficiency of your home be improved? Perform the following energy audit to find out.

The Wind Test

One day when it's windy outside, fasten a sheet of tissue paper onto a hanger with a piece of tape, as shown below. Next, hold the hanger in front of a window at the point where the window meets the wall. Hold the hanger still. If the paper moves, you've found a draft. Note the location of the draft in your *EcoLog.* Check all around the window, making comments about the drafts you find. Then examine all of the other windows, doors, electrical outlets, plumbing pipes, and baseboards that are on the outer walls of your home. Note every place where the tissue moves.

These drafts of air that you've discovered can add 20 to 35 percent to your heating and cooling bills. Fortunately, you can seal these air leaks with weatherstripping and caulk. Weatherstripping is for moving parts, such as doors and window frames. Caulk is for sealing cracks along joints and edges. These materials are relatively inexpensive, can be found at any hardware store, and can save 7 to 20 percent on your heating and cooling bills.

▶This simple device could help you improve the energy efficiency of your home.

FOR MORE INFORMATION

Your local electric company can probably send you a packet of energy- and cost-saving ideas. In addition, your city may sponsor thorough in-house energy audits as well as rebates and loans for improving the energy efficiency of your home. Contact your city's electric utilities conservation department for more information.

Consult your library or bookstore for books on improving your home's energy efficiency. You might find these books helpful.

Consumer Guide to Home Efficiency, 7th ed., by Alex Wilson, Jennifer Thorne, and John Merrill, American Council for an Energy-Efficient Economy. White River Jct., VT: Chelsea Green Pub., 2000.

Energy Efficient Houses, by Fine Homebuilding Magazine. Newtown, CT: Taunton Press, 1993.

ELIMINATING PESTS
NATURALLY

A huge cockroach is crawling across your floor. How will you get rid of it? Don't reach for an expensive store-bought chemical that could possibly contaminate the local water supply or even harm someone in your household. Instead, try a natural remedy!

> **Even the tidiest of homes can be bugged by insect pests. If this happens to your home, fight back—naturally!**

Cockroaches Make a roach trap by putting honey in the bottom of a jar and setting it upright where the pests are most likely to visit. The sweet smell of the honey will lure roaches into the jar, but the stickiness of the substance will make it impossible for them to escape. You could also line the cracks where you think roaches are entering your home with bay leaves. The smell of bay leaves repels roaches. Prevent roaches from entering your home by keeping all food covered and stored and by cleaning dirty dishes. Seal cracks in walls, baseboards, and ducts with caulk so that roaches and other pests can't get in.

Ants Sealing cracks with caulk will also help keep ants out of your home. In the meantime, squeeze fresh lemon or lime juice into the holes or cracks. Then leave the peels where you've seen ants. Scatter mint around your shelves and cabinets, or pour a line of cream of tartar, red chili pepper, salt, paprika, dried peppermint, or talcum powder where ants enter your home. These substances either repel or kill the pests.

Another effective remedy for ridding your home of ants or cockroaches is to sprinkle a mixture of equal parts of boric acid and confectioners' sugar in dry areas where ants and cockroaches are found. The pests will eat the sugar and then die from the effects of the boric acid. Caution: If ingested, boric acid is acutely toxic to pets and small children. Use boric acid only in areas that are out of reach of kids and pets.

Ticks and Fleas If your pet has a problem with ticks or fleas, try feeding the animal brewer's yeast or vitamin B. Also wash your pet regularly with soap and water, then dry the animal and spray an herbal mixture of rosemary and water onto its coat. (You can make the mixture by steeping ½ cup of fresh or dried rosemary in one quart of boiling water. Let the liquid cool, pour it into a pump bottle, and then spray it onto your animal's coat.)

▶ You can help reduce the number of ticks and fleas that bother your pet by bathing it frequently and spraying an herbal mixture on its coat.

You can control the ticks and fleas in your yard by sprinkling the grass with diatomaceous earth, which is available at many nurseries. Diatomaceous earth consists of tiny glasslike skeletons of diatoms (a type of single-celled algae). These organisms scratch the outer layer of an insect's body as it crawls along the ground. Bacteria then enter the insect's body through the open wounds, and the insect dies of disease. Caution: Diatomaceous earth can be harmful to your lungs if inhaled. Wear a protective mask when spreading the substance.

FOR MORE INFORMATION

Your city's environmental and conservation services department (if you have one) may have some other remedies for pests and some recipes for nontoxic household cleaners. Also check your local bookstore or library for books on natural pesticides, organic gardening, and chemical-free homes. You might find these books helpful.

The Good Earth Home and Garden Book, by Casey Kellar. Iola, WI: Krause Publications, 2002.

Natural Pest Control Alternatives to Chemicals for the Home and Garden, by A. Lopez. Austin, TX: Acres USA, 1990.

Ecoskills

Environmental Shopping

Try to count how many products you've used today. It's probably not as easy as you think. In the first few minutes of your day, you may have used a dozen products.

All of those products and their packaging are made from valuable resources. More often than not, once those resources are used, they're tossed in a trash can and eventually hauled to the local landfill.

You can cut back on the amount of waste you send to the landfill and conserve resources in the process. On your next few shopping trips, think about the products you choose. If you're like most Americans, you'll probably be amazed at how many wasteful shopping habits you have. But after a while you'll begin to know instinctively which products are best for you and the environment.

On your next few shopping trips, think about the products you choose. If you're like most Americans, you'll probably be amazed at how many wasteful shopping habits you have.

Your Personal Shopping Guide

Read the information on the following page, and think of a way to reproduce it so that you (and other members of your household) have it handy when you set out on a shopping trip. For example, you may want to copy the questions and answers on the side of a brown paper bag. That way you'll have a shopper's guide, and you'll need one less sack at the checkout stand. Another option is to write your guidelines on the back of an old grocery receipt and then adhere the receipt to the refrigerator with a magnet so that it will be handy for the other shoppers in your household. The options are limitless, so be creative, and try to incorporate recycled items into your design.

Before you create your personal shopping guide, you may want to review the section titled "Reducing Solid Waste" on pages 488 to 492 of your text.

FOR MORE INFORMATION

Consult your local library or a bookstore to find references that will help you with your environmental shopping. You might find one of the following books helpful.

Green Products by Design: Choices for a Cleaner Environment, by Gregory Eyring. Upland, PA: Diane Publications, 1992.

Mother Nature's Shopping List: A Buying Guide for Environmentally Concerned Consumers, by Michael D. Shook. New York: Carol Publishing Group, 1995.

An Environmental Shopper's Guide

Do I really need this product? Can I use something I already have?	Borrow or rent products you don't use often.
Is this a "throwaway" item that is designed to be used once or twice and then thrown away?	Avoid using disposable products whenever possible. Nondisposable alternatives may be more expensive initially, but in the long run they often save you money.
Does this product have more packaging than it really needs?	Look for alternatives with less packaging or wrapping. Purchase products in bulk or in a larger size so that in the long run you use less packaging (and save money!). Buy fresh vegetables and fruit instead of frozen or canned products.
Was this product's container or packaging made with recycled materials?	Choose products that have recycled paper, aluminum, glass, plastic, or other recycled materials in their packaging.
Is this product's container or packaging made from cardboard, aluminum, glass, or another material that I can easily recycle?	Find out which materials you can conveniently recycle, and then buy those sorts of containers. Also, think of ways to reuse old containers rather than throwing them out.
Does this product have bleaches, dyes, or fragrances added to it? Does it contain phosphates? Is it made from a petroleum-based synthetic fabric, such as polyester?	Phosphates and many other chemicals can pollute water sources. Look for natural, organic, and phosphate-free alternatives. When purchasing clothing, choose cotton or wool over synthetic fabrics.
Does the company that makes this product have a good environmental record?	You may have to do a little research to answer this one. Try the references listed on page 598.
Although this product has a "green" label, is it really good for me and the environment?	Don't be deceived by advertising and product labeling; carefully examine the contents of a product before you purchase it.
Do I really need a shopping bag to carry home the items I'm purchasing? If so, will I be more likely to recycle or reuse a plastic shopping bag or a paper one?	If you purchase just one or two items, tell the grocer that you don't need a bag to carry the products. For more items, bring old paper or plastic sacks with you when you go to the store, or use a canvas bag, which will last through many trips.
How much energy do I spend getting to the store?	If possible, ride your bike or walk to the store. If not, condense several short trips into one longer trip for a bigger supply of items.

MAKING YOUR OWN COMPOST HEAP

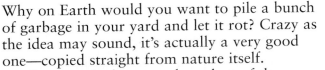

Why on Earth would you want to pile a bunch of garbage in your yard and let it rot? Crazy as the idea may sound, it's actually a very good one—copied straight from nature itself.

Compost is the natural product of the Earth's organic decaying process. When a dead organism decomposes, nutrients are returned to the soil. A compost heap is a collection of organic materials such as leaves, grass, and fruit peelings that will decompose over time to create rich, fertile soil. By making your own compost heap, you can reduce the amount of waste you send to the local landfill and create an excellent natural fertilizer for your garden.

There are many opinions on how to construct the best compost heap—it can be as basic or as fancy as you like. Either way, composting is easy, and it's almost impossible to foul up the process.

A compost heap can be placed just about anywhere in the yard. Either a sunny or a shady spot will be fine. You will want to keep it out of the way of normal activity, however.

Many people choose a spot on a concrete slab or a grassy area and then simply pile their materials there. (See the photo on this page.) This method is easy and effective.

A compost heap contains a mishmash of many different organic materials. Most of your heap will probably consist of grass clippings and leaves. You can also add raw vegetables, other uncooked food scraps, coffee grounds, tea bags, cotton, dust, discarded plants, and weeds. Avoid adding pet manure, cooked foods, and meat of any kind.

> **By making your own compost heap, you can reduce the amount of waste you send to the local landfill and create an excellent natural fertilizer for your garden.**

▶ This is an easy and effective way to make your own compost heap.

Anatomy of a Compost Heap

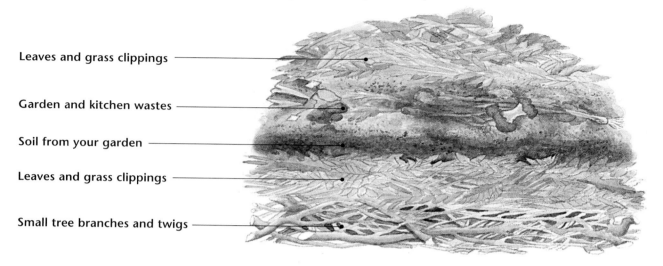

Leaves and grass clippings

Garden and kitchen wastes

Soil from your garden

Leaves and grass clippings

Small tree branches and twigs

If you add raw food wastes, cover them with leaves to keep away flies and to prevent an unpleasant odor.

Your heap will begin to decompose through the action of microorganisms. It's a good idea to shovel a couple of scoops of soil from your yard into the heap. The microorganisms in the soil will immediately begin decomposing the items in the heap.

Turn the heap at least once a month to keep it well aerated and active. Once the organic matter has broken down to the point that no single item is recognizable, it's ready to work into your garden's soil. The entire process can take anywhere from two months to one year, depending on the kinds of materials being decomposed and how often the heap is turned. Composting is more of an art than a science, so be prepared to experiment!

Compost Container

If you choose to contain your compost pile, you will be able to add more materials to a smaller area. You can buy a ready-made container from a hardware store, or you can build one yourself.

If you decide to build one, you may wish to use metal stakes and chicken wire to create a container like the one shown at right. Keep in mind, however, that as long as the container allows air to get in and out, the type of container you choose is limited only by your imagination!

Compost Heap Container

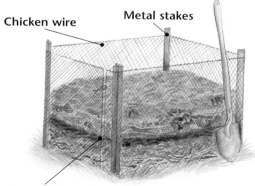

Chicken wire   Metal stakes

Loose wire can be twisted around two sections of chicken wire to create a "door" for easy turning.

▶ You can build this container for your compost heap with a few materials from your local hardware store.

FOR MORE INFORMATION

Consult your library for a manual on composting. You might find one of these helpful.

Let it Rot! The Home Gardener's Guide to Composting, 3d ed., by Stu Campbell. Pownal, VT: Storey Books, 1998.

Rodale Book of Composting, edited by Grace Gershuny and Deborah L. Martin. Emmaus, PA: Rodale Press, 1992.

Creating a Wildlife Garden

Manicured lawns and nonnative vegetation are not part of a natural ecosystem. Although these have been standard in urban and suburban neighborhoods for years, they usually require pesticides, fertilizers, water, and attention just to survive. In addition, they often exclude wildlife by removing some of their natural sources of food, water, and shelter.

To attract wildlife to your home, you simply need to provide native plants and the sorts of water sources and shelters naturally available to the wildlife in your area.

Plants

Plants are probably the most crucial element of your wildlife garden. Whether you have a lot of space for planting a wildflower meadow, a balcony on which you can create a container garden full of native plants, or a few windows to which you can attach boxes full of bright and cheery wildflowers, you will need a variety of native plants. Check with a local nursery, library, or bookstore for recommendations.

To attract wildlife to your home, you simply need to provide native plants and the sorts of water sources and shelters naturally available to the wildlife in your area.

Water

People often overlook the need all animals have for water. Although some animals obtain enough water from the foods they eat, most require additional water for drinking and bathing.

Water sources are easy to provide. Many people purchase hanging or standing birdbaths from a nursery or hardware store. Others create ponds. You can make a simple pond by setting an old trash-can lid upside down in a corner of your yard and filling it with water. Surround your water source with plants, rocks, and other items so that the wildlife can find cover if necessary. In addition, make sure your pond or birdbath is at least partially shallow so that no animal is in danger of drowning, and keep the water clean.

Food and Shelter

Many different kinds of birdhouses and feeders are available at nature stores, hardware shops, and nurseries. Most of these can be hung on a balcony, and some can even be adhered to a window. Or, you could make your own birdhouse or feeder. A milk jug with a large hole cut in its side that is filled with seed and hung from a tree or balcony is an excellent way to feed many birds. If you would like to attract bats to your yard, see pages 606–609 for directions on how to make (or purchase) a bat house.

Woodpiles, rock piles, and brush piles are valuable sources of shelter for wildlife such as lizards and toads that might not frequent your backyard habitat otherwise. The most successful pile is one that incorporates different-sized spaces among the various components. You can make your pile attractive by planting vines in and around it.

Caution: *A shelter like the one described above may also attract poisonous snakes. Find out if any live in your area; if so, you may want to refrain from making a shelter pile.*

FOR MORE INFORMATION

Consult your library or bookstore for books on gardening with plants native to your area, gardening for the wildlife in your area, and xeriscape techniques. You might find these books helpful.

Noah's Garden: Restoring the Ecology of Our Own Back Yards, by Sara Bonnett Stein. Boston: Houghton Mifflin, 1995.

Your Backyard Wildlife Garden: How to Attract and Identify Wildlife in Your Yard, by Marcus Schneck. New York: St. Martin's Press, 1992.

Planting a Tree

In many towns and cities across America today, more trees are dying or being cut down than are being planted. If you live in an urban environment, you can help reverse this disturbing trend by planting a tree.

No matter where you live, you'll undoubtedly find a tree a welcome addition to your neighborhood. If you don't have a yard or can't plant a tree in your yard, consider getting a small tree to put in a pot on your patio or balcony. Or you could participate in a community tree-planting project. Check with a library, nursery, or environmental organization to find out if there are any tree-planting projects in your area.

Finding the Right Tree
There are countless varieties of trees that you can choose from. The following guidelines should help you find the tree that best suits your needs.

Go Native Native trees usually don't require fertilizers, pesticides, or excessive watering. So choose a native tree, and save yourself considerable time, money, and effort.

Long-Living or Quick-Growing? Most people want a tree that will grow quickly to its full height. Unfortunately, many fast-growing trees have a long list of serious problems. As a result, they often do not live as long as slower-growing varieties. So choose a medium- to slow-growing tree, and look forward to watching it mature over the years. (Many slow-growing trees can live more than 100 years!)

Follow the Sun Think about the effect the tree will have on its surroundings. For example, a tree planted within 4.5 m (15 ft.) of your home's south or west side can shade the house in the summer. If the tree is deciduous (sheds its leaves in the fall), sunlight can filter through the branches in the winter and help keep your home warm.

Watch Those Roots and Branches Keep in mind that roots can seriously damage sidewalks, driveways, and sewage systems and that branches can damage shingles, windows, and house siding. Therefore, keep trees that will have large root systems or branches an appropriate distance from anything they might damage.

No matter where you live, you'll undoubtedly find a tree a welcome addition to your neighborhood.

▶ Covering the area around a newly planted tree with compost and mulch will supply the tree with nutrients and retain soil moisture.

Cost Trees sold in plastic 1–5 gallon containers range from $3 on sale to $50 or more for rare or nonnative species.

Seeking Advice The nursery is an obvious place to get advice about trees you could plant, but nearby college agricultural departments, university extension offices, U.S. Forest Service offices, and city government offices can also help you. Sometimes they even have trees for sale, often at their cost.

Planting Your Tree

The following directions will help you give your tree a good start.

1. Dig a square hole *exactly* the same depth as the root ball. The hole should be approximately twice the width of the root ball. Use a ruler or measuring tape to make sure the hole is the right depth and width. The sides of the hole do not have to be smooth. In fact, it is better if they are rough and jagged, as shown in the diagram at right.

2. Now put the tree in the hole, and refill the hole with the soil you took out. You may want to mix the soil with some organic material such as bark mulch or finished compost (*well-decomposed* organic matter). Do not use pure manure, sand, bark, or peat moss. Compact the soil slightly but not too firmly. Water the tree slowly and deeply to help the soil settle into the hole.

3. It's a good idea to cover the area around your newly planted tree with about 1 inch of compost and 3 inches of mulch (shredded hardwood bark, wood chips, or hay). The compost will serve as a natural fertilizer that gives the tree necessary nutrients, and the mulch will help the soil retain moisture.

Caring for Your Tree

The first couple of years of the tree's life are especially important. Don't just plant the tree and forget about it! Water the tree thoroughly when the top 3–4 inches of soil dry out or if the tree's leaves start to wilt.

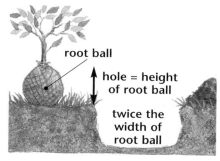

root ball

hole = height of root ball

twice the width of root ball

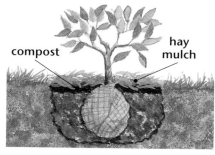

compost

hay mulch

FOR MORE INFORMATION

Global ReLeaf Forests: American Forests, P.O. Box 2000, Washington, D.C. 20013; (202) 955-4000

Tree People: 12601 Mulholland Drive, Beverly Hills, CA 90210-1332; (818) 753-4600 (Ask about their Campus Forester program and their book titled *The Simple Act of Planting a Tree*.)

Building a Bat House

If mosquitoes bother you, you should love bats. A single little brown bat is capable of eating 600 mosquitoes in just one hour. Other bats pollinate flowers and disperse the seeds of the fruit they eat, helping many plants to reproduce. Bats are a vital part of nearly every ecosystem on Earth.

Despite the importance of bats, people have feared and persecuted them for centuries, incorrectly believing them to be vicious creatures that attack humans. This fear is due to misunderstanding. Bats are actually useful creatures that, like most wild animals, generally avoid contact with humans. The chance of catching a disease from a bat is actually very small. If you simply avoid handling bats, you have nothing to fear.

Give a bat a home and keep annoying insects, such as mosquitoes, at bay.

Unfortunately, many bat species in the world today are threatened with extinction. Nearly 40 percent of those living in the United States are on the endangered list or are official candidates for it. You can do your part to help these unique creatures by building a bat house.

Getting Started

To build a bat house for 30 or more bats, collect the following materials. Because you will be working with sharp tools, exercise extreme caution.

Materials

- one 8' long piece of 1" × 8" untreated lumber (for the front and back pieces)*
- one 5' long piece of 2" × 2" untreated lumber (for the ceiling and sides)*
- one 5' long piece of 1" × 4" untreated lumber (for the roof and the posting board for mounting the bat house)*
- one piece of $15\frac{1}{2}$" × 21" fiberglass window screening (Do not use metal screening.)
- nails (approximately 16 8d galvanized nails and 30 6d galvanized nails)
- safety goggles
- staple gun and staples, or approximately 16 upholstery tacks
- metal utility knife
- hammer

* Note: The lumber sizes given are the sizes you will need to ask for at a lumber company. The actual measurements of those boards are slightly smaller, however. The measurements given in the directions and drawings for this activity reflect the actual measurements of those boards. (For example, a 1" × 8" actually measures $\frac{3}{4}$" × $7\frac{3}{4}$".)

- saw
- measuring tape
- silicone caulk (may or may not be necessary)
- small can of exterior latex paint or varnish (light colored or clear if you live in a warm southern climate and dark if you live in a colder northern climate)
- ladder (to mount the bat house)

Building the Bat House

1. Cut the 5' long piece of 1" × 4" lumber into two pieces, one piece $16\frac{1}{2}$" long (roof) and the other 3' long (posting board).

   Caution: *Exercise extreme caution while sawing.*

2. Cut the 1" × 8" × 8' board into six pieces, each measuring $15\frac{3}{4}$" in length. Next, cut one of these pieces lengthwise to make a strip measuring 1" wide; then cut the strip to $12\frac{3}{4}$" long. Be as accurate as possible when sawing.

3. Using 6d nails, nail three of the pieces you cut in step 2 (note: all three should be identical in size) to the posting board, as shown in **Diagram A.** Fit the three boards together as tightly as possible.

4. Using the staple gun or upholstery tacks, tightly secure the fiberglass screening to the three boards you assembled in step 3. (Use the side without the posting board.)

Diagram A

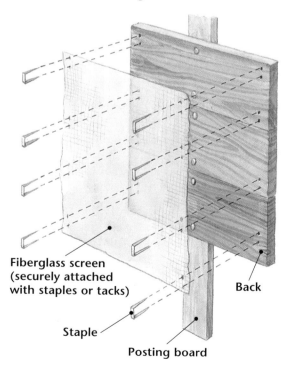

Fiberglass screen (securely attached with staples or tacks)

Back

Staple

Posting board

5. Cut the 2" × 2" × 5' piece into two pieces measuring $21\frac{3}{4}$" long (sides) and one piece measuring $12\frac{3}{4}$" long (ceiling).

6. Construct the internal frame by nailing the sides, ceiling, and entrance restriction (this is the $12\frac{3}{4}$" piece you cut off of one of the six identical boards in step 1) together with 8d nails, as shown in **Diagram B.**

7. Nail the back pieces to the internal frame, using 6d nails.

8. Using 6d nails, secure the front three pieces to the internal frame, allowing a gap of $\frac{1}{2}$" for ventilation between the bottom and middle pieces. (Note: The bottom piece is smaller than the other two.) See **Diagram B.**

9. Nail the roof to the top of the structure, allowing the excess to extend over the front of the bat house. See **Diagram C.**

10. Be sure the house is as draft-free as possible. If there are gaps between the pieces in the top two-thirds of the house, seal those spaces with silicone caulk. Gaps in the lower third of the house will allow ventilation for the bats.

11. Paint or varnish the exterior of the bat house.

Diagram B

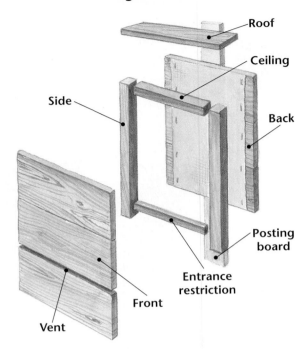

Diagram C

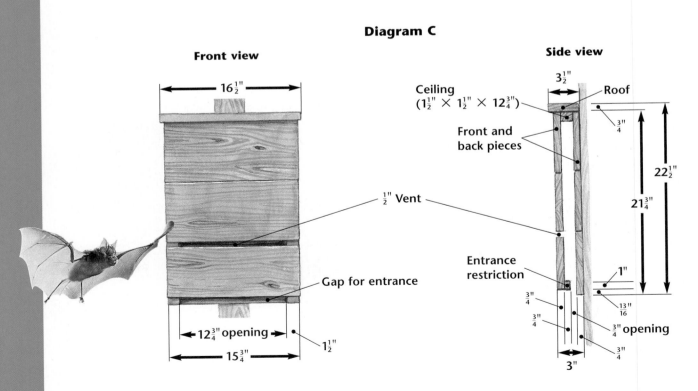

▶ Mount your bat house a minimum of 12 feet from the ground in an area that will receive at least four hours of sunlight a day.

Mounting Your Bat House

Mount your bat house at least 12 feet high (ideally, 15–20 feet or more) on a sturdy pole or on the side of a building away from lights. You can also put the house in a tree as long as the tree's leaves do not shade the bat house too heavily in the summer. If possible, the bat house should be within a mile or so of a water source such as a lake, river, or creek. The amount of sunlight the bat house receives will influence whether bats will inhabit it. All bat houses should receive at least 4 hours of sunlight per day.

If you live in the southern United States, the bat house will need protection from the sun's heat. Position the house so that it receives sunlight in the morning but not in the afternoon. Paint the house white or light brown to reflect the sun's rays.

If you live in the northern part of the country, paint the bat house black or dark brown to retain the sun's heat. Position the house so that it will receive 4–6 hours of sunlight if it is black or 6–12 hours of sunlight if it is painted dark brown. If you live in the central United States, you can paint the house any color, but make sure it receives at least 4 hours of sunlight.

Check your bat house regularly for signs of habitation. If the house is occupied, you can look at the bats with a flashlight, but only for brief periods. If bats don't immediately take up residence, don't get discouraged. It might take a while for them to find your house, especially in the winter (when many species hibernate). Be prepared to experiment with different colors of paint and different locations.

FOR MORE INFORMATION

For more information about how to attract bats to your house or how to build a more elaborate bat house, or to purchase a ready-made bat house, contact:

Bat Conservation International
P.O. Box 162603
Austin, Texas 78716
(512) 327-9721

Ecoskills

FLUSHING LESS WATER

A typical American uses over 100 gallons of water before he or she even leaves for work or school in the morning, and much of that water is wasted. You may wish to review Table 1 on page 277, which shows daily water use in the United States per person.

Many Americans are beginning to change their wasteful practices, however. One simple and inexpensive way you can waste less water is by making a water-displacement device for your toilet's tank. This device takes up space in the tank so that less water is required to fill the tank with every flush. It only takes about 10 minutes to make, and with it you can save 1–2 gallons of water every time you flush. This may not sound like much, but it adds up quickly. Most toilets use 5–7 gallons of water with every flush. If a toilet is flushed an average of eight times per day, it uses around 52 gallons of water per day, or 18,980 gallons per year. If you can save $1\frac{1}{2}$ gallons of water with every flush, you'll save 4,380 gallons of water each year. If just 250 other people take similar measures, over 1 million gallons of water could be saved each year.

Making a Quick and Easy Water Displacer

1. Remove the label from a plastic container. (Milk jugs, juice bottles, and dishwashing soap bottles work well. Be prepared to experiment with different-sized containers.) Drop a few rocks into the container to weigh it own, fill the container with water, and put the lid back on.

2. Place the container in the toilet tank, as shown at left.

3. Be certain that the container doesn't interfere with the flushing mechanism inside the tank.

4. Experiment with different containers. Your goal is to use the largest container that the tank will hold while still maintaining an effective flush.

ONE FINAL IMPORTANT NOTE

The more water you save, the less you pay for. No matter which water-saving device you install, your water bill should be noticeably lower.

SI Conversions

The metric system is used for making measurements in science. The official name of this system is the Système International d'Unités, or International System of Measurements (SI).

SI Units	From SI to English	From English to SI
Length		
kilometer (km) = 1,000 m	1 km = 0.62 mile	1 mile = 1.609 km
meter (m) = 100 cm	1 m = 3.28 feet	1 foot = 0.305 m
centimeter (cm) = 0.01 m	1 cm = 0.394 inch	1 inch = 2.54 cm
millimeter (mm) = 0.001 m	1 mm = 0.039 inch	
micrometer (μm) = 0.000 001 m		
nanometer (nm) = 0.000 000 001 m		
Area		
square kilometer (km^2) = 100 hectares	1 km^2 = 0.386 square mile	1 square mile = 2.590 km^2
hectare (ha) = 10,000 m^2	1 ha = 2.471 acres	1 acre = 0.405 ha
square meter (m^2) = 10,000 cm^2	1 m^2 = 10.765 square feet	1 square foot = 0.093 m^2
square centimeter (cm^2) = 100 mm^2	1 cm^2 = 0.155 square inch	1 square inch = 6.452 cm^2
Volume		
liter (L) = 1,000 mL = 1 dm^3	1 L = 1.06 fluid quarts	1 fluid quart = 0.946 L
milliliter (mL) = 0.001 L = 1 cm^3	1 mL = 0.034 fluid ounce	1 fluid ounce = 29.577 mL
microliter (μL) = 0.000 001 L		
Mass		
kilogram (kg) = 1,000 g	1 kg = 2.205 pounds	1 pound = 0.454 kg
gram (g) = 1,000 mg	1 g = 0.035 ounce	1 ounce = 28.35 g
milligram (mg) = 0.001 g		
microgram (μg) = 0.000 001 g		
Energy		
British Thermal Units (BTU)	1 BTU = 1,055.056 joules	1 joule = 0.00095 BTU

Temperature

°F 0 20 40 60 80 100 120 140 160 180 200 220

°C −20 −10 0 10 20 30 40 50 60 70 80 90 100

Freezing point of water
Room temperature
Normal human body temperature

Conversion of Fahrenheit to Celsius:
$$°C = \frac{5}{9}(°F - 32)$$

Conversion of Celsius to Fahrenheit:
$$°F = \left(\frac{9}{5}°C\right) + 32$$

Mineral Uses

	Mineral	Chemical formula	Identifying characteristics
Metallic Minerals	Chalcopyrite	$CuFeS_2$	brassy color; iridescent tarnish; soft for metal; brittle
	Chromite	$FeCr_2O_4$	iron-black color; weakly magnetic
	Galena	PbS	high density; perfect cleavage in four directions, which forms a cube; low hardness
	Gold	Au	golden color; low hardness; high density; malleable (can be pressed into various forms)
	Ilmenite	$FeTiO_3$	tabular crystals; no cleavage
	Magnetite	Fe_3O_4	8-sided crystals; magnetic
	Uraninite	UO_2	black to steel black color; dull luster; radioactive
Nonmetallic Minerals	Barite	$BaSO_4$	high density for nonmetal
	Borax	$Na_2B_4O_7 \bullet 10H_2O$	low hardness; low density; dissolves in water
	Calcite	$CaCO_3$	perfect cleavage in three directions; low hardness; fizzes in dilute hydrochloric acid
	Diamond	C	extreme hardness; transparency; perfect cleavage in four directions
	Fluorite	CaF_2	cubic or 8-sided crystals; perfect cleavage in four directions
	Gypsum	$CaSO_4 \bullet 2H_2O$	softness; perfect cleavage in one direction and good in two others
	Halite	$NaCl$	low hardness; perfect cleavage in three directions, which forms cubes; salty taste
	Sulfur	S	yellow color; low hardness; poor conductor of heat; odor
	Kaolinite	$Al_2Si_2O_5(OH)_4$	low hardness; white color; noncrystalline
	Quartz	SiO_2	hardness; conchoidal fracture; crystals form six-sided prisms
	Talc	$Mg_3Si_4O_{10}(OH)_2$	very low hardness; massive; perfect cleavage, which forms thin, flexible flakes; soapy or greasy feel

Explanation of Terms

cleavage: the splitting of a mineral along smooth, flat surfaces

fracture: the tendency of a mineral to break along curved or irregular surfaces; conchoidal fracture is a smooth, curved fracture

hardness: a measure of a mineral to resist scratching

luster: the way the surface of a mineral reflects light

Economically important deposits	Important Uses
Chile, USA, Indonesia	power transmission, electrical and electronic products, building wiring, telecommunications equipment, industrial machinery and equipment
South Africa, Kazahkstan, India	production of stainless steel, alloys, metalplating
Australia, China, USA	batteries, ammunition, glass and ceramics, x-ray shielding
South Africa, USA, Australia	computers, communications equipment, spacecraft, jet engines, dentistry, jewelry, coins
Australia, South Africa, Canada	jet engines; missile components; white pigment in paints, toothpaste, and candy
China, Brazil, Australia	steelmaking
Canada, Australia	fuel in nuclear power reactors, manufacture of radioisotopes
China, India, USA	weighting agent in oil well drilling fluids, automobile paint primer, x-ray diagnostic work
Turkey, USA, Russia	glass, soaps and detergents, agriculture, fire retardants, plastics and polymer additives
China, USA, Russia	cement, lime production, crushed stone, glassmaking, chemicals, optics
Australia, Democratic Republic of the Congo, Russia	jewelry, cutting tools, drill bits, computer chip production
China, Mexico, South Africa	hydrofluoric acid, steelmaking, water fluoridation, solvents, glass manufacture, enamels
USA, Iran, Canada	wallboard, building plasters, manufacture of cement
USA, China, Germany	chemical production, human and animal nutrition, highway deicer, water softener
Canada, USA, Russia	sulfuric acid, fertilizers, gunpowder, tires
USA, Uzbehkistan, Czech Republic	glossy paper, whitener and abrasive in toothpaste
USA, Germany, France	glass, computer chips, ceramics, abrasives, water filtration
China, USA, Republic of Korea	ceramics, plastics, paint, paper, rubber, cosmetics

Maps

World Physical Relief

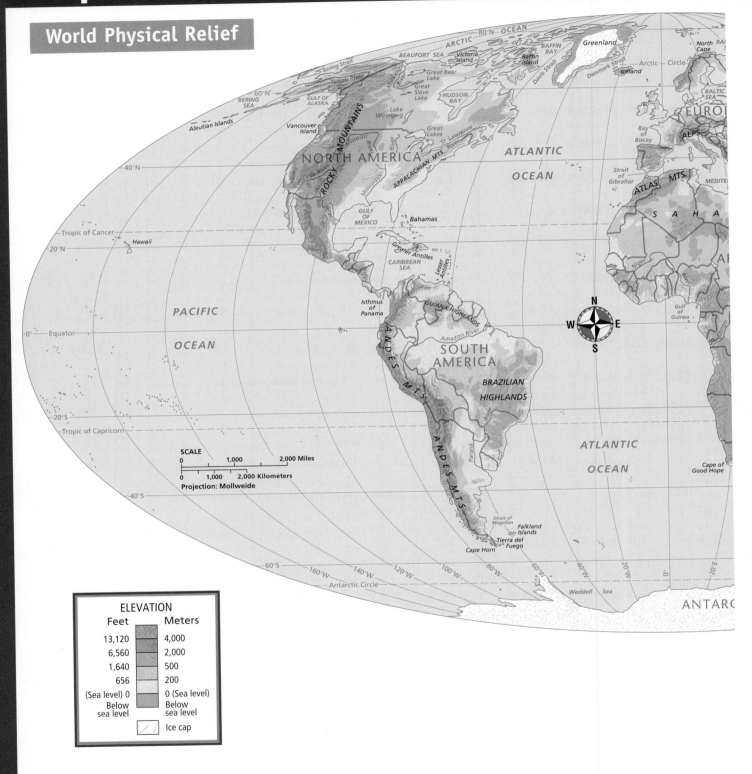

SCALE

0   1,000   2,000 Miles

0   1,000   2,000 Kilometers

Projection: Mollweide

ELEVATION

Feet	Meters
13,120	4,000
6,560	2,000
1,640	500
656	200
(Sea level) 0	0 (Sea level)
Below sea level	Below sea level
	Ice cap

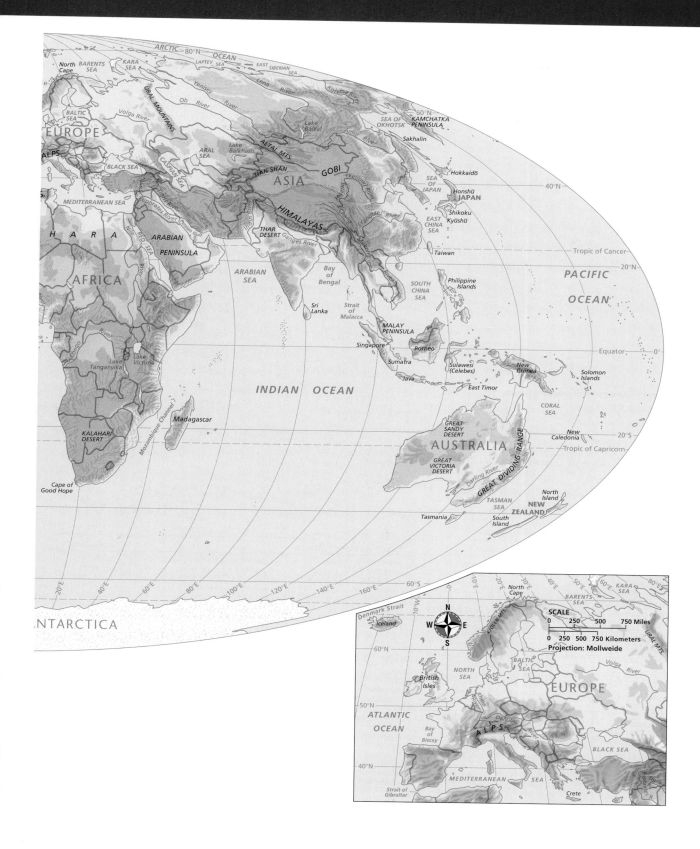

ARCTIC 80°N OCEAN

North Cape
BARENTS SEA
KARA SEA
LAPTEV SEA
EAST SIBERIAN SEA

EUROPE

BALTIC SEA

ALPS

Volga River

URAL MOUNTAINS

Ob River
Yenisey River
Lena River
Kolyma River

Lake Balkal

60°N

KAMCHATKA PENINSULA

SEA OF OKHOTSK

BLACK SEA

CASPIAN SEA

ARAL SEA

Lake Balkhash

ALTAI MTS.

Amur River

Sakhalin

HARA

MEDITERRANEAN SEA

Euphrates River

Tigris River

Nile River

Persian Gulf

ARABIAN PENINSULA

RED SEA

AFRICA

TIAN SHAN

ASIA

GOBI

Yellow River

Hokkaidō

SEA OF JAPAN

40°N

Honshū JAPAN

HIMALAYAS

THAR DESERT

Indus River

Ganges River

Brahmaputra River

Yangtze River

Shikoku
Kyūshū

EAST CHINA SEA

Taiwan

Tropic of Cancer

20°N

PACIFIC

ARABIAN SEA

Bay of Bengal

SOUTH CHINA SEA

Philippine Islands

OCEAN

Sri Lanka

Strait of Malacca

MALAY PENINSULA

Congo River

Lake Tanganyika
Lake Victoria

Singapore

Sumatra

Borneo

Sulawesi (Celebes)

New Guinea

Solomon Islands

Equator 0°

INDIAN OCEAN

Java

East Timor

CORAL SEA

Mozambique Channel

Madagascar

GREAT SANDY DESERT

New Caledonia

20°S

KALAHARI DESERT

AUSTRALIA

GREAT VICTORIA DESERT

GREAT DIVIDING RANGE

Tropic of Capricorn

Darling River

North Island

Cape of Good Hope

TASMAN SEA

NEW ZEALAND

Tasmania

South Island

NTARCTICA

20°E 40°E 60°E 80°E 100°E 120°E 140°E 160°E 60°S

Denmark Strait

Iceland

North Cape

KARA SEA

BARENTS SEA

80°E

NORTH SEA

BALTIC SEA

URAL MTS.

Volga River

British Isles

60°N

ATLANTIC

OCEAN

50°N

Bay of Biscay

Rhine River

Danube River

EUROPE

ALPS

BLACK SEA

40°N

MEDITERRANEAN SEA

Strait of Gibraltar

Crete

Tigris R.

621

Maps

World Climate Regions

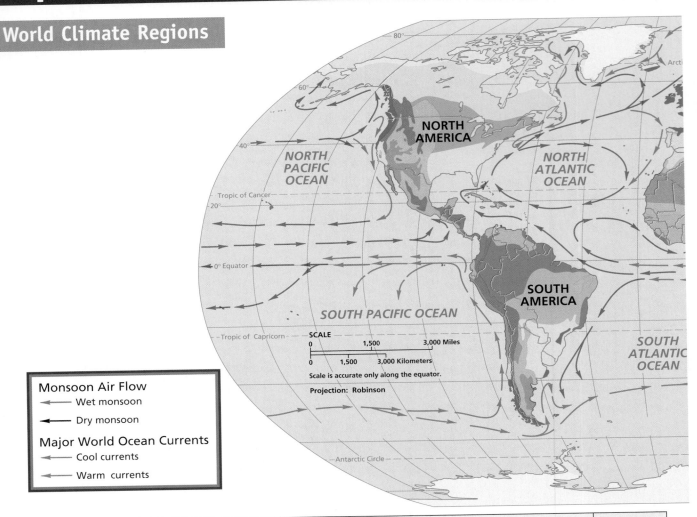

Monsoon Air Flow

← Wet monsoon

← Dry monsoon

Major World Ocean Currents

← Cool currents

← Warm currents

	Climate	Geographic Distribution	Major Weather Patterns	Vegetation
Tropical	**TROPICAL HUMID**	along equator; particularly equatorial South America, Congo Basin in Africa, Southeast Asia	warm and rainy year-round, with rain totaling anywhere from 65 to more than 450 in. (165–1,143 cm) annually; typical temperatures are 90°–95°F (32°–35°C) during the day and 65°–70°F (18°–21°C) at night	tropical rain forest
	TROPICAL WET AND DRY	between humid tropics and deserts; tropical regions of Africa, South and Central America, South and Southeast Asia, Australia	warm all year; distinct rainy and dry seasons; precipitation during the summer of at least 20 in. (51 cm); monsoon influences in some areas, such as South and Southeast Asia; summer temperatures average 90°F (32°C) during the day and 70°F (21°C) at night; typical winter temperatures are 75°–80°F (24°–27°C) during the day and 55°–60°F (13°–16°C) at night	tropical grassland with scattered trees
Dry	**ARID**	centered along 30° latitude; some middle-latitude deserts in interior of large continents and along western coasts; particularly Saharan Africa, Southwest Asia, central and western Australia, southwestern North America	arid; precipitation of less than 10 in. (25 cm) annually; sunny and hot in the tropics and sunny with great temperature ranges in middle latitudes; typical summer temperatures for lower-latitude deserts are 110°–115°F (43°–46°C) during the day and 60°–65°F (16°–18°C) at night, while winter temperatures average 80°F (27°C) during the day and 45°F (7°C) at night; in middle latitudes the hottest month averages 70°F (21°C)	sparse drought-resistant plants; many barren, rocky, or sandy areas
	SEMIARID	generally bordering deserts and interiors of large continents; particularly northern and southern Africa, interior western North America, central and interior Asia and Australia, southern South America	semiarid; about 10–20 in. (25–51 cm) of precipitation annually; hot summers and cooler winters with wide temperature ranges similar to desert temperatures	grassland; few trees
Middle Latitudes	**MEDITERRANEAN**	west coasts in middle latitudes near cool ocean currents; particularly southern Europe, part of Southwest Asia, northwestern Africa, California, southwestern Australia, central Chile, south-western South Africa	dry sunny warm summers and mild wetter winters; precipitation averages 14–35 in. (35–90 cm) annually; typical temperatures are 75°–80°F (24°–27°C) on summer days; the average winter temperature is 50°F (10°C)	scrub woodland and grassland
	HUMID SUBTROPICAL	east coasts in middle latitudes; particularly southeastern United States, eastern Asia, central southern Europe, southeastern parts of South America, South Africa, and Australia	hot humid summers and mild humid winters; precipitation year-round; coastal areas are in the paths of hurricanes and typhoons; precipitation averages 40 in. (102 cm) annually; typical temperatures are 75°–90°F (24°–32°C) in summer and 45°–50°F (7°–10°C) in winter	mixed forest

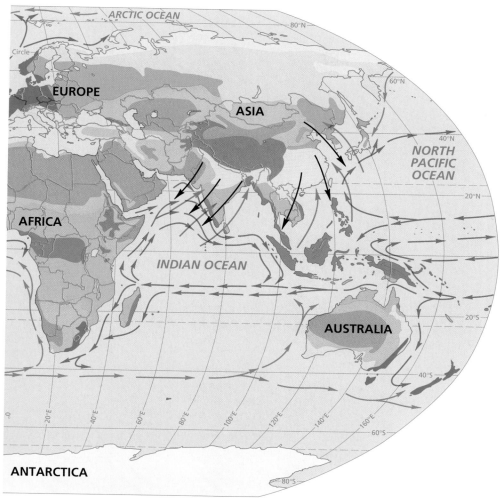

ARCTIC OCEAN

80°N

Circle

EUROPE

60°N

ASIA

40°N

NORTH PACIFIC OCEAN

20°N

AFRICA

INDIAN OCEAN

20°S

AUSTRALIA

40°S

0° 20°E 40°E 60°E 80°E 100°E 120°E 140°E 160°E

60°S

ANTARCTICA

80°S

	Climate	Geographic Distribution	Major Weather Patterns	Vegetation
Middle Latitudes	**MARINE WEST COAST**	west coasts in upper-middle latitudes; particularly northwestern Europe and North America, southwestern South America, central southern South Africa, southeastern Australia, New Zealand	cloudy mild summers and cool rainy winters; strong ocean influence; precipitation averages 20–98 in. (51–250 cm) annually; westerlies bring storms and rain; average temperature in hottest month is usually between 60°F and 70°F (16°–21°C); average temperature in coolest month usually is above 32°F (0°C)	temperate evergreen forest
	HUMID CONTINENTAL	east coasts and interiors of upper-middle latitude continents; particularly northeastern North America, northern and eastern Europe, northeastern Asia	four distinct seasons; long cold winters and short warm summers; precipitation amounts vary, usually 20–50 in. (51–127 cm) or more annually; average summer temperature is 75°F (24°C); average winter temperature is below freezing	mixed forest
High Latitudes	**SUBARCTIC**	higher latitudes of interior and east coasts of continents; particularly northern parts of North America, Europe, and Asia	extremes of temperature; long cold winters and short mild summers; low precipitation amounts all year; precipitation averages 5–15 in. (13–38 cm) in summer; temperatures in warmest month average 60°F (16°C) but can warm to 77°F (25°C); winter temperatures average below 0°F (–18°C)	northern evergreen forest
	TUNDRA	high-latitude coasts; particularly far northern parts of North America, Europe, and Asia, Antarctic Peninsula, subantarctic islands	cold all year; very long cold winters and very short cool summers; low precipitation amounts; precipitation average is 5–15 in. (13–38 cm) annually; warmest month averages less than 50°F (10°C); coolest month averages a little below 0°F (–18°C)	moss, lichens, low shrubs; permafrost bogs in summer
	ICECAP	polar regions; particularly Antarctica, Greenland, Arctic Basin islands	freezing cold; snow and ice year-round; precipitation averages less than 10 in. (25 cm) annually; average temperatures in warmest month do not reach higher than freezing	no vegetation
	HIGHLAND	high mountain regions, particularly western parts of North and South America, eastern parts of Asia and Africa, southern and central Europe and Asia	greatly varied temperatures and precipitation amounts over short distances as elevation changes; prevailing wind patterns can affect rainfall on windward and leeward sides of highland areas	forest to tundra vegetation, depending on elevation

Maps

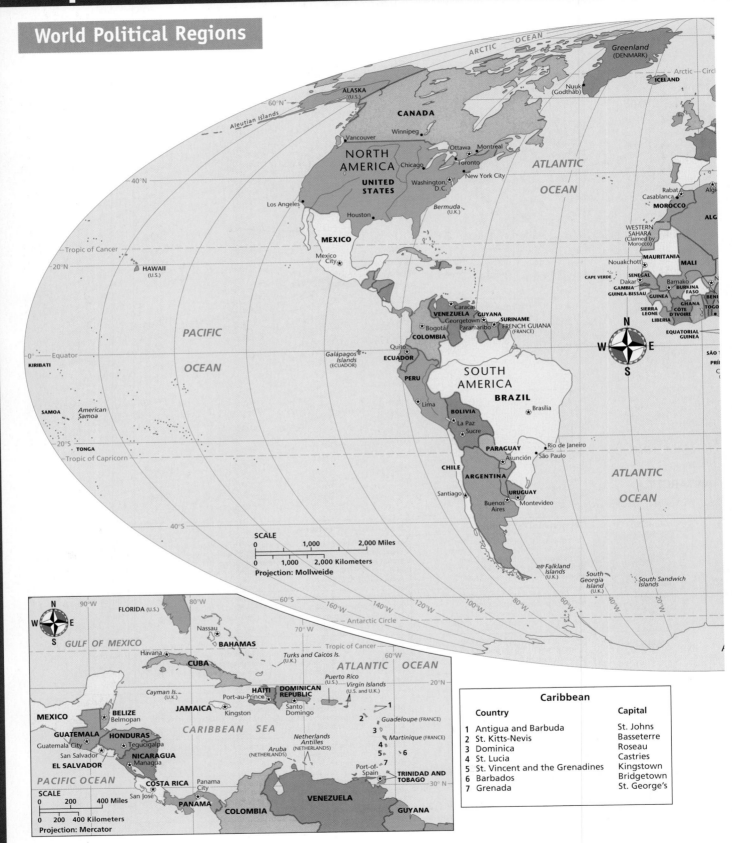

SCALE
0 — 1,000 — 2,000 Miles
0 — 1,000 — 2,000 Kilometers
Projection: Mollweide

SCALE
0 — 200 — 400 Miles
0 — 200 — 400 Kilometers
Projection: Mercator

Caribbean	
Country	Capital
1 Antigua and Barbuda	St. Johns
2 St. Kitts-Nevis	Basseterre
3 Dominica	Roseau
4 St. Lucia	Castries
5 St. Vincent and the Grenadines	Kingstown
6 Barbados	Bridgetown
7 Grenada	St. George's

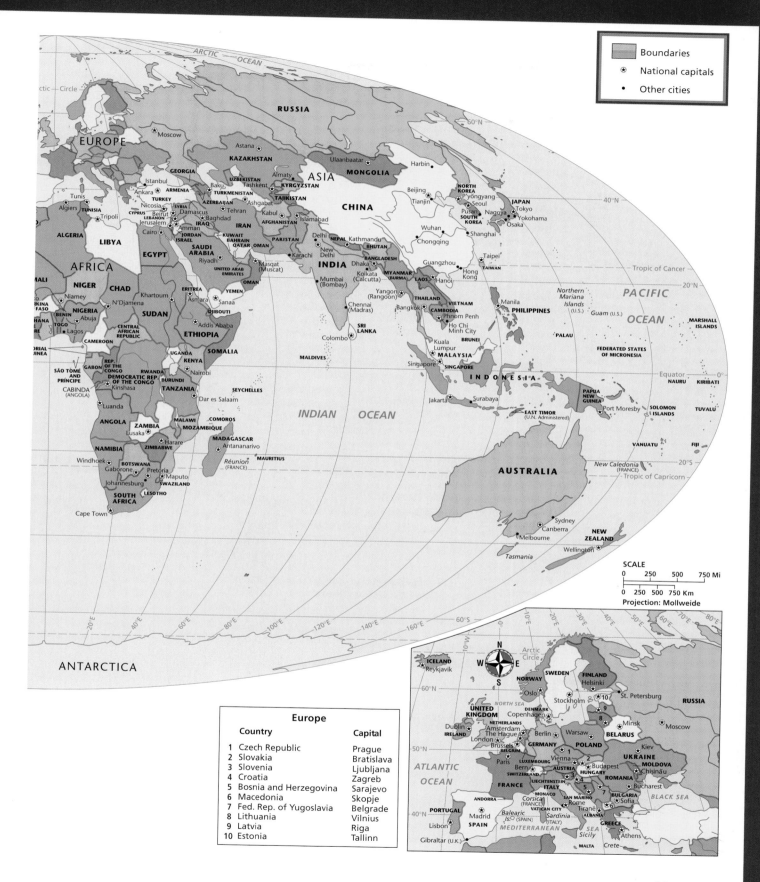

Boundaries

⊛ National capitals

• Other cities

SCALE

0   250   500   750 Mi

0   250   500   750 Km

Projection: Mollweide

Europe	
Country	**Capital**
1 Czech Republic	Prague
2 Slovakia	Bratislava
3 Slovenia	Ljubljana
4 Croatia	Zagreb
5 Bosnia and Herzegovina	Sarajevo
6 Macedonia	Skopje
7 Fed. Rep. of Yugoslavia	Belgrade
8 Lithuania	Vilnius
9 Latvia	Riga
10 Estonia	Tallinn

World Population Density

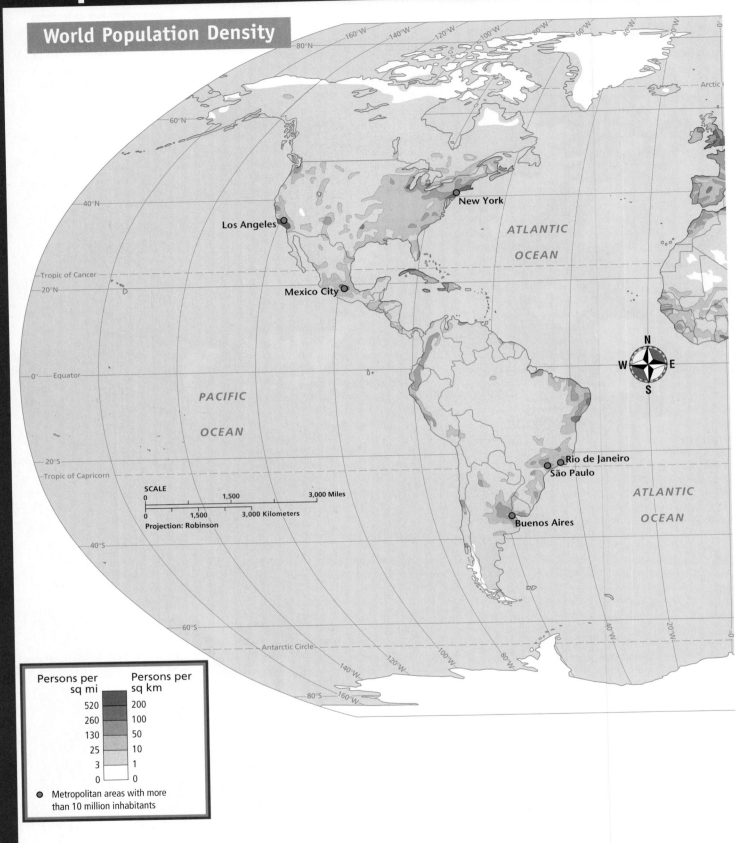

SCALE

| 0 | | 1,500 | | 3,000 Miles |
| 0 | 1,500 | | 3,000 Kilometers | |

Projection: Robinson

Persons per sq mi	Persons per sq km
520	200
260	100
130	50
25	10
3	1
0	0

● Metropolitan areas with more than 10 million inhabitants

ARCTIC OCEAN

Circle

Cairo

Lagos

Karachi

Mumbai (Bombay)

Delhi

Dhaka

Kolkata (Calcutta)

Beijing

Shanghai

Tokyo

Osaka

Manila

Jakarta

INDIAN OCEAN

PACIFIC OCEAN

Tropic of Cancer

Equator

Tropic of Capricorn

Antarctic Circle

Maps

World Carbon Dioxide Emissions Per Person

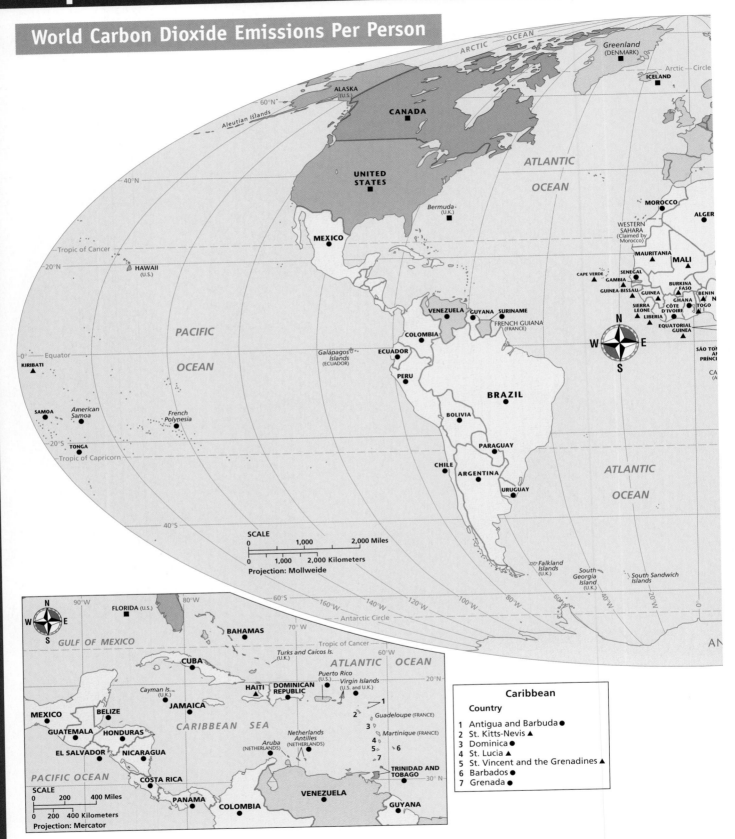

Caribbean

Country

1 Antigua and Barbuda ●
2 St. Kitts-Nevis ▲
3 Dominica ●
4 St. Lucia ▲
5 St. Vincent and the Grenadines ▲
6 Barbados ●
7 Grenada ●

628 Maps

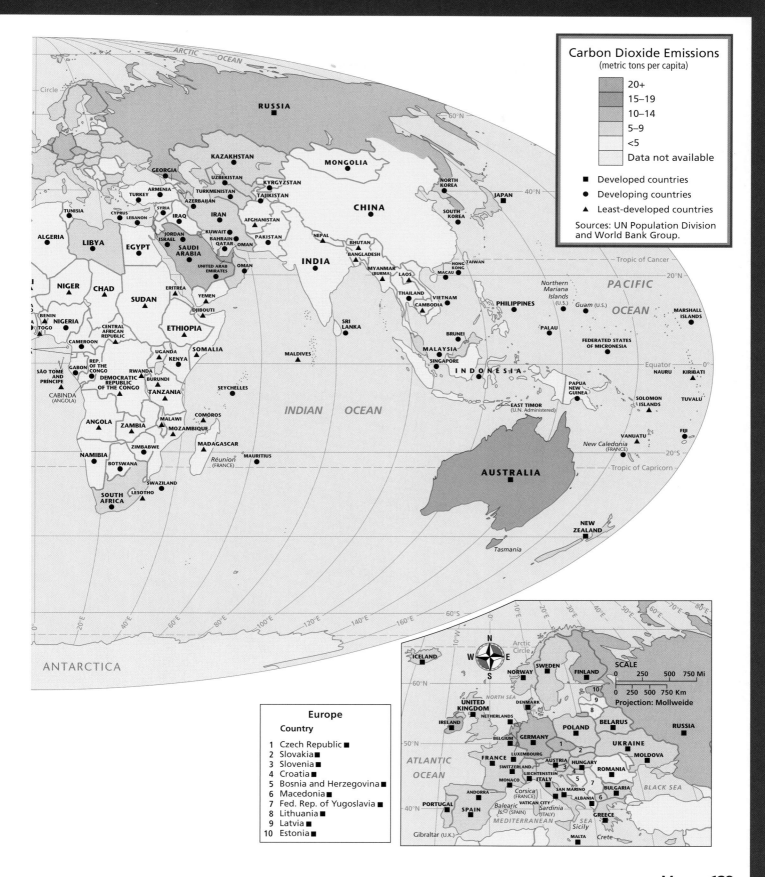

Carbon Dioxide Emissions
(metric tons per capita)

- 20+
- 15–19
- 10–14
- 5–9
- <5
- Data not available

- ■ Developed countries
- ● Developing countries
- ▲ Least-developed countries

Sources: UN Population Division and World Bank Group.

Europe

Country

1 Czech Republic ■
2 Slovakia ■
3 Slovenia ■
4 Croatia ■
5 Bosnia and Herzegovina ■
6 Macedonia ■
7 Fed. Rep. of Yugoslavia ■
8 Lithuania ■
9 Latvia ■
10 Estonia ■

SCALE
0   250   500   750 Mi
0   250   500   750 Km
Projection: Mollweide

Maps

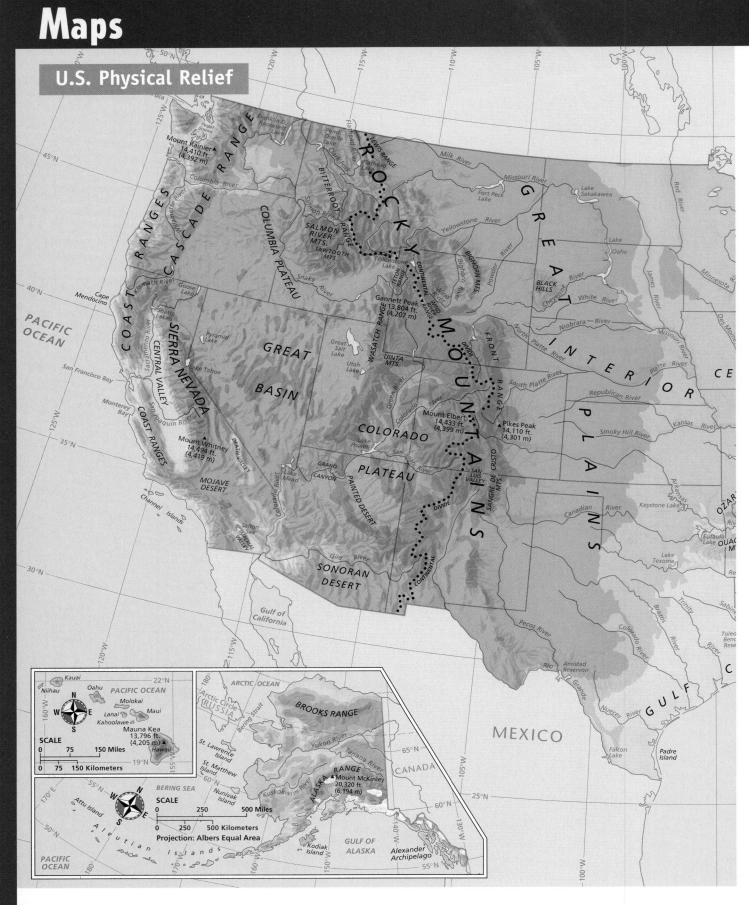

U.S. Physical Relief

PACIFIC OCEAN

Mount Rainier▲
14,410 ft.
(4,392 m)

Cape
Mendocino

COAST RANGES

CASCADE RANGE

COLUMBIA PLATEAU

LEWIS RANGE

BITTERROOT RANGE

SALMON RIVER MTS.

SAWTOOTH MTS.

ROCKY

MOUNTAINS

GREAT INTERIOR PLAINS

BLACK HILLS

CENTRAL VALLEY

SIERRA NEVADA

Lake Tahoe

GREAT BASIN

Great Salt Lake

Utah Lake

WASATCH RANGE

UINTA MTS.

Gannett Peak
13,804 ft.
(4,207 m)

TETON RANGE

WIND RIVER RANGE

CONTINENTAL DIVIDE

FRONT RANGE

San Francisco Bay

Monterey Bay

COAST RANGES

Mount Whitney
14,494 ft.
(4,419 m)

DEATH VALLEY

MOJAVE DESERT

Channel Islands

Salton Sea

IMPERIAL VALLEY

Lake Mead

GRAND CANYON

PAINTED DESERT

COLORADO PLATEAU

Lake Powell

Mount Elbert
14,433 ft.
(4,399 m)

Pikes Peak▲
14,110 ft.
(4,301 m)

SAN LUIS VALLEY

SANGRE DE CRISTO MTS.

DIVIDE

SONORAN DESERT

Gila River

CONTINENTAL

Gulf of California

OZARK

OUACHITA MTS.

GULF

MEXICO

CANADA

PACIFIC OCEAN

ARCTIC OCEAN

RUSSIA

Arctic Circle

Bering Strait

BROOKS RANGE

Yukon River

Tanana River

ALASKA RANGE

Mount McKinley
20,320 ft.
(6,194 m)

St. Lawrence Island

St. Matthew Island

Nunivak Island

Kuskokwim River

BERING SEA

Attu Island

Aleutian Islands

Kodiak Island

GULF OF ALASKA

Alexander Archipelago

SCALE
0   250   500 Miles
0   250   500 Kilometers
Projection: Albers Equal Area

Kauai
Niihau
Oahu
Molokai
Lanai
Kahoolawe
Maui
Mauna Kea
13,796 ft.
(4,205 m)▲
Hawaii

PACIFIC OCEAN

SCALE
0   75   150 Miles
0   75   150 Kilometers

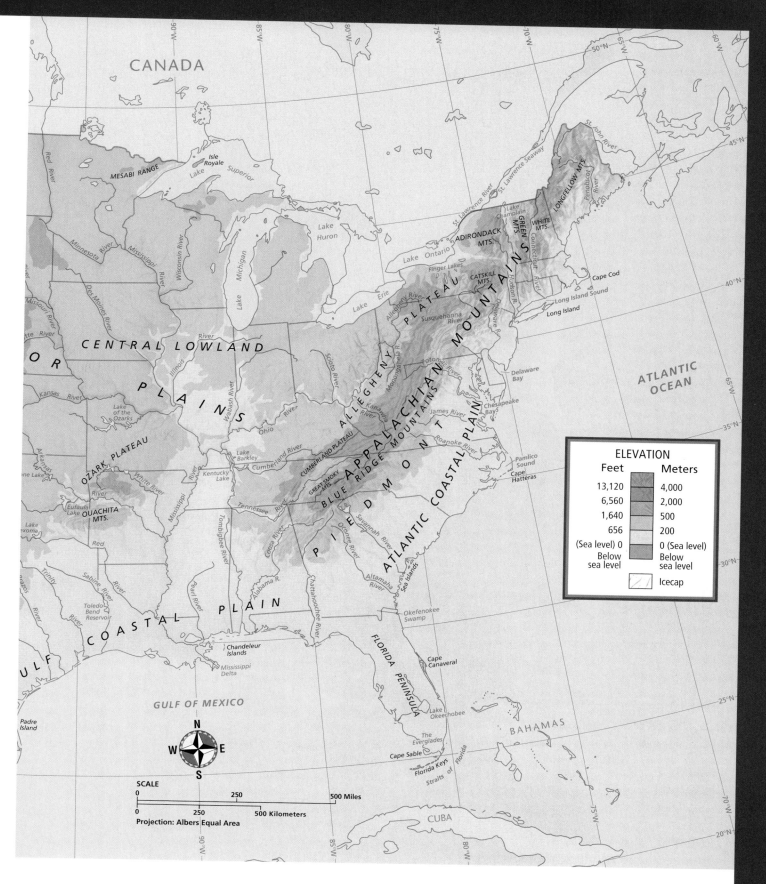

CANADA

MESABI RANGE

Isle Royale

Lake Superior

Lake Huron

Lake Michigan

Lake Ontario

Lake Erie

Finger Lakes

ADIRONDACK MTS.

CATSKILL MTS.

GREEN MTS.

WHITE MTS.

LONGFELLOW MTS.

St. Lawrence Seaway

St. Lawrence River

Lake Champlain

Penobscot R.

St. John River

Connecticut River

Hudson R.

Cape Cod

Long Island Sound

Long Island

ATLANTIC OCEAN

CENTRAL LOWLAND

PLAINS

Red River

Minnesota River

Mississippi River

Des Moines River

Wisconsin River

Illinois River

River

Platte River

Kansas River

Missouri River

Lake of the Ozarks

OZARK PLATEAU

Arkansas River

Eufaula Lake

OUACHITA MTS.

Lake Texoma

White River

Red

Sabine River

Toledo Bend Reservoir

Trinity River

Brazos River

Padre Island

GULF

COASTAL PLAIN

GULF OF MEXICO

Chandeleur Islands

Mississippi Delta

Pearl River

Alabama R.

Tombigbee River

Coosa River

Chattahoochee River

Tennessee River

Cumberland River

Kentucky Lake

Lake Barkley

Ohio River

Scioto River

Wabash River

Allegheny River

ALLEGHENY PLATEAU

Susquehanna River

Delaware R.

Potomac River

Monongahela R.

Kanawha R.

APPALACHIAN MOUNTAINS

BLUE RIDGE MOUNTAINS

GREAT SMOKY MTS.

CUMBERLAND PLATEAU

James River

Roanoke River

PIEDMONT

ATLANTIC COASTAL PLAIN

Delaware Bay

Chesapeake Bay

Pamlico Sound

Cape Hatteras

Savannah River

Oconee River

Altamaha River

Sea Islands

Okefenokee Swamp

FLORIDA PENINSULA

Cape Canaveral

Lake Okeechobee

The Everglades

Cape Sable

Florida Keys

Straits of Florida

BAHAMAS

CUBA

Padre Island

ELEVATION

Feet		Meters
13,120		4,000
6,560		2,000
1,640		500
656		200
(Sea level) 0		0 (Sea level)
Below sea level		Below sea level

Icecap

SCALE

0   250   500 Miles

0   250   500 Kilometers

Projection: Albers Equal Area

N W E S

U.S., Canada, and Mexico Climate Regions

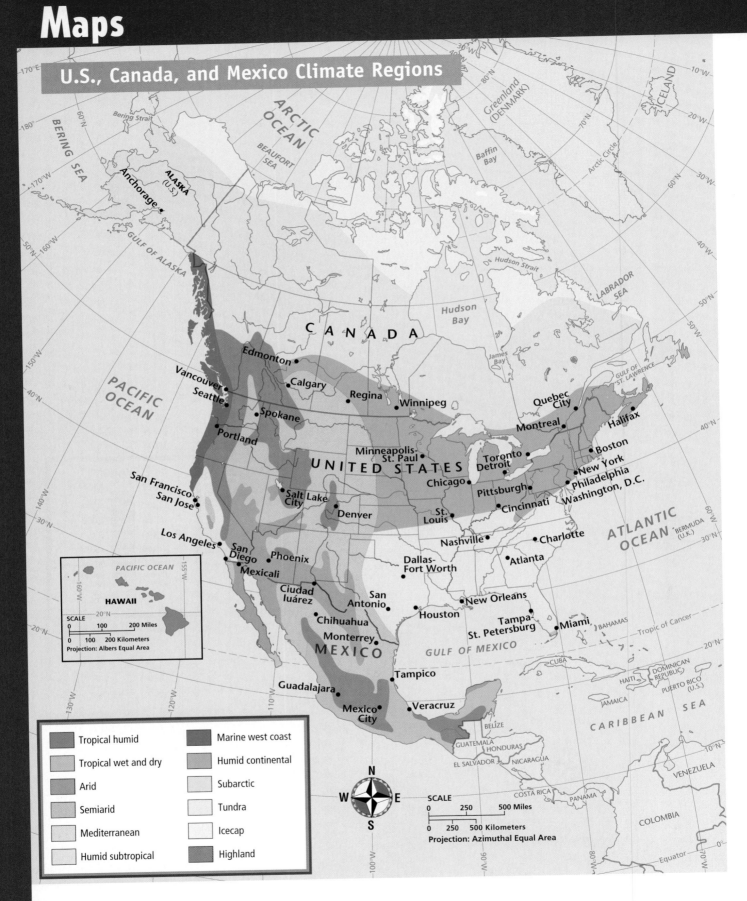

Legend:

- Tropical humid
- Tropical wet and dry
- Arid
- Semiarid
- Mediterranean
- Humid subtropical
- Marine west coast
- Humid continental
- Subarctic
- Tundra
- Icecap
- Highland

HAWAII
PACIFIC OCEAN
SCALE
0   100   200 Miles
0   100   200 Kilometers
Projection: Albers Equal Area

SCALE
0   250   500 Miles
0   250   500 Kilometers
Projection: Azimuthal Equal Area

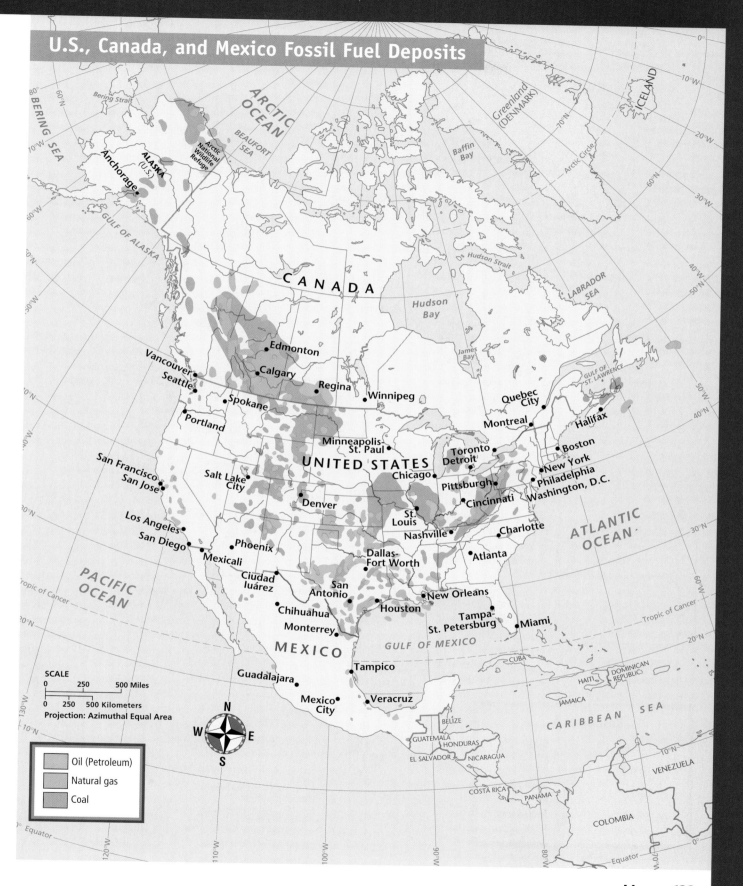

U.S., Canada, and Mexico Fossil Fuel Deposits

Legend:
- Oil (Petroleum)
- Natural gas
- Coal

SCALE
0   250   500 Miles
0   250   500 Kilometers
Projection: Azimuthal Equal Area

Maps

U.S., Canada, and Mexico Mineral and Energy Resources

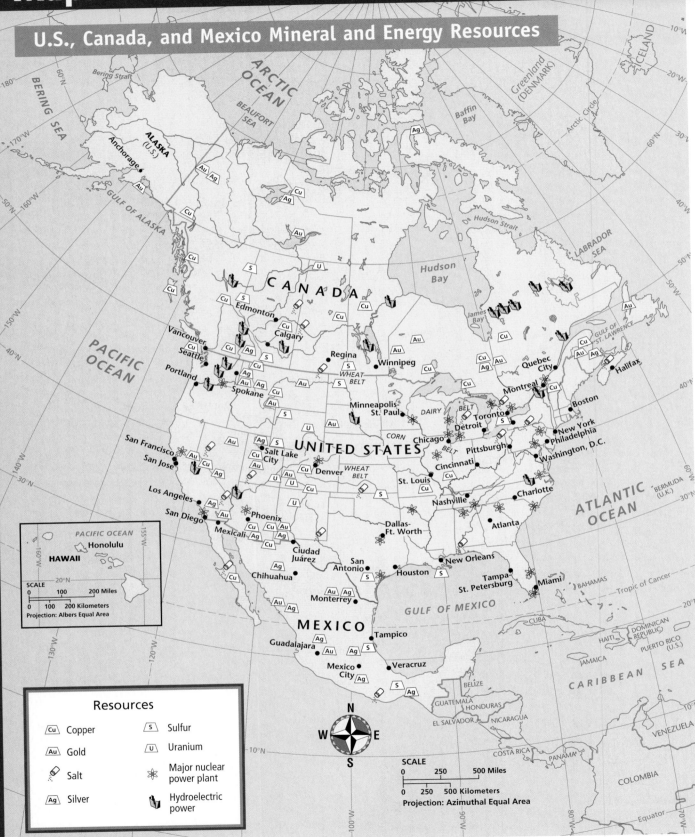

Resources

Cu	Copper	S	Sulfur
Au	Gold	U	Uranium
Salt	Salt	✳	Major nuclear power plant
Ag	Silver	⬛	Hydroelectric power

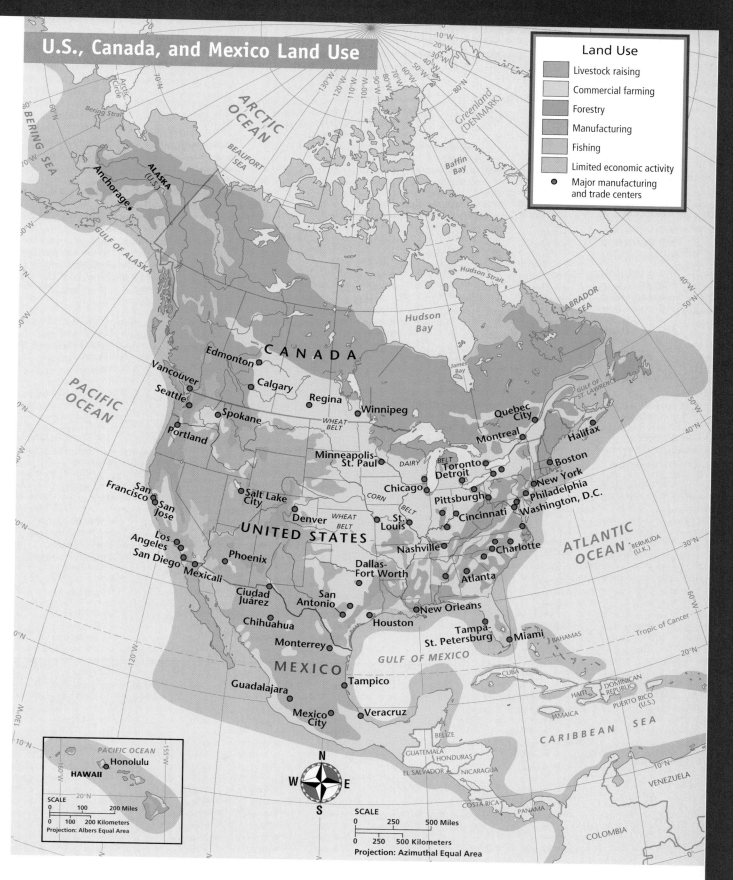

U.S., Canada, and Mexico Land Use

Land Use

- Livestock raising
- Commercial farming
- Forestry
- Manufacturing
- Fishing
- Limited economic activity
- ● Major manufacturing and trade centers

ARCTIC OCEAN

BERING SEA

Bering Strait

Arctic Circle

BEAUFORT SEA

Greenland (DENMARK)

Anchorage

ALASKA (U.S.)

GULF OF ALASKA

PACIFIC OCEAN

Baffin Bay

Hudson Strait

LABRADOR SEA

Hudson Bay

James Bay

CANADA

Edmonton
Vancouver
Calgary
Seattle
Regina
Winnipeg
Spokane
Quebec City
Portland
WHEAT BELT
Montreal
Halifax
GULF OF ST. LAWRENCE

Minneapolis-St. Paul
DAIRY BELT
Toronto
Boston
Detroit
Chicago
New York
San Francisco
Salt Lake City
Pittsburgh
Philadelphia
San Jose
CORN BELT
Cincinnati
Washington, D.C.
Denver
WHEAT BELT
St. Louis
UNITED STATES
Los Angeles
Nashville
Charlotte
San Diego
Phoenix
ATLANTIC OCEAN
BERMUDA (U.K.)
Mexicali
Dallas-Fort Worth
Atlanta
Ciudad Juárez
San Antonio
New Orleans
Chihuahua
Houston
Tampa-St. Petersburg
Miami
Tropic of Cancer
Monterrey
GULF OF MEXICO
BAHAMAS
MEXICO
Tampico
CUBA
Guadalajara
HAITI
DOMINICAN REPUBLIC
PUERTO RICO (U.S.)
Mexico City
Veracruz
JAMAICA
BELIZE
CARIBBEAN SEA
GUATEMALA
HONDURAS
EL SALVADOR
NICARAGUA
VENEZUELA
COSTA RICA
PANAMA
COLOMBIA

PACIFIC OCEAN
Honolulu
HAWAII

SCALE
0   100   200 Miles
0   100   200 Kilometers
Projection: Albers Equal Area

N
W   E
S

SCALE
0   250   500 Miles
0   250   500 Kilometers
Projection: Azimuthal Equal Area

Estimated Time

To give students practice under more realistic testing conditions, allow them 30 minutes to answer all of the questions in this practice test.

 TEST DOCTOR

Question 6 Full-credit answers should include the following points:

- as nonrenewable resources become scarce, their costs rise
- as the cost of a resource rises, people look for alternatives that are cheaper and more readily available

Question 8 Full-credit answers should include the following points:

- the agricultural revolution allowed populations to grow in localized areas
- the larger populations used more resources than the smaller hunter-gatherer groups
- using more resources put more pressure on the environment to replace, or find replacements for, the used resources

Understanding Concepts

Directions (1–4): For *each* question, write on a separate sheet of paper the letter of the correct answer.

1 How do scientists characterize a nonrenewable resource?
- **A.** a resource that is used by humans
- **B.** a resource that can not be replaced
- **C.** a resource that can be replaced relatively quickly
- **D.** a resource that takes more time to replace than to deplete

2 Which of the following is an important foundation of environmental science?
- **F.** ecology
- **G.** economics
- **H.** meteorology
- **I.** political science

3 Which of the following phrases describes the term biodiversity?
- **A.** species that have become extinct
- **B.** the animals that live in an area
- **C.** species that look different from one another
- **D.** the number and variety of species that live in an area

4 Energy from the sun, water, air, wood, and soil are all examples of what kind of energy?
- **F.** ecological energy
- **G.** organic energy
- **H.** renewable energy
- **I.** solar energy

Directions (5–6): For *each* question, write a short response.

5 Pollution is created when societies produce wastes faster than they can be disposed of. Identify the two main types of pollutants.

6 Economic forces influence the use of natural resources. One rule of economics is the law of supply and demand. Analyze how economic forces affect the usage of nonrenewable resources.

Reading Skills

Directions (7–9): Read the passage below. Then answer the questions.

Early hunter-gatherer groups began to collect the seeds of the plants they gathered and to domesticate animals in their environment. This practice of agriculture started in many parts of the world more than 10,000 years ago, and had such a dramatic impact on human societies and their environment that it is often called the agricultural revolution. An area of land could now support up to 500 times as many people by farming than it could by hunting and gathering. As populations grew, they began to concentrate in smaller areas.

As grasslands, forests, and wetlands were replaced with farmland, habitat was destroyed. In addition, much of the converted land was farmed poorly and became infertile. The destruction of farmland had far-reaching environmental effects.

7 The practice of gathering seeds and domesticating some animals led to
- **A.** a reduction in the kinds of food people ate
- **B.** the extension of the prairies as open grassland
- **C.** the disappearance of some large mammal species
- **D.** a growth in population at an unprecedented rate

8 In what ways did the agricultural revolution increase pressure on local environments?

9 What is the ultimate result of the loss of species habitat?
- **F.** the extinction of the species
- **G.** a growth in species population
- **H.** a change in species eating habits
- **I.** the overuse of the remaining resources of the species

636

Answers

1. D
2. F
3. D
4. H
5. The two types of pollutants are biodegradable and nonbiodegradable.
6. Answers will vary. See Test Doctor for detailed scoring rubric.
7. D
8. Answers will vary. See Test Doctor for detailed scoring rubric.
9. F
10. D
11. H
12. B
13. G

Interpreting Graphics

Directions (10–13): **For** *each* **question below, record the correct answer on a separate sheet of paper.**

10 Population growth can result in what ethical environmental problem, addressed by ecologist Garrett Hardin in "The Tragedy of the Commons?"
A. the conflict between water resources and industrial growth
B. the conflict between forest resources and the lumber companies
C. the conflict between political interests and international energy use
D. the conflict between individual interests and the welfare of society

The line graph below shows the world population between the years 1600 and 2000. Use this graph to answer questions 11 and 12.

World Population (1600–2000)

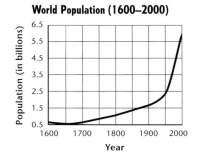

11 What was the total population increase between the years 1600 and 1900?
F. 0.6 billion
G. 0.9 billion
H. 1.0 billion
I. 1.5 billion

12 If the rate of growth from 1900-1950 had been the same as the rate of growth from 1950-2000, what would the world population have been at the end of the century?
A. more than 7 billion
B. more than 10 billion
C. more than 15 billion
D. more than 20 billion

13 Which of the following characterizes the environmental consequences of the current population trend?
F. More people mean more housing construction.
G. The need for food and resources is growing rapidly.
H. The standard of living has risen around the world.
I. There is no connection between population growth and environment.

Test TIP

When reading a graph that shows a change in some variable over time, keep in mind that the steepness and direction of a curve indicate the relative rate of change at a given point in time.

 **TEST DOCTOR**

Question 11 Answer H is correct. You can determine differences in values by identifying each value on the y-axis, which corresponds to a specific value, and then subtracting the smaller values from the larger. Answers F and G are too small, and answer I is much too large.

Question 12 Answer B is correct. To estimate a value, students should draw a best-fit line from a point near where the data ends to the point they are trying to predict. To estimate the value asked in the question, students should estimate the slope of the line from 1950–2000. Then replace the actual line from 1900–1950 with the new line. Identify where the line ends (the population of 1950 under the new conditions). Then apply this line again to the period between 1950 and 2000 to identify what the population in the year 2000 would have been under the new conditions.

Students struggling with this type of question may benefit from practicing producing their own line graphs. For example, ask students to draw two line graphs. The first should have a constant slope throughout. The second should have one slope for the first half, and then double the slope for the second half. Ask student to compare the two graphs.

637

Estimated Time

To give students practice under more realistic testing conditions, allow them 30 minutes to answer all of the questions in this practice test.

Question 3 Answer A is the correct choice. Answers B, C, and D are incorrect because they are all examples of images. Although scientists use maps, diagrams, and satellite images for scientific study, these tools are not physical representations.

Question 5 Full-credit answers should include the following points:

- no; every time a penny is tossed, the probability of its outcome is independent of the tosses that have come before
- the probability of heads or tails is always 50%

Question 6 Full-credit answers should include the following points:

- the mean is a numerical measure for a given population
- the mean allows scientists to compare characteristics of different populations

Understanding Concepts

Directions (1–4): For *each* question, write on a separate sheet of paper the letter of the correct answer.

1 How would a scientist categorize a testable explanation for an observation?
A. a correlation
B. an experiment
C. an hypothesis
D. a prediction

2 What happens when an observation is submitted for peer review?
F. The article is proofread before it is published.
G. A professor gives a lecture based on a published article.
H. The results are looked at closely by other scientific experts.
I. Information on the experimental design is included in published works.

3 Which of the following is an example of a scientist's physical model?
A. a crash-test dummy for a car company
B. a diagram of the structure of an atom
C. a map of Denver, Colorado
D. a satellite image of South America

4 What attribute of a skeptic would contribute to a good scientific mind?
F. willingness to travel
G. an empathetic nature
H. desire to conduct experiments
I. continually questioning observations

Directions (5–6): For *each* question, write a short response.

5 A penny is tossed and comes up heads 7 out of 10 times. Is the probability that it will be heads on the next toss 70%? Why or why not?

6 A mean is the number obtained by adding up the data for a given characteristic of a statistical population, and dividing the sum by the total number of individuals in the given population. Why do scientists calculate the mean of a statistical population?

Reading Skills

Directions (7–9): Read the passage below. Then answer the questions.

We use statistics everyday. Weathermen report the forecast in terms of probabilities, such as "There is a 50 percent chance of rain today." People are constantly guessing at the possibility that something will or will not happen. A guess is one of the ways we express probability.

In scientific terms, risk is the probability of an unwanted outcome. Most people overestimate the risk of dying from sensational causes, such as plane crashes, and underestimate the risk from common causes such as smoking. Likewise, most citizens overestimate the risk of sensational environmental problems such as oil spills, and underestimate the risk of ordinary ones, like ozone depletion. However, when decisions must be made on proposals affecting the environment, it is important that all the benefits and risks of the possible action are calculated.

7 Assess which of the following experts would perceive as having the **highest** risk.
A. the threat of global climate change
B. the radioactivity from the waste of a nuclear power plant
C. the possibility of a tidal wave reaching a highly populated land mass
D. the danger of widespread water pollution

8 Although big oil spills get headlines, what is the **greatest** risk to the world ocean?
F. air pollution
G. offshore drilling
H. runoff from land
I. ships

9 How could a decision-making model be helpful in estimating the benefits and risks of a proposal?
A. It would eliminate uncertainty.
B. It would create a digital image.
C. It would predict the outcome of the decision.
D. It would allow consideration of all the variables.

638

Answers

1. C
2. H
3. A
4. I
5. Answers will vary. See Test Doctor for detailed scoring rubric.
6. Answers will vary. See Test Doctor for detailed scoring rubric.
7. A
8. H
9. D
10. H
11. A
12. F
13. C

Interpreting Graphics

Directions (10–13): For *each* question below, record the correct answer on a separate sheet of paper.

The bar graph below shows the distribution of lengths in a population of dwarf wedge mussels. Use this graph to answer questions 10 through 13.

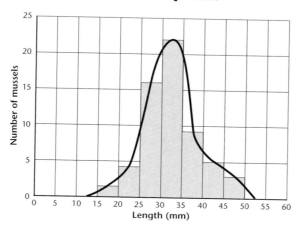

Sizes of Dwarf Wedge Mussels

10 What type of distribution does this bell-shaped curve depict?
 F. asymmetric
 G. correlative
 H. normal
 I. random

11 How many mussels are less than 25 mm in length?
 A. 6
 B. 9
 C. 12
 D. 15

12 Determine the total size of this statistical population of dwarf wedge mussels.
 F. 60
 G. 70
 H. 80
 I. 90

13 What is the **most** likely size predictable for a mussel randomly drawn from this population?
 A. 15–20 mm
 B. 25–30 mm
 C. 30–35 mm
 D. 40–45 mm

TEST DOCTOR

Question 10 Answer H is the correct choice because a bell curve represents a normal distribution. Answer F is incorrect because asymmetric curves are dramatically asymmetrical, while this curve is only slightly asymmetrical. Answer G is incorrect because a correlative curve most resembles a straight line. Answer I is incorrect because random distributions have no distinct shape or pattern.

Question 13 Answer C is correct. Students struggling with this type of question may benefit from practicing calculating probability. For example, ask students what the probability would be of a rolling a 6 on a six-sided die. Then ask students what the probability of rolling a 6 would be if two of the sides on the die were 6.

Test TIP

Probability is the chance of an outcome occurring. The highest probability occurs in the group with the largest number of individuals.

639

Estimated Time

To give students practice under more realistic testing conditions, allow them 30 minutes to answer all of the questions in this practice test.

 TEST DOCTOR

Question 2 Answer H is correct. When tectonic plates collide, Earth's crust becomes thicker, eventually forming mountain ranges. Answer F is incorrect because earthquakes are vibrations of the Earth's crust. Answer G is incorrect because a fault is a break in the Earth's crust. Answer I is incorrect because volcanoes are mountains built from molten rock.

Question 5 Full-credit answers should include the following points:

- the compositional layers of Earth are crust, mantle, and core
- the crust is the outermost and thinnest layer
- the mantle is the middle layer composed of dense iron-rich materials
- the core is the innermost layer and is composed of the densest materials

Question 8 Answer H is correct. Answers F, G, and I are incorrect because even though they describe likely events resulting from a volcanic eruption, they do not affect climate on a global scale.

Understanding Concepts

Directions (1–4): For *each* question, write on a separate sheet of paper the letter of the correct answer.

1 What is the cool, rigid, outermost layer of the Earth?
- A. the asthenosphere
- B. the geosphere
- C. the lithosphere
- D. the mesosphere

2 The collision of tectonic plates creates what geologic feature?
- F. earthquakes
- G. faults
- H. mountains
- I. volcanoes

3 What determines the weather we experience on Earth?
- A. movement of water over land masses
- B. gases trapping heat near Earth's surface
- C. absorption of radiation by the thermosphere
- D. air constantly moving through Earth's atmosphere

4 What is the difference between evaporation and condensation?
- F. Evaporation is the first stage of the water cycle; condensation is the last stage.
- G. Evaporation is the change from water to vapor; condensation is the change from vapor to water.
- H. Evaporation is the process where water is heated by the sea; condensation is the process where water droplets fall from clouds.
- I. Evaporation is the process where water vapor forms droplets; condensation is the process where water vapor forms clouds.

Directions (5–6): For *each* question, write a short response.

5 Differentiate between the three compositional layers of Earth.

6 What is the purpose of the Richter scale?

Reading Skills

Directions (7–9): Read the passage below. Then answer the questions.

A volcano is a mountain built from melted rock that rises from the Earth's interior to the surface. Volcanoes may occur on land or under the sea, where they eventually break the ocean surface as islands. They are often located near tectonic plate boundaries.

Volcanic eruptions can be devastating to local economies, cause great human loss, and affect global climate for several years. In large eruptions, clouds of volcanic ash and sulfur-rich gases may reach the upper atmosphere. In addition, ash that falls to the ground can cause buildings to collapse under its weight, bury crops, and damage the engines of vehicles.

Major volcanic eruptions, such as the eruption of Mount St. Helens, can change Earth's climate for several years.

7 Where are the majority of the world's active volcanoes on land located?
- A. the Amazon rain forest
- B. the Antarctic
- C. the Cape of Good Hope
- D. the Ring of Fire

8 How do volcanic eruptions affect global climate?
- F. The ash buries crops which affects photosynthesis.
- G. The ash mixes with water to produce destructive mudflows.
- H. The ash blocks sunlight which reduces average temperatures.
- I. The ash falls to the ground and produces dust storms.

9 Which of the following is a beneficial byproduct of an active volcano?
- A. new plant life
- B. geothermal energy
- C. melting ice in the polar regions
- D. decrease in seasonal temperature fluctuations

Answers
1. C
2. H
3. D
4. G
5. Answers will vary. See Test Doctor for detailed scoring rubric.
6. The Richter scale is used to quantify the amount of energy released by an earthquake.
7. D
8. H
9. A
10. F
11. D
12. G

Interpreting Graphics

Directions (10–12): For *each* question below, record the correct answer on a separate sheet of paper.

10 Which of the following statements is true?
  F. The world ocean covers 70% of Earth's surface.
  G. The world ocean is the body of water south of Africa.
  H. The world ocean has little effect on Earth's environment.
  I. The world ocean consists of the Atlantic and Pacific Oceans.

The illustration below shows elements in the ocean's water. Use it to answer question 11.

Elements in Ocean Water

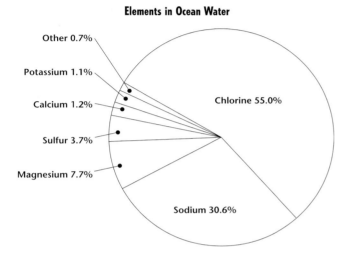

11 What determines the salinity of the ocean's water?
  A. chlorine
  B. chlorine and sulfur
  C. sodium
  D. sodium and chlorine

12 The Atlantic Ocean covers 81,630,000 km^2 of Earth's surface area and the Pacific Ocean covers 165,640,000 km^2. How much more surface area is covered by the Pacific Ocean than the Atlantic Ocean, in millions of km^2?
  F. 72
  G. 84
  H. 96
  I. 108

Test TIP

If you come upon a word you do not know, try to identify its prefix, suffix, or root. Sometimes knowing even one part of the word will help you answer the question.

641

Standardized Test Prep

✚ TEST DOCTOR

Question 11 Answer D is correct. The salinity of the ocean is the concentration of all the salts it contains. Most of the salt in the ocean is sodium chloride, which is made up of the two elements, sodium and chlorine. Answers A and C are incorrect because it is the combination of these elements that creates sodium chloride. Answer B is incorrect because chlorine and sulfur do not combine to create sodium chloride.

Question 12 Answer G is correct. Students struggling with this type of question may benefit from performing simple mathematical operations (such as multiplication, division, and addition) on large numbers. Some students may also benefit from first writing large numbers in scientific notation and then performing the operations.

Estimated Time

To give students practice under more realistic testing conditions, allow them 30 minutes to answer all of the questions in this practice test.

 TEST DOCTOR

Question 5 Full-credit answers should include the following points:

- every organism affects all other organisms in its ecosystem either directly, by eating or being eaten by other organisms, or indirectly by competing for resources
- every organism is dependent on every other organism, so any change to one organism affects the entire ecosystem

Question 6 Full-credit answers should include one of the following points:

- bacteria help to decompose dead organisms
- bacteria return nutrients to the soil
- bacteria recycle mineral nutrients
- some bacteria convert nitrogen into a form plants can use

Question 9 Full-credit answers should include the following points:

The biotic factors would include

- all of the living organisms, such as fish, plants, algae
- all of the dead organisms
- all of the waste products, such as dead leaves

Understanding Concepts

Directions (1–4): For *each* question, write on a separate sheet of paper the letter of the correct answer.

1 What is the term for the area where organisms live together with their physical environment?
- **A.** biome
- **B.** biosphere
- **C.** ecosystem
- **D.** population

2 Which of the following describes the theory of natural selection?
- **F.** Organisms with desired traits are selected for reproduction.
- **G.** Heredity determines which organisms will survive in their environment.
- **H.** Traits are developed in organisms in response to interaction with other organisms.
- **I.** Organisms with strong survival traits are more likely to pass the traits on in reproduction.

3 What inherited trait increases an organism's chance of survival and reproduction in a certain environment?
- **A.** adaptation
- **B.** characteristic
- **C.** evolution
- **D.** resistance

4 What are the six kingdoms of life?
- **F.** Archaebacteria, Eubacteria, Fungi, Protists, Plants, Animals
- **G.** Eubacteria, Fungi, Protists, Plants, Land Animals, Marine Animals
- **H.** Bacteria, Fungi, Plant-like Protists, Animal-like Protists, Plants, Animals
- **I.** Bacteria, Fungi, Protists, Flowering Plants, Non-flowering Plants, Animals

Directions (5–6): For *each* question, write a short response.

5 Everything is nature is connected. Use the concept of interdependence to analyze how an ecosystem works.

6 Describe one of the important roles of bacteria.

Reading Skills

Directions (7–9): **Read the passage below. Then answer the questions.**

Ecosystems are composed of many interconnected parts that often interact in complex ways. People often think of ecosystems as isolated from each other, but ecosystems do not have clear boundaries. Things move from one ecosystem into another.

Ecosystems are made up of both living and nonliving things. Biotic factors are the living and once living parts of an ecosystem, including all the plants and animals. Biotic factors also include dead organisms, dead parts of organisms, such as leaves, and the organisms' waste products. The biotic parts of an ecosystem also interact with the abiotic factors, the nonliving parts of the ecosystem. There are different levels in the ecological organization, from the individual organism to the biosphere.

In order to survive, ecosystems need five basic components. These are energy, mineral nutrients, water, oxygen, and living organisms.

7 What is one example of an abiotic factor?
- **A.** armadillo
- **B.** carnation
- **C.** robin
- **D.** rock

8 What is a species?
- **F.** all of the organisms that live in a land community
- **G.** a group of organisms that can mate to produce fertile offspring
- **H.** one of the basic components an ecosystem needs in order to survive
- **I.** all of the members of a group of organisms that live in the same place

9 Determine the kinds of biotic factors that would be found in an ocean ecosystem.

642

Answers

1. C
2. I
3. A
4. F
5. Answers will vary. See Test Doctor for detailed scoring rubric.
6. Answers will vary. See Test Doctor for detailed scoring rubric.
7. D
8. G
9. Answers will vary. See Test Doctor for detailed scoring rubric.

10. D
11. H
12. D

Interpreting Graphics

Directions (10–12): For *each* question below, record the correct answer on a separate sheet of paper.

The map below shows changes in forest cover in Costa Rica over 40 years. Use this map to answer questions 10 through 12.

Forest Cover in Costa Rica

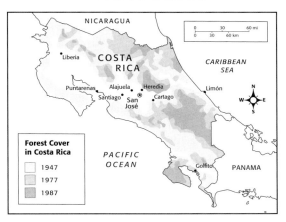

10 Approximately what percentage of Costa Rica was covered by forest in 1947?
  A. 25%
  B. 33%
  C. 50%
  D. 75%

11 What conclusion can be drawn about the forest cover of Costa Rica?
  F. Most of the remaining forests are near cities.
  G. The remaining forests are concentrated along the western coast.
  H. Costa Rica lost more than half of its forest cover in less than 50 years.
  I. Deforestation has accounted for little change in Costa Rica's environment.

12 What can be inferred about organisms adapted to living in trees in Costa Rica?
  A. Organisms that are adapted for living in trees will continue to thrive across the country.
  B. Organisms that are adapted for living in trees will be eliminated from the country's environment.
  C. Organisms that are adapted for living in trees will continue to thrive in areas that used to have forest.
  D. Organisms that are adapted for living in trees will thrive in forested areas but struggle in areas that no longer have trees.

Test TIP

Allow a few minutes at the end of the test-taking period to check for mistakes made in marking answers.

643

Question 11 Answer H is correct. Answer F is incorrect because most remaining forests are far from cities. Answer G is incorrect because the remaining forests are spread out around the country. Answer I is incorrect because wide-scale deforestation can have affects on local and global environments.

Question 12 Answer D is correct. Students struggling with this type of question may benefit from imagining probable consequences of habitat change for a smaller population. For example, have students imagine three communities of tree-dwelling monkeys. In one area, the forests are left as they are. In another, the forests are cleared completely. And in a third are, half of the trees are cut down. Ask students to write out a quick description of how each population will probably change in the year following the habitat change. This specific population can be used to more concretely understand the more general description asked about in the question.

Estimated Time

To give students practice under more realistic testing conditions, allow them 30 minutes to answer all of the questions in this practice test.

✚ TEST DOCTOR

Question 1 Answer B shows the proper sequence. Students choosing answers A or D may have confused producers and consumers. Consumers cannot use energy directly from the sun. Answer C shows decomposers using energy directly from the sun which is incorrect. Decomposers get energy from breaking down dead organisms.

Question 5 Full-credit answers should include the following points:

- the first thing that happens when farmland is abandoned is that native grasses and plants begin to replace the planted crops
- after these plants grow, the animals that feed on them also return to the area
- large trees may also return to the area if they grew there before

Question 8 Full-credit answers should include the following points:

- carbohydrates in organisms are converted into energy
- carbon is released when the organisms die
- the dead organisms form deposits of fossil fuels

Understanding Concepts

Directions (1–4): For *each* question, write on a separate sheet of paper the letter of the correct answer.

1 How does energy move through most ecosystems on Earth?
  A. from the sun to consumers to producers
  B. from the sun to producers to consumers to decomposers
  C. from the sun to decomposers to producers to consumers
  D. from the sun to consumers to producers back to consumers

2 Why is it essential to the understanding of life on Earth to know how energy moves through an ecosystem?
  F. When an animal eats a plant, an energy transfer occurs.
  G. All organisms on Earth require energy for their life processes.
  H. Photosynthesis is needed to convert sugar molecules to energy.
  I. Energy transfer in a food web is more complex than in a food chain.

3 What role do bacteria play during the nitrogen cycle?
  A. Bacteria store nitrogen in wastes.
  B. Bacteria turn nitrogen into phosphates.
  C. Bacteria convert nitrogen into water.
  D. Bacteria turn nitrogen into molecules.

4 What is the process that breaks down food to yield energy called?
  F. cellular digestion
  G. cellular respiration
  H. decomposition
  I. photosynthesis

Directions (5): **Write a short response for the question.**

5 Ecological succession is the gradual process of changes in a community. Describe the succession process that occurs when farmland is abandoned.

Reading Skills

Directions (6–8): **Read the passage below. Then answer the questions.**

Carbon is an essential component of proteins, fats, and carbohydrates, which make up all organisms. The carbon cycle is one of the processes by which materials are reused in an ecosystem. This process cycles carbon between the atmosphere, land, water, and organisms. Carbon enters a short-term cycle in an ecosystem when producers convert carbon dioxide into carbohydrates during photosynthesis. When consumers eat producers, the consumers obtain carbon from the carbohydrates. As the consumers break down the food during cellular respiration, some of the carbon is released back into the atmosphere.

Some carbon enters a long-term cycle. Carbon may be converted into carbonates, which make up the hard parts of bones and shells. Over millions of years, carbonate deposits produce huge formations of limestone rocks. Limestone is one of the largest carbon reservoirs on Earth.

6 During what process do producers convert carbon dioxide into carbohydrates?
  A. Carbon dioxide is converted into carbohydrates during the carbon cycle.
  B. Carbon dioxide is converted into carbohydrates during cellular respiration.
  C. Carbon dioxide is converted into carbohydrates during the eating process.
  D. Carbon dioxide is converted into carbohydrates during photosynthesis.

7 Which of the following groups are producers?
  F. animals
  G. decomposers
  H. herbivores
  I. plants

8 Fossil fuels are essentially stored carbon. What process causes carbon to turn into fossil fuel?

644

Answers

1. B
2. G
3. D
4. G
5. Answers will vary. See Test Doctor for detailed scoring rubric.
6. D
7. I
8. Answers will vary. See Test Doctor for detailed scoring rubric.
9. B
10. F
11. C

Interpreting Graphics

Directions (9–11): For *each* question below, record the correct answer on a separate sheet of paper.

The map below shows carbon dioxide output from the burning of fossil fuels in different regions of the world. Use this map to answer questions 9 and 10.

Carbon Dioxide Output From Fossil Fuels

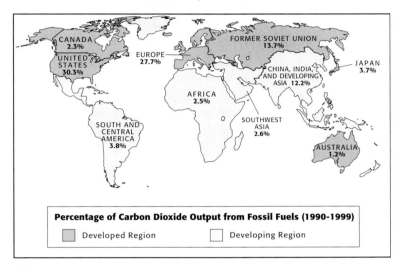

CANADA 2.3%
UNITED STATES 30.3%
EUROPE 27.7%
FORMER SOVIET UNION 13.7%
JAPAN 3.7%
CHINA, INDIA, AND DEVELOPING ASIA 12.2%
AFRICA 2.5%
SOUTH AND CENTRAL AMERICA 3.8%
SOUTHWEST ASIA 2.6%
AUSTRALIA 1.2%

Percentage of Carbon Dioxide Output from Fossil Fuels (1990-1999)

☐ Developed Region   ☐ Developing Region

9 Which continent has the lowest percentage of carbon dioxide output?
A. Asia
B. Australia
C. Europe
D. North America

10 What regions are responsible for the highest percentage of carbon dioxide output?
F. developed regions in the western hemisphere
G. developed regions in the eastern hemisphere
H. developing regions in the western hemisphere
I. developing regions in the eastern hemisphere

11 Which of the following shows an effect on the carbon cycle of the increased burning of fossil fuels?
A. More carbonates remain in fossil fuels.
B. More carbon dioxide is absorbed by organisms.
C. More carbon dioxide is absorbed by the atmosphere.
D. More carbohydrates remain buried deep in the ground

Test TIP

The key helps you interpret the map, and get information about the different types of regions.

 **TEST DOCTOR**

Question 10 Answer F is correct. Answer G incorrectly identifies North America as being a part of the eastern hemisphere. Answers H and I incorrectly identify North America as a developing region.

Question 11 Answer C is correct. Students struggling with this type of question may benefit from practicing organizing information. For example, this type of cause-and-effect question can be solved by drawing a flow chart. The major event is the increase in carbon dioxide being released by the burning of fossil fuels. This predictable effect would cause more carbon dioxide to be absorbed into the atmosphere. Another effect is less carbon being contained in fossil fuel reservoirs as it is released into the atmosphere by the burning of fossil fuels.

Estimated Time

To give students practice under more realistic testing conditions, allow them 30 minutes to answer all of the questions in this practice test.

TEST DOCTOR

Question 5 Full-credit answers should include the following points:

- overgrazing causes increased soil erosion
- increased erosion reduces the overall health of the soils which makes it harder for grasses to grow

Question 6 Full-credit answers should include the following points:

- both desert and tundra have low annual precipitation
- temperatures in the desert are higher than in the tundra

Question 8 Full-credit answers should include the following points:

- deforestation reduces rainfall, making the climate drier
- trees absorb carbon dioxide from the atmosphere
- when there are less trees, more carbon dioxide remains in the atmosphere, which affects temperature

Understanding Concepts

Directions (1–4): For *each* question, write on a separate sheet of paper the letter of the correct answer.

1 Which of the following describes a biome?
A. all the areas on Earth that are life-supporting
B. weather conditions in an area for a specific time period
C. a region characterized by specific climate and organism communities
D. an area where the animal population interacts with its abiotic environment

2 What type of forest has the greatest biodiversity?
F. taiga forest
G. temperate deciduous forest
H. temperate rain forest
I. tropical rain forest

3 What is the diversity of the species in an area dependent on?
A. plant life
B. rainfall
C. sunlight
D. temperature

4 What are the main factors that determine weather?
F. altitude, latitude, precipitation, temperature
G. altitude, latitude, precipitation, vegetation
H. air currents, altitude, temperature, vegetation
I. air currents, precipitation, temperature, vegetation

Directions (5–6): For *each* question, write a short response.

5 A temperate grassland is a biome that is dominated by grasses and that has very few trees. How are temperate grasslands threatened by overgrazing?

6 Compare and contrast the tundra and desert biomes.

Reading Skills

Directions (7–8): Read the passage below. Then answer the questions.

When rain falls on a forest, much of the rain is absorbed by plant roots and transpired into the air as water vapor. Water vapor forms rain clouds. Much of this water will fall as rain somewhere downwind from the forest. Clearing the trees results in deforestation, which can change the climate.

Deforestation led to the disastrous flooding of the Yangtze River in China in 1998. More than 2,000 people died in the floods, and at least 13 million people had to leave their homes. It is estimated that 85 percent of the forest in the Yangtze River basin has been cut down. So the millions of tons of water these trees once absorbed now flows freely down the river and spreads across the fields into towns during the seasonal monsoon rains.

7 How could future flooding on the Yangtze River be avoided?
A. replanting crops
B. rebuilding homes
C. replanting trees
D. rebuilding walls

8 Examine the climate change produced by deforestation.

Directions (9): Read the passage below. Then answer the question.

Tropical rain forests are located in a belt around the Earth near the equator. The climate is ideal for a wide variety of plants and animals. In tropical rain forests, different types of plants grow in different layers. The four main layers above the forest floor are, in order from top to bottom, the emergent layer, the canopy, the epiphytes, and the understory.

9 In which layer will you find most of the animals, and why?

646

Answers

1. C
2. I
3. B
4. F
5. Answers will vary. See Test Doctor for detailed scoring rubric.
6. Answers will vary. See Test Doctor for detailed scoring rubric.
7. C
8. Answers will vary. See Test Doctor for detailed scoring rubric.
9. Most animals live in the canopy because of the abundance of plants.

10. I
11. C
12. G
13. A

Interpreting Graphics

Directions (10–13): For *each* question below, record the correct answer on a separate sheet of paper.

The map below shows the biomes in Africa. Use this map to answer questions 10 through 13.

Biomes of Africa

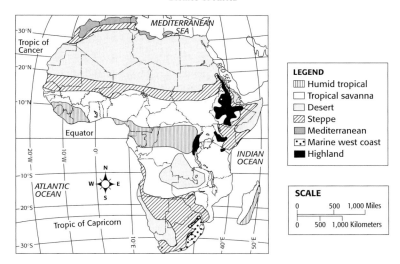

LEGEND
- ▥ Humid tropical
- ☐ Tropical savanna
- ☐ Desert
- ▨ Steppe
- ▦ Mediterranean
- ⠿ Marine west coast
- ■ Highland

SCALE
0   500   1,000 Miles
0   500   1,000 Kilometers

10 What can be inferred about the biomes of Africa?
- **F.** Africa has a large concentration of tropical rain forests.
- **G.** Africa has a limited number of plant and animal communities.
- **H.** Africa has all types of plant life because of the many diverse biomes.
- **I.** Africa has large desert areas that get less than 25.0 centimeters of precipitation a year.

11 Which biome covers the most surface area in Africa?
- **A.** desert
- **B.** highland
- **C.** Mediterranean
- **D.** steppe

12 According to the map, which of the following determines the characteristics of a biome?
- **F.** geographic borders
- **G.** latitude
- **H.** longitude
- **I.** ocean proximity

13 What geographic features are near 10°N, 40°E?
- **A.** mountains
- **B.** plains
- **C.** rivers
- **D.** volcanoes

Test TIP

When several questions refer to the same graph, table or map, answer the questions you are most sure of first.

Standardized Test Prep

✚ TEST DOCTOR

Question 12 Answer G is correct. Students struggling with this type of question may benefit from practicing using the process of elimination. Each incorrect choice can be identified by imagining what a map would look like if each condition was true. For example, if latitude was the determining factor, then biome boundaries would run parallel, or close to parallel, with latitude lines (as they do).

Question 13 Answer A is correct. Students struggling with this type of inference question may benefit from practicing parsing out descriptive titles. For example, the biome type at the given coordinates is described as highlands. Ask students what conclusions they can draw from the specific title highlands. If they are still struggling, tell students to break the word into its two roots, "high" and "lands."

647

Estimated Time

To give students practice under more realistic testing conditions, allow them 30 minutes to answer all of the questions in this practice test.

TEST DOCTOR

Question 6 Full-credit answers should include the following points:

- agricultural pollution, such as pesticides, runs off fields into rivers
- sewage and industrial wastes are often dumped into rivers
- where rivers drain into oceans, this pollution enters the oceans

Question 7 Full-credit answers should include the following points:

- marine ecosystems contain salt water, freshwater ecosystems contain freshwater
- marine ecosystems are located mainly in coastal areas and in the ocean
- freshwater ecosystems include lakes, ponds, rivers, and wetlands

Question 10 Full-credit answers should include the following points:

Scientists probably became aware of the importance of wetlands as

- wetlands began to disappear
- the concentration of pollution in other places increased

Understanding Concepts

Directions (1–5): For *each* question, write on a separate sheet of paper the letter of the correct answer.

1 Organisms living in coastal areas must adapt to what changes?
   A. water level and degree of salinity
   B. water level and amount of sunlight
   C. temperature and availability of oxygen
   D. temperature and availability of nutrients

2 Oil spills, sewage, pesticides, and silt runoff have been linked to the widespread destruction of what kind of marine ecosystem?
   F. coastal wetlands
   G. coral reefs
   H. mangrove swamps
   I. salt marshes

3 Which of the following correctly lists types of organisms in aquatic ecosystems from shallowest to deepest?
   A. plankton, nekton, benthos
   B. plankton, benthos, nekton
   C. benthos, plankton, nekton
   D. benthos, nekton, plankton

4 What is the difference between swamps and marshes?
   F. Marshes attract birds, swamps attract amphibians.
   G. Marshes are freshwater, swamps are saltwater.
   H. Marshes contain non-woody plants, swamps contain woody plants.
   I. Marshes are mostly in the southeast U.S., swamps in the northeast U.S.

5 Which of the following would be considered among the most productive of ecosystems?
   A. barrier island
   B. estuary
   C. river
   D. salt marsh

Directions (6–7): For *each* question, write a short response.

6 Most ocean pollution arises from activity on land. How does land pollution end up in the oceans?

7 Aquatic ecosystems are divided into two types—marine and freshwater. Compare and contrast marine and freshwater ecosystems.

Reading Skills

Directions (8–10): Read the passage below. Then answer the questions.

Wetlands perform several important environmental functions. Some wetlands are used to produce commercially important products, such as cranberries. But wetlands were once considered to be wastelands that provide breeding grounds for insects. Therefore, many have been drained, filled, and cleared for farms or residential and commercial development. For example, the Florida Everglades once covered 8 million acres of south Florida but now it covers less than 2 million acres. From 1982 to 1992 alone, 57 percent of wetlands were converted into land for development. Wetlands are vitally important as habitats for wildlife, and their important role in the environment is now recognized.

8 Which of the following is **not** an environmental function of wetlands?
   F. Wetlands buffer shorelines against erosion.
   G. Wetlands provide the most desirable species of fish.
   H. Wetlands reduce the likelihood of a flood.
   I. Wetlands trap and filter sediments and pollutants.

9 What changes are occurring with regard to wetlands today?
   A. All wetlands are protected by the federal government.
   B. Most states now prohibit the destruction of certain wetlands.
   C. Wetlands have disappeared from the mainland U.S.
   D. Pollution is causing wetlands to lose their ability to purify wastewater.

10 What could be inferred as the cause of scientific awareness of the importance of wetlands?

648

Answers

1. A
2. G
3. A
4. H
5. B
6. Answers will vary. See Test Doctor for detailed scoring rubric.
7. Answers will vary. See Test Doctor for detailed scoring rubric.
8. G
9. B
10. Answers will vary. See Test Doctor for detailed scoring rubric.
11. G
12. A
13. H

Interpreting Graphics

Directions (11–13): **For** *each* question below, record the correct answer on a separate sheet of paper.

The table below shows organisms whose presence or absence can be indicators of water quality. Use this table to answer questions 11 through 13.

Living Water Quality Indicators

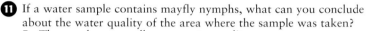

Most-sensitive species Caddisfly larvae, hellgrammites, stonefly larvae, mayfly nymphs, gilled snails, and water penny larvae	mayfly nymph caddisfly larvae
Moderately sensitive species Clams, cranefly larvae, crayfish, damselfly nymphs, dragonfly nymphs, scuds, predacious diving beetle larvae, sowbugs, fishfly larvae, and alderfly larvae	dragonfly nymph predacious diving beetle larva
Tolerant species Aquatic worms, blackfly larvae, leeches, midge larvae, and pouch snails	midge larva leeches

11 If a water sample contains mayfly nymphs, what can you conclude about the water quality of the area where the sample was taken?
  F. The area has generally poor water quality.
  G. The area has generally good water quality.
  H. The water quality cannot be determined from such a sample.
  I. The water in the area has been chemically treated for pollutants.

12 What group of organisms could be seen **only** in a sample taken from unpolluted waters?
  A. leeches and caddisfly larvae
  B. leeches and dragonfly nymphs
  C. leeches and midge larvae
  D. leeches and predacious diving beetle larvae

13 Water Sample A contains only leeches. Water Sample B contains leeches and predacious diving beetle larvae. What comparison can be made between Samples A and B?
  F. Both samples A and B have the same water quality.
  G. Both samples came from the same water source.
  H. Sample A has relatively poorer water quality than sample B.
  I. Sample B has relatively poorer water quality than sample A.

Test TIP

Imagine the increasing levels of pollution that would affect water quality and cause particular species to disappear.

✚ TEST DOCTOR

Question 12 Answer A is the correct choice because caddisfly larvae would not be seen in a sample of polluted waters. Answers B, C, and D could possibly come from an unpolluted area but do not necessarily come from an unpolluted area because similar samples could be draw from relatively polluted water sources.

Question 13 Answer H is correct. Answers F and G are incorrect because the presence of predacious diving beetle larvae implies differing water quality between the two samples. It is very unlikely that one water source could produce samples with different water qualities. Answer I is incorrect because the presence of predacious diving beetle larvae indicates higher water quality.

649

Estimated Time

To give students practice under more realistic testing conditions, allow them 30 minutes to answer all of the questions in this practice test.

TEST DOCTOR

Question 5 Full-credit answers should include the following points:

- density dependent deaths are deaths that occur when individuals of a population are densely packed together
- density independent deaths occur no matter what the density of a population is

Question 6 Full-credit answers should include the following point:

- a species can reduce competition by changing the area in which it competes
- a species can reduce competition by changing the time of day it searches for resources

Understanding Concepts

Directions (1–4): For *each* question, write on a separate sheet of paper the letter of the correct answer.

1 What determines the carrying capacity of an environment?
- **A.** growth rates
- **B.** limiting resources
- **C.** natural selection
- **D.** territorial size

2 Which of the following statements can be made about competition between organisms in a particular ecosystem?
- **F.** Organisms rarely compete with members of their own species.
- **G.** Organisms compete directly when they require the same resources.
- **H.** Organisms only compete when supplies of a resource are unlimited.
- **I.** Organisms only compete for resources when their populations are small.

3 Which of the following describes a species' niche?
- **A.** the unique role the species plays in an ecosystem
- **B.** the physical location where the species can be found on Earth
- **C.** the adaptation of a species population to its physical environment
- **D.** the maximum number of offspring all members of that species can produce

4 Which of the following expressions is used to calculate the change in population size?
- **F.** births plus deaths
- **G.** births plus deaths plus population
- **H.** births minus deaths
- **I.** births minus deaths plus population

Directions (5–6): For *each* question, write a short response.

5 Explain the difference between density dependent and density independent deaths in a population.

6 What are two ways a species can reduce competition when organisms in the same niche use very similar resources?

Reading Skills

Directions (7–9): Read the passage below. Then answer the questions.

A population is all the members of a species living in the same place at the same time. A population is a reproductive group because organisms usually breed with members of their own population. The maximum number of offspring that each member of a population can produce determines the population's reproductive potential. Reproductive potential increases when individuals produce more offspring at a time, reproduce more often, or reproduce earlier in life. Reproducing earlier in life has the greatest effect on reproductive potential. Reproducing early reduces the generation time, the average amount of time it takes a member of a population to reach the age when it reproduces.

Populations sometimes undergo exponential growth, which means they grow faster and faster. In exponential growth, a larger number of individuals is added to the population in each succeeding time period.

7 Which of the following has the shortest generation time?
- **A.** bacteria
- **B.** cattle
- **C.** elephants
- **D.** humans

8 What would reduce a population's reproductive potential?
- **F.** increase the number of survivors
- **G.** decrease the number of predators
- **H.** increase the number of descendants
- **I.** decrease the number of offspring

9 When can exponential growth occur in a population?
- **A.** when the brood size of the population increases
- **B.** when the generation time of a population is shortened
- **C.** when the population has plenty of food and no competition
- **D.** when the dispersion of the population becomes random

650

Answers

1. B
2. G
3. A
4. H
5. Answers will vary. See Test Doctor for detailed scoring rubric.
6. Answers will vary. See Test Doctor for detailed scoring rubric.
7. A
8. I
9. C
10. G
11. C
12. H

Interpreting Graphics

Directions (10–12): For *each* question below, record the correct answer on a separate sheet of paper.

Us the table below to answer question 10.

Types of Interactions Between Two Species

Interaction	Species A	Species B
Competition	Fox	Coyote
Predation	Kit fox	Kangaroo rat
Mutualism	Yucca moth	Yucca seeds
Commensalism	Wren	Cactus

10 Which of the interactions listed in the table is harmful to both species?
- **F.** commensalism
- **G.** competition
- **H.** mutualism
- **I.** predation

Use the illustration below to answer questions 11 and 12.

Carrying Capacity

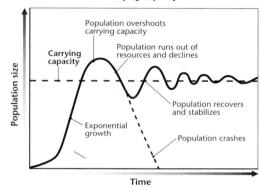

11 What happens to population size between the time it overshoots carrying capacity to when it recovers and stabilizes?
- **A.** It remains stable.
- **B.** It declines steadily.
- **C.** It decreases before it stabilizes.
- **D.** It continues to increase at a steady rate.

12 If the population size was nearly 2,000 when it overshot carrying capacity, and 1,500 when it was at its lowest amount of decline during its recovery, what is the estimated carrying capacity of the population?
- **F.** 1,600
- **G.** 1,750
- **H.** 1,800
- **I.** 1,950

Test TIP

Test questions may not be arranged in order of increasing difficulty. If you are unable to answer a question, mark it and move on to another question.

 **TEST DOCTOR**

Question 11 Answer C is correct. Answer A is incorrect because the population does not remain steady. Answer B is incorrect because the population both increases and decreases as it stabilizes. Answer D is incorrect because the population decreases over time to its carrying capacity.

Question 12 Answer H is correct. The answer is greater than 1750 because the area of the curve above the carrying capacity is greater than the area below the curve. If the areas below and above were equal, than the number would be equidistant between 1500 and 2000, or 1750. Because there is a greater area above the carrying capacity line, the total must be greater than the halfway point of 1750.

651

Estimated Time

To give students practice under more realistic testing conditions, allow them 30 minutes to answer all of the questions in this practice test.

TEST DOCTOR

Question 5 Full-credit answers should include the following points:

- human population growth makes fuelwood more scarce
- more people are using fuelwood faster than it can grow back

Question 6 Full-credit answers should include two of the following points:

- the age structure of a population is how many people are in each age group
- survivorship is the percentage of a population that is likely to survive to a certain age
- fertility rate is the number of babies born each year per 1000 women
- migration is the movement of people from one area to another

Question 8 Full-credit answers should include the following points:

- a community surrounded by farmland must decide whether to continue using land for crops or build new houses
- building houses means less local crops would be available
- using the land for crops means a housing shortage and an increase in home prices

Understanding Concepts

Directions (1–4): For *each* question, write on a separate sheet of paper the letter of the correct answer.

1 How did the human population change in the last 200 years?
   A. increased at a decelerated rate
   B. increased at an accelerated rate
   C. decreased at a decelerated rate
   D. decreased at an accelerated rate

2 Which of the following has the **most** effect on the infant mortality rate?
   F. Parents have a college education.
   G. Parents live in a quiet suburban area.
   H. Parents are in the higher income brackets.
   I. Parents keep a baby fed, clean and warm.

3 Why have populations in most developed countries stopped growing?
   A. In most developed countries, immigration has decreased and emigration increased.
   B. In most developed countries, contagious diseases have reduced life expectancies.
   C. In most developed countries, birth rates have decreased to a level close to the death rate.
   D. In most developed countries, industrial development that created progress has moved to less developed countries.

4 What is the term for the average number of years a person is likely to live?
   F. age structure
   G. demographic transition
   H. life expectancy
   I. migration pattern

Directions (5–6): For *each* question, write a short response.

5 During a period of population growth, the resources most critically affected are fuelwood, water, and arable land. Analyze how the supply of fuelwood is affected by human population growth.

6 Describe two of the factors used to predict human population size.

Reading Skills

Directions (7–8): Read the passage below. Then answer the questions.

Growing populations often make trade-offs between competing uses for land such as agriculture, housing, or natural habitats. Much of the world's population is undergoing urbanization, which means more people are living in cities than in rural areas. In the United States, many cities are becoming large metropolitan areas. People often find work in the cities but want their homes in a nearby suburb. This leads to suburban sprawl, traffic jams, inadequate infrastructure, and the reduction of land for farms, ranches, and wildlife habitat.

7 How does urbanization affect agriculture?
   A. It reduces land available for crops.
   B. It reduces land available for housing.
   C. It reduces the safety of the water supply.
   D. It reduces the variety of foodstuffs.

8 Give an example of a situation where a population must decide how to use limited land resources, and the foreseeable effect of that decision.

Directions (9): Read the passage below. Then answer the question.

The demographic transition model describes how population changes can occur. The model is based on observations of the history of many developed countries. The theory behind the model is that industrial development causes economic and social progress that then affects population growth rates.

9 What are the four stages of the demographic transition model?
   F. agricultural, industrial, transitional, stability
   G. birth rate, growth rate, stability, death rate
   H. low growth, high growth, decreasing growth, zero growth
   I. preindustrial, transitional, industrial, postindustrial

Answers

1. B
2. I
3. C
4. H
5. Human population growth makes fuelwood more scarce because more people are using the wood.
6. Answers will vary. See Test Doctor for detailed scoring rubric.
7. A
8. Answers will vary. See Test Doctor for detailed scoring rubric.
9. I
10. C
11. I
12. C

Interpreting Graphics

Directions (10–12): For *each* question below, record the correct answer on a separate sheet of paper.

The graph below shows population projections for the world at different growth rates. Use this graph to answer questions 10 through 12.

World Population Projection For Different Growth Rates

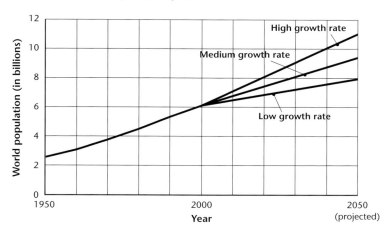

10 Which of the following can be inferred from the medium growth rate predicted?
A. Migration will no longer add to the overall population growth.
B. Life expectancy will decrease more rapidly in less-developed countries.
C. Most countries will have reached replacement level fertility rates by 2050.
D. More-developed countries will have greater growth than less-developed countries.

11 What does the chart project as the most likely outcome for the world population?
F. It will be 12 billion in 2050.
G. It will double by the year 2050.
H. It will increase by 25% by the year 2050.
I. It will be between 8 and 11 billion in 2050.

12 By how many billions of people will the population increase between 1950 and 2005?
A. 2   C. 4
B. 3   D. 5

Test TIP

When solving a math problem, be sure to examine the units involved closely.

✚ TEST DOCTOR

Question 10 Answer C is correct. Answer A is incorrect because migration will continue regardless of growth rates. Answers B and D are incorrect because they would both lead to a higher growth rate rather than a medium growth rate.

Question 11 Answer I is correct. Answers F and G are incorrect because to reach 12 billion or double, the growth rate would have to exceed even the highest probable growth rate shown in the graph. Answer H is incorrect because it assumes that a low growth rate will occur. Low growth is only one of the three possibilities, and therefore has only has a 33% probability. In contrast, the correct answer would be true for all possible growth rates.

Question 12 Answer C is correct. Because the graph does not have smaller units than billions of people, the number of significant figures is one.

Estimated Time

To give students practice under more realistic testing conditions, allow them 30 minutes to answer all of the questions in this practice test.

 TEST DOCTOR

Question 5 Full-credit answers should include the following points:

- they have less political infrastructure to navigate
- they can affect change more easily

Question 6 Full-credit answers should include the following points:

- a threatened species has a declining population and is likely to become endangered if current trends continue
- an endangered species is likely to become extinct if not protected

Question 10 Full-credit answers should include the following points:

- shared seed banks contain germ plasm that can be easily turned into living organisms by planting seeds in fertile ground
- the genetic material in germ-plasm banks is difficult or impossible to transform into living organisms

Understanding Concepts

Directions (1–4): For *each* question, write on a separate sheet of paper the letter of the correct answer.

1 Which of the following phrases describes the term genetic biodiversity?
- **A.** the variety of habitats found in an ecosystem
- **B.** the variety of species present in an ecosystem
- **C.** the differences between populations of species
- **D.** the different genes contained within members of a population

2 What species are critical to the survival of an ecosystem?
- **F.** bottleneck species
- **H.** exotic species
- **G.** endemic species
- **I.** keystone species

3 Which of the following describes a species that is likely to become endangered?
- **A.** insects that have to adapt to an urban environment
- **B.** small mammals that live in urban ecosystems
- **C.** birds that can only survive in rural ecosystems
- **D.** mammals that need an undeveloped habitat to breed successfully

4 Why is international cooperation crucial to securing future biodiversity?
- **F.** Wildlife protection laws vary from country to country.
- **G.** Poaching is the most important reason for a species population decline.
- **H.** Habitat destruction and other causes of extinction cross international borders.
- **I.** Protecting species sometimes conflicts with the interests of human populations.

Directions (5–6): For *each* question, write a short response.

5 Why could private or non-governmental agencies be more effective in protecting species than government agencies?

6 Compare endangered species with threatened species.

Reading Skills

Directions (7–10): Read the passage below. Then answer the questions.

Scientists are developing new methods of conservation in an attempt to preserve species on the verge of extinction. Captive-breeding programs try to restore the population of a species. Another approach is to preserve germ plasm. Germ plasm is any form of genetic material, such as that contained within the reproductive, or germ, cells of plants or animals. Germ-plasm banks store seeds, sperm, eggs, or pure DNA in special controlled environments. Farmers and gardeners also preserve germ plasm when they save and share seeds.

7 What is the risk involved in using captive-breeding programs to preserve species?
- **A.** The species may leave captivity.
- **B.** The species may not reproduce.
- **C.** The species may develop new habits.
- **D.** The species may leave the endangered list.

8 In what type of biome are the most species at risk of becoming extinct?
- **F.** The desert has the most species at risk of becoming extinct.
- **G.** The subtropical savanna has the most species at risk of becoming extinct.
- **H.** The temperate grasslands have the most species at risk of becoming extinct.
- **I.** The tropical rain forests have the most species at risk of becoming extinct.

9 What is the **most** effective way to preserve a species?
- **A.** by watching and protecting their habitats
- **B.** by setting up individual members in a wildlife park
- **C.** by enacting new legislation that prohibits their destruction
- **D.** by promoting public awareness of which species are in danger

10 Contrast the genetic material contained in a germ-plasm bank in a laboratory with the genetic material contained in a shared seed bank.

654

Answers

1. B
2. I
3. D
4. H
5. Answers will vary. See Test Doctor for detailed scoring rubric.
6. Answers will vary. See Test Doctor for detailed scoring rubric.
7. B
8. I
9. A
10. Answers will vary. See Test Doctor for detailed scoring rubric.

11. F
12. B
13. G
14. C

Standardized Test Prep

✚ TEST DOCTOR

Interpreting Graphics

Directions (11–14): For *each* question below, record the correct answer on a separate sheet of paper.

The graph below shows the number of families of marine animals that existed millions of years ago. Use this graph to answer questions 11 and 12.

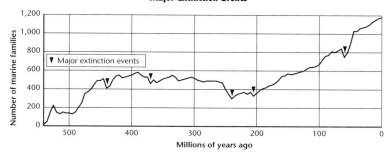

Major Extinction Events

11 How has the biodiversity of marine families changed over the last 500 million years?
F. It has increased.
G. It has decreased slightly.
H. It has remained the same.
I. It has decreased significantly.

12 What is the average number of families of marine organisms lost in a major extinction event?
A. 25   **C.** 100
B. 75   **D.** 150

13 If 90 families were lost in an extinction event that lasted 10 million years, and each family contained 200 species, how many species were lost every 100,000 years during that period?
F. 90   **H.** 200
G. 180   **I.** 360

14 What do we know about the number of individual species currently living on Earth?
A. There are no new species being found.
B. All the species that exist on Earth have been cataloged.
C. About 1.7 million species are known to exist.
D. There are more trees and mammals than there are insects.

Test TIP

Converting the largest numbers to scientific notation may help you simplify your calculations.

Question 11 Answer F is correct. Answer G is incorrect because the change in the number of marine species has been significant. The number of species has gone from less than 50 species 500 million years ago to almost 1200 species today. Answers H and I are incorrect because the overall number of species has increased.

Question 12 Answer B is correct. Students struggling with this type of question may benefit from practicing estimation. For example, this type of determining average problem can be solved quickly by first examining the choices. It is obvious from the answer choices that only an estimate is needed and so, the actual average does not need to be calculated. The answer must be more than 50 (the smallest number of years in an extinction event) and smaller than 100 (the largest number of years in an extinction event). Answer B is the only choice that satisfies both of those criteria.

Estimated Time

To give students practice under more realistic testing conditions, allow them 30 minutes to answer all of the questions in this practice test.

 TEST DOCTOR

Question 5 Full-credit answers should include the following points:

- developing countries use more water resources
- as water resources are depleted, they become more expensive
- people in developing countries will then have less access to clean water and may use polluted water

Question 6 Full-credit answers should include the following points:

- water management projects allow people to live in arid areas
- water management projects are used to generate electricity
- water management projects keep populations from moving to urban areas

Question 9 Full-credit answers should include the following points:

- state legislation varies greatly between states and can be stricter than federal laws
- federal legislation is subject to more limitations, but is often used as the basis for state laws

---

Understanding Concepts

Directions (1–4): For *each* question, write on a separate sheet of paper the letter of the correct answer.

1 Why are the oceans important to the water cycle?
 A. Oceans contain saltwater.
 B. The four major oceans are all joined.
 C. Most of Earth's water is contained in the oceans.
 D. Oceans cover more of Earth's surface area.

2 Where do we find most of the water that is available for human use?
 F. groundwater
 G. the ocean
 H. polar ice caps
 I. rivers and streams

3 Which of the following is an example of a nonpoint-source of pollution?
 A. chemical factory
 B. livestock feedlot
 C. oil spill
 D. wastewater treatment plant

4 How does heat act as a water pollutant?
 F. It slows down the flow of water.
 G. It speeds up chemical reactions.
 H. It increases the nutrients in the water.
 I. It decreases oxygen in the water.

Directions (5–6): For *each* question, write a short response.

5 Economic forces influence how people use natural resources. One rule of economics is the law of supply and demand. This law states that the greater demand for a resource, the more that resource is worth. How do economic forces affect the usage of water resources in developing countries?

6 People often live in areas where water resources are inadequate. Justify the use of water management projects, such as dams, in these areas.

Reading Skills

Directions (7–9): Read the passage below. Then answer the questions.

In 1969 the Cuyahoga River in Ohio was so polluted that the river caught fire and burned for several days. This event was a major factor in the passage of the Clean Water Act of 1972. The stated purpose of the act was to "restore and maintain the chemical, physical, and biological integrity of the nation's waters." Although the goal of the act was not achieved, much progress has been made since the act was passed. The percentage of lakes and rivers that are fit for swimming and fishing has increased by about 30 percent.

Additional water-quality legislation has been enacted, both by the federal government and individual states. Many toxic metals are now removed from wastewater before the water is discharged.

7 What was the goal of the Clean Water Act?
 A. The goal of the Clean Water Act was to clean up the country's major lakes by the turn of the century.
 B. The goal of the Clean Water Act was to make all surface water clean enough for fishing and swimming by 1983.
 C. The goal of the Clean Water Act was to reduce pollution on the Cuyahoga River in five years.
 D. The goal of the Clean Water Act was to remove all toxic metals from wastewater by 1980.

8 What is the term for the accumulation of pollutants at successive levels of the food chain?
 F. biomagnification
 G. containment
 H. sludge
 I. toxicity

9 Legislation has improved water quality in the United States. Analyze the differences between federal and state water-quality legislation.

Answers

1. C
2. F
3. B
4. I
5. Answers will vary. See Test Doctor for detailed scoring rubric.
6. Answers will vary. See Test Doctor for detailed scoring rubric.
7. B
8. F
9. Answers will vary. See Test Doctor for detailed scoring rubric.

10. A
11. I
12. D

Interpreting Graphics

Directions (10–12): For *each* question below, record the correct answer on a separate sheet of paper.

The map below shows the water resources of Canada. Use this map to answer questions 10 through 12.

Canada's Water

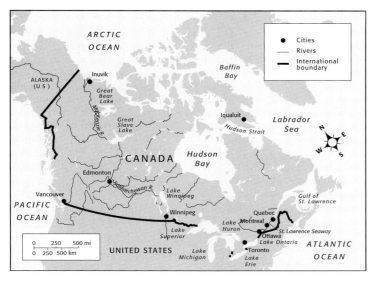

10 What can be inferred about the human population of Canada?
A. Most people live in the southeast part of Canada.
B. The greatest number of people live along the west coast.
C. The Hudson Bay area is the most populous part of Canada.
D. More people live around the Great Lakes than along the Saskatchewan River.

11 Which of the following is the northernmost source of freshwater?
F. Baffin Bay
G. Great Bear Lake
H. Hudson Strait
I. MacKenzie River

12 What is the relationship between cities and water resources?
A. Cities are always located near rivers.
B. Cities are always located near lakes.
C. Cities are never located near salt water.
D. Cities are usually located near water resources.

Test TIP

When analyzing relationships, look for the answer that is not only true but that also best describes the relationship as a whole.

657

Estimated Time

To give students practice under more realistic testing conditions, allow them 30 minutes to answer all of the questions in this practice test.

 TEST DOCTOR

Question 5 Full-credit answers should include the following points:

- acidification changes the balance in soil chemistry
- this change in soil chemistry can cause increased absorption of toxic chemicals and root damage

Question 6 Full-credit answers should include the following points:

- primary pollutants are pollutants that are released directly into the environment
- secondary pollutants form when primary pollutants react with substances in the environment

Question 8 Full-credit answers should include the following points:

- pollution-control devices reduce the amount of pollutants that are released into the air
- this reduces the amount of acid precipitation

Understanding Concepts

Directions (1–4): For *each* question, write on a separate sheet of paper the letter of the correct answer.

1 What is the biggest cause of air pollution?
A. dust particles
B. forest fires
C. human activities
D. volcanic eruptions

2 Which of the following releases the most primary pollutants into the air?
F. electric power plants
G. manufacturing plants
H. mining operations
I. transportation industry

3 Which of the following is a long-term health effect of air pollution?
A. blindness
B. diabetes
C. emphysema
D. hepatitis

4 Which of the following statements is true?
F. Ground-level ozone is harmless to children.
G. Noise pollution occurs at low decibel levels.
H. Light pollution is a direct hazard to human health.
I. Sick-building syndrome is caused by poor air quality.

Directions (5–6): For *each* question, write a short response.

5 Plant communities have adapted over long periods of time to the acidity of the soil in which they grow. Acid precipitation can cause a drop in the pH of soil. Analyze the changes to soil that are caused by acidification.

6 Clean air consists mostly of nitrogen and oxygen gas, as well as very small amounts of argon, carbon dioxide, and water vapor. Substances that pollute the air can be in the form of solids, liquids, or gases. Differentiate between primary and secondary air pollutants.

658

Reading Skills

Directions (7–8): Read the passage below. Then answer the questions.

Power plants that generate our nation's electricity must burn fuel to get the energy they need. They usually burn fossil fuels, but fossil fuels release huge quantities of sulfur dioxide and nitrogen oxide into the air. The Clean Air Act requires many industries to use pollution-control devices, such as scrubbers.

Scrubbers remove some of the more harmful substances that would otherwise pollute the air. A scrubber is a machine that moves gases through a spray of water which dissolves many pollutants. In electrostatic precipitators, gas containing dust particles is blown through a chamber containing an electrical current. An electrical charge is transferred to the dust particles, which causes them to stick to one another and the sides of the chamber. The clean gas is released from the chamber, and the concentrated dust particles can then be collected and removed.

7 Where would you most likely find an electrostatic precipitator being used?
A. car factory
B. chemical plant
C. oil refinery
D. power plant

8 How do pollution-control devices affect acid precipitation?

Directions (9): Read the passage below. Then answer the question.

The Environmental Protection Agency set up an allowance trading system designed to reduce sulfur-dioxide emissions. In this system, one ton of sulfur dioxide emission is equivalent to one allowance. Sulfur dioxide allowances are inexpensive and can be bought from the EPA. There are a limited number of allowances allocated each year.

9 How does a company use the allowance trading system to comply with the Clean Air Act?

Answers

1. C
2. F
3. C
4. I
5. Answers will vary. See Test Doctor for detailed scoring rubric.
6. Answers will vary. See Test Doctor for detailed scoring rubric.
7. D
8. Answers will vary. See Test Doctor for detailed scoring rubric.
9. Answers will vary. See Test Doctor for detailed scoring rubric.

10. H
11. C
12. G
13. C

Interpreting Graphics

Directions (10–13): For *each* question below, record the correct answer on a separate sheet of paper.

The illustration below shows the formation of acid precipitation in the atmosphere. Use it to answer questions 10 through 12.

Formation of Acid Precipitation

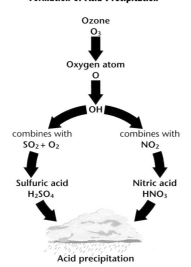

Acid precipitation

10 What is the relationship between sulfur dioxide (SO_2) and sulfuric acid (H_2SO_4)?
F. Sulfur dioxide combines with oxygen to form nitric acid.
G. Sulfur dioxide is the main pollutant that forms nitric acid.
H. Sulfur dioxide is the main pollutant that forms sulfuric acid.
I. Sulfur dioxide combines with nitrogen dioxide to form sulfuric acid.

11 What are the two main polluting components in acid precipitation?
A. ozone and oxygen atoms   **C.** sulfuric acid and nitric acid
B. ozone and water   **D.** sulfuric acid and water

12 What chemical combines with nitrogen dioxide to form nitric acid?
F. H   **H.** O_2
G. OH   **I.** O_3

13 A decrease of one number on the pH scale represents an increase in the concentration of hydronium (OH) ions by a power of 10. How many times more hydronium ions are present in an acid with a pH of 3 than are present in an acid with a pH of 5?
A. 10   **C.** 100
B. 20   **D.** 200

Test TIP

For questions involving chemical reactions, write out the chemical formulas to better visualize the entire reaction.

Question 9 Full-credit answers should include the following points:

• a company uses allowances to avoid exceeding its allowable level of emissions

• allowances let a company determine its most cost-effective way to comply with the law

Question 12 Answer G is correct. The chemical equation of the reaction is

$NO_2 + X \rightarrow HNO_3$. Answer F is incorrect because even though H is added by X, it does not account for the oxygen atom that is added. Answers H and I are incorrect because even though O is added, they do not account for the hydrogen atom that is added.

Question 13 Answer C is correct. The difference between a pH of 5 and a pH of 3 is 2 levels on the pH scale. Each level represents a difference of 10 in the concentration of OH ions. Two levels is a difference of 102 or 100. Answers A and B are too low. Answer D is too high. Answers B and D can also be eliminated quickly because they are not simple powers of ten.

659

Estimated Time

To give students practice under more realistic testing conditions, allow them 30 minutes to answer all of the questions in this practice test.

 TEST DOCTOR

Question 5 Full-credit answers should include the following points:

- without a greenhouse effect on Earth, global temperatures would be much lower
- life on Earth may not have developed
- many life forms could not survive in lower temperatures

Question 7 Full-credit answers should include the following points:

- scientists need information on the relationships of all the different factors that affect Earth's climate
- these relationships can then be programmed into the computer model as equations

Question 8 Full-credit answers should include the following points:

- rainfall follows the warm water moving eastward
- precipitation increases in the southeastern U.S.

Understanding Concepts

Directions (1–4): For *each* question, write on a separate sheet of paper the letter of the correct answer.

1 Which of the following is the most important factor in determining climate?
   A. volcanic activity
   B. nearness to an ocean
   C. distance from the equator
   D. atmospheric circulation patterns

2 What causes the changes of seasons on Earth?
   F. the relatively constant flow of oceanic currents
   G. the tilt of Earth's axis as Earth moves around the sun
   H. the distribution of precipitation influenced by topography
   I. the changes in prevailing wind patterns at different latitudes

3 The development and widespread use of CFCs lead to what scientific discovery?
   A. that CFC molecules were chemically stable
   B. that CFC molecules contain chlorine atoms
   C. that CFC molecules were destroying the ozone layer
   D. that CFC molecules were protecting the ozone layer

4 Why would global warming have an effect on sea levels?
   F. The warmth would cause polar ice caps to melt.
   G. There would be widespread beach erosion.
   H. Rainfall would increase all around the globe.
   I. There would be increased cloud cover blocking the sunlight.

Directions (5): Write a short response for the queston below.

5 The Earth's atmosphere acts like the glass in a greenhouse. Predict what would happen to life on Earth if Earth did not have a greenhouse effect.

Reading Skills

Directions (6–7): Read the passage below. Then answer the questions.

Many scientists believe that the increasing greenhouse gases in our atmosphere will result in increasing average temperatures on Earth. Scientists are currently unable to make accurate predictions about the rate of global warming because climatic patterns are complex and too many variables must be taken into account to be solved even using today's fastest computers. The computer models predict how phenomena such as temperature, rainfall patterns, and sea level, will be affected by carbon dioxide levels, prevailing winds, and other variables.

Computer modeling is complicated by Earth's feedback systems which sometimes make it necessary to use different equations under changing simulated environments.

6 How could faster computers influence the predictions of climate change?
   A. Faster computers could produce more data.
   B. Faster computers could solve more complex equations.
   C. Faster computers could reduce the number of variables needed.
   D. Faster computers could increase the degree of certainty over a prediction.

7 Assess the information needed by scientists to build computer models that can predict climate change.

Directions (8): Read the passage below. Then answer the question.

El Niño is the name given to the periodic change in the location of warm and cold water masses in the Pacific Ocean. During an El Niño, winds in the western Pacific Ocean, which are usually weak, strengthen and push warm water eastward.

8 How does El Niño affect climate in the United States?

660

Answers

1. C
2. G
3. C
4. F
5. Answers will vary. See Test Doctor for detailed scoring rubric.
6. B
7. Answers will vary. See Test Doctor for detailed scoring rubric.
8. Answers will vary. See Test Doctor for detailed scoring rubric.

9. G
10. A
11. G

Interpreting Graphics

Directions (9–11): For *each* question below, record the correct answer on a separate sheet of paper.

The map below shows wind patterns across the Earth. Use this map to answer questions 9 through 11.

Global Wind Patterns

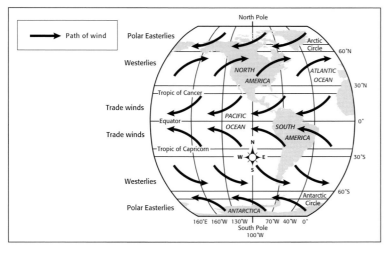

9 Which of the following characteristics has the **largest** effect on wind movement?
F. cardinal direction
G. latitude
H. longitude
I. seasonal change

10 Central America is located in the western hemisphere, linking North and South America. Which way does the wind blow in Central America?
A. northeast to southwest
B. southeast to northwest
C. northwest to southeast
D. southwest to northeast

11 If you were sailing from North America to Europe, in which range of latitudes would you sail?
F. 0° to 30°N
G. 30°N to 60°N
H. 0° to 30°S
I. 30°S to 60°S

Test TIP

Before looking at the answer choices for a question, try to answer the question yourself.

TEST DOCTOR

Question 9 Answer G is correct. Latitude is the distance from the equator measured in degrees north or degrees south. Because of the Earth's rotation, belts of winds are produced in both hemispheres between 0 degrees and 30 degrees, between 30 degrees and 60 degrees, and between 60 degrees and the poles. Choice F refers to the direction in which the wind is blowing. Longitude (Choice H) has no bearing on wind direct ion. Wind currents are one of the causes of seasonal change (Choice I).

Question 10 Choice A is correct. Central America is located in the belt that is between 0 degrees and 30 degrees north of the equator, where the winds blow from the northeast to the southwest.

Question 11 Answer G is correct. Answer F is incorrect because if you sailed in those directions, the winds would take you southwest, which is away from Europe. Answers H and I are incorrect because those latitudes are far away from North America, the intended starting place.

661

Estimated Time

To give students practice under more realistic testing conditions, allow them 30 minutes to answer all of the questions in this practice test.

 **TEST DOCTOR**

Question 6 Full-credit answers should include the following point:

Human activities can make land less productive

- by building structures on it .
- through overuse
- by polluting the soil which reduces the amount of nutrients available to plants

Question 7 Full-credit answers should include the following point:

- if population growth continues, rangelands will need to provide more of the world's food than it does today
- rangeland must be protected from overgrazing so it doesn't become degraded

Question 9 Full-credit answers should include the following points:

- most crops are planted and harvested annually
- trees may take years or even decades to grow to maturity
- unlike other crops, the harvesting of trees can affect the health of nearby trees

Understanding Concepts

Directions (1–5): For *each* question, write on a separate sheet of paper the letter of the correct answer.

1 What is the term for the movement of people from rural areas to cities?
 A. land-use planning
 B. infrastructure
 C. urban sprawl
 D. urbanization

2 Unplanned, rapid urban growth can create what problem?
 F. degradation of the ecosystem
 G. elimination of invasive vegetation
 H. infrastructure that can not adequately support the population
 I. rangeland damaged from overgrazing

3 Which of the following statements generalize population distribution changes over the last 200 years?
 A. Urban areas have doubled in size in 200 years.
 B. More people lived in rural areas 200 years ago.
 C. Population distribution has not significantly changed in 200 years.
 D. Undeveloped countries have seen the most changes in 200 years.

4 Which of the following is an important aspect of land management?
 F. increasing the overall size of the herds
 G. leasing public lands from the federal government
 H. reducing damage to land caused by overgrazing
 I. removing fences from rangeland to allow livestock more grazing area

5 Which of the following is an environmental benefit of open space?
 A. Open space leads to a reduction in traffic flow.
 B. Open space helps filter pollutants from air and water.
 C. Open space means more land is available for planting.
 D. Open space results in lower temperatures in the wintertime.

Directions (6–7): For *each* question, write a short response.

6 As the human population grows, more land resources are needed to support the population. Describe two ways that human activities make rural lands less productive.

7 Analyze the relationship between protecting rangeland and ensuring the world's food supply.

Reading Skills

Directions (8–10): Read the passage below. Then answer the questions.

People use enormous amounts of wood. The worldwide average is 1 800 cm^3 of wood per person each day. However, each person in the United States uses about 3.5 times this amount, the equivalent of cutting down a 30 m tall tree every year.

The timber industry classifies forest lands into three categories—virgin forests, which is forest that has never been cut; native forest, which is forest that is planted and managed; and tree farms, which are areas where trees are planted in rows and harvested like other crops. The two most widely used methods of harvesting trees are clear-cutting and selective cutting. Clear-cutting is the process of removing all trees from an area of land. Clear-cutting large areas destroys wildlife habitat and causes soil erosion. Selective cutting is the process of cutting and removing only middle-aged or mature trees. Selective cutting is more expensive than clear-cutting but selective cutting is usually much less destructive.

8 The forest land classification that can be deduced as the most rare is
 F. deciduous forest
 G. evergreen forest
 H. native forest
 I. virgin forest

9 Compare the harvesting of trees to the harvesting of other kinds of crops.

10 What is the most effective way to help a forest recover from tree harvesting?

Answers

1. D
2. H
3. B
4. H
5. B
6. Answers will vary. See Test Doctor for detailed scoring rubric.
7. Answers will vary. See Test Doctor for detailed scoring rubric.
8. I
9. Answers will vary. See Test Doctor for detailed scoring rubric.
10. The most effective way is by replanting trees, or reforestation.
11. C
12. F
13. B
14. H

Interpreting Graphics

Directions (11–14): For *each* question below, record the correct answer on a separate sheet of paper.

The map below shows land use in the United States. Use this map to answer questions 11 through 14.

Land Use in the United States

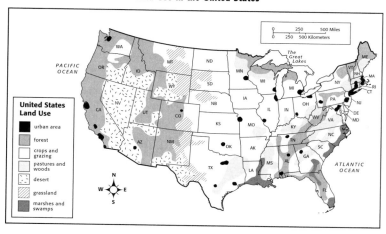

11. Which land-use designation has the **greatest** potential for growth?
    A. forest
    B. desert
    C. urban areas
    D. pasture lands

12. What is the **most** prevalent land use in the continental United States?
    F. crops and grazing
    G. desert
    H. pastures and woods
    I. urban areas

13. Which of the following conclusions is suggested by the map?
    A. Most farmers live west of the Rocky Mountains.
    B. There are more rural lands in the U.S. than there are urban lands.
    C. Manufacturing plants are concentrated along the Mississippi River.
    D. Industries dependent on wood and wood products are located mostly in the Midwest.

14. What percentage of states in the continental United States have some land available for crops and grazing?
    F. 10%
    G. 50%
    H. 80%
    I. 100%

Test TIP

Try to picture in your mind the terrain and buildings that would be present in the various land-use designations.

✚ TEST DOCTOR

Question 11 Answer C is correct. Students struggling with this type of question may benefit from imagining trends or change over time. For example, this type of problem can be solved by imagining what the map may look like in the future if current trends continue.

Question 12 Answer F is correct. Answers G, H and I are incorrect because even though those types of land are predominant in some areas, more land is used for crops and grazing than any other single purpose in the U.S.

Estimated Time

To give students practice under more realistic testing conditions, allow them 30 minutes to answer all of the questions in this practice test.

 TEST DOCTOR

Question 5 Full-credit answers should include the following points:

- resistant pests are more likely to live long enough to reproduce
- they are therefore more likely to pass their genetic abilities onto offspring than are pests without resistance

Question 6 Full-credit answers should include the following points:

- biological pest control is effective for reducing the amount of pesticides used
- biological pest control can help reduce resistance
- biological pest control sometimes requires more planning and overall management of land

Question 8 Full-credit answers should include the following points:

- genetically-modified plants that produce insecticides could speed up the process of resistance
- more insects could potentially be exposed to the particular insecticide

Understanding Concepts

Directions (1–4): For *each* question, write on a separate sheet of paper the letter of the correct answer.

1 Which of the following is a **major** cause of famine in the world today?
  A. food efficiency
  B. improved yield
  C. lack of rainfall
  D. no-till farming

2 What is the main difference between fertile and infertile soil?
  F. Fertile soil supports plant life; infertile soil cannot.
  G. Infertile soil supports plant life; fertile soil cannot.
  H. Fertile soil supports insect life; infertile soil cannot.
  I. Infertile soil supports insect life; fertile soil cannot.

3 What is the eventual result of land degradation?
  A. desertification
  B. fertilization
  C. integration
  D. salinization

4 Which of the following is an effect of soil erosion?
  F. increased crop yields
  G. increase in land fertility
  H. decrease in desertification
  I. decrease in amount of top soil

Directions (5–6): For *each* question, write a short response.

5 Over time, controlling pests with pesticides can make a pest problem worse if pest populations develop resistance. Resistance is the ability to survive exposure to a particular pesticide. Why are subsequent generations of resistant pest populations more likely to be resistant?

6 Evaluate the practice of biological pest control.

Reading Skills

Directions (7–8): Read the passage below. Then answer the questions.

Plant breeding has been used since agriculture began. Farmers select the plants that have the most desirable characteristics. They save seeds from these plants to use in planting the next crop. The selected seeds are more likely to contain the genes for the desired traits. The same result is achieved with genetic engineering, the technology in which genetic material in a living cell is modified for medical or industrial use.

Genetic engineering involves isolating genes from one organisms and implanting them into another. The plants that result from genetic engineering are called genetically modified plants. Sometimes genes introduced into plants are not plant genes. An insecticide gene from a bacterium that produces a chemical that kills plant-eating caterpillars can be implanted into crop plants. Hundreds of gene transfers have now been performed to create many genetically modified crops. In the United States, we now eat and use genetically engineered agricultural products every day.

7 What impact has the technology of genetic engineering had on the agricultural community?
  A. The technology of genetic engineering has increased the efficiency of the agricultural community.
  B. The technology of genetic engineering has decreased the efficiency of the agricultural community.
  C. The technology of genetic engineering has increased cultivation within the agricultural community.
  D. The technology of genetic engineering has decreased cultivation within the agricultural community.

8 What effect could introducing insecticide genes into crops have on insecticide resistance?

664

Answers

1. C
2. F
3. A
4. I
5. Answers will vary. See Test Doctor for detailed scoring rubric.
6. Answers will vary. See Test Doctor for detailed scoring rubric.
7. A
8. Answers will vary. See Test Doctor for detailed scoring rubric.

9. H
10. B
11. G

Interpreting Graphics

Directions (9–11): For *each* question below, record the correct answer on a separate sheet of paper.

The graph below shows world grain production over a 50-year period. Use this graph to answer questions 9 through 11.

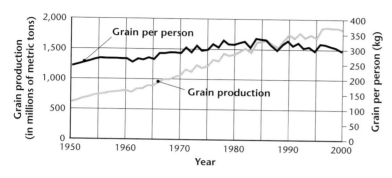

World Grain Production, 1950–2000

9 In what year did the total amount of grain production equal the average amount of grain per person?
F. 1980
G. 1983
H. 1986
I. 1989

10 What is the main reason that, while total grain production has increased each year, the amount of grain per person has been relatively steady?
A. Much of the grain produced could not be distributed efficiently.
B. The world's population has grown faster than grain production.
C. The world's population has remained roughly the same for the last 30 years.
D. Much of the grain produced in the world was used to feed livestock or as seed.

11 If the trend that was in place from 1990-1995 continues through 2005, what will be the average amount of grain available per person?
F. 200 kg
G. 250 kg
H. 300 kg
I. 350 kg

Test TIP

Slow, deep breathing may help you relax. If you suffer from test anxiety, focus on your breathing in order to calm down.

+ TEST DOCTOR

Question 10 Answer B is correct. Answer A is incorrect because even though distribution has become an issue in some areas, it is not the main reason for the reduction in grain availability. Answers C and D are incorrect because they are both untrue statements.

Question 11 Answer G is correct. Answer F is incorrect because it is too low. To reach this amount, the decrease would have to accelerate. Answers H and I are incorrect because they are too high. To reach this amount, the amount of grain available would have to increase dramatically.

665

Estimated Time

To give students practice under more realistic testing conditions, allow them 30 minutes to answer all of the questions in this practice test.

 **TEST DOCTOR**

Question 4 Answer F is correct. Most minerals are compounds, but native elements are not, even though they are considered to be minerals. Answer G is incorrect because gold, silver, and copper are native elements. Answer H is incorrect because solar evaporation is the process used to mine salt. Answer I is incorrect because silica forms most of the minerals that make up the Earth's crust.

Question 5 Full-credit answers should include the following points:

- sedimentation can choke out plants
- this reduces the food supply for some organisms.
- it can damage water quality

Question 6 Full-credit answers should include the following points:

Environmental impacts of mining on the ocean floor are

- increased waste and pollution
- an increase in sediment which could harm fragile ecosystems

---

Understanding Concepts

Directions (1–4): For *each* question, write on a separate sheet of paper the letter of the correct answer.

1 Which of the following statements is true?
  A. Minerals are found both on and beneath Earth's surface.
  B. Minerals are metallic compounds of economic importance.
  C. Minerals are limited resources easily extracted from the ground.
  D. Minerals are unlimited resources extracted from the ground at little cost.

2 Which of these nonmetallic minerals has many uses in the construction industry?
  F. gypsum
  G. quartz
  H. silicon
  I. tourmaline

3 Salt is mined underground using what technique?
  A. longwall mining
  B. open-pit mining
  C. solution mining
  D. surface mining

4 What distinguishes native elements from most other minerals?
  F. Native elements are not compounds.
  G. Native elements are always non-metallic.
  H. Native elements are mined by solar evaporation.
  I. Native elements are essentially the minerals that make up Earth's crust.

Directions (5–6): For *each* question, write a short response.

5 Excess rock from mines is sometimes dumped into large piles. Running water erodes these unprotected dumps, transporting the rocks and sediment into nearby streams. What effect does the resulting sedimentation have on a stream's ecosystem?

6 The ocean floor contains significant mineral resources. Assess the environmental impact of mining on the ocean floor.

Reading Skills

Directions (7–9): Read the passage below. Then answer the questions.

The process of returning land to its original or better condition after mining is completed is called reclamation. In the United States billions of dollars are spent to clean up abandoned mines. The Surface Mining Control and Reclamation Act of 1977 (SMRCA) created a program for the regulation of surface coal mining on public and private lands. The act set standards that would minimize the surface effects of coal mining on the environment. SMRCA also established a fund that is used to reclaim land and water resources that have been adversely affected by past coal-mining activities.

States also have programs to regulate mining activities. Mining companies must obtain permits from state environmental agencies before mining a site.

7 Which of the following is a consequence of not reclaiming previously-mined land?
  A. Unreclaimed land decreases air pollution.
  B. Unreclaimed land decreases soil erosion.
  C. Unreclaimed land increases water quality.
  D. Unreclaimed land increases water pollution.

8 Why is mining one of the most heavily regulated industries in the United States?
  F. Mining is heavily regulated because of its profit potential.
  G. Mining is heavily regulated because of its economic impact.
  H. Mining is heavily regulated because of its environmental impact.
  I. Mining is heavily regulated because of the abundance of small owners.

9 What is the process of mining mineral deposits concentrated on Earth's surface by wind and water called?
  A. hydraulic mining
  B. open-pit mining
  C. placer mining
  D. smelter mining

666

Answers

1. A
2. F
3. C
4. F
5. Answers will vary. See Test Doctor for detailed scoring rubric.
6. Answers will vary. See Test Doctor for detailed scoring rubric.
7. D
8. H
9. C
10. H

11. B
12. F
13. A

Interpreting Graphics

Directions (10–13): For *each* question below, record the correct answer on a separate sheet of paper.

The table below shows the properties for some minerals. Use this table to answer questions 10 through 12.

Properties of Selected Minerals

Mineral	Luster	Color	Streak	Hardness	Specific Gravity
Bauxite (a mixture of several minerals)	varies	brownish yellow, brown, red, gray	varies	varies	varies
Beryl ($Be_3Al_2Si_6O_{18}$)	Vitreous	blue, green, yellow, pink	white	$7\frac{1}{2}$-8	2.6-2.8
Chromite ($FeCr_2O_4$)	Metallic to sub-metallic	Black to brownish black	Dark brown	$5\frac{1}{2}$	4.5-4.8
Cinnabar	Adamantine (diamond-like)	Scarlet red to brownish red	Vermillion (vivid red to red orange)	$2-2\frac{1}{2}$	8.0-8.2
Cuprite (Cu_2O)	Adamantine or submetallic	Red, sometimes very dark red	Brownish red	$3\frac{1}{2}$-4	5.8-6.1

10 If a mineral sample has an adamantine luster, what test could be used to **most** accurately identify the specific mineral?
F. color
G. hardness
H. specific gravity
I. streak

11 Using the table, what conclusion can be reached about beryl?
A. It is a soft mineral.
B. Its color varies greatly.
C. It has a smooth surface.
D. It is seven times heavier than water.

12 What properties can a scientist observe **most** easily in the field?
F. color and luster
G. hardness and streak
H. specific gravity and hardness
I. streak and specific gravity

13 Bauxite is the major ore mineral that composes what element?
A. aluminum
B. copper
C. manganese
D. titanium

Test TIP

Take time to read each question completely, including all of the answer choices. Consider each answer choice before determining which one is correct.

+ *TEST DOCTOR*

Question 10 Answer H is the correct choice because specific gravity is the property that has the largest discrepancy between possible choices. Answers F and I are incorrect because there could be some confusion between red and brown. Answer G is incorrect because even though the hardness may differ, it does not differ by as much as specific gravity.

Question 11 Answer B is correct, the color of beryl varies greatly, from pink to blue to green to yellow. Answer A is incorrect because the hardness of beryl is high, $7\frac{1}{2}$ to 8 on a scale of 1 to 10. Answer C is incorrect because beryl has a vitreous surface that resembles broken china. Answer D is incorrect because the specific gravity of beryl, its weight when compared to the weight of an equal volume of water, is less than 3.

Estimated Time

To give students practice under more realistic testing conditions, allow them 30 minutes to answer all of the questions in this practice test.

TEST DOCTOR

Question 4 Full-credit answers should include the following points:

- many countries outside of the U.S. have a smaller land mass
- many countries outside of the U.S. have developed mass transit systems that require less energy for transporting goods and services than does the use of individual vehicles

Question 5 Full-credit answers should include the following points:

- for safety reasons, nuclear power plants are engineered with many backup systems
- approvals for sites and building take a long time

Question 7 Full-credit answers should include the following points:

- as oil prices increase, industries in developing countries may have to pay more for fuel
- this increase in the price of energy may slow growth

Understanding Concepts

Directions (1–3): For *each* question, write on a separate sheet of paper the letter of the correct answer.

1 Where do coal, oil and natural gas come from?
- **A.** the melting of polar ice packs
- **B.** the remains of organisms
- **C.** the residue of volcanic eruptions
- **D.** the understory of forests

2 When fossil fuels are burned and converted to electricity, how does the total amount of usable energy change?
- **F.** The amount of usable energy remains the same.
- **G.** The amount of usable energy decreases during conversion.
- **H.** The amount of usable energy doubles during conversion.
- **I.** The amount of usable energy depends on how well the power plant is maintained.

3 Which of the following statements describes energy consumption trends today?
- **A.** Developed nations are using less energy per person.
- **B.** Undeveloped nations are using less energy per person.
- **C.** Developed nations are using more energy than undeveloped nations.
- **D.** Undeveloped nations are using more energy than developed nations.

Directions (4–5): For *each* question, write a short response.

4 The United States uses more than 25% of its energy to transport goods and people. This percentage of resources used mostly by trucks and personal vehicles is much higher than energy used for transportation in Japan or Switzerland. Why do these two developed countries use less energy for transportation than the United States?

5 Provide one reason why building and maintaining nuclear power plants is expensive.

Reading Skills

Directions (6–8): Read the passage below. Then answer the questions.

Today, fossil fuels supply about 90% of the energy used in developed countries. Oil production is increasing, but it is increasing much more slowly than it has in the past. Many different factors must be considered when predicting oil production. Oil reserves are oil deposits that can be extracted profitably at current prices using current technology. Some projections suggest that by 2050 world demand for fossil fuels will have doubled, mainly as a result of increased population and industry in developing nations. People in developed societies use much more energy than people in developing countries do. As the demand for the energy resources increases, the cost of fossil fuels will likely increase enough to make other energy sources more attractive.

6 How would you assess the effect of technology on oil prices?
- **F.** Technology raises production and lowers prices.
- **G.** Technology lowers production and raises prices.
- **H.** Technology raises production and raises prices.
- **I.** Technology lowers production and lowers prices.

7 If oil prices continue to increase, what effect would this have on industries in developing countries?

8 Natural gas is a nonrenewable petroleum product. Which of the following statements about this fossil fuel is **false**?
- **A.** Burning natural gas produces fewer pollutants than other fossil fuels.
- **B.** Burning natural gas requires fewer pollution controls.
- **C.** Because of costs, electric power plants can not use natural gas.
- **D.** About 20 percent of the world's nonrenewable energy comes from natural gas.

668

Answers

1. B
2. G
3. C
4. Answers will vary. See Test Doctor for detailed scoring rubric.
5. Answers will vary. See Test Doctor for detailed scoring rubric.
6. F
7. Answers will vary. See Test Doctor for detailed scoring rubric.
8. D
9. I
10. D
11. F
12. D

Interpreting Graphics

Directions (9–12): For *each* question below, record the correct answer on a separate sheet of paper.

The graph below shows past and predicted oil production. Use this graph to answer questions 9 through 12.

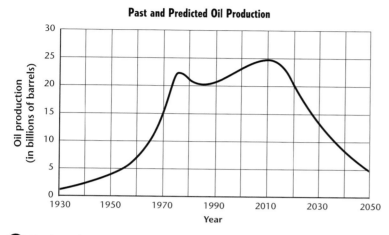

Past and Predicted Oil Production

Question 10 Answer D is correct. Answer A is incorrect because oil production hit its overall low in 1950. Answer B is incorrect because production decreased during the 1980s. Answer C is incorrect because there were times during the 1900s when oil production decreased and so the increase in production was not continuous.

Question 12 Answer D is correct. Answer A is incorrect because even though production increased, its overall increase was less than 50%. Answers B and C are incorrect because production increased.

9 If oil production after 2010 continues at the predicted rate, when will the oil reserves run out?
  F. 2040
  G. 2050
  H. 2060
  I. 2070

10 What can be concluded about oil production from the graph?
  A. Oil production hit its peak in the mid 20th century.
  B. Oil production saw a dramatic increase during the 1980s.
  C. Oil production continued to increase throughout the 1900s.
  D. Oil production more than doubled between 1965 and 1975.

11 What can be inferred about the cost of oil after 2010?
  F. It will increase.
  G. It will decrease steadily.
  H. It will remain unchanged.
  I. It will increase until 2030, then decrease.

12 What is the difference in billions of barrels produced in 1990 compared to 1970?
  A. Production doubled.
  B. Production fell by half.
  C. Production hit a new low.
  D. Production was up six billion barrels.

Test TIP

Sometimes, only a portion of a graph or table is needed to answer a question. Focus only on the necessary information to avoid confusion.

669

Estimated Time

To give students practice under more realistic testing conditions, allow them 30 minutes to answer all of the questions in this practice test.

TEST DOCTOR

Question 6 Full-credit answers should include the following points:

- the potential for wind energy is high
- the windiest spots on Earth are underdeveloped
- as problems of transportation are solved, the number of wind farms will increase

Question 8 Full-credit answers should include the following points:

- thick walls and insulation prevent the trapped heat from escaping quickly
- this keeps the house warm longer

Question 9 Full-credit answers should include two of the following points:

Energy efficiency in the home can be increased by

- replacing old windows
- sealing leaks around doors
- closing doors and turning off lights in unused rooms
- adjusting the thermostat higher in summer, lower in winter
- using hot water only when necessary
- cleaning or replacing air filters

Understanding Concepts

Directions (1–4): For *each* question, write on a separate sheet of paper the letter of the correct answer.

1 What is the ultimate source of all renewable energy?
A. the biosphere
B. the moon
C. the ocean
D. the sun

2 Which of the following is a renewable energy source?
F. coal mine
G. gas pipeline
H. power plant
I. wind farm

3 What is the most important factor in the development and implementation of alternative energy sources?
A. The most important factor is the abundance of the source.
B. The most important factor is its cost effectiveness.
C. The most important factor is whether government approval can be obtained.
D. The most important factor is if the source can gain social acceptance.

4 Why is hydrogen called the fuel of the future?
F. It is very inexpensive to produce.
G. It requires very little energy to produce.
H. It is the most abundant element in the universe.
I. It contains carbon which disperses in the atmosphere when burned.

Directions (5–6): For *each* question, write a short response.

5 Energy efficiency is the percentage of energy put into a system that does useful work. Why is energy efficiency always less than 100%?

6 Estimate the potential of wind power as an energy source.

Reading Skills

Directions (7–8): **Read the passage below. Then answer the questions.**

Ancient cliff dwellings contain elements of the simplest form of solar energy called passive solar heating. Passive solar heating uses the sun's energy to heat something directly. In the Northern Hemisphere, windows facing south receive the most solar energy, so passive solar buildings have large windows that face south. Solar energy enters the windows and warms the house. At night, the heat is released slowly to keep the house warm. Passive solar buildings must be insulated with thick walls and floors.

Passive solar buildings are oriented according to the yearly movement of the sun. In summer, the sun's path is high in the sky and the overhang of the roof shades the building and keeps it cool. If there is reliable winter light, an extremely efficient passive solar heating system can heat a house in very cold weather without using any other source of energy.

7 In hotter climates, in what direction should the largest windows face in order for the house to be as cool as possible?
A. north
B. south
C. east
D. west

8 Why do passive solar houses require thick walls and insulation to be efficient at heating the house in winter?

Directions (9): **Read the passage below. Then answer the question.**

Energy conservation means saving energy. It can occur in many ways, including using energy-efficient devices and wasting less energy. Between 1975 and 1985, conservation made more energy available in the United States than all alternative energy sources combined did.

9 Describe two ways that a family could reduce the amount of energy wasted in their home by increasing energy efficiency.

670

Answers

1. D
2. I
3. B
4. H
5. Because some energy is always lost during conversion from one form to another.
6. Answers will vary. See Test Doctor for detailed scoring rubric.
7. A
8. Answers will vary. See Test Doctor for detailed scoring rubric.
9. Answers will vary. See Test Doctor for detailed scoring rubric.
10. F
11. C
12. F

Interpreting Graphics

Directions (10–12): For *each* question below, record the correct answer on a separate sheet of paper.

The map below shows hydroelectric power generation in the United States. Use this map to answer questions 10 and 11.

Hydroelectric Power

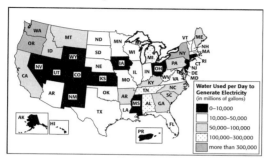

10 What can be inferred about the use of hydroelectric power in the United States?
   F. There are more dams in Oregon than in Kansas.
   G. There is more hydroelectric power used in New Mexico than in Alabama.
   H. The upper Midwest uses more water for hydroelectric power than New England.
   I. The biggest users of water for hydroelectricity are located along the Mississippi River.

11 How many states use 10,000 million gallons of water or less per day?
   A. 5   C. 15
   B. 10   D. 20

Use the diagram below to answer question 12.

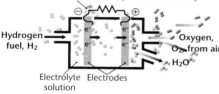

12 What type of electrical generation is depicted in the diagram?
   F. fuel cell
   G. geothermal plant
   H. ocean thermal energy
   I. tidal plant

TEST DOCTOR

Question 10 Answer F is correct. Answer G is incorrect because there is more hydroelectric power used in Alabama than New Mexico. Answer H is incorrect because the upper Midwest uses relatively small amounts of hydroelectric power. Answer I is incorrect because the biggest users of hydroelectric power are in the west and the Mississippi River is in the Midwest.

Question 12 Answer F is correct. The diagram shows how a fuel cell produces electrical energy. Choices G and H produce steam that powers a turbine that generates electricity. A tidal plant (Choice I) works much like a hydroelectric dam.

Test TIP

Carefully read the instructions, the question, and the answer options before choosing an answer.

671

Standardized Test Prep

Estimated Time

To give students practice under more realistic testing conditions, allow them 30 minutes to answer all of the questions in this practice test.

✚ TEST DOCTOR

Question 3 Answer D is correct. Answer A is incorrect because plastic is nonbiodegradable so it increases the amount of solid waste. Answer B is incorrect because packaging is usually thrown away, creating waste. Answer C is incorrect because recycling would reduce the amount of solid waste.

Question 5 Full-credit answers should include the following points:

- compost enriches the soil
- fewer people would need chemical fertilizers
- less pollutants contaminating water means less water pollution

Question 9 Full-credit answers should include the following points:

- any increase in the cost of waste disposal increases the costs of manufacturing
- when costs increase, management looks for more efficient methods
- increased disposal costs mean increased research into reducing waste

Understanding Concepts

Directions (1–4): For *each* question, write on a separate sheet of paper the letter of the correct answer.

1 Which of the following is biodegradable?
- **A.** a nylon jacket
- **B.** a plastic cup
- **C.** a television set
- **D.** a wool sweater

2 What is source reduction?
- **F.** Source reduction is a method by which we can produce less waste.
- **G.** Source reduction is a method of reducing recyclable materials.
- **H.** Source reduction is a process that changes the number of landfills.
- **I.** Source reduction is a process that allows manufacturers to make more products from plastics.

3 What is the relationship between packaging and solid waste?
- **A.** Packaging made of plastic decreases the amount of solid waste.
- **B.** Packaging has little effect on the amount of solid waste produced.
- **C.** Packaging that can be recycled increases the amount of solid waste.
- **D.** Packaging for single-serving items increases the amount of solid waste.

4 Which of the following statements is true?
- **F.** There is little danger of hazardous waste entering groundwater if the waste is disposed of through deep-well injection.
- **G.** During the 1990s the amount of waste generated per person remained almost constant, decreasing the total amount of waste generated.
- **H.** Composting could reduce the amount of waste that restaurants, food-processing plants, and animal feedlots send to landfills.
- **I.** Landfills are the safest way to dispose of solid waste because the materials buried in them decompose quickly.

Directions (5–6): For *each* question, write a short response.

5 How could the widespread use of compost affect water pollution that results from the use of chemical fertilizers?

6 If plastics are made from petroleum, why are plastic products nonbiodegradable?

Reading Skills

Directions (7–9): Read the passage below. Then answer the questions.

Many of the products in use today are produced in modern factories that use thousands of chemicals. Many of these chemicals are classified as hazardous waste, which is any waste that is a risk to the health of living things.

Each year, the United States produces about 252 metric tons of hazardous wastes, and this amount is growing every year. In recent years, many manufacturers have discovered that they can redesign manufacturing methods to produce less or no hazardous waste. For example, some manufacturers who used chemicals to clean metal parts of machines have discovered tat they can use tiny plastic beads instead. The beads act like a sandblaster to clean the parts, can be reused several times, and are not hazardous when disposed. Often, such techniques save the manufacturers money by cutting the cost of materials as well as in cutting the cost of waste disposal.

7 Which of the following items associated with an automobile is considered hazardous waste?
- **A.** back-up light
- **B.** motor oil
- **C.** rubber tire
- **D.** windshield wiper

8 When would chemicals be useful in the handling of hazardous waste?

9 How does the cost of waste disposal affect manufacturing methods?

672

Answers

1. D
2. F
3. D
4. H
5. Answers will vary. See Test Doctor for detailed scoring rubric.
6. Plastics are chemically different from their petroleum sources.
7. B
8. Chemicals are useful when they can change hazardous materials into non-hazardous materials.
9. Answers will vary. See Test Doctor for detailed scoring rubric.
10. I
11. B
12. G
13. B

Standardized Test Prep

Interpreting Graphics

Directions (10–13): For *each* question below, record the correct answer on a separate sheet of paper.

The graph below shows the composition (by weight) of municipal solid wastes in the United States. Use this graph to answer questions 10 through 12.

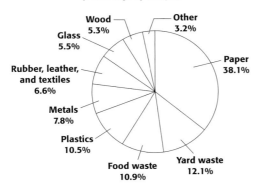

Municipal Solid Waste (Percentage by Weight)

- Wood 5.3%
- Other 3.2%
- Glass 5.5%
- Paper 38.1%
- Rubber, leather, and textiles 6.6%
- Metals 7.8%
- Plastics 10.5%
- Food waste 10.9%
- Yard waste 12.1%

10 If the most recyclable materials are paper, plastics, metal and glass, by what percentage would municipal solid waste be reduced if every person in the country recycled?
F. 38%
G. 43%
H. 54%
I. 62%

11 Which of the following is a type of waste you would likely find under the category of "Other" in this pie graph?
A. banana peels
B. glazed ceramics
C. grass clippings
D. industrial chemicals

12 By how much would the amount of municipal waste be reduced if every household in the country had a compost pile?
F. 12%
G. 23%
H. 35%
I. 41%

13 Where does municipal waste come from?
A. agriculture and mining   C. manufacturing and business
B. households and business   D. mining and households

TEST DOCTOR

Question 10 Answer I is correct. Answer F is incorrect because it only takes into account the amount of paper. Answer G is incorrect because it only accounts for paper and glass. Answer H is incorrect because it only accounts for paper, glass, and plastic.

Question 11 Answer B is correct. Students struggling with this type of question may benefit from imagining the types of items that would be found under each category. For example, food waste would contain items from a kitchen, such as banana peels (Choice A). Grass clippings (Choice C) would be among the items found under yard waste. Choice D involves waste from manufacturing plants and is not represented on this graph.

Test TIP

For multiple-choice questions, try to eliminate any answer choices that are obviously incorrect, and then consider the remaining answer choices.

Estimated Time

To give students practice under more realistic testing conditions, allow them 30 minutes to answer all of the questions in this practice test.

TEST DOCTOR

Question 3 Answer C is correct. Answers A and D are incorrect because naturally occurring pollutants become hazardous when their concentrations are greater than normal levels in the environment. Answer B is incorrect because research has been ongoing for years into how human activities cause pollution.

Question 5 Full-credit answers should include the following points:

Using communal water sources for drinking and bathing

- increases the amount of pollution in the water
- spreads the organisms that cause diseases more readily

Question 8 Full-credit answers should include the following points:

- environmental change changes the habitats of animals
- this puts different animals in close proximity with each other and humans
- by placing animals that have been historically separated by habitat in close proximity with each other creates a situation that can cause cross-species transfers.

Understanding Concepts

Directions (1–4): For *each* question, write on a separate sheet of paper the letter of the correct answer.

1 Which of the following is a naturally occurring pollutant?
  A. pesticides
  B. radon
  C. sewage
  D. vectors

2 What is the study of the spread of disease called?
  F. antibiology
  G. epidemiology
  H. pathogenology
  I. toxicology

3 What is the difference between pollution from natural sources and pollution from human activities?
  A. Naturally occurring pollutants are less toxic.
  B. Pollution from human activities has not been researched.
  C. Pollution from human activities can be predicted and controlled.
  D. Naturally occurring pollutants are only hazardous in small dosages.

4 How has the construction of irrigation canals and dams enabled the spread of infectious disease?
  F. The construction allows viruses to evolve.
  G. Canals and dams provide sites for waste disposal.
  H. Canals and dams provide increased habitats for vectors.
  I. The construction eliminates the natural predators of pathogens.

Directions (5–6): For *each* question, write a short response.

5 Analyze how communal water sources transmit pollution and disease when residents use them for drinking, bathing, and sewage disposal.

6 Identify how naturally-occuring heavy metals can act as pollutants.

674

Reading Skills

Directions (7–9): Read the passage below. Then, answer the questions.

Some of the damage to human health in which the environment plays a role is not caused by toxic chemicals but by organisms that carry disease. Today we have outbreaks of diseases that did not exist one hundred years ago. And diseases that have killed people for centuries are killing as many or more people today. One of the reasons these diseases are now widespread is that we have altered our environment in ways that encourage their spread.

In recent years, scientists have discovered an increasing number of pathogens that have made a cross-species transfer, or have moved from one species to another. When the pathogens invade humans, the pathogens cause serious disease. One example of cross-species transfer occurred in Argentina. Herbicides were sprayed on crops in Argentina. The herbicides killed the native grasses and allowed other plants to invade the farmland. These new plants attracted a species of rodent that feeds on them. The rodents were carrying viruses for a hemorrhagic fever, which infected many of the agricultural workers.

7 In what way do modern ranching practices affect the number of cross-species transfers?
  A. The sizes of herds increase.
  B. Humans have closer contact with the animals.
  C. There is less opportunity to spread viruses.
  D. It reduces the virility of pathogens.

8 Describe the relationship between cross-species transfers and environmental change.

9 How did the Hong Kong flu get its name?
  F. It originated in a Hong Kong zoo.
  G. It was the latest version of the Asian flu.
  H. It was first diagnosed in a Hong Kong hospital.
  I. It was transmitted to humans from ducks bred in Hong Kong for food.

Answers
1. B
2. G
3. C
4. H
5. Answers will vary. See Test Doctor for detailed scoring rubric.
6. Naturally-occurring heavy metals are toxic to humans if ingested in high doses.
7. B
8. Answers will vary. See Test Doctor for detailed scoring rubric.
9. I

10. A
11. I
12. B
13. G

Interpreting Graphics

Directions (10–13): For *each* question below, record the correct answer on a separate sheet of paper.

The graph below shows the environment's contribution to disease in different regions of the world. Use the graph to answer questions 10 and 11.

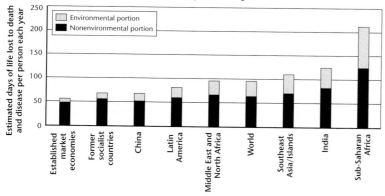

Poor Health by World Region

10 In which of these regions is nonenvironmental pollution the largest percentage of the total?
A. China
B. India
C. Latin America
D. Sub-Saharan Africa

11 In which of these regions are nonenvironmental causes of poor health the smallest percentage of the total?
F. China
G. India
H. Latin America
I. Sub-Saharan Africa

12 What is the main factor in the poor health of undeveloped countries?
A. too few doctors
B. infectious disease
C. industrial pollution
D. more imports than exports

13 The average person in the world will lose 95 days a year to death and disease. In years, approximately how much time will be lost to death and disease by the average person after 20 years?
F. 3
G. 5
H. 7
I. 9

TEST DOCTOR

Question 10 Answer A is correct. Students struggling with this type of question may benefit from extra practice comparing portions and totals. Ask students to draw a similar graph with portions are represented as a part of the whole in a bar graph. Tell students that amounts of one of the portions increased by 50%. Ask students to explain how the portion and the total changed.

Question 13 Answer G is correct. This answer is found by multiplying the average per year by the number of years, (95 X 20 = 1900) then dividing by 365 and rounding off to a whole number (1900/365=5.2).

Test TIP

When making comparisons from a graph, analyze each component separately, comparing the fraction to the total for each component. Then compare the components to each other.

675

Estimated Time

To give students practice under more realistic testing conditions, allow them 30 minutes to answer all of the questions in this practice test.

TEST DOCTOR

Question 4 Full-credit answers should include the following points:

- the ban on unfunded mandates made the federal government more aware of the economic impacts of environmental regulations
- if funds were not available to help pay for needed changes, new rules could not be imposed

Question 5 Full-credit answers should include the following points:

- the slogan, think globally, act locally, means that everyday actions affect global environmental health
- in order to help the global environment, individuals need to live their daily lives in an environmentally responsible manner

Question 8 Full-credit answers should include the following points:

- the work of biologists made the public aware of environmental problems
- this put pressure on the federal government to pass new laws and regulations

Understanding Concepts

Directions (1–3): For *each* question, write on a separate sheet of paper the letter of the correct answer.

1 If you wanted to buy stock in a company with a sustainability policy, what would you look for?
  A. You would look for a company that uses offshore labor.
  B. You would look for a company that uses renewable resources.
  C. You would look for a company that offers high quality products.
  D. You would look for a company that has implemented a global marketing strategy.

2 What kind of information is contained in an Environmental Impact Statement?
  F. An Environmental Impact Statement contains public comment on the proposed project.
  G. An Environmental Impact Statement tells where funds for the project are coming from.
  H. An Environmental Impact Statement predicts the project's expected impact on the environment.
  I. An Environmental Impact Statement contains the total price of all construction materials and labor.

3 How can an individual have an effect on global environmental problems?
  A. Buy things that provide only a single use.
  B. Buy paper goods at the local discount store.
  C. Structure daily activities around convenience.
  D. Purchase vegetables that are organically grown.

Directions (4–5): For *each* question, write a short response.

4 Unfunded mandates are regulations that do not provide funds to implement the regulations. What was one effect of the ban on unfunded mandates in 1995?

5 What does the slogan, think globally, act locally, mean?

676

Reading Skills

Directions (6–8): **Read the passage below. Then answer the questions.**

During the 1960s, biologists such as Paul Ehrlich, Barry Commoner, and Rachel Carson drew public attention to environmental problems such as pollution, rapid population growth, and resource depletion. In the book *Silent Spring,* Rachel Carson argued that resources such as water had to be protected and kept in natural, unpolluted conditions. Partly as a result of this book, Congress passed the Wilderness Act in 1964. This allowed the government to designate some federal lands as wilderness areas.

Also in the 1960s, several environmental disasters made headlines. Air pollution in New York City was blamed for 300 deaths. The bald eagle became endangered as a result of the widespread use of DDT. There was a massive oil spill near Santa Barbara. Lake Erie became so polluted that many of its beaches had to be closed. Eventually, pressure from the public led to new laws and efforts to reduce environmental damage.

6 Predict how a lack of public interest in environmental issues could affect federal legislation.
  F. Congress would hold more committee hearings.
  G. States would take the lead in passing regulations.
  H. New federal laws would be passed more quickly.
  I. The federal government would stop enforcing existing laws.

7 How can citizens be **most** effective at influencing environmental policy?
  A. by volunteering in their local community organizations
  B. by protesting at environmentally polluted sites
  C. by writing directly to the President of the United States
  D. by joining a national environmental lobbying organization

8 How did the work of biologists affect public policy in the 1960s?

Answers

1. B
2. H
3. D
4. Answers will vary. See Test Doctor for detailed scoring rubric.
5. Answers will vary. See Test Doctor for detailed scoring rubric.
6. I
7. A
8. Answers will vary. See Test Doctor for detailed scoring rubric.

9. F
10. A
11. I

Interpreting Graphics

Directions (9–11): For *each* question below, record the correct answer on a separate sheet of paper.

The map below shows the location of national parks in Central America. Use this map to answer questions 9 through 11.

National Parks

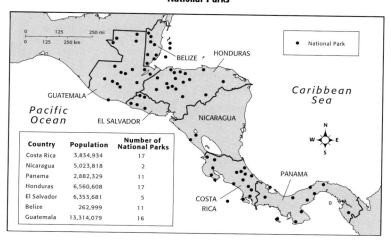

Country	Population	Number of National Parks
Costa Rica	3,834,934	17
Nicaragua	5,023,818	2
Panama	2,882,329	11
Honduras	6,560,608	17
El Salvador	6,353,681	5
Belize	262,999	11
Guatemala	13,314,079	16

9 What percentage of national parks in Guatemala are located within 50 kilometers of the coast?
 F. less than 15%
 G. more than 25%
 H. exactly 50%
 I. almost 75%

10 Which country has the largest number of national parks per capita in Central America?
 A. Belize
 B. El Salvador
 C. Guatemala
 D. Nicaragua

11 How is the number of national parks related to the size of a country?
 F. The bigger countries have more national parks.
 G. The smaller countries have more national parks.
 H. The countries with the most coastline have more national parks.
 I. There is no correlation between a country's size and its national parks.

Test TIP

Per capita means per unit of population.

677

TEST DOCTOR

Question 10 Answer A is correct. This answer is found by comparing the number of parks to the total population of each country to get a per capita (or per person) average. Answers B, C, and D are incorrect because they have smaller areas of national parks per capita than Belize.

Question 11 Answer I is correct. Students struggling with this type of question may benefit from extra practice making comparisons. If they list the largest countries (Choice F), they find Nicaragua with the fewest number of national parks. If they list the countries with the most national parks, they find Honduras, large in population size and area (Choice G). If they check the coastline (Choice H), they find Nicaragua once again, with the fewest number of parks. There is no way to correlate the number of parks with the country size.

Glossary

A

abiotic (ay bie AHT ik) **factor** an environmental factor that is not associated with the activities of living organisms (94)

acid precipitation precipitation, such as rain, sleet, or snow, that contains a high concentration of acids, often because of the pollution of the atmosphere (314)

acid shock the sudden runoff of large amounts of highly acidic water into lakes and streams when snow melts in the spring or when heavy rains follow a drought (316)

active solar heating the gathering of solar energy by collectors that are used to heat water or heat a building (460)

adaptation the process of becoming adapted to an environment; an anatomical, physiological, or behavioral change that improves a population's ability to survive (99)

age structure the classification of members of a population into groups according to age or the distribution of members of a population in terms of age groups (220)

agriculture the raising of crops and livestock for food or for other products that are useful to humans (10)

air pollution the contamination of the atmosphere by wastes from sources such as industrial burning and automobile exhausts (303)

alternative energy energy that does not come from fossil fuels and that is still in development (466)

altitude the height of an object above a reference point, such as sea level or the Earth's surface (145)

angiosperm (AN jee oh SPURM) a flowering plant that produces seeds within a fruit (105)

aquaculture (AK wuh KUHL chur) the raising of aquatic plants and animals for human use or consumption (396)

aquifer a body of rock or sediment that stores groundwater and allows the flow of groundwater (272)

arable land farmland that can be used to grow crops (227, 384)

artificial eutrophication a process that increases the amount of nutrients in a body of water through human activities, such as waste disposal and land drainage (288)

artificial selection the selective breeding of organisms (by humans) for specific desirable characteristics (100)

asbestos any of six silicate minerals that form bundles of minute fibers that are heat resistant, flexible, and durable (312)

asthenosphere the solid, plastic layer of the mantle beneath the lithosphere; made of mantle rock that flows very slowly, which allows tectonic plates to move on top of it (61)

atmosphere a mixture of gases that surrounds a planet, such as Earth (67)

B

bacteria extremely small, single-celled organisms that usually have a cell wall and reproduce by cell division (singular, *bacterium*) (102)

barrier island a long ridge of sand or narrow island that lies parallel to the shore (182)

benthic zone the region near the bottom of a pond, lake, or ocean (174)

benthos the organisms that live at the bottom of the sea or ocean (173)

biodegradable material a material that can be broken down by biological processes (483)

biodiversity the variety of organisms in a given area, the genetic variation within a population, the variety of species in a community, or the variety of communities in an ecosystem (15, 241)

Biodiversity Treaty an international agreement aimed at strengthening national control and preservation of biological resources; associated with the UN Conference on Environment and Development (UNCED or Earth summit) in 1992 (257)

biological pest control the use of certain organisms by humans to eliminate or control pests (391)

biomagnification the accumulation of pollutants at successive levels of the food chain (292)

biomass fuel plant material, manure, or any other organic matter that is used as an energy source (462)

biome a large region characterized by a specific type of climate and certain types of plant and animal communities (143)

biosphere the part of Earth where life exists (80)

biotic factor an environmental factor that is associated with or results from the activities of living organisms (94)

C

canopy the layers of treetops that shade the forest floor (148)

carbon cycle the movement of carbon from the nonliving environment into living things and back (124)

carrying capacity the largest population that an environment can support at any given time (200)

cellular respiration the process by which cells produce energy from carbohydrates; atmospheric oxygen combines with glucose to form water and carbon dioxide (120)

chaparral a type of vegetation that includes broad-leafed evergreen shrubs and that is located in areas with hot, dry summers and mild, wet winters (158)

chlorofluorocarbons hydrocarbons in which some or all of the hydrogen atoms are replaced by chlorine and fluorine; used in coolants for refrigerators and air conditioners and in cleaning solvents; their use is restricted because they destroy ozone molecules in the stratosphere (abbreviation, CFCs) (335)

climate the average weather conditions in an area over a long period of time (144, 327)

climax community a final, stable community in equilibrium with the environment (130)

closed system a system that cannot exchange matter or energy with its surroundings (81)

commensalism a relationship between two organisms in which one organism benefits and the other is unaffected (209)

community a group of various species that live in the same habitat and interact with each other (96)

competition the relationship between two species (or individuals) in which both species (or individuals) attempt to use the same limited resource such that both are negatively affected by the relationship (204)

compost a mixture of decomposing organic matter, such as manure and rotting plants, that is used as fertilizer and soil conditioner (490)

conceptual model a verbal or graphical explanation for how a system works or is organized (43)

condensation the change of state from a gas to a liquid (73)

conduction the transfer of energy as heat through a material (70)

consumer an organism that eats other organisms or organic matter instead of producing its own nutrients or obtaining nutrients from inorganic sources (118)

control group in an experiment, a group that serves as a standard of comparison with another group to which the control group is identical except for one factor (33)

convection the movement of matter due to differences in density that are caused by temperature variations; can result in the transfer of energy as heat (70)

coral reef a limestone ridge found in tropical climates and composed of coral fragments that are deposited around organic remains (183)

core the central part of the Earth below the mantle; also the center of the sun (61)

correlation the linear dependence between two variables (35)

crust the thin and solid outermost layer of the Earth above the mantle (60)

D

dam a structure that is built across a river to control a river's flow (280)

data any pieces of information acquired through observation or experimentation (34)

decibel the most common unit used to measure loudness (abbreviation, dB) (312)

decision-making model a conceptual model that provides a systematic process for making decisions (45)

decomposer an organism that feeds by breaking down organic matter from dead organisms; examples include bacteria and fungi (119)

deep-well injection deep-well disposal of hazardous waste (496)

deforestation the process of clearing forests (366)

demographic transition the general pattern of demographic change from high birth and death rates to low birth and death rates, as observed in the history of more-developed countries (223)

demography the study of the characteristics of populations, especially human populations (219)

Glossary

density the number of individuals of the same species that live in a given unit of area (198)

desalination (DEE SAL uh NAY shun) a process of removing salt from ocean water (283)

desert a region that has little or no vegetation, long periods without rain, and extreme temperatures; usually found in warm climates (160)

desertification the process by which human activities or climatic changes make arid or semiarid areas more desertlike (386)

diet the type and amount of food that a person eats (380)

dispersion in ecology, the pattern of distribution of organisms in a population (198)

distribution the relative arrangement of the members of a statistical population; usually shown in a graph (39)

domesticated describes organisms that have been bred and managed for human use (395)

dose the amount of a harmful substance to which a person is exposed; the quantity of medicine that needs to be taken over a period of time (512)

dose-response curve a graph that shows the relative effect of various doses of a drug or chemical on an organism or organisms (512)

E

ecological footprint a calculation that shows the productive area of Earth needed to support one person in a particular country (19)

ecological succession a gradual process of change and replacement in a community (129)

ecology the study of the interactions of living organisms with one another and with their environment (6)

economics the study of how individuals and groups make decisions about the production, distribution, and consumption of limited resources as the individuals or groups attempt to fulfill their needs and wants (536)

ecosystem (EE koh SIS tuhm) a community of organisms and their abiotic environment (93)

ecosystem services the role that organisms play in creating a healthful environment for humans (357)

ecotourism a form of tourism that supports the conservation and sustainable development of ecologically unique areas (244)

electric generator a device that converts mechanical energy into electrical energy (332)

El Niño (el NEEN yoh) the warm phase of the El Niño–Southern Oscillation; a periodic occurrence in the eastern Pacific Ocean in which the surface-water temperature becomes unusually warm (332)

emergent layer the top foliage layer in a forest where the trees extend above surrounding trees (148)

endangered species a species that has been identified to be in danger of extinction throughout all or a significant part of its range, and that is thus under protection by regulations or conservation measures (245)

Endangered Species Act an act that the U.S. Congress passed in 1973 to protect any plant or animal species in danger of extinction (255)

endemic species a species that is native to a particular place and that is found only there (248)

energy conservation the process of saving energy by reducing energy use and waste (470)

energy efficiency the percentage of energy put into a system that does useful work (468)

Environmental Impact Statement an assessment of the effect of a proposed project or law on the environment (541)

environmental science the study of the air, water, and land surrounding an organism or a community, which ranges from a small area to Earth's entire biosphere; it includes the study of the impact of humans on the environment (5)

epidemiology (EP uh DEE me AHL uh jee) the study of the distribution of diseases in populations and the study of factors that influence the occurrence and spread of disease (513)

epiphyte a plant that uses another plant for support, but not for nourishment (148)

erosion a process in which the materials of the Earth's surface are loosened, dissolved, or worn away and transported from one place to another by a natural agent, such as wind, water, ice, or gravity (66, 386)

estuary an area where fresh water from rivers mixes with salt water from the ocean; the part of a river where the tides meet the river current (179)

eutrophication an increase in the amount of nutrients, such as nitrates, in a marine or aquatic ecosystem (175)

evaporation the change of a substance from a liquid to a gas (73)

evolution a change in the characteristics of a population from one generation to the next; the gradual development of organisms from other organisms since the beginnings of life (97)

exotic species a species that is not native to a particular region (247)

experiment a procedure that is carried out under controlled conditions to discover, demonstrate, or test a fact, theory, or general truth (33)

experimental group in an experiment, a group that is identical to a control group except for one factor and that is compared with the control group (33)

exponential growth logarithmic growth, or growth in which numbers increase by a certain factor in each successive time period (199)

F

famine widespread malnutrition and starvation in an area due to a shortage of food, usually caused by a catastrophic event (379)

fertility rate the number of births (usually per year) per 1,000 women of childbearing age (usually 15 to 44) (221)

food chain the pathway of energy transfer through various stages as a result of the feeding patterns of a series of organisms (122)

food web a diagram that shows the feeding relationships between organisms in an ecosystem (122)

fossil fuel a nonrenewable energy resource formed from the remains of organisms that lived long ago; examples include oil, coal, and natural gas (435)

fresh water water that contains insignificant amounts of salts, as in rivers and lakes (79)

fuel cell a device that produces electricity chemically by combining hydrogen fuel with oxygen from the air (468)

fungus an organism whose cells have nuclei, rigid cell walls, and no chlorophyll and that belongs to the kingdom Fungi (103)

G

gene a segment of DNA that is located in a chromosome and that codes for a specific hereditary trait (242)

genetic engineering a technology in which the genome of a living cell is modified for medical or industrial use (393)

geographic information system an automated system for capturing, storing, retrieving, analyzing, manipulating, and displaying geographic data (abbreviation, GIS) (361)

geosphere the mostly solid, rocky part of the Earth; extends from the center of the core to the surface of the crust (59)

geothermal energy the energy produced by heat within the Earth (464)

germ plasm hereditary material (chromosomes and genes) that is usually contained in the protoplasm of germ cells (253)

global warming a gradual increase in the average global temperature that is due to a higher concentration of gases such as carbon dioxide in the atmosphere (341)

greenhouse effect the warming of the surface and lower atmosphere of Earth that occurs when carbon dioxide, water vapor, and other gases in the air absorb and reradiate infrared radiation (72)

greenhouse gas a gas composed of molecules that absorb and radiate infrared radiation from the sun (339)

groundwater the water that is beneath the Earth's surface (272)

growth rate an expression of the increase in the size of an organism or population over a given period of time (198)

gymnosperm (JIM noh SPURM) a woody vascular seed plant whose seeds are not enclosed by an ovary or fruit (105)

habitat the place where an organism usually lives (96)

habitat conservation plan a land-use plan that attempts to protect threatened or endangered species across a given area by allowing some tradeoffs between harm to the species and additional conservation commitments among cooperating parties (255)

Glossary

hazardous wastes wastes that are a risk to the health of humans or other living organisms (493)

heat island an area in which the air temperature is generally higher than the temperature of surrounding rural areas (360)

host an organism from which a parasite takes food and shelter (519)

hydroelectric energy electrical energy produced by falling water (463)

hypothesis (hie PATH uh sis) a theory or explanation that is based on observations and that can be tested (32)

infrastructure the basic facilities of a country or region, such as roads, bridges, and sewers (225, 359)

invertebrate (in VUHR tuh brit) an animal that does not have a backbone (106)

keystone species a species that is critical to the functioning of the ecosystem in which it lives because it affects the survival and abundance of many other species in its community (242)

Kyoto Protocol an international treaty according to which developed countries that signed the treaty agree to reduce their emissions of carbon dioxide and other gases that may contribute to global warming by 2012 (345)

L

landfill an area of land or an excavation where wastes are placed for permanent disposal (485)

land-use planning a set of policies and activities related to potential uses of land that is put in place before an area is developed (361)

La Niña (la NEEN yah) the cool phase of the El Niño–Southern Oscillation; a periodic occurrence in the eastern Pacific Ocean in which the surface-water temperature becomes unusually cool (332)

latitude the distance north or south from the equator; expressed in degrees (145, 328)

law of supply and demand a law of economics that states that as the demand for a good or service increases, the value of the good or service also increases (17)

leachate a liquid that has passed through solid waste and has extracted dissolved or suspended materials from that waste, such as pesticides in the soil (485)

least developed countries countries that have been identified by the United Nations as showing the fewest signs of development in terms of income, human resources, and economic diversification (228)

life expectancy the average length of time that an individual is expected to live (222)

limiting resource a particular natural resource that, when limited, determines the carrying capacity of an ecosystem for a particular species (201)

lithosphere the solid, outer layer of the Earth that consists of the crust and the rigid upper part of the mantle (61)

littoral zone a shallow zone in a freshwater habitat where light reaches the bottom and nurtures plants (174)

livestock domesticated animals that are raised to be used on a farm or ranch or to be sold for profit (398)

lobbying an attempt to influence the decisions of lawmakers (543)

M

malnutrition a disorder of nutrition that results when a person does not consume enough of each of the nutrients that are needed by the human body (379)

mangrove swamp a tropical or subtropical marine swamp that is characterized by the abundance of low to tall trees, especially mangrove trees (182)

mantle in Earth science, the layer of rock between the Earth's crust and core (61)

mathematical model one or more equations that represent the way a system or process works (44)

mean the number obtained by adding up the data for a given characteristic and dividing this sum by the number of individuals (39)

migration in general, any movement of individuals or populations from one location to another; specifically, a periodic group movement that is characteristic of a given population or species (221)

mineral a natural, usually inorganic solid that has a characteristic chemical composition, an orderly internal structure, and a characteristic set of physical properties (411)

model a pattern, plan, representation, or description designed to show the structure or workings of an object, system, or concept (42)

municipal solid waste waste produced by households and businesses (484)

mutualism a relationship between two species in which both species benefit (208)

N

natural resource any natural material that is used by humans, such as water, petroleum, minerals, forests, and animals (14)

natural selection the process by which individuals that have favorable variations and are better adapted to their environment survive and reproduce more successfully than less well adapted individuals do (97)

nekton all organisms that swim actively in open water, independent of currents (173)

niche (NICH) the unique position occupied by a species, both in terms of its physical use of its habitat and its function within an ecological community (203)

nitrogen cycle the process in which nitrogen circulates among the air, soil, water, plants, and animals in an ecosystem (126)

nitrogen-fixing bacteria bacteria that convert atmospheric nitrogen into ammonia (126)

nonpoint-source pollution pollution that comes from many sources rather than from a single specific site; an example is pollution that reaches a body of water from streets and storm sewers (285)

nuclear energy the energy released by a fission or fusion reaction; the binding energy of the atomic nucleus (444)

nuclear fission the splitting of the nucleus of a large atom into two or more fragments; releases additional neutrons and energy (444)

nuclear fusion the combination of the nuclei of small atoms to form a larger nucleus; releases energy (447)

O

observation the process of obtaining information by using the senses; the information obtained by using the senses (31)

ocean thermal energy conversion the use of temperature differences in ocean water to produce electricity (abbreviation, OTEC) (467)

oil reserves oil deposits that are discovered and are in commercial production (442)

open system a system that can exchange both matter and energy with its surroundings (81)

ore mineral a mineral that contains one or more elements of economic value (412)

organism a living thing; anything that can carry out life processes independently (95)

overgrazing the depletion of vegetation due to the continuous feeding of too many animals (364)

overharvesting catching or removing from a population more organisms than the population can replace (395)

ozone hole a thinning of stratospheric ozone that occurs over the poles during the spring (336)

ozone a gas molecule that is made up of three oxygen atoms (69)

ozone layer the layer of the atmosphere at an altitude of 15 to 40 km in which ozone absorbs ultraviolet solar radiation (335)

P

parasitism a relationship between two species in which one species, the parasite, benefits from the other species, the host, and usually harms the host (208)

particulates (pahr TIK yoo lits) fine particles that are suspended in the atmosphere and that are associated with air pollution (514)

passive solar heating the use of sunlight to heat buildings directly (458)

pathogen a virus, microorganism, or other substance that causes disease; an infectious agent (277, 519)

Glossary

permafrost in arctic regions, the permanently frozen layer of soil or subsoil (162)

permeability the ability of a rock or sediment to let fluids pass through its open spaces or pores (273)

pesticide a poison used to destroy pests, such as insects, rodents, or weeds; examples include insecticides, rodenticides, and herbicides (389)

petroleum a liquid mixture of complex hydrocarbon compounds; used widely as a fuel source (440)

pH a value that is used to express the acidity or alkalinity (basicity) of a system; each whole number on the scale indicates a tenfold change in acidity; a pH of 7 is neutral, a pH of less than 7 is acidic, and a pH of greater than 7 is basic (314)

phosphorus cycle the cyclic movement of phosphorus in different chemical forms from the environment to organisms and then back to the environment (127)

photosynthesis the process by which plants, algae, and some bacteria use sunlight, carbon dioxide, and water to produce carbohydrates and oxygen (117)

pioneer species a species that colonizes an uninhabited area and that starts an ecological cycle in which many other species become established (130)

placer deposit a deposit that contains a valuable mineral that has been concentrated by mechanical action (419)

plankton the mass of mostly microscopic organisms that float or drift freely in the waters of aquatic (freshwater and marine) environments (173)

poaching the illegal harvesting of fish, game, or other species (247)

point-source pollution pollution that comes from a specific site (285)

polar stratospheric cloud a cloud that forms at altitudes of about 21,000 m during the Arctic and Antarctic winter or early spring, when air temperatures drop below −80°C (336)

pollution an undesirable change in the natural environment that is caused by the introduction of substances that are harmful to living organisms or by excessive wastes, heat, noise, or radiation (14)

population a group of organisms of the same species that live in a specific geographical area and interbreed (96, 197)

porosity the percentage of the total volume of a rock or sediment that consists of open spaces (273)

potable suitable for drinking (277)

precipitation any form of water that falls to the Earth's surface from the clouds; includes rain, snow, sleet, and hail (73)

predation an interaction between two species in which one species, the predator, feeds on the other species, the prey (206)

prediction a statement made in advance that expresses the results that will be obtained from testing a hypothesis if the hypothesis is supported; the expected outcome if a hypothesis is accurate (32)

primary pollutant a pollutant that is put directly into the atmosphere by human or natural activity (303)

primary succession succession that begins in an area that previously did not support life (129)

probability the likelihood that a possible future event will occur in any given instance of the event; the mathematical ratio of the number of times one outcome of any event is likely to occur to the number of possible outcomes of the event (40)

producer an organism that can make organic molecules from inorganic molecules; a photosynthetic or chemosynthetic autotroph that serves as the basic food source in an ecosystem (118)

protist an organism that belongs to the kingdom Protista (104)

R

radiation the energy that is transferred as electromagnetic waves, such as visible light and infrared waves (70)

recharge zone an area in which water travels downward to become part of an aquifer (274)

reclamation the process of returning land to its original condition after mining is completed (424)

recycling the process of recovering valuable or useful materials from waste or scrap; the process of reusing some items (489)

reforestation the reestablishment and development of trees in a forest land (367)

renewable energy energy from sources that are constantly being formed (457)

reproductive potential the maximum number of offspring that a given organism can produce (199)

reservoir an artificial body of water that usually forms behind a dam (280)

resistance in biology, the ability of an organism to tolerate a chemical or disease-causing agent (101)

risk the probability of an unwanted outcome (41)

risk assessment the scientific assessment, study, and management of risk; a scientific estimation of the likelihood of negative effects that may result from exposure to a specific hazard (513)

river system a flowing network of rivers and streams draining a river basin (271)

ruminant (ROO muh nuhnt) a cud-chewing mammal that has a three- or four-chambered stomach; examples include sheep, goats, and cattle (398)

rural describes an area of open land that is often used for farming (355)

S

salinity a measure of the amount of dissolved salts in a given amount of liquid (76, 173)

salinization (SAL uh nie ZAY shuhn) the accumulation of salts in soil (388)

salt marsh a maritime habitat characterized by grasses, sedges, and other plants that have adapted to continual, periodic flooding; salt marshes are found primarily throughout the temperate and subarctic regions (182)

sample the group of individuals or events selected to represent a statistical population (40)

savanna a plain full of grasses and scattered trees and shrubs; found in tropical and subtropical habitats and mainly in regions with a dry climate, such as East Africa (155)

secondary pollutant a pollutant that forms in the atmosphere by chemical reaction with primary air pollutants, natural components in the air, or both (303)

secondary succession the process by which one community replaces another community that has been partially or totally destroyed (129)

sick-building syndrome a set of symptoms, such as headache, fatigue, eye irritation, and dizziness, that may affect workers in modern, airtight office buildings; believed to be caused by indoor pollutants (310)

smelting the melting or fusing of ore in order to separate impurities from pure metal (420)

smog urban air pollution composed of a mixture of smoke and fog produced from industrial pollutants and burning fuels (308)

solid waste a discarded solid material, such as garbage, refuse, or sludges (481)

source reduction any change in the design, manufacture, purchase, or use of materials or products to reduce their amount or toxicity before they become municipal solid waste; also the reuse of products or materials (488)

species a group of organisms that are closely related and can mate to produce fertile offspring; also the level of classification below genus and above subspecies (95)

statistics the collection and classification of data that are in the form of numbers (38)

stratosphere the layer of the atmosphere, that lies immediately above the troposphere and extends from about 10 to 50 km above the Earth's surface, in which temperature increases as altitude increases; contains the ozone layer (69)

subsidence the sinking of regions of the ground surface with little or no horizontal movement (423)

subsurface mining a mining method in which ore is extracted from beneath the ground surface (416)

surface impoundment a natural depression or a human-made excavation that serves as a disposal facility that holds an accumulation of wastes (496)

surface mining a mining method in which soil and rocks are removed to reach underlying coal or minerals (417)

surface water all the bodies of fresh water, salt water, ice, and snow that are found above the ground (270)

survivorship the percentage of newborn individuals in a population that can be expected to survive to a given age (220)

sustainability the condition in which human needs are met in such a way that a human population can survive indefinitely (21, 533)

symbiosis a relationship in which two different organisms live in close association with each other (209)

T

taiga a region of evergreen, coniferous forest below the arctic and subarctic tundra regions (153)

tectonic plate a block of lithosphere that consists of the crust and the rigid, outermost part of the mantle; also called lithospheric plate (62)

temperate deciduous forest a forest (or biome) that is characterized by trees that shed their leaves in the fall (152)

Glossary

temperate grassland a community (or biome) that is dominated by grasses, has few trees, and is characterized by cold winters and rainfall that is intermediate between that of a forest and a desert (156)

temperate rain forest a forest community (or biome), characterized by cool, humid weather and abundant rainfall, where tree branches are draped with mosses, tree trunks are covered with lichens, and the forest floor is covered with ferns (151)

temperature inversion the atmospheric condition in which warm air traps cooler air near Earth's surface (308)

thermal pollution a temperature increase in a body of water that is caused by human activity and that has a harmful effect on water quality and on the ability of that body of water to support life (289)

threatened species a species that has been identified to be likely to become endangered in the foreseeable future (245)

topsoil the surface layer of the soil, which is usually richer in organic matter than the subsoil is (385)

toxicology the study of toxic substances, including their nature, effects, detection, methods of treatment, and exposure control (512)

trophic level one of the steps in a food chain or food pyramid; examples include producers and primary, secondary, and tertiary consumers (122)

tropical rain forest a forest or jungle near the equator that is characterized by large amounts of rain and little variation in temperature and that contains the greatest known diversity of organisms on Earth (146)

troposphere the lowest layer of the atmosphere, in which temperature drops at a constant rate as altitude increases; the part of the atmosphere where weather conditions exist (68)

tundra a treeless plain that is located in the Arctic or Antarctic and that is characterized by very low winter temperatures; short, cool summers; and vegetation that consists of grasses, lichens, and perennial herbs (162)

understory a foliage layer that is beneath and shaded by the main canopy of a forest (148)

urban describes an area that contains a city (355)

urbanization an increase in the ratio or density of people living in urban areas rather than in rural areas (227, 358)

urban sprawl the rapid spread of a city into adjoining suburbs and rural areas (359)

value a principle or standard that an individual considers to be important (45)

variable (VER ee uh buhl) a factor that changes in an experiment in order to test a hypothesis (33)

vector in biology, any agent, such as a plasmid or a virus, that can incorporate foreign DNA and transfer that DNA from one organism to another; an intermediate host that transfers a pathogen or a parasite to another organism (520)

vertebrate an animal that has a backbone; includes mammals, birds, reptiles, amphibians, and fish (107)

wastewater water that contains wastes from homes or industry (286)

water cycle the continuous movement of water from the ocean to the atmosphere to the land and back to the ocean (73)

water pollution the introduction into water of waste matter or chemicals that are harmful to organisms living in the water or to those that drink or are exposed to the water (284)

watershed the area of land that is drained by a water system (271)

weather the short-term state of the atmosphere, including temperature, humidity, precipitation, wind, and visibility (327)

wetland an area of land that is periodically underwater or whose soil contains a great deal of moisture (173)

wilderness a region that is not cultivated and that is not inhabited by humans (368)

yield the amount of crops produced per unit area (381)

A

abiotic factor/factor abiótico un factor ambiental que no está asociado con las actividades de l... seres vivos (94)

acid precipitation/precipitació... ...ida pr...e o nie..., que cipitación tal como lluvia, agu... ácido...debido contiene una alta concentra... (314)

...acidez entrada a la contaminación de la a... e agua muy ácida a

acid shock/cambio brus... ieve se derrite en la súbita de grandes can... abundancia después los lagos y arroyos c... primavera o cuand... de una sequía (3...

...amiento solar activo ...olar por medio de

active solar hea... ra calentar agua o un la recopilació... colectores ... el proceso de adaptarse a edificio (...io anatómico, fisiológico o ...ejora la capacidad de super-

adaptati...ación (99)

un a...tura de edades la en ...pos de los miembros de una ...vión de su edad, o bien, la dis- ...niembros de una población en ...os de edad (220)

...icultura cultivar cosechas y criar ...sarlos como alimento o para pro-tos útiles para los seres humanos (10)

/contaminación del aire contami-...a atmósfera con desechos de fuentes o combustión industrial o escapes de ...iles (303)

...tive energy/energía alternativa energía ...que no proviene de los combustibles fósiles y que todavía se encuentra en desarrollo (466)

altitude/altitud la altura de un objeto sobre un punto de referencia, tal como el nivel del mar o la superficie de la Tierra (145)

angiosperm/angiosperma una planta que da flores y que produce semillas dentro de la fruta (105)

aquaculture/acuacultura el cultivo de plantas y animales acuáticos para uso o consumo humano (396)

aquifer/acuífero un cuerpo rocoso o sedimento que almacena agua subterránea y permite que fluya (272)

arable land/tierra arable tierra agrícola que puede usarse para cosechar cultivos (227, 384)

artificial eutrophication/eutrificación artificial un proceso que aumenta la cantidad de nutrientes en una masa de agua debido a actividades humanas, tales como el desecho de residuos y el drenaje de la tierra (288)

artificial selection/selección artificial la repro-ducción selectiva de organismos (por los seres humanos) para obtener características específicas deseables (100)

asbestos/asbesto cualquiera de seis minerales de silicato que forman montones de fibras diminutas que son resistentes al calor, flexibles y resistentes (312)

asthenosphere/astenosfera la capa sólida y plás-tica del manto, que se encuentra debajo de la lito-sfera; está formada por roca del manto que fluye muy lentamente, lo cual permite que las placas tectónicas se muevan en su superficie (61)

atmosphere/atmósfera una mezcla de gases que rodea un planeta, tal como la Tierra (67)

B

bacteria/bacterias organismos extremadamente pequeños, unicelulares, que normalmente tienen pared celular y se reproducen por división celular (102)

barrier islands/isla barrera un largo arrecife de arena o una isla angosta ubicada paralela a la costa (182)

benthic zone/zona béntica la región que se encuentra cerca del fondo de una laguna, lago u océano (174)

benthos/benthos los organismos que viven en el fondo del mar o del océano (173)

biodegradable material/material biodegradable un material que puede descomponerse por medio de procesos biológicos (483)

biodiversity/biodiversidad la variedad de organ-ismos que se encuentran en un área determinada, la variación genética dentro de una población, la variedad de especies en una comunidad o la var-iedad de comunidades en un ecosistema (15, 241)

Spanish Glossary

Biodiversity Treaty/Tratado de la Biodiversidad un acuerdo internacional cuyo objetivo es fortalecer el control y conservación nacional de los recursos biológicos; asociado con la Conferencia de las Naciones Unidas sobre el Medio Ambiente y el Desarrollo (UNCED o Cumbre de la Tierra) en 1992 (257)

biological pest control/control biológico de plagas el uso de ciertos organismos por parte de los seres humanos para eliminar o controlar plagas (391)

biomagnification/bioaumento la acumulación de contaminantes en niveles sucesivos de la cadena alimenticia (292)

biomass fuel/combustible de biomasa material vegetal, abono o cualquier otra materia orgánica que se use como fuente de energía (462)

biome/bioma una región extensa caracterizada por un tipo de clima específico y ciertos tipos de comunidades de plantas y animales (143)

biosphere/biosfera la parte de la Tierra donde existe la vida (80)

biotic factor/factor biótico un factor ambiental que está asociado con las actividades de los seres vivos o que resulta de ellas (94)

C

canopy/dosel vegetal las capas de las copas de los árboles que dan sombra al suelo del bosque (148)

carbon cycle/ciclo del carbono el movimiento del carbono del ambiente sin vida a los seres vivos y de los seres vivos al ambiente (124)

carrying capacity/capacidad de carga la población más grande que un ambiente puede mantener en cualquier momento dado (200)

cellular respiration/respiración celular el proceso por medio del cual las células producen energía a partir de los carbohidratos; el oxígeno atmosférico se combina con la glucosa para formar agua y dióxido de carbono (120)

chaparral/chaparral un tipo de vegetación que incluye arbustos de hoja perenne y ancha, y que se encuentra en áreas donde los veranos son calientes y secos, y los inviernos son templados y húmedos

chlorofluorocarbons/clorofluorocarbonos hidrocarburos en los que algunos o todos los átomos de hidrógeno son reemplazados por cloro y flúor; se usan líquidos refrigerantes para refrigeradores y aireacondicionados y en solventes para limpieza; su uso está restringido porque destruyen las moléculas de ozido de la estratosfera (abreviatura: CFC) (335)

climate/clima las condiciones promedio del tiempo en un área durante un largo período de tiempo (144, 327)

climax community/comunidad clímax una comunidad final y estable en equilibrio con el ambiente (130)

closed system/sistema cerrado una que no puede intercambiar materia con el medio que lo rodea (81)

commensalism/comensalismo una relación entre dos organismos en la que uno se beneficia y el otro no es afectado (209)

community/comunidad un grupo de diferentes especies que viven en el mismo hábitat e interactúan unas con otras (96)

competition/competencia la relación entre dos especies (o individuos) en la que ambas especies (o individuos) intentan usar el mismo recurso limitado, de modo que ambas resultan afectadas negativamente por la relación (204)

compost/composta una mezcla de materia orgánica en descomposición, como por ejemplo estiércol y plantas en estado de putrefacción, que se usa como fertilizante y acondicionador del suelo (490)

conceptual model/modelo conceptual una explicación verbal o gráfica acerca de cómo funciona o está organizado un sistema (43)

condensation/condensación el cambio de estado de gas a líquido (73)

conduction/conducción la transferencia de energía en forma de calor a través de un material (70)

consumer/consumidor un organismo que se alimenta de otros organismos o de materia orgánica, en lugar de producir sus propios nutrientes o de obtenerlos de fuentes inorgánicas (118)

control group/grupo de control en un experimento, un grupo que sirve como estándar de comparación con otro grupo, al cual el grupo de control es idéntico excepto por un factor (33)

Spanish Glossary

A

abiotic factor/factor abiótic un factor ambiental que no está asociado con las actividades de los seres vivos (94)

acid precipitation/precipitación ácida precipitación tal como lluvia, aguanieve o nieve, que contiene una alta concentración de ácidos debido a la contaminación de la atmósfera (314)

acid shock/cambio brusco de la acidez entrada súbita de grandes cantidades de agua muy ácida a los lagos y arroyos cuando la nieve se derrite en la primavera o cuando llueve en abundancia después de una sequía (316)

active solar heating/calentamiento solar activo la recopilación de energía solar por medio de colectores que se usan para calentar agua o un edificio (460)

adaptation/adaptación el proceso de adaptarse a un ambiente; un cambio anatómico, fisiológico o en la conducta que mejora la capacidad de supervivencia de una población (99)

age structure/estructura de edades la clasificación en grupos de los miembros de una población en función de su edad, o bien, la distribución de los miembros de una población en función de grupos de edad (220)

agriculture/agricultura cultivar cosechas y criar ganado para usarlos como alimento o para producir productos útiles para los seres humanos (10)

air pollution/contaminación del aire contaminación de la atmósfera con desechos de fuentes tales como combustión industrial o escapes de automóviles (303)

alternative energy/energía alternativa energía que no proviene de los combustibles fósiles y que todavía se encuentra en desarrollo (466)

altitude/altitud la altura de un objeto sobre un punto de referencia, tal como el nivel del mar o la superficie de la Tierra (145)

angiosperm/angiosperma una planta que da flores y que produce semillas dentro de la fruta (105)

aquaculture/acuacultura el cultivo de plantas y animales acuáticos para uso o consumo humano (396)

aquifer/acuífero un cuerpo rocoso o sedimento que almacena agua subterránea y permite que fluya (272)

arable land/tierra arable tierra agrícola que puede usarse para cosechar cultivos (227, 384)

artificial eutrophication/eutrificación artificial un proceso que aumenta la cantidad de nutrientes en una masa de agua debido a actividades humanas, tales como el desecho de residuos y el drenaje de la tierra (288)

artificial selection/selección artificial la reproducción selectiva de organismos (por los seres humanos) para obtener características específicas deseables (100)

asbestos/asbesto cualquiera de seis minerales de silicato que forman montones de fibras diminutas que son resistentes al calor, flexibles y resistentes (312)

asthenosphere/astenosfera la capa sólida y plástica del manto, que se encuentra debajo de la litosfera; está formada por roca del manto que fluye muy lentamente, lo cual permite que las placas tectónicas se muevan en su superficie (61)

atmosphere/atmósfera una mezcla de gases que rodea un planeta, tal como la Tierra (67)

B

bacteria/bacterias organismos extremadamente pequeños, unicelulares, que normalmente tienen pared celular y se reproducen por división celular (102)

barrier islands/isla barrera un largo arrecife de arena o una isla angosta ubicada paralela a la costa (182)

benthic zone/zona béntica la región que se encuentra cerca del fondo de una laguna, lago u océano (174)

benthos/benthos los organismos que viven en el fondo del mar o del océano (173)

biodegradable material/material biodegradable un material que puede descomponerse por medio de procesos biológicos (483)

biodiversity/biodiversidad la variedad de organismos que se encuentran en un área determinada, la variación genética dentro de una población, la variedad de especies en una comunidad o la variedad de comunidades en un ecosistema (15, 241)

Spanish Glossary

Biodiversity Treaty/Tratado de la Biodiversidad un acuerdo internacional cuyo objetivo es fortalecer el control y conservación nacional de los recursos biológicos; asociado con la Conferencia de las Naciones Unidas sobre el Medio Ambiente y el Desarrollo (UNCED o Cumbre de la Tierra) en 1992 (257)

biological pest control/control biológico de plagas el uso de ciertos organismos por parte de los seres humanos para eliminar o controlar plagas (391)

biomagnification/bioaumento la acumulación de contaminantes en niveles sucesivos de la cadena alimenticia (292)

biomass fuel/combustible de biomasa material vegetal, abono o cualquier otra materia orgánica que se use como fuente de energía (462)

biome/bioma una región extensa caracterizada por un tipo de clima específico y ciertos tipos de comunidades de plantas y animales (143)

biosphere/biosfera la parte de la Tierra donde existe la vida (80)

biotic factor/factor biótico un factor ambiental que está asociado con las actividades de los seres vivos o que resulta de ellas (94)

C

canopy/dosel vegetal las capas de las copas de los árboles que dan sombra al suelo del bosque (148)

carbon cycle/ciclo del carbono el movimiento del carbono del ambiente sin vida a los seres vivos y de los seres vivos al ambiente (124)

carrying capacity/capacidad de carga la población más grande que un ambiente puede sostener en cualquier momento dado (200)

cellular respiration/respiración celular el proceso por medio del cual las células producen energía a partir de los carbohidratos; el oxígeno atmosférico se combina con la glucosa para formar agua y dióxido de carbono (120)

chaparral/chaparral un tipo de vegetación que incluye arbustos de hoja perenne y ancha, y que se ubica en áreas donde los veranos son calientes y secos y los inviernos son templados y húmedos (158)

chlorofluorocarbons/clorofluorocarbonos hidrocarburos en los que algunos o todos los átomos de hidrógeno son reemplazados por cloro y flúor; se usan en líquidos refrigerantes para refrigeradores y aires acondicionados y en solventes para limpieza; su uso está restringido porque destruyen las moléculas de ozono de la estratosfera (abreviatura: CFC) (335)

climate/clima las condiciones promedio del tiempo en un área durante un largo período de tiempo (144, 327)

climax community/comunidad clímax una comunidad final y estable, que está en equilibrio con el ambiente (130)

closed system/sistema cerrado un sistema que no puede intercambiar materia ni energía con el medio que lo rodea (81)

commensalism/comensalismo una relación entre dos organismos en la que uno se beneficia y el otro no es afectado (209)

community/comunidad un grupo de varias especies que viven en el mismo hábitat e interactúan unas con otras (96)

competition/competencia la relación entre dos especies (o individuos) en la que ambas especies (o individuos) intentan usar el mismo recurso limitado, de modo que ambas resultan afectadas negativamente por la relación (204)

compost/composta una mezcla de materia orgánica en descomposición, como por ejemplo, estiércol y plantas en estado de putrefacción, que se usa como fertilizante y acondicionador del suelo (490)

conceptual model/modelo conceptual una explicación verbal o gráfica acerca de cómo funciona o está organizado un sistema (43)

condensation/condensación el cambio de estado de gas a líquido (73)

conduction/conducción la transferencia de energía en forma de calor a través de un material (70)

consumer/consumidor un organismo que se alimenta de otros organismos o de materia orgánica, en lugar de producir sus propios nutrientes o de obtenerlos de fuentes inorgánicas (118)

control group/grupo de control en un experimento, un grupo que sirve como estándar de comparación con otro grupo, al cual el grupo de control es idéntico excepto por un factor (33)

convection/convección el movimiento de la materia debido a diferencias en la densidad que se producen por variaciones en la temperatura; puede resultar en la transferencia de energía en forma de calor (70)

coral reef/arrecife de coral una cumbre de piedra caliza ubicada en climas tropicales, formada por fragmentos de coral depositados alrededor de restos orgánicos (183)

core/núcleo la parte central de la Tierra, debajo del manto; *también,* el centro del Sol (61)

correlation/correlación la dependencia linear entre dos variables (35)

crust/corteza la capa externa, delgada y sólida de la Tierra, que se encuentra sobre el manto (60)

D

dam/presa una estructura que se construye a través de un río para controlar el flujo del río (280)

data/datos cualquier parte de la información que se adquiere por medio de la observación o experimentación (34)

decibel/decibel la unidad más común que se usa para medir el volumen del sonido (abreviatura: dB) (312)

decision-making model/modelo de toma de decisiones un modelo conceptual que brinda un proceso sistemático para tomar decisiones (45)

decomposer/descomponedor un organismo que desintegra la materia orgánica de organismos muertos y se alimenta de ella; entre los ejemplos se encuentran las bacterias y los hongos (119)

deep-well injection/inyección a pozo profundo método de desecho de residuos peligrosos por inyección a pozo (496)

deforestation/deforestación el proceso de talar bosques (366)

demographic transition/transición demográfica el patrón general de cambio demográfico de tasas de nacimiento y mortalidad altas a tasas de nacimiento y mortalidad bajas, tal como se observa en la historia de los países más desarrollados (223)

demography/demografía el estudio de las características de las poblaciones, sobre todo las poblaciones humanas (219)

density/densidad el número de individuos de la misma especie que viven en una unidad superficial determinada (198)

desalination/desalación (o desalinización) un proceso de remoción de sal del agua del océano (283)

desert/desierto una región con poca vegetación o sin vegetación, largos períodos sin lluvia y temperaturas extremas; generalmente se ubica en climas calientes (160)

desertification/desertificación el proceso por medio del cual las actividades humanas o los cambios climáticos hacen que un área árida o semiárida se vuelva más parecida a un desierto (386)

diet/dieta el tipo y cantidad de alimento que come una persona (380)

dispersion/dispersión en ecología, el patrón de distribución de organismos en una población (198)

distribution/distribución la organización relativa de los miembros de una población estadística; normalmente se muestra en una gráfica (39)

domesticated/domesticado término que describe a organismos que han sido reproducidos y criados para uso humano (395)

dose/dosis la cantidad de medicina que se necesita tomar durante un período de tiempo; *también,* la cantidad de una sustancia dañina a la que está expuesta una persona (512)

dose-response curve/curva de dosis-respuesta una gráfica que muestra el efecto relativo de varias dosis de un medicamento o substancia química en un organismo u organismos (512)

E

ecological footprint/huella ecológica un cálculo que muestra el área productiva de la Tierra que se requiere para mantener a una persona en un cierto país (19)

ecological succession/sucesión ecológica un proceso gradual de cambio y sustitución en una comunidad (129)

ecology/ecología el estudio de las interacciones de los seres vivos entre sí mismos y entre sí mismos y su ambiente (6)

economics/economía el estudio de cómo los individuos y grupos toman decisiones acerca de la producción, distribución y consumo de recursos limitados, al mismo tiempo que estos individuos o grupos intentan satisfacer sus necesidades y deseos (536)

ecosystem/ecosistema una comunidad de organismos y su ambiente abiótico (93)

ecosystem services/servicios del ecosistema el papel que juegan los organismos en la creación de un ambiente saludable para los seres humanos (357)

ecotourism/ecoturismo una forma de turismo que apoya la conservación y desarrollo sustentable de áreas ecológicamente únicas (244)

electric generator/descarga eléctrica la liberación de electricidad almacenada en una fuente (332)

El Niño/El Niño la fase caliente de la Oscilación Sureña "El Niño"; un fenómeno periódico que ocurre en el océano Pacífico oriental en el que la temperatura del agua superficial se vuelve más caliente que de costumbre (436)

emergent layer/capa emergente la capa superior de follaje en un bosque, en la que los árboles se extienden sobre los árboles circundantes (148)

endangered species/especie en peligro de extinción una especie que se ha identificado como en peligro de extinción en toda su zona de distribución o en una parte importante de ella, y que, por lo tanto, se encuentra protegida por normas y medidas de conservación (245)

Endangered Species Act/Ley de Especies en Peligro de Extinción una ley que el Congreso de los Estados Unidos emitió en 1973 cuyo fin es proteger las especies de animales o plantes que están en peligro de extinguirse (255)

endemic species/especie endémica una especie que es nativa de un lugar particular y que únicamente se encuentra allí (248)

energy conservation/conservación de energía el proceso de ahorrar energía al reducir el uso y el gasto inútil de energía (470)

energy efficiency/eficiencia energética el porcentaje de energía que se pone en un sistema que realiza un trabajo útil (468)

Environmental Impact Statement/Evaluación del Impacto Ambiental una evaluación del efecto que una propuesta de proyecto o ley tendrá en el ambiente (541)

environmental science/ciencias ambientales el estudio del aire, agua y tierra circundantes en relación con un organismo o comunidad, desde un área pequeña de la Tierra hasta la biosfera completa; incluye el estudio del impacto que los seres humanos tienen en el ambiente (5)

epidemiology/epidemiología el estudio de la distribución de las enfermedades en poblaciones y el estudio de los factores que influyen en la incidencia y propagación de las enfermedades (513)

epiphyte/epifita una planta que utiliza otra planta para sostenerse pero no para alimentarse (148)

erosion/erosión un proceso por medio del cual los materiales de la superficie de la Tierra se aflojan, disuelven o desgastan y son transportados de un lugar a otro por un agente natural, como el viento, el agua, el hielo o la gravedad (66, 386)

estuary/estuario un área donde el agua dulce de los ríos se mezcla con el agua salada del océano; la parte de un río donde las mareas se encuentran con la corriente del río (179)

eutrophication/eutrofización un aumento en la cantidad de nutrientes, tales como nitratos, en un ecosistema marino o acuático (175)

evaporation/evaporación el cambio de una substancia de líquido a gas (73)

evolution/evolución un cambio en las características de una población de una generación a la siguiente; el desarrollo gradual de organismos a partir de otros organismos desde los inicios de la vida (97)

exotic species/especie exótica una especie que no es originaria de una región en particular (247)

experiment/experimento un procedimiento que se lleva a cabo bajo condiciones controladas para descubrir, demostrar o probar un hecho, teoría o verdad general (33)

experimental group/grupo experimental en un experimento, un grupo que es idéntico al grupo de control, excepto por un factor, y que es comparado con el grupo de control (33)

exponential growth/crecimiento exponencial crecimiento logarítmico o crecimiento en el que los números aumentan en función de un cierto factor en cada período de tiempo sucesivo (199)

famine/hambruna desnutrición e inanición generalizadas en un área debido a una escasez de alimento, normalmente causada por un suceso catastrófico (379)

fertility rate/tasa de fertilidad el número de nacimientos (normalmente por año) por cada 1,000 mujeres en edad de procrear (normalmente entre los 15 y 44 años de edad) (221)

food chain/cadena alimenticia la vía de transferencia de energía través de varias etapas, que ocurre como resultado de los patrones de alimentación de una serie de organismos (122)

food web/red alimenticia un diagrama que muestra las relaciones de alimentación entre los organismos de un ecosistema (122)

fossil fuel/combustible fósil un recurso energético no renovable formado a partir de los restos de organismos que vivieron hace mucho tiempo; algunos ejemplos incluyen el petróleo, el carbón y el gas natural (435)

fresh water/agua dulce agua que contiene una cantidad insignificante de sales, como el agua de los ríos y lagos (79)

fuel cell/pila de combustible un aparato que produce electricidad químicamente al combinar combustible de hidrógeno con oxígeno del aire (468)

fungus/hongo un organismo que tiene células con núcleos y pared celular rígida, pero carece de clorofila, perteneciente al reino Fungi (103)

G

gene/gene un segmento de ADN ubicado en un cromosoma, que codifica para un carácter hereditario específico (242)

genetic engineering/ingeniería genética una tecnología en la que el genoma de una célula viva se modifica con fines médicos o industriales (393)

geographic information system/sistema de información geográfica un sistema automatizado que sirve para capturar, almacenar, obtener, analizar, manipular y mostrar datos geográficos (abreviatura: SIG) (361)

geosphere/geosfera la parte principalmente sólida y rocosa de la Tierra; se extiende del centro del núcleo a la superficie de la corteza (59)

geothermal energy/energía geotérmica la energía producida por el calor del interior de la Tierra (464)

germ plasm/plasma germinal material hereditario (cromosomas y genes) que normalmente se encuentra contenido en el protoplasma de las células germinales (253)

global warming/calentamiento global un aumento gradual en las temperaturas globales promedio debido a una concentración más alta de gases (tales como dióxido de carbono) en la atmósfera (341)

greenhouse effect/efecto de invernadero el calentamiento de la superficie terrestre y de la parte más baja de la atmósfera, el cual se produce cuando el dióxido de carbono, el vapor de agua y otros gases del aire absorben radiación infrarroja y la vuelven a irradiar (72)

greenhouse gases/gas de invernadero un gas compuesto de moléculas que absorben radiación infrarroja del Sol y la vuelven a irradiar (339)

groundwater/agua subterránea el agua que está debajo de la superficie de la Tierra (272)

growth rate/tasa de crecimiento una expresión del aumento en el tamaño de un organismo o población a lo largo de un cierto período de tiempo (198)

gymnosperm/gimnosperma una planta leñosa y vascular, la cual produce semillas que no están contenidas en un ovario o fruto (105)

H

habitat/hábitat el lugar donde un organismo vive normalmente (96)

habitat conservation plan/plan de conservación del hábitat un plan de uso de la tierra que tiene como objetivo proteger a las especies amenazadas o en peligro de extinción en un área determinada, permitiendo algunas compensaciones entre el daño a las especies y compromisos adicionales de conservación entre las partes en cooperación (255)

hazardous wastes/residuos peligrosos residuos que son un riesgo para la salud de los seres humano y otros seres vivos (493)

heat island/isla de calor un área en la que la temperatura del aire es generalmente más alta que la temperatura de las áreas rurales circundantes (360)

host/huésped el organismo del cual un parásito obtiene alimento y refugio (519)

hydroelectric energy/energía hidroeléctrica energía eléctrica producida por agua en caída (463)

hypothesis/hipótesis una teoría o explicación basada en observaciones y que se puede probar (32)

Spanish Glossary

infrastructure/infraestructura los servicios básicos de un país o región, tales como caminos, puentes y drenaje (225, 359)

invertebrate/invertebrado un animal que no tiene columna vertebral (106)

keystone species/especie clave una especie que es crítica para el funcionamiento del ecosistema en el que vive porque afecta la supervivencia y abundancia de muchas otras especies en su comunidad (242)

Kyoto Protocol/Protocolo de Kyoto un tratado internacional en función del cual los países desarrollados que lo firmaron acceden a reducir sus emisiones de dióxido de carbono y otros gases que pueden contribuir al calentamiento global para el año 2012 (345)

landfill/entierro de residuos un área de terreno o una excavación donde se colocan los residuos para deshacerse de ellos permanentemente (485)

land-use planning/planeación del uso de tierras un conjunto de políticas y actividades relacionadas con los usos potenciales de la tierra, que se establecen antes de desarrollar un área (361)

La Niña/La Niña la fase fría de la Oscilación Sureña "El Niño"; un fenómeno periódico que ocurre en el océano Pacífico oriental en el que la temperatura del agua superficial se vuelve más fría que de costumbre (332)

latitude/latitud la distancia hacia el norte o hacia el sur del ecuador; se expresa en grados (145, 328)

law of supply and demand/ley de la oferta y la demanda una ley de economía que establece que al aumentar la demanda de un bien o servicio, el valor del bien o servicio también aumenta (17)

leachate/lechado un líquido que ha pasado a través de desechos sólidos y ha extraído materiales disueltos o suspendidos de los desechos, como por ejemplo, pesticidas en el suelo (485)

least developed countries/países menos desarrollados países que la Organización de las Naciones Unidas ha identificado como los que muestran las menores señales de desarrollo en términos de ingresos, recursos humanos y diversificación económica (228)

life expectancy/esperanza de vida la longitud promedio de tiempo que se espera que un individuo viva (222)

limiting resources/recursos limitantes un recurso natural particular que, si está limitado, determina la capacidad de carga de un ecosistema para una especie en particular (201)

lithosphere/litosfera la capa externa y sólida de la Tierra que está formada por la corteza y la parte superior y rígida del manto (61)

littoral zone/zona litoral una zona poco profunda del hábitat de agua dulce donde la luz llega al fondo y nutre a las plantas (174)

livestock/animales de cría animales domesticados que se crían para usarse en una granja o rancho o para ser vendidos con el fin de obtener una ganancia (398)

lobbying/cabildeo un intento de ejercer una influencia en las decisiones de los legisladores (543)

malnutrition/desnutrición un trastorno de nutrición que resulta cuando una persona no consume una cantidad suficiente de cada nutriente que el cuerpo humano necesita (379)

mangrove swamp/manglar un pantano marino tropical o subtropical que se caracteriza por la abundancia de árboles bajos a altos, especialmente árboles de mangle (182)

mantle/manto en las ciencias de la Tierra, la capa de roca que se encuentra entre la corteza terrestre y el núcleo (61)

mathematical model/modelo matemático una o más ecuaciones que representan la forma en que funciona un sistema o proceso (44)

mean/media el número que se obtiene al sumar los datos de una característica determinada y dividir esta suma entre el número de individuos (39)

migration/migración en general, cualquier movimiento de individuos o poblaciones de un lugar a otro; específicamente, un movimiento periódico en grupo que es característico de una población o especie determinada (221)

mineral/mineral un sólido natural, normalmente inorgánico, que tiene una composición química característica, una estructura interna ordenada y propiedades físicas y químicas características (411)

model/modelo un diseño, plan, representación o descripción cuyo objetivo es mostrar la estructura o funcionamiento de un objeto, sistema o concepto (42)

municipal solid waste/desechos sólidos municipales desechos producidos por las casas y negocios (484)

mutualism/mutualismo una relación entre dos especies en la que ambas se benefician (208)

N

natural resource/recurso natural cualquier material natural que es utilizado por los seres humanos, como agua, petróleo, minerales, bosques y animales (14)

natural selection/selección natural el proceso por medio del cual los individuos que tienen condiciones favorables y que están mejor adaptados a su ambiente sobreviven y se reproducen con más éxito que los individuos que no están tan bien adaptados (97)

nekton/necton todos los organismos que nadan activamente en las aguas abiertas, de manera independiente de las corrientes (173)

niche/nicho la posición única que ocupa una especie, tanto en lo que se refiere al uso de su hábitat como en cuanto a su función dentro de una comunidad ecológica (203)

nitrogen cycle/ciclo del nitrógeno el proceso por medio del cual el nitrógeno circula en el aire, suelo, agua, plantas y animales de un ecosistema (126)

nitrogen-fixing bacteria/bacterias fijadoras de nitrógeno bacterias que transforman el nitrógeno atmosférico en amoniaco (126)

nonpoint-source pollution/contaminación no puntual contaminación que proviene de muchas fuentes, en lugar de provenir de un solo sitio específico; un ejemplo es la contaminación que llega a una masa de agua a partir de las calles y los drenajes (285)

nuclear energy/energía nuclear la energía liberada por una reacción de fisión o fusión; la energía de enlace del núcleo atómico (444)

nuclear fission/fisión nuclear la partición del núcleo de un átomo grande en dos o más fragmentos; libera neutrones y energía adicionales (444)

nuclear fusion/fusión nuclear combinación de los núcleos de átomos pequeños para formar un núcleo más grande; libera energía (447)

O

observation/observación el proceso de obtener información por medio de los sentidos; la información que se obtiene al usar los sentidos (31)

ocean thermal energy conversion/conversión de la energía térmica del océano el uso de diferencias en la temperatura del agua del océano para producir electricidad (abreviatura: OTEC, por sus siglas en inglés) (467)

oil reserves/reservas de petróleo depósitos de petróleo que son descubiertos y se encuentran en producción comercial (442)

open system/sistema abierto un sistema que puede intercambiar tanto materia como energía con el medio que lo rodea (81)

ore mineral/mineral metalífero un mineral que contiene uno o más elementos de valor económico (412)

organism/organismo un ser vivo; cualquier cosa que pueda llevar a cabo procesos vitales independientemente (95)

overgrazing/sobrepastoreo el agotamiento de la vegetación debido a la alimentación continua de demasiados animales (364)

overharvesting/sobrecosechar capturar o sustraer de una población más organismos de los que la población puede reemplazar (395)

ozone/ozono una molécula de gas que está formada por tres átomos de oxígeno (69)

ozone hole/agujero en la capa de ozono un adelgazamiento del ozono estratosférico, el cual occure encima de los Polos durante la primavera (336)

ozone layer/capa de ozono la capa de la atmósfera ubicada a una altitud de 15 a 40 km, en la cual el ozono absorbe la radiación solar (335)

Spanish Glossary

parasitism/parasitismo una relación entre dos especies en la que una, el parásito, se beneficia de la otra, el huésped, y normalmente lo daña (208)

particulates/materia particulada partículas finas que se encuentran suspendidas en la atmósfera y que están relacionadas con la contaminación del aire (514)

passive solar heating/calentamiento solar pasivo el uso de la luz solar para calentar edificios directamente (458)

pathogen/patógeno un virus, microorganismo u otra substancia que causa enfermedades; un agente infeccioso (277, 519)

permafrost/permafrost en las regiones árticas, la capa de suelo o subsuelo que se encuentra congelada permanentemente (162)

permeability/permeabilidad la capacidad de una roca o sedimento de permitir que los fluidos pasen a través de sus espacios abiertos o poros (273)

pesticide/pesticida un veneno que se usa para destruir plagas, tales como insectos, roedores o maleza; entre los ejemplos se encuentran los insecticidas, rodenticidas y herbicidas (389)

petroleum/petróleo una mezcla líquida de compuestos hidrocarburos complejos; se usa ampliamente como una fuente de combustible (440)

pH/pH un valor que expresa la acidez o la alcalinidad (basicidad) de un sistema; cada número entero de la escala indica un cambio de 10 veces en la acidez; un pH de 7 es neutro, un pH de menos de 7 es ácido y un pH de más de 7 es básico (314)

phosphorous cycle/ciclo del fósforo el movimiento cíclico del fósforo en diferentes formas químicas del ambiente a los organismos y de regreso al ambiente (127)

photosynthesis/fotosíntesis el proceso por medio del cual las plantas, algas y algunas bacterias utilizan la luz solar, dióxido de carbono y agua para producir carbohidratos y oxígeno (117)

pioneer species/especie pionera una especie que coloniza un área deshabitada y empieza un ciclo ecológico en el cual se establecen muchas otras especies (130)

placer deposit/yacimiento de aluvión un yacimiento que contiene un mineral valioso que se ha concentrado debido a la acción mecánica (419)

plankton/plancton la masa de organismos casi microscópicos que flotan o se encuentran a la deriva en aguas (dulces y marinas) de ambientes acuáticos (173)

poaching/caza furtiva la cosecha ilegal de peces, presas u otras especies (247)

point-source pollution/contaminación puntual contaminación que proviene de un lugar específico (285)

polar stratospheric cloud/nube polar estratosférica una nube que se forma en altitudes de aproximadamente 21,000 m durante el invierno ártico y antártico o al principio de la primavera, cuando la temperatura del aire disminuye a menos de –80 °C (336)

pollution/contaminación un cambio indeseable en el ambiente natural, producido por la introducción de substancias que son dañinas para los organismos vivos o por desechos, calor, ruido o radiación excesivos (14)

population/población un grupo de organismos de la misma especie que viven en un área geográfica específica y se reproducen entre sí (96, 197)

porosity/porosidad el porcentaje del volumen total de una roca o sedimento que está formado por espacios abiertos (273)

potable/potable que puede beberse (277)

precipitation/precipitación cualquier forma de agua que cae de las nubes a la superficie de la Tierra; incluye a la lluvia, nieve, aguanieve y granizo (73)

predation/depredación la interacción entre dos especies en la que una especie, el depredador, se alimenta de la otra especie, la presa (206)

prediction/predicción una afirmación que se hace por anticipado, la cual expresa los resultados que se obtendrán al poner a prueba una hipótesis si ésta es corroborada; el resultado esperado si la hipótesis es correcta (32)

primary pollutant/contaminante primario un contaminante que es colocado directamente en la atmósfera por las actividades humanas o naturales (303)

primary succession/sucesión primaria sucesión que comienza en un área donde previamente no podía existir la vida (129)

probability/probabilidad termino que describe qué tan probable es que ocurra un posible evento futuro en un caso dado del evento; la proporción matemática del número de veces que es posible que ocurra un resultado de cualquier evento respecto al número de resultados posibles del evento (40)

producer/productor un organismo que elabora moléculas orgánicas a partir de moléculas inorgánicas; un autótrofo fotosintético o quimiosintético que funciona como la fuente fundamental de alimento en un ecosistema (118)

protist/protista un organismo que pertenece al reino Protista (104)

R

radiation/radiación la energía que se transfiere en forma de ondas electromagnéticas, tales como las ondas de luz y las infrarrojas (70)

recharge zone/zona de recarga un área en la que el agua se desplaza hacia abajo para convertirse en parte de un acuífero (274)

reclamation/restauración el proceso de hacer que la tierra vuelva a su condición original después de que se terminan las actividades de explotación minera (424)

recycling/reciclar el proceso de recuperar materiales valiosos o útiles de los desechos o de la basura; el proceso de reutilizar algunas cosas (489)

reforestation/reforestación el restablecimiento y desarrollo de los árboles en un bosque (367)

renewable energy/energía renovable energía que proviene de fuentes que se están formando constantemente (457)

reproductive potential/potencial reproductivo el número máximo de crías que puede producir un determinado organismo (199)

reservoir/represa una masa artificial de agua que normalmente se forma detrás de una presa (280)

resistance/resistencia en biología, la capacidad de un organismo de tolerar a un agente químico o causante de enfermedades (101)

risk/riesgo la probabilidad de que se produzca un resultado no deseado (41)

risk assessment/evaluación de riesgos la evaluación, estudio y administración del riesgo por medios científicos; un cálculo científico de la probabilidad de que ocurran efectos negativos debido a la exposición a un peligro específico (513)

river system/sistema fluvial una red de ríos y arroyos en flujo que drenan una cuenca fluvial (271)

ruminant/rumiante un mamífero que mastica los alimentos dos veces, el cual tiene un estómago con tres o cuatro cámaras; entre los ejemplos se encuentran los borregos, cabras y ganado (398)

rural/rural término que describe un área de tierra abierta que a menudo se usa para la labranza (355)

S

salinity/salinidad una medida de la cantidad de sales disueltas en una cantidad determinada de líquido (76, 173)

salinization/salinización la acumulación de sales en el suelo (388)

salt marsh/marisma un hábitat marino que se caracteriza por tener pasto, juncias y otras plantas que se han adaptado a la inundación continua y periódica; las marismas se encuentran principalmente en las regiones templadas y subárticas (182)

sample/muestra el grupo de individuos o sucesos que se seleccionan para representar a una población estadística (40)

savanna/sabana una planicie llena de pastizales y árboles y arbustos dispersos; se encuentra en los hábitats tropicales y subtropicales y, sobre todo, en regiones con un clima seco, como en el este de África (155)

secondary pollutant/contaminante secundario un contaminante que se forma en la atmósfera por medio de una reacción química con contaminantes primarios del aire, componentes naturales del aire o ambos (303)

secondary succession/sucesión secundaria el proceso por medio del cual una comunidad reemplaza a otra, la cual ha sido parcial o totalmente destruida (129)

sick-building syndrome/síndrome del edificio enfermo un conjunto de síntomas, como dolor de cabeza, fatiga, irritación de los ojos y mareo, que puede afectar a las personas que trabajan en edificios modernos que cuentan con ventanas selladas; se cree que es producido por los contaminantes del interior del edificio (310)

Spanish Glossary

smelting/fundir derretir una mena con el fin de separar las impurezas del metal puro (420)

smog/esmog contaminación urbana del aire, compuesta por una mezcla de humo y niebla producida por contaminantes industriales y combustibles (308)

solid waste/desechos sólidos un material sólido desechado, como por ejemplo, basura, residuos o sedimentos (481)

source reduction/reducción de la fuente cualquier cambio en el diseño, manufactura, compra o uso de materiales o productos para reducir su cantidad o toxicidad antes de que se conviertan en desechos sólidos municipales; *también,* la reutilización de productos o materiales (488)

species/especie un grupo de organismos que tienen un parentesco cercano y que pueden aparearse para producir descendencia fértil; *también,* el nivel de clasificación debajo de género y arriba de subespecie (95)

statistics/estadística la recolección y clasificación de datos que encuentran en forma de números (38)

stratosphere/estratosfera la capa de la atmósfera que se encuentra justo encima de la troposfera y se extiende de aproximadamente 10 km hasta 50 km sobre la superficie de la Tierra; ahí, la temperatura aumenta al aumentar la altitud; contiene la capa de ozono (69)

subsidence/hundimiento del terreno el hundimiento de regiones de la superficie del suelo con muy poco o sin ningún movimiento horizontal (423)

subsurface mining/minería subsuperficial un método de explotación de minas en el que la mena se extrae de la parte inferior de la superficie del suelo (416)

surface impoundment/separación superficial una depresión natural o una excavación hecha por el hombre que sirve como vertedero de basura para acumular desechos (496)

surface mining/minería superficial un método de explotación de minas en el que se remueven el suelo y las rocas para llegar al carbón o minerales subyacentes (417)

surface water/agua superficial todas las masas de agua dulce, agua salada, hielo y nieve que se encuentran arriba del suelo (270)

survivorship/supervivencia el porcentaje de individuos recién nacidos de una población que se espera que sobrevivan hasta una edad determinada (220)

sustainability/sustentabilidad la condición en la que se cumple con las necesidades humanas de una forma tal que una población humana pueda sobrevivir indefinidamente (21, 533)

symbiosis/simbiosis una relación en la que dos organismos diferentes viven estrechamente asociados uno con el otro (209)

taiga/taiga una región de bosques siempreverdes de coníferas, ubicado debajo de las regiones árticas y subárticas de tundra (153)

tectonic plate/placa tectónica un bloque de litosfera formado por la corteza y la parte rígida y más externa del manto; también se llama placa litosférica (62)

temperate deciduous forest/bosque caducifolio templado un bosque (o bioma) que se caracteriza por árboles a los que se les caen las hojas en el otoño (152)

temperate grassland/pradera templada una comunidad (o bioma) que está dominada por pastos, tiene pocos árboles y se caracteriza por inviernos fríos y precipitación pluvial que es intermedia entre la de un bosque y la de un desierto (156)

temperate rain forest/selva tropical templada una comunidad de bosque (o bioma) caracterizada por tiempo frío y húmedo y lluvia en abundancia, en la cual las ramas de los árboles están cubiertas por moho, los troncos de los árboles están cubiertos por líquenes y el suelo del bosque está cubierto por helechos (151)

temperature inversion/inversión de la temperatura la condición atmosférica en la que el aire caliente retiene al aire frío cerca de la superficie terrestre (308)

thermal pollution/contaminación térmica un aumento en la temperatura de una masa de agua, producido por las actividades humanas y que tiene un efecto dañino en la calidad del agua y en la capacidad de esa masa de agua para permitir que se desarrolle la vida (289)

threatened species/especie amenazada una especie que se ha identificado como candidata para estar en peligro de extinción en el futuro inmediato (245)

topsoil/capa superior del suelo la capa superficial del suelo, la cual normalmente es más rica en materia orgánica que el subsuelo (385)

toxicology/toxicología el estudio de las substancias tóxicas, incluyendo su naturaleza, efectos, detección, métodos de tratamiento y control de exposición (512)

trophic level/nivel trófico uno de los pasos de la cadena alimenticia o de la pirámide alimenticia; entre los ejemplos se encuentran los productores y los consumidores primarios, secundarios y terciarios (122)

tropical rain forest/selva tropical un bosque o jungla que se encuentra cerca del ecuador y se caracteriza por una gran cantidad de lluvia y poca variación en la temperatura, y que contiene la mayor diversidad de organismos que se conoce en la Tierra (146)

troposphere/troposfera la capa inferior de la atmósfera, en la que la temperatura disminuye a una tasa constante a medida que la altitud aumenta; la parte de la atmósfera donde se dan las condiciones del tiempo (68)

tundra/tundra un llano sin árboles que se ubica en la región ártica o antártica y se caracteriza por temperaturas muy bajas en el invierno, veranos cortos y frescos y vegetación que consiste en pasto, líquenes y hierbas perennes (162)

understory/capa sumergida una capa de follaje que se encuentra debajo de la bóveda principal de un bosque y está cubierta por ella (148)

urban/urbana término que describe a un área que contiene una ciudad (355)

urbanization/urbanización un aumento en la razón o densidad de las personas que viven en áreas urbanas en lugar de en áreas rurales (227, 358)

urban sprawl/derrame urbano la rápida propagación de una ciudad hacia los suburbios adjuntos y áreas rurales (359)

value/valor un principio o norma que un individuo considera importante (45)

variable/variable un factor que se modifica en un experimento con el fin de probar una hipótesis (33)

vector/vector en biología, cualquier agente, como por ejemplo un plásmido o un virus, que tiene la capacidad de incorporar ADN extraño y de transferir ese ADN de un organismo a otro; un huésped intermediario que transfiere un organismo patógeno o un parásito a otro organismo (520)

vertebrate/vertebrado un animal que tiene columna vertebral; incluye a los mamíferos, aves, reptiles, anfibios y peces (107)

wastewater/agua de desecho agua que contiene desechos de los hogares o la industria (286)

water cycle/ciclo del agua el movimiento continuo del agua: del océano a la atmósfera, de la atmósfera a la tierra y de la tierra al océano (73)

water pollution/contaminación del agua la adición al agua de sustancias químicas o materiales de desecho que son dañinos para los organismos que viven en el agua o para aquellos que la beben o que están expuestos a ella (284)

watershed/cuenca hidrográfica el área del terreno que es drenada por un sistema de agua (271)

weather/tiempo el estado de la atmósfera a corto plazo que incluye la temperatura, la humedad, la precipitación, el viento y la visibilidad (327)

wetland/pantano un área de tierra que está periódicamente bajo el agua o cuyo suelo contiene una gran cantidad de humedad (173)

wilderness/área silvestre una región que no ha sido cultivada ni está habitada por seres humanos (368)

Y

yield/rendimiento la cantidad de cosechas producidas por unidad de área (381)

Index

Index

Index

Index

Index

Index

Index

Acknowledgments

Teacher Reviewers continued

David Blinn
Secondary Sciences Teacher
Wrenshall High School
Wrenshall, Minnesota

Bart Bookman
Science Teacher
Stevenson High School
Bronx, New York

Daniel Bugenhagen
Science Teacher
Yutan Community School
Yutan, Nebraska

Robert Chandler
Science Teacher
Soddy-Daisy High School
Soddy-Daisy, Tennessee

Johanna Chase, C.H.E.S.
Health Educator
California State University
Dominquez Hills, California

Cindy Copolo, Ph.D.
Science Specialist
Summit Solutions
Bahama, North Carolina

Linda Culp
Science Teacher
Thorndale High School
Thorndale, Texas

Katherine Cummings
Science Teacher
Currituck County
Currituck, North Carolina

Alonda Droege
Science Teacher
Evergreen High School
Seattle, Washington

Richard Filson
Science Teacher
Edison High School
Stockton, California

Randa Flinn
Science Teacher
Northeast High School
Fort Lauderdale, Florida

Jane Frailey
Science Coordinator
Hononegah High School
Hononegah, Illinois

Art Goldsmith
Biology and Earth Sciences Teacher
Hallandale High School
Hallandale, Florida

Sharon Harris
Science Teacher
Mother of Mercy High School
Cincinnati, Ohio

Carolyn Hayes
Honors Biology and Environmental Science Teacher
Center Grove High School
Greenwood, Indiana

Stacey Jeffress
Environmental Science Teacher
El Dorado High School
El Dorado, Arkansas

Donald R. Kanner
Physics Instructor
Lane Technical High School
Chicago, Illinois

Edward Keller
Science Teacher
Morgantown High School
Morgantown, West Virginia

Kathy LaRoe
Science Teacher
St. Paul School District
St. Paul, Nebraska

Clifford Lerner
Biology Teacher
Keene High School
Keene, New Hampshire

Stewart Lipsky
Science Teacher
Seward Park High School
New York, New York

Mike Lubich
Science Teacher
Mapletown High School
Greensboro, Pennsylvania

Thomas Manerchia
Environmental Science Teacher, Retired
Archmere Academy
Claymont, Delaware

Tammie Niffenegger
Science Chair and Science Teacher
Port Washington High School
Waldo, Wisconsin

Gabriele DeBear Paye
Science and Environmental Technology Lead Teacher
West Roxbury High School
West Roxbury, Massachusetts

Denice Sandefur
Fire Ecology and Science Teacher
Nucla High School
Nucla, Colorado

Jennifer M. Fritz
Science Teacher
North Springs High School
Atlanta, Georgia

Dyanne Semerjibashian, Ph.D.
Science Teacher
Pflugerville High School
Pflugerville, Texas

Bert Sherwood
Science/Health Specialist
Socorro Independent School District
El Paso, Texas

Dan Trockman
Science Teacher
Hopkins High School
Minnetonka, Minnesota

Jim Watson
Science Teacher
Dalton High School
Dalton, Georgia

Staff Credits

Robert Tucek
Executive Editor

Debbie Starr
Managing Editor

Clay Walton
Senior Editor

Editorial Development Team

Bill Burnside

Jen Driscoll

Frieda Gress

Bill Rader

Jim Ratcliffe

Copyeditors

Dawn Marie Spinozza
Copyediting Manager

Anne-Marie De Witt

Simon Key

Jane A. Kirschman

Kira J. Watkins

Editorial Support Staff

Mary Anderson

Soojinn Choi

Jeanne Graham

Shannon Oehler

Stephanie Sanchez

Tanu'e White

Editorial Interns

Kristina Bigelow

Erica Garza

Sarah Ray

Kenneth G. Raymond

Kyle Stock

Audra Teinert

Online Products

Bob Tucek

Wesley M. Bain

Catherine Gallagher

Douglas P. Rutley

Book Design

Kay Selke
Director of Book Design

Media Design

Richard Metzger
Design Director

Chris Smith
Developmental Designer

Cover Design

Kay Selke
Director of Book Design

Publishing Services

Carol Martin
Director

Jeff Robinson
Manager, Ancillary Design

Production

Adriana Bardin Prestwood
Senior Production Coordinator

Eddie Dawson
Senior Production Manager

Technology Services Managers

Laura Likon
Director

Juan Baquera
Manager

JoAnn Stringer
Manager

Senior Technology Services Analysts

Katrina Gnader

Lana Kaupp

Margaret Sanchez

Technology Services Analysts

Sara Buller

Patty Zepeda

eMedia

Melanie Baccus
eMedia Coordinator

Ed Blake
Design Director

Kimberly Cammerata
Design Manager

Lydia Doty
Senior Project Manager

Marsh Flournoy
Technology Project Manager

Dakota Smith
Quality Assurance Analyst

Cathy Kuhles
Technical Assistant

Tara F. Ross
Senior Project Manager

Ken Whiteside
Manager, Application Development

Acknowledgments

Image Credits

COVER IMAGE: butterfly—royalty-free image from Photodisc; leaves—Yasushi Takahashi/Photonica; water droplets—Ken Hiroshi/Photonica.

FRONTMATTER: v(t), E.R. Degginger/Bruce Coleman, Inc.; (tl), Claus Meyer/Minden Pictures; (tml), Fred Bavendam/Minden Pictures; (bml), Yann Layma/Getty Images/Stone; (bl), Luis Vega/The Image Bank/Getty Images ; (b), Anthony Bannister/Photo Researchers, Inc.; vi(ml), Mark Moffett/Minden Pictures; vi(bl), Norbert Wu/Norbert Wu Productions; vii(tl), Earth Imaging/Stone/Getty Images ; vii(ml), S. Hanquet/Peter Arnold, Inc.; vii(bl), Gary Meszaros/Dembinsky Photo Associates; viii(tl), Martin Harvey/Peter Arnold, Inc.; viii(bl), Brandon D. Cole/Corbis; ix(tl), Mark Carwardine/Still Pictures/Peter Arnold, Inc.; ix(ml), Topham/The ImageWorks; ix(bl), John Shaw/Bruce Coleman, Inc.; x(tl), Ralph A. Clevenger/Corbis; x(ml), Chad Ehlers/ImageState; x(bl), NASA/Goddard Space Flight Center/Science Source; xi(ml), Jim Wark/Peter Arnold, Inc.; xi(bl), Grant Heilman/Grant Heilman Photography Inc.; xii(tl), Dan Budnik/Woodfin Camp and Associates; xii(ml), Werner H. Muller/Peter Arnold, Inc.; xiii(ml), Rafael Macia/Photo Researchers, Inc.; xiii(bl), Andrew Rakoczy/Bruce Coleman, Inc.; xiv(tl), Argus Fotoarchiv/Peter Arnold, Inc.; xiv(ml) , Albert Normandin/Masterfile; xx(ml), John Cancalosi/Peter Arnold, Inc.; (tl), Bob Wolf; (tr), Sam Dudgeon/HRW Photo; (br), Karen Allen; xx(bl), NOAA/Department of Commerce/NOAA Central Library; xxi(r), Gerhard Gscheidle/Peter Arnold, Inc.; (l), The Zoological Society of San Diego.

UNIT 1 OPENER: 2, E.R. Degginger/Bruce Coleman, Inc.; 3(t), Mark Moffett/Minden Pictures; (m), Norbert Wu/Norbert Wu Productions; (b), Earth Imaging/Stone/Getty Images.

CHAPTER 1: 4, Mark Moffett/Minden Pictures; 5, Courtesy of Cliff Lerner/HRW Photo; 6(bl), Roland Seitre/Peter Arnold, Inc.; (br), Douglas Faulkner/Photo Researchers, Inc.; 7(br), Matt Meadows/Peter Arnold, Inc.; (bl), K & M Krafft/Photo Researchers, Inc.; 8(br), Courtesy of Gardner Watkins/Study conducted by the students of Dublin Scioto High School and supported by The Columbus Zoo Dublin Scioto H.S. Environmental Club; (bl), AFP/Corbis; 9(b), David Gillison/Peter Arnold, Inc.; (t), NorthWind Picture Archives; 10(b), M. Edwards/Peter Arnold, Inc.;(t), A. Murray/University of Florida/URL:http://plants.ifas.ufl.edu; 11(r), W.M. Weber/Peter Arnold, Inc.; (tl), Lambert/Archive Photos/PictureQuest; 12, NASA; 13, Joel Rodgers/Corbis; 14, Gene Aherns/Bruce Coleman, Inc.; 15(tr), Conor Caffrey/Photo Researchers, Inc.; (br), BIOS/Peter Arnold, Inc.; 18(l), Ric Ergenbright/Corbis; (b), Macduff Everton/Corbis; 20, Terry Farmer/Tony Stone Images; 21, Frank Pedrck/The ImageWorks; 22(t), Courtesy of Gardner Watkins/Study conducted by the students of Dublin Scioto High School and supported by The Columbus Zoo Dublin Scioto H.S. Environmental Club; (b), Ric Ergenbright/Corbis; 26, S. Dudgeon/HRW Photo; 28, Karen Allen; 29, Karen Allen.

CHAPTER 2: 30, Norbert Wu/Norbert Wu Productions; 31, Jeff & Alexa Henry/Peter Arnold, Inc.; 33, Courtesy of Cliff Lerner/HRW Photo; 34, Tek Image/SPL/Photo Researchers, Inc.; 35, Courtesy of U of AK Tree-Ring Lab URL:www.uark.edu/dendro/stahle_bwr.tif; 36, K & K Ammann/Bruce Coleman, Inc.; 37, SPL/PRI/Photo Researchers, Inc.; 38, Cliff Lerner/HRW Photo; 40, Kent Wood/Photo Researchers, Inc.; 41, Jim Olive/Peter Arnold, Inc.; 42, Eye Ubiquitous/ Corbis; 44, Earth Satellite Corp./SPL/Photo Researchers, Inc.; 46(t), G. Lasley/VIREO URL: www.acnatsci.org/vireo; (b), S. Michael Bisceglie/ Animals Animals/Earth Scenes; 47, Matt Bradley/Bruce Coleman, Inc.; 49, Richard Hamilton Smith/Dembinsky Photo Associates; 50(t), Jeff & Alexa Henry/Peter Arnold, Inc.; (m), Earth Satellite Corp./SPL/Photo Researchers, Inc.; (b), Richard Hamilton Smith/Dembinsky Photo Associates; 54, Holt/HRW Photo; 55, Holt/HRW Photo; 57, Merlin D. Tuttle/Bat Conservation International, Inc.

CHAPTER 3: 58, Earth Imaging/Stone/Getty Images ; 63, Jock Montgomery/Bruce Coleman, Inc.; 65, Gary Braasch/Corbis; 66, Dennis Flaherty/Photo Researchers, Inc.; 67, NASA; 69(t), NOAA/Department of Commerce/NOAA Central Library URL: www.photolib.noaa.gov; (br), Chris Madeley/SPL/Photo Researchers, Inc.; 73, Courtesy of Robert Cantor/Christian Grantham URL: www.christianandvince.com gallery/ hawaii/diamondhead.jpg; 74, Peter Ryan/ Scripps/SPL/Photo Researchers, Inc.; 75, K. Crane/WHO1/Visuals Unlimited; 77, Rosentiel School of Marine & Atmospheric Science, U. Of Miami; 79, Bernhard Edmaier/SPL/Photo Researchers, Inc.; 80(b), Provided by the SeaWiFS Project, NASA/Goddard Space Flight Center and ORBIMAGE/ NASA/Seawifs; (bl), Provided by the SeaWiFS Project, NASA/Goddard Space Flight Center and ORBIMAGE/NASA/Seawifs; 81, Reuters NewMedia Inc./Corbis; 82(t), Gary Braasch/Corbis; (m), NOAA/Department of Commerce/NOAA Central Library/URL: www.photolib.noaa.gov; 86, Holt/HRW Photo; 87(t), Sam Dudgeon/Holt, Rinehart and Winston; (m), Sam Dudgeon/Holt, Rinehart and Winston; (b), SamDudgeon/Holt, Rinehart and Winston; 89, PhotoDisc.

UNIT 2 OPENER: 90, Claus Meyer/Minden Pictures; 91(t), S. Hanquet/ Peter Arnold, Inc.; (tc), Gary Meszaros/Dembinsky Photo Associates; (bc), Martin Harvey/Peter Arnold, Inc.; (b), Brandon D. Cole/Corbis.

CHAPTER 4: 92, S. Hanquet/Peter Arnold, Inc.; 93, John Meilcarek/Dembinsky Photo Associates; 94(t, b), J.J. Alcalay/Peter Arnold, Inc.; 95(bl), Scott Smith/Dembinsky Photo Associates; (br), Fred Bruemmer/Peter Arnold, Inc.; 96, Joe McDonald/Bruce Coleman, Inc.; 97, Stephen Dalton/Photo Researchers, Inc.; 98, Tui De Roy/Bruce Coleman, Inc.; 99, Auscape/Parer-Cook/Peter Arnold, Inc.; 100(t), P. La Tourrette/VIREO URL: www.acnatsci.org/vireo; (bc), Tom Brakefield/Bruce Coleman, Inc.; (br), Klein/Hubert/Peter Arnold, Inc.; 103(t), Juergen Berger/SPL/Photo Researchers, Inc.; (bl), Rod Planck/Dembinsky Photo Associates; (br), Steven Mark Needham/ Envision; 104(tl), Lawrence Naylor/Photo Researchers, Inc.; (b), F. Stuart Westmorland/Photo Researchers, Inc.; (tc), Jan Hinsch/ SPL/Dembinsky Photo Associates; 105(t), Daniel Zupanc/Bruce Coleman, Inc.; (b), Kent Foster/Bruce Coleman, Inc.; 106(br), Mark J. Thomas/Dembinsky Photo Associates; (bl), Jim Steinberg/Photo Researchers, Inc.; (bc), Raymond A. Mendez/Animals Animals/Earth Scenes; 107(tl), Ed Reschke/Peter Arnold, Inc.; (tr),Gerard Lacz/Peter Arnold, Inc.; (tc), Klaus Jost/Peter Arnold, Inc.; 108(t), J.J. Alcalay/Peter Arnold, Inc.; (m), P. La Tourrette/VIREO URL: www.acnatsci.org/vireo; (b), Juergen Berger/SPL/Photo Researchers, Inc.; 112, Victoria Smith/ Holt, Rinehart and Winston; 113, Manfred Kage/Peter Arnold, Inc.; 114(bl, br), Lincoln Brower; 115, Lincoln Brower.

CHAPTER 5: 116, Gary Meszaros/Dembinsky Photo Associates; 117(bl), Phillipe Giraud/Sygma; (l), John Durham/SPL/Photo Researchers, Inc.; 118(tl), IFA/Peter Arnold, Inc.; (tc), Van deRostyne/ BIOS/Peter Arnold, Inc.; (tc), Sharon Cummings/ Dembinsky Photo Associates; (tr), Jeff LePore/Dembinsky Photo Associates; (bl), Ron & Valerie Taylor/Bruce Coleman, Inc.; (bc), Alfred Pasieka/SPL/Photo Researchers, Inc.; 119, Marilyn Kazmers/Dembinsky Photo Associates; 120, Harry Engels/Photo Researchers, Inc.; 121, Fritz Polking/Bruce Coleman, Inc.; 125, Ted Spiegel/Corbis; 126(bl), G. R. "Dick" Roberts/ G.R. "Dick" Roberts Photo Library; (m), C.P. Vance/Visuals Unlimited; 128(t), E.R. Degginger/Color-Pic, Inc.; (l), G.R. "Dick" Roberts/G.R. "Dick" Roberts Photo Library; 129, Hans Reinhard/Bruce Coleman, Inc.; 130(b), Stan Osolinski/Dembinsky Photo Associates; (tl), Larry Nielsen/Peter Arnold, Inc.; 131(l), Scott Smith/Dembinsky Photo Associates; (r), Roger W. Archibald/Animals Animals/Earth Scenes; (tr), Patti Murray/Animals Animals/Earth Scenes; 133(l), Scott Smith/ Dembinsky Photo Associates; (r), Norman Owen Tomalin/Bruce Coleman, Inc.; 134(t), Fritz Polking/Bruce Coleman, Inc.; (b), Hans Reinhard/Bruce Coleman, Inc.; (m), G.R. "Dick" Roberts/G.R. "Dick" Roberts Photo Library; 141(b), Stephen Rose/Rainbow; (t), FPG International/Getty Images.

CHAPTER 6: 142, Martin Harvey/Peter Arnold, Inc.; 144, Eastcott/Momatiuk/Animals Animals/Earth Scenes; 146(br), Dr. Morley Read/SPL/Photo Researchers, Inc.; (bl), Patti Murray/Animals Animals/ Earth Scenes; (bc), Michael Fogden/Bruce Coleman, Inc.; 147(bl), Frans Lanting/Minden Pictures; (br), M. Gunther/Peter Arnold, Inc.; 149(r), Mickey Gibson/Animals Animals/Earth Scenes; (bl), E. Woods/ Animals Animals/Earth Scenes; 149(br), Walter H. Hodge/Peter Arnold, Inc.; (l), Alan G. Nelson/Dembinsky Photo Associates; 150, Greg Baker/AP/Wide World; 151(l), Wayne Lawler/Photo Researchers, Inc.; (tr), Jim Steinberg/Photo Researchers, Inc.; 152(bl, bc, br), Scott W. Smith/Animals Animals/Earth Scenes; 153(r), George E. Stewart/ Dembinsky Photo Associates; (tr), Anthony Mercieca/Dembinsky Photo Associates; (inset), Michael F. Sacca/Animals Animals/Earth Scenes; 154(tl), J.J. Alcalay/Peter Arnold, Inc.; (bc, br), S.J. Krasemann/ Peter Arnold, Inc.; (inset), David Cavagnaro/DRK Photo; 156, Tim Davis/ Photo Researchers, Inc.; 157(tl), Joel Bennett/Peter Arnold, Inc.; (tr), Jeff Gnass; 158(t), Dominique Braud/Animals Animals/Earth Scenes; (b), Coco McCoy/Rainbow; 159(t), Bobbi Lane/Stone; (inset), Ernest H. Rogers/Sea Images; 160(b), Jon Mark Stewart/Biological Photo Service; (inset), Frans Lanting/Minden Pictures; 161(bc), Anthony Bannister/Natural History Photographic Agency; (br), C. K. Lorenz/ Photo Researchers, Inc.; 162, Jo Overholt/

AlaskaStock Images; 163(r), Johnny Johnson/Animals Animals/Earth Scenes; (tl), Gerald & Buff Corsi/Visuals Unlimited; 164(m), Dr. Morley Read/SPL/Photo Researchers, Inc.; (b), Tim Davis/Photo Researchers, Inc.; (t), Eastcott/ Momatiuk/Animals Animals/Earth Scenes; 169(t), Adam Jones/ Dembinsky Photo Associates; (b), D. Boone/Corbis; 170(t), Hugh S. Rose/Visuals Unlimited; (b), John W. Warden/West Stock; 171(t), Ken Graham/Tony Stone/Allstock.

CHAPTER 7: 172, Brandon D. Cole/Corbis; 173, Jan-Peter Lahall/Peter Arnold, Inc.; 174(t), Breck P. Kent/Animals Animals/Earth Scenes; (br), Roland Birke/Peter Arnold, Inc.; (bc), A.M. Siegelman/Visuals Unlimited; 175(t), Willard Clay/Dembinsky Photo Associates; (b), Stan Osolinski/ Dembinsky Photo Associates; 176, Dominique Braud/ Dembinsky Photo Associates; 177(t), Brian Miller/Animals Animals/ Earth Scenes; (bl), C.C. Lockwood/Animals Animals/Earth Scenes; (br), Bates Littlehales/Animals Animals/Earth Scenes; 178(t), R. Toms, OSF/ Animals Animals/Earth Scenes; (m), Adam Jones/Photo Researchers, Inc.; 181(tr), Fred Bavendam; (m), Lynda Richardson/ Corbis; 182(t), Fabio Colombini/ Animals Animals/Earth Scenes; (b), Ted Levin/ Animals Animals/Earth Scenes; 183(br), Ron Sefton/Bruce Coleman, Inc.; (m), Norbert Wu; 185(t), Norbert Wu/Peter Arnold, Inc.; (m), Image Life/Corbis/Corbis; 186(t), Breck P. Kent/Animals Animals/Earth Scenes; (b), Image Life/ Corbis; 190, Victoria Smith/HRW Photo; 191, Victoria Smith/HRW Photo; 193(m), Aldo Brando/Peter Arnold, Inc.; (b), Patricia Jordan/Peter Arnold, Inc.

UNIT 3 OPENER: 194, Fred Bavendam/Minden Pictures; 195(t), Mark Carwardine/Still Pictures/Peter Arnold, Inc.; (m), Topham/The ImageWorks; (b), John Shaw/Bruce Coleman, Inc.

CHAPTER 8: 196, Mark Carwardine/Still Pictures/Peter Arnold, Inc.; 197(br), Fred Bavendam/Peter Arnold, Inc.; 197(bl), Stuart Westmoreland/Corbis; 198(tr), Norman Owen Tomalin/Bruce Coleman. Inc.; 198(tl), Michael Fogden/Animals Animals/Earth Scenes; 198(t), David Hughes/Bruce Coleman. Inc.; 200(bl), Bettmann/Corbis; 201(tr), Richard Thom/Visuals Unlimited; 201(tl), Skip Moody/Dembinsky Photo Associates; 202(t), C.C. Lockwood/ Animals Animals/Earth Scenes; 202(l), W. Metzen/Bruce Coleman. Inc.; 203(bl), Beverly Joubert/ National Geographic Society; 203(bc), Tom Brakefield/Tom Brakefield Photography; 203(br), Y. Arthus-Bertand/ Peter Arnold, Inc.; 203(br), Anne Wertheim/Animals Animals/Earth Scenes; 206(tl), Wyman Meizer/Photo Researchers, Inc; 206(r), J.T. Collins/Photo Researchers, Inc; 207(l), Bill Beatty/Visuals Unlimited; 207(r), Joe McDonald/Visuals Unlimited; 208(tr), CNRI/SPL/Photo Researchers, Inc; 208(tl), John Shaw/Bruce Coleman. Inc.; 208(bl), Patti Murray/Animals Animals/ Earth Scenes; 209(t), James Watt/ Animals Animals/Earth Scenes; 210(t), Norman Owen Tomalin/Bruce Coleman. Inc.; 210(b), Wyman Meizer/ Photo Researchers, Inc; 214(b), Dennis Kunkel/Phototake; 215(t), Dr. David P. Frankhauser URL: http://biology.clc.uc.edu/Fankhauser; 216(t), William bernarrd/ Corbis; 217(t), Dan Grandmaison/ Permission of International Wolf Center/URL: www.wolf.org.

CHAPTER 9: 218, Topham/The ImageWorks; 222(tc), Michael Sullivan/ TexaStock; (tl), Marcel & Eva Malherbe/The ImageWorks; (tr), Jeff Greenberg/Rainbow; 224, Sean Sprague/The ImageWorks; 225, Inga Spence/Visuals Unlimited; 226(b), Annie Griffiths Belt/ Corbis; (t), Jean-Leo Dugast/Panos Pictures; 227(bl), Mark E. Gibson/ Visuals Unlimited; (br), NASA/Johnson Space Center/NASA; 229, Jodi Cobb/ National Geographic/Getty Images; 230, Louise Gubb/The ImageWorks; 232(t), Jeff Greenberg/Rainbow; (b), Louise Gubb/The ImageWorks; 239(b), Phil Degginger/Color-Pic, Inc.; (t), Fritz Prenzel/Animals Animals/Earth Scenes.

CHAPTER 10: 240, John Shaw/Bruce Coleman, Inc.; 242(tl), Chip Clark/Smithsonian, NMNH; (bl), Nora Darbyshire/Selmer Ausland/Memories of Deep River URL www.jkcc.com/evje/ trapping.html; (bc), Norbert Wu/Peter Arnold, Inc.; (bc), Jeff Foott/ Bruce Coleman, Inc.; (br), Jeffrey Rotman/Peter Arnold, Inc.; 244(t), Lineair/ Peter Arnold, Inc.; 246, Art Wolfe/Stone/AllStock; 247, M. Timothy O'Keefe/Bruce Coleman, Inc.; 248, Alison Wright/Photo Researchers, Inc.; 249, M. Edwards/Still Pictures/Peter Arnold, Inc.; 250(tl, tr), Tom McHugh/Photo Researchers, Inc.; (br), S. Cordier/Jacana /Photo Researchers, Inc.; (bl), Merlin D. Tuttle/Bat Conservation International, Inc.; (tc), Sally A. Morgan/Corbis; 251(tc, tl), David Dennis/Animals Animals/Earth Scenes; (bl), Bill Lea/Dembinsky Photo Associates; (br), Jeff Foott/Bruce Coleman, Inc.; (tr), Tim Davis/ Davis/ Lynn Images; 252, David Clendenen/U.S. Fish and Wildlife Service;

253(b), Cameramann International; (t), ARS/RDF/Visuals Unlimited; 254, Photo by Rodney North/courtesy of Equal Exchange/ URL www. equalexchange.com; 255, Phillip Roullard/ Phillip Roullard Photography; 256(tl), Robert Caputo/Aurora; (tr), Louise Gubb/J.B. Pictures; 257, Baker/Greenpeace; 258(t), Chip Clark/Smithsonian, NMNH; (m), David Dennis/Animals Animals/Earth Scenes; (b), Phillip Roullard/Phillip Roullard Photography; 262, Sam Dudgeon/Holt, Rinehart and Winston; 263, Sam Dudgeon/Holt, Rinehart and Winston; 264, AP/Wide World; 265, Marc Halevi/Harvard Photographic Services.

UNIT 4 OPENER: 266, Yann Layma/Getty Images/Stone; 267(t), Ralph A. Clevenger/Corbis; (tc), Chad Ehlers/International Stock/ ImageState; (m), NASA/Goddard Space Flight Center/Science Source/Photo Researchers, Inc.; (bc), Jim Wark/Peter Arnold, Inc.; (b), Grant Heilman/Grant Heilman Photography Inc.

CHAPTER 11: 268, Ralph A. Clevenger/Corbis; 270, E.R.I.M./Stone; 273, John Warden/Index Stock Imagery; 278, Jim Zuckerman/Corbis; 279(tr, tl), Grant Heilman/Grant Heilman Photography Inc.; (b), Vanni Archive/ Corbis; 280(br), Michelle Buselle/Getty Images/Stone; (l), Lloyd Cluff/ Corbis; 281, Christi Carter/Grant Heilman Photography Inc.; 282, Dave G. Houser/Corbis; 283, Steve Raymer/National Geographic Society; 284, Thomas Del Brase/Getty Images/Stone; 285(bl), Gary Braasch/Corbis; (b), W. L. McCoy/PictureQuest; (br), Ben Blankenburg/eStock Photo/ PictureQuest; 288(b), Nick Hawkes/ Ecoscene/Corbis; (bl), Siede Preis/ PhotoDisc/PictureQuest; 289, AFP/ Anonio Scorza/Corbis; 292, Bettmann/Corbis; 294(t), E.R.I.M./ Stone; (m), Jim Zuckerman/Corbis; (b), AFP/Anonio Scorza/Corbis; 299, Victoria Smith/HRW Photo; 300(t), IFA/Bruce Coleman, Inc.; (tr), Wolfgang Kaehler/Corbis; 301(bl), Liu Liqun/Corbis; (bc), Keren Su/Corbis.

CHAPTER 12: 302, Chad Ehlers/ International Stock/ImageState; 305, Pictor International/PictureQuest; 306, Sam Dudgeon/Holt, Rinehart and Winston; 307, Michael Newman/PhotoEdit; 309, E. Tobisch/UNEP/Peter Arnold, Inc.; 310, Neal Preston/Corbis; 311, Geoff Tompkins/SPL/Photo Researchers, Inc.; 312(tc), Thomas Ives/Corbis Stock Market; (tl), Dick Blume/The ImageWorks; 313, Gary Braasch/Woodfin Camp and Associates; 315, Simon Fraser/SPL/Photo Researchers, Inc.; 316, David R. Frazier/David R. Frazier Photolibrary; 318(b), Simon Fraser/SPL/Photo Researchers, Inc.; (m), Geoff Tompkins/SPL/Photo Researchers, Inc.; (t), Pictor International/PictureQuest; 322, Victoria Smith/HRW Photo; 324, C. Mayhew & R. Simmon/NASA/GSFC; 325, Pittsburgh Post-Gazette.

CHAPTER 13: 326, NASA/Goddard Space Flight Center/SPL/Photo Researchers, Inc.; 327(bl), Joe Sroka/Dembinsky Photo Associates; (br), Mark J. Thomas/Dembinsky Photo Associates; 330, From Degrees of variation: climate change in Nunavut, Geological Survey of Canada. Misc. Rept. 71, 2001, Natural Resources Canada/Reproduced with the permission of the Minister of Public Works & Govt. Services, 2002; 331(tr), GSFC/NASA; (r), CSIRO/Simon Fraser/SPL/Photo Researchers, Inc.; (l), Michael Sewell/Peter Arnold, Inc.; 332(t), JPL/NASA; 332(b), NASA; 333(br), GSFC/NASA; 336(tr, tc, tl), NASA; 342, Dr. Jeffrey Kiehl/National Center for Atmospheric Research; 343, NASA; 344(b), D. Allan/Animals Animals/Earth Scenes; (t), Terry Brandt/Grant Heilman Photography Inc.; 345, Bruce Brander/Photo Researchers, Inc.; 346(t), Mark J. Thomas/Dembinsky Photo Associates; (m), NASA; (b), Dr. Jeffrey Kiehl/National Center for Atmospheric Research; 350, Victoria Smith/ HRW Photo; 352, Louis Psihoyos/Matrix International; 353, R. Sanders/ Courtesy of Susan Solomon; (tl), NASA/Visuals Unlimited.

CHAPTER 14: 354, Jim Wark/Peter Arnold, Inc.; 356(l), J. Messerschmidt/ Bruce Coleman, Inc.; (b), Hanson Carroll/Peter Arnold, Inc.; 358(bl, bc, br), University of Maryland, Baltimore County/ Courtesy U.S. Geological Survey; 359(t), Brian Brake/Photo Researchers, Inc.; (b), Nicholas DeVore/Bruce Coleman, Inc.; 360(t), Mark Gibson/Index Stock Imagery; (b), Data:LandSat5 thematic Mapper, Data courtesy of C.P. Lo (U. of GA)/NASA/GSFC; 361(br, bc, bl), Reprinted with permission from the City of Seattle, Seattle Public Utilities' IT-Storefront; 362, Morton Beebe/ Corbis; 363, Douglas Peebles/Corbis; 364, E.R. Degginger/Animals Animals/Earth Scenes; 367(m), Tom McHugh/Photo Researchers, Inc.; (t), Peter Arnold/Peter Arnold, Inc.; 369, Scott Smith/Dembinsky Photo Associates; 370(t), Hanson Carroll/Peter Arnold, Inc.; (m), Nicholas DeVore/Bruce Coleman, Inc.; (b), Tom McHugh/Photo Researchers, Inc.; 376/377, Michael Murphy/By Permission of Selah, Bamberger Ranch/ URL: www.bambergerranch.org.

Acknowledgments

Image Credits continued

CHAPTER 15: 378, Grant Heilman/Grant Heilman Photography Inc.; 381(br), Chris Hellier/Corbis; (bl), Alan Bonicatti/Liaison Agency Inc; 382, David Austen/PictureQuest; 383, Grant Heilman/Grant Heilman Photography Inc.; 384, Arthur C. Smith III/Grant Heilman Photography Inc.; 387(b), Larry Lefever/Grant Heilman Photography Inc.; (tr), Alex S. MacLean/Peter Arnold, Inc.; (tl), Melinda Berge/Bruce Coleman, Inc.; 389(tl), Tony Stone/Stone; (tc), Nuridsany/ Perennou/Photo Researchers, Inc.; (tr), Inga Spence/Visuals Unlimited; 390, John Zoiner; 391(tl), Nigel Cattlin/Holt Studios international/ Photo Researchers, Inc.; (tr), Holt Studios International, Ltd.; 394, Patty Melander, permission of the Land Inst. URL www.landinstitute.org; 395, Roland Seitre/Peter Arnold, Inc.; 396, Doug Plummer/Photo Researchers, Inc.; 397, James L. Amos/Peter Arnold, Inc.; 398, Larry Lefever/Grant Heilman Photography Inc.; 399, Inga Spence/Visuals Unlimited; 400(t), Chris Hellier/Corbis; (m), Alex S. MacLean/Peter Arnold, Inc.; (b), Inga Spence/Visuals Unlimited; 405, Victoria Smith/HRW Photo; 406(t), Science VU/ARS/ Visuals Unlimited; (b), Scott Smith/Animals Animals/Earth Scenes; 407(b), Reuters NewMedia, Inc./Corbis; (t), Lynsey Addario/SABA/Corbis.

UNIT 5 OPENER: 408, Luis Vega/The Image Bank/Getty Images; 409(t), Dan Budnik/Woodfin Camp and Associates; (tc), Werner H. Muller/Peter Arnold, Inc.; (bc), Rafael Macia/Photo Researchers, Inc.; (b), Andrew Rakoczy/Bruce Coleman, Inc.

CHAPTER 16: 410, Dan Budnik/Woodfin Camp and Associates; 412(t), E.R. Degginger/Photo Researchers, Inc.; (b), Ken Lucas/Visuals Unlimited; 414(r), Mark A. Schneider/Visuals Unlimited; (b), Randy Jolly/The ImageWorks; 415, Paul A. Souders/Corbis; 416(b), David Barnes/Courtesy of New South Wales, Australia Department of Mineral Resources. New South Wales Mines Department; (t), By kind permission of K+S Aktiengesellschaft; 417(t), Michael Wickes/Bruce Coleman, Inc.; (b), Larsh Bristol/Visuals Unlimited; 418(t), Still Pictures/Peter Arnold, Inc.; (b), Carleton E. Watkins/California Historical Society; 419(tr), James L. Amos/Corbis; (m), Betty Sederquist/Visuals Unlimited; 420, Mireille Vautier/Woodfin Camp and Associates; 421, Phillip Richardson/Corbis; 422, Simon Fraser/SPL/Photo Researchers, Inc.; 423, Dean Purcell/ AP/ Wide World; 424, Courtesy of Dr. Prakash; U of AK, Geophysical Inst. Dr. Anupma Prakash; 425, Leo Touchet/Woodfin Camp and Associates; 426(t), Ken Lucas/Visuals Unlimited; (m), Still Pictures/ Peter Arnold, Inc.; (b), Simon Fraser/SPL/Photo Researchers, Inc.; 430, E.R. Degginger/ Color-Pic, Inc.; 431, Victoria Smith/HRW Photo; 433(b), Arnaud Zajtman/AP/Wide World; (t), Ryan McVay/PhotoDisc, Inc.

CHAPTER 17: 434, Werner H. Muller/Peter Arnold, Inc.; 435(bl), A.J. Copley/Visuals Unlimited; (br), Richard Hutchings/Photo Researchers, Inc.; 436, Bob Burch/Bruce Coleman, Inc.; 439, Leo Touchet/Woodfin Camp and Associates; 441(l), Gary Klinkhammer/College of Oceanic and Atmospheric Sciences; (r), Dr. Ian R. MacDonald/Texas A&M Univ./ AquaPix, LLC; (t), GettyOne.com/Getty Images ; 442, Leonard Lessin/ Peter Arnold, Inc.; 443, Keith Wood/Corbis; 445, George D. Lepp/ Corbis; 446(inset), Courtesy Westinghouse Nuclear Fuel Division/ Westinghouse Nuclear Fuel Division; 446(t), Inga Spence/Visuals Unlimited; 448(t), Keith Wood/Corbis; (b), George D. Lepp/Corbis; 452(br), Holt, Rinehart and Winston/HRW Photo; (bl), Phil Degginger/ Color-Pic, Inc.

CHAPTER 18: 456, Rafael Macia/Photo Researchers, Inc.; 457(br), Gunter Ziesler/Peter Arnold, Inc.; (bl), NASA; 458, Arnout Hyde, Jr./ Dembinsky Photo Associates; 459, Permission of Rocky Mountain Institute URL: www.rmi.org; 461, Schafer & Hill/Peter Arnold, Inc.; 463, Carlos Sanuvo/Bruce Coleman, Inc.; 467, Warren Gretz/NREL/PIX; 470, Francis Dean/The ImageWorks; 472(t), Schafer & Hill/Peter Arnold, Inc.; (b), Francis Dean/The ImageWorks; 479(b), George Retseck; (m), Applied Innovative Technologies, Inc.

CHAPTER 19: 480, Andrew Rakoczy/Bruce Coleman, Inc.; 481, Michelle Bridwell/Frontera Fotos; 482, Mark Sands/SIPA Press, Headquarters; 483, Reuters NewMedia, Inc./TimePix; 484, Ray Pfortner/Peter Arnold, Inc.; 486, Prof. Rathje/The Garbage Project; 488, HRW Photo; 489(bc), Randy Faris/Corbis; (inset), Phil Degginger/ Bruce Coleman, Inc.; (br), Kay Park - Rec. Corp; (bl), Andy Christiansen/HRW Photo; 490, Sam Dudgeon/HRW Photo; 491, Sam Dudgeon/HRW Photo; 492, Roger Ressmeyer/Corbis; 493, Roger Ressmeyer/Corbis; 495, Nancy J. Pierce/

Photo Researchers, Inc.; 496, Ken Sherman/Bruce Coleman, Inc.; 497, Bettmann/Corbis; 498, Lawson Wood/Corbis; 499, Bernd Wittich/ Visuals Unlimited; 500(t), Prof. Rathje/The Garbage Project; (b), Roger Ressmeyer/ Corbis; (m), Phil Degginger/Bruce Coleman, Inc.; 505, Victoria Smith/HRW Photo; 506, Sander/Gamma Liaison.

UNIT 6 OPENER: 508, Anthony Bannister/Photo Researchers, Inc.; 509(t), Argus Fotoarchiv/Peter Arnold, Inc.; (b), Albert Normandin/Masterfile.

CHAPTER 20: 510, Argus Fotoarchiv/Peter Arnold, Inc.; 514(b), Bourseiller/Hoaqui/Photo Researchers, Inc.; (t), John Elk III/Bruce Coleman, Inc.; 515(b), Mark Edwards/Still Pictures/Peter Arnold, Inc.; (r), Peter Turnley/Corbis; 517(m), J. Serrao/Photo Researchers, Inc.; (t), G. DeGrazia/Custom Medical Stock Photo; 518, UNEP/Peter Arnold, Inc.; 520, Jean-Marc Bouju/AP/Wide World; 521(t), George Turner/ Animals Animals/Earth Scenes; (inset), David Scharf/Peter Arnold, Inc.; 523, Brad Rickerby/Bruce Coleman, Inc.; 524(t), Mark Edwards/Still Pictures/Peter Arnold, Inc.; (b), Jean-Marc Bouju/AP/ Wide World; 528, Elvfis Barukcic/Corbis; 529, Dan McCoy/Rainbow; 531(inset), Brian Witte/AP/Wide World; (b), LisaLoucksChristenson/e-Compass Communications.

CHAPTER 21: 532, Albert Normandin/Masterfile; 533, Rommel Pecson/The ImageWorks; 537(r), D. Seifert/UNEP/Peter Arnold, Inc.; (l), R. Sorensen/J. Olsen/NHPA; 538, Gunter Ziesler/Peter Arnold, Inc.; 539, Bettmann/Corbis; 541, C.C. Lockwood/Animals Animals/Earth Scenes; 542(bl), Spencer Grant/PhotoEdit; (br), Larry Kolvoord/The ImageWorks; 543, Louie Balukoff/AP/Wide World; 545(inset), Bob Schutz/AP/Wide World; (bl), Seth Resnick; (br), Pierre Gleizes/AP/Wide World; 547, Rhoda Sidney/The ImageWorks; 548(t), Gunter Ziesler/Peter Arnold, Inc.; (m), Bettmann/ Corbis; (b), Rhoda Sidney/The ImageWorks; 552(t, r), HRW Photo; 553, HRW Photo. 554, Andrew Dunn; 555, Trevor Frey

APPENDIX: 559, Sam Dudgeon/HRW Photo; 587(tl), Ted M. Conde/ Courtesy Niki Espy; (bl), Selvakumar Ramakrishnan; 588, HRW Photo; 589, HRW Photo; 590(br), Louis Psihoyos/Contact Press Images; (bl), Solar Survivor Architecture, Taos; 591(t), Pamela Freund/Solar Survival Architecture, Taos; (b), Pamela Freund/Solar Survival Architecture, Taos; 592, Flavia Castro; 593, Ken Dudzik; 594(tl), Jonathan A. Meyers; (l), Jonathan A. Meyers; 595(t), courtesy Jana L. Walker; (b), Jonathan A. Meyers; 596(tl), Michael Newman/PhotoEdit; (b), Michael Newman/ PhotoEdit; 597, Dian Russell/HRW Photo; 598, Ken Dudzik; 599(t), Alex S. Maclean/Landslides; (b), Ken Dudzik; 600, courtesy Hanna Anderson; 601, Sam Dudgeon/HRW Photo; 603, Bob Wolf; 606, Sam Dudgeon/ HRW Photo; 608(b), Hans Reinhard/Bruce Coleman, Ltd.; (t), Paul S. Conklin; 609(inset), George H. Harrison/ Grant Heilman Photography; (b), Mae Scanlan; 610, Skjold/Photri; 612(t), Merlin D. Tuttle/Bat Conservation International; (b), Donna Hensley/Bat Conservation International; 613(tc, tr), Sam Dudgeon/ HRW Photo; (tl), Stephen Dalton/O.S.F./Animals Animals/Earth Scenes; 614(r), Tony Tlford/O.S.F./ Animals Animals/Earth Scenes; (t), Stephen Dalton/O.S.F./Animals Animals/Earth Scenes; 615(t), Merlin D. Tuttle/ Bat Conservation International; (br), Stephen Dalton/O.S.F./ Animals Animals/Earth Scenes; (bl), Sam Dudgeon/HRW Photo; 616(t), Ken Cole/Animals Animals/Earth Scenes; (b), John Langford/ HRW Photo.

All art created by Function Thru Form except where noted below.

Holt, Rinehart & Winston: 27(r), 84(b), 138(b), 139(t), 140(t), 365(t), 365(m), 516(br). Dr. Jeffrey Kiehl/National Center for Atmospheric Research: 342(tl). MapQuest.com Inc.: 42(b), 89(b), 176(t), 180(b), 238(t), 301(t), 432(t), 530(t). Roberto Osti: 95(t), 122(l), 122(br), 123(t), 136(b), 174(b), 179(b), 184(t), 204(t), 292(tl). Dan Stuckenschneider/Uhl Studios: viii, xi, xii, 59(b), 60(b), 61(bc), 70(b), 72(b), 124(b), 126(b), 127(b), 132(b), 174(b), 179(b), 269(b), 274(t), 276(b), 287(b), 290(b), 298(br), 306(r), 307(br), 308(tl), 328(b), 333(t), 339(b), 385(b), 413(b), 436(b), 444(b), 445(b), 447(b), 455(t), 459(t), 460(t), 463(t), 464(l), 465(br), 467(t), 469(b), 471(b), 474(b), 485(b), 487(t), 497(t), 507(t). Topozone.com: 56. John White/John White Illustration: 204(t).

Answers to Concept Mapping Questions

Science and the Environment

The following pages contain sample answers to all of the concept mapping questions that appear in the Chapter Reviews. Answers may vary because there is more than one way to do a concept map.

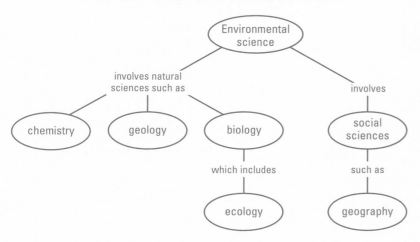

Tools of Environmental Science

The Dynamic Earth

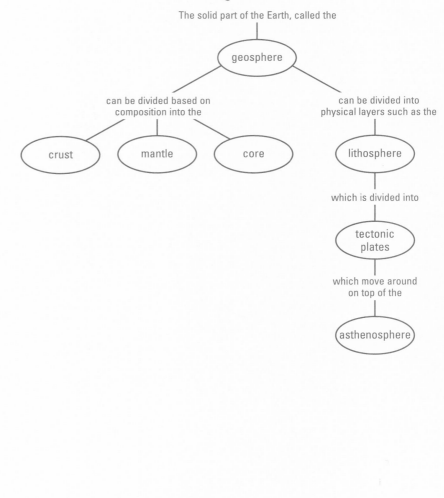

The Organization of Life

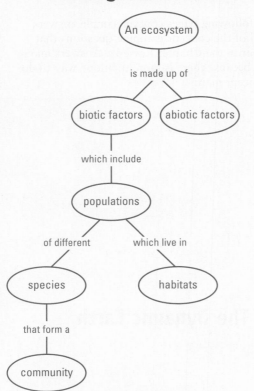

How Ecosystems Work

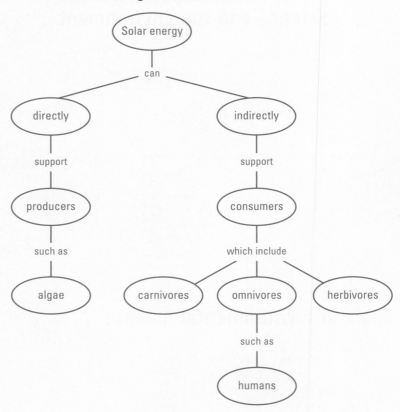

Biomes

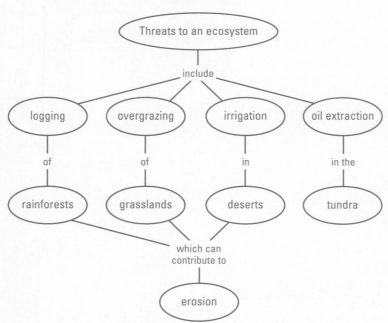

Aquatic Ecosystems

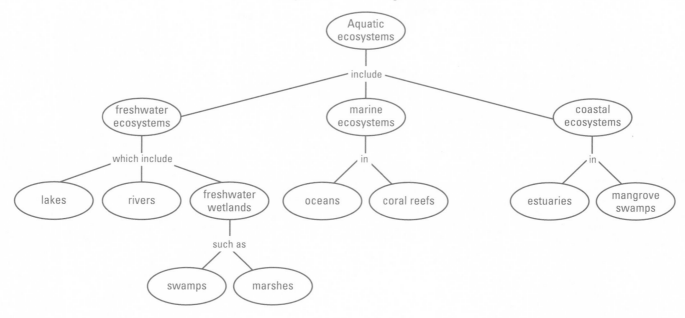

Aquatic ecosystems

— include

freshwater ecosystems — which include — lakes, rivers, freshwater wetlands — such as — swamps, marshes

marine ecosystems — in — oceans, coral reefs

coastal ecosystems — in — estuaries, mangrove swamps

Understanding Populations

Species interactions include

predation — in which a — predator — consumes its — prey

symbiosis — including — parasitism — in which a — parasite — lives on or in its — host

commensalism

mutualism

The Human Population

Demographic transition

is a pattern of increasing

survivorship

leading to

rapid human population growth

which impacts — fuelwood, water, land

until there is a decrease in the — fertility rate

Biodiversity

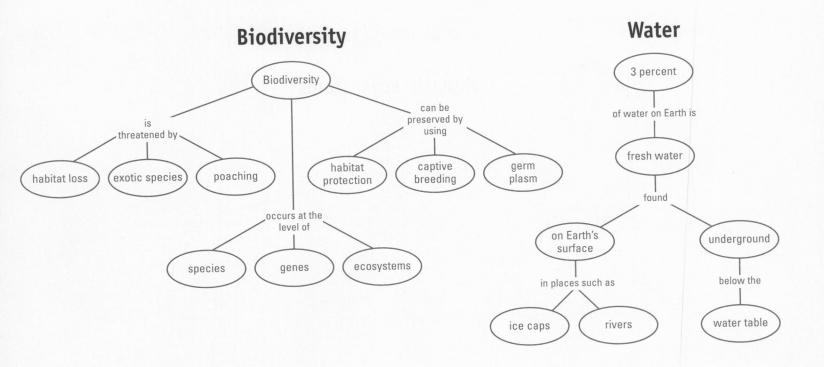

Water

Air

Atmosphere and Climate Change

Land

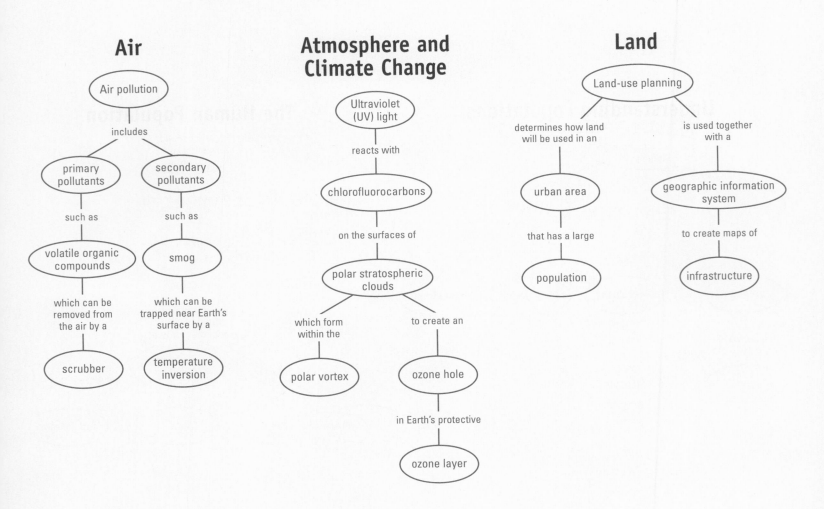

Food and Agriculture

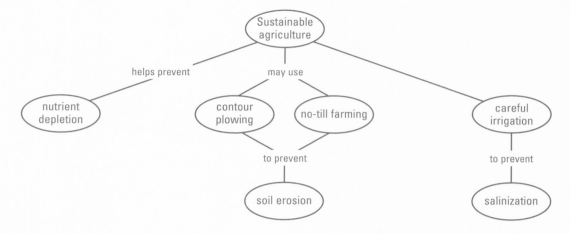

Mining and Mineral Resources

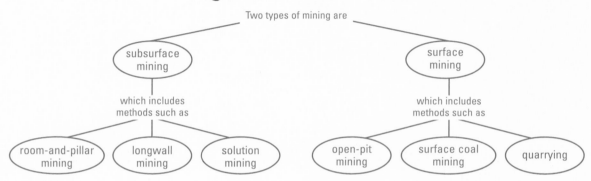

Nonrenewable Energy

Renewable Energy

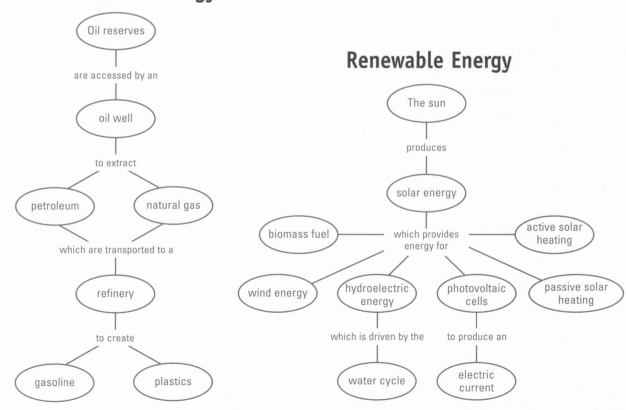

Waste

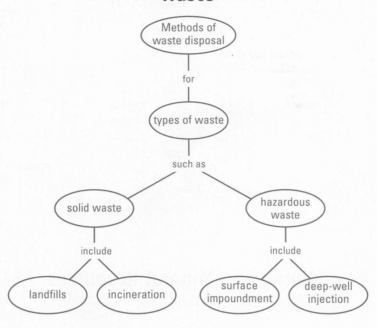

Methods of waste disposal

for

types of waste

such as

solid waste — include — landfills / incineration

hazardous waste — include — surface impoundment / deep-well injection

The Environment and Human Health

Habitat destruction

can expose people to an

animal

which can be a

vector

for a

pathogen

that causes

human disease

Economics, Policy, and the Future

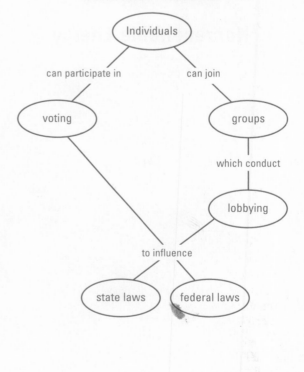

Individuals

can participate in — voting

can join — groups

which conduct — lobbying

to influence — state laws / federal laws